Reproductive Biology and Phylogeny of Snakes

Reproductive Biology and Phylogeny Series
Series Editor: Barrie G. M. Jamieson

Published:

Vol. 1 : Reproductive Biology and Phylogeny of Urodela
(Volume Editor: David M. Sever)

Vol. 2 : Reproductive Biology and Phylogeny of Anura
(Volume Editor: Barrie G. M. Jamieson)

Vol. 3 : Reproductive Biology and Phylogeny of Chondrichthyes
(Volume Editor: William C. Hamlett)

Vol. 4 : Reproductive Biology and Phylogeny of Annelida
(Volume Editors: G. Rouse and F. Pleijel)

Vol. 5 : Reproductive Biology and Phylogeny of Gymnophiona
(Caecilians)
(Volume Editor: Jean-Marie Exbrayat)

Vol. 6 (A and B) : Reproductive Biology and Phylogeny of Birds
(Volume Editor: Barrie G. M. Jamieson)

Vol. 7 : Reproductive Biology and Phylogeny of Cetacea
(Volume Editor: D. Miller)

Vol. 8 (A and B) : Reproductive Biology and Phylogeny of Fishes
(Agnathans and Bony Fishes)
(Volume Editor: Barrie G. M. Jamieson)

Reproductive Biology and Phylogeny of Snakes

Volume edited by
ROBERT D. ALDRIDGE
Department of Biology
Saint Louis University
St. Louis, MO
USA

DAVID M. SEVER
Department of Biological Sciences
Southeastern Louisiana University
Hammond, LA
USA

Volume 9 of Series:
Reproductive Biology and Phylogeny

Series edited by
BARRIE G.M. JAMIESON
School of Integrative Biology
University of Queensland
St. Lucia, Queensland
Australia

THE UNIVERSITY
OF QUEENSLAND
AUSTRALIA

CRC Press
Taylor & Francis Group
an informa business
www.crcpress.com

6000 Broken Sound Parkway, NW
Suite 300, Boca Raton, FL 33487
Schipholweg 107C
2316 XC Leiden, NL
2 Park Square, Milton Park
Abingdon, Oxon OX14 4RN, UK

Science Publishers
Enfield, New Hampshire

Published by Science Publishers, P.O. Box 699, Enfield, NH 03748, USA
An imprint of Edenbridge Ltd., British Channel Islands

E-mail: *info@scipub.net* Website: *www.scipub.net*

Marketed and distributed by:

CRC Press
Taylor & Francis Group
an informa business
www.crcpress.com

6000 Broken Sound Parkway, NW
Suite 300, Boca Raton, FL 33487
Schipholweg 107C
2316 XC Leiden, NL
2 Park Square, Milton Park
Abingdon, Oxon OX14 4RN, UK

ISBN (Series) 978-1-57808-271-1
ISBN (Vol. 9) 978-1-57808-701-3

Library of Congress Cataloging-in-Publication Data

Reproductive biology and phylogeny of snakes/edited by Robert D. Aldridge, David M. Sever.
p. cm. -- (Reproductive biology and phylogeny series ; v. 9)
Includes bibliographical references and index.
ISBN 978-1-57808-701-3 (hardcover)
1. Snakes--Reproduction. 2. Snakes--Phylogeny. I. Aldridge, Robert D.II. Sever, David M.
QL666.O6R42 2011
597.96--dc22

2010039671

Printed in the United States of America

Preface to the Series

This series was founded by the present series editor, Barrie Jamieson, in consultation with Science Publishers, Inc., in 2001. The series bears the title 'Reproductive Biology and Phylogeny' and this title is followed in each volume with the name of the taxonomic group which is the subject of the volume. Each publication has one or more invited volume editors (sometimes the series editor) and a large number of authors of international repute. The level of the taxonomic group which is the subject of each volume varies according, largely, to the amount of information available on the group, the advice of the volume editors, and the interest expressed by the zoological community in the proposed work. The order of publication of taxonomic groups reflects these concerns, and the availability of authors for the various chapters, and does not proceed serially through the animal kingdom in a presumed "ladder of life" sequence. Nevertheless, a second aspect of the series is coverage of the phylogeny and classification of the group, as a necessary framework for an understanding of reproductive biology. It is not claimed that a single volume can, in fact, cover the entire gamut of reproductive topics for a given group but it is believed that the series gives an unsurpassed coverage of reproduction and provides a general text rather than being a mere collection of research papers on the subject. Coverage in different volumes varies in terms of topics, though it is clear from the first volume that the standard is uniformly high. The stress varies from group to group; for instance, modes of external fertilization or vocalization, important in one group, might be inapplicable in another. This is the ninth volume in the series. Previous volumes in the series were devoted to 1. Urodela; 2. Anura; 3. Chondrichthyes: Sharks, Batoids and Chimaeras; 4. Annelida; 5. Gymnophiona (Caecilians); 6 A and B. Birds; 7. Cetacea (whales, dolphins and porpoises); 8 A and B. Fishes (Agnathans and Bony Fishes). My thanks are due to the School of Integrative Biology, University of Queensland, for facilities. I thank my wife, Sheila Jamieson, who has supported me indirectly in so many ways in this work. I am grateful to the publishers, and especially Mr. Raju Primlani, for their friendly support and high standards in producing this series. Sincere thanks must be given to the volume editors and the authors who have freely contributed their chapters in very full schedules. Professors Robert Aldridge and David Sever are particularly to be thanked for conceiving the present volume and for the diligence and outstanding expertise which they brought to its preparation. The editors and publishers are gratified that the enthusiasm and expertise of these contributors has been reflected by the reception of the series by our readers.

THE UNIVERSITY
OF QUEENSLAND
AUSTRALIA
25 June 2010

Barrie G. M. Jamieson
School of Integrative Biology
University of Queensland
Brisbane

Preface to this Volume

> *"To few other animals have been attached so many superstitions and false beliefs as to the snakes. Although most of these beliefs are easily abolished by the scientist, we have to admit, however, that our knowledge of the behaviour of snakes leaves much to be desired. This is perhaps especially true of the biology of reproduction".*
>
> —Volsøe (1944)

We dedicate this volume to the wide variety of snakes in the world and the equally wide variety of dedicated herpetologists who study them. Herpetologists have made some progress in the study of snake reproduction, but the statement of Volsøe (1944) is still valid.

In 2007, we discussed the possibility of organizing a symposium on the Reproductive Biology of Snakes at the Joint Meeting of Ichthyologists and Herpetologists scheduled for July, 2009. We asked many of our colleagues if they would be interested in presenting a paper at the symposium. Nearly all said they would be honored to participate. We sent our request for this symposium to the Herpetologists League, and they approved and provided substantial financial support. The symposium was a huge success, and we thank the Herpetologists League for their moral and financial support.

At the symposium several participants suggested that we publish the papers presented. We contacted Barrie G. M. Jamieson, the editor of the *Reproductive Biology and Phylogeny* series. Barrie liked the idea. Over the course of several months, Barrie reviewed all of the chapters for content and style. We thank Barrie for his advice, patience and skill in helping produce this volume. Without his support this book would not have been possible. We also wish to thank Science Publishers and CRC Press, and particularly Raju Primlani, for the careful production of this volume.

We would also like to thank all of the contributors for their cooperation, professionalism and timely return of their chapters. This has truly been an enjoyable experience for us. We think the volume presents a comprehensive review of all aspects of the reproductive biology and phylogeny of these wonderful and mysterious animals.

This volume benefited from the expertise of the many reviewers of the chapters. We would like to thank the following for their assistance:

Fig. 1 Robert D. Aldridge (left) and David M. Sever at the symposium on the Reproductive Biology of Snakes sponsored by the Herpetologists League at the Joint Meeting of Ichthyologists and Herpetologists in Portland, Oregon in July 2009.

Kraig Adler (Cornell University), Robin M. Andrews (Virginia Tech), Ann C. Burke (Wesleyan University), Don Bradshaw (The University of Western Australia), Gregory P. Brown (University of Sydney), David Cundall (Lehigh University), Thomas Flatt (Veterinärmedizinische Universität Wien), Alex Flemming (University of Stellenbosch), Patrick T. Gregory (University of Victoria), Caleb R. Hickman (Washington University in St. Louis), Benjamin C. Jellen (Saint Louis University), Pilar López (Museo Nacional de Ciencias Naturales), Ignacio Moore (Virginia Tech), Brian Peterson (Thad Cochran National Warmwater Aquaculture Center), Juan M. Pleguezuelos (Universidad de Granada), Lígia Pizzatto (Instituto Butantan), Rick Shine (University of Sydney), Gordon W. Schuett (Georgia State University), Dustin S. Siegel (Saint Louis University), Louis Somma

(Florida Department of Agriculture and Consumer Services), Michael B. Thompson (University of Sydney), Stanley E. Trauth (Arkansas State University), Patrick Weatherhead (University of Illinois) and Sarah Woodley (Duquesne University).

Finally, we wish to thank our wives, Linda Aldridge and Marlis Sever, for their constant support and understanding of our infatuation with the reproduction of snakes.

Volsøe, H. 1944. Structure and seasonal variation of the male reproductive organs of *Vipera berus* (L.). Spolia Zoologica Musei Hauniensis 5: 1-157.

25 June 2010

Robert D. Aldridge
Saint Louis University

David M. Sever
Southeastern Louisiana University

Contents

CHAPTER 1

History of Reproductive Studies on Snakes

Göran Nilson

1.1 INTRODUCTION

Reproductive biology is a central part of many studies of snakes. From a historical perspective, studies by earlier scientists who have contributed to our present knowledge are worth examination. At the same time much of our understanding of the reproductive biology of snakes comes from studies of other reptiles or even other vertebrates. Observations of reproductive behavior or breeding results have been mentioned in much of the older literature, although the observer was not always aware of what occurred. For example, reports of snake balls have been occurring in literature for several hundred years, and this phenomenon of mating aggregations has been well known to farmers and other field people without actually knowing what was happening.

However, it all started much earlier. Among the first published interpretations of snake reproductive biology was that of Herodotus in 440 B.C.E. where in his famous 'History' he wrote: '*So it is also with the vipers...when they are mating in couples and the male is in the very act of emitting his seed, the female, as he does so, seizes him by the neck and, hanging on, never lets him go till she has bitten the neck through. This is how the male dies; but the female pays a kind of recompense, too, to the male. For the children, while still in the womb, take vengeance for their male parent by eating through their mother's insides and so make their entry into the world after eating up her womb. Other snakes, which are not destructive to man, lay eggs and hatch out an infinity of children.*'

This could perhaps be seen as an early start on the discussions about oviparity versus ovoviviparity in snakes, although the contents are somewhat imaginative. Sometime afterwards Aristotle (350 B.C.) made a

Göteborg Natural History Museum, Box 7283, SE 402 35 Göteborg, Sweden short but to some extent generally correct description of the female cloacal anatomy and further discussed oviparity versus ovoviviparity in vipers: '*Serpents as a rule are oviparous, the viper being the only viviparous member of the genus.... The womb of the serpent is long, in keeping with the body, and starting below from a single duct extends continuously on both sides of the spine, so as to give the impression of thus being a separate duct on each side of the spine, until it reaches the midriff, where the eggs are engendered in a row; and these eggs are laid not one by one, but all strung together. And all animals that are viviparous both internally and externally have the womb situated above the stomach, and all the ovipara underneath, near to the loin. Animals that are viviparous externally and internally oviparous present an intermediate arrangement; for the underneath portion of the womb, in which the eggs are, is placed near to the loin, but the part about the orifice is above the gut.*'

In comparatively more recent times Nicolò Leoniceno published a contribution in 1498, with a reprint in 1518, about whether vipers give birth to live young or hatch them from eggs. This was followed by Marco A. Severino in 1650 where he, among other observations, accurately described the anatomy and reproduction of vipers. So, looking back, studies of anatomical configurations in snakes concerning reproductive actions have appeared in literature for many centuries. Edward Tyson publish a study of the Timber Rattlesnake (*Crotalus horridus)* anatomy as early as 1683, with G.-J. Martin Saint Ange (1854) repeating that focus nearly 200 years later on vertebrates in general, including snakes. In addition, shortly afterwards O. Gampert (1866) presented studies of kidney morphology in the Grass Snake (*Natrix natrix)*. Thereby, several of the key mechanisms in reproductive anatomy have long been known to some extent.

From the early statements by Herodotus that '*in Viper females... juveniles eat up their mother from inside...while other snakes lay eggs that hatch*', knowledge has greatly increased. Today we know much about the reproductive biology of snakes and the major patterns and mechanisms that run the breeding cycles and reproductive behaviour although information about some areas concerning specific traits or species groups still remains to be obtained. Many biologists are, or have been during recent times, studying reproductive biology of snakes and as can be seen in this volume a lot is known. However, the gathering of more serious scientific knowledge started with comparatively few individuals who during the last century started up those lines of research that are still productive in this fascinating field of study.

Many researchers have commented on reproduction in snakes during their studies of other aspects of snake natural history. Important general papers on snake biology that contain reproductive and breeding biology sections include papers by DeHaas (1941) on snakes on Java, by Kopstein (1938) on snakes from Malaysia, by Hajime Fukada (1962, 1992) and Koba (1962) on snakes from Japan (mostly *Rhabdophis, Amphiesma, Elaphe* and *Gloydius*), by Razi Dmi'el (1967) on snakes (especially *Spalerosophis*) in Israel, and by Charles C. Carpenter (1952), Lawrence M. Klauber (1956) and Donald

W. Tinkle (1957, 1962) on North American snakes (*Crotalus, Thamnophis*). Raymond Rollinat (1934) wrote a major summary of reproduction in European snakes.

A key person who studied snake reproductive biology in Europe during the last century was Hubert Saint Girons (Fig. 1.1A). He published most of his articles in French although in later years he also published in English. He had a very broad expertise in herpetology and physiology and increased our knowledge on a number of different subjects. Much of his reproductive studies focused on the Asp Viper (*Vipera aspis*) and the European Adder (*Vipera berus*), but vipers in general were his main interest. Other key persons in earlier research concerning reproductive biology of snakes are Henry S. Fitch (Fig. 1.1B), Paul Licht, Hermann Rahn and Harold Fox.

Fig. 1.1 Major historical authorities on the reproductive biology of snakes. **A.** Hubert Saint Girons (1926-2000). **B.** Henry S. Fitch (1909-2009). **C.** R. Wade Fox Jr. (1920-1964). **D.** Frank N. Blanchard (1888-1937).

1.1.1 Primary Reproductive Studies

Henry Fitch, who devoted much of his life to studying the natural history of snakes, contributed enormously to our knowledge of reproductive patterns and strategies in various taxa. He started in the 1940's and devoted more than 50 years (Fitch 1999) to this task. His research covered a great variety of snake taxa such as Gartersnakes (*Thamnophis)* (1940, 1965) and other colubrids like the North American Racer *(Coluber constrictor,* 1963a), the Texas Ratsnake (*Pantherophis obsoletus*, 1963b), the Ring-necked Snake, (*Diadophis punctatus*, 1975) and the Milksnake (*Lampropeltis triangulum*, Fitch and Fleet 1970). Pitvipers (Crotalinae) were included by his monumental study on the ecology of the Copperhead (*Agkistrodon contortrix*, Fitch 1960). In another important study he comprehensively summarized the knowledge of reproductive cycles up to that time (Fitch 1970). Henry Fitch has contributed fundamentally to our understanding of reproductive patterns and mechanisms in snakes and thereby supported all the important studies that lay the foundations of the field from the past to the present.

Considerable knowledge about reproduction in snakes has been produced by reptile breeders and terrarists. Long term captive breeding of snakes in Germany and other countries in central Europe has provided much new information, e.g., Wilhelm Klingelhöffer (1959) has added much new information on snake reproduction and there is a good review of captive breeding by Hans-Günter Petzold (1982; English edition 2008).

1.2 GAMETE PRODUCTION

1.2.1 Female Reproductive Anatomy: Oviducts and Cloaca. Male Urogenital Ducts and Cloacal Anatomy

The first more complete anatomical descriptions of the reproductive organs and the production of eggs and sperm are represented by a few important works. Much of this is summarized in the valuable review of the urogenital system of reptiles by Harold Fox (1977) but prior studies described a number of structures that comprise the reproductive systems in snakes. Helge Volsøe (1944) produced the first complete anatomical description of the reproductive organs of a snake in terms of gross and micro-anatomy in his study of *Vipera berus*. Otherwise, the classical workers with their limited techniques focused mainly on gross anatomy. In the light of more recent studies with much more sophisticated techniques these earlier results are to some extent invalid. Nevertheless they are important in their own right as a platform for current research. The macroscopic structure of the reproductive organs of snakes was first described in detail by Martin St. Ange in 1854 on *Natrix natrix*, but in absence of microscopic technique his studies have no information on microscopic anatomy and histology. As previously mentioned, the anatomy of urogenital organs for a rattlesnake was first described by Tyson in 1683. Otherwise Edward Drinker Cope was probably one of the first Americans who actually examined the gonads and oviducts, whereas Hans Friederich Gadow (1887) is still the only paper cited on cloacal morphology (see **Chapter 10** of this volume).

In Europe, Hans Beuchelt (1936) investigated the reproductive organs in *Natrix natrix* and *Vipera berus*. Without the review in *Biology of the Reptilia* by Raynaud and Pieau (1985), we would have less insight on the development of reproductive organs in squamates.

Clifford H. Pope (1941) with his study of copulatory adjustment in snakes also contributed to this field of knowledge. Wade Fox's (1952, 1956) studies of reproductive systems are other important papers in reproductive anatomy. Wade Fox (Fig. 1.1C) discovered the tubular seminal receptacles in female snakes in 1956. Opinions about the mechanisms behind the expulsion of spermatozoa from these receptacles were proposed by him and later by Hoffman and Wimsatt (1972). The occurrence of intersexuality was studied by Alphonse Hoge and coworkers (1959) on the Golden Lancehead (*Bothrops insularis)* on the island Queimada Grande off the coast of Brazil following up on earlier work by Alfranio do Amaral (1921).

1.2.2 The Testis and Spermatogenesis, Oogenesis

The first more complete descriptions of the production of egg and sperm in snakes were presented in a few important works. The phenomenon of spermatogenesis had been described several times earlier for snakes (e.g., Thatcher 1922). Volsøe (1944) in his study of *Vipera berus* produced, however, a most detailed picture of the complete cycle of spermatogenesis in a snake. Further, Fox (1952) also made an informative study of the seasonal changes in the male reproductive organs in Western Terrestrial Gartersnake (*Thamnophis elegans)* with a detailed description of the cyclic events in the gonads, seminiferous tubules, interstitial cells and spermatogenesis. At the same time Petter-Rousseaux (1953) studied spermatogenesis in the West European Grass Snake (*Natrix natrix helvetica*). Marshall and Woolf (1957) focused on *V. berus* in their study of the activities in the seminiferous tubules in snakes. Somewhat later, Lofts and coworkers (Lofts and Choy 1971, Lofts *et al.* 1966, Tam *et al.* 1969) produced important information about sperm production in the Indian Cobra (*Naja naja*). Brian Lofts (1969, 1977) has contributed considerably to our understanding of spermatogenesis.

In addition, oogenesis and morphology of the ovaries of snakes have been studied a number of times. Tom W. Betz (1963) described these for the Diamond-Backed Watersnake (*Nerodia rhombifer)* during the reproductive cycle. Ballowitz (1901) described gastrulation in the Common Grassnake (*Natrix natrix*) and later (1903) published a study of the early development in the embryogenesis of the adder, *Vipera berus* covering the period up to the closure of the amnion. Later studies by Krull (1906) and Viefhaus (1907) on *N. natrix* provided information on the later stages of embryonic development, focussing on the stages between the neural fold formation and amnion closure.

1.3 BREEDING CYCLES AND PREGNANCY

1.3.1 Male Reproductive Cycles, Female Reproductive Cycles

Presently we know much about how the reproductive cycle runs in tropical and temperate regions, as well as major differences between snake taxa. A larger number of detailed mechanisms affecting the cycles are well elucidated even if there are a number of components which need further study. Much of the knowledge about snakes available to us today has been achieved through research based on the pioneer works that started some half a century ago or more with studies by Blanchard and Blanchard (1941a), Blanchard *et al.* (1979), Rahn (1942), Saint Girons (1947, 1957, 1972, 1982), Fox (1952, 1954) and others. Fitch (1970) in his 'Reproductive Cycles in Lizards and Snakes' put together the knowledge at that time for a considerable portion of the extant snake and lizard families and genera. In addition he reviewed internal and external factors affecting reproductive cycles in squamate reptiles, including snakes, and gathered a considerable amount of information concerning brood size, timing of breeding seasons, and he also discussed ovoviviparity versus oviparity. Saint Girons (1985) summarized the timing of reproductive cycles in snakes and other lepidosaurian reptiles.

In this paper Saint Girons (1985) described patterns of spermatogenesis, vitellogenesis and ovulation as well as the timing of egg laying in oviparous species, and parturition in viviparous species.

1.3.2 Mating Periods

Many studies of reproductive patterns have been performed in temperate regions with an early focus on North America and Europe. The mating periods for most species in temperate regions are a vernal event although the phenomenon of summer and fall mating was discovered early in several groups of snakes. Frank N. Blanchard (Fig. 1.1D) and Frieda C. Blanchard (1941b, 1942) published reports about such patterns in the Eastern Gartersnake (*Thamnophis sirtalis sirtalis).* Fox (1954) also discussed this pattern at an early stage. Subsequently, summer and fall mating was demonstrated for a number of pit vipers as well.

1.3.3 Copulation/Fertilization

Courtship and behavior during mating were discussed by D. Dwight Davis (1936) and Noble (1937), among others, and a variety of different reproductive and courting behaviors in snakes were summarized and further discussed by Charles C. Carpenter (1977). The mating aggregations that over time have been observed in many populations of snakes around the world has been thoroughly and intensively studied in North America, primarily in *Thamnophis*. These studies began with Noble (1937), Blanchard and Blanchard (1941a), Carpenter (1952) and Fox (1955), among others. Reports of aggregations of up to several thousands of individual snakes have appeared and resulted in important series of publications during more recent times. The research on gartersnakes has greatly increased our understanding of mating aggregations.

We now know that the mechanisms behind copulation in snakes are sophisticated, but comparatively little interest has been shown in older literature about this act. Hans Beuchelt (1936) and Clifford H. Pope (1941) initiated the investigations that led to our present understanding of the complex events that characterize snake copulation. Each hemipenis is equipped with two lobes (bifid), each with a sperm transferring branch (the sulcus spermaticus) which fits exactly into the female cloaca, the spines on the hemipenis holding it in the correct position for transferring sperm into the two oviducts in the bilobated cloaca. The first descriptions of snake hemipenes were made by Cope in 1893a, b. The use of snake hemipenis morphology in systematics was pioneered by Herndon G. Dowling and Jay Savage (1960).

1.3.4 Gestation

The study of gestation and placentation in snakes was initiated by H.J. Clausen (1940). Subsequent studies on gestation and placentation are summarized by Yaron (1985) (see also **Chapter 5** of this volume). Some studies, such as Frank Blanchard (1926) on the Northern Ring-necked Snake (*Diadophis punctatus edwardsii*), and the studies by Herman Rahn (1939, 1940a) on *Thanmophis* gestation and placentation are notable. The most comprehensive review of clutch size patterns was produced by Fitch (1970) in his 'Reproductive Cycles in Lizards and Snakes'.

1.3.5 Birth and Early Development

Parental care of offspring in populations of ovoviviparous viperids in temperate regions is an area given much attention in recent times, and evidence of a similar behavior has been documented in tropical oviparous species. The first studies on incubation of eggs in *Python* was made by Lamarre-Picquot (1835, 1842), with detailed follow-up by Valenciennes (1841). Victor H. Hutchison *et al.* (1966) subsequently discussed thermo-regulation in brooding female Indian Pythons (*Python molurus bivittatus*). David Weinland (1857) discovered and described in detail the egg-tooth of snakes. James A. Olivier (1956) described the protection of eggs and nest against predators by the King Cobra (*Ophiophagus hannah*).

1.4 PHYSIOLOGICAL AND COMMUNICATIVE CONTROL OF REPRODUCTION

1.4.1 Physiological Control of Reproduction

The physiological control of reproduction for vertebrates in general has been studied by several authors and sheds light on the situation for snakes. Licht (1984) summarized much of the known literature on reptile reproduction. Hoffman (1970) described functional histology of placentation, and subsequently Saint Girons (1959) also used histological techniques in his studies on physiology of reproduction of vipers.

1.4.2 Endocrine System. Neuroendocrine Control of Reproduction in Snakes. Hormones and Reproduction in Free-ranging Snakes

The endocrine system and the physiological control of the reproductive processes are other fields in which considerable research has been concentrated and where an increasing amount of knowledge has been gathered over the last 50 years or so. In snakes, however, the intensity of investigation increased with Edwin Cieslak's (1945) study on the relation between the pituitary gland and reproductive activities in male Plains Gartersnake (*Thamnophis radix*). Cieslak (1945) found a pituitary cycle in *T. radix* that he felt could be correlated with testicular activity. It includes the key processes that activate a series of physiological and histochemical constraints that are factors in snake reproduction. Much of the information obtained was summarized by Malcolm R. Miller (1959). Work done by biologists including Saint Girons (1959) and Loren H. Hoffman (1970) illustrated the importance of these mechanisms. Saint Girons, together with Gabe (Gabe and Saint Girons, 1962) made a comprehensive study of hormonal activity in *Vipera aspis*, which was later on followed by several other studies by Saint Girons and co-workers (Saint Girons *et al.*, 1993). These endocrine and hormonal activities are induced by pheromones, the way snakes communicate, which in turn have been demonstrated in a number of studies originating from G. Kingsley Noble's research of sense organs involved in the courtship of North American Brownsnakes (*Storeria*), *Thamnophis*, and other species (Noble 1937). Similar pioneer studies were also performed by Goslar (1958) on *Natrix natrix*. A series of papers within the field of endocrinology and hormones was also presented by Bragdon during the same period (1950, 1951, 1952, 1953).

1.5 REPRODUCTIVE BEHAVIOR

As mentioned earlier, the characteristic behavior of snakes in combat has always fascinated people. The mating aggregations and combats in many species have been well known and described in anecdotal and popular items in both Europe, North America and elsewhere. In scientific literature the year 1936 was a starting point for the studies of mating aggregations as well as courtship and mating behavior in several species of snakes by Davis (1936) and in Dekay's Brownsnake (*Storeria dekayi*) by Noble and Clausen (1936). The famous mating aggregations of gartersnakes were first observed by Fox (1955). The combat "dances" in vipers and pit vipers (Viperidae) that have been analyzed and studied a number of times and in detail during the last 30 years or so were on the focus of studies by Shaw

(1948) on *Crotalus* in USA and Thomas (1960) for *Vipera berus* in Europe. Charles M. Bogert and Vincent D. Roth (1966) made a study of male combat in the Pinesnake (*Pituophis melanoleucus*) and much of the different reproductive behaviors in snakes including combat was summarized and further discussed by Charles Carpenter (1977).

1.6 ECDYSIS AND PHEROMONES IN SNAKES

The importance of the ecdysis in communication between snakes is well known today and has during recent times been more fully clarified. Pioneer research in this direction was performed by Goslar (1959) in *Natrix natrix* and by Saint Girons (1980) on European *Vipera*.

With ecdysis, pheromones are released in the near and distant landscape informing other members in the local population about the reproductive status of the carrier. Currently, pheromones are studied by a number of investigators, however, it all started with G. Kingsley Noble in 1937. He was a pioneer within this field of science through his studies on North American natricines.

1.7 THE SEXUAL SEGMENT OF THE KIDNEY

Heinrich Rathke (1839) made the first detailed and illustrated study of the urogenital system of a snake (*Natrix natrix*). The role of the kidney in reproduction has become better understood over time (see **Chapter 11** in this volume), but this was first documented in *N. natrix* in Europe by Gampert (1866) and Heidenhain (1874), who described the variation in the sexual segments. Regaud and Policard (1903) were first to detect that this cyclic variation was a sexual phenomenon only occurring in male snakes. They found a considerable variation in size and structure and supposed that these variations were seasonal in males. Others, including Tribondeau (1903), Zarnik (1910), Cordier (1928) and Herlant (1933) confirmed these results for snakes. Subsequently, Volsøe (1944) and Marshall and Woolf (1957) studied seasonal changes of the sexual elements in the male adder, *Vipera berus*, kidney, followed by a study by Jane E. Bishop (1959) of the sexual segments in males of *Thamnophis sirtalis*. Several of these authors came to a similar conclusion that the secretory activity of these sexual elements in the kidney are involved in the sexual functions of the males. The role of the sexual segments of the kidney in copulatory plug functions has been addressed a number of times and the papers by Volsøe (1944), Fitch (1965) and Devine (1975) could be mentioned as the starting point of this interpretation. In addition, Seshadri (1960) made a study of the structural modification of the cloaca of the Indian Wolf Snake (*Lycodon aulicus*) in relation to urine excretion and the presence of the sexual segment in males. The situation is not clearly cyclic in a more southern species, for instance the Chequered Watersnake (*Xenochrophis piscator*) in India where the diameter of the renal sex segment and height of its epithelial cells remain constant throughout the year (Srivastava and Thapliyal 1965). Takewaki and Hatta (1941) also addressed the role of the kidney in reproduction in their study of the effects of gonadectomy and hypophysectomy in the Tiger Keelback (*Rhabdophis tigrinus*).

1.8 SPERM STORAGE IN MALES AND FEMALES

Today we know that sperm survive a variable length of time in the oviducts of the female after copulation (Rollinant 1934). Spermatozoa can be viable for some weeks before ovulation in *Vipera berus*, which predominantly has a spring mating period (female estrus). Spermatozoa can stay alive for at least six months in North American pit vipers that have a summer-fall mating season, meaning that ovulation and internal fertilization takes place first the following year. A number of important papers have contributed to our understanding of these processes although it was not clear how such fantastic pattern could have evolved. Rahn (1940b, 1942) was one of the first who actually started looking into sperm storage in snakes, which makes him a pioneer, even though other people were conducting studies in this area around the same time. Other important studies were made by Blanchard (1942) in *Thamnophis* and Marion Ludwig and Hermann Rahn (1943) on Prairie Rattlesnakes (*Crotalus viridis*). These were followed by additional and informative studies by Fox (1956) and Hubert Saint Girons (1973) on sperm survival in snakes. A summary can be seen in the excellent review by Harold Fox (1977) in *Biology of the Reptilia*. Some authors such as Haines (1940) observed prolongation of sperm storage over longer periods lasting up to several years. This is extremely interesting, especially as, at least in some cases, parthenogenesis might be involved as an alternative explanation (personal observation). This phenomenon should be further addressed.

1.9 SUMMARY AND CONCLUSIONS

Some species of snakes have been studied intensively over time, and this information forms a foundation of knowledge facilitating present and future in-depth studies of specific questions. Species such as *Vipera berus, V. aspis, Naja naja, Thamnophis sirtalis, Crotalus atrox, C. horridus, C. viridis* and others have been studied for longer periods of time and provide the basic and original knowledge for many aspects of reproductive biology in snakes. Comprehensive ecological studies often contain considerable information on reproduction and even general background for the understanding of reproductive biology for a species. The excellent works of Bernström (1943), Viitanen (1967) and Prestt (1971) on *Vipera berus* have provided information that has led to more detailed questions of the reproductive life and mechanisms involved. These studies are being referred to in a number of more recent zoological studies for that species. In Sweden alone, eight Ph.D. theses on the biology of *V. berus* have appeared during the last 30 years.

Similar patterns can be seen for a number of other species of snakes around the world. By the 1970s at least 30 informative papers on the reproductive biology of North American Gartersnakes (*Thamnophis*) had appeared in scientific publications. The number of publications on reproductive biology on European vipers (*Vipera*) are about equal in the same period of time. These early studies provided the basis for a more scientific approach to satisfy our curiosity on snake reproductive biology. In turn, this has led to greatly increased research efforts leading to the present times.

The history of knowledge of reproduction in snakes goes back several thousands of years, to the times of Herodotus and Aristotle, but a more complete understanding through research is apparent during the last half century or so. Important contributors during this last period were Hubert Saint Girons and Henry Fitch. However, other biologists contributed substantially during this period to our understanding, producing an important source of knowledge for researchers of today including Paul Licht, Lawrence Klauber, Hermann Rahn, Frank and Frida, Blanchard, Harold Fox, Wade Fox and Helge Volsøe. Further important contributors were G. Kingsley Noble, Giacomini, Hoffman, Gadow, Raynaud, Rahn, Cope and others, but also much of the knowledge has come from small studies and as side results while doing herpetological research in other directions. Ecological studies of snakes always contribute in some way to our understanding of reproduction.

1.10 ACKNOWLEDGMENTS

For portraits of F.N. Blanchard, H.S. Fitch and W. Fox, I am grateful to Kraig Adler, Cornell University, Ithaca, New York. I also thank Kraig Adler as well as Robert D. Aldridge, Saint Louis University, St. Louis, Missouri, for their most valuable reviews of the manuscript.

1.11 LITERATURE CITED

Amaral, A. do 1921. Contribuição para conhecimento dos ofídios do Brasil—A. Parte II. Biologia da nova espécie, *Lachesis insularis*. Anexos das Memórias do Instituto de Butantan 1: 39-44.

Aristotle. 350 B. C. E. *The History of Animals* (Translated by D'Arcy Wentworth Thompson). Book III, Part 1. Clarendon Press, Oxford, U.K. 2008. In Nine Webpage Parts.

Ballowitz, E. 1901. Die gastrulation dei der ringelsatter (*Tropidonotus natrix* Boie) bis zum auftreten der falterform der embryonalanlage. Zeitschrift für Wissenschaftliche Zoologie 70: 675-732.

Ballowitz, E. 1903. Entwicklungsgeschichte der kreutzotter (*Pelias berus* Merr.). Teil I. Die entwicklung vom auftreten der ersten furche bis zum schlusse des amnios. Fischer, Jena, Germany. 295 pp.

Bellairs, R., Griffiths, I. and Bellairs, D'A. 1955. Placentation in the adder, *Vipera berus*. Nature 176: 657-658.

Bernström, J. 1943. Till kännedom om huggormen *Vipera berus berus* (Linné). Meddelande Göteborgs Museum Zoologiska Avdelning 103: 1-34.

Betz, T. W. 1963. The gross ovarian morphology of the diamond-backed water snake *Natrix rhombifera* during the reproductive cycle. Copeia 1963: 692-697.

Beuchelt, H. 1936. Bau, funktion und entwicklung der begatubngsorgane der männlichen ringelnatter (*Natrix natrix* L.) und kreuzotter (*Vipera berus* L.). Morphologisches Jahrbuch 78: 445-516.

Bishop, J. E. 1959. A histological and histochemical study of the kidney tubule of the common garter snake, *Thamnophis sirtalis*, with special reference to the sexual segment in the male. Journal of Morphology 104: 307-358.

Blanchard, F. C. 1942. A test of fecundity of the garter snake *Thamnophis sirtalis sirtalis* (Linnaeus) in the year following the year of insemination. Papers of the Michigan Academy of Science, Arts and Letters 28: 313-316.

Blanchard, F. N. 1926. Eggs and young of the eastern ring-neck snake *Diadophis punctatus edwardsii*. Papers of the Michigan Academy of Science, Arts and Letters 7: 279-292, pls. 13-19.

Blanchard, F. N. and Blanchard, F. C. 1941a. Factors determining time of birth in the garter snake *Thamnophis sirtalis sirtalis* (Linnaeus). Papers of the Michigan Academy of Science, Arts and Letters 26: 161-176.

Blanchard, F. N. and Blanchard, F. C. 1941b. The inheritance of melanism in the garter snake *Thamnophis sirtalis sirtalis* (Linnaeus), and some evidence of effective autumn mating. Papers of the Michigan Academy of Science, Arts and Letters 26: 177-193.

Blanchard, F. N. and Blanchard, F. C. 1942. Mating of the garter snake *Thamnophis sirtalis sirtalis* (Linnaeus). Papers of the Michigan Academy of Science, Arts and Letters 27: 215-234.

Blanchard, F. N., Gilreath, M. R. and Blanchard, F. C. 1979. The eastern ringneck snake (*Diadophis punctatus edwardsii*) in northern Michigan (Reptilia, Serpentes, Colubridae). Journal of Herpetology 13: 377-402.

Bogert, C. M. and Roth, V. D. 1966. Ritualistic combat of male gopher snakes, *Pituophis melanoleucus affinis* (Reptilia, Colubridae). American Museum Novitates (2245): 1-27.

Bragdon, D. E. 1950. Hormonal control of the reproductive cycle of ovoviviparous snakes as related to the evolution of viviparity. Virginia Journal of Science 1: 391-392

Bragdon, D. E. 1951. The non-essentiality of the corpora lutea for the maintenance of gestation in certain live bearing snakes. Journal of Experimental Zoology 118: 419-435.

Bragdon, D. E. 1952. Corpus luteum formation and follicular atresia in the common garter snake, *Thamnophis sirtalis*. Journal of Morphology 91: 413-443.

Bragdon, D. E. 1953. A contribution to the surgical anatomy of the water snake *Natrix sipedon sipedon*; the location of the visceral endocrine organs with reference to ventral scutelation. Anatomical Record 117: 145-161.

Carpenter, C. C. 1952. Comparative ecology of the common garter snakes (*Thamnophis s. sirtalis*), the ribbon snake (*Thamnophis s. sauritus*) and Butler's garter snakes (*Thamnophis butleri*) in mixed populations. Ecological Monographs 22: 235-258.

Carpenter, C. C. 1977. A survey of stereotyped reptilian behavioural patterns. Pp. 335-403. In C. Gans and D. W. Tinkle (eds), *Biology of the Reptilia, Ecology and Behaviour A*, Vol. 7. Academic Press, New York.

Cieslak, E. S. 1945. Relations between the reproductive cycle and the pituitary gland in the snake *Thamnophis radix*. Physiological Zoology 18: 299-329.

Clausen, H. J. 1940. Studies on the effect of ovariotomy and hypophysectomy on gestation in snakes. Endocrinology 27: 700-704.

Cope, E. D. 1893a. Prodromus of a new system of the non-venomous snakes. American Naturalist 1893: 477-483.

Cope, E. D. 1893b. The classification of the Ophidia. Transactions of the American Philosophical Society (n.s.) 28 (art. 3): 186-219.

Cordier, R. 1928. Études histophysisologiques sur le tube urinaire des Reptiles. Archives de Biologie 38: 111-171.

Davis, D. D. 1936. Courtship and mating behavior in snakes. Field Museum of Natural History, Zoology Series 20: 257-290.

DeHaas, C. P. J. 1941. Some notes on the biology of snakes and their distribution in two districts of West Java. Treubia, Bogor 18: 327-375.

Devine, M. C. 1975. Copulatory plugs in snakes: enforced chastity. Science 187: 844-845.

Dmi'el, R. (1967). Studies on reproduction, growth, and feeding in the snake *Spalerosophis cliffordi* (Colubridae). Copeia 1967: 332-346.

Dowling, H. G. and Savage, H. M. 1960. A guide to the snake hemipenis: a survey of basic structure and systematic characteristics. Zoologica 45: 17-29.

Fitch, H. S. 1940. A biogeographical study of the ordinoides artenkreis of garter snakes (genus *Thamnophis*). University of California Publications in Zoology 44: 1-150.

Fitch, H. S. 1960. Autecology of the copperhead. University of Kansas Publications, Museum of Natural History 13: 85-288.

Fitch, H. S. 1963a. Natural history of the racer *Coluber constrictor*. University of Kansas Publications, Museum of Natural History 15: 351-468.

Fitch, H. S. 1963b. Natural history of the black rat snake (*Elaphe o. obsoleta*) in Kansas. Copeia 1963: 649-658.

Fitch, H. S. 1965. An ecological study of the garter snake, *Thamnophis sirtalis*. University of Kansas Publications, Museum of Natural History 15: 493-564.

Fitch, H. S. 1970. Reproductive Cycles in Lizards and Snakes. University of Kansas Publications, Museum of Natural History Miscellaneous Publications, No. 52. 1-247.

Fitch, H. S. 1975. A demographic study of the ringneck snake (*Diadophis punctatus*) in Kansas. University of Kansas Publications, Museum of Natural History Miscellaneous Publications 62: 1-53.

Fitch, H. S. 1999. *A Kansas Snake Community: Composition and Changes over 50 Years.* Kreiger Publishing Co., Malibar, Florida. Pp. 178.

Fitch, H. S. and Fleet, R. R. 1970. Natural history of the milk snake (*Lampropeltis triangulum*) in the northeastern Kansas. Herpetologica 26: 387-396.

Fox, H. 1977. The urogenital system of reptiles. Pp. 1-157. In C. Gans and T. S. Parsons (eds), *Biology of the Reptilia*, Vol. 6. Academic Press, New York.

Fox, W. 1952. Seasonal variation in the male reproductive system of Pacific Coast garter snakes. Journal of Morphology 90: 481-554.

Fox, W. 1954. Genetic and environmental variation in the timing of the reproductive cycle of male garter snakes. Journal of Morphology 95: 415-450.

Fox, W. 1955. Mating aggregations of garter snakes. Herpetologica 11: 176.

Fox, W. 1956. Seminal receptacles of snakes. Anatomical Record 124: 519-540.
Fukada, H. 1962. Biological Studies on the snakes IX. Breeding habits of *Agkistrodon halys blomhoffii* (Boie). Bulletin of the Kyoto Gakugei University, Series B, 20: 12-17.
Fukada, H. 1992. *Snake Life History in Kyoto.* Impact Shuppankai Co. Ltd. Tokyo. Pp. 171.
Gabe, M. and Saint Girons, H. 1962. Donées histophysiologiques sur l'élaboration d'hormones sexuelles au cours du sysle reproducteur chez *Vipera aspis* (L.). Acta Anatomica 50 : 22-51.
Gadow, H. 1887. Remarks on the cloaca and the copulatory organs of the Amniota. Philosophical Transactions of the Royal Society (B) 178: 12-37.
Gampert, O. 1866. Ueber die niere von *Tropidonotus natrix* und der Cyprinoiden. Zeitschrift für Wissenschaftliche Zoologie 16: 369-373.
Goslar, H. G. 1958. Über der wirkung verschiedener sexualhormone auf die häutungsgänge det ringelnatter (*Natrix natrix* L.). Dermatologische Wochenschrift 6: 139-146.
Haines, T. P. 1940. Delayed fertilization in *Leptodeira annulata polysticta.* Copeia 1940: 116-118.
Heidenhain, R. 1874. Mikroskopische beiträge zur anatomie und physiologie der nieren. Archiv für Mikroskopische Anatomie 10: 1-50.
Herlant, M. 1933. Recherches histologiques et expérimentales sur les variations cycliques du testicule et des caractères sexuels secondaires chez les reptiles. Archives de Biologie 44: 347-468.
Herodotus, 440 B. C. E. *The History. Book Three: Part 109* (Translated by David Grene 1987). University of Chicago Press, Chicago. Pp. 699.
Hoffman, L. H. 1970. Placentation in the garter snake, *Thamnophis sirtalis.* Journal of Morphology 131: 57-88.
Hoffman, L. H. and Wimsatt, W. A. 1972. Histochemical and electon microscopic observations on the sperm receptacles in the garter snake oviduct. American Journal of Anatomy 134: 71-96.
Hoge, A. R., Belluomini H. E., Schreiber, G. and Penha, A. M. 1959. Sexual abnormaliities in *Bothrops insularis* (Amaral, 1921). Memórias do Instituto Butantan 42/43: 373-496.
Hutchison, V. H., Dowling, H. G. and Vinegar, A. 1966. Thermoregulation in a brooding female Indian python (*Python molurus bivittatus*). Science 151: 694-696.
Klauber, L. M. 1956. *Rattlesnakes.* Vols. I and II. University of California Press, Berkeley and Los Angeles. Pp. 1533.
Klingelhöffer, W. 1959. *Terrarienkunde,* 4. Teil: Schlangen, Schildkröten, Panzerechsen, Reptilienzucht. Stuttgart. Pp. 379.
Kopstein, F. 1938. Ein Beitrag zur Eierkunde Fortpflantzung der Malaiischen Reptilien. Bulletin of the Raffles Museum 14: 81-167.
Krull, J. 1906. Die entwicklung der ringelnatter (*Tropidonotus natrix* Boie) vom ersten austreten des proamnios bis zum schluss des amnios. Zeitschrift für Wissenschaftliche Zoologie 85: 107-155.
Lamarre-Picquot, P. 1835. L'lnstitut 3: 70.
Lamarre-Picquot, P. 1842. Troisème mémoire sur l'incubation et autres phénomènes observés chez les ophidiens (Third report on the incubation and other phenomena observed in the snake house). Les Comptes Rendus de l'Académie des Sciences 14: 164.
Leoniceno, N. 1498. De Tiro, seu Vipera (=1518: De Serpentibus Opus Singulare ac Exactissimum). per Ioannem Antonium iuniorem de Benedictis, Bononiae. Pp. 107.

Licht, P. 1984. Reptiles. Pp. 206-282. In G. E. Lamming (ed.), *Marshall's Physiology of Reproduction*. Churchill Livingston, Edinburgh, U.K.

Lofts, B. 1969. Seasonal cycles in reptilian testes. General and Comparative Endocrinology. Supplement 2: 147-155.

Lofts, B. 1977. Patterns of spermatogenesis and steroidogenesis in male reptiles. Pp. 127-136. In J. H. Calaby and C. H. Tyndale-Biscoe (eds), *Reproduction and Evolution*. Australian Academy of Sciences, Canberra.

Lofts, B. and Choy, L. Y. L. 1971. Steroid synthesis by the seminiferous tubules of the snake *Naja naja*. General and Comparative Endocrinology 17(3): 588-591.

Lofts, B., Phillips, J. G. and Tam, W. H. 1966. Seasonal changes in the testis of the Cobra, *Naja naja* (Linn.). General and Comparative Endocrinology 6(3): 466-475.

Ludwig, M. and Rahn H. 1943. Sperm storage and copulatory adjustment in the prairie rattlesnake. Copeia 1943: 15-18.

Marshall, A. J. and Woolf, F. M. 1957. Seasonal lipid changes in the sexual elements of a male snake, *Vipera berus*. Quarterly Journal of Microscopical Science 98: 89-100.

Martin Saint Ange, G.-J. 1854. *Étude de l'appariel reproducteur dans les cinq classes d'animaux vertébrés, au point de vue anatomique, physiolgique et zoologique*. J.-B. Balliére, Libraire de l'Académe Impériale de Médecine, Paris. Pp. 234.

Miller, M. R. 1959. The endocrine basis for reproductive adaptations in reptiles. Pp. 499-516. In A. Gorbman (ed.), *Symposium on Comparative Endocrinology*. John Wiley and Sons, New York.

Noble, G. K. 1937. The sense organs involved in the courtship of *Storeria*, *Thamnophis* and other snakes. Bulletin of the American Museum of Natural History 73: 673-725.

Noble, G. K. and Clausen H. C. 1936. The aggregation behavior of *Storeria dekayi* and other snakes. Ecological Monographs 6: 269-316.

Olivier, J. A. 1956. Reproduction in the king cobra, *Ophiophagus hannah* Cantor. Zoologica 41: 145-152.

Petter-Rousseaux, A. 1953. Recherches ssur la croissance et le cycle d'activité testiculaire de *Natrix natrix helvetica* (Lacépède). Terre Vie 100: 175-223.

Petzold, H.-G. 2008. *The Lives of Captive Reptiles*. SSAR, Cornell University, Ithaca, New York. Pp. 308.

Pope, C. H. 1941. Copulatory adjustment in snakes. Zoological Series of Field Museum of Natural History 24: 249-252.

Prestt, I. 1971. An ecological study of the viper *Vipera berus* in southern Britain. Journal of Zoology, London 164: 373-418.

Rahn, H. 1939. Structure and function of placenta and corpus luteum in viviparous snakes. Proceedings of the Society for Experimental Biology and Medicine 40: 381-382.

Rahn, H. 1940a. The physiology of gestation in viviparous snakes. Journal of the Colorado-Wyoming Academy of Science 2: 45-46.

Rahn, H. 1940b. Sperm viability in the uterus of the garter snake, *Thamnophis*. Copeia 1940: 109-115.

Rahn, H. 1942. The reproductive cycle of the prairie rattlesnake. Copeia 1942: 233-240.

Rathke, H. 1839. *Entwicklungsgeschichte der Natter (Coluber natrix)*. Königsberg, Germany. Pp. 232.

Raynaud, A. and Pieau, C. 1985. Embryonic development of the genital system. Pp. 149-300. In C. Gans and Frank Billet (eds), *Biology of the Reptilia, Vol. 15 Development B*. Academic Press, New York.

Regaud, C. and Policard, A. 1903. Recherches sur la structure du rein de quelques ophidiens. Archives d'Anatomie Microscopique, Paris 6: 191-282.

Rollinant, R. 1934. *La Vie de Reptiles de la France Centrale*. Delagrave, Paris. Pp. 340.
Saint Girons, H. 1947. Ècologie des Vipères. 1. *Vipera aspis*. Bulletin de la Societe Zoologique de France 72: 158-169.
Saint Girons, H. 1957. Le sycle sexuel chez *Vipera aspis* (L.) dans l'ouest de la France. Bulletin Biologique de la France et de la Belgique 91: 284-350.
Saint Girons, H. 1959. Données histochemiques sur les glucides de l'appareil génital chez les vipères, au cors du cycle reproducteur. Annales de Histochimie 4: 235-243.
Saint Girons, H. 1972. Le cycle sexual de *Vipera aspis* (L.) en montagne. Vie Milieu 23: 309-328.
Saint Girons, H. 1973. Sperm survival and transport in the female genital tract of reptiles. Pp. 105-113. In E. S. E. Hafez and C. G. Thibault (eds), *The Biology of Spermatozoa*. Karger, Basel, Switzerland.
Saint Girons, H. 1980. Le cycle des mues chez les vipères Européennes. Bulletin de la Societe Zoologique de France 105: 551-559.
Saint Girons, H. 1982. Reproductive cycles of male snakes and their relationships with climate and female reproductive cycles. Herpetologica 38: 5-16.
Saint Girons, H. 1985. Comparative data on lepidosaurian reproduction and some time tables. Pp. 35-58. In C. Gans and F. Billet (eds), *Biology of the Reptilia, Vol. 15 Development B*. Academic Press, New York.
Saint Girons, H. and Kramer, E. 1963. Le cycle sexuel chez *Vipera berus* (L.) en montagne. Revue Suisse de Zoologie 70: 15-221.
Saint Girons, H., Bradshaw, S. D. and Bradshaw, F. J. 1993. Sexual activity and plasma levels of sex steroids in the aspic viper *Vipera aspis* L. (Reptilia, Viperidae). General and Comparative Endocrinology 91: 287-297.
Severino, M.A. 1650. Vipera Pythia: Id est, De Viperae Natura, Veneno, Medicina, Demonstrationes, and Experimenta noua. Patavii. (2nd ed.).
Shaw, C. E. 1948. The male combal "dance" of some crotalide snakes. Herpetologica 4: 137-145.
Seshadri, C. 1960. Structural modification of the cloaca of *Lycodon aulicus aulicus* Linn., in relation to urine excretion and the presence of sexual segment in the kidney of male. Proceedings of the National Institute of Science, India 25B: 271-278.
Srivastava, P. C. and Thapliyal, J. P. 1965. The male sexual cycle of the chequered water snake, *Natrix piscator*. Copeia 1965: 410-415.
Takewaki, K. and Hatta, K. 1941. Effect of gonadectomy and hypophysectomy on the kidney and genital tract of a snake, *Natrix tigrina tigrina*. Annotationes Zoologicae Japonenses 20: 4-8.
Tam, W. H., Phillips, J. G. and Lofts, B. 1969. Seasonal changes in the *in vitro* production of testicular androgens by the cobra (*Naja naja* Linn.). General and Comparative Endocrinology 13: 117-125.
Thatcher, L. E. 1922. Spermatogenesis of the garter snake. *Science* 56: 372.
Thomas, S. E. 1960. Kommentkämpfe bei Vipern. Zoologischer Anzeiger 23: 111-116.
Tinkle, D. W. 1957. Ecology, maturation and reproduction of *Thamnophis sauritus proximus*. Ecology 38: 69-77.
Tinkle, D. W. 1962. Reproductive potential and cycles in female *Crotalus atrox* from Northwestern Texas. Copeia 1962: 306-313.
Tribondeau, M. 1903. Recherches anatomiques et histologiques sur le rein des ophidiens. (4° serie des communications) Actes de la Société Linnéenne de Bordeaux 57: 90-105.
Tyson, E. 1683. Anatomy of a rattle-snake. Philosophical Transactions of the Royal Society of London 13: 281-284.

Valenciennes, M. 1841. Observations faites pendant l'incubation d'une femelle du Python a deux raies (*Python bivittatus*, Kuhl) pendant les mois de mai et de juin 1841. Les Comptes Rendus de l'Académie des Sciences, Paris. 13: 126-133.

Viefhaus, T. 1907. Die Entwicklung der Ringelnattere (*Tropidonotus natrix* Boie) nach Ausbildung der Falterform bis zur Erhebung des Proamnios. Zeitschrift für Wissenschaftliche Zoologie 86: 55-99.

Viitanen, P. 1967. Hibernation and seasonal movements of the viper *Vipera berus berus* (L.) in southern Finland. Annales Zoologici Fennici 4: 472-546.

Volsøe, H. 1944. Structure and seasonal variation of the male reproductive organs of *Vipera berus* (L.). Spolia Zoologica Musei Hauniensis 5: 1-157.

Weinland, D. F. 1857. On the egg-tooth of snakes and lizards. Proceedings of the Essex Institute 2: 1-7.

Wharton, C. H. 1966. Reproduction and growth in the cottonmouths, *Agkistrodon piscivorus* Lacépède, of Cedar Keys, Florida. Copeia 1966: 149-161.

Yaron, Z. 1985. Reptilian placentation and gestation: Structure, function, and endocrine control. Pp. 527-603. In C. Gans and F. Billet (eds), *Biology of the Reptilia Vol. 15: Development B*. Academic Press, New York.

Zarnik, B. 1910. Vergleichende studien über den bau der niere von echidna und der reptilienniere. Jenaische Zeitschrift für Naturwissenschaft 46: 113-124.

CHAPTER 2

Evolution and Taxonomy of Snakes

Frank T. Burbrink[1] *and Brian I. Crother*[2]

2.1 INTRODUCTION

This chapter arrives at an interesting and exciting time in the study of snake systematics. The last part of the 20th century and the early part of the 21st century might ultimately be highlighted as the intersection between traditional classifications of snakes based on morphology and those based on molecular data. Classification of organisms has typically and traditionally relied on morphological traits to guide the process, either by phylogenetic methods that attempt to be concordant with evolutionary history or by more arbitrary methods that apply the use of authoritative interpretation of morphology by experts in the field. Given the real possibility of evolutionary convergence among morphological characters in organisms, such as in snakes and other limbless squamates (see Wiens *et al.* 2006), it seems that having a credible understanding of relationships among extant serpents will be through the use of molecular systematics. Another advantage is that molecular systematics can provide thousands to millions of characters as well produce species tree relationships using independently evolving gene estimates free from linkage or convergence. However, there have been important studies using rigorous phylogenetic methods on a large suite of morphological characters scored from extant and extinct snakes (something molecular methods cannot address) that reveal the utility of these characters to address phylogeny (Lee and Scanlon 2002; Lee *et al.* 2007). Therefore, we are not saying that traditional classifications based on morphology are entirely incorrect; in fact many of them still hold up well. However, several studies are revealing that certain traditional groups

[1]Biology Department, 6S-143, 2800 Victory Blvd., College of Staten Island/CUNY, Staten Island, New York 10314 USA
[2]Department of Biology, College of Science and Technology, Southeastern Louisiana University, Hammond, LA 70402 USA

simply cannot be credible given the agreement among independently evolving genes (e.g., the traditional macrostomata, Anilioidea, Colubridae are all likely paraphyletic). Moreover, molecular methods will be more useful at examining relationships at the levels of species, genera, and families. Arriving at a strong consensus with robust trees among studies using unlinked genetic markers has already helped illuminate evolutionary relationships among snake species. These molecular studies inform taxonomy by naming groups that are concordant (i.e., monophyletic) with the evolutionary history of the taxon. These phylogenies ultimately help comparative biologists attain a better understanding of the independent origins of various morphological characteristics, ecologies and behavior.

While we extol the virtues of the current state of molecular systematics and realize how the field will aid the "scholarly snake community" to better comprehend the origins and relationships of snakes, we also realize that our understanding based on a handful of markers is likely to change as snake phylogeneticists lumber into the world of phylogenomics and coalescent based species tree estimation (Edwards 2009). Currently, the only species tree estimation paper that also uses the largest number of genes to date (25 independent loci) has been applied using single representatives of only 21 major snake groups/families (Pyron and Burbrink, unpublished data; but see Wiens *et al.* 2008). In contrast, the densest sampling of snakes for a single phylogenetic project is only 232 species out of ~3,150 described taxa, and using only a single gene (Eckstut *et al.* 2009). Given the decreasing costs for next generation DNA sequencing, it is conceivable that snake systematists will produce phylogenetic trees using thousands of single copy, unlinked genes sampled across the genome for hundreds of species, while properly inferring the species tree given the uncertainty in the gene tree. This again may rapidly change our notions of snake taxonomy and evolutionary relationships. On the other hand, it may show that the information given in number of substitutions and sorting of lineages may never be adequate to resolve some situations. That is, some relationships may simply not be knowable.

This chapter provides a brief overview of the relationships, defining characteristics, and geographic area and dates of origin of all major extant snake groups. Several radical taxonomic changes have been proposed for certain groups in the last decade, leaving little strong consensus about the taxonomy of a given group. For example, several researchers have proposed major changes to the group Colubroidea, yet no single taxonomic scheme has taken hold. We therefore discuss the most conservative aspects of modern snake taxonomy based on published research. This chapter is not meant to be the lexicon of snake taxonomy but rather a fairly detailed introduction to snake systematics primarily based on results from modern studies.

2.1.1 What are Snakes?

We know that snakes are squamates and deeply embedded in the lizard phylogeny. In fact, snakes are simply a very specialized group of extremely diverse limbless lizards. As such, snakes are members of the second most speciose group of living reptile (see Reptile Database: http://www.reptile-database.org/). The evolution of limblessness is quite common in lizards and, including snakes, has evolved independently at least 25 times (Wiens *et al.* 2006). However, no limbless lizard clade is as successful as snakes, with ~3,150 species occurring in nearly every habitat on every continent except Antarctica. Snakes form a monophyletic group and the best available phylogenetic evidence using molecular data, free from morphological convergence due to reduction in character states, suggest that snakes are not related to other limbless lizards like amphisbaenids or dibamids, but rather group with iguanians, lacertiforms and anguimorphs (Townsend *et al.* 2004; Eckstut *et al.* 2009; see Douglas *et al.* 2006 for a contrasting molecular view). The exact placement of snakes within the lizards has yet to be determined, but using multiple independently evolving loci, both Townsend *et al.* (2004) and Vidal and Hedges (2005) demonstrated a close relationship between snakes and anguimorphs, which has also been suggested by other authors (e.g., McDowell and Bogert 1954; Jamieson 1995; Lee 1998; Reynoso 1998; Lee and Caldwell 2000; Eckstut *et al.* 2009). Several studies that include morphological data have claimed a closer relationship between varanids or mosasaurs and snakes (Lee 1997, 1998, 2000; Caldwell 1999; Lee and Caldwell 2000; Lee and Scanlon 2001; Scanlon and Lee 2002; Caldwell and DalSasso 2004) or a group consisting of amphisbaenids, dibamids and snakes. The most recent large scale morphological study, which included fossils, suggested snakes are most closely related to scincoids, the sister to a clade of trogonophids, amphisbaenids, and rhineurids (Conrad 2008). The relationships suggested by these morphological studies have been soundly rejected by those using multiple independently evolving genetic markers, suggesting that convergent evolution or poor character scoring was responsible for these hypothesized relationships.

The early evolutionary history of snakes inferred from the fossil record portrays a fascinating story about the independent evolution of limb reduction in serpents. The earliest identified snake, *Najash rionegrina,* found in Upper Cretaceous deposits in Argentina, was a small terrestrial or burrowing serpent with sacral vertebrae, pelvic elements and hindlimbs (Apesteguia and Zaher 2006). This study conflicts with some theories that suggest snakes (along with their adaptive limb reduction) originated in aquatic habitats, as this earliest snake fossil provides solid evidence for a burrowing/terrestrial origin of snakes. Other Cretaceous fossils, *Pachyrhachis problematicus, Haasiophis terrasanctus,* and *Eupodophis descouensi* are all shallow marine species from Northern Gondwana, found along the Tethyan Coast. These three taxa have hindlimb bones but lack differentiated sacral vertebrae for anchoring pelvic elements (Caldwell and Lee 1997; Tchernov *et al.* 2000; Rage and Escuillié 2000). Moreover, Apesteguia and Zaher (2006) using phylogenetic analyses of morphological data demonstrated that these three fossil taxa do not represent the earliest snakes but are rather nested within the radiation of macrostomatan snakes (see our discussion on Alethinophidia for an alternate view of macrostomatan monophyly). Along with some extant groups (e.g., scolecophidians, boids, pythonids and aniliids), these fossils show that complete limb loss has occurred independently throughout the early evolution of snakes.

All extant snakes share a series of characters including absence of the pectoral girdle and forelimbs. However, remnants of the pelvic girdle are found in various groups including scolecophidians, pythonids, boids, and aniliids. Cloacal spurs appear in boids, pythonids and aniliids (McDowell 1987; Cundall *et al.* 1993). The elongated features of snakes are due to an increase in vertebrae ranging from 120 to 500. Like lizards, all snakes are covered in scutes, with ventral scales extending from the throat to the tail tip, and genitalia are either a single or bilobed organ referred to as a hemipenis. Characters (or character states) unique to snakes include: a supraoccipital that is excluded from the border of the foramen magnum by the exoccipitals, down growths of the parietal bones enclose the ophthalmic branch of the trigeminal nerve which enters the orbit through the optic foramen, the size of the left arterial arch is greater than the right (the reverse is found in most tetrapods), flexible ligamentous connection between dentaries, and a lack of ciliary muscles in the eyes. Many other characters (e.g., characters responsible for increasing gape) appear only as derived conditions in certain groups of snakes (Underwood 1967; McDowell 1987; Pough *et al.* 2004; Vitt and Caldwell 2009).

Although the classification of extant snakes began with Linnaeus in 1758 and received various rearrangements by herpetological luminaries like Duméril (1853), Cope (1894, 1895), Boulenger (1896) and Hoffstetter (1946, 1962), most modern treatments of taxonomy can be traced to Underwood (1967). Since then, numerous studies and lists have been produced attempting to classify snakes. Many of these studies chart the rise of modern computational and molecular systematics (immunological or DNA hybridization). However, our basic treatment of major snake taxonomy in this chapter will primarily be discussed in the context of molecular DNA sequence, character based systematics, while occasionally referring to concordant morphological data.

Among extant snakes, the basal divisions occur between the scolecophidians and the alethinophidians (Rage 1984; Cundall *et al.* 1993; Dessauer *et al.* 1987; Vidal and Hedges 2002; White *et al.* 2005; Burbrink and Pyron 2008; Wiens *et al.* 2008; Eckstut *et al.* 2009), although other classification schemes have been presented (Vidal *et al.* 2009). Outside of the purview of most neontologists are the large number of extinct families and genera of snakes known because of the dedicated work of a few paleoherpetologists. Many of these taxa cannot be confidently placed within the phylogeny of extant serpents because of the absence of various characters or convergence in states, not to mention the obvious complete lack of DNA data. We underscore the importance of these fossils in understanding the area and dates of origins of snakes, as well as morphological changes through time. We also realize that the correct placement of many of these taxa may actually help better understand relationships among extant families. Several of these families, including Palaeopheidae, Dinilysiidae, Nigerophiidae, Lapparentopheidae, Simoliopheidae, Pachyophiidae, Russellopheidae, and several unincorporated genera, *Eupodophis, Goniophis,* likely represent extinct alethinophidians, but the exact position remains contentious (Lydekker 1888; Nopcsa 1923; Romer 1956; Hoffstetter 1961; Rage 1975;1984; McDowell 1987; Holman and Case 1988; Carroll 1988; Holman *et al.* 1991; Averianov 1997; Caldwell and Lee 1997; Khajuria and Prasad 1998; Nessov *et al.* 1998; Zaher 1998; Lee *et al.* 1999; Rage and Werner 1999; Zaher and Rieppel 1999; Caldwell 2000; Lee and Scanlon 2002; Sepkoski 2002; Zaher and Rieppel 2002; Caldwell and Albino 2003; Rieppel and Head 2004; Head *et al.* 2005; Parmley and DeVore 2005; Scanlon 2006; Head *et al.* 2009). A recent study by Apesteguia and Zaher (2006) places *Najash* as the sister taxon to all serpents, Dinilysiidae (*Dinilysia*) as the sister group to all Alethinophidia (including Aniliidae) and Pachyophiidae (represented by *Haasiophis, Pachyrhachis,* and *Eupodophis*) unresolved within a group containing colubroids, pythonids, boids, loxocemids and xenopeltids (Zaher 1998; Zaher and Rieppel 2002). Furthermore, these authors placed the giant extinct snake, *Wonambi,* as sister to the Boidae. This genus along with *Yurlunggur,* are placed within Madtsoiiidae by Scanlon and Lee (2000) and Scanlon (2006). In contrast to Apesteguia and Zaher (2006), Caldwell (2000), Lee and Scanlon (2002) suggest that Madtsoiiidae, *Pachyrhachis, Haasiophis* and *Dinilysia* fall outside the most common ancestor of Scolecophidia and Alethinophidia, are all subtended by the basal nodes in the tree of serpents, and undermine the concept of Macrostomata. We leave the subject of fossil snakes to now focus on the major groups and families within extant snakes.

2.2 SCOLECOPHIDIA

The blind snakes are easily recognized as generally small uniformly scaled snakes, that superficially resemble worms more than they do their sister group, alethinophidian snakes. All scolecophidians retain pelvic elements but display no external limb vestiges. A large number of morphological synapomorphies for this group have been discussed by several authors (McDowell 1987; Rieppel 1988; Cundall *et al.* 1993; Holman 2000; Lee and Scanlon 2002; Vitt and Caldwell 2009). Three families, Anomalepididae, Leptotyphlopidae, and Typhlopidae traditionally represent the scolecophidians. Morphological support for a most recent common ancestor for these three families is large and includes multiple premaxillary foramina, a fenestra for the duct of the Jacobson's organ that opens posteroventrally as well as 27 other characters (Lee and Scanlon 2002). Unfortunately, it is not clear how many of these traits are simply associated with burrowing or simply represent independently evolved states. Therefore, a real possibility is that morphology is overstating support for a monophyletic Scolocophidia. Contrary to the idea that morphology supports monophyly, Cundall and Irish (2008) state "The jaw elements of scolecophidians provide a strong argument in favor of paraphyly." For more information concerning either advanced or primitive morphological characters that separate Scolecophidia from Alethinophidia, see McDowell (1967, 1974, 1987), Cundall *et al.* (1993) and Lee and Scanlon (2002). Recent molecular studies are conflicting with regard to a monophyletic Scolecophidia. Macey and Verma (1997), Vidal and Hedges (2002), Lee *et al.* (2007), suggested they were monophyletic, but Heise *et al.* (1995), Forstner *et al.* (1995), Eckstut *et al.* (2009), Vidal *et al.* (2009) and Wiens *et al.* (2008) all inferred the Scolecophidia to not be monophyletic. Pyron and Burbrink (unpublished data) using species tree methods from 25 loci revealed a sister relationship between Typhlopidae and the remainder of all snakes, with Leptotyphlopidae appearing sister to the group containing Anomalepididae and the Alethinophidia (Fig. 2.1). The combined morphological and molecular analysis of snake relationships in White *et al.* (2005) also inferred a paraphyletic Scolecophidia but Lee *et al.* (2007) indicated that Scolecophidia are monophyletic. Bowing to historical inertia and for ease of discussion, we treat Scolecophidia as monophyletic here, but realize there is considerable uncertainty about this assumption.

The most species rich group of scolecophidians, Typhlopidae, are represented by nine genera and 232 taxa (Reptile Database) and mostly occur in the tropical regions of the world, although two species are found in North America and one in Europe (McDiarmid *et al.* 1999). This group has a toothless dentary as well as 12 other states listed in Lee and Scanlon (2002).

Leptotyphlopidae are found in the tropics and subtropics of Africa and the Americas as well as southwest Asia and are composed of 116 species. They are represented by two genera, although in a rare study on the phylogenetics of any scolecophidian groups using molecular data, Adalsteinsson *et al.* (2009) divided leptotyphlopids into 12 genera. Leptotyphlopidae may be the sister family to the other scolecophidian groups and is distinguished from them by having 11 unique character states, including a toothless maxilla (McDowell 1987; Lee and Scanlon 2002).

The most range restricted group, Anomalepididae, is found in southern Central America and South America. Represented by only 17 species and four genera (Reptile Database; McDowell *et al.* 1999), anomalepidids can be diagnosed by 18 diagnostic character states, including a toothed a maxilla and dentary as well as absence of all pelvic vestiges (see McDowell 1967; McDowell 1987; Lee and Scanlon 2002; Pough *et al.* 2004).

Although the oldest scolecophidian fossils are from the Paleocene (Folie 2006), molecular divergence dating has suggested that the group originated in the early Cretaceous or late Jurassic (Burbrink and Pyron 2008; Vidal *et al.* 2009), a time frame deduced by White *et al.* (2005) based on minimal fossil ages and constrained by phylogeny. Given that the first appearance of a fossil probably underestimates the actual date of origin for the group, it is likely that molecular dating might provide a more realistic estimate of the origin of any group of organisms. The downside to estimating molecular dates of origin is that all of the inferences discussed here assume some very realistic and large quantity of error around the mean date. Yet, it is encouraging that the molecular dates and the deduced dates are similar. However, please consult the original articles where estimated of dates of divergence are concerned.

All scolecophidians are oviparous (although delayed egg deposition is known from *Typhlops squamosus*). The often-introduced typhlopid *Ramphotyphlops braminus* is parthenogenic. Most scolecophidians are fossorial (although some exceptions are known) and consume termites, ants or the eggs and larvae of these prey (Webb *et al.* 2000; Vitt and Caldwell 2009).

2.3 ALETHINOPHIDIA

The remainder of extant snakes belongs to Alethinophidia, and for the most part, these are the serpents with which people are most familiar. They are generally differentiated from the scolecophidians by possessing a well developed squamosal bone that articulates with the quadrate and brain case (absent in Uropeltidae) and lacking or having a small coronoid bone and vertebrae possessing a neural spine (lacking in Uropeltidae). McDowell (1987) provides a detailed review of the anatomy of this group. Fossil records for alethinophidians date to the mid-Cretaceous (Rage and Werner 1999), although the origin of this group has been suggested to have occurred in the late Jurassic or early Cretaceous (White *et al.* 2005). Recent studies using relaxed molecular clocks also indicate they diverged from a common ancestor with scolecophidians around that time (Burbrink and Pyron 2008; Vidal *et al.* 2009).

Based on various morphological studies and combined morphological and molecular phylogenetic analyses (Rieppel 1988; Lee and Scanlon 2002; Lee *et al.* 2007), alethinophidians typically have been divided into Anilioidea (Aniliidae, Cylindrophiidae, Uropeltidae and Anomochilidae) and Macrostomata (Pythonidae, Boidea, Colubroidea and Acrochordidae). However, several molecular and morphological studies have demonstrated this to be in error and conditions that increase gape in macrostomatan genera have either evolved numerous times or have been lost several times, resulting in paraphyletic classifications (Cadle *et al.* 1990; Slowinski and Lawson 2002; Wilcox *et al.* 2002; Lawson *et al.* 2004; Gower *et al.* 2005; Wiens *et al.* 2008; Eckstut *et al.* 2009; Vidal *et al.* 2009).

Molecular phylogenetic studies have shown support for an initial division in Alethinophidia occurring in the later half of the Cretaceous, which sometimes join the Aniliidae and two genera of the Tropidopheidae (*Tropidophis* and *Trachyboa*; the other two genera *Ungaliophis* and *Exiliboa* are related to the Boidea; Wilcox *et al.* 2002; Vidal and Hedges 2002; Lawson *et al.* 2004; Burbrink and Pyron 2008; Wiens *et al.* 2008; Eckstut *et al.* 2009; Vidal *et al.* 2009). The remainder of alethinophidians, the second division, includes

Pythonidae (with the closely related Loxocemidae and Xenopeltidae) and Boidae, which also encompass Ungaliopheidae (*Exiliboa* and *Ungaliophis*), erycine boids and *Calabaria* (Vidal and Hedges 2002; Burbrink and Pyron 2008; Wiens *et al.* 2008; Eckstut *et al.* 2009; Vidal *et al.* 2009). Also, within this second division are the massively diverse caenophidians (Acrochordidae and Colubroidea) as well as Bolyeriidae, Xenophidiidae, Uropeltidae, Cylindrophiidae and Anomochilidae. Pyron and Burbrink (unpublished data) demonstrated, using species tree methods, that the initial division within Alethinophidia divided caenophidians and the remainder of alethinophidians including Aniliidae and Tropidopheidae. This later group was commonly referred to as Henophidia (Fig. 2.1).

In two recent molecular studies using numerous mtDNA and nDNA genes, Bolyeriidae and Uropeltidae (including Cylindrophiidae) are sometimes considered to be removed from the clade containing pythonids, loxocemids, xenopeltids and boids (Vidal *et al.* 2009) or of uncertain position within this second division of alethinophidians, but with a possible sister relationship between the boids and bolyeriids (Wiens *et al.* 2008). Pyron and Burbrink (unpublished data) showed a clade containing pythonids, loxocemids and xenopeltids as the sister group to a clade containing Boidea (and Calabariidae), Bolyeriidae and Uropeltidae (Fig. 2.1). By examination of morphological characters, the extremely rare Xenophidiidae, found only in peninsular Malaysia and Borneo, has been proposed to be closely related to various groups including colubroids, aniliids, tropidopheids or boids (Günther and Manthey 1995; Wallach and Günther 1998). After obtaining a rare, but decayed tissue sample, Lawson *et al.* 2004, demonstrated from only a single gene that xenophidiids are closely related to bolyeriids, which in turn may be related to pythonids or boids. Finally, the remaining group, Anomochilidae, may not actually deserve family ranking. In a study using 12S and 16S DNA sequences, *Anomochilus*, representing the family Anomochilidae, was found to be closely related to Cylindrophiidae, which rendered the genus *Cylindrophis* paraphyletic (Gower *et al.* 2005).

2.3.1 Aniliidae

The South American pipesnakes are a monotypic family composed of a single species, *Anilius scytale* (McDiarmid *et al.* 1999). This species is found throughout tropical northern South America (Greene 1997). This viviparous species superficially resembles the bi-colored coral snakes, however, it lacks a distinctly differentiated head and neck region and has only a single scale covering each eye. Additionally, femurs are present as cloacal spurs and remnants of pelvic elements are found in the musculature of the trunk (McDowell 1987). A large number of morphological characters (~28) appear to diagnose this monotypic family (Underwood 1967; McDowell 1987; Lee and Scanlon 2002; Pough *et al.*, 2004; Vitt and Caldwell 2009). The species is usually smaller than one meter and occurs in tropical forest litter and near water. They are viviparous and generally give birth from 4 to 18 young in

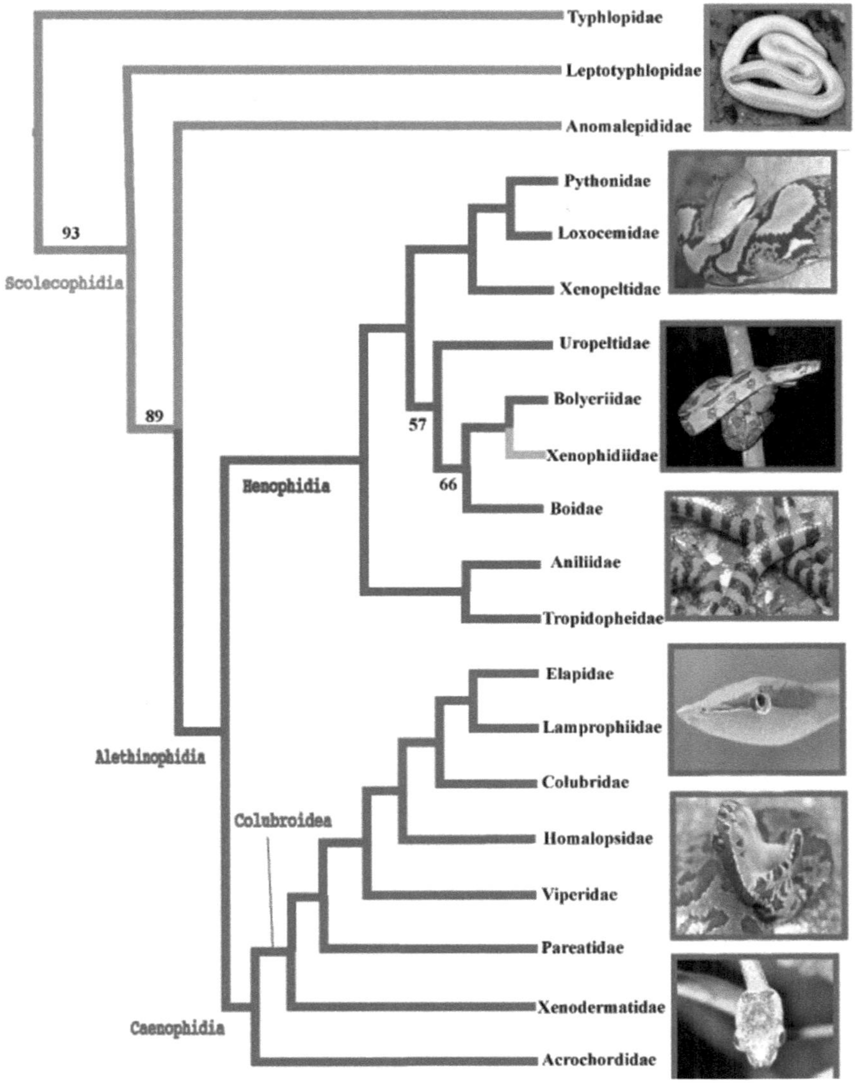

Fig. 2.1 Phylogenetic relationships among snake families and higher level groups using species tree methods (Pyron and Burbrink, unpublished data). Posterior probability support is greater than 95% for all nodes unless indicated otherwise. While Scolecophidia is designated on this tree it was not found to be monophyletic. Taxa illustrated and photo credits from top to bottom: *Leptotyphlops brasiliensis* (Jalapão National Park, Tocantis, Brazil, by Donald Shepard); *Python reticulatus* (Danum Valley, Sabah, Borneo, by Frank Burbrink); *Boa constrictor* (Tortuguero, Costa Rica, by Frank Burbrink); *Anilius scytale* (Cristalino River near Alta Floresta, Mato Grosso, Brazil, by David Shepard); *Oxybelis fulgidus* (Tortuguero, Costa Rica, by Frank Burbrink); *Agkistrodon piscivorus* (Florida, USA, by Frank Burbrink); *Aplopeltura boa* (Danum Valley, Sabah, Borneo, by Frank Burbrink).

Color image of this figure appears in the color plate section at the end of the book.

either the wet or dry season (Martins and Oliveira 1999; Cisneros-Heredia 2005; Maschio *et al.* 2007). Molecular divergence dating indicates that this family likely originated at the K/T boundary (Burbrink and Pyron 2008).

2.3.2 Tropidopheidae

Once considered to have been composed of four genera, Tropidopheidae, now only includes Tropidopheinae and contains only two genera, *Tropidophis* and *Trachyboa*, totaling 23 species (Zaher 1994; Wilcox *et al.* 2002; Lawson *et al.* 2004; Gower *et al.* 2005; Eckstut *et al.* 2009; Reptile Database). These moderate to small snakes are found in the West Indies, Central America and South America. During the late Cretaceous or early Tertiary they diverged from a recent common ancestor with the New World aniliids (Schwartz and Henderson 1991; Tolson and Henderson 1993; Wallach and Günther 1998; Burbrink and Pyron 2008; Vitt and Caldwell 2009). Unlike aniliids, these terrestrial/arboreal snakes share the macrostomatan skull condition and have edentulous premaxillaries. Tropidopheids still retain some pelvic elements. Morphological characters discerning tropidopheines and ungaliopheines (now in Boidae), including parallelization of hyoid horns, are described in Zaher (1994). Tropidopheids primarily feed on lizards and other small vertebrates, are viviparous and are recorded to have two to 12 young (Henderson and Powell 2009). Relationships among ~50% of the species were examined in Wilcox *et al.* (2002).

2.3.3 Uropeltidae

This family, which should also include Cylindrophiidae and the single species of Anomochilidae (Gower 2005), represents a radiation of southern and southeastern Asian non-macrostomatan alethinophidians. Occasionally, the family is considered to be a superfamily composed of Uropeltidae, Cylindrophiidae and Anomochilidae (Reptile Database). However, given that *Cylindrophis* is rendered paraphyletic by *Anomochilus*, the most conservative approach to the classification of this group would be to abandon all separate families except Uropeltidae. If we assume Uropeltidae includes all three families, then it is composed of 10 genera and 62 species. The monophyly of the family for at least three genera (*Cylindrophis*, *Rhinophis*, and *Uropeltis*) was found in Eckstut *et al.* (2009), which supports an early molecular evolution study on this group (Cadle *et al.* 1990). These unusual secretive snakes are distributed in southern India and southeastern Asia. The diet appears to vary in this eclectic group, from earthworms in the uropeltines to larger elongate prey like eels and snakes in *Cylindrophis* (Murphy *et al.* 1999). The closely related *Anomochilus* and *Cylindrophis* still retain pelvic elements with cloacal spurs, whereas all other genera (in the restricted family Uropeltidae) have no limb elements. The stem uropeltids originated in the late Cretaceous or early Tertiary (Burbrink and Pyron 2008; Vidal *et al.* 2009). All members of Uropeltidae appear to be fossorial with the uropeltines possessing biochemical specializations that allow continuous muscle activity for borrowing (Gans *et al.* 1978), and prefer to forage at night on the surface or in loose soil. This habitat preference is suggested for *Anomochilus*, although given the rarity of this species (known from less than one dozen specimens) their lifestyle has yet to be confirmed. All species of uropeltids are viviparous, except *Anomochilus*.

2.3.4 Bolyeriidae

The Mascarene boas found only on Mauritius Island and surrounding islets, contain only two genera and two species, *Bolyeria multocarinata* and *Casarea dussumieri* (McDiarmid *et al.* 1999). A key defining feature in these snakes is their intramaxillary joint. Unique for this enigmatic group, the maxillary is divided into an anterior and posterior section, presumably as an adaptation for feeding on skinks (Bullock 1986; Cundall and Irish 1986; 1989; Wallach and Günther 1998). The stem members of this group probably diverged in the late Cretaceous (Vidal *et al.* 2009). Presumably *Bolyeria* became extinct in the 20th century (Bullock 1986). *Casarea* is apparently oviparous and reproduction in *Bolyeria* is unknown (Cundall and Irish 1989; Vitt and Caldwell 2009).

2.3.5 Xenophidiidae

The family Xenophidiidae is known from only two species, *Xenophidion acanthognathus* and *Xenophidion schaeferi*, found in Sabah, Borneo and Peninsular Malaysia, respectively (Günther and Manthey 1995; Wallach and Günther 1998). Prior to molecular analyses, it had been suggested this family was related to aniliids, tropidopheids, boids, or colubroids. Modern molecular phylogenetic analyses demonstrated that it might be the sister group to Bolyeriidae, or was at least likely to be part of a clade containing bolyeriids, uropeltids, loxocemids, xenopeltids and pythonids (Lawson *et al.* 2004). Dates of origin for this group are unknown, but given their placement among alethinophidians and particularly close relationship with bolyeriids, it is likely that they originated in the later Cretaceous or early Tertiary. Spinejaw snakes are known to live in rainforest habitats and the single female specimen found in Borneo contained large shelled eggs. Presumably the large tooth on the anterior portion of the mandible is used to secure struggling prey, perhaps small vertebrates (Cundall and Irish 1986, 1989; Vitt and Caldwell 2009).

2.3.6 Loxocemidae

The Mexican Burrowing Python (*Loxocemus bicolor*), the only living member of the family, is found from Costa Rica to southeastern Mexico (McDiarmid *et al.* 1999; Reptile Database). Their taxonomic position has been discussed by various authors using morphological data (Haas 1955; Underwood 1967), but based on a review of recent literature on molecular systematics, it is quite clear *L. bicolor* is the sister species to all Old World pythons (Fig. 2.1; Slowinski and Lawson 2002; Wilcox *et al.* 2002; Lawson *et al.*

2004; White *et al.* 2005; Burbrink and Pyron 2008; Wiens *et al.* 2008; Eckstut *et al.* 2009; Vidal *et al.* 2009). However, using combined molecular and morphological data, Lee *et al.* (2007) demonstrated that *Loxocemus* was sister to *Xenopeltis,* and this clade was sister to pythonids. Also demonstrating that they are the sister group to pythonids, Burbrink and Pyron (2008) estimated these two groups shared a common ancestor in the Eocene. Like pythonids, *Loxocemus* has remnant femurs represented as cloacal spurs as well as vestigial pelvic elements (Wilson and Meyer 1985; Savage 2002). *Loxocemus* can attain lengths greater than one meter, though usually they are smaller. They live in tropical and subtropical forests and appear to be fossorial or at least secretive and terrestrial. This species is nocturnal and feeds on small mammals or lizards. *Loxocemus* is oviparous and lays clutches of four large eggs (Odinchenko and Latyshev 1996; Greene 1997; Savage 2002).

2.3.7 Xenopeltidae

Sunbeam snakes are known from two species, *Xenopeltis unicolor* and *X. hainanensis,* known from southern and southeastern China (McDiarmid *et al.* 1999). This family is the sister group to a clade containing pythonids and loxocemids (Slowinski and Lawson 2002; Wilcox *et al.* 2002; Burbrink and Pyron 2008; Wiens *et al.* 2008; Eckstut *et al.* 2009; Vidal *et al.* 2009) and likely diverged from a common ancestor with pythonids and loxocemids in the early Eocene. These snakes have distinctly iridescent scales but lack any pelvic or limb elements. *Xenopeltis unicolor* often occur in rainforests, human modified habitats (e.g., rice fields) and coastal areas. Adults are generally smaller than 1.5 meters and burrow in mud and forage for lizards, snakes and frogs either in the daytime or nighttime. They are oviparous and lay clutches generally smaller than 17 eggs (Cox 1991).

2.3.8 Pythonidae

This family includes nine genera (*Aspidites, Antaresia, Apodora, Bothrochilus, Broghammerus, Leiopython, Liasis, Morelia,* and *Python*) and 38 species found in the Old World, mostly tropical regions (Schleip and O'Shea in review; Reptile Database). Pythonids are generally large snakes with teeth on their premaxillaries (except *Aspidites*; Frazetta 1975) and a low (or lack of) supraoccipital crest (Underwood 1967; Kluge 1991). They have vestigial limb elements (cloacal spurs) and remnants of pelvic elements. Additionally they have no tracheal lung but possess a large left lung. Pythons are oviparous and females usually coil around egg clutches. True brooding is associated with *Python molurus* in order to maintain incubating temperatures by increasing body temperatures (Van Mierop and Barnard 1976, 1978). A combined mtDNA and morphological study on python phylogenetics has demonstrated a basal split that subtends one lineage of Afro-Asian pythons (*P. regius, P. brongersmai, P. sebae* and *P. molurus*) and another which includes the sister species *P. reticulatus* and *P. timorensis* and the remainder of the Indo-Australian species, indicating the genus *Python* is paraphyletic (Rawlings *et al.* 2008). Divergence dates suggest that the stem group likely originated in the early to mid-Tertiary (Noonan and Chippindale 2006; Burbrink and Pyron 2008; Rawlings *et al.* 2008). Additionally, Rawlings *et al.* (2008) demonstrated a four-fold decrease in diversification 45 Ma, with the last speciation events taking place prior to the Pliocene.

2.3.9 Boidae

This family is divided into Boinae and Erycinae. Boinae are composed of 28 species, with two genera occurring in Madagascar (*Acrantophis* and *Sanzinia*), one in southeastern Asia (*Candoia*) and six in the New World tropics (*Boa, Corallus, Epicrates, Eunectes, Exiliboa,* and *Ungaliophis*). Erycinae are composed of 14 species, with one genus in North America (*Charina*) and four genera found in Africa, the Middle East and Europe (*Calabaria, Charina, Eryx* and *Gonglyophis*) (McDiarmid *et al.* 1999; Reptile Database). In contrast to Kluge (1991) molecular studies have all shown that the New World and Madagascar Boinae each form monophyletic groups (Burbrink 2005; Noonan and Chippindale 2006). Burbrink (2005) demonstrated that the New World boines (*Boa, Corallus, Epicrates* and *Eunectes*) are monophyletic, while Noonan and Chippindale (2006) demonstrated that *Calabaria* is the sister species to a clade containing *Acrantophis* and *Sanzinia*, which is in turn sister to a group containing three major geographic radiations; Neotropical (*Corallus, Epicrates, Eunectes* and *Boa*) sister to a Pacific/African/Indian group (*Eryx* and *Candoia*) and a North/Central American group (*Exiliboa, Lichanura* and *Charina*). Eckstut *et al.* (2009) found *Calabaria* as the sister to the rest of the boids and both *Ungaliophis* and *Exiliboa* were nested within the boid clade, as suggested by Zaher (1994), contra Wilcox *et al.* (2002).

Stem members of what we include as Boidea most likely originated in the late Cretaceous (White *et al.* 2005; Noonan and Chippendale 2006; Burbrink and Pyron 2008; Vidal *et al.* 2009). Interestingly, Noonan and Chippindale (2006) demonstrated that all diversification events, even among sister species, within boidae occurred prior to the Neogene. Head *et al.* (2009) discovered one of the earliest boid fossils. *Titanoboa cerrejonesis* was found in deposits 58-60 Ma in the Cerrajon Basin in Colombia, and is expected to have attained the massive size of 13 meters (Head *et al.* 2009).

Boids have edentulous premaxillaries, a coronoid bone, a strongly developed supraoccipital crest, and like pythons have remnant pelvic elements and femurs represented as cloacal spurs. All boids are viviparous and exhibit a large range of litter size (Vitt and Caldwell 2009).

2.4 CAENOPHIDIA

2.4.1 Acrochordidae

This family is composed of a single genus, *Acrochordus*, with three species, *A. arafurae, A. granulatus*, and *A. javanicus*. The filesnakes snakes occur in southern and southeastern Asia as well as Australia. Nearly all modern molecular studies (Lawson *et al.* 2005; Wiens *et al.* 2008; Eckstut *et al.* 2009; Vidal *et al.* 2009; except Kelly *et al.* 2003 who inferred the acrochordids as sister to the Xenodermatidae) have demonstrated that this group represents the sister taxon to the massive superfamily Colubroidea. The placement of Acrochordidae as sister to Colubroidea is supported by a large suite of morphological characters as well (Rieppel 1988, Cundall *et al.* 1993, Scanlon and Lee 2000; Lee and Scanlon 2002). Fossils of putative acrochordids are known from the early Miocene (Head *et al.* 2007) and molecular dates suggest this group originated in the late Cretaceous or early Tertiary (Burbrink and Pyron 2008). This highly aquatic snake is covered with baggy skin in small nonoverlapping, granular scales. No limb or pelvic elements are present (Vitt and Caldwell 2009). These large snakes (ranging from 1.0–2.7 m) are usually found in marine or brackish water and primarily feed on fish (Shine 1986). All species of *Acrochordus* are viviparous and generally give birth from 4 to 40 young in the water. It is thought they reproduce less frequently than other snakes (Shine 1986) and there is evidence that occasionally some females of *A. arafurae* exhibit parthenogenesis (Dubach *et al.* 1997).

2.4.2 Colubroidea

This largest clade of snakes represents 85% of all serpent species and is composed of ~2670 taxa (Reptile Database). This superfamily occurs on every continent (excluding Antarctica) and likely are the most commonly encountered snakes (particularly in North America). It contains all dangerously venomous and medically important snakes and many families have taxa that occupy a wide variety of niches, including arboreal, terrestrial, fossorial, temperate, tropical desert and oceanic habitats (Pough *et al.* 2004). Among a series of characters not possessed by Colubroidea, McDowell (1987) suggested that they are a morphologically distinct superfamily all possessing distinctive rib ends and unique cranioquadrate muscles (Haas 1973; Rieppel 1980). Lee and Scanlon (2002) diagnose this group with only eight morphological characters, including a lack of vomerine flaps, poorly developed or a complete lack of a coronoid process, and intercostal arteries that arise from the dorsal aorta at intervals which span multiple body segments (Wallach and Günther 1998; Pough *et al.* 2004).

A detailed treatment on the taxonomic history of this group is beyond the scope of this chapter. However, we note that a large number of molecular studies that have changed the taxonomy of this group have seen print in the new millennium (Slowinski and Keogh 2000; Kelly *et al.* 2003;

Lawson *et al.* 2005; Vidal *et al.* 2007; Pinou *et al.* 2004; Nagy *et al.*, 2003; Vidal *et al.* 2007; Eckstut *et al.* 2009; Kelly *et al.* 2009; Zaher *et al.* 2009). Some of these have made radical changes to the taxonomy of this group relative to Dowling and Duellman (1978) and Zaher (1999). We take a conservative approach to the taxonomy of colubroid snakes in this chapter and primarily use the classification presented in Lawson *et al.* (2005), which minimizes the number of name changes, such as the retention of Colubroidea as the name for the sister clade to the Acrochordidae [as opposed to, for example, Colubroides (Zaher *et al.* 2009)]. We attempt to strike a balance between long-used taxonomic schemes and molecular phylogenetic estimates. Therefore, given recent evidence presented in Vidal *et al.* (2007), Eckstut *et al.* (2009), Kelly *et al.* (2009), Zaher *et al.* 2009 and Pyron *et al.* (in press), we have made a few modifications from Lawson *et al.* (2005; see Colubridae and Lamprophiidae).

Gone are the days where Colubroidea was nicely divided into four families: Colubridae, Viperidae, Atractaspididae and Elapidae (e.g., Pough *et al.* 2004). All molecular studies have shown that this classification is paraphyletic and in keeping with a Linnaean based hierarchy, this means that other subfamilies have been elevated to familial level. Based on congruence among the multiple studies mentioned previously, we recognize the following seven families and subfamilies (in parentheses): Xenodermatidae, Homalopsidae, Pareatidae, Colubridae (Calamariinae, Colubrinae, Natricinae, Pseudoxenodontinae, and Dipsadinae), Elapidae (Elapinae and Hydrophiinae), Lamprophiidae (Atractaspidinae, Lamprophiinae, Psammophiinae and Pseudoxyrhophiinae), and Viperidae (Azemiopinae, Crotalinae, and Viperinae). This classification limits the proliferation of unnecessary familial ranks but we fully realize that these taxonomic proposals are subject to future testing, like any good scientific hypothesis.

We use the traditional definition of Colubroidea in this chapter. Colubroidea, which is sister to Acrochordidae, includes the families Colubridae, Elapidae, Homalopsidae, Lamprophiidae, Pareatidae, Viperidae, and Xenodermatidae and takes historical precedence (Romer 1956) over other definitions. It is still widely used by systematists, ecologists, conservationists and ethologists (e.g., Dowling and Duellman 1978; Greene 1997; Zaher 1999; Lawson *et al.* 2005; Wiens *et al.* 2008; Vitt and Caldwell 2009). This view is in contrast to Vidal *et al.* (2007) and Zaher *et al.* (2009), who redefine Colubroidea to include only Colubridae (sensu Lawson *et al.* 2005). Concomitantly, these authors elevated the subfamilies of Colubridae to the family level (i.e., Calamariidae, Colubridae, Pseudoxenodontidae, Natricidae, and Dipsadidae [or Xenodontidae]), which required that Colubridae be ranked as a superfamily (Colubroidea). Moreover, along with Pinou *et al.* (2004) they named the node Elapoidea to include Elapidae and Lamprophiidae. Realistically, there is no phylogenetic justification for recognizing these traditional colubrid subfamilies as distinct families, changing the long-standing definition of Colubroidea, or naming Elapoidea to be ranked alongside their newer definition of Colubroidea.

Therefore, we retain the traditional meaning of Colubroidea, and maintain the subfamilies within Colubridae in the remainder of this chapter.

Colubroidea may have diverged from their sister group, Acrochordidae during the early Tertiary. White *et al.* (2005) suggested a late Cretaceous divergence of the Colubroidea. Interestingly, Burbrink and Pyron (2008) showed that an origin of the Colubroidea prior to the K/T boundary at 65 MA is likely when accounting for skewed estimates of divergence dates. This suggests that colubroids survived the cataclysm that effectively ended 76% of life on Earth (Pope *et al.* 1998). There is little doubt that the diversification of families, subfamilies within Colubroidea occurred throughout the Tertiary (Rage 1987; Holman 2000; Burbrink and Pyron 2008; Vidal and Hedges 2009)

2.4.3 Colubridae

This family once contained about 63% of all snake species (Pough *et al.* 2004) but now may contain as few as just over 100 genera (Zaher *et al.* 2009). The former Colubridae is at the heart of most higher-level taxonomic changes within snakes. Of the four original colubroid snake families (e.g., Viperidae, Elapidae, Atractaspididae and Colubridae), it was clear from various molecular phylogenetic studies that Atractaspididae and Elapidae shared a most recent common ancestor with certain colubrid groups (Psammophiinae, Pseudoxyrhophiinae, Lamprophiinae [formerly Boodontinae and Pseudoxyrhophiinae]). Additionally, other colubrid subfamilies (e.g., Pareatinae, Xenodermatinae, and Homalopsinae) fell well outside the traditional Colubridae, which required familial ranking for these groups (e.g., Cadle 1994; Vidal and Hedges 2002; Kelly *et al.* 2003; Nagy *et al.* 2003; Lawson *et al.* 2005; Vidal *et al.* 2007; Eckstut *et al.* 2009; Zaher *et al.* 2009; Pyron *et al.* in press). Many authors have proposed various taxonomic schemes that have yet to be completely accepted (e.g., Kelly *et al.* 2003; Lawson *et al.* 2005; Vidal *et al.* 2007; Zaher *et al.* 2009) but which at least show similar groupings (albeit with different names). The challenge here is to provide a classification that bridges all the recent phylogenies. As noted above, we take a more conservative approach than recent classifications (e.g., Zaher *et al.* 2009), but nonetheless we believe it reflects the congruent features of the recent phylogenies. Therefore, the original Colubridae should be divided into Pareatidae, Homalopsidae, Xenodermatidae, and Lamprophiidae (containing Lamprophiinae, Atractaspidinae, Psammophiinae, and Pseudoxyrhophiinae). We discuss the remainder of Colubridae in this section.

There is clear congruence among phylogenies for the contents of the clades within what we call the Colubridae: Pseudoxenodontinae, Calamariinae, Dipsadinae (formerly Xenodontinae in Lawson *et al.*, 2005), Natricinae, and Colubrinae (Kelly *et al.* 2003; Lawson *et al.* 2005; Vidal *et al.* 2007; Eckstut *et al.* 2009; Kelly *et al.* 2009; Zaher *et al.* 2009). However, the relationships among these groups remain uncertain, with the exception of Calamariinae and Colubrinae. Even this latter relationship is unclear because in some cases Calamariinae render Colubrinae paraphyletic. Even though the content of the Colubridae is reduced, it still is global in scope and exhibits great diversity.

Given these recent taxonomic changes, diagnosing Colubridae using morphological characters has yet to be widely discussed. Zaher *et al.* (2009) diagnose Colubridae (their Colubroidea) exclusively with hemipenal characters and noted that one of these characters, calyces on the hemipenal lobes, has been lost in the natricines.

Natricinae (approximately 33 genera and 207 species) are found on every habitable continent (and Indoaustralian islands) except for South America. They occupy aquatic, mostly freshwater but some brackish and coastal waters, semi-fossorial, and terrestrial habitats. They exhibit a wide diet diversity, including fish and amphibians and some show unusual specialization on prey like slugs, earthworms, and crawfish (a crustacean). Reproductive modes include viviparity (all North American species are viviparous) and oviparity and, interestingly, one taxon, *Tropidonophis mairii* represents one of the only confirmed cases of multi-clutching by a single female during the reproductive season (Brown and Shine 2002). A molecular phylogeny of the entire subfamily using a majority of recognized genera awaits publication, but de Queiroz *et al.* (2002) have examined relationships among and within some New World genera, particularly *Thamnophis*.

Calamariinae is a group of seven genera and 82 species that are distributed in southeastern Asia and the Indonesian-Malaysian islands. Based on their head morphology and other characters as well as field observations it is thought that these snakes are probably fossorial. They are all oviparous (Greene 1997). A molecular phylogeny of these genera has yet to be published and very few specimens are ever included within any molecular phylogeny of Colubroidea.

The smallest clade of colubrids is the Pseudoxenodontinae, composed of two genera (*Plagiopholis* and *Pseudoxenodon*) and 10 species. The species range through southeastern Asia, including one in India, and occupy islands in Indonesia and Malaysia. This subfamily was erected by McDowell (1987). Based on the hemipenes, McDowell (1987) suggested a close relationship with natricines, xenodontines or colubrines, which turned out to be fairly prescient given the subsequent reorganization of Colubridae. Little is known about the natural history of this group, but apparently some of them occur in montane forests and eat frogs and at least one species, *Pseudoxenodon macrops*, is known to be oviparous (Zhao *et al.* 1998).

Dipsadinae (=Dipsadidae in Zaher *et al.* 2009) contains 88 genera and is the most speciose group in all of Colubroidea with almost 700 species. The group is often referred to as Xenodontinae (Bonaparte 1845), however, the name Dipsadinae (Bonaparte 1838) has priority by seven years. The distribution has been considered strictly New World until three recent papers all indicated that the Asian *Thermophis* is nested within Dipsadinae (Guo *et al.* 2009; He *et al.* 2009; Huang *et al.* 2009). Traditionally Dipsadinae has been diagnosed using hemipenal characters and currently this remains the case (Zaher *et al.* 2009). Interestingly, Xenodontinae of Zaher *et al.* (2009) is not diagnosable by any characters, which appears to us to be a good reason to recognize the larger, diagnosable subfamily, Dipsadinae. Furthermore, there are diagnosable subclades within the Dipsadinae (*sensu lato*), as hinted by early immunological studies (Cadle 1984a,b,c) and discussed in Zaher *et al.* (2009), but we wait for these diagnoses. Little is known about the natural history of this group, but apparently some of them occur in montane forests and eat frogs and at least one species, *Pseudoxenodon macrops*, is known to be oviparous (Zhao *et al.* 1998). Such a speciose group, as expected, shows extreme diversity in body form, habitat preferences, and diets, but if one wished to stereotype these snakes morphologically then they could be considered as mostly rear fanged tropical and mild temperate reptile and amphibian feeders (with obvious exceptions like the goo-eaters). The group also contains individuals that might be considered dangerously venomous (e.g., *Conophis lineatus*). This group is predominantly oviparous with some viviparous members. Recent studies have attempted to use molecular phylogenetic methods to infer relationships among some tribes and genera (Crother 1999; Vidal *et al.* 2000; Mulcahy 2007; Zaher *et al.* 2009).

The nominate subfamily Colubrinae is composed of approximately 100 genera and some 650 species. With respect to distribution, the colubrines are global, found on all continents (excluding Antarctica) and in all habitats from the tropics to the deserts and high mountains and high latitudes. There are fossorial, arboreal, aquatic, terrestrial, and even flying (gliding) forms (*Chrysopelea*). Although some genera are dangerously venomous (*Dispholidus* and *Thelotornis*), most species of Colubrinae are not considered harmful. This massive group exhibits an extreme breadth in diet from generalist to peculiar specializations like feeding on tarantulas, scorpions and centipedes. Like dipsadines, colubrines are almost exclusively oviparous with a few notable viviparous taxa (e.g., *Oocatochus rufodostatus*, Ji *et al.* 1997). There have been few attempts at inferring the phylogeny on this massive group, although a few smaller studies exist that have attempted to examine phylogenetic relationships and dates divergence for various subcomponents including Old World and New World ratsnakes/ kingsnakes (e.g., *Elaphe, Coelognathus, Gonyosoma, Rhinechis, Pantherophis, Pituophis,* and *Lampropeltis*, Utiger *et al.* 2002; 2005; Burbrink and Lawson 2007; Pyron and Burbrink 2009), and Old World and New World racers and whipsnakes (e.g., *Coluber, Masticophis, Hemorrhois, Hierophis,* and *Eirenis*; Nagy *et al.* 2004). From Burbrink and Lawson (2007) it is likely that colubrines originated in the Eocene of the Old World.

2.4.4 Xenodermatidae

Composed of six genera and 18 species, this group of colubroids is confined to southern and southeastern Asia and appears to be the sister clade to the rest of the colubroids (Fig. 2.1; Kelly *et al.* 2003; Vidal *et al.* 2007; Eckstut *et al.* 2009; Zaher *et al.* 2009). This obviously required the removal of the group as a subfamily of Colubridae. Other studies (e.g., Lawson *et al.* 2005; Kelly *et al.* 2009) that used *Oxyrhabdium* (instead of *Xenodermus* and *Stoliczkia* like the former studies) found it nested deep in the colubroid clade. Zaher *et al.* (2009) suggested that *Oxyrhabdium* is probably not a member of the Xenodermatidae. Interestingly, *Oxyrhabdium* was only recently added to the family (along with *Xylophis*) by McDowell (1987) because of the shared maxilla–palatine connection among Xenodermatidae, which he even mused as "perhaps merely primitive."

These strange-scaled snakes are unusual in that the scales are almost entirely fused to the skin and have, in some places, large interscalar areas of exposed skin. This contrasts the typical condition where the scales are usually fixed to the underlying skin at one point and the rest of the scale is free and also typical is that the scales are imbricating (although there are other snakes without imbricating scales). These snakes are poorly known but considered to feed primarily on frogs and possibly fish. *Xenodermus* is known to be oviparous (Greene 1997).

2.4.5 Pareatidae

Composed of three genera and 14 species, this group is distributed in southeastern Asia. These snakes are the Old World slug eaters, with probable convergence in jaw and skull characters with New World slug eaters in the family Dipsadinae (e.g., *Dipsas*), such as possessing a reduced preorbital portion of the maxilla and elongate narrow teeth (Cundall and Irish 2008). Molecular phylogenies have inferred the pareatids to be the sister to all the colubroids, exclusive of xenodermatids (Lawson *et al.* 2005; Vidal *et al.* 2007; Eckstut *et al.* 2009; Zaher *et al.* 2009). Like Zaher *et al.* (2009), we do not see the need to erect the name Pareatoidea as a monotypic superfamily as proposed by Vidal *et al.* (2007).

Pareatids are widespread in the tropical and subtropical regions of southeastern Asia and occupy terrestrial and arboreal habitats. They are almost exclusively gastropod feeders except for *Aplopeltura*, which eats lizards. For the gastropod feeders at least, it is thought they do not exhibit ontogenetic changes in diet (Hoso 2007). They are oviparous (Greene 1997) and in one species, Iwasaki's Snail Eater (*Pareas iwasakii*), clutch sizes have ranged from six to 11 (Hoso 2007).

2.4.6 Viperidae

We recognize three subfamilies (following Liem *et al.* 1971), Crotalinae, Viperinae, and Azemiopinae, in this globally (except Australia and Antarctica) distributed group of highly specialized venomous snakes. The most distinctive synapomorphy for Viperidae is the solenoglyph condition, characterized by reduced maxilla each possessing a single modified tooth, which is a hollow hinged fang that is retractable to sit against the roof of the mouth (McDowell 1987). Composed of 28 genera and 81 species, the crotalines are the most diverse group of viperids and found throughout the

New World, and through Asia into southeastern Europe. Viperinae, known from 13 genera and 81 species, are distributed in Africa, Europe, and Asia. The single species of the enigmatic Azemiopinae is restricted to montane regions of Myanmar, Vietnam, and China.

Azemiopinae is monotypic (*Azemiops*), and in some recent molecular phylogenies is inferred to be the sister taxon of a crotaline–viperine clade (Kelly *et al.* 2003 ML tree; Eckstut *et al.* 2009). Castoe and Parkinson (2006) and Zaher *et al.* (2009) indicated that the Azemiopinae is nested within the Viperidae and is the sister clade to the Crotalinae. The preferred tree in Kelly *et al.* (2003) illustrates that *Azemiops* renders the Viperinae paraphyletic. On the other hand, Wüster *et al.* (2008) demonstrated that *Azemiops* was sister to Crotalinae. Currently, there seems to be little consensus about the placement of *Azemiops*.

As one would expect of globally distributed viperids of nearly 300 species, these snakes occupy diverse habitats and exhibit a wide variety of ecologies. They are found in deserts, tropical forests, and freshwater aquatic systems. They are terrestrial, fossorial, and arboreal. Perhaps if there is something shared by all members it is diet: they all appear to feed on vertebrates and be mostly sit and wait predators (but see *Causus*). Crotalines have a pair of specialized heat sensing pits between the nares and eyes that allow them to see prey items based on heat signatures. Viperids exhibit several interesting reproductive strategies. Most crotalines are viviparous but several taxa are oviparous and among viperines both oviparity and ovoviviparity are exhibited. Apparently, even parthenogenesis occurs in viperids (Alemeida-Santos and Salomão 2002). While the general perception of vipers is that of cold-blooded killers, parental care is surprisingly broadly distributed among oviparous and viviparous viperine and crotaline taxa (Greene *et al.* 2002).

The crown group of Viperidae likely originated in the early Tertiary (Wüster *et al.* 2008) and apparently viviparity was a key innovation that coincided with global cooling in the Cenozoic and resulted in rapid adaptive radiation in this group (Lynch 2009). For more information about the biology of these organisms, see several comprehensive texts on the biology of viperids (Campbell and Brodie 1992; Schuett *et al.* 2002; Campbell and Lamar 2004).

2.4.7 Homalopsidae

Eleven genera and 35 species are currently considered homalopsids (Lawson *et al.* 2005; Zaher *et al.* 2009) and they are distributed in southern and southeastern Asia and Australasia (Murphy 2007). The majority of these taxa are grooved rear-fanged aquatic specialists and the key morphological synapomorphies that diagnose this clade are associated with aquatic specialization, such as the dorsal position of the nares and eyes on the head, specialized structures for breathing underwater, and the ability to close the nostrils (Santos-Costa and Hofstadler-Deiques 2002). One taxon, *Brachyorrhos*, is not aquatic, not rear fanged, has lateral eyes, and anterior nares (Murphy 2007). Based on hemipenes, vertebrae and skull characters, McDowell (1987) included it in the group and Zaher *et al.* (2009) followed this suggestion. Lawson *et al.* (2005) treated *Brachyorrhos* as *incertae sedis*. Additionally, *Anoplohydrus* (known from a single specimen collected in Sumatra but lost during the bombing of Dresden in WWII) is supposedly a homalopsid but no phylogenetic study has confirmed its placement (Murphy 2007). Several recent molecular studies are congruent in their inference of the homalopsids as the sister to the remainder of the Colubroidea, exclusive of the xenodermatids, pareatids, and viperids (Kelly *et al.* 2003, Lawson *et al.* 2005; Eckstut *et al.* 2009; Zaher *et al.* 2009, Pyron and Burbrink, unpublished data).

These snakes (except *Brachyorrhos*) inhabit all manner of aquatic environments including freshwater ponds, streams, freshwater wetlands, and agricultural systems (e.g., flooded rice paddies) as well as coastal marine systems like tidal flats, mangrove forests, and estuaries (Gyi 1970; Heatewole 1999). In all of these diverse aquatic habitats homalopsids are mostly associated with mud substrates (Murphy 2007). They feed on amphibians, fish and crustaceans. One of the most interesting observations about the latter food item is that the Crab-eating water snake (*Fordonia leucobalia*) actually dismember the crustaceans as they eat, the only snakes known to do so (Jayne *et al.* 2002). All members are thought to be viviparous and multiple paternities have been documented for two species (Voris *et al.* 2008). A recent study on the phylogenetic history of this group using sequences from a majority of taxa demonstrated that the crown group originated ~22 Ma. Moreover, this phylogeny produced the telling signals that suggested this group diversified in an early explosive burst of speciation (Alfaro *et al.* 2008).

2.4.8 Lamprophiidae

Containing primarily an African radiation of snakes, this group may be sister to Elapidae (Vidal *et al.* 2007; Pyron *et al.* in press). Some studies suggest it might be paraphyletic with regard to elapids (Lawson *et al.* 2005; Kelly *et al.* 2009). Although we use Lamprophiidae to cover the subfamilies Atractaspidinae, Psammophiinae, Pseudoxyrhophiinae, and Lamprophiinae, others have considered these subfamilies (and more) included within Elapidae (Lawson *et al.* 2005) or joined with Elapidae in the superfamily Elapoidea (Pinou *et al.* 2004 [in part]; Vidal *et al.* 2007; Kelly *et al.* 2009; Zaher *et al.* 2009). All of these taxonomic suggestions have yet to be bolstered with a phylogeny inferred using large number of independent loci. Dates and area of origin for Lamprophiidae are hampered by a lack of fossils from Africa. Although the composition and relationships are not identical to a monophyletic Lamprophiidae, it might be inferred from Kelly *et al.* (2009) the family likely originated in late Eocene in Africa.

One of the most distinct subfamilies, Atractaspidinae, contains a dozen genera and 70 species that occupy sub-Saharan Africa. The stem group likely originated in Africa in the late Eocene/Early Oligocene (Kelly *et al.* 2009). Frequently, *Aparallactus* is excluded from this subfamily. For example, Kelly *et al.* (2003) and Vidal *et al.* (2007) did not find those two genera in the same clade (possibly because of sample size issues). However, given evidence in Eckstut *et al.* (2009), Kelly *et al.* (2009) and Pyron *et al.* (in press), we recognize an Atractaspidinae that includes *Aparallactus*. Diagnostic morphological characters that define this subfamily have yet to be found. Hemipenal characters examined in Zaher *et al.* (2009) are variable throughout the groups and the dentition that once marked the subfamily as unique is now confined to a subset of the taxa. Remarkably, this clade contains opistoglyphous, aglyphous, proteroglyphous and solenoglyphous forms! Perhaps, as Zaher *et al.* (2009) stated, the current contents of this clade exist for "...convenience and historical legacy." However, at minimum, several studies are congruent in showing the proteroglyphous *Homoroselaps*, the solenoglyphous *Atractaspis*, and the opistoglyphous/ aglyphous *Aparallactus* to share a most recent common ancestor (Kelly *et al.* 2003; Lawson *et al.* 2005; Vidal *et al.* 2008; Eckstut *et al.* 2009; Kelly *et al.* 2009; Zaher *et al.* 2009). Diet varies among this group from small mammals like shrews and naked mole rats (*Atractaspis*) to elongate vertebrates such as amphisbaenians, and also centipedes. All members are thought to be oviparous (Pough *et al.* 2004).

Lamprophiinae contains 20 genera and 81 species that are distributed through Africa and is much more restricted than the concept in Zaher *et al.* (2009) and more similar to the contents of Kelly *et al.* (2009) and Lawson *et al.* (2005; although Lawson *et al.* 2005 referred to the group as Boodontinae). Vidal *et al.* (2008), Pyron *et al.* (in press) found the lamprophiines to be the sister clade to the pseudoxyrhophiines. To understand the exact generic content of these two groups will require a more thorough sampling of taxa and genes. McDowell (1987) recognized a polyphyletic Boodontinae (sensu Dowling *et al.* 1983) and the molecular data have supported this contention. Essentially, the old Boodontinae has been found to be composed of two groups the lamprophiines, pseudoxyrhophiines and some psammophiines. Obvious diagnostic morphological characters are absent, but Zaher *et al.* (2009) considers the arrangement of spines in transverse rows that form flounce-structures on the hemipenal body indicative of most of the clade. Presumably, snakes in this group are oviparous (e.g., *Lamprophis*, Ford 2001; *Lycophidion*, Greer 1968; *Pseudoboodon*, Spawls 1997). Kelly *et al.* (2009) demonstrated that the stem of this group likely originated in the late Eocene in Africa.

Pseudoxyrhophiinae, as noted above, contains the other portion of the old Boodontinae (*sensu* Dowling *et al.* 1983) and comprises about 20 genera and 80 species (Lawson *et al.* 2005; Zaher *et al.* 2009; Reptile Database). Vidal *et al.* (2008) and Eckstut *et al.* (2009) included 16 genera of this group in their studies and found them to form a monophyletic group. Lawson *et al.* (2005), Kelly *et al.* (2009) and Zaher *et al.* (2009) all considered this group to contain mostly genera from Madagascar plus some mainland taxa (*Amplorhinus*, *Ditypophis*, and *Duberria*). Morphologically, these taxa are diagnosed with the presence of only spinules on the hemipenal lobes

(Zaher 1999), which is apparently also shared with homalopsids. The Madagascar radiation is extremely diverse and may have been colonized more than once. The main radiation of the crown group in Madagascar and Socotra occurred in the late Eocene (Nagy *et al.* 2003; Kelly *et al.* 2009). They feed on vertebrates may be arboreal, fossorial, terrestrial and troglodytic. Most taxa are oviparous but viviparity is known in some species (Pough *et al.* 2004).

Psammophiinae, another primarily African group of snakes (although they reach into southern Europe and western Asia) is composed of six genera and 46 species. The delimitation and monophyly of this subfamily is without dispute (Kelly *et al.* 2008) and the clade is easily diagnosed by a number of morphological characters such as extremely reduced hemipenes and differentiated maxillary and mandibular dentition (Zaher *et al.* 2009). What remains in dispute are their placement within Lamprophiidae. Kelly *et al.* (2003) and Zaher *et al.* (2009) inferred that the psammophiines were the sister to the rest of the lamprophiids and elapids. Lawson *et al.* (2005) found psammophiines to share a more recent common ancestor with pseudoxyrhophiines. Vidal *et al.* (2007) inferred psammophiines as sister to a clade composed of lamprophiines-atractaspidines and later (Vidal *et al.* 2008) inferred they were the sister clade to all the elapids minus the elapines. Eckstut *et al.* (2009) recovered a psammophiine-*Prosymna* clade and Kelly *et al.* (2009) found the psammophiines related to a pseudoxyrhophiine-*Prosymna* clade. Stem origins for this group likely occurred in Africa near the late Eocene (Kelly *et al.* 2009).

This clade of snakes most resemble New World racers or whipsnakes in that they are slender, often big eyed, diurnal serpents that hunt lizards and small mammals (Shine *et al.* 2006). The diet may change ontogenetically with larger individuals taking more mammals. The largest species of the group, *Malpolon monspessulanus*, has a notoriously catholic diet, apparently eating anything it can subdue and swallow, such as lizards, rabbits, snakes, birds, and even tortoises (Shine *et al.* 2006). They are apparently oviparous (Shine *et al.* 2006).

Finally, within Lamprophiidae, Kelly *et al.* (2009) recognized two more groups, Prosymninae and the Pseudaspidinae. The former is monotypic for the genus *Prosymna*, which based on the studies discussed here, is clearly a member of the Lamprophiidae (*sensu lato*) but where it belongs exactly is unclear. In Lawson *et al.* (2005) *Prosymna* could be placed as sister to a number of groups depending on the analysis. The placement of this genus was incongruent in several other studies (Lawson *et al.* 2005; Vidal *et al.* 2008; Eckstut *et al.* 2009; Kelly *et al.* 2009). Pyron *et al.* (in press) demonstrated that they might be sister to Atractaspidinae. Kelly *et al.* (2009) included only two genera in their Pseudaspididae and these were inferred to be monophyletic in all the studies that included them (Lawson *et al.* 2005; Eckstut *et al.* 2009; Kelly *et al.* 2009). Pyron *et al.* (in press) demonstrated that the basal Lamprophiidae node subtends the two genera of Pseudaspididae.

2.4.9 Elapidae

One of the important revelations of modern molecular phylogenetics of snakes was the discovery that the fixed front fanged snakes of the Elapidae (*sensu stricto*) rendered the Colubridae (*sensu lato*) paraphyletic (Kelly *et al.* 2003; Lawson *et al.* 2005) and that several groups of colubrids (lamprophiids, including Atractaspidinae) were more closely related to these proteroglyph snakes than to other snakes without the fixed, front fang condition (Lawson *et al.* 2005). Subsequent to this discovery, as noted above under Colubridae, a number of classification schemes have been put forward to organize this new found set of relationships. Some have considered the families Lamprophiidae and Elapidae combined in the superfamily Elapoidea, which is sister to a redefined Colubroidea (see above; Pinou *et al.* 2004 [in part]; Vidal *et al.* 2007; Kelly *et al.* 2009 and Zaher *et al.* 2009). Because we prefer to retain the broader and more widely used definition of Colubroidea in this chapter (Romer 1956), we are forced to refrain from using Elapoidea.

All elapids have the proteroglyph condition—a fixed, erect fang on each maxillary. The group likely contains two subfamilies, Elapinae and Hydrophiinae (Lawson *et al.* 2005; Castoe *et al.* 2007). Establishing the monophyly of these subfamilies, particularly Elapinae has been somewhat troublesome. Although Elapinae has been recognized in some studies (Lawson *et al.* 2005; Castoe *et al.* 2007), others have not (Zaher *et al.* 2009; Kelly *et al.* 2009). Several studies have demonstrated support for a monophyletic Hydrophiinae (Keogh 1998; Keogh *et al.*, 1998; Scanlon and Lee 2004; Lawson *et al.* 2005; Castoe *et al.* 2007; Sanders *et al.*, 2008). Occasionally, a third subfamily, Laticaudinae is recognized but often is included within Hydrophiinae. Kelly *et al.* (2009) demonstrated that stem elapids likely originated in Asia during the Eocene.

Elapinae is composed of 19 genera and 162 species found throughout the New World, Africa, Asia and Eurasia. Castoe *et al.* (2007) divided Elapinae into two tribes, Calliophiinae containing mostly New World and Old World coralsnakes (i.e., *Micrurus, Micruroides, Sinomicrurus,* and *Calliophis*) and Hemibungarini containing Asian and African cobras, kraits and relatives (e.g., *Naja, Bungarus, Dendroaspsis,* etc.). While dates of origin for this subfamily are not yet known, divergence date estimates have been produced for some groups. For instance, the diversification of spitting cobras in Africa took place during the Mid-Miocene (Wüster *et al.* 2007) and the origin of Calliophiinae occurred in the late Oligocene (Kelly *et al.* 2009). Excluding mambas (*Dendroaspis*) and tree cobras (*Pseudohaje*), most elapines are terrestrial. Many elapids display aposematic coloring and generally feed on vertebrates. Most species are oviparous and clutch size seems to be correlated with body size. The viviparous *Hemachatus* is an exception (Vitt and Caldwell 2009).

The Australo-Melanasian Hydrophiinae is composed of 46 genera and 188 species (Reptile Database) and is generally found in Australasia, the Pacific and Indian Oceans. Diversification within this morphologically very variable group of elapids occurred very recently, within the last 10 Ma, relative to other colubroid groups (Sanders and Lee 2008; Sanders *et al.* 2008).

Terrestrial hydrophiines generally feed on vertebrates and may be either oviparous or viviparous. True seasnakes, referred to as Hydrophiini (Lee *et al.* 2007; Sanders and Lee 2008; Sanders *et al.* 2008) are much more adapted for oceanic life. These adaptations include laterally compressed bodies, paddle-shaped tails and a lack of enlarged ventral scales. All sea snakes are viviparous. In contrast, *Laticauda* (sometimes placed in the subfamily Laticaudinae) lays eggs on land. The sea snake *Pelamis platurus* has one of the largest ranges of any snake, occurring in the Persian Gulf, throughout the Indian Ocean, through the Pacific Ocean in Asia and Australia to Baja California, Central America and South America.

2.5 CONCLUSION

We have attempted to produce a concise introduction to the current state of the taxonomy, systematics, and origins of extant snakes in one compact chapter. Comprehensive chapters like this one are both good and bad. The good first: they identify the most relevant literature and uptodate knowledge on a subject. The bad: they are likely to become woefully out of date and incorrect as new literature on the subject is produced. For snake taxonomy, the bad aspect may not be entirely dismal; like any good science, systematics is likely to change as new evidence is presented and old taxonomic hypotheses are challenged. Unfortunately, many researchers in other fields require that taxonomies remain stable. Imposed stability without regard to new discoveries, however, forces systematics to become religion and not testable science (Crother 2009). Therefore, fluidity in taxonomy should be expected as new data are generated and tested. While many groups discussed will remain taxonomically stable, as they have for decades, we still look forward to the changes in snake systematics as the field enters into the world of phylogenomics.

2.6 LITERATURE CITED

Adalsteinsson, S. A., Branch, W. R., Trape, S., Vitt, L. J. and Hedges, S. B. 2009. Molecular phylogeny, classification, and biogeography of snakes of the Family Leptotyphlopidae (Reptilia, Squamata). Zootaxa 2244: 1-50.

Alfaro, M. A., Karns, D. R., Voris, H. K. and Stuart, B. L. 2008. Phylogeny, evolutionary history, and biogeography of Oriental-Australian rear-fanged water snakes (Colubridae: Homalopsidae) inferred from mitochondrial and nuclear DNA sequences. Molecular Phylogenetics and Evolution 46: 576-593.

Almeida-Santos, S. M. and Salomão, M. D. G. 2002. Reproduction in neotropical pitvipers, with emphasis on species of the genus *Bothrops*. Pp. 445-462. In G. W. Schuett, M. Höggren, M. E. Douglas and H. W. Greene (eds), *Biology of the Vipers*. Eagle Mountain Publishing, Eagle Mountain, Utah.

Apesteguía, S. and Zaher, H. 2006. A Cretaceous terrestrial snake with robust hindlimbs, and a sacrum. Nature 440: 1037-1040.
Averianov, A. O. 1997. Paleogene sea snakes from the eastern part of Tethys. Russian Journal of Herpetology 4: 128-142.
Boulenger, G. A. 1896. *Catalogue of the Snakes in the British Museum*, Vol. 3. Trustees of the British Museum, London, UK. Pp. 727.
Brown, G. P. and Shine, R. 2002. Reproductive ecology of a tropical natricine snake, *Tropidonophis elegans*. Copeia 2001: 508-513.
Bullock, D. J. 1986. The ecology and conservation of reptiles on Round Island and Gunner's Quoin, Mauritius. Biological Conservation 37: 135-156.
Burbrink, F. T. 2005. Inferring the phylogenetic position of *Boa constrictor* among the Boinae. Molecular Phylogenetics and Evolution 34: 167-180.
Burbrink, F. T. and Lawson, R. 2007. How and when did the Old World rat snakes cross into the New World? Molecular Phylogenetics and Evolution 43: 173-189.
Burbrink, F. T. and Pyron, R. A. 2008. The taming of the skew: estimating proper confidence intervals for divergence dates. Systematic Biology 57: 317-328.
Cadle, J. E. 1984a. Molecular systematics of neotropical xenodontine snakes. I. South American xenodontines. Herpetologica 40: 8-20.
Cadle, J. E. 1984b. Molecular systematics of neotropical xenodontine snakes. II. Central American xenodontines. Herpetologica 40: 21-30.
Cadle, J. E. 1984c. Molecular systematics of neotropical xenodontine snakes. III. Overview of xenodontine phylogeny and the history of New World snakes. Copeia 1984: 641-652.
Cadle, J. E. 1994. The colubrid radiation in Africa (Serpentes: Colubridae): phylogenetic relationships and evolutionary patterns based on immunological data. Zoological Journal of the Linnean Society 110: 103-140.
Cadle, J. E., Dessauer, H. C., Gans, C. and Gartside, D. F. 1990. Phylogenetic relationships and molecular evolution in uropeltid snakes (Serpentes: Uropeltidae): allozymes and albumin immunology. Biological Journal Linnean Society 40: 293-320.
Caldwell, M. W. 1999. Squamate phylogeny and the relationships of snakes and mosasauroids. Zoological Journal of the Linnean Society 125: 115-147.
Caldwell, M.W. 2000. On the phylogenetic relationships of *Pachyrhachis* within snakes: a response to Zaher (1998). Journal of Vertebrate Paleontology 20: 187-190.
Caldwell, M. W. and Lee, M. S. Y. 1997. A snake with legs from the marine Cretaceous of the Middle East. Nature 386: 705-709.
Caldwell, M. W. and Albino, A. 2003. Exceptionally preserved skeletons of the Cretaceous snake *Dinilysia patagonica* Woodward, 1901. Journal of Vertebrate Paleontology 22: 861-866.
Caldwell, M. W. and Dal Sasso, C. 2004. Soft-tissue preservation in a 95-million-year-old marine lizard: form, function, and aquatic adaptation. Journal of Vertebrate Paleontology 24: 980-985.
Campbell, J. A. and Brodie, E. D. Jr. (eds). 1992. *Biology of Pitvipers*. Selva, Tyler, Texas. Pp. 467.
Campbell, J. A. and Lamar, W. W. 2004. *The Venomous Reptiles of the Western Hemisphere*, 2 vols. Cornell University Press, Ithaca. Pp. 870.
Carroll, R. L. 1988. *Vertebrate Paleontology and Evolution*. W. H. Freeman and Company, New York. Pp. 698.
Castoe, T. A. and Parkinson, C. L. 2006. Bayesian mixed models and the phylogeny of pitvipers (Viperidae: Serpentes). Molecular Phylogenetics and Evolution 39: 91-110.

Castoe, T. A., Smith, E. N., Brown, R. M. and Parkinson, C. L. 2007. Higher-level phylogeny of Asian and American coralsnakes, their placement within the Elapidae (Squamata), and the systematic affinities of the enigmatic Asian coralsnake *Hemibungarus calligaster* (Wiegmann, 1834). Zoological Journal of the Linnean Society 151: 809-831.

Cisneros-Heredia, D. F. 2005. *Anilius scytale* (Red pipesnake): Reproduction. The Herpetological Bulletin 92: 28-29.

Conrad, J. L. 2008. Phylogeny and systematics of Squamata (Reptilia) based on morphology. Bulletin of the American Museum of Natural History 310: 1-182.

Cope, E. D. 1984. The classification of snakes. American Naturalist 28: 831-844.

Cope, E. D. 1895. The classification of the Ophidia. Transactions of the American Philosophical Society 18: 186-219.

Cox, M. J. 1991. *The Snakes of Thailand and their Husbandry*. Krieger Publishing Company, Florida. Pp. 526.

Crother, B. I. 1999. Phylogenetic relationships among West Indian Xenodontine snakes (Serpentes: Colubridae) with comments on the phylogeny of some mainland xenodontines. Contemporary Herpetology 1999 (2): http://dataserver.calacademy.org/herpetology/herpdocs/index. html.

Crother, B. I. 2009. Are standard names lists taxonomic straightjackets? Herpetologica 65: 129-135.

Cundall, D. and Irish, F. J. 1986. Aspects of loco-motor and feeding behavior in the Round Island boa *Casarea dussumieri*. Dodo, Journal of the Jersey Wildlife Preserve Trust 23: 108-111.

Cundall, D. and Irish, F. J. 1989. The function of the intramaxillary joint in the Round Island boa, *Casarea dussumieri*. Journal of Zoology, London 217: 569-598.

Cundall, D. and Irish, F. J. 2008. The snake skull. Pp. 349-692. In C. Gans, A. S. Gaunt and K. Adler (eds), *Biology of the Reptilia*. Vol. 20 (Morphology H: The Skull of Lepidosauria), Society for the Study of Amphibians and Reptiles, Ithaca, New York.

Cundall, D., Wallach, V. and Rossman, D. A. 1993. The systematic relationships of the snake genus *Anomochilus*. Zoological Journal of the Linnean Society 109: 275-299.

De Queiroz, A., Lawson, R. and Lemos-Espinal, J. A. 2002. Phylogenetic relationships of North American garter snakes (*Thamnophis*) based on four mitochondrial genes: How much DNA sequence is enough. Molecular Phylogenetics and Evolution 22: 315-329.

Dessauer, H. C., Cadle, J. E. and Lawson, R. 1987. Patterns of snake evolution suggested by their proteins. Fieldiana: Zoology 34: 1-34.

Douglas, D. A., Janke, A. and Arnason, U. 2006. A mitogenomic study on the phylogenetic position of snakes. Zoologica Scripta 35: 545-558.

Dowling H. G. and Duellman, W. E. 1978. *Systematic Herpetology: A Synopsis of Families and Higher Categories*. Herpetologogical Information Search Systems Publications, New York. Pp. 188.

Dowling, H. G., Highton, R., Maha, G. C. and Maxson, L. R. 1983. Biochemical evaluation of colubrid snake phylogeny. Journal of Zoology, London 201: 309-329.

Dubach, J., Sajewicz, A. and Pawley, R. 1997. Parthenogenesis in the Arafuran filesnake (*Acrochordus arafurae*). Herpetological Natural History 5: 11-18.

Duméril, A. M. C. 1853. Prodrome de la classification des Reptiles Ophidiens. Mémoires de l'Académie des Sciences de l'Institut de France 23: 399-535.

Eckstut, M. E., Sever, D. M., White, M. E. and Crother, B. I. 2009. Phylogenetic analysis of sperm storage in female squamates. Pp. 185-218. In L. T. Dahnof

(ed.), *Animal Reproduction: New Research Developments*. Nova Science Publishers, Hauppauge, New York.
Edwards, S. V. 2009. Is a new and general theory of molecular systematics emerging? Evolution 63: 1-19.
Folie, A. 2006. Evolution des Amphibiens et squamates de la transition Crétacé-Paléogene en Europe: Les faunes du Maastrichtien du Bassin de Hateg (Roumanie) et du Paléocene du Bassin de Mons (Belgique). Ph.D. Dissertation, Universite libre de Bruxelles. Pp. 271.
Ford, N. B. 2001. Reproduction in the brown house snake, *Lamprophis fuliginosus*, from Tanzania. African Journal of Herpetology 50: 31-34.
Forstner, M. R. J., Davis, S. K. and Arévalo, E. 1995. Support for the hypothesis of anguimorph ancestry for the suborder Serpentes from phylogenetic analysis of mitochondrial DNA sequences. Molecular Phylogenetics and Evolution 4: 93-102.
Frazetta, T. H. 1975. Pattern and instability in the evolving premaxilla of boine snakes. American Zoologist 15: 469-481.
Gans, C., Dessauer, H. C. and Baic, D. 1978. Axial differences in the musculature of uropletid snakes: the freight-train approach to burrowing. Science 199: 189-192.
Gower, D. J., Vidal, N., Spinks, J. N. and McCarthy, C. J. 2005. First inclusion of Anomochilidae in molecular phylogenetics of the major lineages of snakes. Journal of Zoological Systematics and Evolutionary Research 43: 315-320.
Greene, H. W. 1997. *Snakes: The Evolution of Mystery in Nature*. University of California Press, California. Pp. 351.
Greene, H. W., May, P. G., Hardy, D. L. Sr., Sciturro, J. M. and Farrel, T. M. 2002. Parental behavior by vipers. Pp. 179-205. In G. W. Schuett, M. Höggren, M. E. Douglas and H. W. Greene (eds), *Biology of the Vipers*. Eagle Mountain Publishing, Eagle Mountain, Utah.
Greer, A. E. 1968. Mode of reproduction in the squamate faunas of three altitudinally correlated life zones in East Africa. Herpetologica 24: 229-232
Günther, R. and Manthey, U. 1995. *Xenophidion*, a new genus with two new species of snakes from Malaysia (Serpentes, Colubridae). Amphibia-Reptilia 16: 229-240.
Guo, P., Liu, S. Y., Huang, S., He, M., Sun, Z. Y., Feng, J. C. and Zhao, E. -M. 2009. Morphological variation in *Thermophis* Malnate (Serpentes: Colubdridae), with an expanded descrition of *T. zhaoermii*. Zootaxa 1973: 51-60.
Gyi, K. K. 1970. *A Revision of Colubrid Snakes of the Subfamily Homalopsinae*. University of Kansas Publications, Museum of Natural History, 20: 47-223.
Haas, G. 1955. The systematic position of *Loxocemus bicolor* Cope (Ophidia). American Museum Novitates 1748: 1-8.
Hass, G. 1973. Muscles of the jaws and associated structures in the Rhynchocephalia and Squamata. Pp. 285-490. In C. Gans and T. S. Parsons (eds), *Biology of the Reptilia* Vol. 4. Academic Press, London, U.K.
He, M., Feng, J.-C., Liu, S.-Y., Guo, P. and Zhao, E.-M. 2009. The phylogenetic position of *Thermophis* (Serpentes: Colubridae), an endemic snake from the Qinghai-Xizang Plateau, China. Journal of Natural History 43: 479-488.
Head, J. J., Mohabey, D. M. and Wilson, J. A. 2007. *Acrochordus* Hornstedt (Serpentes, Caenophidia) from the Miocene of Gujarat, Western India: Temporal constraints on dispersal of a derived snake. Journal of Vertebrate Paleontology 27: 720-723.
Head, J. J., Holroyd, P. A., Hutchison, J. H. and Ciochon, R. L. 2005. First report of snakes (Serpentes) from the Late Middle Eocene Pondaung Formation, Myanmar. Journal of Vertebrate Paleontology 25: 246-250.

Head, J. J., Bloch, J. I., Hastings, A. K., Bourque, J. R., Cadena, E. A., Herrera, F. A., Polly, P. D. and Jaramillo, C. A. 2009. Giant boid snake from the Palaeocene neotropics reveals hotter past equatorial temperatures. Nature 457: 715-717.

Heatewole, H. 1999. *Sea Snakes*. Krieger Publishing Company, Florida. Pp. 166.

Heise, P. J., Maxson, L. R., Dowling, H. G. and Hedges, S. B. 1995. Higher level snake phylogeny inferred from mitochondrial DNA sequences of 12S rRNA and 16S rRNA genes, Molecular Biology and Evolution 12: 259-265.

Hoffstetter, R. 1946. Remarques sur la classification des Ophidiens et particulierement des Boidae des Mascareignes (Bolyerinae subfam. nov.). Bulletin of the American Museum of Natural History 18: 432-435.

Hoffstetter, R. 1961. Noveaux restes d'un serpent Boidé (*Madtsoia madagascariensis* nov. sp.) dans le Crétacé Supérieur de Madagascar. Bulletin du Muséum National d'Histoire Naturelle, Série 2, Paris 33: 152-160.

Hoffstetter, R. 1962. Revue des recentes acquisitions concernant l'histoire et la systematique des Squamates. Colloques International Centre National de la Recherche Scientifique 104: 243-279.

Holman, J. A. 2000. *Fossil Snakes of North America*. Indiana University Press, Bloomington. Pp. 357.

Holman, J. A. and Case, G. R. 1988. Reptiles from the Eocene Tallahatta Formation of Alabama. Journal of Vertebrate Paleontology 8: 328-333.

Holman, J. A., Dockery, D. T. and Case, G. R. 1991. Paleogene snakes of Mississippi. Mississippi Geology 11: 1-12.

Hoso, M. 2007. Oviposition and hatchling diet of a snail-eating snake *Pareas iwasakii* (Colubridae: Pareatinae). Current Herpetology 26: 41-43.

Huang, S., Liu, S.-Y., Guo, P., Zhang,Y,-P. and Zhao, E.-M. 2009. What are the closest relatives of the hot-spring snakes (Colubridae, *Thermophis*), the relict species endemic to the Tibetan Plateau? Molecular Phylogenetics and Evolution 51: 438-446.

Jamieson, B.G.M. 1995. The ultrastructure of spermatozoa of the Squamata (Reptilia) with phylogenetic considerations. Pp. 359-383. In B. G. M. Jamieson, J. Ausio and J.-L. Justine (eds), *Advances in Spermatozoal Phylogeny and Taxonomy*. vol. 166, Mémoires du Muséum National d'Histoire Naturelle, Paris.

Jayne, B. C., Voris, H. K. and Ng, P. K. L. 2002. Snake circumvents constraints on prey size. Nature 418: 143.

Ji, X., Xie, Y.-Y., Sun, P.-Y. and Zheng, X.-Z. 1997. Sexual dimorphism and female reproduction in a viviparous snake, *Elaphe rufodorsata*. Journal of Herpetology 31: 420-422.

Kelly, C. M. R., Barker, N. P. and Villet, M. H. 2003. Phylogenetics of advanced snakes (Caenophidia) based on four mitochondrial genes. Systematic Biology 52: 439-459.

Kelly, C. M. R., Barker, N. P., Villet, M. H., Broadley, D. G. and Branch, W. R. 2008. The snake family Psammophiidae (Reptilia: Serpentes): phylogenetics and species delimitation in the African sand snakes (*Psammophis* Boie, 1825) and allied genera. Molecular Phylogenetics and Evolution 47: 1045-1060.

Kelly, C. M. R., Barker, N. P., Villet, M. H. and Broadley, D. G. 2009. Phylogeny, biogeography and classification of the snake superfamily Elapoidea: a rapid radiation in the Late Eocene. Cladistics 25: 38-63.

Keogh, J. S. 1998. Molecular phylogeny of elapid snakes and a consideration of their biogeographic history. Biological Journal of the Linnean Society 63: 177-203.

Keogh, J. S., Shine, R. and Donnellan, S. 1998. Phylogenetic relationships of terrestrial Australo-Papuan elapid snakes based on cytochrome *b* and 16S rRNA sequences. Molecular Phylogenetics and Evolution 10: 67-81.

Khajuria, C. K. and Prasad, G. V. R. 1998. Taphonomy of a Late Cretaceous mammal-bearing microvertebrate assemblage from the Deccan inter-trappean beds of Naskal, peninsular India. Palaeogeography, Palaeoclimatology, Palaeoecology 137: 153-172.

Kluge, A. G. 1991. Boine phylogeny and research cycles. Miscellaneous Publications Museum of Zoology, University of Michigan 178: 1-58.

Lawson, R., Slowinski, J. B. and Burbrink, F. T. 2004. A molecular approach to discerning the phylogenetic placement of the enigmatic snake *Xenophidion schaeferi* among the Alethinophidia. Journal of Zoology, London 263: 1-10.

Lawson, R., Slowinski, J. B., Crother, B. I. and Burbrink, F. T. 2005. Phylogeny of the Colubroidea (Serpentes): new evidence from mitochondrial and nuclear genes. Molecular Phylogenetics and Evolution 37: 581-601.

Lee, M. S. Y. 1997. The phylogeny of varanoid lizards and the affinities of snakes. Philosophical Transactions of the Royal Society of London B Biological Sciences 352: 53-91.

Lee, M. S. Y. 1998. Convergent evolution and character correlation in burrowing reptiles: towards a resolution of squamate relationships. Biological Journal of the Linnean Society 65: 369-453.

Lee, M. S. Y. 2000. Soft anatomy, diffuse homoplasy, and the relationships of lizards and snakes. Zoologica Scripta 29: 101-130.

Lee, M. S. Y. and Caldwell, M. W. 2000. *Adriosaurus* and the affinities of mosasaurs, dolichosaurs, and snakes. Journal of Paleontology 74: 915-937.

Lee, M. S. Y. and Scanlon, J. D. 2001. On the lower jaw and intramandibular septum in snakes and anguimorph lizards. Copeia 2001: 531-535.

Lee, M. S. Y. and Scanlon, J. D. 2002. Snake phylogeny based on osteology, soft anatomy, and ecology. Biological Reviews 77: 333-401.

Lee, M.S.Y. and Caldwell, W. and Scanlon, J. D. 1999. A second primitive marine snake: *Pachyophis woodwardi* from the Cretaceous of Bosnia-Herzegovina. Journal of Zoology, London 248: 509-520.

Lee, M. S. Y., Hugall, A. F., Lawson, R. and Scanlon, J. D. 2007. Phylogeny of snakes (Serpentes): combining morphological and molecular data in likelihood, Bayesian, and parsimony analyses. Systematics and Biodiversity 4: 371-389.

Liem, K. F., Marx, H. and Rabb, G. B., 1971. The viperid snake *Azemiops* and its comparative anatomy and phylogenetic position in relation to Viperinae and Crotalinae. Fieldiana Zoologica 59: 67-126.

Linnaeus, C. 1758. Systema Naturae, Ed. X. (Systema naturae per regna tria naturae, secundum classes, ordines, genera, species, cum characteribus, differentiis, synonymis, locis. Tomus I. Editio decima, reformata.) Holmiae. Systema Nat. ed. 10. Pp. i–ii + 1-824.

Lydekker, R. 1888. Notes on Tertiary Lacertilia and Ophidia. Geological Magazine 5: 110-113.

Lynch, V. J. 2009. Live-birth in vipers (Viperidae) is a key innovation and adaptation to global cooling during the Cenozoic. Evolution 63: 2457-2465.

Macey, J. R. and Verma, A. 1997. Homology in phylogenetic analysis: alignment of transfer RNA genes and the phylogenetic position of snakes. Molecular Phylogenetics and Evolution 7: 272-279.

Martins, M. and Oliveira, M. E. 1999. Natural history of the snakes in forests of the Manaus region, Central Amazonia, Brazil. Herpetological Natural History 6: 78-150.

Maschio, G. F., Prudente, A. L. da C., de Lima, A. C. and Feitosa, D. T. 2007. Reproductive Biology of *Anilius scytale* (Linnaeus, 1758) (Serpentes, Aniliidae) from Eastern Amazonia, Brazil. South American Journal of Herpetology 2: 179-183.

McDiarmid, R. W., Campbell, J. A. and Touré, T. A. 1999. *Snake Species of the World, a Taxonomic and Geographic Reference*, Vol. 1. The Herpetologists' League, Washington, D. C. Pp. 512.

McDowell, S. B. 1967. Osteology of the Typhlopidae and Leptotyphlopidae: a critical review. Copeia 1967: 686-692.

McDowell, S. B. 1974. A catalogue of the snakes of New Guinea and the Solomons, with special reference to those in the Bernice P. Bishop Museum, Pt. 1. Scolecophidia. Journal of Herpetology 8: 1-57.

McDowell, S. B. 1987. Systematics. Pp. 3-50. In R. A. Seigel, J. T. Collins and S. S. Novak (eds), *Snakes: Ecology and Evolutionary Biology*. Macmillan Publishing Company, New York.

McDowell, S. B., Jr. and Bogert, C. M. 1954. The systematic position of *Lanthanotus borneensis*, and the affinities of the anguinomorphan lizards. Bulletin of the American Museum of Natural History 105: 1-142.

Mulcahy, D. G. 2007. Molecular systematics of neotropical cat-eyed snakes: a test of the monophyly of Leptodeirini (Colubridae: Dipsadinae) with implications for character evolution and biogeography. Biological Journal of the Linnean Society 92: 483-500.

Murphy, J. C. 2007. *Homalopsid Snakes: Evolution in the Mud*. Krieger Publishing, Florida. Pp. 260.

Murphy, J. C., Voris, H. K., Karns, D. R., Chanard, T. and Suvunrat, K. 1999. The ecology of the water snakes of Ban Tha Hin, Songkhla Province, Thailand. The Natural History Bulletin of the Siam Society 47: 129-147.

Nagy, Z. T., Lawson, R., Joger, U. and Wink, M. 2004. Molecular systematics of racers, whipsnakes and relatives (Reptilia: Colubridae) using mitochondrial and nuclear markers. Journal of Zoological Systematics and Evolutionary Research 42: 223-233.

Nagy, Z. T., Joger, U., Wink, M., Glaw, F. and Vences, M. 2003. Multiple colonization of Madagascar and Socotra by colubrid snakes: evidence from nuclear and mitochondrial gene phylogenies. Proceedings of the Royal Society, Biological Sciences 270: 2613-2621.

Nessov, L. A., Zhegallo, V. I. and Averianov, A. O. 1998. A new locality of Late Cretaceous snakes, mammals and other vertebrates in Africa (western Libya). Annales de Paléontologie 84: 265-275.

Noonan, B. P. and Chippindale, P. T. 2006. Dispersal and vicariance: The complex evolutionary history of boid snakes. Molecular Phylogenetics and Evolution 40: 347-358.

Nopcsa, F. 1923. *Eidolosaurus* und *Pachyophis*, Zwei neue Neocom-Reptilien. Palaeontographica 55: 97-154.

Odinchenko, V. I. and Latyshev, V. A. 1996. Keeping and breeding in captivity the Mexican burrowing python *Loxocemus bicolor* (Cope, 1961) at Moscow Zoo. Russian Journal of Herpetology 3: 95-96.

Parmley, D. and DeVore, M. 2005. Palaeopheid snakes from the Late Eocene Hardie Mine local fauna of central Georgia. Southeastern Naturalist 4: 703-722.

Pinou, T., Vicario, S., Marschner, M. and Caccone, A. 2004. Relict snakes of North America and their relationships within Caenophidia, using likelihood-based Bayesian methods on mitochondrial sequences. Molecular Phylogenetics and Evolution 32: 563-574.

Pope, K. O., D'Hondt, S. L. and Marshall, C.R. 1998. Meteorite impact and the mass extinction of species at the Cretaceous Tertiary boundary. Proceedings of the National Academy of Sciences USA 95: 11028-11029.

Pough, F. H., Andrews, R. M., Cadle, J. E., Crump, M. L., Savitsky, A. H. and Wells, K. D. 2004. *Herpetology*. Pearson Prentice Hall, New Jersey. Pp. 726.

Pyron, R. A. and Burbrink, F. T. 2009. Neogene diversification and taxonomic stability in the snake tribe Lampropeltini (Serpentes: Colubridae). Molecular Phylogenetics and Evolution 52: 524-529.

Pyron, R., Burbrink, F., Colli, G., Kuczynski, C., Montes de Oca, A., Vitt, L. and Wiens, J. J. In Press. The Phylogeny of Advanced Snakes (*Colubroidea*), with Discovery of a New Subfamily and Comparison of Support Methods for Likelihood Trees. Molecular Phylogenetics and Evolution.

Rage, J.-C. 1975. Un Caenophidien primitif (Reptilia, Serpentes) dans l'Eocène inférieur. Compte Rendu Sommaire des Séances de la Société Géologique de France 17: 46-47.

Rage, J.-C. 1984. *Serpentes*. Gustav Fischer, New York. Pp. 80.

Rage, J.-C. 1987. Fossil history. Pp. 49-76. In R. A. Seigel, J. T. Collins and S. S. Novak (eds), *Snakes: Ecology and Evolutionary Biology*. MacMillan, NewYork.

Rage, J.-C. and Werner, C. 1999. Mid-Cretaceous (Cenomanian) snakes from Wadi Abu Hashim, Sudan: The earliest snake assemblage. Palaeontologia Africana 35: 85-110.

Rage, J.-C. and Escuillé, F. 2000. Un nouveau serpent bipe`de du Cénomanien (Crétacé). Implications phylétiques. Comptes Rendus de l'Académie des Sciences Série IIa Earth and Planetary Sciences 330: 513-520.

Rawlings, L. H., Rabosky, D. L., Donnellan, S. C. and Hutchinson, M. N. 2008. Python phylogenetics: inference from morphology and mitochondrial DNA. Biological Journal of the Linnean Society 93: 603-619. Reptile Database: http://www.reptile-database.org.

Reynoso, V.-H. 1998. *Huehuecuetzpalli mixtecus mixtecus* gen. et sp. nov. a basal squamate (Reptilia) from the Early Cretaceous of Tepexide Rodríguez, Central México. Philosophical Transactions of the Royal Society of London Series B Biological Sciences 353: 477-500.

Rieppel, O. 1980. The trigeminal jaw adductors of primitive snakes and their homologies with the lacertilian jaw adductors. Journal of Zoology 190: 447-471.

Rieppel, O. 1988. A review of the origin of snakes. Evolutionary Biology 22: 37-130.

Rieppel, O. and Head, J.J. 2004. New specimens of the fossil snake genus *Eupodophis* Rage and Escuillié, from the Mid-Cretaceous of Lebanon. Memorie della Società Italiana di Scienze Naturali e Museo Civico di Storia Naturale di Milano 23: 1-26

Romer, A. S. 1956. *Osteology of the Reptiles*. University of Chicago Press, Chicago. Pp. 772.

Sanders, K. L. and Lee, M. S. Y. 2008. Molecular evidence for a rapid Late-Miocene radiation of Australasian venomous snakes (Elapidae, Colubroidea). Molecular Phylogenetics and Evolution 46: 1165-1173.

Sanders, K. L., Lee, M. S. Y., Leys, R., Foster, R. and Keogh, S. J. 2008. Molecular phylogeny and divergence dates for Australasian elapids and sea snakes (Hydrophiinae): evidence from seven genes for rapid evolutionary radiations. Journal of Evolutionary Biology 21: 682-695.

Santos-Costa, M. C. dos and Hofstadler-Deiques, C. 2002. The ethmoidal region and cranial adaptations of the neotropical aquatic snake *Helicops infrataeniatus* Jan. 1865 (Serpentes, Colubridae). Amphibia-Reptilia 23: 83-91.

Savage, J. M. 2002. *The Amphibians and Reptiles of Costa Rica. A Herpetofauna between Two Continents, between Two Seas*. University of Chicago Press, Chicago. Pp. 934.

Scanlon, J. D. 2006. Skull of the large non-macrostomatan snake Yurlunggur from the Australian Oligo-Miocene. Nature 439: 839-842.

Scanlon, J. D. and Lee, M. S. Y. 2002. Varanoid like dentition in primitive snakes (Madtsoiidae). Journal of Herpetology 36: 100-106.

Scanlon, J. D. and Lee, M. S. Y. 2004. Phylogeny of Australasian venomous snakes (Colubroidea, Elapidae, Hydrophiinae) based on phenotypic and molecular evidence. Zoologica Scripta 33: 335-366.

Schleip, W. D. and O'Shea, M. In Review. Annotated checklist of the recent and extinct pythons 1 (Serpentes: Pythonidae), with notes on nomenclature, taxonomy, and distribution. ZooKeys.

Schuett, G., Höggren, M., Douglas, M. E. and Greene H. W. (eds). 2002. *Biology of the Vipers*. Eagle Mountain Publishing, Eagle Mountain, Utah. Pp. 580.

Schwartz, A. and Henderson, R. W. 1991. *Amphibians and Reptiles of the West Indies*. University of Florida Press, Florida. Pp. 720.

Sepkoski, J. J. 2002. A compendium of fossil marine animal genera. Bulletins of American Paleontology 363: 1-560.

Shine, R. 1986. Ecology of a low energy specialist: food habits and reproductive biology of the Arafura filesnake (Acrochordidae). Copeia 1986: 424-437.

Shine, R., Branch, W. R., Webb, J. K., Harlow, P. S. and Shine, T. 2006. Sexual dimorphism, reproductive biology, and dietary habits of psammophiine snakes (Colubridae) from southern Africa. Copeia 2006: 650-664.

Slowinski, J. B. and Keogh, J. S. 2000. Phylogenetic relationships of elapid snakes based on cytochrome *b* mtDNA sequences. Molecular Phylogenetics and Evolution 15: 157-164.

Slowinski, J. B. and Lawson, R. 2002. Snake phylogeny: evidence from nuclear and mitochondrial genes. Molecular Phylogenetics and Evolution 23: 194-202.

Spawls, S. 1997. Natural history notes — *Pseudoboodon lemniscatus*; reproduction and size. Herpetological Review 28: 45.

Tchernov, E., Rieppel, O., Zaher, H., Polcyn, M. J. and Jacobs, L. L. 2000. A fossil snake with limbs. Science 287: 2010-2012.

Tolson, P. J. and Henderson, R. W. 1993. *The Natural History of West Indian Boas*. R & A Publishing, London, UK. Pp. 125.

Townsend, T. M., Larson, A., Louis, E. and Macey, J. R. 2004. Molecular phylogenetics of Squamata: The position of snakes, amphisbaenians, and dibamids, and the root of the squamate tree. Systematic Biology 53: 735-757.

Underwood, G. 1967. *A Contribution to the Classification of Snakes*. The British Museum, London. Pp. 179.

Utiger, U., Schatti, B. and Helfenberger, N. 2005. The oriental colubrine genus *Coelognathus* Fitzinger, 1843 and classification of Old and New World racers and ratsnakes (Reptilia, Squamata, Colubridae, Colubrinae). Russian Journal of Herpetology 12: 39-60.

Utiger, U., Helfenberger, N., Schatti, B., Schmidt, C., Ruf, M. and Ziswiler, V. 2002. Molecular systematics and phylogeny of Old and New World rat snakes, *Elaphe* Auct., and related genera (Reptilia, Squamata, Colubridae). Russian J. Herp. 9: 105-124.

Van Mierop, L. H. S. and Barnard, S. M. 1976. Thermoregulation in a brooding female *Python molurus bivattatus* (Serpentes: Boidae). Copeia 1976: 398-401.

Van Mierop, L. H. S. and Barnard, S. M. 1978. Further observations on thermoregulation in the brooding female *Python molurus bivittatus* (Serpentes: Boidae). Copeia 1978: 615-621.

Vidal, N. and Hedges, S. B. 2002. Higher-level relationships of snakes inferred from four nuclear and mitochondrial genes. Comptes Rendus de l'Académie des Sciences, Paris Biologies 325: 977-985.

Vidal, N. and Hedges, S. B. 2005. The phylogeny of squamate reptiles (lizards, snakes, and amphisbaenians) inferred from nizne nuclear protein-coding genes. Comptes Rendus Biologies 328: 1000-1008.

Vidal, N. and Hedges, S. B. 2009. The molecular evolutionary tree of lizards, snakes, and amphisbaenians. Comptes Rendus Biologies 332: 129-139.

Vidal, N., Kindl, S. G., Wong, A. and Hedges, S. B. 2000. Phylogenetic relationships of xenodontine snakes inferred from 12s and 16s ribosomal RNA sequences. Molecular Phylogenetics and Evolution 14: 389–402.

Vidal, N., Rage, J.-C., Couloux, A. and Hedges, S. B. 2009. Snakes (Serpentes). Pp. 390-397. In S. B. Hedges and S. Kumar (eds), *The Timetree of Life*. Oxford University Press, Oxford, U.K.

Vidal, N., Delmas, A.-S., David, P., Cruaud, C., Couloux, A. and Hedges, S. B. 2007. The phylogeny and classification of caenophidian snakes inferred from seven nuclear protein coding genes. Comptes Rendus Biologies 330: 182-187.

Vidal, N., Branch, W. R., Pauwels, O. S. G., Hedges, S. B., Broadley, D. G., Wink, M., Cruaud, C., Joger, U. and Nagy, Z. T. 2008. Dissecting the major African snake radiation: a molecular phylogeny of the Lamprophiidae Fitzinger (Serpentes, Caenophidia). Zootaxa 1945: 51-66.

Vitt, L. J. and Caldwell, J. P. 2009. *Herpetology: An Introductory Biology of Amphibians and Reptiles*. Academic Press, California. Pp. 720.

Voris, H. K., Karns, D. R., Feldheim, K. A., Kechavarzi, B. and Rinehart, M. 2008. Multiple paternity in the Oriental-Australian rear-fanged watersnakes (Homalopsidae). Herpetological Conservation and Biology 3: 88-102.

Wallach, V. and Günther, R. 1998. Visceral anatomy of the Malaysian snake genus *Xenophidion*, including a cladistic analysis and allocation to a new family (Serpentes: Xenophidiidae). Amphibia-Reptilia 19: 385-404.

Webb, J. K., Shiner, R., Branch, W. R. and Harlow, P. S. 2000. Life history strategies in basal snakes: reproduction and dietary habits of the African thread snake *Leptotyphlops scutifrons* (Serpentes: Leptotyphlopidae). Journal of Zoology, London 250: 321-327.

White, M. E., Kelly-Smith, M. and Crother, B. I. 2005. Higher-level snake phylogeny as inferred from 28s ribosomal DNA and morphology. Pp. 156-173. In M. A. Donnelly, B. I. Crother, C. Guyer, M. H. Wake and M. E. White (eds), *Ecology and Evolution in the Tropics*. The University of Chicago Press, Chicago, IL.

Wiens, J. J., Brandley, M. C. and Reeder, T. W. 2006. Why does a trait evolve multiple times within a clade? Repeated evolution of snakelike body form in squamate reptiles. Systematic Biology 60: 123-141.

Wiens, J. J., Kuczynski, C. A., Smith, S. A., Mulcahy, D. G., Sites, J. W., Townsend, T. M. and Reeder, T. W. 2008. Branch length, support, and congruence: testing the phylogenomic approach with 20 nuclear loci in snakes. Systematic Biology 57: 420-431.

Wilcox, T. P., Zwickl, T. A., Heath, T. A. and Hillis, D. M. 2002. Phylogenetic relationships of the dwarf boas and a comparison of Bayesian and bootstrap measures of phylogenetic support. Molecular Phylogenetics and Evolution 23: 361-371.

Wilson, L. D. and Meyer, J. R. 1985. *The Snakes of Honduras*. Special Publications in Biology and Geology, Milwaukee Public Museum, Milwaukee. Pp. 150.

Wüster, W., Peppin, L., Pook, C. E. and Walker, D. E. 2008. Nesting of vipers: phylogeny and historical biogeography of the Viperidae (Squamata: Serpentes). Molecular Phylogenetics and Evolution 49: 445-459.

Wüster, W., Crookes S., Ivan, I., Mané, Y., Pook, C. E., Trape, J.-F. and Broadley, D. G. 2007. The phylogeny of cobras inferred from mitochondrial DNA sequences: evolution of venom spitting and the phylogeography of the African spitting cobras (Serpentes: Elapidae: *Naja nigricollis* complex). Molecular Phylogenetics and Evolution 45: 437-453.

Zaher, H. 1994. Les Tropedopheoidea (Serpentes; Alethinophidia) sont-ils réellement monophylétiques? Arguments en favor de leur polyphylétisme. Comptes Rendus de l'Académie des Sciences Paris 317: 471-478.

Zaher, H. 1998. The phylogenetic position of *Pachyrhachis* within snakes (Squamata, Lepidosauria). Journal of Vertebrate Paleontology 18: 1-3.

Zaher, H. 1999. Hemipenial morphology of the South American xenodontine snakes, with a proposal for a monophyletic Xenodontinae and a reappraisal of colubroid hemipenes. Bulletin of the American Museum of Natural History 240: 1-168.

Zaher, H. and Rieppel, O. 1999. The phylogenetic relationships of *Pachyrhachis problematicus*, and the evolution of limblessness in snakes (Lepidosauria, Squamata). Comptes Rendus de l'Académie des Sciences de Paris, Sciences de la Terre et des Planètes 329: 831-837.

Zaher, H. and Rieppel, O. 2002. On the phylogenetic relationships of the Cretaceous snakes with legs, with special reference to *Pachyrhachis problematicus* (Squamata, Serpentes). Journal of Vertebrate Paleontology 22: 104-109.

Zaher, H., Grazziotin, F. G., Cadle, J. E., Murphy, R. W., Cesar de Moura-Leite, J. and Bonatto, S. L. 2009. Molecular phylogeny of advanced snakes (Serpentes, Caenophidia) with an emphasis on South American xenodontines: a revised classification and descriptions of new taxa. Papéis Avulsos de Zoologia 49(11): 115-153.

Zhao, E., Huang, M., Zong, Y., Jiang, Y., Huang, Q., Zhao, H., Ma, J., Zheng, J., Huang, Z., Wei, G., Yang, D. and Li, D. 1998. *Fauna Sinica. Reptilia*. vol. 3. *Squamata. Serpentes*. Science Press, Beijing. Pp. 522.

CHAPTER 3

The Major Clades of Living Snakes: Morphological Evolution, Molecular Phylogeny, and Divergence Dates

John D. Scanlon[1] *and Michael S. Y. Lee*[2,3]

3.1 INTRODUCTION

Snakes are among the most charismatic and highly-studied organisms (Greene 1997), yet our understanding of their early evolution and phylogeny remains in a state of flux. Extensive anatomical information (e.g., Underwood 1967; McDowell 1974, 1975, 1979), analyzed using quantitative phylogenetic methods (e.g., Kluge 1991; Cundall *et al.* 1993), had led to a broad consensus on relationships among living snakes (Lee and Scanlon 2002; Rieppel *et al.* 2003). The tiny, burrowing, worm-like blindsnakes (scolecophidians) were considered the most basal clade of living snakes, followed by other small burrowing taxa with restricted gapes (pipesnakes and shieldtail snakes). The partly surface-active, and moderate-gaped sunbeam snakes (*Xenopeltis* and *Loxocemus*) were transitional forms, while the typical, generally surface-active and large-gaped snakes (such as pythons, boas, and colubroids) were inferred to represent a single, derived radiation ("core macrostomatans"). This phylogeny implied that snake evolution involved consistent trends towards greater surface activity, increased body size, and enlarged gape (e.g., Underwood 1967; Rodríguez-Robles *et al.* 1999). However, some studies of primitive fossil snakes with large body size and extensive gapes did not support this scenario, although the exact phylogenetic position of these fossils remains debated (e.g., Caldwell 2007; Wilson *et al.* 2010). Most recently, increasingly large molecular sequence datasets have further challenged the traditional scenario (Slowinski and Lawson 2002; Wilcox *et al.* 2002; Lawson *et al.* 2005; Vidal *et al.* 2007a, 2007b; Wiens *et al.* 2008; Burbrink and Crother 2010).

[1]Riversleigh Fossil Centre, Outback at Isa, Marian Street, Mt Isa 4825, Australia
[2]Earth Sciences Section, South Australian Museum North Tce, Adelaide 5000, Australia [Address for correspondence]
[3]School of Earth and Environmental Sciences, University of Adelaide, Adelaide 5005, Australia

Notably, the molecular sequences consistently place the morphologically advanced dwarf boas (tropidophiines) near the base of snakes, and conversely place the primitive sunbeam snakes (*Xenopeltis*) with pythons. Like the fossils, the molecular data imply considerable homoplasy in the evolution of gape size and fossorial habits.

Here, the anatomical evidence is integrated with the extensive molecular sequence data, in a combined analysis of 214 skeletal characters, 20 nuclear genes, 2 mitochondrial genes, plus aligned indels (gaps). Anguimorph lizards are used as outgroups, based on recent molecular and morphological evidence (Lee 2009). In addition to all the major clades of living snakes, we include two enigmatic living taxa (*Anomochilus* and *Xenophidion*), based on all relevant available information (Appendix). The relationships among the major living snake lineages are shown in Figure 3.1. With certain exceptions noted below (e.g., monophyly of blindsnakes), it is generally similar to the recent phylogenies based on large nuclear datasets (e.g., Vidal and Hedges 2007a; Wiens *et al.* 2008; Burbrink and Crother 2010). We first describe the major clades of living snakes, and then briefly discuss important fossil taxa and the likely timeframe for snake diversification.

3.2 THE MAJOR CLADES OF LIVING SNAKES

3.2.1 Blindsnakes: The Subterranean Branch of Living Snakes

The most basal clade of living snakes is the scolecophidians (blindsnakes), which consist of the anomalepidids, leptotyphlopids and typhlopids. They are worm-like reptiles that seldom emerge above ground, and are specialized for rapidly ingesting vast numbers of small subterranean invertebrates (Kley 2001; Cundall and Irish 2008). Scolecophidians include the smallest known snakes, and the only known parthenogenetic snakes. As their common name suggests, their eyes are greatly reduced, containing one (rather than two) types of visual cells (Underwood 1967), often covered by opaque scales. While recent extensive molecular studies have failed to robustly support or refute scolecophidian monophyly (Wiens *et al.* 2008; Vidal *et al.* 2009; Burbrink and Crother 2010), the first two

Fig. 3.1 Phylogenetic relationships between the major clades of snakes, based on analyses of a large data set (20 nuclear genes, 3 mitochondrial genes, indels and morphology/osteology: see Appendix). Both parsimony and Bayesian analyses of the combined data produced this tree. To test the hypothesis that *Anomochilus* is nested within *Cylindrophis* (Gower *et al.* 2005), two species of *Cylindrophis* were included. The 3 numbers at each node denote the following: Parsimony branch (=Bremer) support / Parsimony bootstrap frequency / Likelihood bootstrap frequency (sequence data only). Illustrations are as follows: (1) *Typhlops vermicularis* (blindsnake), (2) *Tropidophis greenwayi* (Neotropical dwarf "boa"), (3) *Cylindrophis ruffus* (asian pipesnake), (4) *Liasis mackloti* (water python), (5) *Eunectes notaeus* (anaconda), (6) *Acrochordus arafurae* (Arafura filesnake), (7) *Crotalus willardi* (ridge-nosed rattlesnake). See Acknowledgments for photo credits.

Color image of this figure appears in the color plate section at the end of the book.

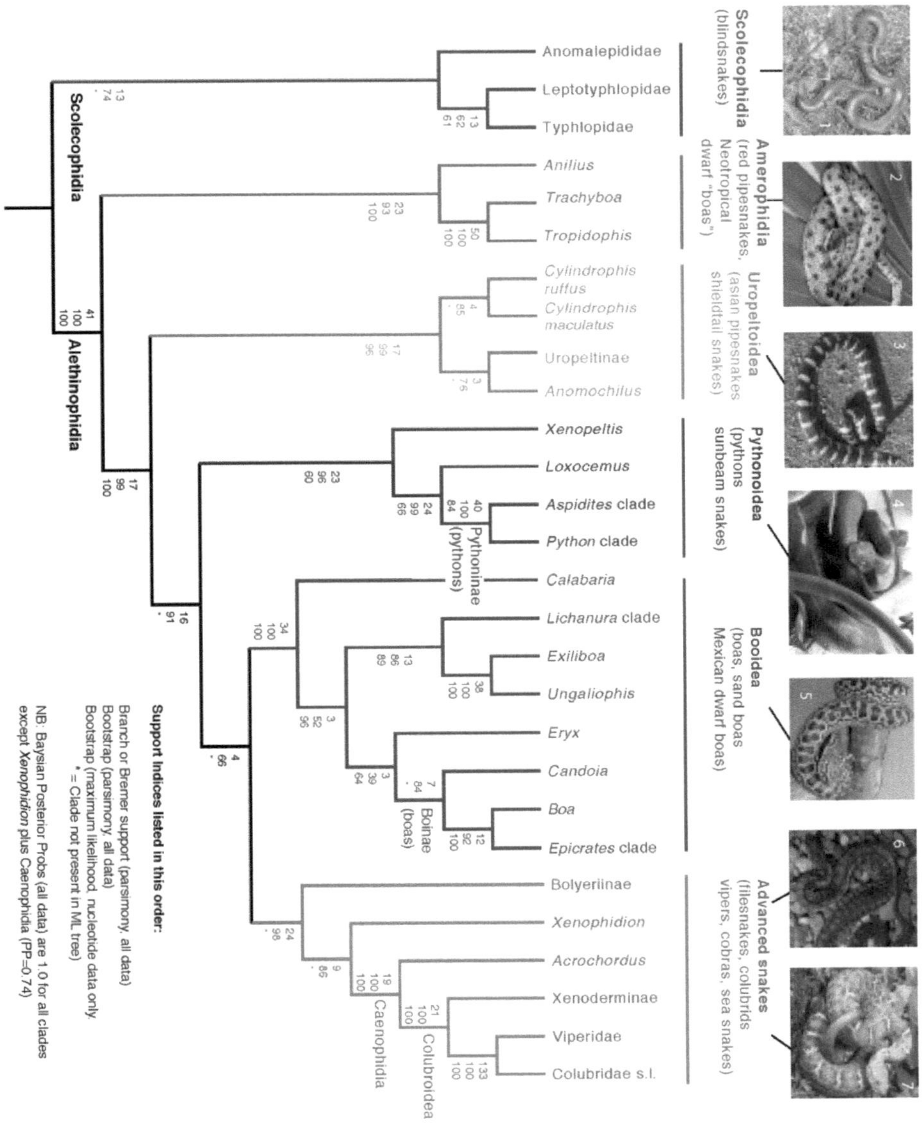
Scolecophidia
Alethinophidia
Anomalepididae
Leptotyphlopidae
Typhlopidae
Anilius
Trachyboa
Tropidophis
Cylindrophis ruffus
Cylindrophis maculatus
Uropeltinae
Anomochilus
Xenopeltis
Loxocemus
Aspidites clade
Python clade
Pythoninae (pythons)
Calabaria
Lichanura clade
Exiliboa
Ungaliophis
Eryx
Candoia
Boa
Epicrates clade
Boinae (boas)
Bolyeriinae
Xenophidion
Acrochordus
Xenoderminae
Viperidae
Colubridae s.l.
Caenophidia
Colubroidea
Scolecophidia (blindsnakes)
Amerophidia (red pipesnakes, Neotropical dwarf "boas")
Uropeltoidea (asian pipesnakes shieldtail snakes)
Pythonoidea (pythons sunbeam snakes)
Booidea (boas, sand boas Mexican dwarf boas)
Advanced snakes (filesnakes, colubrids vipers, cobras, sea snakes)
Support Indices listed in this order:
Branch or Bremer support (parsimony, all data)
Bootstrap (parsimony, all data)
Bootstrap (maximum likelihood, nucleotide data only.
* = Clade not present in ML tree)
NB: Baysian Posterior Probs (all data) are 1.0 for all clades except Xenophidion plus Caenophidia (PP=0.74)

studies have acknowledged that numerous distinctive specializations in all scolecophidians (see below) strongly support monophyly. Here, combining the relatively small morphological dataset with the extensive molecular data restores the traditional picture of scolecophidian monophyly. Anomalepidids are the most basal scolecophidians; this is consistent with presence of (vestigial) supratemporal and ectopterygoid bones in most anomalepidids, and their absence in leptotyphlopids and typhlopids (Rieppel *et al.* 2009).

Many of the distinctive specializations shared by all scolecophidians (Underwood 1967; Kley 2001; Cundall and Irish 2008) are associated with small gape and a diet of numerous small invertebrates. They have reduced or lost several bones of the jaw suspension (supratemporal) and palate (ectopterygoid, pterygoid), and have anteriorly-placed jaw articulations resulting in short jaws housing a greatly reduced number of teeth. Other derived traits are associated with a fossorial lifestyle, such as a consolidated spherical snout, vertebrae totally lacking neural spines but possessing enlarged subcentral foramina, and a short stumpy tail. It has thus been noted that the morphological support for scolecophidian monophyly might be weaker than it appears (Burbrink and Crother 2010). A few other synapomorphies, however, are not obviously correlated with general habitus, leading to greater confidence in monophyly: these include the optic foramen located anteriorly (entirely within the frontal bone), and the frontal bones which overlap the nasals (rather than vice versa).

3.2.2 Basal Alethinophidians and the Evolution of Gape Size

All living snakes apart from blindsnakes form a clade, the alethinophidians, which is robustly supported by both morphology (Underwood 1967; Rieppel 1988; Cundall *et al.* 1993; Lee and Scanlon 2002) and molecules (e.g., Slowinski and Lawson 2002; Wiens *et al.* 2008, Vidal *et al.* 2009; Burbrink and Crother 2010). This clade spans almost the entire range of snake morphologies and ecologies, making generalisations difficult. Unlike scolecophidians, they always feed on fewer, larger prey (often vertebrates). All share several evolutionary novelties absent in scolecophidians. Many are associated with larger prey: they have mobile "prokinetic" snouts (Cundall and Irish 2008), very loose chin connections (Young 1998), a unique arrangement of the jaw adductor muscles (Cundall *et al.* 1993), and often employ constriction (lost in some colubroids including most venomous forms). In the braincase, the olfactory lobes are separated by a median bony wall, there is a distinct ossification around the trigeminal nerve roots (the laterosphenoid), the vidian canal enters the brain cavity, and the sphenoid has a distinct triangular "wing" that projects dorsally (meeting the front of the prootic). They also share some distinctive soft anatomical features, such as four thymus glands (Underwood 1967; Wallach and Gunther 1998), and ventral scales that are at least slightly widened (facilitating above-ground locomotion).

The most basal clade within Alethinophidia is a morphologically disparate group containing the fossorial *Anilius* (red pipe snake), and the terrestrial-to-arboreal tropidophiines: *Tropidophis* (wood boas) and *Trachyboa* (eyelash boas). Their feeding habits are similarly diverse; *Anilius* specializes on narrow, elongate prey (e.g., caecilians, amphisbaenians), but tropidophiines frequently take large-diameter prey (e.g., frogs and lizards). While detailed kinematic analyses have yet to be undertaken, morphology and diet suggest that *Anilius* is small-gaped (non-macrostomatan), while tropidophiines are large-gaped (macrostomatan). Despite such morphological and ecological disparity, the clade nevertheless has a tight neotropical distribution, and has recently been dubbed "Amerophidia" (Vidal *et al.* 2009). Amerophidian monophyly is supported by both mitochondrial (Slowinksi and Lawson 2002) and nuclear (Vidal and Hedges 2007a; Wiens *et al.* 2008; Burbrink and Crother 2010) genes and can be considered robust; however, it remains highly inconsistent with morphology. Tropidophiines exhibit numerous advanced characters across different anatomical systems, which have traditionally placed them with various groups of derived snakes (Underwood 1967). Unlike *Anilius*, tropidophiines have prominent, haemal keel-like hypapophyses on all presacral vertebrae (not just the anterior presacrals; Szyndlar *et al.* 2008), a reduced coronoid bone in the mandible, complex jaw muscles (with a divided intermandibularis anterior), keeled dorsal scales (often associated with arboreality), greatly widened ventral scales (true gastrosteges), and a fully divided hemipenis. If the molecular evidence is correct, all these characters are homoplasious: either primitive in alethinophidians and lost in *Anilius* and uropeltoids (e.g., Vidal *et al.* 2009), or convergent in tropidophiines and true macrostomatans (pythonoids, boids, and advanced snakes). *Anilius* and tropidophiines, however, share at least one unusual derived character: on the anterior palate, the vomers are expanded posteriorly, restricting the bony border of the internal nostrils. This unusual character otherwise occurs only amongst some pythonoids among living snakes. In addition, tropidophiines, like other basal alethinophidians such as *Anilius* and uropeltoids, have a stapes (ear ossicle) that attaches to the dorsal end of the quadrate; in more derived snakes this articulation is typically more ventral (Cundall and Irish 2008). Finally, while the monophyly of Amerophidia is strongly supported by molecular data, its position at the base of Alethinophidia is less robust: one molecular analysis places it with booids and pythonoids (Burbrink and Crother 2010).

The next clade of alethinophidians is both morphologically and geographically homogeneous, containing burrowing, small-jawed forms with an Asian distribution: *Cylindrophis* (Asian pipe snakes), uropeltines (shield-tail snakes), and the rare *Anomochilus* (dwarf pipe snakes). The name Uropeltoidea has been applied to this clade (Vidal *et al.* 2009). All specialize on long-bodied, fossorial prey, such as caecilians and worms (Cundall and Greene 2000). This clade is diagnosed by several evolutionary novelties, such as a well-formed articulation between the snout (premaxilla) and palate (vomer), an occipital condyle (neck articulation) that incorporates the exoccipitals and is spherical (lacking the typical dorsal concavity), and loss of the omohyoideus muscle in the neck region (Lee and Scanlon 2002). Combined morphological and molecular data (Fig. 3.1) robustly place uropeltoids basal to all alethinophidians (except *Anilius* and tropidophiines), but the molecules alone weakly unite them with pythonoids (Vidal *et al.* 2007a) or booids (Wiens *et al.* 2008). The combined dataset also groups *Anomochilus* with uropeltines, consistent with some unusual shared anatomical traits such as a wide, toothless anterior palatine process, a single palatal foramen on the premaxilla, and a distinct flange on the retroarticular process of the lower jaw (Cundall *et al.* 1993; Lee and Scanlon 2002). Morphological analyses previously placed *Anomochilus* at the very base of Alethinophidia (Cundall *et al.* 1993), while the (extremely limited) molecular sequences available nested it within *Cylindrophis* (Gower *et al.* 2005).

In contrast to the strong recent molecular evidence for a basal position of tropidophiines, older analyses based largely on morphological data (reviewed in Lee and Scanlon 2002) have concluded that basal alethinophidians are entirely fossorial, small-gaped forms that feed on elongate burrowing prey (*Anilius, Cylindrophis*, uropeltids and *Anomochilus*). This led to the view that alethinophidian (and thus, snake) evolution was a trend towards increased surface activity, increased gape size, and large-diameter prey, culminating in "macrostomatans" (Greene 1983; Cundall *et al.* 1993). However, the robust molecular evidence placing the partly arboreal, large-gaped tropidophiines near the very base of Alethinophidia (below *Cylindrophis*, uropeltids and *Anomochilus*) makes a very different scenario plausible. Surface-activity, large gape and wide prey might be primitive for snakes, or at least alethinophidians; fossoriality has evolved repeatedly in three basal lineages (blindsnakes, *Anilius*, and uropeltoids) leading to concomitant reduction of gape, and specialization on small or elongate prey (e.g., Vidal *et al.* 2009).

3.2.3 Large Constrictors and Their Allies

The large constrictors (boas and pythons) are not closest relatives; rather, they are each nested within clades of smaller, often fossorial taxa, termed respectively Booidea and Pythonoidea. The shared similarities between boas and pythons (Lee and Scanlon 2002) are thus convergent: extremely flexible tissues between the lower jaws (Cundall and Greene 2000), various skull characters (Lee and Scanlon 2002), and heat sensor pits along the upper and lower lip (variably present: Kluge 1993). Many of these traits probably facilitate predation on large endothermic (warm-blooded) prey, i.e., mammals and birds, and together with novel changes in development (Head and Polly 2007), have probably provided the context for the convergent evolution of gigantism within both boas and pythons.

Booidea comprises boines (boas), erycines (sand boas), ungaliophiines (Central American dwarf boas), and *Calabaria* (burrowing "python"). They are typically wide-gaped, thick-bodied, constricting snakes. The nesting of ungaliophiines within booids has recently been supported by both morphology (Zaher 1994) and molecules (e.g., Noonan and Chippindale 2006; Vidal *et al.* 2007a; Wiens *et al.* 2008). Most booids possess the following three unusual traits in the palate: the maxilla contacts the pterygoid (excluding the palatine from the suborbital fenestra), the maxillary process of the palatine is situated very posteriorly, and the ectopterygoid abuts rather than overlaps the pterygoid. The latter two traits also occur, presumably convergently, in pythons (but not basal pythonoids). *Calabaria* is the most basal booid, and retains egg-laying habits. Live-bearing most probably evolved in the ancestor of the clade consisting of all other booids; the sole egg-laying species of *Eryx* appears to be secondarily oviparous, since neonates lack the egg tooth found in all other oviparous squamates but lost in live-bearing forms (Lynch 2009). There is little evidence for the monophyly of erycines (in any conventional sense), consistent with independent evolution of "sand boas" on different continents (Noonan and Chippindale 2006; Wiens 2008; Lynch 2009). The unexpected clade uniting the New World erycines (*Lichanura* and *Charina*) with ungaliophiines not only has strong molecular support, but is geographically homogenous and also diagnosed by an unusual morphological feature found in ungaliophiines, *Lichanura,* and at least some *Charina*: loss of the coronoid bone on the lower jaw (a rare trait otherwise characterizing caenophidians and some *Tropidophis*).

Pythonoidea is a heterogenous group that includes pythonines (pythons) and their nearest relatives, *Xenopeltis* and *Loxocemus* (sunbeam snakes). This clade is robustly supported by molecular data (Slowinski and Lawson 2002; Wiens *et al.* 2008; Vidal *et al.* 2009; Burbrink and Crother 2010) but contradicts both geography and morphology. Pythons (Afro-Asian-Australasian) and *Xenopeltis* (Asian) are widely disjunct from the neotropical *Loxocemus*. Further, pythons are terrestrial/arboreal forms famous for their prodigious swallowing abilities, whereas *Xenopeltis* and *Loxocemus* are partly fossorial and possess relatively small gapes. Nevertheless, two unusual traits are shared by most pythonoids: presence of teeth on the premaxilla (found elsewhere among living snakes only in *Anilius*), and paired rather than single scales under the tail (found elsewhere only in *Anilius, Bolyeria,* and some colubroids). A fossil loxocemid has also recently been interpreted to be more python-like than the living *Loxocemus,* adding further support to this relationship (Bhullar *et al.* 2009). Hybridization in captivity across highly divergent lineages ("genera") appears to be unusually prevalent in pythons (e.g. Torr 2000), indicating that post-zygotic isolating mechanisms have been slow to develop, although there has been no quantitative analysis of this phenomenon across snakes.

3.2.4 Advanced Snakes: Filesnakes, Colubroids, and Their Relatives

The remaining snakes comprise a clade informally termed "advanced snakes". These include the bolyeriines (Round-Island "boas"), the enigmatic

Xenophidion, acrochordids (filesnakes), and colubroids. The first three lineages are highly specialized, species-poor groups, but the colubroids are by far the most species-rich, diverse and cosmopolitan lineage of snakes. Several evolutionary novelties diagnose all advanced snakes (Groombridge 1979; Lee and Scanlon 2002), many involving the palate region involved in the "pterygoid walk", such as numerous (>9) large teeth on the palatine, and a large flange on the maxilla which articulates with the ectopterygoid (lost in acrochordids).

Bolyeriines are bizarre snakes restricted to Mauritius that possess a divided (two-part) maxilla in the upper jaw, probably an adaptation to grasp slippery-scaled lizards (Cundall and Irish 1989; Maisano and Rieppel 2007). *Bolyeria* is presumed extinct and has no associated molecular data, but molecular sequences of *Casarea* failed to robustly resolve its affinities (Vidal *et al.* 2007a; Wiens *et al.* 2008). In contrast, the combined data here robustly places *Casarea* with advanced snakes, consistent with its possession of the palatal adaptations discussed above. *Xenophidion* is an enigmatic lineage of two species (each known from a single specimen), described very recently (see Wallach and Gunther 1998). The limited morphological data did not conclusively resolve affinities, but successful sequencing of one gene placed it with bolyeriines (Lawson *et al.* 2004; see also Burbrink and Crother 2010). The combined analysis here (Fig. 3.1) is broadly consistent with the molecular data, placing *Xenophidion* near bolyeriines on the stem lineage of advanced snakes.

The bizarre acrochordids (filesnakes) are the nearest relatives of colubroids. The two groups share numerous novelties, such as a long supratemporal and posteriorly inclined quadrate, which together "carry" the jaw articulation backwards permitting long jaws relative to head size. They also have novelties in the nerves of the jaw muscles (Zaher 1994), and in the shape of the bones surrounding Jacobson's organ, the chemosensory structure in the palate which receives particles sampled during tongue-flicking (Underwood 1967; Lee and Scanlon 2002). Acrochordids are stout, granular-skinned snakes which inhabit freshwater and marine ecosystems around southeast Asia and Australasia. They are virtually helpless on land and give birth to live young underwater, like many sea snakes. They are specialized sit-and-wait predators that have extremely low metabolic rates, feeding infrequently and reproducing very slowly (Shine 1986). Their extremely large gape enables them to opportunistically swallow very large prey.

Colubroidea (colubroids) is the most species-rich and widely-distributed snake lineage, and the dominant group of snakes on all continents. They are characterized by advanced adaptations on the roof of the mouth for ratcheting prey down the throat (Cundall 1983; Greene 1997), loss of the right carotid artery and possession of a unique pattern of intercostal arteries (Underwood 1967), novel rib cartilages and axial musculature (e.g., Rieppel 1988), and unusual features of the spermatozoa (Jamieson and Koehler 1994; Tavares-Bastos *et al.* 2008). They also possess physiological novelties that are difficult to quantify. Compared to other snakes, they seem to have more rapid and sophisticated locomotion, which correlates with tendencies to be more diurnal and to use more open habitats (Greene 1997); however, rigorous studies are few. Until recently, the bewildering diversity of colubroids has hindered understanding of their internal phylogenetic relationships. However, large nuclear gene datasets (Vidal *et al* 2007a; Wiens *et al.* 2008; Burbrink and Crother 2010) indicate that the most basal colubroids are the semi-fossorial, amphibious xenodermines (mudsnakes) and terrestrial-to-arboreal pareatines (old-world slug-eaters), a view consistent with earlier mitochondrial analyses (Kelly *et al.* 2003; Lawson *et al.* 2005). Both groups are non-venomous (aglyphous) forms found in southeast Asia, which is also the centre of distribution of the colubroid sister group (acrochordids); this supports the idea of an Asian origin for Colubroids (Greene 1997). Unlike most other colubroids, xenodermines and pareatines retain simple, unmodified posterior maxillary teeth and apparently lack a "dental gland" (Underwood 1997).

Another group of basal colubroids are the highly venomous viperids, which include rattlesnakes (*Crotalus*), adders (*Vipera*), and copperheads (*Agkistrodon*). They are solenoglyphous, possessing short, mobile maxillae bearing fangs which are erected forwards while striking. Because the fangs can be swung backwards when not in use, they can be greatly enlarged. They are generally stout-bodied, sit-and-wait predators, but some arboreal forms are more slender. The venom is often predominantly haemotoxic, damaging the blood circulatory system, muscles, and other tissues and leading to extensive scarring (if survived).

The remaining colubroids form a large and very successful clade ("Colubridae" sensu Wiens *et al.* 2008), making up over 80% of living snake species. The relationships and taxonomy of much of this group remain in a state of flux (e.g., Kelly *et al.* 2003, 2009; Lawson *et al.* 2005; Vidal *et al.* 2007b; Wiens *et al.* 2008; Burbrink and Crother 2010). They include such forms as grass snakes (Natricines), racers, ratsnakes and boomslangs (Colubrinae) and hognose snakes (Xenodontinae); these groups are all discussed in more detail in Burbrink and Crother (2010). Enlarged, fixed rear fangs (opisthoglyphy) occurs widely throughout all these groups, in over 1/3 of species (Greene 1997) and might even be primitive for Colubridae (sensu Wiens *et al.* 2008) as a whole (Underwood 1967; Jackson 2007). Rear fangs are less effective than front fangs, as a deep chewing action is often required for envenomation. Nevertheless, some rear-fanged snakes are highly dangerous, such as boomslangs (*Dispholidus*) and twigsnakes (*Thelotornis*). *Atractaspis* (mole viper) has independently evolved a unique type of solenoglyphy: front fangs which can be erected laterally to "stab" prey via sideways lunges in the confines of burrows (Greene 1997; Deufel and Cundall 2003). In contrast, the elapids, which include cobras (*Naja*), coral snakes (*Micrurus*), mambas (*Dendroaspis*), and taipans (*Oxyuranus*), have fixed fangs at the front of the jaws, termed proteroglyphy. Most elapids are more slender and active than viperids, but many exceptions exist, such as the stocky Australian death "adder" *Acanthophis*. The venom is usually very potently neurotoxic (affecting the nervous system), and the majority of the most deadly snakes are elapids. Living sea snakes, being descended from terrestrial elapids, have inherited proteroglyphy and neurotoxic venom; the two clades of sea snakes (sea kraits *Laticauda* and true sea snakes Hydrophiini) represent separate marine invasions. They have convergently laterally compressed bodies and paddle-like tails, and valves in the nostrils to exclude water. *Laticauda* periodically returns to shore to bask, deposit eggs and drink freshwater (Lillywhite *et al.* 2008), but the Hydrophiini are totally marine, bearing live young underwater (Heatwole 1999). The Hydrophiini are a very young and speciose group, and might be one of the most rapidly radiating groups of reptiles (Sanders *et al.* 2010).

The repeated evolution of venom delivery systems in the rear maxilla (in the upper jaw) of colubroids is striking, especially since these highly advantageous adaptations have never appeared in any other snake or lizard lineage. The restricted distribution of structures for venom secretion and delivery within reptiles is not explained by the hypothesis that (at least incipient) venom systems are broadly distributed across snakes and their lizard relatives, the anguimorphs and iguanians (Fry *et al.* 2006). Rather, the explanation might lie in a unique feature of colubroid development. In colubroids, unlike in all other snakes, the rear maxilla and associated secretory glands develop as a separate coherent unit, permitting their integrated evolution into a fang-and-gland system (Vonk *et al.* 2008). This explanation is supported by evidence that even "front" fangs of proteroglyphs and solenoglyphs are actually modified rear teeth, which have been displaced forward by a shortening of the maxilla during evolution (see review in Jackson 2007). The absence of specialized posterior teeth and "dental glands" in xenodermines and pareatines (Underwood 1997) suggests these basal colubroids had not yet evolved this developmental novelty. *Xenophidion*, a close outgroup to colubroids, is apparently unique among snakes in possessing specialized lower teeth but not upper teeth (Cundall and Irish 2008). This is again consistent with the hypothesis that the developmental module facilitating venom evolution in the upper jaw is a novelty that arose within colubroids.

3.2.5 Primitive Fossil Snakes

Although this chapter has focused on living snakes, some important early fossil snakes require discussion. The Cretaceous pachyophiids are small-headed, laterally compressed, marine snakes such as *Pachyrhachis* (Caldwell and Lee 1997) and *Haasiophis* (Rieppel *et al.* 2003). They might have foraged in crevices and eel-burrows, by analogy with (unrelated) microcephalous living sea snakes (Heatwole 1999). *Najash* (Apesteguía and Zaher 2006) was a large terrestrial predator with a constrictor-type morphology. All these taxa possessed well-developed pelves and hindlimbs, along with adaptations for large gapes. Other early lineages of constrictor-like snakes have also been recently redescribed based on abundant new material: *Dinilysia* (Caldwell and Albino 2002; Caldwell and Calvo 2008), *Sanajeh* (Wilson *et al.* 2010), and madtsoiids, based on late-surviving (up to Pleistocene) forms *Wonambi* (Scanlon 2005a) and *Yurlunggur* (Scanlon 2006). Both *Dinilysia* (Budney *et al.* 2006) and some smaller madtsoiids (Scanlon 1997) appear to have hinged teeth, like several unrelated groups of living snakes and legless lizards that specialize on hard-scaled lizards. The presence or absence of limbs in *Dinilysia*, *Sanajeh*, and madtsoiids cannot yet be confirmed. All the above terrestrial forms appear to have been surface-active predators. Some madtsoiids have tall neural spines on trunk vertebrae, a condition absent in all highly fossorial snakes but consistent with either terrestrial, arboreal or aquatic specialization in extant forms (Johnson 1955). Ontogenetic fusion of posterior braincase elements in *Yurlunggur* and *Menarana*, and fusion of the atlas neural arch and intercentrum in the latter, suggest possible burrowing ancestry (LaDuke *et al.* 2010), but the structure of the snout in *Yurlunggur* (the best-preserved of all these forms) is unspecialized and clearly not adapted for digging behavior (Scanlon 2006). Even the smallest madtsoiids are too large to be totally fossorial, nor do they exhibit any clear specializations for burrowing. A wide range of intermediates between fossoriality and other lifestyles occur in extant snakes (e.g., use of pre-existing burrows, caves, tree-holes, mud, dense vegetation or leaf litter) and would also have been available to early snakes.

While apparently lacking the advanced upper-jaw kinesis characteristic of modern 'macrostomatans', at least some madtsoiids have more elongate mandibles than basal modern snakes and appear capable of swallowing relatively bulky prey; several associations of the medium-sized (~3.5 m) *Sanajeh* with sauropod dinosaur eggs suggest that hatchlings (~0.5 m) were a regular part of its diet (Wilson *et al.* 2010). No direct evidence for reproductive modes in early fossil snakes is yet known, but viviparity has been suggested for two groups. Pachyophiids may have been too aquatically specialized to return to land to lay eggs (Scanlon *et al.* 1999), and Australian Eocene madtsoiids appear to be derived from South American forms, and if so, presumably had ancestors which were viviparous high-latitude forms inhabiting Antarctica (Scanlon 2005b).

The above fossil snakes, where known, retain several primitive, "lizard-like" features indicating they are stem snakes, lying outside the crown-clade of living forms. These include a large "alar process" projecting anteriorly from the braincase, a large pelvis with sacral attachments, sizeable hindlimbs with at least femur, tibia and fibula, a distinct narrow neck region, and V-shaped chevrons that are not fused to the tail vertebrae. However, they also possess long and flexible jaw elements and were clearly adapted for large prey, features that have been sometimes interpreted as indicating a higher (nested) position within snakes. As a result, the affinities of these fossils remain debated but without clear resolution (e.g., Coates and Ruta 2000; Rieppel and Kearney 2001; Caldwell 2007). However, the emerging molecular data weaken the argument that they must be derived snakes because of their macrostomate condition, as it raises the likelihood that macrostomy is primitive in snakes (Vidal *et al.* 2009). If so, the co-existence of large pelves and limbs with large gape in early snakes is not paradoxical, but indeed consistent with a basal position of these fossil forms.

3.3 A REDUCED TIMESCALE FOR SNAKE EVOLUTION

The fossil record of snakes is poorer (in terms of anatomical completeness) than for many other reptile groups, consisting mainly of vertebrae that can be difficult to place phylogenetically. As a result, the tempo of divergence of the major groups of snakes has not been robustly inferred from the stratigraphic record. Molecular dating has recently been applied to the problem, but different analyses have generated often markedly different dates (see Vidal *et al.* 2009 for review). Here, a Bayesian relaxed clock analysis was applied to the combined molecular dataset of 20 nuclear and 3 mitochondrial genes, the largest dataset used for such a study to date. Accurate molecular dating requires multiple robust calibrations, preferably within the group of interest. Previous molecular inferences of divergence dates between the major groups of snakes have (almost necessarily) been calibrated largely with vertebrae that have not been assessed in a quantitative phylogenetic analysis, and/or more complete taxa that are of highly contentious placement (see Section 3.2.5). These issues affected all 10 calibrations in Noonan and Chippindale (2006), and 7 of the 8 calibrations in Vidal and Hedges (2009), which acknowledged these uncertainties. The latter study included one calibration based on a relatively complete fossil of robust phylogenetic relationships (the Miocene pythonine *Morelia riversleighensis*). This calibration, and four similar calibrations (Appendix), are used here to generate estimates of snake divergence times calibrated for the first time with snake fossils that have been placed phylogenetically using explicit derived characters.

The estimated divergence dates among living lineages are shown in Fig. 3.2, and are generally younger and more consistent with the broader fossil record than most recent molecular studies. They are also highly consistent with the lower (recent) part of the age ranges presented by Burbrink and Crother (2010). Snakes diverged from their anguimorph lizard relatives (Lee 2009) in the mid-Jurassic (~172 my). Scolecophidians and

Fig. 3.2 Divergence dates between the major clades of snakes, based on Bayesian relaxed clock analysis of a molecular data set (20 nuclear genes and 3 mitochondrial genes: see Appendix). The numbers at nodes denote the median date estimate, the bars denote the 95% highest posterior density (the narrowest interval that contains 95% of the sampled values). The dark circles numbered 1-5 are the nodal dates used for calibration.

Color image of this figure appears in the color plate section at the end of the book.

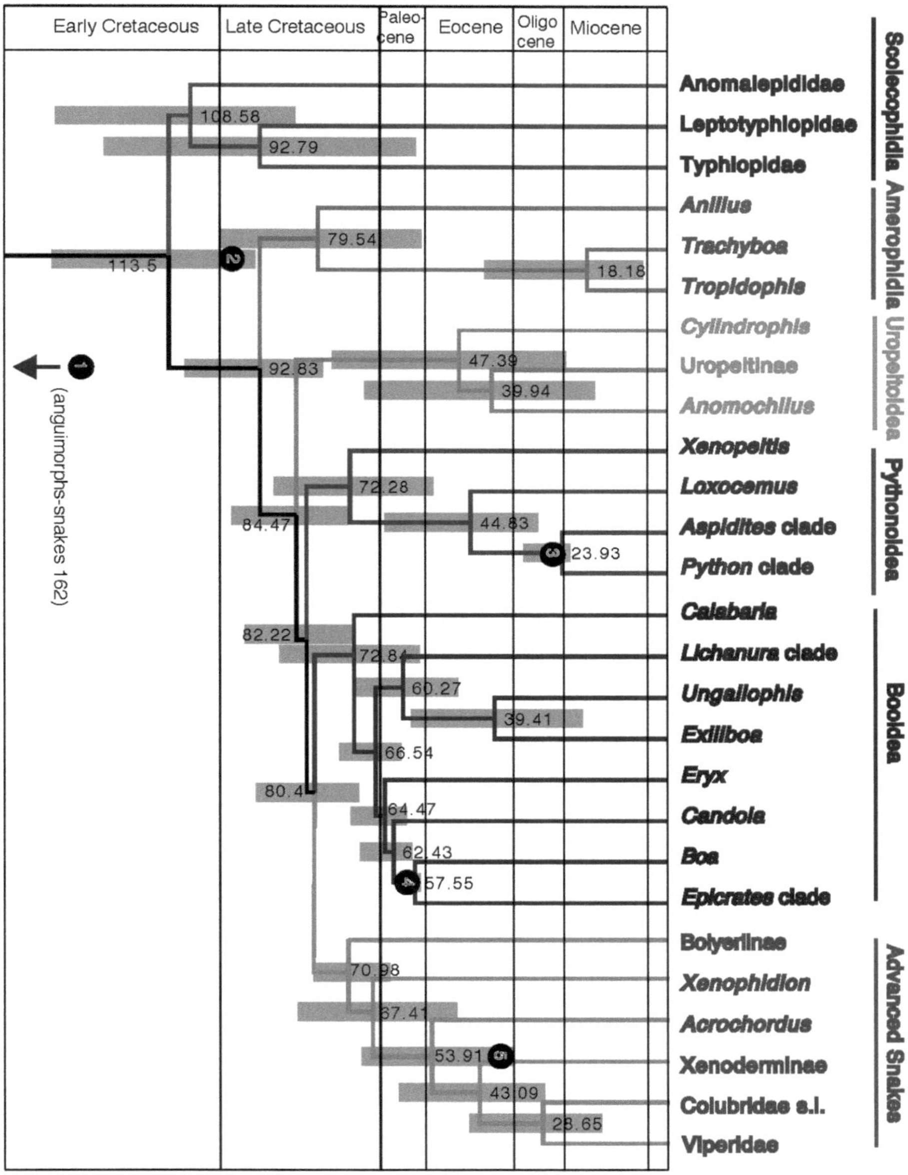

Early Cretaceous
Late Cretaceous
Paleo-cene
Eocene
Oligo cene
Miocene
108.58
92.79
113.5
79.54
18.18
92.83
47.39
39.94
(anguimorphs-snakes 162)
72.28
84.47
44.83
23.93
82.22
72.84
60.27
39.41
66.54
80.4
64.47
62.43
57.55
70.98
67.41
53.91
43.09
28.65
Anomalepididae
Leptotyphlopidae
Typhlopidae
Anilius
Trachyboa
Tropidophis
Cylindrophis
Uropeltinae
Anomochilus
Xenopeltis
Loxocemus
Aspidites clade
Python clade
Calabaria
Lichanura clade
Ungaliophis
Exiliboa
Eryx
Candoia
Boa
Epicrates clade
Bolyeriinae
Xenophidion
Acrochordus
Xenoderminae
Colubridae s.l.
Viperidae
Scolecophidia
Amerophidia
Uropeltoidea
Pythonoidea
Booidea
Advanced Snakes

alethinophidians diverged ~114 my. The three lineages of scolecophidians are each very ancient lineages (>93 my), and the two lineages of "Amerophidians" (*Anilius* and tropidophiines) diverged ~80 my—consistent with their morphological disparity. Thus, the tropidophiines, which superficially appear rather unremarkable booid-like snakes, represent a very ancient lineage that should repay detailed morphological and ecological studies. Uropeltoids, pythonoids, booids, and advanced snakes all diverged from each other in a relatively narrow time window in the late Cretaceous (85-80 my), accounting for why relationships between these lineages have been difficult to resolve with even extensive molecular data. The major booid and advanced snake lineages each radiated around the KT (i.e., K-Pg) boundary 65 mya, a relatively deep age with might account for their geographical disjunction, while consistent with diverse regional radiations of small boa-like snakes that are well represented in the Paleogene of North America and Europe (e.g., Holman 2000; Szyndlar and Rage 2003). The burrowing uropeltoid radiation is a surprisingly recent clade (~47 my), consistent with a restricted distribution in southeast Asia. The exceptionally diverse colubroid radiation is also very recent, and several clades in this radiation have been inferred to exhibit speciation rates that are among the fastest in land vertebrates (Sanders *et al.* 2010). The very recent evolution of higher colubroids, which diverged from viperids ~29 my, is consistent with their absence from Australia until very recently when it approached the Asian plate (Shine 1991); Australia has been continuously isolated from all other landmasses during the last ~50 my. Finally, the current study provides the first estimates for divergence times of two enigmatic snake lineages. *Anomochilus* diverged from uropeltines during the Eocene, while *Xenophidion* has been isolated from other advanced snakes since the uppermost Cretaceous.

This timescale of snake evolution is highly consistent with the broad fossil record of snakes (Rage 1987). While the caveat must be made that the five calibration nodes were based on that record, this does not result in a circular argument, as other studies, which also used fossil calibrations, retrieved divergence dates which are less consistent with stratigraphy (see below). In the discussion below comparing fossil and inferred molecular dates, we only discuss nodes that were free to vary (i.e., uncalibrated). The earliest undisputed fossil snakes are from the earliest late Cretaceous (Evans 2003), consistent with the current inference that the snake stem lineage is no older than 172 my. There are numerous snake fossils in the early to middle Late Cretaceous (100–70 mya), yet none can be unequivocally placed within crown Alethinophidia (see Appendix). While some of the fragmentary fossils have been optimistically assigned to particular modern lineages (e.g., Rage and Werner 1999), these interpretations are questionable given the fragmentary nature of the material and lack of explicit synapomorphies (Head *et al.* 2005). Similarly, the well-preserved marine pachyophiids, and madtsoiid-like taxa, cannot be unambiguously assigned to Alethinophidia (see above). The paucity of crown alethinophidian fossils during the late Mesozoic suggests that this clade either did not yet exist, or at least was not yet very diverse and is thus consistent with the relatively shallow molecular estimate of 93 my for crown Alethinophidia. Within alethinophidians, the late Cretaceous age of the earliest compelling aniliid (~75 mya) and booid (~68 mya) fossils is very consistent with the molecular estimates for their stem ages (~80 my for both clades). Within caenophidians, the earliest viperids are Oligo-Miocene (~22 my), again closely matching the inferred molecular dates (~29 my). While a 'colubrid' has been reported from the late Eocene (Rage *et al.* 1992), its precise relationships are not clear, and a relatively basal position within Colubroidea would be consistent with our timescale. The biggest implied fossil gap involves the scolecophidians, which are known only from the Eocene onwards, yet all three major lineages are inferred to be >90 my in age. The lack of fossil scolecophidians might be explicable based on the low fossilization potential of their tiny delicate skeletons; however, we acknowledge the possibility that some fossils assigned to other taxa, such as the 'aniliid' genus *Coniophis*, may be stem scolecophidians. There is also a large inferred fossil gap for tropidophiines; living forms have a narrow tropical range and relatively low species diversity, and if such characteristics were exhibited throughout their history, this would result in low fossilization potential.

In contrast to the above results, the most recent comprehensive study of snake divergence dates, which used largely different calibrations, is less consistent with the fossil record. The timescale of Vidal and Hedges (2009) implies much larger fossil gaps for certain snake lineages: crown snakes ~160 my (>60 my before the earliest undisputed snake fossils), and caenophidians ~91 my (>50 my ghost lineage).

3.4 PROBLEMS AND PROSPECTS

As in much of the tree of life, the burgeoning wealth of molecular data is substantially refining our understanding of snake evolution. However, even in the genomic age, morphological data can still be critical for phylogenetic reconstruction: here, the (almost universally accepted) monophyly of the blindsnakes continues to defy large molecular datasets, but is restored with the addition of morphological data. This supports the view that all sources of evidence should be considered simultaneously, rather than being pre-judged: an insight forcefully articulated long ago in a study of boine snakes (Kluge 1989). Furthermore, morphological data can provide compelling, independent corroboration of contentious molecular clades; the number of characters are not as important as the fact that they are far removed from DNA. For instance, detailed morphological studies of mammals are revealing novel characters supporting recently identified "molecular clades" such as Afrotheria, a heterogenous group of mammals that superficially have little in common except African distributions (Asher *et al.* 2009). Snake anatomy needs to be similarly reassessed in light of the new molecular results. Numerous derived characters discussed above corroborate two of the unexpected snake clades robustly supported by recent molecular sequences: Booidea (including ungaliophiines) and Pythonoidea (including *Xenopeltis* and *Loxocemus*). However, morphological support for uniting *Anilius* with tropidophiines remains elusive (Maisano and Rieppel 2007). New studies of novel anatomical systems using modern technology (e.g., CT scanning, ultrastructure) might uncover new relevant traits. The few relatively complete fossil snakes also need to be assessed in the context of the new molecular tree. Previous analyses of these fossils have either used exclusively morphological data, or datasets dominated by morphological data and have thus attempted to place these fossils in the context of an incorrect tree of living taxa.

Finally, the disagreement among molecular clock studies of snake divergence times is largely due to the paucity of reliable calibrations. There are relatively few outstanding Tertiary snake fossils, and even fewer have been incorporated into rigorous, numerical phylogenetic analyses. Those that have been analyzed in this fashion (e.g., *Morelia riversleighensis*) are proving to be critically important calibrations for molecular clock studies. However, relationships of other key Tertiary fossils, such as the putative ungaliophiines *Rottophis* (Szyndlar and Böhme 1996) and *Messelophis* (Baszio 2004; Schaal and Baszio 2004; Szyndlar *et al.* 2008), are currently based on largely phenetic comparisons and verbal arguments.

3.5 CONCLUSION

Recent molecular analyses have changed key aspects of our understanding of snake phylogeny and evolution. While many morphological groupings have been upheld, such as the alethinophidians (all living snakes excluding blindsnakes) and caenophidians (filesnakes plus colubroids), others have been convincingly overturned. In particular, the large-gaped tropidophiines are very basal alethinophidians related to *Anilius*, while the relatively primitive *Xenopeltis* and *Loxocemus* are the nearest relatives of pythons. The new molecular phylogeny refutes the widespread view that snake evolution involved gradual elaboration of feeding mechanisms, culminating in an advanced "macrostomatan" clade. Instead, molecular evidence suggests large gape was primitive for snakes (or at least alethinophidians), and reduced repeatedly in many basal, burrowing alethinophidians as well as further elaborated at least twice; this view is consistent with some interpretations of the snake fossil record. Some (but not all) of the new molecular clades exhibit morphological novelties that support their reality. A revised molecular clock analysis of snakes, using five robust fossil calibrations, produces more recent dates for all snake lineages compared to previous molecular clock studies. These shallower dates are much more consistent with the broader fossil record of reptiles, implying far shorter ghost lineages.

3.6 ACKNOWLEDGMENTS

Our research has been supported by the Australian Research Council. We thank the editors and referees for comments and suggestions. Credits for thumbnail photos in Figure 3.1 are as follows: (1) Kiril Kapustin, (2) Matthew Niemiller, (3) W. A. Djatmiko, (4) Tim Vickers, (5) Patrick Jean, (6) S. Macdonald, (7) NBII public domain photograph. Photo 2 copyright Matthew Niemiller and used with permission, all other photos are public domain images used courtesy of Wikimedia under their conditions.

3.7 LITERATURE CITED

[Editorial Note: Some references are cited only in the Appendix]

Apesteguía, S. and Zaher, H. 2006. A Cretaceous terrestrial snake with robust hindlimbs and a sacrum. Nature 440: 1037-1040.

Asher R. J., Bennett, N. and Lehmann, T. 2009. The new framework for understanding placental mammal evolution. Bioessays 31: 853-864.

Baszio S. 2004. *Messelophis variatus* n.gen n.sp., from the Eocene of Messel: a tropidopheine snake with affinities to Erycinae (Boidae). Courier Forschungsinstitut Senckenberg 252: 47-66.

Bhullar, B.-A., Pauly, G., Scanferla, C., Bever, G., and Smith, K.T. 2009. The first fossil Sunbeam Snake and the antiquity of modern snake clades. Journal of Vertebrate Paleontology 29(3): 63A.

Budney, L. A., Caldwell, M. W. and Albino, A. 2006. Unexpected tooth socket histology in the Cretaceous snake *Dinilysia*, with a review of amniote dental attachment tissues. Journal of Vertebrate Paleontology 26: 138-145.

Burbrink, F. T. and Crother, B. 2011. Evolution and taxonomy of snakes. Pp. 19-53. In R. D. Aldridge and D. M. Sever (eds), *Reproductive Biology and Phylogeny of Snakes.* Science Publishers Inc., Enfield, New Hamsphire.

Caldwell, M. W. 2007. Snake phylogeny, origins and evolution: the role, impact, and importance of fossils (1869-2006). Pp. 253-302. In J. S. Anderson and H.-D. Sues (eds), *Major Transitions in Vertebrate Evolution*. Indiana University Press, Bloomington.

Caldwell, M. W. and Lee, M. S. Y. 1997. A snake with legs from the marine Cretaceous of the Middle East. Nature 386: 705-709.

Caldwell, M. W. and Albino, A. 2002. Exceptionally preserved skeletons of the Cretaceous snake *Dinilysia patagonica*, Woodward, 1901. Journal of Vertebrate Paleontology 22: 861-866.

Caldwell, M. W. and Calvo, J. 2008. Details of a new skull and articulated cervical column of *Dinilysia patagonica* Woodward, 1901. Journal of Vertebrate Paleontology 28: 349-362.

Coates, M. and Ruta, M. 2000. Nice snake, shame about the legs. Trends in Ecology and Evolution 15: 503-507.

Conrad, J. L. 2008. Phylogeny and systematics of Squamata (Reptilia) based on morphology. Bulletin of the American Museum of Natural History 310: 1-183.

Cundall, D. 1983. Activity of head muscles during feeding by snakes: A comparative study. American Zoologist 23: 383-396.

Cundall, D. and Irish, F. J. 1989. The function of the intramaxillary joint in the Round Island boa, *Casarea dussumieri.* Journal of Zoology 217: 569-598.

Cundall, D. and Greene, H. W. 2000. Feeding in snakes. Pp. 293-333. In K. Schwenk (ed.), *Feeding: Form, Function and Evolution in Tetrapod Vertebrates*. Academic Press, San Diego.

Cundall, D. and F. Irish. 2008. The snake skull. Pp. 349-692. In C. Gans, A. S. Gaunt and K. Adler (eds), *Biology of the Reptilia, Vol. 20, Morphology H*. Society for the Study of Amphibians and Reptiles, Ithaca, New York.

Cundall, D., Wallach, V. and Rossman, D.A. 1993. The systematic relationships of the snake genus *Anomochilus*. Zoological Journal of the Linnean Society 109: 275-299.

Deufel, A. and Cundall, D. 2003. Feeding in *Atractaspis* (Serpentes: Atractaspididae): A study in conflicting functional constraints. Zoology 106: 43-61.

Drummond, A. J. and Rambaut, A. 2007. *BEAST*: Bayesian evolutionary analysis by sampling trees. BMC Evolutionary Biology 7: e214 (unpaginated).

Drummond, A. J., Ho, S. Y. W., Phillips, M. J. and Rambaut, A. 2006. Relaxed Phylogenetics and Dating with Confidence. PLoS Biology 4: e88 (unpaginated).

Evans, S. E. 2003. At the feet of the dinosaurs: the early history and radiation of lizards. Biological Reviews 78: 513-551.

Fedorov, P. V. and Nessov, L. A. 1992. A lizard from the boundary of the Middle and Late Jurassic of north-east Fergana. Vestnik Sankt-Petersburgskago Universiteta, Seriya 7 (Geologiya, Geografiya) [Bulletin of the St Petersburg University, Series 7, (Geology, Geography)] 3: 9-14.

Fry, B. G., Vidal, N., Norman, J. A., Vonk, F. J., Scheib, H., Ramjan, R., Kuruppu, S., Fung, K., Hedges, S. B., Richardson, M. K., Hodgson, W. C., Ignjatovic, V., Summerhayes, R. and Kochva, E. 2006. Early evolution of the venom system in lizards and snakes. Nature 439: 509-632.

Gardner, J. D. and Cifelli, R. L. 1999. A primitive snake from the Cretaceous of Utah. Special Papers in Paleontology 60: 87-100.

Gower, D. J., Vidal, N., Spinks, J. N. and McCarthy, C. J. 2005. The phylogenetic position of Anomochilidae (Reptilia: Serpentes): first evidence from DNA sequences. Journal of Zoological Systematics and Evolutionary Research 43: 315-320.

Greene, H. W. 1997. *Snakes: The Evolution of Mystery in Nature*. University of California Press, Berkeley. Pp. 351.

Greene, H. W. 1983. Dietary correlates of the origin and radiation of snakes. American Zoologist 23: 431-441.

Groombridge, B. C. 1979. Variations in morphology of the superficial palate of henophidian snakes and some possible systematic implications. Journal of Natural History 13: 447-475.

Head, J. J. and Polly, P. D. 2007. Dissociation of somatic growth from segmentation drives gigantism in snakes. Biology Letters 3: 296-298.

Head, J. J., Mohabey, D. M. and Wilson, J. A. 2007. *Acrochordus hornstedt* (Serpentes, Caenophidia) from the Miocene of Gujarat, Western India: Temporal constraints on dispersal of a derived snake. Journal of Vertebrate Paleontology 27: 720–723.

Head, J. J., Holroyd, P. A., Hutchinson, J. H. and Ciochon, R. L. 2005. First report of snakes (Serpentes) from the Late Middle Eocene Pondaung Formation, Myanmar. Journal of Vertebrate Paleontology 25: 246-250.

Head, J. J., Bloch, J. I., Hastings, A. K., Bourke, J. R., Cadena, E. A., Herrera, F. A., Polly, P. D. and Jaramillo. C. A. 2008. Giant boid snake from the Paleocene neotropics reveals hotter past equatorial temperatures. Nature 457: 715-717.

Heatwole, H. 1999. *Sea Snakes*. University of New South Wales Press, Sydney.

Holman, J. A. 2000. *Fossil Snakes of North America: Origin, Evolution, Distribution, Paleoecology*. Indiana University Press, Bloomington. Pp. 376.

Jackson, K. 2007. The evolution of venom-conducting fangs: Insights from developmental biology. Toxicon 49: 975-981.
Jamieson, B. G. M. and Koehler, L. 1994. The ultrastructure of the spermatozoon of the northern water snake, *Nerodia sipedon* (Colubridae, Serpentes), with phylogenetic considerations. Canadian Journal of Zoology 72: 1648-1652.
Kelly, C. M. R., Barker, N. P. and Villet, M. H. 2003. Phylogenetics of advanced snakes (Caenophidia) based on four mitochondrial genes. Systematic Biology 52: 439-459.
Kelly, C. M. R., Barker, N. P., Villet, M. H. and Broadley, D. G. 2009. Phylogeny, biogeography and classification of the snake superfamily Elapoidea: a rapid radiation in the late Eocene. Cladistics 25: 38-63.
Kley, N. J. 2001. Prey transport mechanisms in blindsnakes and the evolution of unilateral feeding systems in snakes. American Zoologist 41: 1321-1337.
Kluge, A.G. 1991. Boine snake phylogeny and research cycles. Miscellaneous Publications of the Museum of Zoology, University of Michigan 178: 1-58.
Kluge, A.G. 1993. *Aspidites* and the phylogeny of pythonine snakes. Records of the Australian Museum Supplement 19: 1-77.
Kolaczkowski, B. and Thornton J. W. 2009. Long-branch attraction bias and inconsistency in Bayesian phylogenetics. PLoS ONE 4: e7891 e88 (unpaginated).
LaDuke, T.C., Krause, D.W., Scanlon, J.D. and Kley, N.J. 2010. A Late Cretaceous (Maastrichtian) snake assemblage from the Maevarano Formation, Mahajanga Basin, Madagascar. Journal of Vertebrate Paleontology 30: in press.
Lawson, R., Slowinski, J. B. and Burbrink, F. T. 2004. A molecular approach to discerning the phylogenetic placement of the enigmatic snake *Xenophidion schaeferi* among the Alethinophidia. Journal of Zoology 263: 285-294.
Lawson, R., Slowinski, J. B., Crother, B. I. and Burbrink, F. T. 2005. Phylogeny of the Colubroidea (Serpentes): New evidence from mitochondrial and nuclear genes. Molecular Phylogenetics and Evolution 37: 581-601.
Lillywhite, H. B., Leslie S., Babonis, L. S., Sheehy, C. M., III and Tu T.-M. 2008. Sea snakes (*Laticauda* spp.) require fresh drinking water: Implication for the distribution and persistence of populations. Physiological and Biochemical Zoology. 81: 785–796.
Lee, M. S. Y. 2009. Hidden support from unpromising datasets strongly unites snakes and anguimorph lizards. Journal of Evolutionary Biology 22: 1308-1316.
Lee, M. S. Y. and Scanlon, J. D. 2002. Snake phylogeny based on osteology, soft anatomy and ecology. Biological Reviews of the Cambridge Philosophical Society 77: 333-401.
Lynch, V. J. and Wagner, G. P. 2009. Did egg-laying boas break Dollo's Law? Phylogenetic evidence for reversal to oviparity in Sand Boas (*Eryx*: Boidae). Evolution 64: 207-216.
Maisano, J. A. and Rieppel, O. 2007 The Skull of the Round Island Boa, *Casarea dussumieri* Schlegel, Based on High-Resolution X-ray Computed Tomography. Journal of Morphology 268: 371-384.
McDowell, S. B. 1974. A catalogue of the snakes of New Guinea and the Solomons, with special reference to those in the Bernice P. Bishop Museum. Part I. Scolecophidia. Journal of Herpetology 8: 1-57.
McDowell, S. B. 1975. A catalogue of the snakes of New Guinea and the Solomons, with special reference to those in the Bernice P. Bishop Museum. Part II. Anilioidea and Pythoninae. Journal of Herpetology 9: 1-80.
McDowell, S. B. 1979. A catalogue of the snakes of New Guinea and the Solomons, with special reference to those in the Bernice P. Bishop Museum. Part III. Boinae and Acrochordoidea (Reptilia, Serpentes). Journal of Herpetology 13: 1-92.

Noonan, B. P. and Chippindale, P. T. 2006. Dispersal and vicariance: The complex evolutionary history of boid snakes. Molecular Phylogenetics and Evolution 40: 347-358.

Nylander, J. A. A. 2004. *MrModeltest v2.* Evolutionary Biology Centre, Uppsala University. http://www.abc.se/~nylander/mrmodeltest2/mrmodeltest2.html.

Rage, J.-C. 1987. Fossil Record. Pp. 51-76. In R. A. Seigel, J. T. Collins and S. S. Novak (eds), *Snakes: Ecology and Evolutionary Biology.* MacMillan, New York.

Rage, J.-C. and Werner, C. 1999. Mid-Cretaceous (Cenomanian) snakes from Wadi Abu Hashim, Sudan: The earliest snake assemblage. Palaeontologia Africana 35: 85-110.

Rage, J.-C., Folie, A., Rana, R. S., Singh, H., Rose, K. D. and Smith, T. 2008. A diverse snake fauna from the early Eocene of Vastan Lignite Mine, Gujarat, India. *Acta Palaeontologica Polonica* 53: 391-403.

Rambaut, A. and Drummond, A. J. 2007. *Tracer v1.4.* Available from http://beast.bio.ed.ac.uk/Tracer.

Rawlings, L .H., Rabosky, D. L., Donnellan, S. C. and Hutchinson, M.N. 2008. Python phylogenetics: inference from morphology and mitochondrial DNA. Biological Journal of the Linnean Society 93: 603-619.

Rieppel, O. 1988. A review of the origin of snakes. Evolutionary Biology 22: 37-130.

Rieppel, O., Kley, N. J., and Maisano, J. A. 2009. Morphology of the skull of the white-nosed blindsnake, *Liotyphlops albirostris* (Scolecophidia: Anomalepididae). Journal of Morphology 270: 536-557.

Rieppel, O., Zaher, H., Tchernov, E. and Polcyn, M. 2003. The anatomy and relationships of *Haasiophis terrasanctus*, a fossil snake with well-developed hind limbs from the mid-Cretaceous of the Middle East. Journal of Paleontology 77: 536-558.

Rodríguez-Robles, J. A., Bell, C. J. and Greene, H. W. 1999. Gape size and evolution of diet in snakes: feeding ecology of erycine boas. Journal of Zoology 248: 49-58.

Ronquist, F. and Huelsenbeck, J. P. 2003. *MrBayes 3*: Bayesian phylogenetic inference under mixed models. Bioinformatics 19: 1572-1574.

Ronquist, F., Huelsenbeck, J. P. and P. van der Mark. 2005. *MrBayes 3.1 Manual.* Available from mrbayes.scs.fsu.edu/mb3.1_manual.pdf.

Sanders, K.L. and Lee, M. S. Y. 2007. Evaluating molecular clock calibrations using Bayesian analyses with soft and hard bounds. Biology Letters 3: 275-279.

Sanders, K. L. and Lee, M. S. Y. 2008. Molecular evidence for a rapid late-Miocene radiation of Australasian venomous snakes (Elapidae, Colubroidea). Molecular Phylogenetics and Evolution 46: 1165-1173.

Sanders, K. L., Munpuni, and Lee, M. S. Y. 2010. Uncoupling of ecological shift and diversification rate in sea snakes. Journal of Evolutionary Biology 23: 2685-2693.

Scanlon, J. D. 1997. *Nanowana* gen. nov., small madtsoiid snakes from the Miocene of Riversleigh: sympatric species with divergently specialised dentition. Memoirs of the Queensland Museum 41: 393-412.

Scanlon, J. D. 2001. *Montypythonoides*: the Miocene snake *Morelia riversleighensis* (Smith and Plane 1985) and the question of the geographic origin of pythons. Memoirs of the Association of Australasian Palaeontologists 25: 1-35.

Scanlon, J. D. 2003. The basicranial morphology of madtsoiid snakes (Squamata, Ophidia) and the earliest Alethinophidia (Serpentes). Journal of Vertebrate Paleontology 23: 971-976.

Scanlon, J. D. 2005a. Cranial morphology of the Plio-Pleistocene giant madtsoiid snake *Wonambi naracoortensis*. Acta Palaeontologica Polonica 50: 139-180.
Scanlon, J. D. 2005b. Australia's oldest known snakes: *Patagoniophis*, *Alamitophis*, and cf. *Madtsoia* (Squamata: Madtsoiidae) from the Eocene of Queensland. Memoirs of the Queensland Museum 51: 215-235.
Scanlon, J. D. 2006. Skull of the large non-macrostomatan snake *Yurlunggur* from the Australian Oligo-Miocene. Nature 439: 839-842.
Scanlon, J. D., Lee, M. S. Y., Caldwell, M. W. and Shine, R. 1999. The palaeoecology of the primitive snake *Pachyrhachis*. Historical Biology 13: 127-152.
Schaal, S. and Baszio, S. 2004. *Messelophis ermannorum* n.sp., eine neue Zwergboa (Serpentes: Boidae: Tropidopheinae) aus dem Mittel-Eozän von Messel. Courier Forschungsinstitut Senckenberg 252: 67-77.
Shine, R. 1986. Ecology of a low-energy specialist: food habitats and reproductive biology of the Arafura filesnake (Acrochordidae). Copeia 1986: 424-437.
Shine, R. 1991. Strangers in a strange land: ecology of the Australian colubrid snakes. Copeia 1991: 120-131.
Slowinski, J. B. and Lawson, R. 2002. Snake phylogeny: evidence from nuclear and mitochondrial genes. Molecular Phylogenetics and Evolution 24: 194-202.
Sorenson, M. D. and Franzosa, E. A. 2007. *TreeRot, version 3*. Boston University, Boston, MA. http://people.bu.edu/msoren/TreeRot.html.
Stamatakis, A. 2006. RAxML-VI-HPC: Maximum likelihood-based phylogenetic analyses with thousands of taxa and mixed models. Bioinformatics 22: 2688-2690.
Swofford, D. L. 2003. *PAUP*. Phylogenetic Analysis Using Parsimony (*and Other Methods)*. Version 4. Sinauer Associates, Sunderland, Massachusetts. Software, unpaginated.
Szyndlar, Z. and Böhme, W. 1996. Redescription of *Tropidonotus atavus* von Meyer, 1855 from the upper Oligocene of Rott (Germany) and its allocation to *Rottophis* gen. nov. (Serpentes, Boidae). Palaeontographica Abteilung A, 240: 145-161.
Szyndlar, Z., Smith, R. and Rage, J.-C. 2008. A new dwarf boa (Serpentes, Booidea, 'Tropidophiidae') from the Early Oligocene of Belgium: a case of the isolation of Western European snake faunas. Zoological Journal of the Linnean Society 152: 393-406.
Tavares-Bastos, L., Colli, G. R. and Báo, S. N. 2008. The evolution of sperm ultrastructure among Boidae (Serpentes). Zoomorphology 127: 189-202.
Torr, G. 2000. *Pythons of Australia: A Natural History*. UNSW Press, Sydney. Pp. 103.
Underwood, G. 1967. *A Contribution to the Classification of Snakes*. British Museum (Natural History), London. Pp. 179.
Underwood, G. 1997. An overview of venomous snake evolution. Pp. 1-13. In R. S. Thorpe, W. Wuster and A. Malhotra (eds), *Venomous Snakes: Ecology, Evolution and Snakebite*. Oxford University Press, Oxford, U.K.
Vidal, N., Delmas, A.-S. and Hedges, S. B. 2007a. The higher-level relationships of alethinophidian snakes inferred from seven nuclear and mitochondrial genes. Pp. 27-33. In R. W. Henderson and R. Powell (eds), *Biology of the Boas and Pythons*. Eagle Mountain Publishing, Eagle Mountain, Utah.
Vidal, N., Delmas, A.-S., David, P., Cruaud, C., Couloux, A. and Hedges, S. B. 2007b. The phylogeny and classification of caenophidian snakes inferred from seven nuclear protein-coding genes. Comptes Rendes Biologies 330: 182-187.
Vidal, N., Rage, J.-C., Couloux, A. and Hedges, S. B. 2009. Snakes (Serpentes). Pp. 390-397. In S. B. Hedges and S. Kumar (eds), *The Timetree of Life*. Oxford University Press, Oxford, U.K.

Vonk, F. J., Admiraal, J. F., Jackson, K., Reshef, R., de Bakker, M. A. G., Vanderschoot, K., van den Berge, I., van Atten, M., Burgerhout, E., Beck, A., Mirtschin, P. J., Kochva, E., Witte, F., Fry, B. G., Woods, A. and Richardson, M. K. 2008. Evolutionary origin and development of snake fangs. Nature 454: 630-633.

Wallach, V. and Günther, R. 1998. Visceral anatomy of the Malaysian snake genus *Xenophidion*, including a cladistic analysis and allocation to a new family. Amphibia-Reptilia 19: 385-404.

Wiens, J. J., Kuczynski, C. A., Smith, S. A., Mulcahy, D., Sites, J. W. Jr., Townsend, T. M. and Reeder, T. W. 2008. Branch length, support, and congruence: testing the phylogenomic approach with 20 nuclear loci in snakes. Systematic Biology 57: 420-431.

Wilcox, T. P., Zwickl, D. J., Heath, T. A. and Hillis, D. M. Phylogenetic relationships of the dwarf boas and a comparison of Bayesian and bootstrap measures of phylogenetic support. Molecular Phylogenetics and Evolution 25: 361-371.

Wilson, J. A., Mohabey, D. M., Peters, S. E. and Head, J. J. 2010. Predation upon hatchling dinosaurs by a new snake from the Late Cretaceous of India. PLoS Biology 8 (3): e1000322.

Young, B. A. 1998. The comparative morphology of the intermandibular connective tissue in snakes (Reptilia, Squamata). Zoologischer Anzeiger 237: 59-84.

Zaher, H. 1994a. Les Tropidopheoidea (Serpentes: Alethinophidia) sont-ils réellement monophyletique? Arguments en faveur de leur polyphyletisme. Comptes Rendus de l'Academie des Sciences, Série III (Sciences de la Vie) 317: 471-478.

Zaher, H. 1994b. Comments on the evolution of the jaw adductor musculature of snakes. Zoological Journal of the Linnean Society 111: 339-384.

Zaher, H. and Rieppel, O. 1999. The phylogenetic relationships of *Pachyrhachis problematicus*, and the evolution of limblessness in snakes (Lepidosauria, Squamata). Comptes Rendus de l'Academie des Sciences, Sciences de la Terre et des Planètes 329: 831-837.

3.8 APPENDIX: MATERIALS AND METHODS

3.8.1 Morphological and Molecular Data

The full data matrix consists of 30 terminal taxa (28 snakes, with two outgroups Varanidae and Agamidae), scored for 13230 nuclear sites, 1944 mitochondrial sites (1834 alignable), 72 indels and 214 morphological characters as listed below, yielding a total of 15460 characters (15350 alignable). All major snake lineages were represented by both molecular and morphological data, including *Xenophidion* and *Anomochilus*.

3.8.1.1 Molecular sequence data and alignment

Molecular data for the terminal taxa include 20 nuclear genes (Wiens *et al.* 2008) plus mitochondrial Cytochrome b, 12S RNA and 16S RNA. All relevant nuclear sequences used in Wiens *et al.* (2008), and 12S and 16S sequences used in Gower *et al.* (2005) were incorporated. Genbank numbers for additional incorporated sequences are listed below. For each major snake lineage, the same exemplar species was used as in Wiens *et al.* (2008); however, *Candoia carinata* was added due to new data (Vidal and Hedges 2009), and sampling was reduced within Colubroidea, which was represented by three exemplars spanning the basal dichotomy (*Xenodermus javanicus*, *Daboia russelli* and *Coluber constrictor*). In a few cases (indicated with*), sequences for exemplar species were not unavailable and closely related taxa (usually cogeneric) were used instead; this concatenation would not cause problems given the level of this analysis (Wiens *et al.* 2008). All protein-coding loci were readily alignable using reading frame; the new RNA sequences were aligned against the existing alignment in Gower *et al.* (2005) which incorporated information on secondary structure.

The aligned nucleotide data is available on TreeBase (www.treebase.org accession number S2675).

BDNF: *Candoia carinata* FJ433974.1, *Uropeltis melanogaster* FJ433965.1

NTF3: *Candoia carinata* FJ434077, *Ungaliophis continentalis* FJ434081.1, *Stoliczkaia* sp.* FJ434083.1.

RAG-1: *Candoia carinata* AY988065.1, *Uropeltis melanogaster* AY487399.1.

Cytochrome b: *Agama atra** AF355543.1, *Varanus salvator* NC_010974.1, *Acrochordus granulatus* AF217841.1, *Anilius scytale* FJ755180.1, *Aspidites melanocephalus* U69741.1, *Boa constrictor* AB177354.1, *Calabaria reinhardtii* AY099985, *Candoia carinata* AY099984.1, *Casarea dussumieri* U69755.1, *Coluber constrictor* AF217818.1, *Cylindrophis rufus* NC_007401.1, *Daboia russelli* AF471076.1, *Epicrates striatus* U69791.1, *Eryx colubrinus* U69811.1, *Exiliboa placata* AY099989.1, *Leptotyphlops humilis*AB079597.1, *Lichanura (Charina) trivirgata* U69844.1, *Liotyphlops albirostris* AF544672.1, *Loxocemus bicolor* AY099993.1, *Python molurus* AY099983.1, *Tropidophis haetianus* NC_012573.1, *Ungaliophis continentalis* U69870, *Uropeltis phillipsi** AF471034.1, *Xenodermus javanicus* AY425810.1, *Xenopeltis unicolor* NC_007402.1, *Xenophidion schaeferi* AY574279.1.

12S RNA: *Calotes* (Agamidae)* NC 009683, *Varanus salvator* NC_010974.1, *Aspidites melanocephalus* EF545033.1, *Candoia carinata* AF544741.1, *Daboia russelli* NC_011391.1, *Epicrates cenchria** AF368059.1, *Eryx tararicus** AF236681, *Exiliboa placata* AF512742.1, *Leptotyphlops humilis* AB079597, *Lichanura (Charina) trivirgata* AF544749, *Python molurus* EF545038, *Typhlops reticulatus** NC_010971.1, *Ungaliophis continentalis* AF512741, *Xenodermus javanicus* AF544781.

16S RNA: *Agama agama** FJ159562, *Varanus salvator* NC_010974.1, *Aspidites melanocephalus* EF545060, *Candoia carinata* EU419850.1, *Epicrates striatus* AF215273, *Eryx conicus** AF512743, *Exiliboa placata* AF512742.1, *Leptotyphlops humilis* AB079597, *Lichanura (Charina) trivirgata* EU280414, *Liotyphlops albirostris* AF366762, *Python molurus* EF545065, *Typhlops reticulatus** NC_010971.1, *Ungaliophis continentalis* AF512741, *Achalinus** (Xenoderminae) NC_011576.

3.8.1.2 Indels

72 unambiguous gaps across all molecular loci were scored as binary characters (if all gaps were identical in length with no other overlapping gaps), or multistate characters (if there were overlapping gaps of differing lengths). The full indel data matrix is available on TreeBase (accession number S2675).

3.8.1.3 Morphological data

The morphological data consisted of the 214 osteological characters. Characters 24 and 62 are new to this analysis; characters 1-23, 25-61 and 63-214 are discussed in detail in Lee and Scanlon (2002) and Scanlon (2006) numbered respectively in those studies as 1-23, 24-60, and 61-212. The soft anatomical characters in those analyses were mostly too poorly sampled to be scoreable in the smaller terminals used in the current analysis. Multistate characters which formed morphoclines were ordered (2 5 6 11 12 18 22 23 32 35 43 49 50 58 60 63 68 71 72 74 77 80 89 92 98 103 104 106 110 119 123 136 137 139 140 143 148 151 154 159 160 162 164 168 169 184 187 188 194 197 198 202 203 212 214), those which did not form clear morphoclines were left unordered (8 29 30 52 53 55 61 62 82 87 96 107 111 122 138 167 177 200 208). Modifications to the matrix in Scanlon (2006) are listed below. MrBayes only accepts ordered characters if states are ascending (e.g., 0-1-2 not 0-2-1), hence for some multistate characters, the numbers assigned to each state needed to be switched (as discussed below). The morphological matrix with these characters scored for all taxa is available on TreeBase (accession number S2675).

1. Premaxilla. 0: anterior surface convex or straight. 1: anterior surface concave.
2. Ascending process of premaxilla. 0: long and contacting frontals, i.e., extends entire snout-frontal distance. 1: intermediate in length and not contacting frontals. 2: extremely reduced or absent.

3. Ascending process of premaxilla. 0: process transversely expanded, partly roofing external nares. 1: process narrow or spine-like, separating but not roofing external nares.
4. Ascending process of premaxilla. 0: without lateral flange. 1: with lateral flange forming dorsal margin of external naris.
5. Nasal keel (process) of premaxilla. 0: absent. 1: moderately developed, short flange. 2: well developed, long process.
6. Palatal (vomerine) process of premaxilla. 0: extensive overlapping contact with vomer. 1: non-overlapping, point contact with vomer. 2: not in contact with vomer. Inapplicable in some agamids (where process contacts anteromedial flange of maxilla), but others exhibit state 1.
7. Premaxilla-vomer contact. 0: flat overlap. 1: well-defined facet.
8. Premaxillary palatal foramina. 0: paired. 1: single median. 2: multiple. Unordered because of uncertain homologies between the multiple foramina and the median paired foramina.
9. Main body of premaxilla. 0: on anterior end of the snout. 1: on ventral surface of snout.
10. Snout shape. 0: tapering anteriorly in front of orbits. 1: spherical, expanded in front of orbits.
11. Posterior margin of lateral process of premaxilla, in palatal view. 0: oriented anterolaterally. 1: oriented transversely, perpendicular to midline. 2: oriented posterolaterally.
12. Maxilla-premaxilla contact. 0: close, suture or strong abutting contact. 1: close but not abutting, connected by short ligament. 2: loose, widely separated.
13. Anterior (premaxillary) process of maxilla. 0: well developed, forming ventral margin of external naris. 1: poorly developed or absent, maxilla excluded from ventral margin of external naris.
14. Dorsal (ascending or prefrontal) process of maxilla. 0: well developed. 1: poorly developed or absent.
15. Anteromedial maxillary flange. 0: present, small horizontal shelf on medial surface of anterior end of maxilla. 1: absent, anterior end of maxilla without such shelf.
16. Lateral maxillary foramina. 0: present. 1: absent.
17. Maxilla. 0: alveolar (tooth) row oriented longitudinally. 1: alveolar (tooth) row oriented transversely.
18. Maxilla-palatine articulation. 0: located anteriorly, at or in front of anterior orbital margin. 1: located beneath anterior half of orbit. 2: located posteriorly, at same level as centre of orbit or further posterior. Ordered 0-1-2.
19. Palatal process of maxilla. 0: absent, medial margin of maxilla smooth or with (at most) indistinct swelling. 1: present, medial margin of maxilla with distinct process.
20. Palatal process of maxilla. 0: does not approach pterygoid, palatine broadly enters suborbital fenestra. 1: contacts pterygoid, excluding palatine from suborbital fenestra.

21. Palatal process of maxilla. 0: dorsomedial surface pierced by a large foramen. 1: not pierced.
22. Ectopterygoid flange of maxilla. 0: maxilla without distinct posteromedial (ectopterygoid) expansion or flange. 1: maxilla with weak but distinct posteromedial (ectopterygoid) expansion or flange. 2: maxilla with large posteromedial (ectopterygoid) expansion or flange.
23. Posterior extent of maxilla. 0: does not reach middle of orbit. 1: reaches middle of orbit, or slightly further. 2: extends past posterior margin of orbit.
24. Maxilla. Single unit (0); two units (1).
25. Nasal. 0: does not closely approach lateral process of premaxilla. 1: extends anteriorly to almost reach lateral process of premaxilla.
26. Horizontal lamina of nasal. 0: narrow anteriorly, tapering to a point beside premaxilla. 1: horizontal lamina of nasal wide anteriorly, at most tapering only slightly to a blunt anterior end.
27. Horizontal lamina of nasal. 0: posterior width across both nasals wide. 1: posterior width narrow, tapering to a sagittal point.
28. Nasal-frontal contact. 0: horizontal laminae of nasals and frontals in contact. 1: horizontal laminae of nasals and frontals not in contact.
29. Nasal-frontal contact. 0: nasals overlap frontals dorsally. 1: frontals overlap nasals dorsally. 2: clasping junction, nasal fits into anterior groove in frontal and is thus overlapped and underlapped by frontal.
30. Nasal-frontal boundary: 0: concave posteriorly in dorsal view. 1: approximately straight and transverse. 2: convex posteriorly. 3: W-shaped, nasals project posteriorly into embayments in frontals.
31. Descending laminae of nasals. 0: not enlarged (shallow) anteriorly. 1: distinctly enlarged (very deep) anteriorly.
32. Anterior process of prefrontal. 0: poorly developed. 1: moderately developed, triangular flange. 2: greatly elongated process.
33. Anterodorsal lappet of prefrontal. 0: absent or indistinct. 1: distinct process extending along lateral margin of nasal.
34. Prefrontal-nasal contact. 0: prefrontal separated from nasal by fissure continuous with external naris. 1: prefrontal contacts nasal. Character definition modified, combining states 0 and 2, and deleting state 3 which does not occur in the ingroup or proximal outgroups.
35. Prefrontal-maxilla contact. 0: anterior process and ventrolateral margin of prefrontal contact maxilla. 1: anterior process of prefrontal does not contact maxilla and projects freely, only ventrolateral margin of prefrontal contacting maxilla. 2: anterior process and ventrolateral margin of prefrontal do not contact maxilla.
36. Prefrontal-maxilla contact on facial region. 0: tight or interdigitating, relatively rigid. 1: flat or slightly convex surfaces, allowing rocking or sliding motion.

37. Antorbital buttress of prefrontal. 0: lateral foot process (lateral to lacrimal foramen) does not contact palatine. 1: lateral foot process contacts palatine.
38. Antorbital buttress of prefrontal. 0: medial foot process (medial to lacrimal foramen) does not contact maxilla. 1: medial foot process contacts maxilla.
39. Lateral process of palatine. 0: does not reach lateral edge of maxilla. 1: reaches lateral edge of maxilla.
40. Outer orbital (lateral) margin of prefrontal, in lateral view. 0: slants anteroventrally. 1: vertical.
41. Prefrontal lacrimal duct roof. 0: absent. 1: present, a horizontal flange extending anteriorly from lacrimal foramen.
42. Prefrontal-frontal contact in dorsal view. 0: approximately straight. 1: curved, prefrontal fitting into deep embayment in frontal.
43. Prefrontal-frontal contact in dorsal view. 0: oriented approximately parasagittally, prefrontals contact only lateral margins of frontals and are widely separated. 1: oriented anteromedially, prefrontals contact anterolateral margins of frontals and are moderately separated. 2: oriented anteromedially or transversely, prefrontals closely approaching or contacting one another.
44. Prefrontal-frontal contact. 0: prefrontal sutured to or tightly buttressed against frontal. 1: prefrontal moveably articulated to frontal.
45. Antorbital (vertical) buttress of prefrontal. 0: broad, extends medially underneath lateral descending flanges of frontal. 1: narrow, does not extend medially to reach lateral descending flanges of frontal.
46. Lacrimal foramen. 0: bordered ventrally by prefrontal. 1: widely open on ventral edge of prefrontal. States 1 and 2 in Scanlon 2006 combined.
47. Jugal. 0: present. 1: absent.
48. Postorbitofrontal ossification(s) in adults. 0: present, the single ossification in snakes is conventionally termed the postorbital or postorbitofrontal. 1: absent.
49. Postorbitofrontal ossification(s). 0: strongly forked medial margin, anterior and posterior rami tightly clasping frontoparietal suture. 1: weakly forked or straight medial margin overlapping frontoparietal suture. 2: medial margin not forked and without distinct anterior and posterior rami, abutting skull roof laterally.
50. Prefrontal and postorbitofrontal ossification(s). 0: widely separated, frontal broadly enters orbit. 1: narrowly separated, frontal narrowly enters orbit. 2: in contact, frontal excluded from orbit.
51. Lateral process of parietal (at suture with frontal). 0: lateral process distinct. 1: lateral process absent.
52. Posterior orbital margin. 0: complete, closed by postorbital contacting jugal. 1: complete, closed by postorbital contacting ectopterygoid-maxilla unit. 2: incomplete.

53. Frontal shape. 0: frontals tapering anteriorly. 1: frontals rectangular, at most slightly constricted in middle. 2: frontals tapering posteriorly. 3: frontals greatly constricted in middle.
54. Anterior tab of frontal, underlying frontonasal joint. 0: distinct and well-defined. 1: poorly defined or absent.
55. Frontal-parietal contact (dorsal aspect). 0: mostly straight and transverse, slight median notch in frontals at most. 1: Anteriorly concave, i.e., frontals extending posteriorly into broad median embayment in parietals. 2: complex W or M shape.
56. Subolfactory (lateral descending) processes of frontal. 0: not contacting one another ventromedially. 1: meeting ventromedially, below medial descending processes of frontal if present.
57. Medial descending processes of frontal. 0: absent. 1: present.
58. Medial descending processes of frontal. 0: not meeting subolfactory (lateral descending) frontal processes, interolfactory pillar incomplete ventrally. 1: meeting ascending projections of the subolfactoryprocesses at the mesial frontal suture. 2: fused to subolfactory frontal processes, mesial frontal suture obliterated.
59. Mesial frontal suture. 0: on ventral portion of interolfactory pillar. 1: on middle of interolfactory pillar.
60. Length of main body of parietal (i.e., excluding supraorbital or posterior processes). 0: short, at most 40% skull (snout-occiput) length. 1: intermediate, between 40 and 55% of skull length. 2: long, at least 55% of skull length.
61. Suture between frontal and parietal descending flanges. 0: in lateral view, suture between frontal and parietal extends approximately vertically, or slightly anterodorsally. 1: suture greatly inclined anterodorsally, i.e., closer to the horizontal than the vertical. 2: suture curved, extending vertically in its ventral portion and becoming horizontal more dorsally.
62. Optic (=ophthalmic) fenestra. 0: not defined, incomplete ventrally. 1: bordered by frontal, parietal and sphenoid. 2: sphenoid separated from border of foramen by frontal and/or parietal.
63. Optic foramen. 0: posteriorly located, posterior border forming a deep notch in parietal. 1: intermediate position, posterior border formed by straight margin of parietal. 2: anteriorly located, posterior border within frontal.
64. Optic foramen. 0: external opening faces anterolaterally. 1: faces laterally.
65. Anterior (supraorbital) process of parietal. 0: absent or poorly developed. 1: enlarged, extending along at least 40% of lateral margin of frontal.
66. Posterior border of parietal. 0: with distinct median notch. 1: without distinct median notch.
67. Posterior border of parietal. 0: without median projection over supraoccipital. 1: with median projection over supraoccipital.

68. Posterolateral (supratemporal or suspensorial) process of parietal. 0: well developed, posterolateral margin of parietal with a distinct flange. 1: reduced, posterolateral margin of parietal with a triangular corner. 2: absent, posterolateral margin of parietal rounded.
69. Descending flange of parietal. 0: without horizontal crest. 1: with very large horizontal crest, extending from orbital region towards prootic.
70. Descending flange of parietal. 0: does not contact anterior margin of base of basipterygoid process. 1: broadly contacts anterior margin of base of basipterygoid process.
71. Supratemporal. 0: large, quadrate contacts mostly supratemporal. 1: vestigial, quadrate contacts supratemporal and otic capsule broadly. 2: absent, quadrate contacts otic capsule only.
72. Supratemporal. 0: applied to otic capsule with large free posterior projection. 1: applied to otic capsule, with small posterior projection. 2: applied to otic capsule, no posterior projection. Inapplicable in taxa with a greatly reduced supratemporal, or in outgroups where supratemporal is applied to suspensorial ramus of parietal rather than otic capsule.
73. Supratemporal. 0: does not substantially cover dorsolateral surface of prootic. 1: covers almost the entire dorsolateral surface of prootic.
74. Supratemporal. 0: anterior tip well behind anterior margin of prootic. 1: anterior tip slightly behind anterior margin of prootic. 2: anterior tip in line with or in front of anterior margin of prootic.
75. Quadrate. 0: without small ossification (extracolumella or stylohyal) on medial surface, contacting stapes. 1: with such ossification.
76. Dorsoposterior (=suprastapedial) process of quadrate. 0: distinct, large. 1: indistinct, small or absent.
77. Dorsoposterior (suprastapedial) process of quadrate. 0: projects posteroventrally, forming acute angle with quadrate shaft. 1: projects posteriorly, forming approximately a right or slightly obtuse angle with quadrate shaft. 2: projects posterodorsally, forms very obtuse angle with quadrate shaft in lateral view.
78. Length of quadrate shaft (i.e., excluding suprastapedial process). 0: short, maximum length along shaft no more than 25% of snout-occiput length. 1: long, more than 25% of snout-occiput length.
79. Cephalic condyle of quadrate. 0: situated dorsally, approximately level with dorsal margin of prootic. 1: situated ventrally, well below level of dorsal margin of prootic.
80. Quadrate shaft. 0: greatly inclined anteroventrally. 1: slightly inclined anteroventrally. 2: approximately vertical. 3: inclined posteroventrally.
81. Septomaxilla. 0: overlapping lateral process of premaxilla and/or anterior tip of the maxilla. 1: not overlapping lateral process of premaxilla or anterior tip of maxilla.
82. Dorsolateral flange of septomaxilla. 0: blunt, without spine, expansion or calcified ligament. 1: with spine projecting posterolaterally.

2: with posterior expansion projecting posteromedially towards frontal. 3: with calcified ligament.

83. Septomaxilla. 0: maxilla, but not septomaxilla, contributes to posterior border of the external naris. 1: septomaxilla with lateral flange contributing to the posterior border of the external naris.
84. Septomaxilla-frontal contact. 0: posteromedial flange of septomaxilla short, not contacting frontal. 1: posteromedial flange of septomaxilla long, contacting frontal adjacent to midline on lower part of interolfactory pillar.
85. Fenestra for duct of Jacobson's organ. 0: faces ventrally. 1: faces posteroventrally.
86. Vomer. 0: does not enter lateral margin of fenestra for Jacobson's organ. 1: forms posterior part of lateral margin of fenestra for Jacobson's organ.
87. Vomeronasal nerve. 0: does not pierce the ridge on the vomer forming the posterior wall of the vomeronasal organ. 1: pierces ridge *via* a single large foramen (sometimes with one or two additional small foramina). 2: pierces ridge through a cluster of numerous small foramina.
88. Medial fenestra in vomeronasal cupola. 0: posterior ends of sagittal flanges of vomer and septomaxilla with small or no contact, and large intervening fenestra or embayment. 1: posterior ends of sagittal flanges of vomer and septomaxilla in extensive contact, with small or no intervening fenestra.
89. Palatine length (excluding posteromedial process). 0: short anteroposteriorly (much shorter than vomer). 1: intermediate in length anteroposteriorly (as long as vomer). 2: long anteroposteriorly (much longer than vomer).
90. Horizontal (palatal) lamina of vomer. 0: posterior end narrow, tapering to a point; choana wide. 1: posterior end expanded; choana narrow.
91. Vertical (posterior dorsal) lamina of vomer. 0: small or absent. 1: well developed.
92. Palatine-vomer contact. 0: medial (choanal) process of palatine with extensive contact with vomer. 1: tiny point contact. 2: no contact.
93. Palatine-vomer articulation. 0: medial (choanal or vomerine) processes of palatines do not project ventromedially to separate vomers. 1: medial processes of palatines project ventromedially to separate the posterior portions of the vomers. This and the next two characters are inapplicable in outgroups, where the homologous process cannot be clearly identified.
94. Medial (choanal or vomerine) process of palatine. 0: anteroposteriorly broad plate of bone. 1: narrow finger-like process.
95. Medial (choanal or vomerine) process of palatine. 0: without distinct, large anterior flange. 1: with distinct, large plate-like anterior flange, abutting vomer posterolaterally.

96. Anterior process of palatine. 0: freely-projecting anterior process absent, only medial (choanal or vomerine), lateral (maxillary) and posteromedial (pterygoid) processes present. 1: narrow ("dentigerous") process present, in addition to medial, lateral and posteromedial processes. 2: wide horizontal plate present. Unordered.
97. Anterior process of palatine. 0: contacting vomer-septomaxilla complex. 1: not contacting vomer-septomaxilla complex.
98. Palatine-maxilla contact . 0: palatine sutured to maxilla. 1: palatine meets maxilla in a mobile joint. 2: palatine does not contact maxilla.
99. Lateral (maxillary) process of palatine. 0: situated in middle or anterior end of main body of palatine. 1: at posterior end of main body of palatine.
100. Lateral process of palatine. 0: pierced by foramen or constricted notch for palatine nerve (= sphenopalatine branch of facial nerve). 1: lacking foramen.
101. Articulation of palatine with pterygoid. 0: short. 1: long.
102. Pterygoid tooth row curvature. 0: concave medially. 1: straight to slightly convex medially.
103. Ectopterygoid process of pterygoid. 0: well developed, a large rectangular or triangular lateral process. 1: poorly developed, a small rounded lateral flange. 2: absent.
104. Ectopterygoid attachment to pterygoid. 0: anterior to basipterygoid process. 1: lateral to basipterygoid process. 2: posterior to basipterygoid process.
105. Pterygoid quadrate ramus. 0: robust, plate-like. 1: gracile, rod-like.
106. Pterygoid medial margin. 0: with distinct medial spur in region of basicranial articulation. 1: with smooth medial bulge. 2: with straight margin.
107. Pterygoid quadrate ramus. 0: with a shallow groove along ventromedial surface, or no groove. 1: with a very deep groove along ventromedial surface, becoming dorsomedial posteriorly. 2: teardrop-shaped muscle scar with broad anteromedial part on rounded medial lobe, tapering posterolaterally. New state (2) added.
108. Pterygoid quadrate ramus. 0: vertical or oblique sheet. 1: approximately horizontal sheet.
109. Pterygoid quadrate ramus. 0: terminates near jaw joint. 1: projects posteriorly well past jaw joint.
110. Ectopterygoid. 0: large. 1: small. 2: absent. Ordered 0-1-2.
111. Ectopterygoid-pterygoid contact. 0: clasps pterygoid on both dorsal and ventral surfaces. 1: simple overlap on only ventral surface of pterygoid. 2: simple overlap on only dorsal surface of pterygoid. 3: simple contact on only lateral edge only of pterygoid.
112. Ectopterygoid-maxilla contact. 0: posterior tip of the maxilla abuts ectopterygoid. 1: posterior tip of the maxilla is lifted off ectopterygoid and projects freely.

113. Ectopterygoid-maxilla contact. 0: anterior end of ectopterygoid restricted to posteromedial edge of maxilla. 1: ectopterygoid invades significantly the dorsal surface of the maxilla.
114. Ectopterygoid shape. 0: distal end of ectopterygoid with single anterior process projecting dorsally along maxilla. 1: distal end of ectopterygoid with two anterior processes projecting dorsally along maxilla.
115. Lateral edge of the ectopterygoid. 0: straight or slightly curved, lacking distinct angulation. 1: distinctly angulated, a distinct 'corner' present between the anterior (parasagittally oriented) and posterior (posteromedially oriented) portions of the lateral margin.
116. Cultriform process. 0: anterior one-third broad posteriorly and tapering anteriorly. 1: anterior one-third narrow throughout.
117. Interchoanal keel of cultriform process. 0: absent. 1: present, a sagittal flange extending ventrally between the medial processes of the palatines.
118. Parabasisphenoid transverse width immediately behind frontal descending flanges (= posterior orbital region in outgroups). 0: narrow, without concave ventral surface. 1: broad and ventrally concave.
119. Basipterygoid process. 0: prominent, i.e., a pedicel or flange projecting far laterally with distinct distal facet. 1: weak, consisting of a crest or mound without a distinct distal facet. 2: absent.
120. Distal surfaces (facets or crests) of basipterygoid processes. 0: long axes oriented obliquely, or transversely in ventral view. 1: long axes oriented parasagitally in ventral view.
121. Parabasisphenoid (= sphenoid). 0: sphenoid wing absent, no triangular dorsolateral prominence lateral to alar process of dorsum sellae. 1: sphenoid wing present as triangular prominence distinct from alar process, extending up anterior margin of prootic below the trigeminal notch.
122. Ventral surface of parabasisphenoid. 0: smooth posteriorly, lacking keel. 1: with median keel in posterior region, at level of posterior openings of vidian canals. 2: with pair of parasagittal keels.
123. Basioccipital-parabasisphenoid suture. 0: posteriorly positioned, closer to fenestra ovalis than to trigeminal foramen. 1: positioned midway between fenestra ovalis and trigeminal foramen. 2: anteriorly positioned, closer to trigeminal foramen than to fenestra ovalis.
124. Basioccipital. 0: with short posterolateral flanges. 1: with long posterolateral processes.
125. Posterior opening of vidian canal. 0: within basisphenoid, not bordered by prootic. 1: partly bordered by prootic (i.e., on basisphenoid-prootic suture) or entirely within prootic.
126. Vidian canal. 0: does not open intracranially. 1: opens intracranially, emerging on internal surface of sphenoid (primary opening) then emerging externally on sphenoid-parietal suture (secondary opening).

127. Vidian canals. 0: symmetrical. 1: asymmetrical, left larger than right or *vice versa*.
128. Hypophysial pit (sella turcica). 0: without distinct bony anterior boundary. 1: bounded anteriorly by distinct ridge.
129. Cerebral carotid artery. 0: opens into posterior region of hypophysial pit, near posterior transverse wall. 1: opens into middle region of hypophysial pit, well away from posterior transverse wall.
130. Dorsum sellae (crista sellaris). 0: well developed. 1: greatly reduced.
131. Dorsum sellae (crista sellaris). 0: oriented anterodorsally, overhanging the posterior portion of hypophysial pit. 1: oriented dorsally, not overhanging hypophysial pit.
132. Laterosphenoid bridge. 0: absent, V2 and V3 exits of trigeminal foramen confluent. 1: present, fuses to prootic forming vertical bar between exits of V2 and V3.
133. Alar process of prootic. 0: long distinct process projecting anteriorly well past trigeminal (V) foramen. 1: short process not projecting past trigeminal (V) foramen.
134. Trigeminal foramen, anterior margin. 0: closed by parietal at least medially; upper and lower anterior processes of prootic may touch superficially lateral to the parietal. 1: closed by prootic, deep contact or fusion of prootic processes excludes parietal from opening.
135. Exit foramen for the facial (VII) nerve (hyomandibular branch, if distinct). 0: located outside the opening for the mandibular branch of the trigeminal nerve (V3, or V2+V3). 1: located within the opening.
136. Sulcus connecting exit foramen of palatine branch of facial (VII) nerve with posterior opening of vidian canal. 0: weakly recessed, with shallow and smooth margins. 1: deeply recessed, with sharply defined anterior and often also posterior margins. 2: embedded, closed laterally forming a tunnel in prootic.
137. Crista circumfenestralis. 0: juxtastapedial recess bordered by crests which may extend directly laterally but do not converge. 1: juxtastapedial recess surrounded by crests which converge to partly enclose stapedial footplate, much of footplate remains exposed laterally. 2: juxtastapedial recess surrounded by crests which converge to largely enclose stapedial footplate.
138. Supratemporal-supraoccipital contact. 0: supratemporal and supraoccipital separated by dorsal exposures of parietal and exoccipital. 1: supratemporal and supraoccipital separated by dorsal exposures of prootic, parietal and exoccipital. 2: supratemporal and supraoccipital in contact.
139. Paroccipital process. 0: long process. 1: distinct flange. 2: indistinct bump or absent.
140. Supraoccipital. 0: external (dorsoposterior) surface with no, or very weak transverse ridge. 1: external surface with moderate transverse ridge. 2: external surface with very high transverse ridge.

141. Supraoccipital dorsal exposure. 0: long, sagittal dimension more than 50% transverse dimension. 1: short, sagittal dimension less than 50% transverse dimension.
142. Supraoccipital-prootic contact. 0: narrow, less than half supraoccipital-parietal contact. 1: broad, subequal to or as long as supraoccipital-parietal contact.
143. Exoccipital separation dorsal to foramen magnum. 0: exoccipitals widely separated above foramen magnum. 1: exoccipitals with tiny point contact above foramen magnum. 2: exoccipitals in extensive median contact above foramen magnum.
144. Exoccipital separation ventral to foramen magnum. 0: exoccipitals separated below foramen magnum, not in contact along dorsal midline of occipital condyle. 1: exoccipitals in contact below foramen magnum, along dorsal midline of occipital condyle.
145. Occipital condyle. 0: dorsal surface deeply concave, i.e., with deep "fovea dentis". 1: dorsal surface slightly concave at most, i.e., with shallow or no "fovea dentis".
146. Stapedial shaft. 0, straight. 1, angulated.
147. Stapedial shaft. 0, slender and longer than diameter of stapedial footplate. 1, thick and not longer than diameter of footplate.
148. Distal end of stapes. 0, associated with dorsal tip of suprastapedial process of quadrate. 1, associated with ventral end of suprastapedial process and dorsal end of quadrate shaft, i.e., cephalic condyle. 2, associated with middle or ventral half of quadrate shaft.
149. Dentary length. 0: dentary long, more than 40% of main mandible length, i.e., length excluding retroarticular process. 1: dentary short, less than 40% of main mandible length.
150. Mental foramina on lateral surface of dentary. 0: two or more. 1: one.
151. Posterolateral margin of dentary. 0: notch absent, posterolateral margin of dentary straight or slightly concave, dorsoposterior and ventroposterior processes indistinct. 1: with shallow notch, processes short. 2: with deep notch, processes long.
152. Dentary posterior margin. 0: dorsal posterior process does not extend much further than ventral posterior processes. 1: dorsal process extends much further posteriorly than ventral process.
153. Medial shelf of dentary; horizontal flange near splenial-coronoid junction. 0: not exposed medially. 1: exposed medially.
154. Meckel's canal (groove). 0: lacks floor anteriorly, open ventrally anterior to level of anterior inferior alveolar foramen. 1: floored by a horizontal ventral lamina for its full length. 2: enclosed anteriorly, with ventral and medial lamina.
155. Splenial. 0: splenial present as discrete element. 1: splenial not present as discrete element.
156. Splenial size. 0: small, extends no more than 50% of distance between splenial-angular contact and symphysis. 1: large, extends more than 50% of distance.

157. Splenial-angular joint. 0: vertical in medial view. 1: highly oblique in medial view.
158. Foramen within splenial (= anterior mylohyoid foramen). 0: present. 1: absent.
159. Dorsal margin of splenial. 0: deeply notched, posterior region of notch bordered dorsally by anterodorsal spine. 1: moderately notched, posterior region of notch not bordered dorsally by anterodorsal spine. 2: smooth, not notched. Ordered 0-1-2.
160. Splenial-coronoid contact. 0: posterior end of splenial in broad contact with coronoid. 1: posterior end of splenial only just reaches coronoid. 2: posterior end of splenial does not contact coronoid.
161. Splenial lateral exposure. 0: Anterior portion of splenial not exposed laterally. 1: Anterior portion of splenial greatly exposed laterally.
162. Coronoid. 0: coronoid large and distinct. 1: coronoid greatly reduced and sometimes fused to compound. 2: coronoid never present as distinct element.
163. Coronoid. 0: with posteroventral process or expansion. 1: without posteroventral process or expansion.
164. Coronoid lateral exposure. 0: coronoid overlaps lateral surface of surangular and is exposed in lateral view. 1: coronoid does not overlap lateral surface of surangular, but projects dorsally beyond it and is thus well exposed in lateral view. 2: coronoid entirely medial to surangular and is largely covered in lateral view.
165. Coronoid-angular contact. 0: coronoid and angular separated by prearticular, or prearticular portion of compound bone. 1: coronoid contacts angular.
166. Coronoid process. 0: well developed, distinct projection lateral to adductor fossa. 1: poorly developed or absent, smooth rounded crest at most.
167. Surangular eminence. 0: compound postdentary element without dorsal eminence. 1: surangular portion of compound with dorsal crest or process lateral to adductor fossa. 2: prearticular portion of compound with ascending process medial to adductor fossa.
168. Adductor fossa. 0: posterior region exposed medially, prearticular dorsal margin lower than surangular dorsal margin. 1: posterior region exposed dorsally only; prearticular about equal in height to surangular. 2: posterior region exposed laterally only, prearticular higher than surangular.
169. Anterior surangular foramen. 0: situated posteriorly, below apex of coronoid process or more posterior. 1: situated anteriorly, between apex and anterior limit of coronoid process. 2: situated far anteriorly, in front of anterior limit of coronoid process.
170. Lateral crest of compound element, extending anteriorly from articular cotyle along ventrolateral surface of mandible. 0: absent or weak. 1: strongly developed. State descriptions modified from 'absent' and 'present'.

171. Articular-surangular fusion. 0: articular and surangular not fully fused in region of articular facet. 1: articular and surangular fully fused in region of articular facet.
172. Retroarticular process length. 0: long, longer than articular facet. 1: short, not longer than articular facet.
173. Flange on dorsolateral surface of tip of retroarticular process. 0: absent. 1: present.
174. External grooves and ridges on tooth bases. 0: present, surface of bases of mature tooth crowns with vertical ridges and grooves. 1: absent, surface of bases of mature tooth crowns smooth.
175. Premaxillary teeth. 0: present. 1: absent.
176. Premaxillary tooth number. 0: three or more alveoli on each side of the midline. 1: one or two alveoli on each side.
177. Maxillary teeth. 0: nearly uniform in size, with very gradual size changes. 1: distinctly larger near middle of tooth row, smaller anteriorly and posteriorly. 2: distinctly larger near anterior end of tooth row, smaller in middle and posteriorly. 3: distinctly larger posteriorly.
178. Maxillary teeth. 0: nine or more alveoli. 1: eight or fewer alveoli.
179. Dentary teeth. 0: eight or more alveoli. 1: seven or fewer alveoli.
180. Alveoli (in middle of maxilla and dentary). 0: not expanded transversely. 1: wider transversely than anteroposteriorly.
181. Palatine teeth. 0: absent. 1: present.
182. Palatine teeth. 0: nine or more alveoli. 1: eight or fewer alveoli.
183. Pterygoid teeth. 0: present. 1: absent.
184. Pterygoid teeth. 0: twelve or more alveoli. 1: eleven to nine alveoli. 2: eight or fewer alveoli.
185. Median (basihyal) element. 0: present, uniting hyoid cornua. 1: absent.
186. First branchial arch elements. 0: present. 1: absent, replaced by caudal extensions of the lateral edge of the basihyal.
187. Hyoid cornua. 0: diverging sharply posteriorly. 1: diverging only slightly posteriorly. 2: parallel.
188. Number of presacral vertebrae. 0: less than 120. 1: 120-160. 2: 160-200. 3: over 200.
189. Number of caudal vertebrae. 0: more than 20. 1: fewer than 20.
190. Dorsoposterior process on atlas neural arch, overlying axis neural arch. 0: present, well developed. 1: absent or very weak.
191. Second (axis) intercentrum. 0: not fused to anterior region of axis centrum, suturally connected at most. 1: fused to anterior region of axis centrum.
192. Neural spine height. 0: well-developed process. 1: low ridge, or absent.
193. Posterior margin of neural arch. 0: shallowly concave in dorsal view. 1: with deep, V-shaped embayment in dorsal view exposing much of centrum in front of condyle.

194. Zygosphene roof. 0: with deeply concave anterior edge, i.e., deeply notched between zygosphenal facets. 1: with shallowly concave anterior edge, i.e., slightly notched between facets. 2: with straight or slightly sinuous anterior edge, i.e., not uniformly concave.
195. Condyles of mid-trunk vertebrae. 0: oval, sagittal dimension much less than transverse diameter. 1: round, sagittal dimension similar to transverse dimension.
196. Condyles of mid-trunk vertebrae. 0: facing very dorsally, ventral edge (at most) of condyle surface exposed in ventral view. 1: facing posteriorly, or posterodorsally, much of condyle surface exposed in ventral view.
197. Precondylar constriction of centrum. 0: absent or very weak. 1: moderate. 2: strong.
198. Orientation of zygapophyses of mid-trunk vertebrae. 0: steeply inclined medially, 30° or more from the horizontal. 1: moderately inclined medially, between 15° and 30° from the horizontal. 2: not inclined medially, less than 15° from horizontal.
199. Paracotylar foramina (foramen on anterior surface between cotyle and transverse process). 0: present on most or all vertebrae. 1: present on no, or few, vertebrae.
200. Parazygantral foramina (foramen on posterior surface of neural arch, between zygantrum and postzygapophyseal facets when present). 0: absent on all vertebrae. 1: numerous small pits (but no large foramina) in parazygantral area. 2: one (or several) large foramen present on each side.
201. Subcentral foramina. 0: uniform throughout column, small and paired in most vertebrae. 1: irregular, being either small and paired, absent, or single and large, in different vertebrae.
202. Prezygapophyseal process. 0: absent. 1: present as a small process extending slightly laterally from prezygapophyseal facet. 2: present as a prominent process extending laterally or anterolaterally from prezygapophyseal facet.
203. Hypapophyses. 0: present on anterior eight cervicals or fewer. 1: present up to at least cervical ten, but absent in mid- and posterior trunk. 2: present throughout trunk, but poorly developed in posterior trunk. 3: present throughout trunk, well developed throughout.
204. Ventral surface of centra. 0: mid-trunk vertebrae with smooth, transversely convex ventral surface. 1: mid-trunk vertebrae bearing single median haemal keels.
205. Lymphapophyses. 0: No forked free ribs or lymphapophyses. 1: three or more free-ending cloacal vertebrae with lymphapophyses.
206. Caudal vertebrae. 0: with posteroventral projections. 1: without posteroventral projections.
207. Posteroventral elements of caudals. 0: articulate with centrum. 1: fuse with centrum.

208. Posteroventral elements of caudals. 0: distally fused (chevrons). 1: distally separated (haemapophyses). 2: single median element (caudal hypapophyses).
209. Ribs. 0: tuber costae (= tuberculum, tuberculiform process, dorsal process) of rib absent or weakly developed. 1: tuber costae well developed.
210. Ribs. 0: slender throughout body. 1: thickened and heavily ossified (pachyostotic) in middle region of body.
211. Cervical region. 0: present, at least 20 anterior vertebrae bearing ribs distinctly shorter and thinner than ribs of other trunk vertebrae. 1: absent, only the first few anterior vertebrae bear short, thin ribs.
212. Pelvic girdle. 0: three distinct elements present. 1: two distinct elements present. 2: single element present. 3: no elements present.
213. Pelvis. 0: external to sacral or cloacal ribs. 1: internal to sacral or cloacal ribs.
214. Hindlimb. 0: present, with distinct femur, tibia and fibula. 1: vestigial, with one bone (femur) only, and sometimes a single distal spur. 2: absent.

3.8.2 Phylogenetic Analyses: Parsimony, Bayesian Inference and Maximum Likelihood

The phylogenetic tree (Fig. 3.1) is based on combined analysis of the molecular and morphological data. This tree is very similar to the molecular-only tree (see Wiens *et al.* 2008), with differences generally restricted to regions that were poorly resolved in the molecular analysis anyway (e.g. scolecophidian monophyly). Analyses of the morphological data alone yielded a tree very similar to Scanlon and Lee (2002) which appears to be overly influenced by convergence in burrowing traits.

Parsimony analyses and bootstrapping were performed using PAUP* (Swofford 2003), with multistate morphological characters ordered if they formed clear morphoclines (see above). However, analyses with all multistate characters unordered yielded the same tree, as such traits formed a minute proportion of the dataset (<0.36%). Branch (Bremer) support was performed in PAUP* with assistance of TreeRot (Sorenson and Franzosa 2007). All searches employed at least 200 random addition replicates. The most parsimonious tree is shown in Figure 3.1, along with branch support and bootstrapping frequencies.

Bayesian analyses were performed in MrBayes (Ronquist and Huelsenbeck 2003). The "standard" model was used for the indel and morphology datasets; the "variable only" correction for scoring bias was employed. Bayes factors (cutoff >5 difference in log-likelihoods; see Ronquist *et al.* 2005) indicated the rate variation (gamma) parameter was required for the morphology partition (harmonic means 2410.96 vs 2481.23) but not the indel partition (harmonic means 538.43 vs 539.24). A 7-partition scheme for the molecular data (3 nuclear codons, 3 mitochondrial codons,

RNA) was favored over all simpler schemes examined, a very typical result. For example, the log-likelihood for the 7-partition model was –94816.98, while the next most complex 5-partition model (with codons in the short mitochondrial data combined) had a log-likelihood of –95527.09: a difference of only >5 in log-likelihoods is often considered decisive (see Ronquist *et al.* 2005). More elaborate partitioning schemes (e.g., partitioning the data into individual genes as well as codons) had difficulty reaching stationarity even after very long runs (>5 million generations) and could not be evaluated. AIC comparisons using MrModeltest (Nylander 2004), suggested the GTRig model for nuclear codon 1, nuclear codon 2, and RNA, and the GTRg model for all other partitions. This is consistent with the observation that the former three partitions are likely to contain invariant sites due to overall slow evolution, or highly conserved areas.

Two independent MCMC (Markov-Chain Monte Carlo) runs of the final 9-partition model (7 molecular, plus indels and morphology) were performed, each with 6 chains and 4 million steps. The first 50% of trees was discarded as burn-in. Tree proportions were linked across molecular partitions (but absolute tree lengths were unlinked, i.e., allowed to vary using the rate scalar). Both tree proportions and absolute tree length were unlinked across the indel, morphological and combined molecular partitions. Split frequencies (<0.02), and analyses with Tracer, suggested stationarity was reached well before the burn-in of 2 million steps. The majority-rule Bayesian consensus tree was identical to the parsimony tree, and Bayesian posterior probabilities for all clades are shown in Figure 3.1. However, split frequencies between 0.01 and 0.02 indicated stationarity was marginal in this analysis with all taxa (see Ronquist *et al.* 2005), and this persisted when the analysis was repeated. Hence, the analysis was then performed with very poorly known taxa (*Xenophidion*, *Anomochilus* and *Cylindrophis maculatus*) deleted; here, the same run settings almost achieved convergence well before 2 million steps (split frequencies <<0.01), and the relationships between the included taxa were all identical to that in Figure 3.1 with posteriors of 1.0.

Both parsimony and Bayesian analysis have recently been demonstrated to suffer artifacts related to long branch attraction, while maximum likelihood is less susceptible (Kolaczkowski and Thornton 2009). We therefore confirmed that likelihood analyses gave similar results as the parsimony and Bayesian analyses. Likelihood analyses were performed using RaxML (Stamatakis 2006), but only on the molecular data. All the clades strongly supported in the clades in the parsimony and Bayesian analysis (of the combined data) were supported in the ML analysis of molecular data alone.

3.8.3 Molecular Divergence Dating

Divergence dates were inferred in a Bayesian framework using BEAST 1.5.3 (Drummond and Rambaut 2007), which can incorporate uncertainties in calibration, branch lengths and tree topology. Only the nucleotide sequence data could be employed (i.e., indels and morphology excluded). The same partitioned model in the MrBayes analysis (see above) was used, along with a birth-death tree prior and a relaxed clock with lognormally-distributed rate variability. Agamids were assumed to be the furthest outgroup (Lee 2009). To reflect the topology robustly supported by the combined morphological and molecular data, scolecophidians were constrained to be monophyletic, and uropeltoids, pythonoids, bolyeriines, *Anomochilus* and *Xenophidion* were constrained to their positions in that analysis. However, analyses without any topological constraints yielded very similar divergence dates within snakes. Five robust fossil calibrations were used, discussed in detail in the next section.

Two separate MCMC runs were performed, each of 13 million steps, sampling every 1000 steps. Analyses with Tracer (Rambaut and Drummond 2007) suggested stationarity was reached by 3 million steps, leaving 10 million steps for analysis. All parameters had similar and broadly overlapping sampled distributions across both runs; in particular, the estimates for divergence dates for all nodes within snakes were very similar (medians differing by <2 my between runs). Thus, the sampled trees (10,000) across both runs were combined. The maximum posterior probability tree, and divergence dates for each node (with 95% error bars), is shown in Fig. 3.2.

3.8.4 Fossil Calibrations

All 5 calibrations used in the molecular divergence analyses were based on fossils that had been robustly placed using explicit derived characters (synapomorphies). These priors are indicated with numbers 1-5 on Figure 3.2. Lognormal distributions were employed, with the median being the estimated age of the calibration fossils. The younger bound is tight (close to the median) and represented by a "hard" wall; assuming the fossil is correctly assigned, and dated with reasonable accuracy, the real divergence date cannot be much younger. The older bound is loose (further from the median) and represented by a "soft" tapering distribution tail; non-preservation means the real divergence date can always be older than the fossil, but the probability of very deep dates diminishes as older and older fossil deposits yield no fossils of the focal clade. Here, we have set upper bound at the age where earlier deposits occur yielding many small-to-medium land vertebrates, but none are clearly attributable to the focal clade. This date is set as a "soft" 95% upper bound (i.e., allowing a 5% chance the real divergence is still older). These soft upper bounds allow for self-correction of overly young calibration dates. For three calibration nodes, the inferred dates (posteriors) are very similar to the calibration ages (priors). However, for the remaining two calibrated nodes (caenophidians, crown snakes), the posteriors are much older than the priors. This means that the other 3 calibration nodes and the molecular data (branch lengths) strongly suggest that the calibrations for Caenophidia and crown snakes are substantial underestimates (see Sanders and Lee 2007).

1. Anguimorphs-Snakes. Hard lower bound 155, median 162, soft upper bound 200. The earliest undisputed anguimorphs are the dorsetisaurids (Evans 2003; Conrad 2008), first known from the mid-late Jurassic boundary (Fedorov and Nessov 1992); several other anguimorphs occur slightly later. The mid-Jurassic *Parviraptor* has most recently been assigned to the Gekkota (Conrad 2008). Upper bound reflects the observation that no undisputed anguimorphs are known from the rich lower Jurassic record, which preserves numerous other small reptiles.
2. Scolecophidia-Alethinophidia (crown snakes). Hard lower bound 93.6, median 97, soft upper bound 120. A partial snake braincase exhibiting a unique and unreversed alethinophidian character, the sphenoid wing, is present in cranial remains from the upper Wadi Abu Hashim Member of the Wadi Milk Formation (Scanlon 2003), dated as Cenomanian (Rage and Werner 1999). Vertebrae provisionally identified as alethinophidian ("anilioid") also known from the same period (Gardner and Cifelli 1999). Upper bound reflects the observation that no undisputed alethinophidians are known from the lower Cretaceous record, which preserves numerous other small reptiles.
3. *Python-Aspidites* (Pythoninae). Hard lower bound 22, median 25, soft upper bound 55. This calibration was previously used in Vidal *et al.* (2009). The oldest unequivocal fossil pythons are from Riversleigh, Australia, assignable to *Morelia* (e.g., Scanlon 2001 and in prep.). The very loose (old) upper bound reflects (1) the inference that *Morelia* is nested either within the *Python* clade (Kluge 1993) or more likely within the *Aspidites* clade (Rawlings *et al.* 2008), meaning that the split between the *Aspidites* and *Python* clades must be somewhat older and (2) the observation that the Australian fossil record before Riversleigh is relatively poor until the Murgon deposits (Tingamarra Local Fauna) ~55 my. No pythons are known from the Tingamarra LF, which preserves other small and medium-sized vertebrates (e.g., Scanlon 2005b).
4. *Boa-Epicrates* (core Boinae). Hard lower bound 56, median 59, soft upper bound 69. *Titanoboa* (58-60 mya) has been assigned to the *Boa* lineage using derived vertebral traits (Head *et al.* 2009). Upper bound reflects the observation that no undisputed vertebrae attributable to modern boas are known from the rich South American Cretaceous record which preserves other snakes.
5. Colubroidea-*Acrochordus* (Caenophidia). Hard lower bound 35, median 38, soft upper bound 48. This calibration was previously used in Sanders and Lee (2008). The oldest unequivocal colubroids are from the middle Eocene (Head *et al.* 2005), and the *Acrochordus* lineage likely extends to approximately the same time (Head *et al.* 2007). While older vertebrae have been assigned to colubroids or acrochordoids (e.g., Rage *et al.* 2008), these referrals are based on broad similarities rather than clear derived characters and a position outside of crown Caenophidia cannot be excluded.

CHAPTER 4

Oogenesis and Early Embryogenesis

Mary E. White

4.1 INTRODUCTION

In the introduction to their volume on reptilian development, Billett *et al.* (1985) note, "...the study of the developmental anatomy of reptiles reached its zenith between 1850 and 1920." Sadly, this has not changed significantly. Few modern studies in development have focused on reptiles (which for the purposes of this review will not include birds), and those that do concentrate mainly on chelonians. This may be due to the ease of obtaining turtle embryos, along with the observation that eggs of turtles are generally oviposited at much earlier stages of development than are those of squamates (for example, see Shine 1983). Much that we know, or think we know, has been learned from comparisons of snake development to other organisms, particularly birds. There have been few recent morphological studies of snake development, and even fewer molecular investigations.

Hubert provided extensive reviews of the morphology and cytology of oocytes (1985a) and early developmental stages (1985b) of snakes. Significant updates to the body of work have not been made since that time, so I will not focus on cytology. Likewise Blackburn and Flemming (2008) have recently reviewed the morphology, evolution and development of fetal membranes, comparing those in oviparous and viviparous snakes. In **Chapter 5** of this volume, Blackburn and Stewart discuss placentation in viviparous snakes, so these topics will not be covered here. Rather I concentrate on some of what is known, or surmized, about the processes of oogenesis and early development of snakes, focusing particularly on vitellogenesis, gastrulation, axis formation and primordial germ cells.

Department of Biological Sciences, Southeastern Louisiana University, Hammond, LA 70402 USA

4.2 VITELLOGENESIS

Like all amniotes, with the exception of eutherian mammals, snakes have telolecithal (also called mega- or macrolecithal) eggs. While the term telolecithal literally refers to a large concentration of yolk at one end, in reality, the quantity of yolk is so large that it is everywhere except one end—the blastodisc, where the embryo will form.

4.2.1 Reproductive Cycles

The amount of time required to produce these yolk-laden eggs varies among species. Temperate zone snakes generally show seasonal reproduction with many species requiring two to three years for a female to produce mature follicles. In the Diamond-backed Watersnake (*Nerodia rhombifer*), for example, follicular growth occurs slowly over a two-year period, followed by very rapid growth in the spring of the third year (Betz 1963). Bauman and Metter (1977) found similar results in the Northern Watersnake (*N. sipedon*), with little growth in follicles throughout the year but explosive growth just before mating. Aldridge (1979) characterized two different patterns of vitellogenesis in temperate zone snakes. Both groups undergo what he termed primary vitellogenesis, in which the oocytes acquire small amounts of yolk protein, lipids and some calcium, and the follicle grows to approximately 4-6 mm. The second stage, secondary vitellogenesis, involves the rapid growth of the follicle, and its timing differs between the two groups. In Type I snakes, secondary vitellogenesis begins early in the spring and continues until ovulation that summer. In Type II snakes, secondary vitellogenesis begins in the summer following oviposition, pauses during hibernation, and resumes the following spring (Aldridge 1979).

Whereas temperate zone snakes have seasonal and often biennial or triennial female reproductive cycles, tropical snakes show great variation in reproduction. For example, in a study of five species of colubrid snakes from the Yucatan Peninsula, Censky and McCoy (1988) found that two species reproduced annually in the rainy season, one reproduced biannually in the rainy season, one laid two clutches annually in the dry season, and one was aseasonal, showing continuous reproduction. Boids and pythons require several months for vitellogenesis and most do not reproduce annually (Shine 2003; Pizzatto and Marques 2007). Tropical snake reproductive cycles are covered in more detail in **Chapter 12**.

4.2.2 Yolk Biochemistry

The content of yolk has been studied extensively in a number of organisms, but little work has been done on the yolk of snake follicles. Noble (1991) reviewed yolk composition and utilization in "birds and reptiles", but the data are mainly for birds and crocodilians, with a few lizard and turtle examples. He did note that other reptilian eggs in general have a higher percent of yolk in the egg by weight compared to birds, presumably due at least in part to the higher amounts of albumen in bird eggs. Yolk content can vary not only among species, but also due to diet, environmental conditions including temperature, and other factors. Therefore it is incorrect to consider a "standard" yolk composition for a species or group (reviewed in Noble 1991).

A few studies have looked at yolk in specific snakes. Dessauer and Fox (1959) measured a high lipid content (43%) with a much lower phosphoprotein content from a partially purified vitellogenin extract from the serum of the Eastern Ribbon Snake (*Thamnophis sauritus*). In the Common European Adder (*Vipera berus*) yolk has been shown to consist mainly of lipoproteins (Bellairs 1959). Stewart and Castillo (1984) estimated the dry weight of yolk from *Nerodia* eggs and found protein in the highest proportion followed by lipid. However, when comparing the eggs to the internal yolk sacs of newly born young, their results suggest that lipid rather than protein catabolism provides the major energy source for the developing embryo (Stewart and Castillo 1984). Indeed, in his review, Noble (1991) notes that for all reptilian species studied, rapid catabolism of lipid occurs in the later stages of development during the period of maximum growth.

4.2.3 Mechanisms of Vitellogenesis

As with content of yolk, the mechanisms of vitellogenesis are not well known in reptiles in general, and this topic is particularly under-studied in snakes. In lizards, the precursors to yolk proteins are synthesized in the liver and carried to the ovaries through the bloodstream. These precursors are then brought into the oocyte via pinocytosis (reviewed in Hubert 1985a). A similar mechanism is assumed to occur in snakes. Lizard follicles have pyriform cells in the granulosa that form cytoplasmic bridges that are presumed to infuse oocytes with a variety of components including nuclei (Guraya 1969), cytoplasmic organelles (Srivastava 1948; Andreuccetti *et al.* 1978), ribosomes (Taddei 1972) and lipids, glycogen and RNA (Blanc 1971, as reviewed in Hubert 1985a). More recently, Maurizii and coworkers (2009) have shown evidence that vasa, a germ cell-associated protein, is transferred from somatic pyriform cells to developing oocytes in a lacertid lizard. Betz (1963) noted fusion of follicle cells with oocytes in *Nerodia rhombifer*, suggesting a similar process in snakes, but this has not been studied in detail.

4.2.4 Energetics of Vitellogenesis

Whatever the mechanism, clearly a great deal of energy is required for the provisioning of macrolecithal oocytes during vitellogenesis. Drent and Daan (1980) employed economic terms to discuss two strategies for reproductive cycles in birds. They used "capital-based investment" to refer to species in which body reserves are used for producing mature oocytes.

In such cases, the body condition of the animal prior to vitellogenesis is of utmost importance. In contrast, for species using the "interest-based investment" strategy, foraging success prior to ovulation is paramount. Nalleau and Bonnet (1995) noted that both strategies seemed to be used by snakes. Saint Girons (1957) suggested that spring foraging in the European Asp (*Vipera aspis*) was not adequate to provide for the rapid uptake of yolk to the developing follicle, so that these snakes must rely on stored fat. Other studies have detailed the high cost of reproduction as relates to energy stores in *V. aspis* (see Bonnet *et al.* 2002, for example). In Brown Water Pythons (*Liasis fuscus*), reproduction did not occur unless the female had surpassed a threshold level of energy reserves, and reproductive output was directly related to body condition (Madsen and Shine 1999). Dessauer and Fox (1959) observed that in *Thamnophis sauritus*, for example, provisioning of the follicle occurs at the same time as regression of abdominal fat bodies, suggesting that these fat bodies serve as a lipid store for vitellogenesis. However, Garstka *et al.* (1985) noted that while *T. sirtalis* did rely on stored fat during vitellogenesis, it was not nearly to the extent seen in viperids and some other capital breeders. Whittier and Crews (1990) suggested that there is year-to-year variation in the importance of stored reserves versus foraging in *T. s. parietalis*. This is supported by a study in the Rough Greensnake (*Opheodrys aestivus*) in which low fat reserves after an extraordinarily hot and dry summer had no measurable impacts on reproduction the following spring. Foraging success during the period of vitellogenesis was apparently sufficient to fully provision the eggs (Plummer 1983). A similar result was seen in the aquatic Black Swampsnake (*Seminatrix pygaea*); no reduction in reproductive output occurred despite depletion of energy reserves following a severe drought (Winne *et al.* 2006). Naulleau and Bonnet (1995) found that even under normal conditions, the Aesculapean Snake (*Elaphe longissima*) used an "interest-based" strategy. Vitellogenesis in this annual breeder depended on foraging success rather than body reserves. Overall these results suggest that snakes show something of a continuum from those that rely almost entirely on stored reserves (capital breeders), to those that use both stored reserves and foraging, particularly depending on conditions, to those that rely mainly on foraging (interest-based breeders).

4.3 FERTILIZATION AND CLEAVAGE STAGES

4.3.1 Snake Fertilization

In snakes, fertilization is internal and takes place in the upper regions of the oviduct. Males have two intromittent organs called hemipenes that are usually highly ornamented. The hemipenes are generally inverted inside the hemipenal sheath, caudal to the cloaca, but are everted for mating. Many species have hooked or bilobed and bulbous hemipenes, which may be important in anchoring the organ in the female reproductive tract.

Females of most species apparently store sperm, so that mating and fertilization are not necessarily temporally connected (Sever and Hamlett 2002). Sperm storage anatomy is covered in detail in **Chapter 9**.

4.3.2 Early Cleavage Stages

Following fertilization, the early cleavage stages begin while the embryo is still in the oviduct, and thus they have not been well-studied. Zehr (1962) provided a staging series for *Thamnophis sirtalis*, but he only named the earliest cleavage stages without describing the patterns of cleavage. He was able to describe several later developmental stages from each of a number of females using a multiple Caesarean technique. Hubert and Dufaure (1968) gave a more detailed description of early and later stages for *Vipera aspis*, and their staging series is now commonly used for describing snake development. *Thamnophis* and *Vipera* are viviparous snakes, retaining embryos until birth, unlike oviparous snakes that lay their eggs at a fairly late stage (Stage 35-40 according to the Hubert and Dufaure staging), when embryos are considered to be at least halfway through embryonic development (Shine 1983). Early stages of oviparous snake embryogenesis have received even less attention than those of viviparous snakes, though partial series for the Grass Snake (*Natrix natrix*) and Bullsnake (*Pituophis melanoleucus*) have been described (*N. natrix*; Krull 1906; Vielhaus 1907; *P. melanoleucus*; Treadwell 1962). More recently, Jackson (2002) described stages for the Monocled Cobra (*Naja kaouthia*), but his series started at oviposition. Early development in squamates is considered to be similar to that of birds, and thus much of our knowledge comes from comparisons to chick development. As in birds, the huge concentration of yolk leads to discoidal cleavage, in which only a small region of cytoplasm cleaves to form the embryo atop the much larger yolk mass (reviewed in Hubert 1985b; Gilland and Burke 2004). These early meroblastic cleavages result in an embryo that overlies a subgerminal cavity and is continuous with the yolk. The initial cleavages are vertical. Cleavage occurs more easily where there is less yolk, and thus these first furrows produce smaller blastomeres in the central region of the blastodisc, with fewer, larger cells in the marginal zones. The terminology used for birds is often applied to other reptiles, with the area pellucida being the region of the blastodisc over the subgerminal cavity, and the area opaca being the marginal region that directly overlies the yolk.

4.3.3 Blastula Formation

The initial cleavages are followed by a delamination that divides the disc into an upper epiblast that is relatively yolk-free, and a deeper hypoblast with larger cells, many still continuous with the yolk (Hubert 1985b). Incomplete cleavage in the deeper cells and peripheral cells of the blastodisc leads to a yolk syncytial layer. Although early cleavage is generally thought to be similar, by the late blastula stage the epiblast and hypoblast differ somewhat among different groups of reptiles. In all reptiles, the epiblast forms an epithelial layer, but its thickness varies, with some turtles having an epiblast composed of only a single cell layer. In lizards, it is thicker, and may consist of up to four or five layers of cells in some species (Peter 1934, 1938, as reviewed in Gilland and Burke 2004). The snake epiblast is more similar to that of lizards (Hubert 1985b).

As the embryo progresses towards gastrulation, the hypoblast in lizards is continuous, with a thicker region in the central areas underlying the embryonic shield, spreading out to thinner layers in the periphery (Fig. 4.1). In snakes, the region of hypoblast below the embryonic shield is diffuse, with only scattered clumps of cells lying mainly in the subgerminal cavity (Fig. 4.2). This is more similar to the pattern seen in turtles than that seen in lizards (Hubert 1985b) and will be discussed below. However, it is important to note that these observations are based mainly on one species of snake and a few species of lizards. A broader comparative approach might yield somewhat different conclusions.

4.4 GASTRULATION

4.4.1 Overview

Gastrulation is the critical morphogenetic process that leads to the formation of the embryonic germ layers. The early literature contains conflicting reports of germ layer formation in snakes, particularly as regards formation of chordamesoderm and endoderm. Gilland and Burke (2004) attribute these differences to the difficulties of interpreting a dynamic process using static histological sections. When gastrulation in amphibians and chicks was shown to involve massive morphogenetic movements (Gräper 1929; Vogt 1929; Wetzel 1929, as discussed in Gilland and Burke 2004), earlier interpretations of squamate gastrulation that invoked mainly cell proliferation and delamination were deemed to be incorrect. There has also historically been a disagreement as to whether squamate gastrulation involves a blastopore, as in amphibians, or the primitive streak in birds and mammals. All recent studies have stated unequivocally that squamates do not use a primitive streak (Hubert 1985b).

4.4.2 Morphogenetic Movements in Gastrulation

The stages of gastrulation have been studied in the snake, *Vipera aspis* (Hubert 1970, reviewed in Hubert 1985b); however, gastrulation movements have not been tracked using cell marking techniques in squamates. In their review of reptilian gastrulation, Gilland and Burke (2004) noted that morphological analyses suggest that the movements of squamate gastrulation are similar to those of turtles, which have been tracked (Pasteels 1937; Chandrasekharan Nayar 1966, as reviewed in Gilland and Burke 2004). In both, the onset of gastrulation is visible as the posterior

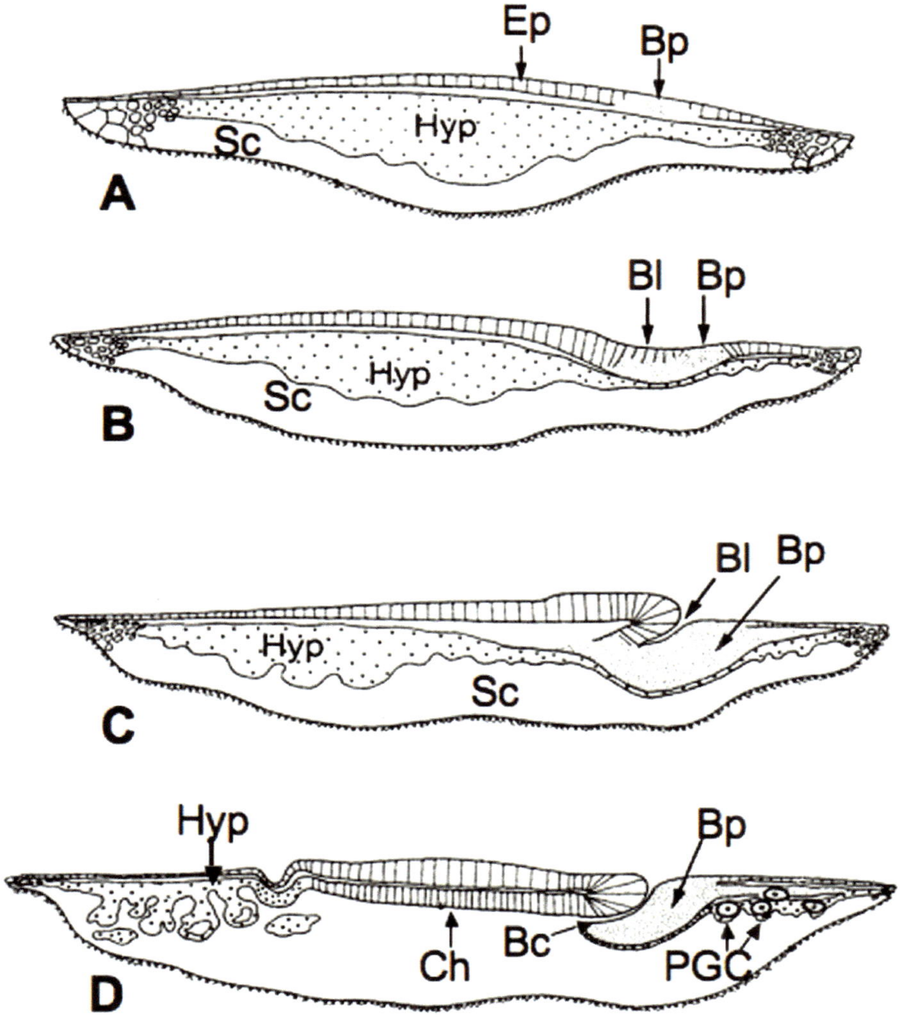

Fig. 4.1 Schematic drawings of stages of gastrulation in the lizard *Lacerta vivipara* shown in sagittal section. The anterior of the embryo is to the left. **A.** A thickening of the posterior region of the epiblast forms the blastoporal plate (Bp). The hypoblast (Hyp) lies beneath the epiblast (Ep) above the subgerminal cavity (Sc). **B.** The blastopore (Bl) forms initially as a small invagination in the anterior region of the blastoporal plate. **C.** The blastopore extends forward and the epiblast begins to involute over the dorsal lip of the blastopore. **D.** The hypoblast has been pushed to the extraembryonic region. The epiblast that has involuted over the dorsal lip of the blastopore forms the chordamesoderm (Ch). The blastoporal canal is well-developed and the blastoporal plate is reduced in size. Primordial germ cells (PGC) are visible in the posterior extraembryonic region. Figure is modified from Gilland, E. H. and Burke, A. C. 2004. Pp. 205-217. In C. D. Stern (ed.). *Gastrulation. From Cells to Embryos,* Fig. 6.

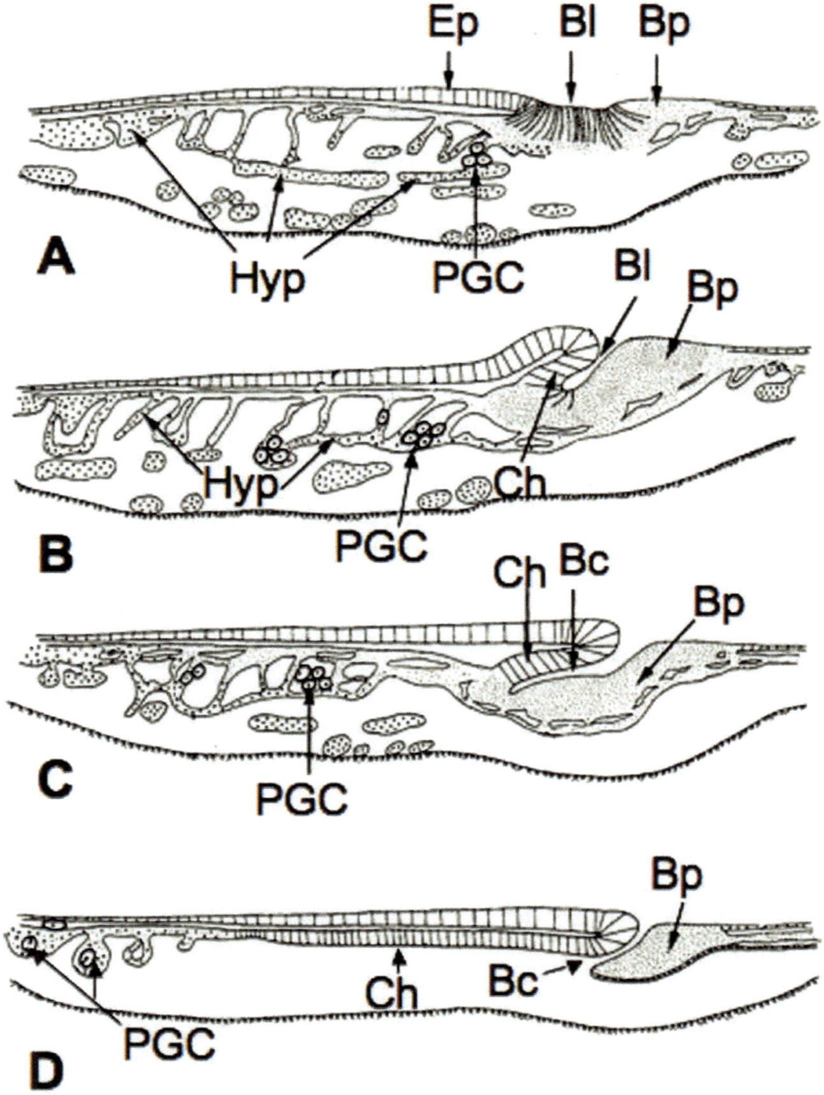

Fig 4.2 Schematic drawings of stages of gastrulation in the snake *Vipera aspis* shown in sagittal section. The anterior of the embryo is to the left. **A.** The blastoporal plate (Bp) forms in the posterior region of the epiblast (Ep). The hypoblast (Hyp) is thin and diffuse. Primordial germ cells (PGC) are visible just anterior to the developing blastopore (Bl). **B.** The blastopore extends to the anterior and the chordamesoderm (Ch) involutes over the dorsal lip of the blastopore. PGCs are moved to the anterior as gastrulation continues. **C.** The chordamesoderm continues to involute over the dorsal lip of the blastopore, pushing PGCs farther to the anterior. **D.** The hypoblast has been pushed to the extraembryonic region in the anterior, and the PGCs lie close to the hypoblast. The chordamesoderm (Ch) is fully extended under the ectoderm. The blastoporal canal is well-developed and the blastoporal plate is reduced in size. Figure is modified from Gilland, E. H. and Burke, A. C. 2004. Pp. 205-217. In C. D. Stern (ed.). *Gastrulation. From Cells to Embryos,* Fig 6.

midline region of the embryonic shield thickens into a small blastoporal plate. As gastrulation progresses, the size of this plate diminishes (Fig. 4.2).

4.4.2.1 Formation of the blastopore

In turtles, the first major cell movements of gastrulation are towards the posterior midline region where the blastopore will form. Cells move in essentially all directions converging on this blastoporal plate, including cells moving from the anterior midline, as well as cells moving in from the posterior and lateral regions (Mitsukuri 1893; Pasteels 1937, reviewed in Bellairs 1991). As the cells converge on the same small area, eventually they must either move inwards or bulge out. In normal embryos, the blastopore forms for inward gastrulation (Fig. 4.3). If inward movement is inhibited experimentally, the embryo will exogastrulate, with material bulging outwards instead.

During gastrulation, the blastopore forms initially as a transverse slit in the more anterior region of the blastoporal plate, but later it becomes crescent-shaped as the blastopore extends first laterally and then caudally and finally medially (Mitsukuri 1893, reviewed in Gilland and Burke 2004).

4.4.2.2 Mesoderm

In turtles, cells have been shown to involute over the blastopore and move forward in the embryo, between the epiblast and the hypoblast. The involuted cells around the midline eventually form mesoderm, with the medial region forming notochord and the paramedial regions forming the somites. The inward and anterior movement of the sheet of cells results in a change in the external appearance of the embryo; lateral thickenings form on either side of the midline as the presumptive mesoderm moves forward. Although the morphological movements have not been followed in snakes, the appearance of the blastopore and the tissue that moves inwards through the blastopore, as well as the external morphology suggest a similar process (Pasteels 1970; Hubert 1985b, reviewed in Gilland and Burke 2004).

4.4.2.3 Formation of the hypoblast and endoderm

The mechanism of forming the presumptive mesoderm during gastrulation is now generally agreed upon, but there has been much debate about formation of the hypoblast and the embryonic endoderm in reptiles. Disagreement may stem from true taxonomic differences, or it may stem from other factors such as narrow taxon sampling and the dearth of cell marking and fate mapping studies. Clearly the appearance of the embryo after a morphogenetic process has occurred does not always reveal the mechanism of that process.

In birds, the formation of the hypoblast involves two separate steps that produce what are often known as the primary and the secondary hypoblast. The primary hypoblast is the result of ingression from the epiblast and consists of islands of cells lying near the floor of the subgerminal cavity.

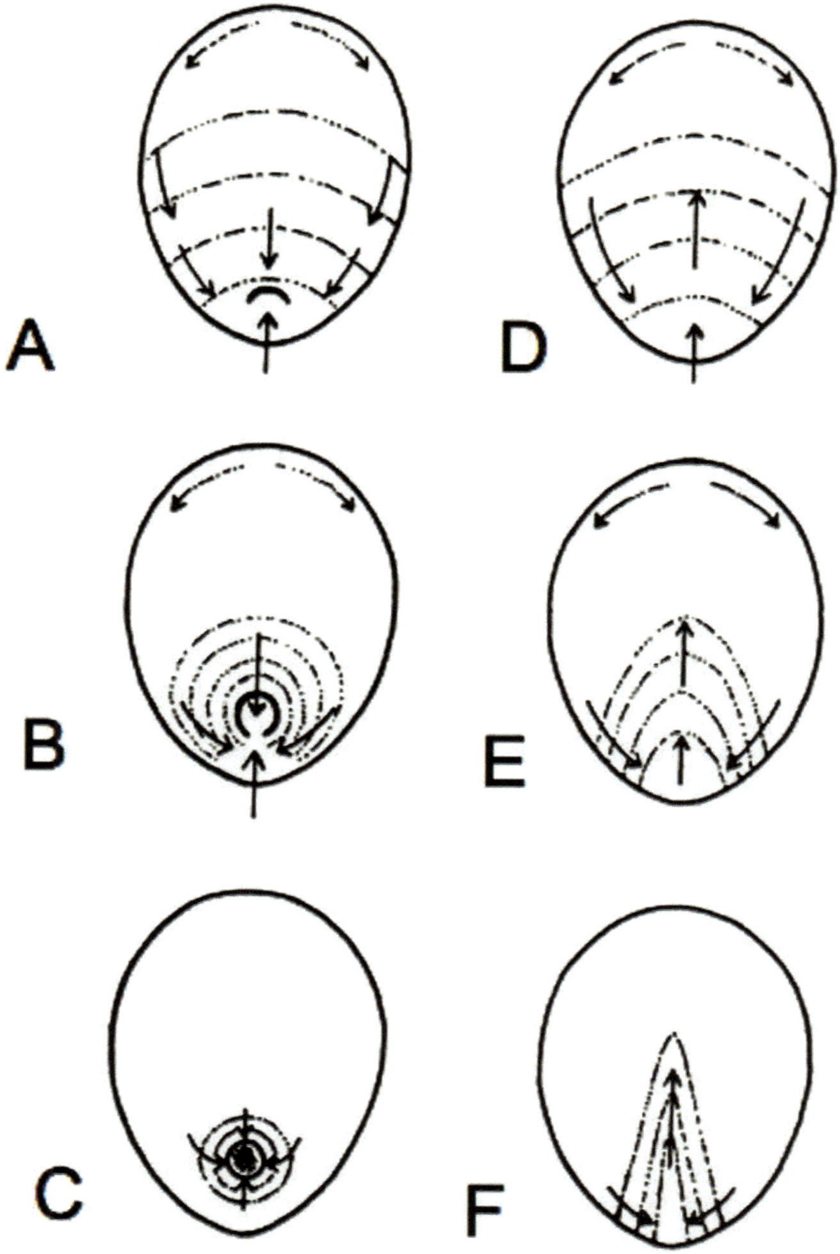

Fig. 4.3 Schematic drawings of gastrulation in the turtle, *Clemmys leprosa* (**A-C**) and the chicken (**D-F**). Arrows denote direction of cell movement. Notice the movement towards the posterior midline of the turtle that is lacking in the chick. Drawings and legend modified from Bellairs, R. 1991. Pp. 371-383. In D. C. Deeming (ed.), *Egg Incubation: Its Effects on Embryonic Development in Birds and Reptiles,* Fig 23.5.

The secondary hypoblast is a result of forward movement of cells from the posterior marginal belt region of the embryo, and these cells join with the primary hypoblast to make a more continuous layer (reviewed in Stern 1990). The hypoblast is later displaced when the definitive or gut endoderm enters from the primitive streak. Early works suggested that the endoderm was the result of delamination from the epiblast, but Rosenquist (1966) labeled and traced cell fates to show that the cells originally being studied were hypoblastic and were pushed aside by the definitive endoderm during gastrulation.

Similar issues have been raised regarding the formation of hypoblast and endoderm in turtles, lizards and snakes. Since cell movements have not been followed in squamates, conclusions are based on comparison to other embryos and on the appearance of different embryonic stages. Pasteels (1937) first suggested that the hypoblast in turtles formed in much the same way as the secondary hypoblast of birds—by forward movement of cells from the posterior marginal region of the embryonic shield. Conversely Peter (1938) argued for delamination of the hypoblast from the epiblast in turtles and squamates. Later, Pasteels (1957) concluded that in turtles both mechanisms occurred—delamination from the epiblast first, followed by forward movement of cells from the blastoporal plate (reviewed in Gilland and Burke 2004). Hubert (1985b) found that the thick lizard hypoblast was formed entirely by delamination of deep cells from the epiblast. Conversely, snake hypoblast, which in appearance is more similar to turtle hypoblast, also forms in a two-step process with delamination from the deep region of the epiblast followed by ingression from the blastoporal plate (Hubert 1970, reviewed in Hubert 1985b). As snake gastrulation proceeds, much of the diffuse hypoblast is pushed to the extraembryonic regions by the cells invaginating through the blastopore (Fig. 4.2D). In snakes, as in birds, some of these gastrular cells will later form the embryonic endoderm.

4.4.3 Blastopore vs. Primitive Streak

The appearance of eggs and early cleavage stages of birds and other reptiles is very similar, so much has been made of the quite different mechanisms of gastrulation, i.e., primitive streak versus blastopore. Bellairs (1991) suggests that the early gastrulation movements seen in turtles and surmised in squamates might give a clue as to the origin of the primitive streak in birds. Recall that in turtles, and presumably in snakes and lizards, cells move in all directions towards the blastoporal plate, including from the anterior midline region (Fig. 4.3A-C). In chick development, similar movements occur from all directions *except* from the anterior midline (Fig. 4.3D,E). In this case, rather than a force from all directions causing the cells to punch inwards and a blastopore to form, movement from the posterior to the anterior would dominate, pushing material along the surface in a trough: the primitive streak (Bellairs 1991). Interestingly, crocodilian embryos have both a blastopore and a primitive streak (reviewed in Ferguson 1985).

Bellairs (1991) suggests that the strength of the movements from anterior to posterior in crocodilians may be intermediate between that in turtles and squamates and that in birds. However, the gastrulation movements of crocodilians are currently unknown.

4.5 ESTABLISHMENT OF BODY AXES

One of the most important events of early development is the establishment of the main body axes, particularly the anterior/posterior and dorsal/ventral axes. As with many other aspects of development, much of the information on squamates has been gleaned from comparison with other vertebrates, particularly birds.

In almost all animals, the dorsal-ventral axis is associated in some way with placement of yolk. Because yolk is denser than cytoplasm, eggs with moderate to large amounts of yolk tend to orient naturally with the yolk side down. In telolecithal embryos, the dorsal side develops on the side farthest from the yolk, and the ventral side develops next to the yolk. Eyal-Giladi and coworkers have extensively studied the importance of gravity in normal development of chick embryos for both dorso-ventral and anterior-posterior axis formation (for review, see Eyal-Giladi 1997, 2001). Although most experiments have used birds and amphibians as model systems, Eyal-Giladi (1997, 2001) argues convincingly that the establishment of bilateral symmetry follows a common theme throughout vertebrate evolution. For anterior-posterior axis formation, maternal determinants located in the egg cytoplasm must move to the proper location to induce the A-P axis. In frogs, this occurs rapidly following fertilization and is influenced by the sperm entry point. A cortical shift driven by a microtubule array moves the determinants, which are located in the vegetal cortical cytoplasm, to the future dorsal, posterior region where they influence subsequent development. This so-called grey crescent region is sufficient to induce a secondary axis when transplanted into the opposite side of another embryo. Although experiments on the space shuttle have shown that normal frog axis formation can occur in a microgravity environment (Souza *et al.* 1995), on earth, gravity can be used to override the grey crescent's signal and to experimentally induce twinning (Gerhart *et al.* 1981).

In telolecithal eggs, the enormous mass of yolk probably precludes the role of microtubules radiating from a sperm entry point in determining the axis of development. Instead, gravity seems to be the chief factor in moving cortical determinants to their proper position in the future posterior region (reviewed in Eyal-Giladi 1997). In birds and other reptiles, the axis is set up *in utero* by the rotation of the egg on its long axis. Indeed as early as 1828 von Baer (as reviewed by Bellairs 1991) recognized that bird embryos generally develop with the A-P axis perpendicular to the long axis of the egg and the embryo's head facing forward when the blunt end of the shell is to the left. Clavert and Zahnd (1955) confirmed the general applicability of von Baer's rule in reptiles. As noted earlier, in most reptiles all of the early developmental stages occur in the oviduct. During the relatively long period of time in the uterus, as the egg rotates on its long axis, there is a continual shifting of the embryo due to the concentration of heavy yolk. This tilts the blastoderm such that one end remains at a higher position than the other. This higher side later forms the posterior end of the embryo. Eyal-Giladi and her colleagues showed that in birds, it is not the rotation per se, but the tilting which produced the A-P axis. When developing embryos were prematurely removed from the uterus and suspended in a saline bath, the region at the upper end of the blasoderm formed the posterior region, with the head at the lower end (Kochav and Eyal-Giladi 1971; Eyal-Giladi and Fabian 1980; reviewed in Eyal-Giladi 1997). If embryos were maintained horizontally so that there were no upper and lower ends, the A-P axis did not form (reviewed in Eyal-Giladi 2001). Gravity acting on the yolk determines the tilting which then directs the determinants to their proper location. Although such experiments have not been performed in non-avian reptiles, the rotation of the egg in the uterus and the direction of embryo formation suggest a similar mechanism (Clavert and Zahnd 1955).

4.6 PRIMORDIAL GERM CELLS

4.6.1 Overview

Primordial germ cells (PGCs) are the founder cells of the germ line. Although germ cells mature in the gonads, in most animals the PGCs are first formed elsewhere and then migrate to the genital ridge later in development. The location of PGCs, and the path and mechanism by which they migrate to the gonadal region can vary markedly among species (reviewed in Hubert 1985a).

PGCs have traditionally been identified by cytological means, including morphological characteristics such as shape and retention of yolk granules, carbohydrate and lipid staining patterns, and with specific carbohydrate antibodies. More recently, molecular markers for germ cell-specific genes, particularly *vasa* and *dazl*, have allowed identification of presumptive PGCs at earlier stages of development.

4.6.2 Location and Migration of PGCs

In describing the location and mechanism of migration of PGCs, Hubert (1985a) differentiates between what he refers to as the bird pattern and the mammalian pattern. In bird embryos just prior to gastrulation, PGCs are located in the central region of the area pellucida on the ventral side (Tsunekawa *et al.* 2000). By early somite stages, the PGCs are found mainly in a germinal crescent that is located anterior to the developing head. This is consistent with an association with the hypoblast as it is pushed forward during gastrulation (Tsunekawa *et al.* 2000). In mammals, the precursors to

PGCs are among the first cells to gastrulate through the primitive streak and can later be found in an extraembryonic crescent closely surrounding the blastoporal plate in the posterior region of the embryo. These alternate locations of PGCs have been associated with the mechanism of migration to the gonadal ridge. In birds the PGCs migrate from the anterior region through the circulatory system to arrive at the gonads. In mammals, on the other hand, PGCs move from the posterior regions interstitially through the gut and dorsal mesentery to the genital ridge.

As with other processes, reptiles show variability in location and migration of primordial germ cells (reviewed in Hubert 1985a). The development of PGCs in reptiles has been studied most extensively in turtles (for examples see Allen 1906; Jordan 1917; Fujimoto *et al.* 1979). Bachvarova *et al.* (2009) analyzed turtle embryos with the germ cell-specific markers *dazl* and *vasa* and found many similarities between turtle and mammalian PGCs in timing of appearance, location and mechanism of migration. They also reviewed the patterns of PGC location and migration in different groups of reptiles, and noted three different patterns, which they refer to as 1) the turtle pattern, 2) the snake pattern, and 3) the circumferential pattern. These patterns correspond to those noted by Hubert (1985a) and are shown in Figure 4.4.

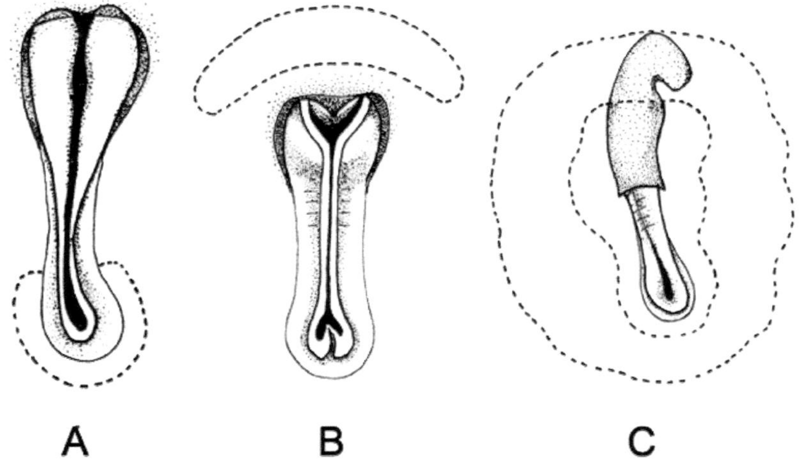

Fig. 4.4 Schematic drawings of the location of primordial germ cells before migration to the genital ridge. The region containing PGCs is shown by stippled lines. **A.** The turtle pattern. Drawing is based on the location in *Lacerta vivipara*. PGCs are in a small crescent in the posterior extraembryonic region of the embryo. **B.** The snake pattern. Drawing is based on location in *Vipera aspis*. PGCs are found in a large crescent in the anterior extraembryonic region. **C.** The circumferential pattern. Drawing is based on the location in the skink *Anguis fragilis*. PGCs are found all around the embryo, though mainly in the anterior. Figure is modified from Hubert, J. 1985a. Pp. 41-74. In C. Gans, F. Billett and P. F. A. Maderson (eds), *Biology of the Reptilia*, Vol. 15, Fig. 4.

4.6.2.1 The turtle pattern

The turtle pattern is similar to the mammalian pattern. Turtle PGCs can first be identified during early gastrulation and are associated with gastrulating mesoderm. Following gastrulation they can be found in a crescent of hypoblast at the posterior region of the embryo (Fig. 4.4A). From this location they migrate to the forming gonads by interstitial movements (Bachvarova *et al.* 2009).

4.6.2.2 The snake pattern

PGCs have been studied in two species of snakes: *Vipera aspis* (Hubert 1976) and *Thamnophis sirtalis* (Risley and Barnett 1941). Both follow the same route, which is referred to as the snake pattern, and is reminiscent of the avian pattern. At the beginning of gastrulation the PGCs are located just ahead of the blastopore, and by early somite stages they form a crescent in the anterior extraembryonic region (Fig. 4.4B). Snake PGCs migrate to the gonads via the vascular system, again in common with birds and contrary to the mechanism seen in turtles.

4.6.2.3 The circumferential pattern

Lizards show at least two different patterns. Several species of lacertid lizards have been shown to follow the turtle pattern (Hubert 1985a). This pattern has also been suggested for some other lizards, but the data are less solid. The circumferential pattern is seen in several species of lizards, including the Grass Top Skink (*Mabuya megalura*), the Slow Worm (*Anguis fragilis*) and the Side Striped Chameleon (*Chamaeleo bitaeniatus*; Pasteels 1953; Hubert 1969, 1971; reviewed in Hubert 1985a). In this pattern, the germ cells are found all around the extraembryonic region surrounding the embryo, though mainly in the anterior (Fig. 4.4C). Migration via vascular channels occurs from both the anterior and the posterior in these lizards. Hubert noted that in snakes and birds, a few germ cells may be found in the posterior region at early somite stages, so there may not be a sharp demarcation between the snake pattern and the circumferential pattern.

4.6.3 Germ Cell Determination Mechanisms

The location of primordial germ cells during development has been used to infer the mechanism of determination of the germ cells (Johnson *et al.* 2003; Bachvarova *et al.* 2009). Among vertebrates, there exist two very different means for determining which cells will become PGCs (reviewed in Extavour and Akam 2003; Johnson *et al.* 2003; Crother *et al.* 2007). In frogs, chicks and some fish, PCGs are determined by maternal molecules that are localized to a region of the oocytes and/or early embryos known as the germ plasm (Komiya *et al.* 1994; Houston and King 2000; Knaut *et al.* 2000, 2002; Tsunekawa *et al.* 2000; Hashimoto *et al.* 2004). Germ plasm is typically localized to a discreet region of the egg cytoplasm and is then differentially distributed to presumptive germ cells during embryogenesis. Evidence from diverse species shows that germ plasm contains germ cell-specific RNA binding proteins and/or the mRNAs that encode them (Houston and King, 2000; Tsunekawa *et al.* 2000; Hashimoto *et al.* 2004). Only cells that inherit this germ plasm later in development will form germ cells. Organisms using this cell-autonomous mechanism are often referred to as predetermined.

In some other vertebrates, including salamanders (Sutasurya and Nieuwkoop, 1974; Maufroid and Capuron 1985; Johnson *et al.* 2001) and mammals (Lawson and Hage 1994; Tam and Zhou 1996), PGCs are not predetermined, but instead must be induced later in development. In these organisms, zygotic signals specify PGCs, unlike the maternal molecules in the predetermined mode. The PGCs are derived from pleuripotent precursor cells that also contribute to somatic mesodermal lineages. This is sometimes referred to as the epigenetic or regulative mode.

Little is known about the mechanisms of determination in non-avian reptiles. Bachvarova *et al.* (2009) give extensive evidence that turtles use the induced mode, much like mammals and salamanders. The first appearance of PGCs at the onset of gastrulation in association with the mesoderm supports induction, as does the apparent absence of germ plasm as detected by *dazl* and *vasa* at earlier stages. Detailed developmental studies such as those done by Bachvarova and colleagues (2009) have not been published for other reptiles. Maurizii and coworkers (2009) looked at the distribution of vasa protein and mRNA during oogenesis in the Italian Wall Lizard (*Podarcis sicula*) but they did not follow through developmental stages. Their results show vasa protein in high concentrations throughout small oogonia. Staining is more diffuse in growing oocytes. As noted earlier, the somatic pyriform cells attached by cytoplasmic bridges provide vasa to growing oocytes, and the location is initially cortical. This cortical location might be consistent with germ plasm, but is also expected of molecules that enter via endocytosis. By later stages of follicle development, the vasa protein appears to be in not only cortical regions, but subcortical areas as well. Since vasa distribution through cleavage has not been studied, it is difficult to make conclusions about whether this represents germ plasm and the predetermined mode. Indeed, the location and migration of PGCs in lacertid lizards suggests they follow the turtle pattern and use the inductive mode of germ cell determination.

In snakes, and in those lizards that show the circumferential pattern, PGCs are found mainly in the anterior extraembryonic regions following gastrulation. Johnson and colleagues (2003) have argued that the environment of signaling molecules in anterior regions would be inconsistent with the induction of germ cells. In mammals and salamanders, PGC induction requires the signaling molecule BMP-4. Anterior regions contain antagonists to BMP-4, suggesting that PGCs located in the anterior could probably not be induced and would therefore be more likely to use the predetermined mode. However, evidence for germ plasm in these species would be necessary to confirm the predetermined mode in snakes and some lizards.

4.7 CONCLUSIONS

It should be clear from this chapter that our knowledge of mechanisms of oogenesis and early development in snakes and other reptiles is incomplete at best. A great deal is known about some topics, such as vitellogenic cycles of snakes, but much less is known about cleavage stages, gastrulation, axis formation and primordial germ cell determination. This chapter has been necessarily comparative because so much of what we "know" is based on comparing the morphological appearance of snake developmental stages with those of related organisms that have been better studied. The development of birds and other reptiles is thought to be conserved in some important aspects, but there are notable differences as well; thus to really understand snake development, many more studies are necessary.

In some cases this chapter has differentiated between mechanisms in snakes versus lizards. For example, the hypoblast is thought to form by delamination alone in lizards and by a two-step process in snakes. When looking at location of primordial germ cells, snakes are considered to show the bird pattern, while lizards have either the mammalian or the circumferential pattern. Comparisons such as these are based on a very few (or even only one) species of each group. Since snakes arguably are one branch of the lizard tree, it seems likely that a much broader comparative approach will shed more light on the diversity of mechanisms found throughout squamates.

4.8 ACKNOWLEDGMENTS

I am particularly grateful to Brian Crother for helpful discussions, support and suggestions. I appreciate Andrew Johnson's assistance with current information on PGCs. I would also like to thank the editors, David Sever and Robert Aldridge, for their patience as I inched my way through this chapter and for all of their work in putting the symposium and volume together.

4.9 LITERATURE CITED

Aldridge, R. D. 1979. Female reproductive cycle of the snakes *Arizona elegans* and *Crotalus viridis*. Herpetologica 35: 256-261.

Allen, B. M. 1906. The origin of the sex cells in *Chrysemys*. Anatomischer Anzeiger 2: 217-236.

Andreuccetti, P., Taddei, C. and Filosa, S. 1978. Intercellular bridges between follicle cells and oocyte during the differentiation of follicular epithelium in *Lacerta sicula* Raf. Journal of Cell Science 33: 341-350.

Bachvarova, R. F., Crother, B. I., Manova, K., Chatfield J., Shoemaker, C. M., Crews, D. P. and Johnson, A. D. 2009. Expression of *dazl* and *vasa* in turtle embryos and ovaries: evidence for inductive specification of germ cells. Evolution and Development 11: 524-533.

Bauman, M. A. and Metter, E. 1977. Reproductive cycle of the northern watersnake *Natrix s. sipedon* (Reptilia, Serpentes, Colubridae). Journal of Herpetology 11: 51-59.

Bellairs, R. 1959. The yolk of the adder (*Vipera berus*). British Journal of Herpetology 2: 155-158.

Bellairs, R. 1991. Overview of early stages of avian and reptilian development. Pp. 371-383. In D. C. Deeming (ed.). *Egg Incubation: Its Effects on Embryonic Development in Birds and Reptiles*. Cambridge University Press, Cambridge, U.K.

Betz, T. W. 1963. The ovarian histology of the diamond-backed watersnake, *Natrix rhombifera*, during the reproductive cycle. Journal of Morphology 113: 245-260.

Billett, F., Gans, C. and Maderson, P. F. A. 1985. Why study reptilian development? Pp. 1-39. In C. Gans, F. Billett and P. F. A. Maderson (eds). *Biology of the Reptilia*, Vol 14. Wiley, New York.

Blackburn, D. G. and Flemming, A. F. 2008. Morphology, development and evolution of fetal membranes and placentation in squamate reptiles. Journal of Experimental Zoology Part B: Molecular and Developmental Evolution 312B: 579-589.

Blanc, C. P. 1971. Études sur les Iguanidae de Madagascar. VI. Observations sur la structure et la signification des cellules folliculeuses ovocytaires chez *Chalarodon madagascariensis* Peters 1854. Annales de l'Université de Madagascar 8: 205-233.

Bonnet, X., Lourdais, O., Shine, R. and Naulleau, G. 2002. Reproduction in snakes (*Vipera aspis*): costs, currencies and complications. Ecology 83: 2124-2135.

Brachet, A. 1914. Recherches sur l'embryologie des reptiles. Acrogenese, Cephalogenese et Cormogenese chez *Chrysemys marginata*. Archives of Biology 29: 501-577.

Censky, E. J. and McCoy, C. J. 1988. Female reproductive ecology of five species of snakes (Reptilia: Colubridae) from the Yucatan Peninsula, Mexico. Biotropica 20: 326-333.

Chandrasekharan Nayar, M. 1966. *In vitro* vital staining of chelonian blastoderms. Indian Journal of Experimental Biology 4: 131-134.

Clavert, J. and Zahnd, J. P. 1955. Sur la déterminisme de la symetrie bilateral et la regle de von Baer chez les Reptiles. Comptes Rendus des Séances et Mémoires de la Société de Biologie Paris 149: 1650-1651.

Crother, B. I., White, M. E. and Johnson, A. D. 2007. Developmental constraint and constraint release: Primordial germ cell determination mechanisms as examples. Journal of Theoretical Biology 248: 322-330.

Dessauer, H. C. and Fox, W. 1959. Changes in ovarian composition with plasma levels of snakes during estrus. American Journal of Physiology 197: 360-366.

Drent, R. H. and Daan, S. 1980. The prudent parent: energetic adjustments in avian breeding. Ardea 68: 225-252.

Extavour, C. G. and Akam, M. 2003. Mechanisms of germ cell specification across the metazoans: epigenesis and preformation. Development 130: 5869-5884.

Eyal-Giladi, H. 1997. Establishment of the axis in chordates: facts and speculations. Development 124: 2285-2296.

Eyal-Giladi, H. 2001. The initial steps leading to axis determination. Pp. 29-42. In S.-L. Ang, R. R. Behringer, H. Sasake, J. S. Altman and C. Coath (eds), *Axis Formation in Vertebrate Embryos. A Comparative Approach.* Human Frontier Science Program, Strasbourg, Germany.

Eyal-Giladi, H. and Fabian, B. C. 1980. Axis determination in uterine chick blastodiscs under changing spatial positions during the sensitive period for polarity. Developmental Biology 77: 228-232.

Ferguson, M. W. J. 1985. Reproductive biology and embryology of the crocodilians. Pp. 329-491. In C. Gans, F. Billett and P. F. A. Maderson (eds). *Biology of the Reptilia,* Vol 14. Wiley, New York.

Fujimoto, T., Ukeshima, A., Miyayama, Y., Horio F. and Ninomiya, E. 1979. Observations of primordial germ cells in the turtle embryo (*Caretta caretta*): light and electron microscopic studies. Development Growth and Differentiation 21: 3-10.

Garstka, W. R., Tokarz, R. R., Diamond, M., Halpert, A. and Crews, D. 1985. Behavioral and physiological control of yolk synthesis and deposition in the female red-sided garter snake (*Thamnophis sirtalis parietalis*). Hormones and Behavior 19: 137-153.

Gerhart, J., Ubbels, G., Black, S., Hara, K. and Kirschner, M. 1981. A reinvestigation of the role of the grey crescent in axis formation in *Xenopus laevis*. Nature 292: 511-516.

Gilland, E. H. and Burke, A. C. 2004. Gastrulation in Reptiles. Pp. 205-217. In: C. D. Stern (ed.), *Gastrulation: From Cells to Embryos.* Cold Spring Harbor Laboratory Press, New York.

Gräper, L. 1929. Die Primitiventwicklung des Hünchens nach stereokinematographischen Untersuchungen, kontrolliert durch vitalen Farbmarkierung und verglichen mit der Entwicklung anderer Wirbeltiere. Wilhelm Roux Archiv für Entwicklungmechanik der Organismen 116: 382-429.

Guraya, S. S. 1969. Follicle cell nuclei as the possible source for the formation of ooplasmic reserve DNA in the garden lizard: a morphological and histochemical study. Acta Embryologiae Experimentalis Italia 1: 91-96.

Hashimoto, Y,, Maegawa, S., Nagai, T., Yamaha, E., Suzuki, H., Yasuda, K. and Inoue, K. 2004. Localized maternal factors are required for germ cell formation. Developmental Biology 268: 152-161.

Houston, D. W. and King, M. L. 2000. Germ plasm and molecular determinants of germ cell fate. Current Topics in Developmental Biology 50: 155-181.

Hubert, J. 1969. Localisation précoce et mode de migration des gonocytes primordiaux chez quelques reptiles. Annales d'Embryology et de Morphogenese 2: 479-494.

Hubert, J. 1970. Développement précoce de l'embryon et localisation extra-embryonaire des gonocytes chez les reptiles. Archives D'Amatomie Microscopique et de Morphologie Experimentale 59: 253-270.

Hubert, J. 1971. Localisation extra-embryonnaire des gonocytes chez l'embryon d'orvet *Anguis fragilis* L. Archives D'Amatomie Microscopique et de Morphologie Experimentale 60: 261-268.

Hubert, J. 1976. Développement à 27ºC d'embryons de *Lacerta vivipara* Jacquin dans des oeufs places in vitro et ayant séjournes in utero ou in vitro à basse temperature. Bulletin de la Societe de Zoologie de France 101: 315-324.

Hubert, J 1985a. The Origin and Development of Oocytes. Pp. 41-74. In C. Gans, F. Billett and P. F. A. Maderson (eds), *Biology of the Reptilia,* Vol 14. Wiley, New York.

Hubert, J. 1985b. Embryology of the Squamata. Pp. 1-34. In C. Gans, F. Billett and P. F. A. Maderson (eds), *Biology of the Reptilia,* Vol. 15. Wiley, New York.

Hubert, J. and Dufaure, J. P. 1968. Table de dévelopment de la vipère aspic: *Vipera aspis*. Bulletin de la Societe de Zoologie de France 93: 135-148.

Jackson, K. 2002. Post-ovipositional development of the monocled cobra, *Naja kaouthia* (Serpentes: Elapidae). Zoology 105:203-214.

Johnson, A. D., Bachvarova, R. F., Drum, M. and Masi, T. 2001. Expression of axolotl *DAZL* RNA, a marker of germ plasm: widespread maternal RNA and onset of expression in germ cells approaching the gonad. Developmental Biology 234: 402-415.

Johnson, A. D., Drum, M., Bachvarova, R. F., Masi, T., White, M. E. and Crother. B. I. 2003. Evolution of predetermined germ cells in vertebrate embryos: Implications for macroevolution. Evolution and Development 5: 414-431.

Jordan, H. E. 1917. The history of the primordial germ cells in the loggerhead turtle embryo. Proceedings of the National Academy of Sciences USA 3: 271-275.

Knaut, H., Pelegri, F., Bohmann, K., Schwarz, H. and Nusslein-Volhard. C. 2000. Zebrafish *vasa* RNA but not its protein is a component of the germ plasm and segregates asymmetrically before germline specification. Journal of Cell Biology 149: 875-888.

Knaut, H., Steinbeisser, H., Schwarz, H. and Nusslein-Volhard, C. 2002. An evolutionary conserved region in the vasa 3'UTR targets RNA translation to the germ cells in the zebrafish. Current Biology 12: 454-466.

Kochav, S. and Eyal-Giladi, H. 1971. Bilateral symmetry in chick embryo determination by gravity. Science 171: 1027-1029.

Komiya, T., Itoh, K., Ikenishi, K. and Furusawa, M. 1994. Isolation and characterization of a novel DEAD box protein family which is specifically expressed in germ cells of *Xenopus laevis*. Developmental Biology 162: 354-363.

Krohmer, R. W. and Aldridge, R. D. 1985. Female reproductive cycle of the lined snake (*Tropidoclonion lineatum*). Herpetologica 41: 39-44.

Krull, J. 1906. Die Entwicklung der Ringelnatter (*Tropidonotus natrix* Boie) vom ersten Aufstreten des Proamnios bis zum Schlusse des Amnios. Zeitschrift für Wissenschaftliche Zoologie 85: 107-155.

Lawson, K. A. and Hage, W. J. 1994. Clonal analysis of the origin of primordial germ cells in the mouse. Ciba Foundation Symposium 182: 68-91.

Madsen, T. and Shine, R. 1999. The adjustment of reproductive threshold to prey abundance in a capital breeder. Journal of Animal Ecology 68: 571-580.

Maufroid, J. P. and Capuron, A. P. 1972. Migration des cellules germinales primodriales chez l'amphibien urodele *Pleurodeles waltii* Michah. Wilhelm Roux's Archives 170: 234-243.

Maurizii, M. G., Cavaliere, V., Gamberi, C., Lasko, P., Garglulo, G. and Taddei, C. 2009. Vasa protein is localized in the germ cells and in the oocyte-associated pyriform cells during early oogenesis in the lizard *Podarcis sicula*. Development, Genes and Evolution 219: 361-367.

Mitsukuri, K. 1893. On the process of gastrulation in Chelonia. Journal of Cell Science of Imperial University Tokyo 6: 227-277.

Naulleau, G. and Bonnet, X. 1995. Reproductive ecology, fat reserves and foraging *Elaphe longissima* mode of two contrasted snake species: *Vipera aspis* (terrestrial, viviparous) and (semi-arboreal, oviparous). Amphibia-Reptilia 16: 37-46.

Noble, R. 1991. Comparative composition and utilisation of yolk lipid by embryonic birds and reptiles. Pp. 17-28. In D. C. Deeming and M. W. J. Ferguson (eds). *Egg Incubation: Its Effects on Embryonic Development in Birds and Reptiles*. Cambridge University Press, Cambridge, U.K.

Pasteels, J. 1937. Étude sur la gastrulation des vertébrés méroblastiques. II. Reptiles. Archives of Biology 48: 105-184.

Pasteels, J. 1953. Contribution à l'étude du développement des Reptiles. I. Origine et migration des gonocytes chez deux Lacertiliens (*Mabuia megalura* et *Chamaeleo bitaeniatus*). Archives of Biology 64: 227-245.

Pasteels, J. 1970. Développement embryonnaire. Pp. 893-971. In P. P. Grasse (ed.), *Traité de Zoologie*. Vol 14. Masson, Paris, France.

Peter, K. 1934. Die Erste Entwicklung des Chamäleon (*Chamaeleo vulgaris*) verglischen mit der Eideschse (Ei, Keimbildung, Furchung, Entodermbildung). Zeitschrift für Anatomie und Entwicklungsgeschichte 103: 147-188.

Peter, K. 1938. Untersuchungen über die Entwicklung des Dotterentoderms. 3. Die Entwicklung des Entoderms bei Reptilien. Zeitschrift für Mikroskopisch-Anatomische Forschung 44: 498-531.

Pizzatto, L. and Marques, O. A. V. 2007. Reproductive ecology of boine snakes with emphasis on Brazilian species and a comparison to pythons. South American Journal of Herpetology 2: 107-122.

Plummer, M. 1983. Annual variation in stored lipids and reproduction in green snakes (*Opheodrys aestivus*). Copeia 1983: 741-745.

Risley, P. L. and Barnett. M. 1941. Origin and migration of primordial germ cells in snake embryo. Anatomical Record 81 (suppl.): 82.

Rosenquist, G. C. 1966. A radioautographic study of labeled grafts in the chick blastoderm: development from primitive streak stages to stage 12. Contributions to Embryology 38: 111-121.

Saint Girons, H. 1957. Le cycle sexuel chez *Vipera aspis* (L.) dans l'ouest de la France. Bulletin Biologique de la France et de la Belgique 91: 284-350.

Sever, D. M. and Hamlett, W. C. 2002. Female sperm storage in reptiles. Journal of Experimental Zoology 292: 187-199.

Shine, R. 1983. Reptilian reproductive modes: The oviparity-viviparity continuum. Herpetologica 39: 1-8

Shine, R. 2003. Reproductive strategies in snakes. Proceedings of the Royal Society of London B270: 995-1004.

Souza, K. A., Black, S. D. and Wassersug, R. J. 1995. Amphibian development in the virtual absence of gravity. Proceedings of the National Academy of Sciences USA 92: 1975-1978.

Srivastava, A. S. 1948. Cytological observations on the oogenesis of certain Indian lizards. I. Infiltrations of cytoplasmic inclusions from the follicle cells into the oocyte. Transactions of the American Microscopy Society 66: 318-327.

Stern, C. D. 1990. The marginal zone and its contribution to the hypoblast and primitive streak of the chick embryo. Development 109: 667-682.

Stewart, J. R. and Castillo, R. E. 1984. Nutritional provision of the yolk of two species of viviparous reptiles. Physiological Zoology 57: 377-383.

Sutasurya, L. A. and Nieuwkoop, P. D. 1974. The induction of the primordial germ cells in the urodeles. Wilhelm Roux's Archives 175: 199-220.

Taddei, C. 1972. The significance of pyriform cells in the ovarian follicle of *Lacerta sicula*. Experimental Cell Research 72: 562-566.

Tam, P. P. and Zhou, S. X. 1996. The allocation of epiblast cells to ectodermal and germ-line lineages is influenced by the position of the cells in the gastrulating mouse embryo. Developmental Biology 178: 124-132.

Treadwell, R. W. 1962. Time and Sequence of Appearance of Certain Gross Structures in *Pituophis melanoleucus sayi* embryos. Herpetologica 18: 120-124.

Tsunekawa, N., Naito, M., Sakai, Y., Nishida, T. and Noce, T. 2000. Isolation of chicken vasa homolog gene and tracing the origin of primordial germ cells. Development 127: 2741-2750.

Viefhaus, T. 1907. Die Entwicklung der Ringelnatter (*Tropidonotus natrix* Boie) nach Ausbildung der Falterform bis zur Erhebung des Proamnios (mit Taf. IV-VI u. 8 im Text). Zeitschrift für Wissenschaftliche Zoologie 86: 55-99.

Vogt, W. 1929. Gestaltungsanalyse am Amphibien keim mit örtlicher Vitalfarung. II. Gastrulation und Mesoderm bildung bei Urodelen und Anuren. Archiv für Entwicklungsmechanik 120: 385-707.

Wetzel, R. 1929. Untersuchungen am Hünchen. Die Entwicklung des Keims während der ersten beiden Bruttage. Archiv für Entwicklungsmechanik 119: 188-320.

Whittier, J. M. and Crews, D. 1990. Body mass and reproduction in female red-sided garter snakes (*Thamnophis sirtalis parietalis*). Herpetologica 46: 219-226.

Winne, C. T ., Willson, J. D. and Gibbons, J. W. 2006. Income breeding allows an aquatic snake *Seminatrix pygaea* to reproduce normally following prolonged drought-induced aestivation. Journal of Animal Ecology 75: 1352-1360.

Zehr, D. R. 1962. Stages in the normal development of the common garter snake, *Thamnophis sirtalis*. Copeia 1962: 322-329.

CHAPTER 5

Viviparity and Placentation in Snakes

Daniel G. Blackburn[1] *and James R. Stewart*[2]

5.1 INTRODUCTION

Viviparity in snakes has been a source of fascination since ancient times and is a particular focus of biological interest today. Questions about how viviparity is accomplished and how and why it has evolved among snakes are significant, complex, and challenging. Such questions must therefore be addressed through wide-ranging empirical studies and theoretical analyses, while being driven by an evolutionary perspective. Accordingly, research on reproduction in live-bearing snakes draws on methods of anatomy, physiology, biochemistry, molecular biology, ecology, systematics, behavior, and phylogenetic analysis (for overview see Blackburn 2000; Thompson and Blackburn 2006; Thompson *et al.* 2010). Such research offers a powerful illustration of the value—and the necessity—of integrative approaches that transcend traditional disciplinary methodologies and perspectives.

From the standpoint of a female snake, viviparity can have the benefits of protecting eggs from predators and the external environment and of modulating conditions of embryo development. However, live-bearing reproduction also entails significant costs and physiological challenges. One major functional difficulty is that developing embryos must be sustained until birth in a reproductive tract that evolved to house eggs for relatively short time periods. Lacking access to the external environment, the developing embryo requires a way to exchange respiratory gases and obtain water; also lacking an eggshell, it needs a source of calcium. Furthermore, the embryo may require (or at least benefit from) nutrients other than those provided in the ovulated yolk. For viviparous squamates,

[1]Department of Biology and Electron Microscopy Facility, Trinity College, Hartford, CT 06106 USA
[2]Department of Biological Sciences, East Tennessee State University, Johnson City, TN 37614 USA

the evolutionary answer to these problems was the development of organs known as placentae.

Snake placentae, like those of viviparous lizards, are formed out of fetal membranes and the maternal oviduct. Such placentae have evolved concomitantly with viviparity in numerous squamate lineages (Blackburn 1992, 2006). Within a given species, different types of placentae have different structural and functional attributes. Furthermore, placentae show striking diversity between squamate species that reflects specializations for different functions (Blackburn 1993; Stewart 1993; Stewart and Thompson 2000; Thompson *et al.* 2004; Thompson and Speake 2006).

We have two major goals in this chapter: (1) to summarize functional and evolutionary aspects of snake viviparity; and (2) to explore the structure, function, and evolution of the placental organs of viviparous snakes. Much of what is known about squamate viviparity comes from studies on lizards. In past reviews, snakes have sometimes been treated as afterthoughts (if mentioned at all) in evaluations of the literature. Nevertheless, viviparous snakes have received considerable empirical study, yielding many data that are available for synthesis. As for snake placentae, most of our knowledge is relatively new and comes from work done on thamnophine snakes in our respective laboratories. The phylogenetic focus of our placental research has the advantage of allowing us to reconstruct a pattern of reproductive evolution in detail, an approach that has proven valuable in research on lizards (Stewart and Thompson 2003, 2009a, b); however, it also precludes the sort of understanding that a broad sampling of diversity would afford. Nevertheless, what we have learned about snake placentation is enlightening and compatible with the growing body of information on lizards. It also provides a basis for comparison to other snake clades, as well as to other lineages of viviparous vertebrates.

Our hope is that this chapter will facilitate an understanding and appreciation of snake viviparity and placentation and spark new research that is broad in methodological and phylogenetic scope. All viviparous squamates share a common oviparous ancestry and biological substrate. However, information derived from studies of lizards should not be assumed to apply directly to snakes; after all, the suborder Sauria is both paraphyletic and diverse. Therefore, ophidian viviparity and placentation deserve consideration in their own right. The diverse clades of viviparous snakes offer untapped potential for an understanding of snake reproduction as well as the viviparous pattern itself.

5.2 VIVIPARITY

Among vertebrates, viviparity is the reproductive pattern in which females retain fertilized eggs in their reproductive tracts and give birth to their young. In viviparous squamate reptiles, the eggs develop to term in the oviduct, resulting in birth of offspring that are (aside from lack of sexual maturity) miniature versions of the adult. In contrast, oviparous squamates lay eggs approximately 2-5 weeks after fertilization (Saint Girons 1964; Tinkle 1967; Clark 1970a; Greene 1997) with embryos that are about 30% of the way through development (Shine 1983a; Blackburn 1995).

5.2.1 Nature and Distribution of Viviparity in Snakes

Viviparity is widespread among snakes; it occurs in 14 nominal families (Table 5.1) and nearly 20% of the species (Blackburn 1985). Nine snake families contain both viviparous and oviparous species (Table 5.1). Furthermore, oviparous and viviparous species are found in several snake genera, including *Amphiesma* (Natricidae), *Aparallactus* (Lamprophiidae), *Coronella* (Colubridae), *Helicops* (Dipsadidae), *Psammophylax* (Lamprophiidae), *Pseudechis* (Elapidae), *Sinonatrix* (Natricidae), *Trimeresurus* (Viperidae), *Typhlops* (Typhlopidae), and *Vipera* (Viperidae) (Blackburn 1985, 1999; Shine 1985).

Table 5.1 Ophidian families that include viviparous species. Taxonomy draws on Vidal *et al.* (2009) and Zaher *et al.* (2009). Common names and distributions refer to viviparous species. Information is summarized in Fitch (1970), Tinkle and Gibbons (1977), Seigel and Fitch (1984), Blackburn (1985), and Shine (1985).

Family	*Common names*	*Distribution*
Aniliidae	pipe snakes	South America
Uropeltidae	shield tailed snakes	South Asia
Acrochordidae	wart snakes, file snakes	Asia, Australia
Typhlopidae[1]	blind snakes	South America
Boidae[1]	boas	Americas, S. Pacific
Tropidophiidae	dwarf boas	tropical Americas
Colubridae[1, 2]	colubrids	cosmopolitan
Dipsadidae[1, 2]	thirst snakes, xenodontines	tropical Americas
Xenopeltidae[1, 2]	keelbacks	South America
Natricidae[1, 2]	garter snakes, water snakes	N. America, East Asia
Homalopsidae[2, 3]	water snakes	India to Australia
Elapidae[1, 4]	seasnakes, swamp snakes, black snakes, spitting cobra	Africa, Australia
Lamprophiidae[1]	centipede eaters, grass snakes	Africa
Viperidae[1]	pit vipers, rattlesnakes	cosmopolitan

[1]also includes oviparous species
[2]traditionally classified in "Colubridae" *sensu lato*
[3]contains some of the snakes traditionally included in the family Elapidae
[4]includes hydrophiine seasnakes, traditionally classified as a separate family

Viviparous snake species are distributed worldwide (Table 5.1) and occupy a variety of environments (Neill 1964; Fitch 1970; Tinkle and Gibbons 1977). For example, viviparous snakes can be aquatic (as in acrochordids, homalopsines, and hydrophiine seasnakes), semi-aquatic (as in thamnophine water snakes and keelbacks of the genus *Helicops*), arboreal (various boines, colubrids, and viperids), fossorial (uropeltids, aniliids, and typhlopids), semi-fossorial (species in the thamnophine genera *Storeria* and *Virginia*), and terrestrial (various colubrids, elapids, and viperids). Viviparous species are disproportionately represented in the squamate fauna of higher latitudes and altitudes (e.g. Tinkle and Gibbons 1977; Shine and Berry 1978; Gregory 2009). Nevertheless, most viviparous species occupy other environments, a fact reflected in the broad ecological and geographical distribution of viviparous snakes.

Gestation lengths vary moderately among viviparous snakes. Durations of 2 to 3 months are common in snakes (e.g., in thamnophines and crotalids) (Fitch 1970; Tinkle and Gibbons 1977) while some of the longest reported gestation lengths (6-8 months) occur among elapid seasnakes (Bergman 1943; Hin *et al.* 1991). This range is equivalent to the total developmental period found in eggs of most oviparous snakes (Greene 1997) and viviparous lizards (see Tinkle and Gibbons 1977). Literature estimates of gestation lengths tend to be approximate, usually being based on surveys of museum specimens captured at different times of year. If synchronous breeding is incorrectly assumed, this approach can lead to discrepant estimates. Figures based on observations of captive animals can also be problematic. Because female snakes can store sperm (Hoffman and Wimsatt 1972; Gist and Jones 1987; Sever and Hamlett 2002; Siegel and Sever 2009), calculations based on the time between mating and parturition may lead to overestimates. A further complication is that gestation length is temperature dependent and therefore reflects maternal behavior and habitat, as well as laboratory temperatures in captive animals. In any case, the span of 2-8 month gestation lengths (with most species lying towards the lower end of the range) demarcates the period during which snake embryos must be carried by the female and sustained in her reproductive tract.

Litter size varies considerably between species of viviparous snakes, and can vary intraspecifically with maternal age, body size, and nutritional state, offspring size, and geographical region (Fitch 1970; Seigel and Fitch 1984, 1985; Seigel *et al.* 1986; Siegel and Ford 1987). In broad taxonomic comparisons, litter size does not correlate with reproductive mode; in fact, some of the smallest and largest litter sizes reported in snakes are for viviparous species. At one extreme of the continuum of viviparous forms lies Shaw's Sea Snake (*Lapemis curtus*), with reported median litter sizes of 2 (Fitch 1970) and 3 embryos (Hin *et al.* 1991). Small litter sizes also occur in such viviparous species as the Western Diamond-Backed Rattlesnake (*Crotalus atrox;* with a reported average of 4.5 offspring and a range of 2-7; Taylor and DeNardo 2005), the Rough Earthsnake (*Virginia striatula;* with a mean of 4.9 offspring and a range of 3-8; Clark 1964), the Southern Copperhead (*Agkistrodon contortrix;* with 5-6 offspring; Fitch 1970), and the File Snake, (*Acrochordus granulatus;* with a mean litter size of about 5; Voris and Glodek 1980; Wangkulangkul *et al.* 2005). At the other extreme lies the viviparous Diamond-backed Watersnake (*Nerodia rhombifer*); in data compiled from 22 broods, litter size averaged 47 (Fitch 1970). Record litter sizes of 70 to more than 100 offspring have been reported among viviparous thamnophines, hydrophiines, and vipers. These include the Fer-de-Lance

(*Bothrops atrox*) with up to 86 embryos (Hirth 1964), the Plains Gartersnake (*Thamnophis radix*) with as many as 92 offspring (Wright and Wright 1957) and Australia's Mainland Tiger Snake (*Notechis scutatus*), in which a case of 109 offspring has been reported (Fitch 1970). No viviparous lizards approach these values.

It may be tempting to infer a causal connection between viviparous production of large litters and maternal defensive abilities, given that female snakes that are venomous (for example) can protect themselves and their embryos from predation (Neill 1964; Fitch 1970). However, the diversity of snakes challenges attempts to find simple correlations of such features. Venomous snakes include those with some of the smallest litter sizes on record (Lemen and Voris 1981), such as those cited above. Likewise, very large clutch sizes (over 90-100 eggs) occur among large non-venomous oviparous snakes, such as some species of *Python* (Pope 1961). Litter size is most likely affected by many attributes including female body size, offspring size, behavior, habitat, and phylogenetic relationships.

In many viviparous snake species, many or most females breed only once every two or more years. This pattern of "low frequency reproduction" (LFR) is scattered widely among viviparous viperids, elapids, and homalopsids (Fitch 1970; Aldridge 1979; Bull and Shine 1979; but see Blem 1982). This pattern has been inferred, for example, in the Chinese Green Tree Viper (*Trimeresurus stejnegeri;* Tsai and Tu 2001), *Crotalus atrox* (Taylor and DeNardo 2005), and the Cottonmouth (*Agkistrodon piscivorus;* Wharton 1966). As an extreme case, in a study of the Timber Rattlesnake (*Crotalus horridus*) near the northern end of its range, most females reproduced at 3-year intervals, and some at 4-year intervals (Brown 1991). Likewise, in northern populations of the Asp Viper (*Vipera aspis*) females reportedly reproduce on a triennial or quadrennial cycle (Saint Girons 1957). In the latter species, frequency of reproduction varies geographically. Low frequency reproduction is not an obligative characteristic of a species but, rather, a facultative response by individuals within populations (see Aldridge 1979; Lordais *et al.* 2002). Thus, it is best viewed as a product of both hereditary capacities and environmental influences.

In contrast to low-frequency reproduction, annual production of more than a single litter has only rarely been reported (Seigel and Ford 1987). The possibility that some viviparous females of a species produce dual clutches has been suggested in the Checkered Gartersnake (*Thamnophis marcianus*) (Ford and Karges 1987) and the Rainbow Water Snake (*Enhydris enhydris*) (Murphy *et al.* 2002) but definitive evidence is lacking in both cases. Dual clutches were reported in the Western Ribbon Snake (*Thamnophis proximus*) and Eastern Ribbon Snake (*Thamnophis sauritus*) (Fitch 1970). However, these accounts were based on two captive-bred females (see Neill 1962; Conant 1965) and evidence is lacking that they correspond to breeding in natural habitats. Such reports may reflect a reproductive plasticity of snakes (see Ford and Seigel 1989; Seigel and Ford 2001) that allows them to maximize reproductive effort as resources permit.

5.2.2 Historical Overview of Research on Snake Viviparity

In the biological literature, references to snake viviparity date to ancient Greece. In his *Historia Animalium* (circa 343 BCE), Aristotle noted that while most snakes lay eggs, a certain "viper" reproduces by giving birth to its young. Aristotle also distinguished between species (such as some sharks and vipers) in which females carry large eggs internally, and those (like mammals) that were said to lack such eggs and to be "internally viviparous." Remarkably, Aristotle had anticipated both of the fundamental reproductive distinctions we make among vertebrates, those based on reproductive products (egg vs. offspring) and the source of nutrients for embryonic development (yolk vs. placenta). However, Aristotle was not the first writer to take note of viviparity in snakes. Yaron (1985) traced mention of snake viviparity to Hebrew writings of 715 BCE. Both he and Boulenger (1913) noted that the word "viper" itself derives from the Latin words *pario* (to produce) and *vivus* (alive), a linguistic origin that implies an ancient past. Nevertheless, recognition of the phenomenon of viviparity in snakes undoubtedly predates the earliest written records, since our pre-literate ancestors on multiple continents surely learned from experience that some snakes carry their developing young internally.

During the 19th through early 20th centuries, scientific publications on snake viviparity fell into four main categories: (a) natural history accounts that documented viviparous habits in various species (e.g., Duméril and Bibron 1834-1854; Boulenger 1913; Serie 1916; Wall 1921; Mole 1924; Amaral 1927; Mell 1929; Smith 1930); (b) analyses of the association of viviparity with certain habitats (Rollinat 1904; Gadow 1910; Mell 1929; Weekes 1933; Sergeev 1940); (c) simple physiological and anatomical studies on the control of gestation (Rahn 1939; Boyd 1940; Clausen 1940; Bragdon 1955); and (d) anatomical studies of the oviduct and placental membranes (Giacomini 1893; Weekes 1929). By the 1960s, viviparity had been widely documented among snakes; however, little was known about how gestation and parturition were accomplished, and detailed placental descriptions were available for only four species (Weekes 1929; Kasturirangan 1951a, b; Parameswaran 1962). Furthermore, no consensus existed on why squamate viviparity and placentation had arisen or the sequence of events through which they evolved (Weekes 1933; Kopstein 1938; Neil 1964; Bauchot 1965; Packard 1966).

The development of life history theory in the 1960s and 1970s provided a major impetus to studies of viviparity in squamates, particularly lizards (Tinkle 1969; Tinkle *et al.* 1970; Vitt and Congdon 1978; Vitt 1981; Vitt and Price 1982). In the ensuing decade, life history theory also was applied to snake reproduction (e.g., Shine 1977; Seigel and Fitch 1984, 1985; Seigel *et al.* 1986; Seigel and Ford 1987). In this context, viviparity came to be viewed as a reproductive strategy that conferred both benefits and costs, as well as an evolutionary product of selective pressures and constraints (Tinkle 1967; Fitch 1970; Packard *et al.* 1977; Tinkle and Gibbons 1977; Shine 1980). The conceptual developments also marked a major shift in research rationales. Traditionally, a major justification for studies on viviparous reptiles was what they might reveal about mammalian reproductive evolution (Giacomini 1891; Kerr 1919; Weekes 1930). From the 1970s onward, viviparous reptiles were seen as worthy of study in their own right and for insights to be gained into the viviparous reproductive pattern itself.

Over the past 30 years, broad interest in the functional and evolutionary questions posed by squamate viviparity has led to research on multiple fronts. Areas of active investigation include reproductive ecology; physiological and behavioral ecology; reproductive anatomy and physiology; placentation and fetal nutrition; reproductive endocrinology; and the evolution of reproductive diversity. Numerous research papers and theoretical analyses are available in these areas (for overviews see Blackburn 2000 and papers in Thompson *et al.* 2010), and the literature continues to grow at a rapid pace.

5.2.3 Viviparity as a Reproductive Strategy

Advantages. Viviparity potentially confers both advantages and disadvantages to an animal, many of which apply directly to snakes (Table 5.2). Such factors have not necessarily operated as selective pressures, since they could well have accrued after viviparity originated. The main overarching benefit is that by retaining eggs in her reproductive tract, a reproducing female can protect them from environmental sources of mortality, including temperature extremes, dehydration, over-hydration, predation, and bacterial and fungal infection (Weekes 1930; Sergeev 1940; Fitch 1970; Tinkle and Gibbons 1977; Shine 1985). A related advantage is that viviparity can allow a terrestrial amniote to occupy habitats that lack suitable sites for deposition of eggs, including aquatic, arboreal, and extremely arid environments. For example, aquatic snakes dwell in locations unsuitable for oviposition. Accordingly, oviparous seasnakes of the genus *Laticauda* must return to land to lay their eggs (Smedley 1930; Saint Girons 1964), whereas viviparous aquatic snakes can give birth in the water (Neill 1964; Lemen and Voris 1981). A third potential benefit to squamates is that the viviparous female can provide placental nutrients to supplement those in the ovulated yolk (Stewart 1989; Stewart *et al.* 1990). A fourth benefit is that as ectotherms, the viviparous females can thermoregulate their eggs behaviorally. A number of researchers have noted this advantage in considering the likelihood that squamate viviparity has evolved preferentially in cold climates; however, the thermoregulatory benefits extend broadly to squamates in warm climates as well (see below).

Considerable evidence for the thermoregulatory advantages of viviparity has emerged from experimental and descriptive studies (e.g., Shine 1983b, 1987a, b, 1995, 2004; Capula *et al.* 1995; Mathies and Andrews 1995; Qualls and Andrews 1999; Andrews 2000; Lordais *et al.* 2004a). While most of this work has been conducted on lizards, it appears to be generally applicable to snakes. One benefit of viviparity is that viviparous females can thermoregulate their developing embryos at optimal developmental temperatures, decreasing development time during the short breeding

Table 5.2 Potential benefits and costs of viviparity in snakes. The following sources are among those offering documentation: Sergeev 1940; Neill 1964; Fitch 1970; Packard *et al.* 1977; Tinkle and Gibbons 1977; Shine and Bull 1979; Blackburn 1982; Shine 1985; Stewart 1989. A given benefit has not necessarily acted as a selective advantage, since benefits may accrue following the evolution of viviparity.

Potential benefits	*Factors and examples*
Protects eggs from environmental sources of mortality	temperature extremes (overheating, freezing) moisture extremes (flooding, dehydration) variable temperatures predators (e.g. insects and vertebrates) bacterial infection fungal attack
Frees females from having to find suitable nest sites	environments that lack suitable sites maternal energy expenditure and exposure risk
Allows use of a range of environments	aquatic habitats arboreal habitats xeric conditions seasonally variable temperatures
Permits maternal thermoregulation of embryos	optimal mean temperatures stable temperatures
Permits maternal oxygen supply to eggs	high altitudes and microenvironments with low oxygen tensions
Allows extra-vitelline nutrient provision by the female	yolk supplementation facultative supply of nutrients
Potential Costs	
Decreased locomotor performance	physical burden of the litter
Anorexia/reduced feeding during pregnancy	abdominal space constraints decreased foraging ability
Decreased maternal activity during pregnancy	altered thermoregulatory behavior decreased locomotor ability
Increased predation on reproducing females	decreased locomotion increased exposure to predators
Metabolic costs	physiological maintenance of the litter physical burden of the litter
Physiological debilitation of reproducing females	physiological costs of pregnancy decreased feeding
Decreased litter size	abdominal space constraints
Decreased offspring size	abdominal space constraints
Decreased clutch frequency	duration of pregnancy

season (e.g., Lordais *et al.* 2004a). Another potential benefit is that viviparous females can protect their eggs from freezing and thus enhance egg survival (Shine and Bull 1979). Yet another possibility is that thermoregulating females in cool or temperate climates enhance offspring quality rather than survival *per se* (Shine 1995). Recent experimental and circumstantial evidence supports the "maternal manipulation hypothesis"—that pregnant females can maintain their embryos at stable (rather than relatively high) developmental temperatures, thereby increasing offspring fitness (Shine 2004; Webb *et al.* 2006; Ji *et al.* 2007; Li *et al.* 2009). Stable temperatures can allow biochemical reactions to proceed at their temperature optima, maximizing speed and coordination of development. Thus, the thermoregulatory benefits of viviparity extend to a variety of environments, including tropical climates (Shine *et al.* 2003). The diversity of thermoregulatory benefits may account for the abundance of viviparous snakes in cold, temperate, and tropical environments, whatever the original selective pressures leading to evolution of their viviparity.

The variety of thermoregulatory benefits of viviparity is reflected in the diversity of patterns found among viviparous squamates. As compared to non-gravid females, pregnant snakes may regulate at relatively higher temperatures (Stewart 1965; Graves and Duvall 1993; Charland 1995; Ladyman *et al.* 2003; Foster *et al.* 2007), less variable temperatures (Osgood 1970), or the same temperatures (Gibson and Falls 1979; Gier *et al.* 1989), depending on the species. Similar variation occurs among lizard species, in which temperature preference during pregnancy may be higher, lower, or similar to non-gravid females; female basking times may be longer or unchanged; and females may either increase or reduce variance in body temperatures (Blackburn 2000; Shine 2006). In terms of its thermal benefits, viviparity offers reproducing squamates a degree of control over embryo development that is exploited in diverse ways under different circumstances.

Disadvantages. Viviparity entails several major disadvantages that reflect the fact that the female snake must carry eggs and fetuses for some months (Table 5.2). First, pregnancy may alter a snake's ability to locomote, forage, and escape predators (Bull and Shine 1979; Shine 1980). For example garter snakes show significant decrements in terrestrial locomotor performance while pregnant (*Thamnophis marcianus:* Seigel *et al.* 1987; *T. ordinoides:* Brodie 1989). In a study of the Northern Death Adder (*Acanthophis praelongus*), females in late pregnancy showed >30% reduction in swimming speeds as compared to non-reproductive females (Webb 2004). Likewise, pregnancy significantly decreased speed of locomotion in the Black Swamp Snake (*Seminatrix pygaea*); after parturition, swimming and crawling speed increased by more than 72% and 59% respectively (Winne and Hopkins 2006). Such impairment has been widely reported in viviparous lizards (Bauwens and Thoen 1981; van Damme *et al.* 1989; Lin *et al.* 2008), and extends to gravid oviparous lizards as well (Miles *et al.* 2000; Shine 2003).

A second, related issue is that female snakes commonly show reduced feeding during pregnancy (Fitch and Twining 1946; Gregory and Stewart 1975; Shine 1980; Krohmer and Aldridge 1985; Gregory and Skebo 1998; Lordais *et al.* 2002; Gignac and Gregory 2005). For example, in *Crotalus atrox* towards the northern part of its range, pregnant females feed infrequently (if ever) during the entire three month gestation period (Keenlyne 1972). Depending on the species, reduction and cessation of feeding may reflect decreased female mobility (through reduced locomotor performance), secretive behavior, inability to accommodate ingested prey (given the large volume of developing eggs in the oviducts: Keenlyne 1972), or a conflict between feeding and gestational thermoregulatory behavior (Gregory *et al.* 1999; Lordais *et al.* 2002). Reduced feeding requires females to fuel activity from lipid reserves and catabolized muscle (Lordais *et al.* 2004b) and can affect future reproduction and even survival (see below).

A third potential disadvantage is the overall physiological cost of pregnancy (Birchard *et al.* 1984a; Ladyman *et al.* 2003; Schultz *et al.* 2008). Shifts in thermal ecology during gestation can increase maternal metabolic rate during the very period that the female has ceased feeding and is depleting her lipid stores. Likewise, altered activity patterns (Lordais *et al.* 2002) can have detrimental effects on female metabolism. Physiological costs of pregnancy have been estimated in a few lizards (Beuchat and Vleck 1990; DeMarco and Guillette 1992; Robert and Thompson 2000) and one viviparous snake (Schultz *et al.* 2008). In *Acanthophis praelongus* the "maintenance cost of pregnancy" (MCP) during the last weeks of pregnancy was measured at 26.4% of total maternal metabolic rate (Schultz *et al.* 2008). Although large, this figure represented a small proportion of total female reproductive effort. Nevertheless, the MCP may impose an important ecological cost given limited female energy stores and their depletion as gestation proceeds.

As a result of these and perhaps other factors, female snakes can be greatly debilitated by pregnancy. A common pattern is for lipid reserves to be greatly diminished during pregnancy, as in the Timber Rattlesnake (*Crotalus horridus*) (Gibbons 1972). In the Rainbow Boa (*Epicrates cenchria maurus*), pregnant females not only deplete lipid reserves but catabolize trunk muscle protein, such that strength and performance are significantly decreased by the time of parturition (Lordais *et al.* 2004b). In the Northern Viper (*Vipera berus*), mortality of reproducing females is very high (40%), largely due to their post-partum emaciation (Madsen and Shine 1992, 1993). In its congener *V. aspis*, most females do not survive even their first attempt at reproduction (Bonnet *et al.* 2002). Even for surviving females of the latter species, anorexia during pregnancy may have serious consequences, because post-partum fat reserves affect the probability and timing of future reproduction (see Bonnet *et al.* 2001). Perhaps for such reasons, in a compilation of squamates with low frequency reproduction (Bull and Shine 1979), most consisted of viviparous species.

Another potential disadvantage of viviparity in snakes is that abdominal space limitations may constrain litter size and offspring size, due to the difficulty of accommodating relatively large eggs in a tubular body of limited diameter. Space constraints are a general problem for snakes regardless of reproductive mode, and multiple reproductive specializations have evolved that minimize them (Blackburn 1998a).

These specializations include (a) elongated eggs, a feature that allows accommodation of relatively large volumes of yolk in the abdomen; (b) an exaggerated oviductal asymmetry that places eggs of the right oviduct cranial to those in the left (instead of side by side, as in most lizards); (c) exaggerated ovarian asymmetry, that similarly prevents overlap (Pizzatto *et al.* 2007); and (d) loss of one oviduct, a feature of scolecophidians (Fox and Dessauer 1962; Robb 1960), some species of *Tantilla* (Clark 1970b; Aldridge and Semlitsch 1992), and two genera of anomalepidids (Robb and Smith 1960). Although these features are not adaptations to viviparity itself (Blackburn 1998a), they offer evidence for disadvantages of gravidity that can be particularly acute for viviparous species. Investigations of interspecific diversity in these features (e.g. Pizzatto *et al.* 2007) might reveal enlightening correlations with reproductive mode.

Relative clutch mass (the ratio of clutch weight to body weight) is significantly lower in viviparous snakes than in oviparous snakes (Seigel and Fitch 1984; also see Seigel *et al.* 1986, 1987). Such data offer a useful indicator of costs to fecundity entailed by viviparity. Clearly the selective pressures leading to viviparity in snakes must be high in order to outweigh the significant disadvantages that this pattern can impose.

5.2.4 Selective Pressures and Influences

Among early researchers, a common approach was to assume that potential benefits of viviparity (as in Table 5.2) had served as selective advantages during its evolution. Several writers deduced relevant selective pressures from the habitats in which viviparous species are currently found. For example, the disproportionate occurrence of live-bearing snakes and lizards in high altitudes in Australia led Weekes (1933) to infer that the colder climates of those environments selected for viviparity. Her conclusions paralleled those of researchers on squamates of other regions (Rollinat 1904; Mell 1929; Sergeev 1940). As Sergeev (1940) noted, viviparous females can thermoregulate their embryos behaviorally at optimal temperatures, ensuring their full development during the relatively short breeding season. Focusing on another potential selective advantage, Sowerby (1930) suggested that semi-aquatic habits of the Red-Backed Ratsnake (*Oocatochus* [*"Elaphe"*] *rufodorsatus*) led to its evolution of viviparity (Shine 1985). Researchers also speculated that other traits such as arboreality, aquatic habitats, cold climates, fossorial habits, and maternal defensive ability have enhanced selection for viviparous reproduction in snakes (Sergeev 1940; Neill 1964; Fitch 1970; Packard *et al.* 1977).

However, ecological and biological features of extant viviparous species do not necessarily indicate those under which this pattern originated (Tinkle 1967; Greene 1970; Tinkle and Gibbons 1977; Shine and Bull 1979). In theory, viviparity in a clade could have preceded, accompanied, or followed the invasion of a habitat in which live-bearing species occur. Similarly, characteristics of particular viviparous snakes may or may not have preceded and led to development of viviparous habits. Only phylogenetic analysis that takes into account character state polarity and chronological sequences through comparison to appropriate outgroups can determine the conditions under which viviparity has evolved in a given lineage (Blackburn 2000). The fact that many origins of squamate viviparity have occurred at relatively low taxonomic levels makes such an analysis possible and practical.

Two independent analyses sought to define the phylogenetic origins of viviparity of snakes and lizards (Blackburn 1985; Shine 1985). These analyses recognized up to 39 potential origins of ophidian viviparity, of which more than 30 were judged as particularly well established (Blackburn 1999). A more recent analysis of boids has eliminated some of these origins while revealing others (Lynch and Wagner 2009). Viviparity clearly has originated many times among snakes, and ongoing refinement of the recognized origins will continue to allow hypotheses to be tested about the conditions of its evolution.

Shine's (1985) review evaluated squamate origins of viviparity in terms of operative ecological conditions, and identified cold climates of high altitudes and latitudes as the most important selective agent leading to live-bearing habits. In contrast, the analysis found little evidence of a preferential evolution of viviparity in aquatic, arboreal, and xeric habitats. Phylogenetic analyses for iguanian lizards (Schulte and Moreno-Roark 2009) and viperid snakes (Lynch 2009) provide further support for an association between origin of viviparity and climatic fluctuations leading to cooler temperature regimes. In addition, origins of viviparity among multiple lineages of iguanian lizards may have been influenced by independent climatic events separated over a considerable historical time frame (Schulte and Moreno-Roark 2009). These conclusions are consistent with strong evidence for the thermoregulatory benefits of viviparity discussed above. Nevertheless, among viviparous snakes are some groups whose reproductive habits cannot readily be traced to a cold climate origin (Neill 1964; Ota *et al.* 1991). Likewise, the diversity of ways that pregnant females modify thermoregulatory behavior (see above) suggests a variety of potential benefits that vary interspecifically, and some of which would apply to tropical species (Webb *et al.* 2006; Ji *et al.* 2007).

Evolution of viviparity should be affected not only by selective pressures but by proto-adaptations ("pre-adaptations") and constraints (Table 5.3). As a group, squamates share several features that have facilitated the evolution of viviparity (Blackburn 1998a). These features include internal fertilization, oviductal egg retention with female thermal modification (Shine 2006; Lourdais *et al.* 2008) (since oviparous females usually retain eggs for ~30% of development), vascularized oviducts (lacking in most teleosts), a vascularized chorioallantois (which permits intrauterine gas exchange), and maternal ability to withstand altered thermoregulation and prolonged anorexia (features lacking in endotherms). Other potential proto-adaptations are discontinuously distributed among squamates, such

Table 5.3 Factors hypothesized to affect the origin of viviparity in squamates. "Proto-adaptations" refer to predisposing factors; those shared by squamates in general are marked with an asterisk. Sources include: Sergeev 1940; Neill 1964; Fitch 1970; Packard *et al.* 1977; Tinkle and Gibbons 1977; Shine and Bull 1979; Bull 1980; Blackburn 1982, 1985; Shine 1983b, 1985, 2006; Andrews and Mathies 2000; Andrews 2002.

Selective pressures	*Proto-adaptations*
Cold climate	Internal fertilization*
Short breeding season	Vascularized oviducts*
Freezing temperatures	Vascular fetal membranes*
Low oxygen tensions	Maternal egg-retention*
Arid habitat	Ectothermy*
Aquatic habitat	Low maternal metabolism
Arboreal habitat	Ability to withstand anorexia*
Predation on eggs	Fetal resistance to hypoxia*
Constraints	Male heterogamety
Loss of fecundity	Facultative egg retention
Decreased maternal survival	Maternal thermophilic behavior
Temperature-dependent sex determination	Thin eggshells
Oxygen availability to oviductal embryos	Maternal brooding behavior
Female heterogamety	Maternal defensive capacity
Highly calcified eggshells	

as venomous capacity (as in viperids, elapids, and colubrids), maternal egg-brooding behavior (as in various colubrids and viperids, as well as scincid lizards; Shine 1988), and physiological capability of embryos to circumvent an hypoxic oviductal environment (Andrews and Mathias 2000; Parker and Andrews 2006). Venomous capacity, or more broadly, defensive ability, can allow pregnant females to avoid disadvantages of viviparity associated with increased susceptibility to predation (Neill 1964; Shine and Bull 1979). Whether maternal brooding facilitates viviparity or represents an alternative means of achieving its advantages has been controversial (Shine and Bull 1979; Shine 1985; de Fraipont *et al.* 1996; Blackburn 1999; Shine and Lee 1999). At present, the proposed relationship can be viewed as unsubstantiated. Costs of viviparity that may have acted as evolutionary constraints are also discontinuously distributed among squamates (Table 5.3). Such constraints include decreased fecundity, and decreased embryonic or maternal survivorship. Other hypothetical constraints, such as calcareous eggshells (Packard *et al.* 1977) and modes of sex determination (Blackburn 1982) have not been supported by subsequent analyses (Blackburn 1985; Stewart *et al.* 2009b; Linville *et al.* 2010).

As discussed, multiple lines of evidence—phylogenetic, descriptive, and experimental—support the hypothesis that viviparity commonly evolves in part due to its overall thermoregulatory benefits, and has frequently evolved in relatively cool or temperate climates. Nevertheless, the precise benefits vary between species, as have operative proto-adaptations, constraints, and consequences. Therefore, while the search for commonalities has yielded evidence of striking patterns of similarity, nevertheless phylogenetic and functional differences are to be expected in groups as diverse as the Serpentes.

5.2.5 Historical Sequences and Patterns

Ideally, detailed reconstructions of the evolution of viviparity and placentation would draw upon representatives of individual clades. The existence of snake genera with both oviparous and viviparous members offers good prospects for such an approach. However, while a few key lizard taxa have been investigated in this way (Heulin *et al.* 1989; Smith and Shine 1997; Qualls and Shine 1998; Stewart and Thompson 2003, 2009; Stewart *et al.* 2004a; Surget-Groba *et al.* 2006), snakes have received relatively little attention in this regard (Lynch 2009; Lynch and Wagner 2009). Nevertheless, a great deal can be gleaned from broad surveys of squamate species that represent multiple lineages.

According to a traditional scenario (summarized by Blackburn 1992, 1995), reproductive evolution in viviparous squamates involved multiple sequential steps: (a) incremental evolutionary increases in oviductal egg retention, accompanying the oviposition of eggs with increasingly more advanced embryos; (b) retention of embryos to term by the female, associated with viviparous reproduction; (c) evolution of a placenta that functioned in gas exchange; (d) incipient placentotrophy, in which placentae supply small quantities of nutrients to the embryo; and (e) substantial placentotrophy, in which placental membranes supply most of the nutrients for development. This scenario distinguishes "lecithotrophy" (in which the yolk provides the organic nutrients for development) from "placentotrophy" (in which nutrients are provided by placental means). Accordingly, the lecithotrophy of oviparous species would be retained upon the origin of viviparity. Incipient placentotrophy and substantive placentotrophy would be viewed as successive sequential modifications.

Tests of predictions based on this scenario have falsified several aspects. Under the modified scenario (Blackburn 1995, 2005, 2006), the evolution of viviparity in squamates occurred simultaneously with placentation and placental provision of small quantities of nutrients. Evidence for this inference comes from the fact that every viviparous squamate (including snakes) that has been appropriately examined has placentae that accomplish respiratory gas exchange and that supply water and (at least) small quantities of nutrients. Given embryonic needs for gas exchange in particular, it is difficult to imagine how viviparity could evolve in squamates without placentae to sustain embryos during their gestation in the female reproductive tract.

Another aspect of the traditional scenario that has been questioned is its assumption that viviparity evolves through continuously incremental increases in egg retention, associated with deposition of eggs with increasingly advanced embryos (Blackburn 1995, 1998b). With rare exceptions (Qualls *et al.* 1995; Smith and Shine 1997; Smith *et al.* 2001), squamate species are bimodally distributed between two patterns: deposition of eggs with early ("limb-bud") stage embryos (typical oviparity) vs. viviparity, with production of fully developed neonates. This distribution has been interpreted as consistent with a punctuated equilibrium model of change, in which the transition from oviparity to viviparity occurs relatively quickly (and at low taxonomic levels) through an evolutionarily unstable intermediate phenotype (Blackburn 1995). This transition is distinct from saltatory change. The punctuated equilibrium scenario has yielded fruitful discussion (Qualls *et al.* 1997; Blackburn 1998b; Gould 2002) and elaboration (Shine and Thompson 2006)

The considerable number of independent origins of viviparity among Squamata indicates that reproductive mode is either highly labile or subject to strong selection pressure under some environmental conditions. In either case, oviparity is advantageous in most habitats because only approximately 20% of squamates are viviparous and the percentage is comparable for both lizards and snakes. The common occurrence of the transition from oviparity to viviparity has stimulated speculation on the potential for reversibility and the methods of analysis required to test for polarity in the transition (de Fraipont *et al.* 1996, 1999; Blackburn 1999; Shine and Lee 1999). Comparisons between conspecific populations and between congeners have revealed few characteristics of reproductive morphology that distinguish oviparous and viviparous reproductive modes (Mulaik 1946; Guillette and Jones 1985; Stewart 1985; Stewart *et al.* 2004a; Heulin *et al.* 2005), increasing the likelihood of evolutionary reversals to oviparity. However, the embryonic stage at parition (oviposition or birth) differs substantially between oviparous and viviparous modes with few exceptions (Qualls *et al.* 1995; Smith and Shine 1997); thus intermediate conditions may be highly unstable (Blackburn 1995, 1998b). In the absence of recognizable phenotypic characters, the best evidence for a reversal to oviparity is provided by comparative phylogenetic analyses. Recent analyses have revealed a few lineages in which reversal to oviparity is a plausible hypothesis (Smith *et al.* 2001; Surget-Groba *et al.* 2006; Lynch and Wagner 2009). One of the most convincing cases is the boid genus *Eryx* in which both phylogenetic and morphological evidence exists for reversal (Lynch and Wagner 2009). If substantiated, such lineages offer an excellent opportunity to understand the evolution of phenotypic traits such as eggshell deposition in the transition between reproductive modes.

In sum, under the modified scenario, reproductive evolution in snakes appears to involve simultaneous development of viviparity, placentation and incipient placentotrophy. One implication is that placentae, like viviparity itself, have evolved convergently in over 100 squamate lineages, including each of 30+ clades of viviparous snakes. To readers unfamiliar with squamate placentae, it may seem puzzling (if not hard to believe) that such structures could evolve so readily and so often. After all, mammal placentae traditionally have been held to epitomize the height of reproductive evolution. However, the mystery is easily resolved by considering what placentae actually are, how they develop, and how they originate evolutionarily.

5.3 PLACENTAE AND PLACENTAL RESEARCH

Placentae, by definition, are organs formed through apposition of embryonic and maternal tissues and that function in physiological exchange (Mossman 1937, 1987). The structural criterion of this broad definition implies nothing about cellular complexity or identity of the tissues. Likewise, under its functional criterion, physiological exchange is interpreted broadly to include respiratory gases as well as inorganic and organic nutrient provision.

The maternal component of reptilian placentae is the lining of the uterine oviduct. The embryonic component of the placenta consists of fetal membranes (chorion, chorioallantois, and yolk sac) (Stewart 1997), structures that line the eggshell in oviparous squamates. The eggshell of oviparous squamates has two components: an inner organic fiber matrix, the shell membrane, overlain by an inorganic layer of calcium carbonate (Packard and Demarco 1991). In contrast, viviparous squamates retain only a greatly reduced shell membrane, which separates fetal and maternal tissues. Therefore, pregnancy in snakes brings fetal membranes very close or even in contact with the oviductal lining, forming the placental structures that sustain the embryo throughout gestation (Blackburn 1992; Stewart 1997).

5.3.1 Research Methods

Our understanding of placental structure and function in squamates stems from studies of placental morphology, physiology, developmental biology, and biochemistry (for reviews see Yaron 1985; Blackburn 1993, 2000; Stewart 1993; Thompson *et al.* 2004; Thompson and Speake 2006). Techniques that have been applied to snakes are summarized in Table 5.4. Anatomical studies traditionally have drawn on microscopic examination of histological sections of developmental series of embryos. These techniques continue to be valuable in revealing placental composition, development, and functional attributes. More recently, electron microscopy (EM) has revealed details of functional morphology at the ultrastructural level. The first placental studies of any squamate to use transmission EM (Hoffman 1970) and scanning EM (Blackburn *et al.* 2002) were done on garter snakes of the genus *Thamnophis.* Ultrastructural techniques have since been applied to other snake species as well (Attaway 2000; Blackburn and Lorenz 2003a, b; Stewart and Brasch 2003; Anderson and Blackburn 2009; Blackburn *et al.* 2009). Histochemical analysis has been used infrequently on snake placentae (Hoffman 1970; Baxter 1987) but holds promise in revealing functional attributes of the placental membranes.

Table 5.4 Viviparous snakes whose placentae have been studied. Methods abbreviations: "comp. analysis" = composition analysis of eggs vs. neonates; "LM" = light microscopy; "SEM" = scanning electron microscopy; "TEM" = transmission electron microscopy. Sources: (1) Attaway 2000; (2) Baxter 1987; (3) Bellairs *et al.* 1955; (4) Blackburn 1998a & unpublished data.; (5) Blackburn and Lorenz 2003a; (6) Blackburn and Lorenz 2003b; (7) Blackburn *et al.* 2002; (8) Blackburn *et al.* 2009; (9) Conaway and Flemming 1960; (10) Hoffman 1970; (11) Jones and Baxter 1991; (12) Kasturirangan 1951a; (13) Kasturirangan 1951b; (14) Parameswaran 1962; (15) Sangha *et al.* 1996; (16) Stewart 1989; (17) Stewart 1992; (18) Stewart and Brasch 2003; (19) Stewart and Castillo 1984; (20) Stewart *et al.* 1990; (21) Weekes 1929.

Taxon	*Methods*	*Sources*
Natricidae		
Thamnophis sirtalis	LM, radioisotope transfer, histochemistry, TEM, SEM	5, 6, 7, 10
T. ordinoides	LM, SEM, comp. analysis	7, 20
T. radix	LM, TEM	5, 6
Virginia striatula	LM, TEM, SEM, comp. analysis	15, 16, 17, 18
Tropidoclonion lineatum	LM, histochemistry	2, 11
Nerodia sipedon	LM, TEM, SEM radioisotope transfer	4, 9
N. rhombifer	LM, comp. analysis	4, 19
N. cyclopion	radioisotope transfer	9
Regina septemvittata	LM, TEM	1
Storeria dekayi	LM, TEM, SEM	8
S. occipitomaculata	LM, TEM (?)[1]	10
Homalopsidae		
Enhydris dussumieri	LM	14
Elapidae		
Austrelaps ramsayi[2]	LM	21
Suta suta[3]	LM	21
Enhydrina schistosa	LM	12
Hydrophis cyanocinctus	LM	13
Viperidae		
Vipera berus	LM[1]	3

[1]These studies contain no corroborative micrographs or drawings.
[2]as *"Denisonia superba"*
[3]as *"Denisonia suta"*

Physiological studies of placental function in snakes have relied on two types of methods. One method is injection of radiolabeled molecules (e.g., sodium, amino acids) into pregnant females to determine what passes across the placenta into embryonic tissues (Conaway and Flemming 1960; Hoffman 1970). This approach offers qualitative information, but usually does not reveal the particular placental tissues involved. A second method is to compare newly ovulated eggs to neonates in terms of wet and dry mass (Clark *et al.* 1955; Clark and Sisken 1956; Shine 1977), on the grounds that any increase in either feature must reflect placental transfer. This approach offers only a very rough measure of placental function but does distinguish species with substantial placental uptake of organic molecules. A gestational decrease in dry mass of the conceptus (yolk + embryo), as occurs in most squamates, does not preclude placental transfer of organic and inorganic molecules, but does indicate that if placental transfer of organic molecules occurs, it does not compensate for metabolic loss (Blackburn 1994; Stewart and Thompson 2000). A third and more sophisticated technique is to compare chemical composition of ovulated eggs to that of sibling neonates or late-term fetuses (Stewart and Castillo 1984; Stewart 1989). Any increase in inorganic molecules during gestation is an indirect estimate of net placental transport because these molecules are not metabolized. Sampling of sibling yolks and neonates allows statistical detection of covariance between transported molecules and of effects due to individual females. Experimental study of *Virginia striatula* suggests that removal of sibling yolks did not affect placental nutritional provision to other embryos (Sangha *et al.* 1996). While the analysis of chemical composition has not been used to quantify transfer of organic molecules in snakes, it has been useful in revealing incipient placentotrophy that involves sodium and calcium. It also permits recognition of obligative vs. facultative patterns of placental provision (Stewart 1989; Stewart *et al.* 1990).

Overall, a battery of anatomical and physiological methods is needed to reveal details of placental structure and function, including details relevant to a reconstruction of placental evolution.

5.3.2 Research Species

Aspects of placentation have been investigated mainly in snakes from three families. Early studies described placental histology, cytology, and development in hydrophiine elapids (Weekes 1929; Kasturirangan 1951a, b) and homalopsids (Parameswaran 1962). In the past two decades snake placental research has focused on North American thamnophines (Natricidae) (Table 5.4), through research in each of our laboratories. Consequently, much of the discussion of snake placentation below focuses on these snakes.

North American thamnophine snakes are a monophyletic group of nine genera and about 50 species (Alfaro and Arnold 2001). These species are combined with Old World forms in the family Natricidae (Zaher *et al.* 2009), within which thamnophines can be considered a subfamily or a tribe. North American thamnophines share a common ancestry with Old World members of the family (Rossman and Eberle 1977; Lawson *et al.* 2005; Zaher *et al.* 2009). Living thamnophines include garter snakes (*Thamnophis*), water snakes (*Nerodia*), swamp snakes (*Seminatrix*), crayfish snakes and queen snakes (*Regina*), brown snakes (*Storeria*), lined snakes (*Tropidoclonion*), and earth snakes (*Virginia*). Many studies have considered thamnophine reproductive ecology and behavior (Seigel and Ford 1987;

Rossman *et al.* 1996; Shine *et al.* 2005a, b) and others have documented aspects of female reproductive anatomy (Hoffman and Wimsatt 1972; Blackburn 1998a; Sever and Ryan 1999; Sever *et al.* 2000; Siegel and Sever 2008) and physiology (e.g., Mead *et al.* 1981; Kleis *et al.* 1986a, b; Whittier *et al.* 1987, 1991; Ingermann *et al.* 1991). A recent molecular phylogeny has documented three major lineages among New World natricids (Alfaro and Arnold 2001): a garter snake clade (*Thamnophis*), a water snake clade (*Nerodia, Tropidoclonion,* and some *Regina*), and a clade of semi-fossorial snakes and their allies (*Virginia, Storeria, Clonophis, Seminatrix,* and some *Regina*).

Placental research on thamnophines has applied a variety of contemporary methods to species throughout the group. Anatomical techniques (such as histology, transmission and scanning EM, and histochemistry) have been used in ten species from six genera (Table 5.4). Physiological techniques (e.g., placental transfer of radioisotopes, chemical composition analysis) have been applied to six species in three genera (Table 5.4). Species studied with these techniques include multiple representatives from each of the three major thamnophine clades (as recognized by Alfaro and Arnold 2001). Furthermore, data on fetal membrane structure and function are now available for a related oviparous snake (the Red Cornsnake *Pantherophis guttatus*: Blackburn *et al.* 2003; Ecay *et al.* 2004; Stewart *et al.* 2004b; Knight and Blackburn 2008). Therefore, results of these studies offer an unprecedented opportunity to reconstruct details of placental structure, function, and evolution.

5.4 PLACENTAL FUNCTIONS IN SNAKES

Placentae of snakes sustain developing embryos by accomplishing gas exchange, provision of water and nutrients and excretion of nitrogenous wastes. Both the maternal and the fetal component of the placentae carry out functions similar to those of their oviparous homologues. In oviparous squamates, the uterine oviduct is that portion of the maternal tract that houses developing eggs before oviposition; it also supplies calcium to the eggshell (Stewart and Ecay 2010 for review) and must be responsible for any gas exchange (Andrews and Mathies 2000; Parker and Andrews 2006) and water provision that occurs. As for the fetal ("extraembryonic") membranes, they accomplish physiological exchange through the eggshell. Specifically, the chorioallantois functions in gas exchange (Andrews and Mathies 2000) and transport of calcium (Ecay *et al.* 2004; Stewart *et al.* 2004b). Tissues derived from the yolk sac are indirectly implicated in water uptake (Weekes 1935; Blackburn and Flemming 2009). These functions are retained and enhanced in viviparous squamates (Blackburn 1993; Stewart 1993).

Viviparity not only extends the time frame in which eggs are sustained in the oviduct but entails an expansion of physiological functions that occur in oviparity. Accordingly, placental functions change over the course of gestation. For example, embryonic needs for oxygen (Vleck and Hoyt 1991; DeMarco 1993; Andrews and Mathies 2000; Robert and Thompson 2000; Parker *et al.* 2004) and calcium (Stewart *et al.* 2004b; Stewart and Ecay 2009; Fregoso *et al.* 2010) are accentuated late in development. A complicating factor is that the fetal membranes that contribute to placentae undergo dramatic developmental changes in accord with the squamate morphogenetic pattern. Identity of the fetal membranes that function in physiological exchange at a given stage partly reflects embryonic needs as well as ancestral morphogenetic patterns of development. Another complication is that diverse placental functions (e.g., for gas exchange and nutrient provision) may require incompatible structural attributes; hence separate regions of a given placenta may be specialized for different functions (Blackburn 1993; Jerez and Ramírez-Pinilla 2001; Stewart and Brasch 2003; Ramirez-Pinilla *et al.* 2006). Furthermore, a given placental function may be accomplished by different mechanisms and structures in different species, even those within a single viviparous clade. Therefore, a full understanding of placental structure and function requires detailed explorations of some clades as well as sampling of species from others.

5.4.1 Respiratory Exchange

Gas exchange is critical for an embryo to survive in the hypoxic oviductal lumen. Although eggs of oviparous squamates begin their development in the oviduct, they are oviposited in early stages when oxygen requirements are low (Clark 1953a; Blackburn 1995). Experimental evidence on lizards suggests that the capacity for maternal-fetal oxygen transfer may be a constraint on the evolution of viviparity (Andrews and Mathies 2000; Andrews 2002; Parker *et al.* 2004; Parker and Andrews 2006). Water provision also appears to be essential, as it is in oviposited eggs of most oviparous squamates (Packard *et al.* 1982; Packard and Packard 1988; Thompson and Speake 2004). Such water allows yolk liquefaction; it also contributes to composition of the embryo, which has a higher concentration of water than does yolk in viviparous species (Table 5.5). Provision of nutrients has several functions and as discussed below, exhibits facultative and obligative components (Stewart 1989; Thompson *et al.* 1999; Stewart and Thompson 2000). Metabolism of nitrogenous wastes is similar in oviparous (*Coluber constrictor*) and viviparous (*Thamnophis sirtalis*) snakes (Clark 1953b; Clark and Sisken 1956). The primary excretory products are soluble molecules, ammonia and urea, which are sequestered in egg compartments in both reproductive modes, but also transferred from embryo to maternal compartments across the placenta in the Common Garter Snake, *Thamnophis sirtalis*.

Transplacental respiratory exchange in viviparous snakes has not been quantified, but can be inferred from indirect lines of evidence. Estimates of the cost of viviparous reproduction in snakes (Birchard *et al.* 1984a; Ladyman *et al.* 2003) and in lizards (DeMarco 1993; Robert and Thompson 2000) reveal that oxygen consumption of gravid females has three

Table 5.5 Composition of recently oviposited/ovulated eggs and hatchlings/neonates in oviparous and viviparous snakes

	Yolk				*Hatchling/Neonate*				
Species	Wet mass (g)	*Dry mass* (g)	*Ash-free dry mass* (g)	*Ash* (mg)	*Wet mass* (g)	*Dry mass* (g)	*Ash-free dry mass* (g)	*Ash* (mg)	*Reference*
Oviparous									
Coluber constrictor	3.6	1.4			4	1			Clark 1953b
Elaphe carinata	34.8	8.4	7.7	687	25.7	6.8	6.0	780	Ji and Du 2001
Elaphe taeniura	25.4	6.6	6.1	502	18.8	5.6	5.0	605	Ji *et al.* 1999
Pantherophis guttatus	7.6	2.4	2.3	130	7.4	2	1.9	150	Stewart *et al.* 2004
Ptyas korros	7	2.1			6	1.6			Ji and Sun 2000
Rhabdophis tigrinus		0.6	0.54	65		0.4	0.30	45	Cai *et al.* 2007
Xenochrophis piscator	2.3	0.6	0.53	69	1.8	0.4	0.36	72	Lu *et al.* 2009
Naja naja	15.9				10.4				Ji *et al.* 1999b
Viviparous									
Nerodia rhombifer	6.7	3.5	3.2	350	10.7	2.7	2.4	340	Stewart and Castillo 1984
Thamnophis ordinoides	1.3	0.6	0.55	46	1.8	0.4	0.39	53	Stewart *et al.* 1990
Virginia striatula	0.28	0.14	0.13	11	0.42	0.1	0.08	12	Stewart 1989
Notechis scutatus	2.7	1.4			5.3	0.9			Shine 1977
Pseudechis porphyriacus	6.6	3.2	2.9	260	20.2	3.2	2.8	400	Shine 1977

components: resting metabolism, embryonic metabolism and pregnancy maintenance metabolism (Birchard *et al.* 1984a). For lizards, energy consumption during development of viviparous embryos is similar to that of oviparous embryos (Robert and Thompson 2000). Thus, measurements of oviparous eggs may provide relevant estimates for embryonic oxygen needs of viviparous species. In oviposited snake eggs, oxygen consumption increases progressively over the course of development (Clark 1953a; Dmi'el 1970; Thompson 1989). Similar data are available for eggs of various lizards (e.g. Thompson and Stewart 1997; Thompson and Russell 1999; Booth *et al.* 2000). Oxygen transport to embryos of viviparous snakes is enhanced by higher blood oxygen affinity in embryos that in at least some species is facilitated by a reduction in oxygen affinity of female blood associated with pregnancy (Birchard *et al.* 1984b; Berner and Ingermann 1988; Ragsdale and Ingermann 1991, 1993; Ingermann 1992; Ragsdale *et al.* 1993). Evidence thus indicates that in various viviparous snakes, fetal needs for respiratory exchange increase during gestation and maternal blood chemistry adjusts to enhance respiratory exchange.

5.4.2 Water Provision

Maternal-fetal transfer of water can be calculated by comparing water content of ovulated eggs with neonates or late-term fetuses. In viviparous snakes, wet mass of the conceptus can double or triple during gestation (Table 5.5; Fig. 5.1A). In contrast, most oviparous hatchlings contain less water than was present in recently ovulated eggs (Table 5.5, Fig. 5.1A). Placental transfer of water is substantial; with regard to embryonic hydration, the uterine environment probably is more stable than the unattended nests of oviparous species.

Viviparous snakes ovulate large yolk-rich eggs that provide most organic nutrients for embryonic development. Accordingly, dry mass of the conceptus commonly decreases during development (Table 5.5, Fig. 5.1B), as occurs in oviparous squamates and most viviparous lizards (Blackburn 1994; Thompson *et al.* 2000). In studies on three thamnophines (Stewart and Castillo 1984; Stewart 1989; Stewart *et al.* 1990), organic content (calculated as ash-free dry mass) decreased by about 24 to 40% during development (Table 5.5, Fig. 5.1C). In *Oocatochus rufodorsatus,* caloric value and lipid content of the conceptus decreases significantly during gestation (Ji 1995). A reduction in dry mass and organic content indicates that fetal nutrition is relatively lecithotrophic but does not preclude a placental supply of nutrients insufficient to compensate for metabolic loss.

5.4.3 Nutrient Provision

Placental transfer of organic nutrients has been suggested for three species of thamnophines. Composition analyses of yolk and embryos/neonates of *Thamnophis sirtalis* (wet mass, dry mass, lipid mass) (Clark *et al.* 1955; Clark and Sisken 1956) and of Graham's Crayfish Snake, *Regina grahami*

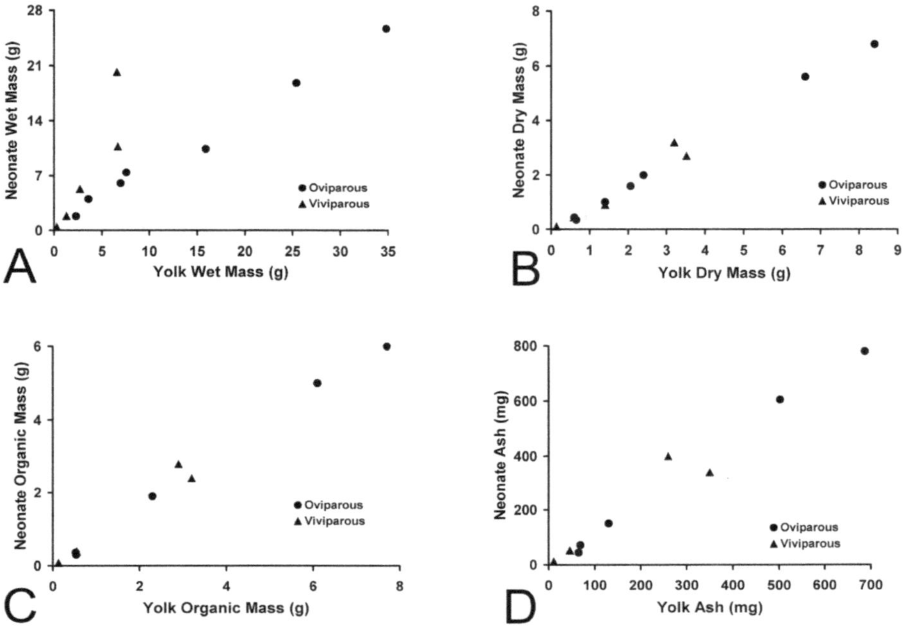

Fig. 5.1 Comparison of the composition of recently oviposited/ovulated eggs and hatchlings/neonates of oviparous and viviparous snakes. **A.** Wet mass. **B.** Dry mass. **C.** Organic content (ash-free dry mass). **D.** Ash content (total inorganics).

(dry mass) (Hall 1969) were interpreted as embryonic uptake of organic nutrients. However, all three studies reported small sample sizes and failed to account for variation among clutches. Analyses of data provided in Clark and Sisken (1956) and Hall (1969) found that the presumed differences were not statistically significant (Shine 1977). Indirect evidence for placental transfer of organic nutrients has been reported for *Virginia striatula*, a predominantly lecithotrophic species. Dry mass of neonates is lower than that of yolk, yet larger females give birth to larger young relative to egg size (Stewart 1989). Placental nutrient transfer may be influenced by female nutritional condition in this species. Direct evidence of net embryonic uptake of organic nutrients via placental transfer is available only for *Thamnophis sirtalis*, in which radiolabeled glycine injected into pregnant females was transferred to embryos via the placentae (Hoffman 1970). Whereas evidence for placental transfer of organic nutrients is sparse, net placental uptake of inorganic nutrients is widespread among snakes.

The developing conceptuses increase in ash content in several snake species (Table 5.5, Fig. 5.1D). Analytical studies have revealed the specific ions that contribute to the increase (Table 5.6). Placental transfer of sodium has been estimated from radioisotopes injected into gravid females, as in *Thamnophis sirtalis* (Hoffman 1970) and the Northern Watersnake, *Nerodia sipedon* (Conaway and Fleming 1960) and from quantitative analysis of yolk and neonates, as in *N. rhombifer* (Stewart and Castillo 1984), *Virginia*

Table 5.6 Placental transfer of radiolabeled molecules and net placental uptake of inorganic ions as indicated by composition analysis for viviparous thamnophines

Species	*Method*	*Molecules*	*Reference*
Nerodia cyclopion	Radioisotopes	Na^+, I^-	Conaway and Fleming 1960
N. sipedon	Radioisotopes	Na^+, I^-	Conaway and Fleming 1960
N. rhombifer	Composition analysis	Na^+, K^+	Stewart and Castillo 1984
Thamnophis sirtalis	Radioisotopes	Na^+, glycine	Hoffman 1970
T. ordinoides	Composition analysis	Na^+, Ca^{2+}, Mg^{2+}	Stewart *et al.* 1990
Virginia striatula	Composition analysis	Na^+, K^+, Ca^{2+}, Mg^{2+}	Stewart 1989

striatula (Stewart 1989) and the Northwestern Gartersnake, *Thamnophis ordinoides* (Stewart *et al.* 1990). Sodium transfer is common and may be a universal attribute of placentation in both snakes and lizards (Thompson *et al.* 2000). Net placental uptake of sodium was correlated with embryonic water uptake in both *V. striatula* and *T. ordinoides* (Stewart 1989; Stewart *et al.* 1990), as would be expected if selective transport of sodium contributed to a favorable osmotic gradient for embryonic water uptake. Placental transport of calcium is also common, but not universal among viviparous snakes (Stewart and Castillo 1984; Stewart 1989; Stewart *et al.* 1990; Fregoso *et al.* 2010). Oviparous squamate embryos typically mobilize calcium from both yolk and eggshell (Packard 1994; Stewart and Ecay 2010). Viviparous snakes also depend on calcium from yolk, but lack a calcareous eggshell; embryos supplement the yolk provision with placental transport of calcium which can contribute 19–53% of neonatal calcium (Stewart 1989; Stewart *et al.* 1990; Fregoso *et al.* 2010).

Viviparous snakes have a complex pattern of embryonic nutrition with both a plesiomorphic (yolk) and a derived (placenta) component. The placenta evolves through association of fetal and maternal tissues and the complexity of the exchange reflects both availability and ease with which various molecules cross tissue barriers. Thus, intrauterine gestation provides a structural relationship that is a minimal barrier to many molecules, i.e., oxygen, carbon dioxide, water, but that is unlikely to transport quantities of other molecules, i.e., organics, in the absence of elaborate specializations. The evolution of viviparity could not occur if extended intrauterine gestation restricted access to substances such as oxygen and water, which are requisite to embryonic development. In contrast to gas and water exchange, placental provision of minerals and organic nutrients may be either facultative or obligative. Some important nutrients, i.e., calcium, may traverse the placenta relatively easily, but may not be readily available in a uterus specialized for oviparous reproduction. Embryonic uptake of these nutrients will be dependent on uterine provisioning in species that lack placental specializations for secretion and transport. Thus, predominantly lecithotrophic viviparous species can supplement yolk nutrients via placental transfer, the magnitude of which will be influenced by maternal physiology. This pattern of embryonic nutrition, in which placental nutrients contribute to, but are not required for, embryonic development is termed facultative placentotrophy and was recognized initially in the snakes *Virginia striatula* and *Thamnophis ordinoides* (Stewart 1989; Stewart *et al.* 1990). In *V. striatula*, placental nutrient provision varies among and between populations and between seasons (Stewart 1989; Sangha *et al.* 1994; Fregoso *et al.* 2010). Facultative placentotrophy may be common among predominantly lecithotrophic squamates generally and may be a definitive pattern of embryonic nutrition for natricid snakes.

Obligatory placentotrophy, defined as placental uptake of nutrients that are required for embryonic development, may also occur in predominantly lecithotrophic species as compensation for deficiencies in yolk and/or eggshell composition. However, this nutritional pattern is most prominent in species with evolutionary reduction in yolk content and morphologically specialized placentae. In contrast to snakes, several lineages of lizards nourish embryos primarily via placental transport and thus exhibit obligatory placentotrophy (Blackburn *et al.* 1984; Thompson *et al.* 1999; Blackburn 2000, 2006; Flemming and Branch 2001; Ramírez-Pinilla 2006; Blackburn and Flemming 2009). However, it would be premature to rule out the possibility that some snakes have evolved extensive placentotrophy, given that fetal nutrition has been studied in few viviparous snakes.

5.5 FETAL MEMBRANE DEVELOPMENT

An understanding of fetal membrane morphogenesis is important because the various types of placentae are defined by their fetal contribution. Snake placentae undergo significant modifications during the months of gestation. Cellular and extracellular components are modified as fetal membranes arise, are transformed, and become replaced. These developmental transformations reflect functional needs of embryos as well as morphogenetic patterns inherited from oviparous ancestors.

Fetal membrane development has been described in eight viviparous snake species representing three families (Table 5.4). The pattern that has emerged is generally consistent, although not identical, with data on various lizards (Stewart and Blackburn 1988; Stewart 1993, 1997) and the one oviparous snake that has been studied (Blackburn *et al.* 2003). One important feature to recognize at the outset is that squamates have a unique pattern of yolk sac development that is unlike that of all other amniotes (Stewart 1997). This pattern is seldom described in textbooks, which incorrectly use the domestic chicken as representative of all Reptilia. The error is unfortunate in that it obscures unusual features of the squamate yolk sac that surely are significant functionally and evolutionarily.

The following general account is based primarily on *Virginia striatula* (Stewart 1990; Stewart and Brasch 2003) and garter snakes of the genus

Thamnophis (Hoffman 1970; Blackburn *et al.* 2002; Blackburn and Lorenz 2003a, b), as supplemented by information on other thamnophines. In addition to the amnion, which does not contribute to any placenta, four distinct fetal membranes are formed. Two of these four membranes persist until the end of development—one is the chorioallantois, and the other is the omphalallantois. The latter complex is derived from the yolk sac but is associated with the allantois.

Early in development, extraembryonic ectoderm, mesoderm, and endoderm spread peripherally from the tiny embryo to cover the dorsal and lateral surface of the yolk mass. These germ layers comprise the trilaminar yolk sac or "choriovitelline membrane" (Table 5.7, Fig. 5.2). This structure is vascularized by its mesodermal component. As the ectoderm and endoderm continue to spread, they eventually enclose the entire vitellus (yolk). However, the expanding mesoderm is diverted into the body of the yolk as a band of "intravitelline mesoderm." As the mesoderm penetrates, it separates a segment of yolk material—the "isolated yolk mass" (or IYM) —from the vitellus proper (Fig. 5.2). The ventral (abembryonic) hemisphere of the isolated yolk mass is thereby bounded externally by an avascular "bilaminar omphalopleure" (bilaminar yolk sac) that consists of ectoderm and endoderm (Table 5.7).

Table 5.7 Major fetal membranes of snakes (exclusive of the amnion). Terminology follows Stewart and Blackburn (1988) and Stewart (1993, 1997). "IYM" = isolated yolk mass.

Fetal membrane	*Components*
Choriovitelline membrane	Ectoderm, mesoderm, endoderm
Chorioallantois	Somatopleure (ectoderm, mesoderm) + allantois
Bilaminar omphalopleure	Ectoderm, endoderm, and IYM or vitellus
Omphalallantois	Bilaminar omphalopleure + allantois

During these early developmental stages, an exocoelom develops in the extraembryonic mesoderm, splitting the choriovitelline membrane. As the exocoelom expands, the choriovitelline membrane accordingly disappears. The allantois penetrates into the exocoelom and contacts the external chorion, forming the "chorioallantois" (Fig. 5.2) in accord with the standard sauropsid pattern. As development proceeds, the chorioallantois spreads to line the entire embryonic hemisphere of the egg (Fig. 5.3). Meanwhile, in the abembryonic hemisphere of the egg, the IYM becomes separated from the vitellus through formation of an exocoelom, the "yolk cleft," lined by the intravitelline mesoderm. The allantois expands into the yolk cleft and thereby comes to line the inside of the IYM (Fig. 5.3). As a result, the ventral pole of the egg is delimited by a tissue complex derived from the avascular bilaminar yolk sac, whose closest blood supply comes from the allantois. Although commonly termed the "omphalallantoic membrane" (or omphalallantois) (Table 5.7) the allantois often is only loosely associated with the yolk sac omphalopleure. Unlike the other two fetal membrane complexes, the omphalallantoic membrane and the chorioallantois persist until the end of gestation.

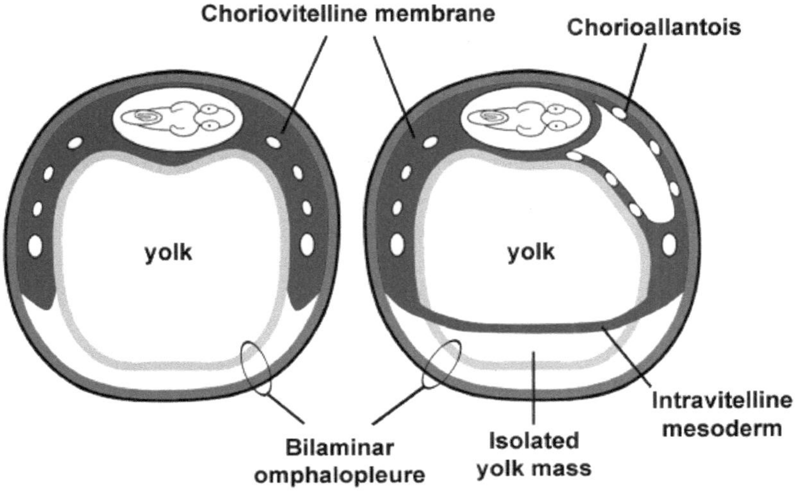

Fig. 5.2 Early development of the fetal membranes in viviparous thamnophine snakes. **Left,** the choriovitelline membrane consists of vascularized mesoderm (red) lying between extraembryonic ectoderm (blue) and endoderm (yellow). A bilaminar omphalopleure (ectoderm plus endoderm) occupies the abembryonic pole. **Right,** the chorioallantois progressively replaces the choriovitelline membrane. Invasion of intravitelline mesoderm into the vitellus cuts off the isolated yolk mass.

Color image of this figure appears in the color plate section at the end of the book.

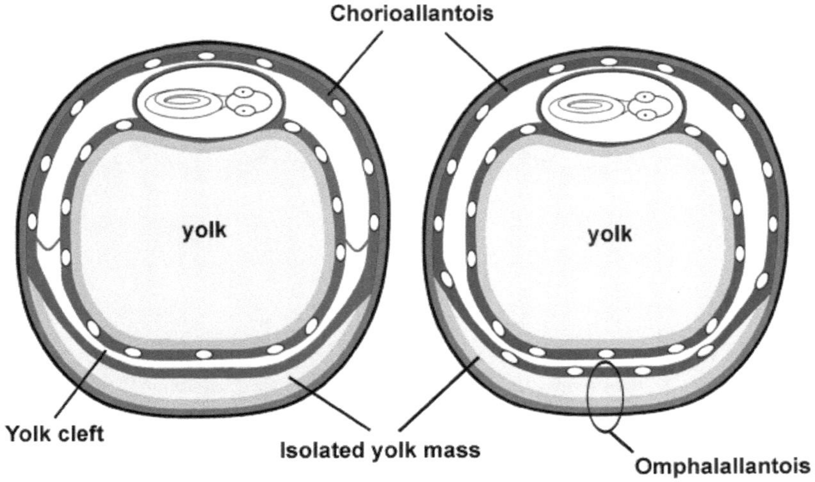

Fig. 5.3 Further development of the fetal membranes in thamnophines. **Left,** chorioallantois expands to line the dorsal hemisphere of the egg. The abembryonic hemisphere is lined by bilaminar omphalopleure and isolated yolk mass. The latter is separated by a yolk cleft from the vitellus (yolk proper). **Right,** expansion of the allantois into the yolk cleft leads to establishment of the omphalallantoic membrane. As in Fig. 5.2, ectoderm is shown in blue, mesoderm in red, and endoderm in yellow.

Color image of this figure appears in the color plate section at the end of the book.

An unusual feature of yolk sac development has been described in *Virginia striatula* (Stewart 1990; Stewart and Brasch 2003). During the period when the chorioallantois is spreading in the dorsal hemisphere, a cavity called the "secondary yolk cleft" (to distinguish it from the main or primary one) splits the IYM into inner and outer portions (Fig. 5.4). When the allantois penetrates the primary yolk cleft, the allantois therefore remains separated from the inside of the IYM by the secondary cleft (Figs. 5.4 and 5.5A). One significant consequence is that throughout the remainder of development, the ventral pole of the egg remains avascular. The closest blood supply to the egg's periphery (the allantoic circulation) is separated by the bilaminar omphalopleure with its IYM as well as by the space of the secondary cleft (Figs. 5.5A,B). Although details of secondary cleft development are available only for *Virginia striatula* (Stewart 1990), evidence suggests that this structure also may form in other thamnophines. A space or cavity separates the allantois from the omphalopleure in species of *Thamnophis* (Hoffman 1970; Blackburn and Lorenz 2003b) and *Regina* (Attaway 2000), a cavity that is lined externally and internally by yolk droplets. These features are understandable if a secondary yolk cleft develops within the original IYM (Blackburn and Lorenz 2003b).

Yolk sac development results in an avascular bilaminar omphalopleure at the abembryonic pole. This omphalopleure consists of ectodermal cells that overlie the isolated yolk mass and yolk endoderm (Figs. 5.5B,C). The allantois either lies on the inner face of the membrane or is separated from the omphalopleure by a (primary or secondary) yolk cleft.

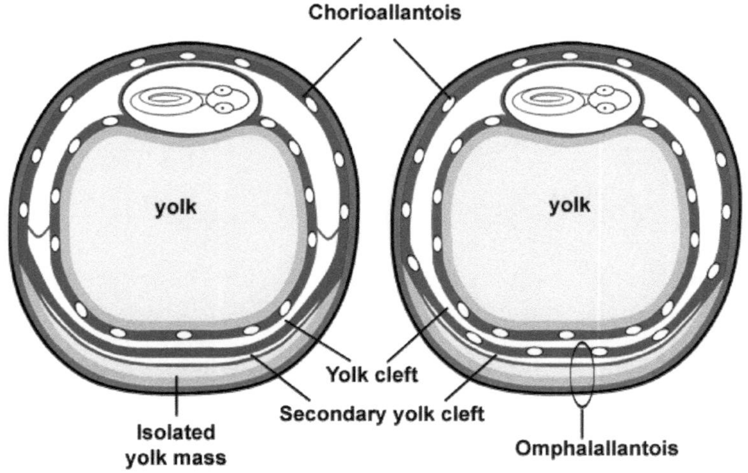

Fig. 5.4 Fetal membrane development in *Virginia striatula*. **Left,** following establishment of the yolk cleft, a secondary yolk cleft splits the isolated yolk mass. **Right,** the omphalallantoic complex forms through expansion of the allantois into the yolk cleft. The allantois does not contact the omphalopleure due to the secondary cleft. Germ layer colors are as shown in Fig. 5.2.

Color image of this figure appears in the color plate section at the end of the book.

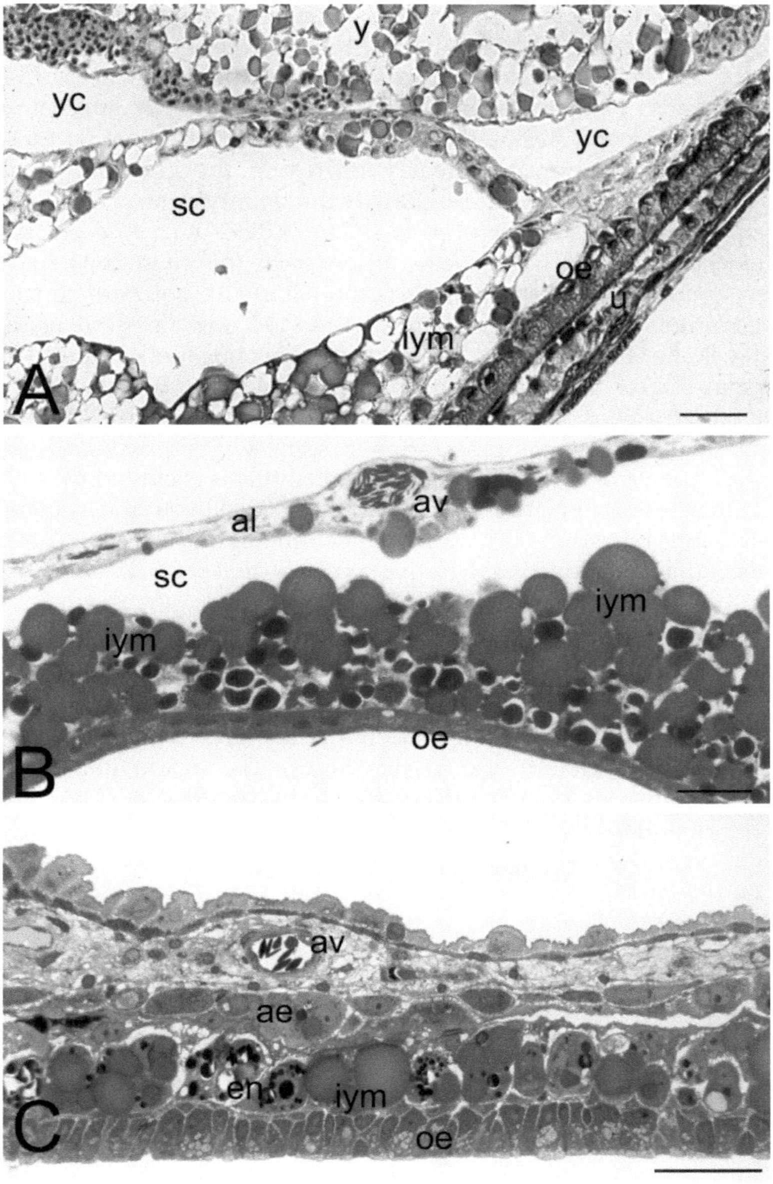

Fig. 5.5 Omphalopleure structure. **A.** *Virginia striatula*, mid- development, showing the secondary yolk cleft (sc). Hematoxylin and eosin. **B.** *Virginia striatula*, late development. Allantois (al) is separated from the isolated yolk mass by remains of the secondary cleft. Toluidine blue. **C.** *Storeria dekayi* at mid development, showing composition of the omphalallantoic complex. Azure II/ methylene blue. ae, allantoic endoderm; av, allantoic blood vessel; en, yolk endoderm; iym, isolated yolk mass; oe, omphalopleure epithelium; u, uterine tissue; y, yolk proper (vitellus); yc, yolk cleft. Scale bars: A-C, 50 μm.

Color image of this figure appears in the color plate section at the end of the book.

How do viviparous thamnophines compare to other snakes? Early light microscopic studies on various hydrophiines (Weekes 1929; Kasturirangan 1951a, b) and Dussumier's Water Snake, the homalopsid *Enhydris dussumieri* (Parameswaran 1962) documented two fetal membranes that persist until the end of development. Although these sources adopt different terminologies, the described membranes clearly represent the chorioallantois and omphalallantoic complex. These same two fetal membranes also develop in an oviparous colubrid, *Pantherophis guttatus* (Blackburn *et al.* 2003; Knight and Blackburn 2008). This species differs from the viviparous snakes late in development, when the yolk sac omphalopleure becomes converted to chorioallantois through regression of the IYM and invasion of allantoic blood vessels. Nevertheless, the overall morphogenetic pattern is very similar to that of the viviparous snakes. In addition, lizards (like snakes) develop both a chorioallantois and a bilaminar yolk sac. However, whereas an omphalallantoic membrane forms in some viviparous lizards (Weekes 1927; Villagran *et al.* 2005), in others, the allantois is excluded from the yolk cleft (Boyd 1942; Stewart 1985; Yaron 1985; Blackburn and Callard 1997; Stewart and Thompson 2000), and no such membrane forms. Whether this difference has any functional consequences is not known.

5.6 PLACENTAL MORPHOLOGY

Each of the four fetal membranes of snakes (excluding the amnion) contributes to a corresponding placenta (Table 5.8), through juxtaposition of the fetal membrane to the uterine lining. The choriovitelline placenta and omphaloplacenta are transitory structures, whereas the chorioallantoic and omphalallantoic placentae persist until the end of development. In Figure 5.6, the topographic positions and extent of the chorioallantoic and omphalallantoic placentae at placental maturity are illustrated for *Virginia striatula*. The general pattern shown applies to other viviparous snakes as well.

Table 5.8 Categories of placentae in viviparous snakes. Terminology follows Stewart and Blackburn (1988), Stewart (1993 1997), and Table 5.7. Two of the placentae persist until birth (asterisks); the other two are transitory.

Placenta	*Components*
Choriovitelline	Choriovitelline membrane + uterine lining
*Chorioallantoic	Chorioallantois + uterine lining
Omphaloplacenta	Bilaminar omphalopleure + uterine lining
*Omphalallantoic	Omphalallantois + uterine lining

In placentae of all four categories, the placental interface is formed by apposition of a fetal membrane to the oviductal (uterine) lining, with only a thin remnant of the shell membrane interposed between them. As revealed by transmission electron microscopy (TEM), the shell membrane of *Thamnophis* (Hoffman 1970; Blackburn and Lorenz 2003a) and *Virginia*

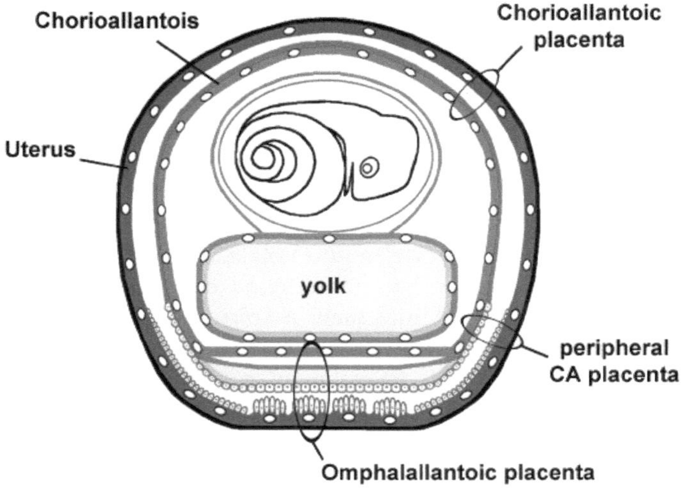

Fig. 5.6 Topography of the placental membranes in *Virginia striatula*. The chorioallantoic placenta surrounds most of the conceptus at placental maturity, and omphalallantoic placenta occupies the ventral pole of the egg. Each placenta is formed through apposition of the corresponding fetal membrane to the uterine lining. A specialized "peripheral region" of chorioallantoic placenta occurs in *V. striatula*, but has not been described in other thamnophines. As in Figs. 5.2 through 5.4, ectoderm is shown in blue, mesoderm in red, and endoderm in yellow.

Color image of this figure appears in the color plate section at the end of the book.

(Stewart and Brasch 2003) consists of two layers, an electron-dense layer adjacent to the embryonic epithelium and a thicker, heterogeneous layer facing the uterine epithelium. The composition of the membrane changes during gestation (Blackburn and Lorenz 2003a) and differs structurally in different placental regions (Stewart and Brasch 2003). Material that accumulates on the uterine face of the shell membrane may represent uterine secretions (Stewart and Brasch 2003). The shell membrane can deteriorate in late gestation, allowing sites of direct contact between the chorioallantois and uterine epithelium (Blackburn and Lorenz 2003a).

Based on studies of thamnophines, each of the four placental types is described and illustrated below. Detailed anatomical descriptions of the placental membranes are available for thamnophine snakes of the genera *Thamnophis* (Hoffman 1970; Blackburn *et al.* 2002; Blackburn and Lorenz 2003a, b), *Virginia* (Stewart 1990; Stewart and Brasch 2003; Anderson and Blackburn 2009), *Storeria* (Blackburn *et al.* 2009), *Nerodia* (Blackburn 1998a; Johnson 2003), *Regina* (Attaway 2000), and *Tropidoclonion* (Baxter 1987; Jones and Baxter 1991).

5.6.1 Choriovitelline Placenta

The choriovitelline placenta consists of the trilaminar yolk sac in apposition to the uterine lining. Given the ephemeral existence of its fetal component (Fig. 5.2; Stewart 1993, 1997), it is not surprising that a choriovitelline placenta has rarely been observed in squamates (Stewart and Thompson 2000). Nevertheless, this placenta has been described in *Virginia* (Stewart 1990; Stewart and Brasch 2003) and *Tropidoclonion* (Baxter 1987), as well as in various lizards (Blackburn and Callard 1997; Stewart and Thompson 1996, 1998, 2000). A choriovitelline placenta presumably forms as a transitory structure in all viviparous squamates (Stewart and Blackburn 1988; Stewart 1993).

The choriovitelline membrane consists of a thin epithelium (ectoderm) overlying a vascularized mesoderm and yolk sac endoderm (Fig. 5.7A). The apposed maternal tissue consists of flattened epithelial cells over uterine capillaries. At the placental interface, a thin shell membrane separates the chorionic and uterine epithelium. The most significant features of the choriovitelline placenta are its vascularity and the close proximity of fetal and maternal capillaries. As the first vascularized placenta to form, the choriovitelline placenta offers a means of maternal-fetal gas exchange early in development, before formation of the chorioallantois (Stewart 1993).

5.6.2 Chorioallantoic Placenta

The chorioallantoic placenta ("allantoplacenta") of snakes consists of the chorioallantois in apposition to the uterine lining. As the only vascularized placenta to persist throughout most of development in viviparous squamates, the chorioallantoic placenta is inferred to function in maternal-fetal gas exchange (Blackburn 1993, 1998a; Stewart and Brasch 2003). In addition, details of the structure of the chorioallantoic placenta suggest a role in nutrient transfer. The close apposition of fetal and maternal blood vessels would allow for hemotrophic nutrient exchange, and cytological features revealed by TEM are indicative of histotrophic transfer (Stewart and Brasch 2003).

As seen with light microscopy, both the chorioallantois and uterus are well vascularized (Fig. 5.7B-D). The uterine and chorionic epithelia form thin layers over the corresponding capillaries, offering a slight barrier to interhemal exchange. The uterine epithelium is so thin that early reports on some thamnophines (Rahn 1939; Conaway and Flemming 1960) inferred that it is eroded at the placental interface. Likewise, an early study on *Thamnophis sirtalis* suggested erosion of the chorionic epithelium (Clark *et al.* 1955). However, ultrastructural examination of various thamnophines confirms that both fetal and maternal epithelia remain intact. Like epithelia at the placental interface, the shell membrane is thin and difficult to see with light microscopy.

Under scanning electron microscopy (SEM), the uterine epithelium is evident as broad, flattened cells with angular borders (Fig 5.8A), interspersed with occasional ciliated cells. The epithelium forms an unbroken lining with no signs of erosion, and overlies a dense network of uterine capillaries (Fig. 5.8B). The external surface of the chorion is also lined by an attenuated epithelium that shows no gaps or eroded regions

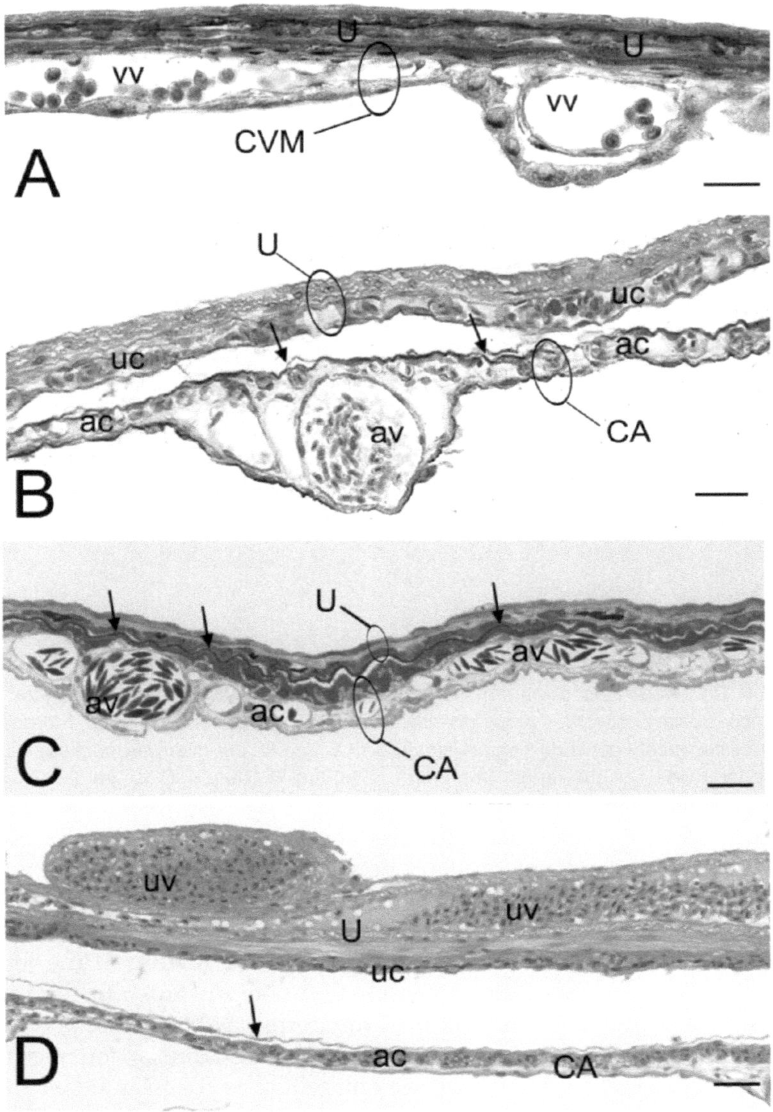

Fig. 5.7 Placental histology in thamnophines. **A.** Choriovitelline placenta, *Virginia striatula,* consisting of choriovitelline membrane (CVM) apposed to the uterus (U); hematoxylin and eosin. **B.** Chorioallantoic placenta, *Thamnophis sirtalis,* consisting of chorioallantois (CA) apposed to the uterus; stained with neutral red and fast green to reveal allantoic capillaries (ac) and uterine capillaries (uc). **C.** Chorioallantoic placenta, *Storeria dekayi.* Azure II/ methylene blue. **D.** Chorioallantoic placenta, *Nerodia sipedon.* Hematoxylin and eosin. In B and D, the chorioallantois and uterus are separated by an artifactual space. av, allantoic vessel; vv, vitelline vessels; arrows, shell membrane at the placental interface. Scale bars: A and C, 25 μm; B, 20 μm; D, 50 μm.

Color image of this figure appears in the color plate section at the end of the book.

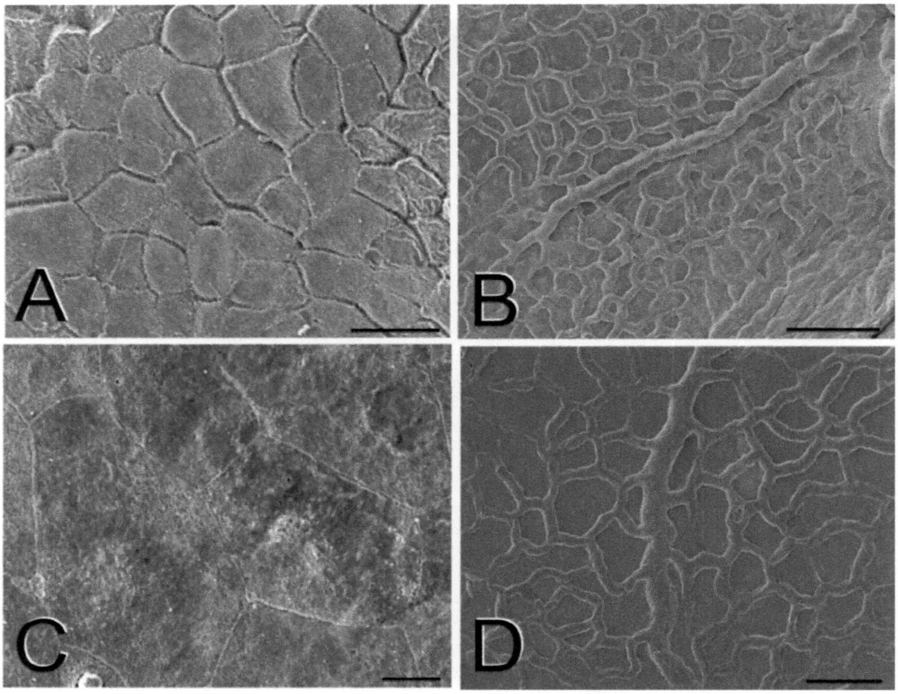

Fig. 5.8 Chorioallantoic placental membranes, SEM. **A.** Uterine lining, *Storeria dekayi*. **B.** Uterine capillary network, *Virginia striatula*. **C.** External surface of the chorion, *Storeria dekayi*. **D.** Allantoic capillary network, *Virginia striatula*. In B and D, the overlying epithelia have been removed to reveal the capillaries. Scale bars: A, 20 μm; B 100 μm, C 10 μm; D, 50 μm.

under SEM (Fig. 5.8C). Beneath the chorionic epithelium, the allantoic vasculature forms a dense capillary network similar to that on the maternal side of the placenta (Fig. 5.8D).

Examination of *Thamnophis* with TEM reveals cytological details of the placental interface (Fig. 5.9). The most prominent feature of the fetal and maternal tissues is small blood vessels that take up much of the width of each membrane. The uterine epithelium forms a thin monolayer of cells that is reduced over the uterine capillaries, through displacement of the cell nuclei and organelles. The chorionic epithelium is represented as a bilayer of flattened cells, and over the chorionic capillaries is similarly attenuated. Both of these epithelial layers progressively thin over the course of development. At the placental interface, the shell membrane forms a thin, irregular barrier and undergoes degeneration in local areas, allowing direct contact of fetal and maternal tissues (Fig. 5.9). Due to epithelial thinning and degeneration of the shell membrane, the diffusion distance between fetal and maternal blood streams in late gestation can be as thin as 2 μm. Although the chorioallantoic placenta appears specialized for gas exchange, ultrastructure of the placental interface suggests an additional role in nutrient transfer (Blackburn and Lorenz 2003a). The uterine and chorionic

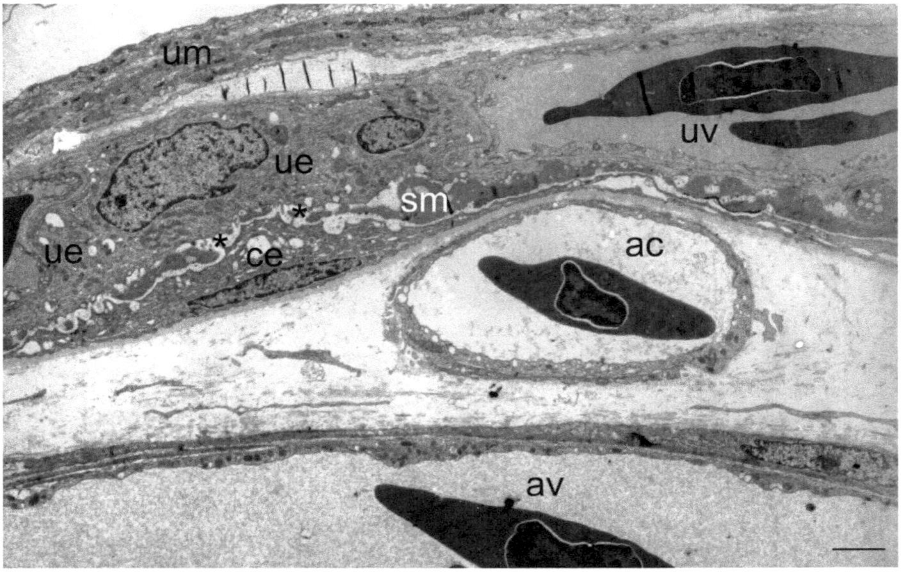

Fig. 5.9 Chorioallantoic placenta in *Thamnophis sirtalis*, TEM. Apposition of uterine epithelium (ue) to chorionic epithelium (ce) forms the placental interface. The thin shell membrane (sm) is undergoing degeneration, allowing contact between maternal and fetal membranes (asterisks). Note the thin diffusion distance between uterine vessels (uv) and allantoic capillaries (ac). av, allantoic vessel; um, uterine muscle. Scale bar: 1 µm.

epithelial cells contain apical vesicles that may indicate histotrophic nutrient transfer. The epithelial cells also are likely to contribute to breakdown of the shell membrane (Blackburn and Lorenz 2003a).

Ultrastructural appearance of the chorioallantoic placenta in *Nerodia sipedon* (Johnson 2003) and *Storeria dekayi* (Blackburn *et al.* 2009) is largely consistent with the above description, as is that of *Virginia striatula* (Fig. 5.10A,B). In *V. striatula*, apical vesicles in squamous uterine epithelial cells adjacent to blood vessels as well as irregularities in the cell membrane and apical vesicles in chorionic epithelial cells indicate that the chorioallantoic placenta functions in nutrient transfer. Thus, as in *Thamnophis*, the presence of attenuated epithelial cells does not preclude histotrophic transfer and this tissue may have multiple functions, i.e., respiratory and nutritive. In addition, the chorioallantoic placenta of *V. striatula* is regionally diversified, unlike descriptions of other thamnophines (Stewart 1990; Stewart and Brasch 2003). Whereas, the structure of the placenta over most of the embryonic hemisphere of the egg is indistinguishable from that of other thamnophines, a peripheral zone of chorioallantoic placentation adjacent to the omphalallantoic placenta differs (Fig. 5.10C,D). Chorionic and uterine epithelial cells of this region are cuboidal and exhibit cytological features that are absent in the remainder of the chorioallantoic placenta (Stewart and Brasch 2003). Uterine and fetal epithelia have characteristics of histotrophic transporting epithelia, but with structural specializations that differ from

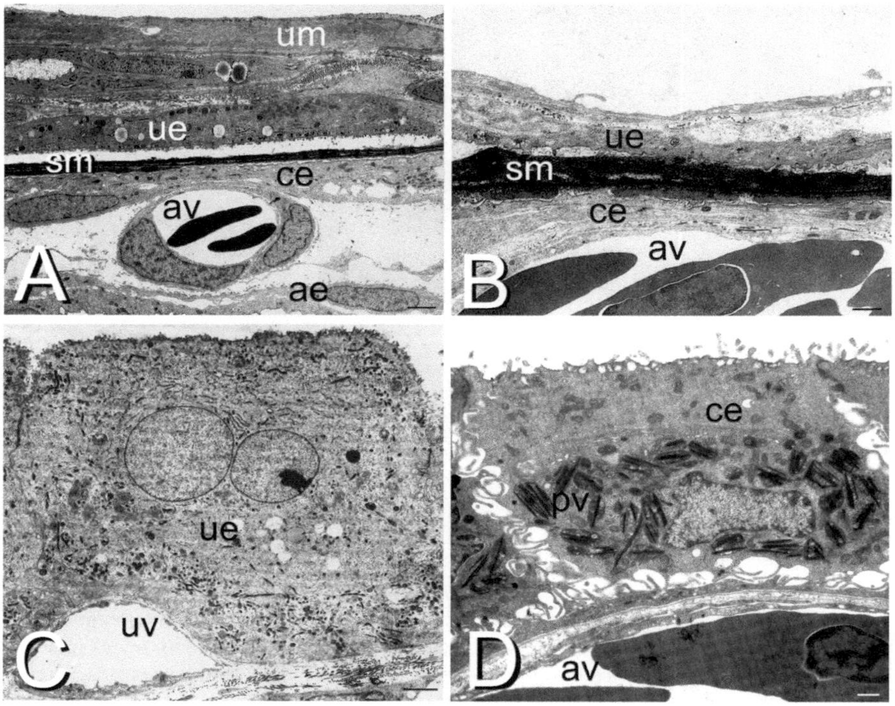

Fig. 5.10 Chorioallantoic placenta in *Virginia striatula*, TEM. **A** and **B.** Generalized region. The chorionic epithelium (ce) is greatly attenuated over the allantoic blood vessels (av); the uterine epithelium (ue) is similarly flattened. **C** and **D.** Specialized (peripheral) region. C shows an enlarged, binucleated cell of the uterine epithelium overlying a uterine blood vessel (uv). D shows the specialized chorionic epithelium superficial to an allantoic vessel. ae, allantoic endoderm; pv, paracrystalline vesicles; sm, shell membrane; um, uterine muscle. Scale bars: A and C, 3 μm; B, 1 μm; D, 0.5 μm.

the squamous epithelial area of chorioallantoic placentation. The two regions are likely specialized for transport of different nutritive molecules. Regional specialization of the chorioallantoic placenta, sometimes termed complex chorioallantoic placentation, occurs in several lineages of lizards, but has not been described in other snakes.

5.6.3 Omphaloplacenta and Omphalallantoic Placenta

Early development in squamates yields an avascular bilaminar yolk sac lined by ectoderm, endoderm, and a substantial isolated yolk mass (IYM) (Figs. 5.3, 5.11). Apposition of the bilaminar omphalopleure to the uterus (and intervening shell membrane) forms the omphaloplacenta (*sensu stricto*). Upon invasion of allantois into the yolk cleft (Fig. 5.3), the complex is termed an omphalallantoic placenta. The two successive stages in fetal membrane development do not appear to be reflected in modifications of cells at the maternal-fetal interface. Therefore, these two placentae shall be discussed together for the sake of simplicity.

Epithelium lining the fetal side of the placenta is elaborately specialized. Surface cells of the omphalopleure are large, cuboidal elements, very unlike the flattened cells of the chorion (Fig. 5.11).

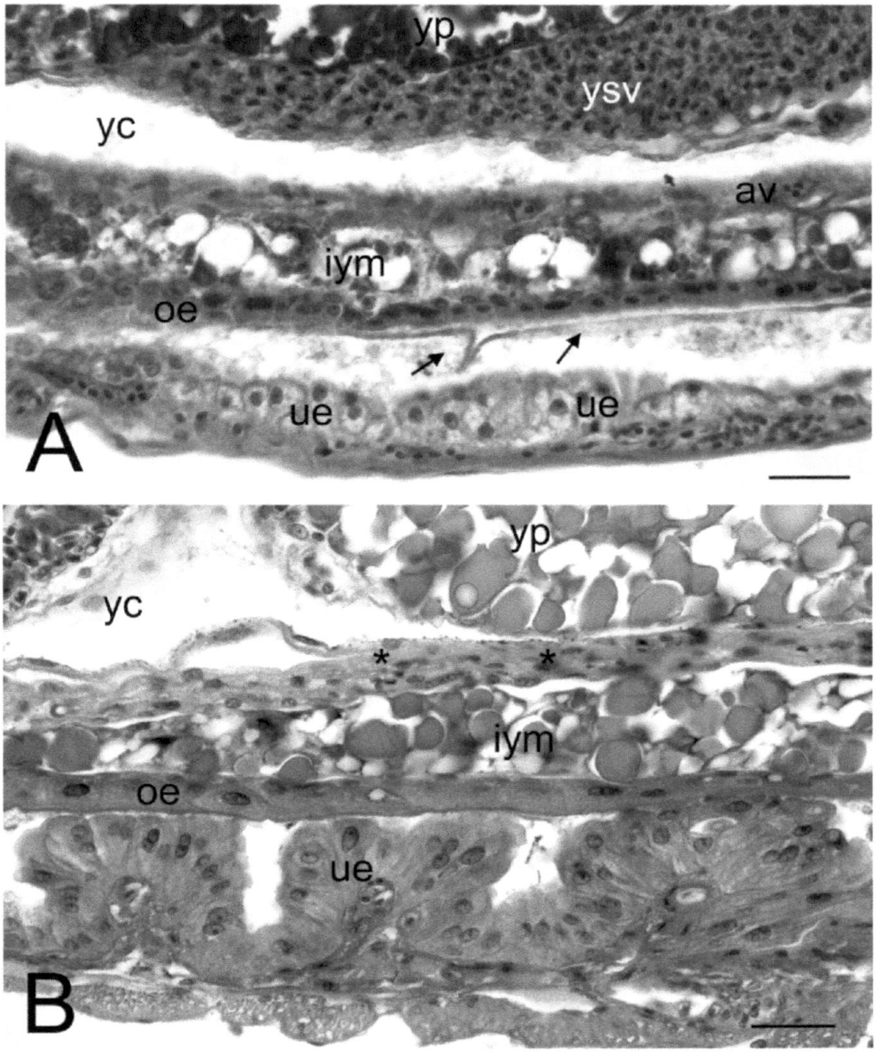

Fig. 5.11 Histology of the omphalallantoic placenta. **A.** *Storeria dekayi.* **B.** *Virginia striatula.* In both, the fetal component consists of enlarged cells of the omphalopleure epithelium (oe), plus isolated yolk mass (iym) and its associated endoderm. The membrane's inner face is lined by allantois (in A) or allantois plus tissue lining the secondary yolk cleft (in B, asterisk). In B, the uterine epithelium is arrayed in pillars extending from the lamina propria. av, allantoic vessels; ue, uterine epithelium; yc, primary yolk cleft; yp, yolk proper (vitellus); ysv, yolk sac vasculature; arrows, shell membrane. Scale bars: A and B, 25 μm.

Color image of this figure appears in the color plate section at the end of the book.

Examination of species of *Thamnophis* and *Storeria* via SEM and TEM has revealed two distinct cell populations (Figs. 5.12A,B) (Blackburn *et al.* 2002; Blackburn and Lorenz 2003a; Blackburn *et al.* 2009). One consists of "brush-border cells" with prominent, irregular microvilli that extend towards the shell membrane and uterus. These cells are rich in mitochondria and Golgi bodies. The other population is formed of "granular cells" which are characterized by very large cytoplasmic droplets that are lipid in nature (Hoffman 1970). These cells can bear short, irregular microvilli (Blackburn and Lorenz 2003b). In *T. sirtalis*, apical cytoplasm of these cells contains small inclusions of sulphated acid mucosubstances (Hoffman 1970). Lateral surfaces of the epithelial cells are sculpted into an extensive network of channels, formed through membranous extensions that interdigitate with

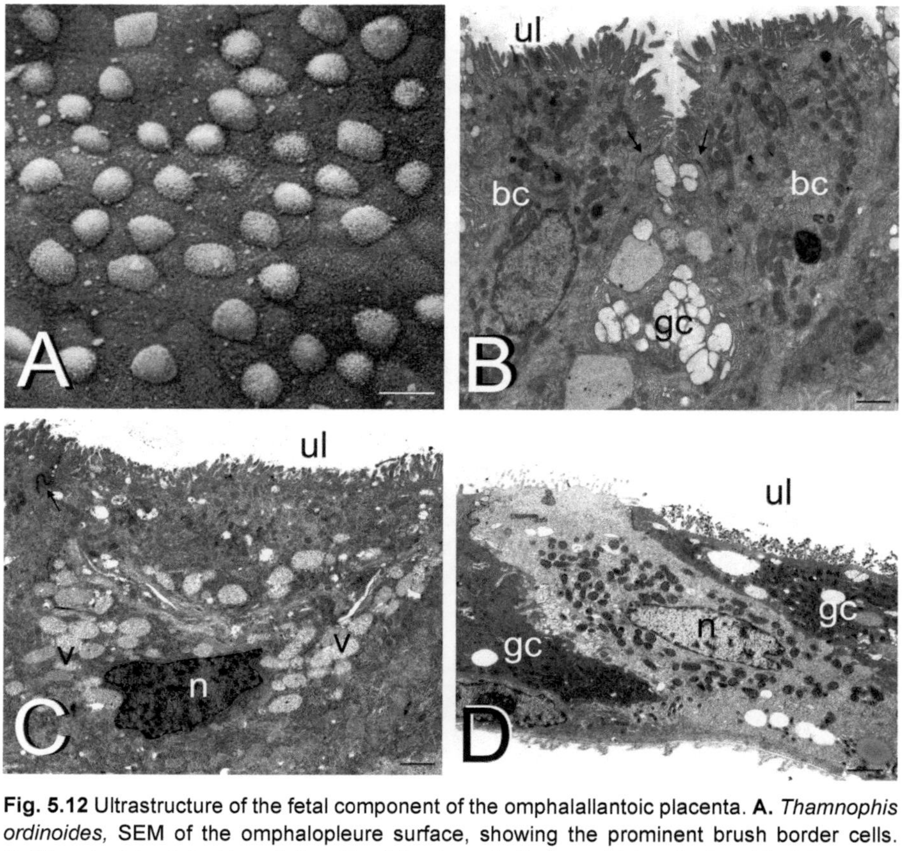

Fig. 5.12 Ultrastructure of the fetal component of the omphalallantoic placenta. **A.** *Thamnophis ordinoides,* SEM of the omphalopleure surface, showing the prominent brush border cells. **B.** *Storeria dekayi* at mid-gestation. Brush border cells (bc) are rich in mitochondria. Intervening granular cells (gc) contain cytoplasmic vesicles, Golgi bodies, and endoplasmic reticulum. **C.** *Virginia striatula,* showing a microvilliated granular cell lining the omphalopleure. **D.** *Virginia striatula,* showing a mitochondria-rich cell, bordered by two granular cells. Microvilli are apparent on both cell types. n, cell nucleus; ul, uterine lumen; v, cytoplasmic vesicles; arrows, apical tight junctions. Scale bars: A,10 µm; B, C, and D, 1.5 µm.

those of the adjacent cells (Blackburn *et al.* 2002; Blackburn *et al.* 2009). Because the cells are linked apically by tight junctions, they are surrounded laterally by an extracellular compartment that lacks continuity with the uterine lumen. The omphalopleure is similar to the above description in *Virginia striatula* (Figs. 5.12C,D) and other thamnophines that have been studied with EM (Baxter 1987; Attaway 2000; Johnson 2003; Stewart and Brasch 2003). Some details may vary between species, but the differences appear to be relatively minor.

The uterine component of these yolk sac placentae is also distinctive. The uterine epithelium consists of tall cells that can exhibit large and abundant cytoplasmic granules or vacuoles, as well as small apical vesicles, mitochondria, ribosomal ER, and Golgi bodies (Fig. 5.13A-C); the cells may also be binucleate. Study of *Thamnophis sirtalis* has revealed that the apical vesicles contain a mucoid secretory product and the large vacuoles contain material like that in the uterine lumen (Hoffman 1970). Their morphology thus indicates a secretory function. Microscopic anatomy of the uterine epithelium is generally similar among thamnophines, including species of *Storeria, Thamnophis, Nerodia, Tropidoclonion, Regina,* and *Virginia* (Hoffman 1970; Baxter 1987; Stewart 1992; Attaway 2000; Blackburn and Lorenz 2003b; Johnson 2003; Stewart and Brasch 2003; Blackburn *et al.* 2009). However, in *Virginia striatula*, the epithelium forms protruding ridges lined by tall columnar epithelial cells (Fig. 5.11B). These cells are binucleate, and their basal plasmalemma is extensively infolded in association with the underlying capillary endothelium (Fig. 5.13D) (Stewart 1992; Stewart and Brasch 2003). These features commonly are found in epithelia specialized for transport. Features of the uterine epithelia likely represent a specialization of *V. striatula*. Alternatively, since this species has been studied in particular detail, the possibility remains that these specializations exist in other thamnophines but have not yet been observed.

5.6.4 Placentation in Other Viviparous Snakes

Placentation has been described in four hydrophiines and one homalopsid (Table 5.4; Weekes 1929; Kasturirangan 1951a, b; Parameswaran 1962). Because these studies are based on light microscopy and lack relevant detail, close comparisons with thamnophines are not feasible. However, placental morphology in these species appears to be consistent with what is known of thamnophines. In each case, morphologically distinct placentae develop from the chorioallantois and omphalallantoic membrane. A shell membrane persists at the placental interface. The chorioallantoic placenta is well-vascularized, and epithelium of both the chorion and uterus are attenuated at the placental interface. In contrast, the epithelia of the omphalallantoic placenta are hypertrophied, and show evidence of maternal secretion and fetal absorption. The similarities with thamnophines are remarkable, given that hydrophiines and homalopsids are derived from separate origins of viviparity (Blackburn 1999; Stewart and Thompson 2000).

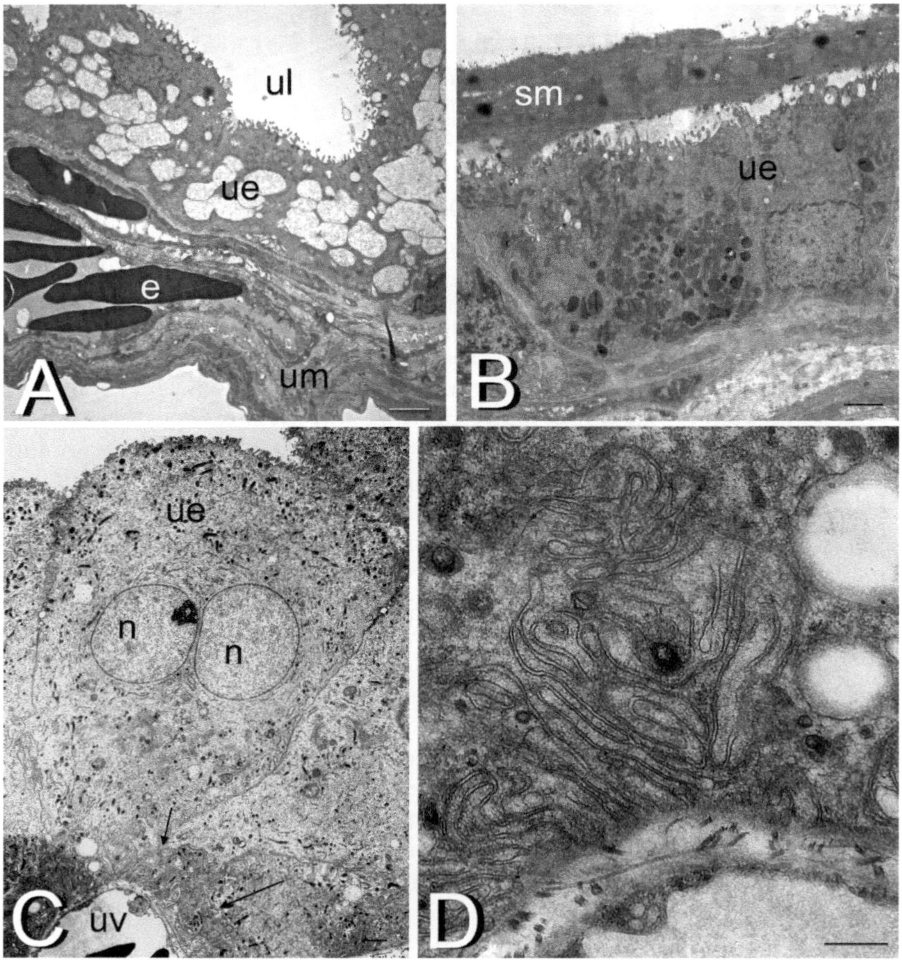

Fig. 5.13 Uterine component of the omphaloplacenta. **A** and **B.** *Storeria dekayi.* In A, the uterine lumen (ul) is lined a cuboidal uterine epithelium (ue); the cells contain large cytoplasmic droplets, evident here as vacuoles. In B, the uterine epithelial cells lack granules but have abundant mitochondria, ribosomal ER, and Golgi bodies. **C.** *Virginia striatula.* Uterine epithelial cells are enlarged and binucleated, and laden with mitochondria and electron dense granules. **D.** *Virginia striatula,* showing a region of C at higher magnification. The basal membrane is highly infolded adjacent to the blood vessel. e, erythrocyte; sm, shell membrane; n, cell nucleus; um, uterine muscle; uv, uterine vessel. Scale bars: A and C, 2.5 μm; B, 1.7 μm; D, 0.5 μm.

5.6.5 Overview of Placental Structure and Function

Structural characteristics of the placentae reflect potentialities that permit integration with information on placental function. Snake placentae vary primarily with regard to three features: (1) character of the fetal and maternal epithelia at the placental interface; (2) presence or absence of a direct fetal blood supply; and (3) presence or absence of a substantial barrier to exchange between maternal and fetal blood streams. They also differ in terms of timing of development and persistence during gestation.

The chorioallantoic placenta of snakes clearly performs the main role in maternal-fetal gas exchange, as suggested for viviparous squamates in general (Weekes 1935; Yaron 1985; Blackburn 1993). This placenta persists from relatively early in development onward, during which time fetal needs for oxygen and removal of carbon dioxide grow to a maximum. The fetal and maternal components of the placenta are well-vascularized and lined by very thin epithelial cells that provide the thinnest of barriers to gas diffusion. In fact, the distance between fetal and maternal blood streams can be reduced to ~2 μm, less than the width of a single erythrocyte. As the only well-vascularized placenta to persist throughout most of development in viviparous snakes, the chorioallantoic placenta is ideally suited for its respiratory function. Characteristics that enhance gas exchange also favor exchange of other molecules. Of particular note are structures in the apical cytoplasm of the thin chorionic epithelial cells that are common features of cells engaged in nutrient transport. Histotrophic delivery of at least some nutrients is compatible with placental respiratory exchange. Additional evidence for a role in nutrient provision for the chorioallantoic placenta is seen in localized epithelia with specializations for nutrient transport and uptake in conjunction with cellular hypertrophy that increases the distance between fetal and maternal vascular systems. Thus, the chorioallantoic placenta in general serves multiple functions, but this placenta can also develop regional specializations for particular functions.

The choriovitelline placenta also shows structural evidence of a respiratory function, given its high vascularity and the thin epithelia at the maternal-fetal interface. Given its structure and the timing of its development, the choriovitelline placenta would be able to meet embryonic respiratory needs very early in development, before such functions are assumed by the chorioallantoic placenta.

In thamnophines, the omphaloplacenta and omphalallantoic placentae are structurally distinct from the other two placental types, and show strong evidence of maternal secretion and fetal absorption. Cells of the uterine epithelium are enlarged and show cytoplasmic machinery indicative of synthetic functions (e.g., ribosomal ER, Golgi bodies), as well as cytoplasmic granules of organic material. Ultrastructural analysis suggests that these cells secrete material into the uterine lumen for uptake by the fetal omphalopleure. Surface cells of the fetal omphalopleure are also enlarged and many bear microvilli. Many of the cells contain large droplets or granules of material, presumably resulting from nutrient uptake. Composition of the transferred nutrients has not been established, nor has organic nutrient provision been quantified. Given that thamnophines are relatively lecithotrophic, placental transfer of organic nutrients is small compared to nutrient provision via the yolk. However, quantity is only one measure of nutrient provision; placental sources might account for provision of specific nutrients that importantly enhance offspring quality. Likewise, morphological adaptations for nutrient provision may allow pregnant females to supplement yolk nutrients facultatively. Such facultative provision has been established for species in three thamnophine genera (Stewart and Castillo 1984; Stewart 1989; Stewart *et al.* 1990; Sangha *et al.* 1996). Ultrastructural examination has revealed probable sites of maternal synthesis and secretion of nutrients as well as their absorption by the omphalopleure.

Another striking feature of the omphaloplacenta and omphalallantoic placenta, one revealed by ultrastructural analysis, suggests a function characteristic of transporting epithelia from a variety of other vertebrate tissues. Epithelial cells of the omphalopleure are separated by elaborate extracellular channels that remain isolated from the uterine lumen by apical tight junctions (Blackburn *et al.* 2002; Blackburn and Lorenz 2003a). Equivalent features have been observed in the chorioallantoic placenta (Stewart and Brasch 2003). This tissue organization is similar to that of epithelia (such as that of the intestine, gall bladder, and kidney) that engage in paracellular transport as one mechanism to move fluid and solutes (Anderson and Van Itallie 1995; Ballard *et al.* 1995). On this basis, we have postulated that one of the functions of the epithelium is sodium-coupled water movement via paracellular channels, as a means of water provision to the embryo (Blackburn *et al.* 2002). This mechanism would contribute to the statistical correlation between placental sodium transfer and water provision that has been demonstrated in *Thamnophis ordinoides* and *Virginia striatula* (Stewart 1989; Stewart *et al.* 1990).

The omphaloplacenta and omphalallantoic placenta are far better suited for histototrophic transfer and water provision than for interhemal exchange. The omphaloplacenta lacks a direct fetal blood supply, because the closest blood vessels lie across the yolk cleft in the vitellus. Similarly, in the omphalallantoic placenta, the fetal (allantoic) blood supply is separated from the omphalopleure by the yolk cleft. Thus, throughout gestation, a substantial barrier lies between maternal and fetal blood streams, formed by the following: enlarged cells of the uterine epithelium, shell membrane, two layers of enlarged omphalopleure cells, IYM and yolk endoderm, plus one or more yolk clefts. In *Thamnophis,* the diffusion distance between fetal and maternal blood streams across the omphalallantoic placenta is on the order of 250-300 μm, over 100 times the interhemal distance across the chorioallantoic placenta (Blackburn and Lorenz 2003b). This large distance precludes efficient interhemal transfer of gases or nutrients, and offers further evidence that these placentae are specialized for other functions, notably histotrophic nutrient transfer.

5.7 PLACENTAL EVOLUTION

5.7.1 The Oviparous Substrate

A major key to understanding placental evolution lies with attributes of the oviduct and fetal membranes in oviparous species. Snake eggs typically are laid at a post-pharyngula stage about 30% of the way through development (Shine 1983a; Blackburn 1995). Before oviposition, the squamate oviduct provides eggs with water (Giersberg 1922; Cordero-López and Morales 1995) and oxygen (Andrews and Mathies 2000; Parker and Andrews 2006). It also deposits fibers of the eggshell (Giersberg 1922; Jacobi 1946; Guillette *et al.* 1989; Heulin *et al.* 2005) as well as eggshell calcium (see Stewart and Ecay 2010 for review).

In preparation for gravidity, the oviparous uterus can undergo several changes (Perkins and Palmer 1996), including an increase in epithelial height, development of the shell glands, and increases in thickness of the uterine connective tissue and musculature (Blackburn 1998a). Epithelium of the gravid oviduct commonly is thinned, perhaps due to stretching of the tissues by the egg. Likewise, in some oviparous lizards, the uterus reportedly increases in vascularity during gravidity (Guillette and Jones 1985; Masson and Guillette 1987; Picariello *et al.* 1989). These features may enhance gas exchange with the young embryo. However, the eggshell would greatly limit any gas exchange between fetal and maternal tissues.

Fetal membranes of oviparous snakes primarily contribute to maintenance of the embryo after oviposition. As the first vascularized fetal membrane to develop, the choriovitelline membrane may contribute to gas exchange in the uterine environment when oxygen requirements are still low. However, the chorioallantois assumes such respiratory functions in the extra-uterine environment (Stewart 1997; Andrews and Mathies 2000). In the Red Cornsnake (*Pantherophis guttatus*), the chorioallantois is beginning to be established at the time of laying, and this membrane expands rapidly to fill the dorsal hemisphere of the egg (Blackburn *et al.* 2003). The timing of its development in this species is consistent with data on oviparous lizards (Stewart and Thompson 1996; Stewart and Florian 2000; Stewart *et al.* 2004a). Thus, by oviposition, the squamate chorioallantois is positioned to begin taking on a major role in gas exchange, a role maintained until the final stages of gestation. This membrane also functions in transport of calcium from the eggshell (Ecay *et al.* 2004; Stewart *et al.* 2004b).

The chorioallantois of oviparous squamates is morphologically specialized for its respiratory functions. In *Pantherophis guttatus,* it is well vascularized and its epithelium thins during development, minimizing the diffusion distance for respiratory gases (Blackburn *et al.* 2003; Knight and Blackburn 2008). Structure of the chorioallantoic membrane in this snake is similar to descriptions of oviparous lizards (Stewart 1985; Stewart and Thompson 1996; Stewart and Florian 2000; Stewart *et al.* 2004a). As for its role in calcium uptake, the chorioallantois mobilizes calcium from the eggshell primarily during a phase of substantial embryonic growth in late gestation (Ecay *et al.* 2004; Stewart *et al.* 2004b).

Functions of the yolk sac omphalopleure in oviparous forms have not been resolved. However, by lining the abembryonic eggshell, the omphalopleure is in a position to take up water from the substrate and perhaps minerals from the eggshell. Epithelial cells of the squamate omphalopleure are enlarged relative to those of the chorioallantois (Stewart and Thompson 1996; Stewart and Florian 2000; Blackburn *et al.* 2003; Stewart *et al.* 2004a). Evidence for their function comes from ultrastructural study of the Red Cornsnake (*Pantherophis guttatus*). In this species, the omphalopleure cells develop surface ridges that increase the surface area (Knight and Blackburn 2008). These ridges disappear as development proceeds, and the membrane becomes converted to chorioallantois. As a result, the entire eggshell is lined by vascularized fetal membrane (Blackburn *et al.* 2003), a transition that probably reflects increased needs for gas exchange. How widespread these features are among oviparous snakes remains to be determined.

5.7.2 Viviparous Species

Viviparity brings fetal membranes into a close structural-functional association with the uterine lining, an association in which both components perform functions like those of their oviparous homologues. Under viviparous conditions, gas exchange functions of the chorioallantois are now carried out in the uterine environment through several features that enhance interhemal exchange. Uterine shell gland activity is decreased (Blackburn 1998a) such that only a thin vestige of the eggshell is deposited. Even this thin barrier is reduced as gestation proceeds (Blackburn and Lorenz 2003a). Fetal and uterine epithelia become attenuated to thin remnants, bringing allantoic and uterine blood vessels into close proximity. Oxygen transfer may be further enhanced by increased oviductal vascularity (Hoffman 1970; Gerrard 1974; Blackburn 1998a; Blackburn and Lorenz 2003a) as well as changes in blood oxygen affinity (Ingermann 1992; Ragsdale and Ingermann 1993; Ragsdale *et al.* 1993).

Putative absorptive functions of the omphalopleure also are extended under viviparous conditions. In thamnophines, the omphalopleuric epithelium develops striking specializations for absorption, potentially for water and minerals (Blackburn *et al.* 2002) as well as organic molecules (Hoffman 1970; Stewart 1992; Blackburn and Lorenz 2003b; Stewart and Brasch 2003). Likewise, the apposed uterine epithelium shows strong evidence of secretory activity. Furthermore, the omphalopleure is retained as a functional component throughout development, instead of being converted to chorioallantois (as in the oviparous *Pantherophis guttatus* (Blackburn *et al.* 2003). This feature ensures maintenance of a placenta with secretory/absorptive characteristics until the end of gestation.

5.7.3 A Model for Evolution of Viviparity and Placentation

Thamnophine snakes share five derived developmental characters that form the basis for a model for the evolution of placentation: 1) a thin shell membrane that lacks an outer calcareous layer, 2) a transitory choriovitelline placenta, 3) a chorioallantoic placenta with thin, squamous uterine and fetal epithelial cells overlying extensive capillary networks, 4) a transitory omphaloplacenta, and 5) an omphalallantoic placenta that persists to parturition. In the scenario that we propose, the evolutionary transformation is initiated by prolonged oviductal egg retention and an associated reduction in eggshell composition. The evolution of placentation is an historical contingency of the resultant relationship between maternal and fetal tissues. Subsequent histological and cytological specializations enhance maternal – fetal exchange and increase functional properties of the placentae.

Evolution of oviparous egg retention (under any of the potential selective pressures) places conflicting functional demands on eggshell structure. The conflict arises because a shell thick enough to protect the egg in the external environment is too thick to allow for gas exchange prior to oviposition (Blackburn 1995). Accordingly, evolution of viviparity accompanies a reduction in shell gland activity, and thus reduction in eggshell thickness, and (potentially) development of other mechanisms of shell reduction, such as phagocytosis by chorionic epithelial cells (Blackburn and Lorenz 2003a).

Sufficient reduction of the shell membrane to accommodate physiological exchange between maternal and fetal tissues yields the structural/functional relationship known as placentation. Squamates (unlike anamniotes) have eggs that are entirely surrounded by fetal membranes available for physiological exchange. Being a very thin tube, the squamate oviduct is greatly stretched by presence of the egg, such that the uterine epithelium lies juxtaposed to these fetal membranes. Thus, the entire egg is surrounded by placental membranes.

Under viviparous conditions, the fetal membranes maintain and extend their original oviparous functions, and both respiratory and nutritive placentae simultaneously evolve. The chorioallantoic placenta accordingly serves as the main respiratory organ of the developing snake embryo. Intrauterine gas exchange across this placenta is enhanced by several specializations—attenuated chorionic and uterine epithelia, increased uterine vascularity, shell membrane degeneration, and changes in oxygen affinity of fetal and maternal blood—features that can differ between ophidian lineages (Blackburn 2000).

Although development in the intrauterine environment entails difficulties in terms of respiration, it also exposes the egg to new potential sources of nutrients. The absorptive capabilities of the omphalopleure are recruited to exploit nutrients arising from the uterine epithelium. The omphaloplacenta and omphalallantoic placenta thereby become sites of histotrophic transfer and provide facultative supplementation of yolk nutrient sources (Stewart 1989). In addition to its prominence as a respiratory organ, the chorioallantoic placenta retains its function in calcium transport in response to secretion from the uterus. Thus, both chorioallantoic and yolk sac placentae arise and provide nutritive functions in response to prolonged oviductal egg retention.

Following the origin of viviparity and placentation, subsequent specializations for nutrient transfer evolve in both chorioallantoic and yolk sac placentae. Prominent innovations in the chorioallantoic placenta include nutrient secretion by most uterine epithelial cells and regional differentiation of the epithelium for enhanced histotrophic transport. In accord with these novel functions, the fetal epithelium develops cellular and tissue specializations for nutrient uptake. The yolk sac placentae evolve further modifications for histotrophic transport and contribute to fetal nutrition throughout gestation. These specializations include specializations for uterine synthesis and secretion of nutrients and for absorption by the fetal omphalopleure.

Figure 5.14 summarizes aspects of placental evolution, as exemplified by viviparous thamnophines. Characteristics of the oviparous corn snake *Pantherophis guttatus* are deemed to be primitive for the group. Most of the relevant corn snake features are found among oviparous lizards and (with the possible exception of the omphalallantoic membrane: Stewart and Thompson 2000) are judged as ancestral for squamates. Most of the

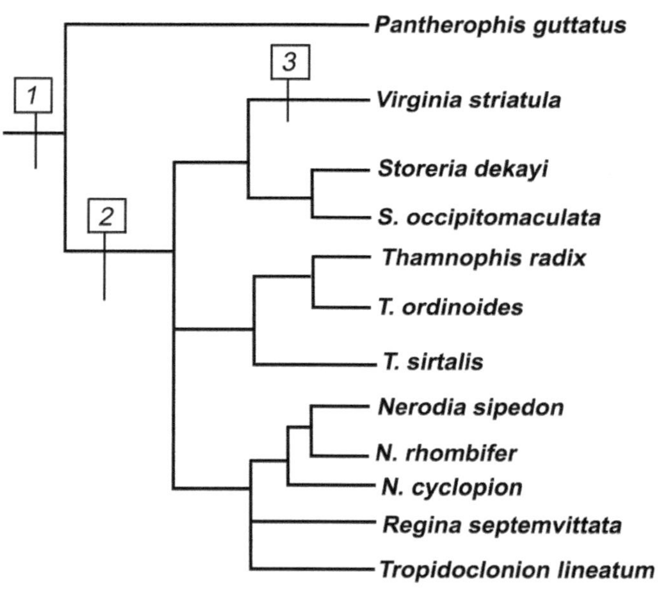

Fig. 5.14 Reproductive evolution in thamnophine snakes. Species include those thamnophines for which placental information is available. Phylogenetic relationships follow Alfaro and Arnold (2001). Reproductive features of nodes 2 and 3 are inferred to have evolved at the indicated positions; those at node 1 are probably ancestral for snakes. Site of evolution of the secondary yolk cleft is uncertain; the structure is found in *Virginia striatula,* but may exist in other species. **Node 1:** oviparity, bilaminar omphalopleure, yolk cleft + isolated yolk mass, allantois in yolk cleft, chorioallantois that functions in respiration, yolk sac omphalopleure that functions in water absorption. **Node 2:** viviparity with incipient placentotrophy, reduced shell membrane, chorioallantoic placenta that is specialized for gas exchange, omphalopleure with specialized absorptive cells, secretory uterine epithelium, omphalallantoic placenta that functions in histotrophic transfer. **Node 3:** omphalallantoic placenta with uterine epithelium aligned on vascular ridges, chorioallantoic placenta with regional variation (specialized peripheral zone).

placental features cited in Figure 5.14 appear to be plesiomorphic for the thamnophine clade. Features that only have been observed in *Virginia striatula* are provisionally inferred to be specializations of that species.

The thamnophine pattern of embryonic nutrition, with primary reliance on yolk nourishment supplemented by placental provision, is common among viviparous squamates. In view of the broad taxonomic distribution of viviparous lecithotrophy, it is unfortunate that we know so little regarding its selective advantages. Given the diversity of feeding and habitat specializations and the breadth of knowledge of their reproductive biology, thamnophine snakes are an ideal taxon to address questions of how pattern of embryonic nutrition influences life history evolution.

Likewise, given our detailed knowledge of thamnophine biology and evolution, these snakes offer a valuable model clade for ongoing and future reproductive studies. Aspects of placentation may well vary between ophidian clades, as do particulars of viviparous reproduction; such is to be expected given the many times this reproductive pattern has evolved among snakes. Nevertheless, research on thamnophines offers an unparalleled opportunity to reconstruct patterns of structure, function, proto-adaptation, and constraint that have operated during the evolution of viviparity.

5.8 SUMMARY

Viviparity and placentation have evolved convergently in numerous lineages of snakes, and are found in 14 families and in species of every utilized habitat. Viviparity can confer significant costs to female snakes, including reduced mobility, decreased feeding behavior, and constraints on litter size and frequency of reproduction. However, viviparity also confers thermal benefits, and permits exploitation of environments where nesting sites are lacking. Circumstantial and experimental evidence indicates that thermal benefits have acted as selective pressures in the origin of snake viviparity.

Placentation evolves simultaneously with viviparity in snakes, in association with a change in composition and thinning of the eggshell and the consequent increased proximity of fetal membranes to the oviductal lining. Physiological studies on North American thamnophines (Natricidae) reveal that placentae are responsible for transfer of oxygen, water, and nutrients from pregnant females to embryos. Histological and ultrastructural studies of thamnophines from 6 genera and 9 species have revealed the morphological basis for these functions. The chorioallantoic placenta is primarily responsible for gas exchange and shows specializations that enhance its functions. Placentae derived from the yolk sac omphalopleure show cellular specializations for maternal nutrient secretion and fetal absorption.

Based on evidence from thamnophines, we detail a model for the evolution of placentation in which the fetal membranes maintain and extend their original oviparous functions. Thus, during prolonged uterine egg retention and viviparity, the choriovitelline membrane and chorioallantois assume functions in intrauterine respiration, and the omphalopleure, in absorption. Subsequent specializations for nutrient transfer evolve in fetal and maternal components of both the chorioallantoic and yolk sac placentae. This model is testable in the many clades of viviparous snakes, and therefore offers a valuable framework for future research on placental function and evolution.

5.9 ACKNOWLEDGMENTS

Research discussed herein has been supported over the years by the National Science Foundation, Howard Hughes Medical Institute, grants from Trinity College, University of Tulsa, and East Tennessee State University, and the Thomas S. Johnson Research Professorship funds. Many students (graduate and undergraduate) at our respective institutions have contributed to the research upon which this chapter has drawn, including Kristie Anderson, Marcus Attaway, Duane Baxter, Richard Castillo, Jessica Chin, David Crotzer, Santiago Fregoso, Greg Gavelis, Amy Johnson, Siobhan Knight, Tim Lishnak, Rachel Lorenz, Shauna McKinney, Jen Petzold, Soni Sangha, Craig Shadrix, Michael Smola, Kera Weaber, and Andy Weisenfeld. Craig Schneider began the snake breeding colony at Trinity College and Jenny Nord has carefully maintained it. Kristie Anderson provided some of the micrographs used in this chapter, and Ann Lehman offered valuable technical advice. Laurie Bonneau and anonymous referees carefully reviewed the manuscript. Thanks also are due to Glenn Shea for advice on the identity of specimens described in Weekes (1929).

5.10 LITERATURE CITED

Adams, S. M., Biazik, J. M., Thompson, M. B. and Murphy, C. R. 2005. Cyto-epitheliochorial placenta of the viviparous lizard *Pseudemoia entrecasteauxii*: a new placental morphotype. Journal of Morphology 264: 264-276.

Aldridge, R. D. 1979. Female reproductive cycles of the snakes *Arizona elegans* and *Crotalus viridis*. Herpetologica 35: 256-261.

Aldridge, R. D. and Semlitsch, R. D. 1992. Female reproductive biology of the southeastern crowned snake (*Tantilla coronata*). Amphibia-Reptilia 13: 209-218.

Alfaro, M. E. and Arnold, S. J. 2001. Molecular systematics and evolution of *Regina* and the thamnophiine snakes. Molecular Phylogenetics and Evolution 21: 408-423.

Amaral, A. D. 1927. Contribuição a biologia dos ofidios brasileiros (reproduccao). Coletanea de trabalhos do Instituto Butantan 2: 185-187.

Anderson, J. M. and Van Itallie, C. M. 1995. Tight junctions and the molecular basis for regulation of paracellular permeability. American Journal of Physiology 269 (Gastrointestional and Liver Physiology 32): G467-G475.

Anderson, K. E. and Blackburn, D. G. 2009. The placental interface in viviparous snakes, as revealed by scanning electron microscopy. American Society of Ichthyologists and Herpetologists, Proceedings of the Annual Meeting, Portland, Oregon.

Andrews, R. M. 2000. Evolution of viviparity in squamate reptiles (*Sceloporus* spp.): a variant of the cold-climate model. Journal of Zoology 250: 243-253.
Andrews, R. M. 2002. Low oxygen: a constraint on the evolution of viviparity in reptiles. Physiological and Biochemical Zoology 75: 145-154.
Andrews, R. M. and Mathies, T. 2000. Natural history of reptilian development: constraints on the evolution of viviparity. Bioscience 50: 227-238.
Attaway, M. B. 2000. Morphology of the late stage placentae in the Queen Snake (*Regina septemvittata*). MSc. Thesis, East Tennessee State University, Johnson City, Tennessee. Pp. 96.
Ballard, S. T., Hunter, J. H. and Taylor, A. E. 1995. Regulation of tight junction permeability during nutrient absorption across the intestinal epithelium. Annual Reviews of Nutrition 15: 35-55.
Bauchot, R. 1965. La placentation chez les reptiles. L' Annėe Biologique 4: 547-575.
Bauwens, D. and Thoen, C. 1981. Escape tactics and vulnerability to predation associated with reproduction in the lizard *Lacerta vivipara*. Journal of Animal Ecology 50: 733-743.
Baxter, D. C. 1987. Placentation in the viviparous lined snake, *Tropidoclonion lineatum:* ontogeny of the extraembryonic membranes and histochemistry of placental tissues. MSc Thesis, University of Tulsa, Tulsa, Oklahoma. Pp. 119.
Bellairs, R., Griffiths, I. and Bellairs, A. A. 1955. Placentation in the adder, *Vipera berus*. Nature 176: 657-658.
Bergman, A. J. 1943. The breeding habits of sea snakes. Copeia 1943: 156-160.
Berner, N. J. and Ingermann, R. L. 1988. Molecular basis of the difference in oxygen affinity between maternal and foetal red blood cells in the viviparous garter snake *Thamnophis elegans*. Journal of Experimental Biology 140: 437-453.
Beuchat, C. A. and Vleck, D. 1990. Metabolic consequences of viviparity in a lizard, *Sceloporus jarrovi*. Physiological Zoology 63: 555-570.
Birchard, G. F., Black, C. P., Schuett, G. W. and Black, V. 1984a. Influence of pregnancy on oxygen consumption, heart rate, and hematology in the garter snake: implications for the "cost of reproduction" in live-bearing reptiles. Comparative Biochemistry and Physiology 77: 519-523.
Blackburn, D. G. 1982. Evolutionary origins of viviparity in the Reptilia. I. Sauria. Amphibia-Reptilia 3: 185-205.
Birchard, G. F., Black, C. P., Schuett, G. W. and Black, V. 1984b. Foetal-maternal blood respiratory properties of an ovoviviparous snake the cottonmouth, *Agkistrodon piscivorus*. Journal of Experimental Biology 108: 247-255.
Blackburn, D. G. 1985. Evolutionary origins of viviparity in the Reptilia. II. Serpentes, Amphisbaenia, and Ichthyosauria. Amphibia-Reptilia 5: 259-291.
Blackburn, D. G. 1992. Convergent evolution of viviparity, matrotrophy, and specializations for fetal nutrition in reptiles and other vertebrates. American Zoologist 32: 313-321.
Blackburn, D. G. 1993. Chorioallantoic placentation in squamate reptiles: structure, function, development, and evolution. Journal of Experimental Zoology 266: 414-430.
Blackburn, D. G. 1994. Standardized criteria for the recognition of developmental nutritional patterns in squamate reptiles. Copeia 1994: 925-935.
Blackburn, D. G. 1995. Saltationist and punctuated equilibrium models for the evolution of viviparity and placentation. Journal of Theoretical Biology 174: 199-216.
Blackburn, D. G. 1998a. Structure, function, and evolution of the oviducts of squamate reptiles, with special reference to viviparity and placentation. Journal of Experimental Zoology 282: 560-617.

Blackburn, D. G. 1998b. Reconstructing the evolution of viviparity and placentation. Journal of Theoretical Biology 192: 183-190.
Blackburn, D. G. 1999. Are viviparity and egg-guarding evolutionarily labile? Herpetologica 55: 556-573.
Blackburn, D. G. 2000. Viviparity: past research, future directions, and appropriate models. Comparative Biochemistry and Physiology—Part A: Molecular and Integrative Physiology 127: 391-409.
Blackburn, D. G. 2006. Squamate reptiles as model organisms for the evolution of viviparity. Herpetological Monographs 20: 131-146.
Blackburn, D. G. and Callard, I. P. 1997. Morphogenesis of the placental membranes in the viviparous, placentotrophic lizard *Chalcides chalcides* (Squamata: Scincidae). Journal of Morphology 232: 35-55.
Blackburn, D. G. and Vitt, L. J. 2002. Specializations of the chorioallantoic placenta in the Brazilian scincid lizard, *Mabuya heathi*: a new placental morphotype for reptiles. Journal of Morphology 254: 121-131.
Blackburn, D. G. and Lorenz, R. 2003a. Placentation in garter snakes. Part II. Transmission EM of the chorioallantoic placenta of *Thamnophis radix* and *T. sirtalis*. Journal of Morphology 256: 171-186.
Blackburn, D. G. and Lorenz, R. 2003b. Placentation in garter snakes. Part III. Transmission EM of the omphallantoic placenta of *Thamnophis radix* and *T. sirtalis*. Journal of Morphology 256: 187-204.
Blackburn, D. G. and Flemming, A. F. 2009. Morphology, development, and evolution of fetal membranes and placentation in squamate reptiles. Journal of Experimental Zoology B. Molecular and Developmental Evolution 312B: 579-589.
Blackburn, D. G., Vitt, L. J. and Beuchat, C. A. 1984. Eutherian-like reproductive specializations in a viviparous reptiles. Proceedings of the National Academy of Sciences, USA, 81: 4860-4863.
Blackburn, D. G., Johnson A. R. and Petzold J. L. 2003. Histology of the extraembryonic membranes of an oviparous snake: towards a reconstruction of basal squamate patterns. Journal of Experimental Zoology 290A: 48-58.
Blackburn, D. G., Stewart, J.R., Baxter, D.C. and Hoffman, L.H. 2002. Placentation in garter snakes. Scanning EM of the placental membranes of *Thamnophis ordinoides* and *T. sirtalis*. Journal of Morphology 252: 263-275.
Blackburn, D. G., Anderson, K. E., Johnson, A. R., Knight, S. R. and Gavelis, G. S. 2009. Histology and ultrastructure of the placental membranes of the viviparous brown snake, *Storeria dekayi* (Colubridae: Natricinae). Journal of Morphology 270: 1137-1154.
Blem, C. R. 1982. Biennial reproduction in snakes: an alternative hypothesis. Copeia 1982: 961-963.
Bonnet, X., Naulleau, G., Shine, R. and Lourdais, O. 2001. Short-term vs. long-term effects of food intake on reproductive output in a viviparous snake, *Vipera aspis*. Oikos 292: 297-308.
Bonnet, X., Lourdais, O., Shine, R. and Naulleau, G. 2002. Reproduction in snakes (*Vipera aspis*): costs, currencies, and complications. Ecology 83: 2124-2135.
Booth, D. T., Thompson, M. B. and Herring, S. 2000. How incubation temperature influences the physiology and growth of embryonic lizards. Journal of Comparative Physiology B 170: 269-276.
Boulenger, G. A. 1913. *The Snakes of Europe.* Methuen, London, U.K. Pp. 151.
Boyd, M. M. M. 1940. The structure of the ovary and formation of the corpus luteum in *Hoplodactylus maculatus*. Quarterly Journal of Microscopic Science 82: 337-376.

Boyd, M. M. M. 1942. The oviduct, foetal membranes, and placentation in *Hoplodactylus maculatus* Gray. Proceedings of the Zoological Society of London, Series A 112: 65-104.
Bragdon, D. E. 1951. The non-essentiality of the corpus luteum for the maintenance of gestation in certain live-bearing snakes. Journal of Experimental Biology 118: 419-435.
Brodie, E. D. III. 1989. Behavioral modification as a means of reducing the cost of reproduction. American Naturalist 134: 225-238.
Brown, W. S. 1991. Female reproductive ecology in a northern population of the timber rattlesnake, *Crotalus horridus*. Herpetologica 47: 101-115.
Bull, J. J. and Shine, R. 1979. Iteroparous animals that skip opportunities for reproduction. American Naturalist 114: 296-316.
Cai, Y., Zhou T. and Ji, X. 2007. Embryonic growth and mobilization of energy and material in oviposited eggs of the red-necked keelback snake, *Rhabdophis tigrinus lateralis*. Comparative Biochemistry and Physiology A 147: 57-63.
Capula, M., Luiselli, L. and Rugiero, L. 1995. Ecological correlates of reproductive mode in reproductively bimodal snakes of the genus *Coronella*. Vie Milieu 45: 167-175.
Charland, M. B. 1995. Thermal consequences of reptilian viviparity: thermoregulation in gravid and nongravid garter snakes (*Thamnophis*). Journal of Herpetology 29: 383-391.
Clark, H. 1953a. Metabolism of the black snake embryo. II. Respiratory exchange. Journal of Experimental Biology 30: 502-505.
Clark, H. 1953b. Metabolism of the black snake embryo. I. Nitrogen excretion. Journal of Experimental Biology 30: 492-501.
Clark, D. R. 1964. Reproduction and sexual dimorphism in a population of the rough earth snake, *Virginia striatula* (Linnaeus). Texas Journal of Science 16: 265-295.
Clark, D. R. 1970a. Ecological study of the worm snake, *Carphophis vermis* (Kennicott). University of Kansas Publications of the Museum of Natural History 19: 85-194.
Clark, D. R. 1970b. Loss of the left oviduct in the colubrid snake genus *Tantilla*. Herpetologica 26: 130-133.
Clark, H. and Sisken, B. F. 1956. Nitrogenous excretion by embryos of the viviparous snake *Thamnophis s. sirtalis* (L.). Journal of Experimental Biology 33: 384-393.
Clark, H., Florio, B. and Hurowitz, R. 1955. Embryonic growth of *Thamnophis s. sirtalis* in relation to fertilization date and placental function. Copeia 1955: 9-13.
Clausen, H. J. 1940. Studies on the effects of ovariotomy and hypophysectomy on gestation in snakes. Endocrinology 27: 700-704.
Conant, R. 1965. Notes on reproduction in two natricine snakes from Mexico. Herpetologica 21: 140-144.
Conaway, C. H. and Fleming, W. R. 1960. Placental transmission of Na^{22} and I^{131} in *Natrix*. Copeia 1960: 360-366.
de Fraipont, M., Clobert, J. and Barbault, R. 1996. The evolution of oviparity with egg-guarding and viviparity in lizards and snakes: a phylogenetic analysis. Evolution 50: 391-400.
de Fraipont, M., Clobert, J. and Barbault, R. 1999. On the evolution of viviparity and egg-guarding in squamate reptiles: a reply to R. Shine and M. S. Y. Lee. Herpetologica 55: 550-555.
DeMarco, V. 1993. Metabolic rates of female viviparous lizards (*Sceloporus jarrovi*) throughout the reproductive cycle: do pregnant lizards adhere to standard allometry? Physiological Zoology 66: 166-180.

DeMarco, V. and Guillette, L. J., Jr. 1992. Physiological cost of pregnancy in a viviparous lizard (*Sceloporus jarrovi*). Journal of Experimental Zoology 262: 383-390.
Dmi'el, R. 1970. Growth and metabolism in snake embryos. Journal of Embryology and Experimental Morphology 23: 761-772.
Duméril, A. M. C. and Bibron, G. 1834 - 1854. *Erpétologie Générale ou Histoire Naturelle Complète des Reptiles.* 10 vols. Roret, Paris.
Ecay, T. W., Stewart, J. R. and Blackburn, D. G. 2004. Expression of calbindin-D28K by yolk sac and chorioallantoic membranes of the corn snake, *Elaphe guttata*. Journal of Experimental Zoology 302B: 517-525.
Fitch, H. S. 1970. Reproductive cycles in lizards and snakes. University of Kansas Museum of Natural History, Miscellaneous Publications 52: 1-247.
Fitch, H. S. and Twining, H. 1946. Feeding habits of the Pacific rattlesnake. Copeia 1946: 64-71.
Flemming, A. F. and Branch, W. R. 2001. Extraordinary case of matrotrophy in the African skink *Eumecia anchietae*. Journal of Morphology 246: 264-287.
Ford, N. B. and Karges, J. P. 1987. Reproduction in the checkered garter snake, *Thamnophis marcianus*, from southern Texas and northeastern Mexico: seasonality and evidence for multiple clutches. Southwestern Naturalist 32: 93-101.
Ford, N. B. and Seigel, R. A. 1989. Phenotypic plasticity in reproductive characteristics: evidence from a viviparous snake. Ecology 70: 1768-1774.
Foster, M., Bissell, K. M., Campa, H., III and Harrison, T. M. 2007. The influence of reproductive status on thermal ecology and vegetation use of female eastern massasauga rattlesnakes (*Sistrurus catenatus catenatus*) in southwestern Michigan. Herpetological Conservation and Biology 4: 48-54.
Fox, W. and Dessauer, H. C. 1962. The single right oviduct and other urogenital structures of female *Typhlops* and *Leptotyphlops*. Copeia 1962: 590-597.
Fregoso, S. P., Stewart, J. R. and Ecay, T. W. 2010. Embryonic mobilization of calcium in a viviparous reptile: evidence for a novel pattern of placental calcium secretion. Comparative Physiology and Biochemistry A, doi:10.1016/j.cbpa.2010.01.014.
Gadow, H. 1910. The effect of altitude upon the distribution of Mexican amphibians and reptiles. Zoologischer Jahrbuch 29: 689-714.
Gerrard, A. M. 1974. Placental transfer of steroids at different stages of development and its possible implications in the sexual differentiation of *Thamnophis radix haydenii* embryos. PhD Thesis, University of Colorado, Boulder. Pp. 81.
Giacomini, E. 1891. Materiali per la storia dello svilluppo del *Seps chalcides* (Cuv.) Bonap. Monitore Zoologico Italiano 2: 179-192, 198-211.
Giacomini, E. 1893. Sull' ovidutto del Sauropsidi. Monitore Zoologico Italiano 4: 202-265.
Gibson, A. R. and Falls, J. B. 1979. Thermal biology of the common garter snake *Thamnophis sirtalis* (L.). I. Temporal variation, environmental effects and sex differences. Oecologia (Berlin) 43: 79-97.
Gibbons, J. W. 1972. Reproduction, growth, and sexual dimorphism in the canebreak rattlesnake (*Crotalus horridus atricaudatus*). Copeia 1972: 222-226.
Gier, P. J., Wallace, R. J. and Ingermann, R. L. 1989. Influence of pregnancy on behavioral thermoregulation in the Northern Pacific rattlesnake *Crotalus viridis oreganus*. Journal of Experimental Biology 145: 465-469.
Giersberg, H. 1922. Untersuchungen über Physiologie und Histologie des Eileiters der Reptilien und Vögel; nebst einem Beitrag zue Fasergenese. Zeitschrift für Wissenschaftliche Zoologie 70: 1-97.
Gignac, A. and Gregory, P. T. 2005. The effects of body size, age, and food intake during pregnancy on reproductive traits of a viviparous snake, *Thamnophis ordinoides*. Ecoscience 12: 236-243.

Girling, J. E. 2002. The reptilian oviduct: a review of structure and function and directions for future research. Journal of Experimental Zoology 293: 141-170.
Gist, D. H. and Jones, J. M. 1987. Storage of sperm in the reptilian oviduct. Scanning Microscopy 1: 1839-1851.
Gould, S. J. 2002. *The Structure of Evolutionary Theory.* Harvard University Press, Cambridge, Massachusetts. Pp. 1433.
Graves, B. M. and Duvall, D. D. 1993. Reproduction, rookery use, and thermoregulation in free-ranging, pregnant *Crotalus v. viridis*. Journal of Herpetology 27: 33-41.
Greene, H. W. 1970. Mode of reproduction in lizards and snakes of the Gomez Farias region, Tamaulipas, Mexico. Copeia 1970: 565-568.
Greene, H. W. 1997. *Snakes: The Evolution of Mystery in Nature.* University of California Press, Berkeley, California. Pp. 366.
Gregory, P. T. 2009. Northern lights and seasonal sex: the reproductive ecology of cool-climate snakes. Herpetologica 65: 1-13.
Gregory, P. T. and Stewart, K. W. 1975. Long distance dispersal and feeding strategy of the red-sided garter snake *(Thamnophis sirtalis parietalis)* in the Interlake of Manitoba. Canadian Journal of Zoology 53: 238-245.
Gregory, P. T. and Skebo, K. M. 1998. Trade-offs between reproductive traits and the influence of food intake during pregnancy in the garter snake, *Thamnophis elegans.* American Naturalist 151: 477-486.
Gregory, P. T., Crampton, L. H. and Skebo, K. M. 1999. Conflicts and interactions among reproduction, thermoregulation and feeding in viviparous reptiles: are gravid snakes anorexic? Journal of Zoology 248: 231-241.
Grosser, O. 1927. *Frühentwicklung Eihautbildung und Placentation des Menschen und der Säugetiere.* J.F. Bergmann, München, Germany. Pp. 454.
Guillette, L. J. Jr. and Jones, R. E. 1985. Ovarian, oviductal, and placental morphology of the reproductively bimodal lizard, *Sceloporus aeneus.* Journal of Morphology 184: 85-98.
Guillette, L. J., Jr., Fox, S. L. and Palmer, B. D. 1989. Oviductal morphology and egg shelling in the oviparous lizards *Crotaphytus collaris* and *Eumeces obsoletus*. Journal of Morphology 201: 145-159.
Hall, R. J. 1969. Ecological observations on Graham's watersnake (*Regina grahami* Baird and Girard). The American Midland Naturalist 81: 156-163.
Heulin, B., Stewart, J. R., Surget-Groba, Y., Bellaud, B., Jouan, F., Lancien, G. and Deunff, J. 2005. Development of the uterine shell glands during the preovulatory and early gestation periods in oviparous and viviparous *Lacerta vivipara.* Journal of Morphology 266: 80-96.
Heulin, B., Arrayago M.-J. and Bea, A. 1989. Expérience d'hybridation entre les souches ovipare et vivipare du lézard *Lacerta vivipara*. Comptes rendus de l'Académie des sciences Paris 308: 341-346.
Hin, K. H., Stuebing, R. B. and Voris, H. K. 1991. Population structure and reproduction in the marine snake, *Lapemis hardwickii* Gray, from the west of coast of Sabah. Sarawak Museum Journal 42: 463-475.
Hirth, H. F. 1964. Observations on the fer-de-lance, *Bothrops atrox,* in coastal Costa Rica. Copeia 1964: 453-454.
Hoffman, L. H. 1970. Placentation in the garter snake, *Thamnophis sirtalis*. Journal of Morphology 131: 57-88.
Hoffman, L. H. and Wimsatt, W. A. 1972. Histochemical and electron microscopic observations on the sperm receptacles in the garter snake oviduct. American Journal of Anatomy 134: 71-96.
Ingermann, R. L. 1992. Maternal-fetal oxygen transfer in lower vertebrates. American Zoologist 32: 322-330.

Ingermann, R. L., Berner, N. J. and Ragsdale, F. R. 1991. Effect of pregnancy and temperature on red cell oxygen - affinity in the viviparous snake *Thamnophis elegans*. Journal of Experimental Biology 156: 399-406.

Jacobi, L. 1936. Ovoviviparie bei einheimischen eidechsen. Zeitschrift für Wissenschaftliche Zoologie 148: 401-464.

Jerez, A. and Ramírez-Pinilla, M. P. 2001. The allantoplacenta of *Mabuya mabouya* (Sauria, Scincidae). Journal of Morphology 249: 132-146.

Ji, X. 1995. Egg and hatchling components in a viviparous snake, *Elaphe rufodorsata*. Journal of Herpetology 29: 298-300.

Ji, X. and Sun, P.-Y. 2000. Embryonic use of energy and post-hatching yolk in the gray rat snake, *Ptyas korros* (Colubridae). Herpetological Journal 10: 13-17.

Ji, X. and Du, W.- G. 2001. The effects of thermal and hydric environments on hatching success, embryonic use of energy and hatchling traits in a colubrid snake, *Elaphe carinata*. Comparative Biochemistry and Physiology A 129: 461-471.

Ji, X., Sun, P.-Y., Fu, S.-Y. and Zhang, H.-S. 1999a. Utilization of energy and material in eggs and post-hatching yolk in an oviparous snake, *Elaphe taeniura*. Asiatic Herpetological Research 8:53-59.

Ji, X., Du, W.-G. and Xu, W.-Q. 1999b. Experimental manipulation of egg size and hatchling size in the cobra, *Naja naja atra* (Elapidae). Netherlands Journal of Zoology 49: 167-175.

Ji, X., Lin, C. X., Lin, L. H., Qiu, Q. B. and Du, Y. 2007. Evolution of viviparity in warm-climate lizards: an experimental test of the maternal manipulation hypothesis. Journal of Evolutionary Biology 20: 1037-1045.

Johnson, A. R. 2003. A comparison of squamate fetal membrane and placental ultra-structure. BSc. Honors Thesis, Trinity College, Hartford, Connecticut. Pp. 115.

Jones, R. E. and Baxter, D. C. 1991. Gestation, with emphasis on corpus luteum biology, placentation, and parturition. Pp. 205-301. In P. K. T. Pang, M. P. Schreibman and R. Jones (eds), *Vertebrate Endocrinology: Fundamentals and Biomedical Implications*, vol. 4, part A. Academic Press, New York.

Kasturirangan, L. R. 1951a. Placentation in the sea-snake, *Enhydrina schistosa* (Daudin). Proceedings of the Indian Academy of Sciences B 34: 1-32.

Kasturirangan, L. R. 1951b. The allantoplacenta of the sea-snake, *Hydrophis cyanocinctus* Daudin. Journal of the Zoological Society of India 3: 277-290.

Keenlyne, K. D. 1972. Sexual differences in feeding habits of *Crotalus horridus horridus*. Journal of Herpetology 6: 234-237.

Kerr, J. G. 1919. *Textbook of Embryology*. Macmillan, London, U.K. Pp. 591.

Kleis-San Francisco, S. M. and Callard, I. P. 1986a. Identification of a putative progesterone receptor in the oviduct of a viviparous watersnake (*Nerodia*). General and Comparative Endocrinology 61: 490-498.

Kleis-San Francisco, S. M. and Callard, I. P. 1986b. Progesterone receptors in the oviduct of a viviparous snake (*Nerodia*): correlations with ovarian function and plasma steroid levels. General and Comparative Endocrinology 63: 220-229.

Knight, S. R. and Blackburn, D. G. 2008. Scanning electron microscopy of the fetal membranes of an oviparous squamate, the corn snake *Pituophis guttatus* (Colubridae). Journal of Morphology 269: 922-934.

Kopstein, F. 1938. Ein beitrag zur eierkunde und zur fortpflanzung der Malaiischen reptilien. Bulletin of the Raffles Museum 14: 81-167.

Krohmer, R. W. and Aldridge, R. D. 1985. Female reproductive cycle of the lined snake *(Tropidoclonion lineatum)*. Herpetologica 41: 39-44.

Ladyman, M., Bonnet, X., Lourdais, O., Bradshaw, D. and Naulleau, G. 2003. Gestation, thermoregulation, and metabolism in a viviparous snake, *Vipera aspis:* evidence for fecundity-independent costs. Physiological and Biochemical Zoology 76: 497-510.

Lawson, R., Slowinski, J. B., Crowther, B. I. and Burbrink, F. T. 2005. Phylogeny of the Colubroidea (Serpentes): New evidence from mitochondrial and nuclear genes. Molecular Phylogenetics and Evolution 37: 581-601.

Lemen, C. A. and Voris, H. K. 1981. A comparison of reproductive strategies among marine snakes. Journal of Animal Ecology 50: 89-101.

Li, H., Qu, Y.-F., Hu, R.-B. and Ji, X. 2009. Evolution of viviparity in cold-climate lizards: testing the maternal manipulation hypothesis. Evolutionary Ecology, published online, doi: 10.1007/s10682-008-9272-2.

Lin, C. -X., Zhang, L. and Ji, X. 2008. Influence of pregnancy on locomotor and feeding performances of the skink, *Mabuya multifasciata:* why do females shift thermal preferences when pregnant? Zoology 111: 188-195.

Linville, B. J., Stewart, J. R., Ecay, T. W., Herbert, J. F., Parker, S. L. and Thompson, M. B. 2010. Placental calcium provision in a lizard with prolonged oviductal egg retention. Journal of Comparative Physiology B 180: 221-227.

Lourdais, O., Bonnet, X. and Doughty, P. 2002. Costs of anorexia during pregnancy in a viviparous snake *(Vipera aspis).* Journal of Experimental Zoology 292: 487-493.

Lourdais, O., Heulin, B. and Shine, R. 2008. Thermoregulation during gravidity in the children's python (*Antaresia childreni*): a test of the preadaptation hypothesis for maternal thermophily in snakes. Biological Journal of the Linnaean Society 93: 499-508.

Lourdais, O., Shine, R., Bonnet, X., Guillon, M. and Naulleau, G. 2004a. Climate affects embryonic development in a viviparous snake, *Vipera aspis.* Oikos 104: 551-560.

Lourdais, O., Brischoux, F., DeNardo, D. and Shine, R. 2004b. Protein catabolism in pregnant snakes *(Epicrates cenchria maurus* Boidae) compromises musculature and performance after reproduction. Journal of Comparative Physiology B 174: 383-391.

Lu, H. -L., Hu, R.-B. and Ji, X. 2009. Embryonic growth and mobilization of energy and material during incubation in the checkered keelback snake, *Xenochrophis piscator*. Comparative Biochemistry and Physiology A 152: 214-218.

Lynch, V. J. 2010. Live-birth in vipers (Viperidae) is a key innovation and adaptation to global cooling during the Cenozoic. Evolution 63: 2457-2465.

Lynch, V. J. and Wagner, G. P. 2010. Did egg-laying boas break Dollo's Law? Phylogenetic evidence for reversal to oviparity in sand boas (*Eryx:* Boidae). Evolution, in press. doi: 10.1111/j.1558-5646.2009.00790.x

Madsen, T. and Shine, R. 1992. Determinants of reproductive success in female adders, *Vipera berus.* Oecologia 92: 40-47.

Madsen, T. and Shine, R. 1993. Costs of reproduction in a population of European adders. Oecologia 94: 488-495.

Masson, G. R. and Guillette, L. J., Jr. 1987. Changes in oviducal vascularity during the reproductive cycle of three oviparous lizards (*Eumeces obsoletus, Sceloporus undulatus* and *Crotaphytus collaris*). Journal of Reproduction and Fertility 80: 361-371.

Mathies, T. and Andrews, R. M. 1995. Thermal and reproductive biology of high and low elevation populations of the lizard *Sceloporus scalaris*: implications for the evolution of viviparity. Oecologia 104: 101-111.

Mead, R. A., Eroschenko, V. P. and Highfill, D. R. 1981. Effects of progesterone and estrogen on the histology of the oviduct of the garter snake, *Thamnophis elegans.* General and Comparative Endocrinology 45: 345-354.

Mell, R. 1929. *Beiträge zur Fauna sinica. IV. Grundzüge einer Ökologie der chinesischen Reptilien und einer herpetologischen Tiergeographie Chinas.* Walter de Gruyter & Co., Liepzig, Germany. Pp. 282.

Miles, D. B., Sinervo, B. and Frankino, W. A. 2000. Reproductive burden, locomotor performance, and the cost of reproduction in free ranging lizards. Evolution 54: 1386-1395.

Mole, R. H. 1924. The Trinidad snakes. Proceedings of the Zoological Society of London 11: 235-278.

Mossman, H. W. 1937. Comparative morphogenesis of the fetal membranes and accessory uterine structures. Carnegie Institute Contributions in Embryology 26: 129-246.

Mossman, H. W. 1987. *Vertebrate Fetal Membranes.* Rutgers University Press, New Brunswick, New Jersey. Pp. 383.

Mulaik, D. D. M. 1946. A comparative study of the urinogenital systems of an oviparous and two ovoviviparous species of the lizard genus *Sceloporus.* Bulletin of the University of Utah 37: 1-24.

Murphy, J. C., Voris, H. K., Stuart, B. L. and Platt, S. G. 2002. Female reproduction in the rainbow water snake, *Enhydris enhydris* (Serpentes, Colubridae, Homalopsinae). Natural History Journal of the Chulalongkorn University 2: 31-37.

Neill, W. T. 1962. The reproductive cycle of snakes in a tropical region, British Honduras. Quarterly Journal of the Florida Academy of Sciences 25: 234-253.

Neill, W. T. 1964. Viviparity in snakes: some ecological and zoogeographical considerations. American Naturalist 98: 35-55.

Osgood, D. W. 1970. Thermoregulation in water snakes studied by telemetry. Copeia 1970: 568-570.

Ota, H., Iwanaga, S., Itoman, K., Nishimura, M. and Mori, A. 1991. Reproductive mode of a natricine snake, *Amphiesma pryeri* (Colubridae: Squamata), from the Ryuku Archipelago, with special reference to the viviparity of *A. p. ishigakiensis.* The Biological Magazine Okinawa 29: 37-43.

Packard, G. C. 1966. The influence of ambient temperature and aridity on modes of reproduction and excretion of amniote vertebrates. American Naturalist 100: 677-682.

Packard, G. C. and Packard, M. J. 1988. The physiological ecology of reptilian eggs and embryos. Pp. 523-605. In C. Gans and R. B. Huey (eds), *Biology of the Reptilia.* vol. 16. A. R. Liss, New York.

Packard, G. C., Tracy, C. R. and Roth, J. J. 1977. The physiological ecology of reptilian eggs and embryos, and the evolution of viviparity within the Class Reptilia. Biological Reviews 52: 71-105.

Packard, M. J. 1994. Patterns of mobilization and deposition of calcium in embryos of oviparous, amniotic vertebrates. Israel Journal of Zoology 40: 481-492.

Packard, M. J. and DeMarco, V. G. 1991. Eggshell structure and formation in eggs of oviparous reptiles. Pp. 53-70. In D. C. Deeming and M. W. J. Ferguson (eds), *Egg Incubation: Its Effects on Embryonic Development in Birds and Reptiles.* Cambridge University Press, Cambridge, U.K.

Packard, M. J., Packard, G. C. and Boardman, T. J. 1982. Structure of eggshells and water relations of reptilian eggs. Herpetologica 38: 136-155.

Parameswaran, K. N. 1962. The foetal membranes and placentation of *Enhydris dussumieri* (Smith). Proceedings of the Indian Academy of Science B 56: 302-327.

Parker, S. L. and Andrews, R. M. 2006. Evolution of viviparity in sceloporine lizards: in utero PO_2 as a developmental constraint during egg retention. Physiological and Biochemical Zoology 79: 581-592.

Parker, S. L., Andrews, R. M. and Mathies, T. 2004. Embryonic responses to variation in oviductal oxygen in the lizard *Sceloporus undulatus* from New Jersey and South Carolina, USA. Biological Journal of the Linnaean Society 83: 289-299.
Perkins, M. J. and Palmer, B. D. 1996. Histology and functional morphology of the oviduct of an oviparous snake, *Diadophis punctatus*. Journal of Morphology 227: 67-79.
Picariello, O., Ciarcia, G. and Angelini, F. 1989. The annual cycle of oviduct in *Tarentola m. mauritanica* L. (Reptilia, Gekkonidae). Amphibia-Reptilia 10: 371-386.
Pizzatto, L., Almeida-Santos, S. M. and Shine, R. 2007. Life-history adaptations to arboreality in snakes. Ecology 88: 359-366.
Pope, C. H. 1961. *The Giant Snakes.* Alfred Knopf, New York. Pp. 290.
Qualls, C. P. and Shine, R. 1998. *Lerista bougainvillii*, a case study for the evolution of viviparity in reptiles. Journal of Evolutionary Biology 11: 63-78.
Qualls, C. P. and Andrews, R. M. 1999. Cold climates and the evolution of viviparity in reptiles: cold incubation temperatures produce poor-quality offspring in the lizard, *Sceloporus virgatus*. Biological Journal of the Linnean Society 67: 353-376.
Qualls, C. P., Andrews, R. M. and Mathies, T. 1997. The evolution of viviparity and placentation revisted [sic]. Journal of Theoretical Biology 185: 129-135.
Qualls, C. P., Shine, R., Donnellan, S. and Hutchinson, M. 1995. The evolution of viviparity within the Australian scincid lizard *Lerista bougainvillii.* Journal of Zoology 237: 13-26.
Ragsdale, F. R. and Ingermann, R. L. 1991. Influence of pregnancy on the oxygen affinity of red cells from the northern Pacific rattlesnake *Crotalus viridis oreganus.* Journal of Experimental Biology 159: 501-505.
Ragsdale, F. R. and Ingermann, R. L. 1993. Biochemical bases for difference in oxygen affinity of maternal and fetal red blood cells of rattlesnake. American Journal of Physiology 264: R481-R486.
Ragsdale, F. R., Imel, K. M., Nilsson, E. E. and Ingermann, R. L. 1993. Pregnancy-associated factors affecting organic phosphate levels and oxygen affinity of garter snake red cells. General and Comparative Endocrinology 91: 181-188.
Rahn, H. 1939. Structure and function of placenta and corpus luteum in viviparous snakes. Proceedings of the Society Experimental Biology and Medicine 40: 381-382.
Ramírez-Pinilla, M. P. 2006. Placental transfer of nutrients during gestation in an Andean population of the highly matrotrophic lizard genus *Mabuya* (Squamata: Scincidae). Herpetological Monographs 20: 194-204.
Robert, K. A. and Thompson, M. B. 2000. Energy consumption by embryos of a viviparous lizard, *Eulamprus tympanum*, during development. Comparative Biochemistry and Physiology A 127: 481-486.
Robb, J. 1960. The internal anatomy of *Typhlops* Schneider (Reptilia). Australian Journal of Zoology 8: 181-216.
Robb, J. and Smith, H. M. 1966. The systematic position of the group of snake genera allied to *Anomalepis*. Natural History Miscellanea of the Chicago Academy of Sciences 184: 1-8.
Rollinat, R. 1904. Observations sur la tendance vers l'ovoviviparité chez quelques sauriens et ophidiens de la France centrale. Mémoires de la Société zoologique de France 17: 30-41.
Rossman, D. A. and Eberle, W. G. 1977. Partition of the genus *Natrix*, with preliminary observations on evolutionary trends on natricine snakes. Herpetologica 33: 34-43.
Rossman, D. A., Ford, N. B. and Seigel, R. A. 1996. *The Garter Snakes: Evolution and Ecology.* University of Oklahoma Press, Norman, Oklahoma. Pp. 332.

Saint Girons, H. 1957. Le cycle sexuel chez *Vipera aspis* (L.) dans l'ouest de la France. Bulletin biologique de la France et de la Belgique 91: 284-350.
Saint Girons, H. 1964. Notes sur L'ecologie et la structure des populations des Laticaudinae (Serpentes, Hydrophiidae) en Novelle Caledonie. Terre Vie 2: 185-214.
Sangha, S., Smola, M. A., McKinney, S. L., Crotzer, D. R., Shadrix, C. A. and Stewart, J. R. 1996. The effect of surgical removal of oviductal eggs on placental function and size of neonates in the viviparous snake *Virginia striatula*. Herpetologica 52: 32-36.
Schulte, J. A. and Moreno-Roark, F. 2009. Live birth in iguanian lizards predates the Pliocene Pleistocene. Biology Letters, in press. doi: 10.1098/rsbl.2009.0707
Schultz, T. J., Webb, J. K. and Christian, K. A. 2008. The physiological cost of pregnancy in a tropical viviparous snake. Copeia 2008: 637-642.
Seigel, R. A. and Fitch, H. S. 1984. Ecological patterns of relative clutch mass in snakes. Oecologia 61: 293-301.
Seigel, R. A. and Fitch, H. S. 1985. Annual variation in reproduction in snakes in a fluctuating environment. Journal of Animal Ecology 54: 497-505.
Seigel, R. A. and Ford, N. B. 1987. Reproductive ecology. Pp. 210-253. In R. A. Seigel, J. T. Collins, and S. S. Novak (eds), *Snakes: Ecology and Evolutionary Biology*. McGraw Hill, New York.
Seigel, R. A. and Ford, N. B. 2001. Phenotypic plasticity in reproductive traits: geographical variation in plasticity in a viviparous snake. Functional Ecology 15: 36-42.
Seigel, R. A., Fitch, H. S. and Ford, N. B. 1986. Variation in relative clutch mass in snakes among and within species. Herpetologica 42: 179-185.
Seigel, R. A., Huggins, M. M. and Ford, N. B. 1987. Reduction in locomotor ability as a cost of reproduction in pregnant snakes. Oecologia 73: 481-485.
Sergeev, A. M. 1940. Research on the viviparity of reptiles. Moscow Society of Naturalists 13: 1-34.
Serie, P. 1916. Ovoviviparidad de una culebra opistoglifa, "*Thamnodynastes nattereri* (Mikan) Gthr." Physis 2: 425.
Sever, D. M. and Ryan, T. J. 1999. Ultrastructure of the reproductive system of the black swamp snake (*Seminatrix pygaea*): Part I. Evidence for oviducal sperm storage. Journal of Morphology 241: 1-18.
Sever, D. M. and Hamlett, W. C. 2002. Female sperm storage in reptiles. Journal of Experimental Zoology 292: 187-199.
Sever, D. M., Ryan, T. J., Morris, T., Patton, D. and Swafford, S. 2000. Ultrastructure of the reproductive system of the black swamp snake (*Seminatrix pygaea*). II. Annual oviducal cycle. Journal Morphology 245: 146-160.
Shine, R. 1977. Reproduction in Australian elapid snakes. II. Female reproductive cycles. Australian Journal of Zoology 25: 655-666.
Shine, R. 1980. "Costs" of reproduction in reptiles. Oecologia 46: 92-100.
Shine, R. 1983a. Reptilian reproductive modes: the oviparity-viviparity continuum. Herpetologica 39: 1-8.
Shine, R. 1983b. Reptilian viviparity in cold climates: testing the assumptions of an evolutionary hypothesis. Oecologia 57: 397-405.
Shine, R. 1985. The evolution of viviparity in reptiles: an ecological analysis. Pp. 605-694. In C. Gans and F. Billet (eds), *Biology of the Reptilia,* vol. 15. John Wiley & Sons, New York.
Shine, R. 1987a. The evolution of viviparity: ecological correlates of reproductive mode within a genus of Australian snakes (*Pseudechis*: Elapidae). Copeia 1987: 551-563.

Shine, R. 1987b. Reproductive mode may determine geographic distributions in Australian venomous snakes (*Pseudechis*, Elapidae). Oecologia 71: 608-612.
Shine, R. 1988. Parental care in reptiles. Pp. 275-329. In C. Gans and R. Huey (eds), *Biology of the Reptilia*, vol. 16. A. R. Liss, New York.
Shine, R. 1995. A new hypothesis for the evolution of viviparity in reptiles. American Naturalist 145: 809-823.
Shine, R. 2003. Effects of pregnancy on locomotor performance: an experimental study on lizards. Oecologia 136: 450-456.
Shine, R. 2004. Does viviparity evolve in cold climate reptiles because pregnant females maintain stable (not high) body temperatures? Evolution 58: 1809-1818.
Shine, R. 2006. Is increased maternal basking an adaptation or a pre-adaptation to viviparity in lizards? Journal of Experimental Zoology 305A: 524-535.
Shine, R. and Berry, J. F. 1978. Climatic correlates of live-bearing in squamate reptiles. Oecologia (Berlin) 33: 261-268.
Shine, R. and Bull, J. J. 1979. The evolution of live-bearing in lizards and snakes. American Naturalist 113: 905-923.
Shine, R. and Lee, M. S. Y. 1999. A reanalysis of the evolution of viviparity and egg-guarding in squamate reptiles. Herpetologica 55: 538-549.
Shine, R. and Thompson, M. B. 2006. Did embryonic responses to incubation conditions drive the evolution of reproductive modes in squamate reptiles? Herpetological Monographs 20: 159-171.
Shine, R., Elphick, M. J. and Barrot, E. G. 2003. Sunny side up: lethally high, not low, nest temperatures may prevent oviparous reptiles from reproducing at high elevations. Biological Journal of the Linnean Society 78: 325-334.
Shine, R., O'Donnell, R., Langkilde, T., Wall, M. D. and Mason, R. T. 2005a. Snakes in search of sex: the relationship between mate-locating ability and mating success in male garter snakes. Animal Behaviour 69: 1251-1258.
Shine, R., Langkilde, T., Wall, M. and Mason, R. T. 2005b. Alternative male mating tactics in garter snakes, *Thamnophis sirtalis parietalis*. Animal Behaviour 70: 387-396.
Siegel, D. S. and Sever, D. M. 2008. Seasonal variation in the oviduct of female *Agkistrodon piscivorus* (Reptilia: Squamata): an ultrastructural investigation. Journal of Morphology 269: 980-997.
Siegel, D. S. and Sever, D. M. 2009. Sperm aggregations in female *Agkistrodon piscivorus* (Reptilia: Squamata): a histological and ultrastructural investigation. Journal of Morphology 269: 189-206.
Smedley, N. 1930. Oviparity in a sea-snake (*Laticauda colubrina*). Nature 126: 312-313.
Smith, M. A. 1930. Ovoviviparity in sea snakes. Nature 126: 568.
Smith, S. A. and Shine, R. 1997. Intraspecific variation in reproductive mode within the scincid lizard *Saiphos equalis*. Australian Journal of Zoology 45: 435-445.
Smith, S. A., Austin, C. C. and Shine, R. 2001. A phylogenetic analysis of variation in reproductive mode within an Australian lizard (*Saiphos equalis*, Scincidae). Biological Journal of the Linnean Society 74: 131-139.
Sowerby, A. de C. 1930. The reptiles and amphibians of the Manchurian region. Pp. 1-41. In *The Naturalists in Manchuria*, vol 4. Tientsin Press, Tientsin, China. [as cited by R. Shine (1985); original not seen].
Stewart, G. R. 1965. Thermal ecology of the garter snakes *Thamnophis sirtalis concinnus* (Hallowell) and *Thamnophis ordinoides* (Baird and Girard). Herpetologica 21: 81-102.
Stewart, J. R. 1985. Placentation in the lizard *Gerrhonotus coeruleus* with a comparison to the extraembryonic membranes of the oviparous *Gerrhonotus multicarinatus* (Sauria, Anguidae). Journal of Morphology 185: 101-114.

Stewart, J. R. 1989. Facultative placentotrophy and the evolution of squamate placentation: quality of eggs and neonates in *Virginia striatula*. American Naturalist 133: 111-137.

Stewart, J. R. 1990. Development of the extraembryonic membranes and histology of the placentae in *Virginia striatula* (Squamata: Serpentes). Journal of Morphology 205: 1-11.

Stewart, J. R. 1992. Placental structure and nutritional provision to embryos in predominantly lecithotrophic viviparous reptiles. American Zoologist 32: 303-312.

Stewart, J. R. 1993. Yolk sac placentation in reptiles: structural innovation in a fundamental vertebrate nutritional system. Journal Experimental Zoology 266: 431-449.

Stewart, J. R. 1997. Morphology and evolution of the egg of oviparous amniotes. Pp. 291-326. In S. S. Sumida and K. L. M. Martin (eds), *Amniote origins*. Academic Press, San Diego.

Stewart, J. R. and Castillo, R. E. 1984. Nutritional provision of the yolk of two species of viviparous reptiles. Physiological Zoology 57: 377-383.

Stewart, J. R. and Blackburn, D. G. 1988. Reptilian placentation: structural diversity and terminology. Copeia 1988: 838-851.

Stewart, J. R. and Thompson, M. B. 1996. Evolution of reptilian placentation: development of extraembryonic membranes of the Australian scincid lizards *Bassiana duperreyi* (oviparous) and *Pseudemoia entrecasteauxii* (viviparous). Journal of Morphology 227: 349-370.

Stewart, J. R. and Thompson, M. B. 1998. Placental ontogeny of the Australian scincid lizards *Niveoscincus coventryi* and *Pseudemoia spenceri*. Journal of Experimental Zoology 282: 535-559.

Stewart, J. R. and Florian, J. D., Jr. 2000. Ontogeny of the extraembryonic membranes of the oviparous lizard, *Eumeces fasciatus* (Squamata: Scincidae). Journal of Morphology 244: 81-107.

Stewart, J. R. and Thompson, M. B. 2000. Evolution of placentation among squamate reptiles: recent research and future directions. Comparative Biochemistry and Physiology A: Molecular and Integrative Physiology 127: 411-431.

Stewart, J. R. and Brasch, K. R. 2003. Ultrastructure of the placentae of the natricine snake, *Virginia striatula* (Reptilia: Squamata). Journal of Morphology 255: 177-201

Stewart, J. R. and Thompson, M. B. 2003. Evolutionary transformations of the fetal membranes of viviparous reptiles: a case study in two lineages. Journal of Experimental Zoology 299A: 13-32.

Stewart, J. R. and Thompson, M. B. 2009a. Parallel evolution of placentation in Australian scincid lizards. Journal of Experimental Zoology. Molecular and Developmental Evolution B 312: 590-602.

Stewart, J. R. and Thompson, M. B. 2009b. Placental ontogeny in Tasmanian snow skinks (genus *Niveoscincus*) (Lacertilia: Scincidae). Journal of Morphology 270: 485-516.

Stewart, J. R. and Ecay, T. W. 2010. Patterns of maternal provision and embryonic mobilization of calcium in oviparous and viviparous squamate reptiles. Herpetological Conservation and Biology (in press).

Stewart, J. R., Heulin, B. and Surget-Groba, Y. 2004a. Extraembryonic membrane development in a reproductively bimodal lizard, *Lacerta (Zootoca) vivipara*. Zoology 107: 289-314.

Stewart, J. R., Ecay, T. W. and Blackburn, D. G. 2004b. Sources and timing of calcium mobilization during embryonic development of the corn snake *Pantherophis guttatus*. Comparative Biochemistry and Physiology 139: 335-341.

Stewart, J. R., Blackburn, D. G., Baxter, D. C. and Hoffman, L. H. 1990. Nutritional provision to the embryos in *Thamnophis ordinoides* (Squamata: Colubridae), a predominantly lecithotrophic placental reptile. Physiological Zoology 63: 722-734.

Stewart, J. R., Ecay, T. W., Garland, C. P., Fregoso, S. P., Price, E. K., Herbert, J. F. and Thompson, M. B. 2009a. Maternal provision and embryonic uptake of calcium in an oviparous and a placentotrophic viviparous Australian lizard (Lacertilia: Scincidae). Comparative Biochemistry and Physiology—Part A: Molecular & Integrative Physiology 153: 202-208.

Stewart, J. R., Ecay, T. W. and Heulin, B. 2009b. Calcium provision to oviparous and viviparous embryos of the reproductively bimodal lizard *Lacerta (Zootoca) vivipara*. Journal of Experimental Biology 212: 2520-2524.

Surget-Groba, Y., Heulin, B., Guillaume, C.-P., Puky, M., Semenov, D., Orlova, V., Kupriyanova, L., Ghira, I. and Smajda, B. 2006. Multiple origins of viviparity, or reversal from viviparity to oviparity? The European common lizard (*Zootoca vivipara*, Lacertidae) and the evolution of parity. Biological Journal of the Linnean Society 87: 1-11.

Taylor, E. N. and DeNardo, D. E. 2005. Reproductive ecology of western Diamond-backed rattlesnakes *(Crotalus atrox)* in the Sonoran Desert. Copeia 2005: 152-158.

Thompson, M. B. 1989. Patterns of metabolism in embryonic reptiles. Respiration Physiology 76:243-256.

Thompson, M. B. and Stewart, J. R. 1997. Embryonic metabolism and growth in lizards of the genus *Eumeces*. Comparative Biochemistry and Physiology A 118: 647-654.

Thompson, M. B. and Russell, K. J. 1999. Embryonic energetics in eggs of two species of Australian skink, *Morethia boulengeri* and *Morethia adelaidensis*. Journal of Herpetology 33: 291-297.

Thompson, M. B. and Speake, B. K. 2004. Egg morphology and composition. Pp. 45-74. In D.C. Deeming (ed.), *Reptilian Incubation: Environment, Evolution and Behaviour*. Nottingham University Press, Nottingham, U.K.

Thompson, M. B. and Blackburn, D. G. 2006. Evolution of viviparity in reptiles: introduction to the symposium. Herpetological Monographs 20: 129-130.

Thompson, M. B. and Speake, B. K. 2006. A review of the evolution of viviparity in lizards: structure, function, and physiology of the placenta. Journal of Comparative Physiology B 176: 179-189.

Thompson, M. B., Stewart J. R. and Speake B. K. 2000. Comparison of nutrient transport across the placenta of lizards differing in placental complexity. Comparative Biochemistry and Physiology A: Molecular and Integrative Physiology 127: 469-479.

Thompson, M. B., Parker, S. L. and Blackburn, D. G. 2010. Reproduction in reptiles from genes to ecology: a retrospective and prospective vision. Herpetological Conservation and Biology (in press).

Thompson, M. B., Adams, S. M., Herbert, J. F., Biazik, J. M. and Murphy, C. R. 2004. Placental function in lizards. International Congress Series 1275: 218-225.

Thompson, M. B., Stewart J. R., Speake B. K., Russell K. J., McCartney R. J. and Surai, P. F. 1999. Placental nutrition in a viviparous lizard (*Pseudemoia pagenstecheri*) with a complex placenta. Journal of Zoology 248: 295-305.

Tinkle, D. W. 1967. The life and demography of the side-blotched lizard, *Uta stansburiana*. Miscellaneous Publications of the Museum of Zoology, University of Michigan 132: 1-182.

Tinkle, D. W. 1969. The concept of reproductive effort and its relation to the evolution of life histories of lizards. American Naturalist 103: 501-516.

Tinkle, D. W. and Gibbons, J. W. 1977. The distribution and evolution of viviparity in reptiles. Miscellaneous Publications of the Museum of Zoology, University of Michigan 154: 1-55.

Tinkle, D. W., Wilbur, H. M. and Tilley, S. G. 1970. Evolutionary strategies in lizard reproduction. Evolution 24: 55-74.

Tsai, T.-S. and Tu, M.-C. 2001. Reproductive cycle of female Chinese green tree vipers, *Trimeresurus stejnegeri stejnegeri,* in northern Taiwan. Herpetologica 57: 157-168.

van Damme, R., Bauwens, D. and Verheyen, R. F. 1989. Effect of relative clutch mass on sprint speed in the lizard *Lacerta vivipara.* Journal of Herpetology 23: 459-461.

Vidal, N., Rage, J.-C., Couloux, A. and Hedges, S. B. 2009. Snakes (Serpentes). Pp. 390-397. In S.B. Hedges and S. Kumar (eds), *The Timetree of Life.* Oxford University Press, Oxford, U.K.

Villagran, M., Mendez, F. R. and Stewart, J. R. 2005. Placentation in the Mexican lizard *Sceloporus mucronatus* (Squamata: Phrynosomatidae). Journal of Morphology 264: 286-297.

Vitt, L. J. 1981. Lizard reproduction: habitat specificity and constraints on relative clutch mass. American Naturalist 117: 506-514.

Vitt, L. J. and Congdon, J. D. 1978. Body shape, reproductive effort, and relative clutch mass in lizards: resolution of a paradox. American Naturalist 112: 595-608.

Vitt, L. J. and Price, H. J. 1982. Ecological and evolutionary determinants of relative clutch mass in lizards. Herpetologica 1982: 237-255.

Voris, H. K. and Glodek, G. S. 1980. Habitat, diet, and reproduction of the file snake, *Acrochordus granulatus,* in the straits of Malacca. Journal of Herpetology 14: 108-111.

Wall, F. 1921. *Ophidia Taprobanica, or the Snakes of Ceylon.* H.R. Cottle, Government Printer, New Delhi, India.

Wangkulangkul, S., Thirakhupt, K. and Voris, H. K. 2005. Sexual size dimorphism and reproductive cycle of the little file snake *Acrochordus.* Science Asia 31: 257-263.

Webb, J. K. 2004. Pregnancy decreases swimming performance of female Northern death adders *(Acanthophis praelongus).* Copeia 2004: 357-363.

Webb, J. K., Shine, R. and Christian, K. A. 2006. The adaptive significance of reptilian viviparity in the tropics: testing the maternal manipulation hypothesis. Evolution 60: 115-122.

Weekes, H. C. 1927. Placentation and other phenomena in the scincid lizard *Lygosoma (Hinulia) quoyi.* Proceedings of the Linnean Society of New South Wales 52: 499-554.

Weekes, H. C. 1929. On placentation in reptiles. I. Proceedings of the Linnean Society of New South Wales 54: 34-60.

Weekes, H. C. 1930. On placentation in reptiles. II. Proceedings of the Linnean Society of New South Wales 55: 550-576.

Weekes, H. C. 1933. On the distribution, habitat and reproductive habits of certain European and Australian snakes and lizards with particular regard to their adoption of viviparity. Proceedings of the Linnean Society of New South Wales 58: 270-274.

Weekes, H. C. 1935. A review of placentation among reptiles, with particular regard to the function and evolution of the placenta. Proceedings of the Zoological Society of London 2: 625-645.

Wharton, C. H. 1966. Reproduction and growth in the cottonmouths, *Agkistrodon piscivorus* Lacépède, of Cedar Keys, Florida. Copeia 1966: 149-161.

Whittier, J. M., Mason, R. T. and Crews, D. 1987. Plasma steroid hormone levels of female red-sided garter snakes, *Thamnophis sirtalis parietalis*: relationship to mating and gestation. General and Comparative Endocrinology 67: 33-43.
Whittier, J. M., West, N. B. and Brenner, R. M. 1991. Immunorecognition of estrogen receptors by monoclonal antibody H222 in reproductive tissues of the red-sided garter snake. General and Comparative Endocrinology 81: 1-6.
Winne, C. T. and Hopkins, W. A. 2006. Influence of sex and reproductive condition on terrestrial and aquatic locomotor performance in the semi-aquatic snake, *Seminatrix pygaea*. Functional Ecology 20: 1054-1061.
Wright, A. H. and Wright, A. A. 1957. *Handbook of Snakes of the United States and Canada*. Comstock Publishing, Cornell University Press, Ithaca, New York. vols. 1 & 2. Pp. 1105.
Yaron, Z. 1985. Reptile placentation and gestation: structure, function, and endocrine control. Pp. 527-603. In C. Gans and F. Billet (eds), *Biology of the Reptilia*, vol. 15. John Wiley & Sons, New York.
Zaher, H., Grazziotin, F. G., Cadle, J. E., Murphy, R. W., de Moura-Leite, J. C. and Bonatto, S. L. 2009. Molecular phylogeny of advanced snakes (Serpentes, Caenophidia) with an emphasis on South American Xenodontines: a revised classification and descriptions of new taxa. Papéis Avulsos de Zoologia 49: 115-153.

CHAPTER 6

The Ophidian Testis, Spermatogenesis, and Mature Spermatozoa

Kevin M. Gribbins[1] *and Justin L. Rheubert*[2]

6.1 INTRODUCTION

This chapter is intended to cover morphological data on the testis, spermatogenesis, ultrastructure of spermiogenesis, and the mature spermatozoon in snakes. Though it would be an immense task in almost all other major vertebrate clades, snakes have largely been ignored in terms of testicular architecture, detailed morphological data on the germ cell development strategy during spermatogenesis, and the ultrastructure during spermatogenesis and/or of the sperm. Most of the recent data (last 10 years) on the testis/spermatogenesis in snakes have focused on reproductive cycles or in some cases the ultrastructure of specific parts of the testis, spermatogenesis, or the mature sperm. In the last 15 years, there has been an effort to provide ultrastructural information for the spermatozoon in several species of snakes, which has led to preliminary constructions of phylogenetic relationships between certain families within the Ophidia (Jamieson 1995; Oliver *et al.* 1996; Tavares-Bastos *et al.* 2008).

This chapter will be restricted to providing overall data on the morphology and ultrastructure of the testis, spermatogenesis, or products of the process of spermatogenesis. Though these data are limited in most regards, they suffice for a preliminary analysis of the comparative biology of the testis and spermatogenic process in snakes. Reproductive or spermatogenic cycles will not be a focus in this chapter, for considerations of these cycles refer to **Chapter 12** in this volume. The information provided here will not be a simple review of what is known but when appropriate (sperm ultrastructure for example) the morphology will be

[1]Department of Biology, Wittenberg University, PO Box 720, Springfield OH 45501-0720
[2]Department of Biological Sciences, Southeastern Louisiana Univeristy, SLU 10736, Hammond LA

placed in an evolutionary perspective. Portraying the morphological characteristics during spermatogenesis and/or of the testis of snakes in this way will provide non-traditional data that will hopefully be useful for cladistic analyses within Ophidia and Reptilia. We hope that this chapter not only sheds some light on what is known of the testis in snakes but, as our major goal, stimulates interest in furthering research efforts so as to better understand the morphology and ultrastructure of the testis, spermatogenesis, and the mature spermatozoa.

6.2 THE TESTIS

Unlike the testes of mammals, those of snakes and other reptiles are intra-abdominal, develop embryonically in close association with the kidneys (Raynaud and Peiau 1985), and are almost always elongated and cylindrical in shape (Fig. 6.1). Each testis is held together and is covered with a thin fibrous tunica albuginea and connected to the body wall via a mesorchium. Because of the restricted width of the tunica (roughly 3-10 µm; Figs. 6.2B and 6.3A), the testis in snakes can be fragile and soft to the touch (especially in nonspermatogenic months), and one can typically visualize the coiled seminiferous tubules just under this connective tissue layer (Fig. 6.2B). However, it should be noted that in times of maximum spermatogenic output the testis can easily double its size, producing an intratesticular pressure that pushes against the tunica albuginea giving the testis added firmness. In most snakes looked at to date, the tunica sends few septa (Figs. 6.2 and 6.3) interiorly to separate the testis into distinct compartments such as lobes. However, leptotyphlopid (Werner and Drook 1967) and typhlopid (Robb 1960) testes show either several distinct lobes made up of several seminiferous tubules or a distinct caudal lobe is observed off of the main body of the testis.

The testes in most snakes show bilateral asymmetry as far as size and distance from the cloaca as noted by Volsøe (1944) and Fox (1952). The right testis is typically longer than the left testis (Fig. 6.1A) and is almost always more rostrally located than the left. In some cases, like the Black Swamp Snake (*Seminatrix pygaea*; Sever 2004) this distance can be substantially different between the right and left testes, or in other species, as in the Cottonmouth (*Agkistrodon piscivorus*), the difference in distance from the cloaca between the testes is less than the actual length of each testis (Fig. 6.1B).

The external color of the testes in snakes ranges between white and cream in most species. The snake testis, as in all vertebrates, is composed of a coiled mass of seminiferous tubules (Fig. 6.2B). The intertubular space between tubules is occupied by interstitial (Leydig) cells, a few lymph and blood vessels, leukocytes, myofibroblasts, and is surrounded by extracellular matrix that consists of many collagen fibers (Fig. 6.3A-C).

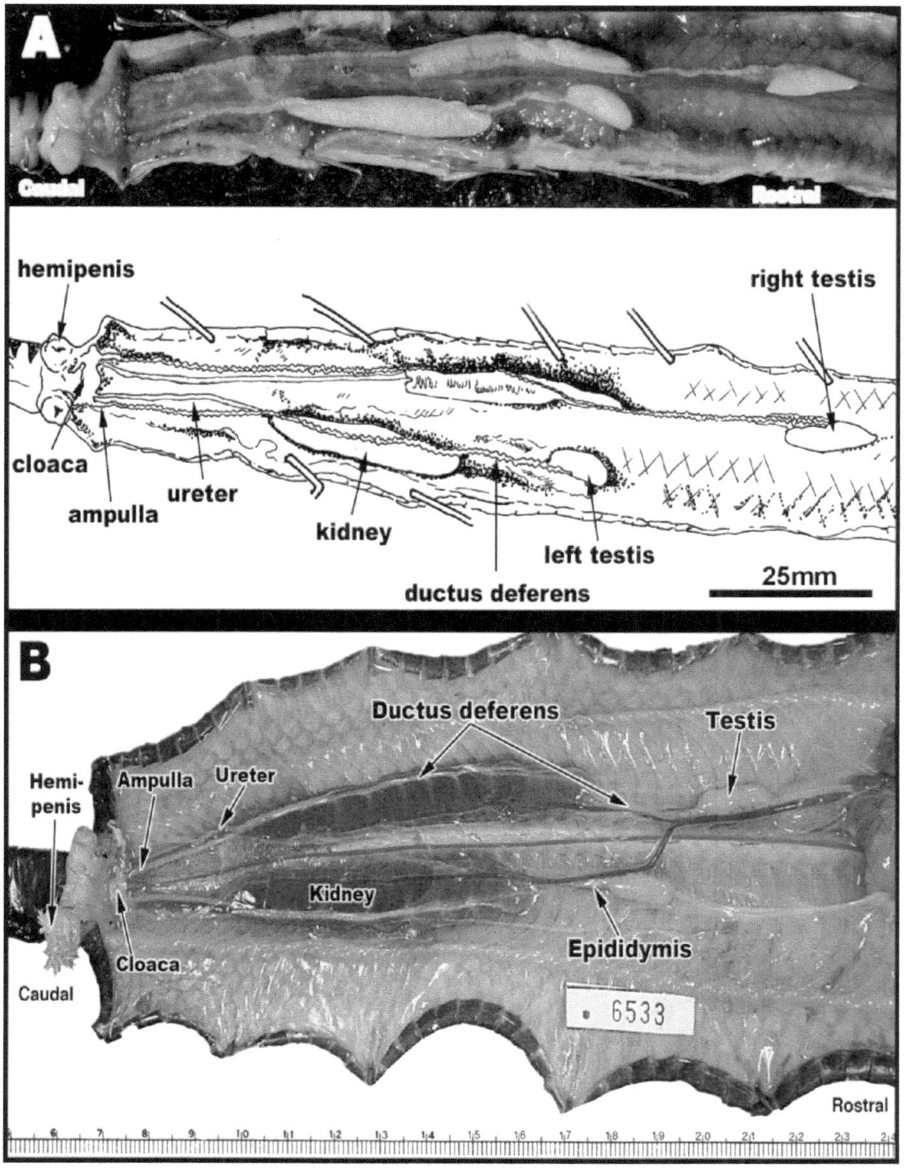

Fig. 6.1 Dissections of the urogenital tracts of *Seminatrix pygaea* (**A**) and of *Agkistrodon piscivorus* (**B**). Scale in mm. Note the location of the testes and excurrent ducts of the reproductive system of each species. Photo A: David M. Sever, Photo B: Dustin S. Siegel.

Color image of this figure appears in the color plate section at the end of the book.

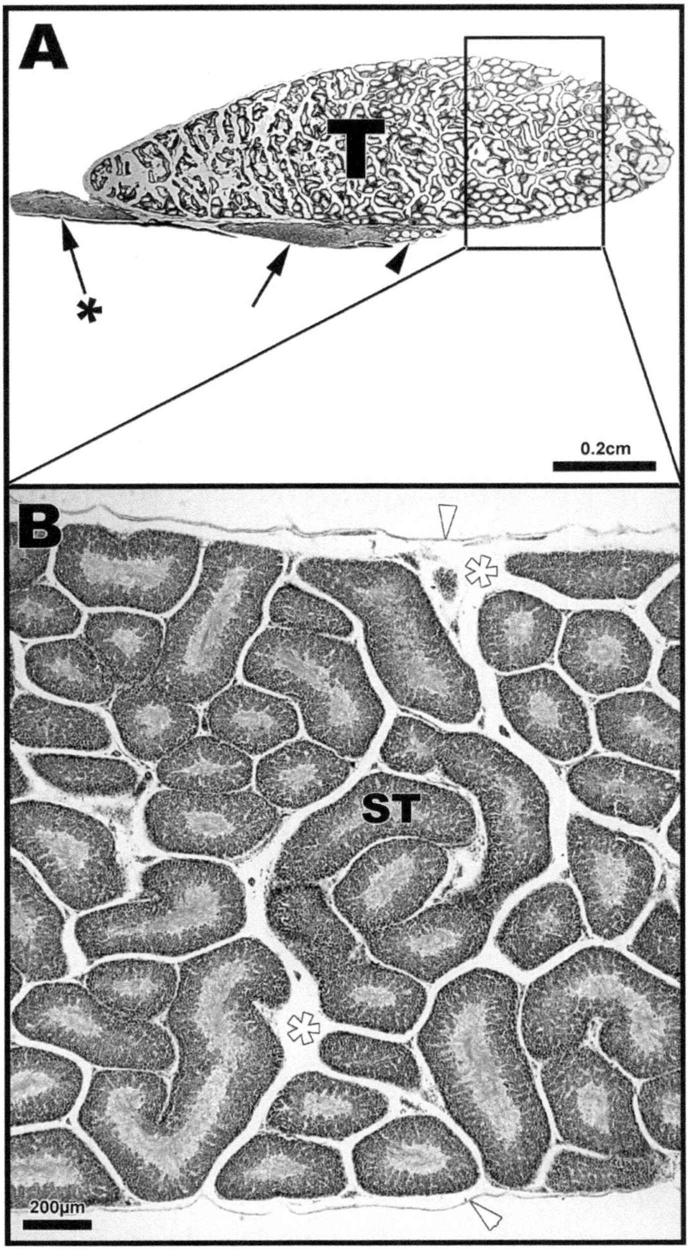

Fig. 6.2 A. Sagittal section of entire testis and anterior excurrent ducts of *Carphophis vermis*. Testis (T), efferent ducts (black arrowhead), rostral epididymis (black arrow), caudal epididymis (*). **B.** Low power view of seminiferous tubules in cross and sagittal sections near the tunica albuginea (white arrowheads) of the testis of *Carphophis vermis*. Seminiferous tubules (ST), Interstitial space (white *). Light micrographs taken from slides in the collection of Robert D. Aldridge.

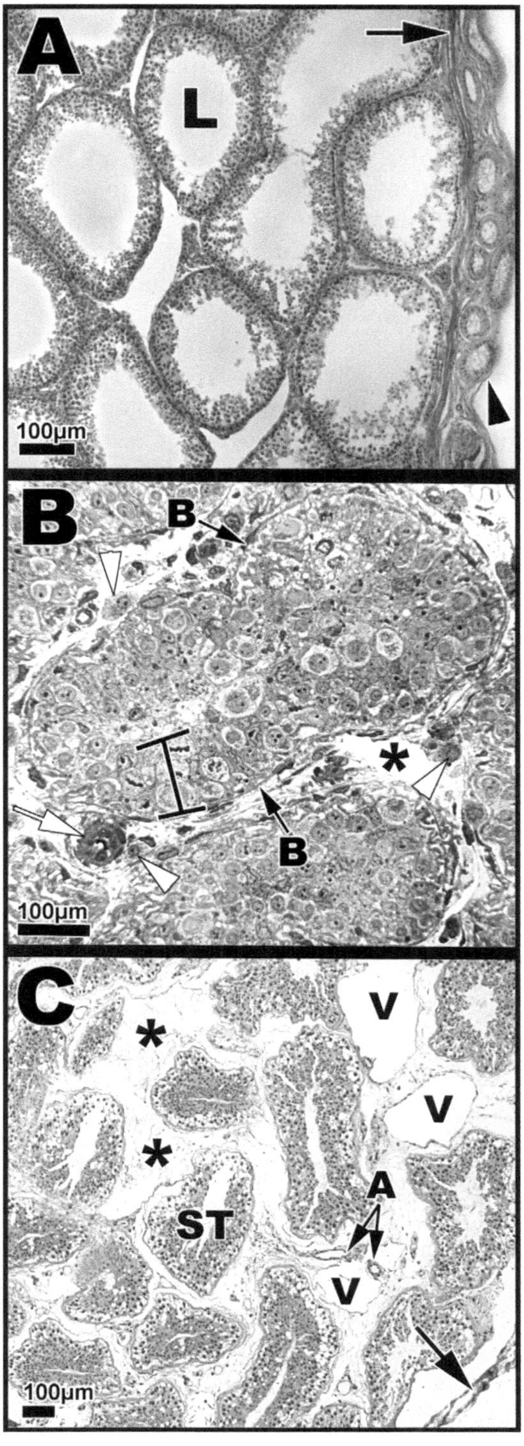

Fig. 6.3 A. Cross-sectional views of seminiferous tubules within the testis of *Carphophis vermis*. Lumen of tubules (L), tunica albuginea (black arrow), anterior excurrent ducts (black arrowhead). Note there are no extensions of the tunica albuginea into the testis proper and between the seminiferous tubules. **B.** Transverse view of two seminiferous tubules within the testis of *Seminatrix pygaea*. The basement membrane is visible (B) and a thick seminiferous epithelium (black line bracket) lines the tubules. Within the interstitial space (*) there are Leydig cells (white arrowheads) in close association to the blood vessels (white arrow). **C.** Scanning view of the seminiferous tubules within the testis of *Seminatrix pygaea*. The interstitial space (*) is thick with many arteries (A) and veins (V). Seminiferous tubules (ST), tunica albuginea (black arrow). Light micrograph A taken from slides in the collection of Robert D. Aldridge.

Color image of this figure appears in the color plate section at the end of the book.

6.2.1 Seminiferous Tubules

The seminiferous tubules of snakes form a tight mass of strongly convoluted tubules surrounded by the tunica albuginea. These tubules are continuous with the anterior ducts (see **Chapter 10**) and mature spermatozoa are emptied from the seminiferous tubules to the efferent ductuli, which drain to the epididymis where spermatozoa will then reside for varying amounts of time before ejaculation. The seminiferous tubules are hollow (Fig. 6.3A-C) and typically have a well-defined lumen that varies in size due to seasonal activities of the testis. The lumen is surrounded by an epithelium called the seminiferous epithelium (Fig. 6.4A), which rests on a thin membrana propria or boundary layer (~between 2-10 μm; Figs. 6.4B, 6.5A,B), and is made up of complex specialized Sertoli cells. The Sertoli cells are highly branched cells that invest and nurture the developing germ cells during spermatogenesis (Fig. 6.4B). The thickness of the seminiferous epithelium again is dependent on the seasonal activity of the testis and on the stage of the spermatogenic cycle of the testis.

6.2.2 Sertoli Cells

The seminiferous epithelium in snakes, as in all amniotes studied to date, has a permanent population of Sertoli (sustentacular) cells. Thus, irrespective of the seasonal activity of the testis, Sertoli cell components are easily identified within the seminiferous epithelium. Thin sections of the testis stained with toluidine blue often reveal the dark staining cytoplasmic processes that wrap around developing germ cells within the seminiferous epithelium (Fig. 6.4B). Because of this close association with the developing germ cells and the support, steroid/lipid synthesis and sequestering ability, and nutritional provisions provided to the germ cells by Sertoli cells (at least in mammals) (Cooksey and Rothwell 1973; Pudney 1993), they are often called "nurse" cells.

Surprisingly in some squamates, the Western Fence Lizard (*Sceloporus occidentalis;* Wilhoft and Quay 1961) and the Tiger Whiptail (*Cnemidorphorous tigris;* Goldberg and Lowe 1966), Sertoli cells are observed being shed to the lumen at the end of the breeding season. But these studies were only performed at the light microscope level and have not been confirmed via electron microscopy. A few snakes do have at least some histological data describing the Sertoli cell and lipid cycle of the seminiferous epithelium at the light microscope level. These studies include the European Viper (*Vipera berus;* Marshall and Woolf 1957), Indian Cobra (*Naja naja;* Lofts *et al.* 1966), *Agkistrodon piscivorus* (Scott *et al.* 1995), the Blackish Blind Snake (*Ramphotyphlops nigrescens;* Shea 2001), and the Black Swamp Snake (*Seminatrix pygaea;* Sever *et al.* 2002). Depending on the species, lipid droplets are typically smallest during active spermatogenesis (particularly during spermiogenesis; Fig. 6.5B). Conversely, during spermatogenic quiescent periods, there is a rapid accumulation of cholesterol rich lipid droplets within the snake seminiferous epithelium (Fig. 6.5A). Once

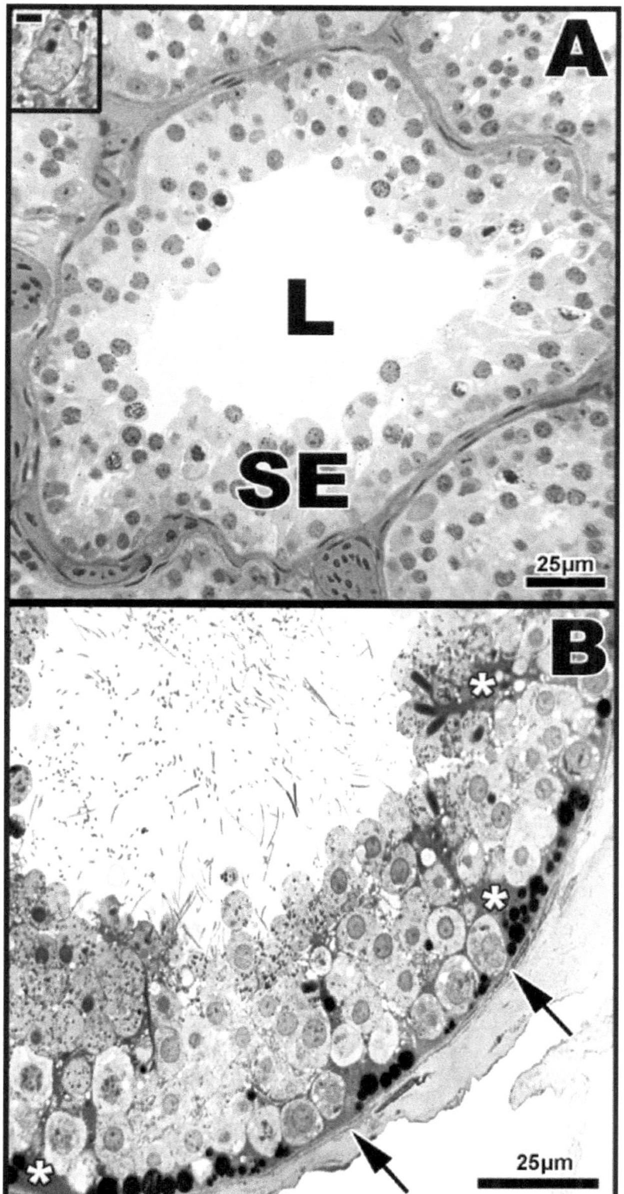

Fig. 6.4 A. A June seminiferous tubule in cross section within the testis of *Seminatrix pygaea*. Note the thick boundary layer around the tubule and the thick seminiferous epithelium (SE) that is made up of Sertoli cells (inset: Sertoli nucleus) that wrap around the developing germ cells. Lumen (L). **B.** The seminiferous epithelium within the July testis of *Seminatrix pygaea*. The dark blue staining Sertoli cell processes (*) are seen investing the lighter staining germ cells. Boundary layer (black arrows). Light micrographs.

Color image of this figure appears in the color plate section at the end of the book.

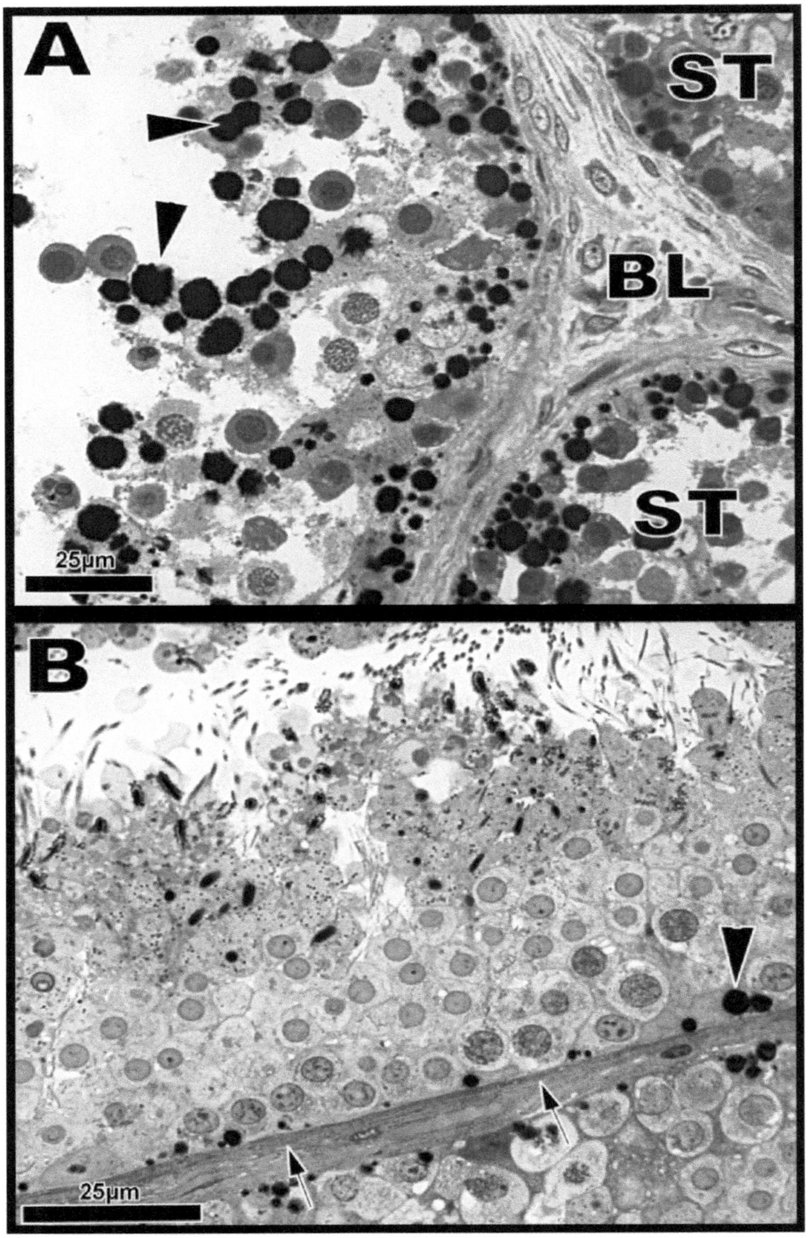

Fig. 6.5 A. September seminiferous tubules within the testis of *Agkistrodon contortrix* show the completion of spermatogenesis and an increase in lipid accumulation or droplets (black arrowheads). Boundary layer and interstitium (BL), seminiferous tubules (ST). **B.** A July seminiferous tubule in *Agkistrodon contortrix* that is spermiogenic and has few lipid droplets (black arrowhead) and a prominent boundary layer (black arrows). Light micrographs provided by Dustin S. Siegel.

spermatogenic recrudescence occurs and spermatogenesis continues through the subsequent phases of mitosis, meiosis, and spermiogenesis, lipid content slowly regresses back to its lowest levels during late spermiogenesis.

There are very few studies that report ultrastructural information on the Sertoli cell in snakes and squamates in general. Two studies describe at least minor details on the Sertoli cell ultrastructure including some information on junctional complexes between germ cells and the Sertoli cell processes within snakes (Hondo *et al.* 1996; Al-Dokhi *et al.* 2004). General trends will be given here with the warning that more ultrastructural information on the Sertoli cell needs to be gathered not only in snakes but also within the paraphyletic Reptilia. Most data presented in this section are unpublished.

Intracellular components and fine structure of the Sertoli cell in the few snakes that have been described are very similar in all species. The irregular nucleus (Fig. 6.6A,B) is usually basally located and has one or two prominent nucleoli. In some instances the Sertoli cell nucleus can appear more apical within the seminiferous epithelium particularly during the late proliferation phase of spermatogenesis. There is typically either diffuse or very large lipid droplets associated with the basal compartment of the Sertoli cells. The droplets are often found juxtapositioned to the nucleus (Fig. 6.6B). The Sertoli cell wraps long cytoplasmic processes around the developing germ cells (Figs. 6.6A, 6.7) and these sustentacular cells sit on a prominent basement membrane (Fig. 6.6A). The cytoplasm associated with both the nucleus and within the cellular processes of Sertoli cells typically contain the same organelles. There is an abundance of small to large lipid inclusions, mitochondria, and well-developed smooth and rough endoplasmic reticulum (Fig. 6.8). There also is typically a prominent Golgi complex with many transport vesicles budding off of it and, depending on the time of year, numerous lysosomes (particularly in the autumn or fall of temperate species) can be found within the Sertoli cell cytoplasm. Bundles of intermediate filaments, actin filaments (especially in association with junctional complexes), and moderate amounts of glycogen granules are represented within the Sertoli cell cytoplasm year round in snakes (Gribbins unpublished).

Relatively detailed information on the ultrastructure of Sertoli/germ cell and Sertoli/Sertoli junctions have been described for only one species of snake, the Japanese Ratsnake (*Elaphe climacophora*; Hondo *et al.* 1996). Adhering junctions (desomsome-like) are seen between spermatogonia, spermatocytes, and most often between spermatids and juxtapositioned Sertoli cells. All three major cell types show the same morphology and ectoplasmic specializations. Like *Elaphe climacophora*, desmosome-like junctions between germ cells and Sertoli cells are common in the testis of *Agkistrodon piscivorus* and the Copperhead (*Agkistrodon contortrix*; Figs. 6.9A,B, 6.10A) (Gribbins unpublished). The desmosome-like junctions in these two pit vipers are very simple in structure and may be described as areas of opposing plasma membrane separated by a 12-22 nm space

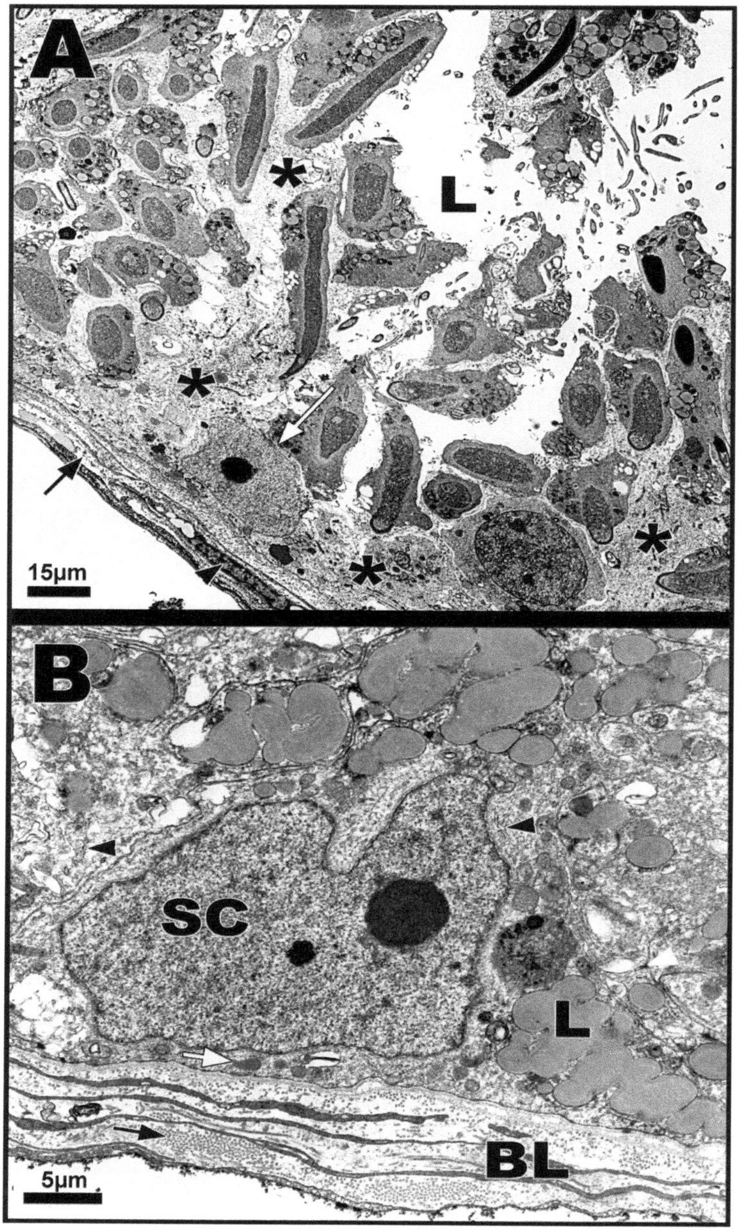

Fig. 6.6 Low (**A**) and high (**B**) power electron micrographs depicting the seminiferous epithelium with a *Thamnophis sauritus* testis. In **A**, Sertoli cells sit on a prominent boundary layer (black arrow) made up of myofibroblasts (black arrowhead) and their nuclei (white arrow) rest on the basement membrane of the seminiferous epithelium. The Sertoli cell processes (*) wrap around developing germ cells. Lumen (L). **B** shows a Sertoli cell nucleus (SC), lipids (L), and Sertoli cytoplasm with mitochondria (white arrow) and abundant ER (black arrowheads) resting on a boundary layer (BL) made up of numerous collagen fibers (black arrow).

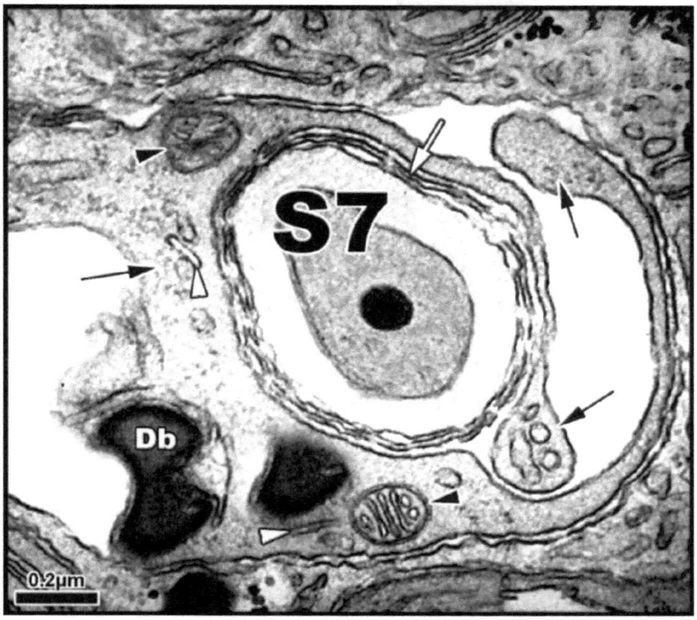

Fig. 6.7 Electron micrographs showing the close association a Sertoli cell process (black arrows) has with a developing spermatid (S7) in the seminiferous epithelium of *Agkistrodon contortrix*. Note that the Sertoli cell produces and surrounds several cell membrane laminae (white arrow) around the spermatid. There are also mitochondria (black arrowheads), smooth ER (white arrowheads), and dense bodies (DB) found within the cytoplasm of the Sertoli cell process.

(Fig. 6.9A inset) (Gribbins unpublished). The junctions are flanked on the cytoplasmic and Sertoli sides of the membranes by thick dark staining subsurface densities (Figs. 6.9A, 6.10A insets). In some cases, the subsurface densities are so thick they appear to cover the intercellular space between the opposing membranes (Fig. 6.10A inset). Smooth endoplasmic reticulum (SER) is associated with the cytoplasmic and Sertoli sides of the membranes (Figs. 6.9B, 6.10A insets). The SER is located between 8 and 40 nm from the plasma membrane and lies either in sagittal or transverse positions to the desmosome-like junction (Gribbins unpublished). Some of the desmosome-like junctions, especially those associated with spermatogonia and spermatids, show an abundance of 7 nm filaments close to the subsurface densities (Fig. 6.9A inset; Gribbins unpublished).

Tight junctions between adjacent Sertoli cells seem to be common within snakes and, in the species mentioned above, appear similar in structure. They are most often basally located and separate spermatogonia and pre-leptotene cells from more advanced spermatocytes and spermatids intercellularly (Fig 6.9B). They are numerous in number and may be found between juxtapositioned Sertoli cells even in the absence of spermatogonia or preleptotene spermatocytes (Fig. 6.10B). Rarely are they found as single

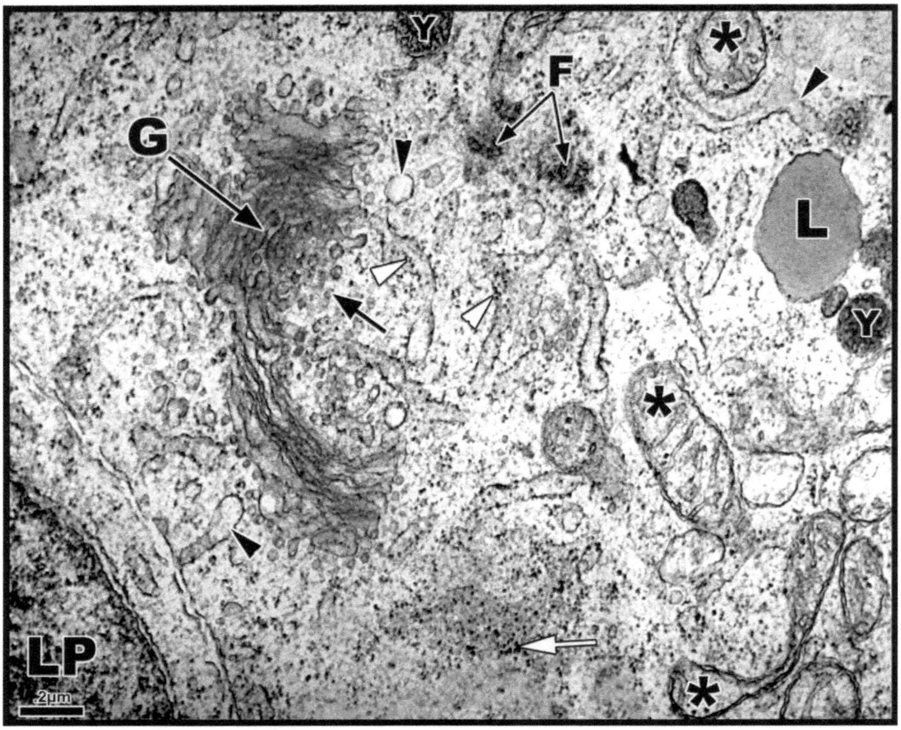

Fig. 6.8 An electron micrograph that represents the cytoplasm within the main body of the Sertoli cell. The represented inclusions are glycogen (white arrow) and lipid droplets (L). The denoted organelles are mitochondria (*), Golgi appratus (G), transport vesicles (black arrow), intermediate filaments (F), smooth ER (black arrowheads), rough ER (white arrowheads), and lysosomes (Y).

junctions between Sertoli cells. Structurally, they appear as focal contacts where opposing Sertoli cell membranes are fused together. Subsurface densities are found on both sides of the opposing membranes but in a more restricted fashion than that seen in desmosome-like junctions (Figs. 6.10B inset, 6.11A). The subsurface densities are concentrated at the focal points of contact between opposing membranes (Fig. 6.11A inset). Often smooth endoplasmic reticulum is found associated with the tight junctions on both sides of the opposing membrane (between 13-45 nm away from the junctional complex) (Fig. 6.11A inset) in *Agkistrodon* (Gribbins unpublished). Sometimes desmosome-like junctions are found close to the tight junctions (Figs. 6.9B, 6.10B insets).

Frequently in *Agkistrodon*, seminiferous epithelium intercellular pathways between Sertoli cells split apical to spermatogonia or spermatocytes forming many invaginations or interdigitations (Figs. 6.9B, 6.10B) between adjacent Sertoli cells. This type of tight junction organization is most often seen during active spermiogenesis within the testis of *A. contortrix* and *A. piscivorus* (Gribbins unpublished). At the point where the Sertoli

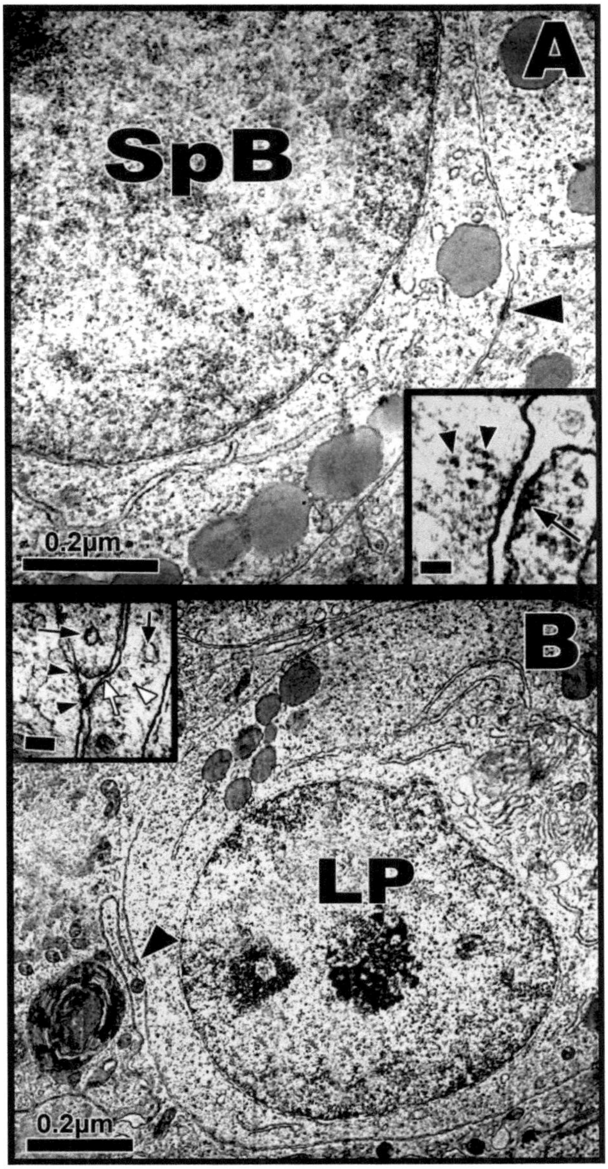

Fig. 6.9 Junctional complexes between Sertoli cells and germ cells within the seminiferous epithelium of *Agkistrodon contortrix*. Bars = 2 µm; inset Bars = 0.2 µm. **A.** Desmosome-like junction (black arrowhead) between a Sertoli cell and spermatogonia B. The inset shows the subsurface densities (black arrow) on both sides of the junction. Also in close association to the junctional complex are groups of 7 nm filaments (inset, black arrowheads). **B.** Shows a combination (black arrowhead) tight (inset, black arrowheads) and desmosome-like junctional (inset, white arrow) complexes within interdigitations of the Sertoli cell near a preleptotene spermatocyte (LP). Inset: Individual 7 nm filaments (white arrow), Smooth ER (black arrows). Electron micrographs.

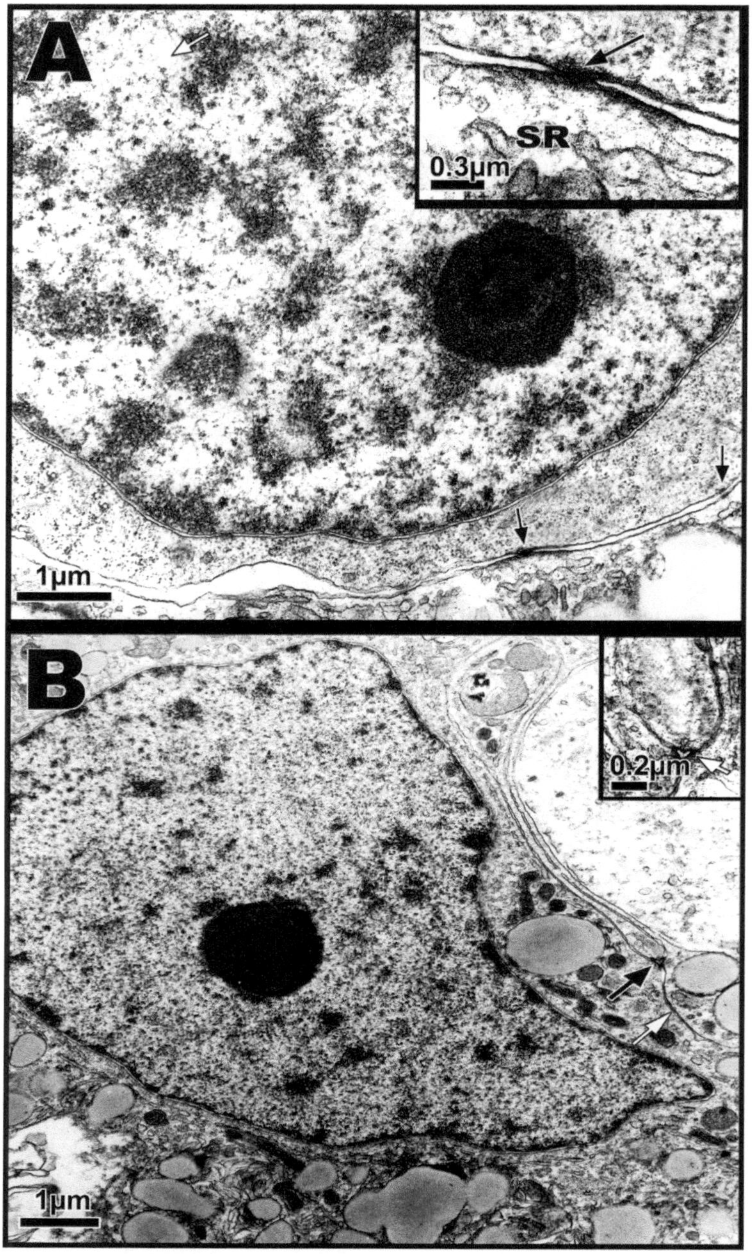

Fig. 6.10 A. Desomosome-like junctions (black arrows) between a spermatogonia and the Sertoli cell within the germinal epithelium of *Agkistrodon piscivorus*. Inset: thick subsurface densities (black arrow), smooth ER (SR). **B.** Interdigitating tight junctions (black arrow) between Sertoli cells within the seminiferous epithelium of *Agkistrodon piscivorus* are positioned just apically to numerous desomsome-like junctions (white arrow). Inset: subsurface densities (white arrows). Electron micrographs.

cell intercellular pathway splits above spermatogonia tight junctions are often found between the three focal meeting points of opposing Sertoli cell membrane (Figs. 6.9, 6.10B inset). These types of tight junctions have been termed interdigitating tight junctions because of their location near invaginations formed by the Sertoli cell processes around developing germ cells (Russell *et al.* 1990). Subsurface densities again are located on either side of opposing membranes but are slightly denser than non-interdigitating tight junctions. Smooth endoplasmic reticulum is also associated with these interdigitating tight junctions (Fig. 6.9B inset).

A blood testis barrier has been demonstrated in squamates (Baccetti *et al.* 1983, Bergmann *et al.* 1984) using electron-dense intercellular tracers. However, no study to date provides data that this barrier in snakes hinders the movement of intercellular tracers between Sertoli cells. Tight junctions have been described in only one species of snake, *Elaphe climacophora* (Hondo *et al.* 1996). Thus, detailed ultrastructural information on lanthanum nitrate (intercellular tracer) penetration in *Seminatrix pygaea* will be provided here (Gribbins unpublished). *S. pygaea* does have a functional blood testis barrier similar to that described in other squamates, birds (Osman *et al.* 1980; Bacettie *et al.* 1983; Bergmann and Schindelmeiser 1987), and mammals (see Pudney 1993). Testis from both active (June, July) and inactive (March) months in the *S. pygaea* were treated with lanthanum nitrate during fixation. In both active and inactive months, tight junctions are intact and show similar morphological characteristics (Fig. 6.11A). These tight junctions resemble those found in *Elaphe climacophora* (Hondo *et al.* 1996) and *Agkistrodon* (Gribbins unpublished). There are focal subsurface densities associated most intensely where the two Sertoli cells membranes meet and abundant SER is observed in close proximity to these occluding junctions. Rarely where only single junctions were seen, three or more areas of the Sertoli cell membrane were often ziplocked together in these junctional complexes.

Lanthanum penetration was halted (Figs. 6.11B, 6.12) above (apically) spermatogonia and before the tracer could reach the level of the spermatocytes and spermatids within *Seminatrix pygaea* seminiferous epithelia (Gribbins unpublished). When spermatogonia were absent (Fig. 6.13B) the tracer would stop just below the spermatocyte layer within the seminiferous epithelium. When spermatogonia and spermatocytes were not present within the seminiferous epithelium the tracer would penetrate just below the developing spermatids (Figs. 6.11B, 6.12). In contrast, the lanthanum nitrate diffused easily through the basement membrane and into intercellular pathways around spermatogonia (Fig. 6.13A) even before active spermatogenesis starts. This tracer, however, never penetrated further apically within these quiescent seminiferous epithelia (Fig. 6.13A white arrow). Thus, as suggested by Pudney (1993) for reptiles overall, the snakes seem to have an intact blood testis barrier before meiosis and spermiogenesis begin. Since the seminiferous epithelium is a permanent component of the testis and intact tight junctions are seen throughout the

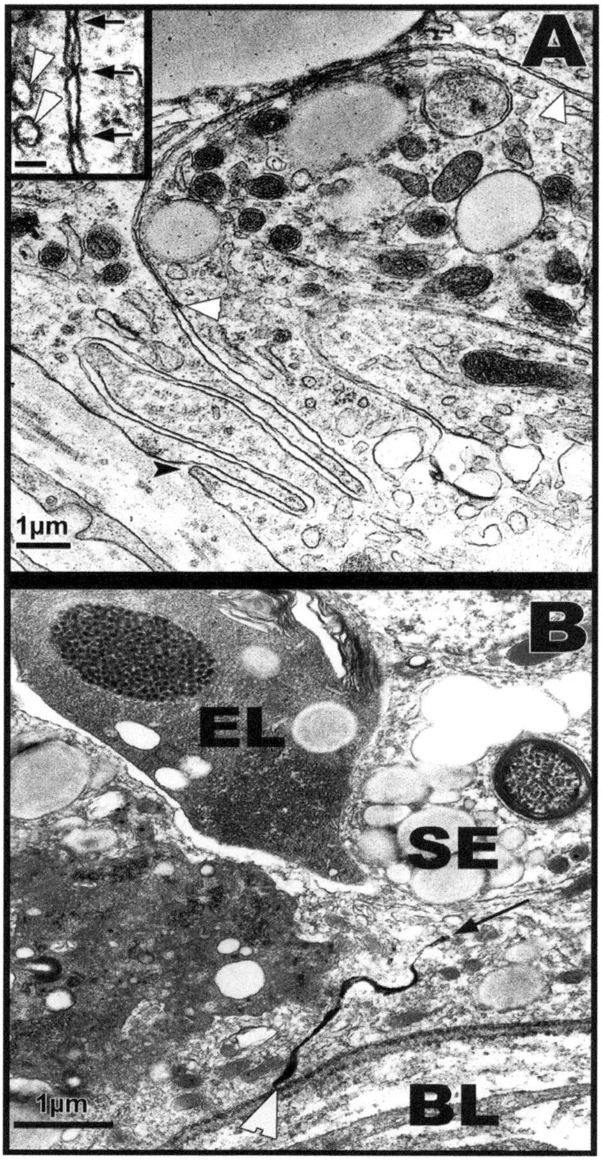

Fig. 6.11 A. Large group of tight junctions (white arrows) near the basement membrane between adjacent Sertoli cells of the seminiferous epithelium in *Seminatrix pygaea*. Entrance to intercellular pathway between Sertoli cells (black arrowhead). Inset: focal subsurface densities (black arrows), smooth ER (white arrowhead). **B.** Lanthanum nitrate penetration within an intercellular pathway (white arrowhead) is stopped (black arrow) just above the boundary layer (BL) in *Seminatrix pygaea*. The location of stoppage occurs in the same area the tight junctions are found within the basal compartment of the seminiferous epithelium. Note the lanthanum does not penetrate up to the more apically located elongating spermatid (EL) within the germinal epithelium (SE). Electron micrographs.

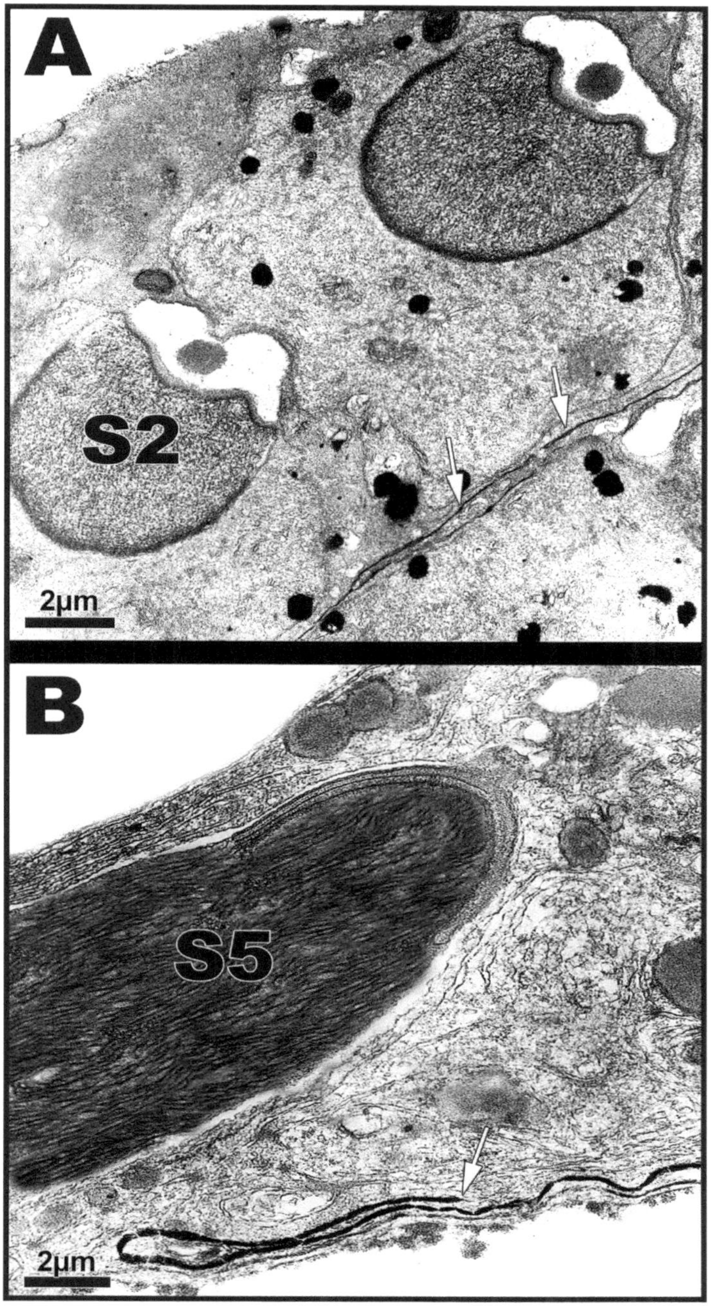

Fig. 6.12 Lanthanum nitrate penetrates freely below round (S2) (**A**, white arrows) and elongating spermatids (S5) **(B)** but never reaches (white arrow in **B**) the intercellular pathways at the level of the spermatids within the seminiferous epithelium of *Seminatrix pygaea*. Bars = 2 μm. Electron micrographs.

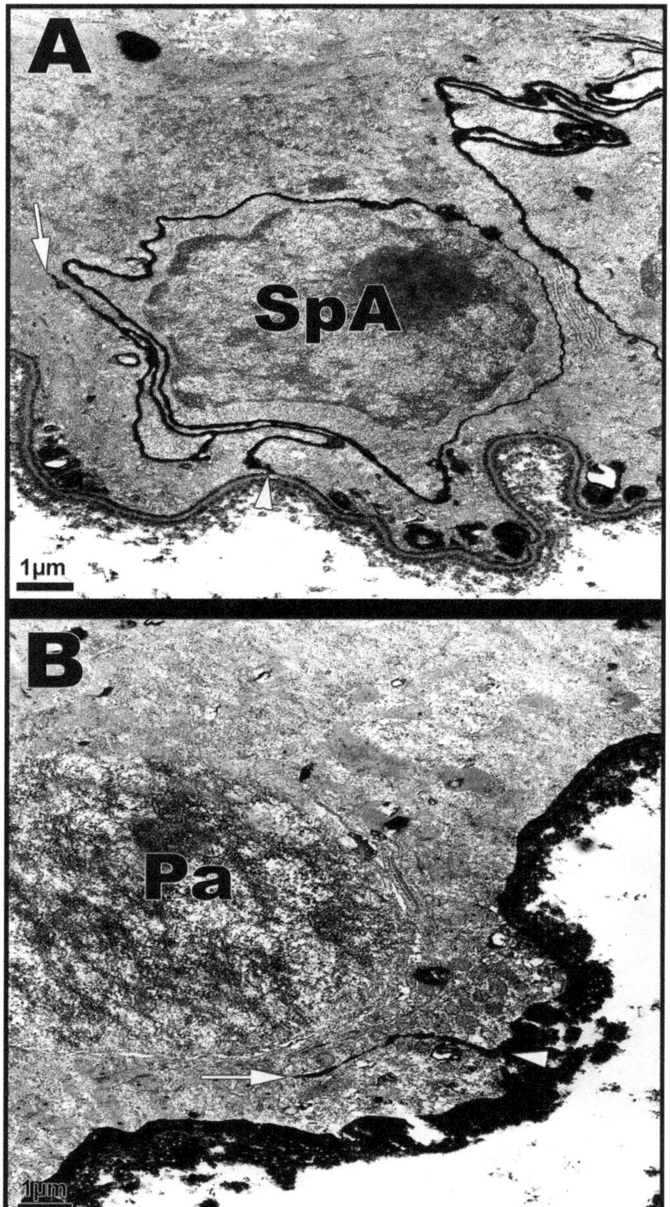

Fig. 6.13 A. Lanthanum nitrate surrounds the developing spermatogonia as it easily moves through the intercellular pathways (white arrowhead) around spermatogonia (SpA) within the non-spermatogenic testis of *Seminatrix pygaea*. Note that the lanthanum is stopped from moving apically between Sertoli cells (white arrow). **B.** Lanthanum nitrate is stopped (white arrow) just below a pachytene spermatocyte (PA) even though spermatogonia are not present within the August seminiferous epithelium of *Seminatrix pygaea*. Origin of Sertoli intercellular pathway (white arrowhead). Electron micrographs.

year in species studied to date, it is tempting to assume that the blood testis barrier is functional year around. This seems to be the case in *S. pygaea*, but again caution is required, as many more species need to be studied to provide clearer evidence that this hypothesis holds true in the majority of snakes and squamates.

6.2.3 Interstitial Tissue of the Testis

The interstitial tissue between seminiferous tubules within snakes is composed of two main layers. The first layer is a compact boundary layer (lamina propria), which is made up of acellular and cellular strata. The other, an inner layer, consists of loose connective tissue, which lies between the boundary layers of adjacent seminiferous tubules and is called the interstitium (Fig. 6.14A,B). In order to see the entire interstitium, one must view the angular edges of where three seminiferous tubules meet. This is especially true in spermatogenically high months when the interstitial tissue is most constricted from view. Typically during spermatogenically active months in snakes, the seminiferous tubules are sandwiched against each other and often the interstitium is absent and the boundary layer is the only subepithelial stratum observed (Fig. 6.5B).

The boundary layer is the subepithelial layer and is stratified into sequential acellular connective tissue and elongated cells and their interconnecting amorphous extracellular matrix. The interstitium when present consists of loosely aggregated Leydig cells, blood vessels, lymphatic vessels, fibroblasts and leukocytes such as marcophages or occasional lymphocytes. One thorough study exists in reptiles in which the authors detail the boundary layer in multiple reptiles including the Grass Snake (*Natrix natrix*; Unsicker and Burnstock 1975). There are no present studies that compile data on the interstitium in snakes. The interstitial and boundary layer data presented here represent new ultrastructural information collected from *Agkistrodon piscivorus*, *A. contortrix* and *Seminatrix pygaea* (Gribbins unpublished) and will be compared to the findings on *Natrix natrix*.

The interstitial tissues of the testis within *Agkistrodon* and *Seminatrix* are very similar in morphology and ultrastructure. The only major difference is the ability to differentiate between fibroblast-like and myoid-like cells in *S. pygaea* during spermatogenically inactive months (March) of the year (Fig. 6.15B). The fibroblasts are more slender in width than myoid cells and contain very elongated nuclei with more heterochromatin associated with the nuclear envelop. In *A. contortrix* and *A. piscivorus* (Fig. 16.15A), it is difficult to differentiate between fibroblast-like cells and myoid cells within the boundary layer. Within *Natrix natrix*, Unsicker and Burnstock (1975) were able to observe distinguishable differences between fibroblasts and myoid cells, with fibroblasts being more prominent in times of late spermiogenesis. The cells within the boundary layer of *Agkistrodon* and even the "myoid" cells of *Seminatrix* showed ultrastructural characteristics

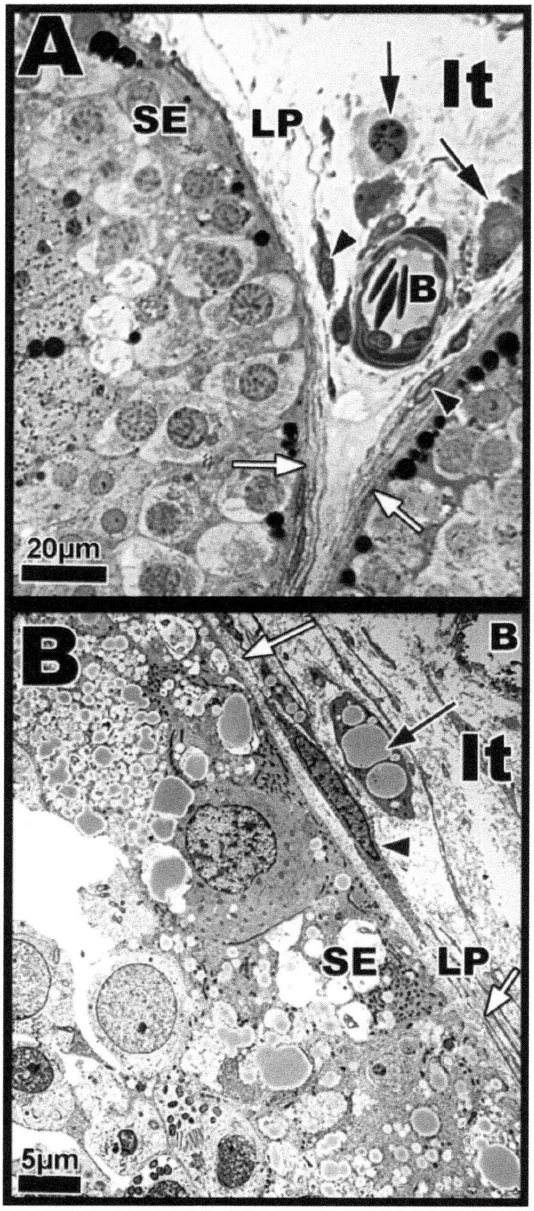

Fig. 6.14 A. Light micrograph showing the interstitial tissue between seminiferous tubules within the June testis of *Agkistrodon contortrix*. **B.** Low power electron micrograph depicting the same area of the interstitial tissue represented in A within the testis of *Agkistrodon piscivorus*. Seminiferous epithelium (SE), basement membrane of the germinal epithelium (white arrows), boundary layer (LP), myofibroblasts (black arrowheads), interstitium (It), Leydig cells (black arrows), blood vessel (B).

Color image of this figure appears in the color plate section at the end of the book.

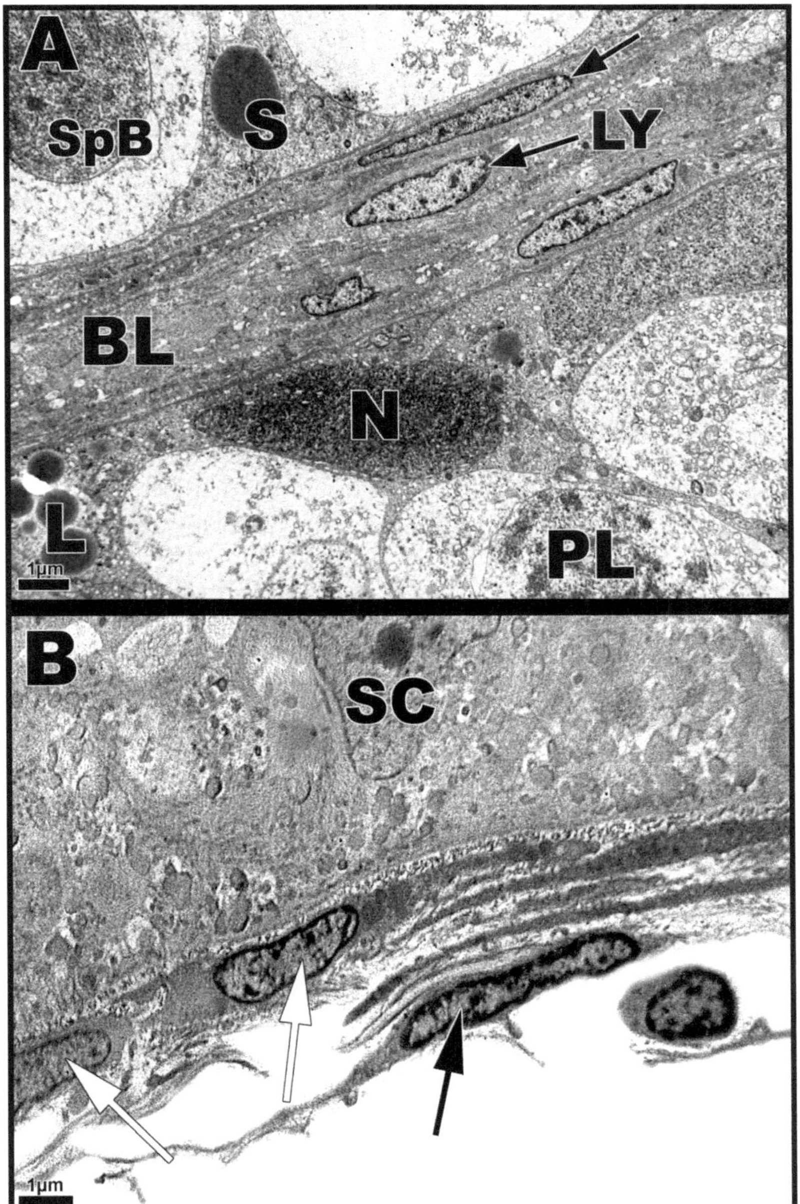

Fig. 6.15 A. The interstitial tissue between two closely associated seminiferous tubules in *Agkistrodon contortrix*. The boundary layer (BL) is the only layer present. It has myofibroblasts (black arrow) and Leydig cells (LY) within a collagen extracellular matrix. Spermatogonia B (SpB), Sertoli cell (S), Sertoli cell nucleus (N), lipids (L), preleptotene spermatocytes (PL). **B.** Boundary layer within the March testis of *Seminatrix pygaea*. Note that two different cell types are observed within the boundary layer: myoid cells (white arrow) and fibroblasts (black arrow). Sertoli cell nucleus (SC). Electron micrographs.

of both fibroblast and myoid cells. Thus, to simplify the discussion here, the term myofibroblasts will be adopted when discussing the cellular components of the boundary layer. The snake myofibroblasts are very similar to the myofibroblasts described in the avian testis (see Aire 2007).

6.2.3.1 The boundary (peritubular) layer

The boundary layers within *Seminatrix* (Figs. 6.15B, 6.16), *Agkistrodon* (Figs. 6.15A, 6.17; Gribbins unpublished), and *Natrix* (Unsicker and Burnstock 1975) contain an inner fibrous layer and an outer cellular layer of myoid-like cells. The fibrous layer directly under the seminiferous epithelium consists of a homogeneous and moderately dense basal lamina juxtapositioned to the basal membrane of the Sertoli cells and an adjacent layer of multi-directional collagen fibrils (Fig. 6.16). In rare cases, the collagen fibrils organize into bundles that loosely resemble mature collagen fibers (Fig. 6.17B).

The mesenchymal myofibroblast layer varies in thickness according to the number of concentric, alternating, and overlapping layers of cells. The number of layers seen within this outer layer of the boundary tissue in *Agkistrodon* (Fig. 6.17) and *Seminatrix* (Fig. 6.16) is typically between 1-5 cells thick, which is consistent with what was reported for *Natrix natrix* (Unsicker and Burnstock 1975). These cells contain elongated nuclei (Figs. 6.15A,B, 6.18) that display uniformly granular chromatin with thick homogeneously staining heterochromatin just inside of the nuclear membrane. One or two prominently staining nucleoli can be visible in most sagittal sections of the nucleus. In some instances, the Leydig cells can be found within the boundary layer (Fig. 6.17A).

The myofibroblast cytoplasm continues past the thickest portion of the cell containing the nucleus as very thin processes that then wrap around the seminiferous tubules. These processes within different cellular layers often overlap each other. These overlaps can possess desmosomes (Figs. 6.17B, 6.19B), or adhering junctions, that are most likely responsible for holding the cells together in an intact layer. The myofibroblasts in *Agkistrodon* and *Seminatrix* (Gribbins unpublished) take on a dark appearance around the nuclei (Fig. 6.18A), which in most cases gives a striated look to their cytoplasm. These dark staining fibers in sagittal and transverse section resemble thin filaments (actin) (Fig. 6.17B) and intermediate filaments (Fig. 6.19B). The actin filaments can appear as clumps or masses or in diffuse loosely organized aggregates in *A. piscivorus* and *A. contortrix*. These thin filaments are most numerous and the myofibroblasts stain the darkest at the climax of spermatogenesis. Also present are dense staining granular plaques (Fig. 6.19A) associated with the cytoplasmic leaflet of the cell membrane. These dense plaques, intermediate filaments, and the presence of actin filaments suggest that these myofibroblasts have the ability to contract. This hypothesized function has been proposed also for birds (Aire 2007) and reptiles (Unsicker and Burnstock 1975). The thin

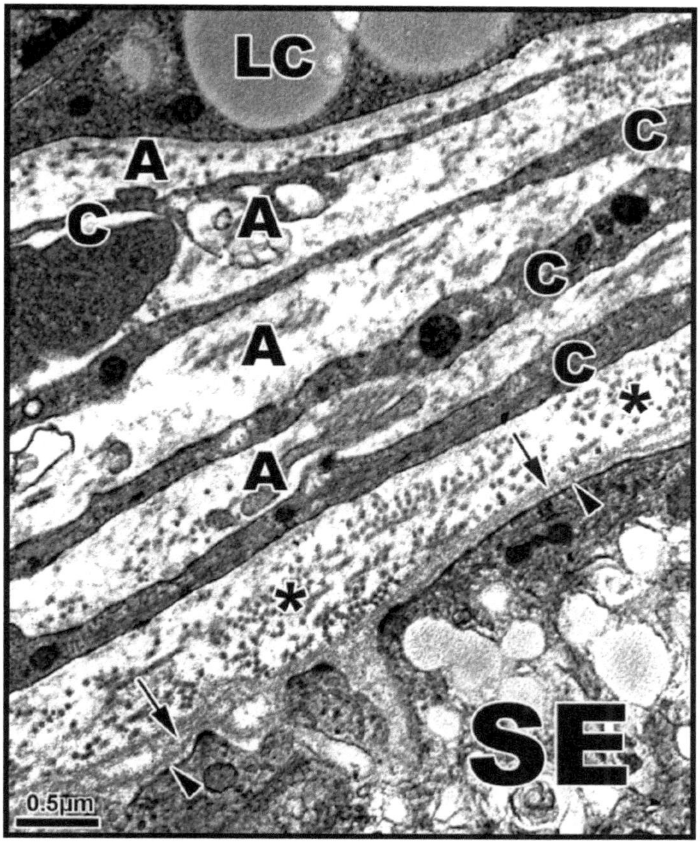

Fig. 6.16 High power electron micrograph of the boundary layer within *Seminatrix pygaea*. This layer begins just under the seminiferous epithelium (SE) as a homogeneous inner basal lamina (black arrow) and a multi-directional layer of collagen (*). Beyond this inner fibrous layer are alternating myofibroblast/cellular (C) and acellular (A) collagen layers. There are typically 1-5 alternating strata within this 2nd layer. Beyond this outer layer is the interstitium, which contains Leydig cells (LC). Basement membrane of the seminiferous epithelium (black arrowheads).

filaments are also most numerous during the climax of spermatogenesis, which intuitively makes sense, as this is when spermatozoa are released to the lumina of the seminiferous tubules. Thus, we propose that these myofibroblasts contract and help propel sperm through the seminiferous tubules to the anterior ducts in the male reproductive tract of snakes. These boundary layer cells also most likely play a role in production of the extracellular matrix of this layer.

The snake myofibroblasts also contain a moderate number of elongated mitochondria, some lipid inclusions, a rare lysosome, and numerous pinocytotic vesicles that are seen both at the cell membrane and within the cytoplasm (Fig. 6.19A,B). There are also short conspicuous distended cisternae of both smooth (Fig. 6.19A,B) and rough endoplasmic reticula

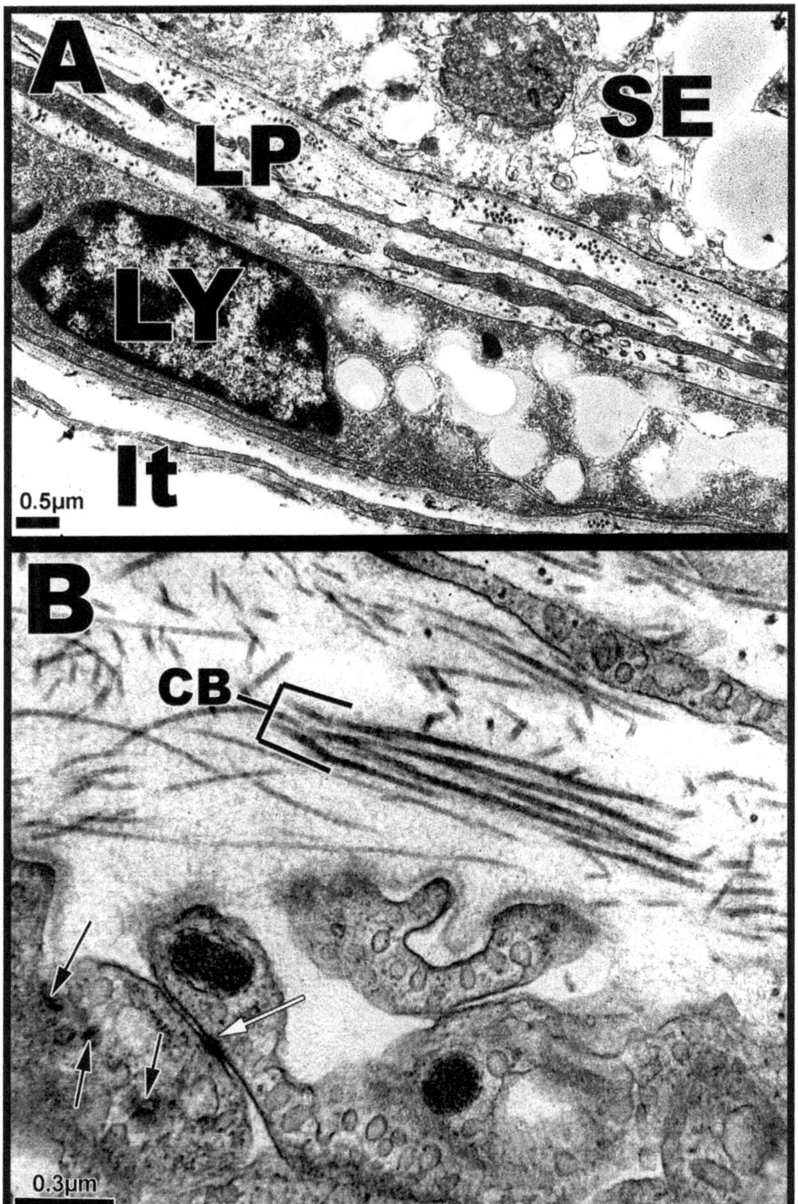

Fig. 6.17 A. The boundary layer (LP) and interstitium within the testis of *Agkistrodon contortrix*. Leydig cells (LY) are found between the boundary layer and the interstitium. Seminiferous epithelium (SE). **B.** High power details of the cellular processes of the myofibroblasts of *Agkistrodon contortrix*. The cytoplasm has bundles of actin filaments (black arrows) and adjacent cellular processes are connected via adhering junctions (white arrow). In rare cases, extracellular collagen fibrils form a bundle similar to a mature collagen fiber (CB). Electron micrographs.

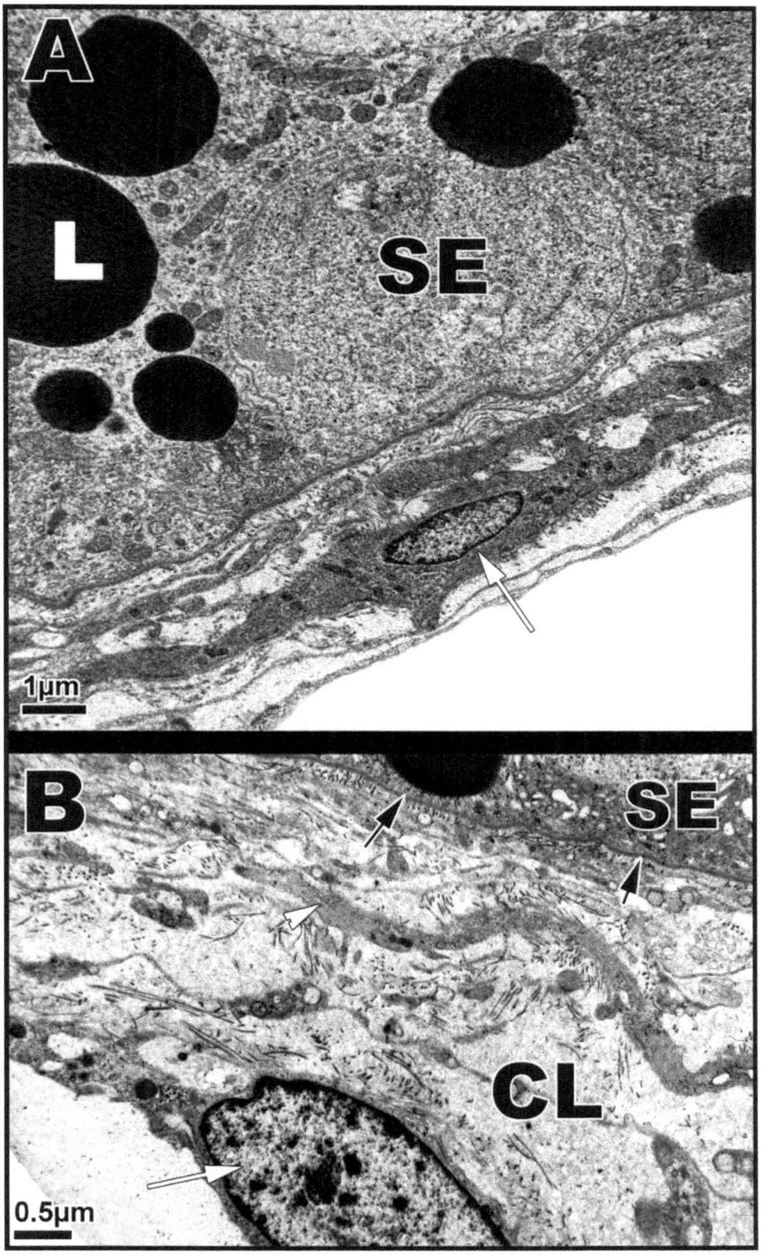

Fig. 6.18 A. *Agkistrodon contortrix* myofibroblast (white arrow) in an August testis exhibiting a dark staining compact cytoplasm packed with thin filament bundles. Lipids (L), seminiferous epithelium (SE). **B.** Myofibroblast (white arrow) in an August testis of *Seminatrix pygaea*. The dark cytoplasm is filled with thin filament bundles. Collagen layer (CL), seminiferous epithelium (SE), myofibroblast cellular process (white arrowhead), basement membrane of the seminiferous epithelium (black arrow). Electron micrographs.

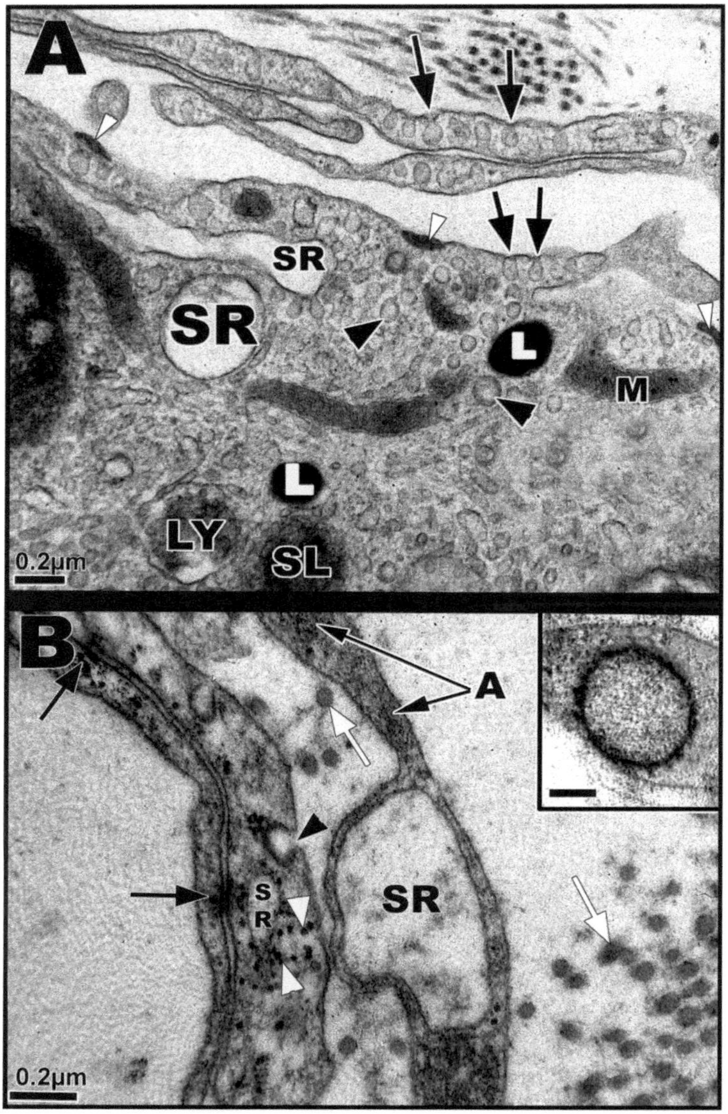

Fig. 6.19 A. *Agkistrodon contortrix* myofibroblast cytoplasm exhibiting dilated smooth ER (SR), lysosomes (LY), secondary lysosomes (SL), lipids (L), and mitochondria (M). There are also many pinocytotic vesicles (black arrowheads) arising from pinocytosis (black arrows) that is occurring all over the cell surface of the myofibroblasts. Just below the plasma membrane there are also dense plaques (white arrows) that have linkages to the intermediate filaments of the cytoskeleton. **B.** *Agkistrodon contortrix* myofibroblast cellular processes. Numerous intermediate filaments (white arrowheads) and actin filaments (A) are seen in the cytoplasm. There is both dilated smooth (SR) and rough (inset) ER within the cytoplasm. Desomosomes anchor adjacent cellular processes together (black arrows) and collagen is abundant in the extracellular space (white arrows). Coated pinocytotic vesicle (black arrowhead). Electron micrographs.

(Fig. 6.19B inset). The snake myofibroblast organelle and cytoplasm composition is very similar to that reported in birds (Aire 2007) and seems to be a combination of the organelles described for the myoid cells and fibroblasts of *Natrix natrix* (Unsicker and Burnstock 1975).

6.2.3.2 The interstitium and leydig cells

The interstitium in snakes, when present, is typically organized as a loose connective tissue (Gribbins unpublished; Fig. 6.20A). It is most prominent in areas where three seminiferous tubules meet forming a wedge of interstitial tissue. The middle portion of the wedge is where most of the interstitium is located (Fig. 6.20A,C). Between two seminiferous tubules there is typically little to no interstitium particularly during spermatogentically active months (Fig. 6.20A). In this case, Leydig cells may be found within the boundary layer (Fig. 6.21) (Gribbins unpublished). The components present within the snake interstitium are a loosely aggregated connective tissue with a number of centrally located blood vessels, lymphatic vessels (rare), a few macrophages, and either single or loosely associated Leydig cells (Fig. 6.22A). The Leydig cells are typically found close to the blood vessels within the interstitium (Fig. 6.22A).

Both in *Seminatrix* and *Agkistrodon*, Leydig cells show a cycle of lipid accumulation and use (Gribbins unpublished). The Leydig cell lipid cycle in snakes follows that reported for other temperate snakes and squamates (Lofts 1968; Guraya, 1973; Unsicker and Burnstock 1975; Callard *et al.* 1978; Engel and Callard 2007). The Leydig cells during spermatogenically inactive times in *S. pygaea* (March) and *A. contortrix* (January/February) have large lipid inclusions within the cytoplasm (Fig. 6.23A,B). In contrast, during active spermatogenesis (*S. pygaea*: July and *A. contortrix*: October), the Leydig cells lack lipid inclusions and appear more granular with multiple mitochondria with tubular cristae (Fig. 6.22B). This indicates that these cells are using the lipids as spermatogenesis progresses and climaxes, presumably for steroid synthesis, which is required for spermatogenesis in reptiles. Leydig cells in the lipid accumulation stage also have very large prominent smooth endoplasmic reticula, fewer tubular mitochondria, and the nucleus is more elongated with no nucleolus and increased heterchromatin associated with the nuclear envelope. In contrast, active Leydig cells have more round nuclei with a prominent nucleolus, many more mitochondria and glycogen granules, and shorter dilated smooth endoplasmic reticula.

6.3 GERM CELL DEVELOPMENT STRATEGY

Although paraphyletic reptiles and their ancestors (basal amniotes), including snakes, hold an important phylogenetic position between amphibians, birds, and mammals, little is still known about the cytological events and germ cell development strategies employed during spermatogenesis in

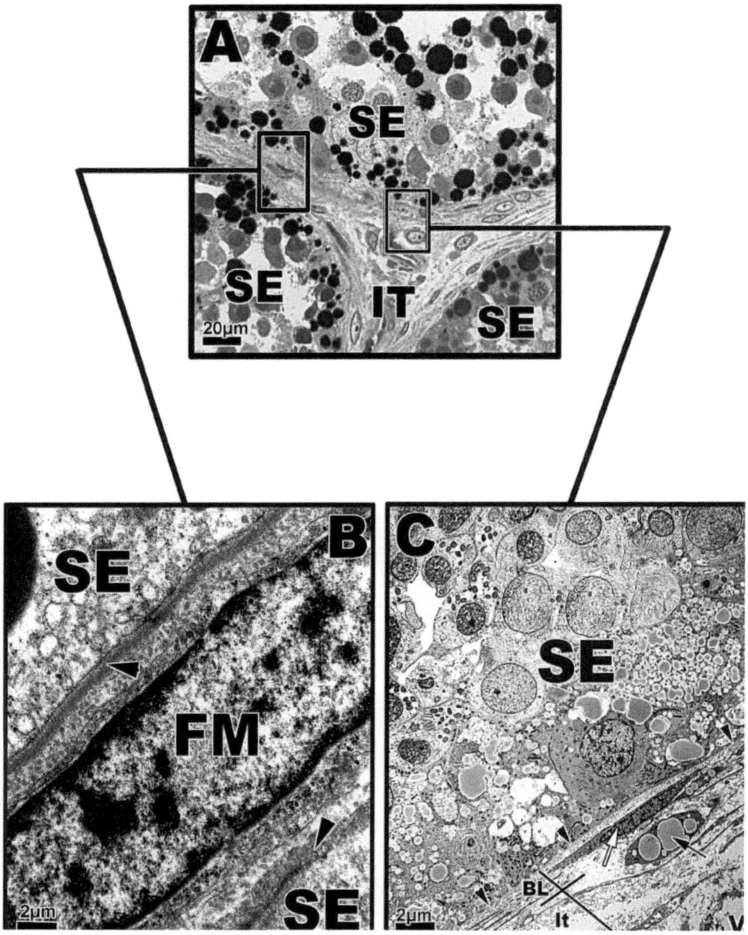

Fig. 6.20 Light (**A**) and electron (**B,C**) micrographs of the boundary layer and interstitium in the testis of *Agkistrodon contortrix*. Typically where three seminiferous tubules meet a wedge of interstitium is seen (A,C) between boundary layers. However, where two tubules meet in juxtaposition there is only room for the boundary layer (B). Seminiferous epithelium (SE), interstitium (IT), myofibroblast (FM, white arrow), boundary layer (BL), Leydig cell (black arrow), basement membrane (black arrowheads), blood vessel (V).

Color image of this figure appears in the color plate section at the end of the book.

these nonmammalian amniotes. Germ cells are in layers within the reptilian seminiferous epithelium; however, evidence suggesting that these layers are consistent like the spatial stages seen in mammals and birds does not exist. When observing germ cells within the testis of most temperate species of snake, one can see that up to five layers of spermatids can exist within the seminiferous epithelium (Fig. 6.24; Gribbins *et al.* 2008).

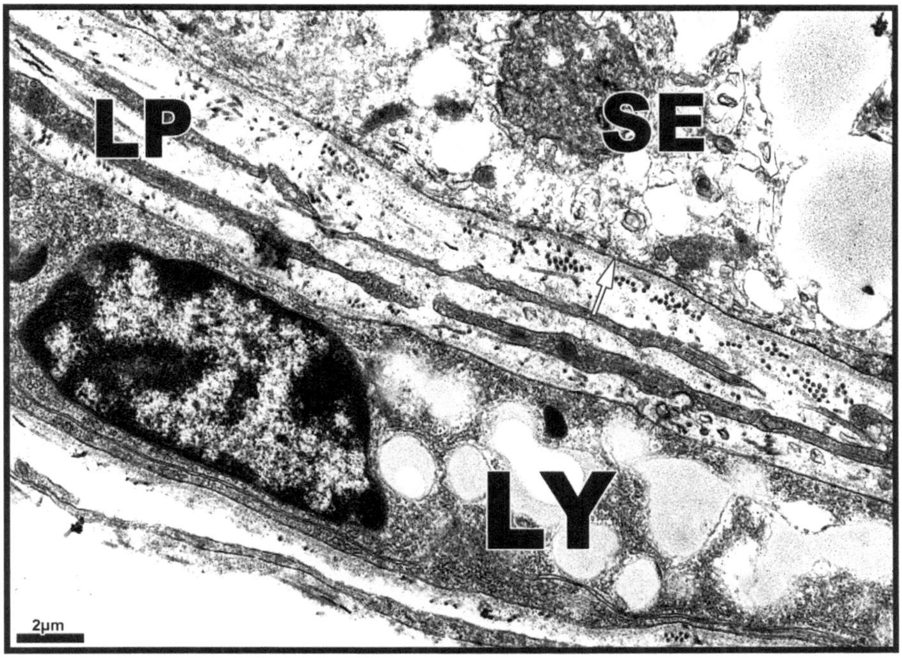

Fig. 6.21 Electron micrograph of the boundary layer (LP) in the July testis of *Seminatrix pygaea*. Leydig cells (LY) are often found in the boundary layer between seminiferous tubules during active spermatogenesis when very little interstitium is seen within the interstitial tissue. Seminiferous epithelium (SE), basement membrane (white arrow).

In all mammals and birds studied to date, germ cells entering the spermatogenic cycle undergo a number of predictable cytological changes during their development (see Roosen-Runge 1972; Russell *et al.* 1990). As new generations of spermatogonia enter the spermatogenic cycle they displace more advanced germ cells centrally toward the lumen. Thus, the seminiferous epithelium at any point in time contains three to five generations of germ cells that are consistently found together. These consistent spatial relationships among germ cells are termed stages and a complete series of stages within the seminiferous epithelium is called the spermatogenic cycle (Russell *et al.* 1990). The spermatogenic cycle of mammals and birds leads to multiple waves of spermiation during spermatogenesis. Seasonally breeding mammals and birds also possess spatial stages within their seminiferous epithelia during their breeding seasons (Roosen-Runge 1972). Though snakes and other squamate testes have attained the amniotic organization of the seminiferous epithelial compartment and possess layers of germ cells within the seminiferous epithelium similar to birds and mammals, they can have up to 8 or 9 generations of germ cells within the germinal epithelium (Fig. 6.24). This is many more representative germ cell generations than in endothermic

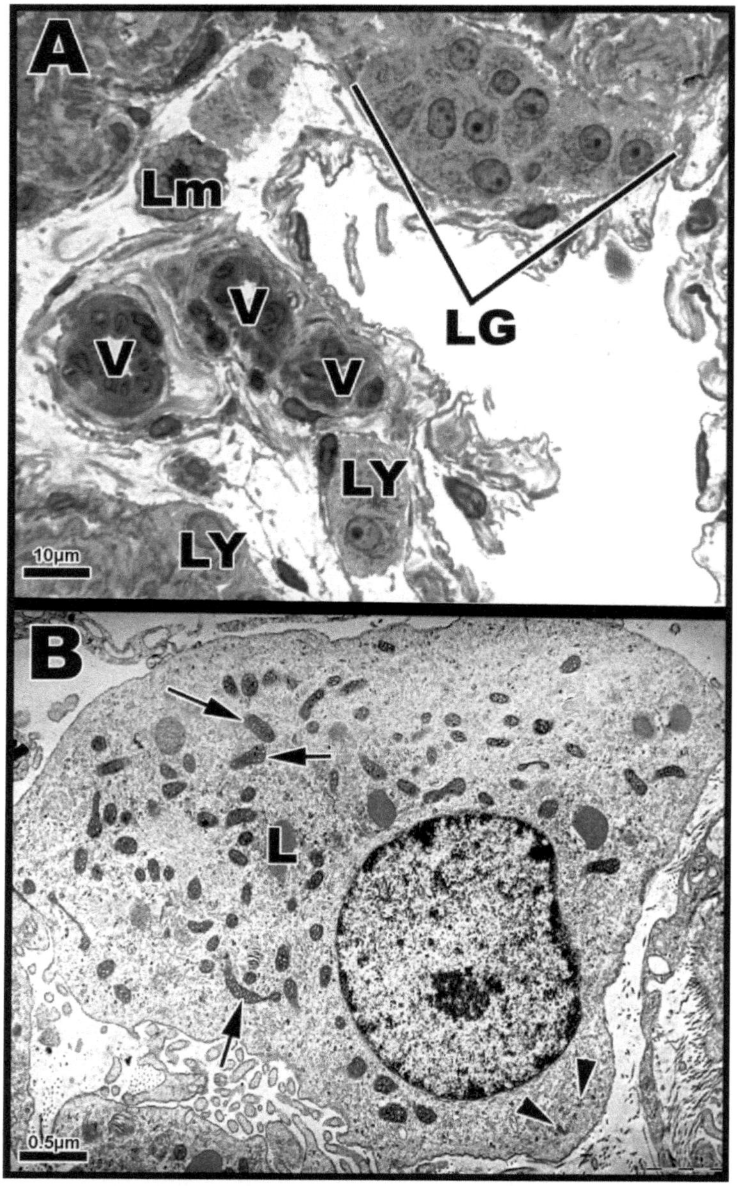

Fig. 6.22 A. Light micrograph of the interstitium in *Agkistrodon contortrix* in August showing single Leydig cells (LY) and a group of Leydig cells (LG). Note, that these testosterone-producing cells are near blood vessels (V). Lymphocyte (LM). Micrograph provided by Dustin Siegel. **B.** High power electron micrograph of a Leydig cell in the August testis of *Agkistrodon contortrix* exhibiting a large prominent nucleolus. Note the lack of large lipid droplets (L) and the numerous mitochondria with tubular cristae (black arrows). There is also scatter glycogen granules (black arrowheads) throughout the cytoplasm.

Color image of this figure appears in the color plate section at the end of the book.

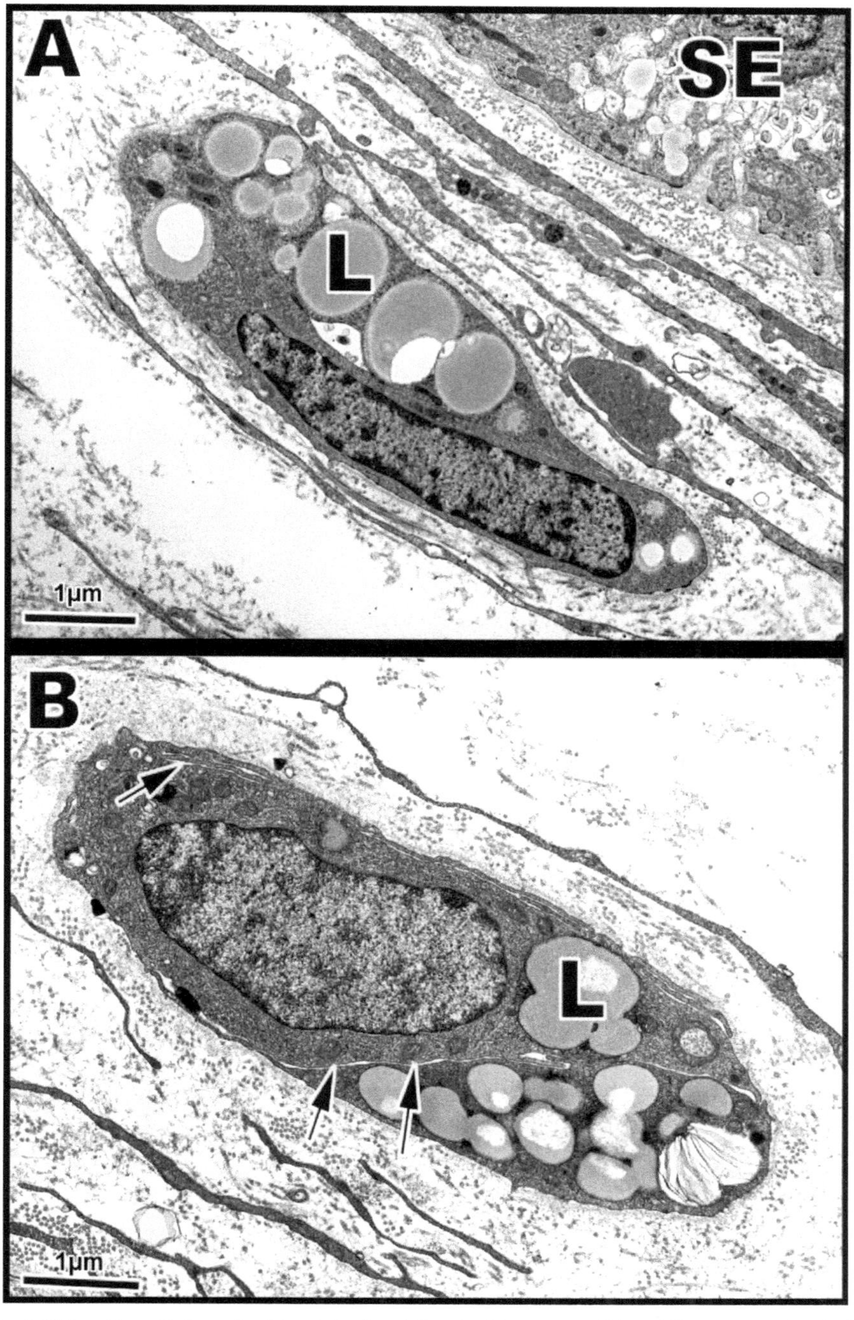

Fig. 6.23 Leydig cells from the March testis of *Seminatrix pygaea* (**A**) and the January testis (**B**) of *Agkistrodon contortrix*. Not the numerous lipid droplets found in these cells, the prominent smooth ER (black arrows), and nuclei lacking nucleoli. Seminiferous epithelium (SE). Electron micrographs.

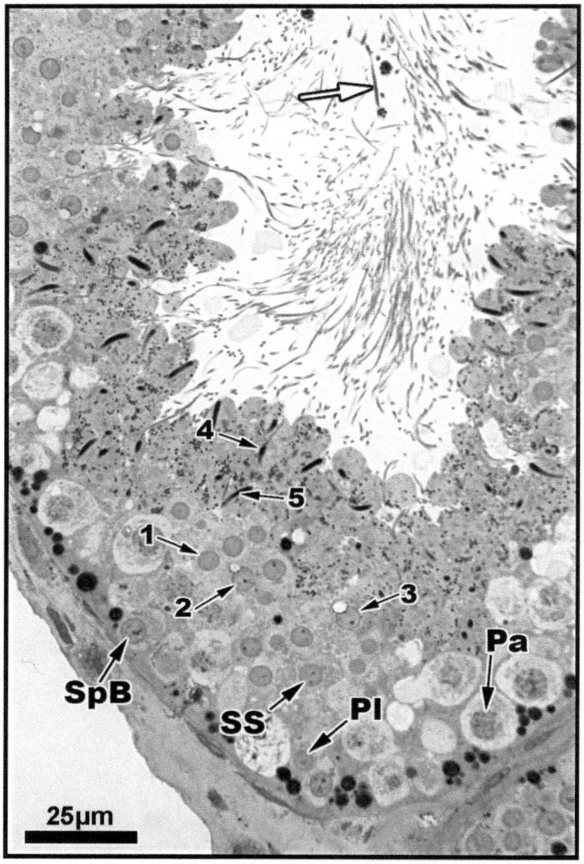

Fig. 6.24 A light micrograph of an August seminiferous tubule in *Agkistrodon contortrix*. There are 9 different cell types found within the germinal epithelium at this time. Five of these cell types are spermatids (1-5). Spermatogonia B (SpB), preleptotene spermatocytes (Pl), Pachytene spermatocytes (Pa), secondary spermatocytes (SS), Steps 1-5 spermatids (1-5), mature spermatozoa (white arrow). Micrograph provided by Dustin Siegel.

amniotes. Thus, it is relevant to study snake spermatogenesis in terms of the germ cell development strategy employed during germ cell development.

Reproductive cycles in snakes have been studied extensively (see **Chapter 12** in this volume). Spermatogenesis in most temperate squamates and snakes is highly seasonal. Classically, two types of seasonal cycles exist in snakes, some produce sperm after mating (postnuptial) and some produce sperm before mating (prenuptial) (Fig. 6.25; Licht 1984). How this seasonality is reflected in germ cell development and subsequent maturation is not known. For example, it is unclear whether germ cells enter the spermatogenic cell cycle continually in snakes during the reproductive season like mammals and birds or whether spermatogenesis is an episodic event like that of anamniotes. Also, no cytological information

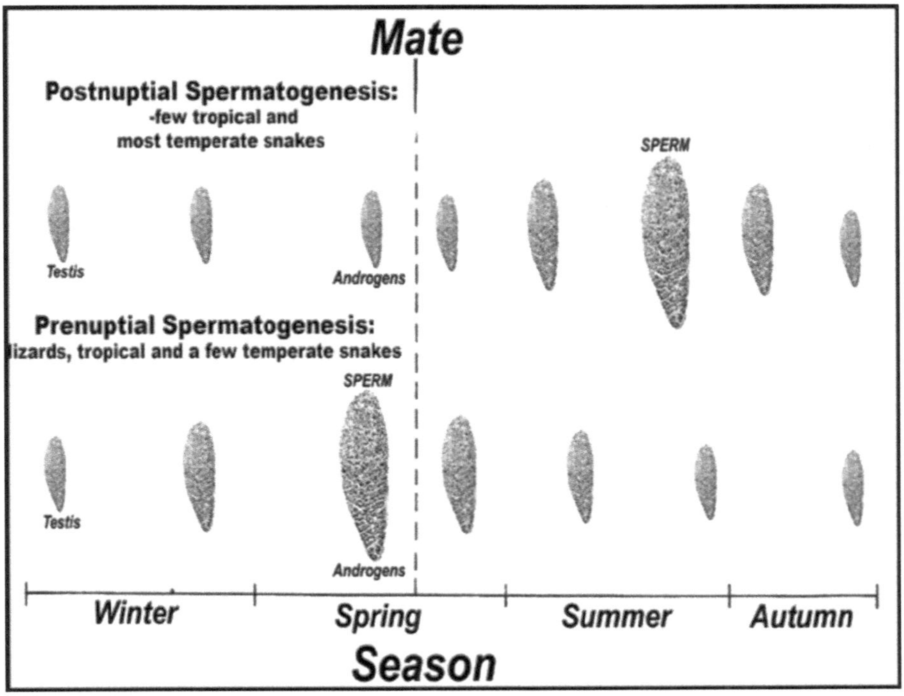

Fig. 6.25 A schematic view that represents the two major seasonal gonadal activities in temperate species of reptiles in relation to spermatogenesis and mating. Sperm production is completed before or after mating. Androgen levels and when sperm are present are also represented within each type of seasonal spermatogenic cycle. Within both types, variations may occur in the rate of recrudescence, spermatogenesis, testis growth, and duration of the breeding season.

confirms that prenuptial and postnuptial germ cell cycles have similar germ cell development strategies within snakes and other reptiles.

Understanding the organization of germ cells within the seminiferous epithelium of reptiles might provide better insight on how the snake testis functions during seasonal cycles. For example, in mammals it is known that FSH stimulates type A spermatogonia during recrudescence of spermatogenesis (Waits and Setchell 1990). FSH has been found in reptiles (Licht 1979; Licht *et al.* 1979) and all recent information suggests that it initiates spermatogenesis in temperate reptiles either before or after mating (Licht *et al.* 1989; Masson and Guillette Jr 2005). However, it is not known whether type A spermatogonia with the ability to divide are present within the seminiferous epithelium at all times of the year or whether the population of dividing spermatogonia is temporal (may be responsible for refractory period for example, see Licht 1984) and only available for FSH activation during certain times of the annual cycle in snakes and other squamates. Understanding the basic cytological events of spermatogenesis in reptiles may help resolve the existing controversies and inconsistencies with regard to refractory periods, temperature effects, and hormonal stimulation within the testicular cycles of reptiles.

Unfortunately, few reproductive biologists in the field of herpetology have paid any attention to the germ cell development strategies during spermatogenesis in snake and other reptiles. Those studies that have been done on reptilian germ cell development strategies do show that no matter the type of reproductive cycle employed during spermatogenesis (pre- or postnuptial, biannual, or continuous) the same temporal germ cell development strategy is utilized by reptiles (Gribbins and Gist 2003; Gribbins *et al.* 2003, 2005, 2007, 2008, 2009; Rheubert *et al.* 2009a; Rheubert *et al.* 2009b). Again it is obvious that more squamate species need to be studied, especially snakes, as the latter show the most diversity as far as reproductive cycle, even within sympatric species (Licht 1984).

In order to determine whether consistent stages or associations are seen between germ cell generations within the snake seminiferous epithelium one must be able to recognize the different morphologies of germ cells as they progress through the phases of spermatogenesis. Three types of cells are found within all vertebrate testes: spermatogonia, spermatocytes, and spermatids. Cytological changes accompanying spermatogenesis are common in the testes of all vertebrates (Volsøe 1944; Hess 1990; Russell *et al.* 1990) and include nuclear changes during mitosis and meiosis and changes to the acrosomal granule and vesicle, changes to the chromatin of the nucleus, and nuclear elongation in spermatids. Using these markers allows for the elucidation of germ cell morphologies as they progress through proliferation, meiosis, and spermiogenesis. Figure 6.26 represents a typical ophidian germ cell development cycle within the testis (from *Seminatrix pygaea*) as described by Gribbins *et al.* (2005). In all reptiles studied to date, germ cell morphologies are remarkably uniform. However, outside of squamates, chelonians have a type of spermatogonia called resting spermatogonia, which are smaller than the two types found within squamates testes. These resting type of spermatogonia are observed only during quiescence in the turtle testis (Gribbins *et al.* 2003). Three species of snakes will be considered here, two from Viperidae: *Agkistrodon piscivorous* (Gribbins *et al.* 2008) and *Agkistrodon contortrix* (Gribbins unpublished), and one from Colubridae: *S. pygaea* (Gribbins *et al.* 2005).

6.3.1 Germ Cell Cycle

The seminiferous epithelium in snakes contains two morphologies of pre-meiotic cells, spermatogonia A and B (Fig. 6.26) during all months of the year. These cells are characterized by nuclei with random clumps of heterochromatin. The major morphological differences between the two types of spermatogonia are that the A type is ovoid in shape with one large nucleolus and B type is more round in shape and usually lacks a prominent nucleolus. Both spermatogonial types are generally found near

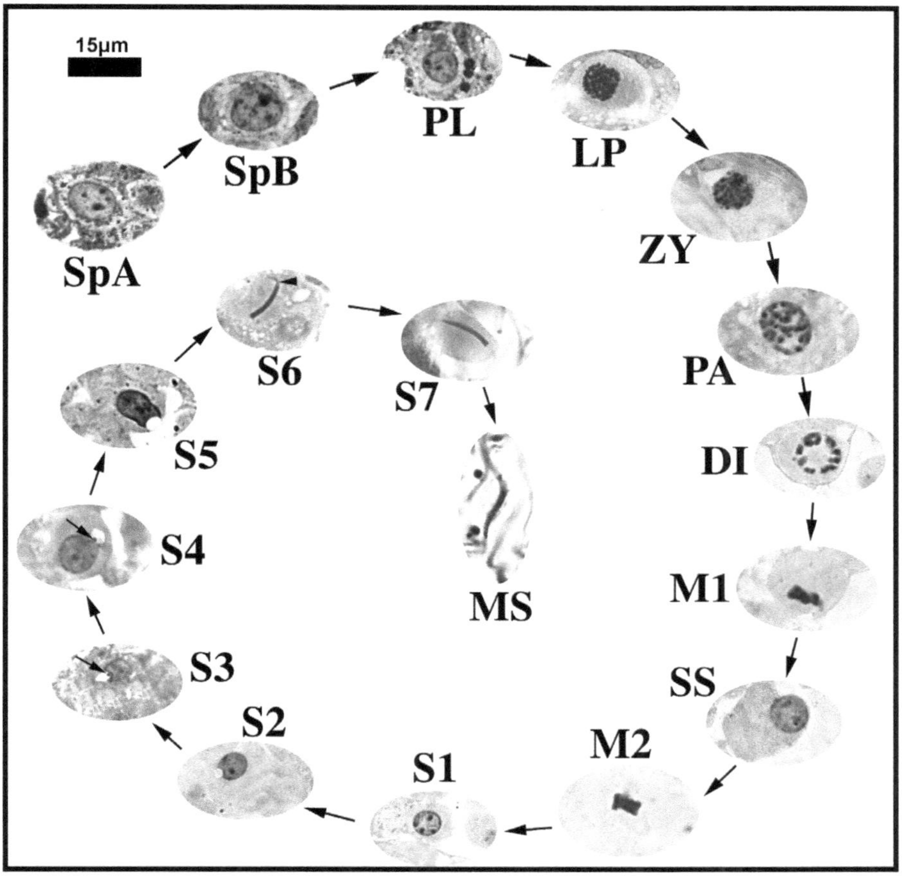

Fig. 6.26 Cell types found within the seminiferous epithelia of *Seminatrix pygaea*. Type A spermatogonia (SpA), type B spermatogonia (SpB), pre-leptotene spermatocytes (PL), leptotene spermatocytes (LP), zygotene spermatocytes (ZY), pachytene spermatocytes (PA), diplotene spermatocytes (DI), meiosis 1 (M1), secondary spermatocytes (SS), meiosis 2 (M2), step 1 spermatid (S1), step 2 spermatid (S2), step 3 spermatid (S3), (black arrow: acrosome granule), step 4 spermatid (S4) (black arrow: acrosome granule), step 5 spermatid (S5), step 6 spermatid (S6) (black arrowhead: apical extension with acrosome), step 7 spermatid (S7), mature sperm (MS). Light micrograph.

Color image of this figure appears in the color plate section at the end of the book.

the basement membrane of the epithelium away from the lumen and associated with the basal compartments formed by Sertoli cells. During the spermatogenic cycle, however, both types of spermatogonia undergo mitosis to maintain the spermatogonial population and many of the B spermatogonia divide to form pre-leptotene spermatocytes. These two types of spermatogonia are typically most active mitotically immediately before the onset of meiosis and spermiogenesis.

Meiotic cells are also very similar between snakes and are characterized by increasing nuclear and cytoplasmic size and the gradual condensation of nuclear chromatin into distinct chromosomes. The majority of spermatogonia B undergo mitotic divisions at different points of the reproductive cycle, within the snakes studied, to produce preleptotene cells. Preleptotene spermatocytes (Fig. 6.26) have nuclei with prominent nucleoli and fine granular chromatin. Preleptotene cells, along with step 1 spermatids (Fig. 6.26), are usually the smallest germ cells (including nucleus and cytoplasm) within the seminiferous epithelium of snakes. Their small size easily distinguishes them from the larger spermatogonia A and B in the basal portion of the seminiferous epithelium.

Leptotene (Fig. 6.26) and zygotene (Fig. 6.26) spermatocytes are slightly larger and stain more intensely than pre-leptotene cells and chromatin fibers pack their nuclei. The major morphological difference between the two is that zygotene cells have more condensed globular chromatin compared to the fine filamentous chromatin found in leptotene cells. They appear together along with pachytene spermatocytes and typically constitute a large portion of the primary spermatocyte population when they are found within the ophidian seminiferous epithelium.

Pachytene cells (Fig. 6.26) will remain in the seminiferous epithelium of most snakes throughout spermatogenesis and are the most common spermatocyte seen within the reptilian testis. These germ cells undergo a substantial size increase over their development and their nuclei contain very thick chromatin fibers that are interspersed with areas of open nucleoplasm. Diakinesis (Fig. 6.26), metaphase I and II (Fig. 6.26), and secondary spermatocytes (Fig. 6.26) are transitional germ cells usually found together within the seminiferous epithelium. Diakinesis cells are characterized by thick fully condensed chromosomal fibers that are interspersed with large open areas of nucleoplasm. The nuclear membrane is indistinguishable and their spoke-like chromosomal fibers are arranged in a circular pattern. Metaphase I and II cells contain a condensed clump of chromosomes that is located on the equatorial plate with no apparent nuclear boundaries. Metaphase II cells are smaller in size than metaphase I cells and visually appear to have half the amount of chromatin when compared to metaphase I cells. Secondary spermatocytes usually are dispersed randomly between metaphase I and II cells. Secondary spermatocytes have lightly stained centrally located nuclei that are typically 15% larger than subsequent step 1 spermatids in ophidian and other reptilian testes.

Spermiogenesis is divided into steps based on the terminology of Russell *et al.* (1990) for mammalian species and includes the development of the acrosome complex, elongation of the nucleus, and condensation of chromatin material. Spermiogenesis is the longest phase of germ cell development. There are typically 7 or 8 recognizable steps of spermiogenesis within the studied North American temperate reptiles including snakes. *Seminatrix pygaea* and the *Agkistrodon* species studied in this chapter have seven recognizable steps of spermiogenesis. The round spermatids

(Fig. 6.26) are the first group of haploid cells that begin to develop the specialized structures of the mature sperm. The round spermatids undergo the development of the acrosome system, which includes the development of the acrosome vesicle and granule (Fig. 6.26) that sit on the apex of the developing sperm nucleus.

Step 1 (Fig. 6.26) and step 2 (Fig. 6.26) spermatids in the temperate ophidian testis mark the beginning of spermiogenesis. Their small size and lightly stained Golgi and small acrosome vesicle that lie in juxtaposition to the nuclear surface characterize these spermatids. Their spherical nuclei are centrally located and contain one or more chromatin bodies and lightly staining diffuse chromatin. The rest of development of the round spermatids involves further contact and growth of the acrosome vesicle, which characteristically forms an indention on the apex of the nucleus (Fig. 6.26).

Once the acrosome system is fully developed, round spermatids undergo a transitional step where the apex of the nucleus begins to elongate (Fig. 6.26). This elongation continues until nuclei are rod shaped and up to 30 μm or more in length (Fig. 6.26). During their development a prominent flagellum is often seen protruding from these spermatids out into the lumen of the seminiferous tubules. The acrosome, in many instances, is still present and is found associated with a thin extension of the apical nucleus (Fig. 6.26), which to date is a unique feature found in some reptiles at least at the light microscopic level. Once elongation is complete, chromatin material is condensed and cytoplasmic material is removed from the developing spermatids to form a more fluid-dynamic cell with a flagellum that is suited for movement through the female reproductive tract. During condensation, it is common in snakes for their nuclei to become slightly curved (Fig. 6.26), which most likely leads to the reptilian spermatozoa characteristic filiform shape. These elongating spermatids are often found in bundles within large columns of seminiferous epithelium and develop together as cohorts of cells. Once spermiogenesis is complete, mature spermatozoa (Fig. 6.26) are shed from the seminiferous epithelium to the lumina of the seminiferous tubules of the testis.

6.3.2 The Mode of Germ Cell Development

Once all germ cell morphologies have been identified for the three phases of spermatogenesis, analysis can be performed to determine whether germ cells are consistently layered together in the ophidian seminiferous epithelium (spatial development) or the germ cells develop as a single population (temporal development). The example we will use first is in *Agkistrodon contortrix* (Gribbins unpublished). In this particular viper, we have samples from only a few months of the year. However, as long as samples that represent at least one wave of spermatogenesis are available then one can determine whether germ cells are developing spatially or temporally. In *A. contortrix*, looking at Figure 6.27, it becomes apparent that the majority of the germ cell population is moving through spermatogenesis as a single cohort.

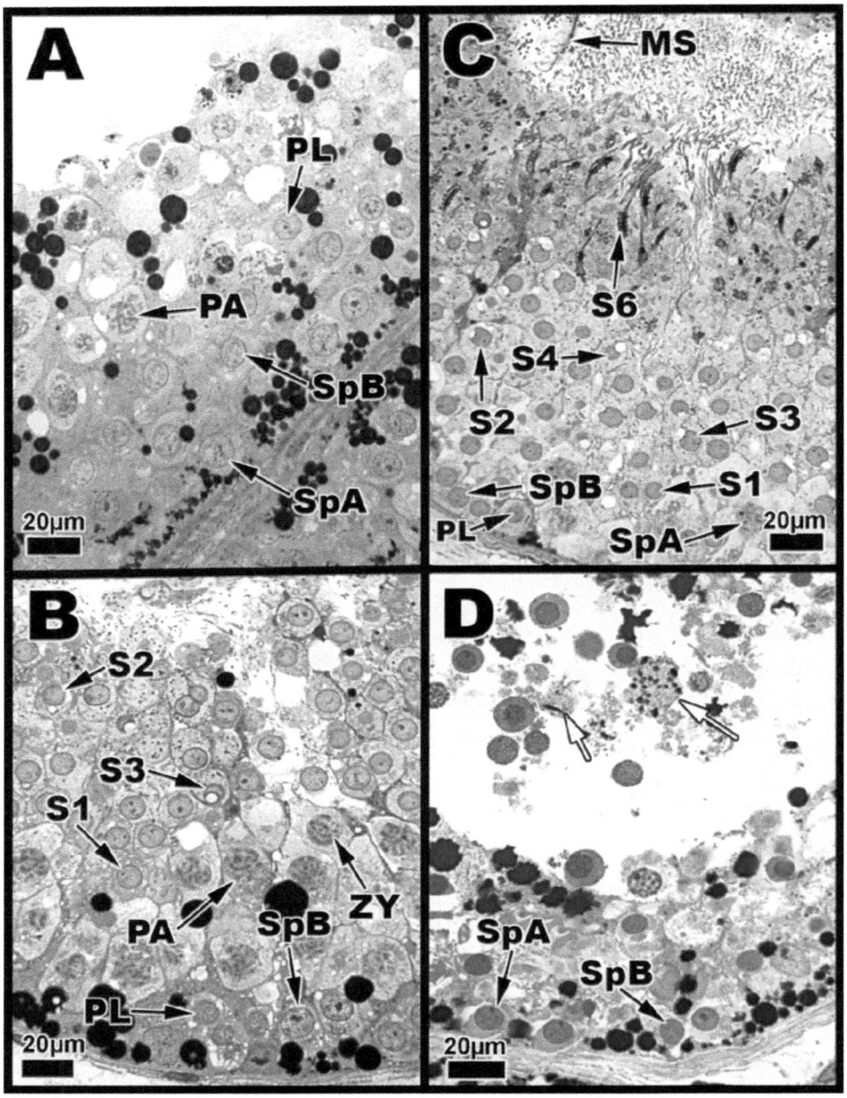

Fig. 6.27 Light micrographs of represented months of seminiferous epithelia within the testis of *Agkistrodon contortrix*. Bars = 20 µm. **A.** May epithelium showing high lipid content and the following cell types: spermatogonia A (SpA) and B (SpB), preleptotene spermatocytes (PL), pachytene spermatocytes (PA). **B.** July epithelium showing fewer lipids and the cell types: spermatogonia B (SpB), preleptotene spermatocytes (PL), zygotene spermatocytes (ZY), pachytene spermatocytes (PA), steps 1-3 spermatids. **C.** Early August epithelium showing almost no lipids and the cell types: spermatogonia A (SpA), spermatogonia B (SpB), preleptotene spermatocytes (PL), steps 1-4, 6 spermatids. **D.** Late August epithelium showing more lipids and the cell types: spermatogonia A (SpA), spermatogonia B (SpB), and shed generations of older cell types (white arrows). Micrographs taken by Dustin Siegel.

Color image of this figure appears in the color plate section at the end of the book.

In May, proliferation is well underway in *Agkistrodon contortrix*. Spermatogonia A and B are common and the early stages of meiosis are also present in the seminiferous epithelium (Fig. 6.27A). Note that no spermatids are seen within the germinal epithelium at this time. By July (Fig. 6.27B), the seminiferous epithelium has thickened and is packed full of early round spermatids. There are still a few spermatocytes, however, proliferation has climaxed and, thus, the number of spermatogonia has decreased substantially when compared to the May seminiferous epithelium. There are no elongating spermatids within the seminiferous epithelium within July testes. Many of the round spermatids have advanced to the elongating spermatid stage of spermiogenesis in early August (Fig. 6.27C). Few mitotic or meiotic cells are present within the seminiferous epithelium and up to 4 or 5 spermatids dominate the seminiferous epithelium. Spermiation has also begun and there are many mature spermatozoa within the seminiferous tubule lumen. By late August to early September (Fig. 6.27D) all the spermatids have completed spermiogenesis and spermiation. Thus, the seminiferous epithelium is very thin and contains mostly spermatogonia A and B. Many of the advanced spermatocytes and spermatids that are left over within the epithelium are being shed to the lumen and are degenerating as evidenced by the large pieces of cellular debris within the seminiferous tubule lumen.

One can easily visualize in this example that germ cells are moving as one major cohort through proliferation (May), Meiosis (July), and spermiogenesis/spermiation (August). This temporal development is very similar to the development seen in anamniotes, and differs greatly from what has been reported for all avian and mammalian species, which practice spatial germ cell development irrespective of their seasonal cycle. Germ cell development in *Agkistrodon contortrix* is very similar to that of the postnuptial Swamp Snake, *Seminatrix pygaea* (Gribbins *et al.* 2005). Though a complete series of samples is not available at this time, a reasonable hypothesis is that the testis of *A. contortrix* also follows this same postnuptial reproductive cycle, which is in corroboration with androgen levels in this species (Schuett *et al.* 1997) and with the fact that most crotalids studied to date follow a postnuptial pattern of male germ cell development (Aldridge and Duvall 2002).

Specifically in *Seminatrix pygaea*, spermatogenic recrudescence begins in March (Gribbins *et al.* 2005) with division of spermatogonia A and B and the appearance of preleptotene spermatocytes (Fig. 6.28A). By June, the majority of the germ cell population has moved into the late events of meiosis and the early round spermatid stage of spermiogenesis (Fig. 6.28B). In July testes, early and late events of spermiogenesis dominate the seminiferous epithelium (Fig. 6.29A). Also, small pockets of spermatozoa are released at this time to the lumina of the seminiferous tubules. The October testis reveals that spermiation is climaxing and what is left of the germ cell population within the seminiferous epithelium are late elongating spermatids. However, spermatogonia A and B are still present resting on

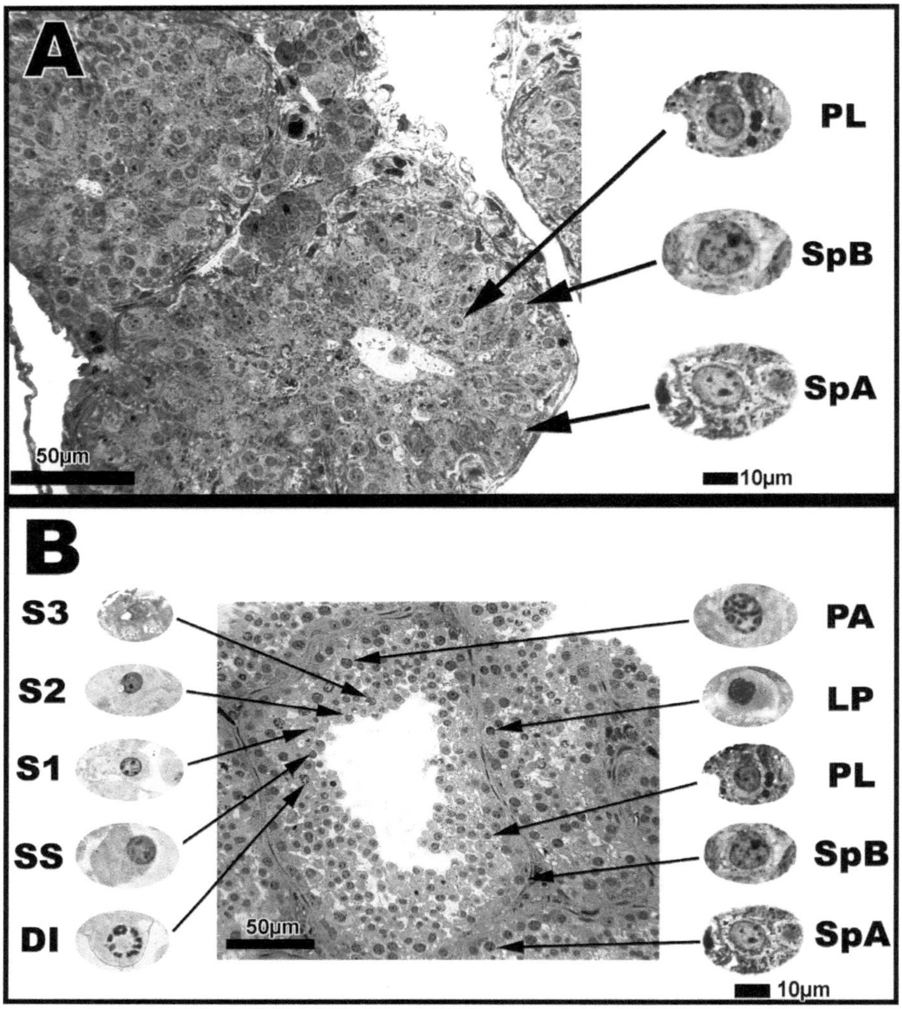

Fig. 6.28 A. A section of March seminiferous epithelium in *Seminatrix pygaea* with higher magnifications of represented cell types. PL, preleptotene; SpB, type B spermatogonia; SpA, type A spermatogonia. **B.** A section of June seminiferous epithelium with higher magnifications of represented cell types. S3, step 3 spermatid; S2, step 2 spermatid; S1, step 1 spermatid; SS, secondary spermatocyte; DI, diplotene; PA, pachytene; LP, leptotene; PL, preleptotene spermatocytes; SpB, type B spermatogonia; SpA, type A spermatogonia. Light micrographs.

Color image of this figure appears in the color plate section at the end of the book.

the basement membrane of the germinal epithelium. *Agkistrodon contortrix* and *S. pygaea* exhibit the same temporal germ cell development observed in other temperate reptilian species (Gribbins *et al.* 2003; Gribbins and Gist 2003; Gribbins *et al.* 2006; Rheubert *et al.* 2009). These reptilian species studied to date are all prenuptial or postnuptial breeders. Thus, it would

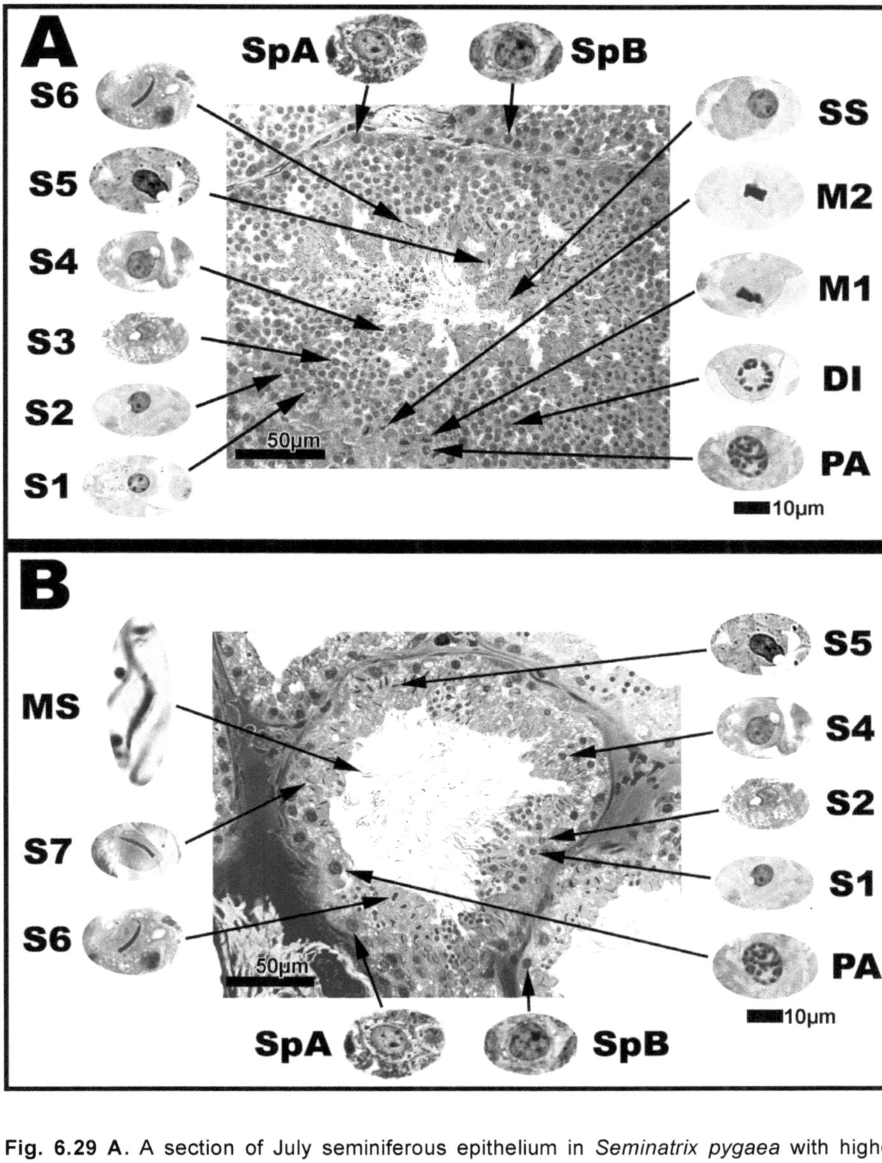

Fig. 6.29 A. A section of July seminiferous epithelium in *Seminatrix pygaea* with higher magnifications of represented cell types. S6, step 6 spermatid; S5, step 5 spermatid; S4, step 4 spermatid; S3, step 3 spermatid; S2, step 2 spermatid; S1, step 1 spermatid; SS, secondary spermatocyte; M2 and M1, meiosis 2 and 1; DI, diplotene spermatocyte; PA, pachytene spermatocytes; SpB, type B spermatogonia; SpA, type A spermatogonia. **B.** A section of October seminiferous epithelium with higher magnifications of represented cell types. MS, mature spermatozoa; S7, step 7 spermatid; S6, step 6 spermatid; S5, step 5 spermatid; S4, step 4 spermatid; S3, step 3 spermatid; S2, step 2 spermatid S1, step 1 spermatid; PA, pachytene spermatocyte; SpB, type B spermatogonia; SpA, type A spermatogonia. Light micrographs.

Color image of this figure appears in the color plate section at the end of the book.

be interesting to determine whether this germ cell development strategy in snakes and other reptiles is similar in other seasonal reproductive cycle types including tropical continuous breeders.

Although no continuous breeding tropical snake to date has been explored for the germ cell development strategy employed during spermatogenesis, a population of *A. piscivorus* at its most southern range limit within the US does practice a different reproductive cycle type (Gribbins *et al.* 2008). This allows the retesting of whether an episodic type of reproductive cycle has an affect on the mode of germ cell development employed during sperm development.

In *Agkistrodon piscivorus*, the testis shows a biannual mode of spermatogenesis (Fig. 6.30) with the first spermatogenic cycle starting in March and ending in June followed by a quiescent period in July and then the second spermatogenic cycle starts again in August and ends in November. Thus, this *A. piscivorus* population from southern Louisiana produces sperm twice, once in the spring and once in the fall within the same calendar year. These data parallel the observations of pairings of male and female *A. piscivorus* being seen throughout the year, except January (Wharton 1966), and sperm in the distal ducts of female cottonmouths from this same population in the spring and fall (Siegel and Sever 2008). The germ cell development strategy within the *A. piscivorus* testis is very similar to the previously described strategies of *A. contortrix* and *Seminatrix pygaea* despite

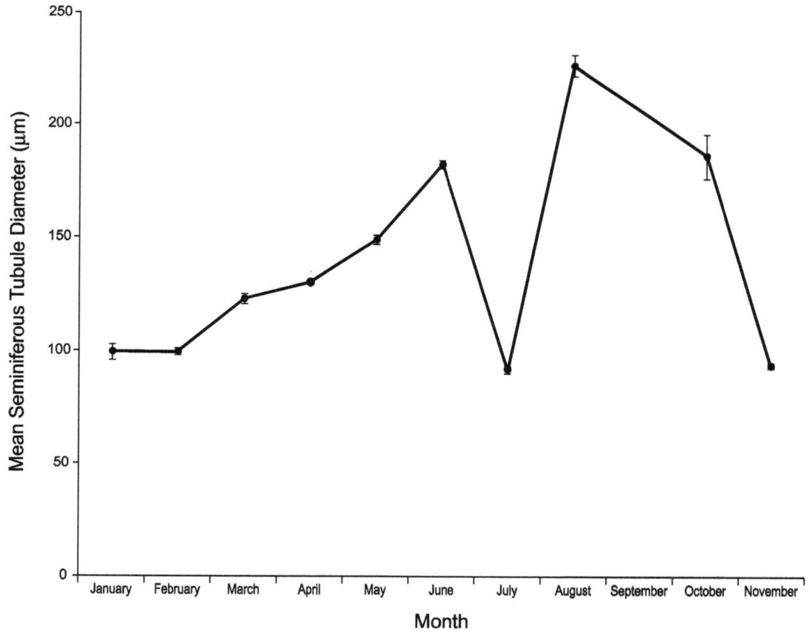

Fig. 6.30 Variation in seminiferous tubule diameter (mean ± 1 SE) during the annual reproductive cycle of the male Cottonmouth, *Agkistrodon piscivorous leucostoma*. Note there are two peaks of spermatogenesis with a quiescent period in July.

its unique spermatogenic cycle. Figure 6.31 represents the first spermatogenic wave seen in *A. piscivorus*, which occurs from March-June. Prior to March the seminiferous tubules are in a quiescent phase of development (January and February, Fig. 6.31A) and spermatogonia A and B are the only major germ cell types present within the seminiferous epithelium.

In March (Fig. 6.31B), spermatogonial proliferation is underway and many spermatogonia have advanced to meiosis and even the early stages of spermiogenesis within the seminiferous epithelium. June samples of testis (Fig. 6.31C) represent the climax of spermiogenesis and an increase in spermiation. Most of the population of germ cells is completing spermiogenesis and entering the lumina of the seminiferous tubules as mature spermatozoa. The lumina increase in size because mature sperm are dumped into the seminiferous tubules and there is an accumulation of generations of elongating spermatids within the seminiferous epithelium. The lack of early meiotic cells and accruing number of elongating spermatids prevents consistent cellular association between germ cell types. By July (Fig. 6.31D), spermiation is complete and the seminiferous tubules have entered their second phase of quiescence, which leads to a dramatic decrease in seminiferous tubule diameter (Fig. 6.30). The only cell types found within the highly vacuolated germinal epithelium are a single row of spermatogonia A and B located against the basement membrane. The lumina of these tubules are void of most spermatozoa and have pieces of the seminiferous epithelium with remnant germ cells (Fig. 6.31D white arrow) from the spring cycle of spermatogenesis.

The second wave of spermatogenesis has begun in the August testes of *Agkistrodon piscivorus* (Fig. 6.32A). The early stages of proliferation and meiosis are similar to the March and April samples but spermiogenic cells are in more advanced stages than the earlier wave. Although all three stages of spermatogenesis are observed in August, no consistent cellular association are formed because of the four to five different spermatids occupying the apical portion of the seminiferous epithelium. Spermatogenesis in the October seminiferous epithelium (Fig. 6.32B) has advanced into spermiogenesis with round and elongating spermatids represented. Spermatocytes have been exhausted and spermatogonia A and B are found near the basement membrane of the seminiferous tubules. Many of the developing spermatids have completed spermiogenesis and are being shed to the lumina of the seminiferous tubules as mature spermatozoa. Like the July sample, the November seminiferous tubules (Fig. 6.32C) are in a state of quiescence with spermatogonia A and B making up the majority of germ cells and lipid rich vacuoles dominating the seminiferous epithelium.

The germ cell development strategy of the cottonmouth is similar to recent studies on other temperate squamates (Gribbins and Gist 2003; Gribbins *et al.* 2005) which, in turn, show a similar temporal germ cell development strategy described here for *Agkistrodon* and *Seminatrix*. This temporal germ cell development differs greatly from the spatial germ

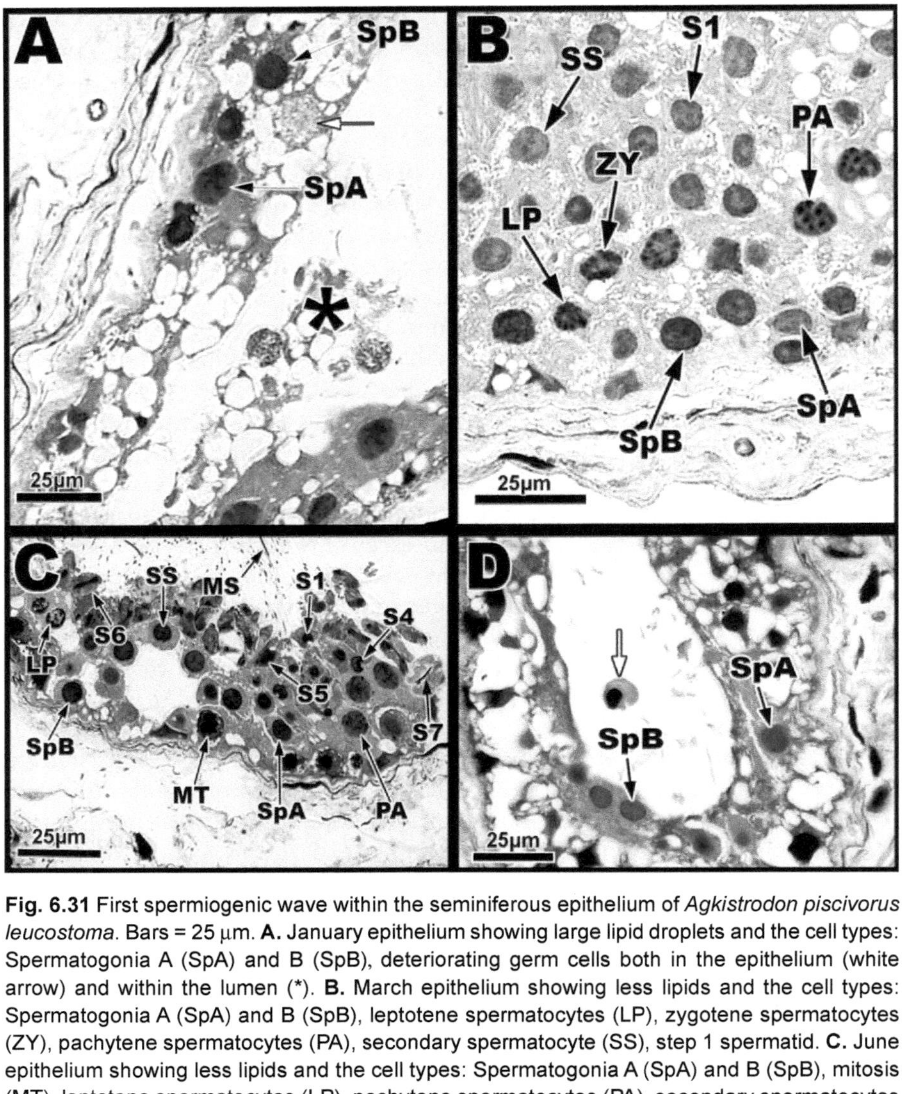

Fig. 6.31 First spermiogenic wave within the seminiferous epithelium of *Agkistrodon piscivorus leucostoma*. Bars = 25 μm. **A.** January epithelium showing large lipid droplets and the cell types: Spermatogonia A (SpA) and B (SpB), deteriorating germ cells both in the epithelium (white arrow) and within the lumen (*). **B.** March epithelium showing less lipids and the cell types: Spermatogonia A (SpA) and B (SpB), leptotene spermatocytes (LP), zygotene spermatocytes (ZY), pachytene spermatocytes (PA), secondary spermatocyte (SS), step 1 spermatid. **C.** June epithelium showing less lipids and the cell types: Spermatogonia A (SpA) and B (SpB), mitosis (MT), leptotene spermatocytes (LP), pachytene spermatocytes (PA), secondary spermatocytes (SS), steps 1, 4-7 spermatids, mature spermatozoa (MS). **D.** July epithelium showing large lipid droplets and the cell types: Spermatogonia A (SpA) and B (SpB), deteriorating germ cells within the lumen (white arrow). Light micrographs.

cell development seen in seasonally and continually breeding birds and mammals (Yamamoto 1967; Roosen-Runge 1972; Tait and Johnson 1982; Tsubota and Kanagawa 1989; Tiba and Kita 1990; Foreman 1997). This episodic spermatogenic development is also very similar to the temporal germ cell development strategy of derived amphibians such as anurans (Lofts 1964; Van Oordt and Brands 1970). Though anuran amphibians have

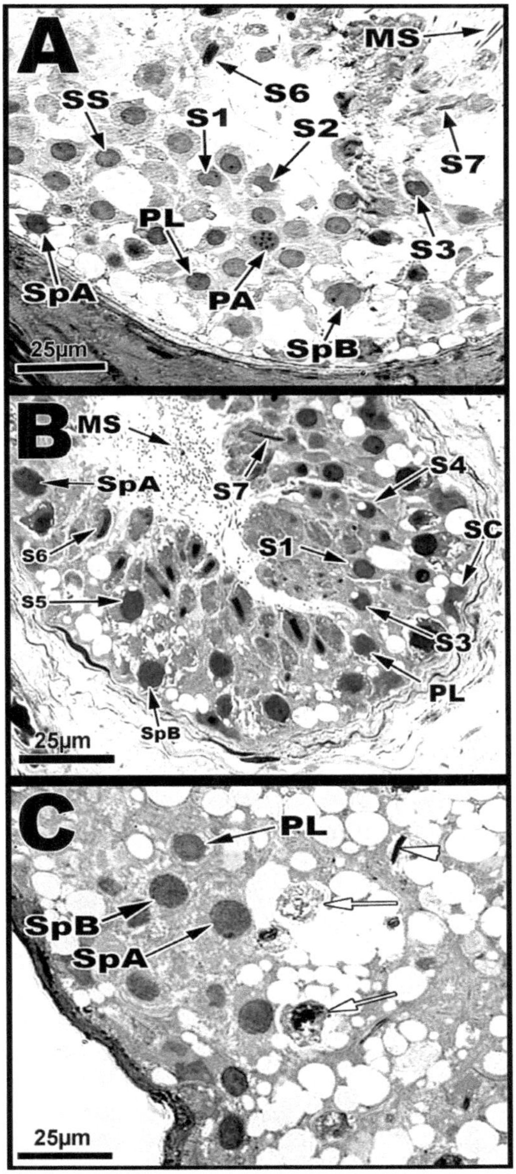

Fig. 6.32 Second spermiogenic wave within the seminiferous epithelium of *Agkistrodon piscivorous leucostoma*. Bars = 25 µm. **A.** August epithelium showing a few large lipid droplets and the cell types: Spermatogonia A (SpA) and B (SpB), preleptotene spermatocytes (PL), secondary spermatocytes (SS), pachytene spermatocytes (PA), steps 1-3, 6, 7 spermatids, mature spermatozoa (MS). **B.** October epithelium showing less lipids and the cell types: Spermatogonia A (SpA) and B (SpB), preleptotene spermatocytes (PL), steps 1, 3-7 spermatids, mature spermatozoa (MS), Sertoli cell nuclei (SC). **C.** November epithelium showing large lipid droplets and the cell types: Spermatogonia A (SpA) and B (SpB), preleptotene spermatocytes and deteriorating germ cells within the lumen (white arrows and arrowheads). Light micrographs.

been called a transitional taxon between the anamniotes and amniotes in terms of testicular organization (Van Oordt 1955), amphibian seminiferous tubules are lined with cysts and do not maintain a permanent epithelium like that of amniotes. The ancestors to the extant reptiles are presumably considered the most primitive amniotes phylogenetically and most likely had testes that were structurally similar to birds and mammals. Thus, reptiles might represent a better transitional intermediary in terms of testicular organization and germ cell development strategy between anurans and the derived amniotic taxa. However, caution must be used when thinking of the evolution of testis structure within reptiles because the extant orders of reptiles are considered paraphyletic (Pough *et al.* 2001). This makes it difficult to make firm conclusions about the evolution of germ cell development strategies within reptiles. As has been stated in earlier publications on squamates (Rheubert *et al.* 2009) the most parsimonious hypothesis for the germ cell development strategy in amniotes would involve retention of the anamniote germ cell development in basal amniotes and modern reptiles and snakes, and an independent origin (convergence) of the spatial development type now seen in modern birds and mammals. In an alternate hypothesis, the spatial development type could have evolved in the ancestor to the Synapsida, but this would require reversal to the state (temporal type) in the crocodilians, lepidosaurians (including snakes), and possibly turtles.

6.4 SPERMIOGENESIS

The last decade has provided a large amount of needed data on the ultrastructure of sperm within Ophidia. These studies have focused on the morphology of the spermatozoa, which is useful for phylogenetic analysis within snakes (see Cunha *et al.* 2008). In contrast, ultrastructural data on spermiogenesis in squamates, particularly snakes, is lacking. The events of spermiogenesis should parallel the final mature structures found in the released spermatozoa (Gribbins *et al.* 2007). Thus, spermiogenesis provides the potential to collect more morphological data that could be combined with spermatozoal ultrastructure and increase the robustness of phylogenetic inferences (Weins 2004). Ontogenic changes in spermatid morphology during spermiogenesis may also lead to detectable morphological characters that are different between closely related species or genera. This would be significant in that spermatozoal ultrastructure presently has only allowed for resolution at the family level in phylogenetic analyses (Jamieson 1995; Teixeira *et al.* 1999).

There are few studies that attempt to at least describe some of the ultrastructural aspects of spermiogenesis within snakes. These include the African Rock Python (*Python sebae*; Boisson and Mattei 1965), the Dark Green Snake (*Coluber viridiflavus*; Saita *et al.* 1988), *Elaphe climacophora* (Hondo *et al.* 1994), and the Arabian Sand Boa (*Eryx jayakari*; Al-Dokhi *et al.* 2005a,b). One elegant study on the entire process of spermiogenesis exists for *P. sebae* (Boisson and Mattei 1966); however, this chapter's entire data set is based on drawings of the different stages of spermatids. These studies also give very little information about the specific sequence of development during the phases of spermiogenesis: acrosome development, elongation, and condensation of the DNA.

There are a few common features observed within all of these studies. There is always a series of microtubules called the manchette associated with elongates (elongating spermatids). The acrosome forms close to the nucleus and then indents its apical surface. There is the presence of an acrosome granule within the acrosome vesicle during some point of the round spermatid stage. The condensation of chromatin in elongating spermatidss always results in filamentous chromatin fibers within the nucleus during the mid stages of spermiogenesis. In most of these previous studies, an annulus is seen migrating distally with the developing axoneme. The acrosome also migrates over the apex of the nucleus during elongation in these snakes, leading to a thinner nuclear rostrum proximally and nuclear shoulders distally. If described, a perforatorium was present in almost all the species looked at to date. Lastly, in most species there is at least a brief description of a subacrosome space between the nucleus and the inner acrosome membrane.

The inconsistencies of the description of spermatids and the lack of relatively complete information for each phase of spermatid development in this series of studies in snakes, lends to the notion that a universal model is needed so that consistent data can be obtained in future studies, especially if these data are to be used in phylogenetic analyses. Thus, a recent study was conducted on the ultrastructure of spermiogenesis within *Agkistrodon piscivorus* (Gribbins *et al.* 2009). This study represents the only comprehensive study of spermiogenesis in a snake. To show the potential of such data for phylogenetic inference, morphological details of the spermatids within the testis of *A. contortrix* will also be provided (Gribbins unpublished) within this section.

The ultrastructure of the spermatids as they traverse spermiogenesis within the testes of *Agkistrodon piscivorus* is similar in most respects to that of other squamates (Gribbins *et al.* 2009). Within squamates, apart from an account of spermatogenesis within *Sphenodon* (Healy and Jamieson 1994), there is only one other major comprehensive study done to date on spermiogenesis, and this was on the Ground Skink (*Scincella lateralis*; Gribbins *et al.* 2007). So data are sorely needed not only in Ophidia but also for squamates in general for spermiogenesis.

During acrosome formation in *Agkistrodon piscivorus* the acrosome vesicle forms from transport vesicles budding from the Golgi apparatus (Fig. 6.33), which accumulate as the acrosome vesicle before contact is made with the nuclear membrane. This is similar to what has been described in *Scincella lateralis* (Gribbins *et al.* 2007). Also, the acrosome granule is seen within this vesicle before nuclear contact (Fig. 6.33A), similar to that of *S. lateralis*; however, the granule is centrally located (rather than basally located). The granule makes its largest growth spurt in this basal position within the acrosome of *A. piscivorus*. Subsequent features of acrosome development (Figs. 6.33B, 6.34A) within *A. piscivorus* are similar to that described for *S. lateralis* and other squamates such as: transport vesicles from the Golgi, prominent subacrosomal space, multilaminar Sertoli cell membranes, and lateral folding (Clark 1967; Da Cruz-Landim and Da Druz-Hofling 1977; Butler and Gabri 1984; Dehlawi *et al.* 1992; Ferreira and Dolder 2002, 2003; Gribbins *et al.* 2007). In both *S. lateralis* and *A. piscivorus* the acrosome granule can be observed prior to the acrosome vesicle making contact with the nuclear membrane, which may suggest this trait is a potential synapomorphy for scleroglossids. This acrosome granule is responsible for the formation of the perforatorium, which is present in all squamates studied to date, including *A. piscivorus* (Ferreira and Dolder 2002; Gribbins *et al.* 2007; Cunha *et al.* 2008).

As early elongation (Fig. 6.35) begins in *Agkistrodon piscivorus*, the acrosome complex (Fig. 6.34B) becomes highly compartmentalized, which is common in Squamata (Healy and Jamieson 1994; Harding *et al.* 1995; Jamieson and Scheltinga 1994; Jamieson *et al.* 1996; Tavares *et al.* 2007). The compartments include the subacrosomal space, perforatorium, acrosomal vesicle, and the outer Sertoli cell laminae (Fig. 6.36), which are all similar to other squamates including *Scincella lateralis* (Gribbins *et al.* 2007). However, stratification occurs to the subacrosome space and an epinuclear lucent zone within *A. piscivorus* spermatids (Figs. 6.36A, 6.38B-D).

During elongation in *Agkistrodon piscivorus*, nuclei move apically and against the cell membranes of the developing spermatids (Fig. 6.35A) as described in *Scincella lateralis* (Gribbins *et al.* 2007). This peripheral location of the spermatid nucleus most likely causes the acrosome to collapse and migrate laterally along the nuclear apex (Fig. 6.35B; Clark 1967; Butler and Gabri 1984) and the relocation of cellular organelles (Sprando and Russell 1988; Lin and Jones 1993, 1997; Soley 1997; Ventela *et al.* 2003). The manchette (Fig. 6.37) can be visualized in both transverse and sagittal sections in elongating spermatids of *A. piscivorus* and is thought to aid in the elongation of the nucleus (Russell *et al.* 1990). Parallel microtubules outnumber the circum-cylindrical fibers in *A. piscivorus*. This absence or limited numbers of circum-cylindrical microtubules is also noted in *S. lateralis* (Gribbins *et al.* 2007), which have thicker bodied spermatozoa than those of other squamate species (Jamieson and Scheltinga 1994). The result of fewer circum-cylindrical fibers may also result in more robust spermatozoa in *A. piscivorus*. Throughout the events of condensation, uniform translucent shoulders (Fig. 6.35A) are observed on either side of the caudally located nuclear fossa where the flagellum attaches to the nuclear body of the developing spermatid. These lucent structures have not been described in any other reptile during spermiogenesis and thus may represent an autapomorphy for *A. piscivorus* or possibly a synapomorphy for the *Agkistrodon* complex if this character is only observed in congeners of this genus.

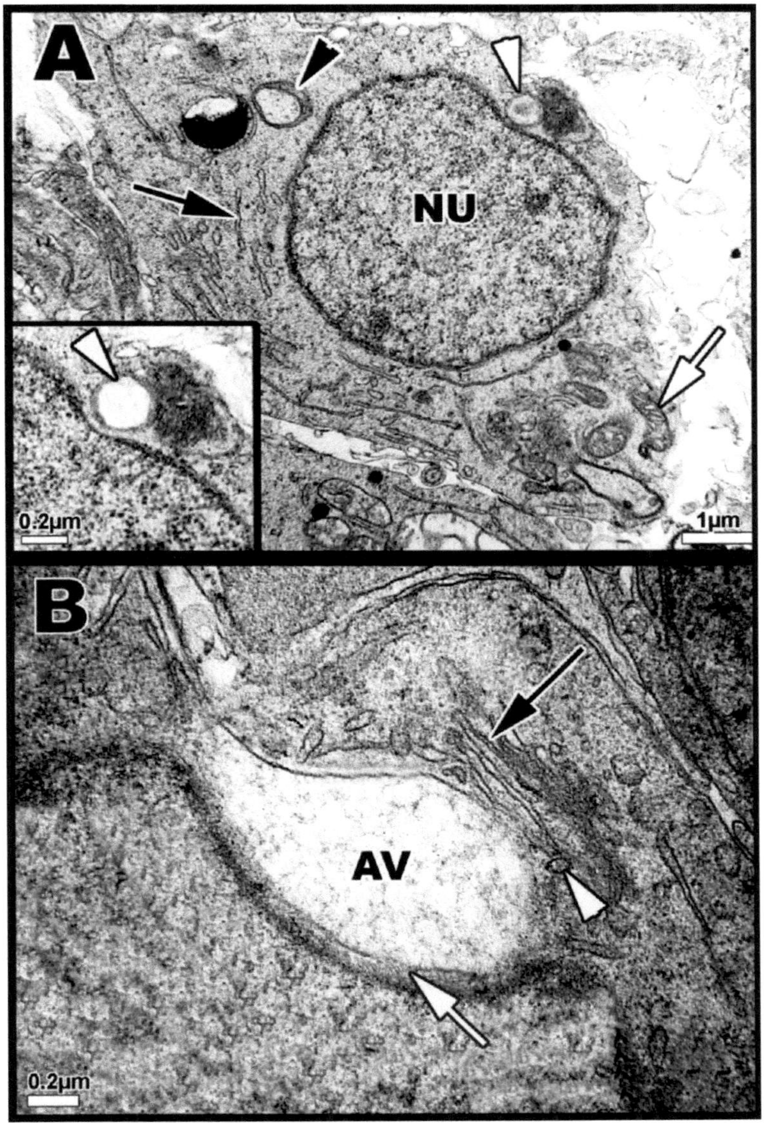

Fig. 6.33 Round spermatids undergoing acrosome development during the early stages of spermiogenesis within the *Agkistrodon piscivorus leucostoma* seminiferous epithelium. **A.** The acrosomal vesicle (white arrowhead) is juxtapositioned to the apical portion of the nucleus (NU). The vesicle is in the early phase of growth. The cytoplasm of the spermatid has numerous mitochondria (white arrow), many layers of endoplasmic reticula (black arrow), and multivesicular bodies (black arrowhead). Inset: Shows that the vesicle (white arrowhead) has not quite made contact with the nuclear membrane. **B.** The Golgi apparatus (black arrow) is prominent and next to the developing acrosome (AV). Transport vesicles (white arrowhead) can be seen budding off of proximal cisterna of the golgi and presumably will merge with the acrosome during its growth phase. A prominent subacrosomal space (white arrow) is also developing between the acrosome membrane and the nuclear membrane.

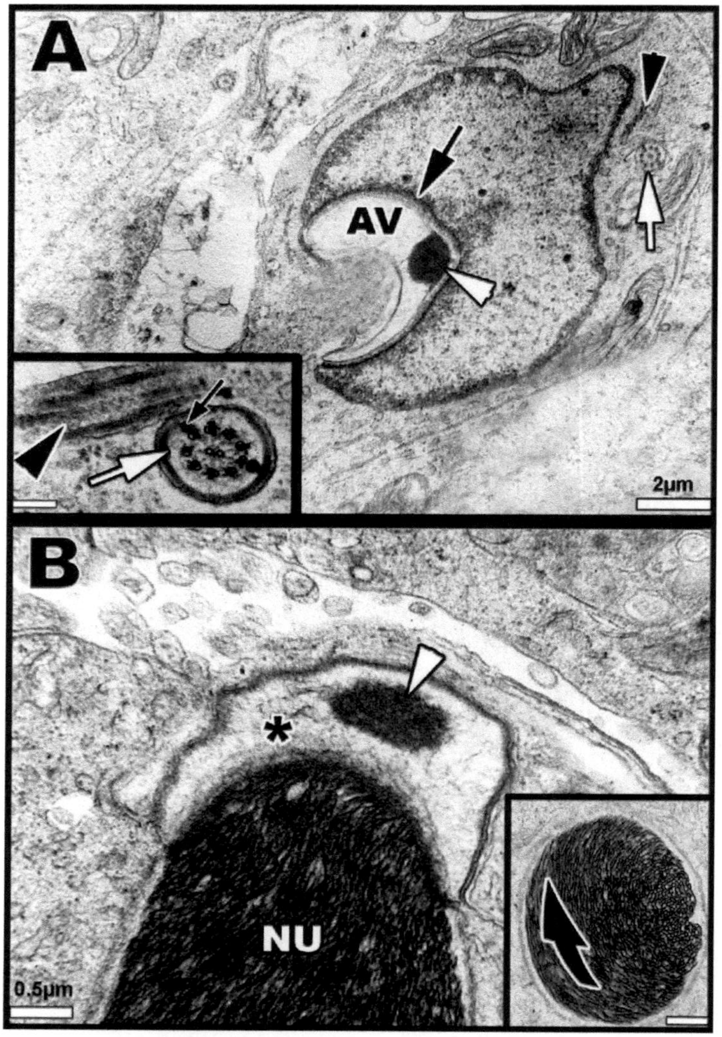

Fig. 6.34 A. Late stage round *Agkistrodon piscivorus leucostoma* spermatid exhibiting a deep indented acrosome (AV) and a prominent acrosomal granule (white arrowhead). An accumulation of dark staining proteins is lining the nuclear membrane side of the subacrosomal space (black arrow). The caudal portion of the nucleus has begun elongation and the proximal centriole (black arrowhead) can be seen in sagittal section and the growing distal neck is shown in transverse section (white arrow) near the caudal end of the nucleus. The insert shows the proximal centriole (black arrowhead) and distal neck (white arrow) in greater detail. The developing flagellum of the neck has two opposing peripheral fibers (black arrow) associated with microtubule doublets 3 and 8. **B.** A middle stage spermatid has an acrosome that begins to flatten and envelop the elongating nucleus (NU) and the acrosomal granule (white arrowhead) migrates from its previous basal position within the vesicle to a more superficial location within the acrosome. The granule starts to break up and become diffuse within the acrosomal vesicle (*).The insert displays the same elongating spermatid step in transverse section and the arrow demonstrates the rotation of the chromatin as it condenses.

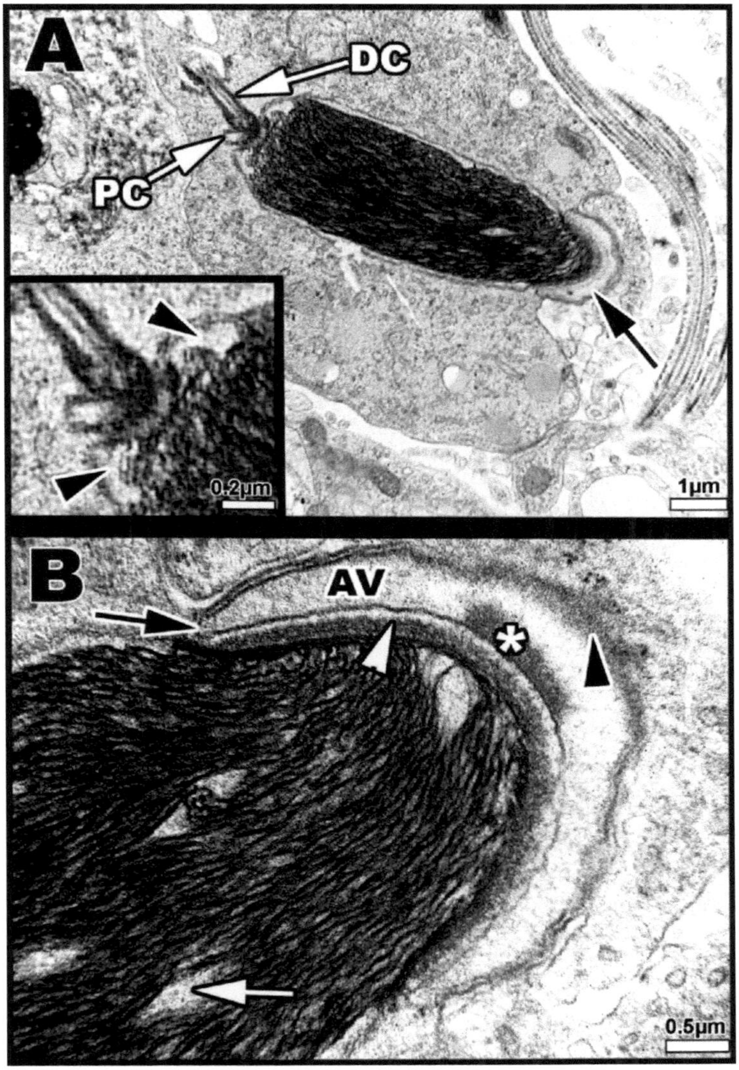

Fig. 6.35 A. The developing flagellum is prominent in sagittal sections of *Agkistrodon piscivorus leucostoma* elongating spermatidss showing the presence of both the proximal (PC) and distal (DC) centrioles. The distal centriole begins to elongating spermatid to form the neck. On the opposite pole of the elongating spermatid is the acrosome vesicle (black arrow). The insert shows the two shoulders (black arrowheads), which are devoid of chromatin and can be seen lateral to the insertion of the flagellum within the flagellar fossa. **B.** A high power view of the acrosome vesicle (AV) of a slightly later staged middle elongating spermatid. The chromatin within the apical nucleus shows spiraling and large open nulceoplasmic spaces (white arrow). The acrosome vesicle shoulders (black arrow) have extended further over the apex of the nucleus. The diffuse acrosome granule (white*) is perfusing within the vesicle (AV). Some of the dense protein material is accumulating underneath the outer acrosome membrane (black arrowhead). Dense protein plaques are also found within the subacrosome space (white arrowhead).

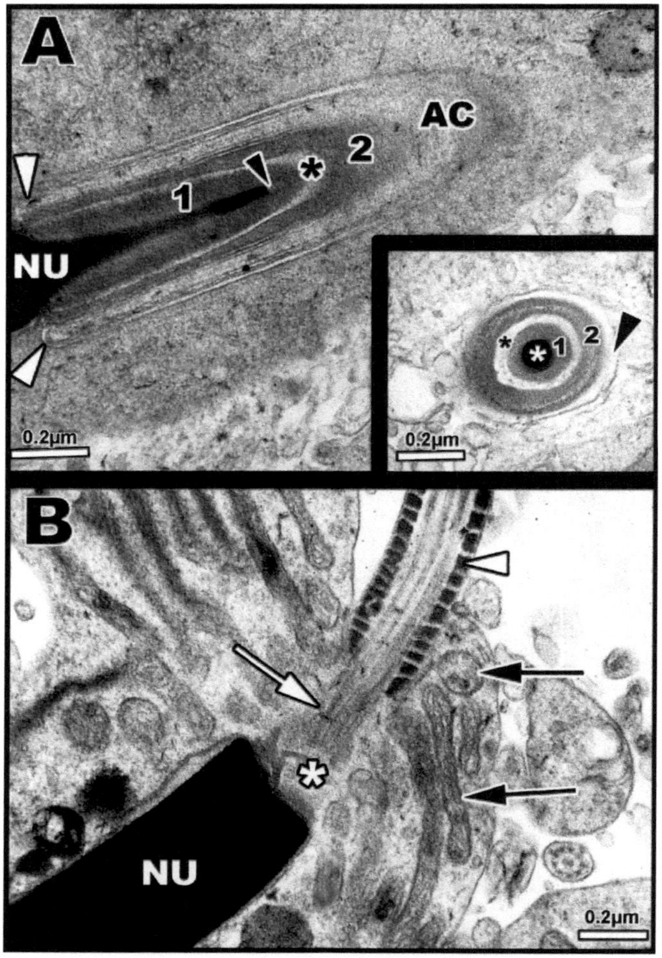

Fig. 6.36 A. A high power view of the acrosome (AC) and the apical nucleus (NU) in a late *Agkistrodon piscivorus leucostoma* elongating spermatid. The acrosome shoulders (white arrowheads) are located where the elongation of the nuclear rostrum (thin nuclear process that extends into the acrosome complex) begins. The rostrum extends into the subacrosomal space and just beyond its tapered tip is a short epinuclear lucent zone (black arrowhead) that will eventually sit caudal to the more rostrally located perforatorium in terminating stages of spermiogenesis. The subacrosomal space is separated into 2 granulated protein layers (1 and 2) by a clear zone (*).The Inset shows the acrosomal complex in transverse section near the tip of the nuclear rostrum (white*). This view confirms the two granulated protein layers (1 and 2) separated by a clear zone (black*) within the subacrosomal space. The acrosome vesicle is also seen in cross section (black arrowhead). **B.** The caudal nucleus (NU) is also uniformly stained with condensed chromatin and the proximal centriole (white*) can be seen in cross section juxtaposition to the distal centriole, which extends to form the neck (white arrow) of the flagellum. The principal piece/midpiece is easily distinguished from the neck by the thick ribs of the fibrous sheath (white arrowhead) that surrounds the microtubules of the flagellum. Mitochondria, both in sagittal and transverse section (black arrows), are accumulating near what will be the midpiece of the mature spermatozoon.

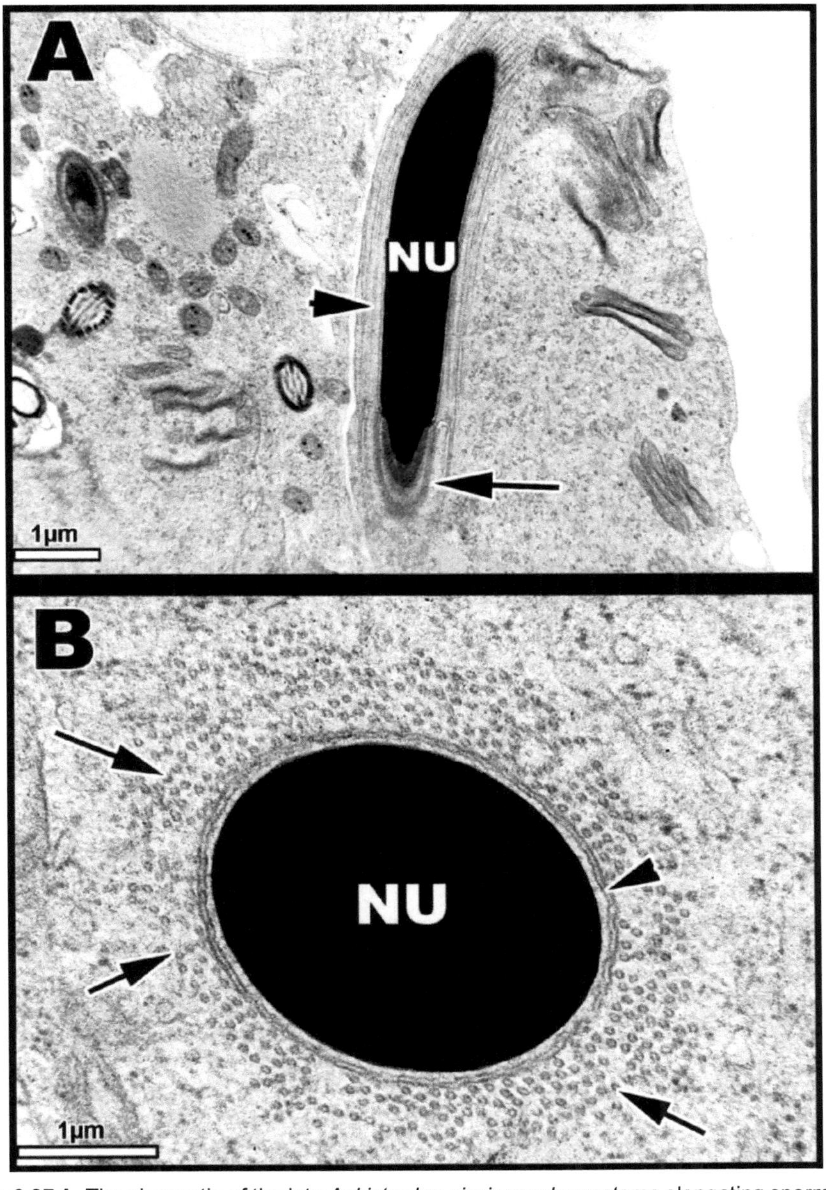

Fig. 6.37 A. The chromatin of the late *Agkistrodon piscivorus leucostoma* elongating spermatid nucleus (NU) lacks open pockets of nucleoplasm and the condensed DNA is uniformly stained. The microtubules composing the manchette (black arrowhead) can be seen in juxtaposition to the sagittally sectioned nucleus. The acrosome complex (black arrow) drapes over the apex of the nucleus, causing a conical shaped cranial nuclear head. **B.** A transverse section through the body of the late elongating spermatid nucleus (NU) showing the uniformly stained DNA and the parallel microtubules of the manchette (black arrows) in cross section. Also note that there is a small group of circum-cylindrical microtubules making up the inner most ring of the manchette (black arrowhead).

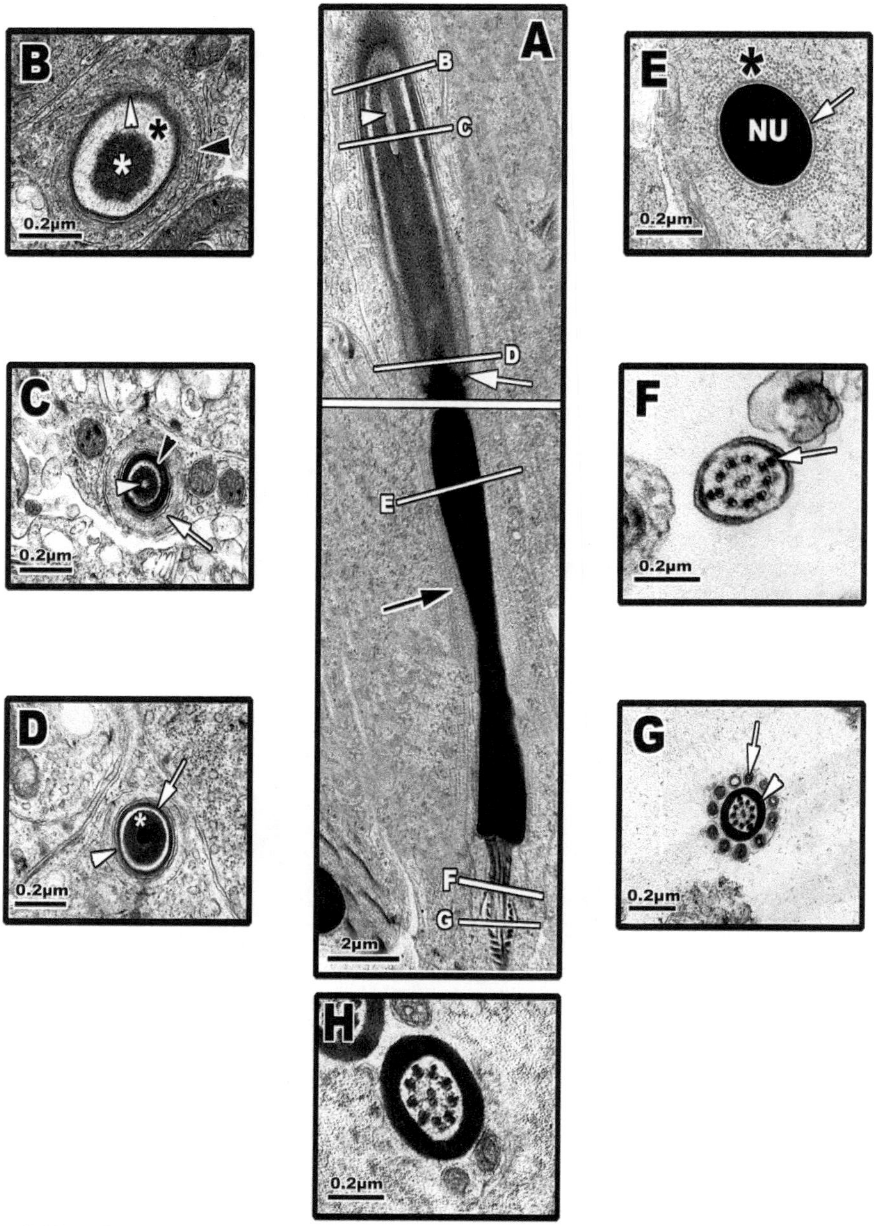

Fig. 6.38 A. Sagittal section of an *Agkistrodon piscivorus leucostoma* elongating spermatid at the end of spermiogenesis with all three major parts of the spermatid visible (acrosome, nucleus, and flagellum; the white line through the middle of A denotes that two separate micrographs were combined to obtain this image). Lines and represented letters show approximately where transverse sections (CS) occurred within spermatids at or near the same stage of

Fig. 6.38 Contd. ...

As the flagellum develops during late elongation (Fig. 6.36B) large numbers of mitochondria become present in the posterior portion of the *Agkistrodon piscivorus* spermatid cytoplasm and are associated with the flagellum, which is consistent with other amniotes (McIntosh and Porter 1967; Lin and Jones 1993; Ferreira and Dolder 2002; Gribbins *et al.* 2007; Cunha *et al.* 2008). The axoneme displays (Fig. 6.38F) the conserved 9+2 microtubule arrangement seen in other amniotes. The previously described enlarged peripheral fibers located near microtubule doublets 3 and 8 (Fig. 6.38F,G; Healy and Jamieson 1994; Ferreira and Dolder 2003; Cunha *et al.* 2008; Tourmente *et al.* 2008), which are considered a synapomorphy for Squamata and Lepidosauria, are also seen within the neck and midpiece axonemes of the flagella of *A. piscivorus*.

The processes of spermiogenesis in *Agkistrodon piscivorus* are very similar in most respects to other ophidians, which have limited data historically, and to other squamates studied to date. This suggests that many aspects of spermiogenesis are highly conserved within ophidian species and within Squamata. Also, much of the described data within this study corroborates the morphology of the mature spermatozoa described for pitvipers, the Tropical Rattlesnake (*Crotalus durissus;* Cunha *et al.* 2008), the Urutu (*Bothrops alternatus*) and the Bolivian Lancehead, (*B. diporus;* Tourmente *et al.* 2008). Presently, these two studies represent the only

... *Fig. 6.38 Contd.*

development as A in order to obtain cross sections B-G. Note the well-developed parallel microtubules of the manchette (black arrow) running alongside the nucleus, the perforatorium (white arrowhead), and the dark staining flange associated with the acrosome shoulders and the nucleus (white arrow). **B.** CS through the rostral subacrosomal space and acrosomal vesicle. Granulated protein layer of subacrosomal space (white*), acrosomal vesicle (black*), protein accumulation on the inside of the outer acrosomal membrane (white arrowhead), multiple Sertoli cell membranes surrounding the acrosome vesicle (black arrowhead). **C.** CS through the subacrosomal space, perforatorium, and acrosomal vesicle. The white arrowhead points to the perforatorium, which is surrounded by the granulated proteins of the subacrosomal space. The white circular region around the subacrosomal space is the acrosomal vesicle, which again has protein accumulations under its outer membrane (black arrowhead). Also note the numerous Sertoli cell membrane layers surrounding the entire acrosomal complex (white arrow). **D.** CS through the nuclear rostrum. The white* labels the subacrosomal space. Within the middle of this space is the conical point of the rostrum in CS. Also present are the protein accumulation under the outer membrane of the acrosome (white arrow) and the Sertoli cell membrane layers (white arrowhead). **E.** CS through the nucleus proper. Nucleus (NU), Manchette (*), inner single circum-cylindrical microtubule layer (white arrow). **F.** CS through the distal neck of the flagellum. Axoneme is nicely represented with 9 outer pairs of microtubules and a single central microtubule doublet, indicating that the section represents the beginning of the elongating flagellum and is most likely the transition point between the distal centriole and the midpiece. Attached enlarged peripheral fibers (white arrow) are seen at microtubule doublets 3 and 8 within the axoneme. **G.** CS through the midpiece. Dense fibrous sheath/ring (white arrowhead), concentric mitochondria (white arrow). **H.** A transverse section of the principal piece that represents the majority of the flagellum caudal to the midpiece. Note there are no enlarged peripheral fibers associated with the axoneme below the midpiece.

morphological data for spermatozoa or spermiogenesis within crotalids. Caution should be taken in making direct comparisons between the morphology of the mature spermatozoa and spermatids as morphological modifications to spermatozoa do occur after spermiogenesis as they pass through the excurrent ducts. Thus, to evaluate whether ultrastructural differences can be discerned between closely related species during spermiogenesis preliminary data will be provided here for spermiogenesis within *A. contortrix*.

The features of spermiogenesis are quite similar in *Agkistrodon contortrix* to those of *A. piscivorus*. This again supports previous studies that suggest the major events of spermiogenesis are highly conserved among squamates and within amniotes. However, there are several key features to *A. contortrix* spermatid ultrastructure that are different from that of *A. piscivorus*. During early acrosome formation (Fig. 6.39A), the acrosome vesicle is seen prior to contact with the nuclear surface similar to that of *A. piscivorus*. However, there is no evidence of the acrosome granule until the acrosome makes contact with the apex of the nucleus (Fig. 6.39B,C) in *A. contortrix* spermatids. A feature of spermiogenesis that seems to be unique to *A. contortrix* and has not been described in any other amniote to date is that two Golgi complexes participate in early acrosome formation (Fig. 6.39B). There is also rough endoplasmic reticulum present in close association with the developing early acrosome vesicle (Fig. 6.39A,C) in *A. contortrix* that is not seen during early spermiogenesis in *A. piscivorus*. The significance of this extra organelle during acrosome formation is not known but it may be an autapomorphy for *A. contortrix*. During mid-elongation (Fig. 6.40), spermiogenesis in *A. contortrix* is again very similar to *A. piscivorus*. There is only one major difference between *A. contortrix* spermatids undergoing early elongation. The nucleus in cross section has at least one endonuclear lacuna (Fig. 6.40A inset) early in elongation.

There is also some variation during late elongation and condensation in the spermatids of *Agkistrodon contortrix*. The manchette is located around the actual acrosome complex (Fig. 6.41A inset), which is absent around *A. piscivorus* acrosomes during elongation. The epinuclear space (Fig. 6.41A) is more pronounced and extends across most of the second subacrosome layer. At the base of the nuclear rostrum in *A. contortrix* elongating spermatidss, there are two distinct lucent spaces (Fig. 6.41A), which are also absent in *A. piscivorus* elongating spermatidss. There is a dense collar associated with the neck of the *A. contortrix* spermatids (Fig. 6.41B) that is again absent in *A. piscivorus* spermatids. In *A. piscivorus* late elongating spermatidss, there are 10-11 concentric mitochondria surrounding the developing midpiece. This number of mitochondria is slightly less (8-9) in *A. contortrix* elongating spermatids (Fig. 6.41B).

Though the above data are preliminary, they offer a potentially powerful data set for phylogenetic analyses in the future, especially in combination with spermatozoal character matrices. Nine character differences exist between two species within the *Agkistrodon* complex

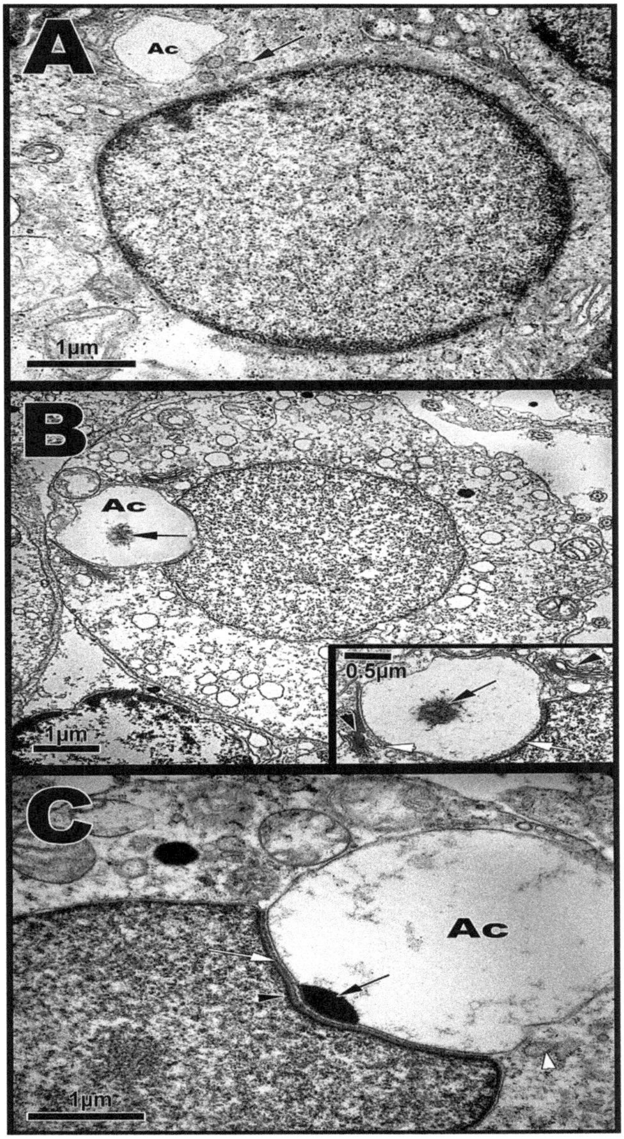

Fig. 6.39 The round spermatid stage of spermiogenesis within the testis of *Agkistrodon contortrix*. **A.** Shows a step 1 spermatid, which has an acrosome (Ac) that is not attached to the nucleus. Note that rough ER (black arrow) is also in close association to the nucleus. **B.** Step 2 spermatid with two prominent Golgi apparati (inset: black arrowheads) associated with the growing acrosome (ac) and acrosome granule (black arrow). Transport vesicles (white arrowhead) are coming off the Golgi and merging with the developing acrosome complex. **C.** Early step 3 spermatid with a large acrosome (Ac) and a basally located acrosome granule (black arrow). There is also rough ER associated with the acrosome (white arrowhead), a visible subacrosome space (white arrow) and dense granular material (black arrow) just under the nuclear membrane.

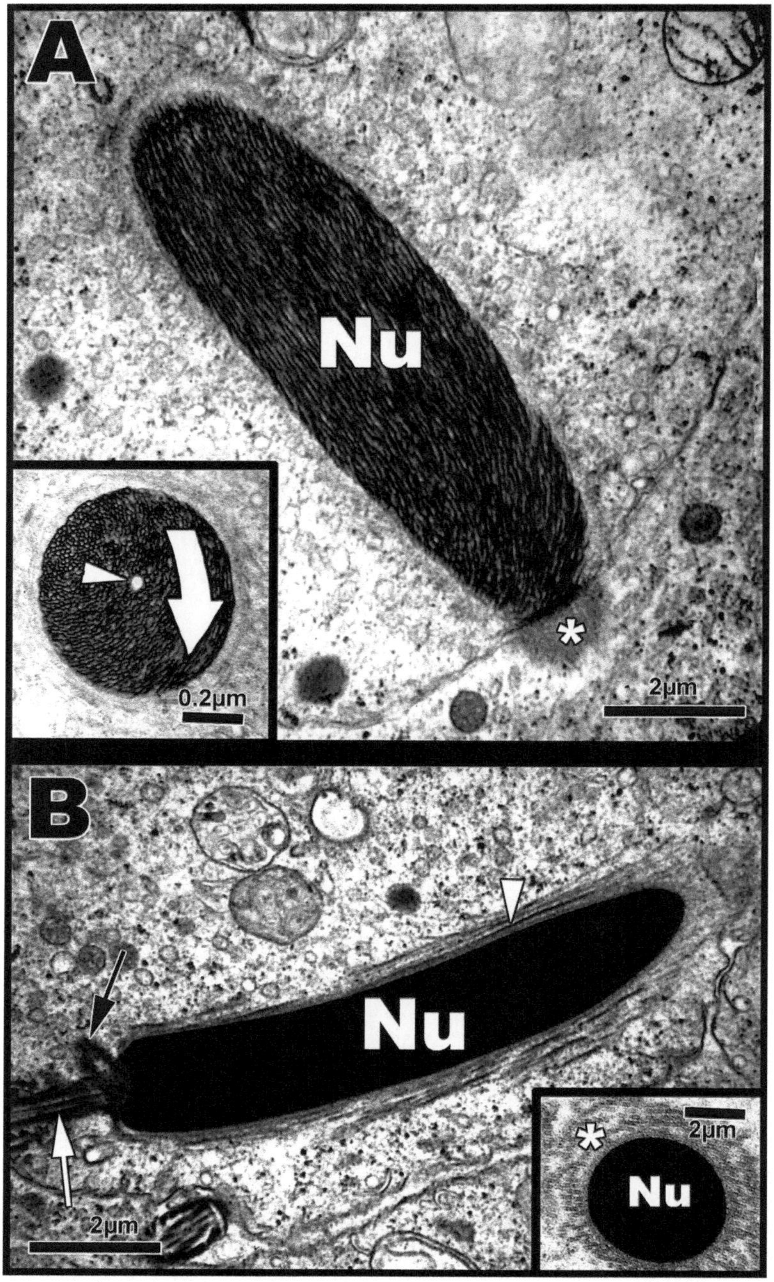

Fig. 6.40 A. A middle elongating spermatid in the testis of *Agkistrodon contortrix*. Note the twisting condensation of the chromatin (inset: white arrow) and the developing acrosome (*) and visible nuclear lacuna (white arrowhead). **B.** Early late elongating spermatid with a uniform nucleus (Nu) and a developing manchette (white arrowhead, inset: *). The proximal (black arrow) and distal (white arrow) centrioles are visible within the nuclear fossa caudally.

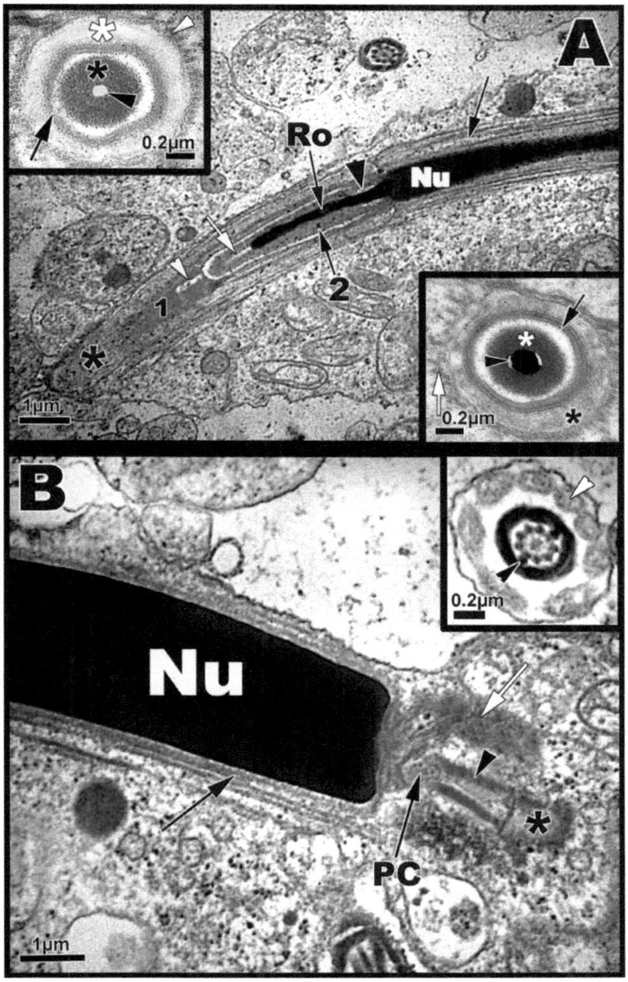

Fig. 6.41 A. A late elongating spermatid apical view of the acrosome within the seminiferous epithelium of *Agkistrodon contortrix*. The acrosome (*) complex surrounds the nuclear (Nu) rostrum (Ro). The complex consists of a subacrosome space that is split into two layers (1/2) via a lucent line. Within layer 2 are two translucent zones: basal rostral zone (black arrowhead) and the epinuclear lucent zone (white arrow). Within layer 1 is the perforatorium (white arrowhead), which is completely prenuclear. The manchette is very prominent and extends up around the acrosome complex (black arrowhead, Insets: white arrow and arrowhead). Upper left inset: acrosome vesicle (white*), layer 1 subacrosome space (black arrow), layer 2 subacrosome space (black*), epinuclear lucent zone (black arrowhead), Lower right inset: acrosome vesicle (black*), layer 1 subacrosome space (black arrow), layer 2 subacrosome space (white*), basal rostral translucent zone (black arrowhead). **B.** Caudal end of the nucleus (Nu) in a late elongating spermatid within the Copperhead testis. The manchette surrounds the nucleus (black arrow). There is a dark staining dense collar (white arrow) around the proximal (PC) and distal (black arrowhead) centrioles. Elongating axoneme of the neck (*). Inset shows the developing midpiece in cross section. Mitochondria (white arrowhead), peripheral fiber 3 and 8 (black arrowhead).

during spermiogenesis. This would provide nine differences that may be important in phylogenetic outcomes if they were mapped onto phylogenies with sperm data. The major obstacle at the present time for such phylogenetic analysis is the lack of spermiogenic data for other closely related snakes within Crotalinae. This is true for all of Ophidia and Squamata and thus does not allow for resolution in determining ancestral conditions. Before definitive support can be given to such spermiogenic ultrastructural differences many more species will have to be studied. Hopefully, the information provided here will stimulate research efforts in the area of spermiogenic and spermatogenic histology and ultrastructure within snakes and other squamate taxa. These data, however, do provide a solid morphological sequence for the development of spermatids during spermiogenesis and may provide baseline ultrastructural information important to histopathological studies regarding spermatogenesis of the semi-aquatic *A. piscivorus*, whose distribution and large numbers make them a potential sentinel species for reproductive toxicant studies.

6.5 THE OPHIDIAN SPERMATOZOON

The largest concentration of ultrastructural information on the ophidian testis is represented by character descriptions of the mature spermatozoa. The first detailed ultrastructural account of an ophidian spermatozoon was completed by Furieri (1965) for the Asp Viper (*Vipera aspis*) in the family Viperidae. Since this original ultrastructural report on ophidian sperm, there have been several other families of snakes studied in terms of overall spermatozoal morphology. The families looked at to date are Viperidae, Colubridae, Elapidae, Boidae, Leptotyphlopidae, Typhlopidae, and Anomalepididae (Table 6.1). There are also several studies that only detail parts of the ophidian spermatozoa and include some of the same families listed above (Austin 1965: Colubridae, Viperidae, Elapidae; Al-Dokhi 2004, 2005a,b: Viperidae). In order to understand the evolution of the ophidian spermatozoon, it is useful to briefly outline the general morphology of the amniotic and squamate spermatozoa.

6.5.1 Plesiomorphies within Amniotic Sperm

Figure 6.42 depicts a hypothesized pleisomorphic spermatozoon of the Amniota (Jamieson 1999). This model sperm is very similar to that of sperm observed in the Chelonia, Crocodylia, and Sphenodontida (Jamieson 2007). Features of this sperm are thought to have been retained from the tetrapod ancestors of amniotes and still can be seen in the aforementioned reptilian clades and to some degree in other taxa of amniotes. Some of the more prominent characters from Figure 6.42 include filiform spermatozoa with cranial acrosomes that are draped over simple subacrosome cones. The acrosome terminates laterally on the apical head of the nucleus and lies on pronounced nuclear shoulders. The perforatorium supports the

Table 6.1 Families of Ophidia for which published spermatozoal ultrastructural data exist.

Family	*Species*	*Author*
Viperidae	*Vipera aspis*	Furieri 1965
	Vipera aspis	Furieri 1970
	Crotallus durissus	Cunha *et al.* 2008
	Bothrops alternatus	Tourmente *et al.* 2008
	Bothrops diporus	Tourmente *et al.* 2008
Colubridae	*Coluber viridiflavus*	Furieri 1970
	Natrix tessellata	
	Natrix natrix	
	Coronella austriaca	
	ELaphe scalaris	Camps and Bergallo 1977
	Nerodia sipedon	Jamieson and Koehler 1994
	Boiga irregularis	Oliver *et al.* 1996
	Stegonotus cucullatus	
Elapidae	*Oxyuranus microlepidotus*	Oliver *et al.* 1996
Boidae	*Aspidites melanocephalus*	Oliver *et al.* 1996
	Boa constrictor occidentalis	Tourmente *et al.* 2006
	Eryx jayakari	Al-Dokhi *et al.* 2007
	Epicrates cenchria	Tavares-Bastos *et al.* 2008
	Boa constrictor amarali	
	Corallus hortulanus	
Leptotyphlopidae	*Leptetyphlops koppsei*	Tavares-Bastos *et al.* 2007
Typhlopidae	*Ramphotyphlops waitii*	Harding *et al.* 1995
	Typhlops reticulatus	Tavares-Bastos *et al.* 2007
Anomalepididae	*Liotyphlops beui*	Tavares-Bastos *et al.* 2007

subacrosome cone and is partly found within the subacrosome and partly embedded into the nuclear head via endonuclear canals, which invest the nucleus almost to its base. The nucleus is elongated and cylindrical in most amniotes studied to date (Jamieson 2007). At its base, the nucleus has a distinct nuclear or implantation fossa that houses the two triplet centrioles. The distal centriole forms the basal body, which elongating spermatids to form the axoneme of the flagellum. The terminus of the midpiece often has an annulus; whether this structure is pleisomorphic or an apomorphic reversal is highly debated (Jamieson 2007). The principal piece of the axoneme is very long and its dense and thick fibrous sheath is easily distinguished from the endpiece, which lacks this sheath.

6.5.2 Sperm Synapomorphies for Squamata

In this chapter, we exclude the ultrastructure of the spermatozoon of *Sphenodon* from the discussion of Squamate sperm as it has been shown (Healy and Jamieson 1992; Jamieson and Healy 1992) to be atypical of the Squamata and to strongly resemble that of turtles and crocodilians.

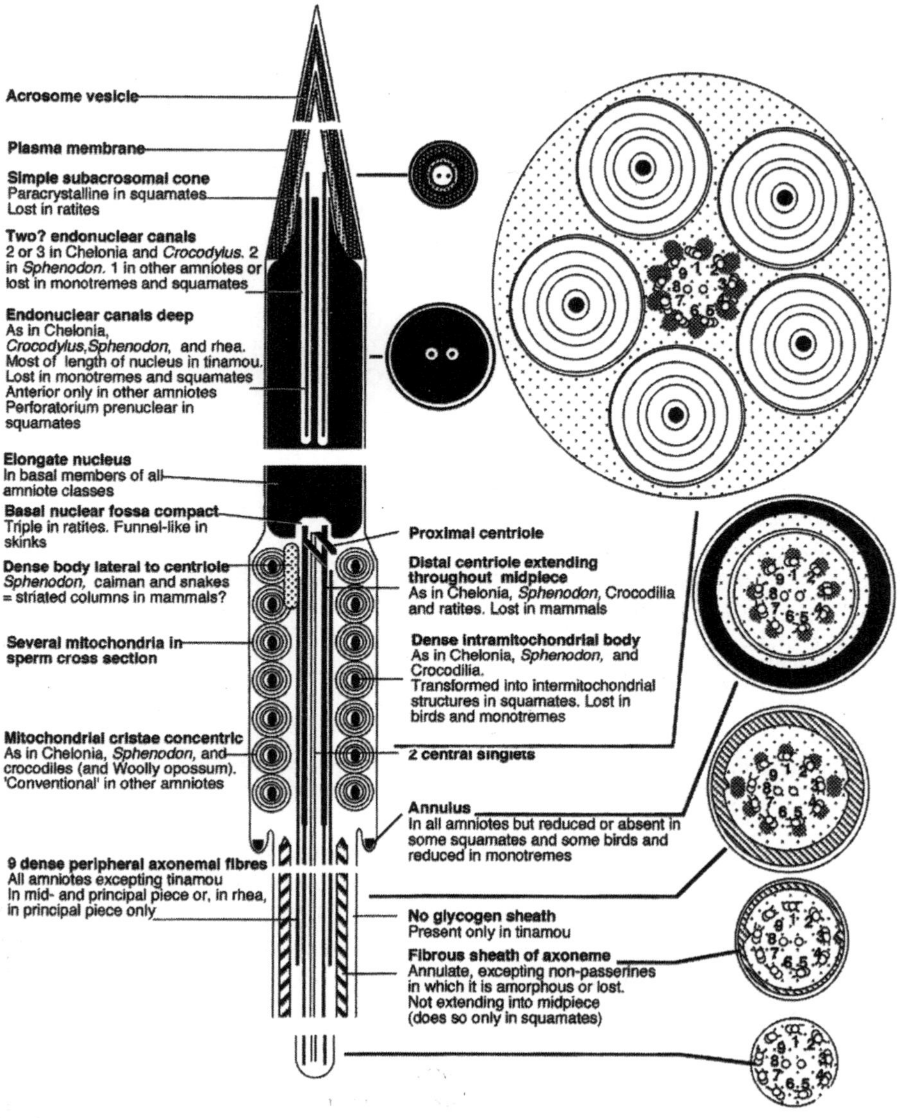

Fig. 6.42 Hypothesized pleisomorphic amniotic spermatozoa. From Jamieson, B. G. M. 2007. Science Publishers, Enfield, New Hampshire, Fig. 8.1.

Figure 6.43 depicts a typical ophidian spermatozoon (Jamieson 1995). It shows many of the synapomorphies that define Squamata; however, there are also several secondary spermatozoal apomorphies that seem to define the Ophidian clade (see Sec. 6.5.3). First let us consider the spermatozoal synapomorphies that to date define Squamata.

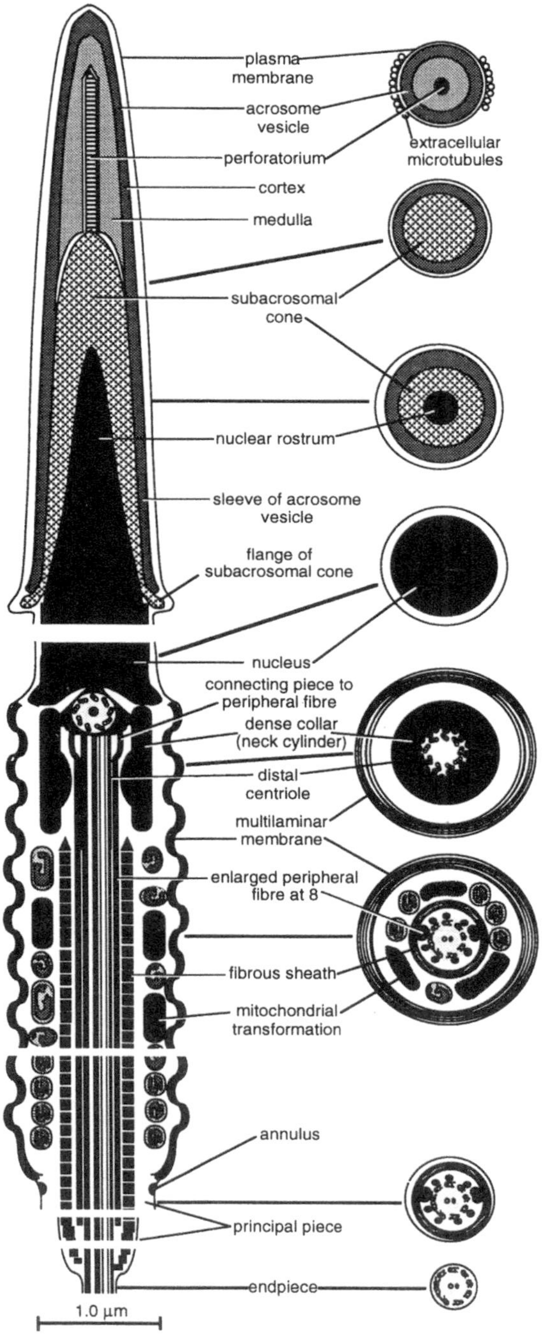

Fig. 6.43 A schematic representation of the known ultrastructural characters of the Serpentes spermatozoon. From Jamieson B. G. M. 1995. Mémoires du Muséum National d'Histoire Naturelle. Tome 166. Editions du Museum, Paris. Fig. 9.

The perforatorium within squamates is completely prenuclear and is confined to the subacrosomal space. Endonuclear canals are absent and a single reduced perforatorium is clearly apomorphic to the multiple canals seen in basal Chelonia and Sphenodontida (Jamieson *et al.* 1996; Oliver *et al.* 1996). The presence of a distinct epinuclear lucent zone is also considered an autapomorphy for Squamata. Intermitochondrial dense bodies within the midpiece are also a probable apomorphy for squamates though they are also poorly developed (possible homoplasy) in some avian species (Jamieson 1999). The subacrosome space or cone has a paracrystalline subarchitecture in squamates (Butler and Gabri 1984; Carcupino *et al.* 1989; Jamieson 1999). Within the midpiece and the principal piece, a thick fibrous sheath, an apparent autapomorphy for squamates, surrounds the axoneme. There are typically nine peripheral fibers associated with the outer microtubule doublets in the midpiece of squamates (3 and 8 are typically enlarge) that then are either reduced or absent in the principal piece (Jamieson and Scheltinga 1993; Jamieson 1995).

6.5.3 Introduction to Ophidian Sperm

Snake sperm possess many of the conserved morphological characters common to the spermatozoa of squamates described above. They are typically filiform in shape, the collapsed acrosome vesicle drapes over a conical subacrosome space, and invests the tapered nuclear apex. The perforatorium (with a poor stopper/knob shaped basal plate) is prenuclear and is exclusive to the subacrosome cone. The midpiece has a fibrous sheath and dense bodies within its concentric mitochondrial layers. In addition to the 9 peripheral fibers that accompany the outer microtubule doublets within the axoneme, fibers 3 and 8 are enlarged and the endpiece lacks both fibers and a sheath. Some species have an epinuclear lucent zone but it is considered poorly developed in snakes (Jamieson 1995).

Multiple studies have been performed detailing the morphological characteristics of spermatozoa in Ophidia (Furieri 1965, 1970; Harding *et al.* 1995; Jamieson 1995; Oliver *et al.* 1996; Cunha *et al.* 2008; Tavares-Bastos *et al.* 2008). However, only a few have put these characters into a phylogenetic context (Jamieson 1995; Oliver *et al.* 1996; Tavares-Bastos *et al.* 2008), and none has included all species on which relevant ultrastructural data exist. Character states for each species are given in Table 6.2. We introduce new observations on ultrastructue of *Seminatrix pygaea* spermatozoa (Figs. 6.44 and 6.45) to represent many of the features of the ophidian sperm and to serve as a basis for comparisons.

Acrosome Complex. The acrosome is an elongating spermatid structure that contains the necessary proteins and enzymes for degradation of oocyte extracellular layers during fertilization. It appears as a roughly cone shaped cap on the apical portion of the spermatozoa (Fig. 6.44 Av). A single unilateral ridge (though rare in snakes) may extend off the lateral portion of the spermatozoon and contains an electron lucent zone

Table 6.2 Ultrastructure characters of ophidian spermatozoa used in preliminary phylogenetic analyses (Sec. 6.5.3)

Character	*States*
1. Acrosome Complex Ridge	0—absent; 1—present
2. Acrosome CS Shape	0—circular; 1—depressed
3. Acrosome Subdivisions	0—absent; 1—present
4. Subacrosome Cone	0—paracrystalline; 1—not paracrystalline
5. Acrosome: Vacuity Subdivision	0—absent; 1—present
6. Epinuclear Lucent Zone	0—absent; 1—present
7. Perforatorium #	0—2; 1—1
8. Perforatorium Tip	0—pointed; 1—rounded
9. Perforatorium Basal Plate	0—absent; 1—present
10. Perforatorium Basal Plate Shape	0—knob—like; 1—stopper-like; 2—N/A
11. Nuclear Lacunae	0—absent; 1—present
12. Neck: Stratified Laminae	0—absent; 1—present
13. Neck: Laminar Projections	0—unilateral; 1—bilateral; 2—N/A
14. Dense Material: Prox. Centriole	0—absent; 1—present
15. Dense Collar (Cylinder)	0—absent; 1—present
16. Midpiece 3 and 8 Fibers	0—grossly enlarged; 1—not grossly enlarged
17. Midpiece: Mito. Cristae	0—concentric; 1—linear
18. Mito. Shape OS	0—columnar; 1—slightly curved; 2—sinuous tubes
19. Mito. Shape LS	0—Trapezoidal; 1—Oval; 2—Columnar sq. ends
20. Mito. Shape CS	0—round/oval; 1— irregular; 2—trapezoidal
21. Midpiece Dense Bodies	0—solid; 1—granular; 2—N/A
22. Dense Bodies CS	0—separated fibrous sheath; 1—juxta fibrous sheath; 2—N/A
23. Midpiece Fibrous Sheath	0—t4; 1—t3; 2—t2; 3—t1; 4—before t1; 5—at annulus level; 6—N/A
24. Midpiece Annulus	0—absent; 1—present
25. Midpiece Annulus Shape	0—scythe; 1—irregular; 2—N/A
26. Principal Piece 3 and 8 Fibers	0—absent; 1—present
27. Multilaminar Membranes	0—absent; 1—present
28. Extracellular Microtubules	0—absent; 1—present

CS, cross section; OS, oblique section; LS, longitudinal

(Fig. 6.44). Cross-sectional views of the acrosome may be compressed into an oval shape, like most outgroups such as Varanidae (Oliver *et al.* 1996), or uncompressed in a circular shape (Fig. 6.44A). The acrosome vesicle (Fig. 6.44) makes up the most exterior portion of the acrosome complex and may be subdivided into two distinct regions, the cortex and medulla, based on electron density (Fig. 6.44). The subacrosomal cone (Fig. 6.44B) lies inside the acrosome vesicle and contains paracrystalline material. An acrosomal vacuity subdivision may be seen, such as in *Ramphotyphlops* (Harding *et al.* 1995), which appears as an electron lucent swirl within the subacrosomal cone. A single perforatorium (Fig. 6.44A) extends within the subacrosomal

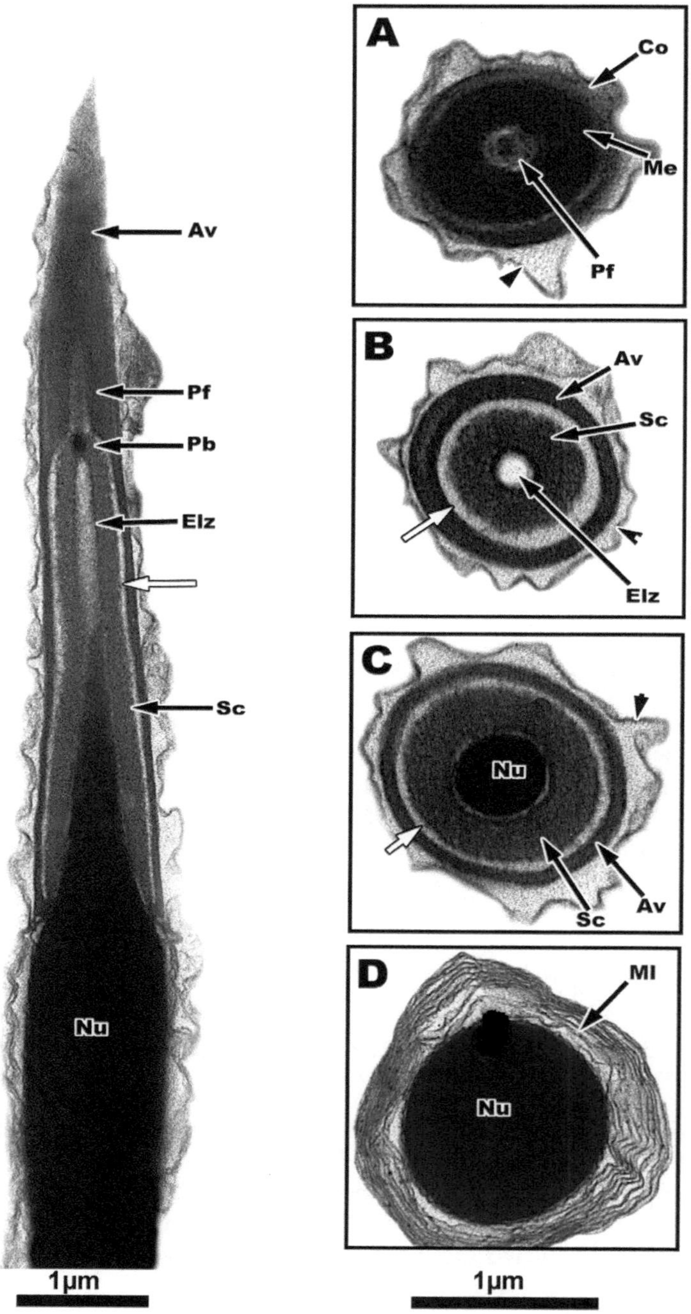

Fig. 6.44 Cranial views of a mature spermatozoon from *Seminatrix pygaea*. Left. Transverse view through the apical portion of a mature spematozoon detailing the acrosome complex.

Fig. 6.44 Contd. ...

cone and can end in a pointed tip, as in *Bothrops alternatus* (Tourmente *et al.* 2008), or end in a rounded tip. Densification of the subacrosomal cone and the caudal perforatorium may or may not be present serving as a perforatorial base plate (Fig. 6.44). The perforatorial base plate can be either stopper-shaped (Fig. 6.44) or knob-like (outgroups; Oliver *et al.* 1996) in shape.

Nucleus. The nucleus is an elongated electron-dense structure (Fig. 6.44) that makes up the body of the spermatozoon and contains the condensed chromatin. The nucleus stains uniformly and may contain nuclear lacunae (Lesser Blind Snake, *Liotyphlops beui;* Tavares-Bastos *et al.* 2007) that appear as electron lucent areas within the center of the nucleus. Apically, the nucleus is tapered into a rostral (Fig. 6.44) shape and extends into the acrosome complex where it may or may not be capped with an epinuclear electron lucent zone (Fig. 6.44). Distally, the nucleus contains an indentation, the nuclear fossa, that houses the proximal centriole (Fig. 6.45A).

Neck Region. The proximal centriole and anterior portion of the distal centriole make up the neck region (Fig. 6.45A) of the spermatozoon. The proximal centriole resides within the nuclear fossa and may contain electron-dense material (Slatey-grey Snake, *Stegonotus cucullatus;* Oliver *et al.* 1996). An electron-dense structure may be observed lateral to the caudal proximal and distal centriole as a dense collar (Fig. 6.45A).

Midpiece. The midpiece (Fig. 6.45) is composed of the proximal region of the axoneme that extends from the distal centriole and the surrounding mitochondria. The axoneme displays the conserved 9+2 microtubule arrangement (Fig. 6.45D) and enlarged peripheral electron-dense fibers (Fig. 6.45D) can be associated with microtubule doublets 3 and 8. Multiple mitochondria with linear cristae occupy the cytoplasm surrounding the distal centriole. In oblique section the mitochondria appear as sinuous tubes (*Stegonotus cucullatus;* Oliver *et al.* 1996) although this is questionable in some species. In longitudinal (sagittal) section, they can appear elongated with rounded ends (Fig. 6.45A) or columnar (*Crotallus durissus;* Cunha *et al.* 2008). In transverse section they appear round/oval (Fig. 6.45D) or irregular in shape (Slender Blind Snake, *Leptotyphlops*

... *Fig. 6.44 Contd.*

AV, acrosome vesicle; Pf, perforatorium; Pb, perforatorial base plate; Elz, epinuclear lucent zone; white arrow, electron lucent band; Sc, subacrosomal cone; Nu, nucleus. **A.** Cross-sectional view through the apical portion of the acrosome complex showing the subdivision of the acrosome into cortex (Co) and medulla (Me) and the perforatorium (Pf) extending into the apical portion of the acrosome. Black arrowhead, plasma membrane. **B.** Cross-sectional view through the medial portion of the acrosome complex detailing the acrosome vesicle (Av), subacrosomal cone (Sc), epinuclear lucent zone (Elz), electron lucent band (white arrow) and plasma membrane (black arrowhead). **C.** Cross-sectional view through basal portion of the acrosome complex showing the acrosome vesicle (Av), subacrosomal cone (Sc), electron lucent band (white arrow) rostrum of the nucleus (Nu), and plasma membrane (black arrowhead). **D.** Cross-sectional view through the nuclear body detailing the homogenous electron density of the nucleus (Nu) and multilaminar membranes (Ml).

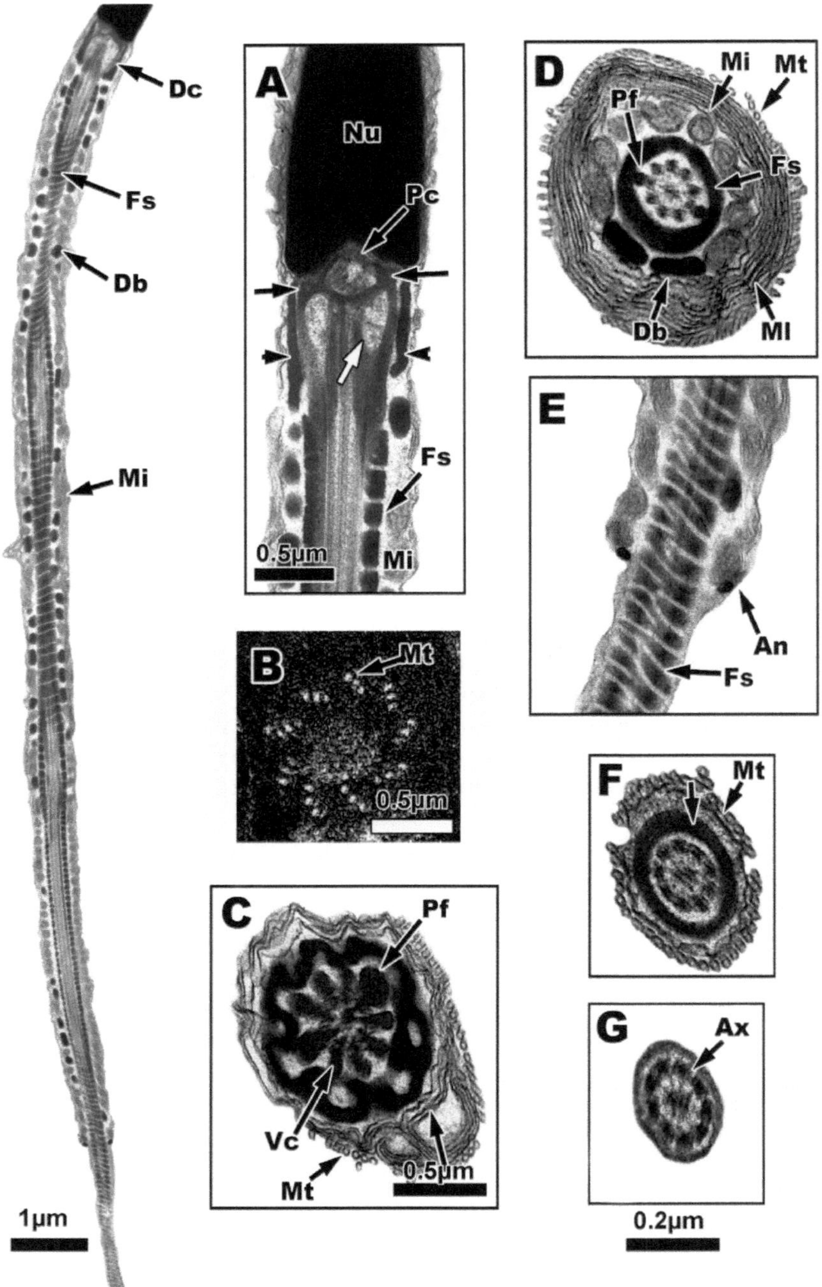

Fig. 6.45 Caudal view of a mature spermatozoon from *Seminiatrix pygaea*. Left. Transverse view of the posterior portion detailing the dense collar (Dc) surrounding the distal centriole, dense bodies (Db) and mitochondria (Mi) of the midpiece, and the fibrous sheath (Fs).

Fig. 6.38 Contd. ...

koppesi; Tavares-Bastos *et al.* 2007). Solid dense bodies may be observed between mitochondria in juxtaposition to the electron-dense blocks making up the fibrous sheath of the midpiece axoneme (Fig. 6.45D). The fibrous sheath may be observed beginning before mitochondrial tier 1 (Northern Water Snake, *Nerodia sipedon*; Jamieson and Koehler 1994), at mitochondrial tier 1 (Fig. 6.45A), or at mitochondrial tier 2 (*Bothrops alternatus*; Tourmente *et al.* 2008). Although other tiers have been suggested for other squamates only these are observed within snakes. The midpiece terminates at an electron-dense ring called the annulus (Fig. 6.45E). The annulus is typically irregular in shape within snakes, as opposed to scythe shape in some other amniotes (Tourmente *et al.* 2008). In all snakes studied to date, the midpiece is extremely long when compared to other squamates, which may be a synapomorphy for Serpentes.

Principal and Endpieces. Most the cytoplasmic material, associated mitochondria, and dense bodies terminate rostrally to the annulus. The fibrous sheath around the axoneme further extends into the principal piece within ophidian sperm (Fig. 6.45F). Peripheral dense fibers next to microtubule doublets 3 and 8 of the principal piece axoneme may (*Stegonotus cucullatus*; Oliver *et al.* 1996) or may not (Fig. 6.45F) be present near the annulus. Extracellularly the spermatozoa may be surrounded by multilaminar membranes and/or microtubules (Fig. 6.45F). The snake endpiece typically lacks the peripheral dense fibers and the fibrous sheath (Fig. 6.45G).

6.5.4 Insight into Ophidian Sperm Evolution

Although previous studies have used sperm characters to reconstruct phylogenetic relationships among the Ophidia (Jamieson 1995; Oliver *et al.* 1996; Tavares-Bastos *et al.* 2008), the purpose of this portion of the chapter is to further elucidate some evolutionary trends seen when sperm characters are mapped onto both morphological and molecular based phylogenies. The reconstruction of phylogenetic relationships based on spermatozoal characters alone is premature because of the low diversity

---

... *Fig. 6.45 Contd.*

A. Enlarged transverse view of the neck region detailing the proximal centriole (Pc) within the nuclear fossa of the nucleus (Nu), bilateral laminar projection (black arrows) extending from the proximal centriole, connecting piece to the distal centriole (white arrow) and dense collar (black arrowhead) extending distally lateral to the distal centriole. **B.** Cross sectional view of the proximal centriole showing the 9+3 microtubule (Mt) arrangement. **C.** Cross-sectional view of the apical portion of the neck detailing the peripheral fibers (Pf) grossly enlarged at microtubules 3 and 8, fibrous sheath (Fs), and extracellular microtubules (Mt). **D.** Cross-sectional view of the midpiece detailing the mitochondria (Mi) and dense bodies (Db) surrounding the fibrous sheath (Fs), peripheral fibers (Pf) associated with doublets 3 and 8 of the axoneme, multilaminar membranes (Ml), and extracellular microtubules (Mt). **E.** Detailed transverse view of the annulus (An) at the base of the midpiece and extension of the fibrous sheath (Fs) into the principal piece. **F.** Cross sectional view of the principal piece detailing the fibrous sheath (Fs) and extracellular microtubules (Mt). **G.** Cross-Sectional view of the endpiece detailing the axoneme (Ax).

of snakes studied to date. However, some unique trends can be seen in spermatozoal character evolution among snakes and will be emphasized here to encourage discussion and hopefully stimulate more research within the field of ophidian spermatozoal ultrastructure.

Characters selected from Vieira *et al.* (2004) were coded for all snake species with representative data by a single author (Justin Rheubert) for standardization, and all character states are listed in Tables 6.2 and 6.3. Character mapping was then performed using MacClade (version 4.08, Maddison and Maddison 2005) onto both a morphological phylogeny (Lee *et al.* 2007), which is the most recent morphological cladogram produced that includes all snake lineages, and a molecular tree (Eckstut *et al.* 2009) that includes the most ophidian taxa sampled to date for a molecular based phylogeny. Internal nodes were determined through MacClade optimization.

Many of the characters used in these analyses are reported as conserved among snakes (see sections 6.5.2, 6.5.3, Tables 6.2, 6.3) and possibly among all squamates although future analyses are needed within both Squamata and Ophidia to justify such hypotheses. These presumed conserved characters can easily be identified in Table 6.3, by the observation that many of the same character states seen in the sister taxon outgroups are observed in many of the species of snake within the table. Because of the limited data set presented here and its poor resolution in the utilized phylogenies, we only discuss nodes with an unequivocal most-parsimonious solution for ancestral state reconstruction. Many of the character states in the two trees presented in this chapter may also be preliminary synapomorphies for certain snake clades and will be mentioned as the trees are discussed. Individual characters will not be examined in detail because all present data can be obtained from Table 6.3 and the sources listed in this table and there is limited space within this chapter.

6.5.4.1 Ophidian sperm character states on a morphology based tree

The beginning of the fibrous sheath in the flagellum is at mitochondrial tier 1 was reconstructed as the ancestral state for Caenophidia and Henophidia (Fig. 6.46 23[3]) but before mitochondrial tier 1 in the scolecophidians (Fig. 6.46 23[4]. A circular acrosome complex (Fig. 6.46 2[0]), stopper shaped perforatorium base plate (Fig. 6.46 10[1]), dense collar in the neck region (Fig. 6.46 15[1]), sinuous tubed mitochondria in oblique section (Fig. 6.46 18[2]), rounded mitochondria in longitudinal section (Fig. 6.46 19[1]), solid dense bodies (Fig. 6.46 21[0]), and multilaminar membranes (Fig. 6.46 27[1]) are resolved as hypothesized synapomorphies for all snakes.

6.5.4.2 Ophidian sperm molecular tree

A ridge on the acrosome surface is present in *Liotyphlops beui* (Fig. 6.47 1[1]). The fibrous sheath beginning before mitochondrial tier 1 is a preliminary synapomorphy for the typhlopid clade (Fig. 6.47 23[4]) as well as a stopper shaped perforatorial base plate (Fig. 6.47 10[1]). The

Table 6.3 Comparative ultrastructure of spermatozoa of Ophidia input data matrix.

	Characters																												
Taxa	1	2	3	4	5	6	7	8	9	10	11	12	13	14	15	16	17	18	19	20	21	22	23	24	25	26	27	28	*Reference*
Boas & Pythons																													
Aspidites melancephalus	0	0	1	0	0	0	1	1	0	2	0	1	1	1	1	0	1	2	1	0	0	1	3	1	1	0	1	1	Oliver *et al.* 1996
Boa constrictor amarali	0	0	0	0	0	1	1	1	1	1	0	1	1	1	1	0	1	?	2	0	0	1	?	1	1	0	1	1	Tavares-Bastos *et al.* 2008
Boa constrictor occidentalis	0	0	?	0	0	0	?	?	0	2	0	1	1	1	1	0	1	2	?	?	0	1	3	?	?	0	1	0	Tourmente *et al.* 2006
Corallus hortulanus	0	0	0	0	0	1	1	1	0	2	0	?	?	0	1	0	1	?	1	0	0	1	?	?	?	0	1	?	Tavares-Bastos *et al.* 2008
Epicrates cenchria	0	0	0	0	0	1	1	1	1	1	0	0	2	1	1	0	1	2	1	0	0	1	3	?	?	0	1	1	Tavares-Bastos *et al.* 2008
Eryx jayakari	0	0	1	0	0	1	1	1	1	1	0	?	?	1	1	0	1	?	2	0	2	2	3	1	1	1	1	1	Al-Dokhi *et al.* 2007
Colubroidea																													
Boiga irregularis	0	0	1	0	0	1	1	1	0	2	0	1	1	0	1	0	1	?	1	0	0	1	4	1	1	0	1	1	Oliver *et al.* 1996
Bothrops alternatus	0	0	1	0	0	1	1	0	0	2	0	1	1	1	1	0	1	2	1	0	2	2	2	?	?	0	0	0	Tourmente *et al.* 2008
Bothrops diporus	0	0	?	0	0	1	?	?	0	2	0	1	1	?	1	0	1	2	1	0	2	2	?	?	?	0	0	0	Tourmente *et al.* 2008
Crotalus durissus	0	0	1	0	0	1	1	1	1	1	1	1	1	1	1	0	1	2	2	0	0	1	3	1	1	1	1	1	Cunha *et al.* 2008
Elaphe scalaris	?	0	?	?	?	?	1	?	?	?	0	1	1	0	1	0	?	2	1	0	?	?	4	?	?	0	1	1	Camps and Bargallo, 1977
Nerodia sipedon	0	0	1	?	0	1	1	1	0	2	0	?	?	1	1	0	1	2	1	1	0	1	4	1	1	1	1	1	Jamieson and Koehler, 1994
Oxyuranus microlepidotus	0	0	1	0	0	1	1	1	0	2	0	0	2	0	1	0	1	2	1	0	0	1	4	1	1	1	1	1	Oliver *et al.* 1996
Seminatrix pygaea	0	0	1	0	0	1	1	1	1	1	0	1	1	0	1	0	1	?	1	0	0	1	3	1	1	0	1	1	This study
Stegonotus cucullatus	0	0	1	0	0	1	1	1	1	1	0	?	?	1	1	0	1	2	2	0	0	1	3	1	1	1	1	1	Oliver *et al.* 1996
Vipera aspis	?	?	?	?	?	?	?	?	?	?	0	?	?	?	?	0	1	2	?	0	?	?	?	?	?	?	?	?	Furieri, 1965
"Scolecophidians"																													
Leptotyphlops koppesi	0	0	1	0	0	1	1	1	1	1	?	1	1	0	1	0	1	2	1	1	0	1	4	1	1	?	1	0	Tavares-Bastos *et al.* 2007
Liotyphlops beui	1	0	1	0	0	1	?	?	?	?	1	?	?	?	?	0	?	?	?	0	0	1	?	?	?	?	?	?	Tavares-Bastos *et al.* 2007
Ramphotyphlops endoterus	0	0	1	0	1	1	1	1	?	?	0	1	1	?	1	0	1	2	1	?	0	1	4	1	1	0	1	1	Harding *et al.* 1995
Ramphotyphlops waitii	0	0	1	0	1	1	1	1	?	?	0	1	1	?	1	0	1	2	1	?	0	1	4	1	1	0	1	1	Harding *et al.* 1995
Typhlops reticulatus	0	1	1	0	0	1	1	1	1	1	1	1	1	1	1	0	1	2	1	0	0	1	4	?	?	?	1	1	Tavares-Bastos *et al.* 2007
Outgroups																													
Iguana iguana	1	1	1	0	0	1	1	0	1	0	0	1	1	1	1	0	1	2	2	2	1	1	2	1	1	1	0	0	Oliver *et al.* 1996
Varanus gouldii	0	1	1	0	0	1	1	1	1	0	0	1	1	1	0	0	1	0	0	0	1	1	2	1	1	0	0	0	Vieira *et al.* 2004

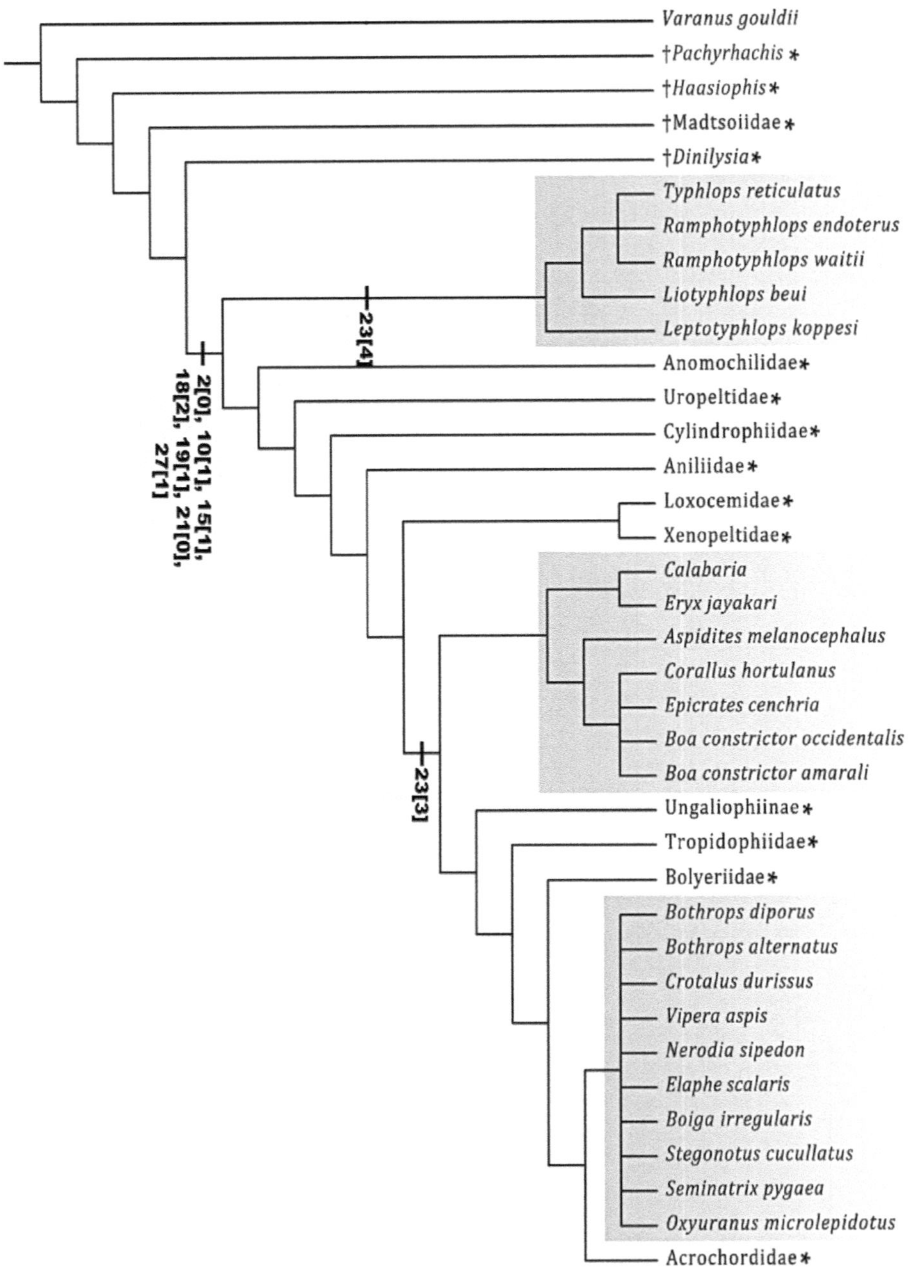

Fig. 6.46 Morphological cladogram. Phylogeny from Lee *et al.* (2007) with sperm ultrastructure character mappings here performed using MacClade. The first number indicates the character (Table 6.2) and the second number in brackers indicates the character state (Table 6.2). (*) taxa with no representative data to date, (+) extinct taxa. Shading of certain lineages outlines the scolecophidians, boids+pythons, and colubrids (top to bottom).

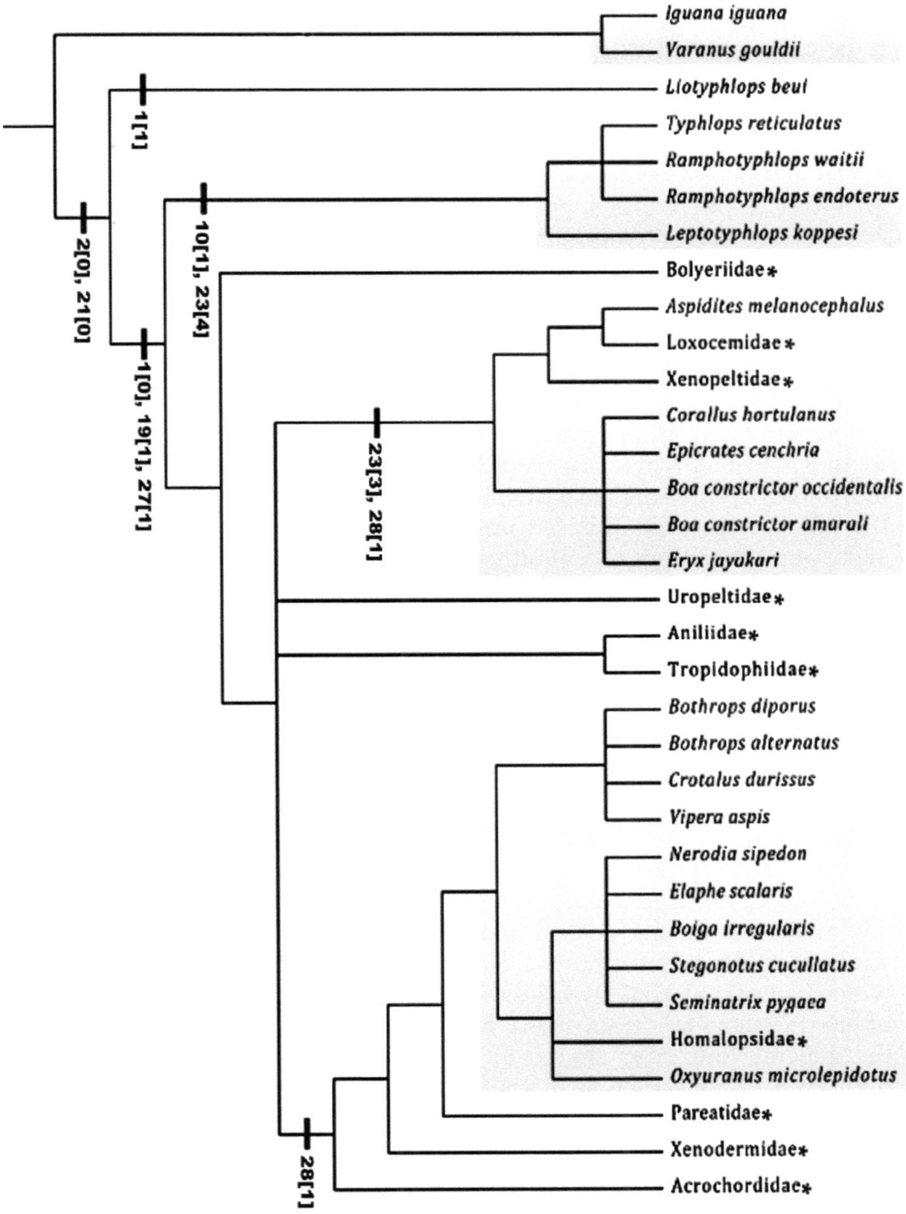

Fig. 6.47 Molecular cladogram. Phylogeny based from Eckstut *et al.* (2009) with sperm ultrastructure character mappings performed by MacClade. The first number indicates the character (Table 6.2) and the second number in brackets indicates the character state (Table 6.2). (*) taxa with no representative data to date, (+) extinct taxa. Shading of certain lineages outlines the scolecophidians, boids+pythons, and colubrids (top to bottom).

beginning of the fibrous sheath at mitochondrial tier 1 (Fig. 6.47 23[3]) and extracellular microtubules (Fig. 6.47 28[1]) are synapomorphies for the boids and possibly xenopeltids and loxcocemids (though more data need to be collected for these two groups). Extracellular microtubules also unite the caenophidians and possibly the colubrids (Fig. 6.47 28[1]) but missing data from a number of key taxa restrict this relationship between boids and colubrids. The absence of an acrosome ridge, (Fig. 6.47 1[0]), a circular acrosomal shape (Fig. 6.47 2[0], rounded mitochondria in longitudinal section (Fig. 6.47 19[1]), solid dense bodies of the midpiece (Fig. 6.47 21[0]), and multilaminar membranes (Fig. 6.47 27[1]) are resolved as ancestral for all snakes.

6.5.4.3 Ophidian sperm conclusions

Very few similarities occur between the morphological and molecular phylogenies. The circular acrosomal shape and solid midpiece dense bodies are the only shared characteristics uniting all snakes. Some of the observed differences such as the perforatorial base plate shape may be attributed to the placement of *Liotyphlops beui* within the scolecophidians in the morphological tree and sister to all other snakes in the molecular tree. Also, the other characters observed uniting snakes may fall out to be possible synapomorphies but no definitive resolution can be made until a further analysis that includes representatives of all squamate clades is conducted (currently underway by authors of this chapter). The morphological tree renders more synapomorphic characters for all snakes, which may be attributed to the outgroups used in the phylogenetic analyses. It is apparent however, from this preliminary study (and a major theme of this chapter), that a higher degree of taxon sampling is needed in order to make definitive conclusions regarding the evolution of sperm characters among snakes and squamates. Without data for certain taxa (i.e., Anomochilidae, Uropeltidae, Cylindrophidae, Bolyeriidae, Acrochordidae) evolutionary trends based on proposed hypotheses cannot be resolved conclusively.

Another striking result of these two trees is that many characters are conserved among snakes and may possibly be conserved among all squamates. This is supported most thoroughly by the present analysis, which includes all snake species with representative data. Even the addition of sperm characters from many new species to the data set did not alter the general conserved characters hypothesized by Jamieson (1995). This high level of conservation of spermatozoal characters in snakes and most likely all squamates also supports the incredibly accurate ultrastructural features proposed by Jamieson *et al.* (1999) for the hypothetical amniote spermatozoon.

6.6 CONCLUSIONS

The authors hope to stimulate the general interests of the reproductive and herpetological communities in the reptilian testis. There is much that needs to be accomplished as far as ultrastructural studies on testicular architecture, particularly the Sertoli cell's general anatomy and function. Spermiogenesis and sperm ultrastructure need to be studied in more taxa within snakes. Though the data to date are very limited, there are many interesting and exciting preliminary results especially with the ontogenic features of spermiogenesis. From the *Agkistrodon* comparison, there were nine character state differences between two closely related species. If this trend holds true for other clades then spermiogenesis could be a powerful tool to help resolve phylogenetic inferences beyond the family level. Spermatozoa data show trends that Barrie G. M. Jamieson has alluded to for more than a decade, in that squamate sperm characters seem to be highly conserved. It is interesting to think that if there is much diversity in the way to produce sperm (i.e., large character differences during spermiogenesis) then why do we not see more differences in spermatozoal ultrastructure? These and other interesting questions need to be answered and this chapter will at least provide some direction to reproductive anatomists interested in pursuing these issues in reptiles.

6.7 ACKNOWLEDGMENTS

K.M. Gribbins wishes to thank Wittenberg University for their support and resources, which aided in the completion of much of the research that appears in this chapter. Without the aid of the university, like maintenance of expensive equipment such as the TEM and SEM, most of my research would be impossible to complete. I would also like to thank David Sever, Robert Aldridge, Dustin Siegel, and Stanley Trauth for their collegial support, as their friendships and shared collecting trips have provided much of the tissue and intellectual stimulation that has produced the data for this chapter. I would also like to acknowledge the tremendous Wittenberg undergraduates (Carrie Happ, Holly McHugh, Jessica Schultz, Jeremy Toffle, Erik Poldemann, Erin Mills, Daniel Jackson, and Marla Anzalone) who have graced my lab space, especially my co-author, Justin Rheubert. Their hard work and dedication to this type of research rivals what is produced at the graduate school level and has resulted in many excellent manuscripts and presentations. Lastly, I could not end without showing the deepest appreciation and adoration for my wife, Sara, who has always supported my efforts and has endured many long nights without my help.

J. L. Rheubert would like to thank both Wittenberg and Southeastern Louisiana Universities for their support and resources. I would also like to thank Dr. Brian Crother and Dr. Kyle Piller for their continuous help in my graduate career thus far. I am indebted to Dr. Kevin Gribbins (co-author) and Dr. David Sever for their continuous support throughout my professional development. Without them I would not be where I am today and my contributions to this growing field of reproductive biology would not have been accomplished. I would also like to thank Dustin

Siegel who has given me a tremendous amount of support and aided in my development and I wish to continue collaborations with him for many years to come. Also to my friends Caleb McMahan (who helped in my phylogenetic understanding), Mike Wilder, Chris Murray, Layla Freeborn, Aaron Geheber, Jennifer Lee, and Mindy Mayon for their continuous support. Last, but certainly not least, I would like to thank my parents Ronald and Diane Rheubert who have supported me in all ways possible. Much of my work, including this chapter, as well as my accomplishments throughout my life would have not been possible without them.

6.8 LITERATURE CITED

Aire, T. A. 2007. Anatomy of the testis and male reproductive tract. Pp. 37-113. In B. G. M. Jamieson (ed.), *Reproductive Biology and Phylogeny of Birds,* vol. 6A. Science Publishers, Enfield, NH.

Al-Dokhi, O. A. 2004. Electron microscopic study of sperm head differentiation in the Arabian Horned Viper *Cerastes cerastes* (Squamata, Reptilia). Journal of Biological Sciences 4: 111-116.

Al-Dokhi, O. A., Al-Onazee, Y. Z. and Mubarak, M. 2004. Light and electron microscopy of the testicular tissue of the snake *Eryx jayakari* (Squamata, Reptilia) with reference to the dividing germ cells. Journal of Biological Sciences 4: 345-351.

Al-Dokhi, O. A., Al-Onazee, Y. Z. and Mubarak, M. 2005a. Ultrastructure of spermiogenesis of the snake *Eryx jayakari* (Squamata, Reptilia): II. Sperm tail differentiation. Indian Journal of Applied and Pure Biology 20: 107-114.

Al-Dokhi, O. A., Al-Onazee, Y. Z. and Mubarak, M. 2005b. Ultrastructure study of the spermiogenesis of the snake *Eryx jayakari* (Squamata, Reptilia): I. Sperm head differentiation. Ultra Scientist of Physical Sciences 17: 1-10.

Al-Dokhi, O.A., Al-Onazee, Y. Z. and Mubarak. M. 2007. Fine structure of the epididymal sperm of the snake *Eryx jayakari* (Squamata, Reptilia). International Journal of Zoological Research 3: 1-13.

Aldridge, R. D. 2002. Evolution of mating season in the pitvipers of North America. Herpetological Monographs 16: 1-25.

Austin, C. R. 1965. Fine structure of the snake tail. Journal of Ultrastructure Research 12: 452-462.

Baccetti, B., Bigliardi, E., Talluri, M. V. and Burrini, A. G. 1983. The Sertoli cell in lizards. Journal of Ultrastructure Research 168: 268-275.

Bergmann, M. and Schindelmeiser, J. 1987. Development of the blood testis barrier in the domestic fowl (*Gallus domesticus*). International Journal of Andrology 10: 481-488.

Bergmann, M., Schindelmeiser, J. and Greven, H. 1984. The blood-testis barrier in vertebrates having different testicular organization. Cell and Tissue Research 238:145-150.

Boisson, C. and Mattei, X. 1965. Sur la spermiogenese de *Python sebae* (Gmelin) etudiee au microscope électronique. Comptes Rendus des Séances de la Societé de Biologia et des ses Filiales 159: 1192-1194.

Boisson, C. and Mattei, X. 1966. La spermiogenese de *Python sebae* (Gmelin) observee au microscope electronique. Annales des Sciences Naturelles. Zoologie et Biologie Animale 8: 363-390.

Butler, R. D. and Gabri, M. S. 1984. Structure and development of the sperm head in the lizard *Podarcis (Lacerta) taurica*. Journal of Ultrastructure Research 88: 261-274.

Camps, J. L. and Bargallo. R. 1977. Espermatogenesis de reptiles. I. Ultraestructura dose 1 espermatozoides de *Elape scalaris* (Schinz). Boletin de la Royal Sociedad Espanola Historia Naturelles 75: 429-446.

Callard, I. P., Callard, G. V., Lance, V., Bolaffi, J. L. and Rosset, J. S. 1978. Testicular regulation in nonmammalian vertebrates. Biology of Reproduction 18: 16-43.

Carcupino, M., Corso, G. and Pala, M. 1989. Spermiogenesis in *Chalcides ocellatus tiligugu* (Gmelin) (Squamata: Scincidae): an electron microscope study. Bollettino di Zoologia 56: 119-124.

Clark, A. Q. 1967. Some aspects of spermiogenesis in a lizard. American Journal of Anatomy 121: 369-400.

Cooksey, E. J. and Rothwell, B. 1973. The ultrastructure of the Sertoli cell and its differentiation in the domestic fowl (*Gallus domesticus*). Journal of Anatomy 114: 329-345.

Cunha, L. D., Tavares-Bastos, L. and Bao S. N. 2008. Ultrastructural description and cytochemical study of the spermatozoa of *Crotalus durissus* (Squamata, Serpentes). Micron 39: 915-925.

Da Cruz-Landim, C. and Da Cruz-Hofling, M. A. 1977. Electron microscope study of lizard spermiogenesis in *Tropidurus torquatus* (Lacertilia). Caryologia 30: 151-162.

Dehlawi, G. Y., Ismail, M. F., Hamdi, S. A. and Jamjoom, M. B. 1992. Ultrastructure of spermiogenesis of a Saudian reptile. The sperm head differentiation in *Agama adramitana*. Archives of Andrology 28: 223-234.

Eckstut, M. E., Sever, D. M., White, M. E. and Crother, B. I. 2009. Phylogenetic analysis of sperm storage in female squamates. Pp. 185-218. In L. T. Dahnof (ed.), *Animal Rreproduction: New Research Developments*. Nova Science Publishers, Inc., Hauppauge, New York.

Engel, K. B. and Callard, G. V. 2007. Endocrinology of Leydig cells in nonmammalian vertebrates. Pp. 207-224. In: A. H. Payne and M. P. Hardy (eds), *Contemporary Endocrinology: The Leydig Cell in Health and Disease*. Humana Press Inc., Totowa, New York.

Ferreira, A. and Dolder, H. 2002. Ultrastructural analysis of spermiogenesis in *Iguana iguana* (Reptilia: Sauria: Iguanidae). European Journal of Morphology 40: 89-99.

Ferreira, A. and Dolder, H. 2003. Sperm ultrastructure and spermatogenesis in the lizard, *Tropidurus itambre*. Biocell 27: 353-362.

Fox, W. 1952. Seasonal variation in the male reproductive system of the Pacific Garter Snake. Journal of Morphology 90: 481-553.

Foreman, D. 1997. Seminiferous tubule stages in the prairie dog (*Cynomys ludovicianus*) during the annual breeding cycle. Anatomical Record 247: 355–367.

Furieri, P. 1965. Prime osservazioni al microscopio elettronico sulla ultrastruttura dello spermatozoo di *Vipera aspis aspis*. Bollettino della Societa Italiana di Biologia Spermimentale 41: 478-480.

Furieri, P. 1970. Sperm morphology in some reptiles: Squamata and Chelonia. Pp. 115-132. In: B. Baccetti (ed.), *Comparative Spermatology*. Academic Press, New York.

Goldberg, S. R. and Lowe, C. H. 1966. The reproductive cycle of the Western Whiptail Lizard (*Cnemidophorous tigris*) in southern Arizona. Journal of Morphology 118: 543-548.

Gribbins, K. and Gist, D. 2003. Cytological evaluation of the germinal epithelium and the germ cell cycle in an introduced population of European Wall Lizards, *Podarcis muralis*. Journal of Morphology 256: 296-306.

Gribbins, K., Gist, D. and Congdon, J. 2003. Cytological evaluation of spermatogenesis in the Red-Eared Slider, *Trachemys scripta*. Journal of Morphology 255: 337-346.

Gribbins, K., Happ, C. S. and Sever, D. M. 2005. Ultrastructure of the reproductive system of the Black Swamp Snake (*Seminatrix pygaea*). V. The temporal germ cell development strategy of the testis. Acta Zoologica 86: 223-230.

Gribbins, K., Elsey, R. and Gist, D. 2006. Cytological evaluation of spermatogenesis in the germinal epithelium of the American Alligator, *Alligator mississippiensis*. Acta Zoologica 87: 59-69.

Gribbins, K., Mills, E. and Sever, D. 2007. The ultrastructure of spermiogenesis within the testis of the ground skink, *Scincella laterale* (Scincidae, Squamata). Journal of Morphology 268: 181-192.

Gribbins, K., Rheubert, J., Collier, M., Siegel, D. and Sever, D. 2008. Histological analysis of spermatogenesis and the germ cell development strategy within the testis of the male Western Cottonmouth Snake, *Agkistrodon piscivorus*. Annals of Anatomy 190: 461-476.

Gribbins, K. M., Rheubert, J. L., Poldemann, E. H., Collier, M. H., Wilson, B. and Wolf, K. 2009. Continuous spermatogenesis and the germ cell development strategy within the testis of the Jamaican Gray Anole, *Anolis lineatopus*. Theriogenology 72: 484-492.

Guraya, S. S. 1973. Histochemical observations on the interstitial (Leydig) cells of the snake. Acta Morphologica Hungarica 21: 1-12.

Harding, H. R., Aplin, K. P., and Mazur, M. 1995. Ultrastructure of spermatozoa of Australian Blindsnakes, *Ramphotyphlops* spp. (Typhlopidae, Squamata): first observations on the mature spermatozoon of scolecophidian snakes. Pp. 385-396. In B. G. M. Jamieson, J. Ausio, and J. L. Justine (eds), *Advances in Spermatozoal Phylogeny and Taxonomy*. vol. 166. Mémoires du Muséum National d'Histoire Naturelle. Editions du Museum, Paris.

Healy, J. M. and Jamieson, B. G. M. 1922. Ultrastructure of the spermatozoon of the tuatara (*Sphenodon punctatus*) and its relevance to the relationships of the Sphenodontida. Phiosophical Transactions of the Royal Society London B 335: 193-205

Healy, J. M. and Jamieson, B. G. M. 1994. The ultrastructure of spermatogenesis and epididymal spermatozoa of the tuatara *Sphenodon punctatus* (Sphenodontidae, Amniota). Philosophical Transactions: Biological Sciences 344: 187-199.

Hess, R. 1990. Quantitative and qualitative characteristics of the stages and transition in the cycle of the rat seminiferous epithelium: light microscopic observations of perfusion-fixed and plastic-embedded testes. Biological Reproduction 43: 525-542.

Hondo, E., Kurohmaru, M., Toriba, M. and Hayashi, Y. 1994. Seasonal changes in spermatogenesis and ultrastructure of developing spermatids in the Japanese Rat Snake, *Elaphe climacophora*. Journal of Veterinary Medical Sciences 56: 836-40.

Hondo, E., Kitamura, N., Toriba, M., Kurohmaru, M. and Hayashi, Y. 1996. Histological study of the seminiferous epithelium in the Japanese Rat Snake, *Elaphe climacophora*: Identification of the spermatogonia. Journal of Veterinary Medical Science 58: 23-29.

Jamieson, B. G. M. 1995. The ultrastructure of spermatozoa of the Squamata (Reptilia) with phylogenetic considerations. Pp. 359-383. In: B. G. M. Jamieson, J. Ausio, and J. L. Justine (eds), *Advances in Spermatozoal Phylogeny and Taxonomy*. vol. 166. Mémoires du Muséum National d'Histoire Naturelle, Editions du Museum, Paris.

Jamieson, B. G. M. 1999. Spermatozoal phylogeny of the Vertebrata. Pp. 303-331. In C. Gagnon (ed.), *The Male Gamete. From Basic Science to Clinical Applications*. Cache River Press, Clearwater, Florida.

Jamieson, B. G. M. 2007. Avian spermatozoa: Structure and phylogeny. Pp. 349-511. In B. G. M. Jamieson (ed.), *Reproductive Biology and Phylogeny of Birds*, Vol. 6A. Science Publishers, Enfield, NH.

Jamieson, B. G. M. and Healy, J. M. 1992. The phylogenetic position of the tuatara, *Sphenodon* (Sphenodontida, Amniota), as indicated by cladistic analysis of the ultrastructure of spermatozoa. Philosophical Transactions of the Royal Society London B 335, 207-219.

Jamieson, B. G. M and Scheltinga, D. M. 1993. The ultrastructure of spermatozoa of *Nangura spinosa* (Scincidae, Reptilia). Memoirs of the Queenlsand Museum 34: 169-179.

Jamieson, B. G. M and Koehler, L. 1994. The ultrastructure of the spermatozoa of the Northern Water Snake, Nerodia sipedon (Colubridae, Serpentes) with phylogenetic considerations. Canadian Journal of Zoology 72: 1648-1652.

Jamieson, B. G. M and Scheltinga, D. M. 1994. The ultrastructure of spermatozoa of the Australian skinks, *Ctenotus taeniolatus*, *Carlia pectoralis*, and *Tiliqua scincoides scincoides* (Scincidae, Reptilia). Memoirs of the Queensland Museum 37: 181-193.

Jamieson, B. G. M., Oliver, S. C. and Scheltinga, D. M. 1996. The ultrastructure of spermatozoa of Squamata. I. Scincidae, Gekkonidae, and Pygonidae (Reptilia). Acta Zoologica 77: 85-100.

Lee, M. S. Y., Hugall, A. F., Lawson, R. and Scanlon, J. D. 2007. Snake phylogeny combining morphological and molecular data in likelihood, Bayesion and parsimony analyses. Systematic Biodiversity 5: 371-389.

Licht, P. 1979. Reproductive endocrinology of reptiles and amphibians: Gonadotropins. Annual Review of Physiology 41: 337-351.

Licht, P. 1984. Reptiles. Pp. 206-282. In: G. E. Lamming (ed.), *Marshall's Physiology of Reproduction*, vol. 1. *Reproductive Cycles of Vertebrates*. Chruchhill Livingstone, New York.

Licht, P., Denver, R. J. and Pavgi, S. 1989. Temperature dependence of in vitro pituitary, testis, and thyroid secretion in a turtle, *Pseudemys scripta*. General Comparative Endocrinology 76: 274-285.

Licht, P., Farmer, S. W., Bona-Gallo, A. and Papkoff, H. 1979. Pituitary gonadotropins in snakes. General and Comparative Endocrinology 39: 34-52.

Lin, M. and Jones, R. C. 1993. Spermiogenesis and spermiation in the Japanese Quail (*Coturnix coturnix japonica*). Journal of Anatomy 183: 525-535.

Lin, M., Harman, A., and Rodger, J. C. 1997. Spermiogenesis and spermiation in a marsupial, the Tammar Wallaby (*Macropus eugenii*). Journal of Anatomy 190: 377-395.

Lofts, B. 1964. Seasonal changes in the functional activity of the interstitial and spermatogenetic tissues of the green frog, *Rana esculenta*. General and Comparative Endocrinology 4: 550-562.

Lofts, B. 1968. Patterns of testicular activity. Pp. 239-245. In: E. J. Barrington and C. B. Jorgensen (eds), *Perspectives in Endocrinology: Hormones in the Lives of Lower Vertebrates*. Academic Press, New York.

Lofts, B., Phillips, J. G. and Tam, W. H. 1966. Seasonal changes in the testis of the cobra, *Naja naja* (Linn). General and Comparative Endocrinology 6: 466-475.

Maddison, D. R. and Maddison, W. P. 2005. *MacClade version 4.08*. Sinauer Associates, Inc., Sunderland, Massachusetts. Unpaginated software.

Marshall, A. J. and Woolf, F. M. 1957. Seasonal lipid changes in the sexual elements of a male snake *Vipera berus*. Quarterly Journal of Microscopy Science 98: 89-100.

Masson, G. R. and Guillette Jr., L. J. 2005. FSH-induced gonadal development in juvenile lizards, *Eumeces obsoletus*. Journal of Experimental Zoology 236: 343-351.

McIntosh, J. R. and Porter, K. R. 1967. Microtubules in the spermatids of the domestic fowl. Journal of Cell Biology 35: 153-173.

Oliver, S. C., Jamieson, B. G. M. and Scheltinga, D. M. 1996. The ultrastructure of the spermatozoa of Squamata. II. Agamidae, Varanidae, Colubridae, Elapidae, and Boidae (Reptilia). Herpetologica 52: 216-241.

Osman, D. I., Ekwall, H. and Ploen, L. 1980. Specialized cell contacts and the blood-testis barrier in the seminiferous tubules of the domestic fowl (*Gallus domesticus*). International Journal of Andrology 3: 553-562.

Pough, F. H., Andrews, R. M., Cadle, J. E., Crump, M. L., Savitzky, A. H. and Wells, K. D. 2001. *Herpetology*, 2nd Edition. Prentice Hall, New Jersey. Pp. 577.

Pudney, J. 1993. Comparative cytology of the non-mammalian vertebrate Sertoli cell. Pp. 611-657. In: L. D. Russell and M. D. Griswold (eds), *The Sertoli Cell*. Cache River Press, Clearwater, Florida.

Raynaud, A. and Pieau, C. 1985. Embryonic development of the genital system. Pp. 149-300. In: C. Gans and F. Billett (eds), *Biology of Reptilia*, vol. 15, *Development B*. John Wiley and Sons, New York.

Rheubert, J. L., McHugh, H., Collier, M. H., Sever, D. M. and Gribbins, K. M. 2009a. Temporal germ cell development strategy during spermatogenesis within the testis of the ground skink, *Scincella laterale* (Sauria: Scincidae). Theriogenology 72: 54-61.

Rheubert, J. L., Poldemann, E. H., Eckstut, M. E., Collier, M. H., Sever, D. M. and Gribbins, K. M. 2009b. Temporal germ cell development strategy during mixed spermatogenesis within the Mediterranean Gecko, *Hemidactylus turcicus* (Reptilia: Gekkonidae). Copeia 2009: 793-800.

Robb, J. 1960. The internal anatomy of *Typhlops schneider* (Reptilia). Australian Journal of Zoology 8: 181-216.

Roosen-Runge, E. C. 1972. *The Process of Spermatogenesis in Animals*. Cambridge University Press, Cambridge, U.K. Pp. 150.

Russell, L. D., Hikim, S. A. P., Ettlin, R. A. and Legg, E. D. 1990. *Histological and Histopathological Evaluation of the Testes*. Cache River Press, Clearwater, Florida. Pp. 275.

Saita, A., Comazzi, M. and Perrotta, E. 1988. Ulteriori osservazioni at M. E. sulla spermiogenesi di un serpente: *Coluber viridiflavus* (Lacepede) in reiferimento ad elementi comparativi nella spermiogenesi dei rettili. Atti dell a Accademia Nazionale dei Lincei. Rendiconti. Classe di Scienze Fisiche, Matematiche e Naturali LXXXII: 137-143.

Schuett, G. W., Harlow, H. J., Rose, J. D., Van Kirk, E. A. and Murdoch, W. J. 1997. Annual cycle of plasma testosterone in male copperheads, *Agkistrodon contortrix* (Serpentes, Viperidae): Relationship to timing of spermatogenesis, mating, and agonistic behavior. General and Comparative Endocrinology 105: 417-427.

Scott, D. E., Fishcer, R. U., Congdon, J. D. and Buss, S. A. 1995. Whole body lipid dynamics and reproduction in the Eastern Cottonmouth, *Agkistrodon piscivorus*. Herpetologica 51: 472-487.

Sever, D. M. 2004. Ultrastructure of the reproductive system of the Black Swamp Snake (*Seminatrix pygaea*). IV. Occurrence of an ampulla ductus deferentis. Journal of Morphology 262: 714-730.

Sever, D. M., Ryan, T. J., Stephens, R. and Hamlett, W. C. 2002. Ultrastructure of the reproductive system of the Black Swamp Snake (*Seminatrix pygaea*). III. Sexual segment of the male kidney. Journal of Morphology 252: 238-254.

Siegel, D. S. and Sever, D. M. 2008. Sperm aggregations in female *Agkistrodon piscivorous* (Reptilia: Squamata): A histological and ultrastructural investigation. Journal of Morphology 269: 189-206.

Shea, G. M. 2001. Spermatogenic cycle, sperm storage, and Sertoli cell size in Scolecophidian (*Ramphotyphlops nigrescens*) from Australia. Journal of Herpetology 35: 85-91.

Soley, J. T. 1997. Nuclear morphogenesis and the role of the manchette during spermiogenesis in the ostrich (*Struthio camelus*). Journal of Anatomy 190: 563-576.

Sprando, R. L. and Russell, L. D. 1988. Spermiogenesis in the red-eared turtle (*Pseudemys scripta*) and the domestic fowl (*Gallus domesticus*): A study of cytoplasmic events including cell volume changes and cytoplasmic elimination. Journal of Morphology 198: 95-118.

Tait, A. J. and Johnson, E. 1982. Spermatogenesis in the Grey Squirrel (*Sciurus carolinensis*) and changes during sexual regression. Journal of Reproduction and Fertility 65: 53-58.

Tavares-Bastos, L., Colli, G. R. and Bao, S. N. 2008. The evolution of sperm ultrastructure among Boidae (Serpentes). Zoomorphology 8: 62-68.

Tavares-Bastos, L., Cunha, G. R., Colli, G. R. and Bao, S. N. 2007. Ultrastructure of spermatozoa of scolecophidian snakes (Lepidosuaria, Squamata). Acta Zoologica 88: 189-197.

Teixeira, R. D., Colli, G. R., Báo, S. N. 1999. The ultrastructure of the spermatozoa of the lizard *Micrablepharus maximiliani* (Squamata, Gymnophthalmidae), with considerations on the use of sperm ultrastructure characters in phylogenetic reconstruction. Acta Zoologica 80: 47–59.

Tibia, T. and Kita, I. 1990. Undifferentiated spermatogonia and their role in seasonally fluctuating spermatogenesis in the ferret, *Mustela putorius furo* (Mammalia). Zoology of Australia and New Zealand 224: 140-155.

Tourmente, M., Giojalas, L. and Chiaraviglio, M. 2008. Sperm ultrastructure of *Bothrops alternatus* and *Bothrops diporus* (Viperidae, Serpentes), and its possible relation to the reproductive features of the species. Zoomorphology 127: 241-248.

Tourmente, M., Cardozo, G., Bertona, M., Guidobaldi, A., Giojalas, L. and Chiaraviglio, M. 2006. The ultrastructure of the spermatozoa of *Boa constrictor occidentalis*, with considerations on its mating system and sperm competition theories. Acta Zoologica 87: 25-32.

Tsubota, T. and Kanagawa, H. 1989. Annual changes in serum testosterone levels and spermatogenesis in the Hokkaido Brown Bear, *Ursus arctos yesoensis*. Journal of Mammalian Society of Japan 14: 11-17.

Unsicker, K. and Burnstock, G. 1975. Myoid cells in the peritublar tissue (lamina propria) of the reptilian testis. Cell and Tissue Research 163: 545-560.

Van Oordt, P. G. W. J. 1955. Regulation of the spermatogenetic cycle in the frog. Memoirs of the Society of Endocrinology 4: 25-38.

Van Oordt, P. G. W. J. and Brands, F. 1970. The Sertoli cell in the testis of the common frog, *Rana temporaria*. Proceedings of the Society of Endocrinology 119th Meeting Journal of Endocrinology 48, Abstract 100.

Ventela, S., Toppari, J. and Parvinen, M. 2003. Intercellular organelle traffic through cytoplasmic bridges in early spermatids of the rat: Mechanism of haploid gene product sharing. Molecular Biology and Cell 14: 2768-2780.

Vieira, G. H. C., Colli, G. R. and Báo, S. N. 2004. The ultrastructure of the spermatozoon of the lizard *Iguana iguana* (Reptilia, Squamata, Iguanidae) and the variability of sperm morphology among iguanian lizards. Journal of Anatomy 204: 451-464.

Volsøe, H. 1944. Structure and seasonal variation of the male reproductive organs of *Vipera berus* (L.). Spolia Zoologica Musei Hauniensis V. Skrifter, Universitetets Zoologiske Museum, København. Pp. 157.

Waits, G. M. H. and Setchell, B. P. 1990. Physiology of the mammalian testis. Pp. 1-105. In: G. E. Lemming (ed.), *Marshall's Physiology of Reproduction*. vol. 2. *Reproduction in the Male*. 4th Edition. Churchill Livingstone, London, U.K.

Weins, J. J. 2004. The role of morphological data in phylogeny reconstruction. Systemic Biology 53: 653-661.

Werner, Y. L. and Drook, K. 1967. The multipartite testis of the snake *Leptotyphlops phillipsi*. Copeia 1967: 159-163.

Wharton, C. H. 1966. Reproduction and growth in the Cottonmouths, *Agkistrodon piscivorus* (Lacepede) of Cedar Keys, Florida. Copeia 1966: 149-161.

Wilhoft, D. C. and Quay, W. B. 1961. Testicular histology and seasonal changes in the lizard, *Sceloporus occidentalis*. Journal of Morphology 108: 95-106.

Yamamoto, S., Tamate, H. and Itikawa, O. 1967. Morphological studies on the sexual maturation in the male Japanese Quail (*Coturnix coturnix jaonica*). Tohuku Jorunal of Agricultural Research 18: 27-37.

CHAPTER 7

Hormones and Reproduction in Free-ranging Snakes

Dale F. DeNardo[1] *and Emily N. Taylor*[2]

7.1 INTRODUCTION: WHAT HORMONES CAN TELL US ABOUT THE REPRODUCTIVE BIOLOGY OF ANIMAL POPULATIONS

The natural history of an organism is rarely discussed without including a description of its reproductive biology. An understanding of reproduction is critical for two primary reasons. First, although reproduction is not essential for the immediate survival of an individual, it is essential for the persistence of the species and is the currency of an individual's fitness. Second, reproduction can come at a sizable energetic cost to organisms, particularly females, and these costs can represent a substantial part of annual expenditures in species where maintenance costs are relatively low (e.g., ectotherms). Therefore, reproduction typically is not a continuous endeavor, and appropriate timing of reproductive effort is critical.

Timing of reproductive events is usually influenced by both internal and external cues. Prior to reproduction, an organism must be in adequate physiological condition and, even when body condition is sufficient, reproduction must be timed to take advantage of environmentally favorable circumstances (e.g., temperature, precipitation, and energy availability). The onset of reproductive activity is signaled by relatively reliable seasonal environmental stimuli such as photoperiod, temperature, or rainfall (Perrins 1970; Wingfield 1980; Perrins and Birkhead 1983; Farner 1986). These cues, which are detected by the organism at various sensory levels, are translated by the endocrine system to hormone signals that communicate the information to the reproductive system (Wingfield 1980, 1983; Wilson and Donham 1988). Hormones also regulate morphological, physiological, and behavioral changes in anticipation of future events (Wingfield 1983;

[1]School of Life Sciences, Arizona State University, Tempe, Arizona
[2]Department of Biological Sciences, California Polytechnic State University, San Luis Obispo, California

Jacobs and Wingfield 2000). Thus, measurements of circulating hormone concentrations are a vital component of understanding the complexity of the timing of behavioral and physiological aspects of reproduction.

Beyond being the primary interface between environmental changes and the onset of reproduction, hormones also modulate reproductive investment based on inputs that continue throughout the reproductive cycle. Changes in climatic patterns, social interactions, energy intake, and health status can all lead to adjustments in life history events including reproductive behavior and investment (Wingfield and Kitaysky 2002). Understanding hormonal responses to such events can elucidate an organism's plasticity to unpredictable situations (Jacobs and Wingfield 2000). Consequently, without an understanding of a species' reproductive endocrinology, it is essentially impossible to adequately understand the mechanisms by which cues, both internal and external, regulate reproduction and how reproductive activity influences behavior and other physiological aspects of an organism.

7.2 SNAKES AS MODELS FOR FIELD ENDOCRINOLOGY

Endocrinology is traditionally a laboratory-based science where hormone concentrations of classical laboratory animals such as mice and rats are manipulated to observe the hormone's effects on behavioral and/or physiological parameters. While laboratory studies are valuable because they reduce or eliminate environmental variation, these studies ignore the complex interactions between animal physiology, behavior, and the environment. Such interactive effects are especially important in understanding reproductive cycles. Therefore, studies in reproductive endocrinology must use field studies to complement laboratory manipulations where conditions are more controlled and simplistic.

Studying animals in their natural habitats can illuminate the ways in which environmental factors such as temperature, photoperiod, resource availability, and social factors affect hormone concentrations and reproduction. However, such studies can be difficult to conduct in the field. For example, many animals are logistically difficult to capture in order to collect blood samples, especially when serial samples are required from the same individuals. However, many species of snakes are easy to capture and monitor in the field, and they exhibit a diversity of reproductive characteristics that make them ideal model organisms for studies of field reproductive endocrinology.

As a group, snakes exhibit dramatic variation in mating systems, life history traits, and other characteristics associated with reproduction. Other chapters in this volume go into detail on this variation, but examples of the diversity of reproductive characteristics in snakes include mating tactics ranging from scramble competition to searching for widely distributed mates, oviparity and viviparity, and income and capital breeding (Seigel and Ford 1987; Duvall *et al.* 1992). However, the roles of hormones in the diverse reproductive patterns observed in the ~3000 snake species are poorly understood and have been studied in a limited number of taxa, particularly garter snakes (*Thamnophis* spp.) and vipers from temperate regions (reviewed in Graham *et al.* 2008; Taylor and DeNardo 2010; see also Krohmer and Lutterschmidt, **Chapter 8** in this volume).

Garter snakes have been heavily studied because they are abundant, resilient to handling and experimental manipulation, and exhibit unique and interesting reproductive behaviors. Vipers have been heavily studied because they are large-bodied and therefore amenable to radiotelemetry, and often occur in high densities at predictable locations (e.g., dens). Also, many vipers are ambush predators and therefore spend substantial amounts of time above ground, which makes them readily available for blood sampling. In contrast, many other species of temperate snakes are small-bodied, fossorial, and rarely encountered; thus they are difficult subjects for field studies of reproductive endocrinology. There are numerous potentially effective study species outside of North America and Europe, but these snakes are rarely studied because scientists are much more heavily concentrated on these two continents.

Most of the studies on hormones and reproduction in free-ranging snakes are descriptive rather than experimental. Most commonly, researchers monitoring a population of snakes via mark-recapture or radiotelemetry draw periodic blood samples and quantify the circulating concentrations of target hormones using radioimmunoassay (RIA) or enzyme-linked immunosorbent assay (ELISA), and relate these results to events in the snakes' reproductive cycles. Hormone manipulations, where plasma hormone concentrations are experimentally increased or decreased, are much less common in field studies of snake reproductive endocrinology.

Snakes exhibit a diversity of reproductive cycles that are described in detail by Saint Girons (1982) and Aldridge *et al.* (2009). Briefly, snakes can be categorized into one of four types of reproductive cycles: 1) the postnuptial (dissociated) or aestival type, in which males undergo spermatogenesis during the summer, sperm are stored over the winter (in the vasa deferentia and sometimes also in female oviducts if fall mating occurs), and the principal mating season occurs in spring; 2) the prenuptial (associated) or vernal type, in which males begin spermatogenesis in the fall and complete it by the following spring or early summer, at which time mating occurs; 3) the mixed type, in which spermatogenesis begins in spring and is completed one year later with one (spring) or two (spring and fall) mating seasons; and 4) the continuous type, in which spermatogenesis and mating behavior occur throughout the year. It is important to recognize that organisms do not always fall neatly into discrete categories assigned by biologists, and that the "type" of reproductive cycle of many species of snakes may therefore vary over time, among populations, and even among and within individuals. We therefore view the categories assigned by Saint Girons (1982) and Aldridge *et al.* (2009) simply as a convenient framework with which to describe the relationship between hormones and reproduction in studies on free-ranging snake species, and point out situations in which species do not appear to fall discretely into the given categories.

7.3 A REVIEW OF THE STUDIES

Field studies on the reproductive endocrinology of snakes have focused on a very small number of species within a narrow taxonomic range for several reasons. First, as stated above, certain species are more amenable to field study due to their catchability, seasonal movement and aggregation patterns, and body size. Second, the vast majority of field research on snake endocrinology has been conducted in temperate regions of North America and Europe. For these reasons, our knowledge of the endocrinological mediation of the relationship between reproduction and environmental conditions in snakes is based almost entirely on studies of garter snakes and vipers from these areas. Because these snakes are viviparous, it follows that relatively few field endocrinological studies have been conducted on oviparous taxa.

The following sections separately review the studies performed on viperids, *Thamnophis*, and other snakes. We focus on field studies, although in many cases complementary laboratory studies have been performed and are cited accordingly. Table 7.1 provides a list of major field studies on the reproductive endocrinology of snakes.

Table 7.1 Studies measuring hormone concentrations of free-ranging snakes. For methods, R = radiotelemetry, CR = capture/recapture, SNE = semi-natural enclosures. For hormones quantified, A = total androgen, B = corticosterone, DHT = dihydrotestosterone, E2 = estradiol, IGF-1 = insulin-like growth factor 1, M = melatonin, P = progesterone, T = testosterone, Th = thyroxine. For data series, S = serial samples on individuals, I = independent samples.

Species	*Methods*	*Hormones quantified*	*Data series*	*Reference*
Acrochordidae				
Acrochordus granulatus	CR	A,E2,P	I	Gorman *et al.* 1981
Colubridae				
Boiga irregularis	CR	B,P,T	I	Mathies *et al.* 2001
B. irregularis	CR	B,E2,P,T	I	Moore *et al.* 2005
B. irregularis	CR	B	I	Waye and Mason 2008
Cerberus rhynchops	CR	A,E2,P	I	Gorman *et al.* 1981
Natrix piscator	CR	M,T	I	Halder and Pandey 1989
Nerodia sipedon	CR	T	I	Weil and Aldridge 1981
Opheodrys aestivus	CR	A	I	Aldridge *et al.* 1990
Thamnophis elegans	CR	B	I	Robert *et al.* 2009
T. elegans	CR	IGF-1	I	Sparkman *et al.* 2009
Thamnophis sirtalis concinnus	CR	B,T	I	Moore *et al.* 2000b
T. s. concinnus	CR	B,T	I	Moore *et al.* 2001
Thamnophis sirtalis parietalis	CR	A,B	I	Krohmer *et al.* 1987

T. s. parietalis	SNE	B,E2,P,T	S	Whittier *et al.* 1987
T. s. parietalis	CR	E2	I	Mendonça and Crews 1989
T. s. parietalis	SNE	E2	I	Mendonça and Crews 1990
T. s. parietalis	SNE	B,T	I	Moore *et al.* 2000a
T. s. parietalis	SNE	B,T	I	Moore and Mason 2001
T. s. parietalis	CR	B,T	I	Moore *et al.* 2001
T. s. parietalis	SNE	A	I	Lutterschmidt *et al.* 2004
T. s. parietalis	SNE	B,T	S	Lutterschmidt and Mason 2005
T. s. parietalis	CR	A,B	I	Cease *et al.* 2007
Elapidae				
Laticauda colubrina	CR	A,E2,P	I	Gorman *et al.* 1981
Viperidae				
Agkistrodon contortrix	R,CR	P,T	S	Smith *et al.* 2010
Agkistrodon piscivorus	CR	A	I	Johnson *et al.* 1982
A. piscivorus	CR	B,T	I	Graham *et al.* 2008
A. piscivorus	CR	B,T	I	Bailey *et al.* 2009
Crotalus atrox	R,CR	B,E2,P,T	I,S	Taylor *et al.* 2004
C. atrox	CR	E2,DHT,T	I	Schuett *et al.* 2005
C. atrox	CR	E2,DHT,T	I	Schuett *et al.* 2006
Crotalus molossus	CR	E2,DHT,T	I	Schuett *et al.* 2005
Crotalus horridus	R	B,E2,T	I	Lutterschmidt *et al.* 2009
Crotalus oreganus	R,CR	B,DHT,E2,P,T	I,S	Lind *et al.* 2010
Crotalus scutulatus	CR	E2,DHT,T	I	Schuett *et al.* 2002
Vipera aspis	SNE	A,Th	S	Naulleau *et al.* 1987
V. aspis	SNE	P	S	Naulleau and Fleury 1990
V. aspis	CR	E2,DHT,P,T	I,S	Saint Girons *et al.* 1993
V. aspis	SNE	E2	S	Bonnet *et al.* 1994
V. aspis	SNE	E2,P	I,S	Bonnet *et al.* 2001
Viridovipera stejnegeri	SNE	E2,P	S	Tsai and Tu 2001

7.3.1 Viperids

Male viperids exhibit inter-specific and sometimes even intra-specific variation in the number of mating periods per year and timing of the mating season relative to gonadal activity. Several studies have shown that androgens (testosterone, T, and/or dihydrotestosterone, DHT) are elevated during the mating season(s), providing strong indication that T stimulates and possibly modulates a range of reproductive behaviors in male viperids. Many viperid species show a single annual mating season, and, where studied, plasma T concentrations are high when spermatogenesis and breeding behaviors occur. In eastern populations of Cottonmouths (*Agkistrodon piscivorus*), T peaks in late summer, at the same time that males show spermiogenic activity, hypertrophy of the sexual segment of the kidney, and breeding behavior (Johnson *et al.* 1982; Graham *et al.* 2008). Similarly, Black-tailed Rattlesnakes (*Crotalus molossus*) from southeastern

Arizona show a single annual peak in T concentrations during their sole summer mating season (Schuett *et al.* 2005), and European Adder (*Vipera berus*) show a single peak in T at the same time as spermiogenesis and mating, although these occur in the spring rather than late summer (Naulleau *et al.* 1987). Timber Rattlesnakes (*Crotalus horridus*) show a single summer mating season but males exhibit no seasonal variation in T concentrations, although the sample size in this study was low (Lutterschmidt *et al.* 2009).

Many snake species, including some viperids, show two annual mating seasons, the first in late summer/fall and a second in the following spring. These mating seasons are separated by winter and summer periods of reproductive inactivity. Western Diamond-backed Rattlesnakes (*Crotalus atrox*), Northern Pacific Rattlesnakes (*Crotalus oreganus*), Mojave Rattlesnakes (*Crotalus scutulatus*), and Aspic Vipers (*Vipera aspis*) mate in both late summer/fall and spring, and males show elevated androgens in both of these seasons (Naulleau *et al.* 1987; Saint Girons *et al.* 1993; Schuett *et al.* 2002, 2005; Taylor *et al.* 2004; Lind *et al.* 2010). Male *C. atrox, C. oreganus*, and *C. scutulatus* show bimodal peaks in circulating DHT that match the timing of the peaks in T (Schuett *et al.* 2002, 2005; Lind *et al.* 2010), although DHT concentrations are much lower than T concentrations. These species show the aestival or postnuptial pattern of spermatogenesis: T concentrations are high during spermatogenesis and mating in the late summer and fall, but are also high during the spring when mating is occurring in the absence of spermatogenesis. Testosterone concentrations are lowest in summer, when mating activity is absent. In the Sonoran Desert, *C. atrox* sampled while basking during the winter exhibit intermediate T concentrations between those of summer (lowest) and of the spring and fall mating seasons (Schuett *et al.* 2006). Interestingly, Copperheads (*Agkistrodon contortrix*) show a single peak in androgen (T) concentrations associated with its single summer mating season at the northern part of its range (Smith *et al.* 2010), but show two annual peaks (late summer/fall and spring) in androgen concentrations associated with its bimodal mating pattern in the southern part of its range (lab study: Schuett *et al.* 1997).

Spermatogenesis, maximal T production, and mating behaviors usually occur simultaneously in snakes with a single annual breeding season, (e.g., the "associated" reproductive pattern, Crews 1984; Whittier and Crews 1987), thereby making it difficult to assess the importance of T in each process. However, studies on species with bimodal mating seasons suggest that T is responsible for mating behaviors, since T is elevated during both mating periods regardless of whether the testes are spermatogenically active or regressed. Some exceptions are evident; for example, male *Vipera aspis* sometimes have very low androgen concentrations during the mating season (Naulleau *et al.* 1987; Saint Girons *et al.* 1993). Species with two mating seasons are intermediate in the associated-dissociated dichotomy of reproductive patterns (Crews 1984; Whittier and Crews 1987), as they have associated reproduction in one season (late summer/fall) and dissociated reproduction in another (spring). The paradigm of associated/dissociated reproduction patterns should be viewed as describing two extremes along a continuum in which many snakes fall somewhere in the middle (Moore and Lindzey 1992; Schuett *et al.* 2006; Benner and Woodley 2007).

In general, studies of the relationship between steroid hormones and reproduction in female viperids show that plasma 17β-estradiol (E2) is elevated during vitellogenesis, reflecting its role in stimulating production of vitellogenin by the liver (Ho *et al.* 1982). Most female vipers initiate vitellogenesis in the fall, become quiescent during the winter, and complete vitellogenesis in the spring (type 2 vitellogenesis, Aldridge 1979). Female *Crotalus oreganus* follow this typical pattern and show increased E2 concentrations in the fall and spring (Lind *et al.* 2010). In contrast, female *C. atrox* in the Sonoran Desert of Arizona show a Type 1 vitellogenic pattern in which they initiate and complete vitellogenesis in the spring (Taylor and DeNardo 2005). Reproductive female *C. atrox* therefore exhibit elevated E2 concentrations from March through June, especially during peak vitellogenesis in April and May (Taylor *et al.* 2004). In non-reproductive female *C. atrox* and *C. oreganus*, plasma concentrations of E2 are low throughout the year (Taylor *et al.* 2004; Lind *et al.* 2010). Lutterschmidt *et al.* (2009) observed low concentrations of E2 in female *C horridus* throughout the year, but the sample sizes, especially of reproductive females, were low. Female *Vipera aspis* in France have the same seasonal mating patterns as *C. atrox* (spring vitellogenesis and late summer parturition), and also show elevated E2 during spring (Saint Girons *et al.* 1993). In Chinese Green Tree Vipers, *Viridovipera* (*Trimeresurus*) *stejnegeri*, E2 concentrations are slightly elevated during vitellogenesis (Tsai and Tu 2001). Almeida-Santos *et al.* (2004) found that E2 peaks during winter vitellogenesis in female Neotropical Rattlesnakes (*Crotalus durissus terrificus*) in southeastern Brazil, and is low throughout the year in non-reproductive females.

Several studies have also measured concentrations of circulating androgens in female vipers. Androgen concentrations of females are always lower than those of males, and are extremely low year-round in non-reproductive female *Crotalus atrox* and *C. oreganus* (Taylor *et al.* 2004; Lind *et al.* 2010). Androgen concentrations increase slightly during spring in reproductive female *C. atrox* and *C. oreganus;* this may occur because T is a precursor to E2 in steroid biosynthesis. Alternatively, the increase in androgens in reproductive females may reflect a functional role of these hormones in stimulating receptivity or other factors related to mating. For example, concentrations of DHT, a non-aromatizable androgen (i.e., cannot be converted to E2), are elevated during spring in female *C. oreganus* (Lind *et al.* 2010). Also, Saint Girons *et al.* (1993) found that female *Vipera aspis* displaying mating behaviors had significantly higher plasma DHT concentrations, but not T concentrations, than females that were not displaying mating behaviors.

Plasma progesterone (P4) is often elevated during gestation in viviparous snakes, reflecting its role in maintenance of pregnancy (Mead *et al.* 1981; reviewed in Custodia-Lora and Callard 2002). Plasma P4 concentrations are very low in non-reproductive female *Crotalus atrox*, and in reproductive females are elevated in May through August, peaking during the gestation period in June-July (Taylor *et al.* 2004; Taylor and DeNardo 2005). Reproductive *Viridovipera* (*Trimeresurus*) *stejnegeri* and *C. durissus terrificus* have dramatically elevated P4 concentrations during gestation (Tsai and Tu 2001; Almeida-Santos *et al.* 2004). Reports of plasma P4 concentrations in reproductive *Vipera aspis* are variable. Saint Girons *et al.* (1993) showed that plasma P4 was highly variable among snakes, with no clear difference between reproductive and non-reproductive females. However, in other studies, plasma P4 concentrations of reproductive female *V. aspis* were elevated in May through August (gestation), decreased in September (after parturition), and rose again during winter (Naulleau and Fleury 1990; Bonnet *et al.* 2001). Bonnet *et al.* (2001) hypothesized that the winter increase in P4, which is especially marked in post-reproductive snakes, may act to block vitellogenesis in emaciated snakes, as P4 may inhibit hepatic synthesis of vitellogenin (Callard *et al.* 1992, 1994).

While corticosterone (CORT, the major glucocorticoid in reptiles) is not typically viewed as a sex steroid hormone, it likely has important roles in reproductive physiology either directing or indirectly via its effects on energy mobilization. Taylor *et al.* (2004) measured CORT concentrations relative to reproduction in free-ranging *Crotalus atrox* and found that CORT concentrations are more variable annually than the other steroid hormones, but increase dramatically in reproductive females in July and August, during late gestation. Plasma CORT concentrations return to baseline along with P4 near the time of parturition in reproductive females. Lutterschmidt *et al.* (2009) found that both female and male *C. horridus* had lower CORT concentrations in summer than in spring or fall, but that reproductive females had higher CORT concentrations than non-reproductive females. In sum, these results demonstrate a relationship between gestation and high CORT concentrations, yet the functional role of CORT remains uncertain. CORT may be elevated in female vipers to aid in energy mobilization during reproduction, when metabolic rate is elevated (Beaupre and Duvall 1998; Beaupre 2002). Alternately, CORT may play a role in regulating development of the offspring, as is seen with glucocorticoids in mammals (Ballard and Ballard 1972). Distinguishing these roles for CORT will require manipulative experiments under controlled laboratory conditions.

In addition to potentially playing a supportive role for either the female or offspring during reproduction, CORT may have an inhibitory effect on reproduction via its role in the stress response. CORT concentrations tend to rise as a result of stressful stimuli such as handling by humans. Bailey *et al.* (2009) showed that *Agkistrodon piscivorus* subjected to 30 minutes of confinement stress in a bag exhibited higher CORT concentrations than snakes bled immediately upon capture. The stress response did not differ between male and female snakes. Lutterschmidt *et al.* (2009) also found that CORT concentrations increased in *Crotalus horridus* subjected to one hour of confinement stress. Reproductive and post-parturient females had a greater stress-induced increase in CORT concentrations than non-reproductive females, suggesting that reproductive state modulates the hypothalamo-pituitary-adrenal axis. There was a significant relationship between baseline CORT and T concentrations in male *C. horridus*, where snakes with higher CORT concentrations had lower T concentrations. CORT and T often exhibit a reciprocal relationship because stress may inhibit reproduction (Greenberg and Wingfield 1987; Moore and Jessop 2003). Interestingly, however, confinement stress led to a slight increase in T concentrations in male *C. horridus*.

7.3.2 The Red-sided Garter Snake (*Thamnophis sirtalis parietalis*)

The Red-sided Garter Snake (*Thamnophis sirtalis parietalis*) is by far the best-studied snake species in terms of reproductive endocrinology. These snakes overwinter in communal dens at high latitudes (e.g., Manitoba, Canada), and in the spring they emerge from the dens and aggregate by the thousands to breed. Researchers are therefore able to obtain huge sample sizes for their studies. In addition, male snakes are so keen to breed that they completely ignore researchers, even when placed in experimental semi-natural enclosures in the field or returned to the laboratory, allowing easy observation of natural behaviors. These characteristics have made *T. s. parietalis* model organisms not just for reproductive endocrinology but also for studies of behavior, sexual selection, chemical ecology, and more. The literature on *T. s. parietalis* is extensive, and noteworthy review articles on their reproductive physiology include Garstka *et al.* (1982), Krohmer *et al.* (1987), Krohmer (2004), and Krohmer and Lutterschmidt (**Chapter 8** in this volume).

In *Thamnophis sirtalis parietalis*, concentrations of circulating hormones have been quantified in free-living snakes, in snakes exposed to experimental conditions in semi-natural enclosures, and in the laboratory. Many of these studies have gone beyond describing seasonal hormone profiles to actually performing experimental manipulations to examine the functional role of hormones in reproductive physiology and behavior. The reproductive endocrinology of *T. s. parietalis* is discussed in detail in Krohmer and Lutterschmidt (**Chapter 8**, this volume) and Taylor and DeNardo (2010), so in this chapter we will focus only on several key field studies on reproductive endocrinology in this species.

Male *Thamnophis sirtalis parietalis* emerge from communal hibernacula in spring and initiate courtship behavior that lasts several weeks (Crews and Garstka 1982). During spring mating, males utilize sperm produced the previous summer and stored through the winter in the vasa deferentia (Krohmer *et al.* 1987); they therefore exhibit the postnuptial, dissociated reproductive tactic (Crews 1976). However, other studies show that *T. s. parietalis* also mate in the late summer and fall (Mendonça and Crews 1989; Whittier and Crews 1989). It was initially reported that plasma androgen concentrations are low in the spring when mating occurs (Camazine *et al.* 1980), again highlighting the postnuptial, dissociated pattern of reproduction in this species. However, later studies demonstrated that androgen concentrations are, in fact, elevated upon emergence from hibernation but drop rapidly after emergence (Krohmer *et al.* 1987; Moore *et al.* 2000b; Cease *et al.* 2007). The sexual segment of the kidney, which is known to be stimulated by androgens in squamate reptiles (Bishop 1959), is hypertrophied in late summer and spring but regressed during the summer when androgen concentrations are low (Krohmer *et al.* 1987). Since there is now evidence for both spring and late summer mating, and elevated T during both fall and emergence from hibernation in spring, it is clear that this species does not exhibit a strictly dissociated pattern of reproduction. However, some laboratory studies have questioned the dependence of mating behavior on T in male *T. s. parietalis* (Camazine *et al.* 1980; Crews 1984; Crews *et al.* 1984). Courtship behavior persists despite castration, adrenalectomy, or hypophysectomy, and treatment with hypothalamic hormones, gonadotropins, arginine vasotocin, or sex steroid hormones does not stimulate courtship behavior in adult males (Garstka *et al.* 1982; Crews *et al.* 1984). Nonetheless, it is possible that prior exposure to androgens organizes brain regions involved in reproduction but the onset of reproductive behavior does not occur until initiated by environmental cues at a later time (Crews 1991). Alternatively, steroids may be synthesized in the brain, bypassing the influence of circulating hormones on reproductive behavior (Soma 2006). Further studies, perhaps incorporating the use of steroid receptor antagonists, are needed to more effectively evaluate the role of T in regulating reproductive behavior in *T. s. parietalis* and other snake taxa.

Several studies have examined the relationship between reproduction and the stress response in male *Thamnophis sirtalis parietalis*, as well as the closely related *T. s. concinnus* (Moore *et al.* 2000a, Moore *et al.* 2001; Lutterschmidt and Mason 2005; Cease *et al.* 2007). Baseline CORT concentrations in *T. s. parietalis* are highly variable from year to year, and are generally highest during the spring breeding season, lowest in summer, and intermediate in fall (when breeding can also occur, Moore *et al.* 2001). Male *T. s. parietalis* show variable hormonal responses to confinement stress in cloth bags. When CORT concentrations are low (e.g., in summer), males exhibited a stress-induced increase in CORT and decrease in T (Moore *et al.* 2001). During the main breeding season in spring, Moore and colleagues (2001) and Lutterschmidt and Mason (2005) observed no stress-induced change in CORT or T concentrations. If male *T. s. parietalis* suppress the stress response during their brief spring breeding season, they may avoid the negative effect that the stress response may have on their ability to reproduce. In support of this, Cease and colleagues (2007) found that actively courting males do not show a CORT stress response, but males already dispersing from the den site do exhibit an increase in CORT in response to stress. However, unlike the studies above, Moore *et al.* (2000a) observed a stress-induced increase in CORT and decrease in T in actively courting males. The rationale behind this discrepancy among studies is unclear at this time and warrants further study.

Male *Thamnophis s. concinnus* exhibit a stress-induced increase in CORT throughout the year. This subspecies has an extended breeding season when compared to *T. s. parietalis*; thus it may not face the same pressure to suppress the stress response during breeding. During spring, T concentrations decreased with stress but during summer and fall they actually increased. Moore *et al.* (2000b) found that throughout their annual cycle, baseline CORT and T concentrations were positively correlated. The differences in the stress-induced changes in CORT and T observed in the two subspecies (Moore *et al.* 2001) highlight the fact that the traditional negative relationship between stress and reproduction (e.g., Greenberg and Wingfield 1987; Moore and Jessop 2003) is not always true. Interestingly, while males continue to display courtship behaviors even when stressed (Moore *et al.* 2000a), injection of exogenous CORT suppresses courtship behavior in a dose-dependent manner but does not affect circulating androgen concentrations (Moore and Mason 2001; Lutterschmidt *et al.* 2004). Also, treatment with exogenous melatonin suppresses courtship behavior but does not affect androgen concentrations (Lutterschmidt *et al.* 2004; Lutterschmidt and Mason 2005). This suggests that any suppression of reproductive behavior by CORT is not due to suppression of androgens, but via another mechanism.

Very few studies have been performed in the field examining the relationship between hormones and reproduction in female *Thamnophis sirtalis parietalis*. However, extensive laboratory studies have been performed on females of this species, and we refer readers to Taylor and DeNardo (2010) and Krohmer and Lutterschmidt (**Chapter 8** in this volume) for reviews of this information. The few field studies conducted have shown that female *T. sirtalis parietalis* exhibit a rather unusual relationship between reproductive events and hormones when compared to other snake species. Spring mating occurs when E2 concentrations are low (Garstka *et al.* 1982), but Mendonça and Crews (1996) have shown through ovariectomy and hormone replacement therapy that even low E2 concentrations appear to be important in making female snakes attractive and receptive to males. Additionally, the physical act of mating induces a surge in E2 in females, but plasma E2 concentrations are not necessarily elevated during vitellogenesis (Garstka *et al.* 1985; Whittier *et al.* 1987; Whittier and Crews 1989; Mendonça and Crews 1990). Like many other snakes studied, plasma T is elevated during vitellogenesis (Whittier *et al.* 1987). Interestingly, female *T. s. parietalis* do not show elevated P4 during gestation (Whittier *et al.* 1987), which is in contrast to other snakes and vertebrates, in general.

7.3.3 Other Snakes

While the vast majority of research on reproduction and hormones in free-ranging snakes has focused on temperate vipers and *Thamnophis sirtalis parietalis*, circulating hormone concentrations of several other species of free-ranging snakes have been quantified. Like the more commonly studied species, most of these are terrestrial, temperate, North American species in which reproduction is highly seasonal. Both Northern Water Snakes (*Nerodia sipedon*) and Rough Greensnakes (*Opheodrys aestivus*) breed only in spring, but males show elevated T during the spring breeding season and during the late summer postnuptial spermatogenic period (Weil and Aldridge 1981; Aldridge *et al.* 1990). Male *T. s. concinnus* have peak T concentrations during spermatogenesis in the fall, and levels are much lower during the spring breeding season (Moore *et al.* 2000b).

A noteworthy exception to the preponderance of studies on temperate, terrestrial snakes is the study by Gorman and colleagues (1981) that describes the seasonal steroid hormone concentrations in relation to reproductive events in three marine snakes (Filesnake, *Acrochordus granulatus;* Dog-faced Water Snake, *Cerberus rhynchops;* and Banded Sea Krait, *Laticauda colubrina*) from the Philippines. These three species show varying degrees of seasonality in their reproductive patterns. *Acrochordus granulatus* is the most seasonal, with mating activity, spermatogenesis, and peak T concentrations occurring in the fall, shortly after vitellogenesis in females. In contrast, *C. rhynchops* and especially *L. colubrina* show spermatogenesis throughout the year. Although spermatogenesis and T concentrations in *C. rhynchops* peak in the fall, there is evidence for year-round gonadal activity. In *L. colubrina,* males show no seasonal trends in spermatogenic activity or T concentrations, suggesting that they most closely follow the continuous pattern of reproduction (Saint Girons 1982). These data suggest that species inhabiting more thermally constant environments, such as the ocean and the tropics, may show year-round reproduction. However, the paucity of data on tropical snakes highlights the need for research in this area. Plasma E2 and T concentrations were elevated during vitellogenesis in the two viviparous species (*A. granulatus* and *C. rhynchops*), and in the oviparous *L. colubrina* E2 and T concentrations increased progressively through vitellogenesis and gravidity. Similarly, P4 concentrations were highest during pregnancy in the two viviparous species and increased progressively during gravidity in *L. colubrina.*

Halder and Pandey (1989) measured concentrations of T and melatonin in Indian Chequered Water Snakes (*Natrix piscator*), a tropical, aquatic snake, throughout the year to examine the relationship between pineal and testicular activity. They found that T concentrations were elevated during the fall, when maximal weights of the testes, vasa deferentia, and kidneys were observed. Plasma melatonin concentrations and pineal weight, in contrast, were lowest during the fall and peaked in the spring, suggesting that melatonin may play a role in inhibiting the reproductive axis.

Western Terrestrial Garter Snakes (*Thamnophis elegans*) exist as two ecotypes in a population near Eagle Lake, California: snakes from meadow habitats are slow-growing, long-lived, and have low annual reproductive output, whereas snakes from lakeshore habitats are fast-growing, shorter-lived, and have higher annual reproductive output. Robert *et al.* (2009) measured plasma CORT concentrations in free-living, pregnant meadow and lakeshore snakes. They found that CORT concentrations were higher in the meadow snakes, possibly because meadow snakes experienced lower food availability than lakeshore snakes. Sparkman *et al.* (2009) measured plasma concentrations of insulin-like growth factor-1 (IGF-1) in this species. They found that IGF-1 concentration was positively associated with body size in lakeshore snakes, but that the relationship between IGF-1 concentrations and size in meadow snakes depended on resource availability. Furthermore, females that gave birth to larger litters had higher concentrations of IGF-1.

Introduced Brown Tree Snakes (*Boiga irregularis*) in Guam have been the subject of several studies on reproductive and stress hormones. Moore *et al.* (2005) measured concentrations of sex steroids and CORT in free-ranging and captive *B. irregularis* throughout a two-year period. Most of the free-ranging snakes they captured were non-reproductive, had lower body condition, and had higher CORT concentrations than captive snakes, suggesting that they were in negative energy balance. Plasma T was low throughout the year in free-ranging males and exhibited a spring peak in captive males. Plasma E2 and P4 were low in free-ranging females throughout the year compared to captive females. Waye and Mason (2008) compared CORT concentrations and body condition of the snakes in the Moore *et al.* (2005) study (collected 1991-1993) to those of snakes collected in 2003. They found that by 2003, body condition had improved and CORT concentrations had dropped, suggesting that the snakes were no longer chronically stressed by energy limitations. Waye and Mason (2008) suggest that introduction of new prey species, including frogs, may have contributed to the increased body condition and decreased CORT concentrations in *B. irregularis*. To investigate the effect of capture and confinement stress on *B. irregularis*, Mathies *et al.* (2001) compared concentrations of CORT and sex steroids in *B. irregularis* captured and held in traps for various durations with snakes bled immediately upon capture. Almost all of the snakes they captured were non-reproductive, therefore concentrations of T in males and P4 in females were low, in concordance with the results of Moore *et al.* (2005). They found that females had CORT concentrations about twice those of males. Confinement in cloth bags led to increased concentrations of CORT, similar to the studies of *Agkistrodon piscivorus* (Bailey *et al.* 2009) and *Crotalus horridus* (Lutterschmidt *et al.* 2009) described in Section 8.3.1. Snakes captured in a trap and held overnight showed increased concentrations of CORT relative to snakes bled immediately upon capture, but CORT concentrations returned to baseline in snakes held for three nights in a trap, suggesting that the snakes may acclimate to the traps.

7.4 SUGGESTED DIRECTIONS FOR FUTURE RESEARCH

To establish a comprehensive understanding of how various hormones regulate reproductive physiology and behavior among snakes in natural conditions, there is dire need for additional work. First, greater amounts of comparative data are needed. Snakes are a diverse group consisting of approximately 3,000 species in 18 families (Integrated Taxonomic Information System 2010). Yet, only 36 field studies provide 40 species accounts of reproductive endocrinology (Table 7.1). Worse, those 40 accounts examined only 18 species in four families (two families are represented by studies on a single species). In fact, 21 of the 40 accounts (53%) are limited to two genera – *Thamnophis* and *Crotalus*.

All accounts of field endocrinology in snakes are limited to the more derived Caenophidia (Lee *et al.* 2007). While the Caenophidia contain the vast majority of extant snake species, the lack of any data from more primitive lineages prohibits any consideration of derived versus primitive patterns in reproductive endocrinology. Even within the speciose families of the Caenophidia, accounts are sparse. While the Viperidae and Elapidae each consist of over 200 species, field reports of reproductive endocrinology are limited to nine species within the Viperidae and only a single species in the Elapidae. Despite most accounts being of species within the Colubridae, this enormously diverse family (approximately 2000 species) is represented by merely seven species.

Equally concerning to the phylogenetic voids in our understanding of snake reproductive endocrinology under natural conditions is the limited coverage of the varied natural history traits among snakes. Without examining diverse snakes under many environmental conditions, we will not be able to assess the universality of findings or what phylogenetic, morphological, physiological, or environmental characteristics influence endocrinological patterns. *Thamnophis sirtalis* has been the most intensively examined species (11 of the 36 publications), and the results from these studies provide us the greatest insight into the reproductive endocrinology of snakes under natural conditions. Unfortunately, the atypical traits of this species (e.g., high latitude distribution, communal denning, formation of mating balls immediately upon female emergence) put into question the broad applicability of these results to snakes in general.

Even looking beyond the *Thamnophis sirtalis* data, there remains considerable bias in that the preponderance of studies has been conducted on temperate terrestrial species, particularly in North America. Despite snakes being widely distributed on most continents except Antarctica, there are no endocrinological field studies from South America, Africa, or Australia. Current literature is dominated by studies of viviparous species, with only four of the 18 species examined thus far being oviparous. In addition, the vast majority of field studies focus on male rather than female snakes. Males are often easier to study because they are more abundant and/or catchable in the field, but the male bias observed in studies may also reflect a greater inherent interest in the reproductive biology of male snakes on the part of biologists. Indeed, the categories to which snakes are assigned when studying their reproductive physiology (i.e., associated/ dissociated, prenuptial/postnuptial, etc.) are often based solely on males, ignoring females. Aldridge *et al.* (2009) review studies on reproductive cycles of male and female colubrid snakes, and provide suggestions for terminology that bridge male and female reproductive patterns (i.e., pre- and post-ovulatory spermatogenesis). There is an obvious need for studies examining hormones and reproduction in both males and females of a given species so that hormonal correlates of the reproductive cycles of both sexes, and their interrelationships, can be examined.

There is a bright side. Of the 36 studies conducted, 23 (64%) of them have been published in the past ten years. Continued development of field technologies and interest in field reproductive endocrinology will hopefully lead to a continuation of the explosive growth in literature on the subject. Because of both financial and manpower limitations, study planning should be well-planned to maximize its value. While additional descriptive studies on endocrinological patterns in free-ranging snakes will make significant contributions, there would be greater benefit from studies that take an organized comparative approach. For example, great insight could come from a study that examines the reproductive endocrinology of a community of snakes that represent varied reproductive modes, activity patterns, and phylogeny. While such a study would be challenging, current technologies make it feasible. An alternate comparative approach could examine multiple viviparous and oviparous species across diverse phylogenies to ascertain the extent to which reproductive mode influences hormonal profiles. Yet another comparative approach of value would be to examine the reproductive endocrinology of multiple populations of a species that is distributed widely across habitats, since ecologically based intra-specific differences in both reproductive cycles and hormone profiles have been documented (e.g., *Agkistrodon contortrix*: Schuett *et al.* 1997, Smith *et al.* 2009; *Thamnophis elegans*: Robert *et al.* 2009). Regardless of which comparative approach is taken, studies involving comparisons of more than two species or two populations are preferable (Garland and Adolph 1994).

Also essential to strengthening our understanding of snake reproductive endocrinology are manipulative studies that link hormonal profiles to reproductive states and behaviors, explore the functional roles of these hormones, and investigate the impact of environmental and physiological conditions on hormone profiles. Endocrine organ removal and hormone replacement therapy have been used to explore the chemical nature of physiology even before the discovery of hormones, and these manipulative techniques were vital to the emergence of the field of reproductive endocrinology in the early 20th century (Borell 1985). Not surprisingly, these techniques have been used in snakes, but relatively sporadically (e.g., Fraenkel *et al.* 1940; Bragdon 1951; Crews 1976; Camazine *et al.* 1980; Garstka *et al.* 1982; Crews *et al.* 1984, 1993; Padgoankar and Samuel

1993; Mendonça and Crews 1996). Furthermore, neither gonadectomy nor hormone treatment have been employed in long-term studies of free-ranging snakes. With the development of radiotelemetry, serially locating and processing free-ranging snakes is now relatively common and the incorporation of hormone manipulation into such studies would not only expand our understanding of reproductive endocrinology, but likely also our understanding of costs of reproduction, sex-based differences in growth and activity, and other vital ecophysiological traits.

7.5 SUMMARY

While there is still much to be learned about the reproductive endocrinology of snakes, especially in their natural environments, our knowledge of this field has grown considerably over the last decade. Here, we make an attempt to synthesize the existing data across taxa to identify consistencies and discrepancies within snakes in general. Additionally, since snakes have diverse reproductive cycles and reproductive modes, we can compare results from various studies to make some preliminary assessments regarding the function of hormones in reproduction despite a paucity of data from manipulative experiments.

In the vast majority of species studied, elevated plasma T concentrations in males are associated with reproductive activity, regardless of the number of mating seasons or the time of year that the mating season occurs. Species that show an associated reproductive cycle where spermatogenesis and mating occur together have a single peak in plasma T that coincides with this reproductive period (e.g., *Acrochordus granulatus*: Gorman *et al.* 1981; *Vipera berus*: Naulleau and Fleury 1987; *Agkistrodon piscivorus*: Graham *et al.* 2008). In species where breeding seems to occur year-round, plasma T remains elevated throughout the year (e.g., *Cerberus rhynchops, Laticauda colubrina*: Gorman *et al.* 1981). The most revealing descriptive data for the role of T in male snakes comes from those species where mating occurs twice a year but spermatogenesis occurs only during one of those two mating periods. In these species, plasma T is elevated during both mating seasons (e.g., *Vipera aspis*: Saint Girons *et al.* 1993; *Crotalus atrox*: Taylor *et al.* 2004; *Crotalus scutulatus*: Schuett *et al.* 2005; *Crotalus oreganus:* Lind *et al.* 2010). Together, these data from various species that have diverse reproductive cycles strongly suggest that T is critical for reproductive activity, especially mating behavior. However, the most studied species, *Thamnophis sirtalis*, does not follow this generality. In this species, T peaks when spermatogenesis is occurring in fall. While T is initially high at spring emergence, plasma concentrations decrease through the mating season (Krohmer *et al.* 1987; Moore *et al.* 2000b; Cease *et al.* 2007). In fact, the removal of T via castration does not inhibit male courtship behavior (Garstka *et al.* 1982; Crews *et al.* 1984). The reason for this inconsistency in T cycles between *T. sirtalis* and most other species exist is uncertain, but it emphasizes the need for caution when broadly applying results derived from a single species.

In females, the general steroid profiles are similar to those of other vertebrates. Peaks in plasma E2 concentrations are associated with vitellogenesis regardless of whether vitellogenesis occurs over a single season (e.g., *Acrochordus granulatus*, *Cerberus rhynchops*: Gorman *et al.* 1981; *Viper aspis*: St. Girons *et al.* 1993; *Crotalus atrox*: Taylor *et al.* 2004) or two seasons (e.g., *Crotalus oreganus*: Lind *et al.* 2010). Consistently, P4 peaks post-ovulation regardless of whether the species is viviparous (e.g., *A. granulatus*, *Cerberus rhynchops*: Gorman *et al.* 1981; *Viridovipera stejnegeri*: Tsai and Tu 2001; *C. atrox*: Taylor *et al.* 2004) or oviparous (e.g., *Laticauda colubrina*: Gorman *et al.* 1981). However, as with the T profiles, results from female *Thamnophis sirtalis parietalis* are not consistent with those from other snakes. While mating induces a surge in E2 in females, plasma E2 concentrations are not necessarily elevated during vitellogenesis (Garstka *et al.* 1985; Mendonça and Crews 1990). Additionally, female *T. s. parietalis* do not show elevated P4 during gestation (Whittier *et al.* 1987).

Among all the steroids hormones, CORT shows the fewest consistencies in relationship to snake reproduction. Like many species, snakes typically show elevated CORT associated with stress (e.g., *Boiga irregularis*, Mathies *et al.* 2001; *Agkistrodon piscivorus*: Bailey *et al.* 2009; *Crotalus horridus*: Lutterschmidt *et al.* 2009), but this stress response is inhibited during reproduction in one subspecies of *Thamnophis sirtalis* but not another (Moore *et al.* 2001). The role of CORT in snake reproduction has simply been attributed to mobilizing energy stores during this period of high metabolic demand but low energy intake (Beaupre and Duvall 1998; Beaupre 2002). However, *C. atrox* show elevated plasma CORT concentrations associated with late gestation followed by a sudden decrease in CORT concentrations at parturition, without food intake (Taylor *et al.* 2004). This result leaves open the possibility of a direct role of CORT in embryonic development prior to parturition.

While general patterns are beginning to develop, the reproductive endocrinology of snakes essentially remains in its infancy and is in great need of further study, particularly strategically designed comparative studies intended to find causal links between hormone profiles and phylogenetic, ecologic, and physiologic characteristics. Also needed are manipulative studies to better elucidate the function of the various hormones associated with reproductive events.

7.6 ACKNOWLEDGMENTS

We thank the organizers of the Snake Reproduction symposium at the 2009 Joint Meeting of Ichthyologists and Herpetologists in Portland, Oregon, as well as the editors of this volume, for enlisting a diverse group of authors to present a thorough amalgamation of the current knowledge on the reproductive biology and phylogeny of snakes. We also thank Gordon W. Schuett and two anonymous reviewers for their comments on earlier versions of this chapter.

7.7 LITERATURE CITED

Aldridge, R.D. 1979. Female reproductive cycles of the snakes *Arizona elegans* and *Crotalus viridis*. Herpetologica 35: 256-261.

Aldridge, R. D., Greenhaw, J. J. and Plummer, M. V. 1990. The male reproductive cycle of the rough green snake (*Opheodrys aestivus*). Amphibia–Reptilia 11: 165-172.

Aldridge, R. D., Goldberg, S. R., Wisniewski, S. S., Bufalino, A. P. and Dillman, C. B. 2009. The reproductive cycle and estrus in the colubrid snakes of temperate North America, Contemporary Herpetology 2009: 1-31.

Almeida-Santos, S. M., Abdalla, F. M. F., Silveira, P. F., Yamanouye, N., Breno, M. C. and Salomão, M. G. 2004. Reproductive cycle of the neotropical *Crotalus durissus terrificus*: I. Seasonal levels and interplay between steroid hormones and vasotocinase. General and Comparative Endocrinology 139: 143-150.

Bailey, F. C., Cobb, V. A., Rainwater, T. R., Worrall, T. and Klukowski, M. 2009. Adrenocortical effects of human encounters on free-ranging cottonmouths (*Agkistrodon piscivorus*). Journal of Herpetology 43: 260-266.

Ballard, P. L. and Ballard, R. A. 1972. Glucocorticoid receptors and the role of glucocorticoids in fetal lung development. Proceedings of the National Academy of Science 69: 2668-2672.

Benner, S. L. and Woodley, S. K. 2007. The reproductive pattern of male dusky salamanders (genus *Desmognathus*) is neither associated nor dissociated. Hormones and Behavior 51: 542-547.

Bragdon, D. E. 1951. The non-essentiality of the corpora lutea for the maintenance of gestation in certain live-bearing snakes. Journal of Experimental Zoology 118: 419-435.

Beaupre, S. J. 2002. Modeling time-energy allocation in vipers: individual responses to environmental variation and implications for populations. Pp. 463-481. In G. W. Schuett, M. Hoggren, M. E. Douglas and H. W. Greene (eds), *Biology of the Vipers*. Eagle Mountain Publishing, Eagle Mountain, Utah.

Beaupre, S. J. and Duvall, D. 1998. Variation in oxygen consumption of the western diamondback rattlesnake (*Crotalus atrox*): implications for sexual size dimorphism. Journal of Comparative Physiology B 168: 497-506.

Bishop, J. E. 1959. A histological and histochemical study of the kidney tubule of the common garter snake, *Thamnophis sirtalis*, with special reference to the sexual segment in the male. Journal of Morphology 104: 307-358.

Bonnet, X., Naulleau, G. and Mauget, R. 1994. The influence of body condition on 17-beta estradiol levels in relation to vitellogenesis in female *Vipera aspis* (Reptilia, Viperidae). General and Comparative Endocrinology 93: 424-37.

Bonnet, X., Naulleau, G., Bradshaw, D. and Shine, R. 2001. Changes in plasma progesterone in relation to vitellogenesis and gestation in the viviparous snake *Vipera aspis*. General and Comparative Endocrinology 121: 84-94.

Borell, M. 1985. Organotherapy and the emergence of reproductive endocrinology. Journal of the History of Biology 18: 1-30.

Callard, I. P., Fileti, A. L., Perez, L. E., Sobrera, L. A., Giannoukos, G., Klosterman, P. T. and McCracken, J. A. 1992. Role of the corpus luteum and progesterone in the evolution of vertebrate viviparity. American Zoologist 32: 264-275.

Callard, I. P., Giannoukos, G., Charnock-Jones, D. S., Benson, S. and Paolucci, M. 1994. Hormone regulation of vitellogenin genes and the evolution of viviparity. Pp. 325-332. In K. G. Davey, R. E. Peter and S. S. Tobe (eds), *Perspectives in Comparative Endocrinology*. National Research Council of Canada, Ottawa.

Camazine, B., Garstka, W., Tokarz, R. and Crews, D. 1980. Effects of castration and androgen replacement on male courtship behavior in the red-sided garter snake (*Thamnophis sirtalis parietalis*). Hormones and Behavior 14: 358-372.
Cease, A. J., Lutterschmidt, D. I. and Mason, R. T. 2007. Corticosterone and the transition from courtship behavior to dispersal in male red-sided garter snakes (*Thamnophis sirtalis parietalis*). General and Comparative Endocrinology 150: 124-131.
Crews, D. 1976. Hormonal control of male courtship behavior and female attractivity in the garter snake (*Thamnophis sirtalis parietalis*). Hormones and Behavior 7: 451-460.
Crews, D. 1984. Gamete production, sex hormone secretion, and mating behavior uncoupled. Hormones and Behavior 18: 22-28.
Crews, D. 1991. Trans-seasonal action of androgen in the control of spring courtship behavior in male red-sided garter snakes Proceedings of the National Academy of Sciences 88: 3545-3548.
Crews, D. and Garstka, W. 1982. The ecological physiology of a garter snake. Scientific American 247: 158-168.
Crews, D., Camazine, B., Diamond, M., Mason, R., Tokarz, R. R. and Garstka, W. R. 1984. Hormonal independence of courtship behavior in the male garter snake. Hormones and Behavior 18: 29-41.
Crews, D., Robker, R. and Mendonça, M. 1993. Seasonal fluctuations in brain nuclei in the red-sided garter snake and their hormonal control. Journal of Neuroscience 13: 5356-5364.
Custodia-Lora, N. and Callard, I. P. 2002. Progesterone and progesterone receptors in reptiles. General and Comparative Endocrinology 127: 1-7.
Duvall, D., Arnold, S. J. and Schuett, G. W. 1992. Pitviper mating systems: Ecological potential, sexual selection, and microevolution. Pp. 321–336. In J. A. Campbell and E. D. Brodie, Jr. (eds), *Biology of the Pitvipers*. Selva, Tyler, Texas.
Farner, D. S. 1987. Generation and regulation of annual cycles in migratory passerine birds. American Zoologist 26: 493-501.
Fraenkel, L., Martins, T. and Mello, R. F. 1940. Studies on the pregnancy of viviparous snakes. Endocrinology 27: 836-837.
Garland, T., Jr. and Adolph, S. C. 1994. Why not to do two-species comparative studies: limitations on inferring adaptation. Physiological Zoology 67: 797-828.
Garstka, W. R., Camazine, B. and Crews, D. 1982. Interactions of behavior and physiology during the annual reproductive cycle of the red-sided garter snake (*Thamnophis sirtalis parietalis*). Herpetologica 38: 104-123.
Garstka, W. R., Tokartz, R. R., Diamond, M., Halpert, A. and Crews, D. 1985. Behavioral and physiological control of yolk synthesis and deposition in the female red-sided garter snake (*Thamnophis sirtalis parietalis*). Hormones and Behavior 19: 137-153.
Gorman, G. C., Licht, P. and McCollum, F. 1981. Annual reproductive patterns in three species of marine snakes from the central Philippines. Journal of Herpetology 15: 335-354.
Graham, S. P., Earley, R. L., Hoss, S. K., Schuett, G. W. and Grober, M. S. 2008. The reproductive biology of male cottonmouths (*Agkistrodon piscivorus*): Do plasma steroid hormones predict the mating season? General and Comparative Endocrinology 159: 226-235.
Greenberg, N., and Wingfield, J. C. 1987. Stress and reproduction: Reciprocal relationships. Pp. 461–503. In D. O. Norris and R. E. Jones (eds), *Hormones and Reproduction in Fishes, Amphibians, and Reptiles*. Plenum Press, New York.

Haldar, C. and Pandey, R. 1989. Effect of pinealectomy on annual testicular cycle of Indian chequered water snake, *Natrix piscator*. General and Comparative Endocrinology 76: 214-222.

Ho, S. M., Kleis-San Francisco, S., McPherson, R., Heiserman, G. J. and Callard, I. P. 1982. Regulation of vitellogenesis in reptiles. Herpetologica 38: 40-50.

Integrated Taxonomic Information System (www.itis.gov/index.html). Retrieved 6 February 2010.

Jacobs, J. D. and Wingfield, J. C. 2000. Endocrine control of life cycle stages: A constraint on response to the environment? The Condor 102: 35-51.

Johnson, L. F., Jacob, J. S. and Torrance, P. 1982. Annual testicular and androgenic cycles of the cottonmouth (*Agkistrodon piscivorus*) in Alabama. Herpetologica 38: 16-25.

Krohmer, R. W. 2004. The male red-sided garter snake (*Thamnophis sirtalis parietalis*): reproductive pattern and behavior. Institute of Laboratory Animal Resources Journal 45: 54-74.

Krohmer, R. W., Grassman, M. and Crews, D. 1987. Annual reproductive cycle in the male red-sided garter snake, *Thamnophis sirtalis parietalis*: Field and laboratory studies. General and Comparative Endocrinology 68: 64-75.

Lee, M. S. Y., Hugall, A. F., Lawson, R. and Scanlon, J. D. 2007. Phylogeny of snakes (Serpentes): combining morphological and molecular data in likelihood Bayesian and parsimony analyses. Systematics and Biodiversity 5: 371-389.

Lind, C. M., Husak, J. F., Eikenaar, C., Moore, I. and Taylor, E. N. 2010. The relationship between plasma steroid hormone concentrations and the reproductive cycle in the Northern Pacific Rattlesnake, *Crotalus oreganus*. General and Comparative Endocrinology 166: 590-599.

Lutterschmidt, D. I. and Mason, R. T. 2005. A serotonin receptor antagonist, but not melatonin, modulates hormonal responses to capture stress in two populations of garter snakes (*Thamnophis sirtalis parietalis* and *Thamnophis sirtalis concinnus*). General and Comparative Endocrinology 141: 259-270.

Lutterschmidt, D. I., LeMaster, M. P. and Mason, R. T. 2004. Effects of melatonin on the behavioral and hormonal responses of red-sided garter snakes (*Thamnophis sirtalis parietalis*) to exogenous corticosterone. Hormones and Behavior 46: 692-702.

Lutterschmidt, W. I., Lutterschmidt, D. I., Mason, R. T. and Reinert, H. K. 2009. Seasonal variation in hormonal responses of timber rattlesnakes (*Crotalus horridus*) to reproductive and environmental stressors. Journal of Comparative Physiology B 179: 747–757.

Mathies, T., Felix, T. A. and Lance, V. A. 2001. Effects of trapping and subsequent short-term confinement stress on plasma corticosterone in the brown treesnake (*Boiga irregularis*) on Guam. General and Comparative Endocrinology 124: 106-114.

Mead, R. A., Eroschenko, V. P., and Highfill, D. R. 1981. Effects of progesterone and estrogen on the histology of the oviduct of the garter snake, *Thamnophis elegans*. General and Comparative Endocrinology 45: 345-354.

Mendonça, M. T. and Crews, D. 1989. Effect of fall mating on ovarian development in the red-sided garter snake. American Journal of Physiology: Integrative and Comparative Physiology 257: 1548-1550.

Mendonça, M. T. and Crews, D. 1990. Mating-induced ovarian recrudescence in the red-sided garter snake. Journal of Comparative Physiology A 166: 629-632.

Mendonça, M. T. and Crews, D. 1996. Effects of ovariectomy and estrogen replacement on attractivity and receptivity in the red-sided garter snake (*Thamnophis sirtalis parietalis*). Journal of Comparative Physiology A 178: 373-381.

Moore, I. T. and Mason, R. T. 2001. Behavioral and hormonal responses to corticosterone in the male red-sided garter snake, *Thamnophis sirtalis parietalis*. Physiology and Behavior 72: 669-674.
Moore, I. T. and Jessop, T. S. 2003. Stress, reproduction, and adrenocortical modulation in amphibians and reptiles. Hormones and Behavior 43: 39-47.
Moore, I. T., LeMaster, M. P. and Mason, R. T. 2000a. Behavioural and hormonal responses to capture stress in the male red-sided garter snake, *Thamnophis sirtalis parietalis*. Animal Behaviour 59: 529-534.
Moore, I. T., Lerner, J. P., Lerner, D. T. and Mason, R. T. 2000b. Relationships between annual cycles of testosterone, corticosterone, and body condition in male red-spotted garter snakes, *Thamnophis sirtalis concinnus*. Physiological and Biochemical Zoology 73: 307-312.
Moore, I. T., Greene, M. J. and Mason, R. T. 2001. Environmental and seasonal adaptations of the adrenocortical and gonadal responses to capture stress in two populations of the male garter snake, *Thamnophis sirtalis*. Journal of Experimental Zoology 289: 99-108.
Moore, I. T., Greene, M. J., Lerner, D. T., Asher, C. E., Krohmer, R. W., Hess, D. L., Whittier, J. and Mason. R. T. 2005. Physiological evidence for reproductive suppression in the introduced population of brown tree snakes (*Boiga irregularis*) on Guam. Biological Conservation 121: 91-98.
Moore, M. C. and Lindzey, J. 1992. The physiological basis of sexual behaviour in male reptiles. Pp. 70-113. In C. Gans and D. Crews (eds), *Biology of the Reptilia*, Vol. 18. University of Chicago Press, Chicago, Illinois.
Naulleau, G. and Fleury, F. 1990. Changes in plasma progesterone in female *Vipera aspis* L. (Reptilia, Viperidae) during the sexual cycle in pregnant and nonpregnant females. General and Comparative Endocrinology 78: 433-443.
Naulleau, G., Fleury, F. and Boissin, J. 1987. Annual cycle in plasma testosterone and thyroxine in the male aspic viper *Vipera aspis* L. (Reptilia, Viperidae) in relation to the sexual cycle and hibernation. General and Comparative Endocrinology 65: 254-263.
Padgoankar, A. S. and Samuel, J. 1993. Effects of administration of follicle stimulating hormone (FSH) on the regressed testes of the snake *Acrochordus granulatus* (Schneider). Journal of Advanced Zoology 14: 44-47.
Perrins, C. M. 1970. The timing of birds breeding seasons. Ibis 112: 242-255.
Perrins, C. M. and Birkhead, T. R. 1983. *Avian Ecology*. Blackie, London, U.K. Pp. 221.
Robert, K. A., Vleck, C. and Bronikowski, A. M. 2009. The effects of maternal corticosterone levels on offspring behavior in fast- and slow-growth garter snakes (*Thamnophis elegans*). Hormones and Behavior 55: 24-32.
Saint Girons, H. 1982. Reproductive cycles of male snakes and their relationships with climate and female reproductive cycles. Herpetologica 38: 5-16.
Saint Girons, H., Bradshaw, D. and Bradshaw, F. J. 1993. Sexual activity and plasma levels of sex steroids in the aspic viper *Vipera aspis* L. (Reptilia, Viperidae). General and Comparative Endocrinology 91: 287-297.
Schuett, G. W., Harlow, H. J., Rose, J. D., Van Kirk, E. A. and Murdoch, W. J. 1997. Annual cycle of plasma testosterone in male copperheads, *Agkistrodon contortrix* (Serpentes, Viperidae): Relationship to timing of spermatogenesis, mating, and agonistic behavior. General and Comparative Endocrinology 105: 417-424.
Schuett, G. W., Carlisle, S. L., Holycross, A. T., O'Leile, J. K., Hardy, D. L. Sr., Van Kirk, E. A. and Murdoch, W. J. 2002. Mating system of male Mojave rattlesnakes (*Crotalus scutulatus*): seasonal timing of mating, agonistic behavior, spermatogenesis, sexual segment of the kidney, and plasma sex steroids. Pp. 515-532. In

G. W. Schuett, M. Hoggren, M. E. Douglas and H. W. Greene (eds), *Biology of the Vipers*. Eagle Mountain Publishing, Eagle Mountain, Utah.

Schuett, G. W., Hardy, D. L., Sr., Greene, H. W., Earley, R. L., Grober, M. S., Van Kirk, E. A. and Murdoch, W. J. 2005. Sympatric rattlesnakes with contrasting mating systems show differences in seasonal patterns of plasma sex steroids. Animal Behaviour 70: 257-267.

Schuett, G. W., Repp, R. A., Taylor, E. N., DeNardo, D. F., Early, R. L., Van Kirk, E. A. and Murdoch, W. J. 2006. Winter profile of plasma sex steroid levels in free-living male western diamond-backed rattlesnakes, *Crotalus atrox* (Serpentes: Viperidae). General and Comparative Endocrinology 149: 72-80.

Seigel, R. A. and Ford, N. B. 1987. Reproductive ecology. Pp. 210–252. In R. A. Seigel, J. T. Collins and S. S. Novak (eds), *Snakes: Ecology and Evolutionary Biology*. McGraw-Hill, New York.

Smith, C. F., Schuett, G. W. and Schwenk, K. 2010. Relationship of plasma sex steroids to the mating season of copperheads at the north-eastern extreme of their range. Journal of Zoology. 280: 362-370.

Soma, K. K. 2006. Testosterone and aggression: Berthold, birds and beyond. Journal of Neuroendocrinology 18: 543-551.

Sparkman, A. M., Vleck, C. M. and Bronikowski, A. M. 2009. Evolutionary ecology of endocrine-mediated life-history variation in the garter snake *Thamnophis elegans*. Ecology 90: 720-728.

Taylor, E. N. and DeNardo, D. F. 2005. Reproductive ecology of Western Diamond-backed Rattlesnakes (*Crotalus atrox*) in the Sonoran Desert. Copeia 2005: 152-158.

Taylor, E. N. and DeNardo, D. F. 2010. Hormones and reproductive cycles in snakes. Pp. 355-372. In D. O. Norris and K. H. Lopez (eds), *Hormones and Reproduction in Vertebrates*, Vol. 3, *Reptiles*. Academic Press, San Diego, California.

Taylor, E. N., DeNardo, D. F. and Jennings, D. H. 2004. Seasonal steroid hormone levels and their relation to reproduction in the Western Diamond-backed Rattlesnake, *Crotalus atrox* (Serpentes: Viperidae). General and Comparative Endocrinology 136: 328-337.

Tsai, T. and Tu, M. 2001. Reproductive cycle of female Chinese green tree vipers, *Trimeresurus stejnegeri stejnegeri*, in northern Taiwan. Herpetologica 57: 157-168.

Waye , H. L. and Mason, R. T. 2008. A combination of body condition measurements is more informative than conventional condition indices: Temporal variation in body condition and corticosterone in brown tree snakes (*Boiga irregularis*). General and Comparative Endocrinology 155: 607-612.

Weil, M. R. and Aldridge, R. D. 1981. Seasonal androgenesis in the male water snake *Nerodia sipedon*. General and Comparative Endocrinology 44: 44-53.

Whittier, J. M. and Crews, D. 1987. Seasonal reproduction: Patterns and control. Pp. 385-409. In D. O. Norris and R. E. Jones (eds.), *Hormones and Reproduction in Fishes, Amphibians, and Reptiles*. Plenum Press, New York.

Whittier, J. M. and Crews, D. 1989. Mating increases plasma levels of prostaglandin F2 alpha in female garter snakes. Prostaglandins 37: 359-367.

Whittier, J. M., Mason, R. T. and Crews, D. 1987. Plasma steroid hormone levels of female red-sided garter snakes *Thamnophis sirtalis parietalis*: Relationship to mating and gestation. General and Comparative Endocrinology 67: 33-43.

Wilson, F. E. and Donham, R. S. 1988. Daylength and control of seasonal reproduction in male birds. Pp. 101-120. In M. H. Stetson (ed.), *Processing of Environmental Information in Vertebrates*. Springer-Verlag, Berlin, Germany.

Wingfield, J. C. 1980. Fine temporal adjustments of reproductive function. Pp. 367-389. In A. Epple and M. H. Stetson (eds), *Avian Endocrinology*. Academic Press, New York.
Wingfield, J. C. 1983. Environmental and endocrine control of reproduction: An ecological approach. Pp. 265-288. In S. I. Mikami, K. Homma, and M. Wada (eds), *Avian Endocrinology: Environmental and Ecological Perspectives*. Springer-Verlag, Berlin, Germany.
Wingfield, J. C. and Kitaysky, A. S. 2002. Endocrine responses to unpredictable environmental events: Stress or anti-stress hormones? Integrative and Comparative Biology 42: 600–609.

CHAPTER 8

Environmental and Neuroendocrine Control of Reproduction in Snakes

Randolph W. Krohmer[1] *and Deborah I. Lutterschmidt*[2]

8.1 INTRODUCTION

There is a vast literature on the neural and hormonal regulation of reproduction in mammals and birds (for review, see Becker *et al.* 2002; Pfaff *et al.* 2002; Nelson 2005). In comparison, there are relatively few studies investigating the neuroendocrine control of reproduction in reptiles, and of these studies, the majority have been conducted in lizards and turtles. Unfortunately for this chapter, there is a paucity of data on the neuroendocrine regulation of reproduction in ophidians. Thus, to provide a more complete context for understanding the neuroendocrinology of snake reproduction, we will include, where appropriate, a comparative analysis that incorporates some of the existing literature on other vertebrate species. Because many aspects of reproductive control are likely conserved across taxonomic groups, particularly in ectothermic vertebrates, we hope that this comparative analysis provides a much needed synthesis of our current understanding of how snake reproduction is regulated by interactions between environmental and neuroendocrine signals.

Figure 8.1 depicts our working conceptual model of the neuroendocrine regulation of reproduction. Consequently, this model serves as an outline for the chapter's content. We first address the environmental cues known to be involved in the neuroendocrine control of reproduction in snakes. We then review the influence of environmental cues on neuroendocrine signaling followed by the impact of these neuroendocrine signals on reproductive physiology and behavior. The latter portion of the chapter reviews the literature addressing the role of neural pathways, neuropeptides, and neuromodulators in regulating snake reproduction.

[1]Department of Biological Sciences, Saint Xavier University, 3700 W. 103rd Street, Chicago, IL 60655, U.S.A.
[2]Department of Biology, Portland State University, P.O. Box 751, Portland, OR, 97207, U.S.A.

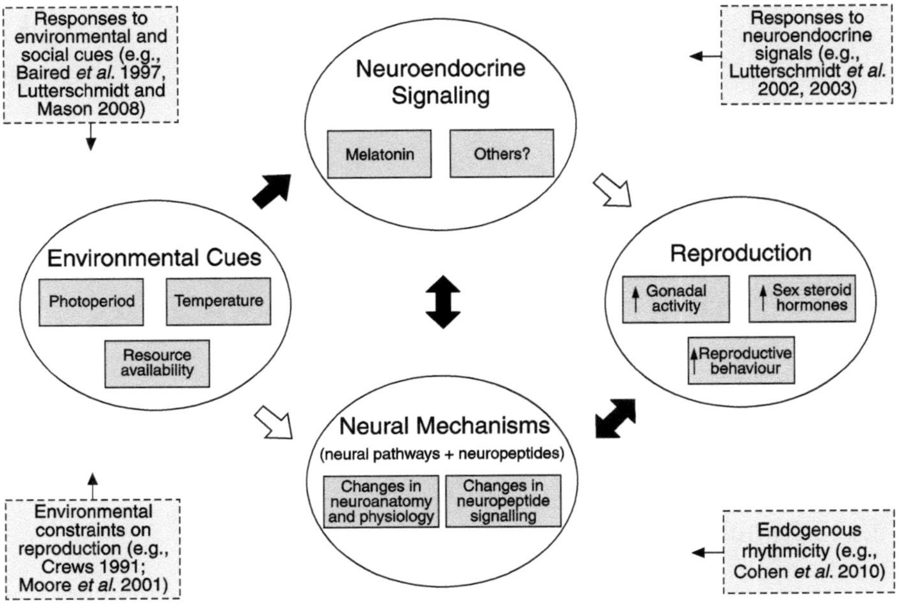

Fig. 8.1 Conceptual model illustrating the neuroendocrine regulation of reproduction. Importantly, the model includes specific examples of how variation in neuroendocrine mechanisms might contribute to evolutionary differences in the timing and patterns of reproductive function both within snakes and among vertebrates (see boxes with hashed borders). Thick black arrows denote primary neuroendocrine pathways. White (open) arrows highlight potential direct effects of environmental cues on neuroanatomy and physiology and of neuroendocrine signaling on reproduction (e.g., photoperiodic induction of reproduction in birds and receptor-mediated effects of melatonin in the gonads, respectively). Bidirectional arrows indicate that changes in brain-behavior relationships are mutually reinforcing (Wilczynski *et al.* 2005). Note that while both resource availability and energy reserves are known to be potent modulators of reproduction in snakes (see review in Shine 2003), there are no experimental studies addressing the neuroendocrine mechanisms by which these extrinsic factors are relayed to and modulate the reproductive axis.

An important feature of Figure 8.1 is that it indicates where variation in neuroendocrine mechanisms may have contributed to evolutionary differences in the timing and patterns of reproductive function both within snakes and among vertebrates. For example, although plasma melatonin is elevated during the scotophase (i.e., dark phase of the photoperiod) in all vertebrates studied, some nocturnal ectotherms do not respond to this neuroendocrine signal in the same manner as diurnal ectotherms (Lutterschmidt *et al.* 2002, 2003). This is just one example of how differences in the sensitivity to neuroendocrine signals may underlie variation in life history characteristics. Unfortunately, the literature on the neuroendocrinology of snake reproduction is so limited that this review is necessarily biased toward studies of one particular snake population, Red-sided Garter Snakes (*Thamnophis sirtalis parietalis*) in

Manitoba, Canada. We emphasize here and throughout the chapter the importance of a comparative approach to understanding reproductive regulatory mechanisms, particularly with regard to variation in life history characteristics. Future studies directed at understanding the neuroendocrine mechanisms underlying differences in the timing of reproduction in sympatric species, for example, will provide a great deal of insight into the evolution of reproductive regulation.

This chapter focuses exclusively on the neural and neuroendocrine mechanisms thought to play a role in regulating reproduction. A description of the endocrine correlates of reproduction is beyond the scope of this chapter but can be found in **Chapter 7** of this text (DeNardo and Taylor 2010). Obviously there will be areas of continuity between the neuroendocrine regulation of reproduction described here and the endocrine control described in **Chapter 7** (DeNardo and Taylor 2010). To avoid redundancy, we will not discuss the role of endocrine factors (e.g., plasma sex steroid hormones, glucocorticoids) in regulating reproduction. Of course, as the distinction between endocrinology and neuroendocrinology becomes increasingly blurred, it will become concomitantly more difficult to make such distinctions between regulatory factors. It is our hope that the combination of these chapters reviewing the endocrine and neuroendocrine control of snake reproduction will stimulate future research into this under-explored area of regulatory biology.

8.1.1 Patterns of Reproductive Activity

Of the relatively few studies examining the neuroendocrine control of reproduction in ophidians, the vast majority have focused on one species, the Common Garter Snake (*Thamnophis sirtalis*). Within this species, the available studies are further limited almost exclusively to one geographic population of snakes: *Thamnophis sirtalis parietalis* in Manitoba, Canada. Because of the extreme environmental constraints this population has adapted to in this habitat, there are some aspects of its reproductive biology that may differ from other snakes. For example, in most snake species studied to date, increased sex steroid hormones are associated with seasonal activation of reproductive function (e.g., **Chapter 7**, DeNardo and Taylor 2010). In contrast, *T. s. parietalis* is one of the best-studied examples of a dissociated reproductive pattern, in which mating behavior does not coincide with peak gonadal activity (reviewed in Woolley *et al.* 2004). Because this chapter is necessarily biased toward studies of *T. s. parietalis*, we feel it is important to briefly highlight the differences in the reproductive pattern of *T. s. parietalis* and review the evidence that that this population exhibits dissociated reproduction. Furthermore, we provide a synthesis of these studies with the most current data and hypotheses regarding how reproduction is regulated in *T. s. parietalis*. Although continuing to advance our understanding of reproductive regulation in this extremely well-studied model system is necessary, future research comparing the reproductive regulatory mechanisms across taxa with different life history strategies will provide a valuable context with which the neuroendocrine control of reproduction may be better understood.

Associated reproductive cycles. The majority of seasonally breeding vertebrates exhibit an associated reproductive pattern, mating at a time when gonads are maximally active and sex steroid hormone levels are elevated (Crews 1984). In these species, castration eliminates sexual activity while reimplantation of the testes into the body cavity or administration of exogenous sex steroids by injection or implantation directly into the brain extends the breeding season or initiates reproductive activity in non-courting individuals (Crews 1982). In reptiles, the activational role of sex steroid hormones has been most extensively studied in the Green Anole (*Anolis carolinensis*). In *A. carolinensis*, rising ambient temperatures induce emergence from winter dormancy and initiate gonadal development (Crews 1975). As spermatogenesis and steroidogenesis progress, the concentration of circulating androgens increase to their highest levels, resulting in males establishing territories and vigorously courting receptive females. Consequently, there is a close temporal association between peak gonadal activity (spermatogenesis and steroidogenesis) and courtship behavior and mating in most seasonally breeding reptiles (e.g., **Chapter 7,** DeNardo and Taylor 2010).

Dissociated reproductive cycles. A few vertebrate species, including *Thamnophis sirtalis parietalis*, mate at a time when the gonads are quiescent and circulating levels of sex steroid hormones are reportedly low (see reviews in Licht 1984 and Woolley *et al.* 2004). The initiation of courtship behavior and mating in adult male *T. s. parietalis* has been reported to be independent of testicular or pituitary hormone control (Camazine *et al.* 1980). Castration of *T. s. parietalis* does not eliminate sexual activity (Crews 1984; Crews *et al.* 1984). In addition, systemic administration of sex steroids, hypothalamic or pituitary hormones, or implantation of sex steroid hormones directly into the anterior hypothalamus-preoptic area (AHPOA) does not extend the breeding season nor will it initiate reproductive behaviors in non-courting individuals (Camazine, *et al.* 1980; Crews *et al.* 1984; Friedman and Crews 1985a; reviewed in Woolley *et al.* 2004). Moreover, males continue to exhibit courtship behavior for up to three years following castration or hypophysectomy (Crews 1991). Thus, the seasonal activation of sexual behavior in male *T. s. parietalis* appears to be dissociated from changes in gonadal development as well as maximal sex steroid hormone synthesis.

Thamnophis sirtalis parietalis is the most extensively studied species exhibiting this unusual reproductive pattern. The range of *T. s. parietalis* extends farther north than any other reptile in the western hemisphere (Logier and Toner 1961). This species has been able to adapt successfully to extreme environmental conditions, surviving winter temperatures that reach –40°C and snow cover that can remain for as long as nine months (Aleksiuk 1970, 1971; Aleksiuk and Stewart 1971). The environment inhabited by northern populations of *T. s. parietalis* in Manitoba, Canada offers very restricted periods of activity each year. Immediately upon emergence from winter dormancy, *T. s. parietalis* displays intense breeding activity lasting approximately 4-5 weeks. During this time, the testes of *T. s. parietalis* are completely regressed and spermatogenesis is not initiated until the end of the breeding season (Fig. 8.2; Crews *et al.* 1984).

Although mating behavior of *Thamnophis sirtalis parietalis* does not coincide with peak gonadal activity, this does not necessarily preclude a role for sex steroid hormones in regulating reproductive behavior. Similar to the suggestions of Crews (1991) and Saint Girons *et al.* (1993), our current hypothesis is that elevated androgen concentrations during the fall and winter dormancy period, in combination with low temperature exposure, play a role in inducing the changes in neuroanatomy and neurophysiology necessary to elicit reproductive behavior in the spring (Krohmer *et al.* 1987; Krohmer 2004; Lutterschmidt 2006; Lutterschmidt and Mason 2009).

Initial field studies in northern populations of *Thamnophis sirtalis parietalis* indicated that circulating androgens were very low or absent

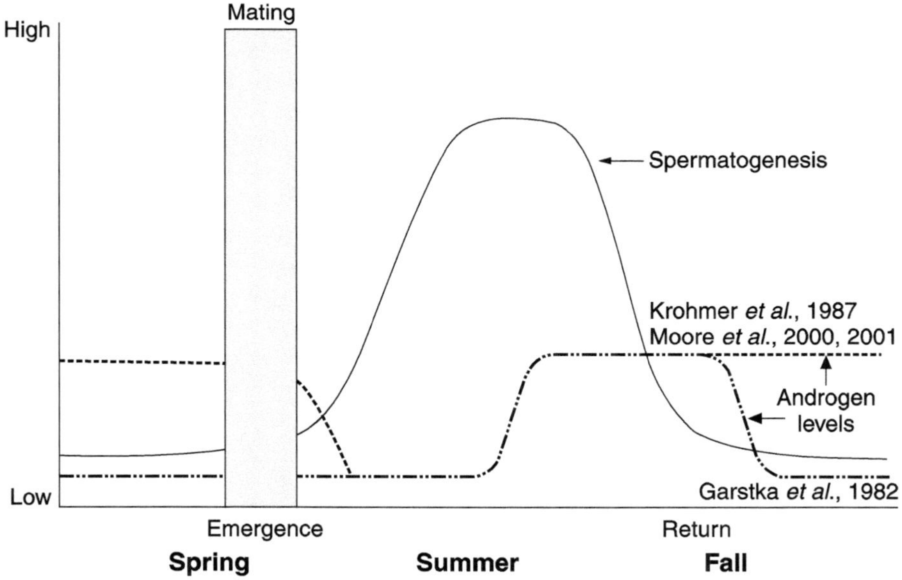

Fig. 8.2 Seasonal pattern of reproductive activity, spermatogenesis and steroidogenesis in male *Thamnophis sirtalis parietalis*, a species exhibiting a dissociated reproductive pattern. Courtship behavior and mating is initiated upon emergence from winter dormancy when gonadal activity is quiescent. Androgen levels were initially reported to be low as the animals enter dormancy and remain low at the initiation of the breeding season (Garstka *et al.* 1982). Subsequently, spermatogenesis and steroidogenesis are not initiated until the end of the breeding season. Later studies have shown that androgen levels, elevated as the animals enter low temperature dormancy, remain elevated throughout dormancy and can be elevated upon emergence and the initiation of courtship and mating (Krohmer *et al.* 1987; Moore *et al.* 2000, 2001). Drawing by L. Kostovich.

upon emergence from winter dormancy (Fig. 8.2; Crews and Garstka 1982; Garstka *et al.* 1982). However, subsequent investigations found that circulating androgens, elevated in the fall prior to winter dormancy, remain elevated throughout low temperature dormancy (LTD; Krohmer *et al.* 1987; Lutterschmidt and Mason 2009) and are not basal upon emergence in the spring (Krohmer *et al.* 1987; Moore *et al.* 2000, 2001; Lutterschmidt and Mason 2005). More recently, Lutterschmidt and Mason (2009) demonstrated that androgen concentrations are elevated during the fall and decline significantly faster in snakes hibernated at warmer temperatures (i.e., 10 versus 5°C; Fig. 8.3).

Thus, androgen concentrations appear to decline during winter dormancy via metabolic clearance. These results indicate that the observed annual variation in spring androgen levels may be related to variation in environmental conditions, particularly temperature profiles (Lutterschmidt and Mason 2009).

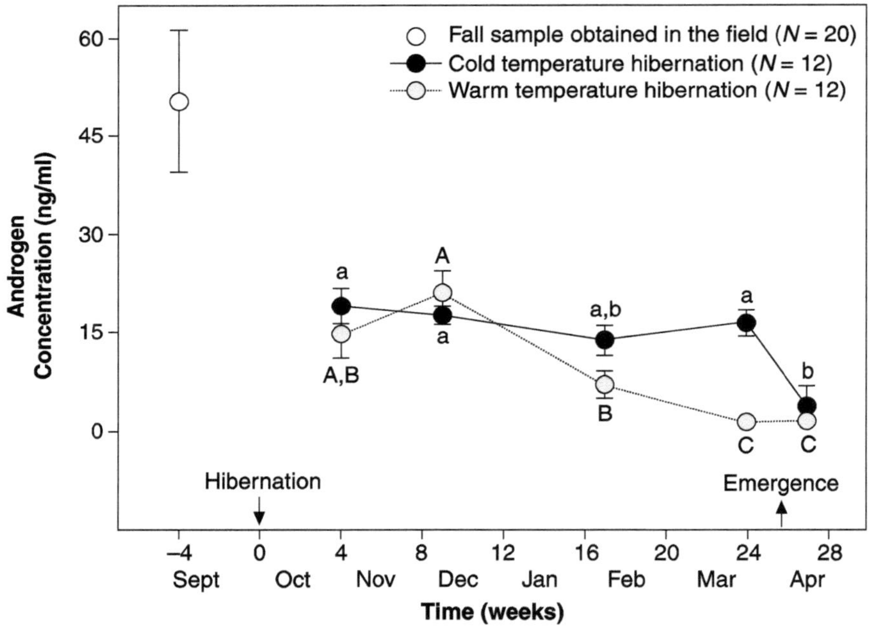

Fig. 8.3 Influence of elevated hibernation temperatures on androgen concentrations of male *Thamnophis sirtalis parietalis* during winter dormancy. Fall pre-hibernation steroid concentrations were determined from a randomly-selected subset of males (n = 20) sampled immediately upon capture in the field. Each subsequent data point is the mean hormone concentration ± 1 s.e.m. of snakes in response to cold (5°C; n = 12) and warm (10°C; n = 12) hibernation temperatures. Within the cold temperature hibernation treatment (black symbols), differences among sampling periods are indicated by lower-case letters near the error bars, while those within the warm temperature hibernation treatment (grey symbols) are indicated by capital letters (results from Tukey's multiple comparisons test). Induction of hibernation and spring emergence are indicated by arrows along the abscissa. Modified from Lutterschmidt, D. I. and Mason, R. T. 2009. Journal of Experimental Biology 212: 3108-3118, Fig. 4A.

In summary, we propose that the reproductive behaviour of *Thamnophis sirtalis parietalis* is merely temporally dissociated from peak gonadal steroid synthesis and gametogenesis. While the reproductive pattern of *T. s. parietalis* is indeed unusual relative to other reptiles, additional studies of this snake population might reveal novel neuroendocrine mechanisms underlying behavior. Moreover, studies examining reproductive regulation in extreme habitats may provide insight into the neuroendocrine mechanisms that govern all seasonal breeders, again highlighting the importance of comparative studies for understanding the evolution of regulatory mechanisms.

8.2 ENVIRONMENTAL CUES AND NEUROENDOCRINE CONTROL OF REPRODUCTION

In most vertebrates, the neuroendocrine control of reproductive physiology and behavior operates in concert with environmental control. Seasonal breeders must track changing environmental signals so that physiological and behavioral processes may be synchronized with optimal environmental conditions. Such "timekeeping" optimizes both individual survival as well as reproductive fitness. Thus, it is important to understand how environmental cues are transduced into neuroendocrine signals and how these signals in turn integrate an animal's reproductive physiology and behavior with its environment.

Understanding how a particular physiological and/or behavioral parameter is regulated will require a broad perspective that examines how environmental cues and neuroendocrine signals interact to regulate different (and often temporally isolated) life history phases. For example, many seasonally breeding ectotherms from diverse environments undergo periods of prolonged winter dormancy prior to spring mating. Despite this fact, relatively little is known about the environmental and hormonal mechanisms regulating entrance into or emergence from winter dormancy. Most intriguing is that in species where reproduction occurs immediately following spring emergence, the concomitant changes in neurophysiology and behavior that accompany reproduction are likely to occur during winter dormancy. Thus, significant changes in reproductive physiology and behavior may occur during this "dormancy" period, and the seasonal control of reproduction is therefore likely linked to the environmental and hormonal mechanisms that control the timing of winter dormancy.

For these reasons, we provide here a brief synopsis of the studies investigating the role of environmental cues, particularly temperature, in regulating spring emergence and seasonal reproduction in snakes. We then discuss how these environmental cues may be transduced into neuroendocrine signals and the influence of these signals on reproductive physiology and behavior. Finally, we present the evidence for geographic variation in the transduction of environmental cues into neuroendocrine signals and discuss the significance of such variation to the stability of animal populations, particularly with regard to environmental perturbations such as global climate change.

8.2.1 Environmental Cues

One of the most reliable environmental cues thought to function in regulating seasonality is photoperiod. Unlike other environmental signals (e.g., temperature and humidity) that can vary dramatically both within seasons and among years, changes in photoperiod length accurately and reliably reflect changing environmental seasons. Many ectotherms inhabiting north-temperate climates, however, undergo periods of prolonged winter dormancy prior to spring breeding. Animals that occupy underground hibernacula during winter dormancy are not exposed, or receive little exposure, to changing photoperiodic conditions (e.g., Whittier *et al.* 1987; Grobman 1990). Thus, photoperiod is likely not a critical factor in synchronizing spring emergence from over-wintering locations. For example, photoperiod prior to and during winter dormancy has no influence on the initiation and timing of reproductive behavior of *Thamnophis sirtalis parietalis* upon spring emergence (Nelson *et al.* 1987; Whittier *et al.* 1987). However, future studies comparing the influence of photoperiod and temperature on reproduction across different snake taxa and life histories (e.g., temperate versus tropical habitats) are needed.

In many ectothermic species, increases in ambient and ground temperatures during spring are thought to play a role in initiating emergence from winter dormancy and subsequent reproductive behavior (e.g., Hawley and Aleksiuk 1975, 1976; Jacob and Painter 1980; Crews and Garstka 1982; Licht 1984; Whittier *et al.* 1987; Macartney *et al.* 1989; Crawford 1991). For example, emergence from winter dormancy in the Eastern Box Turtle (*Terrapene carolina)* and the Ornate Box Turtle (*T. ornata)* occurred after subsurface ground temperatures increased for several consecutive days (Grobman 1990). Sexton and Marion (1981) demonstrated that emergence of Prairie Rattlesnakes (*Crotalus viridis*) from winter hibernacula is regulated by a reversing thermal gradient within natural dens. Emergence of Northern Pacific Rattlesnakes (*Crotalus oreganus oreganus*) also occurred as hibernaculum temperatures increased (Macartney *et al.* 1989). Etheridge *et al.* (1983) further demonstrated experimentally that increasing ambient temperatures stimulated emergence of the Six-lined Racerunner (*Cnemidophorus sexlineatus*) from winter dormancy. However, some ectothermic species (especially those inhabiting extreme northern latitudes) can occupy underground dens at depths where ground temperatures do not change significantly prior to spring emergence. For example, *Thamnophis sirtalis parietalis* in Alberta and Manitoba, Canada emerged at body temperatures as low as 0.5°C, indicating that increases in ground temperatures are not necessary for spring emergence in this species (Macartney *et al.* 1989; Lutterschmidt *et al.* 2006). In addition, body temperatures of *T. s. parietalis* measured in Lutterschmidt *et al.* (2006) increased more rapidly than ground temperatures during the period of spring emergence, suggesting that spring emergence of *T. s. parietalis* (and perhaps other ectothermic vertebrates) may be regulated, in part, by an endogenous circannual cycle that is modulated by environmental temperatures (Lutterschmidt *et al.* 2006).

Because of the close temporal relationship between spring emergence and reproduction in many ectothermic vertebrates, temperature also likely plays an important role in regulating reproduction. While this has not been extensively studied across snake taxa (e.g., see Fig. 8.1), temperature appears to be the primary environmental cue regulating reproduction in reptiles (Licht 1972, 1984; Duvall *et al.* 1982; Whittier *et al.* 1987). Increased environmental temperatures initiated gonadal development and steroidogenesis in *Anolis carolinensis* (Crews 1975). Mating behavior of *Thamnophis sirtalis parietalis* was also induced by increased environmental temperatures following winter dormancy (Ross and Crews 1978; Garstka *et al.* 1982; Bona-Gallo and Licht 1983; Krohmer and Crews 1987a; Whittier *et al.* 1987). Both male and female *T. s. parietalis* required a period of low temperature conditions to initiate sexual behavior upon return to warm temperatures (Camazine *et al.* 1980; Bona-Gallo and Licht 1983). This period of low temperature exposure must be at least 4 wks in duration to elicit courtship behavior (Garstka *et al.* 1982). Furthermore, the proportion of males exhibiting courtship behavior increased as the length of low temperature exposure increased (Fig. 8.4; Garstka *et al.* 1982). If low temperature exposure is required to "reset" the seasonal clock controlling the timing of reproduction, then what is transducing low temperature exposure?

8.2.2 The Pineal Gland and Melatonin: Neuroendocrine Transducers of Environmental Stimuli

The pineal gland and its major secretory product, melatonin (*N*-acetyl-5-methoxytryptamine), are the primary neuroendocrine transducers of environmental stimuli in vertebrates (e.g., Axelrod 1974; Binkley *et al.* 1978; Cassone 1990; Cassone and Natesan 1997). Because melatonin production is directly influenced by environmental cues such as photoperiod and temperature, melatonin rhythms form the neuroendocrine interface between extrinsic cues and intrinsic physiology and behavior. Although detectable levels of melatonin are present in the plasma during daylight, large increases in melatonin are observed during darkness (e.g., Underwood 1985a; Vivien-Roels *et al.* 1988; Moyer *et al.* 1995; Bolliet *et al.* 1996; García-Allegue *et al.* 2001). Nocturnal increases in plasma melatonin concentrations result primarily from increased synthesis and secretion of melatonin from the pineal gland. Pinealectomy abolished circadian rhythms of plasma melatonin in Desert Iguanas (*Dipsosaurus dorsalis*) and Ruin Lizards (*Podarcis sicula*) (Janik and Menaker 1990; Foà *et al.* 1992). However, extrapineal melatonin synthesis occurs at several sites in the body,

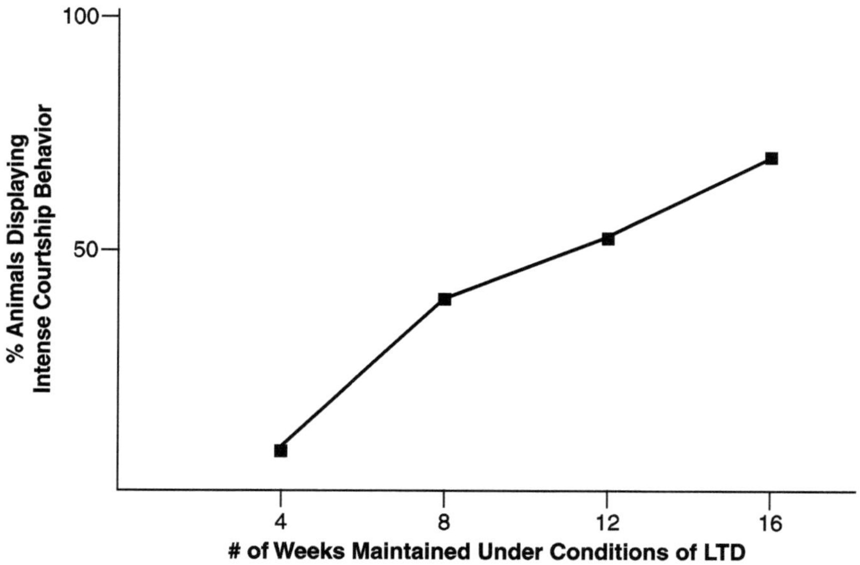

Fig. 8.4 Exposure to low temperatures (4±1.5°C) during dormancy increased the percent of animals expressing intense courtship behavior upon spring emergence in male *Thamnophis sirtalis parietalis* (N=20/group). Redrawn from Garstka, W. R., *et al.* 1982. Herpetologica 38: 104-123, Fig. 9.

including the Harderian gland (or Harder's lacrimal gland), retina, and intestine (Gern and Ralph 1979; Ralph 1980; Gern and Karn 1983; Hadley 1996; Norris 1997). Thus, detectable levels of circulating melatonin are often still present even after pinealectomy. For example, pinealectomy of neotenic Tiger Salamanders (*Ambystoma tigrinum*) reduced melatonin levels during scotophase by only 55% (Gern and Norris 1979). Similarly, photophasic and scotophasic plasma melatonin levels did not differ significantly between pinealectomized and sham-operated *Thamnophis sirtalis parietalis* (Mendonça *et al.* 1996a). Thus, it must be noted that extrapineal melatonin synthesis may significantly contribute to baseline levels of plasma melatonin in some species, with the pineal gland contributing to this baseline level in an additive manner during scotophase (Gern and Norris 1979).

Influence of environmental cues on melatonin rhythms. To accurately and reliably reflect prevailing environmental conditions, the synthesis and secretion of melatonin must be sensitive to external factors. Diel melatonin rhythms are in fact modulated by a number of extrinsic cues, including photoperiod phase and duration, light intensity, and temperature (Gern and Norris 1979; Birks and Ewing 1986; Wilson *et al.* 1986; Firth and Kennaway 1987, 1989; Underwood and Calaban 1987; Firth *et al.* 1989; Underwood and Hyde 1989; Rawding and Hutchison 1992; Reiter 1992, 1993; Hyde and Underwood 1993; Tilden and Hutchison 1993; Falcón *et al.* 1994; Lerchl *et al.* 1998; García-Allegue *et al.* 2001).

Because melatonin production increases during the scotophase, melatonin cycles provide a neuroendocrine index of time of day. Likewise, changes in the duration of the melatonin signal reflect annual changes in the photophase:scotophase ratio and therefore seasonal changes in day length. The duration of increased melatonin synthesis is implicated in the timing of seasonal reproductive cycles in some species (Bittman *et al.* 1983; Carter and Goldman 1983). Lastly, while photoperiod influences the phase and duration of the melatonin cycle, environmental temperature modulates its amplitude (Tilden and Hutchison 1993; García-Allegue *et al.* 2001). In Diamondback Water Snakes (*Nerodia rhombifer*), cold and warm temperature extremes decreased the amplitude of the melatonin cycle (Fig. 8.5; Tilden and Hutchison 1993). Thus, photoperiod and temperature interact to influence diel melatonin rhythms. This relationship has been observed in *N. rhombifer* as well as the European Sea Bass (*Dicentrarchus labrax*), Mudpuppy (*Necturus maculosus*), Three-toed Box Turtle (*Terrapene carolina triunguis*), Marbled Gecko (*Christinus marmoratus*), and *Anolis carolinensis* (Underwood 1985*a*; Vivien-Roels *et al.* 1988; Rawding and Hutchison 1992; Tilden and Hutchison 1993; Moyer *et al.* 1995; García-Allegue *et al.* 2001). Interactions among the amplitude, duration, and phase of the melatonin cycle therefore ultimately determine the effects of this neuroendocrine signal on an animal's physiology and behavior.

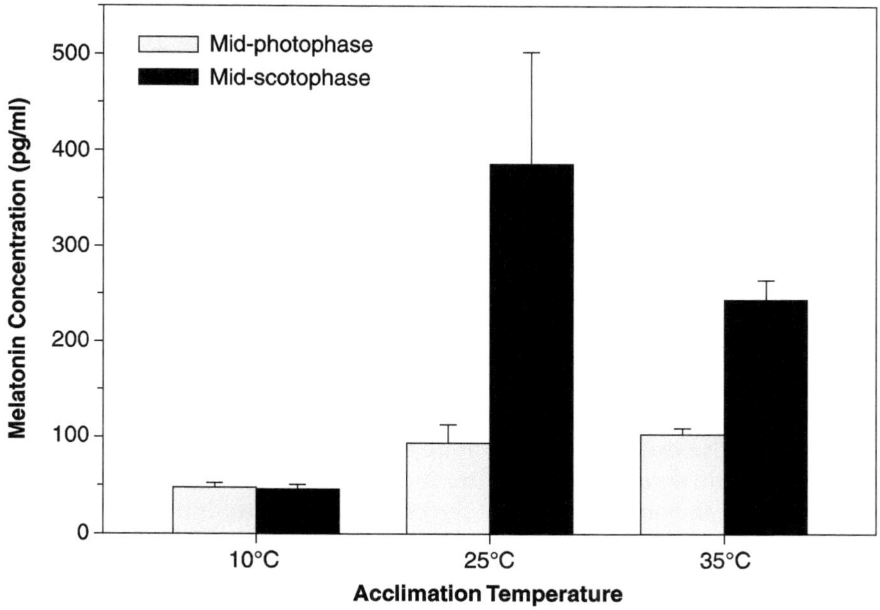

Fig. 8.5 Effects of acclimation temperature on mid-photophase and mid-scotophase serum melatonin in *Nerodia rhombifera*. Snakes were acclimated to 10, 25, or 35°C with a 12:12 h L:D photoperiod for 2 weeks. Each bar is the mean hormone concentration + 1 SEM for five animals. Redrawn from Tilden, A. R. and Hutchison, V. H. 1993. General and Comparative Endocrinology 92: 347-354, Fig. 5.

Lutterschmidt and Mason (2009) investigated the endocrine mechanisms by which temperature induces seasonal reproductive behavior in *Thamnophis sirtalis parietalis*. Because *T. s. parietalis* requires a period of low temperature exposure to elicit reproductive behavior (Camazine *et al.* 1980; Bona-Gallo and Licht 1983), it is an excellent model system for investigating the mechanisms by which temperature influences reproductive parameters. Elevated hibernation temperatures (i.e., 10°C vs. 5°C) significantly increased overall melatonin concentrations of snakes (Fig. 8.6; Lutterschmidt and Mason 2009).

These results demonstrated that environmental temperature, in the absence of changing photoperiodic cues, is sufficient in modulating diel melatonin rhythms (as well as androgen concentrations) of *T. s. parietalis* during winter dormancy. Importantly, the percent body mass loss of snakes did not differ significantly between the hibernation temperature treatments, suggesting that the observed changes in hormone profiles were indeed temperature-induced and not simply an indirect result of significant changes in the energy balance of snakes in the group hibernated at a higher temperature (Lutterschmidt and Mason 2009). Further, the temperature-induced differences in melatonin rhythms between the 5°C and 10°C treatment groups persisted even after both groups were again acclimated to 10°C, indicating that cold temperature exposure has a lasting influence on melatonin rhythms (Lutterschmidt and Mason 2009).

Although inconclusive, it is worth noting that the amplitude of the melatonin cycle appeared to increase in response to prolonged low temperature conditions (Fig. 8.6; Lutterschmidt and Mason 2009). Peak melatonin concentrations of snakes were higher after exposure to 5°C for 12 weeks compared to those of snakes following 4 weeks of 5°C exposure (but note the large variation in melatonin concentrations and the lack of statistical significance; Fig. 8.6B). It is noteworthy that this trend was also observed in diel melatonin cycles during a preliminary hibernation study, in which a melatonin cycle was observed only in *Thamnophis sirtalis parietalis* acclimated to low temperature conditions (5°C vs. 15°C; D. I. Lutterschmidt, unpublished data). These results are intriguing because previous studies in *T. s. parietalis* have shown that the period of low temperature exposure must be greater than 4 weeks in duration to elicit courtship behavior following return to high temperature conditions (Bona-Gallo and Licht 1983). Thus, the duration of cold temperature exposure may be transduced by changes in the amplitude of the melatonin rhythm. This hypothesis is supported by the observation that the cold temperature-induced differences in melatonin rhythms persisted even after snakes in the cold temperature hibernation treatment were returned to 10°C (Lutterschmidt and Mason 2009). Further studies are necessary to determine whether the duration of cold-temperature exposure is indeed coded by changes in the amplitude of the melatonin rhythm.

Influence of melatonin on reproduction. Melatonin's ability to transduce environmental information into appropriate neuroendocrine

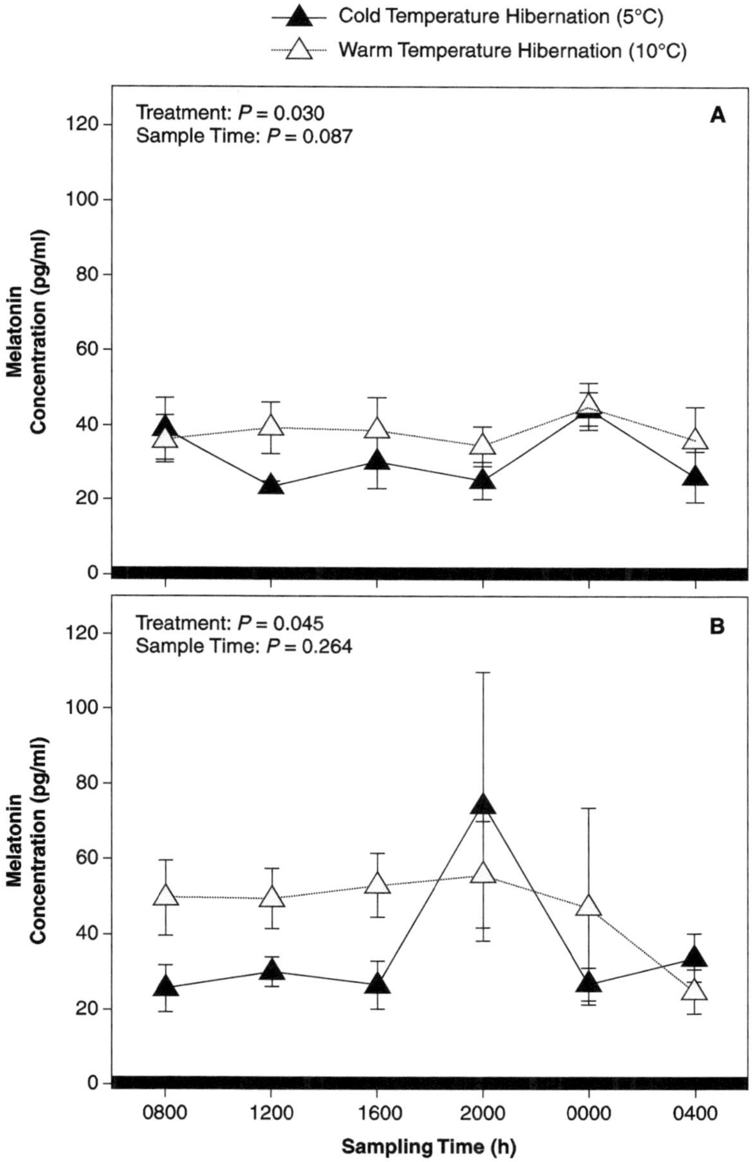

Fig. 8.6 Influence of elevated hibernation temperatures on diel melatonin rhythms of male *Thamnophis sirtalis parietalis*. **A.** Melatonin cycles following 4 weeks of 5ºC exposure. **B.** Melatonin cycles following 12 weeks of 5ºC exposure. Animals were maintained in constant darkness. Each data point is the mean melatonin concentration of 7 independent snakes (± 1 s.e.m.) randomly selected from a total of 42 animals in each treatment group. Main effects of temperature treatment and sampling time are listed in the top left corner of each panel (statistical values from two-way ANOVAs). Melatonin rhythms of snakes in the two treatment groups did not differ significantly prior to temperature manipulation. Modified from Lutterschmidt, D. I. and Mason, R. T. 2009. Journal of Experimental Biology 212: 3108-3118, Figs. 2B and 3A.

signals plays an important role in synchronizing physiology and behavior with optimal environmental conditions. One of the best-studied functions of the pineal gland and melatonin is their role in regulating seasonal reproductive cycles. Although extensive experiments have been done in both birds and mammals, there are fewer studies investigating the relationship between the pineal complex and reproductive function in other vertebrates (reviewed in Turek and Van Cauter 1994; Mayer *et al.* 1997).

The most extensively studied model organism used for investigating the photoperiodicity of reproduction in ectothermic vertebrates is *Anolis carolinensis* (Mayer *et al.* 1997). There is clear evidence for an inhibitory role of the pineal gland and melatonin in the control of reproduction in this species (Levey 1973; Mayer *et al.* 1997; Underwood 1981a,b, 1985a,b). For example, pinealectomy of male anoles stimulated testicular growth and spermatogenesis (Underwood 1985b), while pinealectomy of female anoles stimulated follicular development and increased ovarian and oviduct weights (Levey 1973). Melatonin treatment abolished the effects of pinealectomy on reproduction in female anoles (Levey 1973). In other experiments, treatment of pinealectomized and intact anoles with melatonin had either no effect or inhibited reproductive parameters depending upon the photoperiod and season tested (Underwood 1981a, 1985b).

Antigonadotropic effects of melatonin and the pineal gland also occur in other reptiles, including the Indian Garden Lizard (*Calotes versicolor*) and the Zebra-tailed Lizard (*Callisaurus draconoides*) (Haldar and Thapliyal 1977; Packard and Packard 1977; Thapliyal and Misra nee Haldar 1979; Haldar-Misra and Thapliyal 1981a,b; Vivien-Roels and Pévet 1983; reviewed in Mayer *et al.* 1997). In male Hermann's Tortoises (*Testudo hermanni*), subcutaneous melatonin implants decreased gonadosomatic index and testosterone concentrations (Vivien-Roels and Arendt 1983; Vivien-Roels and Pévet 1983). In studies conducted on two lizard species (*Anolis carolinensis* and *Sceloporus occidentalis*), removal of the pineal gland blocked reproductive responsiveness to photoperiod (Underwood 1981a, 1985a,b). Subsequently, these animals became reproductively active in the non-breeding season while exposed to inhibitory photoperiods (Underwood 1985b).

Melatonin's inhibitory effects on reproduction, in combination with its seasonal rhythm of secretion, are thought to modulate and synchronize reproductive behavior with environmental cues (Underwood 1981a, 1985b; Rismiller and Heldmaier 1987; Crews *et al.* 1988; Mendonça *et al.* 1995). However, the effects of pinealectomy and/or melatonin treatment on reproductive parameters vary among species, experimental protocols, and seasons, making it difficult to draw conclusions about the role of the pineal gland in reproduction (de Vlaming 1975; de Vlaming and Vodicnik 1977; Haldar and Thapliyal 1977; Vodicnik *et al.* 1978; Thapliyal and Misra nee Haldar 1979; Hontela and Peter 1980; Underwood 1981a, 1985b; Chanda and Biswas 1982, 1992; Garg 1988; Haldar and Pandey 1988; Nayak and Singh 1988). For example, the effects of pinealectomy on testis size in the Indian Chequered Water Snake (*Natrix piscator*) are influenced by humidity; pinealectomy decreased testicular size in high humidity (which normally stimulates testicular growth) but increased testicular size in both low and moderate humidity (Haldar and Pandey 1989a,b). In addition, treatment of Indian chequered water snakes with either melatonin or 5-methoxytryptamine (an indoleamine structurally related to melatonin) suppressed testicular function of active testes, but did not influence inactive testes (Haldar and Pandey 1988). The effects of pinealectomy in Black-spined Toads (*Bufo melanostictus*) also vary among this species' seasonal activity patterns: pinealectomy of toads during the hibernation phase stimulated testicular maturation, while pinealectomy during the breeding season did not influence testicular function (Chanda and Biswas 1984).

Similarly, previous studies in *Thamnophis sirtalis parietalis* demonstrated that pinealectomy prior to winter dormancy inhibited male courtship behavior upon spring emergence (Nelson *et al.* 1987; Crews *et al.* 1988; Mendonça *et al.* 1996a). In contrast, pinealectomy following spring emergence had no effect on the expression of male courtship behavior (Nelson *et al.* 1987; Mendonça *et al.* 1996a). These results suggest the pineal gland is necessary for transducing environmental stimuli during winter dormancy. Whether the observed influence of pinealectomy on reproductive behavior is related to changes in melatonin signaling requires further study. For example, treatment of pinealectomized snakes with melatonin did not influence the effects of pinealectomy on courtship behavior (Mendonça *et al.* 1996a). Similarly, treatment of intact snakes with melatonin-filled silastic capsules during winter dormancy did not significantly influence either male courtship behavior or sex steroid concentrations of male or female snakes (Lutterschmidt 2006). However, the arrhythmic and supra-physiological melatonin concentrations (e.g. approximately 3 ng/ml) resulting from the silastic capsule hormone implants must be considered. Implant-based manipulations of melatonin concentrations not only mask endogenous melatonin rhythms, but such high melatonin levels may cause down-regulation of melatonin receptors in the neural areas controlling reproductive behavior (e.g., Krohmer and Crews 1987a). Further studies using alternative techniques to manipulate melatonin signaling are needed to appropriately evaluate the role of diel melatonin rhythms during winter dormancy on reproductive behavior in *T. s. parietalis* as well as other snake species.

In contrast to melatonin's role in the seasonal regulation of reproduction, melatonin also acts as a synchronizer of daily activity patterns. Diel melatonin rhythms during spring emergence most likely play a role in synchronizing reproductive behavior (and activity) with the appropriate time of day. Indeed, disrupted melatonin cycles, with higher melatonin secretion occurring during the photophase, were observed in *T. s. parietalis* that failed to exhibit courtship behavior during the spring breeding season (Fig. 8.7A; Mendonça *et al.* 1996b). Pinealectomy significantly increased courtship behavior of previously noncourting male snakes (Fig. 8.7B; Mendonça *et al.* 1996b), again suggesting that melatonin synchronizes reproductive behavior with daily activity patterns during the spring

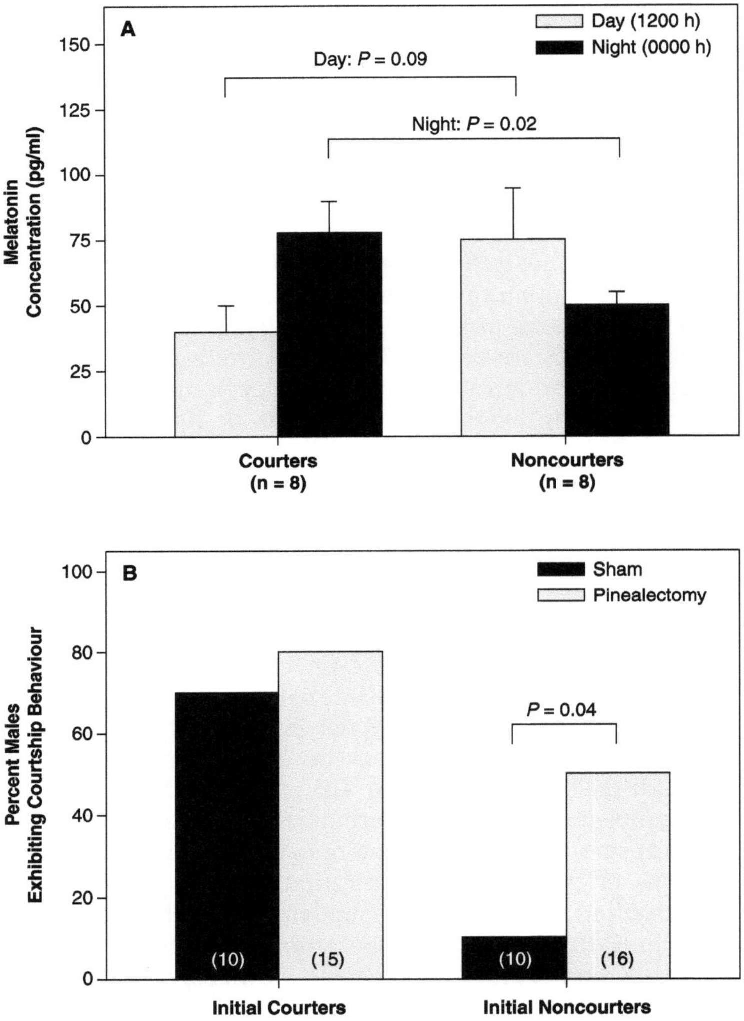

Fig. 8.7 A. Plasma melatonin concentrations during the day (1200 h) and night (0000 h) in courting and noncourting male *Thamnophis sirtalis parietalis* 3 days after emergence. Mean melatonin concentrations of courters were significantly different from those of noncourters at 0000 h but not at 1200 h (results from t-tests). **B.** Effect of pinealectomy on the reproductive behavior of courting and noncourting male *T. s. parietalis*. Depicted is the percentage of initial courters and noncourters that exhibited courtship behavior post-surgery. Redrawn from Mendonça, M. T., *et al.* 1996b. Hormones and Behavior 30: 176-185, Figs. 2 and 4.

mating season. Besides temperature, pinealectomy is the only other known stimulus capable of inducing reproductive behavior in nonreproductive *T. s. parietalis* (Mendonça *et al.* 1996b). In addition, Lutterschmidt *et al.* (2004) demonstrated that acute melatonin treatment (0.03 and 0.3 mg kg^{-1} body mass) significantly inhibited courtship behavior of male snakes during spring (Fig. 8.8). These studies indicate that reproductive behavior

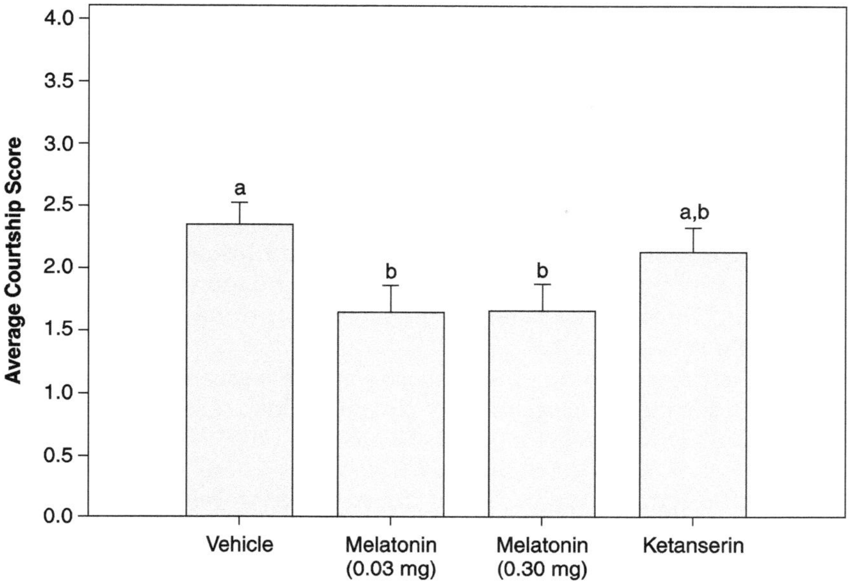

Fig. 8.8 Average courtship scores of male *Thamnophis sirtalis parietalis* following treatment with vehicle, melatonin (0.03 or 0.30 mg), or ketanserin, a serotonergic type 2A receptor antagonist. Standard errors (+ 1) are shown by the vertical lines; $n = 32$ in each treatment group. Statistically significant differences among treatment groups are indicated by letters above the horizontal lines. Ketanserin treatment was included in this experiment to test the hypothesis that melatonin modulates reproductive behavior via antagonism of serotonin receptors. Note that ketanserin treatment did not mimic the effects of melatonin on courtship behavior, suggesting that melatonin specifically modulates reproductive behavior of male snakes. Modified from Lutterschmidt, D. I., LeMaster, M. P. and Mason, R. T. 2004. Hormones and Behavior 46: 692-702, Fig. 1.

of male *T. s. parietalis* is sensitive to melatonin signaling during the spring mating season.

Collectively, the available evidence supports the hypothesis that the pineal gland plays a role in the environmental induction of reproduction. However, current studies indicate that the pineal gland and its secretory product function primarily in synchronizing physiological and behavioral parameters with changing environmental stimuli, rather than strictly inhibiting any one particular parameter (e.g. Lutterschmidt *et al.* 2002, 2003). Pinealectomy therefore merely abolishes the controlling influences of environmental stimuli, which in turn desynchronizes an animal's physiology and behavior with environmental cues. Melatonin treatment would be expected to restore pineal function only if melatonin administration closely mimics the natural melatonin cycle normally determined by interacting environmental factors. Thus, pinealectomy and melatonin treatment exhibit both positive and negative influences on reproduction depending on when experiments are conducted in relation to a species' unique natural history traits.

8.2.3 Geographic Variation in the Transduction of Environmental Cues

The transduction of environmental cues into a rhythmic neuroendocrine signal is central to an animal's timekeeping system. One question that requires further investigation is whether these neuroendocrine rhythms vary among taxa inhabiting different environments. For example, does the neuroendocrine pathway transduce environmental cues similarly in all species, regardless of phylogeny and evolutionary history? In other words, is variation in melatonin rhythms among populations environmentally induced or does this variation reflect evolutionary differences in timekeeping systems?

There is precedence for these questions in observations of geographic variation in the molecular genetics of circadian clock cells in both *Drosophila* and *Arabidopsis* (Michael *et al.* 2003; Khare *et al.* 2005). Furthermore, Firth *et al.* (1989) demonstrated significant differences in melatonin rhythmicity between cold-temperature adapted Tuatara (*Sphenodon punctatus*) and a population of desert-adapted Sleepy Lizards (*Tiliqua rugosa*). However, whether the neuroendocrine interface between the environment and an organism's physiology and behavior is genetically and/or phylogenetically constrained is not clear.

Using a common garden approach, Lutterschmidt and Mason (2008) investigated possible geographic variation in timekeeping systems by comparing 24-h melatonin rhythms and reproductive behavior among 3 populations of garter snakes that were maintained in the laboratory under identical environmental regimes. The snake populations were chosen based on their distinct latitudinal and longitudinal distributions and very different life histories: *Thamnophis sirtalis parietalis* from Manitoba, Canada, Red-spotted Garter Snakes (*Thamnophis sirtalis concinnus*) from western Oregon, and Eastern Garter Snakes (*Thamnophis sirtalis sirtalis*) from southern Florida (Fig. 8.9A). If physiological timekeeping systems evolved to ensure that individuals are temporally well adapted to a particular environment, then one would expect to observe geographic differences in the transduction of environmental cues into neuroendocrine signals. Indeed, melatonin cycles differed significantly among the three snake populations in a majority of the sampling periods (Fig. 8.9B; Lutterschmidt and Mason 2008). Population differences were observed across a wide range of acclimatization conditions and were themselves plastic (i.e., one snake population was not consistently different from the others). Changes in male courtship behaviour (Fig. 8.9C) and androgen concentrations (Fig. 8.9D) during emergence also varied significantly among populations, despite the fact that all snake populations were acclimatized to identical environmental conditions (Lutterschmidt and Mason 2008). These data support the hypothesis that endogenous timekeeping systems have evolved in the presence of unique environmental conditions and indicate that individual animal populations will respond uniquely to a given suite of environmental stimuli. Traditional common garden rearing experiments are needed to determine if the observed geographic variation in timekeeping systems results from inherent genetic differences among these populations or different environmental conditions during key developmental stages (Lutterschmidt and Mason 2008).

If timekeeping systems are indeed evolutionarily constrained, then rapid changes in climate predictability, such as those associated with global climate change (e.g., IPCC 2001), could have a disproportionately large influence on reproductive fitness. For example, disrupted breeding cycles and consistently earlier breeding are occurring in birds and amphibians and have been associated with global climate change (e.g., Forchhammer *et al.* 1998; Brown *et al.* 1999; Dunn and Winkler 1999; Moss *et al.* 2001). Further, Root *et al.* (2003) demonstrated that in more than 80% of the species showing temperature-related shifts, the changes are consistent with and predicted by the species' physiological constraints.

While the results of Lutterschmidt and Mason (2008) suggest that animal populations will respond uniquely to changing environmental cues (and environmental perturbations), additional studies are necessary to determine

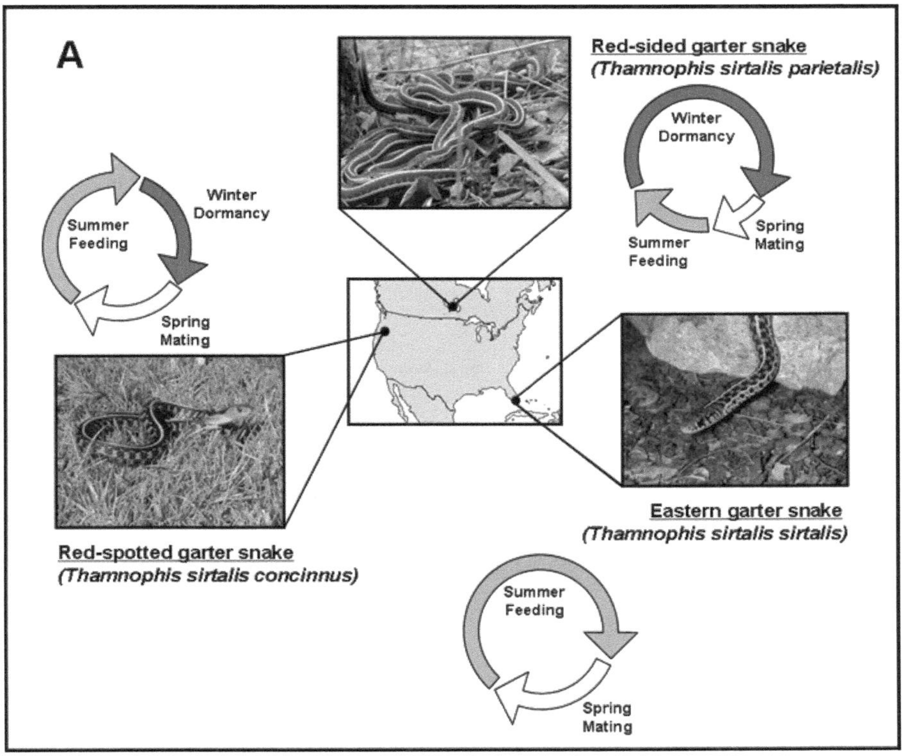

Color image of this figure appears in the color plate section at the end of the book.

Fig. 8.9 Contd. ...

... *Fig. 8.9 Contd.*

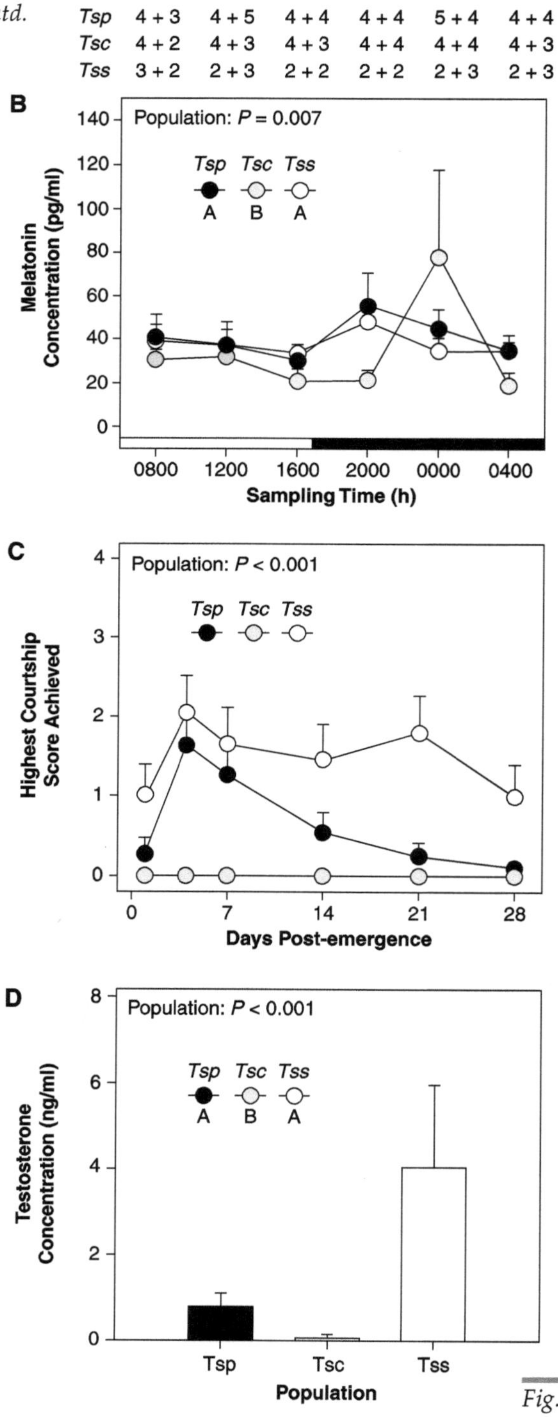

Fig. 8.9 Contd. ...

the degree of plasticity in timekeeping systems within populations. For example, high phenotypic plasticity in one or more traits may be adaptive in highly variable environments, which may allow some populations to adjust more readily, and perhaps favorably, to environmental perturbations (e.g., Visser 2008). Comparative phylogenetic analyses of the biochemistry underlying the transduction of environmental cues would also provide insight into the evolution of circadian and circannual rhythmicity. For example, the geographic variation in melatonin rhythms observed in Lutterschmidt and Mason (2008) might be due to different isoforms of the enzymes regulating melatonin synthesis (e.g., *N*-acetyltransferase, hydroxyindole-*O*-methyltransferase). Such data regarding plasticity in timekeeping systems both within and among animal populations are necessary for understanding the potential impacts of environmental perturbations on seasonal rhythms in physiology and behavior.

8.3 NEURAL PATHWAYS REGULATING REPRODUCTION

The anterior hypothalamus-preoptic area (AHPOA) is a major integrative region for coordination of internal and external stimuli regulating

... *Fig. 8.9 Contd.*

Fig. 8.9 Geographic variation in reproductive physiology and behavior among three populations of garter snakes housed in a common garden regime. **A.** *Thamnophis sirtalis parietalis* (Tsp) is the most northerly living reptile in North America and is found in extremely high numbers throughout south central Manitoba, Canada. These northern-latitude populations of snakes undergo a period of winter dormancy for approximately 8 months each year followed by an attenuated mating season lasting approximately 4 weeks. In contrast, the midlatitude *T. sirtalis concinnus* (Tsc) of western Oregon has an extended breeding season that lasts 10–12 weeks from March through May. Although *T. sirtalis concinnus* do exhibit periods of winter dormancy, they can be active during 10–12 months of the year given appropriate environmental conditions. Similarly, *T. sirtalis sirtalis* (Tss) in the southern subtropical regions of Florida may be active during most of the year, since environmental conditions are extremely mild during winter months. **B.** Plasma melatonin concentrations (pg/ml) measured every 4 h for one 24-h period in each of the three garter snake populations. Each data point is the mean hormone concentration + SE. Sample sizes at each sampling time are shown above the x-axis for each snake population (shown as males + females). Snakes were acclimatized to an 11:13 h L:D photoperiod and a 25:20°C thermoperiod. Black bars above the abscissa indicate the period of scotophase. **C.** Courtship scores of male garter snakes following winter dormancy in the laboratory. Reproductive behavior of *T. sirtalis parietalis* (Tsp; n = 30), *T. s. concinnus* (Tsc; n = 24), and *T. s. sirtalis* (Tss; n = 20) was scored using an ethogram of male courtship behavior. Each data point is the mean highest courtship score achieved by male snakes + SE. **D.** Androgen concentrations (ng/ml) of male snakes during the period of reproductive behavior. Each bar is the mean hormone concentration + SE of 8 male snakes in each population. The main effects of snake population are listed in the top left corner of each panel. Statistically significant differences among snake populations are indicated by letters below the legend (from multiple comparisons tests). Differences in courtship behavior among snake populations in panel C are not shown due to a significant interaction between days post-emergence and population. **B** and **C** modified from Lutterschmidt, D. I. and Mason, R. T. 2008. Physiological and Biochemical Zoology 81: 810-825, Figs. 1 and 2.

reproductive behavior in male vertebrates (Crews and Silver 1985; Ingle and Crews 1985; Baum 2004). Afferents of the major sensory systems important to sexual behavior converge in the AHPOA. In addition, the integrity of the AHPOA is essential for the initiation and maintenance of sexual behavior. Subsequently, lesions of the AHPOA during the breeding season resulted in the rapid extinction of courtship behavior and mating in a wide variety of species (Brookhart and Dey 1941; Barfield 1965; Hart and Leedy 1985; Leedy and Hart 1986; de Jonge *et al.* 1989).

As in other vertebrate species, an intact AHPOA is essential for the display of reproductive behaviors in male *Thamnophis sirtalis parietalis*. Lesions placed in the AHPOA of actively-courting males extinguished all courtship behavior (Friedman and Crews 1985b), while males receiving lesions in the AHPOA prior to entering winter dormancy failed to initiate courtship behavior upon emergence (Krohmer and Crews 1987a). In contrast, bilateral lesions placed in the nucleus sphericus (NS) or septum (S) prior to low temperature dormancy, reduced the time required to initiate intense courtship behavior and increased the total number of days males exhibited moderate to intense courtship (Krohmer and Crews 1987b). These data suggest that input from the NS or S normally inhibit sexual behaviors.

These data led Krohmer and Crews (1987b) to hypothesize that lesions of the NS affected the efferent outflow through the bed nucleus of the stria terminalis (Bst) to the AHPOA. As in other vertebrates, the NS is the major zone receiving efferents from the accessory olfactory bulb (AOB). In turn, the NS sends efferents through the Bst, the preoptic area (POA) and the ventromedial nucleus of the hypothalamus (VMH). The VMH is the major target of efferents from the NS (similar to the mammalian amygdaloid region as both receive efferents from the accessory olfactory bulb (Halpern and Silfen 1974; Halpern and Kubie 1984).

In the only other studies assessing the role of the amygdala (AMG) in the control of reproductive behavior in reptiles, bilateral lesions placed in the AMG of *Anolis carolinensis* (Crews 1979; Greenberg *et al.* 1984) and the western fence lizard, *Sceloporus occidentalis* (Tarr 1977) abolished sexual behavior as well as reduced aggressive behavior. Lesions in the S of castrated, androgen treated, or intact *Anolis carolinensis* resulted in the reduction of courtship activity as well as male-male aggressive behavior in both of these groups (Crews 1979).

In addition to being a major integrative area for stimuli affecting sexual behavior, the AHPOA has been reported to play an important role in temperature regulation/perception in vertebrates (Satinoff 1978, 1983; Szymusiak and Satinoff 1982). In *Thamnophis sirtalis parietalis* thermoregulatory behavior appears to be associated more specifically with the anterior portion of the POA and not the AHPOA as a whole. Krohmer and Crews (1987a) reported that male *T. s. parietalis* that received lesions confined primarily to the anterior portion of the POA exhibited atypical thermoregulatory behavior (Fig. 8.10). Animals receiving lesions

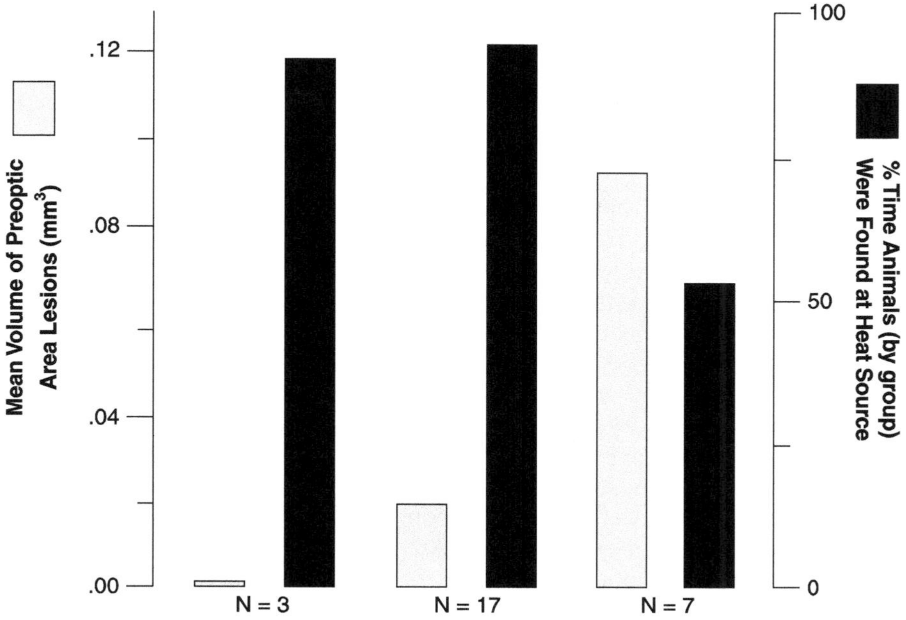

Fig. 8.10 Comparison of the volume of POA lesions and deficits in thermoregulatory behavior. Redrawn from Krohmer, R. W. and Crews, D. 1987a. Behavioral Neuroscience 101: 228-236, Fig. 7.

to the AHPOA, centered at the interface between the POA and anterior hypothalamus, moved to and remained under a heat source for several hours each day, while animals with lesions centered in the anterior portion of the POA (leaving the anterior hypothalamus essentially intact) regularly failed to approach the heat source. Initially, the animals with lesions centered in the anterior POA failed to court attractive females, however, after a period of forced warming, most of these animals exhibited normal courtship behavior (Krohmer and Crews 1987a).

As described previously, the only known requirement for the initiation of courtship behavior and mating in *Thamnophis sirtalis parietalis* is an extended period of low temperature dormancy (Aleksiuk and Gregory 1974; Garstka *et al.* 1982). Furthermore, the sensitivity of the neural pathways, to as yet unidentified internal and/or external cues, that regulate courtship behavior may be increased by extending the time maintained in low temperature dormancy (Garstka *et al.* 1982). In addition, this sensitivity may be "reset" by exposure to short periods of low temperature dormancy during or at the end of the breeding season (Krohmer and Crews 1989).

In *Thamnophis sirtalis parietalis*, an intact POA is critical for the integration of temperature information (Krohmer and Crews 1985a). In addition, removal of the pineal gland disrupts specific components of a temperature detection system, preventing the snake from recognizing warm temperatures upon emergence (Nelson *et al.* 1987) and preventing temperature-dependent vernal mating behavior in *Thamnophis sirtalis parietalis*. As a result, the pineal gland appears to mediate courtship behavior via a neural pathway with the POA that is independent of photoperiodic cues (Nelson *et al.* 1987).

8.3.1 Sex Steroid Concentrating Nuclei

The brains of all vertebrate species studied to date have been found to contain sex steroid concentrating regions (nuclei), and although the species examined are quite diverse, there is a notable uniformity in the location of these sex steroid concentrating nuclei. Sex steroid concentrating nuclei have been identified within neural pathways regulating reproductive behavior (Crews and Silver 1985) and have been found to play a vital role in the regulation of species-typical behaviors (Ingle and Crews 1985). Neural regions such as the POA, VMH, S and AMG are essential for the assimilation of sensory and motor events associated with intrinsic behaviors. In many vertebrate species, these limbic nuclei are sexually dimorphic in size and fluctuate in volume depending on their reproductive status (Gorski *et al.* 1978; Madeira *et al.* 1999; Simerly 2002). While there is a large literature on the role of hormones on specific brain regions in mammals (Dohler *et al.* 1984; Tobet *et al.* 1986; Ulibarri and Yahr 1988) and birds (Panzica *et al.* 1991; Ball and Balthazart 2002), only a few studies have examined sex steroid concentrating nuclei in reptiles (snakes: Halpern 1980; Halpern *et al.* 1982; lizards: Crews *et al.* 1993; Wade *et al.* 1993; O'Bryant and Wade 2002; Beck *et al.* 2008).

Garstka *et al.* (1982) reported that *Thamnophis sirtalis parietalis* initiates courtship and mating immediately upon emergence from winter dormancy when the gonads are quiescent and circulating sex steroid hormones are low. Although the level of sex steroid hormones has now been found to be elevated upon emergence (Fig. 8.2; Krohmer *et al.* 1987; Moore *et al.* 2000; 2001; Lutterschmidt *et al.* 2004; Lutterschmidt and Mason 2009), initiation of courtship behavior still appears to be independent of direct hormonal control. Interestingly, the brains of *T. sirtalis parietalis* have been found to exhibit a pattern of sex steroid concentrating nuclei similar to other species (Halpern *et al.* 1982). Although nuclei in the garter snake brain respond to both testosterone and estrogen, the number of hormone-accumulating cells and density of labeling was greater in response to estrogen compared to testosterone administration (Halpern *et al.* 1982; Kohmer 2004). While the distribution of hormone-concentrating cells following administration of the two hormones was similar, several regions that contained labeled cells after administration of estradiol did not contain labeled cells following treatment with testosterone. Halpern *et al.* (1982) also reported no consistent differences in the intensity or pattern of labeling in the brains of males and females receiving the same hormone treatment. Furthermore, in the garter snake telencephalon, sex steroid hormone-accumulating cells were found in the AMG, S, POA, and hypothalamus (Hy) (Halpern *et al.* 1982). Two regions of the snake's AMG contained high concentrations of labeled cells. These were the ventral amygdaloid nucleus and the NS. Collectively, the S, POA and hypothalamic structures contained numerous, densely labeled cells, comprising some of the most heavily labeled cells in the garter snake brain (Halpern *et al.* 1982).

Sex steroid hormone concentrating nuclei may be sexually dimorphic and seasonally regulated. In the brain of *Thamnophis sirtalis parietalis*, Crews *et al.* (1993) found sex steroid concentrating nuclei in the POA, VMH, external nucleus of the optic tract, and the medial forebrain bundle. In this study, comparison of animals by season revealed sexually dimorphic differences in two areas. In animals maintained in LTD, the POA was significantly smaller in females compared to males, while in the non-LTD group; the NS was significantly smaller in females compared to males. Crews *et al.* (1993) also found no significant differences in the size of brain areas in male *T. s. parietalis* when compared either by season or by hormonal manipulation. However, in a recent study, it was found that the sex steroid concentrating nucleus in the POA of *T. s. parietalis* did undergo significant changes in response to both seasonal and hormonal manipulation (Baleckaitis and Krohmer, unpublished data; reviewed in Krohmer 2004). Furthermore, this study found that sex steroid concentrating neurons in the POA exhibited greater hypertrophy when treated with estrogen compared to animals treated with testosterone or a placebo. These findings corroborate those of Halpern *et al.* (1982) who reported that estrogen had a greater effect on the hypertrophy of sex steroid concentrating neurons in the POA than did testosterone. In addition, animals treated with either estradiol or testosterone maintained under conditions of LTD for 12 wks exhibited greater hypertrophy of the sex steroid concentrating neurons in the POA compared to their counterparts maintained under warm conditions (Fig. 8.11; Krohmer 2004).

8.3.2 Effect of Lesions and Intracranial Steroid Implantation on Reproductive Physiology

Lesions centered in the AHPOA caused no apparent changes in the reproductive physiology of male *Thamnophis sirtalis parietalis*. Friedman and Crews (1985b) reported that neither the gonadosomatic index (GSI) or level of circulating androgens varied between animals receiving lesions or sham-lesions in the AHPOA. Furthermore, the reproductive condition of the testes of all animals was found to be similar, containing transforming spermatids and mature sperm with only a few sperm in the epididymides indicating spermiation had just been initiated. In a later study, animals receiving lesions or sham-lesions in the AHPOA prior to LTD revealed no significant difference between treatment groups in the GSI of testes removed at either four or eight weeks post emergence (Krohmer and Crews 1987a). In addition, no significant differences were observed in the spermatogenic stage of lesioned or sham-lesioned animals. At four

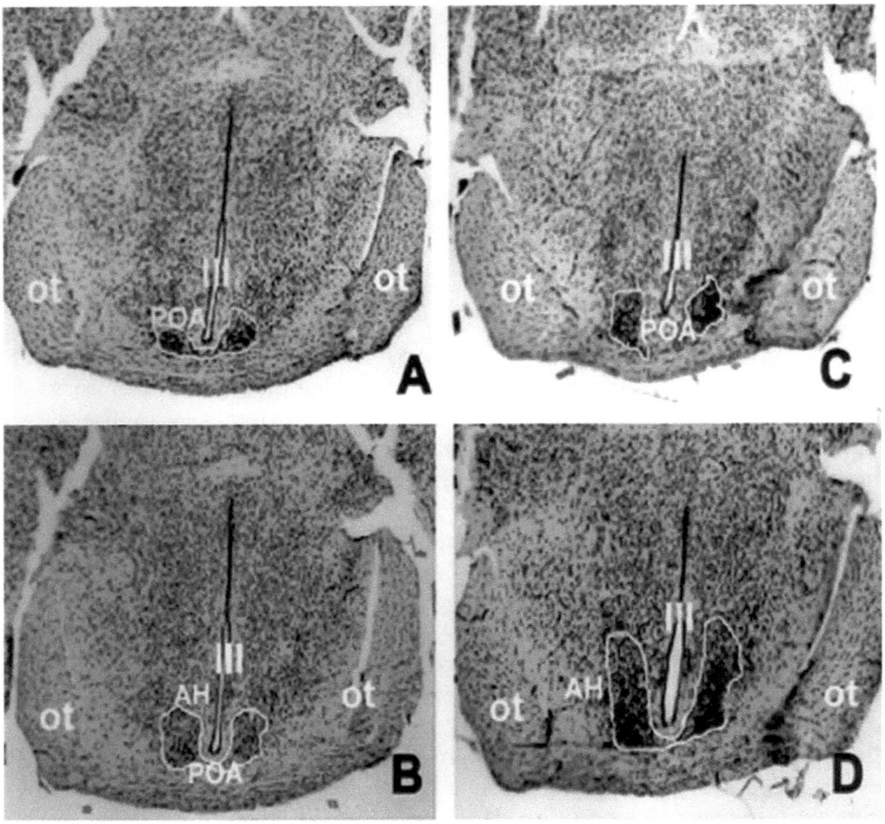

Fig. 8.11 Photomicrographs of cross sections through the POA of male *Thamnophis sirtalis parietalis* after various hormonal and temperature regimens. Tissues were stained with luxol fast blue and basic fuchsin. Outlined regions depict the extent of the sex steroid concentrating nucleus within the POA. **A.** Animals implanted with time release tablets of testosterone (T) (Innovative Research, Inc.) and maintained under warm, spring-like conditions for 12 weeks. **B.** Animals implanted with time release T and maintained in LTD for 12 weeks. Although some hypertrophy is apparent in animals maintained under ambient conditions, a prolonged period of LTD appeared to intensify the response of the sex steroid concentrating neurons to T. **C.** Animals implanted with time release 17β estradiol (17β E) (Innovative Research, Inc.) and maintained under warm, spring-like conditions for 12 weeks. **D.** Animals implanted with time release 17β E and maintained in LTD for 12 weeks. The amount of hypertrophy exhibited in the animals maintained under warm conditions was not significantly different from either of the T implanted groups. However, animals receiving 17β E in combination with LTD exhibited significantly greater hypertrophy of the sex steroid concentrating neurons within the POA. From Krohmer, R. W. 2004. Institute for Laboratory Animal Research 45: 65-74, Fig. 6.

weeks post-emergence, primary and secondary spermatocytes were the dominant cell type; while at eight weeks; mature sperm were the dominant cell type although spermiation had not yet been initiated (Krohmer and Crews 1987a). While no significant difference between treatment groups was observed in the level of circulating androgens at either four or eight weeks post-emergence, there was a significant reduction in circulating androgens between four and eight weeks post-emergence (Krohmer and Crews 1987a). Finally, examination of the renal sexual segment (RSS) of the kidney (Saint Girons 1982) at eight weeks post-emergence, revealed no significant difference in tubular diameter or epithelial height between treatment groups.

In contrast, bilateral lesions in the S or NS did affect the reproductive physiology of male *Thamnophis sirtalis parietalis* (Krohmer and Crews 1987b). The concentration of circulating androgens in animals with lesions in the NS was significantly higher than levels measured in animals receiving septal or sham lesions. While there was no measurable difference in GSI among treatment groups, the diameter and epithelial height of the seminiferous tubules and RSS, as well as the quantity of sexual granules in the RSS epithelia (Bishop 1952) was greater in both lesioned groups compared to sham-lesioned animals (Krohmer and Crews 1987b). Hypertrophy of the RSS and production of sexual granules in the epithelium have been shown to be dependent on the level of circulating androgens (Bishop 1952; Panda and Thapliyal 1964). Subsequently, increased quantities of sexual granules would reflect the elevated level of androgens found in the circulation of animals receiving lesions in the S and NS (Krohmer and Crews 1987b).

In *Anolis carolinensis* the testes of animals receiving lesions in the S appeared normal compared to control animals, suggesting that the S is not involved in the control of pituitary function (Crews 1979). These data support the findings of Nance *et al.* (1974, 1975) who found no evidence for an effect of lesions in the S on the pattern of gonadotropin release in both male and female rats.

Comparative data on the effect of intracranial steroid implantations on reproductive physiology is relatively scarce. Most studies involving intracranial implantation of steroids have removed the testes to negate the effect of systemic androgens on areas of the brain other than the one being studied. However, since garter snakes will court regardless of the presence or absence of the testes, the study by Friedman and Crews (1985b) utilized intact animals. Animals receiving androgen implants in the AHPOA had a lower (although not significant) GSI than animals receiving implants in other areas or in sham and nonsurgical control groups. Friedman and Crews (1985b) found that circulating androgens were significantly elevated in animals receiving implants in the AHPOA compared to animals with implants outside the AHPOA and controls. This increase is most likely not caused by leakage into the general circulation since animals receiving implants in other regions of the brain did not exhibit elevated androgen levels. These data suggest that neurons in the AHPOA are part of a positive feedback loop in which elevated androgen levels stimulate additional gonadotropin release (Friedman and Crews 1985b). The testes of snakes receiving androgen implants in the AHPOA were more developed than those with implants outside the AHPOA or controls. The testes of AHPOA implanted animals contained mature sperm and few spermatids and the epididymides were hypertrophied and packed with sperm. In contrast, the testes of animals with implants outside the AHPOA and controls contained transforming spermatids and spermatozoa and the epididymides were less developed and had few or no sperm (Friedman and Crews 1985b).

While lesions or implantation of sex steroid hormones in the AHPOA had no effect on the reproductive physiology of male *Thamnophis sirtalis parietalis*, bilateral lesions in the S or NS did have a profound effect. Animals receiving lesions in the S or NS exhibited earlier and greater development of the gonads, circulating androgen levels and the RSS. These data would appear to agree with the assumption presented in Section 8.3 that found lesions in the S or NS "disinhibited" courtship behavior by initiating courtship sooner and extending the courtship season.

8.4 NEUROPEPTIDES AND NEUROMODULATORS REGULATING REPRODUCTION

Available evidence in snakes (and ectotherms in general) indicates that reproductive function is regulated by interactions between environmental cues and the neural pathways controlling reproduction. However, we know very little about which neuropeptides and neural factors orchestrate seasonal changes in brain-hormone-behavior relationships. Here we review the evidence supporting a role for gonadotropin-releasing hormone, aromatase, and nitric oxide in snake reproduction. It is likely that a number of other neural signals are also involved in seasonal reproduction, including arginine vasotocin/vasopressin, kisspeptin, corticotropin releasing factor, mesotocin/oxytocin, thyroid hormones, leptin, orexin (hypocretin) neuropeptide Y, gonadotropin-inhibitory hormone, prolactin, and neurotrophic factors. For many of these potential regulators of seasonal reproduction, we know nothing about their neuroanatomical distribution or function in snakes. For the few neural signals that have been examined, we have only limited descriptive data about their presence and/or distribution in the snake brain. Mechanistic approaches to understanding how these neural signals influence seasonal changes in brain-hormone-behavior relationships are lacking, and future studies using an experimental approach are necessary for understanding the importance of these neural signals in integrating environmental control of reproduction with underlying neural circuits.

8.4.1 Gonadotropin-releasing Hormone

Traditionally, the hypothalamic gonadotropin-releasing hormone (GnRH) system has been considered to be the pinnacle of hierarchical control of pituitary and gonadal function. In snakes, data addressing the presence, distribution, and function of GnRH are very limited and mostly descriptive in nature. Studies of GnRH function in snakes are further limited because the structures of snake gonadotropins (e.g., follicle stimulating hormone, luteinizing hormone) are unknown. Sherwood and Whittier (1988) and Smith *et al.* (1997) demonstrated *via* high performance liquid chromatography and immunohistochemistry, respectively, that *Thamnophis sirtalis parietalis* have a form of GnRH most closely resembling the chicken-I GnRH form. In contrast, Licht *et al.* (1984) showed that neither chicken GnRH nor mammalian GnRH produced significant changes in plasma hormones in the Spectacled Cobra (*Naja naja*). Other immunohistochemical studies in *Elaphe climacophora, E. quadrivirgata,, E. conspicillata,* and *Rhabdophis tigrinus* also support the presence of GnRH in the snake brain, although the antiserum used in these studies was generated against the mammalian form of GnRH (Nozaki *et al.* 1984). Similar to the GnRH distribution in amphibians, birds, and mammals, immunoreactive GnRH neurons were found in the terminal nerve ganglion, nucleus of the diagonal band of Broca, medial preoptic area, hypothalamus, median eminence, and infundibulum (Nozaki *et al.* 1984; Muske 1993; Smith *et al.* 1997).

Only one study has investigated the influence of the GnRH system on reproductive behavior in snakes. Smith and Mason (1997) demonstrated that courtship behavior of male *Thamnophis sirtalis parietalis* was not influenced by intracerebroventricular administration of either chicken-I or chicken-II GnRH (although it should be noted that GnRH administration was not pulsatile in nature). The courtship parameters examined included the latency to court and the total amount of time spent courting when males were exposed to both unmated, attractive females and preparations of the female sex attractiveness pheromone (Smith and Mason 1997). In contrast, intracerebroventricular injections of D-Phe^{2,6}, Pro^{3}-GnRH, an antagonist of mammalian GnRH, significantly decreased the latency to court unmated females and significantly increased time spent courting the female attractiveness pheromone. These seemingly contradictory findings may be the result of unknown actions of D-Phe^{2,6}, Pro^{3}-GnRH on chicken-I GnRH receptor in *T. s. parietalis* (Smith and Mason 1997). Alternatively, the authors also suggest that the influence of D-Phe^{2,6}, Pro^{3}-GnRH on male courtship behavior may result from the effects of this peptide on another GnRH receptor subtype, perhaps that associated with the neuromodulatory population of GnRH neurons (e.g., Muske 1993). Interestingly, D-Phe^{2,6}, Pro^{3}-GnRH administration was not capable of inducing reproductive behavior of male snakes during the nonbreeding season (Smith and Mason 1997), indicating that the sensitivity and/or activity of the neural GnRH system changes seasonally.

The upstream mechanisms governing how the GnRH system receives, integrates, and responds to environmental cues have yet to be characterized. Although it is well established that episodic secretion of GnRH from the hypothalamus is required for normal gonadotropin release and reproductive function in mammals, the molecular mechanisms underlying the synchronous release of GnRH are unknown (e.g., Chappell *et al.* 2003). In addition, while the importance of photoperiod and melatonin in reproduction have been well established in several seasonally breeding animals (e.g., Turek and Van Cauter 1994; Mayer *et al.* 1997; Goldman *et al.* 2004), little is known about how environmentally-derived cues are integrated into a suite of synchronized reproductive responses.

Recent evidence suggests seasonal changes in GnRH signaling may be regulated by the neuropeptide kisspetin, an RFamide peptide that exerts profound and long-term effects on reproductive physiology and behavior in mammals (reviewed in Dungan *et al.* 2006; Roa and Tena-Sempere 2007). Kisspeptin directly depolarized GnRH neurons (Han *et al.* 2005) and induced significant, dose-dependent increases in the release of the gonadotropins luteinizing hormone and follicle stimulating hormone (reviewed in Dungan *et al.* 2006; Greives *et al.* 2007). Kisspeptin may also participate in coordinating seasonal reproductive events, as photoperiod modulated the number of kisspeptin-immunoreactive cells as well as the expression of kisspeptin mRNA (Revel *et al.* 2006; Greives *et al.* 2007; reviewed in Clarke *et al.* 2009 and Smith 2008, 2009).

Similar studies of the biological role of kisspeptin in reproduction in ectotherms are limited. The presence of kisspeptin receptor GPR54 has been documented in fish and amphibians (Parhar *et al.* 2004; Mohamed *et al.* 2007; Nocillado *et al.* 2007; reviewed by Roa and Tena-Sempere 2007). There is growing evidence that the kisspeptin/GPR54 system is at least partially regulated by sex steroids and likely plays a role in development and reproduction in fish (reviewed in Elizur 2009). In *Anolis carolinensis*, kisspeptin-like immunoreactive neurons were found in the preoptic area, with vesiculated fibers traveling through the paraventricular zone of the hypothalamus and preoptic area and extending into the rostral telencephalon (Dunham *et al.* 2009). No sex differences were observed in kisspeptin immunoreactivity between male and female anoles (Dunham *et al.* 2009). The documented presence of kisspeptin RFamide peptide and/or kisspeptin receptor GPR54 in at least one species of fish, amphibian, and reptile suggests that this reproductive regulatory signal is evolutionarily conserved. Whether this neuropeptide functions similarly in regulating seasonal changes in the reproductive axis of ectotherms requires further study. The potential role of other RFamide peptides such as prolactin releasing factor (also known as tuberalin and vasoactive intestinal peptide; e.g., Stirland *et al.* 2001; Johnston *et al.* 2003) and gonadotropin-inhibitory hormone (e.g., Kriegsfeld 2006; Bentley *et al.* 2009) in the neuroendocrine control of snake reproduction also warrants future attention.

8.4.2 Aromatase

Aromatase, the enzyme that catalyzes the transformation of androgens into estrogens was initially reported in the brains of rats and humans more than 30 years ago (Naftolin *et al.* 1975). Subsequent studies found the aromatase enzyme to be present in the brains of all major vertebrate groups examined (Callard *et al.* 1978a,b; Callard 1983).

Numerous studies have demonstrated that testosterone, or estrogenic metabolites, aromatized in the brain from circulating testosterone, are critical for the neonatal development of the neural pathways that will ultimately control reproductive behaviors in the adult male. In the rat and ferret, sex steroid hormones organized sexually dimorphic nuclei in the male brain (Gorski 1973; Cherry *et al.* 1990, respectively). In both the rat and ferret, organization of neural pathways depended primarily on estrogens aromatized from circulating testosterone in the brain, opposed to the direct action of androgens (George and Ojeda 1982; MacLusky *et al.* 1985; Krohmer and Baum 1989; Weaver and Baum 1991).

In adult animals, aromatization of circulating androgens appears to play an essential role in the activation of male courtship behavior and mating (Vagell and McGinnis 1997). Treatment with aromatase inhibitors significantly decreased or completely blocked the effects of systemic treatment of castrates with testosterone (Balthazart *et al.* 1990; Bonsall *et al.* 1992).

Utilizing an antibody raised against quail recombinant aromatase (Foidart *et al.* 1995a), two distinct types of aromatase-immunoreactive (ARO-ir) cells were identified in male *Thamnophis sirtalis parietalis* forebrain (Krohmer *et al.* 2002). Large type I cells exhibited intense immunostaining in both the cell body and cellular processes while smaller type II cells exhibited only weak to moderate immunostaining in the cell body with little or no staining in the processes (Fig. 8.12; Krohmer *et al.* 2002). The types of ARO-ir cells identified in the garter snake correspond to descriptions of ARO-ir cells in studies of other vertebrate species (Shinoda *et al.* 1994; Balthazart *et al.* 1996a; 2003).

In the forebrain of male *Thamnophis sirtalis parietalis* the distribution of ARO-ir cells was found to vary both regionally and seasonally. In fall-collected animals, large numbers of type II cells could be identified

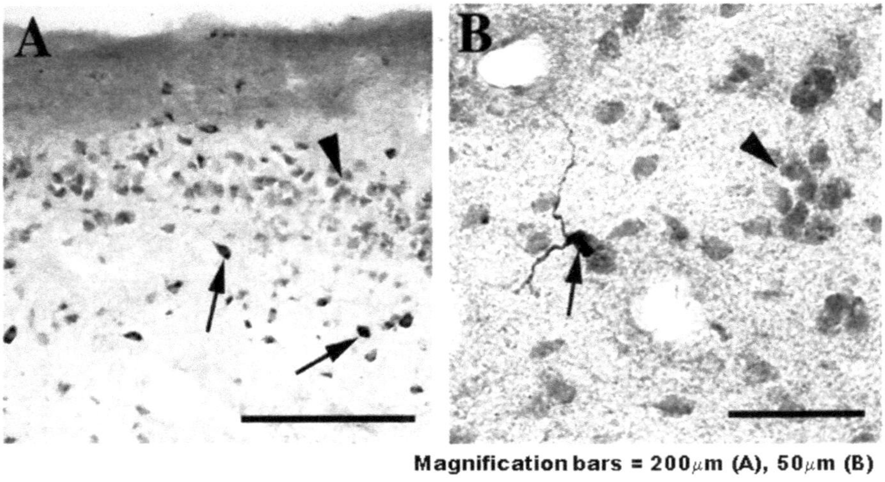

Fig. 8.12 Photomicrographs of the two types of ARO-ir cells in the brain of male *Thamnophis sirtalis parietalis*. **A** and **B**. Sections through the POA, which contain both types of immunoreactive cells. Type I ARO-ir cells (arrows) have larger, darker staining perikarya and visible processes. Type II ARO-ir cells (arrowheads) have smaller perikarya and few or no visible processes.

in all regions of the forebrain, whereas a much smaller number of type I cells were concentrated primarily in regions found to be critical for the control and maintenance of reproductive behaviors (Fig. 8.13; Friedman and Crews 1985a; Krohmer and Crews 1987a,b; Krohmer *et al.* 2002). In actively courting animals collected in the spring, the number of type II cells were drastically reduced and found primarily in those regions critical for the control of reproductive behaviors while the greatest concentration of type I cells could be identified in the olfactory region (Fig. 8.14; Krohmer, unpublished data).

The distribution of type I immunoreactive cells found in the brain of *Thamnophis sirtalis parietalis* corresponded to regions of elevated aromatase activity and/or areas of dense populations of ARO-ir cells in birds and mammals (Roselli *et al.* 1985; Panzica *et al.* 1994; Foidart *et al.* 1995b; Wagner and Morrell 1996; Balthazart *et al.* 1996a,b; Veney and Rissman 1998). To date, no direct comparison of types and distribution of ARO-ir structures are available for any other reptile.

Although aromatase activity (AA) has been documented in all classes of vertebrates (Callard *et al.* 1977; 1978a,b), relatively few studies have been conducted on reptiles. In fact, the vast majority of AA studies in reptiles have been conducted on lizards, animals exhibiting an associated reproductive pattern (Gobbetti *et al.* 1994; Wade 1997). In *Anolis carolinensis* testosterone was reported to facilitate sexual behavior in males (Winkler and Wade 1998; O'Bryant and Wade 2002) while conversion of testosterone to estrogens aids receptivity in females (Winkler and Wade 1998). Moreover, Rosen and Wade (2001) found that courtship in male *A. carolinensis* was partially mediated by 5α-reductase, the enzyme that converts testosterone to dihydrotestosterone. Nevertheless, AA was greater in breeding males compared to non-breeding males while there was no sex or seasonal variation in 5α-reductase activity.

Callard *et al.* (1978a) found AA in the forebrain of female Brown Watersnakes (*Natrix taxispilota*) but no activity was detected in the mid- or hindbrain. However, this study did not examine specific regions within the forebrain or examine possible seasonal variations. Aromatase activity in the forebrain of male *Thamnophis sirtalis parietalis* has been found to vary regionally as well as seasonally (Krohmer *et al.* 2010). In the spring, actively courting males have significantly higher AA in the olfactory region compared to the AHPOA, S or NS. However, in animals collected in the fall as they are returning to the dens prior to LTD, AA was significantly greater in the AHPOA and S compared to the O and NS (Fig. 8.15). These findings support earlier studies that have suggested the ability of target cells in the brain to metabolize androgens can be modified depending on hormonal status and/or environmental factors (Hutchison 1974a,b; Hutchison *et al.* 1986). Consequently, regional and seasonal differences in AA in the brain of *T. s. parietalis* may play a critical role in the priming of specific neural pathways that control stereotypical behavior patterns at the appropriate time.

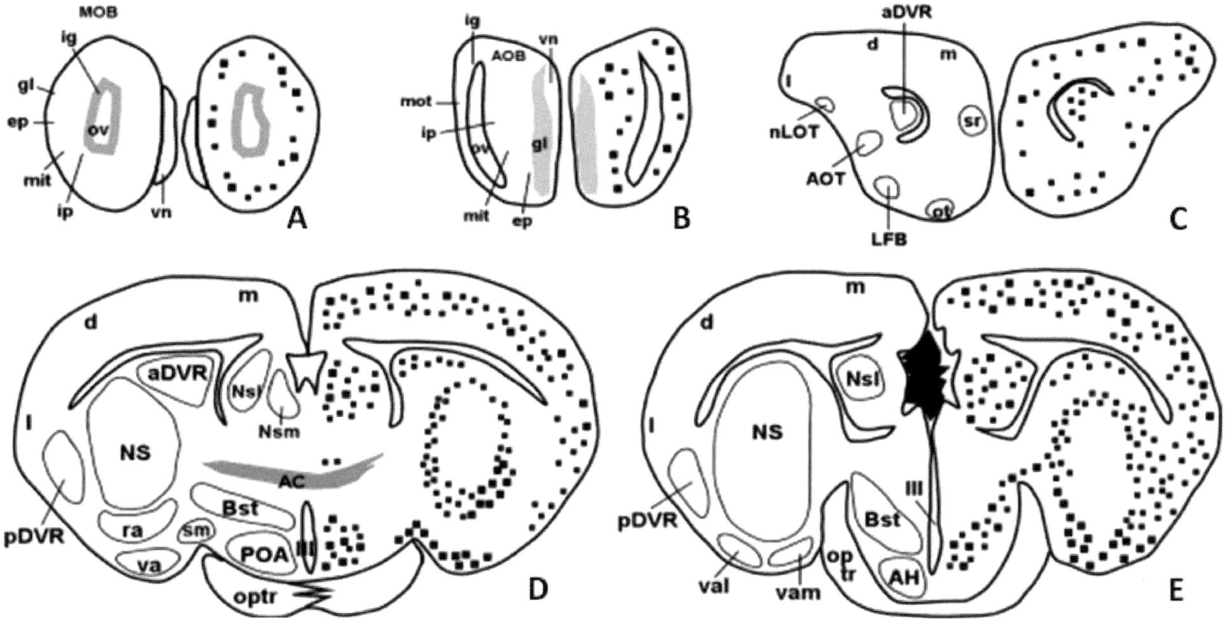

Fig. 8.13 Schematic drawings of coronal sections through the forebrain of fall collected male *Thamnophis sirtalis parietalis* illustrating the distribution of ARO-ir structures in a rostral to caudal order (**A** to **E**). The location of ARO-ir perikarya (dots) is illustrated on the right half of the drawings and structure names shown on the left half. The density of the symbols has been adjusted to give a qualitative estimate of the number of immunoreactive structures. Large dots represent densely stained cells usually showing a few visible processes (type I); small dots represent weakly immunoreactive cells in which no stained processes are visible (type II). III, third ventricle; aDVR, anterior dorsal ventricular ridge; AHPOA, anterior hypothalamus-preoptic area; AC, anterior commissure; AOB, accessory olfactory bulb; AOT, accessory olfactory tract; Bnst, bed nucleus of the stria terminalis; d, dorsal cortex; l, lateral cortex; LFB, lateral forebrain bundle; m, medial cortex; MOB, main olfactory bulb; nLOT, nucleus of the lateral olfactory tract; NS, nucleus sphericus; Nsl, lateral septal nucleus Nsm, medial septal nucleus; optr, optic tract; olfactory tubercle; pDVR, posterior dorsal ventricular ridge; POA, preoptic area; ra, rostral amygdaloid nucleus; sm, strial medullaris; sr, rostral septal nucleus; va, ventral amygdaloid nucleus.

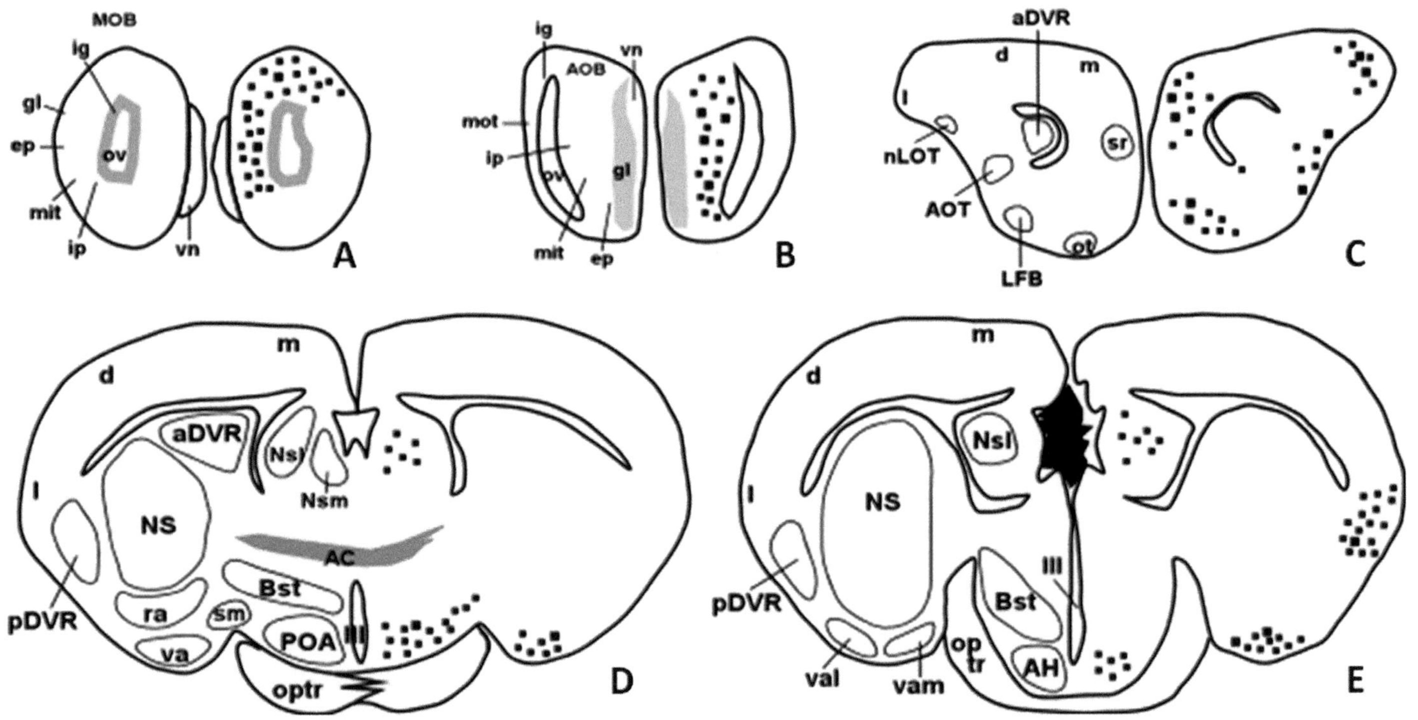

Fig. 8.14 Schematic drawings of coronal sections through the forebrain of spring collected *Thamnophis sirtalis parietalis* illustrating the distribution of ARO-ir structures in a rostral to caudal order (**A** to **E**). See legend of Fig. 8.13 for additional comments.

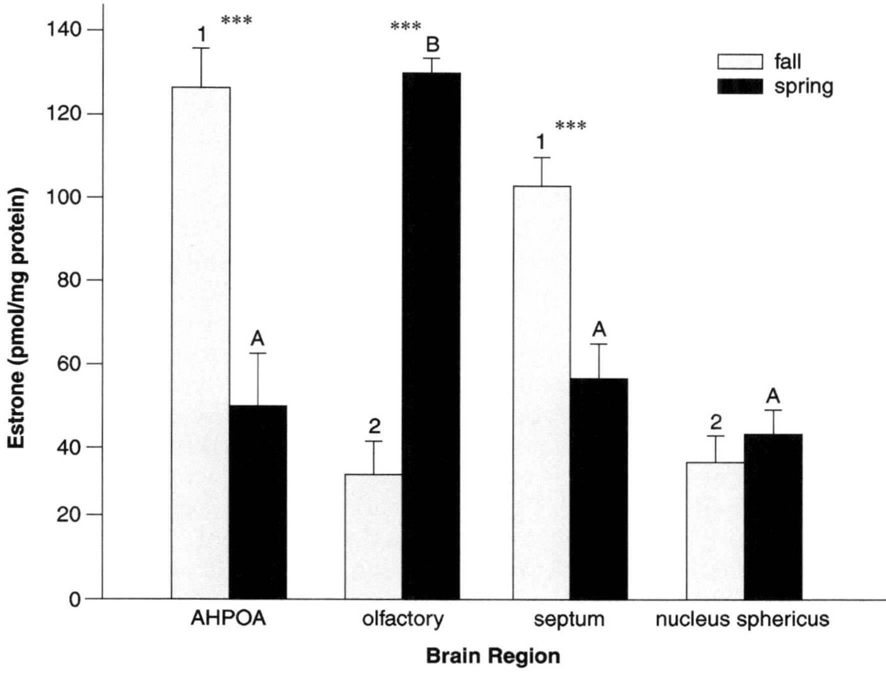

Fig. 8.15. Estrone production (pmol/mg protein; mean + 1 SEM) of punch microdissected brain areas of male *Thamnophis sirtalis parietalis* in spring (black bars) and fall (grey bars). Asterisks indicate significant differences between spring and fall estrone concentrations within each brain region. Capital letters represent statistical differences in estrone production among regions during the spring, while the numbers represent statistical differences among regions within the fall. Estrone concentrations were significantly higher during the spring in the olfactory region, while both the AHPOA and septum produced significantly more estrone in the fall. In the spring, estrone levels were highest in the olfactory region. In the fall, estrone levels were significantly higher in the AHPOA and septum. **Redrawn** from Krohmer *et al.* 2010. Hormones and Behavior 58: 485-492,, Fig. 4.

The elevated AA observed in the S and AHPOA of male snakes collected in the fall (Krohmer *et al.* 2010), in combination with the known importance of the AHPOA in regulating reproduction (Friedman and Crews 1985a; Krohmer and Crews 1987a), suggests that the conversion of androgens to estrogens in the POA in fall animals might be essential to the expression reproductive behaviors in the spring. Although mating behavior in *Thamnophis sirtalis parietalis* does not coincide with peak gonadal activity, this does not automatically rule out a role for sex steroid hormones in regulating reproductive behavior. Furthermore, because reproduction occurs immediately following spring emergence, the concomitant changes in neurophysiology and behavior that accompany reproduction are likely to occur during winter dormancy (Lutterschmidt and Mason 2009).

To date, the only cue found to initiate courtship and mating in *Thamnophis sirtalis parietalis* is a period of LTD. In addition, it has been demonstrated that the duration of time in LTD also plays an important role (Fig. 8.4; Garstka *et al.* 1982). But what is the function of LTD? Recently, it was found that the aromatase enzyme in the brains of *T. s. parietalis* remained active at LTD temperatures (5±1.5ºC; Krohmer, unpublished data) In this study, an aromatase assay was run under conditions of LTD for 4, 8, 12 and 16 weeks, corresponding to the time periods used in Garstka *et al.* (1982). Estrogen accumulation in the S/AHPOA homogenate increased significantly between 4 and 16 weeks, corresponding with the observed increase in courtship intensity reported by Garstka (1982) (Fig. 8.16). Thus, in a dissociated breeder, sex steroid hormones may be playing a role during LTD to prime the nervous system for the expression of mating behavior upon emergence in the spring. The seasonal variation in AA found in the olfactory region (O) of *Thamnophis sirtalis parietalis* (Krohmer *et al.* 2010) may offer a mechanism that could explain trailing behavior in garter snakes (Ford 1981, 1982). These studies found that in garter snakes, the capability of males to trail conspecific females is altered during the active season (Ford 1981; O'Donnell *et al.* 2004). During the spring mating season, when male garter snakes exhibit strong trailing behavior (Ford 1981, 1982; LeMaster and Mason 2001; O'Donnell *et al.* 2004), AA in the O is also elevated. In contrast, trailing behavior decreased in the summer and fall, coinciding with periods of low AA in the O (Ford 1981, 1982). Therefore, seasonal changes in AA in the O might mediate seasonal changes in trailing behavior.

Further research is needed to examine how the conversion of androgens to estrogens during fall and the prolonged period of LTD might affect the

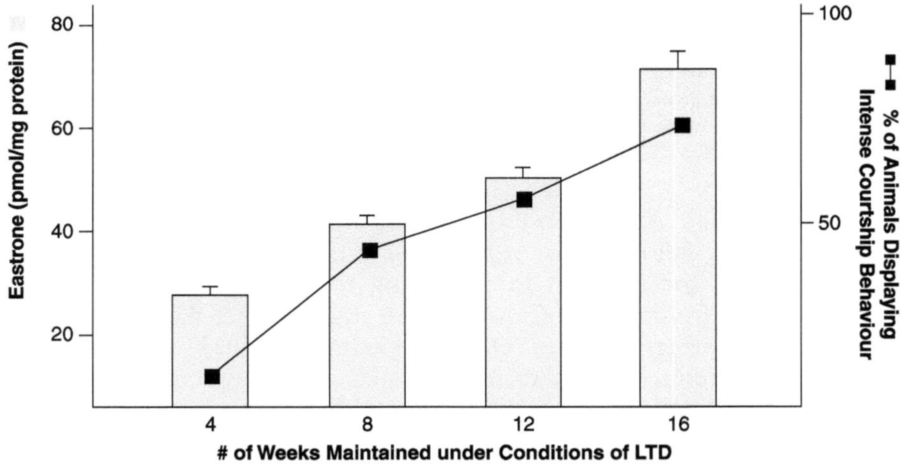

Fig. 8.16 Effect of low temperature dormancy on aromatase activity and courtship intensity. AA, measured as estrone produced (pmol/mg protein) (n=8/group) increased significantly ($F_{3,127}$=52.91, $p<0.001$) from 4 to 16 weeks (gray bars). Intensity of courtship behavior also increased as length of time in LTD increased. These data suggest that estrogens, aromatized from androgens in the brain, may play a critical role in the "set up" of the neural pathways controlling courtship behavior and mating in the male *Thamnophis sirtalis parietalis*.

neuroanatomy and neurophysiology controlling reproductive behaviors. In addition, studies addressing the influence of melatonin signaling on aromatase expression and activity would provide information regarding the neuroendocrine mechanisms regulating neural aromatase activity. These studies would advance our understanding of how reproduction is regulated in animals exhibiting a dissociated reproductive strategy.

8.4.3 Nitric Oxide

Nitric oxide (NO), first identified as a smooth muscle regulator, may also act as a neurotransmitter (Nelson *et al.* 1997). Since NO has a half-life of less than five seconds many studies have examined NO indirectly by assessing its synthetic pathways. The NO producing enzyme, nitric oxid synthase (NOS) is broadly distributed in the mammalian and avian brain, particularly in steroid-sensitive areas that are implicated in the control of reproductive behavior such as the POA and ventromedial nucleus of the hypothalamus (VMH).

The expression of NOS in the POA appears to be controlled by sex steroids. Castration decreases NOS immunoreactivity (NOS-ir) in the medial preoptic nucleus of male hamsters and rats (Hadeishi and Wood 1996; Du and Hull 1999). However, this effect can be reversed by administration of exogenous testosterone. Furthermore, the effects of testosterone on NO can be mimicked by its estrogenic metabolites but not by non-aromatizable androgens. Subsequently, many NOS-ir neurons have been found to contain both androgen and estrogen receptors.

In male *Thamnophis sirtalis parietalis*, NOS-ir neurons could be identified throughout the forebrain. However, NOS activity appeared to be greater in areas previously involved in the initiation and maintenance of reproductive behavior (Krohmer, unpublished data). In addition, there was an apparent co-localization of NOS and ARO enzyme within the pathways (and possibly the same cells) regulating courtship and mating in the male garter snake (Figs. 8.17, 8.18; Krohmer, unpublished data).

These data agree with studies on the Japanese quail (Panzica *et al.* 2000; Balthazart *et al.* 2003), in which the distribution of NOS was found to overlap with steroid sensitive cells in the medial POA, the Bst and the VMH (Panzica *et al.* 1994). Moreover, there was a dense collection of NOS-ir cells in areas known to contain large populations of ARO expressing neurons. In the quail, it appeared that NOS and ARO enzymes were co-localized, but did not exist within the same cells (Balthazart *et al.* 2003).

8.4.4 Other Neuropeptides and Neuromodulators

Neurohypophyseal factors: Oxytocin. Like most vertebrates, snakes possess two main neurohypophyseal factors. The neurohypophyseal hormones arginine vasopressin and oxytocin are part of a peptide superfamily that resulted from a duplication of the gene encoding the precursor molecule and subsequent mutations to produce the vasopressin-like and oxytocin-

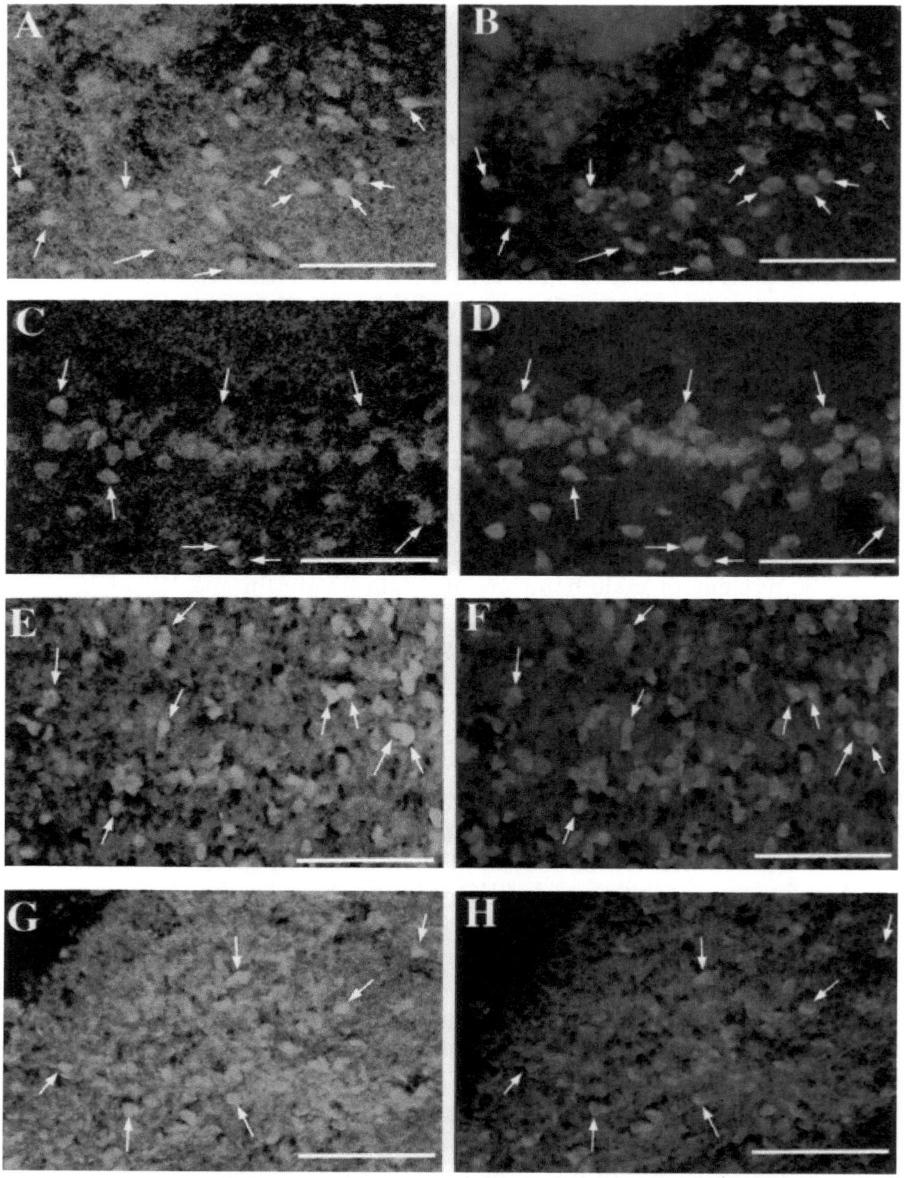

Fig. 8.17 Photomicrographs of male *Thamnophis sirtalis parietalis* brain. Sections simultaneously stained by double label immunocytochemistry for NOS (FITC green label) and aromatase (RITC red label) illustrating the anatomical relationships between these two antigens. Panels located at the same level in the left (NOS) and right (ARO) columns illustrate the same section visualized with different filters. **A-B.** Main olfactory bulb. **C-D.** Medial cortex. **E-F.** Lateral cortex. **G-H.** Septum. Arrows point to examples of cells showing clear co-localization of NOS and aromatase. Magnification bars = 100 μm.

Color image of this figure appears in the color plate section at the end of the book.

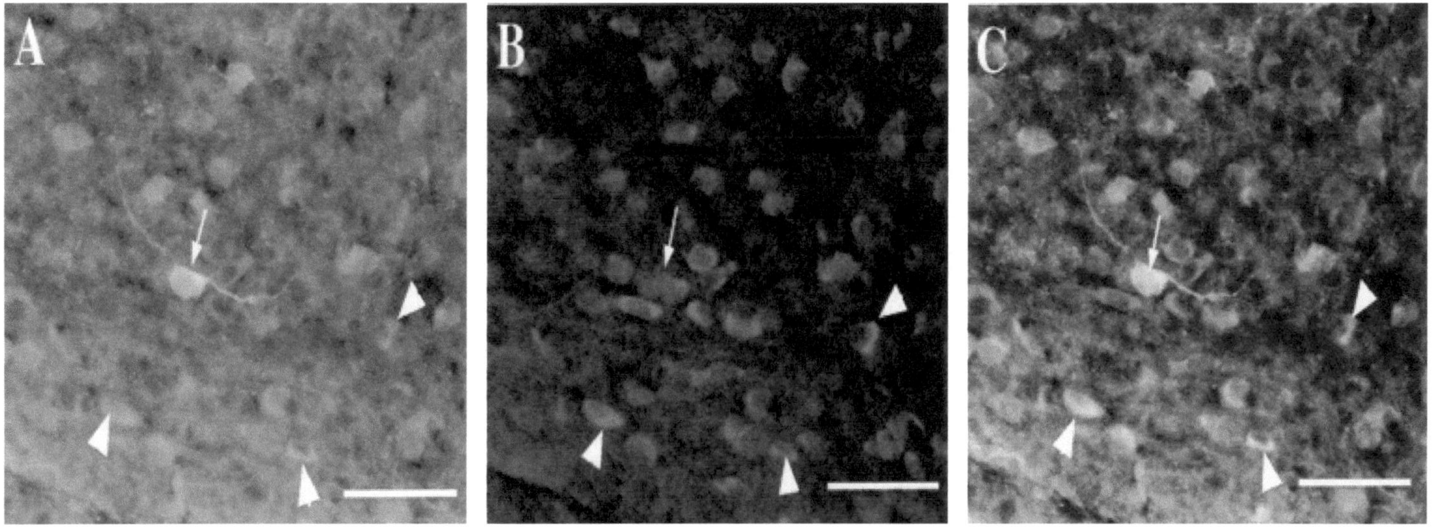

Fig. 8.18 Photomicrographs of the male *Thamnophis sirtalis parietalis* brain showing a section in the preoptic area stained with NOS (FITC green label) and aromatase (RITC red label) illustrating the colocalization between these two antigens. **A.** FITC filters illustrating NOS-immunoreactivity in one type I cell (arrow) and type II cell (arrow heads). **B.** Same section viewed under RITC filter showing ARO-ir cells in the same region. Arrow/Arrowheads point to cells illustrating the co-localization. **C.** Computer reconstruction of the two immunoreactive signals illustrating the co-localization of the NOS (green) and aromatase (red) immunoreactivity. Magnification bar = 50 µm.

Color image of this figure appears in the color plate section at the end of the book.

like lineages (Sawyer 1977; Acher *et al.* 1995; Venkatesh *et al.* 1997; Hoyle 1998). In amphibians, reptiles, birds, and marsupials the oxytocin-like peptide is reportedly mesotocin (see review in Acher *et al.* 1995). However, there is some conflicting evidence regarding the identity of the oxytocin-like peptide in snakes. Follett (1967) and Pickering (1967) reported the presence of oxytocin in hypophysial extracts of the snakes *Naja naja* and *Tropidonotus natrix*. Perez-Figares *et al.* (1995) demonstrated the presence of oxytocin-like immunoreactivity in the hypothalamus and neurohypophysis of the snake *Natrix maura*. However, most evidence indicates that the oxytocin-like peptide in snakes is mesotocin (*Bothrops jararaca*: Silveira *et al.* 2002; Lazari *et al.* 2006; *Elaphe quadrivirgata*: Acher *et al.* 1969; *Naja naja*: Pickering 1967; Acher *et al.* 1969, *Natrix maura*: Fernández-Llebrez *et al.* 1988; Mancera *et al.* 1991; Andrades *et al.* 1994; *Vipera aspis*: Acher *et al.* 1968, 1969). If snakes do express oxytocin peptide, it may simply be a result of random genetic drift (e.g., Acher 1996). For example, some marsupials have both mesotocin and oxytocin, but the presence of oxytocin in this case is thought to be the result of a neutral mutation rather than selective evolution (Bathgate *et al.* 1995). Comparative studies evaluating the effects of mesotocin and oxytocin would not only provide information about the functional roles of these neuropeptides in snakes but would also provide experimental evidence as to whether snakes possess a functional oxytocin signaling system.

Neuropeptides associated with metabolic regulation and energy balance. Seasonal reproduction in animals is accompanied by both environmental and energetic challenges. Appropriate timing of gamete production, mating, gestation, and birth are crucial to reproductive fitness, and both males and females must have adequate energy stores to sustain these events. Neural factors related to energy balance regulation that may influence the control of reproduction include corticotropin-releasing factor, leptin, orexin (hypocretin), and neuropeptide Y, especially in light of the known importance of body condition and body fat reserves to snake reproduction (e.g., Bonnet *et al.* 1994, 2001, 2002; Bonnet and Naulleau 1996; Aubret *et al.* 2002; Taylor *et al.* 2005). While both resource availability and energy reserves are known to be potent modulators of reproduction in snakes (see review in Shine 2003), we have little empirical evidence regarding the neuroendocrine mechanisms by which these extrinsic factors are relayed to and modulate the reproductive axis.

In *Thamnophis sirtalis parietalis*, intracerebroventricular administration of neuropeptide Y to male snakes inhibited courtship behavior and induced feeding behavior (Morris and Crews 1990). These effects of neuropeptide Y administration were transient, as snakes resumed pre-surgery courtship levels and refused food 24 h after surgery (Morris and Crews 1990). During the spring mating season, male *T. s. parietalis* show neuropeptide Y immunoreactivity throughout the brain, although the cell bodies are small and sparsely distributed (Lutterschmidt, unpublished data). Whether seasonal changes in neuropeptide Y signaling contribute to the seasonal transition between courtship and foraging behavior is unknown.

No studies have examined the presence or function of leptin or orexin in snakes, an obvious area for future research.

Because the hypothalamus-pituitary-adrenal (HPA) axis functions in energy allostasis, it provides a mechanism by which information about an individual's energy balance can be incorporated into "decisions" regarding reproductive physiology and behavior, parental care, territorial aggression, etc. Corticotropin-releasing factor, the hypothalamic neurohormone that regulates the HPA axis, may therefore be an important neuromodulator of snake reproduction. In the only study to examine corticotropin-releasing factor in snakes, Mancera *et al.* (1991) found immunoreactive neurons and nerve fibers in the brain of *Natrix maura*. Immunoreactive structures were located in the dorsal cortex, *nucleus accumbens*, amygdala, subfornical organ, *lamina terminalis*, nucleus of the paraventricular organ, nucleus of the oculomotor nerve, nucleus of the trigeminal nerve, and reticular formation. In addition, corticotropin-releasing factor co-localized with both arginine vasotocin and mesotocin in some neurons of the paraventricular nucleus (Mancera *et al.* 1991).

Thyroid hormones also may be an important regulator of seasonal reproduction in snakes, particularly with respect to temperature-induced reproductive activity. As in most vertebrates, thyroid hormones regulate metabolism in snakes (Thapliyal *et al.* 1975; Wong *et al.* 1975; Chiu and Woo 1988; Etheridge 1993). However, thyroid hormones are also sensitive to temperature cues (e.g., Turner and Tipton 1972a,b) and may play a role in regulating reproduction (e.g., Wong and Chiu 1974; Nilson 1982; Naulleau *et al.* 1987). Chiu and Lynn (1971) reported that treatment of hypophysectomized male Glossy Snakes (*Arizona elegans*) with thyroid stimulating hormone (TSH) stimulated and restored spermatogenesis [although the effect of TSH in this case may result from cross-reactivity of the hormone with gonadotropin receptors (Chiu and Lynn 1971)]. Several studies have also documented seasonal variation in the thyroid gland and/or thyroid hormones in snakes, particularly in relation to cycles of hibernation and reproduction (Wong and Chiu 1974; Bona-Gallo *et al.* 1980; Garstka *et al.* 1982; Nilson 1982; Fleury and Naulleau 1987; Naulleau *et al.* 1987; Chiu and Lam 1994). Lastly, Mohanty and Naik (1997) identified TSH immunoreactive cells in the pituitary *pars distalis* as well as the hypothalamus and median eminence of the Rat Snake, *Ptyas mucosus*.

Recent evidence suggests that photoperiodic induction of seasonal breeding in birds and mammals is regulated by changes in localized thyroid hormone metabolism in the basal hypothalamus (e.g., Hanon *et al.* 2008; Nakao *et al.* 2008). Importantly, thyroid hormones regulate both neural development and neuroplasticity and affect the expression of many genes involved in synapse formation (e.g., Zhang *et al.* 2008). Thus, thyroid hormone-induced neuroplasticity within the hypothalamus underlies seasonal changes in the reproductive axis of birds and mammals. Whether seasonal neuroplasticity contributes to rhythms in reproductive function in snakes and other ectothermic vertebrates requires investigation.

Neurogenesis. There is increasing evidence that seasonal rhythms in physiology and behavior may represent seasonal cycles in brain development (see review in Kaslin *et al.* 2008). For example, many developmental processes in the brain vary seasonally, including cell proliferation, cell death, and neuron migration (e.g., Morgan *et al.* 2006; Ebling and Barrett 2008). Indeed, the first demonstration that adult neurogenesis could occur was in the neural circuitry controlling seasonal changes in song production in birds (see review in Nottebohm 2004). Preliminary studies using bromodeoxyuridine (BrdU) immunoreactive labeling indicated that the brain of adult male *Thamnophis sirtalis parietalis* also shows cell proliferation and migration during the spring mating season (Lutterschmidt, unpublished data). Although seasonal cycles in neurogenesis are pervasive across non-mammalian vertebrates, in most instances the functional significance of such changes in the adult brain is unknown.

8.5 CONCLUSIONS AND FUTURE DIRECTIONS

Collectively, the available data examining reproductive regulatory mechanisms in snakes indicate that, as in many other ectothermic vertebrates, environmental temperature plays a central role in regulating reproductive physiology and behavior. An important challenge is now to understand how interacting photoperiod and temperature cues influence the neural and neuroendocrine signals that regulate reproduction. A number of studies have begun to address these issues. For example, we now know that cold temperature exposure, in the absence of changing photoperiod cues, is sufficient in modulating diel melatonin rhythms of garter snakes (Lutterschmidt and Mason 2008, 2009). Melatonin in turn modulates reproductive behavior (Lutterschmidt *et al.* 2004) and can influence neuropeptide signaling in the brain (e.g., arginine vasotocin; Lutterschmidt, unpublished data). We also know that snakes, like other vertebrates, exhibit seasonal changes in neural aromatase activity (Krohmer *et al.* 2010). Further research is necessary to examine how changes in neuroendocrine signaling during the fall and winter dormancy period, in combination with low temperature exposure, function in inducing the changes in neuroanatomy and neurophysiology necessary to elicit spring reproductive behavior. Such studies would significantly advance our understanding of how reproduction is regulated in animals exhibiting different reproductive strategies.

A number of studies have also demonstrated that snakes express many of the same neural and neuroendocrine regulatory factors observed in other vertebrates: gonadotropin-releasing hormone, arginine vasotocin, aromatase, nitric oxide, mesotocin, corticotropin releasing factor, neuropeptide Y, and thyroid hormones. However, there is a lack of a mechanistic approach to understanding how these neural signals influence seasonal changes in brain-hormone-behavior relationships. In addition, much of our current knowledge is limited to studies of one population, *Thamnophis sirtalis parietalis* in Manitoba, Canada.

Future research comparing the reproductive regulatory mechanisms across snake taxa with different life history strategies would provide a valuable context with which the neuroendocrine control of reproduction may be better understood. Studies addressing the molecular biology underlying these reproductive regulatory mechanisms are also needed. Such studies would not only greatly advance our understanding of the neuroendocrine regulation of reproduction, but they would also significantly improve our ability to integrate multiple levels of reproductive regulation—from the level of individual genes to behavior to the selection pressures that produce geographic variation in reproductive patterns and the timing of breeding. This level of integration will prove invaluable not only for our basic understanding of how environmental and neuroendocrine factors are integrated into a coordinated reproductive response, but also for informing conservation efforts regarding the influence of environmental perturbations such as global climate change on the reproductive biology of vertebrates.

8.6 ACKNOWLEDGMENTS

We would like to thank the following people for their helpful feedback and discussions of various aspects of this review: Emma J. Coddington, Leslie A. Dunham, Chris R. Friesen, Michael P. LeMaster, M. Rockwell Parker, Scarlett J. Salem, Walter Wilczynski, and two anonymous reviewers. R. W. K. is funded, in part, by the Center for Educational Practice and the Dean's Fund, Saint Xavier University.

8.7 LITERATURE CITED

Acher, R. 1996. Molecular evolution of fish neurohypophysial hormones: Neutral and selective evolutionary mechanisms. General and Comparative Endocrinology 102: 157–172.

Acher, R., Chauvet, J. and Chauvet, M. T. 1968. Neurohypophysial hormones of reptiles: Isolation of mesotocin and vasotocin of the viper *Vipera aspis*. Biochimica et Biophysica Acta 154: 255-257.

Acher, R., Chauvet, J. and Chauvet, M. T. 1969. The neurohypophysial hormones of reptiles: Comparison of the viper, cobra, and elaphe active principles. General and Comparative Endocrinology 13: 357–360.

Acher, R., Chauvet, J. and Chauvet, M. T. 1995. Man and the chimaera. Selective versus neutral oxytocin evolution. Advances in Experimental Medicine and Biology 395: 615-627.

Aleksiuk, M. 1970. The effects of in vivo light and temperature acclimation on in vitro responses of heart rate to temperature in a cold climate reptile, *Thamnophis sirtalis parietalis*. Canadian Journal of Zoology 48: 1155-1161.

Aleksiuk, M. 1971. Temperature dependent shifts in the metabolism of a cold climate reptile, *Thamnophis sirtalis parietalis*. Comparative Biochemistry and Physiology 39A: 495-503.

Aleksiuk, M. and Stewart, K. W. 1971. Seasonal changes in the body composition of the garter snake (*Thamnophis sirtalis parietalis*) at northern latitudes. Ecology 52: 485-490.

Aleksiuk, M. and Gregory, P. T. 1974. Regulation of seasonal mating behavior in *Thamnophis sirtalis parietalis*. Copeia 1976: 681-689.

Andrades, J. A., Pérez, J. and Fernández-Llebrez, P. 1994. Combined use of immunocytochemistry and lectin histochemistry for the study of the hypothalamic neurosecretory system of the snake *Natrix maura* (L.). Annals of Anatomy 176: 259-261.

Aubret, F., Bonnet, X., Shine, R. and Lourdais, O. 2002. Fat is sexy for females but not males: The influence of body reserves on reproduction in snakes (*Vipera aspis*). Hormones and Behavior 42: 135-147.

Axelrod, J. 1974. The pineal gland: a neurochemical transducer. Science 184: 1341-1348.

Baird, T. A., Fox, S. F. and McCoy, J. K. 1997. Population differences in the roles of size and coloration in intra- and intersexual selection in the collared lizard, *Crotaphytus collaris*: Influence of habitat and social organization. Behavioral Ecology 8: 506-517.

Ball, G. F. and Balthazart, J. 2002. Neuroendocrine mechanisms regulating reproductive cycles and reproductive behavior in birds. Pp. 649-798. In D. W. Pfaff, A. P. Arnold, A. M. Etgen, S. E. Fahrbach and R. Y. Rubin (eds), *Hormones, Brain and Behavior*, vol 2. Academic Press. San Diego, California.

Balthazart J., Panzica G. C. and Krohmer R. W. 2003. Anatomical relationships between aromatase-immunoreactive neurons and nitric oxide synthase as evidenced by NOS immunohistochemistry or NADPH diaphorase histochemistry in the quail forebrain. Journal of Chemical Neuroanatomy 25: 39-51.

Balthazart, J., Tlemcani, O. and Harada, N. 1996a. Localization of testosterone sensitive and sexually dimorphic aromatase-immunoreactive cells in the quail preoptic area. Journal of Chemical Neuroanatomy 11: 147-171.

Balthazart, J., Absil, P., Foidart, A., Houbart, M., Harada, N. and Ball, G. F. 1996b. Distribution of aromatase immunoreactive cells in the forebrain of zebra finches (*Taeniopygia guttata*): implications for the neural action of steroids and nuclear definition in the avian hypothalamus. Journal of Neurobiology 31: 129-148.

Balthazart, J., Foidart, A., Surelmont, C., Vockel, A. and Harada, N. 1990. Distribution of aromatase in the brain of the Japanese quail, ringdove, and zebra finch: an immunocytochemical study. Journal of Comparative Neurology 301: 276-288.

Barfield, R. J. 1965. Effects of preoptic lesions on the sexual behavior of male domestic fowl. American Zoologist 5: 686-687.

Bathgate, R. A., Parry, L. J., Fletcher, T. P., Shaw, G., Renfree, M. B., Gemmell, R. T. and Sernia, C. 1995. Comparative aspects of oxytocin-like hormones in marsupials. Advances in Experimental Medicine and Biology 395: 639-655.

Baum, M. J. 2004. Neuroendocrinology of sexual behavior in the male. Pp. 153-204. In J. B. Becker, S. M. Breedlove, D. Crews and M. M. McCarthy (eds), *Behavioral Endocrinology*, 2nd ed. The MIT Press, MA.

Beck, L. A. and Wade, J. 2009a. Effects of estradiol, sex and season on estrogen receptor alpha mRNA expression and forebrain morphology in adult green anole lizards. Neuroscience 160: 577-586.

Beck, L. A. and Wade, J. 2009b. Morphology and estrogen receptor alpha mRNA expression in the developing green anole forebrain. Journal of Experimental Zoology 311: 162-171.

Beck, L. A., O'Bryant, E. L. and Wade, J. 2008. Sex and seasonal differences in morphology of limbic forebrain nuclei in the green anole lizard. Brain Research 1227: 68-75.
Becker, J. B., Breedlove, S. M., Crews, D. and McCarthy M. M. (eds). 2002. *Behavioral Endocrinology*, 2nd ed. The MIT Press, Cambridge, Massachusetts. Pp. 776.
Bentley, G. E., Ubuka, T., McGuire, N. L., Calisi, R., Perfito, N., Kriegsfeld, L. J., Wingfield, J. C. and Tsutsui, K. 2009. Gonadotrophin-inhibitory hormone: a multifunctional neuropeptide. Journal of Neuroendocrinology 21: 276-281.
Binkley, S., Riebman, J. and Reilly, K. 1978. The pineal gland: A biological clock in vitro. Science 202: 1198-1201.
Birks, E. K. and Ewing, R. D. 1986. Seasonal changes in pineal melatonin content and hydroxyindole-*O*-methyltransferase activity in juvenile chinook salmon, *Oncorhynchus tshawytscha*. General and Comparative Endocrinology 64: 91-98.
Bishop, J. E. 1959. A histological and histochemical study of the kidney tubule of the common garter snake, *Thamnophis sirtalis*, with special reference to the sexual segment in the male. Journal of Morphology 104: 307-358.
Bittman, E. L., Demsey, R. J. and Karsch, F. J. 1983. Pineal melatonin secretion drives the reproductive response to daylength in the ewe. Endocrinology 113: 2276-2283.
Bolliet, V., Ali, M. A., Lapointe, F.-J. and Falcon, J. 1996. Rhythmic melatonin secretion in different teleost species: An *in vitro* study. Journal of Comparative Physiology 165B: 677-683.
Bona-Gallo, A. and Licht, P. 1983. Effect of temperature on sexual receptivity and ovarian recrudescence in the garter snake, *Thamnophis sirtalis parietalis*. Herpetologica 39: 173-182.
Bona-Gallo, A., Licht, P., MacKenzie, D. S. and Lofts, B. 1980. Annual cycles in levels of pituitary and plasma gonadotropin, gonadal steroids, and thyroid activity in the Chinese cobra (*Naja naja*). General and Comparative Endocrinology 42: 477-493.
Bonnet, X. and Naulleau, G. 1996. Are body reserves important for reproduction in male dark green snakes (*Coluber viridiflavus*)? Herpetologica 52: 137-146.
Bonnet, X., Naulleau, G. and Mauget, R. 1994. The influence of body condition on 17b-estradiol levels in relation to vitellogenesis in female *Vipera aspis* (Reptilia, Viperidae). General and Comparative Endocrinology 93: 424-437.
Bonnet, X., Naulleau, G., Shine, R. and Lourdais, O. 2001. Short-term versus long-term effects of food intake on reproductive output in a viviparous snake (*Vipera aspis*). Oikos 92: 297-308.
Bonnet, X., Lourdais O., Shine R. and Naulleau G. 2002. Reproduction in a typical capital breeder: Costs, currencies and complications in the *Aspic viper*. Ecology 83: 2124-2135.
Bonsall, R .W., Clancy, A. N. and Michael, R. P. 1992. Effects of the nonsteroidal aromatase inhibitor, fadrazole, on sexual behavior in male rats. Hormones and Behavior 26: 240-254.
Brookhart, J. M. and Dey, F. L. 1941. Reduction of sexual behavior in male guinea pigs by hypothalamic lesions. American Journal of Physiology 133: 551-554.
Brown, J. L., Li, S-H. and Bhagabati, N. 1999. Long term trend toward earlier breeding in an American bird: A response to global warming? Proceedings of the National Academy of Sciences USA 96: 5565-5569.
Callard, G. V. 1983. Androgen and estrogen actions in the vertebrate brain. American Zoolologist 23: 607-620.

Callard, G. V., Petro, Z. and Ryan, K. J. 1977. Identification of aromatase in the reptilian brain. Endocrinology 100: 1214-1218.

Callard, G. V., Petro, Z. and Ryan, K. J. 1978a. Phylogenetic distribution of aromatase and other androgen-converting enzymes in the central nervous system. Endocrinology 103: 2283-2290.

Callard, G. V., Petro, Z. and Ryan, K. J. 1978b. Conversion of androgen to estrogen and other steroids in the vertebrate brain. American Zoologist 18: 511-523.

Camazine, B., Garstka, W., Tokarz, R. R. and Crews, D. 1980. The effects of castration and androgen replacement therapy on male courtship behavior in the red-sided garter snake (*Thamnophis sirtalis parietalis*). Hormones and Behavior 14: 358-372.

Carter, D. S. and Goldman, B. D. 1983. Antigonadal effects of timed melatonin infusion in pinealectomized male Djungarian hamsters (*Phodopus sungorus sungorus*): Duration is the critical parameter. Endocrinology 113: 1261-1267.

Cassone, V. M. 1990. Melatonin: Time in a bottle. Oxford Reviews of Reproductive Biology 12: 319-367.

Cassone, V. M. and Natesan, A. K. 1997. Time and time again: The phylogeny of melatonin as a transducer of biological time. Journal of Biological Rhythms 12: 489-497.

Chanda, S. and Biswas, N. M. 1982. Effect of morning and evening injections of melatonin on the testis of toad (*Bufo melanosticus*). Endocrinology Japan 29: 483-485.

Chanda, S. and Biswas, N. M. 1984. Influence of pineal gland on spermatogenesis of toad (*Bufo melanosticus*). Hormone Research 19: 200-204.

Chanda, S. and Biswas, N. M. 1992. Effect of melatonin on the testis of pinealectomized and sham operated toads during the breeding season. Comparative Physiology and Ecology 17: 1-5.

Chappell, P. E., White, R. S. and Mellon, P. L. 2003. Circadian gene expression regulates pulsatile gonadotropin-releasing hormone (GnRH) secretory patterns in the hypothalamic GnRH-secreting GT1-7 cell line. Journal of Neuroscience 23: 11202-11213.

Cherry, J. A., Basham, M. E., Weaver, C. E., Krohmer, R. W. and Baum, M. J. 1990. Ontogeny of the sexually dimorphic male nucleus in the preoptic/anterior hypothalamus of ferrets and its manipulation by gonadal steroids. Journal of Neurobiology 21: 844-847.

Chiu, K. W. and Lynn, W. G. 1971. The hypophysis-gonad relation in the glossy snake, *Arizona elegans*. Herpetologica 27: 303-307.

Chiu, K. W. and Woo, N. Y. S. 1988. Metabolic effects of thyroid hormones at low temperature in the snake. Journal of Thermal Biology 13: 179-184.

Chiu, K. W. and Lam, K. Y. 1994. Plasma T3 and T4 levels in a snake, *Elaphe taeniura*. Comparative Biochemistry and Physiology A 107: 107-112.

Clarke, I. J., Smith, J. T., Caraty, A., Goodman, R. L. and Lehman, M. N. 2009. Kisspeptin and seasonality in sheep. Peptides 30: 154-163.

Cohen, R., Smale, L. and Kronfeld-Schor, N. 2010. Masking and temporal niche switches in spiny mice. Journal of Biological Rhythms 25: 47-52.

Crawford, K. M. 1991. The winter environment of painted turtles, *Chrysemys picta*: Temperature, dissolved oxygen, and potential cues for emergence. Canadian Journal of Zoology 69: 2493-2498.

Crews, D. 1975. Psychobiology of reptilian reproduction. Science 189: 1059-1065.

Crews, D. 1979. The neuroendocrinology of lizard reproduction. Biology of Reproduction 20: 51-73.

Crews, D. 1982. The ecological physiology of a garter snake. Scientific American 244: 158-168.
Crews, D. 1984. Gamete production, sex hormone secretion, and mating behavior uncoupled. Hormones and Behavior 18: 22-28.
Crews, D. 1991. Tran-seasonal action of androgen in the control of spring courtship in male red-sided garter snake. Proceedings of the National Academy of Sciences USA 88: 3545-3548.
Crews, D. and Garstka, W. 1982. The ecological physiology of reproduction in the Canadian red-sided garter snake. Scientific American 247: 158-168.
Crews, D. and Silver, R. 1985. Reproductive physiology and behavior interactions in non-mammalian vertebrates. Pp. 101-182. In N. Adler, D. Pfaff and R. W. Goy (eds), *Handbook of Behavioral Neurobiology*, vol. 8. Plenum Press, New York.
Crews, D., Hingorani, V. and Nelson, R. J. 1988. Role of the pineal gland in the control of annual reproductive behavioral and physiological cycles in the red-sided garter snake (*Thamnophis sirtalis parietalis*). Journal of Biological Rhythms 3: 293-302.
Crews, D., Robker, R. and Mendonça, M. 1993. Seasonal fluctuations in brain nuclei in the red-sided garter snake and their hormonal control. Journal of Neuroscience 13: 5356-5364.
Crews, D., Camazine, B., Diamond, M., Mason, R., Tokarz, R. R. and Garstka, W. R. 1984. Hormonal independence of courtship behavior in the male garter snake. Hormones and Behavior 18: 29-41.
de Jonge, F. H., Lowerse, A. L., Ooms, M. P., Evers, P., Endert, E. and Van De Poll, N. E. 1989. Lesions of the SDN-POA inhibit sexual behavior of male wistar rats. Brain Research Bulletin 23: 483-492.
de Vlaming, V. L. 1975. Effects of pinealectomy on gonadal activity in the cyprinid teleost, *Notemigonus crysoleucas*. General and Comparative Endocrinology 26: 36-49.
de Vlaming, V. L. and Vodicnik, M. J. 1977. Effect of pinealectomy on pituitary gonadotrophs, pituitary gonadotropin potency and hypothalamic gonadotrophin releasing activity in *Notemigonus crysoleucas*. Journal of Fish Biology 10: 73-86.
Dohler, K. D., Coquelin, A., Davis, F., Hines, M., Shryne, J. E. and Gorski, R. A. 1984. Pre- and postnatal influences on testosterone propionate and diethylstilbestrol on differentiation of sexually dimorphic nucleus of the preoptic area of male and female rats. Brain Research 302: 291-295.
Du, J. F. and Hull, E. M. 1999. Effects of testosterone on neuronal nitric oxide synthase and tyrosine hydroxylase. Brain Research 836: 90-98.
Dungan, H. M., Clifton, D. K. and Steiner, R. A. 2006. Minireview: Kisspeptin neurons as central processors in the regulation of gonadotropin-releasing hormone secretion. Endocrinology 147: 1154-1158.
Dunham, L. A., Lutterschmidt, D. I. and Wilczynski, W. 2009. Kisspeptin-like immunoreactive neuron distribution in the green anole (*Anolis carolinensis*). Brain, Behavior and Evolution 73: 129-137.
Dunn, P. O. and Winkler, D. W. 1999. Climate change has affected the breeding date of tree swallows throughout North America. Proceedings of the Royal Society of London 266: 247.
Duvall, D., Guillette, L. J., Jr. and Jones, R. E. 1982. Environmental control of reptilian reproductive cycles. Pp. 201-231. In C. Gans and F. H. Pough (eds), *Biology of the Reptilia*, vol. 13. Academic Press, London.
Ebling, F. J. P. and Barrett P. 2008. The regulation of seasonal changes in food intake and body weight. Journal of Neuroendocrinology 20: 827-833.

Elizur, A. 2009. The Kiss1/GPR54 system in fish. Peptides 30: 164-170.
Etheridge, K. 1993. Thyroxine-induced changes in metabolic rate and cytochrome oxidase activity in *Thamnophis sirtalis*: Effects of nutritional status. General and Comparative Endocrinology 91: 66-73.
Etheridge, K., Wit, L. C. and Sellers, J. C. 1983. Hibernation in the lizard *Cnemidophorus sexlineatus* (Lacertilia: Teiidae). Copeia 1983: 206-214.
Falcón, J., Bolliet, V., Ravault, J. P., Chesneau, D., Ali, M. A. and Collin, J.-P. 1994. Rhythmic secretion of melatonin by the superfused pike pineal organ: thermo- and photoperiod interaction. Neuroendocrinology 60: 535-543.
Firth, B. T. and Kennaway, D. J. 1987. Melatonin content of the pineal, parietal eye and blood plasma of the lizard, *Trachydosaurus rugosus*: Effects of constant and fluctuating temperature. Brain Research 404: 313-318.
Firth, B. T. and Kennaway, D. J. 1989. Thermoperiod and photoperiod interact to affect the phase of the plasma melatonin rhythm in the lizard, *Tiliqua rugosa*. Neuroscience Letters 106: 125-130.
Firth, B. T., Thompson, M. B., Kennaway, D. J. and Belan, I. 1989. Thermal sensitivity of reptilian melatonin rhythms: "cold" tuatara vs. "warm" skink. American Journal of Physiology 256: R1160-R1163.
Fleury, F. and Naulleau, G. 1987. Relations between hibernation and the resumption of endocrine, testicular and thyroid activities, in *Vipera aspis* L. (Reptilia, Viperidae). General and Comparative Endocrinology 68: 271-277.
Foà, A., Janik, D. and Minutini, L. 1992. Circadian rhythms and plasma melatonin in the ruin lizard, *Podarcis sicula*: Effects of pinealectomy. Journal of Pineal Research 12: 109-113.
Foidart, A., Harada, N. and Balthazart, J. 1995a. Aromatase-immunoreactive cells are present in mouse brain areas that are known to express high levels of aromatase activity. Cell Tissue Research 280: 561-574.
Foidart, A., Reid, J., Absil, P., Yoshimura, N., Harada N. and Balthazart, J. 1995b. Critical re-examination of the distribution of aromatase-immunoreactive cells in the quail forebrain using antibodies raised against human placental aromatase and against the recombinant quail, mouse or human enzyme. Journal of Chemical Neuroanatomy 8: 267-282.
Follett, B. K. 1967. Neurohypophysial hormones of marine turtles and of grass snake. Journal of Endocrinology 39: 293-294.
Forchhammer, M. C., Post, E. and Stenseth, N. C. 1998. Breeding phenology and climate. Nature 391: 29-30.
Ford, N. B. 1981. Seasonality of pheromone trailing behavior in two species of garter snake, *Thamnophis* (Colubridae). Southwestern Naturalist 26: 385-388.
Ford, N. B. 1982. Species specificity of sex pheromone trails of sympatric and allopatric garter snakes (*Thamnophis*). Copeia 1982: 10-13.
Friedman, D. and Crews, D. 1985a. Role of the anterior hypothalamus-preoptic area in the regulation of courtship behavior in the male Canadian red-sided garter snake, *Thamnophis sirtalis parietalis*: Lesion experiments. Behavioral Neuroscience 99: 942-949.
Friedman, D. and Crews, D. 1985b. Role of the anterior hypothalamus-preoptic area in the regulation of courtship behavior in the male Canadian red-sided garter snake, *Thamnophis sirtalis parietalis:* Intracranial implantation experiments. Hormones and Behavior 19: 122-136.
García-Allegue, R., Madrid, J. A. and Sánchez-Vázquez, F. J. 2001. Melatonin rhythms in European sea bass plasma and eye: influence of seasonal photoperiod and water temperature. Journal of Pineal Research 31: 68-75.

Garg, S. K. 1988. Pinealectomy, ovarian activity and vitellogenin levels in the catfish *Herteropneustes fossilis* (Bloch), exposed to various combinations of photoperiods. Reproduction, Nutrition, and Development 28A: 265-274.
Garstka, W. R., Camazine, B. and Crews, D. 1982. Interactions of behavior and physiology during the annual reproductive cycle of the red-sided garter snake (*Thamnophis sirtalis parietalis*). Herpetologica 38: 104-123.
George, F. W. and Ojeda, S. R. 1982. Changes in aromatase activity in the rat brain during embryonic, neonatal and infantile development. Endocrinology 111: 522-529.
Gern, W. A. and Norris, D. O. 1979. Plasma melatonin in the neotenic tiger salamander (*Ambystoma trigrinum*): Effects of photoperiod and pinealectomy. General and Comparative Endocrinology 38: 393-398.
Gern, W. A. and Ralph, C. L. 1979. Melatonin synthesis by the retina. Science 204: 183-184.
Gern, W. A. and Karn, C. M. 1983. Evolution of melatonin's functions and effects. Pineal Research Reviews 1: 49-90.
Gillespie, J. M. A., Roy, D., Cui, H. and Belsham, D. D. 2004. Repression of gonadotropin-releasing hormone (GnRH) gene expression by melatonin may involve transcription factors COUP-TFI and C/EBP beta binding at the GnRH enhancer. Neuroendocrinology 79: 63-72.
Gobbetti, A., Zerani, M., DiForce, M. M. and Botte, V. 1994. Relationships among GnRH, substance P, prostaglandins, sex steroids and aromatase activity in the brain of the male lizard *Podarcis sicula sicula* during reproduction. Journal of Reproduction and Fertility 101: 523-528.
Goldman, B. D., Gwinner, E., Karsch, F. J., Saunders, D., Zucker, I. and Ball, G. F. 2004. Circannual rhythms and photoperiodism. Pp. 107-142. In J. C. Dunlap, J. J. Lorosand and P. J. DeCoursey (eds), *Chronobiology: Biological Timekeeping.* Sinauer Associates, Inc., Sunderland, Massachusetts.
Gorski, R. A., Shryne, J. H. and Southam, A. M. 1978. Evidence for a morphological sex difference within the medial preoptic area of the rat brain. Brain Research 143: 333-346.
Greives, T. J., Mason, A. O., Scotti, M.-A. L., Levine, J., Ketterson, E. D., Kriegsfeld, L. J. and Demas, G. E. 2007. Environmental control of kisspeptin: Implications for seasonal reproduction. Endocrinology 148: 1158-1166.
Greenberg, N., Scott, M. and Crews, D. 1984. Role of the amygdala in the reproductive and aggressive behavior of the lizard, *Anolis carolinensis*. Physiology and Behavior 32: 147-151.
Grobman, A. B. 1990. The effect of soil temperatures on emergence from hibernation of *Terrapene carolina* and *T. ornata*. American Midland Naturalist 124: 366-371.
Hadeishi, Y. and Wood, R. I. 1996. Nitric oxide synthase in mating behavior circuitry of male Syrian hamster brain. Journal of Neurobiology 30: 480-492.
Hadley, M. E. 1996. *Endocrinology*, 4th ed. Prentice Hall Publishers, Upper Saddle River, New Jersey. Pp. 608.
Haldar, C. and Thapliyal, J. P. 1977. Effects of pinealectomy on the annual testicular cycle of *Calotes versicolor*. General and Comparative Endocrinology 32: 395-399.
Haldar, C. and Pandey, R. 1988. Effect of melatonin and 5-methoxytryptamine administration on the testis and pineal gland activity of the freshwater snake, *Natrix piscator*. Archives of Anatomy, Histology and Embryology 71: 85-95.
Haldar, C. and Pandey, R. 1989a. Effect of pinealectomy on the testicular response of the freshwater snake, *Natrix piscator*, to different environmental factors. Canadian Journal of Zoology 67: 2352-2357.

Haldar, C. and Pandey, R. 1989b. Effect of pinealectomy on annual testicular cycle of Indian checkered snake, *Natrix piscator*. General and Comparative Endocrinology 76: 214-222.

Haldar-Misra, C. and Thapliyal, J. P. 1981a. Effect of melatonin on the testes and the renal sex segment in the garden-lizard, *Calotes versicolor*. Canadian Journal of Zoology 59: 70-74.

Haldar-Misra, C. and Thapliyal, J. P. 1981b. Responses of reptilian gonad to melatonin. Neuroendocrinology 33: 328-332.

Halpern, M. 1980. The telencephalon of snakes. Pp. 257-295. In S. O. E. Ebbesson (ed.), *Comparative Neurology of the Telencephalon*. Plenum Press, New York.

Halpern, M. and Silfen, R. 1974. The efferent connections of the nucleus sphericus in the garter snake, *Thamnophis sirtalis*. Anatomical Record 178: 368.

Halpern, M. and Kubie, J. L. 1984. The role of the ophidian vomeronasal system in the species-typical behavior. Trends in Neuroscience 7: 472-477.

Halpern, M., Morrell, J. I. and Pfaff, D. W. 1982. Cellular [3H] testosterone localization in the brains of garter snakes: An autoradiographic study. General and Comparative Endocrinology 46: 211-224.

Han, S. K., Gottsch, M. L., Lee, K. J., Popa, S. M., Smith, J. T., Jakawich, S. K., Clifton, D. K., Steiner, R. A. and Herbison, A. E. 2005. Activation of gonadotropin-releasing hormone neurons by kisspeptin as a neuroendocrine switch for the onset of puberty. Journal of Neuroscience 25: 11349-11356.

Hanon, E. A., Lincoln, G. A., Fustin, J. M., Dardente, H., Masson-Pévet, M., Morgan, P. J. and Hazlerigg, D. G. 2008. Ancestral TSH mechanism signals summer in a photoperiodic mammal. Current Biology 18: 1147-1152.

Hart, B. L. and Leedy, M. G. 1985. Neurological basis of sexual behavior: a comparative study. Pp. 373-410. In N. Adler, D. Pfaff and R. W. Goy (eds) *Handbook of Behavioral Neurobiology*, vol. 7. Plenum Press, New York.

Hawley, A. W. L. and Aleksiuk, M. 1975. Thermal regulation of spring mating behavior in the red-sided garter snake (*Thamnophis sirtalis parietalis*). Canadian Journal of Zoology 53: 768-776.

Hawley, A. W. L. and Aleksiuk, M. 1976. The influence of photoperiod and temperature on seasonal testicular recrudescence in the red-sided garter snake (*Thamnophis sirtalis parietalis*). Comparative Biochemistry and Physiology 53A: 215-221.

Hontela, A. and Peter, R. E. 1980. Effects of pinealectomy, blinding, and sexual condition on serum gonadotropin levels in the goldfish. General and Comparative Endocrinology 40: 168-179.

Hoyle, C. H. 1998. Neuropeptide families: Evolutionary perspectives. Regulatory Peptides 73: 1-33.

Hutchison, J. B. 1974a. Effects of photoperiod on the decline in behavioral responsiveness to intra-hypothalamic androgen in doves (*Streptopelia risoria*). Journal of Endocrinology 63: 583-584.

Hutchison, J. B. 1974b. Post-castration decline in behavioral responsiveness to intra-hypothalamic androgen in doves. Brain Research 81: 169-181.

Hutchison, J. B., Steimer, Th. and Jaggard, P. 1986. Effects of photoperiod on formation of oestradiol-17b in the dove brain. Journal of Endocrinology. 109: 371-377.

Hyde, L. L. and Underwood, H. 1993. Effects of nightbreak, T-cycle, and resonance lighting schedules on the pineal melatonin rhythm of the *Anolis carolinensis*: Correlations with the reproductive response. Journal of Pineal Research 15: 70-80.

Ingle, D. and Crews, D. 1985. Vertebrate neuroethology: definitions and paradigms. Annual Review of Neuroscience 8: 457-494.

Intergovernmental Panel on Climate Change (IPCC). 2001. Climate change 2001: In M. Manning and C. Nobre (eds), *Impacts, Adaptation, and Vulnerability*. Intergovernmental Panel on Climate Change Technical Summary, 56 Pp.

Jacob, S. J. and Painter, C. W. 1980. Overwinter thermal ecology of *Crotalus viridis* in the north-central plains of New Mexico. Copeia 1980: 799-805.

Janik, D. and Menaker, M. 1990. Circadian locomotor rhythms in the desert iguana. I. The role of the eyes and the pineal. Journal of Comparative Physiology 166A: 803-810.

Johnston, J. D., Cagampang, F. R. A., Stirland, J. A., Carr, A. J. F., White, M. R. H., Davis, J. R. E. and Loudon, A. S. I. 2003. Evidence for an endogenous *Per1* and *ICER*-independent seasonal timer in the hamster pituitary gland. Federation of American Societies for Experimental Biology Journal 17: 810-815.

Kaslin, J., Ganz, J. and Brand, M. 2008. Proliferation, neurogenesis and regeneration in the non-mammalian vertebrate brain. Philosophical Transactions of the Royal Society B 363: 101-122.

Khare, P. V., Satralkar, M. K., Vanlalnghaka, C., Keny, V. L., Kasture, M. S., Shivagaje, A. J., Barnabas, R. J. and Joshi, D. S. 2005. Altitudinal variation in the circadian rhythm of oviposition in *Drosophila ananassae*. Chronobiology International 22: 45-57.

Kriegsfeld, L. J. 2006. Driving reproduction: RFamide peptides behind the wheel. Hormones and Behavior 50: 655-666.

Krohmer, R. W. 2004. The male red-sided grater snake (*Thamnophis sirtalis parietalis*): reproductive pattern and behavior. Institute for Laboratory Animal Research 45: 65-74.

Krohmer, R. W. and Crews, D. 1987a. Temperature activation of courtship behavior in the male red-sided garter snake (*Thamnophis sirtalis parietalis*): role of the anterior hypothalamus-preoptic area. Behavioral Neuroscience 101: 228-236.

Krohmer, R. W. and Crews, D. 1987b. Facilitation of courtship behavior in the male red-sided garter snake (*Thamnophis sirtalis parietalis*) following lesions of the septum or nucleus sphericus. Physiology Behavior 40: 759-765.

Krohmer, R. W. and Baum, M. J. 1989. Effect of sex, intrauterine position and androgen manipulation on the development of brain aromatase activity in fetal ferrets. Journal of Neuroendocrinology 1: 265-271.

Krohmer, R. W. and Crews, D. 1989. Control of length of courtship season in the red-sided garter snake, *Thamnophis sirtalis parietalis*: the role of temperature. Canadian Journal of Zoology 67: 987-993.

Krohmer, R. W., Grassman, M. and Crews, D. 1987. Annual reproductive cycle in the male red-sided garter snake, *Thamnophis sirtalis parietalis*: field and laboratory studies. General and Comparative Endocrinology 68: 64-75.

Krohmer, R., Boyle, M. H., Lutterschmidt, D. I. and Mason, R. T. 2010. Seasonal aromatase activity in the brain of the red-sided garter snake. Hormones and Behavior, 58: 485-492.

Krohmer, R. W., Bieganski, G. J., Baleckaitis, D. D., Harada, N. and Balthazart, J. 2002. Distribution of aromatase immunoreactivity in the forebrain of red-sided garter snakes at the beginning of the winter dormancy. Journal of Chemical Neuroanatomy 23: 59-71.

Lazari, M. F. M., Alponti, R. F., Freitas, T. A., Breno, M. C., da Conceicao, I. M. and Silveira, P. F. 2006. Absence of oxytocin in the central nervous system of the snake *Bothrops jararaca*. Journal of Comparative Physiology B 176: 821-830.

Leedy, M. G. and Hart, B. L. 1986. Medial preoptic-anterior hypothalamic lesions in prepubertal male cats: effects on juvenile and adult sociosexual behaviors. Physiology and Behavior 36: 501-506.

LeMaster, M. P. and Mason, R. T. 2001. Evidence for a female sex pheromone mediating male trailing behavior in the red-sided garter snake, *Thamnophis sirtalis parietalis*. Chemoecology 11: 149-152.

Lerchl, A., Zachmann, A., Ali, M. A. and Reiter, R. J. 1998. The effects of pulsing magnetic fields on pineal melatonin synthesis in a teleost fish (brook trout, *Salvelinus fontinalis*). Neuroscience Letters 256: 171-173.

Levey, I. L. 1973. Effects of pinealectomy and melatonin injections at different seasons on ovarian activity in the lizard *Anolis carolinensis*. Journal of Experimental Zoology 185: 169-174.

Licht, P. 1972. Environmental physiology of reptilian breeding cycles: Role of temperature. General and Comparative Endocrinology Suppl. 3: 477-488.

Licht, P. 1984. Seasonal cycles in reptilian reproductive physiology. Pp. 206-282. In E. Lamming (ed.), *Marshall's Physiology of Reproduction*, vol. 1, 4th ed. Churchhill Livingstone, New York.

Logier, E. B. S. and Toner, G. C. 1961. *Check List of the Amphibians and Reptiles of Canada and Alaska*. Life Science Division of the Royal Ontario Museum Contribution 53. University of Toronto Press, Toronto. Pp. 92.

Lutterschmidt, D. I. 2006. Chronobiology of Garter Snakes: Environmental and Hormonal Cues Mediating Hibernation and Reproduction. Ph.D. thesis, Oregon State University, Corvalis, Oregon.

Lutterschmidt, D. I. and Mason, R. T. 2005. A serotonin receptor antagonist, but not melatonin, modulates hormonal responses to capture stress in two populations of garter snakes (*Thamnophis sirtalis parietalis* and *Thamnophis sirtalis concinnus*). General and Comparative Endocrinology 141: 259-270.

Lutterschmidt, D. I. and Mason, R. T. 2008. Geographic variation in timekeeping systems of three populations of garter snakes (*Thamnophis sirtalis*) in a common garden. Physiological and Biochemical Zoology 81: 810-825.

Lutterschmidt, D. I. and Mason, R. T. 2009. Endocrine mechanisms mediating temperature-induced reproductive behavior in red-sided garter snakes (*Thamnophis sirtalis parietalis*). Journal of Experimental Biology 212: 3108-3118.

Lutterschmidt, D. I., Lutterschmidt, W. I. and Hutchison, V. H. 2003. Melatonin and thermoregulation in ectothermic vertebrates: A review. Canadian Journal of Zoology 81: 1-13.

Lutterschmidt, D. I., LeMaster, M. P. and Mason, R. T. 2004. Effects of melatonin on the behavioral and hormonal responses of red-sided garter snakes (*Thamnophis sirtalis parietalis*) to exogenous corticosterone. Hormones and Behavior 46: 692-702.

Lutterschmidt, D. I., LeMaster, M. P. and Mason, R. T. 2006. Minimal over-wintering temperatures of red-sided garter snakes (*Thamnophis sirtalis parietalis*): A possible cue for emergence? Canadian Journal of Zoology 84: 771-777.

Lutterschmidt, D. I., Lutterschmidt, W. I., Ford, N. B. and Hutchison, V. H. 2002. Behavioral thermoregulation and the role of melatonin in a nocturnal snake. Hormones and Behavior 41: 41-50.

Macartney, J. M., Larsen, K. W. and Gregory, P. T. 1989. Body temperatures and movements of hibernating snakes (*Crotalus* and *Thamnophis*) and thermal gradients of natural hibernacula. Canadian Journal of Zoology 67: 108-114.

MacLusky, N. J., Philip, A., Hurlburt, C. and Naftolin, F. 1985. Estrogen formation in the developing rat brain: sex differences in aromatase activity during early post-natal life. Psychoneuroendocrinology 10: 355–361.

Madeira, M. D., Leal, S. and Paula-Barbosa, M. M. 1999. Stereological evaluation and Golgi study of the sexual dimorphisms in volume, cell numbers, and cell size in the medial preoptic nucleus of the rat. Journal of Neurocytology 28: 131-148.

Malpaux, B., Thiery, J. C. and Chemineau, P. 1999. Melatonin and the seasonal control of reproduction. Reproduction Nutrition Development 39: 355-366.

Mancera, J. M., López Avalos M. D., Pérez-Figares, J. M. and Fernández-Llebrez, P. 1991. The distribution of corticotrophin-releasing factor-immunoreactive neurons and nerve fibers in the brain of the snake, *Natrix maura*. Coexistence with arginine vasotocin and mesotocin. Cell and Tissue Research 264: 539-548.

Mayer, I., Bornestaf, C. and Borg, B. 1997. Melatonin in non-mammalian vertebrates: Physiological role in reproduction? Comparative Biochemistry and Physiology 118A: 515-531.

Mendonça, M. T., Tousignant, A. J. and Crews, D. 1995. Seasonal changes and annual variability in daily plasma melatonin in the red-sided garter snake (*Thamnophis sirtalis parietalis*). General and Comparative Endocrinology 100: 226-237.

Mendonça, M. T., Tousignant, A. J. and Crews, D. 1996a. Pinealectomy, melatonin, and courtship behavior in male red-sided garter snake (*Thamnophis sirtalis parietalis*). Journal of Experimental Zoology 274: 63-74.

Mendonça, M. T., Tousignant, A. J. and D. Crews. 1996b. Courting and noncourting male red-sided garter snake, *Thamnophis sirtalis parietalis*: plasma melatonin levels and the effects of pinealectomy. Hormone and Behavior 30: 176-185.

Michael, T. P., Salomé, P. A., Yu, H. J., Spencer, T. R., Sharp, E. L., McPeek, M. A., Alonso, J. M., Ecker, J. R. and McClung, C. R. 2003. Enhanced fitness conferred by naturally occurring variation in the circadian clock. Science 302: 1049-1053.

Mohamed, J. S., Benninghoff, A. D., Holt, G. J. and Khan, I. A. 2007. Developmental expression of the G protein-coupled receptor 54 and three GnRH mRNAs in the teleost fish cobia. Journal of Molecular Endocrinology 38: 235-244.

Mohanty, K. C. and Naik, D. R. 1997. Immunohistochemistry and tinctorial affinity of adenohypophysial cells of the rat snake *Ptyas mucosus* (Colubridae). General and Comparative Endocrinology 105: 302-313.

Moore, I. T., LeMaster, M. P. and Mason, R. T. 2000. Behavioral and hormonal responses to capture stress in the male red-sided garter snake *Thamnophis sirtalis parietalis*. Animal Behavior 59: 529-534.

Moore, I. T., Green, M. J. and Mason, R. T. 2001. Environmental and seasonal adaptations of the adrenocortical and gonadal response to capture stress in two populations of the male garter snake, *Thamnophis sirtalis*. Journal of Experimental Zoology 289: 99-108.

Morgan, P. J., Ross, A. W., Mercer, J. G. and Barrett, P. 2006. What can we learn from seasonal animals about the regulation of energy balance? Progress in Brain Research 153: 325-337.

Morris, Y. A. and Crews, D. 1990. The effects of exogenous neuropeptide Y on feeding and sexual behavior in the red-sided garter snake (*Thamnophis sirtalis parietalis*). Brain Research 530: 339-341.

Moss, R., Oswald, J. and Baines, D. 2001. Climate change and breeding success: decline of the capercaillie in Scotland. Journal of Animal Ecology 70: 47-61.

Moyer, R. W., Firth, B. T. and Kennaway, D. J. 1995. Effect of constant temperatures, darkness and light on the secretion of melatonin by pineal explants and retinas in the gecko *Christinus marmoratus*. Brain Research 675: 345-348.

Muske, L. E. 1993. Evolution of gonadotropin releasing hormone (GnRH) neuronal systems. Brain, Behavior and Evolution 42: 215-230.

Naftolin, F., Ryan, K. J., Davies, I. J., Takaoka, Y. and Wolin, L. 1975. The formation of estrogens by central neuroendocrine tissues. Recent Proceedings in Hormone Research. 31: 295-319.

Nakao, N., Ono, H., Yamamura, T., Anraku, T., Takagi, T., Higashi, K., Yasuo, S., Katou, Y., Kageyama, S., Uno, Y., Kasukawa, T., Iigo, M., Sharp, P. J., Iwasawa, A., Suzuki, Y., Sugano, S., Niimi, T., Mizutani, M., Namikawa, T., Ebihara, S., Ueda, H. R. and Yoshimura, T. 2008. Thyrotrophin in the pars tuberalis triggers photoperiodic response. Nature 452: 317-322.

Nance, D. M., Shryne, J. and Gorski, R. A. 1974. Septal lesions: Effects on lordosis behavior and pattern of gonadotropin release. Hormones and Behavior 5: 73-81.

Nance, D. M., Shryne, J. and Gorski, R. A. 1975. Effect of septal lesions on behavioral sensitivity of female rats to gonadal hormones. Hormone and Behavior 6: 59-64.

Naulleau, G., Fleury, F. and Boissin, J. 1987. Annual cycle in plasma testosterone and thyroxine in the male aspic viper *Vipera aspis* L. (Reptilia, Viperidae) in relation to the sexual cycle and hibernation. General and Comparative Endocrinology 65: 254-263.

Nayak, P. K. and Singh, T. P. 1988. Effect of pinealectomy on testosterone, estradiol-17b, estrone, and 17b-hydroxyprogesterone levels during the annual reproductive cycle in the freshwater catfish, *Clarias batrachus*. Journal of Pineal Research 5: 419-426.

Nelson, R. J. 2005. *An Introduction to Behavioral Endocrinology*, 3rd ed. Sinauer, Sunderland, Massachusetts. Pp. 822.

Nelson, R .J., Mason, R. T., Krohmer, R. W. and Crews, D. 1987. Pinealectomy blocks vernal courtship behavior in the red-sided garter snake. Physiology and Behavior 39: 231-233.

Nelson, R. J., Demas, G. E., Huang, P. L., Fishman, M. C., Dawson, V. L., Dawson, T. M. and Snyder, S. H. 1995. Behavioural abnormalities in male mice lacking neuronal nitric oxide synthase. Nature 378: 383-386.

Nilson, G. 1982. Thyroid activity and experimental evidence for its role in reproduction in the Adder *Vipera berus*. General and Comparative Endocrinology 47: 148-158.

Nocillado, J. N., Levavi-Sivan, B., Carrick, F. and Elizur, A. 2007. Temporal expression of G-protein-coupled receptor 54 (GPR54), gonadotropin-releasing hormones (GnRH), and dopamine receptor D2 (*drd2*) in pubertal female grey mullet, *Mugil cephalus*. General and Comparative Endocrinology 150: 278-287.

Norris, D. O. 1997. *Vertebrate Endocrinology*. Academic Press, San Diego, California. Pp. 634.

Nottebohm, F. 2004. The road we travelled: discovery, choreography, and significance of brain replaceable neurons. Annals of the New York Academy of Sciences 1016: 628-658.

Nozaki, M., Tsukahara, T. and Kobayashi, H. 1984. Neuronal systems producing LHRH in vertebrates. Pp. 3-27. In K. Ochiai (ed.) *Endocrine Correlates of Reproduction*. Japan Scientific Societies Press, Tokyo and Springer-Verlag, Berlin.

O'Bryant, E. L. and Wade, J. 2002. Seasonal and sexual dimorphisms in the green anole forebrain. Hormones and Behavior 41: 384-395.

O'Donnell, R. P., Shine, R. and Mason, R. T. 2004. Seasonal anorexia in the male red-sided garter snake, *Thamnophis sirtalis parietalis*. Behavioral Ecology and Sociobiology 56: 413-419.

Packard, M. J. and Packard, G. C. 1977. Antigonadotrophic effects of melatonin in male lizards (*Callisaurus draconoides*). Experientia 33: 1665-1666.

Pandha, S. K. and Thapliyal, J. P. 1964. Effect of male hormone on the renal sex segment of castrated males of the lizard *Caloties versicolor*. Copeia 1964: 579-581.

Panzica, G., Viglietti-Panzica, C., Sanchez, F., Sante, P. and Balthazart, J. 1991. Effects of testosterone on a selected neuronal population within the preoptic sexually dimorphic nucleus of the Japanese quail. Journal of Comparative Neurology 303: 443-456.

Panzica, G. C., Verzé, L., Viglietti-Panzica, C. and Rissman, E. F. 2000. Modifications of NOS-immunoreactive system in estrogen receptor alpha knock-out male mice. Society of Neuroscience Abstracts 26: 1275.

Panzica, G. C., Arévalo, R., Sánchez, F., Alonso, J. R., Aste, N., Viglietti-Panzica, C., Aijón, J. and Vázquez, R. 1994. Topographical distribution of reduced nicotinamide adenine dinucleotide phosphate-diaphorase in the brain of the Japanese quail. Journal of Comparative Neurology 342: 97-114.

Parhar, I. S., Ogawa, S. and Sukama, Y. 2004. Laser-captured single digoxigenin-labeled neurons of gonadotropin-releasing hormone types reveal a novel G protein-coupled receptor (GPR54) during maturation in cichlid fish. Endocrinology 145: 3613-3618.

Perez-Figares, J. M., Mancera, J. M., Rodriguez, E. M., Nualart, F. and Fernandez-Llebrez, P. 1995. Presence of an oxytocin-like peptide in the hypothalamus and neurohypophysis of a turtle (*Mauremys caspica*) and a snake (*Natrix maura*). Cell and Tissue Research 279: 75-84.

Pfaff, D. W., Arnold, A. P., Etgen, A. M., Fahrbach, S. E. and Rubin, R. T. (eds). 2002. *Hormones, Brain and Behavior*, Vols 1-5. Academic Press, Los Angeles, California.

Pickering, B. T. 1967. The neurohypophysial hormones of a reptile species, the cobra (*Naja naja*). Journal of Endocrinology 39: 285-293.

Ralph, C. L. 1980. Melatonin production by extrapineal tissues. Pp. 35-46. In N. Birau and W. Schloot (eds), *Melatonin: Current Status and Perspectives*. Pergamon Press, Oxford, U.K.

Rawding, R. S. and Hutchison, V. H. 1992. Influence of temperature and photoperiod on plasma melatonin in the mudpuppy, *Necturus maculosus*. General and Comparative Endocrinology 88: 364-374.

Reiter, R. J. 1992. Changes in circadian melatonin synthesis in the pineal gland of animals exposed to extremely low frequency electromagnetic radiation: a summary of observations and speculation on their implications. Pp. 13-27. In M. C. Moore-Ede, S. S. Campbell and R. J. Reiter (eds), *Electromagnetic Fields and Circadian Rhythmicity*. Birkhäuser, Boston, Massachusetts.

Reiter, R. J. 1993. Electromagnetic fields and melatonin production. Biomedicine and Pharmacotherapy 47: 439-444.

Revel, F. G., Saboureau, M., Masson-Pévet, M., Pévet, P., Mikkelsen, J. D. and Simonneaux, V. 2006. Kisspeptin mediates the photoperiodic control of reproduction in hamsters. Current Biology 16: 1730-1735.

Rismiller, P. D. and Heldmaier, G. 1987. Melatonin and photoperiod affect body temperature in the lizard *Lacerta viridis*. Journal of Thermal Biology 12: 131-134.

Roa, J. and Tena-Sempere, M. 2007. KiSS-1 system and reproduction: Comparative aspects and roles in the control of female gonadotropic axis in mammals. General and Comparative Endocrinology 153: 132-140.

Root, T. L., Price, J. T., Hall, K. R., Schneider, S. H., Rosenzweig, C. and Pounds, J. A. 2003. Fingerprints of global warming on wild animals and plants. Nature 421: 57-60.

Roselli, C. E., Horton, L. E. and Resko, J. A. 1985. Distribution and regulation of aromatase activity in the rat hypothalamus and limbic system. Endocrinology 117: 2471-2477.

Rosen, G. J. and Wade, J. 2001. Androgen metabolism in the brain of the green anole lizard (*Anolis carolinensis*): Effects of sex and season. General and Comparative Endocrinology 122: 40-47.

Ross, P., Jr. and Crews, D. 1978. Stimuli influencing mating behavior in the garter snake, *Thamnophis radix*. Behavioral Ecology and Sociobiology 4: 133-142.

Roy, D., Angelini, N. L., Fujieda, H., Brown, G. M. and Belsham, D. D. 2001. Cyclical regulation of GnRH gene expression in GT1-7 GnRH-secreting neurons by melatonin. Endocrinology 142: 4711-4720.

Saint Girons, H. 1982. Reproductive cycles of male snakes and their relationships with climate and female reproductive cycles. Herpetologica 38: 5-16.

Saint Girons, H., Bradshaw, S. D. and Bradshaw, F. J. 1993. Sexual activity and plasma levels of sex steroids in the Aspic viper *Vipera aspis* L. (Reptilia, Viperidae). General and Comparative Endocrinology 91: 287-297.

Satinoff, E. 1978. Neural organization and evolution of thermoregulation in mammals. Science 201: 16-22.

Satinoff, E. 1983. A reevaluation of the concept of the homeostatic organization of temperature regulation. Pp. 443-472. In E. P. Satinoff and P. Teitlebaum (eds), *Handbook of Behavioral Neurobiology*, vol 6. *Motivation*. Plenum Press, New York.

Sawyer, W. H. 1977. Evolution of neurohypophyseal hormones and their receptors. Federation Proceedings 36: 1842-1847.

Sexton, O. W. and Marion, K. R. 1981. Experimental analysis of movements by prairie rattlesnakes, *Crotalus viridis*, during hibernation. Oecologia 51: 37-41.

Sherwood, N. M. and Whittier, J. M. 1988. Gonadotropin-releasing hormone from brains of reptiles: turtles (*Pseudemys scripta*) and snakes (*Thamnophis sirtalis parietalis*). General and Comparative Endocrinology 69: 319-327.

Shine, R. 2003. Reproductive strategies in snakes. Proceedings of the Royal Society of London B 270: 995-1004.

Shinoda, K., Nagano, M. and Osawa, Y. 1994. Neural aromatase expression in preoptic, strial, and amygdaloid regions during late prenatal and early post-natal development in the rat. Journal of Comparative Neurology 343: 113-129.

Simerly, R. B. 2002. Wired for reproduction: Organization and development of sexually dimorphic circuits in the mammalian forebrain. Annual Review of Neuroscience 25: 507-536.

Smith, J. T. 2008. Kisspeptin signaling in the brain: Steroid regulation in the rodent and ewe. Brain Research Reviews 57: 288-298.

Smith, J. T. 2009. Sex steroid control of hypothalamic *Kiss1* expression in sheep and rodents: Comparative aspects. Peptides 30: 94-102.

Smith, M. T. and Mason, R. T. 1997. Gonadotropin antagonist modulates courtship behavior in male red-sided garter snake, *Thamnophis sirtalis parietalis*. Physiology and Behavior 61: 137-143.

Smith, M. T., Moore, F. L. and Mason, R. T. 1998. Neuroanatomical distribution of chicken-I gonadotropin-releasing hormone (cGnRH-I) in the brain of the male red-sided garter snake. Brain, Behavior and Evolution 49: 137-148.

Stirland, J. A., Johnston, J. D., Cagampang, F. R., Morgan, P. J., Castro, M. G., White, M. R., Davis, J. R. and Loudon, A. S. 2001. Photoperiodic regulation of prolactin gene expression in the Syrian hamster by a pars tuberalis-derived factor. Journal of Neuroendocrinology 13: 147-157.

Szymusiak, R. and Satinoff, E. 1982. Acute thermoregulatory effects of unilateral electrolytic lesions of the medial and lateral preoptic area in rats. Physiology and Behavior 28: 161-170.

Tarr, R. S. 1977. The role of the amygdala in the intraspecific aggressive behavior of the iquanid lizard, *Sceloporus occidentalis*. Physiology and Behavior 18: 1153-1158.

Taylor, E. N., Malawy, M. A., Browning, D. M., Lemar, S. V. and DeNardo, D. F. 2005. Effects of food supplementation on the physiological ecology of female western diamond-backed rattlesnakes (*Crotalus atrox*). Oecologia 144: 206-213.

Thapliyal, J. P. and Misra nee Haldar, C. 1979. Effect of pinealectomy on the photoperiodic gonadal response of the Indian garden lizard, *Calotes versicolor*. General and Comparative Endocrinology 39: 79-86.

Thapliyal, J. P., Kumar, D. S. and Oommen, O. V. 1975. Variations in thyroid activity and respiratory rate during a 24-hr period and role of testosterone and thyroxine on the oxidative metabolism of the water snake, *Natrix piscator*. General and Comparative Endocrinology 26: 100-106.

Tilden, A. R. and Hutchison, V. H. 1993. Influence of photoperiod and temperature on serum melatonin in the diamondback water snake, *Nerodia rhombifera*. General and Comparative Endocrinology 92: 347-354.

Tobet, S. A., Zahniser, D. J. and Baum, M. J. 1986. Differentiation in male ferrets of a sexually dimorphic nucleus in the preoptic/anterior hypothalamic area requires prenatal estrogen. Neuroendocrinology 44: 299-308.

Turek, F. W. and Van Cauter, E. 1994. Rhythms in reproduction. Pp. 487-550. In E. Knobil and J. D. Neil (eds), *Physiology of Reproduction*. Raven Press, New York.

Turner, J. E. and Tipton, S. R. 1972a. Metabolic response to temperature acclimation and T4 in the water snake. General and Comparative Endocrinology 18: 98-101.

Turner, J. E. and Tipton, S. R. 1972b. Environmental temperature and thyroid function in the green water snake, *Natrix cyclopion*. General and Comparative Endocrinology 18: 195-197.

Ulibarri, C. and Yarh, P. 1988. Role of androgens in sexual differentiation of brain structure, scent marking, and gonadotropin secretion in gerbils. Behavioral and Neural Biology 49: 27-44.

Underwood, H. 1981a. Circadian organization in the lizard *Sceloporus occidentalis*: the effects of pinealectomy, blinding, and melatonin. Journal of Comparative Physiology A 141: 537-547.

Underwood, H. 1981b. Effects of pinealectomy and melatonin on the photoperiodic gonadal response of the male lizard *Anolis carolinensis*. The Journal of Experimental Zoology 217: 417-422.

Underwood, H. 1985a. Pineal melatonin rhythms in the lizard *Anolis carolinensis*: Effects of light and temperature cycles. Journal of Comparative Physiology A 157: 57-65.

Underwood, H. 1985b. Annual testicular cycle of the lizard *Anolis carolinensis*: effects of pinealectomy and melatonin. The Journal of Experimental Zoology 233: 235-242.

Underwood, H. and Calaban, M. 1987. Pineal melatonin rhythms in the lizard *Anolis carolinensis*. I. Response to light and temperature cycles. Journal of Biological Rhythms 2: 179-193.

Underwood, H. and Hyde, L. L. 1989. The effect of daylength on the pineal melatonin rhythm of the lizard *Anolis carolinensis*. Comparative Biochemistry and Physiology 94A: 53-56.

Vagell, M. E. and McGinnis, M. Y. 1997. The role of aromatization in the restoration of male rat reproductive behavior. Journal of Neuroendocrinology 9: 415-421.

Veney, S. L. and Rissman, E. F. 1998. Co-localization of estrogen receptor and aromatase enzyme immunoreactivities in adult musk shrew brain. Hormones and Behavior 33: 151-162.

Venkatesh, B., Si-Hoe, S. L., Murphy, D. and Brenner, S. 1997. Transgenic rats reveal functional conservation of regulatory controls between the Fugi

isotocin and oxytocin genes. Proceedings of the National Academy of Sciences USA 94: 12462-12466.
Visser, M. E. 2008. Keeping up with a warming world; assessing the rate of adaptation to climate change. Proceedings of the Royal Society, Series B 275: 649-659.
Vivien-Roels, B. and Arendt, J. 1983. How does the indoleamine production of the pineal gland respond to variations of the environment in a nonmammalian vertebrate, *Testudo hermanni* Gmelin? Psychoneuroendocrinology 8: 327-332.
Vivien-Roels, B. and Pévet, P. 1983. The pineal gland and the synchronization of reproductive cycles with variations of the environmental climatic conditions, with special reference to temperature. Pineal Research Reviews 1: 91-143.
Vivien-Roels, B., Pévet, P. and Claustrat, B. 1988. Pineal and circulating melatonin rhythms in the box turtle, *Terrapene carolina triunguis*: Effect of photoperiod, light pulse, and environmental temperature. General and Comparative Endocrinology 9: 163-173.
Vodicnik, M. J., Kral, V. L., de Vlaming, M. J. and Crim, L. W. 1978. The effects of pinealectomy on pituitary and plasma gonadotropin levels in *Carassius auratus* exposed to various photoperiod-temperature regimes. Journal of Fish Biology 12: 187-196.
Wade, J. 1997. Androgen metabolism in the brain of the green anole lizard (*Anolis carolinensis*). General and Comparative Endocrinology 106: 127-137.
Wade, J., Huang, J-M. and Crews, D. 1993. Hormonal control of sex differences in the brain, behavior and accessory sex structures of whiptail lizards (*Cnemidophorus* species). Journal of Neuroendocrinology 5: 81-93.
Wagner, C. K. and Morrell, J. L. 1996. Distribution and steroid hormone regulation of aromatase mRNA expression in the forebrain of adult male and female rats: a cellular-level analysis using in situ hybridization. Journal of Comparative Neurology 370: 71-84.
Weaver, C. E, Baum M. J. 1991. Differential regulation of brain aromatase by androgen in fetal ferrets. Endocrinology 128:1247-1253.
Whittier, J. M., Mason, R. T., Crews D. and Licht, P. 1987. Role of light and temperature in the regulation of reproduction in the red-sided garter snake, *Thamnophis sirtalis parietalis*. Canadian Journal of Zoology 65: 2090-2096.
Wilczynski, W., Lynch, K. S. and O'Bryant, E. L. 2005. Current research in amphibians: studies integrating endocrinology, behavior, and neurobiology. Hormones and Behavior 48: 440-450.
Wilson, B. W., Chess, E. K. and Anderson, L. E. 1986. 60-Hz electric-field effects on pineal melatonin rhythms: Time course for onset and recovery. Bioelectromagnetics 7: 239-242.
Winkler, S. M. and Wade, J. 1998. Aromatase activity and regulation of sexual behaviors in the green anole lizard. Physiology and Behavior 64: 723-731.
Wong, K. L. and Chiu, K. W. 1974. The snake thyroid gland. I. Seasonal variation of thyroidal and serum iodoamino acids. General and Comparative Endocrinology 23: 63-70.
Wong, K. L., Chiu, K. W. and Wong, C. C. 1975. The effect of thyroid hormones on the oxygen consumption of the liver tissue in the squamate reptiles. Comparative Biochemistry and Physiology B 52: 355-358.
Woolley, S. C., Sakata, J. T. and Crews, D. 2004. Evolutionary insights into the regulation of courtship behavior in male amphibians and reptiles. Physiology and Behavior 83: 347-360.
Zhang, H. M., Su, Q. and Luo, M. 2008. Thyroid hormone regulates the expression of SNAP-25 during rat brain development. Molecular and Cellular Biochemistry 307: 169-175.

CHAPTER 9

Female Reproductive Anatomy: Cloaca, Oviduct, and Sperm Storage

Dustin S. Siegel[1], *Aurélien Miralles*[2], *Ryan E. Chabarria*[3] *and Robert D. Aldridge*[1]

9.1 OVERVIEW

The following is a comprehensive review of the literature on the reproductive anatomy of female snakes, particularly the cloaca and paired oviducts. Historical literature is brought together with recent and new findings to provide consistent nomenclature for the female reproductive tract. Literature on sperm storage and transport in the female reproductive tract of snakes is also reviewed. All snake taxonomy follows Zaher *et al.* (2009) for the Caenophidia and Vidal and Hedges (2002) for the Henophidia and Scolecophidia, unless otherwise noted.

9.2 THE CLOACA

9.2.1 Overview

As reported by Gadow (1887), "the cloaca is defined as a chamber at the terminal portion of the rectum, into which open the rectum, the ureters, and the genital tubes." Unterhössel (1902) provided a more traditional embryological definition of the cloaca as he referred to the cloaca as the open space in which the rectum, allantois, and urinary and genital ducts flow (following Balfour 1881). These definitions only encompass the cloaca of amniotes, and only some anamniotes and, thus, Gerhardt (1937) defined the cloaca as a common chamber that receives excretory ducts from at least

[1]Department of Biology, Saint Louis University, St. Louis, MO 63103, USA
[2]Technische Universität Braunschweig, Zoologisches Institut, Spielmannstr. 8, D-38106 Braunschweig, Germany
[3]Department of Life Sciences-Marine Biology, Texas A&M University-Corpus Christi, Corpus Christi, TX 78412, USA

two different organ systems, so as to create a cloaca definition that includes all vertebrates. Considering this review focuses only on female snakes, any of these definitions would accurately describe the common chamber for genital, fecal, and urinary material exit/entrance.

Few studies exist that describe the histology of the female snake cloaca; however, a number of families are represented for comparison (see Table 9.1 for species and families of snakes utilized for different types of studies on snake reproductive anatomy). Families investigated include the Colubridae, Dipsadidae, Elapidae, Leptotyphlopidae, Natricidae, Pythonidae, Typhlopidae, and Viperidae. Unterhössel (1902) described the cloaca of the European Grass Snake (*Natrix natrix*) in an effort to evaluate the gross regions depicted by Gadow (1887). Of the Scolecophidia, the Texas Blind Snake (*Leptotyphlops dulcis*), Western Thread Snake (*L. humilis*), Flowerpot Snake (*Ramphotyphlops braminus*), Angola Blind Snake, (*Typhlops angolensis*), and one unknown species of Typhlopidae were investigated by Fox and Dessauer (1962). The most in-depth histological description of the female cloaca was conducted by Gabe and Saint-Girons (1965), which also utilized a scolecophidian, the Spotted Blind Snake (*Typhlops punctatus*), and the Banded Krait (*Bungarus fasciatus*), Desert Horned Viper (*Cerastes cerastes*), Smooth Snake (*Coronella austriaca*), False Smooth Snake (*Macroprotodon cucullatus*), Carpet Python (*Morelia spilotes*), and Asp Viper (*Vipera aspis*) of the Caenophidia. To make comparisons between cloacal morphology and urine excretion Seshadri (1959) described the cloacal anatomy of the Indian Wolf Snake (*Lycodon aulicus*). Uribe *et al.* (1998) described the histology of the cloaca of the Lined Tolucan Ground Snake (*Conopsis lineata*) and are the only investigators to study seasonal variation in the histology of the cloaca. In an attempt to reconstruct the phylogenetic relationships of Squamata, Sánchez-Martínez *et al.* (2007) described the cloacal histology of one member of an unknown ground snake (*Atractus* sp.).

Gabe and Saint-Girons (1965) included a description of the glandular masses associated with the cloaca. Whiting (1969) expanded on the data on female cloacal glands by adding gland histological data from 30 species representing the families Acrochordidae, Boidae, Colubridae, Dipsadidae, Elapidae, Homalopsidae, Natricidae, Typhlopidae, and Viperidae. Species investigated included the Javan File Snake (*Acrochordus javanicus*), Long-nosed Tree Snake (*Ahaetulla nasuta*), Gold-ringed Cat Snake (*Boiga dendrophila*), Indian Krait (*Bungarus caeruleus*), *B. fasciatus*, Mussurana (*Clelia clelia*), Kirtland's Snake (*Clonophis kirtlandii*), the Emerald Tree Boa (*Corallus caninus*), Timber Rattlesnake (*Crotalus horridus*), Green Tree Snake (*Dendrelaphis punctulatus*), Loo-choo Big-tooth Snake (*Dinodon semicarinata*), Boomslang (*Dispholidus typus*), Saw-Scaled Viper (*Echis carinatus*), Rainbow Water Snake (*Enhydris enhydris*), Tentacle Snake (*Erpeton tentaculatum*), MacMahon's Desert Viper (*Eristicophis macmahoni*), Puff-faced Water Snake (*Homalopsis buccata*), Egyptian Cobra (*Naja haje*), Dice Snake (*Natrix tessellata*), Northern Water Snake (*Nerodia sipedon*), Western Rat Snake (*Pantherophis obsoletus*), Neuwied's False Boa (*Pseudoboa neuwiedii*), Chinese

Table 9.1 Taxa and focus of structural studies on the female reproductive tract of snakes

Taxa	*Cloaca*	*Cloacal glands*	*Oviducts*	*Sperm storage*	*References*
Leptotyphlopidae					
Leptotyphlops bicolor		x			Young *et al.* 1999
Leptotyphlops dulcis	x		x	x	Fox and Dessauer 1962
Leptotyphlops humilis	x		x	x	Fox and Dessauer 1962
Typhlopidae					
Ramphotyphlops braminus	x	x	x	x	Fox and Dessauer 1962; Young *et al.* 1999
Typhlops angolensis	x		x	x	Fox and Dessauer 1962
Typhlops muelleri		x			Young *et al.* 1999
Typhlops punctatus	x				Gabe and Saint-Girons 1965
Typhlops sp.	x		x	x	Fox and Dessauer 1962
Pythonidae					
Morelia spilotes	x				Gabe and Saint-Girons 1965
Python curtus	x				This chapter
Erycidae					
Gongylophis conicus	x				Das 1959
Boidae					
Acrantophis madagascariensis	x				Gadow 1887
Corallus caninus		x			Whiting 1969
Epicrates sp.			x		Blackburn 1998
Acrochordidae					
Acrochordus javanicus		x			Whiting 1969

Table 9.1 Contd. ...

... *Table 9.1 Contd.*

Viperidae					
Agkistrodon contortrix			x		Blackburn 1998
Agkistrodon piscivorus			x	x	Blackburn 1998; Siegel and Sever 2006, 2008a,b, Siegel *et al.* 2009b
Cerastes cerastes	x	x			Gabe and Saint-Girons 1965; Whiting 1969
Crotalus atrox		x			Young *et al.* 1999
Crotalus durissus			x		Almeida-Santos and Salomão 1997
Crotalus horridus		x			Whiting 1969
Crotalus viridis			x	x	Ludwig and Rahn 1943
Cryptelytrops albolabris	x				Pope 1941
Daboia russelii		x			Young *et al.* 1999
Echis carinatus		x			Whiting 1969; Young *et al.* 1999
Eristicophis macmahoni		x			Whiting 1969
Lachesis muta	x				This chapter
Macrovipera lebetina			x		Korneva 1973
Protobothrops sp.			x		Yokoyama and Yoshida 1994
Sistrurus miliarius		x			Whiting 1969
Vipera berus		x			Whiting 1969
Vipera aspis	x	x	x	x	Giacomini 1893; Van Den Broek 1933; Saint-Girons 1957; Gabe and Saint-Girons 1965
Viridovipera stejnergeri	x				Pope 1941
Homalopsidae					
Enhydris enhydris		x			Whiting 1969
Erpeton tentaculatum		x			Whiting 1969
Homalopsis buccata		x			Whiting 1969

Elapidae				
Bungarus caeruleus		x		Whiting 1969; Young *et al.* 1999
Bungarus fasciatus	x	x		Gabe and Saint-Girons 1965; Whiting 1969
Enhydrina schistose			x	Kasturirangan 1951a
Hemachatus haemachatus		x		Young *et al.* 1999
Hydrophis cyanocinctus			x	Kasturirangan 1951b
Naja haje		x		Whiting 1969
Naja kaouthia		x		Young *et al.* 1999
Pseudolaticauda semifasciata		x		Whiting 1969
Lamprophiidae				
Liopholidophis sexlineatus	x			This chapter
Calamariidae				
Calamaria lumbricoidea	x			Inger and Marx 1962
Colubridae				
Ahaetulla nasuta		x		Whiting 1969
Boiga dendrophila		x		Whiting 1969
Boiga irregularis			x	Bull *et al.* 1997
Conopsis lineata	x			Uribe *et al.* 1998
Coronella austriaca	x	x	x	Giacomini 1893; Gabe and Saint-Girons 1965
Dendrelaphis punctulatus		x		Whiting 1969
Dinodon semicarianata		x		Whiting 1969; Young *et al.* 1999
Dispholidus typus		x		Whiting 1969
Dolichophis jugularis			x	Sacchi 1888; Giacomini 1893

Table 9.1 Contd. ...

... *Table 9.1* Contd.

Elaphe quatuorlineata			x		Giacomini 1893
Gonyosoma oxycephalum		x			Young *et al.* 1999
Lampropeltis triangulum		x			Young *et al.* 1999
Lampropeltis getula		x			Price and LaPointe 1981
Lycodon aulicus	x				Seshadri 1959
Macroprotodon cucullatus	x	x			Gabe and Saint-Girons 1965
Pantherophis obsoletus		x			Whiting 1969
Pituophis melanoleucus		x			Young *et al.* 1999
Ptyas mucosus	x				Das 1959
Tantilla coronata			x	x	Aldridge 1992
Thelotornis kirtlandii		x			Whiting 1969
Colubridae *inserta sedis*					
Grayia smithii	x				This chapter
Natricidae					
Clonophis kirtlandii		x			Whiting 1969
Natrix natrix	x		x		Gadow 1887; Giacomini 1893; Unterhössel 1902; Giersberg 1922
Natrix tessellata		x			Whiting 1969; Young *et al.* 1999
Nerodia sipedon		x	x	x	Whiting 1969; Bauman and Metter 1977; Blackburn 1998; Young *et al.* 1999
Rhabdophis tigrinus			x		Takewaki and Hatta 1941
Seminatrix pygaea			x	x	Sever and Ryan 1999; Sever *et al.* 2000
Thamnophis elegans			x	x	Fox 1956; Mead *et al.* 1981
Thamnophis proximus		x			Whiting 1969

Thamnophis radix		x			Kissner *et al.* 1998; Young *et al.* 1999
Thamnophis sauritus		x			Whiting 1969; Young *et al.* 1999
Thamnophis sirtalis		x	x	x	Gibson 1934; Rahn 1940; Seaman 1949; Fox 1956; Whiting 1969; Hoffman 1970; Hoffman and Wimsatt 1972; Halpert *et al.* 1982; Blackburn 1998
Tropidoclonion lineatum		x			Young *et al.* 1999
Dipsadidae					
Atractus sp.	x				Sánchez-Martínez *et al.* 2007
Clelia clelia		x			Whiting 1969; Young *et al.* 1999
Coniophanes fissidens	x		x		This chapter
Diadophis punctatus		x	x	x	Perkins and Palmer 1996; Young *et al.* 1999
Heterodon platyrhinos	x				Edgren 1943
Liophis poecilogyrus	x				Pope 1941
Pseudoboa neuwiedii		x			Whiting 1969

Sea Snake (*Pseudolaticauda semifasciata*), Pygmy Rattlesnake (*Sistrurus miliarius*), Western Ribbon Snake (*Thamnophis proximus*), Eastern Ribbon Snake (*T. sauritus*), Common Garter Snake (*T. sirtalis*), Forest Vine Snake (*Thelotornis kirtlandii*), an unknown blindsnake (*Typhlops* sp.), and the Adder (*Vipera berus*). Whiting (1969) used these data to conclude scolecophidians were more closely related to the alethinophidians than to lizards by using a subjective approach (basically the presence of "scent glands" in snakes and absence in lizards).

Young *et al.* (1999) conducted an investigation similar to that of Whiting (1969). However, they utilized a more objective phenetic analysis to determine the evolutionary history of only snakes, from just the morphology of the scent gland. They added 12 species to those studied by Whiting (1969): Western Diamond-backed Rattlesnake (*Crotalus atrox*), Russell's Viper (*Daboia russelii*), Ringneck Snake (*Diadophis punctatus*), Red-tailed Green Rat Snake (*Gonyosoma oxycephalum*), Ring-necked Spitting Cobra (*Hemachatus haemachatus*), Scarlet Kingsnake (*Lampropeltis triangulum*), Two-colored Blind Snake (*Leptotyphlops bicolor*), Monocled Cobra (*Naja kaouthia*), Eastern Pine Snake (*Pituophis melanoleucus*), *Ramphotyphlops braminus*, Plains Garter Snake (*Thamnophis radix*), Lined Snake (*Tropidoclonion lineatum*), and Müller's Blind Snake (*Typhlops muelleri*). The cluster analysis performed by Young *et al.* (1999) revealed an evolutionary topology of snakes unlike any proposed before, with all major groups being paraphyletic. Price and LaPointe (1981) also studied the histology of the scent gland in the Common Kingsnake (*Lampropeltis getula*) in an effort to elucidate gland/muscle association and scent gland function. Kissner *et al.* (1998) studied the sexual dimorphism in structure of scent glands in *Thamnophis radix*, while Kissner *et al.* (2000) investigated sexual dimorphism in secretion odor intensity of the scent gland in the Black Ratsnake (*Pantherophis obsoleta*) and *Thamnophis sirtalis*.

Pope (1941) was the first investigator to describe the correlation between female cloacal morphology and hemipenis morphology (an idea hypothesized by Cope 1898), in which he utilized the White-lipped Tree Viper (*Cryptelytrops albolabris*) and Chinese Green Tree Viper (*Viridovipera stejnegeri*) of the Viperidae. Inger and Marx (1962) described the gross anatomy of the female cloaca in the Variable Reed Snake (*Calamaria lumbricoidea*) of the Calamariidae from different geographic locations, the only study to consider geographic affects on anatomical variation. Other studies on gross morphology have been conducted on the female cloaca of the Madagascar Ground Boa (*Acrantophis madagascariensis*; Gadow 1887), Rough-tailed Sand Boa (*Gongylophis conicus*; Das 1959), *Natrix natrix* (Gadow 1887), and the Oriental Ratsnake (*Ptyas mucosus*; Das 1959), along with the gross description of the cloacae from many taxa of snakes by Cope (1898).

The interaction of the male hemipenis with the cloaca in female snakes was examined by the fixation of male and female snakes in copulo. Pope (1941) described this in the Amazon Ground Snake (*Liophis poecilogyrus*;

Dipsadidae) while Edgren (1953) investigated gross copulatory adjustment in the Eastern Hognose Snake (*Heterodon platirhinos*) with somewhat conflicting results.

The following review focuses on the four distinct features of the snake cloaca: 1) regionality as depicted by Gadow (1887), 2) the cloacal/ oviducal junction, 3) glands associated with the cloaca, and 4) copulatory adjustment. New gross examination of four species representing the Pythonidae, Viperidae, Colubridae *inserta sedis,* and Lamprophiidae are utilized to investigate cloacal morphology in more detail. Species include Smith's African Watersnake (*Grayia smithii,* Field Museum of Natural History #191089), South American Bushmaster (*Lachesis muta;* Field Museum of Natural History #33691), Six-lined Water Snake (*Liopholidophis sexlineatus;* Field Museum of Natural History #75612), and the Blood Python (*Python curtus;* Field Museum of Natural History #71586). In addition, complete transverse and sagittal sections were made from the cloaca of the Yellowbelly Snake (*Coniophanes fissidens;* Field Museum of Natural History #20465 and #20474) to better visualize the histological regions depicted in historical accounts. Histological techniques were identical to those used by Siegel and Sever (2008a).

9.2.2 Gross Morphology

Grossly, the cloaca can be divided into three to four regions depending on the species (Fig. 9.1): a proctodaeum (region a), urodaeum (region b), a cranial urodaeal extension (not found in *Python curtus;* region b^1), and a two-region coprodaeal complex (regions c and d). The vent opens up into the proctodaeum (region a) ventrally, which can be distinguished caudally by low longitudinal folds and keratinized covering (Fig. 9.1A-C; folds were stretched out in *P. curtus* during mounting). The proctodaeum transitions into the urodaeum (region b), at which point no specific junction or sphincter can be observed; however, the longitudinal folds increase in size (Fig. 9.1A-C). Two openings on the urinary papilla projecting from the dorsal wall can be observed at the transition of low folds to deep folds (e.g., the transition from the proctodaeum to the urodaeum (Fig. 9.1A-E). A sphincter-like duct (region c) branches off of the urodaeum ventrally and communicates with the intestine (region d) ventrally (Fig. 9.1C). This group of structures make up the coprodaeal complex.

In *Python curtus* two papillae project caudally from the ventro-cranial wall of the urodaeum where the oviducts insert into the cloaca (Fig. 9.1A,D). In *Grayia smithii, Lachesis muta,* and *Liopholidophis sexlineatus* the insertions of the oviducts are not demarcated by papillae (Fig. 9.1E-G). In *G. smithii* and *L. muta* the large deep folds of the urodaeum terminate at a transverse septum (Fig. 9.1E,F), while in *L. sexlineatus* no septum is observed (Fig. 9.1G).

Cranial to the transverse septum longitudinal folds persist cranially in *Lachesis muta* (Fig. 9.1E). These folds decrease in size cranially, and are replaced by transverse ridges of folds (Fig. 9.1E). In *L. muta,* immediately

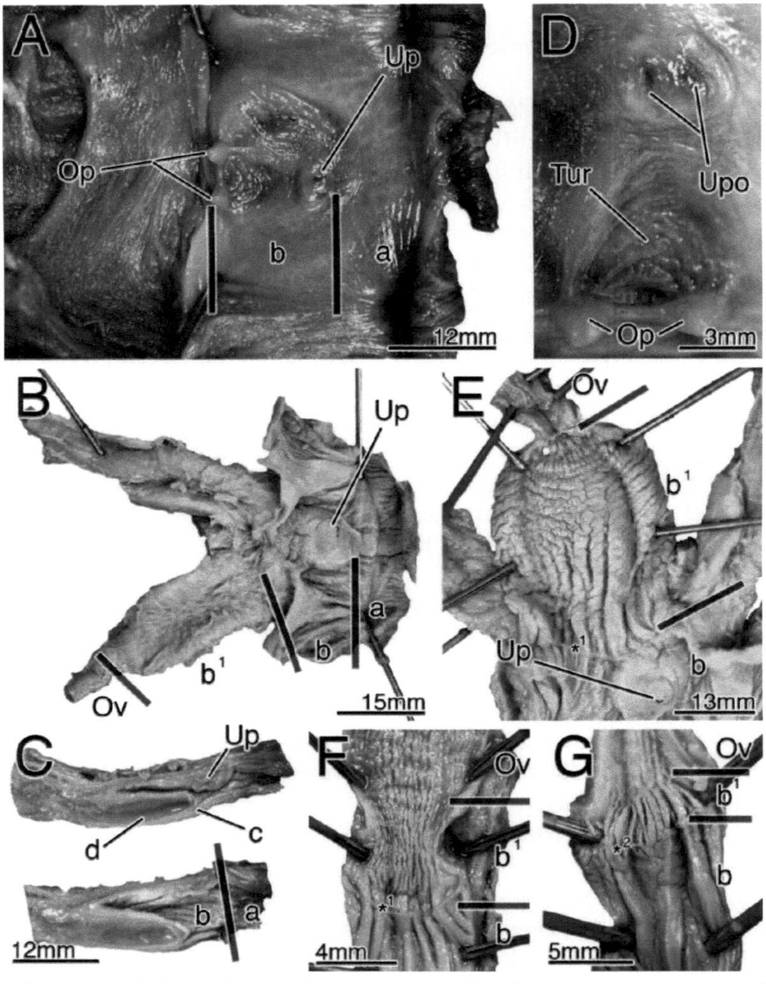

Fig. 9.1 Gross morphology of the snake cloaca, note that the opaque bars represent major division between the proctodaeum (a), urodaeum (b), and oviducts (Ov). **A.** Ventral view (anterior is to the left) of the cloaca of *Python curtus* exhibiting cranially inserting oviducal papilla (Op; note that cloaca is stretched cranially and therefore papillae appear to be originating dorsally) and dorsally projecting urinary papillae (Up). **B.** Ventral view (anterior is to the left) of *Lachesis muta,* note the enlarged vaginal pouch region (b^1) and dorsally projecting urinary papilla (Up). **C.** Sagittal view (anterior is to the left) of the cloaca of *Grayia smithii,* note the orientation of the urinary papillae (Up), urodaeal sphincter (c), and posterior intestine (d; top) compared to the urodaem and proctodaeum (bottom). **D.** Higher magnification of **A** (anterior is up), note the common urinary papilla with two ureteric openings (Upo) and the caudally projecting oviducal papillae (Op) formed from the insertion of the oviduct into the cranial terminal urodaeum (Tur). **E.** Ventral view (anterior is up) of *Lachesis,* note the transverse septum (black asterisk[1]) that separates the urodaeum from the pouch (b^1). **F.** Ventral view (anterior is up) of *Grayia,* note the transverse septum (black asterisk[1]) that separates the urodaeum from the pouch (b^1). **G.** Ventral view (anterior is up) of *Liopholidophis sexlineatus,* note the transition from the urodaeum to the pouch (b^1) without a well defined transverse septum (black asterisk[2]).

cranial to the transverse septum, the cloaca bifurcates dramatically forming two large pouches (region b^1; Fig. 9.1B,E). These pouches terminate at a sphincter-like region cranially, the oviduct (Fig. 9.1B,E).

Cranial to the transverse septum in *Grayia smithii*, long narrow and shallow longitudinal folds persist to the oviduct (Fig. 9.1F). A similar bifurcation of the cloaca occurs cranial to the transverse septum as seen in *Lachesis muta*; however, the two pouches (region b^1; Fig. 9.1F) formed are not enlarged as prominently as in *L. muta*. A similar occurrence forms in the cranial portion of the cloaca of *Liopholidophis sexlineatus*; however, a transverse septum does not delineate the division between the large longitudinal folds and narrower, shallow folds (Fig. 9.1G). The development of the pouch (region b^1) in *L. sexlineatus* is even less compared to *G. smithii*. In *L. sexlineatus* the narrow folds terminate at the oviduct (Fig. 9.1G), as is observed in *G. smithii* and *L. muta*.

9.2.3 Proctodaeum (Region a)

Limited variation has been reported in proctodaeal morphology/histology between taxa, though descriptions of the proctodaeum are relatively few. Figures 9.2 (sagittal sections of the cloaca) and 9.3 (transverse sections of the cloaca) have been constructed from *Coniophanes fissidens* to highlight the histological regions of the snake cloaca. In *C. fissidens* and *Morelia spilotes* the vent of the snake opens into a "U" shaped chamber (in transversal section) with the convex end of the "U" facing ventral (see Fig. in XVII-4 in Gabe and Saint-Girons 1965; Fig. 9.3A, this chapter). The "U" shape is formed by a dorsal projection from the wall of the proctodaeum of which ducts from two dorsal glands open laterally (see Fig. XVII-4 in Gabe and Saint-Girons 1965; Fig. 9.3A, this chapter). Although the only detailed historical proctodaeal description comes from *M. spilotes*, Gabe and Saint-Girons (1965) state that this condition is also present in many members of the Colubroides (represented by Fig. XVIII-5 for *Coronella austriaca*, Gabe and Saint-Girons 1965); however, no detailed analysis of the scolecophidian structure is given. From Fig. XVL-3 (Gabe and Saint-Girons 1965), it appears that the proctodaeal structure differs in *Typhlops punctatus* with the proctodaeum being relatively dorso-ventrally flattened, with dorso-lateral branches leading to remnant hemipenal pouches. It is unclear if the proctodaeum takes on the traditional "U" shape as described for the Henophidia and Caenophidia.

The proctodaeum is characterized by a stratified squamous epithelium (Fox and Dessauer 1962; Gabe and Saint-Girons 1965; Uribe *et al.* 1998; Sánchez-Martínez *et al.* 2007) lining a thin lamina propria encompassed proportionately by a thin layer of circular smooth muscle (Sánchez-Martínez *et al.* 2007; Fig. 9.3A). This smooth muscle is surrounded by bundles of longitudinal muscle making up the trunk musculature of the snake body (Sánchez-Martínez *et al.* 2007; Fig. 9.3A). The epithelium is keratinized, and was described as a Malphigian-keratinized epithelium by Gabe and

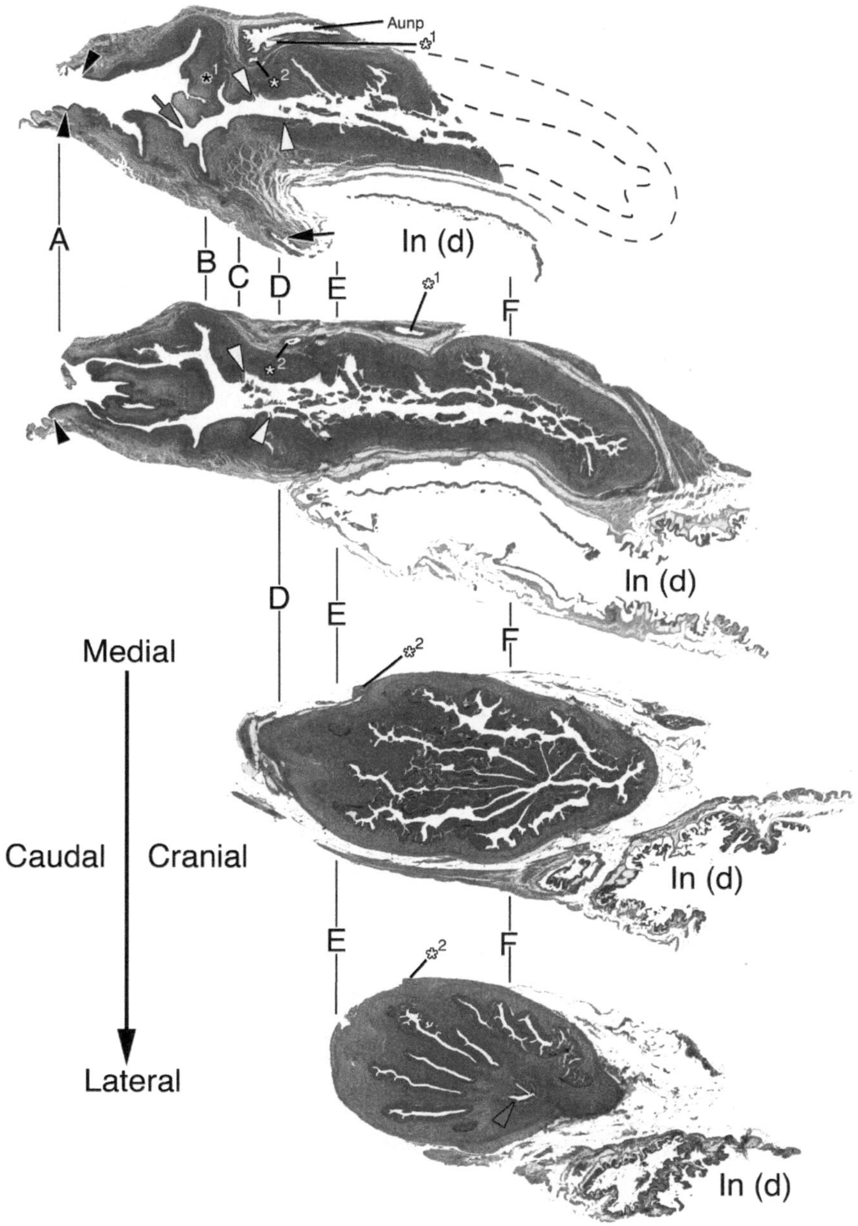

Fig. 9.2 Sagittal sections from the cloaca of *Coniophanes fissidens* (medial to lateral; Hematoxylin and Eosin). A, B, C, D, E, and F correspond with transverse sections of the same letters in Fig. 9.3. The partial top section is most medial exhibiting the transition from

Fig. 9.2 Contd. ...

Saint-Girons (1965) because of the similarity of this epithelium to typical deep layers of the skin. The basal layers of the proctodaeal epithelium are more cubic in appearance, and become increasingly flattened apically (Gabe and Saint-Girons 1965). Two cell layers deep to the superficial keratinized layer, cells are present in different stages of keratinization in *Macroprotodon cucullatus* (Gabe and Saint-Girons 1965). Superficial to the apical squamous layer, desquamating cells are observed in the lumen of the proctodaeum, similar to the observation of shedding scales or skin (see Figs. 59-60 in Gabe and Saint-Girons 1965).

The thickness of the stratified epithelium surrounding the lumen of the proctodaeum appears to vary between the scolecophidians, henophidians, and caenophidians (Gabe and Saint-Girons 1965). For example, *Typhlops punctatus* (Scolecophidia) has a relatively thick proctodaeal lining of 40-50 μm, while colubroid snakes (*Coronella austriaca* and *Macroprotodon cucullatus*; Caenophidia) have a relatively thin proctodaeal lining between 20 and 25 μm thick. *Morelia spilotes* (Henophidia) has a proctodaeal epithelium between that of the scolecophidians and caenophidians approximately 30-40 μm thick.

Gabe and Saint-Girons (1965) describe the anterior extremity of the proctodaeum as transitioning into a stratified cuboidal epithelium (mucous epithelium; Fig. 9.2 [top two sections]). This epithelium represents the transition from the skin-like keratinized epithelium to the typical epithelium of the urodaeum. In lizards, invaginations of epithelium into the lamina propria (forming glands) can be observed in this transitional region; however, no snake exhibits glands in this region (Fox and Dessauer 1962; Gabe and Saint-Girons 1965; Sánchez-Martínez *et al.* 2007).

9.2.4 Urodaeum (Region b)

General structure. The urodaeum has a larger chamber than the proctodaeum (Figs. 9.2 [top sections] and 9.3B-F), and is the most variable

... *Fig. 9.2 Contd.*

the proctodaeum (region a) to the urodaeum (region b; black arrowheads), urodaeum to the extended pouch (region b^1; white arrowheads), the opening of the urodaeal sphincter (region c; grey arrow), the urodaeal sphincter (black arrow) that communicates the posterior intestine (region d; In) with the urodaeum, the ampullary papilla (black asterisks), and the ducts that communicate with the urodaeum through the ampulla urinary papilla (Aunp, ampulla urinary papilla; White asterisk[1], ureter; White asterisks[2], Wolffian duct). The section below shows a section slightly lateral to the former where the ampullary papilla and urodaeal sphincter can no longer be observed in section. Below (immediately above bottom section), the proctodaeum and urodaeum cannot be observed in this section highlighting the enlargement of the cranial aspects of the cloaca. Note the highly folded epithelium signifying the expansion ability of the cranial pouch of the cloaca (region b^1), and the presence of the Wolffian duct as it follows the cranial pouch laterally. The bottom section shows the oviduct (grey arrowhead) communicating with the cloaca at a lateral position. Note that the insertion of the oviduct is not at the cranial extremity of the cloaca and, thus, the pouch terminates in a blind cranial cavity.

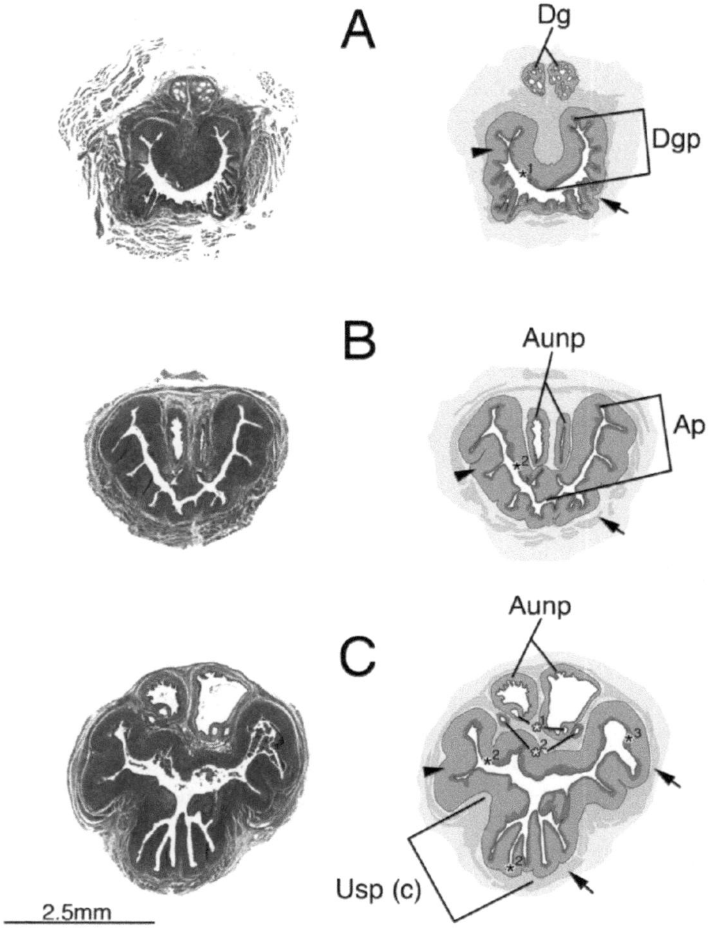

Fig. 9.3 Transversal sections from the cloaca of female *Coniophanes fissidens* (caudal to cranial; Hematoxylin and Eosin). **A.** At the level of the proctodaeum (region a): note the epithelium is stratified keratinized squamous throughout the proctodaeum (black asterisks[1]), also note the dorsal gland papilla (Dgp) projecting off of the dorsal proctodaeal walls through which ducts from the dorsal glands (Dg) travel to communicate with the proctodaeum. **B.** At the level of the caudal region of the urodaeum (region b), note the epithelium is a stratified cuboidal/columnar mucous epithelium (black asterisks[2]), also note the dorsal projection is now the ampullary papilla (Ap) through which the ampulla urinary papilla (Aunp) travel to communicate with the urodaeum. **C.** At the level of the mid-caudal region of the urodaeum (region b-b[1]), note the epithelium is a bistratified cuboidal/columnar mucous epithelium (black asterisk[2]) covering the left lateral portion of the urodaeum and the urodaeal sphincter (Usp [c]), whereas, the epithelium is transitioning to simple tall columnar (black asterisk[3]) dorsally and on the right lateral side of the urodaeum (beginning of region b[1]), also note two ducts are now observed that emptied into the ampulla urinary papilla (Aunp) more caudally, the ureters (white asterisk) and the Wolffian ducts (white asterisk[2]). **D.** At the level of the mid-cranial region of the urodaeum (region b-b[1]), note the epithelium is similar to that of **C**; however the

Fig. 9.3 Contd. ...

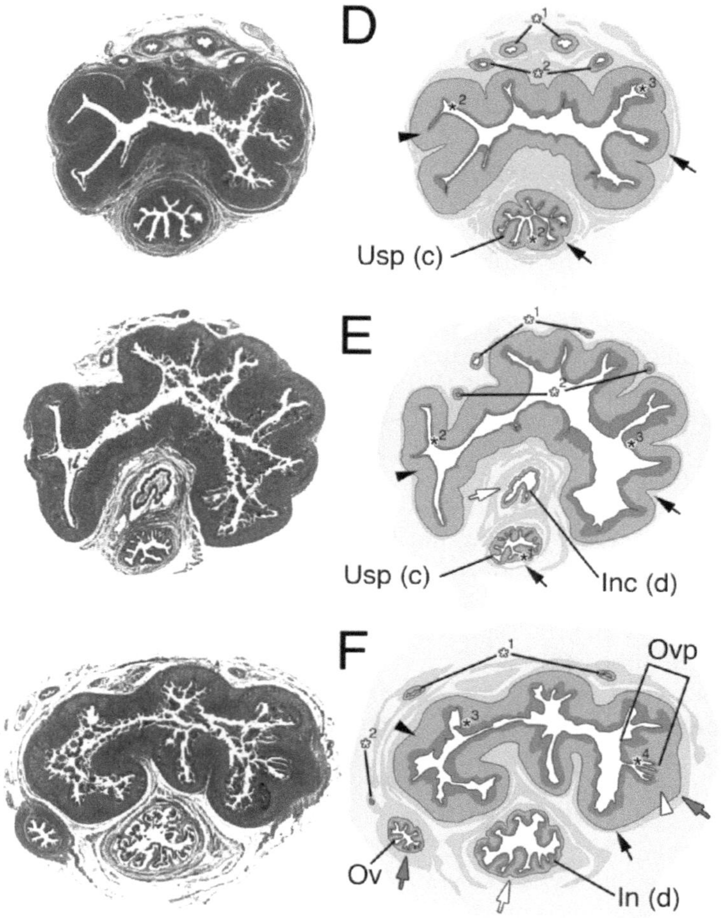

... *Fig. 9.3 Contd.*

simple tall columnar epithelium (black asterisks^{3}) is gradually taking over the entire lining of the urodaeum, also note the urodaeal sphincter (Usp [c]) has now completely branched off of the urodaeum and the ampulla urinary papilla has terminated, leaving only the ureters (white asterisks^{1}) and Wolffian ducts (white asterisks^{2}) observed. **E.** At the level of the cranial region of the urodaeum (region b-b^{1}), note the epithelium is identical to that of **D**, also note the posterior intestinal caecum (Inc [d]) can now be observed medial to the urodaeum and urodaeal sphincter (Usp [c]). **F.** At the level of the oviducal/cloacal junction (region b^{1}), note the simple tall columnar epithelium (black asterisks^{3}) has now taken over the entire lining of the cloaca, also note the intestine (In [c]) cranial to the intestinal sphincter communication is observed, and the left oviduct (Ov) has already inserted into the cloaca, whereas, the right oviduct is found inserting into the cloaca via a small oviducal papilla (Ovp); the epithelium of the oviduct is simple columnar and highly ciliated (black asterisks^{4}). Black arrow, stereotypical circular muscle layer of the cloaca; Black arrowhead, lamina propria; Gray arrow, stereotypical longitudinal and circular muscle layers of the oviduct; White arrow, stereotypical longitudinal and circular muscle layers of intestine; White arrowhead, lamina propria of the oviduct.

region of the female snake cloaca. The urodaeum is non-bifurcated in *Coniophanes fissidens*, *Macroprotodon cucullatus*, *Morelia spilotes*, and *Typhlops punctatus*. However, the urodaeum is longer on the right side of *T. punctatus* (which has only the right oviduct), *C. fissidens* (Fig. 9.3F), and *M. cucullatus* (Gabe and Saint-Girons 1965; this chapter). In contrast, the urodaeum is bifurcated in *Atractus* sp., *Bungarus fasciatus*, *Cerastes cerastes*, *Coronella austriaca*, and *Vipera aspis* (by fusion of the dorsal and ventral urodaeal walls, Gabe and Saint-Girons 1965; Sánchez-Martínez *et al.* 2007). The caudal portion of the urodaeum carries the typical "U" shape as described in the proctodaeum because of the ventrally projecting urinary papilla formed from the dorsal wall of the urodaeum (Fig. 9.3B).

As stated previously, the urodaeum is a continuation of the proctodaeum cranially with a bistratified (*Bungarus fasciatus*, *Typhlops angolensis*, and *T. punctatus*) to stratified (*Atractus* sp., *Cerastes cerastes*, *Coniophanes fissidens*, *Conopsis lineata*, *Coronella austriaca*, *Lycodon aulicus*, *Macroprotodon cucullatus*, *Morelia spilotes*, and *Vipera aspis*) mucous epithelium (Seshadri 1959; Fox and Dessauer 1962; Gabe and Saint-Girons 1965; Uribe *et al.* 1998; Sánchez-Martínez *et al.* 2007). Gabe and Saint-Girons (1965) also note that at the region where the coprodaeal complex communicates with the urodaeum (see section 9.2.5 for description of coprodaeal complex; Figs. 9.2 [top section] and 9.3C), the epithelium of the urodaeum is slightly thickened (e.g., more stratified). They term this the "région anale." The superficial epithelial cells of the urodaeum have been described as cuboidal to columnar (Gabe and Saint-Girons 1965; Sánchez-Martínez *et al.* 2007); however, this could be relative to the reproductive stage of each snake species. Secretions have been reported as positive for neutral carbohydrates (periodic acid Schiff's) and acidic mucoid substances (alcian blue; Gabe and Saint-Girons 1965; Uribe *et al.* 1998). The lamina propria surrounding the urodaeum is slightly thickened compared to the proctodaeum, and smooth circular muscle encompasses the mucosa (Sánchez-Martínez *et al.* 2007; Fig. 9.3B-F). The urodaeum is also highly folded, indicating the ability of expansion (possibly during copulation).

In the Colubroides investigated, the anterior urodaeum (Sánchez-Martínez *et al.* 2007; "diverticolo della cloaca," Giacomini 1893; "vagina," Cope 1898; "l'orifice génital" and "poche vaginale," Gabe and Saint-Girons 1965), which makes up 1/3 (*Conopsis lineata*, *Coronella austriaca*, and *Macroprotodon cucullatus*) to 2/3 (*Atractus* sp., *Bungarus fasciatus*, *Cerastes cerastes*, *Coniophanes fissidens*, and *Vipera aspis*) of the entire length of the urodaeum, transitions into a pseudostratified (*M. cucullatus*) to simple (*Atractus* sp., *B. fasciatus*, *C. cerastes*, *C. fissidens*, *C. lineata*, *C. austriaca*, and *V. aspis*) tall columnar epithelium (Gabe and Saint-Girons 1965; Uribe *et al.* 1998; Sánchez-Martínez *et al.* 2007; Figs. 9.2 [all sections] and 9.3C-F). This epithelium begins cranial to the transverse septum visualized by the gross morphology in section 9.2.2 and, thus, is the epithelium that lines the cranial pouch(es) (region b^1; Fig. 9.1E-G). The tall columnar epithelium is uniformly secretory (*Atractus* sp., Sánchez-Martínez *et al.* 2007; *C. lineata*, Uribe *et al.*

1998), or predominantly secretory with sporadic apically positioned ciliated cells (*B. fasciatus*, *C. cerastes*, *C. fissidens*, *C. austriaca*, and *V. aspis*; Gabe and Saint-Girons 1965; this chapter). The secretory cells react positively to histochemical stains for neutral carbohydrates (periodic acid-Schiff's) and acidic mucoid substances (alcian blue; Gabe and Saint-Girons 1965; Uribe *et al.* 1998). In *C. lineata*, crypt-like glands form in the epithelium during the height of mating activity (Uribe 1998). This region of the urodaeum makes up the majority of the bifurcation when observed in the cloaca, and is grossly enlarged in the elapids and viperids studied (Giacomini 1893; Gabe and Saint-Girons 1965; see Fig. 9.1B,E). As seen in *C. fissidens*, this region is also highly enlarged laterally compared to the more caudal regions of the cloaca, and increased lateral longitudinal folds indicate its greater ability in expanding (Figs. 9.2 [compare top sections with bottom sections] and 9.3 [compare caudal sections with cranial sections; A to F]). In *C. austriaca* and *M. cucullatus*, the tall columnar epithelium only surrounds the lumen of the dorsal portion of the urodaeum (Gabe and Saint-Girons 1965), whereas in *C. fissidens* this epithelium lines the entire cavity of the cranial extremities of the cloaca (Figs. 9.2 [all sections] and 9.3F). Giacomini (1893) first identified this region communicating the oviduct with the urodaeum in the colubrids *C. austriaca*, the Large Whip Snake (*Dolichophis jugularis*), and the Four-lined Ratsnake (*Elaphe quatuorlineata*), and the viperid *V. aspis*. Giacomini's descriptions add little to the detailed analysis above and basically state that the terminal portion of the oviduct opens into a large diverticulum, which has a mucous epithelium, and is most distinct in *V. aspis*.

The anterior urodaeum appears to be equivalent to the vaginal pouches depicted by Siegel and Sever (2008b) as part of the posterior oviduct in the Cottonmouth (*Agkistrodon piscivorus*) as evidenced by the exact same epithelial structure and histochemical reaction (see Fig. 9.7A,C,E). Other authors (Ludwig and Rahn 1943; Fox 1956; Aldridge 1992; Blackburn 1998) have also depicted this region as a derivative of the oviduct (Müllerian duct origin), and this controversy will be discussed in section 9.2.9. Under the assumption that this region was part of the oviduct, Siegel and Sever (2008b) described the ultrastructure of the epithelium of these cells. The epithelial cells are packed with apically positioned secretory vacuoles filled with an electron-dense material, with a highly electron-dense inner core. Lipid droplets can be seen in the basal position of the cells around the basally positioned nuclei, and it appears that a filamentous material is secreted onto the apical surface of the epithelium by a merocrine type secretion.

Urinary papilla/ae. From the dorsal wall of the urodaeum, the ureters communicate with the cloaca via ducts leading ventrally through one or two urinary papilla/ae (Gabe and Saint-Girons 1965; Figs. 9.2 [top section] and 9.3B). Immediately caudal to the genital orifice in *Typhlops punctatus*, the ureters pass through two large symmetrical papillae (Gabe and Saint-Girons 1965). Although not discussed in the text by Gabe and Saint-Girons (1965), remnant Wolffian ducts also travel through this papillae and may communicate with the urodaeum separately from the ureters (see Fig. 11 in Fox and Dessauer 1962). More caudal to the genital tubercles in *Morelia spilotes* the ureters communicate with the lumen of the urodaeum through two large urinary papillae (Gabe and Saint-Girons 1965). The Wolffian ducts are very enlarged in *M. spilotes;* however, Gabe and Saint-Girons (1965) do not discuss their origin or insertion. The authors do note that a material is present in the lumen of the Wolffian ducts that could be sperm (perhaps from a recent mating) or secretions from the Wolffian duct epithelium.

Representative high-resolution microscopy was conducted on the Wolffian ducts of a pregnant *Thamnophis sirtalis* from the spring to assess the origin of secretory material in the lumen of the Wolffian ducts (Siegel, unpublished; Fig. 9.4A-C). Cloacal material from this specimen was processed in the same manner as tissues for transmission electron microscopy in Siegel and Sever (2008a). Light microscopy micrographs of plastic sections at 1 μm reveal that the material residing in the lumen of the Wolffian duct is being secreted from the pseudostratified columnar epithelium of the Wolffian duct mucosa (Fig. 9.4A). Transmission electron microscopy of 70 nm sections reveals small electron dense secretory granules, larger condensing vacuoles, and floccular lipoidal material not constrained by a membrane in the apical regions of the cells (Fig. 9.4B). The basal regions possess euchromatic nuclei (Fig. 9.4B), perinuclear small mitochondria and profiles of rough endoplasmic reticulum. Mature electron dense secretory granules are transported to the apical surface of the cell and fuse with the cell membrane to release their product in a merocrine manner (Fig. 9.4C). Unorganized floccular lipoidal material diffuses through the apical cell membrane and appears to make up the majority of material found in the lumen of the Wolffian duct in *T. sirtalis.* (Fig. 9.4C). Basal cells tend to have more heterochromatic nuclei, possess many small mitochondria and have rough endoplasmic reticulum filling the entire cytoplasm. The production of material in the Wolffian duct does not appear to be equivalent to what is being synthesized in the ductus deferens of males because material production in the ductus deferens is scant, if any at all (Volsøe 1944; Sever 2004; Siegel *et al.,* 2009b; Sever 2010). Seasonal variation and histochemistry of this structure is desired to aid in determining the function of this odd organ.

The ureters communicate with the urodaeum in *Bungarus fasciatus, Coronella austriaca, Macroprotodon cucullatus* (see Fig. XVIII-4 in Gabe and Saint-Girons 1965), and *Lycodon aulicus* (Seshadri 1959) through one ventrally projecting urinary papilla. This papilla is far caudal to that of the genital tubercle in *M. cucullatus,* and the genital orifice in *C. austriaca* (no genital tubercle in *C. austriaca;* Gabe and Saint-Girons 1965). The posterior portions of the ureters, which communicate with the urodaeum, are often dilated into ampullae in these taxa (Gabe and Saint-Girons 1965). It is unclear from Sánchez-Martínez *et al.* (2007), whether one or two urinary papillae are present in *Atractus* sp. The viperids studied (*Cerastes cerates* and *Vipera aspis*) have two symmetrical urinary papillae similar to the condition

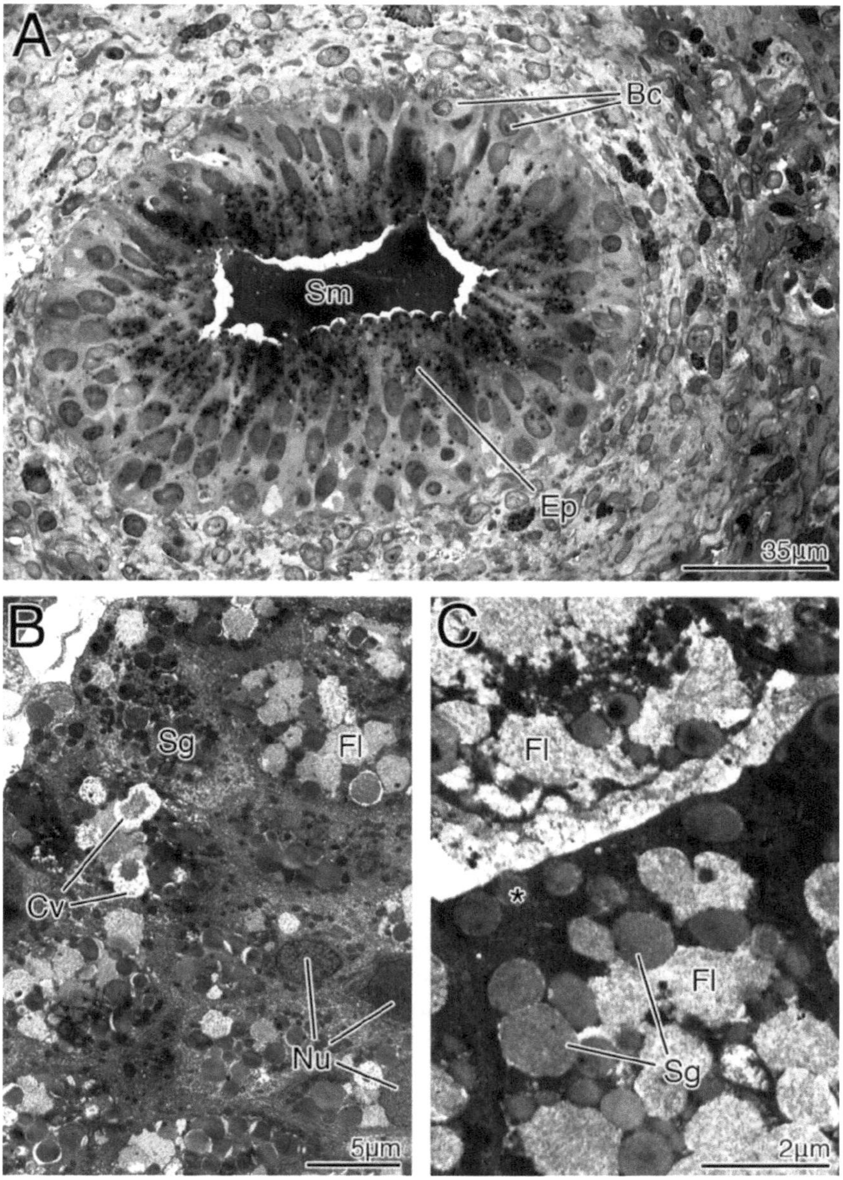

Fig. 9.4 The Wolffian duct of *Thamnophis sirtalis*. **A.** Transverse section of the Wolffian duct, note pseudostratified tall columnar epithelium and lumen filled with dense secretory material (Toluidine blue). **B.** TEM of the epithelial cells of the Wolffian duct showing numerous secretory complexes with varying densities of secretory material (Uranyl Acetate and Lead Citrate). **C.** TEM of the apical border of the Wolffian duct epithelium revealing merocrine type secretory mode (asterisk) of small electron dense granules and simple diffusion of floccular lipoidal material across the apical plasma membrane (uranyl acetate and lead citrate). Bc, basal cells; Cv, condensing vacuoles; Ep, superficial epithelial cells; Fl, floccular lipoidal material; Nu, nuclei; Sg, secretory granules; Sm, secretory material.

found in the scolecophidians and henophidians (Gabe and Saint-Girons 1965). Wolffian ducts are present running through the papillae parallel to the ureters in all of the above taxa (Gabe and Saint-Girons 1965), except *Atractus* sp., of which these ducts were not discussed (Sánchez-Martínez *et al.* 2007). No snake investigated possesses a urinary bladder (Cope 1898; Seshadri 1959; Fox and Dessauer 1962; Gabe and Saint-Girons 1965; Sánchez-Martínez *et al.* 2007).

The histology of the urinary papilla of *Coniophanes fissidens* is identical to that described for *Coronella austriaca, Bungarus fasciatus, Lycodon aulicus,* and *Macroprotodon cucullatus* (Seshadri 1959; Gabe and Saint-Girons 1965) with one urinary papilla projecting from the dorsal urodaeal wall (Fig. 9.3B). However, a novel network of ducts is present that were either missed in previous investigations or autapomorphic to *C. fissidens*. While moving cranially through the posterior half of the urodaeum, two ducts (duct 1 and duct 2) branch obliquely off of what was termed the ureter (duct 3) in previous studies instead of just one (Wolffian duct; Fig. 9.3C). Following these ducts cranially, duct 3 terminates (labeled Aunp in Fig. 9.3B,C), and ducts 1 and 2 continue cranially (Fig. 9.3C to Fig. 9.3D). Duct 1 (most lateral duct) is the Wolffian duct, while duct 2 travels all the way to the collecting ducts of the kidney (medially to the Wolffian duct and duct 3; Fig. 9.3C). Thus, this duct is the ureter. Therefore, duct 3, which communicates with the urodaeum through the urinary papilla, is not the ureter; it is a common duct that the Wolffian duct and ureter communicate with (labeled Aunp in Fig. 9.3B,C). This duct is undoubtedly equivalent with the "ampulla urogenital papilla" described in male members of the Colubroidea (see **Chapter 10**). Thus, we feel that this blind duct is probably not autapomorphic to *C. fissidens* and may be a synapomorphy for the Colubroidea, as it was not found in male viperids. We term this duct in females the "ampulla urinary papilla" due to the fact that this duct does not carry genital material as it does in males. We also utilize the term ampullary papilla for the dorsal projection off of the urodaeal wall through which this complex network of ducts travels, but only in female snakes where the three-duct system is present (at this time only *C. fissidens*). The papilla retains the nomenclature urinary papilla in snakes that do not possess the three-duct network.

9.2.5 Coprodaeum (Regions c and d)

Overview. As with most studies on reproductive morphology, the transition from gross morphological examination to microscopic examination can cause problems in terms of nomenclature. This nomenclatural problem has not evaded the coprodaeal complex. Although regional definitions from gross morphology are imperative for orientation, statements similar to, "rectal or innermost cloacal chamber [(in reference to the coprodaeum by Gadow 1887)]" leave much desired in terms of cloacal comparisons across taxa. Unaware of the fact that the intestine communicates with the urodaeum through a narrow channel (Unterhössel 1902; Seshadri 1959; Gabe and Saint-Girons 1965; region c), Gadow (1887) termed the posterior portion of the intestine the coprodaeum in squamates (Fig. 9.5A). Gadow (1887) described the coprodaeum as the cranial, intestinal chamber of the cloaca, separated from the urodaeum by a fold (Fig. 9.5A). Apparently this fold was supposed to indicate the transition from the primitive cloaca (formed from an epiblastic and hypoblastic tissue) to the hindgut (coprodaeum; Balfour 1881). Thus, the demarcation between the coprodaeum and the "primitive cloaca" was considered indistinct.

Seshadri (1959) followed the lead of Gadow (1887), as he described the posterior intestine (region d) as the coprodaeum; however, Seshadri (1959) also described a sphincter (first described by Unterhössel 1902; region c) that communicates the posterior intestine (his coprodaeum) to the urodaeum. Subsequent authors have termed this sphincter the coprodaeum in squamates (Regamey 1935; Trauth *et al.* 1987; Sánchez-Martínez *et al.* 2007), as it undoubtedly is derived from cloacal tissue (for review see Raynaud and Pieau 1985), whereas, the posterior intestine (Seshadri's coprodaeum) is derived from the mid-gut (for review see Raynaud and Pieau 1985). In developmental studies, the sphincter was termed the isthmus, which adjoins the urodaeum and posterior intestine (for review see Raynaud and Pieau 1985). The isthmus and posterior intestine has been termed the anlage of the coprodaeum (for review see Raynaud and Pieau 1985). It must be presented that Gadow (1887) was aware that he was terming the posterior intestine the coprodaeum in snakes, as he depicts this structure as secondarily assuming cloacal functions due to the fact that it was not of cloacal origin. However, he was inaccurate on how the intestine communicated with the urodaeum, as Gadow believed the intestine was continuous with the urodaeum (separated only by a fold; Fig. 9.5A).

Gabe and Saint-Girons (1965) avoided the confusing coprodaeal description of Gadow (1887), and decided on new terminology (even though in earlier work on lizards Saint-Girons [1962b] followed the terminology of Gadow [1887], and did not recognize the sphincter as a distinct structure). Gabe and Saint-Girons (1965) termed the "sphincter" the "sphincter anale" but do not refer to a coprodaeum in their extensive description of the squamate cloaca, except when referring to the work of Gadow (1887). Gabe and Saint-Girons (1965) refer to the region Gadow (1887) termed the coprodaeum as the intestine. We feel that Gabe and Saint-Girons (1965) have led future investigators in the direction of abandoning the term "coprodaeum", as the original "coprodaeum" was described inaccurately due to only gross examination. In fact, there is no intestinal chamber that makes up the cranial extremity of the urodaeum (Fig. 9.1E; Fig. 9.5B). As Gabe and Saint-Girons (1965) discuss, the sphincter leading to the posterior intestine is a ventral branching of the urodaeum (maintains identical epithelium to that of the urodaeum and single layer of muscle; Fig. 9.3C-E) and, thus, no part of the main cloacal chamber is composed of intestinally derived tissue (Fig. 9.5B). Our histological analysis of *Coniophanes fissidens* and gross observation from *Grayia smithii* (Fig. 9.5B) *Lachesis muta, Liopholidophis sexlineatus,* and *Python curtus* support this claim and, thus, Gadow's original drawing from 1887 has been redrawn more accurately here (Fig. 9.5C). This drawing is more consistent with the gross drawings of the cloaca by Gabe and Saint-Girons (see Fig. I in Gabe and Saint-Girons 1965). So as to not ignore the historical precedence established

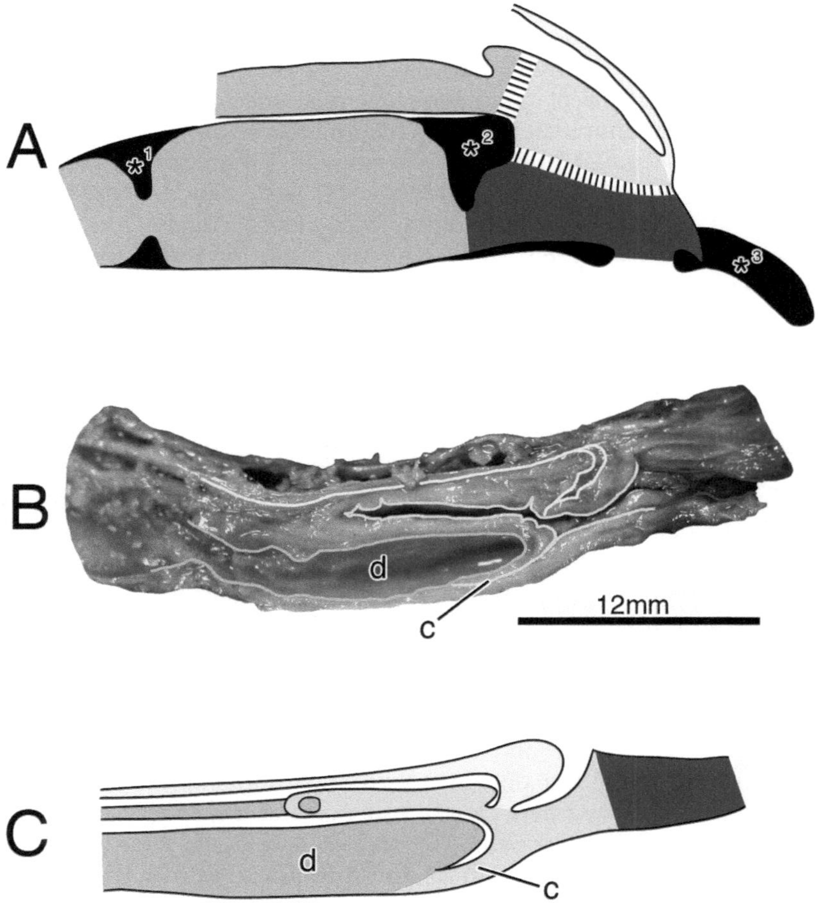

Fig. 9.5 Overview of the cloacal regions. **A.** Cloaca as represented by Gadow (1887), redrawn from Figure 19 in Gadow (1887) of a sagittal view through the midline of *Natrix natrix*. **B.** Outline of the cloacal chambers of *Grayia smithii* from this investigation depicting the general cloacal bauplan for snakes. **C.** Contemporary view of cloacal regionality presented by this review. Blue, proctodaeum (region a); Green, urodaeum (region b); Orange, oviduct; Pink, coprodaeum (region c)/intestine (region d); Asterisks[1], "fold that separates the coprodaeum from the intestine" (Gadow 1887); Asterisks[2], "fold that separates the urodaeum from the coprodaeum" (Gadow 1887); Asterisks[3], anal gland.

Color image of this figure appears in the color plate section at the end of the book.

by Gadow (1887), here the posterior intestine and urodaeal sphincter are collectively termed the "coprodaeal complex," and each will be compared in different taxa individually as such. Keep in mind, this basically means that the coprodaeal complex is not a term for any anatomical structure, but only for a functional complex of structures (see Table 9.2 for a review on the nomenclature of the coprodaeal complex).

Urodaeal sphincter (region c). The sphincter is a ventral projection from the urodaeum (Figs. 9.2 [top section] and 9.3C) that communicates with the posterior intestine. The sphincter has a very thick musculature, mostly composed of circular muscle and occasional bundles of striated muscle fibers (Sánchez-Martínez *et al.* 2007; Fig. 9.3C-E). In *Typhlops angolensis*, the sphincter communicates with the urodaeum at its most caudal position, on the ventral side of the urodaeum (Fox and Dessauer 1962). The posterior portion of the sphincter has an epithelium similar to that of the urodaeum (stratified columnar/cuboidal) and transitions to a simple columnar epithelium at its cranial extremity, indicating the transition of the sphincter to the posterior intestine (Fox and Dessauer 1962). The sphincter attaches to the posterior intestine at the terminal portion of the posterior intestine. This appears to be how the sphincter communicates with the intestine in *Morelia spilotes* as well (Gabe and Saint-Girons 1965).

In *Atractus* sp., *Bungarus fasciatus*, *Cerastes cerastes*, *Coniophanes fissidens*, *Coronella austriaca*, *Macroprotodon cucullatus*, and *Vipera aspis* a different sphincter/posterior intestine communication is observed. Whereas, the sphincter still communicates with the urodaeum at the caudal portion of the ventral side of the urodaeum, the sphincter does not communicate with the posterior intestine at its caudal terminal end. Instead, the sphincter travels ventro-cranial to the intestine and attaches to the posterior intestine on its ventral side (Gabe and Saint-Girons 1965; Sánchez-Martínez *et al.* 2007; Figs. 9.2 [top section] and 9.3D,E), forming a caudally projecting intestinal caecum (Fig. 9.3E). Although this is depicted in the cross sectional drawings of Gabe and Saint-Girons (1965, Fig. XVII-2, Fig. XVIII-3, and Fig.

Table 9.2 Historical nomenclature of the coprodaeal complex

Investigator(s)	*Coprodaeal complex*	
	c*	d*
Gadow 1887	Not discussed	Coprodaeum
Regamey 1935[1]	Coprodaeum	Intestine
Seshadri 1959	Sphincter	Coprodaeum
Gabe and Saint-Girons 1965	Sphincter anale	Intestine
Reynaud and Pieau 1985	Isthmus	Coprodaeum
Trauth *et al.* 1987[2]	Coprodaeum	Intestine
Sanchez-Martinez *et al.* 2007	Coprodaeum	Intestine
Terminology used here	Sphincter	Posterior intestine

[1]Nomenclature from an investigation on a lizard (Lacertidae), *Lacerta agilis*.
[2]Nomenclature from an investigation on a lizard (Scincidae), *Plestiodon laticeps*.
*represent female cloacal regions delineated in the text.

XIX-3), this is not repetitive of the their sagittal drawings overviewing the female snake cloacal complex in their Fig. I-5, which we therefore assume are slightly inaccurate (see Fig. 9.5B,C). The epithelium lining the muscular sphincter of *Atractus* sp., *B. fasciatus*, *C. cerastes*, *C. fissidens*, *C. austriaca*, and *M. cucullatus* were each similar to their own urodaeal epithelium, with a transition to intestinal epithelium (increased occurrence of goblet cells) at the communication with the intestine (Gabe and Saint-Girons 1965; Sánchez-Martínez *et al.* 2007).

It is important to note that in only one species of snake investigated, *Cerastes cerastes*, a ventral projection off of the urodaeal sphincter was observed (Gabe and Saint-Girons 1965). This was similar to the vestigial urinary bladder observed in some lizards, termed the "bladder stalk" by Mulaik (1946), and carried the same epithelial lining of the urodaeum (Gabe and Saint-Girons 1965). Thus, the conclusion that snakes have completely lost a urinary bladder-like structure may be premature.

Posterior intestine (region d). The posterior intestine in snakes is characterized by series of transverse septa that create distinct chambers (Seshadri 1959). Seshadri (1959) reveals up to five, cranially to caudally, aligned chambers in *Lycodon aulicus*, which were referred to as coprodaeal chambers in male vipers (Volsøe 1944).

The lamina propria of the posterior intestine is very thin, and has a typical intestinal muscularis, quite thickened in snakes (Seshadri 1959; Sánchez-Martínez *et al.* 2007). The posterior intestine possesses a pseudostratified epithelium with a superficial layer of tall columnar mucous secreting cells in *Typhlops punctatus* (Gabe and Saint-Girons 1965). *Morelia spilotes* possesses a similar posterior intestine; however, the epithelium is simple. In *Bungarus fasciatus*, *Cerastes cerastes*, *Coronella austriaca*, *Macroprotodon cucullatus*, and *Vipera aspis* the epithelium of the posterior intestine is similar to *Typhlops punctatus;* however, two types of secretory cells are observed. Some cells are tall columnar and produce neutral carbohydrates and acidic mucides, while others are more caliciform in appearance and produce only an acidic mucous type secretion (Gabe and Saint-Girons 1965). In *Lycodon aulicus*, the epithelium of the posterior intestine is composed of tall columnar cells, with the cells reaching their greatest height in the most cranial chamber of the posterior intestine (Seshadri 1959).

Seshadri (1959) describes the storage and processing of "urinary pellets" in the posterior intestine of the coprodaeal complex, a phenomenon that may have arisen because of the lack of a functioning urinary bladder in snakes (Gabe and Saint-Girons 1965). Apparently, urine, which is composed of 80% uric acid (Seshadri 1959), is passed from the urinary papillae (which lay directly dorsal to the urodaeal sphincter; Fig. 9.5B,C), through the urodaeal sphincter, and cranially through the transverse septa of the posterior intestine while water is absorbed from the urine. It is unclear if lizards, of which some have a functioning urinary bladder (Beuchat 1986), possess a chambered posterior intestine created by transverse septa.

9.2.6 Cloacal/Oviducal Junction (region b or b^1)

The posterior portion of the oviduct (characterized by a heavily muscularized duct with a simple to pseudostratified columnar, predominantly ciliated, epithelium; discussed in detail in section 9.3.3) inserts laterally into the urodaeal chamber covered by a stratified mucous epithelium in *Typhlops angolensis* and *T. punctatus* without the formation of a genital tubercle (Fox and Dessauer 1962; Gabe and Saint-Girons 1965). In *Morelia spilotes*, the oviducts insert into this urodaeal chamber ventro-laterally with the formation of large genital tubercles (Gabe and Saint-Girons 1965). In all other snakes investigated (*Atractus* sp., *Bungarus fasciatus, Cerastes cerastes, Coniophanes fissidens, Conopsis lineata, Coronella austriaca, Macroprotodon cucullatus*, and *Vipera aspis*), the oviducts communicate with a simple to pseudostratified columnar segment of the cloaca only found in the Colubroides (Fig 9.2 [bottom section] and Fig. 9.3F; Giacomini 1893; Gabe and Saint-Girons 1965; Uribe *et al.* 1998; Sánchez-Martínez *et al.* 2007). The former condition appears to be the pleisiomorphic condition as this is the condition found in lizards and Tuatara (*Sphenodon punctatus*; Gabe and Saint-Girons 1965; Trauth *et al.* 1987). In *C. fissidens* and *M. cucullatus*, the oviducts insert into the tall columnar epithelium region via small genital tubercles (Gabe and Saint-Girons 1965; Figs. 9.2 [bottom section] and 9.3F). In all other Colubroides, the simple tall columnar region transitions into the oviducts without genital tubercles (Gabe and Saint-Girons 1965; Sánchez-Martínez *et al.* 2007). The gross morphology presented here (Fig. 9.1A-G) confirms the previous histological findings. Although not discussed in previous investigations on Colubroides, the oviducts of *C. fissidens* insert medial/obliquely forming a blind cranial pouch of the cloaca.

9.2.7 Cloacal Glands

We limit the discussion of cloacal glands to their presence and location, with few remarks on histology. We discuss only the glands that have been identified in female snakes. Although some works provide cloacal gland histology, it is often unclear which sex is being discussed. We refer readers interested in cloacal gland histology to Gabe and Saint-Girons (1965), Whiting (1969), and Young *et al.* (1999).

Cloacal glands are somewhat enigmatic organs, with the function of their secretions still undetermined. Cloacal gland function even drew speculation from Charles Darwin (from Whiting 1969) on the Red Bow-fingered Gecko (*Cyrtodactylus rubidus*), where he states that these glands, "probably serve to emit an odour." Our knowledge of cloacal gland function in snakes has not proceeded far beyond this point. Whiting (1969) summarized other hypotheses for cloacal gland function. Researchers believed that the glands were of the same derivation and function as the prostate gland and/or seminal vesicles, but their presence in females contradicts the functional hypothesis. Others predicted that these glands may be involved in displaying mating readiness, and/or opposite conspecific sex determination (Cooper and Trauth 1992); however, more recent investigations refute this hypothesis and suggest these types of pheromonal signals are actually produced in the skin (Garstka and Crews 1981). Of the hypotheses presented that have not been refuted, the possibility of these glands functioning to deter predators still carries weight, although only one study has indirectly confirmed this hypothesis by refuting pheromonal communication (Price and LaPointe 1981). It has also been proposed that the cloacal glands simply serve to lubricate the cloaca for hemipenis insertion, but even the father of this hypothesis (Beuchelt 1936) doubts that this is the sole function of cloacal glands. More recent studies hypothesize that cloacal glands (particularly the scent gland) function in interspecific response and intraspecific alarm cues (for review see Young *et al.* 1999).

Four cloacal gland types have been described in female snakes: 1) the median cloacal gland, 2) the ventral glands, 3) the dorsal gland/s, and 4) the scent glands (also referred to as anal glands). Although, other than the scent gland, it is unclear from previous descriptions to what degree the glands investigated are homologous. It is clear; however, that all of the glands described communicate with the proctodaeum via one or two ducts.

The median cloacal gland has only been reported in the scolecophidians, specifically *Typhlops punctatus* and *Typhlops* sp. (Gabe and Saint-Girons 1965; Whiting 1969). It is found immediately ventral to the scent gland and is comprised of many tubules that open into the dorsal side of the proctodaeum via one common duct. While Gabe and Saint-Girons (1965) depict this gland as being ovoid in shape, Whiting (1969) described this gland as triangular, with the apex of the triangle directed dorsal. The epithelium of the tubules is made up of a simple cuboidal epithelium that produces a secretion that stains positive with the periodic acid-Schiff's procedure indicative of neutral carbohydrates (Gabe and Saint-Girons 1965; Whiting 1969).

The presence of glands lying ventral to the cloaca, appropriately termed ventral glands, is somewhat controversial. While they have been identified in two investigations utilizing *Coronella austriaca* (Disselhorst 1904) and *Vipera aspis* (Beuchelt 1936), ventral glands have not been identified in any other study (besides a note in Gabe and Saint-Girons 1964); even those that utilized *C. austriaca* and *V. aspis* (Gabe and Saint-Girons 1965; Whiting 1969). Ventral gland histology and anatomy are depicted as identical to dorsal glands, which are discussed in the following paragraph.

The dorsal glands may be paired and dorso-lateral or lateral in position (most common in snakes; Fig. 9.3A), or single median (Gabe and Saint-Girons 1965; Whiting 1969). Paired lateral and single median glands are present simultaneously in *Bungarus fasciatus*, *Enhydris enhydris*, *Thamnophis proximus*, *T. sauritus*, and *T. sirtalis* (Whiting 1969). The single median gland may be the result of fusion of the paired glands during development in *Thamnophis* evidenced by two ducts leading from the gland to the proctodaeum (Whiting 1969). Similar to the ventral glands, the dorsal gland/s is/are described as "tubulous" or "tubuloacinous," although the species that possess each type are not provided (Whiting 1969). Whiting (1969) described the epithelial lining as stratified columnar (*Dinodon semicarianata* and *Nerodia sipedon*) or simple columnar (*Dendrelaphis punctulatus*, and *Enhydris enhydris*), although the sex of these species description was not given. In *Morelia spilotes*, two poorly developed glands are described located in a dorsal position in relation to the midline of the proctodaeum (Gabe and Saint-Girons 1965). A large duct from each gland, with a similar epithelium to that of the proctodaeum communicates the dorsal glands and proctodaeum. The cells of the glandular epithelium are ovoid in appearance and secrete a glycoprotein mixture (Gabe and Saint-Girons 1965). A small dorsal cloacal mass forms the dorsal gland in *Macroprotodon cucullatus* with an epithelium similar to that of *M. spilotes*, while the dorsal glands in *Coronella austriaca* are reduced to two small tubes that sink into the median-dorsal region of the proctodaeum (Gabe and Saint-Girons 1965). *Bungarus fasciatus* has similar dorsal glands as *M. cucullatus*; however, a bistratified prismatic epithelium is present. In *Vipera aspis* two dorsal glands are present that are surrounded by the same serosa and have a simple cuboidal epithelium similar to that of *C. austriaca*, *M. cucullatus*, and *M. spilotes* (Gabe and Saint-Girons 1965). Dorsal glands in *Cerastes cerates* are identical to *V. aspis*, except that the two glands are clearly separated by their own serosa (Gabe and Saint-Girons 1965).

The work by Young *et al.* (1999) provides the most accurate comparative description of the scent glands. All snakes investigated have paired scent glands that are located posterior to the cloaca, embedded in the tail (Whiting 1969; Young *et al.* 1999). Grossly, these glands are large, white to yellow in appearance, and have a thick serosa that connects to the trunk musculature of snakes (Young *et al.* 1999). Four layers are described in the scent glands of snakes by Young *et al.* (1999): 1) the outer serosa, 2) the epithelium, 3) the sloughed epithelium (a product of holocrine secretion in these glands), and 4) cellular debris, which forms "concentric rings" around an "inner core of amorphous secretory material" (Young *et al.* 1999) in the lumen of the glands. In *Daboia russelii*, *Leptotyphlops bicolor*, and *Pituophis melanoleucus* the scent glands are bilobed cranially and caudally (Young *et al.* 1999). A septum was observed dividing the scent gland in *Bungarus caeruleus*, *Crotalus atrox*, *Diadophis punctatus*, *Gonyosoma oxycephalum*, *P. melanoleucus*, *Thamnophis radix*, and *Typhlops muelleri* (Young *et al.* 1999). The epithelium of the scent glands is always stratified (see Table 1 in Young *et al.* 1999 for number of cell layers making up the epithelium in each taxon investigated) and ranges from squamous to columnar (Young *et al.* 1999). Variation was also observed in the contours of the epithelium and the staining reaction of the epithelium (see Table 1 in Young *et al.* 1999 for distribution amongst taxa). Descriptions of the morphology of the scent glands by Gabe and Saint-Girons (1965) are similar to those of Young *et al.* (1999). The scent gland, and scent gland ducts are encompassed by striated muscle, which is indicative of the fast expulsion of the material (Price and LaPointe 1981). Also of note, females appear to have slightly larger scent glands than males, and also appear to be more malodorous (Kissner *et al.* 1998; Kissner *et al.* 2000).

9.2.8 Copulatory Adjustment

The cloaca serves as an exit for urinary and fecal waste (Gadow 1887) and as the vagina in which the male copulatory organ, the hemipenis, resides during copulation (Pope 1941; Edgren 1953).

Cope (1898) noted that male species with particularly spiny hemipenes have con-specific females with particularly thick walled cloacae. Pope (1941) confirmed this phenomenon in *Cryptelytrops albolabris* and *Viridovipera stejnegeri*. The male hemipenis of the former is long with few spines, while the male hemipenis of the latter is short and very spiny. The conspecific female of the former had a highly bilobed cloaca with a very thin wall, while the conspecific female of the latter had a short cloaca with a very thick wall. "The correlation is obvious enough," as stated by Pope (1941). Pope used this correlation to hypothesize that the differences in structures could be isolating mechanisms in these sympatric species.

Pope (1941) also stated that the hemipenis of *Liophis poecilogyrus* abuts the cloaca where the oviducts open; "thus, each sulcus makes a direct connection with an oviduct." Considering this species is a member of the Colubroides, we feel it is possible that the hemipenis in Colubroides insert into the region with a tall columnar epithelium as hypothesized by Ludwig and Rahn (1943) for *Crotalus viridis*. Ludwig and Rahn (1943); however, were not the first to hypothesize that hemipenis enter the enlarged "pouches." Giacomini (1893) made clear that the large orifices at the cranial end of cloaca of *Vipera aspis* created ample room for a copulatory organ.

Edgren (1953) reported a different finding than Pope (1941) from two *Heterodon platyrhinos* captured *in coitus*. Edgren (1953) stated that the hemipenis only inserts into the posterior ¾ of the cloaca and that the sulci do not abut snuggly against the openings of the oviducts. He found that the anterior ¼ of the cloaca, possibly the area of simple tall columnar epithelium, is filled with sperm and secretions that possibly form a copulatory plug. Citing the work of Beuchelt (1936) on hemipenis erectile physiology, Edgren (1953) believed that the hemipenis takes on the shape of the female cloaca and, thus, differences in preserved structural complexity of the hemipenis may not act as an isolating mechanism for sympatric species (as long as spines do not cause a fatality during copulation). We concur with Pope (1941) that because of varying hemipenal and female cloacal morphology, that the "problem [of copulatory adjustment in snakes] is barely opened," and much more insight from diverse species is necessary to make any conclusions in terms of copulatory adjustment in diverse snake taxa.

9.2.9 Giacomini's "Diverticulum" (Region b^1)

All Colubroides studied have a distinct region of tall, primarily secretory, columnar cells that communicate the urodaeum with the oviducts (see section 9.2.4). Giacomini (1893) was the first to notice this distinctive region histologically. Shortly after, Cope (1898) described this region grossly and termed it a common bifurcated vagina because this region enlarges at the caudal end of the posterior oviduct and then combines and communicates with the urodaeum (also confirmed from the gross morphology in this review; however, no transverse septum demarcation was observed by Giacomini (1893) or Cope (1898) between the urodaeum and diverticulum; see Fig. 9.1E,F). Cope (1893) noted that this region was not bifurcated in the Boidae, Erycidae, and Pythonidae (Cope's "Peropoda"). However, Cope (1898) did not realize in his description of the vagina that an equivalent histological region to the bifurcated vagina in the Colubroides was not present in the Peropoda, and subsequently was not found to be present in the scolecophidians (Fox and Dessauer 1962; Gabe and Saint-Girons 1965). Thus, Giacomini's diverticulum may be a synapomorphy for the Colubroides.

The embryonic origin of Giacomini's diverticulum is controversial. Some investigators suggest that this region is derived from the Müllerian duct (see Saint-Girons 1957; Blackburn 1998; Siegel and Sever 2008b), while others suggest this region is part of the cloaca (Giacomini 1893; Gabe and Saint-Girons 1965; Sánchez-Martínez *et al.* 2007). The origin of this region has even confused the same author working on the same species; e.g. Saint-Girons (1957) described Giacomini's diverticulum in *Vipera aspis* along with the oviduct (implicating Müllerian duct origin), while in his co-authored work with Manfred Gabe (Gabe and Saint-Girons 1965), described this region as a bifurcation of the cranial extremity of the urodaeum (implicating cloaca origin). The extensive review on development of the urogenital system by Raynaud and Pieau (1985) unfortunately does not give any insight into this controversy and, thus, no developmental data exist describing the origin of Giacomini's diverticulum. We hypothesize that because genital tubercles protrude through the cranial wall of the urodaeum of the scolecophidians and henophidians, and small tubercles protrude through the cranial extremities of Giacomini's diverticulum in *Coniophanes fissidens* (Figs. 9.2 [bottom section] and 9.3F) and *Macroprotodon cucullatus* (Gabe and Saint-Girons 1965), that Giacomini's diverticulum represents a specialized extension of the urodaeum. In other species of Colubroides investigated the tubercles are lost, concurrent with a dramatic increase in size of Giacomini's diverticulum, especially in Viperidae and Elapidae (whether these events occur in a phylogenetically conserved pattern or independently has not been determined; Gabe and Saint-Girons 1965; Sánchez-Martínez *et al.* 2007).

9.2.10 Conclusions

Grossly and histologically, the cloaca can be divided into three to four regions depending on the species. In the Scolecophidia, Henophidia, and Caenophidia the posterior proctodaeum (region a) has a stratified keratinized squamous epithelium, similar to the epidermis of the skin.

Cranial to the proctodaeum the epithelium becomes a bistratified to stratified mucous epithelium. This region is termed the urodaeum (region b) in all taxa studied. The demarcation of the proctodaeum and urodaeum is not easily observed, however, Seshadri (1959) states that, "the position of [the] urinary aperture [(urinary/ampullary papilla(ae))] seems to indicate the boundary of the two chambers." It is also important to note that the urodaeum and proctodaeum are completely divided during early development (for review see Raynaud and Pieau 1985). Ventrally, projecting off of the posterior urodaeum, is the coprodaeal complex. This complex is composed of the urodaeal sphincter (region c), which adjoins the urodaeum and the posterior intestine (region d), which possibly functions in uric acid storage in some species (*Lycodon aulicus*, Seshadri 1959; Brown House Snake, *Lamprophis fuliginosus*, personal observation). However, the idea of an anatomically defined "coprodaeum" as depicted by Gadow (1887) is suspect. Cranial to, and separated from the urodaeum by a transverse septum, a transition in longitudinal fold morphology, or both, is an area of unknown origin with a simple to pseudostratified tall columnar epithelium, found only in the Colubroides (Giacomini's diverticulum; region b^1). The oviducts insert into the anterior portion of the urodaeum via tubercles, or Giacomini's diverticulum via tubercles or a continuous transition. The former being the pleisiomorphic condition in snakes.

Dorsally projecting off of the posterior urodaeum is/are the urinary papilla/ae. Through this/these papilla/ae the ureters and Wolffian ducts communicate with the urodaeum. However, data from *Coniophanes fissidens* reveal that accessory ducts (ampulla urinary papilla), possibly equivalent to the ampulla urogenital papilla of males, communicate directly with the urodaeum, while the Wolffian ducts and ureters communicate with these accessory ducts.

Conflicting results are present in terms of the association between the male hemipenis and the female cloaca. More studies on the gross and histological anatomy of mating snakes may reveal significant variation in copulatory adjustment in snakes.

Although their function is unclear, at least three types of extra-cloacal glands (possibly four) empty into the proctodaeum of female snakes. These are: 1) median cloacal glands (scolecophidians only), 2) dorsal gland/s, 3) scent glands, and 4) possibly ventral glands. More detailed work is necessary to understand the diversity of the median cloacal glands, dorsal gland/s, and ventral glands; however, in depth analysis of the scent gland (Young *et al.* 1999) reveals little variation in a diverse sampling of taxa.

9.3 THE OVIDUCTS

9.3.1 Overview

A minor controversy that must be addressed before discussing the oviducts in snakes is that of the preference of the term "oviducal" versus "oviductal."

Smith *et al.* (1989) stated that there is "no overwhelming preponderance of usage or dictionary consensus favoring one term over the other," but they go on to conclude that, "'oviductal' is overwhelmingly preferable to 'oviducal.'" Their decision was based solely on the fact that "ducal" was in reference to a duke, while "duct" was a derivative of the Latin word "ductus," which directly gave rise to the word oviduct. Smith *et al.* (1989), however, noted that oviducal (first known usage 1839) had historical precedence over oviductal (first known usage 1860). We suggest subsequent authors use whichever terminology they choose, and we utilize the word "oviducal" in this investigation because of the historic precedence.

Two recent reviews exist on the morphology and function of the snake oviduct (Blackburn 1998; Girling 2002). Blackburn (1998) focused mainly on comparative morphology and function of the squamate oviduct, while Girling (2002) compared the oviducal regions of all reptiles. Blackburn (1998) utilized histological information from 21 species belonging to the families Colubridae, Dipsadidae, Elapidae, Leptotyphlopidae, Natricidae, Typhlopidae, and Viperidae. Species utilized in the review included *Coronella austriaca* (Giacomini 1893), Prairie Rattlesnake (*Crotalus viridis;* Ludwig and Rahn 1943), *Diadophis punctatus* (Perkins and Palmer 1996), *Dolichophis jugularis* (Sacchi 1888; Giacomini 1893), *Elaphe quatuorlineata* (Giacomini 1893), Common Seasnake (*Enhydrina schistose;* Kasturirangan 1951a), Annulated Seasnake (*Hydrophis cyanocinctus;* Kasturirangan 1951b), *Leptotyphlops dulcis* (Fox and Dessauer 1962), *L. humilis* (Fox and Dessauer 1962), Levantine Viper (*Macrovipera lebetina*; Korneva 1973), *Natrix natrix* (Giacomini 1893, Giersberg 1922), *Nerodia sipedon* (Bauman and Metter 1977, Blackburn 1998), Habu (*Protobothrops flavoviridus;* Yokoyama and Yoshida 1994), Tiger Keelback (*Rhabdophis tigrinus;* Takewaki and Hatta 1941), Western Terrestrial Garter Snake (*Thamnophis elegans;* Fox 1956, Mead *et al.* 1981), *T. sirtalis* (Fox 1956, Hoffman 1970, Hoffman and Wimsatt 1972, Halpert *et al.* 1982), *Ramphotyphlops braminus* (Fox and Dessauer 1962), Southeastern Crowned Snake (*Tantilla coronata;* Aldridge 1992), *Typhlops angolensis* (Fox and Dessauer 1962), an unknown *Typhlops* (Fox and Dessauer 1962), and *Vipera aspis* (Giacomini 1893, Van Den Broek 1933, Saint-Girons 1957). Girling (2002) added information to the historical data with work on a natricid, the Black Swamp Snake (*Seminatrix pygaea;* Sever and Ryan 1999; Sever *et al.* 2000).

Several studies were not included in the reviews by Blackburn (1998) and Girling (2002) and these include the following: two dissertations on Natricidae by Seaman (1949; unavailable for this review also) and Gibson (1934; *Thamnophis sirtalis*), and histological work on the posterior oviduct of a viperid, the Tropical Rattlesnake (*Crotalus durissus;* Almeida-Santos and Salomão 1997). Also missing from the reviews are histological studies on the Brown Treesnake (*Boiga irregularis;* Bull *et al.* 1997) and *Nerodia sipedon* (Bauman and Metter 1977). This review will include the above studies plus recent studies on the reproductive morphology of snakes that include works by Siegel and Sever (2006, 2008a,b; *Agkistrodon piscivorus*) and Siegel *et al.* (2009b; *A. piscivorus*).

Blackburn (1998) also reviewed and provided new gross morphological description from the Copperhead (*Agkistrodon contortrix*), *Agkistrodon piscivorus*, a boa (*Epicrates* sp.), *Nerodia sipedon* and *Thamnophis sirtalis*, along with a review of the gross morphological features of the oviduct from other taxa of snakes. Siegel *et al.* (2009b) also provided detailed gross anatomical description of the oviduct of *A. piscivorus*.

As noted by Blackburn (1998), a complete description of the reproductive system from a single species of female squamate is currently lacking, although expansion of the work on *Agkistrodon piscivorus*, *Diadophis punctatus*, *Seminatrix pygaea*, *Tantilla coronata*, *Thamnophis sirtalis*, or *Vipera aspis*, from which much data already exists, would bring us closer to this point in snakes (particularly data on ovarian histology/ultrastructure). The following includes a review of the gross morphology and microanatomy of the oviducts in snakes from the above mentioned literature with new histological description of the entire oviduct of vitellogenic *Coniophanes fissidens* (Dipsadidae; Field Museum of Natural History #20465 and #20474). This review focuses on the oviducal morphology of snakes. Histological techniques follow Siegel and Sever (2008a). Histological slides of the entire oviduct of vitellogenic *A. piscivorus* (Viperidae; Southeastern Louisiana University Vertebrate Museum #6451) were obtained to illustrate the oviducal regions of a viviparous snake. Dissection of *A. piscivorus* and *Nerodia sipedon* was utilized to depict oviducal regions grossly, and show the variation that can occur in region length between species.

9.3.2 Gross Morphology

The gross description of the oviduct follows Hoffman (1970) and Blackburn (1998) unless otherwise noted. The female reproductive tract consists of two paired oviducts encompassed in a thin visceral pleuroperitoneum and suspended into the coelomic cavity via a dorsal mesentery. The right oviduct is longer than the left. Three to five oviducal regions are apparent: the posterior oviduct (Fig. 9.6 region 2), which has the smallest diameter of the oviduct and is heavily muscularized, the middle oviduct (Fig. 9.6 region 3), which is the longest oviducal region and is more flaccid and is often slightly folded, and the anterior oviduct (Fig. 9.6 regions 4 and 5), which is highly folded, often translucent, and opens up into the coelomic cavity via a "funnel-shaped" ostium (Perkins and Palmer 1996). The anterior oviduct is sometimes divided into two gross regions; a caudal highly folded opaque region (Fig. 9.6 region 4), and a cranial less folded translucent region (Figure 9.6 region 5; Ludwig and Rahn 1943). Another region, more caudal than the posterior oviduct described above has also been identified in vipers (Fig. 9.6 region b^1 left; Ludwig and Rahn 1943; Saint-Girons 1957, 1962a; Siegel and Sever 2008a); however, this region is equivalent to Giacomini's diverticulum as described in section 9.2.9. Giacomini's diverticulum is present in all Colubroides studied (Fig. 9.6 region b^1 left and right) and may not be of Müllerian duct origin (see following section, 9.3.3). All of the gross regions depicted above correspond to histologically unique oviducal regions and are described below. From the subsequent review, many terms for different oviducal regions are presented, and these terms have been tabulated in Table 9.3. Equivalent histological regions are placed in the same row to accurately demonstrate

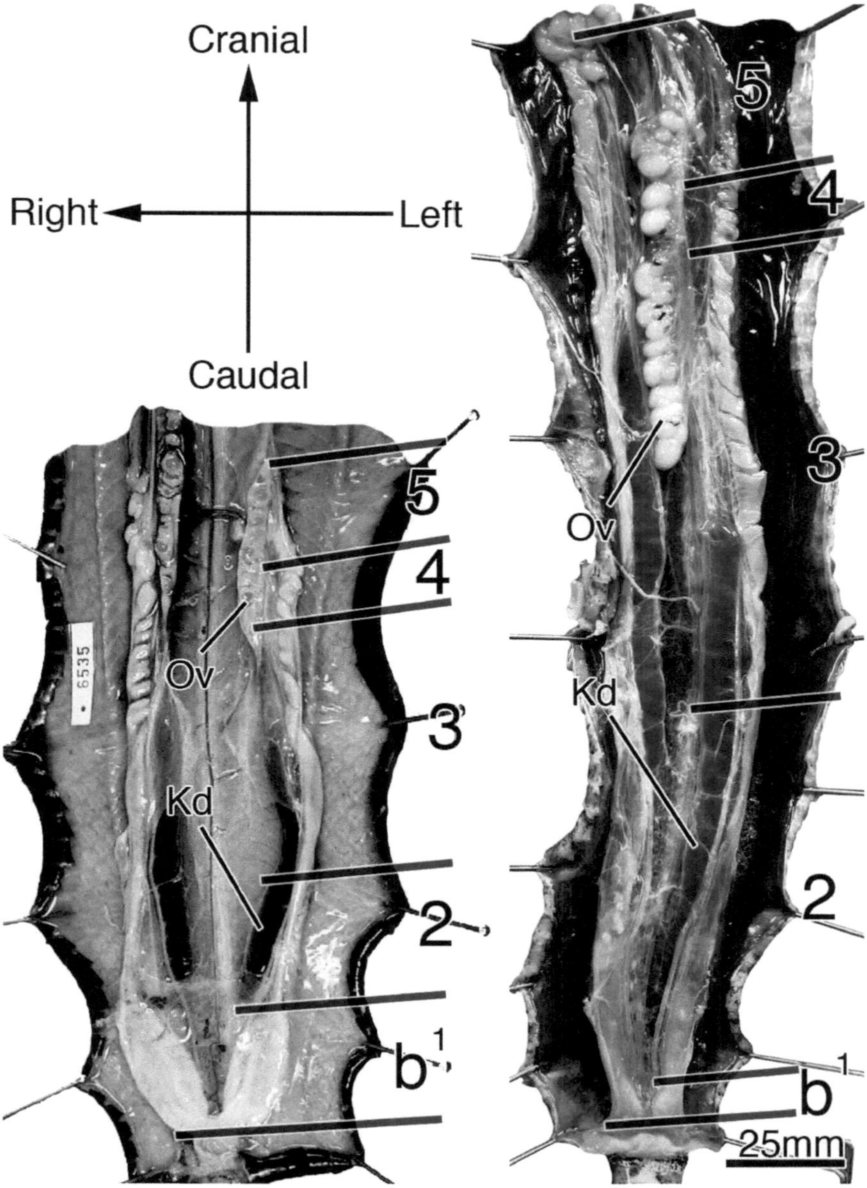

Fig. 9.6 Gross morphology of the urogenital system of *Agkistrodon piscivorus* and *Nerodia sipedon*. **Left.** *Agkistrodon piscivorus*. **Right.** *Nerodia sipedon*. Regions b[1],2,3,4,5 correspond to regions of same number referred to in text and in Table 9.2. Kd, kidney; Ov, ovary.

Table 9.3 Historical nomenclature of the oviducal regions

Investigator(s)	*Family*	*Historical nomenclature of the oviduct (caudal → cranial)*				
		b[1]	2	3	4[1]	5
Sacchi 1888	Colubridae	Not discussed	Partie terminale	Partie inférieure	Partie albumifère	Portion supérieure
Giacomini 1893	Viperidae/ Colubridae/ Natricidae	Diverticolo della cloaca	Porzione terminale	Uterus	Tuba	Imbuto (funnel)
Cope 1898	Variety	Vagina	------------Oviduct------------			
Giersberg 1922	Natricidae	Not discussed	Vagina	Uterus	Tube[2]	Trichters
Ludwig and Rahn 1943	Viperidae	Vaginal pouch[3]	Anterior vagina[3]	Uterus	Tuba[4]	Infundibulum
Kasturirangan 1951	Elapidae	Not discussed	Caudal tubular region	Cranial tubular region[5]	Tubular part of the funnel	Funnel region
Fox 1956	Natricidae	Posterior vagina (pouch)	Anterior vagina	Uterine segment	Seminal receptacles	Infundibulum
Saint-Girons 1957,62a	Viperidae	Poche vaginale	------Uterus------		Fausses glandes de la trompe/tuba	Infundibulum
Fox and Dessauer 1962	Typhlopidae/ Leptotyphlopidae	Not present	Vaginal pouch	Posterior uterus	Seminal receptacles[6]	Infundibulum
Gabe and Saint-Girons 1965	Typhlopidae/ Pythonidae	Not present	Tube vaginal	---------Not discussed---------		

Gabe and Saint-Girons 1965	Colubridae	Poche vaginale/ Orifice génital	Tube Vaginal	----------------------------Not discussed----------------------------		
Hoffman 1970; Hoffman and Wimsatt 1972[7]	Natricidae	------------------Vagina-----------------		Uterus	Sperm receptacles	Infundibulum
Saint-Girons 1973	Synthesis	Not discussed	Vaginal tube	Uterus	Tuba	Infundibulum
Bauman and Metter 1977	Natricidae	-----Vagina or vaginal pouch[8]------		Uterus	Seminal receptacles	Infundibulum
Fox 1977	Synthesis	---Non-glandular vaginal pouch[9]--		Uterus	Seminal receptacles	Infundibulum
Mead et al. 1981	Natricidae	------------Not discussed------------		Uterus	----------------Not discussed----------------	
Halpert et al. 1982	Natricidae	Vagina (A)	Vagina (B)	Uterus (C)	Sperm storage tubules (D)	Infundibulum (D)
Aldridge 1992	Colubridae	Vagina	Posterior uterus	Anterior uterus	Seminal receptacles	Infundibulum
Perkins and Palmer 1996	Colubridae/ Dipsadidae	Posterior vagina	Vagina	Uterus	Tube	Infundibulum
Almeida-Santos and Salomão 1997	Viperidae	Vagina	---------------Uterus[10]--------------		-------------------Oviduct[10]-------------------	
Uribe et al. 1998	Colubridae	Oviductal/ Cloacal junction	Vagina	---------------------------Not discussed--------------------------		
Blackburn 1998	Synthesis	Vaginal pouch	Vagina	Uterus	Posterior infundibulum (tube)	Infundibulum
Sever et al. 2000	Natricidae	Not discussed	Vagina	Uterus	Sperm storage tubules	Infundibulum

Fig. 9.2 Contd. ...

... *Fig. 9.2 Contd.*

Sánchez-Martínez *et al.* 2007	Dipsadidae	Anterior urodaeum	Vagina	Not discussed	Not discussed	Not discussed
Siegel and Sever 2008a,b	Viperidae	Vaginal pouch	Non-glandular uterus	Glandular uterus	Sperm storage tubules	Infundibulum
Proposed terminology	**Ophidia**	**Pouch**	**Non-glandular uterus**	**Glandular uterus**	**Posterior infundibulum**	**Anterior infundibulum**

[1]Cells with sperm storage structure terminology instead of oviducal region definitions do not possess unique terminology for region 4. They simply indicate the terminology for the sperm storage structures and that the structures are in the posterior infundibulum.

[2]Although it is typically cited that Giersberg (1922) described a tube in squamates (region 4), he states there are three regions to the squamate oviduct: 1) funnel with tube 2) uterus, and 3) vagina. Thus, Giersberg was more accurately describing the tube as the posterior portion of the infundibulum in squamates.

[3]Ludwig and Rahn (1943) state that regions 1 and 2 are histologically similar and, thus, the vagina can be divided into an anterior vagina (also called the coiled tube) and a posterior vaginal pouch.

[4]Although Ludwig and Rahn (1943) claimed the four regions they described (considering the vaginal pouch and anterior vagina were grouped together as one region) were based on histological analysis, there is no evidence in their manuscript that histology was conducted cranial to the region 3 (uterus).

[5]Even though Kasturirangan described region 3 as the cranial tubular region, he states in his text that this region functions as the uterus.

[6]Although termed sperm seminal receptacles, Fox and Dessauer (1962) never found sperm in the posterior infundibulum of the Scolecophidia.

[7]Hoffman 1970 was only concerned with the histology of region 3 (uterus), while Hoffman and Wimsatt (1972) were concerned only with region 4 (posterior infundibulum). Thus, their regional descriptions may be from literature review or only gross examination.

[8]Bauman and Metter (1982) group regions 1 and 2 together, although, they do describe the stereotypical variation between the two (e.g. the vagina constricts to a muscular tube cranially).

[9]Similar to Bauman and Metter (1982), Fox (1977) groups regions 1 and 2 together, but then goes on to describe the stereotypical variation between the two regions.

[10]There is no evidence Almeida-Santos and Salomão (1997) conducted histology cranial to region 2.

equivalency of region, even when termed dissimilar by different authors. Numbers corresponding to the regions depicted in Fig. 9.6 are on the top row as a quick reference guide for coordination between nomenclature and location of region.

Smaller snakes with constrained body sizes often exhibit the loss of an oviduct. In the Scolecophidia, this includes the complete loss of the left oviduct (Fox and Dessauer 1962). In *Tantilla*, function of the left oviduct is lost (Clark 1970); however, it persists as a small vestigial organ terminating 2-3 millimeters cranial to the cloaca (Aldridge 1992). Interestingly, although not a functional oviduct, all the distinct regions of the vestigial left oviduct are present in *T. coronata* (Aldridge 1992).

9.3.3 Posterior Oviduct (Region 2)

The oviducts are derived from the Müllerian ducts (paramesonephric ducts), which develop as growths from the dorsal coelomic mesothelium cranial to caudal (for review see Raynaud and Pieau 1985). After the caudal portion inserts into the cloaca, a lumen develops through the growth caudal to cranial (for review see Raynaud and Pieau 1985). Controversy about where the actual oviduct begins and where the cloaca ends is discussed briefly above (section 9.3.2) and in section 9.2.9. Basically, the controversy is whether the tall columnar epithelial layer described above in the cloaca (depicted in Fig. 9.7A,C,E) of Colubroides (e.g., Giacomini's diverticulum; region b^1) is part of the oviduct or the cloaca. We hypothesized that this distinct region was part of the cloaca, and is described in section 9.2.4. Consequently, we begin our discussion of the micro-anatomy of the oviducts with the region of the oviduct that communicates with the urodaeum or Giacomini's diverticulum cranially (Fig. 9.7B,D,F). Some authors considered Giacomini's diverticulum as an expanded posterior region of what we are considering the posterior oviduct, with a slightly higher epithelium (Fox 1956; Fox 1977; Halpert *et al.* 1982; Perkins and Palmer 1996). However, it is clear from histology and histochemistry of these two regions (Fig. 9.7) that this is not the case. The mucosa of Giacomini's diverticulum has a very tall lamina propria that is filled with mast cells (Fig. 9.7A,C,E), has an epithelium that produces neutral carbohydrates and acidic mucous (Fig. 9.7B), and is sparsely ciliated. What we call the posterior oviduct possesses none of these characteristics as described in detail below.

Inserting into the urodaeum of the cloaca, or inserting into Giacomini's diverticulum in Colubroides (see section 9.2.6), is a highly muscularized, almost sphincter-like region (Sánchez-Martínez *et al.* 2007) of the oviduct. This region has been identified in all snake taxa, and is highly consistent in morphology between taxa. Although consistent in morphology, different nomenclature has been utilized for this region of the oviduct by multiple authors. This region has been most consistently termed the vagina (Giersberg 1922; Hoffman and Wimsatt 1972; Bauman and Metter 1977; Perkins and Palmer 1996; Uribe *et al.* 1998; Blackburn 1998; Sever *et al.*

2000; Sánchez-Martínez *et al.* 2007), but has also been termed the "partie terminale" (Sacchi 1888), "porzione terminale" (Giacomini 1893), anterior vagina (Ludwig and Rahn 1943; Fox 1956; Halpert *et al.* 1982), vaginal pouch (Fox and Dessauer 1962; Bauman and Metter 1977), vaginal tube/"le

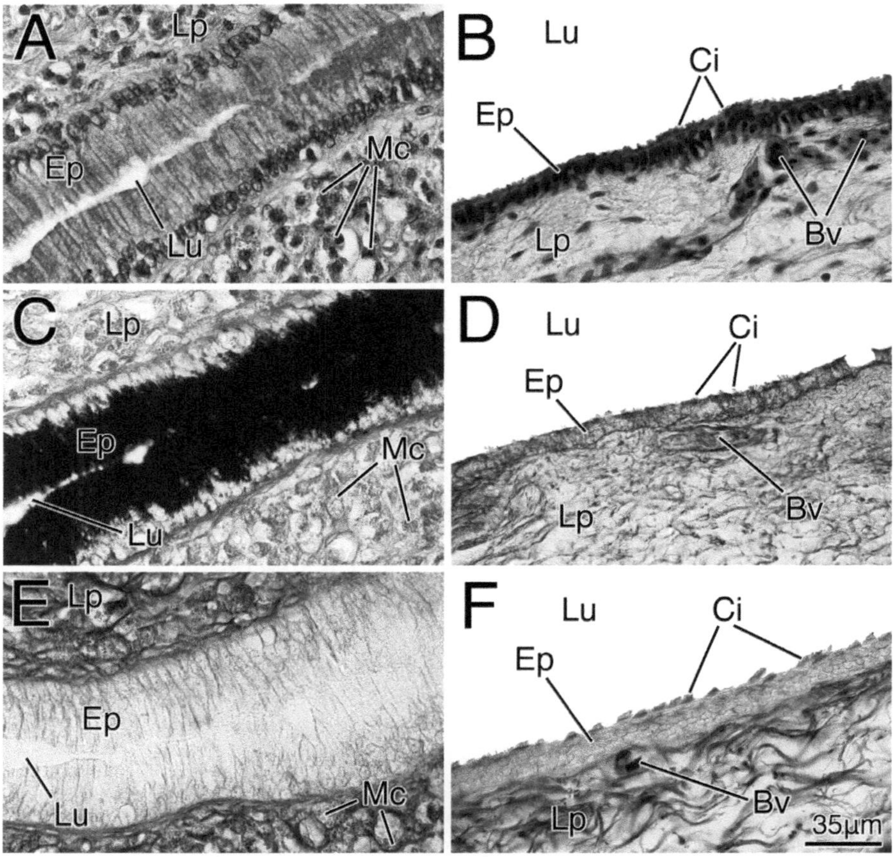

Fig. 9.7 Histology and histochemistry of Giacomini's diverticulum (b^1) and posterior oviduct (oviduct region 2) of *Agkistrodon piscivorus*. **A.** General histology of diverticulum, note tall columnar epithelium and high density of mast cells in the lamina propria (Hematoxylin and Eosin). **B.** General histology of posterior oviduct, note shorter epithelium than in **A** (Hematoxylin and Eosin). **C.** Histochemistry of diverticulum showing a positive reaction for neutral carbohydrates and acidic mucoid substances (Periodic Acid Schiff's and Alcian Blue). **D.** Histochemistry of posterior oviduct showing scant positive reaction for acidic mucoid substances in the apices of the epithelial cells of this region (Periodic Acid Schiff's and Alcian Blue). **E.** Histochemistry of diverticulum showing a negative reaction for protein material in the epithelium, including cilia (Bromophenol Blue). **F.** Histochemistry of posterior oviduct showing a positive reaction for proteins caused by the high density of cilia protruding through the apical cell membrane of many epithelial cells (Bromophenol Blue). Bv, blood vessels; Ci, cilia; Ep, epithelium; Lp, lamina propria; Lu, lumen; Mc, mast cells.

Color image of this figure appears in the color plate section at the end of the book.

conduit vaginal" (Saint-Girons 1957, 1962a, 1975), anterior portion of the non-glandular vaginal pouch (Fox 1977), posterior uterus (Aldridge 1992), uterus (Almeida-Santos and Salamão 1997), and most recently the non-glandular uterus (Siegel and Sever 2006, 2008a,b).

In all taxa studied, including *Coniophanes fissidens* in this chapter, the posterior oviduct is characterized by a thick layer of longitudinal muscle thrown into longitudinal folds and a proportionally thick layer of circular muscle deep to the longitudinal layer (Fox 1956; Saint-Girons 1957, 1962a; Fox and Dessauer 1962; Bauman and Metter 1977; Perkins and Palmer 1996; Sever *et al.* 1999; Sánchez-Martínez *et al.* 2007; Siegel and Sever 2008a). The mucosa is shallower than in the mid-oviduct cranially and cloaca caudally, with deep longitudinal furrows created by lumenally projecting rugae of the lamina propria, and covered by a simple columnar epithelium composed primarily of ciliated cells (Fig. 9.7B,F). Numerous secretory cells are observed interspersed among the ciliated cells in *C. fissidens* and the majority of other taxa studied (Sacchi 1888; Giacomini 1893; Giersberg 1922; Ludwig and Rahn 1943; Fox 1956; Saint-Girons 1957, 1962a; Gabe and Saint-Girons 1965; Hoffman and Wimsatt 1972; Bauman and Metter 1977; Halpert *et al.* 1982; Almeida-Santos and Salamão 1997; Uribe 1998; Sever *et al.* 2000; Sánchez-Martínez *et al.* 2007; Siegel and Sever 2008a,b). The epithelium lining the lumen is pseudostratified in *Tantilla coronata* (Aldridge 1992), and uniformly ciliated in the Scolecophidia (Fox and Dessauer 1962) and *T. coronata* (Aldridge 1992). Saint-Girons (1957) noted that the epithelium changes from simple to pseudostratified during the height of reproductive activity in *Vipera aspis* and states the condition in *Cerastes cerastes* is identical (Saint-Girons 1962a). When present, secretory cells of the epithelium serve mainly to secrete an acidic mucous (Saint-Girons 1975; Bauman and Metter 1977; Perkins and Palmer 1992; Sever *et al.* 2000; Siegel and Sever 2008b; Fig. 9.7D), although positive histochemical reactions for neutral carbohydrates have also been observed in *Diadophis punctatus* (Perkins and Palmer 1996), *Seminatrix pygaea* (Sever *et al.* 2000), and members of the genus *Thamnophis* (Fox 1956; Halpert *et al.* 1982). Although not mentioned in any other study, blood vessels lying immediately beneath the basal lamina are common in the posterior oviduct as they are in the middle oviduct (see section 9.3.4). Mast cells are found in the lamina propria of all regions of the snake oviduct; however, they are at their highest concentration in the posterior oviduct (Saint-Girons 1957; Perkins and Palmer 1996; Uribe *et al.* 1998; Siegel and Sever 2008b), although not as high as in Giacomini's diverticulum (see Fig. 9.7).

The epithelium of this oviducal region has been described ultrastructurally in *Agkistrodon piscivorus* (Siegel and Sever 2008b) and *Seminatrix pygaea* (Sever *et al.* 2000). As confirmed with light microscopy (Fig. 9.7B,D,F), ciliated cells with elongate cilia are found alternating with secretory cells (Sever *et al.* 2000; Siegel and Sever 2008b). These ciliated cells possess numerous mitochondria associated with basal bodies anchoring the cilia into the apical region of each ciliated cell (Sever *et al.* 2000; Siegel and Sever 2008b). Secretory cells have basal euchromatic nuclei and apical secretory vacuoles with the occasional eccentric dense core in both *A. piscivorus* (Siegel and Sever 2008b) and *S. pygaea* (Sever *et al.* 2000). While secretory vacuoles are consistently flocculent in *S. pygaea* (Sever *et al.* 2000), they exhibit varying degrees of electron densities in *A. piscivorus* (Siegel and Sever 2008b). The most conspicuous organelles are mitochondria and Golgi complexes in *A. piscivorus* (Siegel and Sever 2008b) and rough endoplasmic reticulum and Golgi complexes in *S. pygaea* (Sever *et al.* 2000). Sever *et al.* (2000) also note small invaginations into the lamina propria of the cranial portion of the posterior oviduct (described as simple tubular glands), although these invaginations do not vary histologically or ultrastructurally from the epithelium lining the lumen of the posterior oviduct.

During the height of reproductive activity the synthesis and release of secretory products, via an apocrine type secretion, increase in secretory cells found in the epithelium of this oviducal region (Sever *et al.* 2000; Siegel and Sever 2008b). Sperm can also be found in unordered masses in the deep longitudinal furrows and the narrow central lumen during the mating seasons of *A. piscivorus* (Siegel and Sever 2006, 2008a), *Boiga irregularis* (Bull *et al.* 1997), *Cerastes cerastes* (Saint-Girons 1962a), *Conopsis lineata* (Uribe *et al.* 1998), *Crotalus durissus* (Almeida-Santos and Salamão 1997), *Crotalus viridis* (Ludwig and Rah 1943), *Seminatrix pygaea* (Sever and Ryan 1999), *Tantilla coronata* (Aldridge 1992), members of the genus *Thamnophis* (Fox 1956; Halpert *et al.* 1982), and *Vipera aspis* (Saint-Girons 1957). Sever and Ryan (1999) note that sperm are also found in the small invaginations in the cranial region of the posterior oviduct in *S. pygaea*, and in these crevices sperm are slightly more organized. Sperm storage and transport will be discussed in more detail in section 9.4.2.

9.3.4 Middle Oviduct (Region 3)

The middle oviduct shows the most histological diversity in the oviduct. This region is responsible for the shelling of eggs (Hoffman 1970; Palmer *et al.* 1993) and the incubation of embryos in live bearing squamates (Giacomini 1893; Hoffman 1970; Blackburn 1998) and, thus, some of the level of diversity can be accounted for by the fact that squamates have evolved viviparity independently numerous times (Blackburn 1999). Subsequently, the morphology of the middle oviduct is representative of multiple parity modes in snakes.

The middle oviduct has most commonly been termed the uterus (Giacomini 1893; Giersberg 1922; Gibson 1934; Ludwig and Rahn 1943; Saint-Girons 1957, 1962a; Hoffman 1970; Hoffman and Wimsatt 1972; Saint-Girons 1975; Bauman and Metter 1977; Fox 1977; Mead *et al.* 1981; Perkins and Palmer 1996; Blackburn 1998; Sever *et al.* 2000), but has also been termed the "partie inférieure" (Sacchi 1888), cranial tubular region (Kasturirangan 1951a,b), uterine portion/segment (Takewaki and Hatta 1941; Fox 1956), anterior uterus (as the posterior portion of the oviduct was termed the posterior uterus, Aldridge 1992), and most recently the glandular uterus (Siegel and Sever 2008a,b). Almeida-Santos and Salomão (1997) term the posterior oviduct and middle oviduct collectively as the uterus. Fox and Dessauer (1962) describe two regions in the middle oviduct of the scolecophidia (the posterior uterus and anterior uterus). It is of note, in *Coniophanes fissidens* and other taxa studied, no other investigator/s has described two regions in the middle oviduct; however, Fox (1956) did note some regional variation in the oviducts of members of the genus *Thamnophis*. Regional differences have also been noted in the middle oviduct that correspond to incubation chambers (Giacomini 1893; Gibson 1934; Hoffman 1970; Mead *et al.* 1981; Perkins and Palmer 1996; Blackburn 1998).

While the two muscular layers (outer longitudinal and inner circular) are thinner in the middle oviduct compared to the posterior oviduct, and the lamina propria is slightly thicker (Gibson 1934; Kasturirangan 1951a,b; Saint-Girons 1957, 1962a; Bauman and Metter 1977; Mead *et al.* 1981; Perkins and Palmer 1996; Blackburn 1998; Sever *et al.* 2000; Siegel and Sever 2008a), the middle oviduct can most accurately be identified by the presence of the numerous glandular invaginations into lamina propria (Fig. 9.8A-D). In snakes, the middle oviduct makes up the majority of the entire oviduct structure. Few studies investigate the micro-anatomy of the mid-oviduct in reproductive and non-reproductive condition. However, the ones that do note striking differences in glandular activity, height of epithelial cells, and epithelial cell structure (ciliated versus non-ciliated cells composition) between snakes in different reproductive condition (see Giacomini 1893; Giersberg 1922; Saint-Girons 1957, 1962a; Hoffman 1970; Mead *et al.* 1981; Perkins and Palmer 1996; Blackburn 1998; Siegel and Sever 2008b).

Histological changes in the middle oviduct corresponding to embryo development have been addressed briefly in viviparous snakes. In general, during embryo development, the epithelium of the middle oviduct decreases in height and the lamina propria becomes highly vascularized immediately deep to the epithelium. Considering placentation in viviparous snakes is the area of expertise of other authors, we refer readers to Blackburn (1998) for historical review of this subject and **Chapter 5** in this text for more recent information on mid-oviduct morphology during embryogenesis.

The epithelium lining the lumen is hypothesized to provide the acid mucopolysaccharide component of the shell membrane (Hoffman 1970). Histologically the middle oviduct is not much differentiated from that of the posterior oviduct, and fairly consistent along the entire length of the middle oviduct (Gibson 1934) except during gravidity where ciliated cells seem to decrease in number in the enlarged incubation chambers (Kasturirangan 1951a,b; Hoffman 1970; Blackburn 1998). However, this could result from poor resolution as ciliated cells have been observed in the incubation chambers of *Thamnophis sirtalis* with electron microscopy (for review see Blackburn 1998) and *Nerodia sipedon* with light microscopy (Mead *et al.* 1981). Giersberg (1922) and Saint-Girons (1957, 1962a) note a decrease in ciliated cells concurrent with advancement in reproductive condition, but do not discuss this occurrence in terms of incubation chambers. In general, as in *Coniophanes fissidens*, the epithelium lining the lumen is simple cuboidal/columnar with secretory and ciliated cells (Giacomini 1893; Gibson 1934; Fox 1956; Saint-Girons 1957, 1962a; Fox and Dessauer 1962; Hoffman 1970; Bauman and Metter 1977; Mead *et al.* 1981; Perkins and Palmer 1996; Blackburn 1998; Sever *et al.* 2000; Siegel and Sever 2008a,b; Fig. 9.8C). The secretory cells have electron-lucent secretory vacuoles apically (*Agkistrodon piscivorus* and *Seminatrix pygaea*; Sever *et al.* 2000; Siegel and Sever 2008b) and lipid droplets basally (*S. pygaea*; Sever *et al.* 2000), and the most conspicuous organelles are those of mitochondria (*A. piscivorus* and *S. pygaea*; Sever *et al.* 2000; Siegel and Sever 2008b) and Golgi complexes (*A. piscivorus*; Siegel and Sever 2008b). As in the posterior oviduct, the product synthesized by the epithelium lining the lumen of the mid-oviduct is an acidic mucoid secretion in *A. piscivorus* and *Thamnophis* species (Fox 1956; Siegel and Sever 2008b); however, the epithelium of the middle oviduct has also been shown to produce neutral carbohydrates as well in *Diadophis punctatus*, *S. pygaea*, and *Thamnophis sirtalis* (Hoffman 1970; Perkins and Palmer 1996; Sever *et al.* 2000).

Glandular invaginations into the lamina propria have multiple morphological descriptions depending on the species: 1) tubular (*Cerastes cerates*, Saint-Girons 1962a; *Coniophanes fissidens*, this chapter; *Crotalus viridis*, Ludwig and Rahn 1943; *Natrix natrix*, Giersberg 1922; *Vipera aspis*, Saint-Girons 1957; Fig. 9.8A,C), 2) complex club or flask shaped with a bulbous terminal region or simple/branched tubulo-alveolar (*Thamnophis sirtalis*, Gibson 1934; Fox 1956; Mead *et al.* 1981; branched tubulo-acinar by Hoffman 1970), 3) compound alveolar (*Tantilla coronata*, Aldridge 1992), and 4) simple to branched tubular (*Agkistrodon piscivorus*, Siegel and Sever 2008b; scolecophidians, Fox and Dessauer 1962; *Seminatrix pygaea*, Sever *et al.* 2000; Fig. 9.8B,D). During the height of reproductive activity, secretory granules can be observed filling the epithelium of the uterine glands in many species (*A. piscivorus*, Siegel and Sever 2008b; *C. cerastes*, Saint-Girons 1962a; *C. fissidens*, this chapter; *C. austriaca*, Giacomini 1893; *D. punctatus*, Perkins and Palmer 1996; *D. jugularis*, Sacchi 1888, Giacomini 1893; *Elaphe quatuorlineata*, Giacomini 1893; *N. natrix*, Giacomini 1893, Giersberg 1922; *Rhabdophis tigrinus*, Takewaki and Hatta 1941; Scolecophidia, Fox and Dessauer 1962; *Thamnophis* species, Fox 1956, Hoffman 1970, Mead *et al.* 1981; *Vipera aspis*, Saint-Girons 1957; Fig. 9.8C). These glands are thought to provide the pseudokeratin component of the eggs shell membrane (Hoffman 1970). The ducts leading to the terminal portion of the glands are composed of epithelium similar to that of the epithelium lining the lumen of the middle oviduct in most species, although, ciliated cells are absent (Giersberg 1922; Fox 1956; Saint-Girons 1957, 1962a; Hoffman 1970; Blackburn 1998). However, in *C. fissidens* and *A. piscivorus* (Siegel and Sever 2008b) it appears that there is no epithelial variation along the lengths of the uterine glands.

Giacomini (1893) noted that gland size and amount of granulation during peak activity was less in viviparous snakes compared to oviparous snakes, which has led to the hypothesis that viviparous snakes have decreased activity in middle oviduct glands (Blackburn 1998). Giersberg (1922) confirmed this hypothesis with comparison to viviparous lizards. Glands of the middle oviduct also appeared to be denser in the lamina propria of oviparous snakes than in viviparous snakes (Giacomini 1893; Blackburn 1998; compare gland densities between Fig. 9.8 A,C oviparous and B,D viviparous). However, gland activity and product synthesis was variable in different viviparous species. While the middle oviduct glands were not differentiated from the epithelium lining the lumen of the mid-oviduct in *Seminatrix pygaea* (Sever *et al.* 2000), the terminal portion of glands in the middle oviduct of *Agkistrodon piscivorus* (Fig. 9.8D) and *Thamnophis sirtalis* are specialized and produce dense proteins (Hoffman 1970; Siegel and Sever 2008b). During the height of reproductive activity, the cells making up the tubular glands in *A. piscivorus* are also filled with

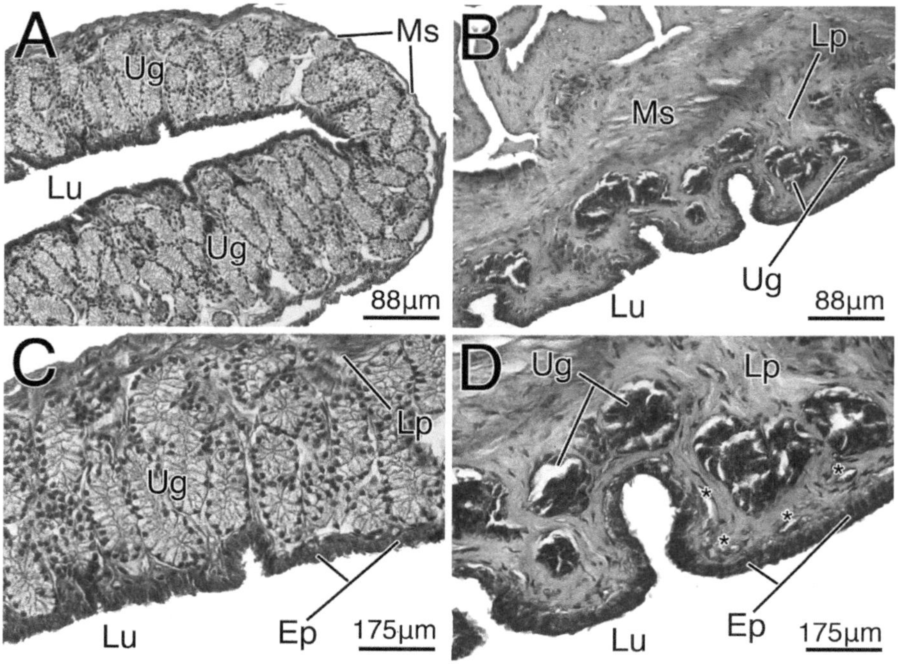

Fig. 9.8 General histology of the middle oviduct (region 3) from an oviparous and a viviparous snake (Hematoxylin and Eosin). **A,B.** Low magnification of the uterus of an oviparous snake (*Coniophanes fissidens*; **A**) and viviparous snake (*Agkistrodon piscivorus*; **B**) demonstrating the higher gland density in the lamina propria of the oviparous species. **C,D.** High magnification of **A** and **B** respectively, highlighting the characteristics of the epithelium lining the lumen and uterine glands of an oviparous and viviparous snake. Note the blood vessels (Asterisks) in the lamina propria of the viviparous snake (D) just beneath the epithelium lining the lumen. Ep, epithelium lining the lumen; Lu, lumen; Lp, lamina propria; Ms, muscularis; Ug, uterine glands.

abundant synthetic organelles (Siegel and Sever 2008b). Blood vessels also appear to be denser in the lamina propria of viviparous species in comparison to oviparous species (Fig. 9.8D).

Regional differences in the middle oviduct were only noted in the scolecophidians and *Thamnophis* species (Fox 1956; Fox and Dessauer 1962). Whereas, highly granulated glands in the posterior 2/3 of the mid-oviduct fill the lamina in *Leptotyphlops*, glands in the anterior 1/3 are sparser, agranulated and only produce mucous (Fox and Dessauer 1962). This variation was less pronounced in *Typhlops* species. Fox and Dessauer (1962) believed the anterior 1/3 of the mid-oviduct in the scolecophidians was equivalent to the albumen-producing region in other oviparous amniotes. They also hypothesized that this region could represent an unusually long tube, as seen in other reptiles and therefore may not be a part of the middle oviduct (Fox and Dessauer 1962). Even less variation was noted in members of the genus *Thamnophis* and only included lighter staining glands cranially with more wide lumina compared to the more caudally positioned glands (Fox 1956).

9.3.5 Anterior Oviduct (Regions 4 and 5)

Overview. The anterior oviduct lies directly cranial to the middle oviduct and has been, in most investigations, divided into two regions. The posterior region has been termed the "partie albumifère" (Sacchi 1888), tuba/tube (Giacomini 1893; Giersberg 1922; Ludwig and Rahn 1943; Saint-Girons 1957, 1962a, 1975), sacculated segment (Seaman 1949), oviduct (Almeida-Santos and Salamão 1997), or posterior infundibulum (Blackburn 1998; Siegel and Sever 2008a). The posterior region possesses structures for sperm storage known as seminal receptacles (Fox 1956; Fox and Dessauer 1962; Fox 1977; Aldridge 1992), sperm storage tubules (Halpert *et al.* 1982; Sever and Ryan 1999; Sever *et al.* 2000; Siegel and Sever 2008a), sperm receptacles (Hoffman and Wimsatt 1972), or "fausses glandes de la trompe" (Saint-Girons 1957, 1962a). All snakes investigated possess sperm storage tubules except *Boiga irregularis* (Bull *et al.* 1997), although Bauman and Metter (1977) imply that structures with typical sperm storage receptacle morphology do not serve as sperm storage structures in *Nerodia sipedon* also. Bauman and Metter (1977) do not provide data on the time of the year or number of specimens examined, and do not provide an alternative function. Siegel (unpublished data) has observed sperm in the seminal receptacles in *N. sipedon* collected in the spring in Missouri.

The cranial region of the anterior oviduct was termed the infundibulum by most investigators; however, other terms have been utilized by Sacchi (1888; "portion supérieure"), Giacomini (1893; "imbuto"), Giersberg (1922; "trichters"), and Kasturirangan (1951a,b; funnel region). Almeida-Santos and Salomão (1997) term the entire caudal and cranial infundibulum the "oviduct." The defining characteristics of the infundibulum are termination of mid-oviducal glands, decrease in size of the mucosa and muscularis, and abundant folding grossly and histologically (Giacomini 1893; Saint-Girons

1957, 1962a; Fox and Dessauer 1962; Bauman and Metter 1977; Halpert *et al.* 1982; Aldridge 1992; Perkins and Palmer 1996; Blackburn 1998; Sever *et al.* 2000; Siegel and Sever 2008a). We will first review the infundibulum "proper" and then review the morphology of the sperm storage receptacle region.

Infundibulum (region 5). The epithelium lining the lumen of the infundibulum proper has been reported as simple squamous in *Agkistrodon piscivorus* (Siegel and Sever 2008b; Fig. 9.9C), simple cuboidal to columnar in *Coniophanes fissidens* (this chapter; Fig. 9.9A,B) and *Nerodia sipedon* (Blackburn 1998), simple squamous to columnar in *Seminatrix pygaea* (Sever *et al.* 2000), or simple columnar in the Scolecophidia and *Tantilla coronata* (Fox and Dessauer 1962; Aldridge 1992). The epithelial cells have been depicted as uniformly ciliated in *Diadophis punctatus* (Perkins and Palmer 1996), Scolecophidia (Fox and Dessauer 1962), and *T. coronata* (Aldridge 1992), uniformly secretory in *Cerastes cerastes* and *Vipera aspis* (Saint-Girons 1957, 1962a; although Giacomini [1893] depicts the epithelium as alternating ciliated and secretory in *V. aspis*), and are alternating ciliated and secretory in *A. piscivorus* (Siegel and Sever 2008b), *C. fissidens* (this study; Fig. 9.9A,B) *Coronella austriaca* (Giacomini 1893), *Dolichophis jugularis* (Giacomini 1893), *Elaphe quatuorlineata* (Giacomini 1893), *Natrix natrix* (Giacomini 1893), *S. pygaea* (Sever *et al.* 2000), and *Thamnophis* species (Fox 1956).

The secretory cells produce eosinophilic granules in *Protobothrops flavoviridus* (Yokoyama and Yoshida 1994) or primarily lipoidal bodies in *Agkistrodon piscivorus* (Siegel and Sever 2008b) and *Seminatrix pygaea* (Sever *et al.* 2000). Siegel and Sever (2008b) describe two types of lipoidal material secreted into the infundibular lumen in *A. piscivorus*: one of organized droplets and another composed of unorganized floccular material.

Only Siegel and Sever (2008b) describe the small epithelial invaginations into the lamina propria of the infundibulum in *Agkistrodon piscivorus* as simple tubular glands (as observed in Fig. 9.9C); the epithelium lining these small invaginations does not vary from that of the epithelium lining the lumen. However, Fox (1956) also notes that secretory cells are more common at the base of the mucosal folds in the infundibulum and, thus, some regional differentiation of epithelial cell composition could be present in some taxa. Through examination of the caudal portions of the infundibulum just cranial to the sperm storage structures (described below) simple tubular glands exist in both *A. piscivorus* and *Coniophanes fissidens* (Fig. 9.9A). However, whereas these glands terminate when moving cranially towards the ostium in *C. fissidens* (Fig. 9.9B), they are consistently observed all the way to the ostium in *A. piscivorus* (Fig. 9.9E). As noted, these invaginations do not appear to be differentiated along their lengths in either *A. piscivorus* or *C. fissidens*, so the descriptions of these as multi-cellular glands may be unwarranted; however, undifferentiated invaginations have been termed glands in other species (e.g. *S. pygaea*; see Sever *et al.* 2000), and we feel that these infundibular structures should be described as glands also.

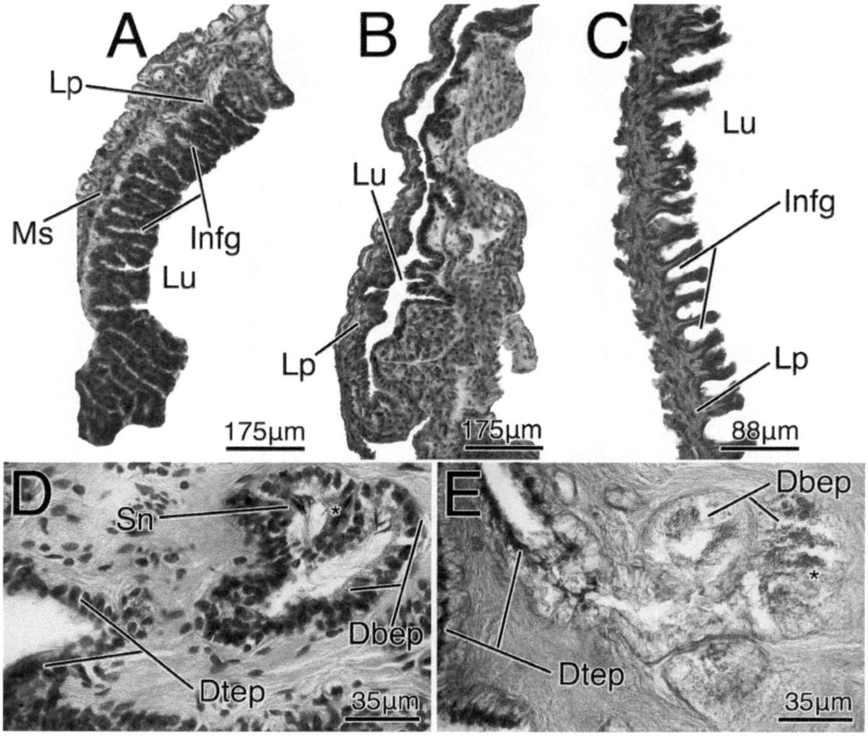

Fig. 9.9 General histology of the anterior oviduct (regions 4 and 5). **A.** Caudal portion of the anterior infundibulum (region 5) of *Coniophanes fissidens*, note tubular gland-like invaginations into the lamina propria (Hematoxylin and Eosin). **B.** Cranial portion of the anterior infundibulum (region 5) of *C. fissidens*, note the absence of tubular gland-like invaginations (Hematoxylin and Eosin). **C.** Cranial portion of the anterior infundibulum (region 5) of *Agkistrodon piscivorus*, note the tubular gland-like invaginations continuing to the ostium (Hematoxylin and Eosin). **D.** Posterior infundibulum (region 4) of *A. piscivorus* showing a branched tubular sperm storage structure (Hematoxylin and Eosin). **E.** Branched tubular sperm storage structure (region 4) in *A. piscivorus*, note a positive reaction for neutral carbohydrates in the distal bulb and a positive reaction for acidic mucoid substances in the gland duct epithelium (Periodic Acid Schiff's and Alcian Blue). Dbep, epithelium of distal bulb of gland; Dtep, epithelium of gland duct; Infg, infundibular gland; Lp, lamina propria; Lu, lumen; Ms, muscularis, Sn, sperm nuclei.

Color image of this figure appears in the color plate section at the end of the book.

Ultrastructurally, the ciliated cells of the infundibulum are no different from those of the more caudal regions of the oviduct. In *Diadophis punctatus* the nuclei take a central position in the secretory cells, while the secretory cells possess basal euchromatic nuclei (Sever *et al.* 2000; Siegel and Sever 2008b). The most conspicuous organelles in the secretory cells of the infundibulum in *Agkistrodon piscivorus* and *Seminatrix pygaea* are mitochondria and smooth endoplasmic reticulum (Sever *et al.* 2000; Siegel and Sever 2008b); however, rough endoplasmic reticulum is found in the secretory cells of *S. pygaea* as well (Sever *et al.* 2000). Little to no seasonal variation has ever been reported in the infundibulum proper (Perkins and Palmer 1996; Sever *et al.* 2000; Siegel and Sever 2008b).

The muscularis seems to take on a somewhat odd confirmation in the infundibulum as it was noted that the continuous layer of longitudinal and circular muscle fibers may not be continuous in the cranial extremity of the infundibulum in *Diadophis punctatus* and *Nerodia sipedon* (Perkins and Palmer 1996; Blackburn 1998). Blackburn (1998) notes from personal observation (apparently from *N. sipedon*) that the posterior circular muscle fibers join with longitudinal fibers cranially.

Sperm storage receptacles (region 4). Specialized glands for storing sperm are thought to be located in the infundibulum because of the presence of infundibular glands intermixed with discrete sperm storage receptacles in the caudal extremities of the anterior oviduct. Infundibular glands are basically undifferentiated invaginations into the lamina propria (see above), while sperm storage receptacles take on a variety of morphologies. Sperm storage receptacles are most dense at the corners of the mucosal folds in the posterior infundibulum, and the lamina propria is slightly thicker in these regions due to the bulging sperm storage receptacles (Fox 1956).

Sperm storage receptacles have three morphologies: 1) branched tubular in *Agkistrodon piscivorus* (Siegel and Sever 2008a; Fig. 9.9D), *Cerastes cerastes* (Saint-Girons 1962a), *Seminatrix pygaea* (Sever and Ryan 1999), and *Vipera aspis* (Saint-Girons 1957), 2) simple to branched alveolar in *Coniophanes fissidens* (this study), *Diadophis punctatus* (Perkins and Palmer 1996), and the Scolecophidia, (although sperm were never found in the receptacles of the Scolecophidia; Fox and Dessauer 1962), and 3) branched to compound alveolar in *Thamnophis* species (Fox 1956). Giacomini (1893) described the glands of the tuba present in *Vipera aspis* as "tubuli ghiandolari (tubular glands);" however, his description was really of the glands identified by Saint-Girons in the same species approximately 60 years later as "fausses glandes de la trompe." Fox (1956) described the complexity of sperm storage receptacles in *Thamnophis* members as up to six alveoli opening into a common ciliated duct and up to five of those ciliated ducts opening into a common ciliated duct, which in turn opens to the lumen of the infundibulum.

The terminal portion of the receptacles, or the alveolus region when present, is composed entirely of secretory cells in all taxa studied (Fig. 9.9D), while the ducts leading to the terminal portion are uniformly ciliated and stain positive for acidic mucous (Fox 1956; Fox and Dessauer 1962; Hoffman and Wimsatt 1972; Perkins and Palmer 1996; Siegel and Sever 2008a,b; Fig. 9.9E) except for in *Cerastes cerastes* (Saint-Girons 1962a), *Seminatrix pygaea* (Sever and Ryan 1999), and *Vipera aspis* (Saint-Girons 1957). In *C. cerastes* (Saint-Girons 1962a) and *V. aspis* (Saint-Girons 1957) sperm receptacles were described as uniformly ciliated whereas in *S. pygaea*, sperm receptacles were undifferentiated from the epithelium lining the lumen of the infundibulum (alternating ciliated and secretory cells; Sever and Ryan 1999). Ciliated cells at the opening of some sperm receptacles (when differentiation along a receptacle is present) are elongate and coarse in *Agkistrodon piscivorus* (Siegel and Sever 2008a) and *Thamnophis* species (Fox 1956), while they are finer in the scolecophidians studied (Fox and Dessauer 1962).

The secretory cells of the sperm receptacles, when present, stain positive for neutral carbohydrates in *Agkistrodon piscivorus* (Siegel and Sever 2008a; Fig. 9.9E), *Diadophis punctatus* (Perkins and Palmer 1996), *Seminatrix pygaea* (Sever and Ryan 1999), *Tantilla coronata* (Aldridge 1992) and *Thamnophis sirtalis* (Hoffman and Wimsatt 1972), and proteins in *A. piscivorus* (Siegel and Sever 2008a) and *S. pygaea* (Sever and Ryan 1999). Ultrastructurally, apical secretory vacuoles in *S. pygaea* are filled with a flocculent material with an electron dense core (Sever and Ryan 1999), while vacuoles in *A. piscivorus* are electron lucent in the fall, but then fill with an electron dense material in the spring subsequent to hibernation (Siegel and Sever 2008a,b). The presence of lipid droplets in the basal portions of secretory cells of the sperm receptacles has also been noted in *A. piscivorus* (Siegel and Sever 2008a), *S. pygaea* (Sever and Ryan 1999), and *T. sirtalis* (Hoffman and Wimsatt 1972). The most conspicuous organelles have been reported as mitochondria, Golgi complexes, and rough endoplasmic reticulum in *A. piscivorus* (Siegel and Sever 2008a) and *T. sirtalis* (Hoffman and Wimsatt 1972). The synthesis of secretory material is concurrent with vitellogenesis and culminates at the timing of ovulation in all taxa studied throughout the reproductive and non-reproductive seasons (Hoffman and Wimsatt 1972; Perkins and Palmer 1996; Sever and Ryan 1999; Sever *et al.* 2000; Siegel and Sever 2008a,b).

Sperm nuclei abut the terminal secretory portion of the sperm receptacles when present in *Agkistrodon piscivorus* (Siegel and Sever 2008a; Fig. 9.9D), *Diadophis punctatus* (Perkins and Palmer 1996), and members of the genus *Thamnophis* (Fox 1956; Hoffman and Wimsatt 1972; Halpert *et al.* 1982). The ducts leading to the terminal portion are filled with undulating parallel arrays of sperm tails. In *Seminatrix pygaea*, sperm are rather unordered, which could be the result of the lack of specialization along the sperm storage receptacle; however, sperm nuclei tend to be oriented towards the opening of the sperm storage receptacles (Sever and Ryan 1999). Sperm were found embedded between secretory cells in the receptacles of *A. piscivorus* (Siegel and Sever 2008a) and were also observed penetrating the epithelium in *A. piscivorus* (Siegel and Sever 2008a) and *Thamnophis* species (Fox 1956; Hoffman and Wimsatt 1972). Only Hoffman and Wimsatt (1972) note the possibility of resorption of sperm by the secretory cells of the seminal receptacle epithelium in *T. sirtalis*.

9.4 SPERM STORAGE AND TRANSPORT

9.4.1 Overview

From the above anatomical reviews, two major sites of sperm aggregations in the snake oviduct exist, and are identified as the posterior oviduct (region 2) and posterior infundibulum (region 4; Rahn 1940; Ludwig and Rahn 1946; Fox 1956; Hoffman and Wimsatt 1972; Halpert *et al.* 1982; Aldridge 1992; Perkins and Palmer 1996; Almeida-Santos and Salomão 1997, 2002; Sever and Ryan 1999; Siegel and Sever 2006, 2008a). Thus, sperm aggregations occur in a caudal and cranial location in the oviducts of snakes, and these regions (regions 2 and 4) will be the only regions discussed in this section unless otherwise noted. Literature on sperm storage in the posterior oviduct and posterior infundibulum will be reviewed below in detail, with comments on the evolution of sperm aggregations and sperm storage tubule location in snakes.

9.4.2 Posterior Sperm Storage (Region 2)

By probing the orifice of the uterus of wild *Thamnophis sirtalis* from May, Rahn (1940) found evidence of motile spermatozoa *in utero*. Of the 44 females examined, 38 had motile sperm from the swabbing method employed, and of those with sperm 13 gave birth. One female possessed motile spermatozoa 23 days after initial swabbing. Approximately two weeks later, that individual had embryos *in utero*. Thus, Rahn (1940) concluded that sperm could survive for at least one month in the female uterus of *T. sirtalis*. It is unclear from what part of the middle oviduct the sperm were residing during probing due to the lack of histological evidence; however, probes were inserted up to 15 cm into the female reproductive tract. Thus, motile sperm may have been residing as far anterior as the infundibulum, the middle oviduct (region 3), the posterior oviduct, and/or anterior cloaca (region b or b^1).

Using histological techniques Ludwig and Rahn (1943) found an aggregation of spermatozoa in the posterior oviduct (what he called the anterior vagina; region 2) in *Crotalus viridis*. No sperm were ever found in the large "vaginal pouch" chamber (equivalent to Giacomini's diverticulum 1893; region b^1) or in the middle oviduct (region 3). Thus, the conclusion was made that long-term sperm storage exists in the lumen and/or furrows of the posterior oviduct. Ludwig and Rahn (1943) did not examine the infundibular region of the oviduct.

Almeida-Santos and Salamão (1997) took the observations from Ludwig and Rahn (1943) one step further by defining the proximate cause of sperm storage in the posterior oviduct (termed uterus in their investigation) as "utero-muscular twisting (UMT)" by utilizing female *Crotalus durissus*. Basically, after mating, the posterior oviduct contracts and becomes twisted and, thus, holds sperm in the posterior oviduct. At the time of ovulation the oviduct relaxes, sperm move cranially, and fertilization occurs somewhere in the oviduct where the caudally moving ovulated eggs and cranially moving sperm meet. These conclusions were made from histological micrographs showing a slight spiraling of the posterior oviduct with sperm in the furrows and no sperm in the furrows in a non-reproductive snake exhibiting no spiraling of the histological section.

Almeida-Santos and Salomão (2002) examined the gross oviducal anatomy of two species of *Agkistrodon*, 13 species of *Bothrops*, and two species of *Crotalus*, and concluded that all reproductively active North American viperids possess UMT and, thus, they believed UMT may be the ancestral means of sperm storage for all vipers.

Analysis of seminal fluid physiology in *Crotalus durissus* revealed that seminal fluid physiology changed after sperm and seminal fluid entered the oviduct (Marinho *et al.* 2009). A conclusion was drawn that the posterior oviduct may also be secreting a seminal fluid to help maintain the viability of sperm in the posterior oviduct (Marinho *et al.* 2009). The investigation by Marinho *et al.* (2009) was in support of Halpert *et al.* (1982) who believed that the epithelium of the posterior oviduct in *Thamnophis sirtalis* shed after mating and acted to facilitate sperm cranial into the oviduct. Sever and Ryan (1999) also noted epithelial cells in the lumen of the posterior oviduct in *Seminatrix pygaea*. This cellular material was hypothesized to originate from the male urogenital system based on the fact that too much material was present for it to be completely produced by the female; however, the authors stated that the source of the material could not be definitively determined (Sever and Ryan 1999). Siegel and Sever (2008a) found similar cellular association with sperm in *Agkistrodon piscivorus*; however, the cellular material stained identical to material from the male reproductive tract (a glycoprotein mixture similar to secretions from the sexual segment of the kidney; Sever *et al.* 2002; Sever *et al.* 2008; Siegel *et al.* 2009a), and not the epithelium lining the lumen of the posterior oviduct (composed of only acidic mucous-like secretions; see Fig. 9.7D). Uribe *et al.* (1998) noted eosinophilic and periodic acid-Schiff's positive cellular material caudal to the posterior oviduct (region 2) associated with sperm (in Giacomini's diverticulum; region b^1) and, thus, we feel this adds support to the hypothesis presented by Siegel and Sever (2008a) that the majority of the material associated with sperm in the posterior oviduct is of male origin.

Other studies also noted the occurrence of sperm in the posterior portion of the snake oviduct (Fox 1956; Saint-Girons 1957, 1962a, 1975; Halpert *et al.* 1982; Aldridge 1992; Perkins and Palmer 1996; Bull *et al.* 1997; Uribe *et al.* 1998; Sever and Ryan 1999; Siegel and Sever 2006, 2008a). However, in these studies sperm resided in the furrows and/or the lumen of the posterior oviduct for as little as a few weeks, to as long as over hibernation, until subsequent migration to specialized infundibular glands in the posterior infundibulum. No specific structures (e.g. multicellular glands or derived epithelium in the furrows) in the posterior portion of the oviduct have ever been described that resemble sperm receptacles in the posterior oviduct of lizards (for review see Sever and Hamlett 2002; Sever and Hopkins 2005). Sever and Ryan (1999) refer to glands in the posterior oviduct of *Seminatrix pygaea*; however, it is clear from this investigation, and a subsequent study on *S. pygaea* (Sever *et al.* 2000), that these glands are highly unspecialized and not differentiated from the epithelium lining the lumen of the posterior oviduct. Thus, these glands act as small extensions of the furrows formed by the longitudinal folds of the posterior oviduct.

The fertilization ability of sperm that reside in the posterior oviduct to the time of ovulation has never been reported; however, Siegel and Sever (2008a) showed that sperm degrade in the posterior oviduct shortly before ovulation in *Agkistrodon piscivorus*.

9.4.3 Anterior Sperm Storage (Region 4)

Fox (1956) found prolonged sperm aggregations in the posterior oviduct "puzzling" because in his *Thamnophis* species, and in all subsequent studies on snakes in which sperm storage or transport was studied, specialized sperm storage tubules were found in the posterior infundibulum (see section 9.3.5). The only caveat to this statement are two studies claiming no sperm storage structures are found in the infundibulum of *Boiga irregularis* (Bull *et al.* 1997) and *Nerodia sipedon* (Bauman and Metter 1977), and one study in which the sperm storage tubules were not that specialized from that of the epithelium lining the lumen of the posterior infundibulum in *Seminatrix pygaea* (Sever and Ryan 1999). In *Seminatrix*, the only specialization comes from a scant production of Periodic Acid-Schiff's positive material in the structurally undifferentiated glands compared to the epithelium lining the lumen of the infundibulum.

In all species with infundibular sperm receptacles sperm fill posterior sperm receptacles shortly after mating, starting with the more caudally located glands first (Fox 1956; Hoffman and Wimsatt 1972). It could be hypothesized that sperm are attracted to the specialized neutral carbohydrate secretions of the seminal receptacles in the posterior infundibulum (see Fig. 9.9E), evidenced by sperm orienting with their nuclei abutting the terminal secretory portion of the glands (see Fig. 9.9D). Hoffman and Wimsatt (1972) believed that the orientation of sperm in the seminal receptacles was evidence for ciliary action forcing sperm into the glands, and Fox (1956) stated "cilia in the ducts [...] may be instrumental in aiding the sperm to reach the receptacles." Although the seminal receptacles stained positive with the periodic acid-Schiff's procedure in *Seminatrix pygaea*, secretory cells were not aggregated at the terminal portion of the glands and, thus, sperm were oriented more with their nuclei facing the opening of the glands (Sever and Ryan 1999). It is unlikely that the neutral carbohydrate secretion acts to maintain sperm viability due to the fact that sperm can remain in sperm storage tubules for many months after the secretory activity in sperm storage receptacles ceases (Siegel and Sever 2008a,b), and in the posterior oviduct with no such secretion (Saint-Girons 1975). However, Hoffman and Wimsatt (1972) hypothesized that the lipids produced in seminal receptacles may aid in sperm viability. Fox (1956) and Aldridge (1992) believed that sperm receptacles only functioned to physically protect sperm from getting swept caudally in the oviduct by the first eggs ovulated, and data on *Agkistrodon piscivorus* (Siegel and

Sever 2008a,b) and *Tantilla coronata* (Aldridge 1992) confirm that sperm in the seminal receptacles are the only sperm remaining in the entire oviduct after ovulation.

The fate of sperm in the sperm receptacles is either fertilization of an egg or evacuation. Sperm are not forced out of sperm receptacles in snakes by contractile force because sperm receptacles in snakes are not associated with any type of myoepithelial border or any other type of contractile element (Hoffman and Wimsatt 1972; Blackburn 1998; Sever and Ryan 1999). Thus, sperm appear to come into contact with eggs by pressure from the ovulated macrolecithal eggs forcing a sperm-ovum association while the eggs are squeezed through the narrow infundibulum (Fox 1956; Hoffman and Wimsatt 1972; Sever and Ryan 1999; Siegel and Sever 2008a).

Evacuation of sperm that do not fertilize eggs is an enigmatic event. Halpert *et al.* (1982) described sperm moving out of sperm storage receptacles from fall matings (long-term sperm storage) to make way for fresh sperm from spring matings (short term sperm storage) in *Thamnophis sirtalis*. Sperm exiting from glands was also observed to be degrading; however, no other study claims to identify the degradation of sperm in the posterior infundibulum or the evacuation of sperm from sperm receptacles. In *Diadophis punctatus* sperm disappear from sperm receptacles immediately after ovulation (Perkins and Palmer 1996), and no sperm were found in seminal receptacles of *Cerastes cerastes* and *Vipera aspis* at least two months after ovulation (for review see Saint-Girons 1975). Siegel and Sever (2008a,b) and Aldridge (1992) noted that sperm were present in receptacles for months after ovulation in *Agkistrodon piscivorus* and *Tantilla coronata* respectively, and as long as post-partum in *A. piscivorus*, but disappeared before the start of the subsequent mating season (Siegel and Sever 2008a,b). The proximate cause for sperm disappearance was not determined in *A. piscivorus*, *D. punctatus*, or *T. coronata*, although Aldridge (1992) states that sperm "apparently had been resorbed" in *T. coronata*. Sperm were found in a fecal sample of the Massasauga Rattlesnake (*Sistrurus catenatus*) during late pregnancy, when mating had long ceased, which led B. C. Jellen, D. S. Siegel, and R. D. Aldridge to conclude that sperm may be swept down the oviduct and evacuated from the reproductive tract intermittently with defecation after fertilization (Jellen *et al.*, in review).

9.4.4 Conclusions on Oviducal Sperm Storage

Sperm aggregations. The above review of sperm aggregations in the oviducts of snakes creates a quandary between anterior and posterior sperm storage because there are overlapping cases depicted in the same species. Rahn (1940) described the sperm storage location in *Thamnophis sirtalis* as uterine, whereas all subsequent studies depicted it as infundibular with sperm residing in the posterior oviduct for some period of time before cranial migration (Fox 1956; Hoffman and Wimsatt 1972; Halpert *et al.* 1982). Almeida-Santos and Salomão (2002) described sperm storage in *Agkistrodon piscivorus* occurring in the posterior oviduct by means of

UMT, while subsequent studies depicted sperm storage in this species as infundibular with sperm residing in the posterior oviduct for some time before cranial migration (Siegel and Sever 2006, 2008a). Considering sperm aggregations in the posterior infundibulum are concurrent with specialized gland activity that varies in secretory activity directly with the mating season and/or sperm presence (Hoffman and Wimsatt 1972; Halpert *et al.* 1982; Aldridge 1992; Sever and Ryan 1999; Siegel and Sever 2008a), it would appear that sperm are viably and orderly stored in the posterior infundibulum in snakes (at least in the species that have been determined to store sperm). We feel that sperm receptacles in the posterior infundibulum do not provide materials that help to sustain sperm viability due to the fact that sperm remain in sperm receptacles for many months after they become quiescent in *A. piscivorus* (Siegel and Sever 2008a,b). Thus, these structures may only be aiding in attracting sperm. Subsequently, these tubules may add mechanical protection from sperm being forced out of the oviduct by earlier ovulated eggs (Fox 1956; Aldridge 1992) and, thus, increase the fertilization rate of later ovulated eggs.

Sperm are not associated with any specialized structures in the posterior oviduct of snakes and are commonly found in the lumen or in deep furrows (with glandular extensions; Sever and Ryan 1999). No specialized secretions concurrent with mating activity or sperm presence have been described in the posterior oviduct. We propose that because sperm may reside in this area of the oviduct for prolong time periods, the entire oviduct acts to maintain the physiological processes associated with sperm viability. Saint-Girons (1975) intimated this idea and predicted that the mucous secretion of the epithelium lining the lumen of the oviduct could serve to maintain sperm viability in the oviduct in lizards (Saint-Girons 1962b). The aggregate nature of sperm in the posterior oviduct compared to the mid oviduct (region 3) is possibly a combination of the large amount of sperm deposited in this region during mating, the bottleneck created by the decrease in luminal size between the cloaca and oviducts, or transitional zone (in Colubroides) and oviducts (Siegel and Sever 2008a), or possibly some mechanical oviducal contraction causing a coiling in the posterior oviduct (Ludwig and Rahn 1943; Almeida-Santos and Salomão 1997, 2002).

When considering the location of sperm aggregations in the oviduct of female snakes, two locations should be considered: the posterior oviduct, in which sperm aggregations are a function of mating activity and oviducal anatomy/physiology, and specialized sperm receptacles in the posterior infundibulum, in which sperm aggregations are a function of attraction of sperm into tubular glands in the posterior infundibulum. The fertilization capability and dynamics of sperm between the two regions is unknown; however, Saint-Girons (1975) hypothesized that sperm in the posterior infundibulum may be for fertilization of ova within one mating season, while sperm in the posterior oviduct may be for fertilization of ova across reproductive seasons. We reject this hypothesis due to the fact that sperm in the posterior oviduct disappear following ovulation (Aldridge 1992; Siegel and Sever 2008a). Another hypothesis seems to be that sperm are present in the posterior oviduct in the fall through hibernation and, thus, the posterior site of long-term sperm storage is the posterior oviduct (Fox 1956; Perkins and Palmer 1996). Subsequently, sperm move to anterior sperm receptacles in the spring at the onset of vitellogenesis (Perkins and Palmer 1996). However, this is clearly not the case in North American crotalines as vitellogenesis begins in the fall (for review see Aldridge and Duvall 2002). Interestingly, in North American crotalines, sperm appear in the infundibular receptacles in the fall (Siegel and Sever 2008a), and this could be thought of as support for the hypothesis by Perkins and Palmer (1996) that vitellogenesis triggers the cranial migration of sperm. Thus, the hormones responsible for vitellogenesis may trigger secretory activity in sperm receptacles, which may subsequently be invaluable in attracting sperm, which would explain why sperm seem to persist in the posterior oviduct in non-viperid Colubroides throughout hibernation.

A working hypothesis is that sperm enter the posterior oviduct after mating, and subsequently are attracted to the posterior infundibular sperm storage receptacles by the secretions produced in their epithelium where the sperm have a better chance of fertilizing eggs. This "better chance" is due to the fact that ovulated eggs actually pass the sperm storage receptacles while they move posterior to the middle oviduct for egg-shelling or incubation (Fox 1956). Some eggs never pass through the posterior oviduct until parturition. A certain, unknown, percentage of sperm never migrate cranially to the infundibular sperm storage receptacles and these sperm eventually degrade naturally as observed in Siegel and Sever (2008a) in *Agkistrodon piscivorus*.

Evolution of sperm storage location. Two studies have traced the evolution of sperm storage receptacle location in squamates, with similar results. Sever and Hamlett (2002) and Eckstut *et al.* (2009) describe sperm storage receptacle location in snakes as infundibular in *Thamnophis sirtalis* and infundibular and "vaginal (posterior oviduct)" in *Seminatrix pygaea*. Eckstut *et al.* (2009) adds to their analysis with the addition of only infundibular sperm storage receptacles found in *Agkistrodon piscivorus* (Siegel and Sever 2008a). The ancestral mode of sperm storage receptacle location in snakes was found to be infundibular (based on outgroup comparison). However, these studies lack data as they only analyze sperm storage receptacles and not the presence of prolonged sperm aggregations, which would more accurately elucidate the evolution of sperm storage location in snakes, and squamates in general. It is apparent that *Sphenodon punctatus* stores sperm; however, no sperm storage receptacles, or sperm, have been identified in the entire reproductive tract of this sister group to all extant squamates (Gabe and Saint-Girons 1964). Saint-Girons (1975) hypothesized that *S. punctatus* stored sperm in the lumen (or possibly furrows) of the posterior portions of the oviduct until ovulation. Thus, sperm stored in the oviducal lumen or in posterior furrows would be the plesiomorphic condition in the Lepidosauria and, thus, sperm storage structures evolved secondarily, and possibly independently in the major squamate lineages (for review on sperm storage structure diversity in squamates see Eckstut *et al.* 2009).

It appears that the ancestor of snakes possessed infundibular sperm storage structures (Sever and Hamlett 2002; Eckstut *et al.* 2009), and in general, all snakes have similar sperm storage structures (see section 9.4.3). However, it appears that all snake taxa have also retained the pleisiomorphic condition of sperm storage in Lepidosauria, as all snakes examined throughout a reproductive season possess sperm in the posterior oviduct for extended periods of time. *Seminatrix pygaea* possesses the most undifferentiated sperm storage structures identified to date with simple glandular invaginations that are unspecialized from the epithelium lining the lumen of the infundibulum. Colubrid and natracine snakes (e.g., *Diadophis punctatus* and *Thamnophis sirtalis* respectively) have the most specialized sperm storage structures with many large complex alveolar glands, wheras, viperids seem to have an intermediate structure (e.g. simple differentiating tubular glands). However, it also appears that some snakes have lost anterior sperm storage, as these structures were not functionally identified in a colubrid (*Boiga irregularis*).

In conclusion all investigated snakes, except *Boiga irregularis* (Bull *et al.* 1977), possess specialized sperm storage receptacles in the posterior infundibulum, these sperm storage receptacles possess some morphological variation (see section 7.4.3), and sperm reside in the unspecialized posterior oviduct for a prolonged period of time (possibly because of the retention of the ancestral condition observed in the Tuatara *Sphenodon punctatus*; Saint-Girons 1975) before migration to infundibular sperm storage receptacles in most taxa examined.

9.5 CONCLUSIONS

9.5.1 Nomenclature

Overview. As observed from Tables 9.2 and 9.3 and the review above, nomenclature has not been consistent across investigations and, thus, not consistent across species. We feel that the most appropriate way to term regions of the female reproductive tract is by homology, so that accurate inter-specific comparison can be made. The function of many oviducal regions is unknown and, thus, terming regions based on "proposed" function carries little weight in comparative studies. However, as long as homologous regions are termed identically, evolutionary comparisons can be made more easily. In the following we suggest terms that easily reflect historical precedence and/or morphologically distinct regions in the female reproductive tract.

Cloaca. We begin our nomenclature proposal with the most posterior portion of the female reproductive tract, the cloaca. The proposed nomenclature is presented in Table 9.4. The opening of the cloaca (vent) opens dorsally into a stratified keratinized region most often referred to as the proctodaeum (region a). Although historically significant, this term is anatomically inaccurate. The proctodaeum in snakes is not the posterior portion of the gastrointestinal tract, and is actually the common entrance and exit of reproductive organs and fecal and urinary waste respectively. However, this inaccuracy is consistent with all vertebrate taxa (Gadow 1887) and so we retain the usage of the term proctodaeum.

Table 9.4 Proposed terminology of the cloaca

Historical	*Suggested*
Proctodaeum (Gadow 1887)	Proctodaeum (region a)
Urodaeum (Gadow 1887)	Urodaeum (region b and b^1)
Coprodaeum (Gadow 1887)	Urodaeal sphincter (region c) and posterior intestine (region d)
Vagina (Giersberg 1922)	Posterior oviduct (region 2)

The proctodaeum transitions inconspicuously to a stratified/bistratified mucous epithelium typically termed the urodaeum (region b). As with the proctodaeum, this term carries great historic value, and is the common chamber of the urinary and reproductive ducts. Thus, we retain the terminology urodaeum for this region. However, as observed in Colubroides, the oviduct may insert into an apomorphic cranial pouch-like region (b^1), which carries a simple to pseudostratified tall columnar, primarily secretory epithelium. The origin of this region (whether cloacal tissue or oviducal tissue) is presently unknown (hypothesized to be cloacal derived here; see section 9.2.9), and this region appears to be distinct, but sometimes continuous, with the traditional urodaeum (region b) and oviduct. This region has been commonly referred to as the vagina or vaginal pouch, which we choose to abandon, as it has not been confirmed that a copulatory organ enters this region in all snakes (though this highly folded region appears desirably expandable for a copulatory organ; see Figs. 9.2 [bottom sections] and 9.3F). Also, even though Pope (1941) suggests that a copulatory organ reaches this region, terming this region the vagina or vaginal pouch would then carry the connotation that scolecophidians and henophidians do not have a vagina. We also urge against the terminology of Giacomini ("diverticolo"; 1893), although utilized throughout this text, as this region is continuous with the oviduct and cloaca and, thus, does not really form a diverticulum. We choose the term "pouch" for this region so to not propose developmental or functional implications.

From the posterior portion of the urodaeum projects a sphincter-like region (region c) ventrally that communicates the intestine (region d) with the urodaeum, and carries a transitioning epithelium similar to the urodaeum caudally and the posterior intestine cranially. We term this region the "urodaeal sphincter." We recognize that this region is the most posterior region of the gastrointestinal tract that carries only fecal waste (besides the transfer of uric acid material into the posterior intestine) and, thus, this is the region that most appropriately should have been called the proctodaeum. However, as discussed above, historical precedence has already caused the proctodaeum to be used elsewhere. Coprodaeum was used for describing the intestinal structures attaching to the cloaca; however, we suggest abandoning this term, in lieu of the fact that the term coprodaeum has been given to the urodeal sphincter and the posterior intestine.

The goal of the nomenclatural review above was to aid in future comparison of female cloacal structures between species of snakes; however, we would like to note that the idea of a "cloaca" as depicted by Gadow (1887) and his contemporaries is quite a conundrum, as emphasized by Gabe and Saint-Girons (1965). As reported by Gadow (1887), one definition of the cloaca is "a chamber at the terminal portion of the rectum, into which open the rectum, the ureters, and the genital tubes." After reading the review above, it is clear in female snakes that this definition fits the definition of what historically has been termed the urodaeum. Considering that the uriniferous ducts and urodaeal sphincter open into the posterior extremity of the urodaeum, removal of what has been traditionally termed the proctodaeum (epiblastic invagination; Gadow 1887) causes the common exit/entrance for waste/reproductive materials to disappear. Could not then cloacae be defined by the presence of the proctodaeum? Thus, maybe what has been traditionally termed the proctodaeum should really be considered the cloaca? However, without a comparative study of all vertebrate cloacae, we do not advocate a complete change of cloacal terminology in all vertebrates from investigation of only female snakes.

Oviduct. As Blackburn (1998) states, the naming of the posterior end of the oviduct (Fig. 9.6 region 2) as the vagina is rather unfortunate. Developmentally, this region is derived from the Müllerian duct, and this region does not house a copulatory organ during copulation. Thus, the term vagina is inappropriate from a developmental and functional standpoint. Considering the epithelium lining the lumen of this oviducal region is similar to that of the middle oviduct, which is commonly termed the uterus, we term this region after the nomenclature presented by Siegel and Sever (2006, 2008a,b) the "non-glandular uterus." To recognize the distinction between this region and the middle oviduct (typically termed the uterus; Fig. 9.6; region 3), we term the portion of the oviduct that possesses the uterine glands the "glandular uterus" after Siegel and Sever (2006, 2008a,b).

As it has been noted that sperm storage receptacles are present in the posterior infundibulum of snakes (Fig. 9.6 region 4), we term the region of sperm storage in snakes as the "posterior infundibulum." The usage of "tuba" should be avoided due to the fact that homology between the posterior infundibulum in snakes and the tuba in other amniotes has not been confirmed. If confirmed in subsequent work, the term tuba is appropriate. This leaves the final, and most cranial region of the oviduct

(Fig. 9.6 region 5), which we term the "anterior infundibulum" in order to distinguish this region from the sperm storage region of the posterior infundibulum.

9.5.2 Future Directions

From the above review, it is clear that variation exists in the reproductive tract of female snakes. However, the extent of this variation is unknown due to the lack of investigations on diverse snake families. It is also unknown if the variation in reproductive anatomy in female snakes is a product of phylogeny, environment, or co-evolution with male reproductive morphology. Interestingly, because of the multiple functions of the female cloaca (excretion of fecal and urinary waste and acceptor of a copulatory organ), variation in anatomy in different regions could be due to the different functions of the parts. Utilization of gross examination of museum specimens may add insights to these questions; however, basic histological examination of representative specimens is needed to make primary hypotheses of homology of different regions of the female reproductive tract.

Other than extensive work in phylogenetic systematics utilizing male hemipenis structure, few studies have tested the utility of female reproductive anatomy in the reconstruction of the evolutionary history of snakes. In fact, only one study (Young *et al.* 1999) utilized cloacal structures (cloacal scent glands) to reconstruct a snake phylogeny, and this investigation utilized phenetics in contrast to a more accepted approach of assessing character polarity and shared derived characters (e.g., a cladistics approach). Thus, the capability of female reproductive structures in reconstructing a phylogeny in snakes is basically unknown. As stated before, histological examination is necessary to assure that sound primary hypotheses are being made for different regions of the oviduct across taxa in subsequent analyses.

Besides the use of female reproductive anatomy in comparative studies, other questions, particularly those dealing with development and function, are also of interest. Other than few studies on the uterus and sperm storage receptacles, very little is known about the function of the female reproductive tract in snakes. Except for the extensive work on the development of the female reproductive tract in *Natrix tessellata* (for review see Raynaud and Pieau 1985), little data exist on the development of the urogenital system in snakes. In the studies by Raynaud and Pieau (1985), correlation of reproductive tract development to the adult structures is not presented. These types of studies are necessary to better assess the homology of reproductive tract regions and, thus, would greatly aid in the nomenclature of the regions in the female reproductive tract of snakes and, furthermore, increase the utility of reproductive morphology in comparative studies. For example, whether or not the pouch (region b^1) is considered oviducal or cloacal derived changes the view of the female reproductive tract dramatically. If the pouch in Colubroides is considered cloacal, it appears that members of Colubroides have a bifurcated cloaca, however, if considered oviducal members of Colubroides have a common oviducal opening into the cloaca.

Whatever the direction of research on the female reproductive tract in snakes proceeds, it is clear that there are abundant interesting questions and testable hypotheses awaiting future investigators. We hope this review and synthesis of the cloaca, oviducts, and sperm storage structures will aid future investigators in all their endeavors, and we strongly suggest utilizing the above nomenclature for no other reason than to maintain consistency in terming homologous regions of the female reproductive tract in snakes.

9.6 ACKNOWLEDGMENTS

The following individuals and institutions were invaluable in the completion of this review: Julie A. Wiese for laboratory assistance; Brian I. Crother for helpful discussion on squamate systematics; David M. Sever, Stanley E. Trauth, and Kevin M. Gribbins for helpful discussions on squamate reproductive morphology; Eric R. Pianka and Dan G. Blackburn for assistance in gathering literature; Saint Louis University; Southeastern Louisiana University for microscope access; Centre National de la Recherche Scientifique; Texas A&M University-Corpus Christi; The Field Museum of Natural History for permission to examine and dissect specimens. Alan Resetar for helpful assistance at the Field Museum of Natural History.

9.7 LITERATURE CITED

Aldridge, R. D. 1992. Oviductal anatomy and seasonal sperm storage in the southeastern crowned snake (*Tantilla coronata*). Copeia 1992: 1103-1106.

Almeida-Santos, S. M. and Salomão, M. G. 1997. Long-term sperm storage in the female neotropical rattlesnake *Crotalus durissus terrificus* (Viperidae: Crotalinae). Japanese Journal of Herpetology 17: 46-52.

Almeida-Santos, S. M. and Salomão, M. 2002. Reproduction in neotropical pitvipers, with emphasis on species of the genus *Bothrops*. Pp. 445-462. In G. W. Schuett, M. Höggren, M. E. Douglas and H. W. Greene (eds), *Biology of the Vipers*. Eagle Mountain Publishing, Eagle Mountain, Utah.

Balfour, F. M. 1881. *A Treatise on Comparative Embryology, Volume II*. Cambridge University Press, Massachusetts. Pp. 655.

Bauman, M. A. and Metter, D. E. 1977. Reproductive cycles of the northern watersnake, *Natrix s. sipedon* (Reptilia, Serpentes, Colubridae). Journal of Herpetology 11: 51-59.

Beuchat, C. A. 1986. Phylogenetic distribution of the urinary bladder in lizards. Copeia 1986: 512-517.

Beuchelt, H. 1936. Bau, funktion und entwicklung der begattungsorgane der männlichen ringelnatter (*Natrix natrix* L.) und kreuzotter (*Vipera berus* L.). Morphologisches Jahrbuch 78: 445-516.

Blackburn, D. G. 1998. Structure, function, and evolution of the oviducts of squamate reptiles, with special reference to viviparity and placentation. Journal of Experimental Zoology 282: 560-617.

Blackburn, D. G. 1999. Viviparity and oviparity: evolution and reproductive strategies. Pp. 994-1003. In T. E. Knobil and J. D. Neill (eds), *Encyclopaedia of Reproduction.* Academic Press, New York.

Broek Van Den, A. J. P. 1933. Gonaden und susführungsgänge. Pp. 1-154. In L. Bolk, E. Göppert, E. Kallius and W. Lubosh (eds), *Handbuch der Vergleichenden Anatomie der Wirbelthiere,* Vol. 6. Urban & Schwarzenberg, Berlin, Germany.

Bull, K. H., Mason, R. T. and Whittier, J. 1997. Seasonal testicular development and sperm storage in tropical and subtropical populations of the brown tree snake (*Boiga irregularis*). Australian Journal of Zoology 45: 479-488.

Clark, D. R. 1970. Loss of the left oviduct in the colubroid snake genus *Tantilla.* Herpetologica 26: 130-133.

Cope, E. D. 1898. The crocodilians, lizards and snakes in North America. Report of the U. S. National Museum 1890: 153-1270.

Cooper, W. E. and Trauth, S. E. 1992. Discrimination of conspecific male and female cloacal chemical stimuli by males and possession of a probably pheromone gland by females in a cordylid lizard, *Gerrhosaurus nigrolineatus.* Herpetologica 48: 229-236.

Das, S. M. 1959. A comparative functional study of the urino-genital system in *Uromastyx hardwickii* Gray (sand lizard), *Ptyas mucosus* Linné (rat snake) and *Eryx conicus* Boulenger (sand boa). Proceedings of the National Academy of Science, India 30: 59-78.

Disselhorst, R. 1904. Ausführapparat und anhangsdrüsen der männlichen geschlechtsorgane. Pp. 1-386. In A. Oppel (ed.), *Lehrbuch der Vergleichenden Mikroskopischen Anatomie der Wirbeltiere,* Vol. 12. Fischer, Jenna, Germany.

Duguy, R. and Saint-Girons, H. 1966. Cycle annuel d'activité et reproduction de la Couleuvre vipérine, *Natrix maura* (L.). Terre et Vie 1966: 423-457.

Eckstut, M. E., Sever, D. M., White, M. E. and Crother, B. I. 2009. Phylogenetic analysis of sperm storage in female squamates. Pp. 185-215. In L. T. Dahnof (ed.), *Animal Reproduction: New Research Developments.* Nova Science Publishers, Inc., Hauppauge, New York.

Edgren, R. A. 1953. Copulatory adjustment in snakes and its evolutionary implications. Copeia 1953: 162-164.

Fox, H. 1977. The urogenital system of reptiles. Pp. 1-157. In C. Gans (ed.), *Biology of the Reptilia,* Vol. 6. Academic Press, New York.

Fox, W. 1956. Seminal receptacles of snakes. Anatomical Record 124: 519-540.

Fox, W. and Dessauer H. C. 1962. The single right oviduct and other urogenital structures of female *Typhlops* and *Leptotyphlops.* Copeia 1962: 590-597.

Gabe, M. and Saint-Girons, H. 1964. *Histologie de Sphenodon punctatus.* Centre National de la Recherche Scientifique, Paris. Pp. 148.

Gabe, M. and Saint-Girons, H. 1965. Contribution à la morphologie comparée du cloaque et des glandes épidermoïdes de la région cloacale chez les lépidosauriens. Mémoires du Muséum National d'Histoire Naturelle XXXIII: 149-332.

Gadow, H. 1887. Remarks on the cloaca and on the copulatory organs of the Amniota. Philosophical Transactions of the Royal Society of London B 178: 5-37.

Garstka, W. R. and Crews, D. 1981. Female sex pheromone in the skin and circulation of a garter snake. Science 214: 681-683.

Gerhardt, E. 1937. Kloake und begattungsorgane. Pp. 253-348. In L. Bolk, E. Göppert, E. Kallius, and W. Lubosch (eds), *Handbuch der Vergleichenden Anatomie der Wirbeltiere,* Vol. 6. Urban and Schwarzenberg, Berlin, Germany.

Giacomini, E. 1893. Sull'ovidutto del Sauropsidi. Monitore Zoologico Italiano 4: 202-265.

Gibson, W. W. 1934. The uterus and embryonic membranes of the common garter snake (*Thamnophis s. sirtalis*) with special reference to placentation. Ph.D. thesis, University of Iowa, Iowa City, Iowa.
Giersberg, H. 1922. Untersuchungen über physiologie und histologie des eileiters der reptilien und vögel; nebst einem beitrag zue fasergenese. Zeitchrift für Wissenshchaftliche Zoologie 70: 1-97.
Girling, J. E. 2002. The reptilian oviduct: a review of structure and function and direction for future research. Journal of Experimental Zoology 293: 141-170.
Halpert, A. P., Gartska, W. R. and Crews, D. 1982. Sperm transport and storage and its relation to the annual cycle of the female red-sided garter snake, *Thamnophis sirtalis parietalis*. Journal of Morphology 174: 149-159.
Hoffman, L. H. 1970. Placentation in the garter snake, *Thamnophis sirtalis*. Journal of Morphology 131: 57-88.
Hoffman, L. H. and Wimsatt, W. A. 1972. Histochemical and electron microscopic observations on the sperm receptacles in the garter snake oviduct. American Journal of Anatomy 134: 71-96.
Inger, R. F. and Marx, H. 1962. Variation of hemipenis and cloaca in the colubrid snake *Calamaria lumbricoidea*. Systematic Zoology 11: 32-38.
Kasturirangan, L. R. 1951a. Placentation in the sea snake *Enhydrina schistose* (Daudin). Proceedings of the Indian Academic Society 34: 1-32.
Kasturirangan, L. R. 1951b. The allantoplacenta of the sea snake *Hydrophis cyanocintus* Daudin. Journal of the Zoological Society of India 2: 277-290.
Kissner, K. J., Blouin-Demers, G. and Weatherhead, P. J. 2000. Sexual dimorphism in malodorousness of musk secretions of snakes. Journal of Herpetology 34: 491-493.
Kissner, K. J., Secoy, D. M. and Forbes, M. R. 1998. Sexual dimorphism in size of cloacal glands of the garter snake, *Thamnophis radix haydeni*. Journal of Herpetology 32: 268-270.
Korneva, L. G. 1973. Morphology of female reproductive system in *Vipera lebetina* with special reference to its activity. Zoologicheskii Zhurnal 52: 85-93.
Lawson, R., Slowinski, J. B., Crother, B. I. and Burbrink, F. T. 2005. Phylogeny of the Colubroidea (Serpentes): New evidence from mitochondrial and nuclear genes. Molecular Phylogenetics and Evolution 37: 581-601.
Ludwig, M. and Rahn, H. 1943. Sperm storage and copulatory adjustment in the prairie rattlesnake. Copeia 1943: 15-18.
Marinho, C. E., Almeida-Santos, S. M., Yamasaki, S. C. and Silveira, P. F. 2009. Peptidase activities in the semen from the ductus deferens and uterus of the neotropical rattlesnake *Crotalus durisus terrificus*. Journal of Comparative Physiology B 179: 635-642.
Mead, R. A., Eroschenko, V. P. and Highfill, D. R. 1981. Effects of progesterone and estrogen on the histology of the oviduct of the garter snake, *Thamnophis elegans*. General and Comparative Endocrinology 45: 345-354.
Mulaik, G. 1946. A comparative study of the urogenital systems of an oviparous and two ovoviviparous species of the lizard genus *Sceloporus*. Bulletin of the University of Utah Biology Series 37: 3-24.
Palmer, B. D., Demarco, V. G. and Guillette, L. 1993. Oviductal morphology and eggshell formation in the lizard, *Sceloporus woodi*. Journal of Morphology 217: 205-217.
Perkins, J. M. and Palmer, B. D. 1996. Histology and functional morphology of the oviduct of an oviparous snake, *Diadophis punctatus*. Journal of Morphology 227: 67-79.

Pope, C. H. 1941. Copulatory adjustment in snakes. Zoological Series of Field Museum of Natural History 24: 249-252.
Price, A. H. and LaPointe, J. L. 1981. Structure-functional aspects of the scent gland in *Lampropeltis getulus splendida*. Copeia 1981: 138-146.
Rahn, H. 1940. Sperm viability in the uterus of the garter snake, *Thamnophis*. Copeia 1940: 107-115.
Rahn, H. 1942. The reproductive cycle of the prairie rattler. Copeia 1942: 233-240.
Raynaud, A. and Pieau, C. 1985. Embryonic development of the genital system. Pp. 149- 300. In C. Gans and F. Billett (eds), *Biology of the Reptilia*, Vol. 15B. John Wiley and Sons, New York.
Regamey, J. 1935. Les caractères sexuels du lézard (*Lacerta agilis* L.). Revue Suisse de Zoologie 42: 87-168.
Sacchi, M. 1888. Contribution à l'histologie de l'oviducte des sauropsides. Archives Italiennes de Biologie 9: 267-284.
Saint-Girons, H. 1957. Le cycle sexuel chez *Vipera aspis* dans l'ouest de la France. Bulletin Biologique de la France et de la Belgique 91: 284-350.
Saint-Girons, H. 1962a. Le cycle reproducteur de la vipère à cornes, *Cerastes cerastes* (L.), dans la nature et en captivité. Bulletin de la Société Zoologique de France 87: 41-51.
Saint-Girons, H. 1962b. Présence de recéptacles séminaux chez les caméléons. Beaufortia 9: 165-172.
Saint-Girons, H. 1975. Sperm survival and transport in the female genital tract of reptiles. Proceedings of the INSERM International Symposium, Nouzilly (1973), Pp. 105-113. In E. S. E. Hafez and C. G. Thibault (eds), *The Biology of the Spermatozoa*. Karger, Basel, Switzerland.
Sánchez-Martínez, P. M., Ramírez-Pimilla, M. P. and Miranda-Esquivel, D. R. 2007. Comparative histology of the vaginal-cloacal region in Squamata and its phylogenetic implications. Acta Zoologica 88: 289-307.
Seaman, A. R. 1949. The histology of the oviduct of the common garter snake, *Thamnophis ordinatus ordinatus*, throughout the reproductive cycle. Ph.D. thesis, Cornell University, Ithaca, New York.
Seshadri, C. 1959. Structural modification of the cloaca of *Lycodon aulicus aulicus* Linn., in relation to urine excretion and the presence of sexual segment in the kidney of male. Proceedings of the National Institute of Science, India 25B: 271-278.
Sever, D. M. 2004. Ultrastructure of the reproductive system of the black swamp snake (*Seminatrix pygaea*). IV. Occurrence of an ampullae ductus deferentis. Journal of Morphology 262: 714-730.
Sever, D. M. 2010. Ultrastructure of the reproductive system of the black swamp snake (*Seminatrix pygaea*). VI. Anterior testicular ducts and their nomenclature. Journal of Morphology 271: 104-115.
Sever, D. M. and Ryan, T. J. 1999. Ultrastructure of the reproductive system of the black swamp snake (*Seminatrix pygaea*). I. Evidence for oviducal sperm storage. Journal of Morphology 241: 1-18.
Sever, D. M. and Hamlett, W. C. 2002. Female sperm storage in reptiles. Journal of Experimental Zoology 292: 187-199.
Sever, D. M. and Hopkins, W. A. 2005. Oviductal sperm storage in the ground skin *Scincella lateralae* Holbrook (Reptilia: Scincidae). Journal of Experimental Zoology 301A: 599-611.
Sever, D. M., Ryan, T. J., Morris, T., Patton, D. and Swafford, S. 2000. Ultrastructure of the reproductive system of the black swamp snake (*Seminatrix pygaea*): Part II. The annual oviducal cycle. Journal of Morphology 245: 146-160.

Sever, D. M., Stevens, R. A., Ryan, T. J. and Hamlett, W. C. 2002. Ultrastructure of the reproductive system of the black swamp snake (*Seminatrix pygaea*). III. Sexual segment of the male kidney. Journal of Morphology 252: 238-254.

Sever, D. M., Siegel, D. S., Bagwill, A., Eckstut, M. E., Alexander, L., Camus, A. and Morgan, C. 2008. Renal sexual segment of the cottonmouth snake, *Agkistrodon piscivorus* (Reptilia, Squamata, Viperidae). Journal of Morphology 269: 640-653.

Siegel, D. S. and Sever, D. M. 2006. Utero-muscular twisting and sperm storage in viperids. Herpetological Conservation and Biology 1: 87-92.

Siegel, D. S. and Sever, D. M. 2008a. Sperm aggregations in female *Agkistrodon piscivorus* (Reptilia: Squamata): A histological and ultrastructural investigation. Journal of Morphology 269: 189-206.

Siegel, D. S. and Sever, D. M. 2008b. Seasonal variation in the oviduct of female *Agkistrodon piscivorus* (Reptilia: Squamata): An ultrastructural investigation. Journal of Morphology 269: 980-997.

Siegel, D. S., Aldridge, R. D., Clark, C. S., Poldemann, E. H. and Gribbins, K. M. 2009a. Stress and reproduction in the brown treesnake (*Boiga irregularis*) with notes on the ultrastructure of the sexual segment of the kidney in squamates. Canadian Journal of Zoology 87: 1138-1146.

Siegel, D. S., Sever, D. M., Rheubert, J. L. and Gribbins, K. M. 2009b. Reproductive biology of *Agkistrodon piscivorus* (Squamata, Ophidia, Viperidae, Crotalinae). Herpetological Monographs 23: 74-107.

Smith, H. M., Preston, M. J. and Jones, R. E. 1989. Oviductal, not oviducal. The Anatomical Record 223: 446-447.

Takewaki, K. and Hatta, K. 1941. Effect of gonadectomy and hypophysectomy on the kidney and genital tract of a snake, *Natrix tigrina tigrina*. Annotationes Zoologicae Japonenses 20: 4-8.

Trauth, S. E., Cooper, W. E., Vitt, L. J. and Perrill, S. A. 1987. Cloacal anatomy of the broad-headed skink, *Eumeces laticeps*, with a description of a female pheromonal gland. Herpetologica 43: 458-466.

Unterhössel, P. 1902. Die eideschsen und schlangen. Pp. 541-581. In A. Fleischmann (ed.), *Morhpologische Studien über Kloake und Phallus der Amnioten*. Morphologisches Jahrbuch 30: 539-675.

Uribe, M. C. A., González-Porter, G., Palmer, B. D. and Guillette, L. J. Jr. 1998. Cyclic histological changes of the oviductal-cloacal junction in the viviparous snake *Toluca lineata*. Journal of Morphology 237: 91-100.

Vidal, N. and Hedges, S. B. 2002. Higher-level relationship of snakes inferred from four nuclear and mitochondrial genes. Comptes Rendus Biologies 325: 977-985.

Volsøe, H. 1944. Structure and seasonal variation of the male reproductive organs of *Vipera berus* (L.). Spolia Zoologica Musei Hauniensis 5:1–157.

Whiting, A. M. 1969. Squamate cloacal glands: morphology, histology and yhistochemistry. Ph.D. thesis, Pennsylvania State University, State College, Pennsylvania.

Yokoyama, F. and Yoshida, H. 1994. The reproductive cycle of the female habu, *Trimeresurus flavoviridis*. Journal of Herpetology 28: 54-59.

Young, B. A., Marsit, C. and Meltzer, K. 1999. Comparative morphology of the cloacal scent gland in snakes (Serpentes: Reptilia). The Anatomical Record 256: 127-138.

Zaher, H., Grazziotin, R. W., Cadle, J. E., Murphy, R. W., de Moura-Leite, J. C. and Bonatto S. L. 2009. Molecular phylogeny of advanced snakes (Serpentes, Caenophidia) with an emphasis on South American xenodontines: a revised classification and description of new taxa. Papéis Avulsos de Zoologia 49: 115-153.

CHAPTER 10

Male Urogenital Ducts and Cloacal Anatomy

Stanley E. Trauth[1] *and David M. Sever*[2]

10.1 INTRODUCTION

From seminiferous tubules of the testis, snakes and other squamates possess a system of efferent ducts, at least some of which function not only in the passage of sperm but also in sperm storage and as accessory sex glands. The seminiferous tubules connect to the rete testis, which passes sperm sequentially to the ductuli efferentes, the ductus epididymis, and the ductus deferens. The sexual segment of the kidney, described in chapter 11, is a male accessory sex gland that passes secretions into the ureter. The ductus deferens merges with its ipsilateral ureter in viperid snakes, whereas the ductus deferens empties posterior to its ipsilateral ureter and into its ipsilateral ampulla urogenital papilla in colubroid snakes. The above description represents a new interpretation of the alignment and termination of the urogenital ducts in viperid and colubroid snakes and will be illustrated and discussed in detail later in this chapter. Eventually, all products within the adjoined urogenital ducts exit via a single urogenital papilla or paired urogenital papillae into the cloaca. Depending on taxon, species may have ampullae ductus deferentia, ampullae ureters, and/or ampullae urogenital papillae. In this chapter, we review the literature on these structures, provide much new information, and indicate directions for future research.

[1]Department of Biological Sciences, Arkansas State University, State University, Arkansas 72467 USA
[2]Department of Biological Sciences, Southeastern Louisiana University, Hammond, Louisiana 70402 USA

10.2 PROXIMAL EFFERENT DUCTS

10.2.1 Overview

Volsøe (1944) indicates that the testes of the Common European Adder (*Vipera berus*), as well as the rest of the urogenital system, are retroperitoneal, but we have found a distinct mesorchium in the species we have examined. Posterior to the testes, the ductus deferens is supported by mesentery until reaching the dorsal surface of the kidney, where the serosae of the kidney and ductus deferens are united but distinct. Posterior to the kidney, the ductus deferens is suspended by the same mesentery as the ureter until reaching the cloacal wall.

Until recently, the ducts leading from the seminiferous tubules of snakes were called the ductuli efferentes and these connected to the ductuli epididymides, which had a proximal portion that was non-ciliated (ductuli epididymides I) and a distal portion with ciliated epithelium (ducutli epidymides II) connecting to the ductus epididymis (Volsøe 1944; Fox 1952; Saint Girons 1958; Fox 1977).

Sever (2010) demonstrated the proximal testicular ducts of the Black Swamp Snake (*Seminatrix pygaea*) were histologically and ultrastructurally similar to those of other amniotes that have been studied with these techniques, including a lizard (Akbarsha *et al.* 2007), a turtle (Holmes and Gist 2004), a crocodilian (Guerrero *et al.* 2004), birds (Aire 2007), and mammals (Hess 2002). Sever (2010) adopted the nomenclature used for these other amniotes so that the ductuli efferentes and ductuli epididymides I become known as the rete testis, and the ductuli epididymides II is called the ductuli efferentes (Table 10.1; Jones 1998).

Most of the observations below concern *Seminatrix pygaea* from Sever (2004, 2010) and results on the Cottonmouth (*Agkistrodon piscivorus*) recently published by Siegel *et al.* (2009), but we also present some new findings

Table 10.1 Nomenclature for the proximal efferent ducts of snakes since Volsøe (1944) and general characteristics of the epithelial lining

Sever (2010) name	*Jones (1998) name*	*Volsøe (1944) name*	*Epithelium*	*Absorption*	*Secretion*
Rete testis	Rete testis	Ductuli efferentes Ductuli epididymides I	Simple squamous/ cuboidal, non-ciliated	Yes	Yes
Ductuli efferentes	Ductuli efferentes	Ductuli epididymides II	Simple cuboidal, ciliated and non-ciliated cells	Yes	Yes
Ductus epididymis	Ductus epididymis	Ductus epididymis	Non-ciliated, pseudostratified	Yes	Yes
Ductus deferens	Ductus epididymis	Ductus deferens	Non-ciliated, pseudostratifed	Yes	Yes/No

on *A. piscivorus*. Light micrographs of the proximal testicular ducts of the two species are illustrated in Figs. 10.1-10.2, and ultrastructure is shown in Figs. 10.3-10.10.

Distal ends of seminiferous tubules as well as the anterior testicular ducts and the adrenal gland are encased in a modification of the serosa on the dorsal surface of the testis called the epididymal sheath (Siegel *et al.* 2009). Intratesticular portions of the rete testis branch off the seminiferous tubules within the capsule of the testis (Fig. 10.1A-C), whereas extratesticular portions extend into the epididymal sheath (Figs. 10.1E, 10.2A,C,D). In *Agkistrodon piscivorus*, some rete tubules do not have intratesticular portions as their seminiferous tubules extend into the epididymal sheath (Fig. 10.2B-D). The ductuli efferentes are entirely extratesticular within the epididymal sheath (Figs. 10.1A,D,E, 10.2A-C).

Rete testis branch off along the entire border of the testis, although the ductus epididymis extends only along the posterior two-thirds of the testis in *Agkistrodon piscivorus* and *Seminatrix pygaea*. The proximal portions of the rete testis widen after branching off the seminiferous tubules (Fig. 10.1C), and divide into smaller branches before passing into the ductuli efferentes. The ductuli efferentes are small, highly branched and convoluted (Figs. 10.1D,E, 10.2B,C).

The rete testis lacks ciliated epithelium whereas the ductuli efferentes possess both ciliated and non-ciliated cells, making the two regions easy to distinguish, even with light microscopy. Indeed, the ductuli efferentes are the only ducts in the male snake reproductive tract with ciliated cells (Table 10.1).

The rete testis, ductuli efferentes, and ductus epididymis of *Agkistrodon piscivorus* and *Seminatrix pygaea* generally give positive histochemical reactions for neutral carbohydrates (periodic-acid/Schiff's reagent: PAS positive) and proteins (bromphenol blue positive), although the carbohydrate reaction in the ductus epididymis is not as intense as in other regions in *S. pygaea* (Fig. 10.1E,F). PAS positive granules are conspicuous in the ductuli efferentes of *A. piscivorous* throughout the year (Siegel *et al.* 2009).

10.2.2 Ultrastructure of Proximal Testicular Ducts

In *Seminatrix pygaea*, the cuboidal epithelial cells of the rete testis have large, indented euchromatic nuclei, and cells are separated by narrow intercellular canaliculi which are labrinythine apically (Fig. 10.3). Short microvilli are present, and mitochondria are elongate and electron dense. Numerous small vesicles occur along the luminal border and a flocculent material occurs in the supranuclear area of most epithelial cells. Although some vesicles may be involved in absorption, Sever (2010) found evidence that exocytosis is also occurring by an apocrine process. Chromatin material becomes more condensed in spring and summer, when active spermatogenesis is occurring, with spermiation commencing in July (Gribbins *et al.* 2005).

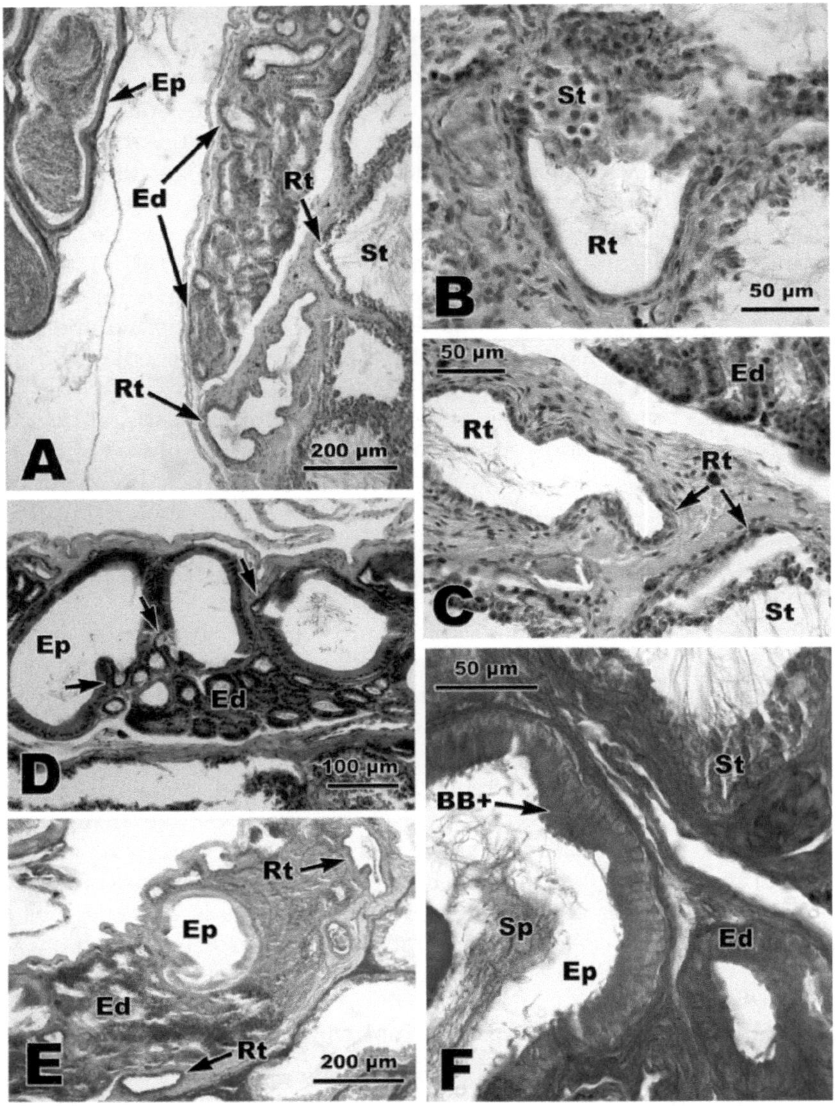

Fig. 10.1 Light micrographs of proximal testicular ducts of *Seminatrix pygaea*. **A.** Overview of relationships among seminiferous tubules, rete testis, ductuli efferentes, and ductus epididymis. **B,C.** Intratesticular rete testis evaginating from seminiferous tubules. **D.** Ductuli efferentes merging with ductus epididymis in extratesticular epididymal sheath (arrows). A-D stained with hematoxylin and eosin. **E.** PAS/AB histochemistry of extratesticular rete testis, ductuli efferentes, and ductus epididymis. PAS positive reactions indicating neutral carbohydrates are illustrated by intensity of the red stain. **F.** Bromphenol blue histochemistry of ductuli efferentes and ductus epididymis. Positive (BB+) reactions indicating proteins are illustrated by the intensity of the blue stain. Ed, ductuli efferentes; Ep, ductus epididymis; Rt, rete testis; Sp, sperm; St, seminiferous tubules.

Color image of this figure appears in the color plate section at the end of the book.

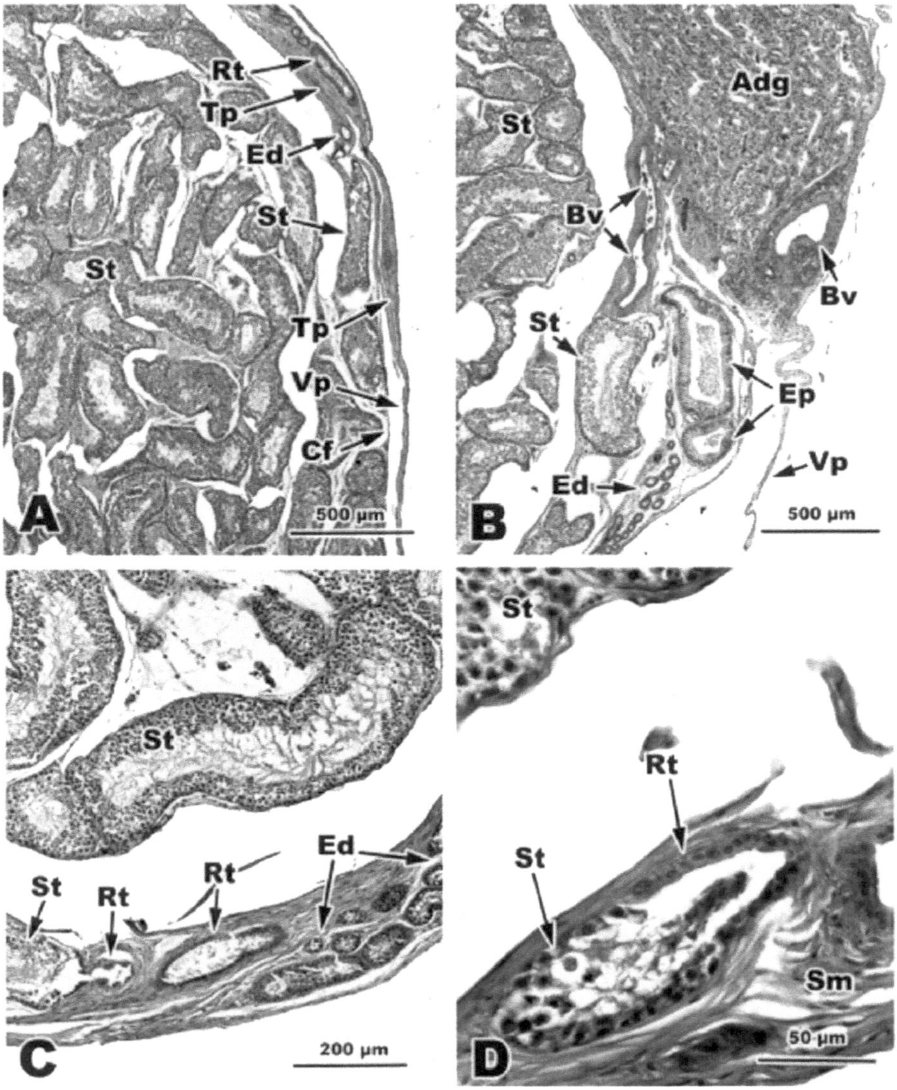

Fig. 10.2 Light micrographs of proximal testicular ducts of *Agkistrodon piscivorus*, stained with hematoxylin and eosin. **A.** Overview of relationships among seminiferous tubules, rete testis, ductuli efferentes, and ductus epididymis. **B.** Extratesticular seminiferous tubules in the epididymal sheath with ductuli efferentes, ductus epididymis, and adrenal gland. **C,D.** Extratesticular rete testis evaginating from seminferous tubule. Adg, adrenal gland; Bv, blood vessel; Cf, collagen fibers; Ed, ductuli efferentes; Ep, ductus epididymis; Rt, rete testis; Sm, smooth muscle; St, seminiferous tubule; Tp, tunica propria; Vp, visceral pleuroperitoneum.

Color image of this figure appears in the color plate section at the end of the book.

The ductuli efferentes have both ciliated and secretory cells (Figs. 10.4, 10.5). Lumina are narrower than in the rete testis, and cilia seem to fill the space (Fig. 10.5). The supranuclear flocculent material is conspicuous in many secretory cells, and apocrine blebs cleave off into the lumen (see Sever 2010). Like the rete testis, intercellular canaliculi are narrow.

At the base of microvilli, coated pits form larger vesicles, endosomes (Fig. 10.6). Electron-dense lysosomes occur in the supra-nuclear regions, and the rough endoplasmic reticulum is well-developed. Seasonal variation again includes an increase of chromatin material in spring and summer, and also an increase in lysosomal activity.

The ductus epididymis is pseudostratified with non-ciliated, columnar principal cells and scattered basal cells (Fig. 10.7). Some of the columnar principal cells are wider than others, but a distinct "narrow cell" as in mammals was not recognized by Sever (2010), because all of the columnar cells seem similar in cytology. Microvilli are short and the elongate "stereocilia" observed in mammals are absent (Figs. 10.7, 10.10). The cytoplasm of the principal cells is characterized by basal, oval nuclei with peripheral chromatin and nucleoli, numerous mitochondria (especially basally), narrow intercellular canaliculi, enlarged cisternae of endoplasmic reticulum, numerous electron-dense primary lysosomes (especially apically), conspicuous apocrine blebs, and an abundance of small vesicles (Figs. 10.7-10.10). The basal cells do not appear metabolically active, and contain only a few scattered mitochondria.

The flocculent material seen in the ductuli efferentes is not as abundant in the ductus epididymis, but the supranuclear Golgi complexes bud off vesicles which contain a product similar in density to the flocculent material. The vesicles empty their contents in the apical cytoplasm to form the apocrine blebs (Fig. 10.10). Definite coated vesicles have not yet been reported, but, as noted above, lysosomes are common, and Sever (2010) illustrated vesicular inclusions in sparsely granulated membranous structures associated with rough endoplasmic reticulum. The only seasonal variation noted in *Seminatrix pygaea* is an increase in chromatic material in the nuclei and increased granular material in the lumina of post-breeding individuals from summer.

10.3 DUCTUS DEFERENS AND AMPULLA DUCTUS DEFERENTIS

10.3.1 Ductus Deferens

Review. In squamate reptiles, the ductus deferens is usually described as rather uniform in structure along its length and is the acknowledged organ of sperm storage (Fox 1952). Most anatomical studies on the ductus deferentia of squamates have been limited to gross morphology and light microscopy. The exceptions are ultrastructure studies by Sever (2004) on the snake *Seminatrix pygaea* and by Akbarsha *et al.* (2005) on the Fan-throated Lizard (*Sitana ponticeriana*), and several electron micrographs of the ductus

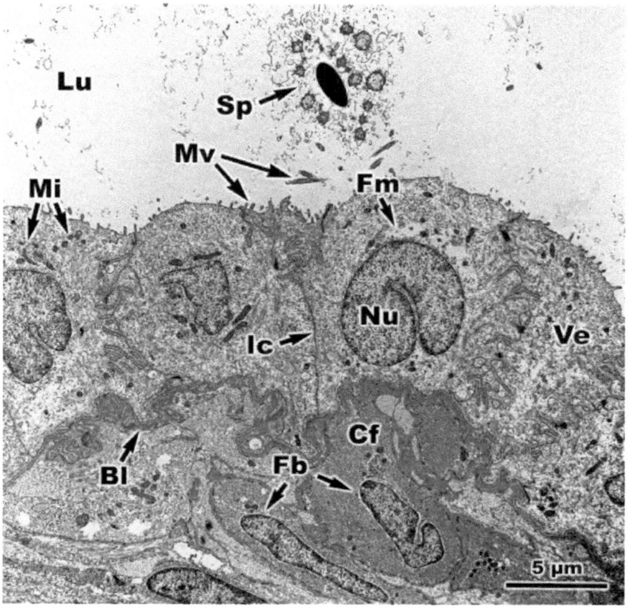

Fig. 10.3 Transmission electron micrograph of the rete testis of *Seminatrix pygaea*. Bl, basal lamina; Cf, collagen fibers; Fb, fibroblast nuclei; Fm, flocculent material; Ic, intercellular canaliculi; Lu, lumen; Mi, mitochondria; Mv, microvilli; Nu, nucleus; Sp, sperm; Ve, vesicles.

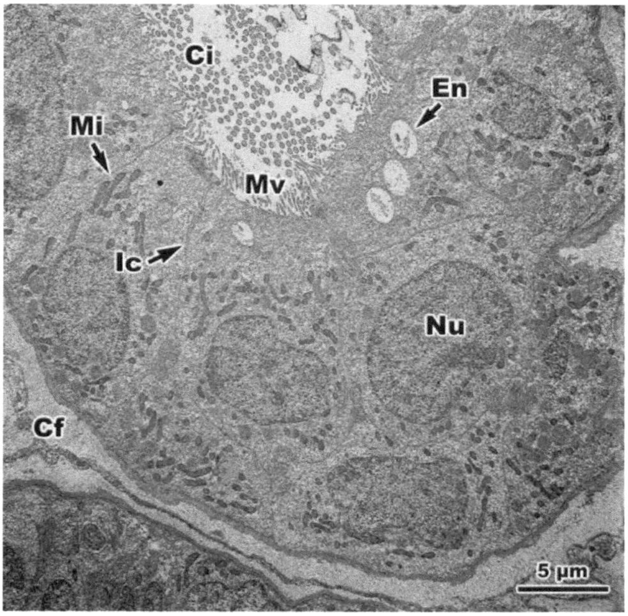

Fig. 10.4 Transmission electron micrograph of the ductuli efferentes of *Seminatrix pygaea*. Cf, collagen fibers; Ci, cilia; En, endocytic vacuole; Ic, intercellular canaliculi; Mi, mitochondria; Mv, microvilli.

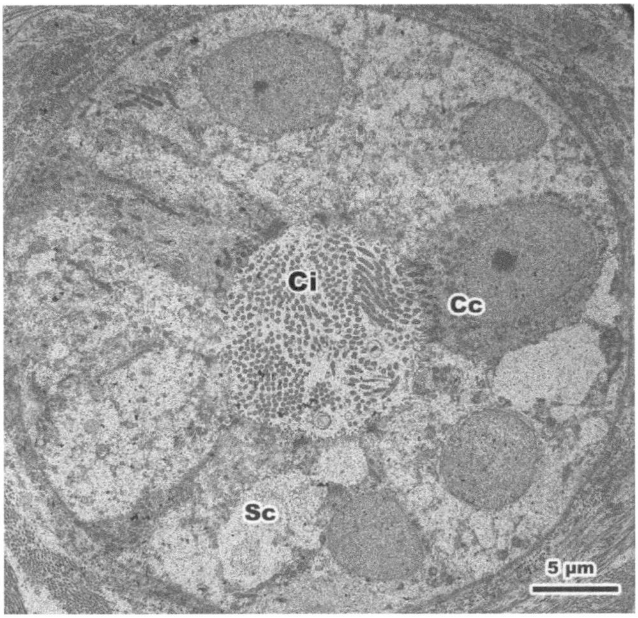

Fig. 10.5 Transmission electron micrograph of the ductuli efferentes of *Agkistrodon piscivorus*. Cc, ciliated cell; Ci, cilia; Sc, secretory cell.

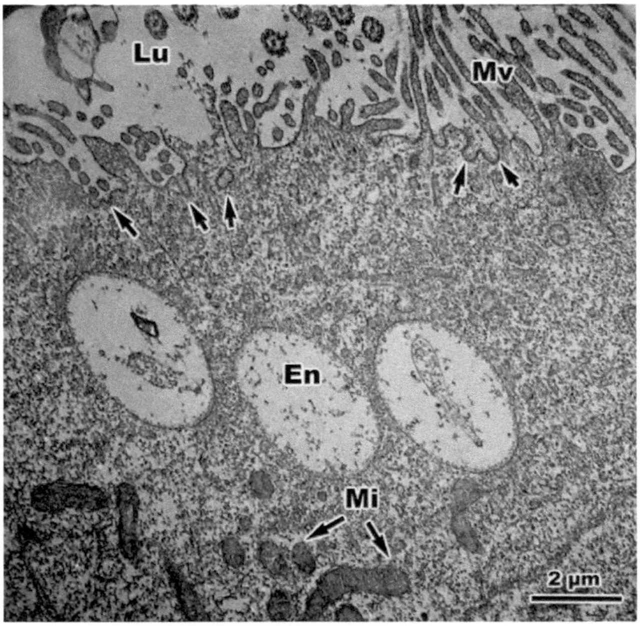

Fig. 10.6 Transmission electron micrograph of the apical cytoplasm of the ductuli efferentes of *Seminatrix pygaea*. Unlabeled arrows indicate coated vesicles. En, endocytic vacuoles; Lu, lumen; Mi, mitochondria; MV, microvilli.

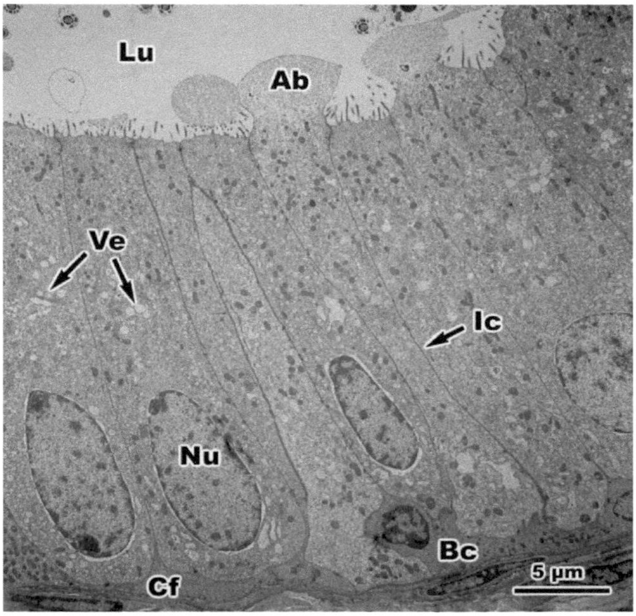

Fig. 10.7 Transmission electron micrograph of the ductus epididymis of *Seminatrix pygaea*. Ab, apocrine bleb; Bc, basal cell; Cf, collagen fibers; Ic, intercellular canaliculi; Lu, lumen; Nu, nucleus; Ve, vesicles.

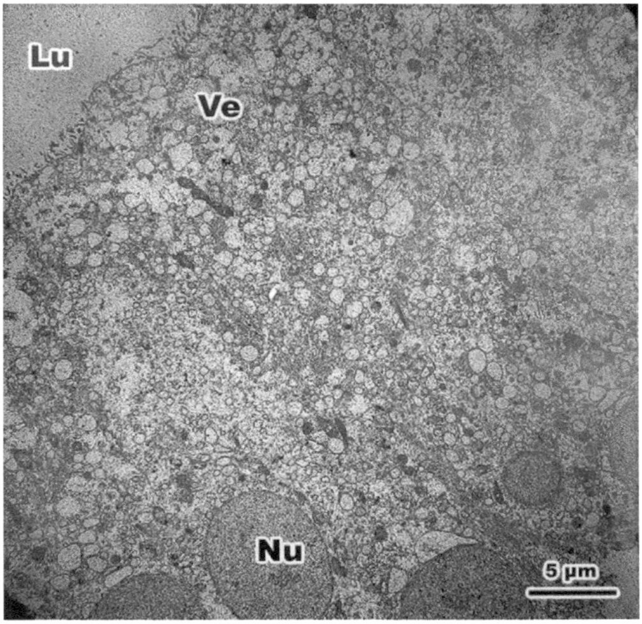

Fig. 10.8 Transmission electron micrograph of the apical cytoplasm of the ductus epididymis of *Agkistrodon piscivorus*. Lu, lumen; Nu, nucleus; Ve, vesicles.

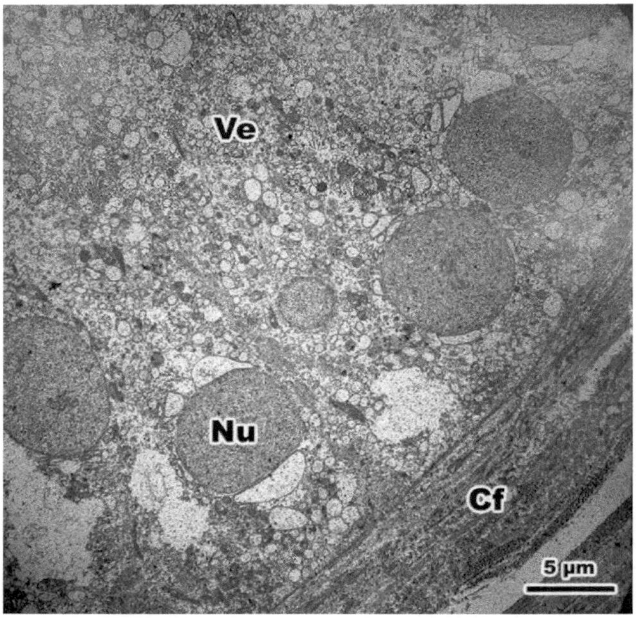

Fig. 10.9 Transmission electron micrograph of the basal cytoplasm of the ductus epididymis of *Agkistrodon piscivorus*. Cf, collagen fibers; Nu, nucleus; Ve, vesicles.

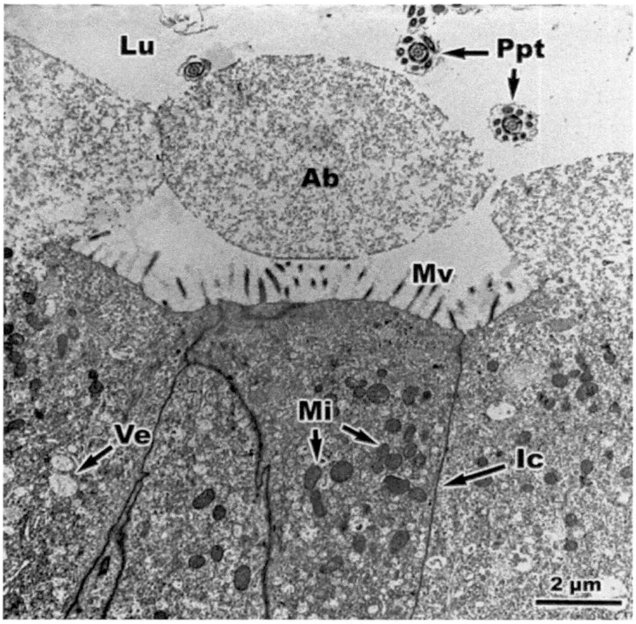

Fig. 10.10 Transmission electron micrograph of the apical cytoplasm of the ductus epididymis of *Seminatrix pygaea*. Ap, apocrine bleb; Ic, intercellular canaliculus; Lu, lumen; Mi, mitochondria; Mv, microvilli; Ppt, principal piece of the tail; Ve, vesicles.

deferens of *Agkistrodon piscivorus* illustrated by Siegel *et al.* (2009). With light microscopy, neither Volsøe (1944) in *Vipera* nor Fox (1952) in *Thamnophis* sp. reported secretory activity in the ductus deferens. Sever (2004) found that the ductus deferens of *S. pygaea* reacted negatively to histochemical tests for carbohydrates and proteins, but Siegel *et al.* (2009) found some positive reactions with these tests for *A. piscivorus*, as indicated below.

The ductus deferens is formed into short tight loops (not coils) as it passes caudally. The demarcation between epididymis and ductus deferens is gradual, with the lumen of the ductus becoming wider, and the epithelium becoming lower. Like the epididymis, the ductus deferens is pseudostratified (Figs. 10.11, 10.12). The epithelium of the principal cells varies from cuboidal to low columnar, and the rather scant cytoplasm is basophilic. As in the epididymis, apocrine blebs are present (Fig. 10.11B). Golgi bodies, rough endoplasmic reticulum, and vesicles are numerous in the cytoplasm (Figs. 10.10C, 10.12A,B).

Sever (2004) in *Seminatrix pygaea* and Siegel *et al.* (2009) in *Agkistrodon piscivorus* reported presence of copious sperm in the ductus deferentia of specimens collected throughout the year. Secretory material associated with the sperm mass in *A. piscivorus* is eosinophilic, bromphenol blue positive (for proteins), and PAS positive (for neutral carbohydrates). These reactions are most intense in areas around the border of the sperm mass that lack sperm, and the secretory material is globular (Siegel *et al.* 2009). Scattered reactions to alcian blue (glycosaminoglycans) occur around the sperm mass in specimens collected in all seasons, and numerous vacuoles are found between the apical cytoplasm and the luminal secretory material in some specimens. Like *S. pygaea*, much of the cytoplasm of the ductus deferens seems non-reactive to histochemical tests in *A. piscivorus*, but in all seasons PAS positive, alcian blue positive, and bromophenol blue positive material is located in the apical cytoplasm. Transmission electron microscopy reveals the presence of secretory vacuoles containing a diffuse material, characteristic of carbohydrates (Fig. 10.13).

New observations. We examined the distal urogenital tracts of several North American colubroid species and provide new micro-anatomical information below on the ductus deferens of adult snakes. Among the representative snakes investigated, we report on the histological and ultrastructural features of the following species: Northern Scarletsnake (*Cemophora coccinea copei*), Western Diamond-backed Rattlesnake (*Crotalus atrox*), Prairie Ringneck Snake (*Diadophis punctatus arnyi*), Western Mudsnake (*Farancia abacura reinwardtii*), Texas Ratsnake (*Pantherophis obsoletus*), Western Pygmy Rattlesnake (*Sistrurus miliarius streckeri*), and the Eastern Gartersnake (*Thamnophis sirtalis sirtalis*). Segments of the urogenital tracts in each species were extracted from a region between 5-10 mm anterior to the cloacal cavity.

We found ductal morphologies in two of these snakes congruent with our previous descriptions above for *Seminatrix pygaea*. Among the species whose ductal morphology was most similar to the ductus deferens of

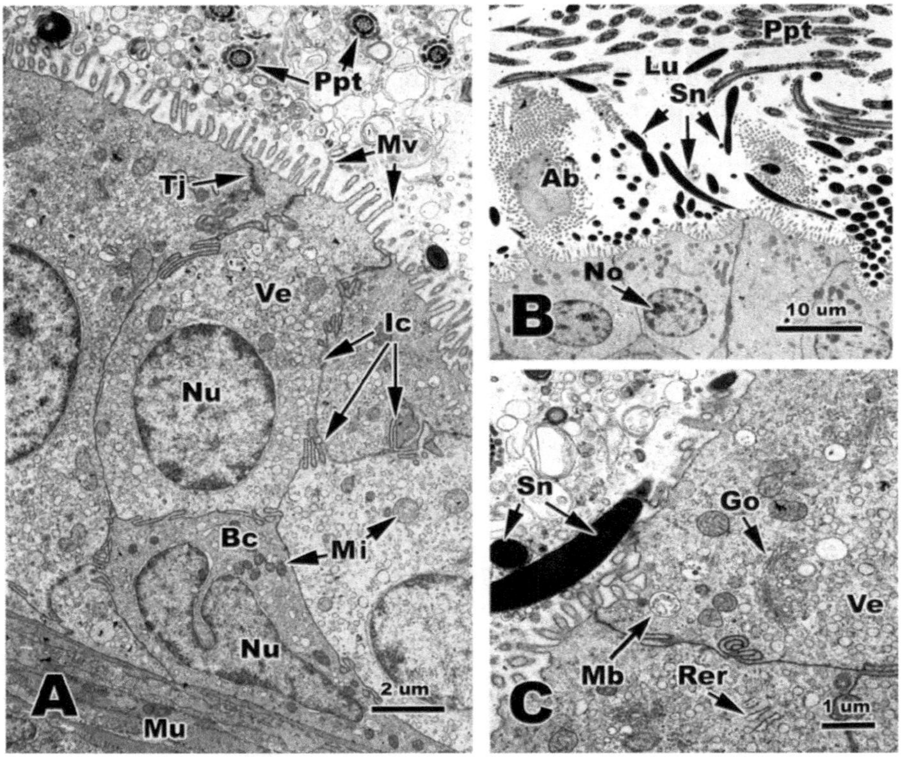

Fig. 10.11 Transmission electron micrographs of the ductus deferens of *Seminatrix pygaea*. **A.** Overview of epithelium. **B,C.** Apical cytoplasm. Ab, apocrine bleb; Bc, basal cell; Go, Golgi bodies; Ic, intercellular canaliculi; Lu, lumen; Mb, multivesicular body; Mi, mitochondria; Mu, muscularis; Mv, microvilli; No, nucleolus; Nu, nucleus; Ppt, principal piece of the tail; Rer, rough endoplasmic reticulum; Sn, sperm nuclei; Tj, tight junction; Ve, vesicles.

S. pygaea was the micro-anatomy found in *Diadophis punctatus arnyi*, which exhibits a pseudostratified, columnar epithelium consisting of a mixture of tall-to-medium height principal cells, narrow cells, and low basal cells (Fig. 10.14D). The epithelial cell cytoplasms exhibit a combination of basophilic and eosinophilic staining properties in *D. punctatus arnyi*. *Sistrurus miliarius streckeri* also possesses a ductal pseudostratification, but, in addition, revealed patches of simple columnar cells (Fig. 10.14F).

Crotalus atrox has a low columnar epithelium (Fig. 10.14G,H) different from that of *Sistrurus miliarius* and more similar to the ductus deferens of *Cemophora coccinea copei* and *Thamnophis sirtalis sirtalis*, which represent the simplest form anatomically and histologically. In the latter two species, the duct is highly looped, greatly distended, and exhibits a thin epithelium with short or flattened cells similar to that illustrated by Fox (1952) in the Coast Gartersnake (*T. elegans terrestris*). The low columnar cells of the epithelial lining stain basophilically in both *C. coccinea copei* and *T. s. sirtalis* (Fig. 10.14A-C) as well as in *C. atrox* (Fig. 10.14G,H). One notable difference

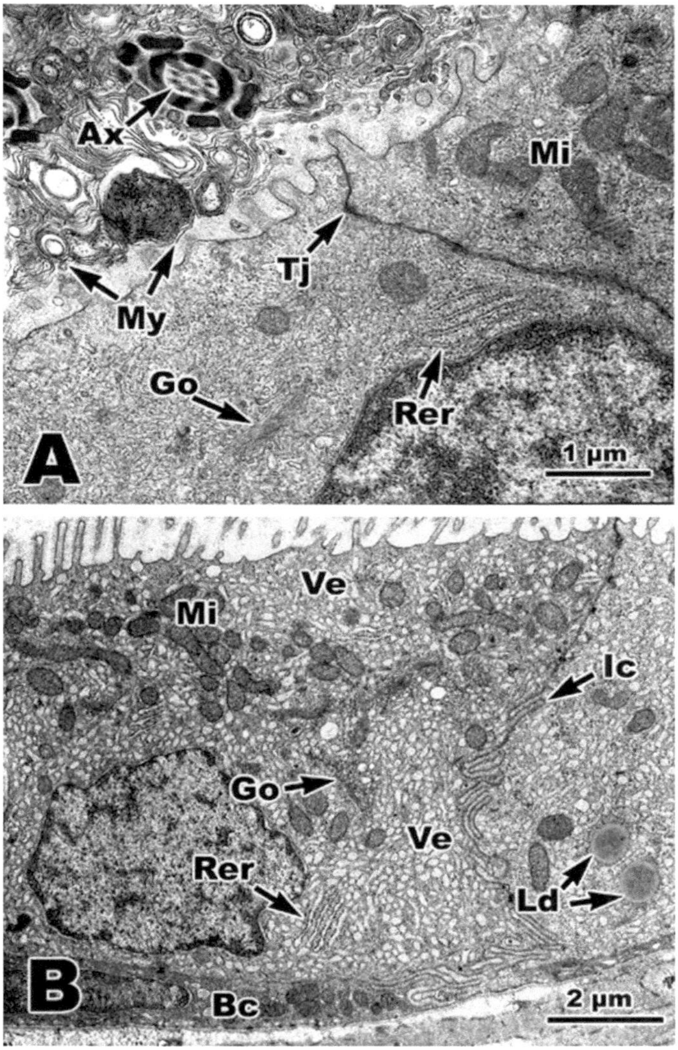

Fig. 10.12 Transmission electron micrographs of the ductus deferens of *Seminatrix pygaea*. **A.** Apical cytoplasm. **B.** Epithelium with numerous vesicles. Ax, axoneme; Bc, basal cell; Go, Golgi bodies; Ic, intercellular canaliculus; Ld, lipid droplets; Mi, mitochondria; My, myelinic figures; Rer, rough endoplasmic reticulum; Tj, tight junction; Ve, vesicles.

is in the thickness of the ductal muscularis, which is greater in diameter in *T. s. sirtalis* than in *C. coccinea copei* (Fig. 10.14A-C). Not illustrated is the relatively thick ductal muscularis in *C. atrox*, which is similar to the relative thickness of the ductal wall of minute typhlopids (Fox 1965). Basophilic ductal epithelia also occur in both viperids, *C. atrox* and *S. miliarius streckeri* (Fig. 10.14F-H). We found an eosinophilic epithelium in *Farancia abacura*

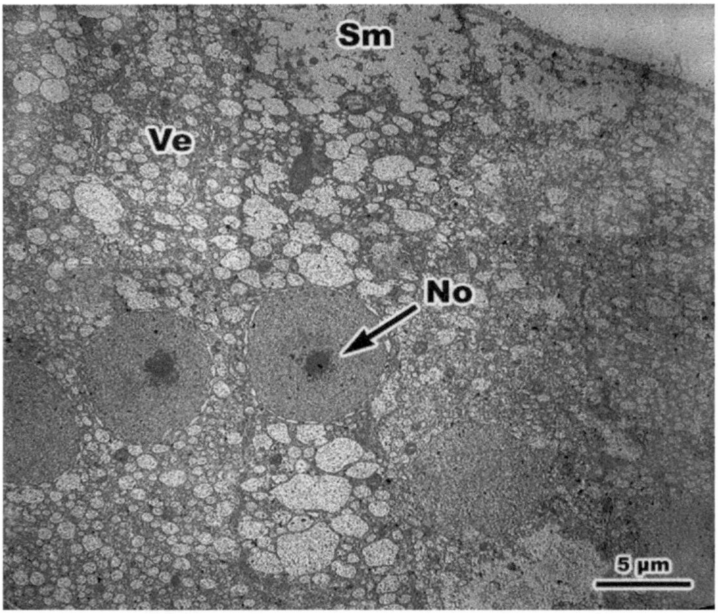

Fig. 10.13 Transmission electron micrograph of the ductus deferens of *Agkistrodon piscivorus*. No, nucleolus; Sm, secretory material; Ve, vesicles.

reinwardtii, and the epithelium has large intercellular spaces along the basal lamina of the ductal epithelium (Fig. 10.14E).

The secretory nature of the ductus deferens displayed in Fig. 10.14 is unclear, primarily because we utilized no specific histological staining tests for carbohydrates and proteins in these species. We suspect, however, that some secretory activity may occur in *Diadophis punctatus arnyi* and *Farancia abacura reinwardtii* based upon the eosinophilic staining properties of their ductal epithelia using general cytological stains.

The ductus deferens of *Pantherophis obsoletus* provided an opportunity to investigate, in some detail, both the ductal morphology and secretory activity of a colubrine species. The ductal secretory activity of a reproductively active individual is illustrated in Fig. 10.15. Large apocrine cytoplasmic blebs dominate the ductal epithelial lining in this species. These cytoplasmic blebs exhibit eosinophilia (Fig. 10.15A) and ultrastructurally appear to be either passively released by cleaving the apical region of cells or are pinched off *en masse* as membrane-bound globular blebs (Fig. 10.15B). These apocrine blebs are similar to those noted in the ductus epididymis and ductus deferens of *Seminatrix pygaea* (Figs. 10.7-10.11). The secretory epithelium of *P. obsoletus* exhibits a low columnar epithelium and lacks stereocilia. The apical cytoplasm as well as the apical blebs found in *P. obsoletus* contain secretory vacuoles (Fig. 10.15B) similar to the description of those found in the viperid, *Agkistrodon piscivorus* (Fig. 10.13).

10.3.2 Ampulla Ductus Deferentis

Review. In many mammals, the distal end of the ductus deferens differs histologically from more proximal portions and is called the ampulla. In contrast to the rest of the ductus deferens, the epithelium of the ampulla is deeply folded, and tortuous out-pocketings that arise between the folds extend into the surrounding muscularis and possess secretory cells (Setchell *et al.* 1994). The mammalian ampulla is morphologically similar to the seminal vesicles (Riva *et al.* 1982) and has been variously implicated in maturation, nourishment, storage, and phagocytosis of sperm (Cooper and Hamilton 1977; Bergerson *et al.* 1994). Among squamates, the occurrence of an ampulla has been documented histologically in two species of lizards, the Oriental Garden Lizard (*Calotes versicolor*; Akbarsha and Meeran 1995) and *Sitana ponticeriana* (Akbarsha *et al.* 2005), and two species of snake, *Seminatrix pygaea* (Sever 2004) and *Agkistrodon piscivorus* (Siegel *et al.* 2009).

Sever (2004) found that the distal end of the ductus deferens of the snake *Seminatrix pygaea* differs from more proximal portions of the genital tract by possessing a highly fluctuated epithelium. As noted above, *S. pygaea* stores sperm throughout the ductus deferens during the entire year, and this includes the ampullary portion. Largely on the basis of negative histochemical tests, Sever (2004) reported that the epithelium of the ampulla of *S. pygaea* is not secretory but noted that the numerous small apical vesicles could function in fluid absorption. In addition to the highly folded epithelium, clusters of sperm nuclei are more intimately associated with the apical ampullary epithelium than elsewhere in the ductus deferens, but no evidence of phagocytosis of sperm was found. In contrast, the ampulla of mammals is glandular and phagocytic. Sever (2004) concluded that the only character shared by *S. pygaea* and mammals is the folded epithelium.

Akbarsha *et al.* (2005) subsequently conducted an ultrastructural study on the ampulla of *Sitana ponticeriana* (Agamidae). The ampulla of *S. ponticeriana* is differentiated into storage and glandular portions. The epithelium of the storage portion is like that in the ductal portion of the ductus deferens, although wider in diameter. The cells of the ductus deferens and storage region of the ampulla consist of principal cells (involved in endocytosis and phagocytosis of dead sperm), dark cells (characterized by numerous coated apical pits and membrane bound vesicles in the cytoplasm), and basal cells (which appear inactive). The principal cells of the storage region are broader than tall. The glandular portion consists of dark and light principal cells and foamy cells, all of which are tall and form villous folds. All three types are secretory, with the products likely different from one another. No evidence of phagocytic activity was found in the glandular portion.

Akbarsha *et al.* (2005) suggested that the ampulla of agamid lizards is a composite gland of the ampulla ductus deferentis and seminal vesicles of mammals. They proposed that within the Squamata, the ampulla differs

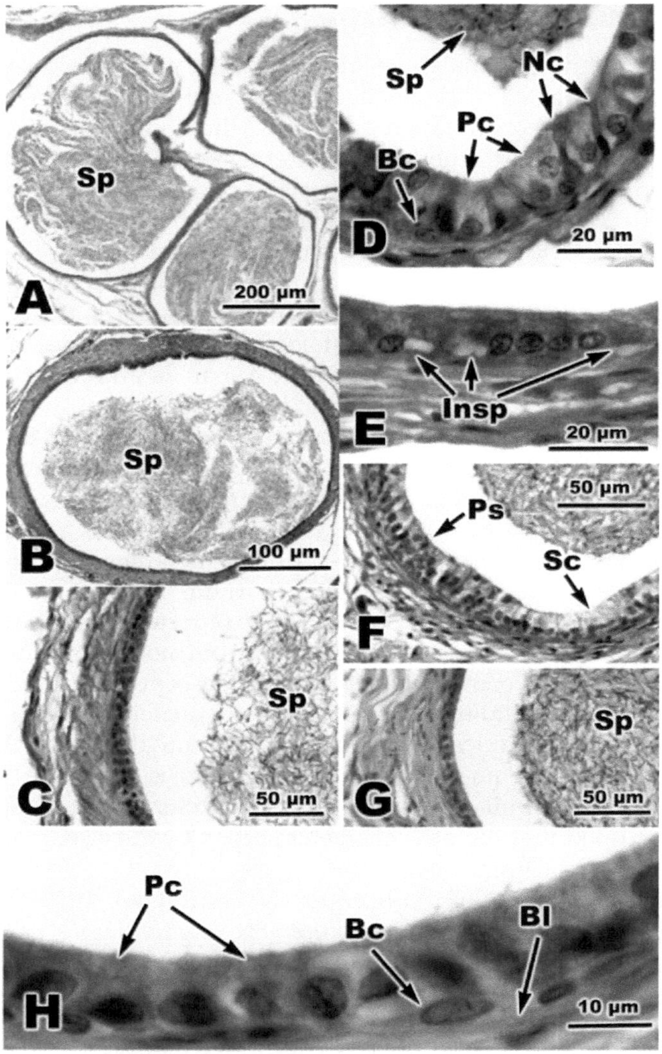

Fig.10.14 Light micrographs of the distal region of the ductus deferens and its secretory epithelium in representative male North American colubroid snakes, stained with hematoxylin and eosin. **A.** *Cemophora coccinea copei*, transverse section revealing a convoluted region of the ductus deferens, which are packed with spermatozoa (Sp). The secretory epithelium is reduced and rests upon a thin ductal wall. **B.** *Thamnophis sirtalis sirtalis*, transverse section of the ductus deferens reveals a thickened muscular wall; a mass of spermatozoa lies within its lumen. **C.** Higher magnification of C showing a thin layer of low columnar cells. **D.** *Diadophis punctatus arnyi*, portion of secretory epithelium showing principal cells (Pc), narrow cells (Nc), and basal cells (Bc). **E.** *Farancia abacura reinwardtii*, portion of the wall of the ductus deferens revealing low columnar cells. Intercellular spaces (Insp) interspersed along basal lamina. **F.** *Sistrurus miliarius streckeri*, showing regional variation of the ductus deferens with area of simple columnar cells (Sc) versus region of pseudostratified columnar cells (Ps).

Fig. 10.14 Contd. ...

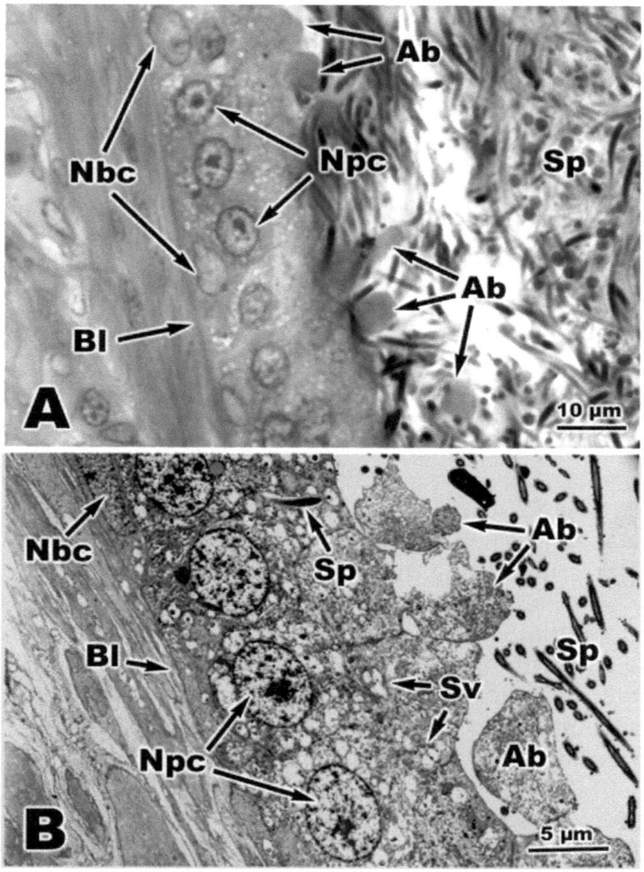

Fig. 10.15 The ductus deferens and its secretory epithelium in *Pantherophis obsoletus*. **A.** Light micrograph of the low columnar epithelium revealing numerous cytoplasmic apocrine blebs (Ab) lying adjacent to epithelial cells and within the ductal lumen. Principal cells with their oval-shaped nuclei (Npc) outnumber basal cells which exhibit oblong nuclei (Nbc). Spermatozoa are crowded along the epithelial lining. Ladd multiple stain of plastic-embedded tissue. **B.** Transmission electron micrograph of epithelium shown in A. Apocrine blebs can be seen detaching from the apical surfaces of the principal cells. Numerous secretory vacuoles (Sv) can be seen at several levels within these cells. Note sperm embedded within cytoplasm of principal cell. Ab, apocrine bleb; Bl, basal lamina; Nbc, nuclei of basal cell; Npc, nuclei of principal cell; Sp, sperm; Sv, secretory vacuoles.

Color image of this figure appears in the color plate section at the end of the book.

... Fig. 10.14 Contd.

G. *Crotalus atrox*, low simple columnar epithelium rests upon a thickened ductal wall (similar to B). **H.** Higher magnification of G. Epithelium greatly reduced in height as in A; basal cells (Bc) reside along the basal lamina (Bl). Narrow cells, which are common in D, were not observed. Bc, basal cells; Bl, basal lamina; Insp, intercellular space; Nc, narrow cells; Pc, principal cells; Ps, pseudostratified; Sc, simple columnar; Sp, sperm.

Color image of this figure appears in the color plate section at the end of the book.

in structure and function. Akbarsha *et al.* (2005) found additional cell types (dark and foamy cells) in the ampulla of *Sitana ponticeriana* not noted by Sever (2004) in *Seminatrix pygaea*. More importantly, three cell types in the lizard were secretory in the glandular portion of the ampulla, and no secretory activity was reported for *S. pygaea.*

Siegel *et al.* (2009) reported that the ampulla of *Agkistrodon piscivorus* is composed of a portion suspended in the posterior part of the pleuroperitoneal coelom alongside the ureter and a segment that passes through the cloacal walls to join with the ureter. Externally, the coelomic ampulla is discerned by a straightening of the loops from the ductus deferens. Internally, the ampulla is characterized by irregular folded epithelial walls until the final intramural portion when the epithelium becomes regular again as the junction with the ureter is approached. Sperm occur in the ampulla of *A. piscivorus* throughout the year, although sperm are scant in October and November specimens.

Siegel *et al.* (2009) reported that the cytoplasm of the anterior portion of the ampulla of *Agkistron piscivorus* is characterized by numerous vacuoles, and the vacuoles also occur between the apices of the epithelial cells and the secretory material encasing the sperm mass. Throughout the year, luminal material is eosinophilic and intensely PAS positive, and small PAS positive granules are scattered throughout the epithelium. Positive reactions with alcian blue and with bromphenol blue are also observed in the cytoplasm and luminal material, but these reactions are not as strong and pervasive as those for PAS. Siegel *et al.* (2009) present electron micrographs that show the presence of apical secretory vacuoles containing a diffuse substance and a dense matrix containing sperm in the lumen.

Thus, in one snake, *Seminatrix pygaea*, no secretory activity is reported for the ampulla (Sever 2004), whereas in another, *Agkistrodon piscivorus*, ample evidence for production of a carbohydrate secretion exists (Siegel *et al.* 2009). It is likely, as in the proximal testicular ducts, that the numerous vesicles in the ampullary cytoplasm of *S. pygaea* do not merely function in endocytosis of luminal fluids. Instead the vesicles may be indicative of a constitutive secretory pathway in which the secretory product is not stored in granules but is continually produced and released through these vesicles. Still, the absence of positive histochemical tests indicate that the ampulla of *S. pygaea* may have less secretory activity than that found in *A. piscivorus*, and some of the species described below.

New observations. We report below on new histological and ultrastructural information pertaining to the ampulla ductus deferentis in several adult North American colubrine snakes; these include the following species (Figs. 10.16-10.19): *Cemophora coccinea copei*, Southern Black Racer (*Coluber constrictor priapus*), Speckled Kingsnake (*Lampropeltis holbrooki*), Mississippi Green Watersnake (*Nerodia cyclopion*), Rough Greensnake (*Opheodrys aestivus*), *Pantherophis obsoletus*, Midland Brownsnake (*Storeria dekayi wrightorum*), Flat-headed Snake (*Tantilla gracilis*), and Rough Earthsnake (*Virginia striatula*).

As in *Seminatrix pygaea*, diagnostic morphological characteristics of the ampulla found in seven of the eight snake species listed above are the highly fluctuating epithelium (Fig. 10.16) and a greatly enlarged ampullary diameter (a diameter which is greater than that of the associated ductus deferens). We found that the complexity of the epithelial folding of the ampulla varies sharply among the species examined. For example, a simple set of 4-6 low, non-villous folds occur in the ampulla of *Storeria dekayi wrightorum* (Fig. 10.17A), whereas *Nerodia cyclopion* possessed the most complex arrangement of ampullary folds (Fig. 10.16G). In *S. dekayi*, the region possessing the ampullary folds is restricted to a very narrow segment of the ampulla (less that 100 μm in total length), and this portion of the ductal region appears rudimentary in overall structure. For instance, this ampulla contains poorly developed ampullary crypts between the folds. In addition, a low pseudostratified, columnar epithelium lines the ampullary crypts and the entire ampullary luminal lining.

The comparative micro-anatomy of the internal ampullary wall (including folds, irregular projections, and crypts) between two species, *Coluber constrictor priapus* and *Opheodrys aestivus* (Fig. 10.16C,F), reveals a great similarity in ampullary duct morphology. This condition represents a higher level of anatomical complexity over that seen in *Storeria dekayi wrightorum*. In the ampulla of these two species, the ductal epithelium exhibits mostly low villous or non-villous folds, which are separated by weakly-defined ampullary crypts (Fig. 10.17B). These folds are generally less numerous and reduced in hillock height when compared to the folds found by Sever (2004) in *Seminatrix pygaea*. The low pseudostratified columnar epithelium of *C. constrictor priapus* did, however, show some evidence of secretory activity as the apical portions of these hillock cells are eosinophilic when stained with hematoxylin and eosin.

A comparison of another pair of species, *Tantilla gracilis* and *Virginia striatula*, shows near identical ductal morphologies with one another (Fig. 10.16D,E). Both species possess relatively tall epithelial hillocks or projections and well developed ampullary crypts. None of the crypts houses any sperm aggregates, although some sperm appear captured within apical cells (Fig. 10.17D). The apical ampullary epithelia (Fig. 10.17D,E) of these two species are more similar to that of *Seminatrix pygaea* than to the micro-anatomies of either *Coluber constrictor priapus* or *Opheodrys aestivus*. Unlike the longitudinally-directed ampullary duct morphology of *V. striatula* (and all the other species mentioned thus far), the ampulla of *T. gracilis* appears as a looped cluster of ducts, and they contain oblong plate-like esosinophilic masses within their lumina (Fig. 10.16D).

A third level of micro-anatomical ampullary complexity occurs in *Cemophora coccinea copei* (Fig. 10.16A). In this species, the ampullary apical epithelium exhibits mostly tall to medium-sized hillocks (irregular epithelial projections) that are festooned (Fig. 10.17F), an indication of possible endocytic activity. Apical cell nuclei, however, appear to be undergoing pyknosis, and cytoplasmic fragmentation occurs, resulting in

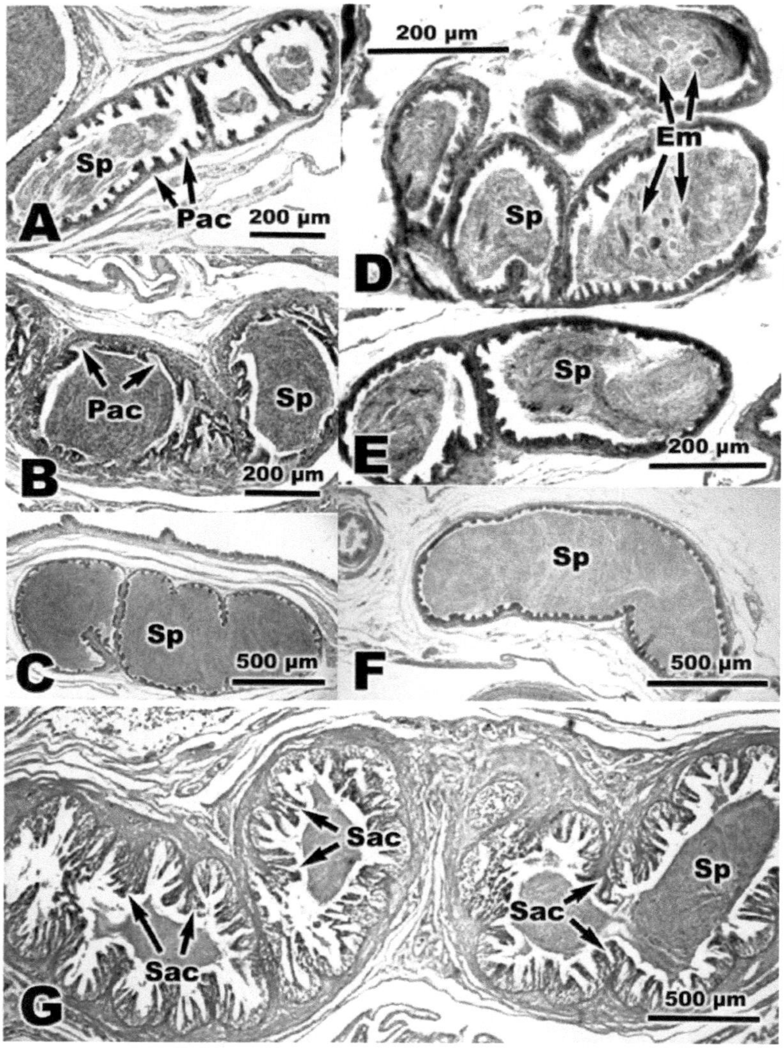

Fig. 10.16 Light micrographs of the ampulla ductus deferentis in adult male North American colubrine snakes, stained with hematoxylin and eosin. All histosections contain densely packed masses of spermatozoa. **A.** *Cemophora coccinea* copei, ampullary epithelial folds exhibit a repetitively festooned appearance with primary ampullary crypts (Pac). **B.** *Lampropeltis holbrooki.* **C.** *Coluber constrictor priapus,* expansive ampullary duct with reduced folds and crypts. **D.** *Tantilla gracilis*, showing multiple loops of the ampulla ductus deferentis. Dense eosinophilic masses (Em) are embedded among spermatozoa. **E.** *Virginia striatula*, with ampulla ductus deferentis similar to morphology found in D. **F.** *Opheodrys aestivus*, ampullary epithelium similar to morphology found in C. **G.** *Nerodia cyclopion*, ampullary epithelial lining highly convoluted and highly branched exhibiting primary and secondary (Sac) ampullary crypts. Em, eosinophilic masses; Pac, primary ampullary crypts; Sac, secondary ampullary crypts; Sp, sperm.

Color image of this figure appears in the color plate section at the end of the book.

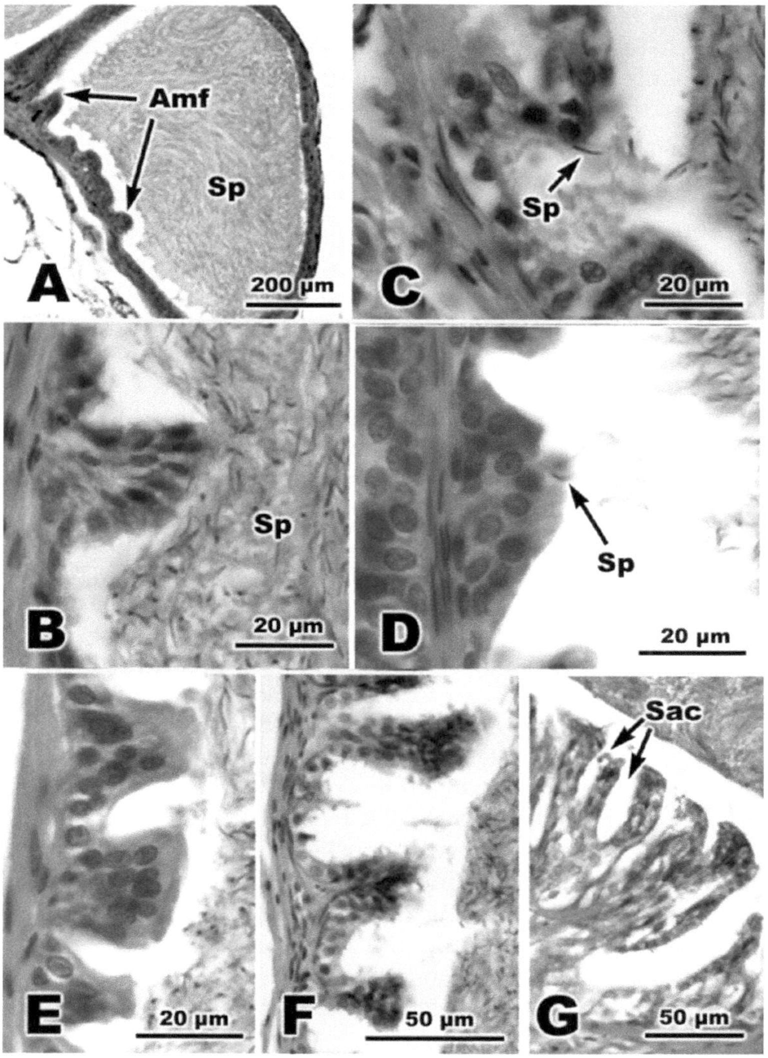

Fig. 10.17 Light micrographs of the ductal epithelium of the ampulla ductus deferentis in adult male colubrine snakes (B-G histosections from Fig. 10.16), stained with hematoxylin and eosin. **A.** *Storeria dekayi*, ductal epithelium has low to flattened columnar cell arrangement with few villous internal ampullary folds (Amf). **B.** *Coluber constrictor priapus*, ductal epithelium elevated, but without festooned appearance. **C.** *Lampropeltis holbrooki*, ampullary crypt exhibiting sperm cell (Sp) embedded within epithelial cell. **D.** *Virginia striatula*, portion of ductal epithelium exhibiting pyramidal-shaped cell layers flanked by narrow ampullary crypts. Note sperm cell embedded with epithelial cell (similar to C). **E.** *Tantilla gracilis*, the ductal epithelium comprised truncated cell clusters residing between narrow ampullary crypts. **F.** *Cemophora coccinea copei*, laterally expanded ampullary crypts lie between tall columns of the ductal epithelium. **G.** *Nerodia cyclopion*, tentacle-like branching pattern of the ductal epithelium showing secondary ampullary crypt. Amf, ampullary folds; Sac, secondary ampullary crypts; Sp, sperm.

Color image of this figure appears in the color plate section at the end of the book.

cell surfaces which lack uniformity (i.e., exhibited uneven boundaries). A basophilia characterizes these distal hillock cell layers, and one could also argue that some exocytic secretory pathway is actively taking place.

The fourth level of ampullary complexity occurs in *Lampropeltis holbrooki* (Fig. 10.16B). In this species, the ampullary folds are wider and taller than in any of the previous species; thus, the ampullary crypts are also enlarged. The ductal wall in *L. holbrooki* (Fig. 10.16B) is thicker than in any of the previous species. Unlike the ampullary crypts of the storage region in *Seminatrix pygaea*, few sperm are found within crypts, although sperm are embedded into cells along the apical hillocks (Fig. 10.17C).

The most complex ampullary micro-anatomy is found in *Nerodia cyclopion*. The entire ampullary lining has numerous large primary ampullary folds (Fig. 10.16G), which form a repetitive cup-like pattern along the ampullary ductal wall. Interspersed along the primary ampullary folds are individual as well as unique, tentacle-like arrangements of secondary ampullary folds, the latter characterized by secondary ampullary crypts (Fig. 10.17G). The apical epithelial cell regions of the secondary ampullary folds shows a mild basophilia, but not as intense as that in the apical cell clusters of *Cemophora coccinea copei* and *Lampropeltis holbrooki*. No spermatozoa are apparent within any of the primary or secondary ampullary crypts in *N. cyclopion*.

As was the case with our examination of the ductus deferens of snakes aforementioned, we were able to examine the ampulla ductus deferentis of *Pantherophis obsoletus* in greater micro-anatomical detail than other snake species. The ampulla of *P. obsoletus* represents another degree of complexity (differing from all other species) by exhibiting a gland-like appearance. The ampulla consists of a lobate cluster of ducts (Fig. 10.18A) along with a multitude of blind outpocketings or pouches that terminate as ampullary crypts. These ampullary crypts also differ from the other species examined by containing dense aggregates of sperm cells whose nuclei occur in bundles within the crypts (Fig. 10.18B). Transmission electron microscopy reveals, however, that the sperm heads are not embedded into the epithelial cell lining of the crypts (Fig. 10.19). This same observation was mentioned by Sever (2004) for *Seminatrix pygaea*. Cytologically, principal cells within crypts contain scattered patches of rough endoplasmic reticulum and mitochondria along the apical surfaces of their nuclei (Fig. 10.19); also, no secretory vacuoles are evident within these cells. A comparison with epithelial cells of the ductus deferens (Fig. 10.15) shows striking differences. Secretory vacuoles that dominate cytoplasmic volume in the secretory epithelium of the ductus deferens are absent from the ampullary epithelium. Although the ductal epithelium within crypts remains pseudostratified columnar throughout most of the ampulla, the irregular height of the principal cells within crypts as well as the presence of apocrine blebs within the luminal region of the crypts point to the possibility of secretory activity.

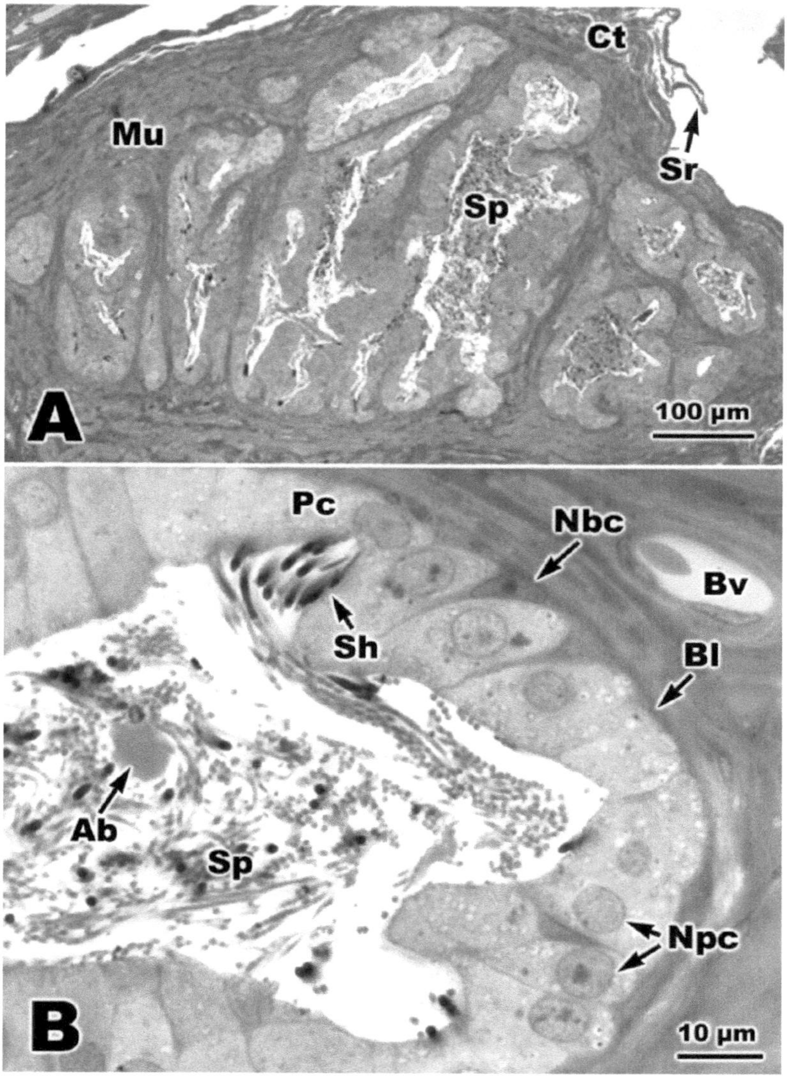

Fig. 10.18 Light micrographs of the ampulla ductus deferentis of an adult *Pantherophis obsoletus*, stained with the Ladd multiple stain. **A.** Section through several coils of the ampulla revealing a tall-to-irregular columnar epithelium; spermatozoa fill the lumina of the coils. The body of the ampulla is surrounded by a muscularis (Mu) and a connective tissue sheath (Ct). A serosa lines the external surface of the ampulla. **B.** Higher magnification of a crypt with the irregular columnar epithelium shown in A. A cluster of sperm heads (Sh) can be seen lying within a depressed space between principal cells (Pc), which contain mostly round nuclei (Npc). A basal cell and its nucleus (Nbc) appear tightly squeezed between two principal cells. Ab, apocrine bleb; Bl, basal lamina; Bv, blood vessel; Ct, connective tissue; Mu, muscularis; Nbc, nuclei of basal cells; Npc, nuclei of principal cells; Pc, principal cell; Sh, sperm heads; Sp, sperm; Sr, serosa.

Color image of this figure appears in the color plate section at the end of the book.

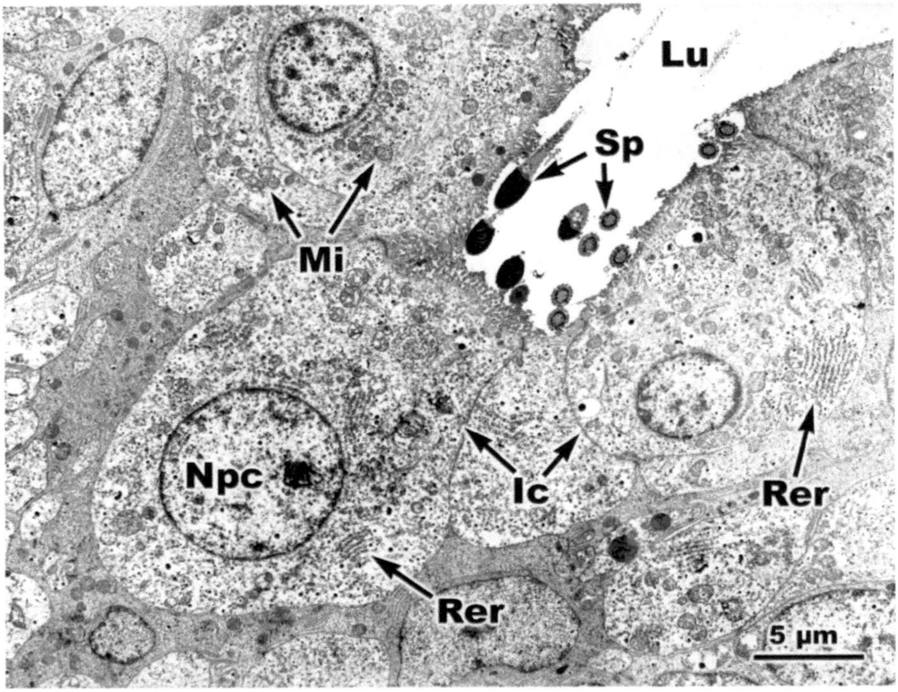

Fig. 10.19 Transmission electron micrograph of an epithelial crypt of the ampulla ductus deferentis (shown in Fig. 10.18) of *Pantherophis obsoletus*. The borders of several principal cells surround a cluster of spermatozoa. Ic, intercellular canaliculi; Lu, lumen; Mi, mitochondria; Npc, nucleus of principal cell; Rer, rough endoplasmic reticulum; Sp, sperm.

10.3.3. Ampulla Urogenital Papilla and Ampulla Ureter

Review. In this section, we recognize the significance of a unique anatomical illustration provided by Martin Saint Ange (1854). We have redrawn it in Fig. 10.20 in order to help clarify our explanations below about snake distal urogenital tract anatomy. Depicted within this figure (two-headed arrow) is a cutaway internal view of the distal expansion of the ureter of a colubrine snake in which Martin Saint Ange inserted tracts of both the ureter proper (as the small anterior tube) and the internal ductus deferens (the small posterior tube) configured from its lateral position. Henceforth, we shall refer to this expanded portion housing the two urogenital ducts as the ampulla urogenital papilla, a new terminology applied to this urogenital region of male colubrine snakes.

Much uncertainty has prevailed in the snake literature dealing with the distal reproductive anatomy of males, especially in reference to the lower urogenital tracts (e.g., see Sever 2004). A clear understanding of distal urogenital tract morphology has been hampered primarily by arguments and descriptions by Volsøe (1944), who equated the distal urogenital tract morphology (and terminology) he had employed for *Vipera*, a viperid species, with that of the tract morphology of a colubrine, the Grass Snake (*Tropidonotus* = *Natrix natrix*) as illustrated by Martin Saint Ange (Fig. 10.20B). In fact, both anatomies are descriptively correct and illustrate two fundamentally different and distinct distal urogenital duct morphologies found in males. Volsøe (1944) added additional obfuscations to understanding this region by denying the presence of an ampulla ductus deferentis (discussed previously) and by explaining away other pertinent incongruities in tract morphologies as anatomical misinterpretations by previous authors. In the following, we provide new descriptive as well as comparative morphological information regarding the micro-anatomy of the ampulla urogenital papilla, which is morphologically similar in all colubrine snakes investigated thus far. We, then, compare this micro-anatomy to its functional anatomical counterpart, the ampulla ureter, which is found in all viperid snakes (see section 10.3.4). We also provide ultrastructural evidence for secretory activity of the epithelium in both the ampulla urogenital papilla and the ampulla ureter.

New observations. The ampulla urogenital papilla (pl., ampullae urogenital papillae) is one of two small complimentary blind pouches, each representing the terminal repository of products released by the male urogenital tracts in North American colubrine snakes. The two pouches increase in volume as they advance posteriorly and reach their maximum diameters within the urogenital papilla. Each normally receives urinary and reproductive products, which include nitrogenous wastes (uric acid), sperm, and the epithelial secretions from the linings of the ureter and ductus deferens. Their overall sizes are similar to one another. They become greatly inflated by the presence of sperm or uric acid, but this increased volume is not sufficient to consider them important in sperm or uric acid storage. The expanded length of an individual pouch varies according to snake body size. We measured the distance from the anteriormost edge to the level of their orifices on the urogenital papilla. For example, the longest ampulla urogenital papilla occurs in the Eastern Coachwhip, *Coluber flagellum flagellum* (5.50 mm), and smallest is in *Diadophis punctatus arnyi* (0.60 mm). The following are other total lengths measured (in decreasing order of size in mm): Eastern Hog-nosed Snake, *Heterodon platirhinos* (5.02), *Farancia abacura reinwardtii* (4.29), *Pantherophis obsoletus* (3.85), *C. constrictor priapus* (3.66), *Thamnophis sirtalis sirtalis* (2.86), Red Milksnake, *Lampropeltis triangulum syspila* (2.76), *L. holbrooki* (2.55), *Cemophora coccinea copei* (2.45), *Opheodrys aestivus* (2.36), *Nerodia cyclopion* (2.19), Orange-striped Ribbonsnake, *T. proximus proximus* (1.93), *Tantilla gracilis* (1.20), *Sonora semiannulata* (1.14), *Virginia striatula* (1.09), Graham's Crayfish Snake, *Regina grahamii* (0.99), *Storeria dekayi wrightorum* (0.99), Smooth Earthsnake, *V. valeriae* (0.83), and Western Wormsnake, *Carphophis vermis* (0.70).

Although macroscopic images and SEM images (external ampullary folds) reveal the general region occupied by the ampulla urogenital papilla (Figs. 10.20E, 10.21), this structure is best delineated microscopically through transverse serial histosections beginning at the level of the anteriormost

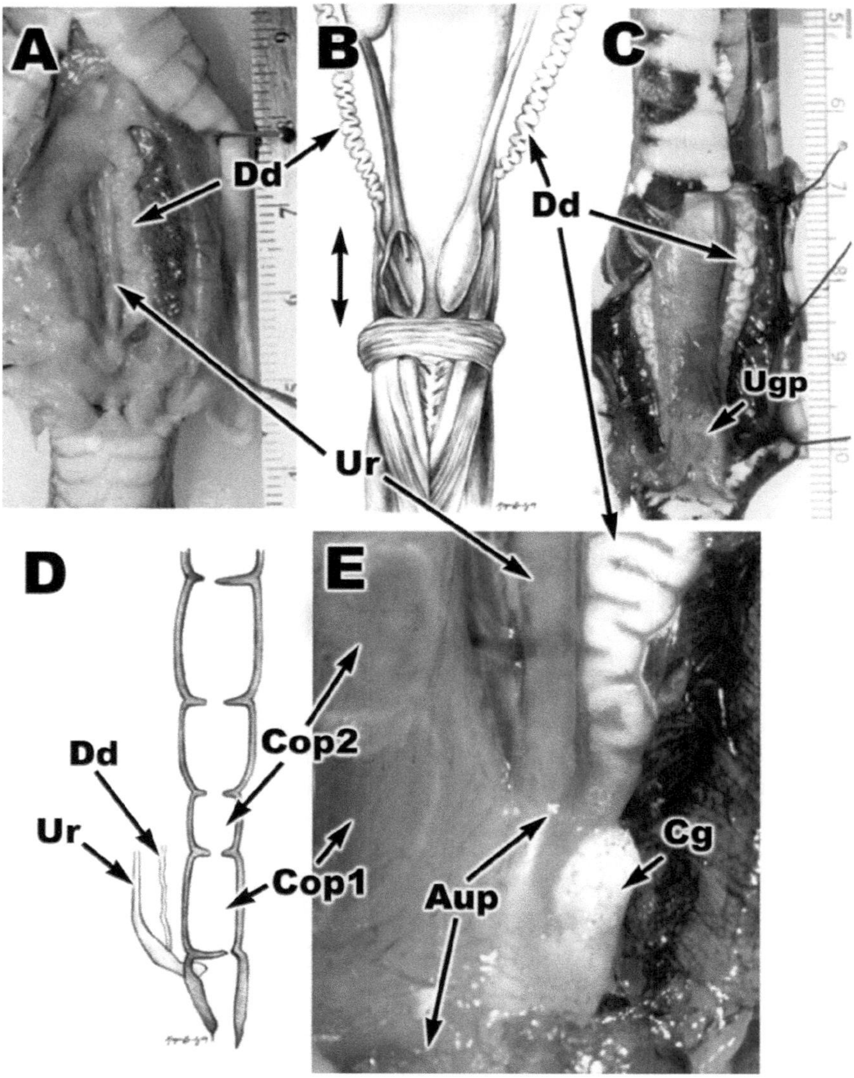

Fig. 10.20 Composite illustration of the distal urogenital anatomy of adult male North American colubrine snakes matched to anatomical diagrams (**B** and **D**) redrawn from Martin Saint Ange (1854) and Volsøe (1944), respectively. Two-headed arrow in B encompasses an entire ampulla urogenital papilla. Ventral views of *Lampropeltis calligaster, Nerodia fasciata confluens*, and *Thamnophis proximus proximus* (**A**, **C**, and **E**) reveal positional relationships among the anatomical structures. Aup, ampulla urogenital papilla; Cg, cloacal gland; Cop1, first coprodaeal chamber; Cop2, second coprodaeal chamber; Dd, ductus deferens; Ugp, urogenital papilla (Ugp); Ur, ureter. **B** is adapted from Martin Saint Ange, G.-J. 1854. J.-B. Balliére, Libraire de l'Académe Impériale de Médecine, Paris, Plate X, Fig. 2. **D** is adapted from Volsøe, H. 1944. Spolia Zoologica Musei Hauniensis V. Skrifter, Universitetets Zoologiske Museum, København, Text Fig. 10.

Color image of this figure appears in the color plate section at the end of the book.

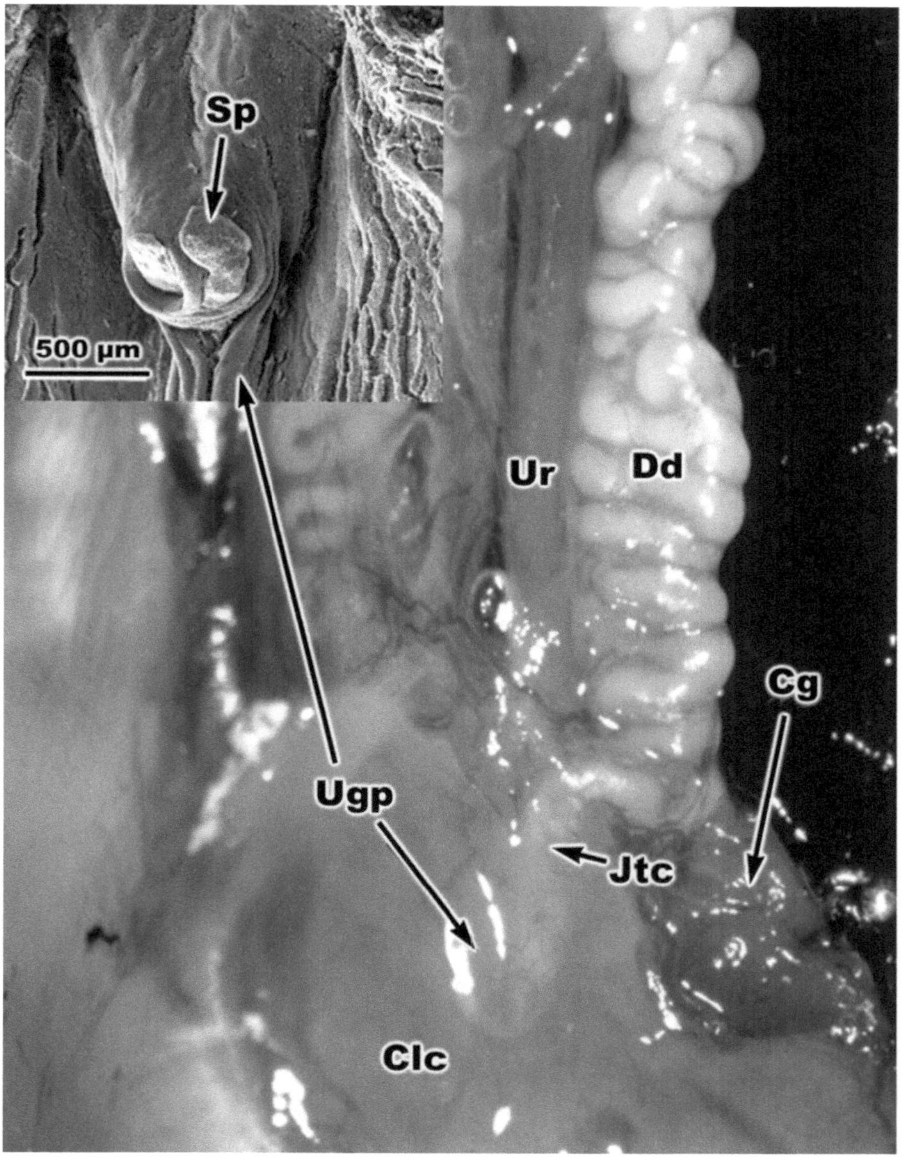

Fig. 10.21 Macroscopic view of the distal urogenital anatomy of a male *Opheodrys aestivus*, a North American colubrine snake, showing urogenital ducts and associated structures. Scanning electron micrograph (inset) reveals sperm extruding from the urogenital papilla in this species. Clc, cloacal chamber; Cg, cloacal gland; Dd, ductus deferens; Jtc, ductus deferens merging with ampulla urogenital papilla; Sp, sperm; Ugp, urogenital papilla; Ur, ureter.

Color image of this figure appears in the color plate section at the end of the book.

limit of the cloacal gland and then proceeding posteriorly toward the orifices of the urogenital papilla. We have chosen a representative colubrine, *Thamnophis proximus proximus* (same specimen measured above), for our descriptive treatment of this structure (Figs. 10.22, 10.23, 10.24).

In Fig. 10.22A, the ureters and the ductus deferentia can be seen in a horizontal plane with each ductus deferens being located lateral to its respective ureter. Both sets of ducts are positioned dorsal to the coprodaeal chamber. At the level viewed in Figs. 10.22B to 10.22D, each ductus deferens changes position as it begins a gradual progression in a ventromedial direction so as to eventually lie nearly ventral to their respective ureter (as shown in Fig. 10.22D-F). Immediately following the plane level of Fig. 10.22D, the anteriormost region of one ampulla urogenital papilla can be seen as a small pouch breaking through the dorsal wall of the muscularis of one ureter (Fig. 10.22E,F). Eventually, both ampullae can be seen in this area, and they continue to enlarge and push the ureters downward until they reside in a near midventral position beneath their respective ampullary cavities (Fig. 10.23A). Soon thereafter, a ureter can be seen opening directly into the cavity of its ampulla (Fig. 10.23B); eventually, both ureters open into the ampullary cavities (Fig. 10.24A). The ductus deferentia move toward the ventral surface of the ampullary cavities and empty into them at a point nearly identical to where the ureters open (Fig. 10.24A,B). The ampullary cavities now expand into the cloacal cavity, leaving a bulging, outward curved depression on the surface topography of the anterior dorsal recess. These bulges are best seen using SEM and are the external ampullary folds (Fig. 10.32). Eventually, the ampullae reach maximum width in the plane level of the orifices of the urogenital papilla (Fig. 10.24D). Posterior to the plane of the orifices, the ampullae quickly disappear.

The degree of folding and the relative thickness of the epithelial lining of the ampulla urogenital papilla, as observed within the urogenital papilla, are highly variable among the different species of snakes (Fig. 10.25). One of the simplest design morphologies can be seen in *Cemophora coccinea copei* (Fig. 10.25A), which exhibits a relatively smooth, unfurrowed epithelial lining when the ampullae are packed with sperm (as in Fig. 10.25A). A thin incomplete midsagittal ampullary wall separates the two ampullary cavities and, thus, allows their contents to combine at the apical region near the common urogenital orifice, which this species possesses. The thin lips of the urogenital papilla are also apparent here. This peculiar bellows configuration found in *C. coccinea copei* was not observed in other species. A low, pseudostratified columnar epithelium lines the two ampullary cavities.

The midsagittal partition separating the two ampullary cavities is complete in *Sonora semiannulata* (Fig. 10.25B) as well as the remaining species examined. This species exhibits thickened ampullary walls, but they still retain a smooth, unfurrowed epithelial surface similar to that of *Cemophora coccinea copei*. The columnar epithelium is 4-5 cell layers in thickness.

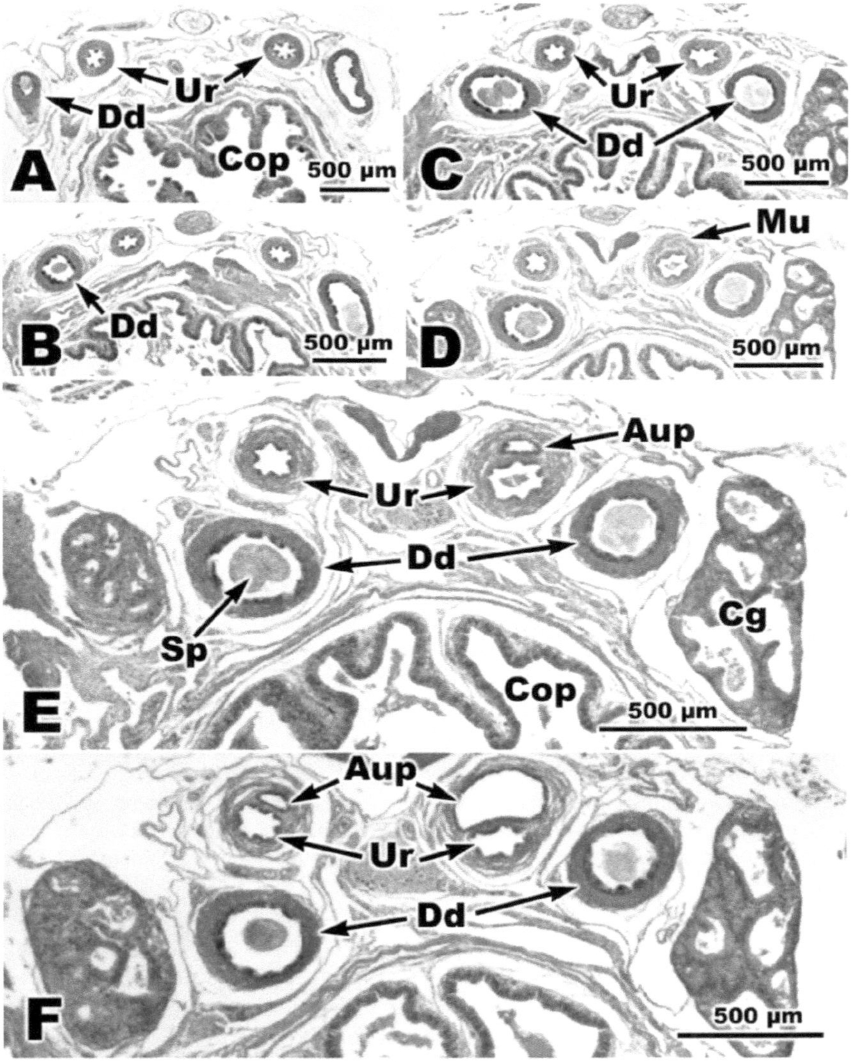

Fig. 10.22 Light micrographs of anterior-to-posterior serial histosections (Figs. 10.22-10.24) of approximately 5.0 mm of urogenital tissue revealing distal urogenital ducts and the anterior cloacal anatomy of *Thamnophis proximus proximus*, stained with hematoxylin and eosin. **A.** Anteriormost section showing the relationship between the ureter (Ur), the ductus deferens (Dd), which is situated laterally, and the first coprodaeal chamber (Cop). **B.** Section posterior to A showing gradual enlargement and medial displacement of the left ductus deferens **C.** Section revealing the near pairing of each ureter and its respective ductus deferens. **D.** Dorsomedial enlargement of the ureter muscularis indicates the anteriormost extension of the ampulla urogenital papilla (Aup); the ductus deferens now resides ventral to the left ureter. **E.** The ampulla urogenital papilla appears as a small pouch dorsal to its ipsilateral ureter. **F.** Bilateral ampullae rapidly expand their anatomical position dorsal to their respective ureters. Aup, ampulla urogenital papilla; Cg, cloacal gland; Cop, coprodaeal chamber; Dd, ductus deferens; Mu, muscularis; Sp, sperm; Ur, ureter.

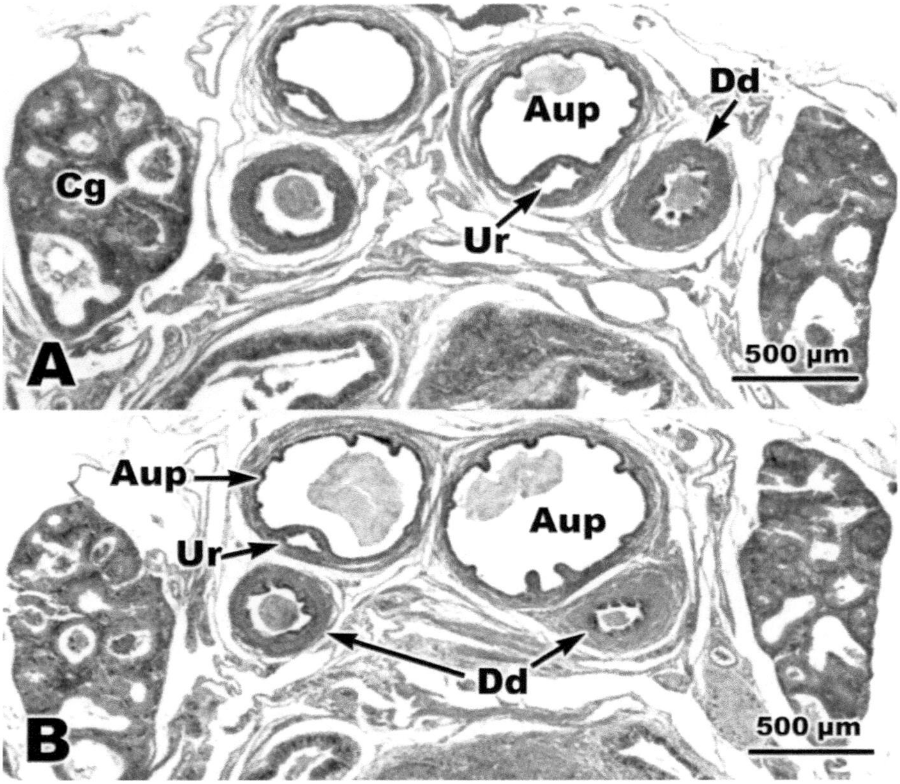

Fig. 10.23 *Thamnophis p. proximus*, continuation of Fig. 10.22. **A.** The ampullae urogenital papillae extend, displace, and constrict the ureters. **B.** The right ureter opens into its ipsilateral ampulla. Aup, ampulla urogenital papilla; Cg, cloacal gland; Dd, ductus deferens; Ur, ureter.

The epithelial lining of the ampullary cavities, as seen in *Farancia abacura reinwardtii* (Fig. 10.25C), is variable in thickness and consists of a mixture of simple columnar and low pseudostratified columnar. Little folding of the ampullary wall is evident. Although this species possesses two distinct urogenital papillae, the ampullary cavities show a thin midventral wall, a partition just prior to the orifices. This ventral partition broadens to become part of the posterior papillary ridge in this species.

In *Diadophis punctatus arnyi* (Fig. 10.25D), the ampullary walls are laterally indented near the level of the orifices by the presence of spherical muscle masses within the lamina propria. These muscle masses are not evident in other species examined. The low pseudostratified columnar epithelial lining varies little throughout the ampullary cavities, but the lining itself shows some simple folds.

The ampullary cavities of *Carphophis vermis* (Fig. 10.25E) are unusual in several anatomical features. Both cavities exhibit additional partitions of the internal ampullary walls. In addition, these well-defined internally

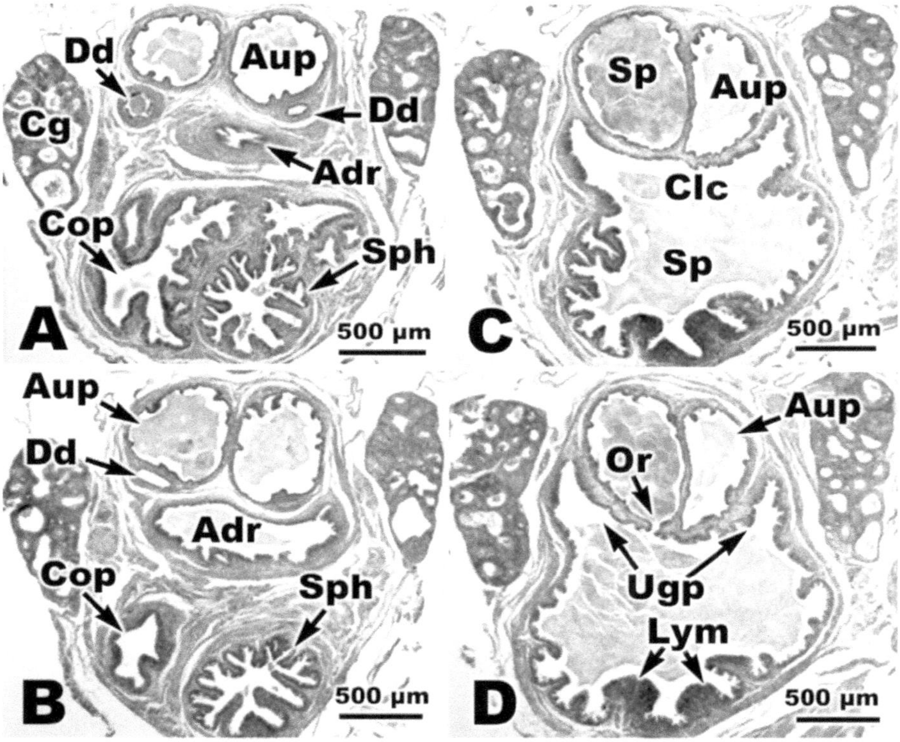

Fig. 10.24 *Thamnophis p. proximus*, continuation of Fig. 10.23. **A.** Spermatozoa fill the greatly enlarged ampullae urogenital papillae. The anterior dorsal recess (Adr) now appears beneath the ampullae urogenital papillae. The coprodaeal chamber is now flanked by the posterior coprodaeal sphincter (Sph), which separates the coprodaeum from the cloacal chamber (Clc). **B.** As the ampullae and the dorsal recess continue to enlarge, the ductus deferens on both sides near the point of confluence with the ampullae. The coprodaeal chamber becomes a blind pouch lateral to the posterior sphincter. **C.** The dorsal recess merges with the cloacal cavity to become an expanded anterior region of the cloaca proper. **D.** Sperm exit through the left orifice (Or) of the urogenital papilla (Ugp). Enlarged lymphoid tissue masses reside in the ventral wall of the cloacal cavity. Adr, anterior dorsal recess; Aup, ampulla urogenital papilla; Cg, cloacal gland; Clc, cloacal chamber; Cop, coprodaeal chamber; Lym, lymphoid masses; Or, orifice; Sp, sperm; Sph, sphincter muscle; Ugp, urogenital papilla.

thickened wall partitions escort each ductus deferens from a dorsal position to a ventral position at a point near its opening with each orifice of the urogenital papilla. A more complex version of this structural microanatomy dealing with the ductus deferens is found in *Regina grahamii* (Fig. 10.25I). The ampullary epithelium remains low, pseudostratified columnar. The tall stature of the urogenital papilla (the extent to which it projects into the cloacal cavity) of *C. vermis* can also be observed in Fig. 10.25E.

In the species examined in Fig. 10.25A-E, the ampullary lining can be characterized as being relative smooth and unfurrowed. In *Lampropeltis holbrooki* (Fig. 10.25F), as well as the three species to be discussed hereafter,

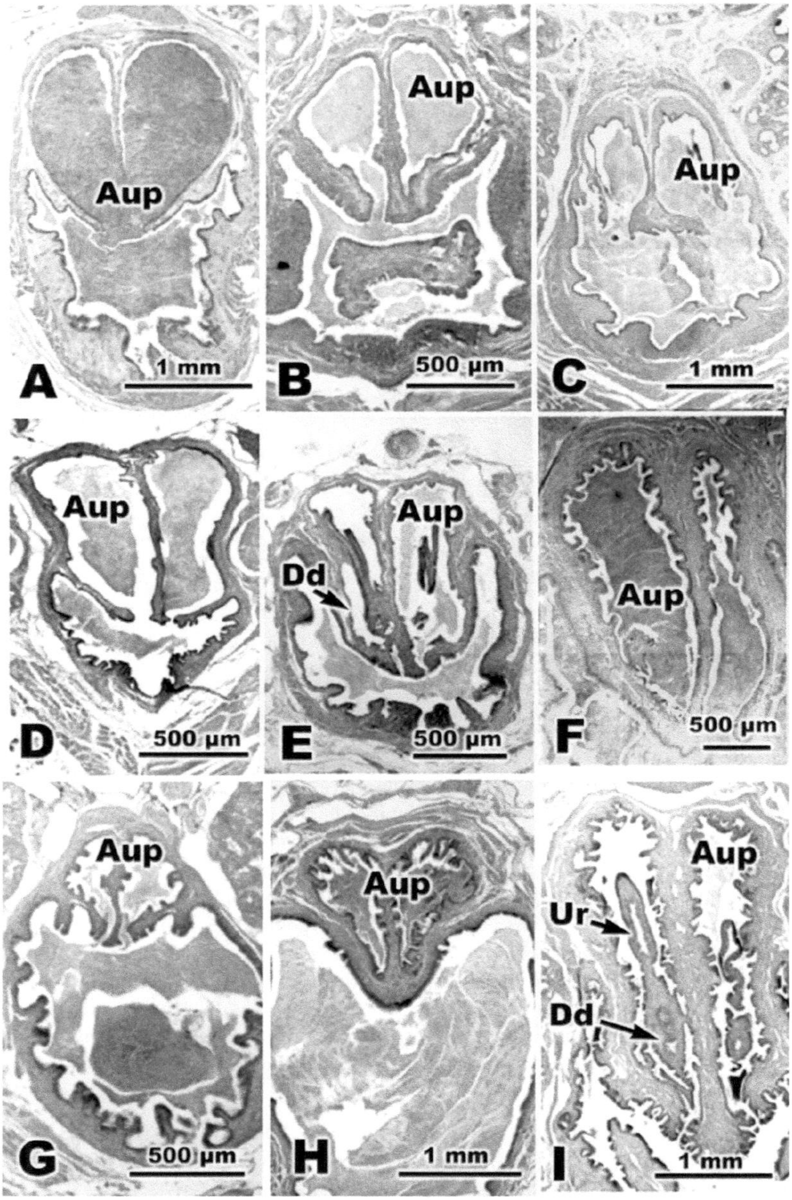

Fig. 10.25 Light micrographs of the ampulla urogenital papilla (Aup) and urogenital papillae in representative species of male North American colubrine snakes, stained with hematoxylin and eosin. All, except H, show the release of sperm through one (A) or two (B-I) urogenital orifices. Species are ranked from A though I according to the degree of folding of the epithelial lining of the Aup (see explanation in text). **A.** *Cemophora coccinea copei.* **B.** *Sonora semiannulata.* **C.** *Farancia abacura reinwardtii.* **D.** *Diadophis punctatus arnyi.* **E.** *Carphophis vermis.* **F.** *Lampropeltis holbrooki.* **G.** *Storeria dekayi wrightorum.* **H.** *Opheodrys aestivus.* **I.** *Regina grahamii.* Aup, ampulla urogenital papilla; Dd, ductus deferens; Ur, ureter.

an extensive folding of the internal ampullary wall is evident. A thickened midsagittal ampullary wall partition as well as relatively thick lateral ampullary walls characterizes the ampulla urogenital papillae in these species. Surface patches of low and tall pseudostratified columnar epithelial regions line the ampullary cavities in *L. holbrooki.*

Both *Storeria dekayi wrightorum* and *Opheodrys aestivus* possess similar ampullary cavities, each being greatly folded internally on all walls of the ampulla urogenital papilla. Two differences, observed in *O. aestivus* and not in *S. d. wrightorum*, are the spherical bulging of the dorsal ampullary cavities and the presence of primarily simple columnar epithelium, whereas *S. dekayi wrightorum* exhibits a low-to-tall pseudostratified columnar epithelium. Possibly the most strikingly unusual micro-structuring within the ampulla urogenital papilla is found in *Regina grahamii* (Fig. 10.25I). In this species, both ampullary cavities exhibit extensive internal surface folding, and the folding is so profound that additional small subcavities appear along most of the ampullary walls. Furthermore, each ureter and ductus deferens becomes enveloped by a thin longitudinally directed, vertical tissue column. The ureter resides in a dorsal position, whereas the ductus deferens resides in a ventral position. Both sets of urogenital ducts dump their products into their respective ampullary cavities at approximately the same plane level with the ureter opening slightly prior to the ductus deferens. Only *Carphophis vermis* and *Pantherophis obsoletus* (see below) examined thus far exhibit a similar urogenital micro-anatomy.

We focused, here again, some detailed histochemical and ultrastructural attention on the ampulla urogenital papilla and the associated urogenital ducts of *Pantherophis obsoletus*. In a histochemical analysis, we found that the lining of the ampulla urogenital papilla and the ureter in this species is highly secretory (Fig. 10.26). The epithelial lining of both structures is PAS positive, a strong indication of the presence of neutral carbohydrates (Fig. 10.26C,F). Upon using alcian blue to test for glycosaminoglycans, we found a strong positive reaction in the epithelial lining of the ampulla urogenital papilla (Fig. 10.26D,E). As noted by Sever (2004) for *Seminatrix pygaea*, the ductus deferens did not react positively with either PAS or with alcian blue. On the other hand, in the region of the merging of the ductus deferens and the ampulla urogenital papilla, a strong PAS positive stain occurs (Fig. 10.26F) within an enclosed sperm aggregate (Fig. 10.26F), indicating that the epithelial secretions within the ampulla possibly move up into the ductus deferens some distance.

The secretory nature of the epithelial lining of the ampulla urogenital papilla of *Pantherophis obsoletus* is further supported by ultrastructural studies. We first noted that in preparation for TEM, the pseudostratified columnar epithelium of both the ampulla and the ureter tested positive for carbohydrates using Ladd multiple stain (Fig. 10.27A,B) on plastic-embedded sections. The supranuclear surfaces of these cells were enriched by secretory material in cytoplasmic vesicles that stain dark red. Upon investigation by TEM, we found numerous secretory vesicles exhibiting a

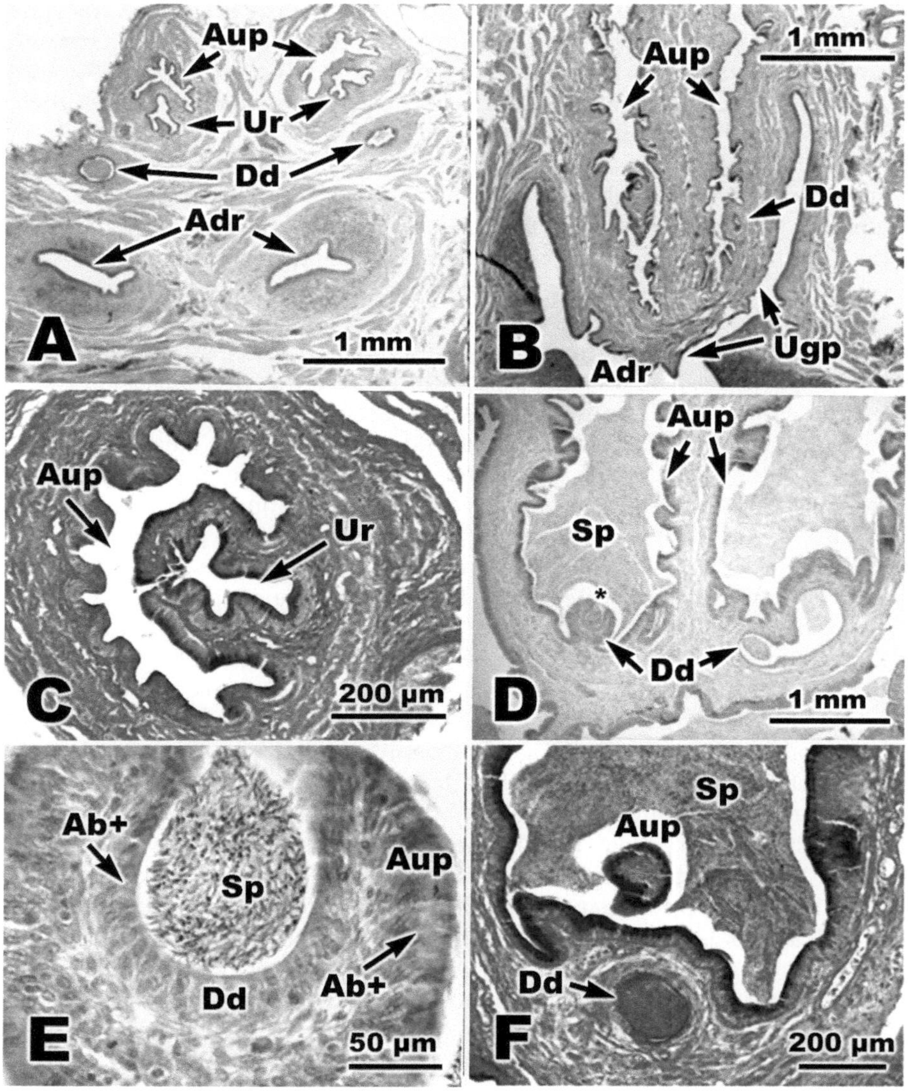

Fig. 10.26 Light micrographs (transverse histosections) of the distal urogenital anatomy of *Pantherophis obsoletus* revealing relationships among the ureter, ductus deferens, and the ampulla urogenital papilla as well as the staining properties of their secretory epithelia. **A.** Section showing urogenital anatomy prior to the merging of the ureter and the ampulla urogenital papilla. Paired extensions of the dorsal recess lie beneath the ampulla urogenital papilla-ureter-ductus deferens complexes. **B.** Section through the urogenital papilla (ventral half) in a non-reproductive adult showing the near merging of the ductus deferens with the ampulla urogenital papilla. Paired extensions of the dorsal recess surround the urogenital papilla. A and B both stained with hematoxylin and eosin. **C.** Section showing the merging of the ureter with the ampulla urogenital papilla. The epithelia of both structures exhibit a strong

Fig. 10.26 Contd. ...

variety of differing electron densities within this surface region. Some of these products appear to be released by the process of exocytosis, whereas others are possibly retained within the cells to perform cell maintenance duties. We also observed what could be phagocytosis of sperm within the epithelial lining (Fig. 10.27B).

10.3.4. Ampulla Ureter

Review. The ampulla ureter is the expanded distal segment of the ureter in male viperid snakes (Volsøe 1944). It is the functional equivalent to the ampulla urogenital papilla of colubrine snakes discussed previously. In Fig. 10.28, we have redrawn an illustration of the ampulla ureter as depicted by Volsøe (1944) and juxtaposed it with a macroscopic view of this structure from the Southern Copperhead (*Agkistrodon contortrix*) in order to better illuminate the morphology of this structure. Below, we also provide new urogenital information on the micro-anatomy of the ampulla ureter in two crotaline species, *Sistrurus miliarius* and *A. contortrix*, using light and electron microscopy.

New observations. The following Figs. 10.28-10.31 illustrate the anatomical design of the ampulla ureter and show the basic differences between this structure and the ampulla urogenital papilla of colubrine species as previously seen in Figs. 10.20-10.27. For example, the distal urogenital ducts of the colubrine species (*Thamnophis proximus proximus*) and a crotaline species (*Sistrurus miliarius streckeri*) are essentially identical at the histological planes shown in Fig. 10.22A and Fig. 10.29A, respectively.

The relative sizes (diameters) of the ureter and the ductus deferens in each species are equally similar to one another. Following a posterior histological progression, the ductus deferens does not enlarge in *Sistrurus milarius streckeri* as it does in *Thamnophis proximus proximus* (Fig. 10.22E), but instead becomes smaller than the ureter as each ductus deferens moves medially (Fig. 10.29B,C). What becomes strikingly apparent in *S. m. streckeri* in Fig. 10.29E is the enlargement of the ureters through a thickening of

... Fig. 10.26 Contd.

PAS positive stain. **D.** Section through the urogenital papilla (ventral half) in a reproductive adult showing the near merging of the ductus deferens with the ampulla urogenital papilla (see asterisk). Note the secretory epithelium of the each ampulla urogenital papilla has reacted positively with alcian blue stain (pH = 2.5-2.8). **E.** Higher magnification of region designated by the asterisk in D. The epithelial surface of the ductus deferens shows a slight positive reaction with alcian blue compared to the more intense (darker blue) reaction seen on the epithelial surface of the ampulla urogenital papilli. **F.** Ventral regions of an ampulla urogenital papilla (similar to view in D) reveals a strong PAS positive stain in the epithelium as evidenced by the intensity of the red stain. Ab+, alcian blue positive reaction; Adr, anterior dorsal recess; Aup, ampulla urogenital papilla; Dd, ductus deferens; Sp, sperm; Ugp, urogenital papilla; Ur, ureter.

Color image of this figure appears in the color plate section at the end of the book.

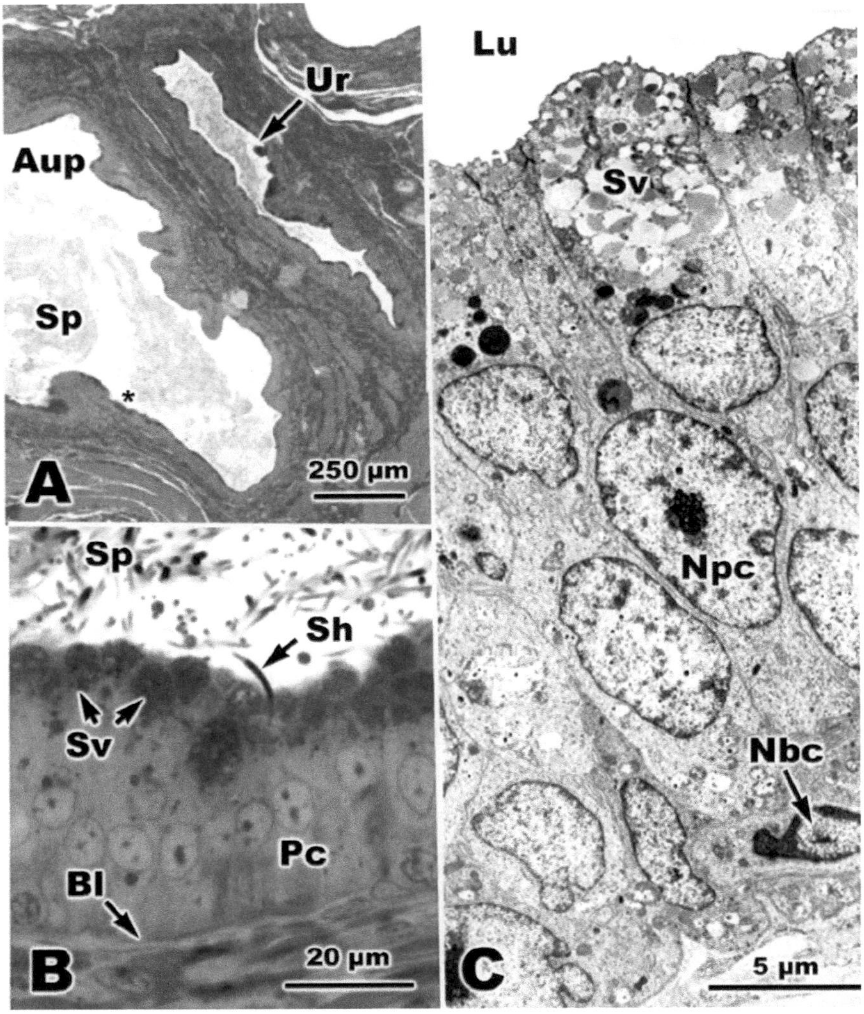

Fig. 10.27 Light and transmission electron micrographs revealing the secretory epithelium of the ampulla urogenital papilla of an adult male *Pantherophis obsoletus*. **A.** Plastic thick section (ca. 1 µm) showing the relative position of the ampulla urogenital papilla and the ureter (see Fig. 10.38A and C for comparison). Sperm are scattered within the lumina of both structures. Asterisk indicates the approximate region shown at higher magnification in B. **B.** Section through the pseudostratified columnar epithelium of the ampulla urogenital papilla as shown in A revealing intense staining of apical secretory vesicles. Sperm are abundant near epithelial surface including a sperm cell whose head is embedded within epithelium. A and B stained with Ladd multiple stain. **C.** Transmission electron micrograph showing a portion of the secretory epithelium. Note clusters of clear and opaque secretory vacuoles near the surface of the epithelial cells. Aup, ampulla urogenital papilla; Bl, basal lamina; Lu, lumen; Nbc, nucleus of basal cell; Npc, nucleus of principal cell; Pc, principal cells; Sh, sperm head; Sp, sperm; Sv, secretory vacuoles; Ur, ureter.

Color image of this figure appears in the color plate section at the end of the book.

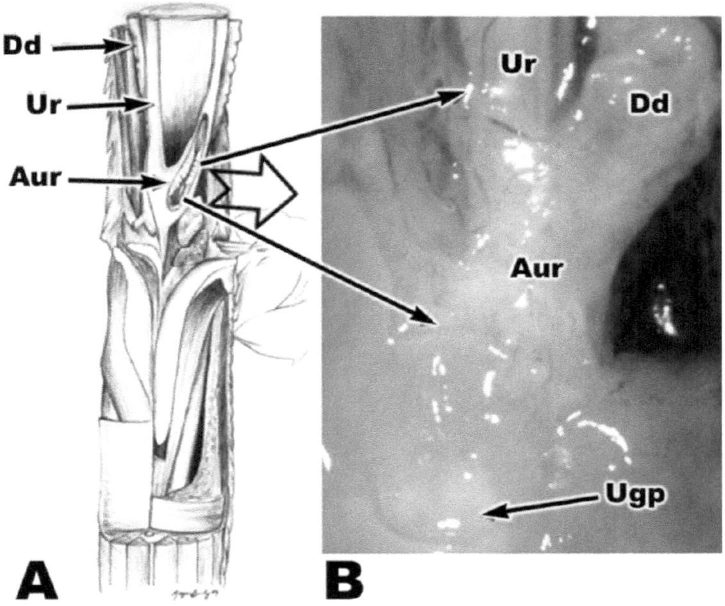

Fig. 10.28 Composite diagram including an illustration of the adult male distal urogenital anatomy of a European viperine snake, *Vipera berus* (**A**; redrawn from Volsøe 1944) matched alongside a North American crotaline snake, *Agkistrodon contortrix* (**B**).The presence of an ampulla ureter (Aur) represents a fundamental anatomical difference between viperine urogenital morphology and the colubrine urogenital morphology as shown in Fig. 10.20. Aur, ampulla ureter; Dd, ductus deferens; Ugp, urogenital papilla; Ur, ureter. **A** is adapted from Volsøe, H. 1944. Spolia Zoologica Musei Hauniensis V. Skrifter, Universitetets Zoologiske Museum, København, Fig. 11.

Color image of this figure appears in the color plate section at the end of the book.

the ureterine muscularis along with a commitment expansion of their lumina. In contrast, the ureterine muscularis of *T. p. proximus* enlarges because of the development and swelling of the ampulla urogenital papilla embedded within its dorsomedial wall (Fig. 10.22E,F). In *S. m. streckeri,* however, no similar development occurs. Instead, the two enlarging ureters move medially and become enveloped within a common muscularis in conjunction with the wall of the newly emerging anterior dorsal recess (Fig. 10.29E,F). At this point, each ureter can now justifiably be called an ampulla ureter. At the same time that this enlargement occurs, each ductus deferens also moves medially to become a small component of a singular medial urogenital complex (Fig. 10.29F). Each ampulla ureter now characteristically exhibits a highly convoluted internal lining. In a more posterior plane, as seen in Fig. 10.29G, the ampullae begin to lengthen dorsoventrally; in Fig. 10.29H, the anterior dorsal recess merges with the cavity of the coprodaeum. The medial urogenital complex has now also incorporated the muscularis of the coprodaeum to become a singular mass,

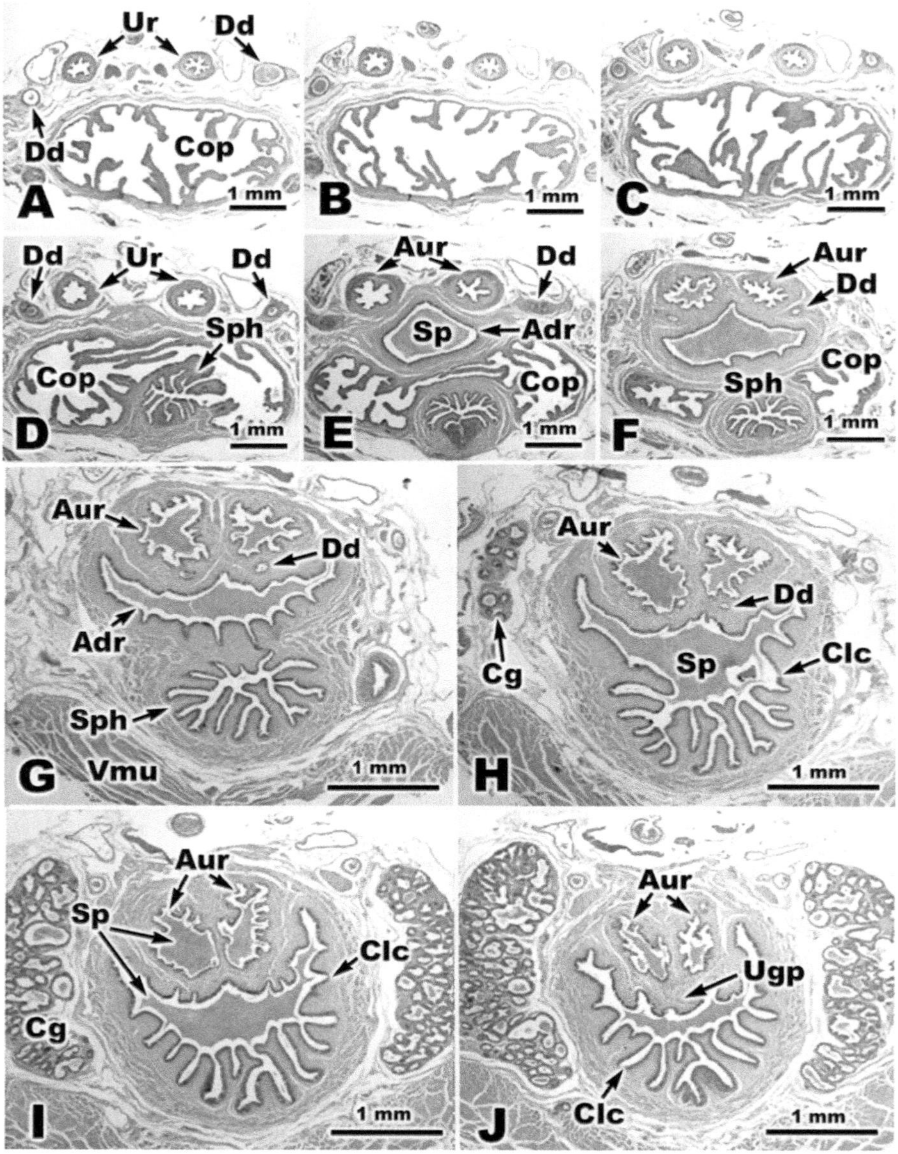

Fig. 10.29 Light micrographs of selected anterior-to-posterior serial histosections of approximately 7.0 mm of urogenital tissue revealing distal urogenital ducts and the anterior cloacal chamber of *Sistrurus miliarius streckeri*, a North American crotaline snake, stained with hematoxylin and eosin. **A.** Anteriormost section of series showing the relationship between each ureter (Ur), ductus deferens (Dd), each lying lateral to its respective ureter, and the first coprodaeal chamber (Cop). **B.** Section posterior to A showing gradual decrease in diameter of the right ductus deferens. **C.** Section revealing a gradual enlargement of the ureters; each

Fig. 10.29 Contd. ...

the cloacal complex. Cloacal glands begin to appear within this histological plane and reside lateral to the cloacal complex. At this point, each ductus deferens has lost most of its muscularis and has become a minute ventral duct lying beneath the ampullae. They soon merge with the ampullary cavities (Fig. 10.29H). The ventral surfaces of each ampulla protrude in the form of individual folds into the cloacal cavity and are referred to as the external ampullary folds of the urogenital papilla (Fig. 10.29I). Finally, the ampullae narrow as they near the urogenital orifice (in this species a single orifice; Fig. 10.29J).

For an evaluation of the secretory activity of the ampulla ureter, we examined this structure in a reproductively active individual of the crotaline species, *Agkistrodon contortrix* (Figs. 10.30, 10.31). The internal walls of the ampullae of *A. contortrix* lack the characteristic folding seen in *Sistrurus miliarius streckeri*. The epithelial lining of the ampullae exhibits a highly secretory, tall, thick pseudostratified columnar epithelium. Spherical cytoplasmic blebs (Fig. 10.30B) form a distinctive layer along the apical surfaces of the epithelial cells. These cytoplasmic blebs show little affinity for histological stains. Under transmission electron microscopy (Fig. 10.31), the apical blebs appear to contain amorphous materials released as an apocrine-packaged unit housed within a compartment created by a growing free cytoplasmic membrane, hence, the term apocrine blebs is appropriate as in more anterior efferent ducts. The epithelial cells appear to accumulate materials within numerous secretory vesicles in a region just below the apical bleb (Fig. 10.31A). These membrane-bound secretory vesicles appear to swell, lose their integrity, and disintegrate along a well-defined linear boundary line. In general, secretory vesicles remain on the nuclear side of the boundary line, whereas secretory materials occupy the luminal side. As the membrane coats of the secretory vesicles are lost, an expansion of the

---

... Fig. 10.29 Contd.

ductus deferens remains situated lateral to its respective ureter. **D.** The muscularis and the lumen of both ureters have continued their enlargement; the coprodaeal sphincter (Sph) appears along the floor of the coprodaeum. **E.** Each enlarged ureter, now nearly twice the diameter as seen in A, is becoming medially displaced and is also transitioning into an ampulla ureter (see text for explanation); the anterior dorsal recess (Adr) appears medial and dorsal to the coprodaeal chamber. **F.** The ampullae ureters are now encompassed by a common muscularis envelope and are continuing their medial displacement and enlargement; sperm fill the lumen of the dorsal recess. Blind pouches of the coprodaeal cavity now lie on either side of the coprodaeal sphincter. **G.** Each ductus deferens now resides ventral to its respective ipsilateral ampulla ureter; the dorsal recess has expanded laterally. **H.** The left ductus deferens nears merging with its ipsilateral ampulla ureter; the lumen of the coprodaeal sphincter has merged with the dorsal recess creating an expansive anterior cloacal chamber (Clc). **I.** The ampullae reach maximum volume; spermatozoa fill the anterior cloacal chamber. **J.** Each lumen of the individual ampulla ureter elongates as both descend into the tip of the urogenital papilla. Adr, anterior dorsal recess; Aur, ampulla ureter; Cg, cloacal gland; Clc, cloacal chamber; Cop, coprodaeal chamber; Dd, ductus deferens; Sp, sperm; Sph, sphincter; Ugp, urogenital papilla; Ur, ureter; Vmu, ventral musculature.

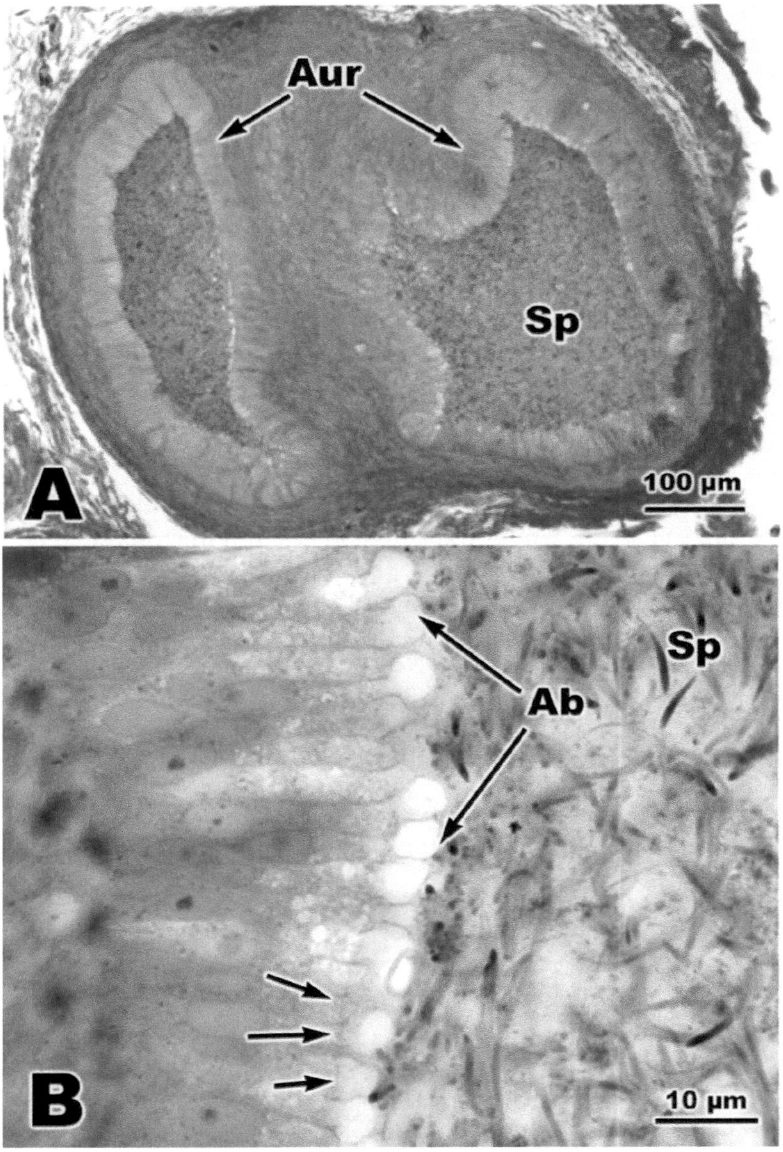

Fig. 10.30 Light micrographs of the ampulla ureter and its secretory epithelium of an adult male *Agkistrodon contortrix*, a North American crotaline snake, stained with Ladd multiple stain. **A.** Section from plastic embedded tissue (ca. 1 mm in thickness) showing the tall pseudostratified columnar epithelium (pale band) lining the lumina of the ampullae. Spermatozoa completely fill luminal cavities. **B.** Portion of secretory epithelium from A exhibiting the presence of apocrine blebs (Ab) being liberated from numerous principal cells. A series of arrows (lower left) follow a faint, but distinctive cytoplasmic demarcation line, which lies at the base of these cytoplasmic blebs. Ab, apocrine blebs; Aur, ampulla ureter; Sp, sperm.

Color image of this figure appears in the color plate section at the end of the book.

plasma membrane accommodates the festooning accumulation of cellular products along the free surface. Apocrine blebs appear to be squeezed away from the parent cell to become a luminal bleb (Fig. 10.31C,D). Thus, the eccrine process is similar to that described for the ductus epididymis and ductus deferens. Sperm lie in close proximity to the margins of these cellular and luminal blebs.

10.4. UROGENITAL PAPILLA AND CLOACA

Review. In male snakes, spermatozoa transversing through the ductus deferens and urine transversing through the ureter exit directly into the cloacal urodaeum via a common pore at the terminal end of a distinctive anatomical structure known as the urogenital papilla (Fig. 10.32). Snakes either possess a single, medial urogenital papilla or paired, bilateral urogenital papillae. In either case, the urogenital papilla projects ventrally from the dorsal wall of the anterior cloacal chamber (Seshadri 1959; Bellairs 1970; Fox 1977; Lombardi 1998). Little anatomical information is available regarding this structure. No authors have incorporated species-specific illustrations, descriptions, and details of the urogenital papilla when discussing the distal urogenital tracts of individual snake species in conjunction with its urogenital anatomy (e.g., W. Fox 1952; H. Fox 1977). Instead, some authors have redrawn the basic anatomical sketches provided by Volsøe (1944) for *Vipera berus* and by Gabe and Saint Girons (1965) for *Vipera aspis*. Surprisingly, contemporary illustrations featuring male snake gross internal anatomy often omit the region occupied by the urogenital papilla entirely (e.g., Shine 1991; Vitt and Caldwell 2009). Gabe and Saint Girons (1965) provided a histological cross section of the urogenital papilla in a male Banded Krait (*Bungarus fasciatus*), making this drawing the only micro-anatomical depiction of the male urogenital papilla in the snake literature until the present study.

Illustrated descriptions of the co-joining of the ureter and the ductus deferens appear as infrequently as do depictions of the urogenital papilla in snake anatomical literature. The only exceptions include an illustration by Martin Saint Ange (1854) for a European natricine (*Natrix natrix*) and the aforementioned viperid species drawn by Volsøe (1944) and by Gabe and Saint Girons (1965).

New observations. Below, we provide descriptive coverage of the urogenital papilla by revealing its macroscopic, microscopic, and ultrastructural anatomy from a diverse group of male North American colubroid snakes. We utilized 39 species and subspecies from four subfamilies (Crotalinae, Xenodontinae, Natricinae, Colubrinae) and follow the classification provided by Vitt and Caldwell (2009); these representative species include the following: *Agkistrodon cortortrix*, *A. piscivorus*, *Crotalus atrox*, Timber Rattlesnake (*C. horridus*), Prairie Rattlesnake (*C. viridis*), *Sistrurus miliarius streckeri*, *Carphophis vermis*, *Cemophora coccinea copei*, Eastern Yellow-bellied Racer (*Coluber constrictor flaviventris*), *C. c. priapus*,

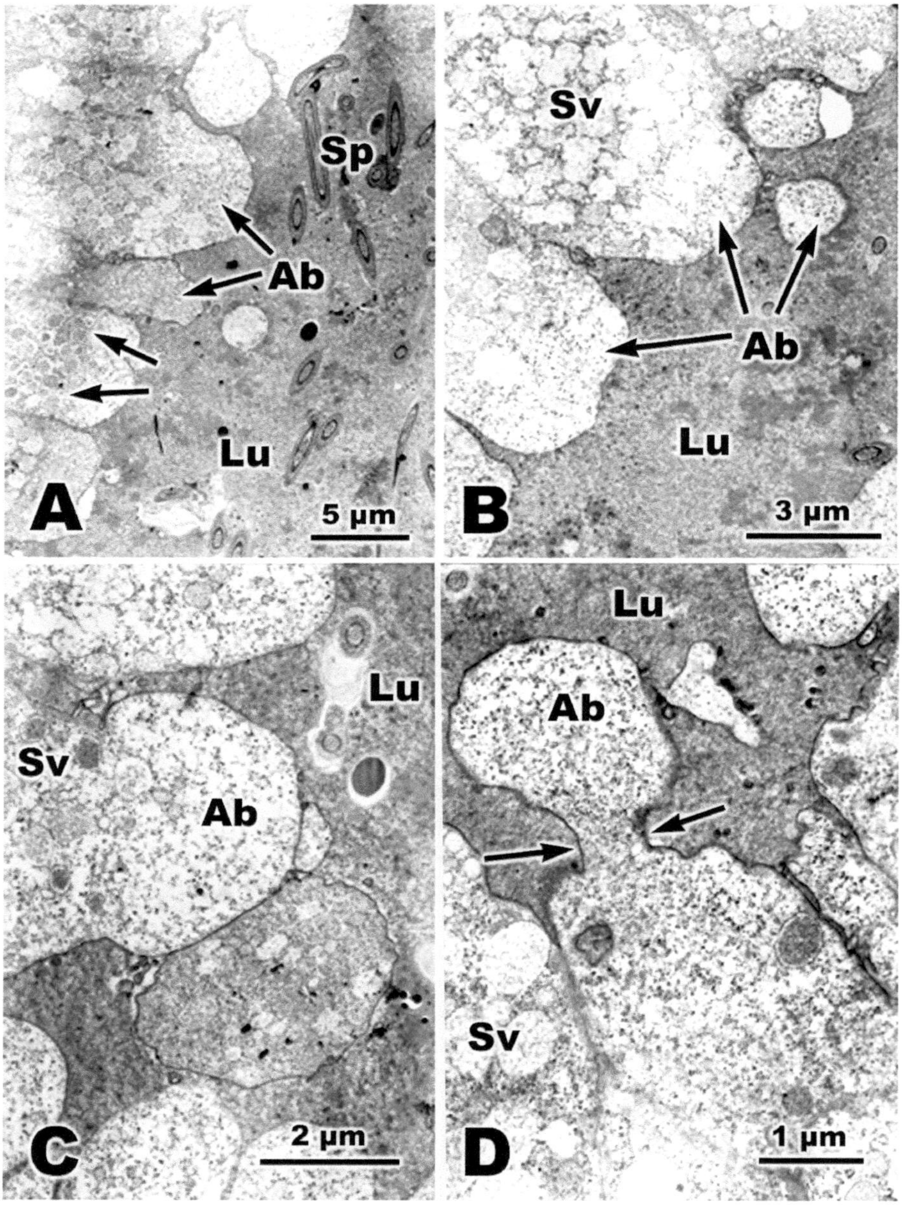

Fig. 10.31 Transmission electron micrographs of the apocrine bleb region of the secretory epithelium of *Agkistrodon contortrix* illustrated in Fig. 10.30. **A.** Arrows point to interface region between cytoplasmic secretory vacuoles and their dissociation to become secretory material of apocrine blebs (Ab). **B,C.** Higher magnification of region of apocrine blebs showing transition from circular secretory vesicles to mostly amorphous material of apocrine blebs and breakdown of vesicular membranes. **D.** Opposing arrows show neck of an apocrine bleb just prior to release into lumen. Ab, apocrine blebs; Lu, lumen; Sp, sperm.

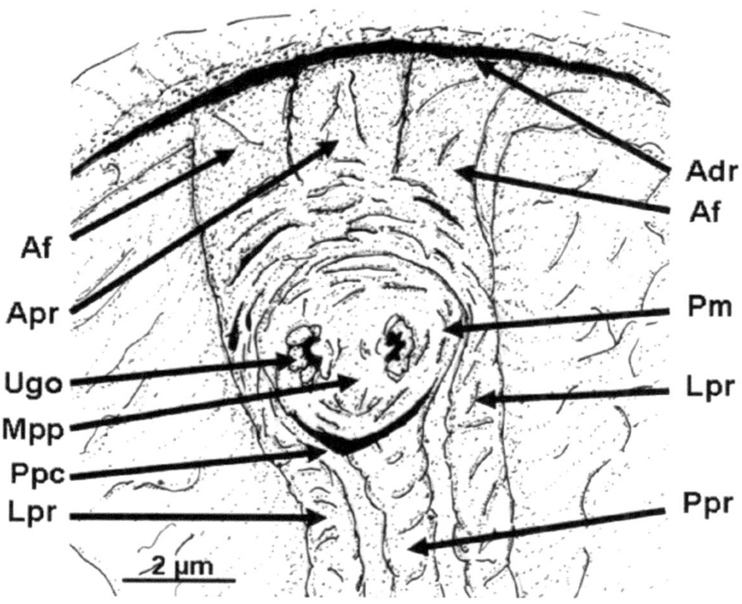

Fig. 10.32 Generalized illustration of a composite urogenital papilla (posterioventral view) including its surrounding tissues of a male North American colubrine snake. Abbreviations: Adr, anterior dorsal recess; Af, ampullary folds (paired); Apr, anterior papillary ridge; Lpr, lateral papillary ridge (paired); Mpp, medial papillary prominence; Pm, papillary mound; Ppc, posterior papillary recess; Ppr, posterior papillary ridge; Ugo, fleshy lips surround the paired urogenital orifices. Drawing by S. E. Trauth.

Black-masked Racer (*C. c. latrunculus*), *C. flagellum flagellum*, Desert Striped Whipsnake (*C. taeniatus taeniatus*), *Diadophis punctatus arnyi*, *Pantherophis obsoletus*, *Farancia abacura reinwardtii*, *Heterodon platirhinos*, Prairie Kingsnake (*Lampropeltis calligaster calligaster*), *L. holbrooki*, *L. triangulum syspila*, Yellow-bellied Watersnake (*Nerodia erythrogaster flavigaster*), Broad Banded Watersnake (*N. fasciata confluens*), Northern Diamond-backed Watersnake (*N. rhombifer rhombifer*), Midland Watersnake (*N. sipedon pleuralis*), *N. cyclopion*, *Opheodrys aestivus*, Bullsnake (*Pituophis catenifer sayi*), Long-nosed Snake (*Rhinocheilus lecontei*), Queensnake (*Regina septemvittata*), *Regina grahamii*, Texas Patch-nosed Snake (*Salvadora grahamiae lineata*), Variable Groundsnake (*Sonora semiannulata semiannulata*), *Storeria dekayi wrightorum*, Northern Red-bellied Snake (*Storeria occipitomaculata occipitomaculata*), *Tantilla gracilis*, *Thamnophis proximus proximus*, *T. sirtalis sirtalis*, *Virginia striatula*, and Western Smooth Earthsnake (*Virginia valeriae elegans*).

A variety of micro-anatomies exists for the urogenital papilla within our sample of crotaline species. For example, two closely-related species of *Agkistrodon* (Fig. 10.33A-D) possess sharply contrasting morphologies. In *A. contortrix*, the urogenital orifices open individually, project posteriorly, and are narrowly separated on two small, but distinct, papillary mounds (Fig. 10.33A,B). This bilobate condition is reinforced by the structure of the ampullary folds, as they extend anteriorly and diagonally away from the individual papillary mounds. In *A. piscivorus*, however, the urogenital orifices lie atop a single well-defined, papillary mound, and the ampullary folds lie side by side anteriorly (Fig. 10.33C). This particular bi-lobate morphology was not observed in any other snake species in our sample.

Within the genus *Crotalus*, congruent morphologies exist between *C. atrox* and *C. horridus* (Fig. 10.33E,F), but neither morphology is similar to that of *C. viridis* (Fig. 10.34A-C). One striking feature of the urogenital papilla in *C. atrox* and *C. horridus* is the flat, platform-like, papillary mound. The urogenital orifices project in a postereolateral direction. The papillary mound is supported by a posterior papillary ridge as well as the two anterior ampullary folds. In *C. viridis*, a round papillary mound has no posterior papillary ridge (Fig. 10.34A,B). Two urogenital orifices, which open ventrally, are grooved, chevron shaped pits with mostly non-fleshy lips (Fig. 10.34C).

The urogenital papilla of another crotaline species, *Sistrurus miliarius streckeri*, is distinctly different from the other viperids described above as well as from all other species examined. It is the only species that exhibits co-joined urogenital orifices (Fig. 10.34D,E). The papillary mound in this species is conical and possesses circular grooves. The mound summit terminates into a pinnacle, and a large anterior papillary ridge exists. No posterior and lateral papillary ridges occur in this species.

In *Carphophis vermis*, the urogenital papilla possesses fleshly-lipped and relatively large, thinly separated orifices (Fig. 10.35A). The ampullary folds can be viewed between two, pin-hole artifacts in the cloacal wall. The lateral papillary ridges support the papillary mound, and a narrow posterior papillary recess is seen beneath the urogenital orifices.

The papillary mound in *Cemophora coccinea* is shaped in the form of a tugboat (Fig. 10.35B,C) and lies atop an oblong papillary base. The mound possesses several concentric grooves near its basal region. Posteriorly, the mound exhibits two large, cup-shaped orifices (right orifice occluded by sperm mass). A distinct posterior papillary ridge supports the papillary base, which lacks supportive lateral papillary ridges.

Three subspecies of racers and a coachwhip (genus *Coluber;* i.e., *C. constrictor flaviventris*, *C. c. priapus*, and *C. c. latrunculus*, Figs. 10.35D-F; 10.36A and *C. flagellum flagellum*, Fig. 10.36B) exhibit very similar papillary morphologies. These similarities include tall, circular, papillary mounds with fleshy-lipped, grooved orifices spaced widely apart atop a smooth papillary summit. The lips appear slightly raised and protrude away from the summit, yielding a bulging appearance in *C. constrictor priapus* (Fig. 10.35F). Within *C. taeniatus* a very different morphology exists (Fig. 10.36C). This species possesses a distinctive, blunt spear-shaped, papillary mound, posteriolaterally projecting orifices, and a deep posterior papillary recess. The recess is flanked by a moderately sized lateral papillary

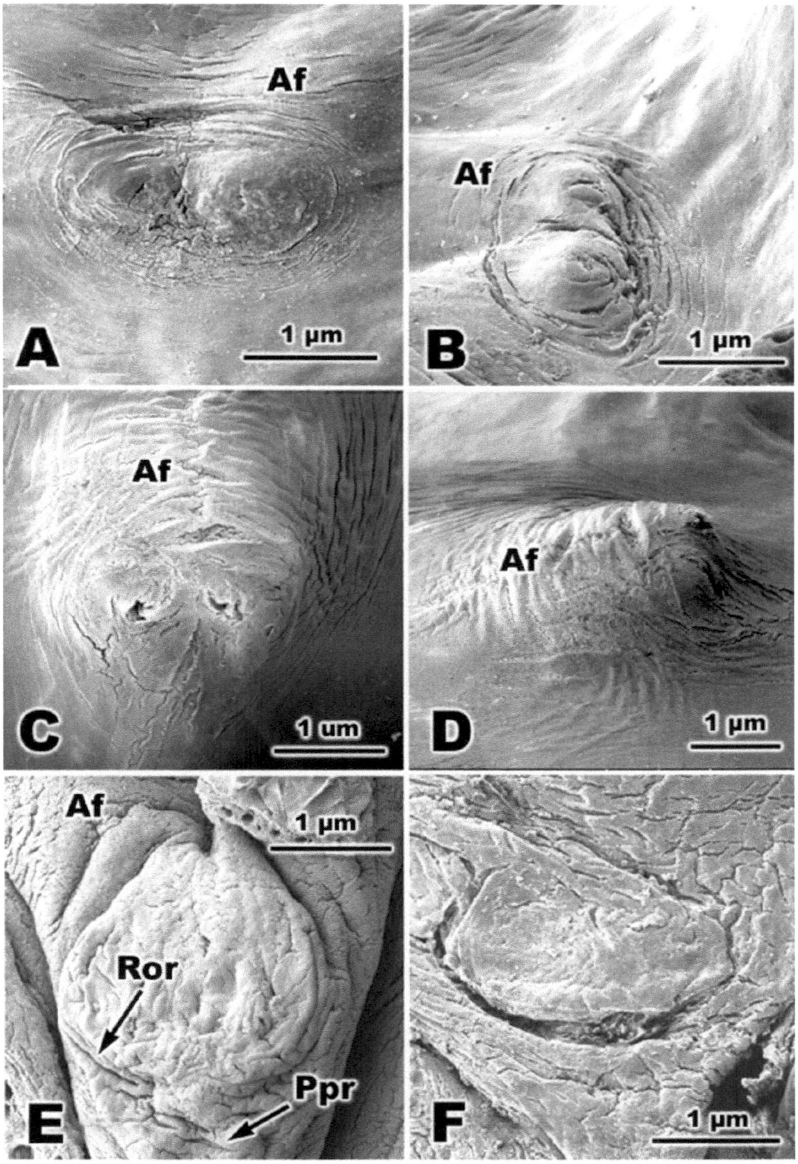

Fig. 10.33 Scanning electron micrographs of the urogenital papilla of representative North American male crotaline snakes. **A.** *Agkistrodon contortrix*, posterioventral view of bi-lobate papillae. **B.** *A. contortrix*, right ventrolateral view of A. **C.** *Agkistrodon piscivorus*, ventral view; urogenital orifices clearly visible. **D.** *A. piscivorus*, right ventrolateral view. An elevated, anterior papillary ridge (Af) leads to the urogenital orifices. **E.** *Crotalus atrox*, ventral view; urogenital orifices project posteriolaterally away from a depressed spatulate-like, urogenital papilla, which lies atop a posterior papillary ridge (Ppr); Right orifice is indicated (Ror). **F.** *Crotalus horridus*, ventral view and anatomical parts similar to E. Af, ampullary fold; Ppr, posterior papillary ridge; Ror, right orifice.

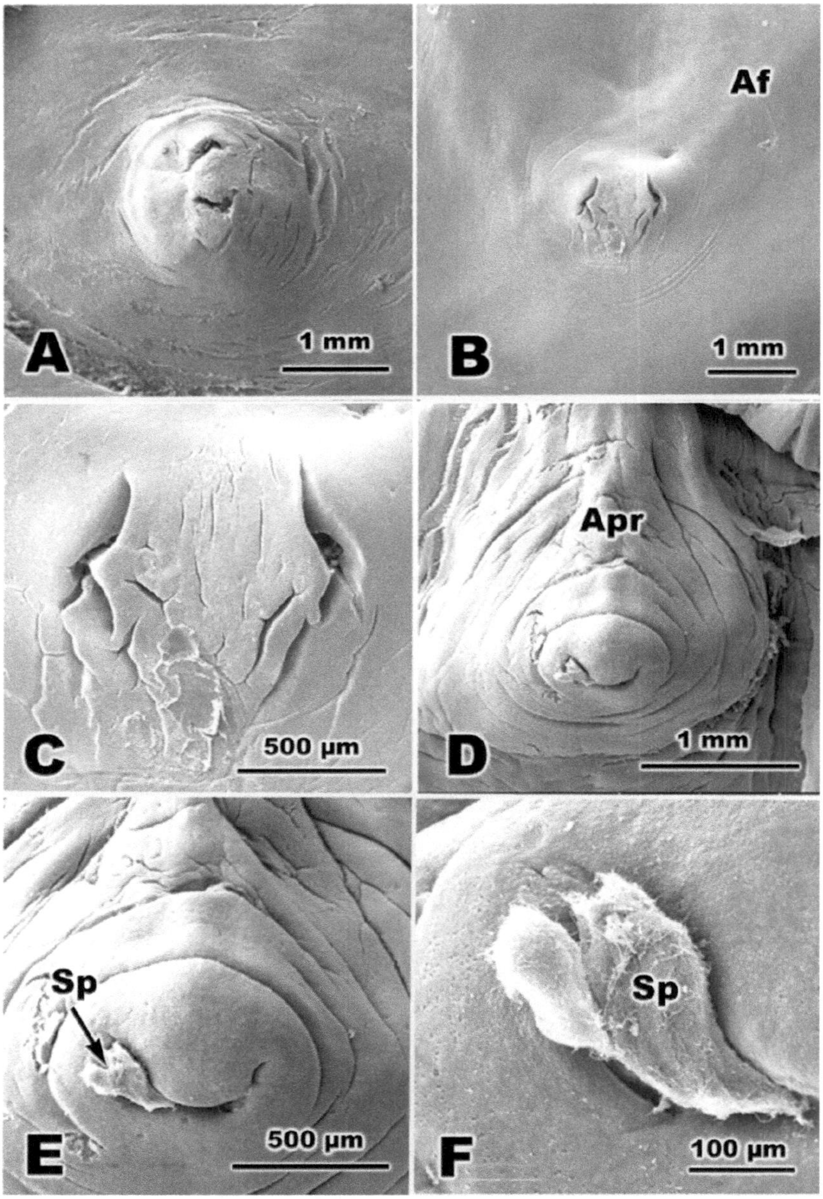

Fig. 10.34 Scanning electron micrographs of the urogenital papilla of representative North American male crotaline snakes. **A.** *Crotalus viridis*, right ventrolateral view. **B-C.** Same species as A, but showing ventral view of opposing, chevron-shaped urogenital orifices (B) and higher magnification (C). **D-F.** *Sistrurus miliarius streckeri*, ventral view of pinnacle-like papillary mound (Apr) and a single urogenital orifice exuding sperm. **E.** Higher magnification of D showing concentric grooves circumscribing pinnacle. **F.** Higher magnification of E showing C-shape urogenital orifice and sperm aggregation. Af, ampullary fold; Apr, anterior papillary ridge; Sp, sperm.

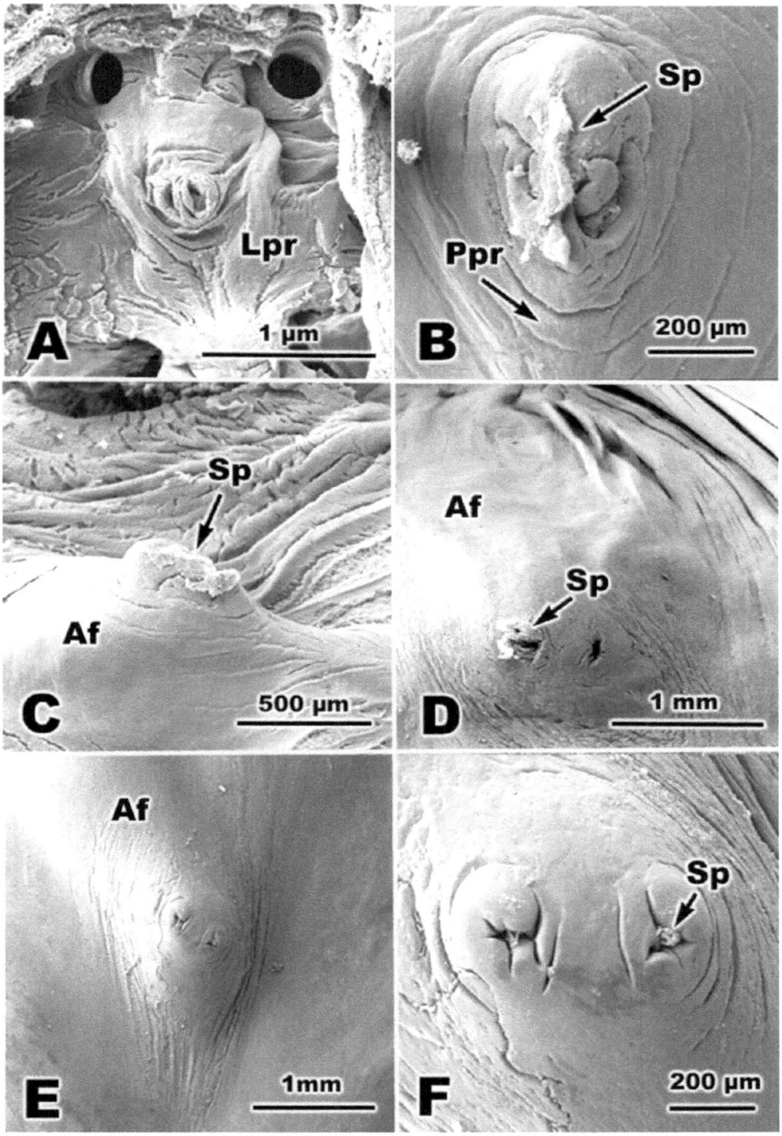

Fig. 10.35 Scanning electron micrographs of the urogenital papilla of representative North American male colubrine colubroid snakes. **A.** *Carphophis vermis*, ventral view showing paired half-moon shaped orifices, a small posterior papillary recess, and lateral papillary ridges (Lpr). **B-C.** *Cemophora coccinea copei*, ventral and right ventrolateral views, respectively, showing elevated tugboat-shaped papillary mound with sperm masses exuding from right orifice. **D.** *Coluber constrictor flaviventris*, ventral view of round papillary stalk with widely spaced orifices; sperm are aggregated at right orifice. **E-F.** *Coluber constrictor priapus*, ventral view of contorted papillary mound (E) not supported by well-defined anterior and posterior papillary ridges; orifices possess fleshy grooved lips (F) with left orifice exuding sperm. Af, ampullary fold; Lpr, lateral papillary ridge; Ppr, posterior papillary ridge; Sp, sperm.

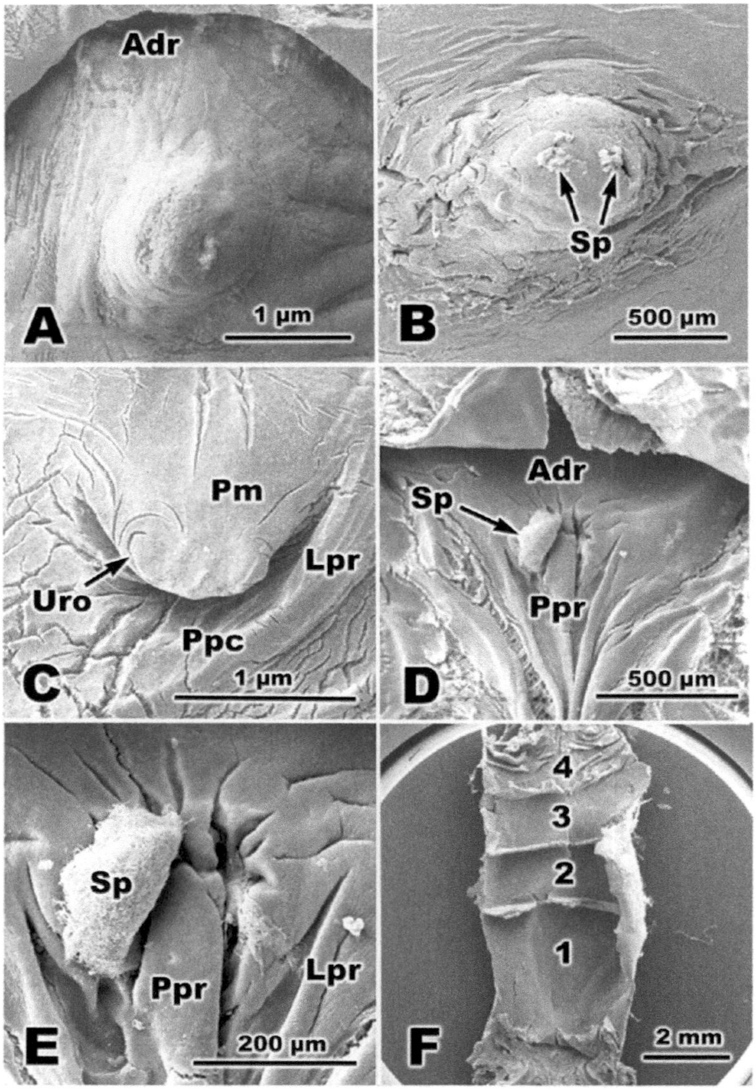

Fig. 10.36 Scanning electron micrographs of the urogenital papilla of representative North American male colubrine colubroid snakes. **A.** *Coluber constrictor latrunculus*, ventral view of circular papillary mound with orifices inconspicuous due to occlusion by sperm; anterior dorsal recess in background. **B.** *Coluber flagellum flagellum*, ventral view of papillary mound similar to A with sperm aggregates occluding orifices. **C.** *Coluber taeniatus*, ventral view of tongue-like papilla with posteriolateral orifices and deep posterior papillary recess. Left urogenital orifice (Uro) is indicated. **D-F.** *Diadophis punctatus arnyi*, ventral view of mostly flat, highly grooved papillary mound with large sperm aggregate occluding right orifice (E). An expansive anterior dorsal recess is evident in D. Coprodaeal chambers (see Fig. 10.20) in F, numbered 1-4, extend anteriorly from cloacal region (as viewed in D). Adr, anterior dorsal recess; Lpr, lateral papillary ridge; Pm, papillary mound; Ppc, posterior papillary recess; Ppr, posterior papillary ridge; Sp, sperm; Uro, urogenital orifice.

ridge on the left size. Recent taxonomic changes removed both *C. flagellum* and *C. taeniatus* from their former genus *Masticophis* (Utiger *et al.* 2005). The inclusion, however, of *M. taeniatus* into *Coluber* could be disputed based upon morphological homologies pertaining to the urogenital papilla among these similar species.

We found the simplest papillary mound in the urogenital papilla of *Diadophis punctatus arnyi* (Fig. 10.36D-F). The mound is not raised above a basal level. Several grooves radiate from the papillary orifices (Fig. 10.36E); these non-fleshy grooves, therefore, preclude the presence of well-defined circular or cup-shaped papillary lips as is the cases with many species in our sample. This species does, however, possess a relatively, large posterior papillary ridge that extends anteriorly through the midline of the mound to a point between the orifices and two lateral papillary ridges. We show the anatomical relationship among coprodaeal chambers and the urogenital papilla in this species (Fig. 10.36F; see also Fig. 10.20D,E).

In *Pantherophis obsoletus* (Fig. 10.37A-C), the urogenital papilla is a robust structure containing well-seated pits and grooves in the papillary summit, well-developed anterior and posterior papillary ridges, a posterior papillary recess, and two cup-shaped urogenital orifices. Anterior lateral ridges, which support the papillary mount, also occur; these prominent ridges are unique to this species.

Among the two xenodontine snakes examined (*Farancia abacura reinwardtii* and *Heterodon platirhinos*), we observed two distinct morphologies. For example, *F. abacura reinwardtii* is the only species among all snakes examined that possesses two distinct urogenital papillae and two distinct posterior papillary recesses (Fig. 10.37D-F). Reproductive and non-reproductive specimens provide contrasting, but congruent, micro-anatomies in this species. In the reproductive individual (Fig. 10.37D,E), no distinct papillary mound exists, but rather bulbous distal segments of the ampullary folds take the place of what should be typical papillary mounds. The folds are highly rugose and closely aligned beside one another. Grooved, fleshy lips bulge around distinctly separate orifices, and spermatozoa can be seen exuding from the left orifice (Fig. 10.37D). In the non-reproductive individual (Fig. 10.37F) the ampullary folds are not rugose, and the bulb-like expansions of the ampullary folds are regressed. In both specimens, the posterior papillary ridge can be seen penetrating far anteriorly and, thus, separating the ampullary folds into two distinct entities. In *H. platirhinos* (Fig. 10.38A), a single urogenital papilla exists. The ampullary folds merge into a single unit and provide support to the papillary mound. The papillary summit and its pair of orifices are not well differentiated in the micrograph.

Marked differences occur between two of the three *Lampropeltis* species (*L. calligaster* and *L. holbrooki*) with the third species (*L. triangulum*) examined (Fig. 10.38B-F). First, both *L. calligaster* and *L. holbrooki* exhibit highly grooved and oblong urogenital orifices, although individually distinct lips are not recognizable in *L. holbrooki*. The species differ from

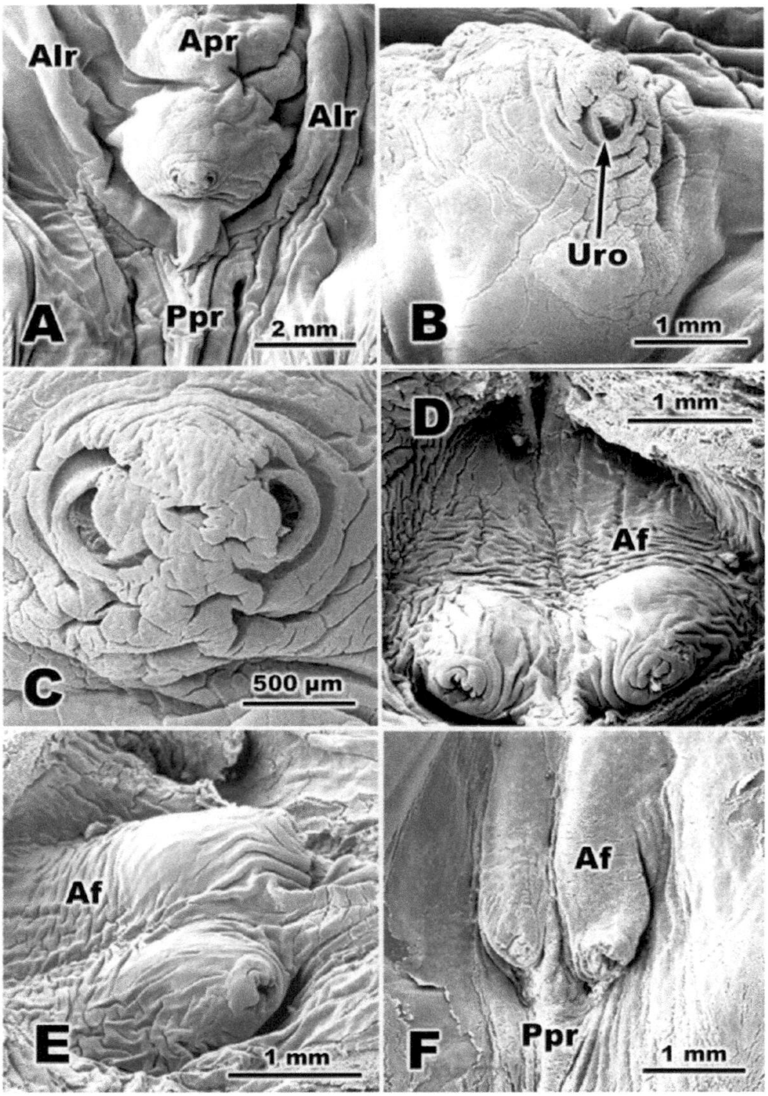

Fig. 10.37 Scanning electron micrographs of the urogenital papilla of representative North American male colubrine colubroid snakes. **A-C.** *Pantherophis obsoletus*, ventral view of pyramidal-shaped papilla with paired (cup-shaped) lateral orifices, well-developed anterior (Apr) and posterior papillary ridges (Ppr), and a small posterior papillary recess. Anterior lateral ridges (Alr) are well developed. Crescent-shaped grooves surround the urogenital orifices. **D-F.** *Farancia abacura reinwardtii*, ventral view of paired urogenital papillae exhibiting bulbous distal segments of the ampullary folds. Fleshy lips surround orifices which project posteriorly, and left orifice in D contains sperm. **E.** Right lateral view of distinct bulbous distal segments seen in D. **F.** Non-reproductive male shows deflated distal segments of papillary mounds (E), and the posterior papillary ridge extends anteriorly between individual papillae. Left and right ampullary folds are distinct. Af, ampullary fold; Alr, anterior lateral ridges; Apr, anterior papillary ridge; Ppc, posterior papillary recess; Ppr, posterior papillary ridge, Uro, urogenital orifice.

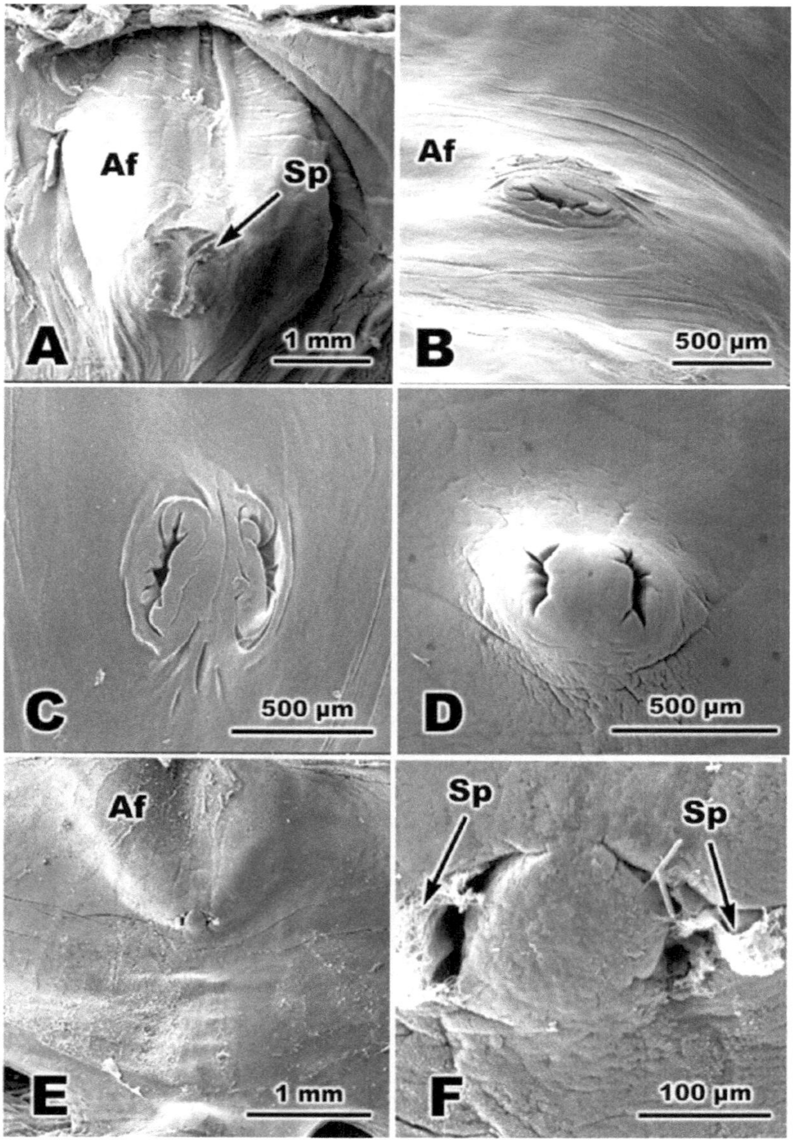

Fig. 10.38 Scanning electron micrographs of the urogenital papilla of representative North American male colubrine colubroid snakes. **A.** *Heterodon platirhinos*, ventral view showing greatly expanded anterior ampullary folds; urogenital orifices mostly obscured and/or occluded by sperm masses. **B-C.** *Lampropeltis calligaster*, right ventrolateral view (B) and ventral view (C) of papillary mound (possibly in a regressed condition?) showing pair of large, grooved, fleshy-lipped orifices. **D.** *Lampropeltis holbrooki*, ventral view of small, nipple-like papilla with pair of large, highly grooved orifices. **E-F.** *Lampropeltis triangulum syspila*, ventral view of papillary region showing no individual papillary mound (E); elevated ampullary folds are evident; F shows crescent-shaped orifices (filled with sperm) separated by a small tissue partition. Af, ampullary fold; Sp, sperm.

one another, however, by the presence of a well-defined, circular papillary mound, which *L. holbrooki* possesses (Fig. 10.38D) and *L. calligaster* generally lacks (Fig. 10.38B,C). The major difference between these two species and *L. triangulum* is the lack of grooved lips within the latter. In *L. triangulum*, the urogenital orifices are also roughly C-shaped slits, and they possess a prominent spherical partition between them (Fig. 10.38F). This micro-structure is not a papillary mound nor is it papilla-like in configuration. Broad ampullary folds are distinctive in *L. triangulum*, but are generally absent in the other two species. All three species lack ridge attachments to the papillary mound (or just the papillary lips) as well as any vestige of a posterior papillary recess.

The natricine watersnakes (Figs 10.39; 10.40A,B) exhibit an array of synapomorphic character states that they share with conspecifics, but at the same time, they also possess distinctive features of their urogenital papillae that are found in other snakes. For instance, four of five species of *Nerodia* (*N. erythrogaster, N. fasciata, N. rhombifer*, and *N. sipedon*) examined possess the following morphologies: 1) the presence of a papillary mound with concentric papillary grooves; 2), the presence of grooved, fleshy papillary lips, and 3) the presence of conspicuous posterior and/or lateral papillary ridges. One synapomorphic character, the presence of a distinctive medial papillary prominence consisting of minute tissue spheres (best observed in Fig. 10.39D, F), is shared by at least *N. rhombifer* and *N. sipedon*. Otherwise, all features mentioned above, except for number 2, are absent from the urogenital papilla of a fifth species of watersnake, *N. cyclopion* (Fig. 10.40B). Alfaro and Arnold (2001) placed *N. cyclopion* and the closely related Florida Green Watersnake (*N. floridana*) in paraphyly with all other *Nerodia*. Our micro-anatomical data, therefore, lends support to their phylogeny of thamnophiine snakes.

Several snake species represent a morphological assortment of character states previously mentioned, and some add unique features of their own to the urogenital papilla micro-anatomy. For example, the urogenital papilla of *Opheodrys aestivus* (Fig. 10.40C,D) exhibits circular urogenital orifices along with smooth, enlarged and elevated, non-grooved lips. The circular orifices are unique to this species. In addition, the lateral papillary ridges are enlarged in this species and, apparently, exclude the development of a posterior papillary ridge. In *Pituophis catenifer sayi*, however, all three posterior ridges are approximately of equal width as they lead away from the papillary mound (Fig. 10.40E). *Pituophis c. sayi* also has a pronounced medial papillary prominence between fleshy, grooved orifices, although the prominence lacks the tissue configuration described earlier for the two *Nerodia* species. Finally, the anterior ampullary folds in *O. aestivus* are somewhat similar to the condition found in *Rhinocheilus lecontei* (Fig. 10.40F).

A morphological disparity exists between the urogenital papillae of two species of *Regina* (Fig. 10.41A,B). In *R. septemvittata*, the orifices are grooved, half-moon shaped openings that lie to the side of the papillary mound.

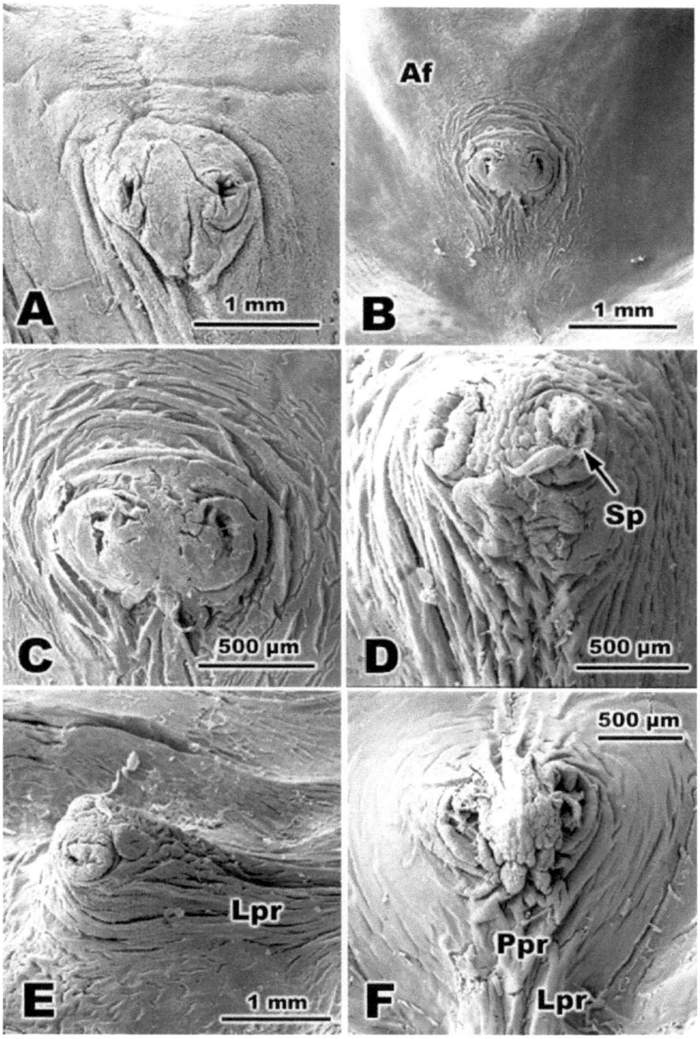

Fig. 10.39 Scanning electron micrographs of the urogenital papilla of representative North American male natricine colubroid snakes. **A.** *Nerodia erythrogaster flavigaster*, ventral view of oblong, smooth papillary mound revealing widely spaced, grooved, fleshy-lipped orifices; a shallow posterior papillary recess is present. **B-C.** *Nerodia fasciata confluens,* ventral view similar to A, but showing instead a circular papillary mound. Ventally directed, paired orifices possess robust lips separated by midventral tissue tuffs (C). Circular grooves surround entire mound, with the exception of the region occupied by the posterior papillary ridge. **D-E.** *Nerodia rhombifer rhombifer*, ventral view of papillary mound characterized by a midventral longitudinal series of 6-8 rows of minute tissue tuffs; fleshy-lipped orifices point ventrally. Sperm mass occludes left orifice (D); right ventrolateral view of D showing lateral papillary ridges extending posteriorly; no posterior papillary ridge present. **F.** *Nerodia sipedon pleuralis* (see also Fig. 10.30A), ventral view showing very similar micro-anatomy as found in D, except that a posterior papillary ridge is present and flanked by lateral papillary ridges. Af, ampullary fold; Lpr, lateral papillary ridge; Ppr, posterior papillary ridge; Sp, sperm.

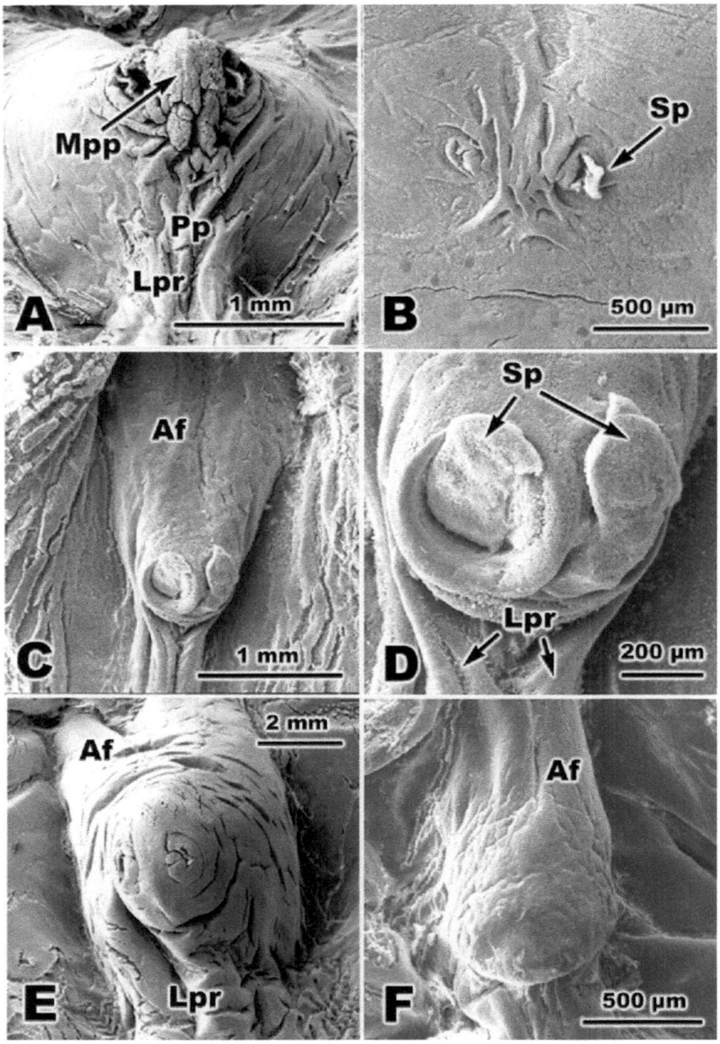

Fig. 10.40 Scanning electron micrographs of the urogenital papilla of representative North American male natricine and colubrine colubroid snakes. **A.** *Nerodia sipedon* (same specimen as Fig. 10.29F), postereoventral view showing elevated position of the midventral rows of papillary tissue tuffs extending beyond the papillary mound; posterior papillary ridge reduced in size and flanked by lateral papillary ridges. Medial papillary prominence (Mpp) exhibiting minute tissue spheres. **B.** *Nerodia cyclopion*, ventral view of flattened mound with no anterior or posterior papillary ridges; midventral tissue tuffs absent; sperm mass resides at mouth of left orifice. **C-D.** *Opheodrys aestivus*, ventral view of papilla (C) showing circular orifices occluded with sperm and fleshy, non-grooved lips; posterior papillary ridge well developed; a small posterior papillary recess present. **E.** *Pituophis catenifer sayi*, ventral view of papillary mound exhibiting fleshy-lipped, crescent-shaped orifices; a small posterior papillary recess present. A well-developed posterior papillary ridge is flanked by lateral papillary ridges. **F.** *Rhinocheilus lecontei*, ventral view of papillary mound with occluded orifices. Af, ampullary fold; Lpr, lateral papillary ridge; Mpp, medial papillary prominence; Pp, papillary prominence; Sp, sperm.

The papillary mound is well supported by all three posterior papillary ridges. In addition, this species has a deep posterior papillary recess and a broad medial papillary prominence. In contrast, *R. grahamii* exhibits posterior-projecting orifices, a narrow medial papillary prominence, and lacks a papillary mound and posterior ridge components. The two species share other features such as grooved ampullary folds, a deep posterior papillary recess, and fleshy lips. Alfaro and Arnold (2001) found these two species to be closely related, whereas the differing papillary morphologies we found in these two species remain equivocal and require further study.

Two colubrine species possessed strikingly different papillary micro-anatomies compared to all other species examined. The papillary mound of *Salvadora grahamiae* (Fig. 10.41C) appears as a tall, laterally compressed pyramid with minute urogenital orifices residing atop a narrow papillary summit. The papillary mound lacks grooves. The ampullary folds project ventrally in the form of distinct ridges. The posterior papillary ridge is absent; instead, the two lateral papillary ridges stand as tall, narrow sheets of tissue. *Sonora semiannulata* (Fig. 10.41D-F) exhibits a conical, slightly grooved, papillary mound that projects posterioventrally. A smooth, pointed, knob-like medial papillary prominence characterizes the papillary summit. The orifices open laterally as distinctive slit-like configurations located at the base of the prominence. No conspicuous posterior ridges and no posterior papillary recess are evident. Well-developed ampullary folds are present in this individual, which was in active reproductive condition.

Two small, closely related, thamnophiine species, *Storeria dekayi wrightorum* and *S. o. occipitomaculata*, share several papillary traits, and, at the same time have features conspicuously different from one another (Fig. 10.42A-D). The grooved, oblong, fleshy-lipped urogenital orifices of both species are clearly similar to the lips of natricine species (Figs. 10.39; 10.40), their allied thamnophiines (Alfaro and Arnold 2001). Both species also exhibit grooved walls in their papillary mounds. Circular grooves exist in *S. d. wrightorum*, whereas mostly diagonal ones are found in *S. o. occipitomaculata*. Both species also exhibit small posterior papillary recesses as well as posterior papillary ridges, although *S. o. occipitomaculata* appears to lack lateral papillary ridges or all three posterior ridges are fused into one structure. The two species differ, however, in the construction of their papillary mounds. *Storeria d. wrightorum* exhibits a somewhat laterally compressed mound base that appears linear in relation to the ampullary folds and the posterior papillary ridges. On the other hand, *S. o. occipitomaculata* possesses a widened, flaring papillary base, which is particularly evident in Fig. 10.42C.

The smallest colubrine species examined, *Tantilla gracilis*, exhibits a disproportionally large urogenital papilla in comparison to its anterior dorsal recess (Fig. 10.42E,F). This species also exhibits a slightly enlarged medial papillary prominence which separates two slit-like orifices on its papillary summit. The base of the papillary mound is circular; lateral papillary ridges provide support to the papillary mound.

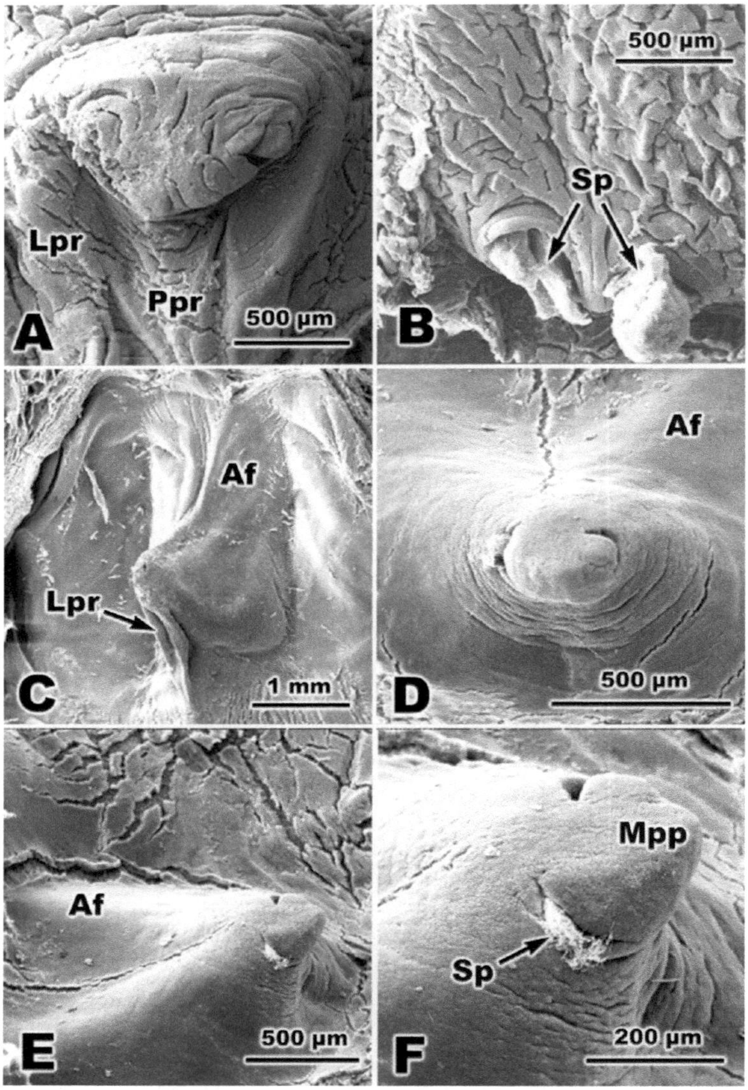

Fig. 10.41 Scanning electron micrographs of the urogenital papilla of representative North American male colubrine colubroid snakes. **A.** *Regina septemvittata*, ventral view of papillary mound characterized by several surface grooves and indentations between posteriolaterally directed orifices. A well-developed posterior papillary recess, posterior papillary ridge, and paired lateral papillary ridges are present. **B.** *Regina grahamii*, ventral view of paired, fleshy-lipped orifices full of sperm; papillary mound flat with highly grooved and pleated surface. **C.** *Salvadora grahamiae,* left ventrolateral view of tall laterally compressed papilla well supported by anterior and posterior papillary ridges. **D-F**. *Sonora semiannulata*, ventral view of conical papilla containing a fleshy medial papillary prominence; laterally directed, paired orifices with non-fleshy lips are present; sperm are existing right orifice; no anterior or posterior papillary ridges present. Af, ampullary fold; Lpr, lateral papillary ridge; Mpp, medial papillary prominence; Ppr, posterior papillary ridge; Sp, sperm.

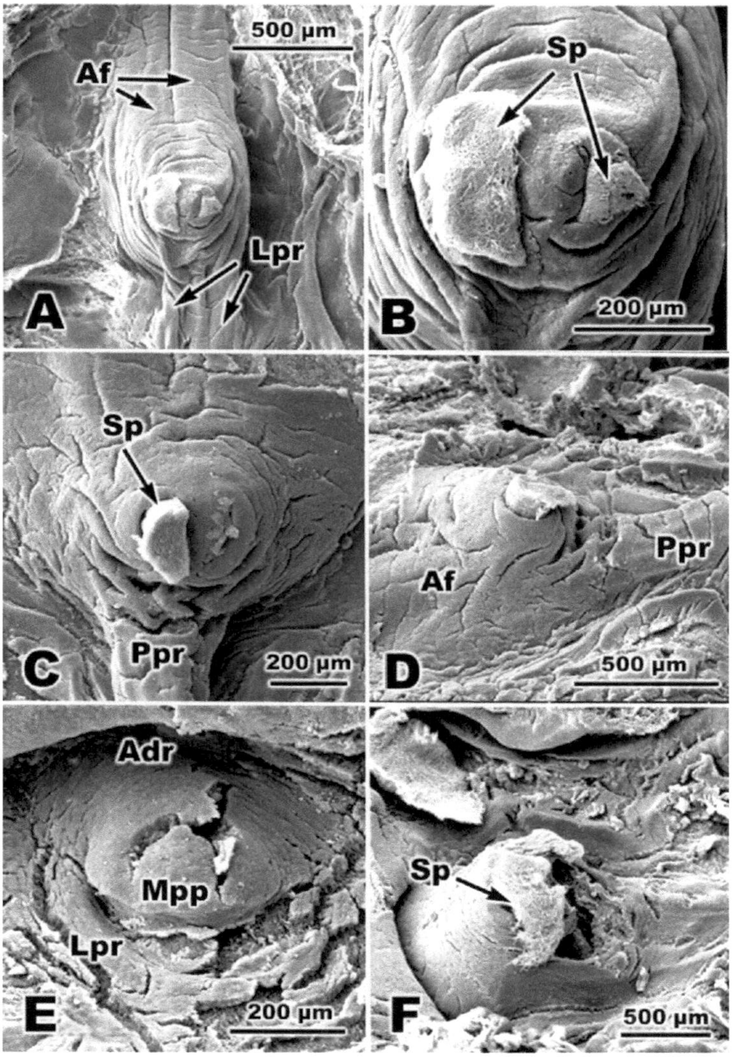

Fig. 10.42 Scanning electron micrographs of the urogenital papilla of representative North American male colubrine colubroid snakes. **A-B.** *Storeria dekayi wrightorum*, ventral view of papilla featuring a laterally compressed papillary mound with large, fleshy-lipped, cup-shaped orifices occluded by sperm masses; posterior papillary recess small; posterior and lateral papillary ridges present. Several circular grooves surround papillary summit. **C-D.** *Storeria occipitomaculata occipitomaculata*, ventral view of papillary summit (C); mound conical with orifices similar to anatomy found in A-B. Right ventrolateral view (D) of single, large posterior papillary ridge; posterior papillary recess small. **E-F.** *Tantilla gracilis*, posteriolateral view of small conical, papillary mound with two slit-like orifices with no fleshy lips (E). Right ventrolateral view (F) of different individual from one shown in E revealing papilla in close proximity to anterior dorsal recess (Adr). Sperm fill the orifices. Papillae of both E and F lack circular grooves. Adr, anterior dorsal recess; Af, ampullary fold; Lpr, lateral papillary ridge; Mpp, medial papillary prominence; Ppr, posterior papillary ridge; Sp, sperm.

Four thamnophiine species (two large-bodied: *Thamnophis proximus proximus* and *T. sirtalis sirtalis* and two diminutive species: *Virginia striatula* and *V. valeriae*) provide additional papillary peculiarities. Although both *T. p. proximus* and *T. s. sirtalis* (Fig. 10.43A,B) exhibit fleshy-lipped orifices as is found among other related species mentioned above, the papillary mound is prominent in *T. p. proximus* and it is nearly non-existent in *T. s. sirtalis*. Both species also exhibit numerous grooves and depressions within their papillary structures, but *T. s. sirtalis* has the most furrowed ampullary folds of all species examined. In addition, this species has dramatically shortened lateral papillary ridges, a very narrow posterior papillary ridge, and a posterior papillary recess.

Distinctive papillary morphologies are found in the two species of *Virginia*. A most peculiar inverted bowl-like papillary base supports a small, circular, recessed papillary mound in *V. striatula* (Fig. 10.43C,D). Extremely large, irregular, cup-shaped and fleshy-lipped orifices dominate the surface of its papillary summit. On either side of the mound are tall posterior papillary ridges that attach firmly to the oval papillary base and not to the papillary mound, *per se*. In addition, two short anterior papillary ridges attach directly to the papillary summit and extend anteriorly for a short distance (Fig. 10.43D). The oval papillary base, indeed, represents an extreme morphology for the paired ampullary folds, which in this specimen, are undoubtedly filled with sperm. Finally, the micro-structure of the urogenital papilla in *V. valeriae* is similar to that of *V. striatula*, except for the lack of well-developed fleshy lips surrounding the orifices. Both species exhibit posterior papillary recesses, although this region is regressed in *V. valeriae* even though both are reproductive individuals. These two species exhibit the most extreme variation in papillary micro-anatomy compared to the basic urogenital papilla morphology shown in Fig. 10.32.

10.5 CONCLUSIONS AND DIRECTIONS FOR FUTURE RESEARCH

The only reproductive tract characters that have been coded for use in recent phylogenetic analyses concern morphology of sperm, hemipenes, and/or cloacal glands (Jamieson 1995; Oliver *et al.* 1996; Young *et al.* 1999; Lee 2000; Sánchez-Martínez 2007). The only broadly based comparative study of the accessory duct system of squamates is that of Dufaure and Saint Girons (1984) who examined histology of the epididymides of 89 species of squamates, representing 72 genera and 18 families, including 25 species of snakes. They found 5 types of secretory activity in the main duct of the epididymis. Type 1 consists of large secretory granules with a chromophilic central core surrounded by a chromophobic vacuole, and this type characterizes the Lacertidae. Types 2-4 show decreasing size and density of secretory products, and Type 5 demonstrates no secretory activity (Table 10.2). Interestingly, type 5 characterizes all snakes that were examined. Dufaure and Saint Girons (1984) noted that differences especially between types 3 and 4 are not considerable, and that electron

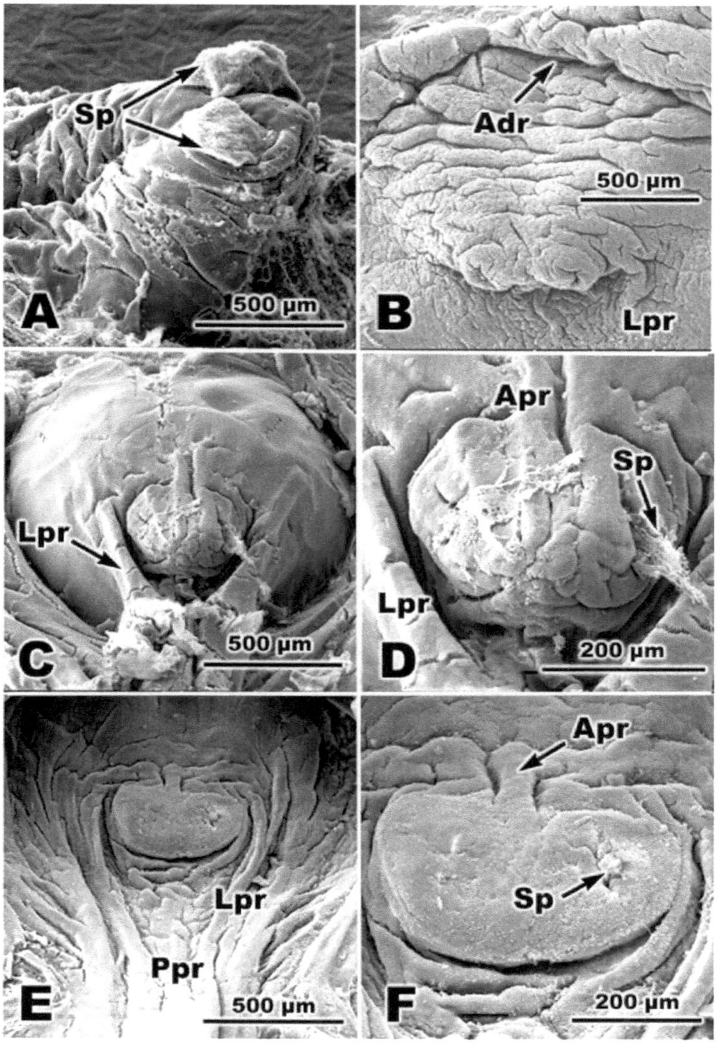

Fig. 10.43 Scanning electron micrographs of the urogenital papilla of representative North American male thamnophiine colubroid snakes. **A.** *Thamnophis proximus proximus*, right ventrolateral view of papillary mound with smooth, fleshy-lipped orifices filled with sperm. Numerous circular papillary grooves surround papilla. **B.** *Thamnophis sirtalis sirtalis*, ventral view of papillary mound exhibiting numerous transverse folds and ridges; anterior dorsal recess conspicuous. Orifices with fleshy lips and are cup-shaped in appearance. **C-D.** *Virginia striatula*, ventral view of papillary mound which is recessed into a cavity between the large lateral papillary ridges and resides atop bulbous basilar region. Two small, anterior papillary ridges are attached to papillary mound (C). Higher magnification of C in D reveals fleshy-lipped and grooved orifices exuding sperm. **E-F.** *Virginia valeriae*, ventral view of papillary mound similar to that found in C, but no bulbous basilar region. Flat, hemispheric papillary summit with two groove-lipped orifices. Large lateral papillary ridges present; single anterior papillary ridge; posterior papillary recess large. Sperm are exiting left orifice. Adr, anterior dorsal recess; Apr, anterior papillary ridge; Lpr, lateral papillary ridge; Ppr, posterior papillary ridge; Sp, sperm.

Table 10.2 States of secretory activity in the epididymides of squamates as described by Dufaure and Saint Girons (1984). They described 5 "Types" and an intermediate stage, which are coded as 6 different character states.

Character State	*Families*	*Description*
1 (Type 1)	Lacertidae	numerous large vacuoles 10-12 μm dia with a central dense core
2 (Type 2)	Agamidae	clusters of apical dense granules 1 μm dia
3 (Type 3)	Iguanidae, some Gekkonidae	finer granules than Type 2 and not clustered
4 (Intermediate)	Anguidae, Amphisbaenidae	in between Types 3 and 4, granulation apparent in dense cytoplasm
5 (Type 4)	Chameleonidae, Iguanidae, Teiidae, Polychrotidae, Scincidae, some Gekkonidae	granules not distinct but cytoplasm dense and heterogeneous; secretory product in lumen
6 (Type 5)	Serpentes	secretory activity completely absent

microscopy is needed to fully resolve the differences in secretory activity in the epididymis among squamates.

When mapped upon a phylogeny from Lee (2005) based upon morphology and a phylogeny from Vidal and Hedges (2005) inferred from nine nuclear protein-coding genes, the results are shown on the cladograms illustrated in Fig. 10.44. Relationships shown among taxa are limited to those groups examined by Dufaure and Saint Girons (1984). Despite the differences in the morphological and molecular phyletic hypotheses, the mapping results are consistent in the two cladograms. The ancestral state is 5 (Type 4), a state in which secretory granules are not present but the cytoplasm appears active as it is dense and heterogenous. Dufaure and Saint Girons (1984) also indicate that even when secretory granules are not present, some secretory product appears associated with sperm in the lumen of the epididymis, and this may have come from the epididymal epithelium. The most interesting conclusion, however, is that snakes lost all secretory activity whereas lizards either retained the ancestral condition or evolved increased secretory activity. Thus, basal snakes went in a different evolutionary direction from lizards.

The interesting question that follows is why this variation in secretory activity exists. The occurrence of an epididymis as a sperm conduit between the testis and ductus deferens is most parsimoniously considered an ancestral character for amniotes (or else it independently evolved numerous times). Thus, the epididymis is a very ancient character that has been evolving separately in major clades for hundreds of millions of years. We know that some functional differences occur in the epididymides of these groups (for example, in mammals the epididymis is the site for sperm storage and maturation whereas in squamates the ductus deferens is the major sperm storage area), but much of the basic anatomy seems to

Fig. 10.44 Epididymal secretions from most copious (1) to none (6) based upon data from Dufaure and Saint Girons (1984) and mapped on phyletic hypotheses using McClade version 4.1. **A.** A phyletic hypothesis based upon morphological characters from Lee (2005). **B.** A phyletic hypothesis based upon molecular characters by Vidal and Hedges (2005).

be conserved. If the epididymal secretions in some lizards play a role in formation of the seminal fluid, capacitation of sperm, etc., those functions must occur elsewhere in snakes.

As noted here and by Sever (2010), however, secretory activity is apparent in the two snakes, *Seminatrix pygaea* and *Agkistrodon piscivorus*, in which the ductus epididymis has been examined by electron microscopy. *Seminatrix pygaea* and *A. piscivorus* possess a constitutive secretory pathway, similar to that of the proximal efferent ducts of mammals (Hoffer *et al.* 1973), in which the product is transported to the surface in small vesicles that Dufaure and Saint Girons (1984) would not have observed with light microscopy. Thus, the product is not concentrated or stored in granules while waiting for a neural or hormonal stimulus, but released as it is produced. In the case of *S. pygaea*, production seems to be year-around, although somewhat reduced in the summer. The product is both PAS positive and bromphenol blue positive, and therefore contains glycoprotein.

Studies on lizards demonstrate that secretions of the epididymis may contain lipid (Haider 1987), proteins (Depeiges and Dufaure 1980), and/or glycoproteins (Manimekalai and Akbarsha 1992). Labate *et al.* (1997) found a variety of glycoconjugates using lectin histochemistry in the ductuli efferentes of the European Wall Lizard (*Podarcis sicula*), and the binding pattern varied seasonally. In the same species, Desantis *et al.* (2002) found that secretory cells along the length of the epididymis have 12 lectins binding to the secretions, with some regionalization. Although the function of the glycoconjugates in the epididymal secretory granules was not determined, perhaps they produce an environment for sperm storage and/or maturation, as known in mammals (Acott and Hoskins 1981).

We have examined the male urogenital ducts and cloacae of only a small subset of the 2800+ extant species of snakes. We do not currently have enough comparative material to enable hypotheses of ancestral and derived traits of the squamate reproductive tract and the use of such data in phylogenetic analyses. The primary direction for future research is obvious: examine more species from a wide range of taxa with special attention to cytological variation and secretory activity of the reproductive ducts. Then, the challenge is to place these data in a phylogenetic context.

Our conclusions, therefore, at this time are modest. (1) The proximal testicular ducts in snakes are probably similar in structure and function to those of other amniotes. We need to look for regionalization of cell types and functions in the ductus epididymis, as well known in mammals and reported in some lizards. (2) The ductus deferens stores sperm in snakes, but how long sperm can remain viable is unknown. Also, secretory activity of the ductus deferens varies among the species examined so far, and the significance of this variation needs resolution. (3) Colubrines examined so far possess an ampulla urogenital papilla, whereas viperids have an ampulla ureter. We need to discover whether these differences are consistent in other taxa in these groups and discern the ancestral condition for Serpentes, including the examination of non-colubroid snakes. (4) The urogenital papilla shows wide variation among the taxa examined so far. Some of this variation is concordant with phylogenetic hypotheses for these groups, but again, snakes from more basal taxa require examination.

10.6 ACKNOWLEDGMENTS

DMS acknowledges with much appreciation support from National Science Foundation grant DEB-0809831. We also thank Layla Freeborn and Justin Rheubert for their input into this manuscript. SET sincerely thanks the many snake collectors, especially Richard Baxter, Matt Connior, Richard F. Hoyer, Phillip Jordan, Chris T. McAllister, Jonathan Stanley, and Benjamin Wheeler, who provided live specimens. Megan Bundy is thanked for the line drawings in Figs. 10.20 and 10.28. Kevin Gribbins is appreciated for his insightful advice during the early preparation of this manuscript. Finally, we thank Barrie G. M. Jamieson and Dustin S. Siegel for their critical reviews of the chapter.

10.7 LITERATURE CITED

Acott, T. S. and Hoskins, D. D. 1981. Bovine sperm forward motility protein: binding to epididymal spermatozoa. Biology of Reproduction 24: 234-240.

Aire, T. A. 2007. Anatomy of the testis and male reproductive tract. Pp. 37-113. In B. G. M. Jamieson (ed.), *Reproductive Biology and Phylogeny of Birds*, Vol. 6A. Science Publishers, Enfield, New Hampshire.

Akbarsha, M. A. and Meeran, M. M. 1995. Occurrence of ampulla in the ductus deferens of the Indian garden lizard *Calotes versicolor* Daudin. Journal of Morphology 225: 261-268.

Akbarsha, M. A., Kadalmani, B. and Tamilarasan, V. 2007. Efferent ductules of the fan-throated lizard *Sitana ponticeriana* Cuvier; light and transmission electron microscopy study. Acta Zoologica (Stockholm) 88: 265-274.

Akbarsha, M. A., Tamilarasan, V., Kadalmani, B. and Daisy, P. 2005. Ultrastructural evidence for secretion the epithelium of ampulla ductus deferentis of the Fan-Throated Lizard, *Sitana ponticeriana* Cuvier. Journal of Morphology 266: 94-111.

Alfaro, M. E. and Arnold, S. J. 2001. Molecular systematics and evolution of *Regina* and the thamnophiine snakes. Molecular Phylogenetics and Evolution 21: 408-423.

Bellairs, A. 1970. *The Life of Reptiles*. Volumes 1 and 2. Universe Books, New York. Pp. 590.

Bergerson, W., Amselgruber, W., Sinowatz, F. and Bergerson, M. 1994. Morphological evidence of sperm maturation in the ampulla ductus deferentis of the bull. Cell and Tissue Research 275: 537-541.

Cooper, T. G. and Hamilton, D. W. 1977. Phagocytosis of spermatozoa in the terminal region and gland of the vas deferens of the rat. American Journal of Anatomy 150: 247-268.

Depeiges, A. and Dufaure, J. P. 1980. Major proteins secreted by the epididymis of *Lacerta vivipara*. Isolation and characterization by electrophoresis of the central core. Biochimica et Biophysica Acta 628: 109-115.

Desantis, S., Labate, M., Labate, G. M. and Cirillo, F. 2002. Evidence of regional differences in the lectin histochemistry along the ductus epididymis of the lizard, *Podarcis sicula* Raf. Histochemical Journal 34: 123-130.

Dufaure, J. and Saint Girons, H. 1984. Histologie compare de l'epididyme et de ses secretions chez les reptiles (lezards et serpents). Archives d'anatomie microscopique et de morphologie expérimentale 73 : 15-26.

Fox, H. 1977. The urinogenital system of reptiles. Pp. 1-157. In C. Gans and T. S. Parsons, (eds), *Biology of the Reptilia*, Vol. 6. Academic Press, New York.

Fox, W. 1952. Seasonal variation in the male reproductive system of Pacific coast garter snakes. Journal of Morphology 90: 481-553.

Fox, W. 1965. A comparison of the male urogenital systems of blind snakes, Leptotyphlopidae and Typhlopidae. Herpetologica 21: 241-256.

Gabe, M. and Saint Girons, H. 1965. Contribution à la morphologie comparée du cloaque et des glandes épidermoïdes de la region cloacale chez les lépidosauriens. Mémoires du Muséum National d'Histoire Naturelle. Séries A Zoologie 33: 149-292.

Gribbins, K. M., Happ, C. S. and Sever, D. M. 2005. Ultrastructure of the reproductive system of the black swamp snake (*Seminatrix pygaea*). V. The temporal germ cell development of the testis. Acta Zoologica (Stockholm) 86: 223-230.

Guerrero, S. M., Calderón, M. L., Pérez, G. R. and Pinilla, M. P. R. 2004. Morphology of the male reproductive duct system of *Caiman crocodiles* (Crocodylia, Alligatoridae). Annals of Anatomy 186: 235-245.

Haider, S. and Rai, U. 1987. Epididymis of the Indian wall lizard (*Hemidactylus flaviviridis*) during the sexual cycle and in response to mammalian pituitary gonadotropins and testosterone. Journal of Morphology 191: 151-160.

Hess, R. A. 2002. The efferent ductules: structure and functions. Pp. 49-80. In B. Robaire and B. T. Hinton (eds), *The Epididymis: From Molecules to Clinical Practice*. Kluwer Academic/Plenum Press, New York.

Hoffer, A. P., Hamilton, D. W. and Fawcett, D. W. 1973. The ultrastructure of the principal cells and intraepithelial leuckocytes in the initial segment of the rat epididymis. Anatomical Record 175: 169-202.

Holmes, H. J. and Gist, D. H. 2004. Excurrent duct system of the male turtle *Chrysemys picta*. Journal of Morphology 261: 312-322.

Jamieson, B. G. M. 1995. The ultrastructure of spermatozoa of the Squamata (Reptilia) with phylogenetic considerations. Pp. 359-383. In B. G. M. Jamieson, J. Ausio and J.-L. Justin (eds), *Advances in Spermatozoal Phylogeny and Taxonomy.* Mémoires du Muséum National d'Histoire Naturelle 166, Paris.

Jones, R. C. 1998. Evolution of the vertebrate epididymis. Journal of Reproduction and Fertilization Supplement 53: 163-181.

Labate, M., Desantis, S. and Corriero, A. 1997. Glycoconjugates during the annual sexual cycle in lizard epididymal ductuli efferentes: a histochemical study. European Journal of Histochemistry 41: 47-56.

Lee, M. S. 2000. Soft anatomy, diffuse homoplasy, and the relationships of lizards and snakes. Zoologica Scripta 29: 101-130.

Lee, M. S. Y. 2005. Squamate phylogeny, taxon sampling, and data congruence. Organisms, Diversity & Evolution 5: 25-45.

Lombardi, J. 1998. *Comparative Vertebrate Reproduction*. Kluwer Academic Publishers, Boston, Massachusetts. Pp. 469.

Manimekalai, M. and Akbarsha, M. A. 1992. Secretion of glycoprotein granules in the epididymis of the agamid lizard *Calotes versicolor* (Daudin) is region-specific. Biology of Structural Morphogenesis 4: 96-101.

Martin Saint Ange, G.-J. 1854. Étude de l'appariel reproducteur dans les cinq classes d'animaux vertébrés, au point de vue anatomique, physiolgique et zoologique. J.-B. Balliére, Libraire de l'Académe Impériale de Médecine, Paris. Pp. 234.

Oliver, S. C., Jamieson, B. G. M. and Scheltinga, D. M. 1966. The ultrastructure of spermatozoa of Squamata. II. Agamidae, Varanidae, Colubridae, Elapidae, and Boidae (Reptilia). Herpetologica 52: 216-241.

Riva, A., Testa-Riva, F., Usai, E. and Cossu, M. 1982. The ampulla ductus deferentis in man, as viewed by SEM and TEM. Archives of Andrology 8: 157-164.

Sánchez-Martínez, P. M., Ramirez-Pinilla, M. P. and Miranda-Esquivel, D. R. 2007. Comparative histology of the vaginal-cloacal region in Squamata and its phylogenetic implications. Acta Zoologica (Stockholm) 88: 289-307.

Seshardi, C. 1959. Structural modification of the cloaca of *Lycodon aulicus aulicus*, Linn., in relation to urine excretion and the presence of sexual segment in the kidney of male. Proceedings of the National Institute of Science of India B 25: B: 271-278.

Setchell, B. P., Maddocks, S. and Brooks, D. E. 1994. Anatomy, vasculature, innervation, and fluids of the male reproductive tract. Pp. 1063-1175. In E. Knobil and J. D. Neill (eds), *The Physiology of Reproduction*, 2nd ed., Raven Press, New York.

Sever, D. M. 2004. Ultrastructure of the reproductive system of the black swamp snake (*Seminatrix pygaea*). IV. Occurrence of an ampulla ductus deferentis. Journal of Morphology 262: 714-730.

Sever, D. M. 2010. Ultrastructure of the reproductive system of the black swamp snake (*Seminatrix pygaea*). IV. The proximal testicular ducts. Journal of Morphology 271: 104-115.

Shine, R. 1991. *Australian Snakes: A Natural History*. Cornell University Press, Ithaca, New York. Pp. 223.

Siegel, D. S., Sever, D. M., Rheubert, J. L. and Gribbins, K. M. 2009. Reproductive biology of *Agkistrodon piscivorus* Lacépède (Squamata, Serpentes, Viperidae, Crotalinae). Herpetological Monographs 23: 74-107.

Vidal, N. and Hedges, S. B. 2005. The phylogeny of squamate reptiles (lizards, snakes, and amphisbaenians) inferred from nine nuclear protein-coding genes. Comptes Rendus Biologies 328: 1000-1008.

Vitt, L. J. and J. P. Caldwell. 2009. *Herpetology*, 3rd ed., Elsevier Inc., Burlington, Massachusetts. Pp. 697.

Volsøe, H. 1944. Structure and seasonal variation of the male reproductive organs of *Vipera berus* (L.). Spolia Zoologica Musei Hauniensis 5: 1-157.

Young, B. A., Marsit, C. and Meltzer, K. 1999. Comparative morphology of the cloacal scent gland in snakes (Serpentes: Reptilia). Anatomical Record 256: 127-138.

CHAPTER 11

The Sexual Segment of the Kidney

Robert D. Aldridge[1], *Benjamin C. Jellen*[1], *Dustin S. Siegel*[1] *and Samantha S. Wisniewski*[1, 2]

11.1 INTRODUCTION

Differences between males and females are generally divided into two categories: primary sexual characteristics, which include the gonads, efferent ducts and copulatory organs; and secondary sexual characteristics, which include all other physical and behavioral differences. Secondary sexual characteristics are generally thought to be brought about by sexual selection. A common type of sexual selection in snakes is precopulatory intrasexual selection, in which the trait or behavior provides an advantage to the male in competition with other males (i.e., body size, Shine 1978). Within snakes, secondary sexual characteristics include: differences in adult size (Shine 1994), color (Northern Viper, *Vipera berus*, Shine and Madsen 1994; Rock Rattlesnake, *Crotalus lepidus klauberi*, Jacob and Altenbach 1977), body proportions (*V. berus*, Madsen 1988), special ornamentations (nasal extension, *Langaha*, Greene 1997; spurs of boids, Carpenter *et al.* 1978) and behaviors (male-male combat, Shine 1994; mate searching behaviors, Duvall and Schuett 1997; Shine *et al.* 2005). The majority of these differences are developed to assist males in prenuptial mating endeavors. However, not all sexual dimorphism is a result of sexual selection. Shine and Madsen (1994) suggest that the brightly colored male *V. berus* may be a result of natural selection acting selectively on males during the mating season. They suggest that the bright color of males causes a "flicker-fusion" effect in which moving males, while searching for females, become difficult for predators to focus on and to assess the direction and velocity of the viper's movement. Lindell and Forsman (1996) found support for this hypothesis in a field study of *V. berus* on islands east of Sweden. They reported that

[1]Department of Zoology, Saint Louis University, St. Louis, Missouri 63103, USA
[2]Caesar Kleberg Wildlife Research Institute, Texas A&M University - Kingsville, Kingsville, Texas 78363, USA

the survival rate of brightly colored males was significantly higher than melanistic males.

One of the most prominent secondary sexual structures in squamates, the sexual segment of the kidney (SSK), may have evolved via natural selection, to enhance the survival of sperm, but also as a result of postcopulatory intrasexual competition, to preclude the female from mating with subsequent males. In either case, the SSK appears to have evolved primarily to assist in postcopulatory reproductive endeavors. This chapter describes the SSK and its role in the reproduction of snakes.

11.1.1 Sexual Role of the Kidney in Vertebrates

In the evolution of the vertebrates, the testis has a history of usurping parts of the kidney nephron and ducts to transport sperm to the copulatory organs (Romer and Parsons 1986). However, the development of an additional (secondary) reproductive function has occurred only three times in vertebrates; in gasterosteid fish, in some urodeles and in squamates. In gasterosteids (sticklebacks), the highly secretory region of the ventral kidney ducts produces a protein called "spiggin" that is the major constituent of the foam nests of stickleback fishes (Courrier 1922; Craig-Bennet 1931; Jakobsson *et al.* 1999). In urodeles, a secondary sexual portion of the kidney has been described by Aron (1924) in the kidney of the male Alpine Newt (*Ichthyosaura alpestris*) of the Salamandridae. The presence of a SSK in the Central Newt (*Notophthalmus viridescens*) has also been discovered (Siegel and Aldridge, unpublished observations). The function and how taxonomically widespread the SSK in salamanders is unknown. The presence of the SSK in squamates, specifically snakes, is the subject of this chapter. In the three taxa above (fish, salamanders and squamates), the SSK is androgen dependent and appears to have evolved independently (Siegel and Aldridge, unpublished data).

11.1.2 Terms for the Sexual Segment

A variety of terms (and abbreviations) have been used by various authors to refer to the sexual segment of the kidney in snakes. These terms include "renal sexual segment" (Deb and Sarkar 1963) and its abbreviation RSS (Krohmer *et al.* 1987) and "kidney sexual segment" and its abbreviation KSS (Mathies *et al.* 2010). Regaud and Policard (1903) used the phrase "segment sexuel" in their original description of this structure in snakes. Volsøe (1944) also used this phrase to discuss the sexual segment. Saint Girons used the French phrase "segment sexual du rein" in his first paper on this topic (Saint Girons 1957) and continued to use this phrase in all of his studies on this structure. Numerous investigators continued to use sexual segment in their publications (Fox 1952, 1954; Bishop 1959; Platt 1969; Licht and Pearson 1969 and many others). However, several papers by Licht (Licht and Donaldson 1969; Licht and Tsui 1975) used the phrase renal sexual segment. The most recent investigations on the morphology of the sexual segment have utilized both renal sexual segment (Sever *et al.* 2002; Sever *et al.* 2008) and sexual segment of the kidney (Siegel *et al.* 2009). In this chapter we use the phrase sexual segment of the kidney (SSK) because of the historical precedence of this term.

11.1.3 The Sexual Segment in Lizards

Information on the SSK and its secretions in the reproduction of lizards is limited. Regaud and Policard (1903) were the first to name the squamate SSK the "sex segment" and describe its presence in lizards. Reiss (1923) showed that the SSK in two species of *Lacerta* exhibited a seasonal cycle. Later studies by Herlant (1933), Regamey (1935) and Misra *et al.* (1966) established the link between androgens and SSK development.

As in snakes, the SSK is primarily developed in males and is a modified portion of the nephron and ureter (Sever and Hopkins 2005). The specific portions of the nephron that develop into the SSK varies in different taxa (see Fox 1977). Development of the SSK in female lizards has been described in two taxa (*Cnemidophorus,* Del Conte 1972; Del Conte and Tamayo 1973; *Scincella lateralis,* Sever and Hopkins 2005). In both of these, the SSK in females was in the same location as in males, but was considerably smaller. In these studies, the authors suggested that females had a low level of natural androgens which caused the development (Del Conte and Tamayo 1973; Sever and Hopkins 2005). No function was proposed in either species.

Most species of lizards exhibit pre-nuptial spermatogenesis, a pattern in which spermatozoa are produced prior to and during the mating season. In these species, the hypertrophy and recrudescence of the SSK is synchronous with androgen secretion and spermatogenic activity (Sever and Hopkins 2005).

The products and possible functions of the SSK secretions in lizards are similar those of snakes. In some taxa, SSK secretions form a copulatory plug (*Lacerta,* Moreira and Birkhead 2003), while in others the secretions do not form a copulatory plug (the Dragon Lizard, *Ctenophorus,* Olsson 2001). Whether a plug forms or not, the SSK secretions do not appear to affect subsequent matings by the female. Olsson (2001) reported that SSK secretions do not form a copulatory plug and do not affect the ability of the female to mate immediately following the first mating.

Studies on the Iberian Rock Lizard (*Lacerta monticola*) showed that the SSK secretions form a hard copulatory plug that adheres firmly to the female's cloaca following copulation and occludes both oviducal openings (Moreira and Birkhead 2003). Subsequent behavioral tests on this lizard showed that the plug does not prevent subsequent matings nor does it reduce the female's attractiveness (Moreira and Birkhead 2003). Moriera *et al.* (2006) presented evidence that the copulatory plug may contain chemicals which convey information about the male's identity and dominance and serve as a type of intraspecific communication.

Cuellar *et al.* (1972) published the only experimental study on the functions of SSK secretions. This *in vitro* study examined the role of SSK secretions on sperm motility in the Green Anole (*Anolis carolinensis*). The SSK secretions were obtained by macerating the kidney. They found that the motility of sperm, after 24 min, was significantly greater when mixed with SSK secretions compared to female kidney extracts and saline. However, because sperm are viable days to weeks following mating, the relevance of these data is questionable.

11.2 OVERVIEW OF THE SEXUAL SEGMENT IN SNAKES

The SSK is a specialized portion of the nephron, whose location is described as terminal or preterminal in various squamates (Fox 1977). In many species the SSK tubules are visible macroscopically and make up a major part of the mass of the kidney (Fig. 11.1). In snakes, the SSK is present in all males (Fox 1977) but can be stimulated by androgens in females (Krohmer *et al.* 2004). The SSK is androgen dependent, with ontogenic development (Krohmer 2004) and seasonal hypertrophy (Bishop 1959; Fox 1977) dependent on elevated plasma testosterone levels. The SSK secretions are transferred to the female during copulation (Fox 1977). The magnitude of the seasonal changes in the hypertrophy of the SSK differs in lizards and snakes. In lizards during the non-mating season, the SSK is hardly distinguishable from the uriniferous tubules (Fox 1977) whereas, in snakes, the SSK often varies in diameter and/or granule composition but the SSK is always developed. In most snakes studied, the seasonal hypertrophy of the SSK corresponds with the mating season. The major exception to this is found in some species of snakes in which the SSK is also hypertrophied in the non-mating season. For example, in the Northern Watersnake (*Nerodia sipedon*),

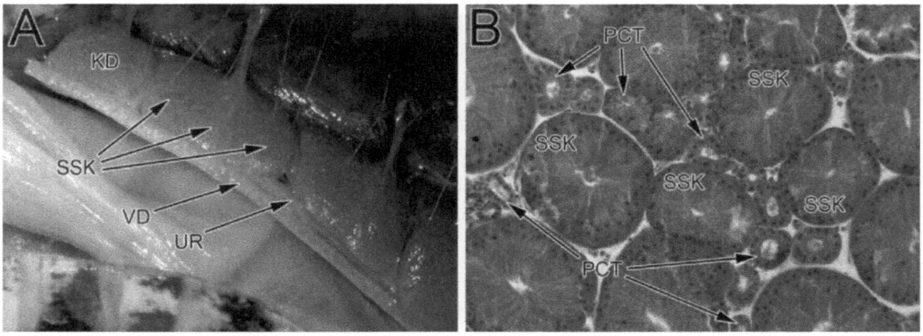

Fig. 11.1 A. Macroscopic view of the sexual segment of the kidney tubules (SSK), ureter (UR) and vas deferens (VD) of the Massasauga (*Sistrurus catenatus*). **B.** Microscopic view of the sexual segment of the kidney tubules in the kidney (SSK) and more proximal convoluted tubules (PCT) in the Red Cornsnake (*Pantherophis guttatus*).

Color image of this figure appears in the color plate section at the end of the book.

the SSK is hypertrophied in the late summer and fall, although mating at this time has not been observed (Weil and Aldridge 1981). Weil and Aldridge (1981) suggested that the hypertrophy in the fall prepares the SSK for immediate use in the spring. The mass of the SSK portion of the kidney is determined not only by the diameter of the tubules but also the length of the SSK portion of the nephron. Because the tubules are convoluted in the kidney, increased length is reflected in an increase number of tubules in the cross section. The epithelial lining of the SSK tubule consists of a simple columnar epithelium surrounding a small lumen. Differences in the diameters of the SSK tubules are determined primarily by the height of the secretory cells. It appears that the length of the nephron, that potentially can develop into the SSK, is probably determined ontogenetically, whereas the actual number of tubules that develop into the SSK and the epithelial height (thus diameter) are determined by the activational effect of plasma androgens. Saint Girons (1957) was the first to measure the gross changes in kidney mass during the reproductive cycle. In *Vipera aspis*, he found that the kidney mass, expressed as the percent of body mass, increased in the spring, during the mating season, then dropped rapidly in the summer and began to enlarge again in the late summer and fall.

11.3 HISTORICAL DESCRIPTION OF THE SEXUAL SEGMENT

Detailed accounts of the early studies on the SSK in snakes are reviewed by Herlant (1933), Volsøe (1944), Bishop (1959), Saint Girons (1972) and Fox (1977) and only highlights will be repeated here. The SSK was first described by Gampert (1866) in the Grass Snake (*Natrix natrix*). In this study of the histology of the kidney, Gampert (1866) noted that parts of the distal nephron were enlarged. Heidenhain (1874) examined the same species and described the numerous granular inclusions the cytoplasm of the SSK. Regaud and Policard (1903) studied *N. natrix* and the Green Whip Snake (*Zamenis viridiflavus*) (=*Hierophis viridiflavus*) and noted that the SSK was restricted to males. Regaud and Policard (1903) postulated that the SSK had a reproductive function and named the structure the sexual segment (of the kidney). Cordier (1928) examined the SSK of three species (*N. natrix*; Viperine Water Snake, *N. maura*; Eastern Garter Snake, *Thamnophis sirtalis*) from captive populations. He concluded that the SSK did not vary seasonally. However, as Bishop (1959) points out, the lack of seasonality that Cordier (1928) observed may have been due to the limited sampling times. In her study, Bishop (1959) found that the SSK of *T. sirtalis* varied seasonally. The most comprehensive study of the seasonality of the SSK was on the Northern Viper (*Vipera berus*) by Volsøe (1944). In this study Volsøe (1944) reported that the SSK of *V. berus* varied seasonally. However, these differences were not very large, and perhaps would have not been apparent without large sample sizes. Volsøe (1944) also hypothesized functions of the SSK secretions (discussed in section 11.6). More recent studies on the SSK have focused on seasonal changes in the SSK in various species and in the morphology, composition and function of the SSK. These are discussed in the relevant sections of this chapter.

11.4 ANATOMY OF THE SEXUAL SEGMENT

11.4.1 Gross Anatomy of the Sexual Segment

The snake kidney is a longitudinally elongate structure, broken into lobes transversely in large species, appears red in color, and lies retroperitoneal on the ventral wall of the body cavity (Fig. 11.2). Blood is supplied to the kidney medially via renal arteries branching from the dorsal aorta. The kidney is emptied via paired renal veins that lie on the medial edge of the kidney and empty into the common posterior vena cava. A portal system travels through the kidney (lateral to medial) and serves as an alternate pathway to drain the blood from the posterior extremities (Fig. 11.1; the other being the ventral abdominal vein not shown).

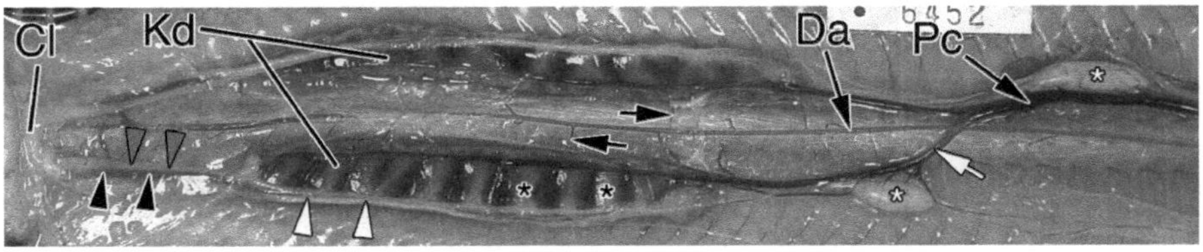

Fig. 11.2 Macroscopic view of the urogenital system of reproductively active *Agkistrodon piscivorus*. Black arrows, renal arteries; black arrowheads, renal portal vein; black asterisks, sexual segment portion of the kidney; Da, dorsal aorta; Grey arrowheads, ureter; Kd, kidney; Pc, postcava; white arrowheads, ductus deferens; white arrow, renal vein; white asterisks, testes.

Color image of this figure appears in the color plate section at the end of the book.

In reproductively active male snakes, the lateral edges of the kidney lobes lose their reddish appearance and take on a creamy yellow color that often appears white after fixation. High magnification of the lateral edges reveals that this color change is due to thickened portions of the nephron tubules aggregated along the lateral edges of the kidney. These thickened portions of the nephron tubules represent the sexual segment of the kidney grossly. The ureter exits the kidney at its lateral extremity and travels caudally to the cloaca alongside the ductus deferens from the testis (Fig. 11.2). The urogenital ducts and cloaca of male snakes are discussed in detail in **Chapter 10**.

11.4.2 Histology and Histochemistry of the Nephron and Sexual Segment

Several studies have been conducted on the chemical nature of the SSK granules. Regaud and Policard (1903) examined five species of snakes (the Smooth Snake, *Coronella austriaca; Natrix natrix, N. viridiflavus; Zamenis viridiflavus; Vipera aspis*) and found that the SSK material stained with mucicarmine indicating that mucus was present. Bishop (1959) stained the SSK of *Thamnophis sirtalis* with a variety of histochemicals and reported that the SSK contained mucopolysaccharides, glycogen, lipids and proteins. Bishop (1959) identified acid phosphatase in cytoplasm, but not granules, in the SSK. Burtner *et al.* (1965) examined the SSK of the Eastern Diamond-backed Rattlesnake (*Crotalus adamanteus*) and found that the SSK granules bind acid dyes strongly and contain lipids and neutral glyco- or mucoproteins. Kühnel and Krisch (1974) examined *N. natrix* and reported that the SSK was positive for phospholipids, but lacked neutral and acid mucosubstances. Kühnel and Krisch (1974) added that the SSK also contained hydrolases and oxydoreductases that show enzymatic activities associated with the glycolytic pathway. Weil (1984) also studied *T. sirtalis* and reported that granules stained most intensely with periodic acid-Schiff, Sudan black B, Oil red O indicating the presence of neutral carbohydrates and lipids. Weil (1984) added that the staining intensity varied during the season, with peak staining in the spring and later in the summer.

Recent detailed studies that described the histology and histochemistry of the SSK were conducted on the Brown Treesnake (*Boiga irregularis*, Siegel *et al.* 2009), Cottonmouth (*Agkistrodon piscivorus*, Sever *et al.* 2008), *Nerodia sipedon* (Krohmer 2004), and *Seminatrix pygaea* (Sever *et al.* 2002). The following description is based on these studies.

The Bowman's capsule marks the beginning of the snake nephron (Fig. 11.3). The capsule consists of a complex network of capillaries surrounded by a one cell-layer thick squamous parietal epithelium (capsular epithelium). A thin squamous epithelium (visceral epithelium) lines the capillaries. Six distinct, successive, regions can be observed traveling away from the Bowman's capsule in the snake nephron: 1) the neck region, 2) the proximal convoluted tubule, 3) the distal convoluted tubule, 4) the sexual segment of the kidney, 5) the post-terminal region, and 6) the collecting duct. Descriptions of these regions below are based on reproductively active *Agkistrodon piscivorus* (Sever *et al.* 2008) and *Boiga irregularis* (Siegel *et al.* 2009).

The neck is the smallest tubule in diameter and is lined by low cuboidal epithelial cells with basally located, irregularly shaped heterochromatic nuclei (Fig. 11.3). The proximal portion of this region is continuous with the squamous epithelium of the Bowman's capsule and communicates the Bowman's capsule with the proximal convoluted tubule. The epithelium of this region is primarily ciliated (Fig. 11.3) and the cytoplasm of the epithelial cells stains intensely basophilic. Cilia are very elongated and fill the lumen of this tubule.

The proximal convoluted tubule possesses an increased diameter compared to the neck region, and is the longest region of the snake nephron. Simple columnar epithelial cells with large round nuclei line this region (Fig. 11.3). These nuclei are located in a basal position and are generally euchromatic with dense central nucleoli. The cytoplasm of the

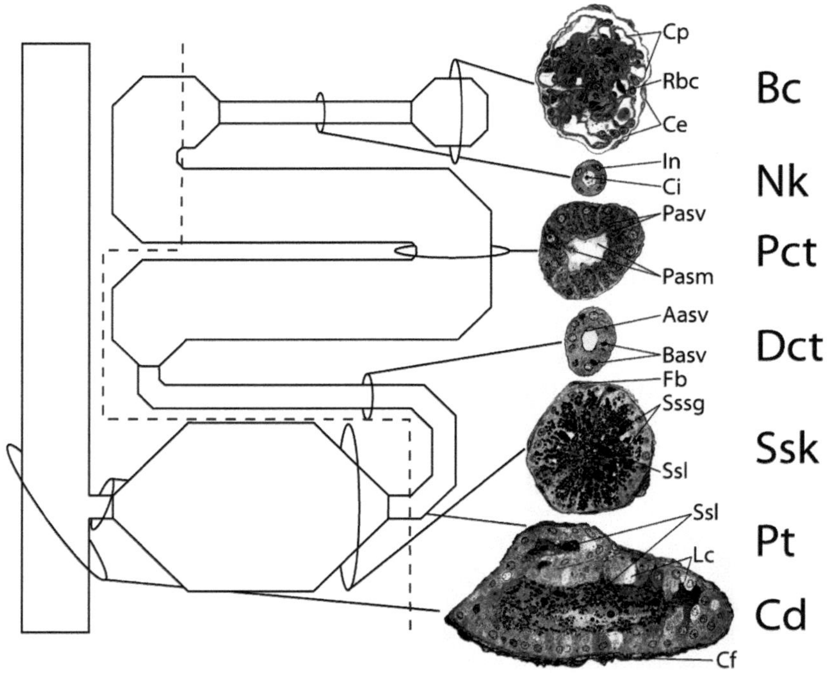

Fig. 11.3 Overview of the histology of the snake nephron as observed in reproductively active Brown Treesnake (*Boiga irregularis*). Left of red dotted line delineates structures found in the cortex. Right of red dotted line delineates structures found in the medulla. Drawing is not to scale. Aasv, apical alcian blue positive secretory vesicles; Basv, basal alcian blue positive secretory vesicles; Bc, Bowman's capsule; Cd, collecting duct; Ce, capsular epithelium; Ci, cilia; Cp, capillary; Dct, distal convoluted tubule; Fb, fibroblast; In, irregular shaped nuclei; Lc, light cells; Nk, neck region; Pasm, periodic acid-Schiff's positive secretory material; Pasv, periodic acid-Schiff's positive secretory vesicle; Pct, proximal convoluted tubule; Pt, post-terminal portion of the nephron; Rbc, red blood cell; Ssk, sexual segment of the kidney; Ssl, sexual segment granules in the lumen; Sssg, sexual segment of the kidney secretory granules.

epithelial cells is basophilic, but less intensely than in the neck region. The cytoplasm is also filled with small vesicles that stain positive for neutral carbohydrates (Fig. 11.4A), but negative for glycosaminoglycans with the periodic acid-Schiff's/alcian blue procedure. There is a weak reaction to the bromophenol blue stain indicative of some protein component of the vesiles (Fig. 11.4B). The epithelial cells are non-ciliated except in the region where the neck transitions to the proximal convoluted tubule. Although the majority of luminal secretory material is a diffuse material, it appears that intact vesicles are also released into the lumen, which is indicative of merocrine and apocrine mode of secretion respectively.

The proximal convoluted tubule narrows and transitions to the distal convoluted tubule, which is lined by a single layer of cuboidal epithelial cells (Fig. 11.3). Like those of the proximal convoluted tubule, the epithelial cells of this region are non-ciliated and slightly basophilic. In contrast to the proximal convoluted tubule, the cytoplasm of the distal convoluted tubule is filled with aggregates of basal secretory vesicles and possesses a thin layer of secretory vesicles apically (Fig. 11.3). The secretory vesicles stain positive for glycosaminoglycans, but not for neutral carbohydrates with the periodic acid-Schiff's/alcian blue procedure or proteins with the bromophenol blue stain (Fig. 11.4A,B). These vesicles appear to be released primarily in a merocrine manner, as intact vesicles are never observed in the lumen. Like the proximal convoluted tubule, the nuclei of the epithelial cells in the distal convoluted tubule are generally euchromatic with centrally located nucleoli (Fig. 11.3).

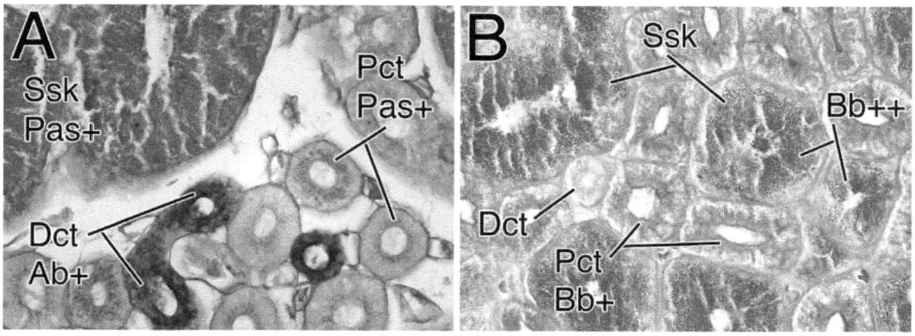

Fig. 11.4 Histochemistry of the snake nephron as observed in reproductively active *Agkistrodon piscivorus* and *Boiga irregularis*. **A.** Periodic acid-Schiff's and alcian blue (Ab) reaction in *A. piscivorus*. **B.** Bromophenol blue reaction in *B. irregularis*. Ab+, alcian blue positive; Bb+, scantly bromophenol blue (Bb) positive; Bb++, intensely bromophenol blue positive; Dct, distal convoluted tubules; Pas+, periodic acid-Schiff's positive; Pct, proximal convoluted tubules; Ssk, sexual segment of the kidney.

Color image of this figure appears in the color plate section at the end of the book.

The distal convoluted tubule enlarges dramatically as it transitions into the SSK. This region has the highest tubular diameter and tall columnar epithelial cells line the tubules (Figs. 11.1B, 11.3, 11.4A,B). In contrast to the loose connective tissue in the interstitium of the more proximal regions, regular dense collagen fibers and fibroblasts surround the SSK tubules (Fig. 11.3). The epithelial cells possess basal nuclei, which appear euchromatic with central nucleoli, and the apices of the cells are filled with intensely eosinophilic secretory granules (filling approximately 4/5 of the total cell; Figs. 11.1B, 11.3). The secretory granules stain intensely for neutral carbohydrates and negatively for glycosaminoglycans with the periodic acid-Schiff's/alcian blue procedure (Fig. 11.4A), and positive for proteins with a bromophenol blue stain (Fig. 11.4B), indicative of a neutral glycoprotein secretion. In reproductively active snakes, secretory granules are so numerous that other cytoplasmic properties, and the delineation between epithelial cells, are often obscured. The epithelium is non-ciliated and the lumen is filled with intact secretory granules indicative of an apocrine mode of secretion (Fig. 11.3).

The SSK empties into the narrow post terminal portion which in turns opens into the histologically identical, but highly enlarged collecting duct (many post terminal tubules open into one collecting duct), and thus these terminal regions will be discussed together. In general the epithelium of the post terminal portion of the nephron and collecting duct is simple columnar. These regions of the nephrogenic tubule are encompassed in collagen fibers and fibroblasts (Fig. 11.3), and a thin layer of smooth muscle at the junction of the collecting duct and ureter. Individual epithelial cells commonly possess basal euchromatic nuclei with central nucleoli, although small irregular shaped heterochromatic nuclei are found in some epithelial cells (Fig. 11.3). The majority of epithelial cells possess a dark cytoplasm, however, light cells can occasionally be found interspersed (Fig. 11.3). The epithelium is non-ciliated, and the post terminal portion and collecting duct do not function in secretion (Fig. 11.3). However, the lumina of the post terminal portion and collecting duct are filled with dense secretory granules and other cellular contents from the highly active SSK (Fig. 11.3). The collecting duct terminates at the ureter, a region discussed in more detail in **Chapter 10**.

11.4.3 Ultrastructure of the Sexual Segment

Since the original light microscope description by Gampert (1866), five studies have been published on the ultrastructure of the SSK in snakes. These studies were conducted on *Agkistrodon piscivorus* (Sever *et al.* 2008), *Boiga irregularis* (Siegel *et al.*, 2009), *Natrix natrix* (Kühnel and Kirsch 1974), *Nerodia sipedon* (Krohmer 2004) and *Seminatrix pygaea* (Sever *et al.* 2002). These five studies will be utilized in the following discussion on the ultrastructure of the sexual segment.

The epithelial cells of the SSK lie against a dense basal lamina. Deep to the basal lamina, a tunica propria exists which is composed of mostly collagen fibers and the occasional fibroblast (Sever *et al.* 2002; Sever *et al.* 2008). Kühnel and Krisch (1974) state that the SSK is surrounded by elongate contractile cells (myoepthelium) in *Natrix natrix*, however, no other study observed this cell type encompassing the SSK, and Sever *et al.* (2002) questioned the presence of contractile fibers in Kühnel and Krisch's Fig. 2. However, Fox (1965) noted a layer of smooth muscle encompassing the SSK in *Typhlops vermicularis* with light microscopy. He described the muscular layer of the SSK as a thin muscular sheath of circular or oblique muscle fibers, differing from that of the muscularis of the ureters, which had a thick layer of circular muscle surrounded internally and externally by thin layers of longitudinal fibers. Thus, it is possible that in some snakes the SSK is modified to actively contract and expel its products into the ureters.

Two types of nerve fibers were observed embedded in Schwann cells and passing between the basal lamina and "myoepthelium" in *Natrix natrix*: 1) fibers with dense synaptic vesicles and many neurotubules with a central filament, and 2) fibers with few synaptic vesicles and many membrane bound granules of varying sizes (Kühnel and Krisch 1974). Neither of these fibers was ever actually observed penetrating the basal lamina, so direct innervation with the SSK epithelium is questionable. Sever *et al.* (2002) also note the occurrence of unmyelinated axons in the tunica propria in *Seminatrix pygaea*, and Fox (1965) reported few large autonomic ganglion cells surrounding the SSK in *Typhlops vermicularis*.

In *Natrix natrix* intercellular caniculi are distinct, narrow to labyrinthine, and interdigitate basally (Kühnel and Krisch 1974). This is similar to *Seminatrix pygaea* (Sever *et al.* 2002) and *Boiga irregularis* (Siegel *et al.* 2009). In *Agkistrodon piscivorus* large vacuoles are found continuous with a labyrinthine caniculus, which often is found interdigitating between adjacent epithelial cell membranes (Sever *et al.* 2008). Zonulae adherentes fuse adjacent cell membranes in *N. natrix* (Kühnel and Krisch 1974), while maculae adherentes fuse the membranes in *Boiga irregularis* (Siegel *et al.* 2009). Due to the breakdown of the apical epithelial membrane in *B. irregularis* during secretion (see section 11.4.4), apical cohesive units were not observed. Sever *et al.* (2002) state that many junctional complexes are found adhering secretory cells in *S. pygaea*, however, only apical zonulae occludentes are observable in their figures. In *A. piscivorus* tight junctions are observed apically, but no connections exist basally (Sever *et al.* 2008).

The synthetic machinery of the SSK epithelium is situated basally around the nucleus in all taxa (Kühnel and Krisch 1974; Sever *et al.* 2008; Siegel *et al.* 2009) except *Seminatrix pygaea*, in which the nucleus is located centrally. Light microscopy analysis of Graham's Crayfish Snake (*Regina grahamii*; unpublished data) indicates a similar condition, suggesting that central nuclei may be a synapomorphy for the semi-fossorial natricines. The cellular machinery of the SSK includes smooth endoplasmic reticulum in *Natrix natrix* (Kühnel and Krisch 1974) and rough endoplasmic reticulum in *Agkistrodon piscivorus* (Sever *et al.* 2008), *Boiga irregularis* (Siegel *et al.* 2009), and *S. pygaea* (Sever *et al.* 2002), both of which are found in association with the cis region of supra- or perinuclear Golgi complexes. In turn, the trans region of the Golgi is associated with large condensing vacuoles (termed Golgi vacuoles by Kühnel and Krisch 1974). Kühnel and Krisch (1974) also noted dense material within the Golgi cisternae in *N. natrix*, whereas, in *B. irrigularis*, dense material is found within the cisternae of the rough endoplasmic reticulum (Siegel *et al.* 2009). The cisternae of the rough endoplasmic reticulum in *B. irregularis* are also distended. Sever and Hopkins (2005) attributed this distended condition to protein precursor formation before association with carbohydrates in the Golgi of the Ground Skink (*Scincella laterale*) SSK. This distended condition is similar to that observed in the widened cisternae of the thyroid gland. Although mitochondria are found most commonly in a supranuclear region, they can be found throughout the SSK epithelial cells (Sever *et al.* 2008). In *A. piscivorus* dark granules are interspersed among the cristae of the mitochondria (Sever *et al.* 2008).

Three types of secretory granules were reported in the apical regions of the SSK epithelial cells in *Natrix natrix*: 1) electron-dense, 2) granular, and 3) vesicular or foamy inclusions (Kühnel and Krisch 1974). However, studies on *Boiga irregularis*, *Nerodia sipedon*, and *Seminatrix pygaea* reported only one type of secretory granule, which varies morphologically during granular condensation (Sever *et al.* 2002; Krohmer 2004; Siegel *et al.* in review), which in turn, is associated with circulating androgen levels (Krohmer 2004). Krohmer (2004) believed that granule development was the cause of the different granule types found in *N. sipedon*, and this idea was supported in the work of Siegel *et al.* (2009) on *B. irregularis*. In *Agkistrodon piscivorus*, two types of secretory granules were observed abutting the large vacuoles of the intercellular canaliculi: 1) large electron-dense granules filling the entire cellular region apical to the nucleus and 2) small diffuse granules located only apically immediately deep to the apical plasma membrane (Sever *et al.* 2008). In general, secretory granules are situated only apical to the nucleus, however, in *Seminatrix pygaea*, which possesses a central nucleus, secretory granules are also concentrated basal to the nucleus (Sever *et al.* 2002).

The surface of the SSK epithelial cells are covered with numerous short and slender microvilli (Kühnel and Krisch 1974; Sever *et al.* 2002; Sever *et al.* 2008; Siegel *et al.* 2009). Kühnel and Krisch (1974) also state that some cilia are present on the surface of the SSK epithelial cells. However, this has not been supported by any other study on SSK ultrastructure, and Kühnel and Krisch (1974) do not provide a supporting micrograph. Thus, we question this occurrence.

The cellular make-up and components of the SSK in lizards are very similar to those of snakes. However, the intercellular canaliculi are indistinct in *Scincella laterale* (Sever and Hopkins 2005) and secretory granules often have two components: a dense inner core with a lucent collar in *Scincella laterale* (Sever and Hopkins 2005), the Rainbow Lizard (*Cnemidophorus lemniscatus*, Del Conte and Tamayo 1973), and the Italian Wall Lizard (*Podarcis sicula*, Furieri and Lanzavecchia 1959).

11.4.4 Seasonality of the Cytology of the Sexual Segment

While quantitative differences in SSK activity may be difficult to observe, qualitative differences reported from seasonal histological and histochemical analyses reveal clear seasonal variation in SSK activity. During the reproductive season(s) the number of granules observed in the SSK epithelium increases dramatically, causing a hypertrophy in the epithelium, and a decrease in luminal volume (Krohmer and Aldridge 1985; Krohmer *et al.* 1986; Clesson *et al.* 2002; Sever *et al.* 2002; Sever *et al.* 2008). The dense aggregation of granules is also associated with increased staining intensity with general histological procedures (Krohmer and Aldridge 1985; Krohmer *et al.* 1987; Clesson *et al.* 2002) and different reactions with histochemical and enzymatic tests (Reddy *et al.* 1972; Kühnel and Krisch 1974; Weil 1984; Sever *et al.* 2002; Sever *et al.* 2008). These histochemical and enzymatic studies revealed seasonal differences associated with lipid, protein, and carbohydrate content in the epithelial cells of the SSK throughout the reproductive and non-reproductive seasons.

The cellular processes described in the epithelium of the SSK in snakes with seasonal SSK cycles were from *Agkistrodon piscivorus* (Sever *et al.* 2008) and *Seminatrix pygaea* (Sever *et al.* 2002), while Siegel *et al.* (2009) describe the cellular processes in a snake with a continuous reproductive cycle, *Boiga irregularis* (Mathies *et al.* 2010). These three species will be utilized for detailed review as they include members of three different families (Natricidae, Colubridae, and Viperidae), were analyzed with the same histochemical and ultrastructural techniques, and were sampled throughout the year to confirm seasonality of SSK activity.

In *Seminatrix pygaea,* peak secretory activity of the SSK occurs in the spring (March; representing the highest tubular diameter and a lumina at their narrowest). Secretory activity is reduced in late spring/early summer (June; representing the lowest tubular diameter), and resumes again in the fall (October; representing tubular diameter almost as high as in the spring). During peak months, PAS+ (Periodic Acid-Schiff) and BB+ (Bromphenol Blue) secretory granules fill the apical and basal portions of the SSK epithelial cells. In times of decreased secretory activity, the BB+ and PAS+ granules are only aggregated towards the apical ends. March specimens were the only specimens exhibiting luminal material staining identically to the granules.

At the peak of SSK hypertrophy (spring) in *Seminatrix pygaea* (Sever *et al.* 2002), the majority of secretory granules are electron-dense (spring), the nuclei are euchromatic, synthetic organelles are not observed, and apocrine blebs of secretory material are found commonly pinching off from the apical membranes of the epithelial cells making up the SSK (Sever *et al.* 2002). During the late spring/summer (peak atrophy), uniformly electron-dense secretory granules disappear and are replaced by condensing vacuoles and secretory granules of varying densities. Heterochromatic nuclei, perinuclear rough endoplasmic reticulum (RER) profiles, Golgi complexes, and basal elongate mitochondria appear at this time. Few apocrine blebs are observed at the surfaces of the epithelial cells. As the summer proceeds into the fall (October), the sexual segment begins to take on the appearance of the kidney in the spring. Thus, during the summer the ultrastructure is indicative of material synthesis and the spring and fall months are indicative of secretory material release.

The seasonal cycle of the SSK in *Agkistrodon piscivorus* (Sever *et al.* 2008) is almost identical to that of *Seminatrix pygaea,* however, the lack of mature secretory granules during atrophy is even more pronounced. In *A. piscivorus,* there are no secretory granules present in epithelial cells during the mid-summer (June-July). Smaller secretory vesicles not found in *S. pygaea* were also missing during this time. However, like *S. pygaea,* condensing vacuoles are present supranuclear, nuclei become heterochromatic, and synthetic organelles such as RER and large mitochondria are conspicuous throughout the SSK epithelial cells. During times of maximum hypertrophy (February-

May and August-November), the epithelial cells of the SSK possess large uniformly electron-dense granules and smaller apically positioned secretory vesicles. During these months the large granules are observed be secreted by an apocrine type secretion while the smaller vesicles may show merocrine secretion. The pattern of histochemical reaction (e.g. PAS/AB and BB) is identical to that of *S. pygaea*. Thus, like *S. pygaea*, two stages are seen, material synthesis and secretory stages.

In *Boiga irregularis* (Mathies *et al.* 2010), which possess continuous spermatogenesis and SSK hypertrophy, no epithelial cell is filled with entirely uniform secretory granules (i.e., early, mid, and mature secretory granules) can always be observed, with mature granules aggregating towards the apical ends of the epithelial cells. Supranuclear condensing vacuoles are abundant, along with synthetic organelles like perinuclear RER and Golgi bodies. Concentrations of synthetic machinery vary between cells. Nuclei are either found as euchromatic and round, or heterochromatic and irregular. Apocrine type secretion occurs at the apical ends of the majority of epithelial cells. The SSK epithelium is PAS+ and BB+. Thus, it appears that in *B. irregularis* synthesis and secretion are occurring at the same time, and can be in different stages depending on the cell.

In those species with seasonal SSK activity (*Seminatrix pygaea*, *Agkistrodon piscivorus*) the SSK tubules have a synthesis phase during the inactive season and secretory phase during the active (hypertrophied) season. However, in species with a continuous SSK cycle (i.e., *Boiga irregularis*), individual epithelial cells occur in either the synthesis or the secretory phase.

In order to create a consistent terminology for describing the patterns of the SSK, we suggest using the terms developed for the testis of amniotes. In the testis, two strategies are described, temporal and spatial. Temporal sperm cell development refers to spermatogenesis occurring uniformly throughout the seminiferous tubule, whereas, spatial sperm cell development refers to the occurrence of different stages spermatogenesis in different portions of the seminiferous tubules (see Gribbins *et al.* 2008).

Thus, we term the SSK cycles in snakes as either 1) a temporal SSK strategy, in which synthesis and secretory phases are separated into non-reproductive and reproductive seasons respectively (i.e., *Seminatrix pygaea*, *Agkistrodon piscivorus*), or 2) a spatial SSK strategy, in which synthesis and secretory phases occur simultaneously in different epithelial cells in the same portion of the SSK (i.e., *Boiga irregularis*).

11.4.5 Secretory Process

The secretory process of the SSK in snakes is controversial. Kühnel and Krisch (1974) were first to describe the process in *Natrix natrix* as apical membrane protrusion into the lumen caused by granule transportation to the apical surface, and then subsequent pinching off of granules with cytoplasmic components attached after membrane rupture. Krohmer (2004) believed that the electron-dense granules may convert to more lucent granules and then be released by a merocrine mode of secretion in *Nerodia sipedon*, although, this was not actually observed. Krohmer (2004) also hypothesized that simple diffusion of material across the apical plasma membrane could be contributing to the secretion process.

Interestingly, while Kühnel and Krisch (1974) describe an apocrine-like process (e.g., cytoplasm attached to intact vacuoles after membrane rupture releases the vacuoles), they also described granular and apical membrane fusion that is merocrine-like. Sever *et al.* (2008) described apocrine secretion of the large electron-dense vacuoles observed in *Agkistrodon piscivorus* in light of intact vacuoles in the SSK lumen, but state that lucent material in the small apical granules may be released by a merocrine mode of secretion. Further, Sever *et al.* (2008) state that the small granules may be converted forms of the large electron-dense granules, and thus suggesting a similar secretion scenario as hypothesized by Krohmer (2004).

In *Seminatrix pygaea* (Sever *et al.* 2002) and *Boiga irregularis* (Siegel *et al.* 2009) a merocrine mode of secretion was never observed. Sever *et al.* (2002) described blebbing of the apical membrane in *S. pygaea* similar to that observed by Kühnel and Krisch (1974), while Siegel *et al.* (2009) only describe apical membrane rupture, with no blebbing (described as an atypical apocrine mode of secretion).

Studies on lizards only add to the confusion of secretory process. In *Cnemidophorus lemniscatus* (Del Conte and Tamayo 1973) the process of secretion was described as holocrine because large portions of cytoplasm encompassed in a plasma membrane with nuclei and secretory granules were observed in the SSK lumen. However, studies on the Crimean Wall Lizard (*Podarcis taurica*, Gabri 1983) and *Scincella laterale* (Sever and Hopkins 2005) described an apocrine secretion process with cytoplasm associated with intact granules in the lumen of the SSK.

The processes involved in SSK secretion in squamates cannot be told from examination of only a few taxa. More studies are needed to determine if different secretory processes (holocrine, merocrine, apocrine, or diffusion) are due to actual variation between taxa, fixation artifacts, or different subjective opinions of the investigators involved in the studies.

11.4.6 Relationship between Sexual Segment Diameter and Species Snout-Vent Length

To determine if the diameters of the SSK tubules of snakes varied with species length, we analyzed the relationship by linear regression. The regression analysis from 27 studies representing 23 species and six families (Table 11.1) indicates that mean tubule diameter of the SSK is statistically independent of SVL ($F_{1,24} = 1.1$, $P = 0.30$, $r^2 = 0.046$). Separate regression analysis of the Colubridae ($F_{1,4} = 0.93$, $P = 0.41$, $r^2 = 0.24$), Natricidae ($F_{1,7} = 4.23$, $P = 0.08$, $r^2 = 0.38$) and Viperidae ($F_{1,7} = 1.1$, $P = 0.75$, $r^2 = 0.02$), also showed that SSK diameter did not covary with SVL. Saint Girons (1972) came to similar conclusions for squamates in general.

Table 11.1 Family, species, mean snout-vent length (SVL), mean tubule diameter of the sexual segment of the kidney (SSK), maximum diameter of the SSK, and mean sexual segment epithelial height (SSEH) in the Ophidia. SVL obtained from authors or from field guides

Family *Species*	*SVL* (mm)	*Mean diameter SSK* (µm)	*Max diameter SSK* (µm)	*SSEH* (µm)	*Reference*
Colubridae					
Arizona elegans	588	165	225	NA	Aldridge *et al.* 1990
Boiga irregularis[1]	1143	79	103	32	Aldridge *et al.* In Press
Boiga irregularis[2]	1202	92	120	44	Aldridge *et al.* In Press
Opheodrys aestivus	356	118	136	NA	Aldridge *et al.* 1990
Tantilla coronata	185	106	139	NA	Aldridge and Semlitsch 1992
Leptotyphlopidae					
Leptotyphlops humilis	30	150	200	77	Fox 1965
Natricidae					
Natrix natrix	500	170	178	65	Duguy and Saint Girons 1966
Nerodia rhombifer	600[3]	125	173	NA	Aldridge *et al.* 1995
Nerodia sipedon	450	130	174	NA	Weil and Aldridge 1981
Regina septemvittata	450	170	190	48	Minesky 1985
Seminatrix pygaea	258	167	210	80	Sever *et al.* 2002
Thamnophis elegans	550[3]	118	NA	49	Fox 1952, 1954
Thamnophis sirtalis	500[3]	150	NA	70	Bishop 1959
Thamnophis sirtalis	500[3]	155	NA	70	Krohmer *et al.* 1987
Thamnophis sirtalis	500[3]	NA	70	50	Fox 1954
Tropidoclonion lineatum	240	155	170	60	Krohmer and Aldridge 1985
Typhlopidae					
Typhlops vermicularis	30	80	80	22	Fox 1965
Viperidae					
Agkistrodon piscivorus	629	164	184	58	Sever *et al.* 2008
Crotalus horridus	1000	179	210	NA	Aldridge and Brown 1995
Crotalus oreganus	930	189	215	NA	Aldridge 2002
Crotalus viridis	874	203	251	NA	Aldridge 1993
Sistrurus catenatus	565	225	250	80	Aldridge *et al.* 2008
Trimeresurus stejnegeri	>370	142	150	NA	Tsui and Tu 2000
Vipera aspis	650[3]	180	200	65	Saint Girons 1957
Vipera berus	600[3]	170	170	NA	Saint Girons and Kramer 1963
Vipera berus	600	140	184	60	Volsøe 1944
Xenodontidae					
Heterodon nasicus	450	135	NA	60	Platt 1969

[1]Guam population
[2]Tropical Australasia population
[3]Snout-vent length estimated from the literature.

11.4.7 Comparison of Sexual Segment Mass and Testis Mass

The SSK in snakes is a prominent structure in the kidney that is visible macroscopically and well as microscopically. In order to quantify the relationship between the mass of the SSK and mass of the testis, we measured the mass of the kidney and testis from 11 species representing five families of snakes (Table 11.2). To determine if the SSK varied along the length of the kidney, we sectioned an anterior, middle and posterior portion of the kidney and measured the percent of the cross sectional area occupied by the SSK using ImageJ analysis software (National Institutes of Health, http://rsb.info.nih.gov/ij/index.html). In all of the species examined the cross sectional area occupied by the SSK varied little among the sections. Thus, the percent area occupied by the SSK multiplied by the kidney mass was used to determine the SSK mass.

Table 11.2 Family, species, snout-vent length (SVL), kidney mass, percent sexual segment of the kidney (SSK) and its corresponding mass, and testis mass of male Ophidia measured during the spring (sp) and summer (su)

Family *Species*	*SVL* (mm)	*Kidney mass* (g)	*% SSK*	*SSK mass* (g)	*Testis mass* (g)
Colubridae					
Masticophis flagellum (sp)[1]	1370	2.06	60	1.24	0.67
Masticophis flagellum (su)	1412	3.95	60	2.37	3.46
Sonora semiannulata (sp)[2]	240	0.040	60	0.024	0.072
Sonora semiannulata (su)	208	0.044	60	0.026	0.040
Elapidae					
Micrurus fulvius (sp)[4]	570	0.33	70	0.23	0.24
Micrurus fulvius (su)	498	0.07	70	0.05	0.03
Natricidae					
Nerodia sipedon (sp)[1]	574	1.39	70	1.03	0.27
Nerodia sipedon (su)	592	0.53	70	0.39	0.90
Regina grahamii (sp)[1]	490	0.30	60	0.18	0.052
Regina grahamii (su)	490	0.41	60	0.25	0.170
Thamnophis proximus (sp)[1]	534	0.222	60	0.133	0.046
Thamnophis proximus (su)	406	0.128	60	0.076	0.104
Viperidae					
Agkistrodon contortrix (sp)[2]	680	1.32	80	1.06	0.20
Agkistrodon contortrix (su)	700	1.95	75	1.46	0.31
Crotalus horridus (sp)[3]	905	2.36	70	1.65	0.106
Crotalus horridus (su)	815	2.58	70	1.81	0.358
Sistrurus miliarius (sp)[3]	390	0.28	70	0.196	0.026
Sistrurus miliarius (su)	329	0.29	70	0.203	0.044
Xenodontidae					
Carphophis vermis (sp)[2]	197	0.042	60	0.025	0.004
Carphophis vermis (su)	215	0.062	70	0.043	0.010
Heterodon nasicus (sp)[2]	407	0.73	60	0.44	0.10
Heterodon nasicus (su)	403	0.39	60	0.23	0.55

[1]Species has post-ovulatory spermatogenesis, spring mating season.
[2]Species has post-ovulatory spermatogenesis, spring and summer mating season.
[3]Species has post-ovulatory spermatogenesis, summer mating season.
[4]Species has pre-ovulatory spermatogenesis.

In the majority of temperate zone snakes the testis is spermatogenic and at its maximum mass, in the summer (Aldridge and Duvall 2002; Aldridge *et al.* 2009). However, in many of these snakes, the mating season occurs in the spring which often coincides with maximum SSK hypertrophy. In the Western Wormsnake (*Carphophis vermis)* and the Viperidae, the SSK mass (in the spring) exceeded the testis mass by 4 to 6 times. In the Natricidae, all of which have a spring mating season and summer spermatogenesis, the spring SSK mass exceeds the summer testis mass. In *Masticophis*, *Sonora* and *Heterodon*, the SSK mass is slightly less that the testis mass.

Assuming that mass reflects the amount of energy devoted to development and maintenance of these structures, these data suggest that the development of the SSK represents a substantial energetic "cost" to the male that may be equal to, and in some species greater than, the energetic cost of the development of the testis. These data support our contention that the SSK is an important, and energetically expensive, structure vitally important to reproductive biology of the Ophidia.

11.5 FUNCTION OF SEXUAL SEGMENT SECRETIONS

Early workers speculated on the function of the SSK secretions. Regaud and Policard (1903) speculated that the SSK was related to the sexual functions of the males. They suggested that the SSK secretion is dissolved in the urine and utilized by the spermatozoa in the cloaca. Volsøe (1944) objected to their conclusion, stating "the mingling with the semen cannot take place in the cloaca, because the semen is probably not kept in the cloaca, but only passes through the seminal groove at the moment of copulation." Volsøe (1944) added that if SSK secretions are really mixed with the semen, then this process must occur in the ampulla of the ureter, in which the semen is known to accumulate in the mating period. Volsøe (1944) concluded that in most other vertebrates special mechanisms, either morphological or physiological, prevent the mingling of urine and semen. Cordier (1928) also suggested that SSK secretions were associated with reproduction and emphasized that the SSK had nothing to do with the urinary function of the kidney. Volsøe (1944) suggested several functions for SSK secretions:

1) the secretions help to secure the separation of the semen from the urine by temporarily blocking the renal tubules or the ureter during copulation;
2) the secretions may be expelled after the semen thereby securing the total emptying of the ampulla and seminal grooves;
3) the secretions may form a plug which blocks the female's cloaca or oviduct to insure retention of the semen (i.e., resembling the vaginal plug found in some mammals); and
4) the secretions are dissolved in the urine the secretion might give this excretion a characteristic scent, thus enabling the male to leave a trail of scent.

Volsøe (1944) immediately discounted his last speculation by stating that during during copulation it is the male that tracks the female, which has no scent gland (i.e., SSK), and that the anal glands, which produce a very strongly smelling secretion, are more likely to serve as scent glands.

Bishop (1959) added to the list of potential functions of SSK secretions:

1) the secretions serve as a nutrient medium for sperm (she added that there is a long time between mating and ovulation, i.e., 1-4 months);
2) the presence of acid phosphatase in the secretion suggests an activating function analogous to that of the prostatic secretions in mammals;
3) the secretions increase the viscosity of the seminal fluids thus enabling it to remain in the sulcus spermaticus during mating; and
4) the secretions could serve as a lubricant to assist the flow of seminal fluid through the sulci of the hemipenes.

Bishop (1959) suggested that 3 and 4 above seem logical because only squamates possess a hemipenis with sulcus spermaticus.

Nilson and Andrén (1982) reported that adders (*Vipera berus*) do not have a true copulatory plug. They suggested that the SSK secretions, which were present in the posterior uterus and coprodaeum (functional vagina in snakes) of recently mated snakes, functioned to contract the sphincter muscles of the uterus. This contraction prevented the sperm of subsequent copulations from entering the uterus. Nilson and Andrén (1982) added that dominant males guarded reproductive females and chased away rival males, thus enhancing the effectiveness of the SSK secretions to prevent subsequent males from mating with this female. The effectiveness of the SSK secretions in preventing subsequent matings was questioned by Stille *et al.* (1986) who reported that litters of *V. berus* could be fathered by at least two males, with both matings occurring in the same mating season. Andrén and Nilson (1987) replied to Stille *et al.* (1986) stating that the greater frequency of first mating progeny compared to second mating progeny suggests that the SSK secretions are effective in preventing or reducing subsequent matings.

Almeida-Santos and Salomão (1997) suggested another role of the SSK secretions in the South American Rattlesnake (*Crotalus durissus*). They proposed that SSK secretions induced a "utero-muscular twisting" (UMT) which functioned primarily to maintain sperm in the posterior oviduct during the winter. They suggest that ovulation reduces the UMT and sperm then travel towards the anterior oviduct for fertilization. Almeida-Santos and Salomão (1997) added that UMT does not preclude the possibility of intraseasonal multiple copulation, sperm competition, and multiple paternity. Siegel and Sever (2006) questioned the uterine twisting proposed by Almeida-Santos and Salomão (1997) stating that the continuous dorsal mesentery that encloses blood vessels and nerves passing to the uterus may prevent the uterine twisting. Siegel and Sever (2006) found no evidence of UMT in the vipers they examined (*Agkistrodon piscivorus*; the Pigmy

Rattlesnake, *Sistrurus miliarius; Crotalus durissus*) and they questioned the hypothesis that UMT could be an ancestral mode for sperm storage in the group (Yamanouye *et al.* 2004).

Recent work in our laboratory on the African Brown House Snake (*Lamprophis fuliginosus*), suggests that SSK secretions may serve as a non-coagulated copulatory plug (Wisniewski 2007). The viscous SSK secretions remain in the cloaca for hours and do not physically block the vagina (cloaca) from other males but may create a viscous barrier, which serves to reduce the likelihood or speed of transmission of sperm (from subsequent matings) from reaching the oviduct.

11.6 SEASONAL CYCLES OF THE SEXUAL SEGMENT

11.6.1 Seasonal Cycles of the Sexual Segment in Snakes

In most temperate zone snakes the diameter of the SSK show seasonal changes. The most typical pattern in the north temperate region consists of an enlarged SSK in the spring, corresponding to the spring mating season. The diameter decreases rapidly during or following the mating season to a seasonal low in June and July. During June and July the seminiferous tubules undergo recrudescence and spermatogenesis is beginning. The SSK begins to hypertrophy in July and, in many species, reaches its peak diameter in August-September. This summer/fall increase occurs in species that have a summer mating season (i.e., Southeastern Crowned Snake, *Tantilla coronata*; Aldridge and Semlitsch 1992), as well as those species that do not (i.e., *Nerodia sipedon*; Weil and Aldridge 1981, 1985; see Aldridge *et al.* 2009 for review of mating seasons). The SSK appears to remain relatively developed during winter hibernation until its spring hypertrophy. This pattern, with minor variations, has been describe for the *Agkistrodon piscivorus* (Johnson *et al.* 1982) the Glossy Snake (*Arizona elegans*, Aldridge 2001), the Plains Hognose Snake (*Heterodon nasicus*, Platt 1969), *Nerodia sipedon* (Bauman and Metter 1977; Weil and Aldridge 1981, 1985), the Rough Greensnake, (*Opheodrys aestivus*, Aldridge *et al.* 1990), *Tantilla coronata* (Aldridge and Semlitsch 1992), *Thamnophis sirtalis* (Bishop 1959; Krohmer *et al.* 1987), the Lined Snake (*Tropidoclonion lineatum*, Krohmer and Aldridge 1985), and *Vipera berus* (Saint Girons and Kramer 1963).

The seasonal differences in the size of the SSK may be large. Saint Girons and Kramer (1963) described changes in the diameter of the SSK in *Vipera berus*. They reported that the SSK was 170 μm in diameter in May during mating season, decreased to 80 μm in June, and subsequently increased to 130 μm in July. Similar seasonal changes in SSK diameter were described for *V. aspis* (Saint Girons 1957)

Not all species, however, show seasonal variation in the diameter of the SSK. In temperate zone populations (25 18 N; 83 1 E) of *Natrix piscator* (Srivastava and Thapliyal 1965) the SSK is smaller than other species (~50 mμ) and does not show a seasonal cycle. Saint Girons and Pfeffer

(1971) also examined the SSK in *Natrix piscator* (from Cambodia) and stated that their results were considerably different from that reported by Srivastava and Thapliyal (1965). Saint Girons and Pfeffer (1971) reported that the diameter of the SSK was small in February (45 mμ), when the testes were involuted, and increased in November (140 mμ), when the testes were spermatogenic. Saint Girons and Pfeffer (1971) noted this difference between these studies of this species and stated they were perplexed by the difference. Saint Girons and Pfeffer (1971) also noted that the period of SSK hypertrophy occurred while the testes were hypertrophied but noted that hypertrophy of the SSK is much shorter than the period of spermatogenesis in these species.

In the temperate zone Timber Rattlesnake (*Crotalus horridus*), Aldridge and Brown (1995) did not find a seasonal difference in the diameter of the SSK. They postulated that this was due to the small sample size. However, these authors did report that the diameter of the SSK increased with the size and age of the snake up to eight years of age.

Graham *et al.* (2008) found no seasonal differences in the SSK diameter in *Agkistrodon piscivorus*. They did note that the diameter of the SSK and the testis was correlated, however, they added that the correlations disappeared when they employed a conservative statistical correction. Graham *et al.* (2008) suggested "that SSK-testicular associations may be characteristic of pitvipers that exhibit a single mating season in late summer and fall. Graham *et al.* (2008) added that the lack of seasonality of the SSK in *A. piscivorus* was surprising because the hypertrophy of the SSK consistently predicts the mating season in all snake species previously examined.

In tropical populations of *Boiga irregularis*, the SSK does not exhibit a seasonal cycle (Mathies *et al.* 2010), however, more temperate populations of this species exhibit a seasonal cycle (Bull *et al.* 1997). Mathies *et al.* (2010) also noted that the SSK increased in size with increasing male size. They reported that the SSK diameters increase in snakes until the snakes reach about 1300 mm in length, at which time the SSK tubules reach maturation.

Another major exceptions to the typical pattern described above was described for *Tropidoclonion lineatum* (Krohmer and Aldridge 1985) and the Prairie Rattlesnake (*Crotalus viridis*, Aldridge 1993). In these species the SSK had a single peak in summer coinciding with the mating season. The SSK diameters were low in the spring and early summer. It is noteworthy that these species mate primarily in the summer (Aldridge and Duvall 2002; Aldridge *et al.* 2009).

Differences in the SSK cycles do not appear to be related to phylogeny. Saint Girons (1976) examined the seasonality of SSK development in six species of *Vipera* in Europe. He found that two different patterns were apparent among the six species. In three species (*V. berus, V. ursinii, V. ammodytes*) mating occurs only in the spring. In these species the SSK is hypertrophied primarily in the spring. Whereas, in the other species (*V. aspis, V. seoanei, V. latastei*) mating occurs in the summer/fall and following spring. The SSK in these species is hypertrophied in the summer/

fall and spring. Saint Girons (1976) suggested that these patterns may be useful in determining the phylogenetic relationships between these species. However, a recent phylogenetic analysis (Garrigues *et al.* 2005) which included all of the above species of *Vipera*, shows that the two most closely related species (*V. berus* and *V. seoanei*) have different mating and SSK patterns, suggesting that mating patterns and SSK hypertrophy may be independent of phylogenetic relationships.

11.6.2 Seasonal Cycles of the Sexual Segment in Squamates

We suggest that the differences between lizards and snakes in the seasonal development of the SSK may be a result of different reproductive patterns and associated sperm storage requirements. The majority of lizards have pre-nuptial spermatogenesis (sperm are produced immediately before ovulation) whereas snakes have primarily post-nuptial spermatogenesis sperm are produced following ovulation). The majority of lizards produce sperm and mate during or immediately following sperm production. Sperm remaining in the vas deferens (and epididymis) degrade following the mating season (Goldberg and Lowe 1966; Goldberg 1971; Goldberg and Lowe 1997). The post-nuptial pattern in snakes requires that sperm be stored in the vas deferens until the spring mating season. If elevated plasma androgen levels are required for the maintenance of the epithelium of the vas deferens and sperm storage in snakes, as suggested by Aldridge and Arackal (2005), perhaps the continuous development of the SSK is a secondary result of the need for long-term sperm storage.

Support for the contention that hypertrophy of the SSK in snakes is related to the storage of sperm is presented by Saint Girons and Pfeffer (1971) in their study of four species of snakes (Chequered Watersnake, *Natrix piscator;* Puff-faced Watersnake, *Homalopsis buccata;* Rainbow Watersnake, *Enhydris enhydris;* Green Cat Snake, *Boiga cyanea*) from Cambodia. Saint Girons and Pfeffer (1971) found that the SSK these snakes decreased dramatically following the mating season, to approximately one quarter to one third of their maximum diameter (both maximum diameter and seasonal variation varied among species). Saint Girons and Pfeffer (1971) also reported that the vas deferens is empty following the mating season. We observed from their illustrations that, although the SSK tubules are small, they are distinguishable from uriniferous tubules.

11.7 COPULATORY PLUGS IN SNAKES

The first description of what appears to be a copulatory plug in snakes was by Edgren (1953) in an Eastern Hog-nosed Snake (*Heterodon platirhinos*) preserved *in copulo*. Edgren (1953) reported that the hemipenis occupied the posterior three-fourth of the cloaca and that the anterior fourth of the cloaca was filled with a clotted substance that contained spermatozoa. Fitch (1965) used the term copulatory plug in snakes to describe SSK secretions in recently mated *Thamnophis sirtalis* and was the first to report that females with a copulatory plug were not courted by sexually aroused males. Devine (1975) observed copulatory plugs in *T. sirtalis,* Butler's Gartersnake (*Thamnophis butleri*) and the Brown Watersnake (*Nerodia taxispilota*). Devine (1975) described the plug as gelatinous material that formed a tight seal in the anterior end of the urodaeum which prevented seminal fluid from escaping from the oviduct. Devine (1975) also examined the chemical nature of the plug and concluded that it consisted of secretions similar to those of the SSK.

The primary function of the copulatory plug is to act as a physical barrier to prevent copulation of the female by successive males (Devine 1975). Under laboratory conditions, Devine (1984) reported that the plug lasted from two days (at 21°C) to two weeks (at 4°C) and that the plug could be an alternative to mate guarding behavior. In the species that he studied, Devine (1975) stated that the snakes mated upon emergence from the den and then immediately dispersed and that the short-lived copulatory plug, which reduced the female's attractiveness, might be effective in reducing sperm competition.

Devine (1975) postulated that if the copulatory plug evolved primarily to prevent subsequent matings, rather than merely function to retain sperm, it should be present in species that do not have other forms of mate protection such as male combat. Among the eight species (see below) which have been reported to have copulatory plugs, only *Coluber constrictor* is known to have male combat behaviors (Devine 1984). However, it may also be that the majority of snakes have not been examined for the presence of copulatory plugs.

The role of the copulatory plug in snakes is limited. As Shine *et al.* (2000) states, much of the research on copulatory plugs in snakes has been done on one population of a single species (*Thamnophis sirtalis parietalis*). Shine *et al.* (2000) also posed the question, why do gartersnakes have mating plugs whereas most species of snakes do not? We suggest that the absence of data on the copulatory plug (or other types of secretions of the SSK) may be due to the failure of researchers to look for the SSK material in the female. The SSK is present in all snakes that have been examined (Fox 1977; Aldridge and Duvall 2002), however, our search of the literature found that copulatory plugs have only been described in eight colubrids; *Coluber constrictor*; the Rough Earthsnake *(Virginia striatula),* Devine 1975; the Eastern Hog-nosed Snake (*Heterodon platirhinos*), Edgren 1953; the Brown Watersnake (*Nerodia taxispilota*), Herrington 1989; *Thamnophis sirtalis parietalis,* see Shine *et al.* 2000; *Thamnophis butleri,* the Eastern Ribbonsnake (*T. sauritus*), Devine 1975 and Klemens 1993; the Lined Tolucan Ground Snake (*Toluca lineata*), Uribe *et al.* 1998 and one boid (the Anaconda, *Eunectes murinus,* Rivas 2000).

In some snakes, the SSK secretions do not appear to form a distinct copulatory plug. For example, in the viperid *Vipera berus,* Nilson and Andrén (1982) reported that SSK secretions were deposited in the posterior uterus of females during copulation. These secretions did not form a copulatory plug, but instead, caused the contraction of the sphincter muscles at the caudal end of the oviduct (see section 11.5). They added this contraction might act as a mechanical barrier preventing sperm, from subsequent matings, from entering the uterus. Nilson and Andrén (1982) also reported that there was no clear trend suggesting that SSK secretions affected the female's attractivity. Attractive females, experimentally treated with SSK secretions remained attractive to males. In *Lamprophis fuliginosus* (Wisniewski 2007), the copulatory plug (material) consisted of a viscous, semitransparent liquid, containing sperm. Mating females were immediately (within 15 min) non-attractive to males and remained non-attractive for several days. Similar to the findings of Nilson and Andrén (1982), the non-attractiveness was not due to SSK secretions because attractive females, when treated with SSK secretions from mated females, remained attractive to males.

Several investigators suggested that the production of a mating plug is a strategy males use to reduce sperm competition by preventing the female from remating (Devine 1975, 1977; Ross and Crews 1977). Using *Thamnophis sirtalis* as a model, Shine *et al.* (2000) designed experiments to determine if the mating plug functioned as a physical block or as source of pheromonal cues that reduce female attractivity. Shine *et al.* (2000) concluded that the copulatory plug functioned primarily as a physical block to delay re-mating by the female and added that fluids associated with the plug, and not the plug *per se*, were responsible for the decrease in female attractiveness. Further studies on this species by O'Donnell *et al.* (2004) confirm that the copulatory fluids, and not the plug, reduce the rival male's interest in the female.

Copulatory plug do not necessarily prevent subsequent copulations by the female. In *Thamnophis sirtalis*, Shine *et al.* (2000) reported that approximately 7% of the females examined in the wild had multiple plugs present in the cloaca. Similarly, Whittier and Crews (1986) reported a substantial number of mated female *T. s. parietalis* (35%) with copulatory plugs, re-mated within 2 h of the initial mating.

The frequency of copulatory plug formation following mating was examined by Shine *et al.* (2000) in *Thamnophis sirtalis parietalis*. They found that copulatory plugs were present in approximately 90% of matings, both in the field and in experiments. Shine *et al.* (2000) suggested that the male's prior mating history might affect plug formation. In repeated matings the frequency of plug formation dropped from 93% (first mating) to 77% (subsequent matings), however, the results were not statistically significant.

11.8 ENDOCRINE CONTROL OF THE SEXUAL SEGMENT

Takewaki and Hatta (1941) were the first to study the role of the testis and hypophysis on SSK development in snakes. Following castration of male

Natrix natrix, they reported that the SSK appeared normal for 14-38 days. At 41 days post-operative, the SSK tubules were reduced in diameter but the epithelial cells were still filled with secretions. At 67 days, the SSK granules decreased especially at the outer periphery of the cells and at 69-101 days the SSK tubules were involuted but still contained many granules. It was not until 157 days that epithelial cells were completely devoid of granules. They noted at this time that the vas deferens were normal and contained sperm.

In the snakes that were hypophysectomized the results on the SSK was similar to those castrated. Substantial regression of the SSK did not occur until day 79.

Takewaki and Fukuda (1935a,b) examined the effects of castration on the SSK in lizards and reported that SSK regression occurred sooner (at day 45) than in snakes and that sperm disappeared from the vas deferens in lizards at this time. Takewaki and Hatta (1941) suggested that the slowness of regression of the SSK and presence of sperm in the vas deferens in snakes, may explain the lack of SSK involution in snakes as is seen in lizards.

Bishop (1959) was unaware of the Takewaki and Hatta (1941) paper and designed experiments to determine the effects of castration and testosterone replacement on the SSK on *Thamnophis sirtalis*. Her results were similar to Takewaki and Hatta (1941). Bishop (1959) also injected an intact female *T. sirtalis* with testosterone and reported that the SSK in the female underwent hypertrophy and developed intracellular granules similar to those of males.

Krohmer *et al.* (2004) examined the effects of exogenous testosterone and 17β-estradiol on the kidneys of immature natricine snakes (*Thamnophis sirtalis parietalis, Nerodia sipedon*, and the Diamondback Watersnake, *Nerodia rhombifer rhombifer*). They reported that both hormones had effects on the SSK, although some differences in responses occurred in different species. In all species testosterone and 17β-estradiol stimulated granule development in both sexes at 16 and 23 days, however, the effects of testosterone were greater. Krohmer *et al.* (2004) proposed that the lack of an SSK response to 17β-estradiol in adult females may be due to the lower concentration of circulating sex hormones in females or an inhibitory effect of the intact female reproductive system.

Krohmer (2004) studied the effects of castration, hypophysectomy and testosterone administration on adult male *Thamnophis sirtalis*. At 14-28 days post-surgery androgen levels had decreased in hypophysectomized snakes but not in castrated snakes when compared to initial measurements. Similar to other studies presented here, the SSK (diameter and epithelial height) did not regress in the treated snakes. Krohmer (2004) reported that the adrenal glands of castrated snakes were enlarged compared to control and hypophysectomized snakes and suggested that the continued hypertrophy of the SSK at 14-28 days may have been due to testosterone production by the adrenal gland. At 60 days, castrate snakes showed a complete reduction in granules. Krohmer (2004) added that administration of testosterone to castrates stimulated granule formation within 7 days.

Haldar and Pandey (1989) reported that pinealectomy in *Natrix piscator* resulted in regression of the SSK. They added that the reduction in the SSK was probably the result of disruption of the hypothalamico-hypophyseal axis (Haldar and Pandey 1989)

The common observation in all of these studies is that SSK regression occurs relatively slowly following castration or hypophysectomy, however, administration of testosterone stimulated granule development relatively rapidly.

11.9 DUAL FUNCTION OF THE KIDNEY

The kidneys of squamates have two major functions: 1) maintaining homeostasis and 2) producing fluids that contribute to the ejaculate. To maintain homeostasis, the kidneys function to remove nitrogenous wastes, primarily uric acid, from the body. Uric acid makes up to 80% of the nitrogenous wastes (Khalil 1948 a,b). Because snakes lack a urinary bladder (Fox 1977) snakes often store uric acid in the ureter. The reproductive function is to synthesize materials in the sexual segment region and excrete these fluids with sperm in the ejaculate. The SSK secretions are transferred to the female during copulation. The dual function of the kidney raises the question of how snakes prevent secretions of uric acid mixing with SSK secretions in the ejaculate. Snakes have solved this problem in two ways: 1) at least one species, *Lamprophis fuliginosus*, the males routinely excrete uric acid when they encounter an attractive female (Aldridge *et al.* 2005) and 2) some species transfer uric acid into the intestine for storage (i.e., the Indian Wolf Snake, *Lycodon aulicus*; Seshadri 1959a), *Nerodia sipedon* (Aldridge, unpublished data) and *Lamprophis fuliginosus* (Aldridge, unpublished data). The storage of uric acid in the intestine of male and female monitor lizards (Seshadri 1959b), a taxon related to basal snakes (Sanders and Lee 2008), suggests that the storage of uric acid in the intestine may be a primitive trait in the Ophidia.

11.10 CONCLUSIONS AND FUTURE RESEARCH

The SSK is a prominent secondary sexual structure in the Ophidia. In many snakes the maximum mass of the SSK is equal to or exceeds the maximum mass of the testis. The SSK is under the control of androgens from the testis, adrenal or both. The secretory granules present in the SSK vary seasonally, from primarily mucus in the non-mating seasons to proteinaceous during periods when the SSK is hypertrophied.

Despite the conspicuous size and ubiquitous presence of the SSK in squamates, the functions of its secretions remain largely unknown. Experiments on the role of SSK secretions in the ejaculate have been hampered by the fact that the SSK is part of an essential organ (kidney) and removal of these structures results in death. A number of anatomical studies, however, suggest that the right and left reproductive tracts function independently in squamates (Dowling and Savage 1960; Crews 1978). Experimental studies also suggest that the tracts operate independently. In lizards, which generally alternate hemipenal use in repeated copulations (Tokarz 1988), the quantity of sperm is reduced if males mate repeatedly with only a single hemipenis (Tokarz and Slowinski 1990). Similar results have been observed in snakes. Zweifel (1997) reported that snakes alternate hemipenal usage when matings occur with in three days. Shine *et al.* (2000) also observed alternating hemipenal use in the *Thamnophis sirtalis* and suggested that the reason may be to prevent depletion of SSK secretions.

We suggest that with surgical removal of one kidney, the role of SSK secretions on the fertilizing capacity of sperm could be examined experimentally in captive snakes. If females are mated successively with intact and nephrectomized males, the effects of the SSK secretions could be determined using molecular techniques to identify paternity from the intact and manipulated males. Studies such as these may confirm and/or refute the hypotheses on the role of SSK secretions in the Ophidia.

11.11 ACKNOWLEDGMENTS

We thank Laurie Vitt, Sam Nobel Oklahoma Museum of Natural History and Linda Trueb, University of Kansas Natural History Museum for access to specimens used in this study. We thank Alyssa S. Abrams for assistance in collecting and measuring tissue samples. We also thank Donald Shepard for his assistance in the Sam Nobel Museum.

11.12 LITERATURE CITED

Aldridge, R. D. 1993. Male reproductive anatomy and seasonal occurrence of mating and combat behavior of the rattlesnake *Crotalus v. viridis*. Journal of Herpetology 27: 481-484.

Aldridge, R. D. 2001. Reproductive anatomy, mating season and cost of reproduction in the glossy snake (*Arizona elegans*). Amphibia-Reptilia 22: 243-249.

Aldridge, R. D. 2002. The link between mating season and male reproductive anatomy in the rattlesnakes *Crotalus viridis oreganus* and *C. v. helleri*. Journal of Herpetology 36: 295-300.

Aldridge, R. D. and Semlitsch, R. D. 1992. Male reproductive biology of the southeastern crowned snake (*Tantilla coronata*). Amphibia-Reptilia 13: 219-225.

Aldridge, R. D. and Brown, W. S. 1995. Male reproductive cycle, age at maturity, and cost of reproduction in the timber rattlesnake (*Crotalus horridus*). Journal of Herpetology 29: 399-407.

Aldridge, R. D. and Duvall, D. 2002. The evolution of the mating season in the pitvipers of North America. Herpetological Monographs 16: 1-25.

Aldridge, R. D. and Arackal, A. A. 2005. Reproductive biology and stress of captivity in male brown treesnakes (*Boiga irregularis*) on Guam. Australian Journal of Zoology 53: 1-8.

Aldridge, R. D., Greenhaw, J. J. and Plummer, M. W. 1990. The male reproductive cycle of the rough green snake (*Opheodrys aestivus*). Amphibia-Reptilia 11: 165-172.

Aldridge, R. D., Flanagan, W. P. and Swarthout, J. T. 1995. The reproductive biology of the diamondback water snake (*Nerodia rhombifer*) from Veracruz, Mexico, with comparisons of tropical and temperate snakes. Herpetologica 51: 182-192.

Aldridge, R. D., Bufalino, A. P., Robison, C., Salgado, C. and Khayyat, P. 2005. Courtship behavior and evacuation of the urinary ducts in captive brown house snakes (*Lamprophis fuliginosus*). Amphibia-Reptilia 26: 576-582.

Aldridge, R. D., Goldberg, S. R., Wisniewski, S. S., Bufalino, A. P. and Dillman, C. B. 2009. The reproductive cycle and estrus in the colubrid snakes of temperate North America. Contemporary Herpetology 2009: 1-31.

Aldridge, R. D., Jellen, B. C., Allender, M. C., Dreslik, M. J., Shepard, D.B., Cox, J. M. and Phillips, C. A. 2008. Reproductive biology of the massasauga (*Sistrurus catenatus*) from South-central Illinois. Pp. 403-412. In W. K. Hayes, K. R. Beaman, M. D. Cardwell and S. P. Bush (eds), *The Biology of the Rattlesnakes*, Loma Linda University Press, Loma Linda, California.

Aldridge, R. D., Siegel, D. S., Bufalino, A. P., Wisniewski, S. S. and Jellen, B. C. in press. A multi-year comparison of the male reproductive biology of the brown treesnake (*Boiga irregularis*) on Guam. Australian Journal of Zoology.

Almeida-Santos, S. M. and Salomão, M. G. 1997. Long-term sperm storage in the female Neotropical Rattlesnake *Crotalus durissus terrificus* (Viperidae: Crotalinae). Japanese Journal of Herpetology 17: 46-52.

Andrén, C. and G. Nilson. 1987. The copulatory plug of the Adder, *Vipera berus*: does it keep sperm in or out? Oikos 49: 230-232.

Aron, M. 1924. Recherches morphologiques et expérimentales sur le déterminisme des caractères sexuels mâles chez les Urodèles. Archives de Biologie Morphologie Experimentale et Comparée 34: 1-166.

Bauman, M. A. and Metter, D. E. 1977. Reproductive cycle of the northern watersnake, *Natrix s. sipedon* (Reptilia, Serpentes, Colubridae). Journal of Herpetology 11: 51-59.

Bishop, J. E. 1959. A histological and histo-chemical study of the kidney tubule of the common garter snake, *Thamnophis sirtalis*, with special reference to the sexual segment in the male. Journal of Morphology 104: 307-358.

Bull, K. H., Mason, R. T. and Whittier, J. 1997. Seasonal testicular development and sperm storage in tropical and subtropical populations of the brown tree snake (*Boiga irregularis*). Australian Journal of Zoology 45: 479-488.

Burtner, H. J., Floyd, A. D. and Longley, J. B. 1965. Histochemistry of the "sexual segment" granules of the male rattlesnake kidney. Journal of Morphology 116: 189-196.

Carpenter, C. C., Murphy, J. B. and Mitchell, L. A. 1978. Combat bouts with spur use in the Madagascan boa (*Sanzinia madagascariensis*). Herpetologica 34: 207-212.

Clesson, D., Bautista, A., Baleckaitis, D. D. and Krohmer, R. W. 2002. Reproductive biology of male eastern garter snakes (*Thamnophis sirtalis sirtalis*) from a denning population in central Wisconsin. American Midland Naturalist 147: 376-386.

Cordier, R. 1928. ltudes histophysiologiques sur le tube urinaire des reptiles. Archives de Biologie 38: 111-71.

Crews, D. 1978. Hemipenile preference: stimulus control of male mounting behavior in the lizard *Anolis carolinensis*. Science 199: 195-196.

Cuellar, H. S., Roth, J. J., Fawcett, J. D. and Jones, R. E. 1972. Evidence for sperm sustenance by secretions of the renal sexual segment of male lizards, *Anolis carolinensis*. Herpetologica 28: 53-57.

Courrier, R. 1922. Etude préliminaire du déterminisme des caractères sexuels secondaires chez les poisons. Archives d'Anatomie d'Histologie et d'Embryologie Experimentale 1: 115-144.

Craig-Bennett, A. 1931. The reproductive cycle of the three-spined stickleback, *Gasterosteus aculeatus*, Linn. Philolsophical Transactions of the Royal Society London B. 219: 197-279.

Deb, C. and Sarkar, C. 1963. Histochemistry of renal sex segment in garden lizard, *Calotes versicolor*. Proceedings of the National Institute of Sciences of India 29: 197-202

Del Conte, E. 1972. Granular secretion in the kidney sexual segments of female lizards, *Cnemidophorus l. lemniscatus* (Sauria, Teiidae). Journal of Morphology 137: 181-192.

Del Conte, E. and Tamayo, J. G. 1973. Ultrastructure of the sexual segments of the kidneys in male and female lizards, *Cnemidophorus l. lemniscatus* (L.). Zeitschrift für Zellforschung und Mikroskopische Anatomie 144: 325-327.

Devine, M. C. 1975. Copulatory plugs in snakes; enforced chastity. Science 187: 844-845.

Devine, M. C. 1977. Copulatory plugs, restricted mating opportunities and reproductive competition among male garter snakes. Nature 267: 345-346.

Devine, M. C. 1984. Potential for sperm competition in reptiles: behavioral and physiological consequences. Pp. 509–521. In R. L. Smith (ed.), *Sperm Competition and the Evolution of Animal Mating Systems*. Academic Press, Orlando, Florida, U.S.A.

Dowling, H. G. and Savage, J. M. 1960. A guide to the snake hemipenis: a survey of basic structure and systematic characteristics. Zoologica 45: 17-28.

Duguy, R. and Saint Girons, H. 1966. Cycle annuel d'activite et reproduction de la coleuvre viperine *Natrix maura* (L.), d'après les notes manuscrites de Rollinat et des observations personelles. Terre et la Vie 113: 423-457.

Duvall, D. and Schuett, G. W. 1997. Straight-line movement and competitive mate searching in prairie rattlesnakes, *Crotalus viridis viridis*. Animal Behavior 54: 329–334.

Edgren, R. A. 1953. Copulatory adjustment in snakes and its evolutionary implications. Copeia 1953: 162-164.

Fitch, H. S. 1965. An ecological study of the garter snake, *Thamnophis sirtalis*. University of Kansas Publications, Museum of Natural History 15: 493-564.

Fox, W. 1952. Seasonal variation in the male reproductive system of Pacific Coast garter snakes. Journal of Morphology 90: 481-554.

Fox, W. 1954. Genetic and environmental variation in the timing of the reproductive cycle of male garter snakes. Journal of Morphology 95: 415-450.

Fox, W. 1965. A comparison of the male urogenital systems of blind snakes, Leptotyphlopidae and Typhlopidae. Herpetologica 21: 241-256.

Fox, H. 1977. The urogenital system of reptiles. Pp. 1-157. In C. Gans and T. S. Parsons (eds), *Biology of the Reptilia*, vol. 6. Academic Press, New York.

Furieri, P. and Lanzavecchia, G. 1959. Le secrezione dell' epididimo e del rene sessuale nei rettili. Studio al microscopio elettronico. Archivio Italiano di Anatomia e di Embriologia 64: 357–379.

Gampert, O. 1866. Ueber die niere von *Tropidonotus natrix* und der Cyprinoiden. Zeitschrift für Wissenschaftliche Zoologie 16: 369–373.

Garrigues, T., Dauga, C., Ferquel, E., Choumet, V. and Failloux, A-B. 2005. Molecular phylogeny of *Vipera* Laurenti, 1768 and the related genera *Macrovipera* (Reuss, 1927) and *Daboia* (Gray, 1842), with comments about neurotoxic *Vipera aspis aspis* populations. Molecular Phylogenetics and Evolution 35: 35-47.

Goldberg, S. R. 1971. Reproductive cycle of the ovoviviparous iguanid lizard *Sceloporus jarrovi* Cope. Herpetologica 27: 123-131.

Goldberg, S. R. and Lowe, C. H. 1966. The reproductive cycle of the western whiptail lizard (*Cnemidophorus tigris*) in southern Arizona. Journal of Morphology 118: 543-548.

Goldberg, S. R. and Lowe, C. H. 1997. Reproductive cycle of the gila monster, *Heloderma suspectum*, in Southern Arizona. Journal of Herpetology 31: 161-166.

Graham, S. P., Earley, R. L., Hoss, S. K., Schuett, G. W. and Grober, M. S. 2008. The reproductive biology of male cottonmouths (*Agkistrodon piscivorus*): Do plasma steroid hormones predict the mating season? General and Comparative Endocrinology 159: 226-235

Greene, H. W. 1997. *Snakes, the Evolution of Mystery in Nature*. University of California Press, Berkeley and Los Angeles, California. Pp. 351.

Haldar, C. and Pandey, R. 1989. Effect of pinealectomy on annual testicular cycle of Indian chequered water snake, *Natrix piscator*. General and Comparative Endocrinology 76: 214-222.

Heidenhain, R. 1874. Mikroskopische beiträge zur anatomie und physiologie der nieren. Archiv für Mikroskopische Anatomie 10: 1-50.

Herlant, M. 1933. Recherches histologiques et expérimentales sur les variations cycliques du testicule et des caractéres sexuels secondaires chez les reptiles. Archives de Biologie 44: 347-468.

Jacob, J. S. and Altenbach, J. S. 1977. Sexual color dimorphism in *Crotalus lepidus klauberi* Gloyd (Reptilia, Serpentes, Viperidae). Journal of Herpetology 11: 81-84.

Jakobsson, S., Borg, B., Haux, C. and Hyllner, S. J. 1999. An 11-ketotestosterone induced kidney-secreted protein: the nest building glue from the male three-spined stickleback, *Gasterosteus aculeatus*. Fish Physiology and Biochemistry 20: 79-85.

Johnson, L. F., Jacob, J. S. and Torrance, P. 1982. Annual testicular and androgenic cycles of the cottonmouth (*Agkistrodon piscivorus*) in Alabama. Herpetologica 38: 16–25.

Khalil, F. 1948a. Excretion in reptiles. II. Nitrogen constituents of the urinary concretions of the oviparous snake *Zamens diadema*, Schlegel. Journal of Biological Chemistry 172: 101-103.

Khalil, F. 1948b. Excretion in reptiles. III. Nitrogen constituents of the urinary concretions of the viviparous snake *Eryx thebaicus*, Reuss. Journal of Biological Chemistry 172: 105-106.

Klemens, M. W. 1993. *Amphibians and Reptiles of Connecticut and Adjacent Regions*. State Geological and Natural History Survey of Connecticut, Bulletin Pp. 318.

Krohmer, R. W. 1986. Effects of mammalian gonadotropins (FSH and LH) on testicular development in the immature water snake, *Nerodia sipedon*. General and Comparative Endocrinology 64: 330–338.

Krohmer, R. W. 2004. Variation in seasonal ultrastructure of sexual granules in the renal sexual segment of the northern water snake, *Nerodia sipedon sipedon*. Journal of Morphology 261: 70–80.

Krohmer, R. W. and Aldridge, R. D. 1985. Male reproductive cycle of the lined snake (*Tropidoclonion lineatum*). Herpetologica 41: 33-38.

Krohmer, R. W., Grassman, M. and Crews, D. 1987. Annual reproductive cycle in the male red-sided gartersnake, *Thamnophis sirtalis parietalis*: field and laboratory studies. General and Comparative Endocrinology 68: 64-75.
Krohmer, R. W., Martinez, D. and Mason, R. T. 2004. Development of the renal sexual segment in immature snakes: effect of sex steroid hormones. Comparative Biochemistry and Physiology, Part A 139: 55-64.
Kühnel, W. and Krisch, B. 1974. On the sexual segment of the kidney in the snake (*Natrix natrix*). Cell and Tissue Research 148: 412-429.
Licht, P. and Donaldson, E. M. 1969. Gonadotropic activity of salmon pituitary extract in the male lizard (*Anolis carolinensis*). Biology of Reproduction 1: 307-314.
Licht, P. and Pearson, A. K. 1969. Effects of mammalian gonadotropin (FSH and LH) on the testes of the lizard *Anolis carolinensis*. General and Comparative Endocrinology 13: 367-381.
Licht, P. and Tsui, H. W. 1975. Evidence for the intrinsic activity of ovine FSH on spermatogenesis, ovarian growth, steroidogenesis and ovulation in lizards. Biology of Reproduction 12: 346-350.
Lindell, L. E. and Forsman A. 1996. Sexual dichromatism in snakes: support for the flicker-fusion hypothesis. Canadian Journal of Zoology 74: 2254-2256.
Mathies, T., Cruz, J. A., Lance, V. A., Miller, L. A. and Savidge, J. A. 2010. Reproductive biology of male brown treesnakes (*Boiga irregularis*) on Guam. Journal of Herpetology 44: 209-221.
Madsen, T. 1988. Reproductive success, mortality and sexual size dimorphism in the adder, *Vipera berus*. Holarctic Ecology 11: 77-80.
Minesky, J. J. 1985. The reproductive biology of the queen snake *Regina septemvittata* in Pennsylvania. Master's Thesis, Saint Louis University, St. Louis, Missouri.
Misra, U. K., Sanyal, M. K. and Prasad, M. R. N. 1965. Phospholipids of the sexual segment of the kidney of the Indian house lizard, *Hemidactylus flaviviridis* Rüppell. Life Sciences 4: 159-166.
Moreira, P. L. and Birkhead, T. R. 2003. Copulatory plugs in the Iberian rock lizard do not prevent insemination by rival males. Functional Ecology 17: 796-802.
Moriera, P. L., López, P. and Martín, J. 2006. Femoral secretions and copulatory plugs convey chemical information about male identity and dominance status in Iberian rock lizards (*Lacerta monticola*). Behavioral Ecology and Sociobiology 60: 166-174.
Nilson, G. and Andrén, C. 1982. Function of renal sex secretion and male hierarchy in the adder, *Vipera berus*. Hormones and Behavior 16: 404-413.
O'Donnell, R. P., Ford, N. B., Shine, R. and Mason, R. T. 2004. Male red-sided garter snakes, *Thamnophis sirtalis parietalis*, determine female mating status from pheromone trails. Animal Behaviour 68: 677-683.
Olsson, M. 2001. 'Voyeurism' prolongs copulation in the dragon lizard *Ctenophorus fordi*. Behavioral Ecology and Sociobiology 50: 378-381.
Platt, D. R. 1969. Natural history of the hognose snakes *Heterodon platyrhinos* and *Heterodon nasicus*. University of Kansas Publications, Museum of Natural History 18: 253-420.
Reddy, P. R. K., Prasad, M. R. N. and Misra, U. K. 1972. Dynamics of phospholipids metabolism in the sexual segment and kidney of the Indian house lizard, *Hemidactylus flaviviridis* Ruppell. Comparative Biochemistry and Physiology 43B: 9-20.
Regaud, C. and Policard, A. 1903. Recherches sur la structure du rein de quelques ophidians. Archives d'Anatomie Microscopique et de Morphologie Experimentale 6: 191-282.

Regamey, 1935. Les caractères sexuels du lezard (*Lacerta agilis* L.). Revue Suisse de Zoologie 42: 87-168.

Reiss, P. 1923. Le cycle testiculaire du lezard. Comptes Rendus de l'Académie des Sciences Paris, 88: 445-447.

Rivas, J. A. 2000. Life history of the green anaconda (*Eunectes murinus*) with emphasis on its reproductive biology. Ph.D. thesis, University of Tennessee, Knoxville, Tennessee.

Romer, A. S. and Parsons, T. S. 1986. *The Vertebrate Body*, Sixth Edition. Saunders College Publishing, Philadelphia, Pennsylvania. Pp. 679.

Ross, P. Jr. and D. Crews. 1977. Influence of the seminal plug on mating behaviour in the garter snake. Nature 267: 344-345.

Saint Girons, H. 1957. Le cycle sexuel chez *Vipera aspis* (L.) dans l'ouest de la France. Bulletin Biologique de la France et de la Belgique 91: 284-350.

Saint Girons, H. 1972. Morphologie comparée du segment sexuel du rein des Squamates (Reptilia). Archives d'Anatomie Microscopique et de Morphologie Expérimentale 61: 243-266.

Saint Girons, H. 1976. Les differents types de cycles sexuels des males chez les *Viperes europeennes*. Comptes Rendus de l'Académie des Sciences 282: 1017-1019.

Saint Girons, H. and Kramer, E. 1963. Le cycle sexuel chez *Vipera berus* (L.) en montagne. Revue Suisse de Zoologie 70: 191-221.

Saint Girons, H. and Pfeffer, P. 1971. Le cycle sexuel des serpents du Cambodge. Annales Des Sciences Naturelles Comprenant la Zoologie Series 12e 13: 543-572.

Sanders, K. L. and Lee, M. S. Y. 2008. Molecular evidence for a rapid late-Miocene radiation of Australasian venomous snakes (Elapidae, Colubroidea). Molecular Phylogenetics and Evolution 46: 1180-1188.

Seshadri, C. 1959a. Structural modification of the cloaca of *Lycodon aulicus aulicus* Linn., in relation to urine excretion and the presence of sexual segment in the kidney of male. Proceedings of the National Institute of Science, India 25B: 271-278.

Seshadri, C. 1959b. Functional morphology of the cloaca of *Varanus monitor* (Linnaeus) in relation to water economy. Proceedings of the National Institute of Sciences of India 25 B (2): 101-106.

Sever, D. M. and Hopkins, W. A. 2005. Renal sexual segment of the ground skink, *Scincella laterale* (Reptilia, Squamata, Scincidae). Journal of Morphology 266: 46-59.

Sever, D. M., Ryan, T. J., Stephens, R. and Hamlett, W. C. 2002. Ultrastructure of the reproductive system of the black swamp snake (*Seminatrix pygaea*). III. Sexual segment of the male kidney. Journal of Morphology 252: 238-254.

Sever, D. M., Siegel, D. S., Bagwill, A., Eckstut, M. E., Alexander, L., Camus, A. and Morgan, C. 2008. Renal sexual segment of the cottonmouth snake, *Agkistrodon piscivorus* (Reptilia, Squamata, Viperidae). Journal of Morphology 269: 640-653.

Shine, R. 1978. Sexual size dimorphism and male combat in snakes. Oecologia 33: 269-277.

Shine, R. 1994. Sexual size dimorphism in snakes revisited. Copeia 1994: 326-346.

Shine, R. and Madsen, T. 1994. Sexual dichromatism in snakes of the genus *Vipera*: A review and a new evolutionary hypothesis. Journal of Herpetology 28: 114-117.

Shine, R., Olsson, M. M. and Mason, R. T. 2000. Chastity belts in gartersnakes: the functional significance of mating plugs. Biological Journal of the Linnean Society 70: 377-390.

Shine, R., O'Donnell, R. P., Langkilde, T., Wall, M. D. and Mason, R. T. 2005. Snakes in search of sex: the relation between mate-locating ability and mating success in male garter snakes. Animal Behaviour 69: 1251-1258.

Siegel, D. S. and Sever, D. M. 2006. Utero-muscular twisting and sperm storage in viperids. Herpetological Conservation and Biology 1: 87-92.

Siegel, D. S., Aldridge, R. D., Clark, C. S., Poldemann, E. H. and Gribbins, K. M. 2009. Stress and reproduction in the brown treesnake (*Boiga irregularis*) with notes on the ultrastructure of the sexual segment of the kidney in squamates. Canadian Journal of Zoology 87: 1138-1146.

Srivastava, P. C. and Thapliyal, J. P. 1965. The male sexual cycle of the chequered water snake, *Natrix piscator*. Copeia 1965: 410-415.

Stille, B., Madsen, T. and Niklasson, M. 1986. Multiple paternity in the adder, *Vipera berus*. Oikos 47: 173-175.

Takewaki, K. and Fukuda, S. 1935a. Effect of gonadectomy and testicular transplantation on the kidney and epididymis of a lizard, *Takydromus tachydromoides*. Journal of the Faculty Science Tokyo Imperial University, Section IV, Zoology 1: 63-76.

Takewaki, K. and Fukuda, S. 1935b. Viability of epididymal spermatozoa in unilaterally and bilaterally castrated lizards. Journal of the Faculty Science Tokyo Imperial University, Section IV, Zoology 1: 77-82.

Takewaki, K. and Hatta, K. 1941. Effect of gonadectomy and hypophysectomy on the kidney and genital tract of a snake, *Natrix tigrina tigrina*. Annotationes Zoologicae Japonenses 20: 4-8.

Tokarz, R. R. 1988. Copulatory behaviour of the lizard *Anolis sagrei*: alternation of hemipenis use. Animal Behaviour 36: 1518-1524.

Tokarz, R. R. and Slowinski, J. B. 1990. Alternation of hemipenis use as a behavioural means of increasing sperm transfer in the lizard, *Anolis sagrei*. Animal Behaviour 40: 374-379.

Tsai, T-S. and Tu, M-C. 2000. Reproductive cycle of male Chinese green tree viper, *Trimeresurus stejnegeri stejnegeri*, in northern Taiwan. Journal of Herpetology 34: 424-430.

Uribe, M. C., González-Porter, G., Palmer, B. D. and L. J. Guillette Jr. 1988. Cyclic histological changes of the oviductal-cloacal junction in the viviparous snake *Toluca lineata*. Journal of Morphology 237: 91-100.

Volsøe, H. 1944. Structure and seasonal variation of the male reproductive organs of *Vipera berus* (L.). Spolia Zoologica Musei Hauniensis 5: 1-157.

Weil, M. R. 1984. Seasonal histochemistry of the renal sexual segment in male common water snakes, *Nerodia sipedon* (L.). Canadian Journal of Zoology 62: 1737-1740.

Weil, M. R. and Aldridge, R. D. 1981. Seasonal androgenesis in the male water snake, *Nerodia sipedon*. General and Comparative Endocrinology 44: 44-53.

Weil, M. R. and Aldridge R. D. 1985. Seasonal histochemistry of the renal sexual segment in the male northern water snake (*Nerodia sipedon*). Journal of Experimental Zoology 210: 327-332.

Whittier, J. M. and Crews, D. 1986. Ovarian development in red-sided garter snakes, *Thamnophis sirtalis parietalis*: relationship to mating. General and Comparative Endocrinology 61: 5-12.

Wisniewski, S. S. 2007. The role of the copulatory plug in *Lamprophis fuliginosus* and the contribution of the SSK in *Nerodia sipedon*. Master's Thesis, Saint Louis University, St. Louis, Missouri.

Yamanouye, N., Silveira, P. F., Abdalla, F. M., Almeida Santos, S. M., Breno, M. C. and Salomão, M. G. 2004. Reproductive cycle of the neotropical *Crotalus durissus terrificus*: II. Establishment and maintenance of the uterine muscular twisting, a strategy for long-term sperm storage. General and Comparative Endocrinology 139: 151-157.

Zweifel, R. G. 1997. Alternating use of hemipenes in the kingsnake, *Lampropeltis getula*. Journal of Herpetology 31: 459-461.

CHAPTER 12

Reproductive Cycles of Tropical Snakes

Tom Mathies

12.1 INTRODUCTON

Snakes are among the most successful of vertebrates, as indicated by their nearly world-wide distribution and occupancy of a wide range of latitudes, altitudes, and habitats. Given their wide distribution and diverse evolutionary lineages, it is not surprising that considerable variation in patterns of reproductive cyclicity exist. Historically, as seen in other areas of early biological inquiry, those species inhabiting temperate zones have received an inequitable degree of attention compared to those in sub-tropical and tropical zones. It was not until the mid-1960s to early 1970s that this condition was considered sufficiently alleviated by some (e.g., Fitch 1970), but numerically, at least, the disparity between studies on temperate and tropical zones persisted into the 1980s (Seigel and Ford 1987) and indeed to present. In the last 20 years however, there has been a marked increase in studies from certain understudied regions, most notably, South America and Australia. It is largely those additions that make the present chapter possible.

The substantial body of work on ophidian reproduction has shown that reproductive cycles of temperate zone species are uniformly seasonal and highly synchronous among individuals (Licht and Gorman 1970; Shine 1985; Duvall *et al.* 1982). Perhaps less anticipated, the same is true for most subtropical and tropical species (Fitch 1982; Vitt and Vangilder 1983). But it is also apparent that many species of tropical snakes are reproductive over extended periods and some apparently reproduce year-round (Saint Girons 1982). This chapter focuses on species with 'extended' or 'aseasonal' reproductive cycles. These species are still poorly understood, not only

United States Department of Agriculture, Wildlife Services, National Wildlife Research Center, 4101 LaPorte Avenue, Fort Collins, CO 80521, USA from ecological and ultimate (evolutionary) perspectives, but also from the simply mechanistic standpoint of the dynamics of reproductive processes of individuals. This chapter first addresses the types of cyclicity present at the level of the individual and how these determine the type of cyclicity manifest at the level of the population. A revised classification system with standardization of the terminology is offered. For the most part this chapter excludes discussion of steroid cycles and neuroendocrine control of reproduction because these areas are covered in detail in **Chapters 7** and **8** and because information on species with aseasonal cycles is still too scant to permit general conclusions.

12.2 VARIATION IN CYCLICITY OF REPRODUCTION

Reproductive processes of individuals may follow three patterns:

1. Discontinuous cyclical, where gonads or accessory organs become reproductively quiescent for some period during the year. This pattern is well-documented and widespread in both temperate and tropical zones.
2. Continuous cyclical, where gonads or accessory organs do not become completely quiescent, but show reduced activity for some period of the year. This pattern has been inferred but not conclusively verified in snakes.
3. Acyclical, where gonads and accessory organs exhibit essentially constant levels of activity throughout the year. This pattern has also been inferred but not conclusively verified in snakes.

More than one form of cyclicity may be operant at the level of the individual. Depending on species or population, variation in cyclicity may differ between the sexes or among the reproductive organs of a single individual. For example, in the Chequered Water Snake (*Natrix piscator*) in India, recrudescence and regression of spermatogenesis are strongly seasonal whereas activity of Leydig cells and the sexual segment of the kidney are acyclic (Srivastava and Thapliyal 1965). Thus one term may not fully describe 'reproductive cyclicity' at the level of the individual. With respect to assessing cyclicity type for populations in later sections, however, this chapter considers temporal variation in either gonadal activity, accessory organ activity, or steriodogenic activity, grounds for rejecting an aseasonal designation in favor of seasonal.

Cyclicity of reproduction at the level of the population has long been of interest to biologists and contemporary conservation efforts often depend on such information (e.g., Valdujo *et al.* 2002; Fitzgerald *et al.* 2004; Brooks *et al.* 2009). But documenting cyclicity at the level of the population and elucidating its underlying causes is not always straightforward. The form of cyclicity documented at the population level, typically by collecting samples at regular intervals, may not necessarily permit conclusions on the underlying cycles of individuals The cyclicity of individuals and its temporal expression by individuals throughout the course of a reproductive period produces the overall type of cyclicity observable at the population level. Cyclicity at the population level, as derived from cyclicities of individuals of either sex, can be viewed as two types, seasonal or aseasonal (Fig. 12.1).

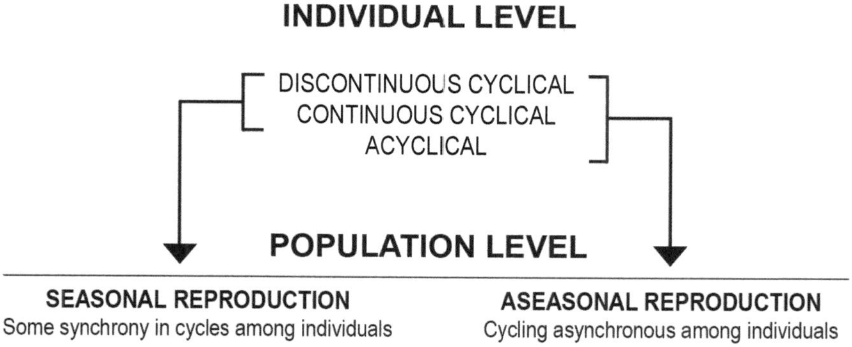

Fig. 12.1 The two types of reproductive cyclicity at the population level (seasonal and aseasonal) and their derivations from the cycles of individuals as mediated by reproductive synchrony among individuals. Discontinuous cyclical: gonads or accessory organs of individuals become reproductively quiescent for some period during the year. Continuous cyclical: gonads or accessory organs of individuals show a reduction in activity for some period of the year. Acyclical: gonads and accessory organs of individuals exhibit essentially constant levels of activity throughout the year.

The vast majority of snake species investigated to date exhibit seasonal reproduction, which by definition, requires at least some degree of synchrony of gonadal activity among individuals within sex. In practice then, seasonal reproduction is evidenced simply by a disproportionate number of individuals in a population exhibiting similar stages of the reproductive process at a given time of year. All available data indicate snake species inhabiting cool or cold higher latitudes exhibit seasonal reproduction (Fitch 1970; Shine 1977a,b; Aldridge 1979), and the degree of reproductive synchrony among individuals has been shown to increase with increasing latitude (Shine 1980a,b). Most species in subtropical, tropical, and equatorial regions also exhibit seasonal reproduction, although the extent of synchrony among individuals may not be nearly as marked. Far fewer species appear to exhibit aseasonal reproduction, and without exception, all are confined to tropical and equatorial areas. Aseasonal reproduction is evidenced by equable proportions of individuals exhibiting similar stages of the reproductive process in every month of the year.

Others have defined aseasonal reproduction somewhat differently, using the terms 'continuous', 'acyclical', and 'aseasonal' interchangeably when referring to either individuals or populations (Licht 1984). As has been narrowly defined by Licht (1984), 'continuous' or aseasonal reproduction requires that rates of reproductive processes in all individuals in a population are constant throughout the year. Some, however, have used the definition more loosely, positing aseasonal reproduction simply on the finding of some reproductive individuals in all or most months (e.g., Fitch 1970; Saint Girons 1971). Because of the considerable distance between these two views, there has been disagreement over whether any snake (or reptile) species exhibits 'truly' aseasonal reproduction (Licht 1984; Callard and Kleis 1987). The narrow view where there is no temporal variation in the rates of reproductive processes in individuals or populations obviously accommodate very few, if any, ectotherms, and thus subsume nearly all species to the alternate classification, 'seasonal'. Adherence to this narrow view renders this useful dichotomous classification system of seasonal *vs.* aseasonal quite unbalanced, leading to obfuscation of the underlying patterns and processes and hence fruitful discussion. The alternate view, such as that allowed by Saint Girons (1971) and Fitch (1970), whereby demonstration of equitable proportions of individuals exhibiting similar stages of the reproductive process in every month is not even attempted is also clearly unsatisfactory, but only for methodological rather than conceptual reasons. The framework laid out above for identifying seasonal versus aseasonal reproduction lies between these two views, providing classifications for cycle types of individuals, a more workable definition for aseasonal reproduction, and a methodological criterion for assigning type of cyclicity at the population level.

Although assessment of the type of cyclicity at the population level can oftentimes be accomplished using standard population sampling methods alongside gonadal and steroidogenic examinations, the resultant quantitative and qualitative data may reveal little about the underlying cycles of individuals, depending on 1) the type of individual cycles operant and 2) the extent of synchrony of reproductive processes among individuals. For example, in a population with seasonal reproduction where testicular cycles of individuals are discontinuous cyclic (i.e., testis are in complete regression for some time period) and highly synchronized among individuals, examination of samples of moderate sizes collected regularly throughout the period of reproduction will adequately reveal the phenology of the spermatogenesis. Seasonal reproduction with tight synchrony among individuals is thus easily demonstrated. Similarly, in a population where the testicular cycle of individuals is acyclic, samples of moderate sizes collected regularly throughout the year may adequately reveal aseasonality at the population level and inference of acyclicity at the individual level, as judged by the absence of individuals with testes showing complete regression. Problems in inferring the cycles of individuals arise in cases where individual cyclicity is discontinuous or continuous but pronouncedly asynchronous among individuals. The problem arises from what is effectively an attempt to sample a 'moving target'; unless sample sizes collected at each time interval are substantial, the variation in stage of reproductive cycle present among individuals at any sample time may not be adequately captured. The greater the asynchrony among individuals, the greater the sample sizes needed. If for example, in two consecutive sample periods all individuals collected happen to have testes that are either regressed or in regression, it might be falsely concluded that the population exhibits seasonal reproduction. Just as likely, such data might go unpublished due to their 'inconclusiveness'. This is unfortunate because it is for these species, all of which are either sub-tropical or tropical, that the least data exist.

Regardless of the degree of reproductive synchronicity among individuals, quantitative analyses of gravimetric and morphometric changes in gonadal and accessory sexual organs together with detailed investigation of gametogenesis, steroidogenic activity, extent of sperm storage by both sexes, are often required to adequately characterize and interpret reproductive cycles at the individual and population levels. In a later section problems and practicalities for assessing cyclicities of individuals are discussed. The following two sections on the sexes focus on species-specific examples of aseasonal reproduction at the population level, and where possible, attention to the underlying cyclicities of individuals. Material for this chapter comes from an extensive review of the literature and is based on published studies on reproductive cycles of tropical snakes from peer-reviewed journals and academic dissertations. Over 135 sources provided quantitative data on monthly reproductive activity and many of these sources providing data for more than one species.

12.2.1 Male

A pervasive shortcoming among studies investigating reproductive biology of snakes is inadequate investigation of male reproductive state. Testis and accessory ducts are often examined only macroscopically, and then only to determine maturity, not reproductive condition. For example, from a representative sample of the literature on central and South American species, considering only those studies where direct gonadal examinations on both sexes were performed, 4.4% of studies did not present any data on testicular condition (2 of 45 studies), 49% simply judged whether testis were 'enlarged' or deferent or efferent ducts were 'opaque' (i.e., sensu stricto Shine 1977a; Shine 1980a,c), 31% determined testis mass or size, but only 15.5% employed histological methods to examine testicular condition. This bias is due in part to the greater importance placed on the female for population persistence, but probably more to the comparative ease at which female reproductive condition can be assessed (macroscopic assessment of follicular state and whether ovigerous).

In contrast to the female, an adequate understanding of cyclicity in the male requires assessment of several features of the urogenital system with histological examination of the testis being key; plasma levels of sex steroids are not necessarily indicative of recrudescence of spermatogenesis and spermatozoa production and impart little information in cases where levels are found seasonally invariant (e.g., Mathies *et al.* 2010).

Of the techniques commonly used to assess male gonadal condition, only histological examination of the testis and accessory ducts provides unequivocal proof of spermatogenic activity. It has been stated that rates of gonadal processes cannot be judged from histological preparations because of their 'static' nature (see Licht 1984). However, the spatial extent of spermiogenesis within tubules does in fact, provide such a measure, albeit only when there is seasonal variability. Measures of testicular size or mass are often used to infer testicular activity but with the same shortcoming that potential variation in activity cannot be ruled out when these features are temporally invariant. The same is true in cases where measures of features of the accessory ducts do not vary. For example, in the Green Vine Snake (*Oxybelis fulgidus*) it was assumed that because the mean diameter of the vas deferens did vary among months that males did not store sperm (Scartozzoni *et al.* 2009). Although this could be true, one could just as easily conclude the opposite. Cyclicity may occur within the vas deferens even when the amount of semen within does not vary seasonally; male Brazilian Rattlesnakes (*Crotalus durissus terrificus*) retain sperm in the vas deferens year-round with no apparent seasonal variation in semen volume, but highest sperm counts are observed just prior to the mating season (Almeida-Santos *et al.* 2004).

Conceptualizations of cycles for individuals as they contribute to seasonality of reproduction of populations are shown in Figure 12.2. In individuals that are 'discontinuous cyclical', complete testicular regression occurs during some part of the year; 'continuous cyclical' applies where the extent of spermatogenesis within the seminiferous tubules varies during the year, but complete regression does not occur; 'acyclical' applies where there is little to no temporal variation in the extent of spermatogenesis within the seminiferous tubules throughout the year. An example of use of the terminology for delineating the cyclicity of a population is as follows: in a population where monthly sampling reveals some individuals in each sample with regressed testes, but a relatively high proportion of individuals with recrudescent testes during a particular period, that population could be said to exhibit seasonal-semi-synchronous reproduction where the underlying cycles of individuals are discontinuous-cyclical (see Fig. 12.2).

The following considerations of whether a study has demonstrated aseasonal reproduction, rely, at minimum, on data presented for gonadal activity (direct or indirect; e.g., histological examination of the spermatogenesis or temporal invariance in testis mass, respectively) fully expecting that future data from histological investigations will invalidate many of the assessments based on indirect data. Further, because the majority of reproductive studies make use of museum specimens, which are often collected opportunistically, sample sizes are often low or nonexistent in some months. Nevertheless, a number of such studies were included when the data seemed reasonably strong, but again, expecting that further data may lead to reassessments. It should also be noted that, because reproductive processes in males of most species examined are cyclic

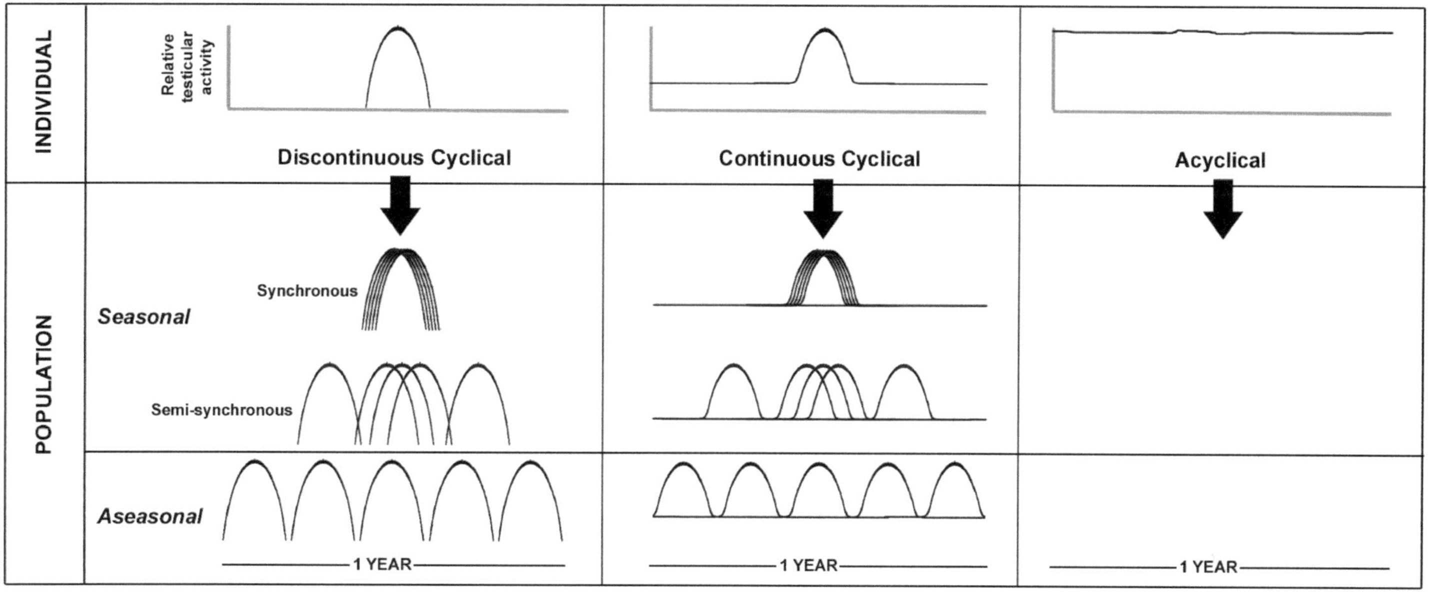

Fig. 12.2 Schematic representation of the variation in testicular cycles of individuals and how inter-individual synchrony in cycles determines type of reproductive cycle at the population level (seasonal or aseasonal). Cycles of individuals: Discontinuous cyclical; testis or accessory organs of individuals become reproductively quiescent for some period during the year. Continuous cyclical; testis or accessory organs of individuals show a reduction in activity for some period of the year. Acyclical; testis and accessory organs of individuals exhibit essentially constant levels of activity throughout the year. Population-level cycles: Synchronous; cycles of individuals in a population progress in close synchrony. Semi-synchronous; cycles of individuals tend to be more coincident at a particular time of year than another, identifiable as a peak period of reproduction. Aseasonal; when cycles of individuals are either discontinuous cyclical or continuous cyclical, the proportion of individuals in reproductive condition is essentially constant throughout year; when cycles of individuals are acyclical, every individual within a population is continuously reproductive.

(i.e., spermatogenesis, secretory activity of the kidney sexual segment, steroidal secretion as judged by plasma levels: Saint Girons and Pfeffer 1971; Saint Girons 1982; Shine 1977a; Butler 1993; Tsai and Tu 2000; Pizzatto *et al.* 2007; Aldridge *et al.* 2009), the more reproductive features that are measured, the more likely that a designation of 'acyclic' will be falsified. Demonstrating acyclity in the male is comparatively tedious, requiring a reasonably comprehensive survey of the male urogenital system. For these reasons a number of studies where the authors have unintentionally illustrated some of the above problems were highlighted.

Since reproductive cycles of male snakes were reviewed over two decades ago when data for tropical species were relatively few (Licht 1984; Siegel and Ford 1987), there have been many subsequent studies claiming documentation of aseasonal reproduction in males. In the present survey 31 cases of putative aseasonal reproduction were identified (Table 12.1). Inspection of these cases, however, reveals that few studies were based on histological examination of the testis, and for those that were, sample sizes were often small or sometimes nonexistent for some months (e.g., Saint Girons and Pfeffer, 1971; Goldberg 2003, 2004b, 2007). That said, inclusion of these latter studies is testament to the information-providing power of histological examination even in the face of small sample sizes. For example, small monthly sample sizes ($N \approx 3$) for a tropical lizard exhibiting year-round spermatogenesis were sufficient to detect population-level fluctuations in seminiferous tubule diameters and epithelial heights associated with fluctuations in the numbers of elongating spermatids and spermiation events (Gribbins *et al.* 2009). Similarly, in males of two species of tropical bolitoglossine salamanders found to be spermatogenic year-round, monthly sample sizes of 1-3 males were sufficient to demonstrate significant monthly variation in levels of spermatozoa in the seminiferous tubules that were not apparent from the external appearance of the testis (Chan 2003).

It is the studies conducted on tropical snakes in the 1960s to 1980s that employed adequate sample sizes, histological investigation, and other metrics of gonadal activity, that still stand as the exemplars providing the greatest insights into the underlying cycles of individuals and how they determine cyclicity at the population level. Namely, Gorman *et al.* (1981) found that males of the Banded Sea Krait (*Laticauda colubrina*) were spermatogenic throughout the year and testis mass and plasma testosterone levels were relatively invariant with no consistent pattern among months. Individuals are thus apparently acyclic and the reproduction in the population was evidently aseasonal. Males of a syntopic species, the Dog-faced Watersnake (*Cereberus rhynchops*), were also spermatogenic year-round, but there were associated peaks in seminiferous tubule height, testis mass, and plasma testosterone in September to November suggesting an increased rate of testicular activity during that period. Testicular activity in individuals thus appears to be continuous cyclical and the population is seasonal. Note that in the latter species, in the absence of histological examination it might have been incorrectly surmised that the testis underwent complete regression. In another early study providing similarly insightful observations, Berry and Lim (1967) found that in the Puff-faced Watersnake (*Homalopsis buccata*) some males in every sample taken throughout the year were spermatogenic, but the proportions of males in this condition varied significantly among months. Testis mass, however, did not vary among months. The cycle of individual males was thus inferably discontinuous cyclic, and because of the disproportionate numbers of spermatogenic males in some months, the population cycle was seasonal. Here, note that in the absence of histological data it might have been incorrectly surmised that cycles of individuals were acyclic and the population cycle was aseasonal! The larger implication of this finding is that it calls into question studies lacking histological examination where features of the urogenital system were found to be invariant among sample periods. Although enlargement of the testis is generally correlated with spermatogenesis in temperate zone snakes (Volsøe 1944, and for review see Fox 1977), this is not necessarily the case in tropical species where the magnitude of excursion in testis mass may be much less marked (e.g., less than 2-fold: *Cerberus rhynchops*, Gorman *et al.* 1981; Guibe's Flame Snake, *Oxyrhopus guibei*, Pizzatto and Marquez 2002), even in some temperate zone species (Blackish Blind Snake, *Ramphotyplops nigrescens*, Shea 2001; Mojave Rattlesnake, *Crotalus scutulatus*, Schuett *et al.* 2002). In Table 12.1, for example, there are 18 cases where testis size or mass did not vary among sample periods and population was thus judged aseasonal. Presence of sperm within the efferent ducts (or epididymis) and particularly activity within the sexual segment of the kidney were infrequently examined.

A survey of the literature revealed no cases of aseasonal reproduction as narrowly defined by Licht (1985). However, reproduction by males in the Guam population of the Brown Treesnake (*Boiga irregularis*) nearly meets the criterion; an apparently constant rate of gonadal activity (e.g., spermatogenesis, seminiferous tubule epithelial height, testis mass) and steroidogenic activity (testosterone) was exhibited by all males in all months, but features of the kidney sexual segment (e.g., tubule diameter) varied significantly among months (Mathies *et al.* 2010). Thus, whereas the cycle of individuals could be judged continuous cyclical (kidney sexual segment was secretory in all individuals), reproduction at the population level would be judged seasonal as defined herein. Note the substantial number of features that were investigated in order to detect putative evidence of seasonality. It remains to be determined whether such subtleties in reproductive physiology are associated with seasonality of mating. Estrus in temperate zone colubrids has been investigated (Aldridge *et al.* 2009) but such information is lacking for most tropical species.

Other investigations on the reproductive biology of *Boiga irregularis* have provided additional insights into the potential diversity in cycles in tropical snakes. Studies on this species in a subtropical part of its natural range revealed that testis of males in those populations undergo seasonal

Table 12.1 Summary of the conditions of testis and accessory structures of 31 species of tropical snakes with putative aseasonal reproduction. O, oviparous, V, viviparous. Category 'Spermatogenic Each Sample Period' based on histological investigation of seminiferous tubules of testis. Designations of individual cycle type are based on data presented by authors, and may be contradictory to conclusions stated in the original paper. NV indicates no variation and NE indicates not examined.

				Variation in feature during year (e.g., size, mass, sperm presence, secretory activity)					
Family, Subfamily	*Species*	*Parity mode*	*Spermatogenic each sample period*	*Testis*	*Efferent ducts*	*Vas deferens*	*Kidney sexual segment*	*Individual cycle type*	*References*
Boidae	*Corallus hortulaus*	V	NE	NV[1]	NE	NV[2]	NE	Unknown	Pizzatto and Marques 2007
Boidae	*Epicrates cenchria cenchria*	V	NE	NV[1]	NE	NV[2]	NE	Unknown	Pizzatto and Marques 2007
Colubridae	*Boiruna maculata*	O	NE	NV[1]	NE	NV[2]	NE	Unknown	Pizzatto 2005
Colubridae	*Dendrelaphis pictus*	O	All males	NV[3]	NE	NE	NV[2]	Acyclic	Saint Girons and Pfeffer 1971
Colubridae	*Dendrophidion vinitor*	O	All males	NE	NE	NV[4]	NV[5]	Acyclic	Goldberg 2003
Colubridae	*Dipsas catesbyi*	O	NE	NV[6]	NE	NV[2]	NE	Unknown	Alves *et al.* 2005
Colubridae	*Dipsas neivai*	O	NE	NV[6]	NE	NV[2]	NE	Unknown	Alves *et al.* 2005
Colubridae	*Oligodon taeniatus*	O	All males	NV[3]	NE	NE	NV[2]	Acyclic	Saint Girons and Pfeffer 1971
Colubridae	*Erythrolamprus bizona*	O	All males	NE	NE	NV[4]	NV[5]	Acyclic	Goldberg 2004a
Colubridae	*Geophis godmani*	O	All males	NE	NE	NE	NE	Acyclic	Goldberg 2007a
Colubridae	*Leptodeira annulata*	O	NE	NV[6]	NE	NV[2]	NE	Unknown	Pizzatto *et al.* 2008a
Colubridae	*Liophis milaris*	O	NE	NV	NE	NE	NE	Unknown	Pizzatto 2003
Colubridae	*Mastigodryas bifossatus*	O	NE	NV[1]	NE	NE	NE	Unknown	Leite *et al.* 2009
Colubridae	*Mastigodryas melanolomus*	O	All males	NE	NE	NV[4]	NE	Acyclic	Goldberg 2006a
Colubridae	*Ninia maculata*	O	All males	NE	NE	NV[4]	NV[5]	Acyclic	Goldberg 2004b
Colubridae	*Ptyas korros*	O	All males	NV[3]	NE	NE	NV[2]	Acyclic	Saint Girons and Pfeffer 1971
Colubridae	*Rhadinea decorata*	O	All males	NE	NE	NV[4]	NE	Acyclic	Goldberg 2007b
Colubridae	*Sibynomorphus mikanii*	O	NE	NV[6]	NE	NV[2]	NE	Unknown	Pizzatto *et al.* 2008a
Colubridae	*Sibynomorphus neuwiedi*	O	NE	NV[6]	NE	NV[2]	NE	Unknown	Pizzatto *et al.* 2008a

Colubridae	*Sibynomorphus ventrimaculatus*	O	NE	NV[6]	NE	NV[2]	NE	Unknown	Pizzatto *et al.* 2008a
Colubridae	*Waglerophis merremii*	O	NE	NV[1]	NV	NV[2]	NE	Unknown	Pizzatto *et al.* 2008b
Colubridae	*Xenodon neuwiedii*	O	NE	NV[1]	NE	NV[2]	NE	Unknown	Pizzatto *et al.* 2008b
Elapidae	*Cacophis squmulosus*	O	NE	NE	NV[7]	NE	NE	Unknown	Shine 1980a
Elapidae	*Cacophis harriettae*	O	NE	NE	NV[7]	NE	NE	Unknown	Shine 1980a
Elapidae	*Cacophis krefftii*	O	NE	NE	NV[7]	NE	NE	Unknown	Shine 1980a
Elapidae	*Micrurus nigrocinctus*	O	All but one male	NE	NE	NE	NE	Unknown	Goldberg 2004c
Hydrophidae	*Laticauda colubrina*	O	All males	NV[6]	NV[4]	NE	NE	Acyclic	Gorman *et al.* 1981
Hydrophidae	*Laticauda semifasciata*	O	NE	NV[6]	NE	NV[4]	NE	Unknown	Tu *et al.* 1990
Viperidae	*Causus maculatus*	O	NE	NE	NV[7]	NE	NE	Unknown	Ineich *et al.* 2006
Viperidae	*Causus lichtensteinii*	O	NE	NE	NV[7]	NE	NE	Unknown	Ineich *et al.* 2006
Viperidae	*Causus resimus*	O	NE	NE	NV[7]	NE	NE	Unknown	Ineich *et al.* 2006

[1]Volume.
[2]Diameter.
[3]Seminiferous tubule diameter.
[4]Sperm present.
[5]Sexual segment secretory.
[6]Length.
[7]Presence of sperm indirectly assessed based on exterior appearance of efferent ducts (sensu Shine 1980a)
[8]Mass.

recrudescence and regression whereas in a tropical population there are always some spermatogenic males in all months (Bull *et al.* 1997). Whether such plasticity in reproductive cyclicity is unique to this species, or whether it is more widely present among other tropical species, awaits further inquiry. The timing of testicular cycles also can vary substantially between populations of the same species. In Costa Rica, testicular cycles (as judged by change in testis mass) of the Terciopelo (*Bothrops asper*) in Atlantic and Pacific populations at equivalent latitudes were out of phase and the Atlantic population showed greater excursions in testis mass than the Pacific population (Solórzano and Cerdas 1989).

Ignoring for the moment the more tenuous designations of acyclicity in Table 12.1, aseasonal reproduction occurs in both oviparous and viviparous species, and is present in Boidae, Colubridae, Elapidae, and Viperidae. The majority of cases of acyclicity, however, are within Colubridae. Whether the preponderance within Colubridae reflects a real predisposition for aseasonality in this taxon or whether it is merely commensurate with the disproportionate number of species in this polyphyletic group (Heise *et al.* 1995) is unclear.

Given the relative wealth of recent studies on species that seem to exhibit aseasonal reproduction, it speaks to the insufficiencies of the methodologies employed that unequivocal (histological) evidence for aseasonal reproduction still only exists for one species, *Laticauda colubrine* (Gorman *et al.* 1981).

12.2.2 Female

"For these six species wherein the largest number of eggs was found in the oviducts ranged from the beginning of July to the end of November, whence it may be concluded that the 'season', if we may speak of any 'season', is not very pronounced. On the other hand it cannot be denied that annually there seems to be a time of increased propagation." A quote from C.P.J. De Haas (1941), commenting on a the six most common species of snakes from a collection of snakes acquired over a two year period (34 species, 3509 snakes total) from two plantations in different districts of west Java.

Aseasonal reproduction. The scope and import of the work by De Haas (1941) is remarkable by the standard of any day, and his findings aptly illustrate the inherent difficulty in detecting subtle, but biologically important, trends in timing of reproduction in species of tropical snakes with extended periods of reproduction. This subtlety is due, in part, to fundamental differences between the reproductive processes of females and males. Production of gametes by individual females, unlike the spermatogenesis by males, cannot be acyclic; follicles recruited to begin vitellogenesis enter into and complete vitellogenesis as a discreet cohort (Callard and Kleis 1987). Females of all snake species thus exhibit discontinuous cyclic reproduction. Females of those species that produce multiple clutches (evidence presented below) where vitellogenesis of a cohort of follicles is initiated while the female is still ovigerous could be considered continuous cyclic, but only if there is no period of ovarian quiescence between bouts of consecutive clutches. No such cases are known or implied. Because individual females do not exhibit acyclic reproduction, truly aseasonal reproduction at the population level is therefore less likely to occur in females than males. Where it does, 'continuity' at the population level rests tenuously on overlap of the relatively brief cycles of individuals.

The criteria used to judge whether a species exhibited aseasonal reproduction, like those for males, were purposefully non-conservative, and it is expected that further studies will show that many of these assessments were too liberal. Some cases that might seem to implicate aseasonality are discussed here along with several others where seasonality was demonstrated because they are instructive for illustrating the difficulties in demonstrating aseasonal reproduction. For example, De Haas (1941) noted for one species, *Elapoidis fusca*, that although ovigerous females were collected in every month of the year, the percentage of females that were ovigerous varied seasonally. Reproduction at the population level was thus seasonal. If not for his adequately high sample sizes such variation might have been undetectable and an aseasonal cycle might have been incorrectly inferred. With even smaller monthly samples, reproduction in some months might have gone undetected and a seasonal cycle might have been correctly inferred, but without the realization that reproduction was actually occurring year-round.

What metric of the female reproductive process is most reliable for judging aseasonality (or seasonality)? For the individual, the best metric is one indicating that reproduction would have been completed within or near the period of the study when the individual was acquired. For oviparous species, gravidity is obviously the preferred metric, as completion of reproduction (oviposition) is relatively imminent, but also because the length of time a female is ovigerous may not be nearly as variable as the time it is vitellogenic; the duration eggs are retained in the oviducts is largely conserved across species as developing embryos of the majority of squamates proceed to a common developmental stage at oviposition (embryo Stage 30, Shine 1983; see also Dufaure and Hubert 1961). Further, whereas rates of vitellogenesis seem to be temperature-permissive, rates of embryogenesis are temperature-dependent. Thus, observation of vitellogenic follicles may not be a reliable indicator of impending ovulation. For example, for female Neuwied's Lanceheads (*Bothrops neuwiedi pubescens*) in Brazil, there were similar size distributions of vitellogenic follicles in most months of the year with no apparent overall trend in increasing follicle size, but pregnant females were observed only October to March (Hartmann *et al.* 2004). The apparent asynchrony in follicular development together with seasonal synchrony in parturition indicates rates of vitellogenesis varied substantially among individuals. Licht (1984) stressed that observation of vitellogenic follicles in an individual yields only a 'static' view of gametogenesis; i.e., the rate of vitellogenesis cannot be assumed to be constant. Rates of follicular development, like all other aspects of reptilian physiology, are influenced by temperature, but for species that are 'income breeders' (vitellogenesis supported more through recent food consumption than mobilization of stores from the corpora adiposa; see Drent and Daan 1980; Reading 2004), rates of vitellogenesis can be lengthy and variable (e.g., Cree *et al.* 1991). Indeed, in species that produce multiple clutches per season and have extended or aseasonal reproduction we should expect rates of vitellogenesis to be variable. This follows from the rationale that multiple clutching is tenable only through an income breeding mechanism of vitellogenesis (Ineich *et al.* 2006) which is in turn mediated by a food supply that may be temporally variable. Species that are capital breeders (Drent and Daan 1980) have a much greater potential to support relatively constant rates of vitellogenesis because all necessary stores have been sequestered in the corpora adiposa (e.g., Chinese Cobra, *Naja naja*, Lance and Lofts 1978; Australian Death Adder, *Acanthophis antarcticus*, Shine 1980c).

For viviparous species, pregnancy, the homologue of gravidity, is similarly the best metric for assessing aseasonality. However, because viviparous species necessarily retain the embryos within the oviducts much longer than oviparous species, progression of reproduction should be documented throughout this period. Typically embryos are examined and assigned a stage of embryonic development (for staging schemes for colubrid and viperid embryos, see Zehr 1962, and Hubert and Dufaure 1968, respectively), or modifications thereof (e.g., Brooks *et al.* 2009). Crude assessments as merely pregnant or not-pregnant may give the false appearance that females are initiating reproduction year-round (c.f., *Helicops infrataeniatus*, Schmidt de Aguir and Di-Bernardo 2005). Almost all studies on seasonality of reproduction in snakes present data on monthly occurrence of hatchlings or neonates as evidence for extended or aseasonal reproduction. But again, caution is advised as acceptance of an aseasonal designation would require equable counts in every month of the year.

How good is the evidence for aseasonal reproduction in females? Since the last comprehensive review on this subject a little over two decades ago (Licht 1984), there have been a substantial number of studies purporting aseasonal reproduction. Many have had to rely on museum specimens secured over large geographical areas (e.g., Zug 1979; Cottone *et al.* 2009) and many decades (e.g., Pizzatto 2005). Nearly all have relied on observation of females with vitellogenic follicles; relatively few have been able to document ovigerous females in all or most months of the year (Table 12.2). The scarcity of ovigerous females (oviparous) in collections is most certainly due to the brief time eggs reside in the oviducts relative to the duration of vitellogenesis, but perhaps also to the reduction in feeding rates (e.g., Solórzano and Cedras 1989; Kofron 1990; but see Pinto and Fernandes 2004) and thus activity of the females while ovigerous. For

Table 12.2 Summary of the condition and monthly occurrence of ovarian follicles and eggs of 26 species of tropical snakes with putative aseasonal reproduction. To facilitate comparisons among species, months for northern hemisphere species have been inverted. O, oviparous, V, viviparous, □ = ovigerous, ● = vitellogenic follicles, ○ = non-vitellogenic follicles; each symbol type indicates at least one female of that condition reported for that month. Dash indicates data not reported. Data were pooled by month in the few cases where data for more than one year were presented.

			Month													
Family	*Species*	*Parity*	J	F	M	A	M	J	J	A	S	O	N	D	*Multiple clutches*	*References*
Colubridae	*Boiga irregularis Guam form*	O	-	-	-	-	-	-	-	-	-	-	-	-	None Reported	Savidge *et al.* 2007
			●	●	●	●	●	●	●	●	●	●	●	●		
			○	○	○	○	○	○	○	○	○	○	○	○		
Colubridae	*Calamaria multipunctata*	O	□	□	□	□	□	□	□	□	□	□	□	□	None Reported	De Haas 1941
			-	-	-	-	-	-	-	-	-	-	-	-		
			-	-	-	-	-	-	-	-	-	-	-	-		
Colubridae	*Calamaria lumbricoidea*	O	-	□	-	□	□	□	□	□	□	-	□	□	None Reported	De Haas 1941
			-	-	-	-	-	-	-	-	-	-	-	-		
			-	-	-	-	-	-	-	-	-	-	-	-		
Colubridae	*Clelia plumbea*	O	-	-	-	-	-	-	-	-	-	-	-	-	None Reported	Pizzatto 2005
			●	●	-	-	●	●	-	●	●	-	●	-		
			○	-	○	○	-	-	○	-	-	○	○	○		
Colubridae	*Dipsas catesbyi*	O	□	-	□	-	□	□	□	-	-	-	-	□	None Reported	Alves *et al.* 2005
			-	-	●	●	●	-	-	●	●	-	●	●		
			-	○	○	○	○	-	○	-	○	○	○	○		
Colubridae	*Dipsas neivai*	O	□	□	□	□	□	-	□	□	-	□	□	□	None Reported	Porto and Fernandes 1996
			-	●	-	●	-	●	●	●	●	●	-	●		
			-	-	-	-	-	-	-	-	-	-	-	-		
Colubridae	*Dipsas neivai*	O	-	-	-	-	□	-	□	□	□	□	□	□	None Reported	Alves *et al.* 2005
			●	●	●	●	●	●	●	●	●	●	●	●		
			○	○	○	○	○	○	-	○	○	○	○	-		
Colubridae	*Dendrophidion dendrophis*	O	□	□	□	-	-	-	□	-	-	□	-	-	None Reported	Prudente *et al.* 2007
			-	-	-	●	●	●	●	●	-	-	-	●		
			-	○	-	-	○	○	○	-	-	○	○	○		

Table 12.2 Contd. ...

... *Table 12.2* *Contd.*

Colubridae	*Dendrophidian vinitor*	O	-	-	-	-	-	□	-	□	-	-	□	-	None Reported	Stafford 2003
			-	●	-	-	-	-	●	-	●	●	-	●		
			-	-	-	-	○	-	-	-	-	-	-	-		
Colubridae	*Elaphe triaspis*	O	-	□	□	-	-	-	-	-	-	-	-	-	None Reported	Censky and McCoy 1988
			●	●	●	●	●	-	●	●	-	●	●	●		
			-	-	-	-	-	-	-	-	○	○	○	-		
Colubridae	*Erythrolamprus bizona*	O	-	-	-	-	-	-	□	□	-	---	-	-	None Reported	Goldberg 2004
			-	-	●	●	●	●	●	●	●		-	●		
			○	-	-	○	-	○	○	-	○		○	○		
Colubridae	*Gongylosoma baliodeira*	O	-	□	□	□	□	□	□	□	□	□	□	□	None Reported	De Haas 1941
			-	-	-	-	-	-	-	-	-	-	-	-		
			-	-	-	-	-	-	-	-	-	-	-	-		
Colubridae	*Leptodeira annulata*	O	□	□	□	-	-	-	□	□	□	□	□	□	None Reported	Fitch 1970
			-	-	-	-	-	-	-	-	-	-	-	-		
			○	○	○	-	-	○	○	○	○	○	○	○		
Colubridae	*Leptophis ahaetulla*	O	□	□	□	-	□	□	□	□	-	-	□	□	None Reported	Oliver 1947
			-	-	-	-	-	-	-	-	-	-	-	-		
			○	○	○	○	○	○	○	○	-	-	○	○		
Colubridae	*Liophis poecilogyrus poecilogyrus*	O	-	□	□	□	□	□	-	□	□	□	-	□	Yes	Pinto and Fernandes 2004
			●	●	●	●	●	●	●	●	●	●	●	●		
			○	○	○	○	○	○	○	○	○	○	○	○		
Colubridae	*Rhabdophis chrysargos*	O	□	□	□	□	□	□	□	□	□	□	□	□	None Reported	De Haas 1941
			-	-	-	-	-	-	-	-	-	-	-	-		
			-	-	-	-	-	-	-	-	-	-	-	-		
Colubridae	*Rhabdophis subminiatus*	O	□	-	-	-	□	□	□	□	□	□	□	□	None Reported	De Haas 1941
			-	-	-	-	-	-	-	-	-	-	-	-		
			-	-	-	-	-	-	-	-	-	-	-	-		
Colubridae	*Rhabdophis vittata*	O	□	□	□	□	□	□	□	□	□	□	□	□	None Reported	Kopstein 1938
			-	-	-	-	-	-	-	-	-	-	-	-		
			-	-	-	-	-	-	-	-	-	-	-	-		

Colubridae	*Tantilla melanocephala*	O	-	-	□	□	□	□	□	□	□	□	□	-	None Reported	Santos-Costa *et al.* 2006
			●	●	●	●	●	●	●	●	●	●	●	●		
			-	-	-	-	-	-	-	-	-	-	-	-		
Colubridae	*Xenochrophis vittata*	O	□	□	□	□	□	□	□	□	□	□	□	□	None Reported	Kopstein 1938
			-	-	-	-	-	-	-	-	-	-	-	-	for wild-caught	
			-	-	-	-	-	-	-	-	-	-	-	-		
Colubridae	*Xenodon neuwiedii*	O	-	□	-	□	-	-	□	□	-	□	□	-	None Reported	Pizzatto *et al.* 2008b
			●	●	●	●	●	●	●	●	●	●	●	●		
			○	○	○	○	○	○	-	-	-	-	○	○		
Elapidae	*Furina* sp.	?	-	-	-	-	-	-	-	-	-	-	-	-	None Reported	Shine 1981
			-	-	●	-	●	●	●	●		●	●	-		
			-	-	○	○	○	○	○	○	○	○	-	○		
Elapidae	*Vermicella annulata*	O	-	□	-	-	-	-	-	-	-	-	-	-	None Reported	Shine 1980d
			-	●	●	-	-	●	-	-	-	●	●	●		
			○	○	○	○	○	○	○	○	○	○	○	○		
Homalopsidae	*Enhydris enhydris*	V	□	□	-	-	-	-	□	□	□	□	□	□	None Reported	Brooks *et al.* 2009
			●	●	●	-	-	●	●	●	●	●	●	●		
			○	○	○	-	-	○	○	○	○	○	○	○		
Homalopsidae	*Enhydris longicauda*	V	□	-	-	-	-	-	□	□	□	-	-	-	None Reported	Brooks *et al.* 2009
			●	●	●	-	●	●	●	●	●	-	-	-		
			○	○	○	-	○	○	○	○	○	-	-	-		
Hydrophidae	*Laticauda colubrina*	O	-	-	-	-	-	-	-	-	-	-	-	-	None Reported	Gorman *et al.* 1981
			-	-	●	●	●	●	-	●	●	●	●	-		
			○	-	○	○	○	○	○	○	○	○	○	-		
Viperidae	*Causus maculatus*	O	-	-	-	-	-	-	-	-	-	-	-	-	Yes	Ineich *et al.* 2006
			●	●	●	-	●	●	●	●	●	●	●	●		
			-	-	-	-	-	-	-	-	-	-	-	-		

example, despite large numbers of vitellogenic brown treesnakes (*Boiga irregularis*) collected by Savidge *et al.* (2007), collecting efforts overall resulted in only one ovigerous female.

A major shortcoming in the studies presented in Table 12.2 is that no study investigated, or was able to investigate, whether mean vitellogenic follicle size, the proportion of females vitellogenic, or the proportion of females ovigerous varied significantly among months. This is more an observation than criticism; some studies were conducted prior to routine employment of statistical analysis whereas some of the later instances were due to limited numbers of specimens available in museum collections.

The survey of the literature presented here yielded only 26 species where data would suggest aseasonal reproduction by females. This is slightly less than findings for males (31 species). Given that far fewer studies investigate the male than female, this suggests that aseasonality is less common in females than males, as would be predicted given the rationale presented near the beginning of this section. Further, although males of many species store sperm in the vas deferens for long periods (Shine 1977a), gonads of males might be expected to remain active over longer periods than those of females so as to increase breeding success. Finally, compared to the female, reproduction in the male is thought to require less energy (but see Bonnet and Naulleau 1996).

Shine (1991) and colleagues (Brown and Shine 2002, 2006) in their detailed studies on a tropical Australian colubrid, the Keelback (*Tropidonophis mairii*) aptly (though unintentionally) illustrated the potential problem of inferring aseasonality based on the monthly size distributions of vitellogenic follicles. They also convincingly showed that even the presence of ovigerous females in each month of the year can lead to a false determination of aseasonality, as defined herein. Shine (1991) presented data showing that distributions of sizes of vitellogenic follicles were similar in each month (Fig. 12.3a). It was concluded, and not incorrectly, that "...*Tropidonophis* reproduced virtually year-round." Follow-up studies of Brown and Shine (2002, 2006), however, showed that the percentage of ovigerous females varied over the year, increasing to maximums (ca 85%) in June or July and then decreasing to essentially zero from December to March (Fig. 12.3b). Female *T. mairii* containing enlarged vitellogenic follicles in November-December thus apparently 'delay' ovulation until at least March-April, apparently through reduction in the rate of vitellogenesis. The results of these studies are instructive and illustrate: 1) the inherent problems in inferring extended or aseasonal reproduction based on presence and size of vitellogenic follicles, and 2) the importance of obtaining sample sizes of ovigerous females large enough to detect of monthly differences in the fraction of the population that is reproducing.

The possibility that individual females produce multiple consecutive clutches in a single season further calls into question how many species have a truly aseasonal reproductive cycle. Until recently there was little evidence for production of more than one clutch in a single season for

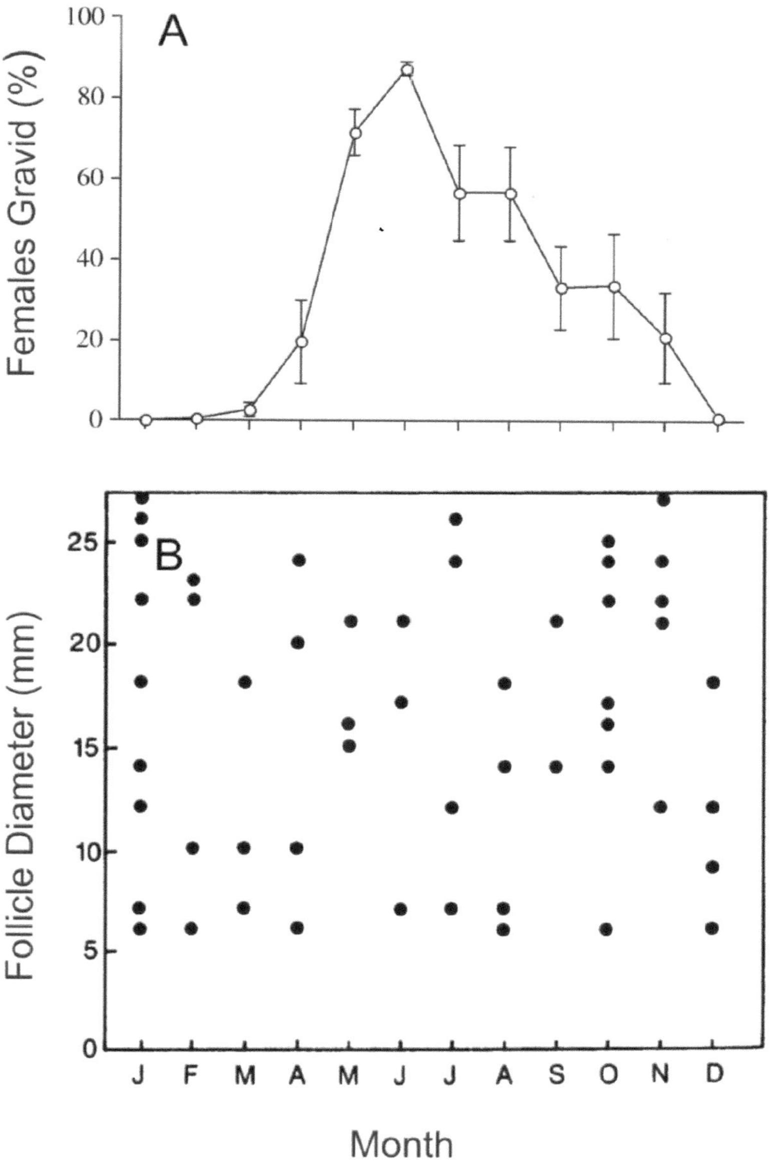

Fig. 12.3 *Tropidonophis mairii* (Colubridae). **A.** Seasonal distribution of mean percentage of females ovigerous (total N = 809 females). Note that reproduction is seasonal, with most females ovigerous in June. From Brown, G. P. and Shine, R. 2006, Fig 1a. **B.** Seasonal distributions of maximum sizes of vitellogenic follicles. Non-vitellogenic follicles (< 5 mm diameter) omitted because of large number of observations in all months. Note that similar size distributions of follicles in all months of the year suggest (falsely) reproduction is aseasonal. Rates of vitellogenesis presumably vary among individuals; such variation would account for observed variation in percent ovigerous shown in upper panel. From Shine, R. 1991, Copeia 1991: 120-131, Fig 2b.

snakes but, in tropical colubrids at least, this phenomenon has become well documented (see Table 12.2). At the time Fitch (1970, 1982), Licht (1984) and Siegel and Ford (1987) surveyed the literature, the only evidence for multiple clutches in snakes came from wild-caught females that had been brought into captivity. Although many such cases were known, and indeed some quite remarkable (e.g., Common Night Adder, *Causus rhombeatus*, seven clutches produced between April and October; Woodward 1933) these observations were necessarily discounted because of potential permissive effects of captive conditions (e.g., warm equable temperatures and steady food supply: see Siegel and Ford 1987). But production of at least two consecutive clutches in the field, as evidenced by oviductal eggs together with enlarged vitellogenic follicles within a single female, is now well-documented for a number of species of tropical snakes (Bacold 1983; Vitt 1983; Marques 1996a; Stafford 2003; Pinto and Fernandes 2004; Balestrin and Di-Berardo 2005; Schmidt de Aguir and Di-Bernardo 2005; Goldberg 2006b; Ineich *et al.*, 2006; Marques and Muriel 2007).

Because production of multiple clutches was unknown at the time of the reviews above (e.g., Saint Girons and Pfeffer 1971), observations of vitellogenic or ovigerous females over many months of the year (i.e., more than typically observed for highly synchronous temperate zone species) were often considered suggestive of aseasonality. Otherwise it is reasonable to expect that individuals producing one clutch a season would reproduce in synchrony, timing reproduction to biotic or abiotic factors maximizing reproductive success, much in the same way as temperate zone snakes. However, if we accept that many tropical species of snakes typically produce multiple clutches a season, then it is inherently impossible for an individual to exactly time each clutch to the time of year (regardless of whether or not annually variable) when conditions best enhance reproductive success. Thus, it might be predicted that each individual would center its 'cohort' of clutches on the time of year optimizing overall reproductive success. If there is only one such time each year, then the timing of bouts of multiple clutching would vary somewhat among individuals, but the highest frequency of reproductive activity would be centered on the optimal time. A caveat to this scenario, however, is that in geographic areas where there are two such favorable periods per year, individuals of species producing two clutches might be expected to reproduce in synchrony with reproductive activity being bimodal (e.g., Rainbow Watersnake, *Enhydris enhydris*: Saint Girons and Pfeffer 1971). Evidence for this scenario and factors known to affect reproductive success in tropical snakes with seasonal reproduction are discussed in the next section. If such a scenario proves more common than thought, then researchers must first rule out this pattern of 'extended' reproduction prior to consideration of aseasonality.

The above-mentioned studies of Shine (1991) and Brown and Shine (2002) on *Tropidonophis mairii* collectively suggest that any snake species with production of two clutches, even those in fairly close succession (mean clutch interval was 69 days) along with the two attendant periods of vitellogenesis, can produce the type of pattern revealed by traditional monthly sampling that might seem to indicate aseasonal reproduction. Ford and Karges (1987) surmised that as far north as northeastern Mexico and southern Texas, female Checkered Garter Snakes (*Thamnophis marcianus*) often produce two clutches a season. However, had they not examined their data in terms of percent females vitellogenic, the span of months wherein females were vitellogenic (10 months) might have suggested aseasonality. Reproductive cycles like those documented for *Tropidonophis mairii* and *Thamnophis marcianus* thus need to be considered and the data given critical evaluation before assuming aseasonality.

As in the male, female cyclicity type may differ between populations of the same species. In Australia, vitellogenesis in female *Boiga irregularis* is seasonal and highly synchronous (Shine 1991; Whittier and Limpus 1996), whereas in the Guam population vitellogenic follicles of all sizes were well represented in all months of the year, strongly implicating aseasonality (Savidge *et al.* 2007). Similarly, reproduction by female Banded Cat-eyed Snakes (*Leptodeira annulata*) in Brazil is seasonal based on the distribution of vitellogenic and ovigerous females (Pizzatto *et al.* 2008a) whereas in Amazonian Peru, Fitch (1970) recorded ovigerous females in all but three months of the year, suggesting that reproduction may be aseasonal in that region. In subtropical eastern Brazil, reproduction by female Black-headed Snakes (*Tantilla melanocephala*) was seasonal (Marques and Puorto 1998) whereas in Amazonian Brazil it appeared to be aseasonal (Santos-Costa *et al.* 2006). With the exception of *Boiga*, it remains to be determined whether these variations in cyclicity type (aseasonal *vs.* seasonal) within species are more apparent than real.

There is no doubt, however, that the timing (synchrony) and rates of reproductive cycles of individuals of some tropical snake species can vary considerably among populations, even over limited geographical areas. Female Military Ground Snakes (*Liophis miliaris*) in a coastal population (Mata Atlântica do sul da Bahia) located between 13° S and 18° S in Brazil where mean temperature and rainfall were relatively equable among months exhibited size distributions of vitellogenic follicles year-round, but ovulation did not occur until October (Pizzatto 2003). In two of the more southerly populations studied where seasonal fluctuations in temperature and rainfall were more pronounced, progression of vitellogenesis appeared to be more synchronized among individuals and ovulation also commenced in October. In the marine snakes, Shaw's Sea Snake (*Lapemis curtis*, sensu Gritis and Voris 1990) and the Elegant Sea Snake (*Hydrophis elegans*) studied at the same locality on the northern Australian continental shelf (with presumably equable water temperatures), the former exhibits an apparently much more rapid rate of vitellogenesis than the latter (Ward 2001).

'Extended' Reproduction. In this chapter 'extended' reproduction is defined simply as a lengthy period of time over which females are ovigerous or exhibit similar size distributions of vitellogenic follicles compared to temperate zone colubrids. The term could also be applied to male cycles. This term has been used previously (cf., Seigel and Ford) but a definition has not been adequately circumscribed. Adoption of this term for cases where reproduction is neither seasonally synchronous nor definitively aseasonal is proposed here. Thus, it is somewhat artificial in that the designation may simply be a neutral category for some species until aseasonal reproduction is adequately shown. Recognize, however, that for species that truly do exhibit extended reproduction, the basic pattern is still seasonal. The term is not intended for cases where the rate of vitellogenesis is slow (i.e., vitellogenic females in most months), but where ovulation is fairly synchronous and annual (e.g., *Bothrops neuwiedi pubescens*: Hartmann *et al.* 2004). Nor is it intended for those viviparous species where ovulation is synchronous and annual but females are pregnant during most other times of the year (e.g., False Coral Snake, *Anilius scytale*; Maschio *et al.* 2007). The term 'continuous' reproduction should no longer be used for either sex as it implies rates of all reproductive processes in an individual remain virtually constant throughout its adult lifetime (sensu Licht 1984), which is likely untenable in any vertebrate. Likewise for populations, it is unlikely that the rate of reproduction in any population is truly continuous throughout the year.

Extended reproduction, as defined herein, is now documented in many species. Just a few examples follow. For these species, periods over which vitellogenic females were acquired ranged from 6 to 10 months and periods over which ovigerous females were acquired ranged from 4 to 7 months. All inhabit tropical zones and are oviparous colubrids: Aesculapian False Coral Snake (*Erythrolamprus aesculapii*, Marques 1996a); Colombian Earth Snake (*Geophis brachycephalus*, Sasa 1993); Hoffmann's Earth Snake (*Geophis hoffmanni*, Goldberg 2006b); Jaeger's Ground Snake (*Liophis jaegeri*, Frota 2005); Lined Ground Snake (*Liophis lineatus)* and Goldbauch-Buntnatter (*Liophis poecilogyrus*, Vitt 1983); Military Ground Snake (*Liophis semiaureus*, Bonfiglio 2007); Crown Ground snake (*Liophis viridis*, Vitt 1983); Tanganyika Water Snake (*Lycodonomorphus bicolor*, Madsen and Osterkamp 1982); Rio Tropical Racer (*Mastigodryas bifossatus*, Marques and Muriel 2007); Diamond-backed Watersnake (*Nerodia rhombifer*, Aldridge *et al.* 1995); Green Vine Snake (*Oxybelis fulgidus*, Scartozzoni *et al.* 2009); Guibe's Flame Snake (*Oxyrhopus guibei*, Pizzatto and Marques 2002); Brazilian Green Racer (*Philodryas aestivus*), Paraguay Racer (*P. nattereri*), and Lichtenstein's Green racer (*P. olfersii*, Fowler *et al.* 1998); Sao Paulo False Coral Snake (*Simophis rhinostoma*, Jordão and Bizerra 1995).

12.3 BASES FOR VARIATIONS IN CYCLICITY

Seasonal reproduction in the seasonally wet–dry tropics. A longstanding generalization for seasonality of reproduction by snakes inhabiting regions of the tropics where there is a distinct 'wet' and 'dry' season is that reproduction is concentrated in the wet season (Kopstein 1938; Duellman 1958; Fitch 1970, 1982; Saint Girons and Pfeffer 1971; Angelini and Picariello

1975). The underlying bases for this view are that most organisms display recognizable seasonal peaks of reproductive activity and a wet-dry season is usually the most conspicuous variation in climate at low latitudes. Seigel and Ford (1987) in their review of the available literature on seasonality of reproduction in tropical and sub-tropical snakes summarized findings for females of 19 species (their Table 8-1), concluding that the reproductive period for all 19 species was 'associated with a wet season'. Since that time, reproductive data for many other species has accumulated that also point to this relationship. The survey presented here of these more recent studies was not all-encompassing and yielded an unintentional bias toward Neotropical species in general, and Neotropical vipers in particular (Table 12.3). General trends, however, are apparent: vitellogenesis occurs during the drier months, oviposition occurs in either the dry or wet seasons, and hatching or parturition occurs primarily in the wet season. The greatest variation among species is seen in the seasonality of oviposition, which is discussed later in this section. All species of Neotropical vipers studied to date uniformly give birth during the wet season. Exceptions to this pattern are seen in viviparous species that are highly aquatic, e.g., the Green Anaconda (*Eunectes murinus*) (Rivas 1999) where young might be expected to fare better when water levels in their habitat are reduced and resources are more concentrated. Females of another highly aquatic snake, the Arafura Filesnake (*Acrochordus arafurae*), similarly give birth at the end of the wet season (Shine 1986).

Although the majority studies on reproduction in tropical snakes have attempted to correlate the period of reproduction with seasonality in rainfall, the number of species investigated to date are probably too few not to expect greater diversity in the seasonal timing of reproductive events among species than that given in Table 12.3. In contrast to the situation for snakes, there is considerable data for associations between reproduction and wet–dry seasons for tropical lizards for which a wide diversity in the timing of reproductive events has been documented (Fitch 1982; Licht 1984; James and Craig 1985) including oviposition in the dry season (James and Craig 1985).

Aseasonal reproduction in the aseasonal tropics. Other studies conducted on tropical species of squamates during the 1960s and 1970s where reproduction was judged aseasonal led to the prevailing view that aseasonal reproductive cycles were associated with regions where monthly variations in temperatures and rainfall are equable (q.v., James and Shine 1985, and earlier references for studies on lizards therein). This generalization had been applied to tropical snakes with presumed aseasonal reproduction (Duellman 1958, 1978). However, quantitative evidence supporting this view has been meager. Information on seasonality of rainfall provided in some the studies listed in Table 12.2 now make a limited assessment of this generalization possible. Do the species of snakes with aseasonal reproduction in Table 12.2 occur in areas where rainfall is similar across months, or are they found in regions with distinct wet–dry seasons? Information on seasonality of rainfall for the areas where studies

Table 12.3 Summary of the timing of vitellogenesis, oviposition, and hatching or parturition, with season (wet or dry) in 44 species of tropical snakes.

			Season			
Family	*Species*	*Parity mode*	*Vitellogenesis*	*Oviposition*	*Hatching or parturition*	*References*
Aniliidae	*Anilius scytale*	V	Dry-Wet		Wet	Maschio *et al.* 2007
Boidae	*Boa constrictor occidentalis*	V	Dry		Wet	Bertona and Chiaraviglio 2003
Boidae	*Eunectes murinus*	V	Wet[1]		Dry	Rivas 1999
Acrochordidae	*Acrochordus arafurae*	V	Dry		Wet	Shine 1986
Acrochordidae	*Acrochordus granulatus*	V	Dry		Wet	Wangkulangkul *et al.* 2005
Homalopsidae	*Enhydris longicauda*	V	Dry		Dry-Wet	Brooks *et al.* 2009
Homalopsidae	*Homalophis buccatta*	V	Dry		Dry-Wet	Brooks *et al.* 2009
Homalopsidae	*Enhydris bocourti*	V	Dry		Dry-Wet	Brooks *et al.* 2009
Homalopsidae	*Erpeton tentaculatus*	V	Dry		Dry-Wet	Brooks *et al.* 2009
Colubridae	*Chironius bicarinatus*	O	Dry	Wet	Not Reported	Marques *et al.* 2009
Colubridae	*Dipsas albifrons*	O	Wet	Wet	Wet-Dry	Hartmann *et al.* 2002
Colubridae	*Duberria lutrix*	V	Wet-Dry		Wet	Kofron 1990
Colubridae	*Erythrolamprus aesculapii*	O	Both seasons	Wet	Dry	Marques 1996a
Colubridae	*Helicops leopardinus*	V	Not Reported		Wet	Ávila *et al.* 2006
Colubridae	*Liophis lineatus*	O	Dry-Wet	Dry-Wet	Wet	Vitt 1983
Colubridae	*Liophis poecilogyrus*	O	Dry-Wet	Dry	Wet	Vitt 1983
Colubridae	*Liophis miliaris*	O	Wet	Wet	Wet-Dry	Pizzatto 2003[2]
Colubridae	*Liophis viridis*	O	Dry-Wet	Dry	Wet	Vitt 1983
Colubridae	*Mastigodryas bifossatus*	O	Dry	Wet	Wet	Marques and Muriel 2007
Colubridae	*Oxybelis fulgidus*	O	Dry	Wet	Wet	Scartozzoni *et al.* 2009
Colubridae	*Oxyrhophus guibei*	O	Dry-Wet	Wet; some in Dry	Wet-Dry	Pizzatto and Marques 2002
Colubridae	*Philodryas aestivus*	O	Dry-Wet	Wet	Not Reported	Fowler *et al.* 1998
Colubridae	*Philodryas nattereri*	O	Dry-Wet	Wet	Not Reported	Fowler *et al.* 1998
Colubridae	*Philodryas olfersii*	O	Dry-Wet	Wet	Not Reported	Fowler *et al.* 1998

Colubridae	*Philodryas patagoniensis*	O	Wet	Wet	Not Reported	Fowler *et al.* 1998
Colubridae	*Psammophis phillipsi*	O	Wet-Dry	Dry	Wet	Butler 1993
Colubridae	*Psammophis phillipsi*	O	Not Reported	Dry	Wet	Akani *et al.* 2002
Colubridae	*Pseudablabes agassizii*	O	Dry-Wet	Wet?	Not Reported	Marques *et al.* 2006
Colubridae	*Sibon sanniola*	O	Dry-Wet	Wet	Wet	Kofron 1983
Colubridae	*Symphimus mayae*	O	Dry	Wet	Wet	Stafford 2005
Colubridae	*Tomodon dorsatus*	V	Wet		Dry-Wet	Bizerra *et al.* 2005
Colubridae	*Tropidonophis mairii*	O	Both seasons	Both but mainly Dry	Wet	Shine 1991, Brown and Shine 2006
Colubridae	*Waglerophis merremii*	O	Dry-Wet	Insufficient data	Wet	Vitt 1983
Elapidae	*Demansia vestigiata*	O	Dry	Late Dry	Wet	Fearn and Trembath 2009
Elapidae	*Micrurus corallinus*	O	Early Wet	Wet	Wet-Dry	Marques 1996b
Lamprophiidae	*Liopholidophis sexlineatus*	V	Not Reported		Wet	Cadle, 2009
Viperidae	*Bothrops asper. Atlantic versant*	V	Dry		Wet	Solórzano and Cedras 1989
Viperidae	*Bothrops asper. Pacific versant*	V	Wet-Dry		Wet	Solórzano and Cedras 1989
Viperidae	*Bothrops mattogrossensis*	V	Dry		Wet	Vitt 1983; Monteiro *et al.* 2006
Viperidae	*Bothrops moojeni*	V	Dry		Wet	Nogueira *et al.* 2003
Viperidae	*Bothrops neuweiedi pauloensis*	V	Dry		Wet	Valdujo *et al.* 2002
Viperidae	*Bothrops neuwiedi pubescens*	V	Wet [3]		Wet [3] (warmest months)	Hartmann *et al.* 2004
Viperidae	*Porthidium picadoi*	V	Not Reported		Wet	Solórzano 1990
Viperidae	*Porthidium yucatanicum*	V	Wet-Dry		Wet	McCoy and Censky 1992
Viperidae	*Trimeresurus stejnegeri stejnegeri*	V	Dry		Wet	Tsai and Tu 2001

Dry-Wet = feature initiated in dry season, continues into wet season.
Wet-Dry = feature initiated in wet season, continues into dry season.
[1] inferred from data presented.
[2] three populations studied in three regions between 20° S and 30° S, Brazil.
[3] no dry season.

were conducted was available for 17 species of snakes given in Table 12.2, and all species inhabited regions with a marked wet and dry season. Thus there is as yet no evidence to support the generalization that aseasonal reproduction occurs mainly in areas where rainfall is equable throughout the year.

Proximate and ultimate bases for variations in reproductive cyclicity. Despite the extensive number of studies conducted to date on reproductive cycles of tropical snakes, virtually none have attempted to quantitatively demonstrate physiological bases mediating synchronization of cycles with proximate (exogenous) cues. This difficulty is due largely to the significant problems in discriminating among the effects of climatic variables, their interactions, and their governing effects on prey abundance. An alternative approach has been to use focal species or assemblages to test assumptions of hypotheses for the proximate and ultimate factors mediating seasonality of reproduction (Brown *et al.* 2002; Brown and Shine 2006). Brown and Shine (2006) examined hypotheses for the evolutionary determinants of reproductive seasonality in *Tropidonophis mairii*, a common species on one of their study areas in northern topical Australia. Findings of this seminal work did not support hypotheses invoking biotic factors for timing of oviposition, which in this species, coincided with the wet season; timing did not minimize egg predation or maximize food availability or survival for hatchings. Instead timing coincided with cessation of rains when soil moisture content (abiotic) apparently favored embryogenesis, as judged by enhanced hatch rates and larger hatchlings, while avoiding earlier wetter conditions when eggs may become waterlogged and die. Potential linkages between seasonal rainfall and the fractions of females reproducing in a given year have been intensively investigated in two other snakes in this area, the Water Python (*Liasis fuscus*) and the Arafura File Snake (*Acrochordus arafurae*). Despite their phylogenetic and ecological dissimilarities, it is the duration of rainfall occurring in previous seasons that determined prey abundance, feeding rates, and hence female fat stores in both species. And because both species are capital breeders, only the fraction of females in the population having sufficient fat reserves to produce a clutch or litter initiated vitellogenesis in a given year (Madsen and Shine 2000; Madsen *et al.* 2006; Shine and Brown 2008).

The comparative approach has also provided insights into the bases of reproductive cycles. Seigel and Ford (1987) in their review of studies on assemblages of tropical snakes (Vitt and Vangilder 1983; Vitt 1983) and lizards (James and Shine 1985) at single localities concurred with those authors that the diversity in reproductive cycles observed at single locations supported the idea that phylogeny and biogeographical histories contributed appreciably to the observed variation in cyclicity among species (see also Censky and McCoy 1988; Vitt 1992). This view is quite different from an earlier view that reproductive patterns of tropical snakes are 'adapted' to local climatic and biotic regimes, constrained only by species-specific traits (e.g., Zug *et al.* 1979). The idea that phylogeny supersedes adaptation at the local level (Vitt 1992) is intriguing and is in need of further investigation, but the potentially confounding effects of plasticity (facultative) of reproductive cycles at the species level (Shine 2002) also require consideration.

Considering the various methodological approaches available and the substantial number of studies conducted on reproductive cycles of tropical snakes to date, it is remarkable that only one study (Brown and Shine 2006) has explicitly examined hypotheses for the evolutionary determinants of the timing between reproductive events and seasonality of rainfall. Findings of the various studies given in Table 12.3 are mixed with respect to the timing of oviposition. Seven of the studies, because oviposition was registered in the dry season, are in line with the findings of Brown and Shine (2006) that abiotic factors (i.e., levels of nest moisture promoting hatching success and hatchling quality) more directly influence reproductive success and hence timing of reproduction than biotic factors (e.g., foraging ecology and seasonal variation in resource availability, sensu Vitt 1987). Conversely, the fourteen cases in Table 12.3 where oviposition was registered in the wet season implicate biotic factors as being most important. However, extending inferences from such coarse characterizations of rainfall as 'wet' or 'dry' to the hydric conditions eggs actually experience in a nest is of limited value. Hatching success and hatchling 'quality' in squamates can depend on a complex interplay among several environmental factors experienced during incubation (e.g., Andrews *et al.* 2000; Warner and Andrews 2002; Brown and Shine, 2004, 2005; Shine and Brown 2008). Elucidation of potential abiotic determinants for timing of oviposition in tropical snakes has received little attention and will require detailed measurements of soil water potential, relative humidity within nest cavity, and diel soil and air temperatures in the field and under controlled conditions in the laboratory.

Selective forces operating on the seasonal timing for production of young may act differentially on oviparous species than on viviparous species. Of the 21 viviparous species in Table 12.3, 20 species give birth in the wet season or the transition thereto. Because embryos of placental species do not normally experience physiologically 'wet' or 'dry' conditions, at least not in the same sense as do those in a nest, increased resources for neonates during the wet season (Janzen and Schoener 1968) or other factors affecting their survival would likely be the primary drivers behind timing of parturition. Indeed, viviparous species may offer a 'simpler' model than oviparous species to test putative environmental determinants for the timing of release of offspring into the environment as abiotic factors affecting eggs in the nest are absent.

12.4 CONCLUSIONS

"How difficult is the study of this kind of work is shown by this publication. In spite of observations on some thousands of snakes over a period of more than three years, some of the problems are still quite or partly unsolved." This quote is the concluding remark of Kopstein (1938), on his studies on the reproductive biology snakes in Malaysia based on substantial sample sizes.

In temperate zone species where there is tight synchrony in the cycles of individuals, examinations of monthly samples will reveal reproductive cycles at both the population and individual levels. There has therefore never been a particularly great need to sample individual animals over time. This method is obviously equally well-suited for characterizing cycles of sub-tropical and tropical species where reproduction is synchronous among individuals. In contrast, and as lamented by Kopstein above, in those tropical species in which the cycles of individuals are extended or asynchronous, monthly sampling may yield data pointing to a certain type of individual cycle but the findings are likely to remain equivocal. Because the underlying cycles of individuals are unknown, causal bases of the overall pattern apparent at the population level are often unclear. Characterization of cycles of individuals of species with extended or aseasonal reproduction will require repeated sampling of individuals. Testicular cycles of males in populations where reproduction appears aseasonal continue to remain virtually unknown, a condition due almost entirely to failure to employ histological methods (but see body of work by Goldberg). Similarly, almost nothing is known about the steroid cycles in either sex. Species have now been identified that would serve as excellent models for investigating endocrine bases of acyclic spermatogenesis in males and multiple clutching in females. Examples for males of species that are common include *Liophis milaris* (Pizzatto 2003) and *Causus maculatus* (Ineich et al. 2006). *Oxyrhophus guibei*, a common Brazilian species, is interesting in that males exhibit only a small decrease in testis volume from February to April and the diameter of the vas deferens is invariant throughout the year (Pizzatto and Marques 2002). For females, *Tropidonophis mairii* (Brown and Shine 2006) would be the ideal candidate.

There seems to have always been confusion in evaluations of evidence for aseasonal reproduction, due in part to inadequate sample sizes and combining of specimens from geographically distant areas, but also to lack of a standardized terminology. For example, Seigel and Ford (1987) in their review of the literature judged that only one species could rightly be considered aseasonal; on the findings of Berry and Lim (1967) concerning *Homalopsis buccata* they stated: "...continuous breeding appears to be real in at least some instances (*H. bucatta*)." Berry and Lim (1967), however, clearly showed that, although there were reproductive individuals of both sexes throughout the year, reproduction in both sexes was unmistakably seasonal; the proportion of reproductive females (vitellogenic and pregnant females pooled) varied significantly among months. Moreover, for males, the presence of individuals in each sample with testis in full regression indicated the testicular cycle was discontinuous cyclic, not acyclic or continuous cyclic as documented for *Laticauda colubrina* or *Cerberus rhynchops*, respectively (Gorman *et al.* 1981). Standardization of the terminology for characterizing cycles on individual and population levels will facilitate advancements in understanding these patterns.

Aseasonal reproduction seems to be relatively uncommon, particularly in the female. Many of the 'aseasonal' species given in Table 12.2 will undoubtedly be proven to have extended periods of reproduction (i.e., seasonal) with peaks of reproductive activity when studies on those species are conducted at single locations over a limited time periods. In cases where aseasonality cannot be reasonably demonstrated the more conservative designation of 'extended' reproduction should be used. Limited asynchrony among individuals in bouts of multiple clutching may prove to be the more common underlying mechanism of acyclicity than complete asynchrony among individuals that reproduce only once in a season.

12.5 PROBLEMS, PRACTICALITIES, AND FUTURE DIRECTIONS

The arguments presented here against attempting to characterize the phenology of reproduction of species or populations with extended or aseasonal cycles using only the traditional sampling method should lead the way to approaches that are more directed. The traditional method where individuals are collected and terminally sampled at regular intervals throughout the year should more properly be viewed as a tool for assaying among species or populations for those exhibiting extended or aseasonal reproduction. Having then identified such cases, follow-on work applying other sampling techniques has great potential to provide new insights into the nature of cyclicities of individuals, the physiologies supporting these cycles, as well as underlying proximate and ultimate causes. The only way to determine with certainty the cyclicities of individuals and how their dynamics collectively manifest at the population level is to monitor the histories of individuals through time (Vitt and Seigel 1985). Mark-recapture procedures are labor intensive, but when applied to populations with extended or apparent aseasonal reproduction, they offer perhaps the only means of clarifying these issues, as well as resolving other issues such as multiple clutching, reproductive effort, age at sexual maturation, chronologies and rates of vitellogenesis, chronologies and rates of spermatogenesis, and periods of estrus. Although costs of construction and maintenance may be substantial, snake 'populations' can be enclosed using fencing, and all individuals therein marked, such as the 5 ha area of jungle on Guam used for intensive study of *Boiga irregularis* (Rodda *et al.* 2007). Recent efforts characterizing snake assemblages containing species with extended or aseasonal reproduction are helping to clarify practicalities and logistics for directed reproductive studies on these species (Sawaya 2003; Bernarde *et al.* 2006; Sawaya *et al.* 2008). A well-recognized challenge, however, is those species of interest that are rare or difficult to detect (e.g., genera *Cleia* and *Boiruna*; q.v., Pizzatto 2005). Rare species of snakes tend to occur in the tropics (Myers 2003, and references therein; Luiselli 2006) and may have limited geographic distributions (Dunn 1949) thus making their study particularly difficult.

Although monitoring individuals through time provides the opportunity to collect otherwise unobtainable data, non-invasive techniques for repeated sampling individuals are limited, particularly for the male. Presence and motility of sperm in the vas deferens can be monitored through collection of semen (Zacariotti *et al.* 2007). Bertona and Chiaraviglio (2003) used ultrasound scanning to measure follicle and testis diameters of field-active Argentine Boa Constrictors (*Boa constrictor occidentalis*), a technology that might also be effective at least in the female of smaller snake species. Females can be abdominally palped (Brown and Shine 2002) for vitellogenic follicles or oviductal eggs and both can oftentimes be accurately counted (sacrificing a series for validation of method). However, small vitellogenic follicles may not be detectable (Fearn and Trembath 2009). In catch and release studies of a tropical lizard, Ayala and Spain (1975) used small samples of blood and a vital stain technique to identify not only onset and duration of vitellogenesis, but also cases where vitellogenesis had apparently slowed or ceased. Such a technique, if transferable to snakes, might be used to monitor rates of vitellogenesis in individuals. Vitellogenin content of blood plasma can be assayed using laser densitometer (Cree *et al.* 1991) and rates of synthesis can be determined using an isotopic method (Craik 1978). Repeated blood sampling of individuals for plasma levels of sex steroids is routine and will reveal seasonality (but not necessarily aseasonality) of reproduction (e.g., female *Trimeresurus stejnegeri stejnegeri*, Tsai and Tu 2001; female *Crotalus durissus terrificus*, Almeida-Santos *et al.* 2004; *C. atrox*, Taylor *et al.* 2004). However, care should be taken in all the above procedures to minimize potential confounding effects of handling on the subjects under investigation; capture can increase plasma levels of stress hormones (e.g., corticosterone: *Boiga irregularis*, Mathies *et al.* 2001) which has been implicated in induction of testicular regression (*B. irregularis*; Aldridge and Arackal 2005; Siegel *et al.* 2009) which is normally acyclic in this population (Mathies *et al.* 2010).

As previously mentioned, more attention to the male reproductive cycle is needed. However, adoption of the practice of assessing whether males undergoing spermatogenesis based solely on whether the testis appear 'turgid' or efferent ducts appear 'opaque and thickened' (e.g., Shine *et al.* 1995; Shine *et al.* 1996; Keogh *et al.* 2000; Ineich *et al.* 2006; Cottone and Bauer 2009) should be considered carefully. Although this meristic for testicular activity may be justified where cycles of individual are discontinuous cyclic and synchronous within the population (i.e., changes are marked; Shine 1986; Slip and Shine 1988), application to populations with extended or aseasonal reproduction could yield potentially misleading or equivocal results, particularly when sample sizes are low in some months (e.g., Keogh *et al.* 2000). Small upward or downward 'excursions' in 'turgidity', resulting simply from sampling bias, could engender false invocation of 'seasonality'. Shine (1977a), in a study on eight species of Australian elapids, validated his methods for inferring spermatogenesis and reproductive maturity based on the outward appearances of the testis and vas deferens by examination macerated preparations of fresh material under light microscopy. Outward appearances of the testis and vas deferens are reasonable meristics for inferring sexual maturity in other species (as was presumably the only intention of Shine 1977a), but not necessarily active spermatogenesis. For *Boiga irregularis* in southeastern Queensland, Whittier and Limpus (1996) in their examination of histological preparations from spermatogenically active and inactive males noted that "...testicular volume is not a reliable indicator of the inseminating capacity of the male..." Nor is necessarily the exterior appearance of the epididymis and efferent ducts; although the epididymis and efferent ducts are not generally thought to function in the maturation or storage of sperm (Jones 1998; Sever *et al.* 2002), in the Australian scolecophidon, *Ramphotyphlops nigrescens*, the epididymis apparently serves as site for sperm storage well after testis became post-spermatogenic (Shea 2001). Long-term sperm storage in the epididymis has also been noted in the Chinese Watersnake (*Enhydris chinensis*, Meixi and Fuying 1989). If such meristics are employed for inferring spermatogenesis, they should be validated for the species under study by subjecting subsamples from the various sampling periods to histological examination (e.g., Almeida-Santos *et al.* 2006). Commercial histological laboratories or collaborations with colleagues with such capabilities are now quite accessible and can process materials at a cost well worth the wealth of definitive information provided.

Of overarching importance is a need to return to the practice of reporting the number of reproductive individuals together with number non-reproductive individuals each sampling period (q.v., De Haas 1941; Berry and Lim 1967). Inclusion of numbers of non-reproductive individuals observed is critical for reducing potential bias caused when sampling effort varies among sample times. Without such information, a robust assessment of whether the reproductive pattern of a population is actually seasonal (*vs.* giving the *appearance* of aseasonal) is not possible. For example, similar size distributions of vitellogenic follicles were documented in every month of the year in the Guam population of *Boiga irregularis* (Savidge *et al.* 2007), but because monthly proportions of vitellogenic to non-vitellogenic females were not presented, the most definitive measure for aseasonality could not be assessed. Similarly, had Shine (1991) reported numbers of female *Tropidonophis mairii* containing non-vitellogenic follicles (only vitellogenic graphically presented, see Fig. 12.3b) the proportions vitellogenic could have been subjected to statistical analysis for investigation of possible seasonality.

Over the last two decades investigations in tropical and subtropical regions have yielded a relative wealth of species or populations where reproduction is extended or potentially aseasonal. Regardless of their true nature, the opportunities to conduct longer-term and more detailed studies on these comparatively unstudied patterns of reproductive cyclicity hold much promise for filling what are still major gaps in our knowledge of the biology of tropical snakes. Nearly all the species with extended of aseasonal reproduction identified by workers in the last century are of Asian or Malaysian origin (Kopstein 1938; De Haas 1941; Saint Girons and Pfeffer 1971). Unfortunately, few subsequent reproductive studies on those taxa have being conducted. Workers in Brazil, on the other hand, have in recent years added immensely our knowledge of reproduction and ecology of Neotropical snakes and their active research program is to be commended and encouraged. Many new opportunities and avenues are available for understanding the reproductive biology of tropical snakes and are only awaiting more directed inquiry.

12.6 ACKNOWLEDGMENTS

I would like to thank Laurie Zuckerman for preparing the figures, Marilyn Howell for securing the more difficult to obtain literature, Melissa Jewth for assembling and formatting the Literature Cited section, and Janet Mathies for the judicious editing. Comments from two anonymous reviewers greatly improved the quality of this work.

12.7 LITERATURE CITED

Akani, G. C., Eniang, E. A., Ekpo, I. J., Angelici, F. M. and Luiselli, L. 2002. Thermal and reproductive ecology of the snake *Psammophis phillipsi* from the rainforest region of southern Nigeria. Herpetological Journal 12: 63-67.

Aldridge, R. D. 1979. Female reproductive cycles of the snakes *Arizona elegans* and *Crotalis viridis*. Herpetologica 35: 256-261.

Aldridge, R. D. and Arackal, A. A. 2005. Reproductive biology and stress of captivity in male brown treesnakes (*Boiga irregularis*) on Guam. Australian Journal of Zoology 53: 249-256.

Aldridge, R. D., Flanagan, W. P. and Swarthout, J. T. 1995. Biology of the water snake *Nerodia rhombifer* from Veracruz, Mexico. Herpetologica 51: 182-192.

Aldridge, R. D., Goldberg, S. R., Wisniewski, S. S., Bufalino, A. P. and Dillman, C. B. 2009. The reproductive cycle and estrus in the colubrid snakes of temperate North America. Contemporary Herpetology 2009: 1-31.

Almeida-Santos, S. M., Pizzatto, L. and Marques, O. A. V. 2006. Intra-sex synchrony and inter-sex coordination in the reproductive timing of the Atlantic coral snake *Micrurus corallinus* (Elapidae) in Brazil. The Herpetological Journal 16: 371-376.

Almeida-Santos, S. M., Abdalla, F. M. F., Silveira, P. F., Yamanouye, N., Breno, M. C. and Salomão, M. G. 2004. Reproductive cycle of the Neotropical *Crotalus durissus terrificus*: I. Seasonal levels and interplay between steroid hormones and vasotocinase. General and Comparative Endocrinology 139: 143-150.

Alves, F. Q., Argolo A. J. S. and Jim, J. 2005. Biologia reprodutiva de *Dipsas neivai* Amaral e *D. catesbyi* (Sentzen) (Serpentes, Colubridae) no sudeste da Bahia, Brasil. Revista Brasileira de Zoologia 22: 573-579.

Andrews, R. M., Mathies, T. and Warner, D. 2000. Effect of incubation temperature on morphology, growth, and survival of juvenile *Sceloporus undulatus*. Herpetological Monographs 14: 420-431.

Angelini, F. and Picariello, O. 1975. The course of spermatogenesis in reptilia. Accademia de Scienze Fisiche e Matematiche, series 3a, 9: 62-107.
Ávila, R. W., Ferreira, V. L. and Arruda, J. A. O. 2006. Natural history of the South American water snake *Helicops leopardinus* (Colubridae: Hydropsini) in the pantanal, Brazil. Journal of Herpetology 40: 274-279.
Ayala, S. C. and Spain, J. L. 1975. Annual oogenesis in the lizard *Anolis auratus* determined by blood smear technique. Copeia 1975: 138-141.
Bacold, P. T. 1983. Reproductive biology of two sea snakes of the genus *Laticauda* from central Philippines. Phillipene Scientist 20: 39-56.
Balestrin, R. L. and Di-Bernardo, M. 2005. Reproductive biology of *Atractus reticulatus* (Boulenger, 1885) (Serpentes, Colubridae) in southern Brazil. Herpetological Journal 15: 195-199.
Bernarde, P. S and Abe, A. S. 2006. A snake community at Espigão do Oeste, Rondônia, southwestern Amazon, Brazil. South American Journal of Herpetology 1: 102-113.
Bertona, M. and Chiaraviglio, M. 2003. Reproductive biology, mating aggregations, and sexual dimorphism of the Argentine boa constrictor (*Boa constrictor occidentalis*). Journal of Herpetology 37: 510-516.
Berry, P. Y. and Lim, G. S. 1967. The breeding pattern of the puff-faced water snake, *Homalopsis buccata* Boulenger. Copeia 1967: 307-313.
Bizerra, A., Marques, O. A. V. and Sazima, I. 2005. Reproduction and feeding in the colubrid snake *Tomadon dorsatus* from south-eastern Brazil. Amphibia-Reptilia 26: 33-38.
Bonfiglio, F. 2007. Biologia reprodutiva e dieta de *Liophis semiaureus* (Serpentes—Colubridae) no Rio Grande do Sul, Brazil. Dissertação de mestrado, Pontifícia Universidade Católica do Rio Grande do Sul, Rio Grande do Sul.
Bonnet, X. and Naulleau, G. 1996. Are body reserves important for reproduction in male dark green snakes (Colubridae: *Coluber viridiflavus*)? Herpetologica 52: 137-146.
Brooks, S. E., Allison, E. H., Gill, J. A. and Reynolds, J. D. 2009. Reproductive and trophic ecology of an assemblage of aquatic and semi-aquatic snakes in Tonle Sap, Cambodia. Copeia 2009: 7-20.
Brown, G. P. and Shine, R. 2002. Reproductive ecology of a tropical natricine snake, *Tropidonophis mairii* (Colubridae). Journal of Zoology 258: 63-72.
Brown, G. P. and Shine, R. 2004. Maternal nest-site choice and offspring fitness in a tropical snake (*Tropidonophis mairii*, Colubridae). Ecology 85: 1627-1634.
Brown, G. P. and Shine, R. 2005. Do changing moisture levels during incubation influence phenotypic traits of hatchling snakes (*Tropidonophis mairii*, Colubridae)? Physiological and Biochemical Zoology 78: 524-530.
Brown, G. P. and Shine, R. 2006. Why do most tropical animals reproduce seasonally? Testing hypotheses on an Australian snake. Ecology 87: 133-143.
Brown, G. P., Shine, R. and Madsen, T. 2002. Responses of three sympatric snake species to tropical seasonality in northern Australia. Journal of Tropical Ecology 18: 549-568.
Bull, K. H., Mason, R. T. and Whittier, J. 1997. Seasonal testicular development and sperm storage in tropical and subtropical populations of the brown tree snake (*Boiga irregularis*). Australian Journal of Zoology 45: 479-488.
Butler, J. A. 1993. Seasonal reproduction in the African olive grass snake, *Psammophis phillipsi* (Serpentes: Colubridae). Journal of Herpetology 27: 144-148.
Cadle, J. E. 2009. Sexual dimorphism and reproductive biology in the Malagasy snake genus *Liopholidophis* (Lamprophiidae: Pseudoxyrhophiinae). Proceedings of the California Academy of Sciences 60: 461-502.

Callard, I. P. and Kleis, S. M. 1987. Reproduction in reptiles. Pp. 187–205. In I. Chester-Jones, P. M. Ingleton and J. G. Phillips (eds), *Fundamentals of Comparative Vertebrate Endocrinology*. Plenum, New York.
Censky, E. J. and McCoy, C. J. 1988. Female reproductive cycles of five species of snakes (Reptilia: Colubridae) from the Yucatan peninsula, Mexico. Biotropica 20: 326-333.
Chan, L. M. 2003. Seasonality, microhabitat and cryptic variation in tropical salamander reproductive cycles. Biological Journal of the Linnean Society 78: 489-496.
Craik, J. C. 1978. Kinetic studies of vitellogenin metabolism in the elasmobranch *Scyliorhinus canicula* L. Comparative Biochemistry and Physiology A 61: 355-361.
Cree, A., Guillette, L. J. Jr., Brown, M. A., Chambers, G. K., Cockrem, J. F. and Newton, J. D. 1991. Slow estradiol-induced vitellogenesis in the tuatara, *Spenodon punctatus*. Physiological Zoology 64: 1234-1251.
Cottone, A. M. and Bauer, A. M. 2009. Sexual size dimorphism, diet, and reproductive biology of the Afro-Asian sand snake, *Psammophis schokari* (Psammophiidae). Amphibia-Reptilia 30: 331-340.
DeHass, C. P. J. 1941. Some notes on the biology of snakes and their distribution in two districts of west Java. Treubia 18: 327-375.
Drent, R. H. and Daan, S. 1980. The prudent parent: Energetic adjustments in avian breeding. Pp. 225-252. In H. Klomp and J. W. Woldendorp (eds), *The Integrated Study of Bird Populations*. North-Holland, Amsterdam, The Netherlands.
Duellman, W. E. 1958. A monographic study of the colubrid snake genus *Leptodeira*. Bulletin of the American Museum of Natural History 114: 1-152.
Duellman, W. E. 1978. *The Biology of an Equatorial Herpetofauna*. University of Kansas Museum of Natural History Miscellaneous Publications No. 65, Pp. 252.
Dufaure, J. P. and Hubert, J. 1961. Table de developpement du lezard vivipare: *Lacerta* (Zootoca) *vivipara* Jaquin. Archives d'Anatomie Microscopique et de Morphologie Experimentale 50: 309-328.
Dunn, E. R. 1949. Abundance of some Panamanian snakes. Ecology 30: 39-57.
Duvall, D., Guillette, L. J., Jr. and Jones, R. E. 1982. Environmental control of reptilian reproductive cycles. Pp. 201-231. In C. Gans and F. H. Pough (eds), *Biology of the Reptilia*, vol. 13. Academic Press, London, U.K.
Fearn, S. and Trembath, D. F. 2009. Body size, reproduction and growth in a population of black whip snakes (*Demansia vestigiata*) (Serpentes: Elapidae) in tropical Australia. Australian Journal of Zoology 57: 49-54.
Fitch, H. S. 1970. Reproductive cycles of lizards and snakes. University of Kansas, Museum of Natural History Miscellaneous Publications. 52: 1-247.
Fitch, H. S. 1982. Reproductive Cycles in Tropical Reptiles. Occasional Papers of the Museum of Natural History University of Kansas 96: 1-53.
Fitzgerald, M., Shine, R. and Lemckert, F. 2004. Life history attributes of the threatened Australian snake (Stephen's banded snake *Hoplocephalus stephensii*, Elapidae). Biological Conservation 119: 121-128.
Ford, N. B. and Karges, J. P. 1987. Reproduction in the checkered garter snake, *Thamnophis marcianus*, from southern Texas and northeastern Mexico: seasonality and evidence for multiple clutches. The Southwestern Naturalist 32: 93-101.
Fox, H. 1977. The urogenital system in reptiles. Pp. 1-157. In C. Gans (ed.), *Biology of the Reptilia*. vol. 6. Academic Press, New York.
Fowler, I. R., Salomão, M. G. and Jordão, R. S. 1998. A description of the female reproductive cycle in four species from the Neotropical colubrid snake *Philodryas* (Colubriae, Xenodontinae). The Snake 28: 71-78.

Frota, J. G. 2005. Biologia reprodutiva e dieta de *Liophis jaegeri jaegeri* (Günther, 1858) (Serpentes, Colubridae, Xenodontinae). Dissertação de mestrado, Pontifícia Universidade Católica do Rio Grande do Sul, Rio Grande do Sul.

Goldberg, S. R. 2003. Reproduction in four species of *Dendrophidion* from Costa Rica (Sepentes: Colubridae). Transactions of the Illinois State Academy of Science 96: 295-300.

Goldberg, S. R. 2004a. Notes on reproduction in the false coral snakes, *Erythrolamprus bizona*, and *Erythrolamprus minus*. The Texas Journal of Science 56: 171-174.

Goldberg, S. R. 2004b. Reproduction in the coffee snake, *Ninia maculata* (Serpentes: Colubridae), from Costa Rica. The Texas Journal of Science 56: 81-84.

Goldberg, S. R. 2004c. Notes on reproduction in the Central American coral snake, *Micrurus nigrocinctus* (Serpentes: Elapidae) from Costa Rica. Caribbean Journal of Science 40: 420-422.

Goldberg, S. R. 2006a. Reproductive cycle of the salmon-bellied racer, *Mastigodryas melanolomus* (Serpentes, Colubridae), from Costa Rica. Phyllomedusa 5: 145-148.

Goldberg, S. R. 2006b. *Geophis hoffmanni* (Hoffmann's earth snake). Reproduction. Herpetological Review 37: 351.

Goldberg, S. R. 2007a. Note on the testicular cycle of Godman's earth snake, *Geophis godmani* (Serpentes: Colubridae). Bulletin of the Chicago Herpetological Society 42: 7-8.

Goldberg, S. R. 2007b. Notes on reproduction of the adorned graceful brown snake, *Rhadinea decorata* (Serpentes, Colubridae), from Costa Rica. Phyllomedusa 6: 151-153.

Gorman, G. C., Licht, P. and McCollum, F. 1981. Annual reproductive patterns in three species of marine snakes from the central Philippines. Journal of Herpetology 15: 335-354.

Gribbins, K. M., Rheubert, J. L., Poldemann, E. H., Collier, M. H., Wilson, B. and Wolf, K. 2009. Continuous spermatogenesis and the germ cell development strategy within the testis of the Jamaican gray anole, *Anolis lineatopus*. Theriogenology 72: 484-492.

Gritis, P. and Voris, H. K. 1990. Variability and significance of parietal and ventral scales in the marine snakes of the genus *Lapemis* (Serpentes: Hydrophiidae), with comments on the occurrence of spiny scales in the genus. Fieldiana Zoology 56: 1-13.

Hartmann, M. T., Marques, O. A. V. and Almeida-Santos, S. M. 2004. Reproductive biology of the southern Brazilian pitviper *Bothrops neuwiedi pubescens* (Serpentes, Viperidae). Amphibia-Reptilia 25: 77-85.

Hartmann, M. T., Grande, M. L., Costa Gondim, M. J., Mendes, M. C. and Marques, O. A. V. 2002. Reproduction and activity of the snail-eating snake, *Dipsas albifrons* (Colubridae), in the southern Atlantic forest in Brazil. Studies on Neotropical Fauna and Environment 37: 111-114.

Heise, P. J., Maxson, L. R., Dowling, H. G. and Hedges, S. B. 1995. Higher-level snake phylogeny inferred from mitochondrial DNA sequences of 12s rRNA and 16s rRNA genes. Molecular Biology and Evolution 12: 259-265.

Hubert, J. and Dufaure, J. P. 1968. Table de developpement de la vipere aspic: *Vipera aspis*. Bulletin de la Socie´te´ Zoologique de France 93: 135-148.

Ineich, I., Bonnet, X., Shine, R., Shine, T., Brischoux, F., Lebreton, M. and Chirio, L. 2006. What, if anything, is a 'typical' viper? Biological attributes of basal viperid snakes (genus *Causus* Wagler, 1830). Biological Journal of the Linnean Society 89: 575-588.

James, C. and Shine, R. 1985. The seasonal timing of reproduction: A tropical-temperate comparison in Australian lizards. Oecologia 67: 464-474.
Janzen, D. H. and Schoener T. W. 1968. Differences in insect abundance and diversity between wetter and drier sites during a tropical dry season. Ecology 49: 96-110.
Jones, R. C. 1998. Evolution of the vertebrate epididymis. Journal of Reproduction and Fertility Supplement 53: 163-181.
Jordão, R. S. and Bizerra, A. F. 1995. Reprodução, dimorfismo sexual e atividade de *Simophis rhinostoma* (Serpentes, Colubridae). Revista Brasileira de Biologia 56: 507-512.
Keogh, J. S., Branch, W. R. and Shine, R. 2000. Feeding ecology, reproduction and sexual dimorphism in the colubrid snake *Crotaphopeltis hotamboeia* in southern Africa. African Journal of Herpetology 49: 129-137.
Kofron, C. P. 1983. Female reproductive cycle of the Neotropical snail-eating snake *Sibon sanniola* in northern Yucatan, Mexico. Copeia 1983: 963-969.
Kofron, C. P. 1990. Female reproductive cycle of the African snail-eating snake, *Duberria lutrix*. Amphibia-Reptilia 11: 15-21.
Kopstein, F. 1938. Ein beitrag zur eierkunde und zur fortpflanzung der Malaiischen reptilien. Bulletin of the Raffles Museum 14: 81-167.
Lance, V. and Lofts, B. 1978. Studies on the annual reproductive cycle of the female cobra, *Naja naja*. Journal of Morphology 157: 161-180.
Leite, P. T., Nunes, F. S., Kaefer, I. L. and Cechin S. Z. 2009. Reproductive biology of the swamp racer *Mastigodryas bifossatus* (Serpentes: Colubridae) in subtropical Brazil. Zoologia 26: 12-18.
Licht, P. 1984. Reptiles. Pp. 206-282. In G. E. Lamming (ed.), *Marshall's Physiology of Reproduction*. Churchill Livingstone, Edinburgh, U.K.
Licht, P., and Gorman, G. C. 1970. Reproductive and fat cycles in Caribbean *Anolis* lizards. University of California Publications Zoology 95: 1-52.
Luiselli, L. 2006. Testing hypotheses on the ecological patterns of rarity using a novel model of study: snake communities worldwide. Web Ecology 6: 44-58.
Madsen, T. and Osterkamp, M. 1982. Notes on the biology of the fish-eating snake *Lycodonomorphus bicolor* in Lake Tanganyica. Journal of Herpetology 16: 185-188.
Madsen, T. and Shine, R. 2000. Rain, fish and snakes: climatically driven population dynamics of Arafura filesnakes in tropical Australia. Oecologia 124: 208-215.
Madsen, T., Ujvari, B., Shine, R., Olsson, M. and Loma, J. 2006. Rain, rats and pythons: climate-driven population dynamics of predators and prey in tropical Australia. Australian Journal of Ecology 31: 30-37.
Marques, O. A.V. 1996a. Biologia reprodutiva da cobra-coral *Erythrolamprus aesculapii* Linnaeus (Colubridae), no sudeste do Brasil. Revista Brasileira de Zoologia 13: 747-753.
Marques, O. A. V. 1996b. Growth of the coral snake, *Micrurus corallinus* (Elapidae), in southeastern Atlantic forest in Brazil. Amphibia-Reptilia 17: 277-285.
Marques, O. A.V. and Puorto, G. 1998. Feeding, reproduction and growth in the crowned snake. Amphibia-Reptilia 19: 311-318.
Marques, O. A.V. and Muriel, A. P. 2007. Reproductive biology and food habits of the swamp racer *Mastigodryas bifossatus* from southeastern South America. Herpetological Journal 17: 104-109.
Marques, O. A.V., Sawaya, R. J., Stender-Olivera, F. and Franca, F. G. R. 2006. Ecology of the colubrid snake *Pseudablabes agassizii* in southeastern South America. Herpetological Journal 16: 37-45.
Marques, O. A. V., Almeida-Santos, S., Rodrigues, M. and Camargo, R. 2009. Mating and reproductive cycle in the Neotropical colubrid snake *Chironius bicarinatus*. South American Journal of Herpetology 4: 76-80.

Maschio, G. F., Prudente, A. L. C., Lima, A. C. and Feitosa, D. T. 2007. Reproductive biology of *Anilius scytale* (Linnaeus, 1758) (Serpentes, Anilidae) from eastern Amazonia, Brazil. South American Journal of Herpetology 2: 179-183.

Mathies, T., Felix, T. A. and Lance, V. A. 2001. Effects of trapping and subsequent short-term confinement stress on plasma corticosterone in the brown tree-snake (*Boiga irregularis*) on Guam. General and Comparative Endocrinology 124: 106-114.

Mathies, T., Cruz, J. A., Lance, V. A. and Savidge, J. A. 2010. Reproductive biology of male brown treesnakes (*Boiga irregularis*) on Guam. Journal of Herpetology 44: 209-221.

McCoy, C. J. and Censky, E. J. 1992. Biology of the Yucatan hognosed viper, *Porthidium yucatanicum*. Pp. 217-222. In J. A. Campbell, E. D. Brodie, Jr. (eds), *Biology of the Pitvipers*. Selva, Tyler, Texas.

Meixi, W. and Fuying, F. 1989. The reproductive pattern of the snake, *Enhydris chinensis* (Gray). Acta Zoologica Sinica 35: 82-87.

Monteiro, C., Montgomery, C. D., Spina, F., Sawaya, R. J. and Martins, M. 2006. Feeding, reproduction, and morphology of *Bothrops mattogrossensis* (Serpentes, Viperidae, Crotalinae) in the Brazilian Pantanal. Journal of Herpetology 40: 408-413.

Myers, C. W. 2003. Rare snakes—five new species from eastern Panama: Reviews of northern *Atractus* and southern *Geophis* (Colubridae: Dipsadinae). American Museum Novitates 3391: 1-47.

Nogueira, C., Sawaya, R. J. and Martins, M. 2003. Ecology of the pitviper, *Bothrops moojeni*, in the Brazilian Cerrado. Journal of Herpetology 37: 653-659.

Oliver, J. A. 1947. The seasonal incidence of snakes. American Museum Novitates 1363: 1-14.

Pinto, R. R. and Fernandes, R. 2004. Reproductive biology and diet of *Liophis poecilogyrus poecilogyrus* (Serpentes, Colubridae) from southeastern Brazil. Phyllomedusa 3: 9-14.

Pizzatto, L. 2003. Reprodução de *Liophis miliaris* (Serpentes: Colubridae) no Brasil: influência históricia e variações geográficas. Dissertação de Mestrado, Instituto de Biologia, Universidade Estadual de Campinas, Campinas, São Paulo, Brasil.

Pizzatto, L. 2005. Body size, reproductive biology and abundance of the rare pseudoboini snakes genera *Clelia* and *Boiruna* (Serpentes, Colubridae) in Brazil. Phyllomedusa 4: 111-122.

Pizzatto, L. and Marques, O. A. V. 2002. Reproductive biology of the false coral snake *Oxyrhopus guibei* (Colubridae) from southeastern Brazil. Amphibia-Reptilia 23: 495-504.

Pizzatto, L. and Marques, O. A. V. 2007. Reproductive ecology of boine snakes with emphasis on Brazilian species and a comparison to pythons. South American Journal of Herpetology 2: 107-133.

Pizzatto, L., Almeida-Santos, S. M. and Marques, O. A. V. 2007. Biologia reprodutiva de serpentes Brasileiras. Pp. 201-221. In L. B. Nacimento and M. E. Oliveira (eds), Herpetologia no Brasil II. Sociedade Brasileira de Herpetologia, Belo Horizonte, Minas Gerais, Brazil.

Pizzatto, L., Cantor, M., Lima de Oliveira, J., Marques, O. A. V., Capovilla, V. and Martins, M. 2008a. Reproductive ecology of dipsadine snakes, with emphasis on South American species. Herpetologica 64: 168-179.

Pizzatto, L., Jordão, R. S. and Marques, O. A. V. 2008b. Overview of reproductive strategies in Xenodontini (Serpentes: Colubridae: Xenodontinae) with new data for *Xenodon neuwiedii* and *Waglerophis merremii*. Journal of Herpetology 42: 153-162.

Porto, M. and Fernandes, R. 1996. Variation and natural history of the snail-eating snake *Dipsas neivai* (Colubridae: Xenodontinae). Journal of Herpetology 30: 269-271.

Prudente, A. L. C., Maschio, G. F., Yamashina, C. E. and Santos-Costa, M. C. 2007. Morphology, reproductive biology and diet of *Dendrophidion dendrophis* (Schlegel, 1837) (Serpentes, Colubridae) in Brazilian Amazon. South American Journal of Herpetology 2: 53-58.

Reading, C. J. 2004. The influence of body condition and prey availability on female breeding success in the smooth snake (*Coronella austriaca* Laurenti). Journal of Zoology 264: 61-67.

Rivas, J. A. 2000. The life history of the green anaconda (*Eunectes murinus*), with emphasis on its reproductive biology. Ph.D. thesis, University of Tennessee, Knoxville, Tennessee

Rodda, G.H., Savidge, J. A., Tyrrell, C. L., Christy, M. T. and Ellingson, A. R. 2007. Size bias in visual searches and trapping of brown treesnakes on Guam. Journal of Wildlife Management 71: 656-661.

Saint Girons, H. 1992. Reproductive cycles of male snakes and their relationships with climate and female reproductive cycles. Herpetologica 38: 5-16.

Saint Girons, H. and Pfeffer P. 1971. Le cycle sexuel des serpents du Cambodge. Annales des Sciences Naturelles, Zoologie 13: 543-572.

Santos-Costa, M. C., Prudente, A. L. C. and Di-Bernardo, M. 2006. Reproductive biology of *Tantilla melanocephala* (Linnaeus, 1758) (Serpentes, Colubridae) from eastern Amazonia, Brazil. Journal of Herpetology 40: 553-556.

Sasa, M. 1993. Distribution and reproduction of the grey earth snake *Geophis brachycephalus* (Serpentes: Colubridae) in Costa Rica. Revista de Biologia Tropical 41: 295-297.

Savidge, J. A., Qualls, F. J. and Rodda, G. H. 2007. Reproductive biology of the brown tree snake, *Boiga irregularis* (Reptilia: Colubridae), during colonization of Guam and comparison with that in their native range. Pacific Science 61: 191-199.

Sawaya, P. S. 2003. História natural e ecologia das serpentes de Cerrado da região de Itirapina, SP. 145 p. Tese (Doutorado, Ecologia), Instituto de Biologia, Universidade Estadual de Campinas.

Sawaya, P. S. Marques, O. A. V. and Martins, M. 2008. Composition and natural history of a cerrado snake assemblage at Itirapina, São Paulo state, southeastern Brazil. Biota Neotropica 8: 127-149.

Scartozzoni, R. R., Salomão, M. and Almeida-Santos, S. M. 2009. Natural history of the vine snake *Oxybelis fulgidus* (Serpentes, Colubridae) from Brazil. South American Journal of Herpetology 4: 81-89.

Schuett, G. W., Carlisle, S., Holycross, A., O'Leile, J., Hardy, D., Van Kirk, E. and Murdoch, W. 2002. Mating system of male Mojave rattlesnakes (*Crotalus scutulatus*): seasonal timing of mating, agonistic behavior, spermatogenesis, sexual segment of the kidney, and plasma sex steroids. Pp. 515-532. In G. W. Schuett, M. Hoggren, M. Douglas and H. Greene (eds), *Biology of the Vipers*. Eagle Mountain Publishing, Eagle Mountain, Utah.

Seigel, R. A. and Ford, N. B. 1987. Reproductive ecology. Pp. 210–252. In R. A. Seigel, J.T. Collins and S. S. Novak (eds), *Snakes: Ecology and Evolutionary Biology*. McGraw-Hill Publishing Company, New York.

Siegel, D. S., Aldridge, R. D., Clark, C. S., Poldemann, E. H. and Gribbins, K. M. 2009. Stress and reproduction in *Boiga irregularis* with notes on the ultrastructure of the sexual segment of the kidney in squamates. Canadian Journal of Zoology 87: 1138-1146.

Sever, D. M., Stevens, R. A., Ryan, T. J. and Hamlett, W. C. 2002. Ultrastructure of the reproductive system of the black swamp snake (*Seminatrix pygaea*): III. Sexual segment of the male kidney. Journal of Morphology 252: 238-254.

Schmidt de Aguiar, L. F. and Di-Bernardo, M. 2005. Reproduction of the water snake *Helicops infrataeniatus* (Colubridae) in southern Brazil. Amphibia-Reptilia 26: 527-533.

Shine, R. 1977a. Reproduction in Australian elapid snakes. I. Testicular cycles. Australian Journal of Zoology 25: 647-653.

Shine, R. 1977b. Reproduction in Australian elapid snakes. II. Female reproductive cycles. Australian Journal of Zoology 25: 655-666.

Shine, R. 1980a. Comparative ecology of three Australian snake species of the genus *Cacophis* (Serpentes: Elapidae). Copeia 1980: 831-838.

Shine, R. 1980b. Ecology of eastern Australian whipsnakes of the genus *Demansia*. Journal of Herpetology 14: 381-389.

Shine, R. 1980c. Ecology of the Australian death adder *Acanthophis antarcticus* (Elapidae): evidence for convergence with the Viperidae. Herpetologica 36: 281-289.

Shine, R. 1980d. Reproduction, feeding and growth in the Australian burrowing snake *Vermicella annulata*. Journal of Herpetology 14: 71-77.

Shine, R. 1983. Reptilian reproductive modes: the oviparity-viviparity continuum. Herpetologica 39: 1-8.

Shine, R. 1985. The reproductive biology of Australian reptiles: search for general patterns. Pp. 297–303. In G. Grigg, R. Shine and H. Ehmann (eds), *Biology of Australian Frogs and Reptiles*. Surrey Beatty and Sons, Chipping Norton, Australia.

Shine, R. 1986. Ecology of a low-energy specialist: food habits and reproductive biology of the arafura filesnake (Acrochordidae). Copeia 1986: 424-437.

Shine, R. 1991. Strangers in a strange land: ecology of the Australian colubrid snakes. Copeia 1991: 120-131.

Shine, R. 2003. Reproductive strategies in snakes. Proceedings of the Royal Society of London B 270: 995-1004.

Shine, R. and Brown, G. B. 2008. Adapting to the unpredictable: reproductive biology of vertebrates in the Australian wet-dry tropics. Philosophical Transactions of the Royal Society 363: 363-373.

Shine, R., Harlow, P., Keogh, J. S. and Boeadi. 1995. Biology and commercial utilization of acrochodid snakes, with special reference to kurung (*Acrochordus javanicus*). Journal of Herpetology 29: 352-360.

Shine, R., Harlow, P. S., Branch, W, R. and Webb, J. K. 1996. Life on the lowest branch: Sexual dimorphism, diet, and reproductive biology of an African twig snake, *Thelotornis capensis* (Serpentes, Colubridae). Copeia 1996: 290-299.

Shea, G. M. 2001. Spermatogenic cycle, sperm storage, and sertoli cell size in a scolecophidian (*Rhamphotyphlops nigrescens*) from Australia. Journal of Herpetology 35: 85-91.

Slip, D. J. and Shine, R. 1988. The reproductive biology and mating system of diamond pythons, *Morelia spilota* (Serpentes: Boidae). Herpetologica 4: 396-404.

Solórzano, A. 1990. Reproduction in the pit viper *Porthidium picadoi* Dunn (Serpentes: Viperidae) in Costa Rica. Copeia 1990: 1154-1157.

Solórzano, A. and Cerdas, L. 1989. Reproductive biology and distribution of the terciopelo, *Bothrops asper* Garman (Serpentes: Viperidae), in Costa Rica. Herpetologica 45: 444-450.

Srivastava, P. C. and Thapliyal, J. P. 1965. The male sexual cycle of the chequered water snake, *Natrix piscator*. Copeia 1965: 410-415.

Stafford, P. J. 2003. Trophic ecology and reproduction in three species of Neotropical forest racers (*Dendrophidion*; Colubridae). Herpetological Journal 13: 101-111.
Stafford, P. J. 2005. Diet and reproductive ecology of the Yucatán cricket-eating snake Symphimus mayae (Colubridae). Journal of Zoology 265: 301-310.
Taylor, E. M., DeNardo, D. F. and Jennings, D. H. 2004. Seasonal steroid hormone levels and their relation to reproduction in the western diamond-backed rattlesnake, *Crotalus atrox* (Serpentes: Viperidae). General and Comparative Endocrinology 136: 328-337.
Tsai, T. S. and Tu, M. C. 2001. Reproductive cycle of female Chinese green tree vipers, *Trimeresurus stejnegeri stejnegeri*, in northern Taiwan. Herpetologica 57: 157-168.
Tu, M. C., Fong, S. C. and Lue, K. Y. 1990. Reproductive biology of the sea snake, *Laticauda semifasciata*, in Taiwan. Journal of Herpetology 24: 119-126.
Valdujo, P. H., Nogueira, C. and Martins, M. 2002. Ecology of *Bothrops neuwiedi pauloensis* (Serpentes: Viperidae: Crotalinae) in the Brazilian Cerrado. Journal of Herpetology 36: 169-176.
Vitt, L. J. 1983. Ecology of an anuran-eating guild of terrestrial tropical snakes. Herpetologica 39: 52-66.
Vitt, L. J. 1992. Diversity of reproductive strategies among Brazilian lizards and snakes: the significance of lineage and adaptation. Pp. 135-149. In W. C. Hamlett (ed.), *Reproductive Biology of South American Vertebrates*. Springer-Verlag, New York.
Vitt, L. J. and Vangilder, L. D. 1983. Ecology of a snake community in northeastern Brazil. Amphibia-Reptilia 4: 273-296.
Vitt, L. J. and Seigel, R. A. 1985. Life history traits of lizards and snakes. The American Midland Naturalist 125: 480-484.
Volsøe, H. 1944. Structure and seasonal variation of the male reproductive organs of *Vipera berus* (L). Spolia Zoologica Musei Hauniensis 5: 7-157.
Wangkulangkul, S., Thirakhupt, K. and Voris, H. 2005. Sexual size dimorphism and reproductive cycle of the little file snake *Acrochordus granulatus* in Phangnga Bay, Thailand. Science Asia 31: 257-263.
Ward, T. M. 2001. Age structures and reproductive patterns of two species of sea snake, *Lapemis hardwickii* Grey (1836) and *Hydrophis elegans* (Grey 1842), incidentally captured by prawn trawlers in northern Australia. Marine Freshwater Research 52: 193-203.
Warner, D. A. and Andrews, R. M. 2002. Nest-site selection in relation to temperature and moisture by the lizard *Sceloporus undulatus*. Herpetologica 58: 399-407.
Whittier, J. M. and Limpus, D. 1996. Reproductive patterns of a biologically invasive species: the brown tree snake (*Boiga irregularis*) in eastern Australia. Journal of Zoology 283: 591-597.
Woodward, S. F. 1933. A few notes on the persistence of active spermatozoa in the African night-adder, *Causus rhombeatus*. Proceedings of the London Zoological Society 1933: 189-190.
Zacariotti, R. L., Grego, K. F., Fernandes, W., Sant'Anna, S. S. and Barros Vaz Guimarães, M. A. 2007. Semen collection and evaluation in free-ranging Brazilian rattlesnakes (*Crotalis durissus terrificus*). Zoo Biology 26: 155-160.
Zehr, D. R. 1962. Stages in the normal development of the common garter snake, *Thamnophis sirtalis sirtalis*. Copeia 1962: 322-329.
Zug, G. R., Hedges, S. B. and Sunkel, S. 1979. Variation in reproductive parameters of three Neotropical snakes, *Coniophanes fissidens*, *Dipsas catesbyi*, and *Imantodes cenchoa*. Smithsonian Contributions to Zoology 300: 1-20.

CHAPTER 13

Pheromones in Snakes: History, Patterns and Future Research Directions

M. Rockwell Parker and Robert T. Mason

13.1 INTRODUCTION

The term pheromone, first coined by Karlson and Lüscher (1959), describes a chemical or semiochemical produced by one individual that effects a change in the physiology or behavior of conspecifics. Taken in the animal communication context, a pheromone is a signal produced by a signaler with the effect of modifying the receiver in some way. Karlson and Lüscher's definition allows researchers to ascribe to unidentified substances the term pheromone because one can evaluate the ability of chemical cues to affect receiver behavior without analytical isolation and characterization of putative semiochemicals, or chemicals with signal function. This chapter will focus on identified pheromones in snakes as well as systems where observations of pheromone-based behaviors have been made. Because this volume is focused on reproduction, we will focus this chapter on reproductive pheromones instead of putative aggregation pheromones (none of which has been identified to date).

13.2 METHODS FOR ASSESSING PHEROMONE USAGE IN OPHIDIAN REPRODUCTION

Most studies invoking the use of pheromones in snake reproduction have done so based on behavioral observations. Several types of bioassays for chemical investigation exist, all of which center on tongue-flicking, the proximate mechanism by which chemical cues can be discriminated via the vomeronsasal (VN) system ("vomerolfaction"; Cooper and Burghardt 1990). During tongue-flicking, snakes (and other squamate reptiles) protrude the tongue and collect chemical cues from the environment, typically from the

Department of Zoology, Oregon State University, Corvallis, OR 97331, USA substrate (Gove 1979; Halpern 1992). The cues are then delivered to the vomeronsasal organ (VNO) in the roof of the mouth where they are sensed, though the precise mechanism of delivery is contentious and remains unresolved (Young 1993).

A functional VN system is required to enable pheromone-based behaviors, which was first demonstrated experimentally by researchers such as Kahmann (1932). Also, male snakes will not court female snakes if the VN nerves are cut, but they do court females following olfactory nerve lesion (Kubie *et al.* 1978). It has further been shown via electrophysiological experiments that the VN system is the primary sensory system responding to chemical cues collected by the tongue of snakes, and olfaction appears to play a minimal or negligible role (Inouchi *et al.* 1993) (Although see section 13.3.2). Whether the VN system evolved specifically for detection of pheromones is an unresolved question, but it is clearly responsible for more than just pheromone reception. For example, garter snakes detect their earthworm prey by tongue-flicking chemoattractants that are detected exclusively by the VN system (Jiang *et al.* 1990; Wang *et al.* 1997). To date, many researchers have developed consistent, reliable methods for assessing the use of the VN system in investigations of mating behavior and conspecific discrimination as well as prey detection (Baxi *et al.* 2006). The quality and unambiguity of such bioassays, however, is critical in the level of inference that can be made about observed behaviors.

13.2.1 Tongue-flick Rates

Many researchers quantify tongue-flick rates to assess individual interest in chemical cues or putative pheromones (Cooper 1998). This method is rapid, repeatable, and inexpensive. Typically, tongue-flick rates are counted once the substance or chemical cue of interest has been placed on a cotton swab and the swab is presented to the focal animal. This assay can be problematic if proper controls are not used for determining background tongue-flick rates. Often, control stimuli are volatile organic solvents or compounds, such as cologne, that are most likely not detected by, or in some cases actually suppress the VN system for which the tongue-flick assay was developed. In many studies water is used, which is an inert, biologically irrelevant chemical control stimulus. Given that the VN system is most responsive to nonvolatile cues, nonvolatile stimuli like oils or lipids serve as better controls for the biological substance of interest (used first in G. K. Noble's studies; Noble and Clausen 1936; Noble 1937). Further, tongue-flick rates primarily reveal to the researcher whether or not the focal animal is interested in the presented cue relative to other stimuli. If biological stimuli are paired during swab tests, discrimination by the focal animal can be accurately assessed. But, without further corroborating behaviors, examination of tongue-flicking behavior alone falls short of capturing the biological meaning of the presented cue. Recognizing this, investigators have included additional behaviors quantified along with the tongue-flick assay, such as incorporating swab attack into the computation of the score in behavioral tests involving perception of prey or feeding cues (Burghardt 1967, 1969). To date, there have been no reports of any corroborating behaviors in conjunction with swab tests examining reproductive chemical cues or putative pheromones in snakes, though recent work in other reptiles has (e.g., Martín *et al.* 2007). Although tongue-flick rates provide adequate metrics in the chemical ecology of foraging and prey selection, they currently provide minimal support for building a case for invoking pheromone use compared to other methods. This is clearly an area where new paradigms are needed in order to make progress in elucidating the role of pheromones in snakes and other reptiles.

13.2.2 Trailing Experiments

Trailing experiments serve as excellent bioassays for testing for putative pheromones. By responding to chemical cues on the substrate, the focal animal actively chooses to either approach (attractants) or avoid (repellents) the source of the semiochemicals as they actively move along the trail. Several researchers have utilized trailing apparatuses for assessing the role of isolated chemical cues in eliciting or maintaining reproductive behavior, typically in males (e.g., Noble and Clausen 1936; Noble, 1937; Gehlbach *et al.* 1971; Ford 1986; Greene *et al.* 2001). As noted by Ford (1995), researchers studying snake reproductive behavior in the laboratory face challenges that necessitate creativity in the design of experimental apparatuses. In published studies on snake chemosensory behavior, many experimental setups have been used and modified, such as Y-mazes made of Plexiglas (Parker and Kardong 2005), open Y-mazes with pegs for allowing taxis along the trail (Ford and Low 1984; LeMaster and Mason 2001a, see Fig. 13.1), and rectangular arenas with a washable or replaceable substrate for creation of random trails (Noble 1937). Only one study has specifically tested pheromone trailing behavior in the field under natural conditions with bioassays using isolated, known chemical cues (LeMaster *et al.* 2001). There are, however, a number of detailed observations from radiotelemetry studies of individual males trailing female chemical trails (e.g., Duvall *et al.* 1985; Slip and Shine 1988; Cardwell 2008). Field experiments have revealed that pheromone trailing by males is a complex process in nature, especially when numerous males trail a single female, which appears to disrupt her pheromone trail in the natural environment and forces males to rely on other cues (e.g., movement) for relocation (Shine *et al.* 2005b). Recent laboratory experiments have highlighted the usefulness of female chemical cues in male snake trailing behavior (e.g., Fornasiero *et al.* 2007).

Trailing experiments, when applicable, are better bioassays for assessing the role of semiochemicals in chemosensory behaviors than are tongue-flick tests because they require persistent interest in the cue to enable sustained and accurate trailing which is mediated by high tongue-flick rates. Moreover, trailing experiments eliminate all other cues such as researcher

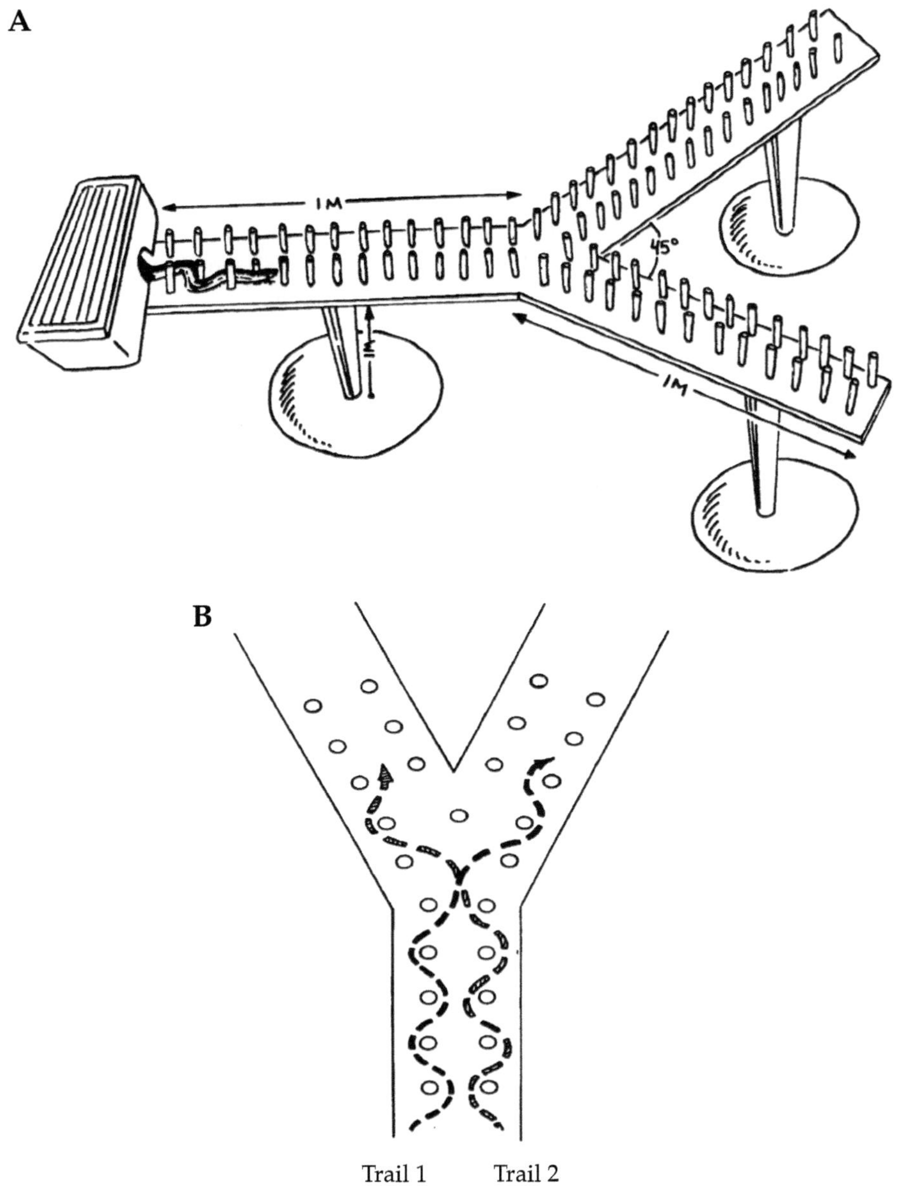

Fig. 13.1 A. Example of a Y-maze trailing apparatus. The snake begins in a holding box and is allowed to explore the trailing surface, using the pegs to locomote through the maze. Choice is scored once the animal passes some marked point at the end of one of the arms, typically the last set of pegs. **B.** Method for creation of the scent trail(s). Trails are woven between the pegs then crossed at the junction of the Y to present the trailing animal with both trails and force a definite choice at the junction. Note: this apparatus also works well for prey trailing experiments. Image reproduced from LeMaster, M. P. and Mason, R. T. 2001. Chemoecology 11: 149-152, Fig. 1.

interaction and other animal behaviors from the source individual. The inability of the focal animal to accurately follow a chemical trail from the source animal also informs the researcher about the relevance, or lack thereof, of any pheromone cues in mate location behavior (in the case of putative pheromones). Accurate trail following behavior demonstrates reception of pheromone cues because the animal is motivated to find the source through the act of trailing itself. The drawback to trailing experiments, however, is that they only assess the role of the chemical cues in mate location behavior, not necessarily in the maintenance of the male-female interaction. We will later discuss the role that trailing behavior plays in the general sequence of snake reproductive behaviors described most recently by Gillingham (1987).

13.2.3 Sexual Behavior

The ultimate category of experimentation for determing the role of chemical cues or pheromones in the reproductive behavior of snakes is sexual behavior, which can be easily observed and measured using staged courtship trials or, in the optimal case, by observing courtship occurring naturally in the field. Courtship behavior in most snakes is an unambiguous, quantifiable process, as has been clearly delineated in at least one species, the Red-Sided Garter Snake (*Thamnophis sirtalis parietalis*; Crews *et al.* 1984; Moore *et al.* 2000). Direct bioassays of behavioral responses to chemical cues or pheromones allow the researcher to quantify data and draw inferences and conclusions that go beyond choice and interest, the two main types of data gathered in trailing and tongue-flick tests, respectively. By placing sexually motivated males in arenas with reproductively active females, female sexual attractiveness can be assessed. In studies from the Crews and Mason laboratories, female and other focal animal attractiveness is quantified based on either the individual courtship scores of males (a point scale based on either half or whole-integer increments, see Table 13.1) or the number of males (proportion or total count) courting the focal animal over a set period of time using field arenas (see Figure 13.2). Such bioassays not only reveal the intensity of male interest in the pheromone cues present on the female's skin but also allow for an assessment of the persistence of female-directed behavior. The drawback to these bioassays, however, is that

Table 13.1 Ethogram of male courtship behavior in the Red-sided Garter Snake, *Thamnophis Sirtalis Parietalis*, redrawn from Moore *et al.* 2000

Courtship score	*Description of behavior*
1.0	Male investigates female, increased tongue-flick rate
2.0	Male chin rubs female with rapid tongue-flicks
3.0	Male aligns body with female
4.0	Male actively tail searches and attempts cloacal apposition and copulation with female; possible caudocephalic waves
5.0	Male copulates with female

other cues are present to males, such as visual and behavioral information from the female, that may (and certainly do) modify male behavior. To address this, Mason and collaborators presented isolated skin lipid chemical cues, and eventually individual synthesized chemical compounds onto filter papers or paper towels and presented them to courting males in the field. Quantification of male sexual behaviors observed only in a reproductive context to these isolated chemical cues in the absence of any other cues provided unambiguous evidence for the identification of the first reptilian pheromone, a series of long-chain saturated and monounsaturated methyl ketones (Mason *et al.* 1989, 1990).

Fig. 13.2 Examples of outdoor arenas for observing snake mating behavior. The arenas measure 1 m^3 and have jersey mesh wind panels to prevent flapping. The vinyl arenas are tied to metal stakes to anchor them. In the top right picture, a mating ball of males containing a single female can be seen on the floor of each arena, demonstrating that typical mating behavior occurs in the arenas and enables detailed, repeatable observation of courtship and mating under natural conditions in the field.

Although laboratory and field experiments are useful for testing hypotheses on the role and relative importance of pheromones in snake reproduction, the most illuminating data come from field studies of snake behavior. Indeed, a number of studies on the ecology of snakes have revealed the importance of chemical communication in all major groups of snakes. We now describe the history of such critical natural history observations and highlight species that are promising for their obvious reliance on sex pheromones in reproduction.

13.3 HISTORY OF REPRODUCTIVE CHEMICAL ECOLOGY STUDIES IN SNAKES

Previous work has extensively reviewed the role of chemical communication in snakes and all reptiles (Mason 1992; Mason *et al.* 1998). Here, we examine the history of key studies and observations of the reproductive behaviors of snakes, most of which have focused on the abilities of males to follow female pheromone trails. A culmination of studies in one species, the Common Garter Snake (*Thamnophis sirtalis*), ultimately resulted in the identification, characterization and synthesis of the only known reptilian pheromone, the sexual attractiveness pheromone of *Thamnophis sirtalis parietalis* (Mason *et al.* 1989, 1990).

Natural history observations sparked the first studies into the role of pheromones in the reproductive ecology of snakes, particularly the behavior of males as they follow the trails of females in the environment. The list of such observations is extensive, but we will focus in this section on explicit tests of the role that female trails and chemical cues play in coordinating reproduction and reproductive behaviors. The pioneering studies were conducted in the European Asp (*Vipera aspis*) where males follow female trails in the laboratory and exhibit high tongue-flick rates to and maintain close proximity with the substrate traversed by females (Baumann 1929). It was further shown that the vomeronsasal organ (VNO) was critical for enabling both female and male identification and subsequent initiation of sex-directed behavior in *Vipera* (Andrén 1982; Andrén 1986). Thus, it can be concluded that this genus uses pheromones for mate recognition, initiation of courtship and combat behavior. It is important to note that the courtship behavior of vipers is highly conserved, with striking similarities existing between vipers and natricines (see Carpenter and Ferguson 1977). Recent work has demonstrated strong sexual dimorphism in the tongue morphologies of Copperheads (*Agkistrodon contortrix*; Smith *et al.* 2008). Copperheads and Cottonmouths (*Agkistrodon piscivorus*) exhibit stereotypical natricine-like courtship behavior with the expression and perception of pheromones being suggested (see below; Martin 1984; Schuett and Gillingham 1988). So the sexual dimorphism in tongue morphology in this group would at the very least facilitate conspecific trailing, may suggest reliance on pheromones during courtship and possibly combat behavior, and certainly implicates a major role for chemical communication in reproduction in *Agkistrodon*, and possibly all pit vipers. It is clear that further experimental work should be conducted on the reproductive biology and chemical ecology of pit vipers.

The earliest studies in the Common European Adder (*Vipera berus*) were conducted before the studies of Noble and colleagues on *Thamnophis*, and it is clear that Noble followed the methods of Baumann and Kahmann (Kahmann 1932) in the design of both of his initial laboratory studies on colubrid reproductive behavior (Noble and Clausen 1936; Noble 1937). A greater volume of information is available on the reproductive behavior and ecology of *Thamnophis* than any other group of snakes due in part to the early and pioneering work of G. K. Noble. His initial study examined the role of chemical trails in the aggregation behavior and reproductive behavior of the Brown Snake (*Storeria dekayi*, Noble and Clausen 1936). The major findings from that study were that male *Storeria* do not trail cloacal secretions from females but that the cues enabling mate location resided instead on the skin of females. It is interesting to note that Noble and Clausen's conclusions were quite unambiguous, yet one can still find investigators after this time addressing the same question and reaching the same conclusion. In a subsequent set of experiments (Noble 1937), it was clearly demonstrated that the skin of both *Thamnophis* and *Storeria* females was the sole source of a chemical attractiveness cue that enabled males to recognize the sex of conspecifics and initiate courtship behavior. Further experiments using skin tubes from sacrificed females slipped over the bodies of males showed that courtship behavior was elicited purely by the cues present on the skin of snakes (Gillingham and Dickinson 1980). Later experiments established that female *Thamnophis* chemical trails are species-specific and that males' abilities to follow female trails are seasonal, although the latter is not consistent for all species (Ford 1981, 1982). An interesting finding in Ford's studies was that species-specificity of female chemical trails is variable, with males of some species showing greater accuracy and a stronger preference for conspecific female trails (Ford 1982). Further testing of other *Thamnophis* species provided strong inference for the role that pheromones may have played in the radiation of this speciose group of snakes in North America (Ford and O'Bleness 1986). All of these findings are likely due to subtle compositional differences in the pheromone blends of these species, which can be determined using established methods (Mason *et al.* 1989; Mason 1992). Such studies in species of *Thamnophis* other than *T. sirtalis* have not been conducted to date, and this information would help to further illuminate the evolution of pheromonal communication in this well-studied genus of snakes.

13.3.1 The Manitoba Populations of *Thamnophis sirtalis parietalis*

Thamnophis sirtalis parietalis has become a model snake species for studies on pheromonal communication in large part because of the intensity of research on this species' reproductive ecology over the past 40+ years. Every spring, *T. s. parietalis* engage in impressive displays of mating behavior in the northern extent of their range in Manitoba, Canada, as both sexes emerge from large, communal hibernacula in the tens of thousands. At this time, male courtship behavior is quite stereotyped and explicit ethograms for scoring male courtship behavior have been created and used repeatedly in numerous studies for assessing the attractiveness of females and experimental animals in the field (Crews *et al.* 1984; Moore *et al.* 2000). As is the case with most snakes, courtship behavior begins with males orienting toward and tongue-flicking to locate the female. Once tongue contact is made with the female, the male rubs his chin along the female's dorsum while rapidly tongue-flicking to detect female pheromone and, thus, insure contact is maintained with the female. The male then aligns his body with the female and orients his head in the same direction as hers, which quickly proceeds to the initiation of caudocephalic waves. The last phase of courtship is tail-searching behavior, where the male attempts to align his cloaca with the female's and copulate.

13.3.1.1 Sex pheromone

The sexual attractiveness pheromone of female *Thamnophis sirtalis parietalis* was subsequently isolated, identified, synthesized, and tested in field bioassays to complete the characterization process (Mason *et al.* 1989, 1990). The female sexual attractiveness pheromone was identified as a blend of 17 homologous long-chain saturated and unsaturated methyl ketones (C29-C37; molecular weights 394 to 532 Da, see Fig. 13.3), which can be extracted from the skin of euthanized or live snakes by full body immersion in n-hexane, a nonpolar solvent that has also been used in other species of snakes to extract skin lipids for bioassays (Mason *et al.* 1989; Greene and Mason 1998; Cressman *et al.* unpubl. data).

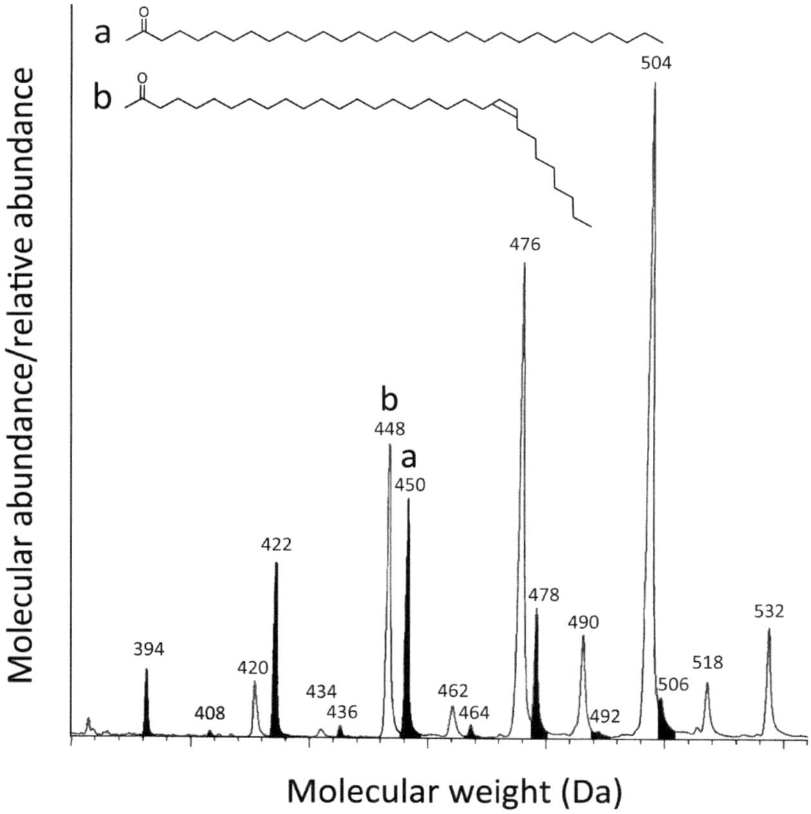

Fig. 13.3 Individual chromatogram for pheromone from a female Red-Sided Garter Snake (*Thamnophis sirtalis parietalis*). Each peak represents a single methyl ketone (black peaks=saturated methyl ketones, white peaks=monounsaturated methyl ketones), numbers above peaks represent molecular weights (Da). **A** and **B** are the methyl ketones weighing 450 and 448 Da, respectively, and their chemical structures are drawn in the top left corner.

The process of extraction yields a complex mixture of skin lipids which must be further fractionated and the chemical constituents isolated by column chromatography (Mason *et al.* 1989). Purified pheromone samples can then be analyzed using small aliquots (1 µl) of pheromone:hexane mixtures (1 mg:1 ml) injected on a fused-silicon glass capillary column following a standard protocol for gas chromatography/mass spectrometry (see Parker and Mason 2009 for the most recent explanation of these methods; see Fig. 13.4 for a depiction of several individual chromatograms).

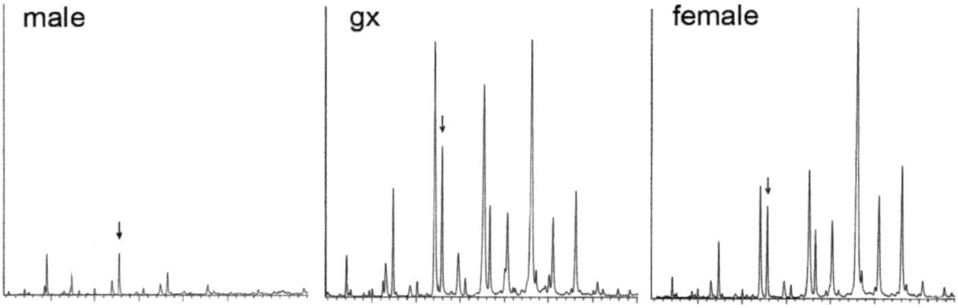

Fig. 13.4 Individual chromatograms of pheromone blends from intact male (left), castrated male (middle), and intact female (right) Red-sided Garter Snakes (*Thamnophis sirtalis parietalis*). The chromatograms represent pheromone blends that are in order from unattractive to most attractive (L to R). The X-axis is retention time on the GC column, and the Y-axis is molecular abundance. These chromatograms have been scaled by using an internal standard (methyl stearate, see LeMaster *et al.* 2008 for methods). For reference, the arrow in each chromatogram indicates the 450 Da methyl ketone peak.

The pheromone of female *Thamnophis sirtalis parietalis* encodes information about condition, age, reproductive state, species, and season (Mason *et al.* 1987; LeMaster and Mason 2001b; LeMaster and Mason 2002; Shine *et al.* 2003a; LeMaster and Mason 2003; O'Donnell *et al.* 2004). Perhaps most intriguing is that the pheromone blend becomes dominated by the longest chain, unsaturated methyl ketones with increasing length and body condition (LeMaster and Mason 2002). Males prefer to court and mate with longer females over shorter females, and they prefer females in better body condition. LeMaster and Mason (2002) demonstrated that males are able to discriminate among variable female body conditions based solely on information provided by female skin lipid pheromone cues (Shine *et al.* 2003a). In this species, female fecundity increases with snout-to-vent length (SVL) and body condition, which suggests that the female sexual attractiveness pheromone functions as an honest signal in this system. Although several laboratory experiments with numerous *Thamnophis* species have clearly demonstrated the ability of pheromones to facilitate mate location, few studies have shown that pheromone trails are actually used in the field by wild males for finding mates during the breeding season. LeMaster *et al.* (2001; LeMaster and Mason 2001a), however, successfully showed that male *Thamnophis sirtalis parietalis*, but not females, use pheromone trails from conspecifics to locate mates and that this ability is seasonal, occurring only during the breeding season. Finally, variation in the presence and abundance of the methyl ketone components of the pheromone blend provide information on population differences within *T. s. parietalis*. Males choose to court and mate with females from their own dens, and females choose to mate with males from their own dens (LeMaster and Mason 2003). Apparently there are features or aspects associated with individual dens and their inhabitants that exert selective pressures to avoid outbreeding within populations. Collectively, the study of the sexual attractiveness pheromone in *T. s. parietalis* has elucidated the critical importance of sex pheromones in the fundamental coordination of reproduction at multiple levels (individual, population, species) in this utilitarian snake model system.

13.3.1.2 Inhibitory pheromone(s)

Although much is known about the sexual attractiveness pheromone of *Thamnophis sirtalis parietalis*, less is known about the pheromone associated with the copulatory plug. Following successful mating, a male deposits a copulatory plug in the female's cloaca, preventing her from immediately mating again by physically blocking her urogenital opening but also by rendering the female transiently unattractive (Devine 1975, 1977; Ross and Crews 1977). Although much has been hypothesized on the role of copulatory plugs in the evolution of sperm competition in reptiles (e.g., Devine 1975; Shine *et al.* 2000b), experimental evidence indicates that the plug of garter snakes, or the fluids associated with mating, have physiological effects on the female that inhibit receptivity and attractivity (e.g., Ross and Crews 1977; Whittier *et al.* 1985; Mendonca and Crews 2001; O'Donnell *et al.* 2004). It is still a matter of controversy whether the inhibitory pheromone is produced in the copulatory plug, is derived from the male's ejaculate, the female's cloaca, or a combination of the two. In the most recent work, Shine *et al.* (2000b) demonstrated that the copulatory fluids from mating males contained the inhibitory pheromone that renders the female transiently unattractive. The actual plug itself did not possess pheromonal properties. Rather, the plug appears to serve as a physical barrier to subsequent matings and possibly as a means to prevent the leakage of sperm from the female's cloaca.

Chemical isolation and identification of the inhibitory pheromone of the copulatory plug has only been partially completed. Mason *et al.* (1989, 1990) identified squalene as a component of the male sex recognition system in *Thamnophis*. Shine *et al.* (2005a) used squalene in field tests of female attractivity and were able to render sexually attractive females transiently unattractive, approximating what is observed in newly mated females. Interestingly, parallel findings have also come from studying the role of female cloacal secretions in the mating behavior of male Brown Tree Snakes (*Boiga irregularis*). Rather than copulatory plug compounds, female cloacal secretions have been shown to decrease male courtship intensity and duration (Greene and Mason 2000; Greene *et al.* 2003). This phenomenon is analogous to the conflicting chemical stimuli delivered by male plethodontid salamanders during courtship (PRF vs. PMF; Rollmann *et al.* 1999; Houck *et al.* 2007). Male *B. irregularis*, unlike male *Thamnophis*, are much larger than the females, and female production and expression of inhibitory pheromones serves to fend off unwanted male courtship. The fact that female *B. irregularis* actively solicit courtship from dominant males supports the evolution of this form of mating system (Greene and Mason 1998). Identification of the nature of these compounds in snakes would uncover the evolutionary trajectory of inhibitory compounds on the courtship behavior of not only snakes but in other terrestrial tetrapods.

13.3.1.3 She-males

One of the more peculiar aspects to the Manitoba garter snake mating system is the existence of "she-males" (male garter snakes that produce female pheromone and are courted as if they were females; Mason and Crews 1985, 1986). Although male garter snakes have been shown to exhibit male-oriented courtship in the laboratory (e.g., Noble 1937, Vagvolgyi and Halpern 1983), the Manitoba populations of *Thamnophis sirtalis parietalis* are the only ones in which she-males are consistently observed. Originally, it was suggested that she-males gained a competitive mating advantage in mating balls, where numerous males attempt to copulate with a single female (Aleksiuk and Gregory 1974; Mason and Crews 1985, 1986). By smelling like a female, she-males could then distract other males in the mating ball and thus gain a better position in which to copulate with the female over competitors (Mason and Crews 1985; 1986). Later work showed that in pairwise choice tests, male garter snakes do not distinguish between female and she-male chemical trails, suggesting that the chemical composition of the pheromone produced by she-males and females is similar enough to prevent discrimination (LeMaster and Mason 2001a). However, more recent work has shown that most males, upon emergence, smell similar to and are briefly courted as if they are females (Shine *et al.* 2000a, 2001). These most recent studies conclude that the advantage to smelling like a female is related to thermoregulation and predation. Cold she-males emerging at near-freezing body temperatures receive a thermal energy benefit via heat transfer by conduction from courting males (Shine *et al.* 2001). Warmer males that emerged earlier transfer some of their body heat to these newly emerged she-males. In addition, due to the she-males' very low body temperatures at emergence, they are more vulnerable to predators because they are unable to move very quickly. By attracting male courtship upon the she-male's emergence, she-males are covered by courting males which reduces their exposure to predation. By transferring warmth from the courting males they are also able to warm up faster and initiate courtship behavior sooner. Overall, however, the intensity of courtship received by she-males is slight in comparison to that received by large females, especially when large females and she-males are present in the same mating ball or experimental trial (Shine *et al.* 2000a; Shine *et al.* 2003b).

In biochemical analyses, the amount and quality of female pheromone on the skin of she-males diminishes with the amount of time spent at the den, suggesting that most, if not all males express female pheromone immediately upon emergence (LeMaster *et al.* 2008). Earlier, it was demonstrated that the skin of she-males exhibited higher aromatase activity than the skin of normal males, suggesting that localized formation of estrogens in the skin contributed to the feminization of she-male skin lipids (Mason and Crews 1986). Current work in the Mason laboratory has demonstrated that aromatase activity is central to the feminization of the skin and to the production and expression of the female sexual attractiveness pheromone. This suggests that the ability of some males in the population to persist in their attractiveness may be the result of a female-organized skin, perhaps occurring during development (Parker and Mason, unpubl. data). The working hypothesis is that the organization of the skin as a specific pheromone-producing organ occurs during development, most likely directed by steroid hormones experienced during gestation. Additional hormonal manipulations and experimental evidence are needed to fully support this model and to elucidate how it fits into the activation-organization scheme of vertebrate reproductive development (Phoenix *et al.* 1959).

13.3.2 Pheromones and Other Sensory Modalities

Studies of more ecologically unique snakes have provided valuable insights into how the properties of sex pheromones affect their persistence and accessibility in nature. An excellent example comes from a set of studies on the mating behavior of sea snakes. On land, Banded Sea Kraits (genus *Laticauda*) exhibit stereotypical courtship behavior involving chin-pressing, rapid tongue-flicks, and body alignment as males attempt to maintain contact with and court females (Shine *et al.* 2002). The skin lipid profiles of two species of Banded Sea Kraits from that study, *Laticauda colubrina* and *L. frontalis*, are compositionally different, but in the ocean, such cues are only accessible to the snakes over very short distances (Shine 2005). Thus, the properties of the putative pheromones in these *Laticauda* prevent mate location over long distances, which appears to be accomplished instead by visual cues (Shine 2005). Once on land however, where sea kraits aggregate to mate in groups of hundreds of individuals, female skin-derived sexual attractiveness pheromones play a remarkably similar role to that found in garter snakes. Interestingly, in studies of the Turtle-Headed Sea Snake (*Emydocephalus annulatus*), skin lipid pheromones play only a minor role in mate choice (Shine 2005). Because these sea snakes are entirely aquatic, skin lipid pheromones would be of little use in locating potential mates at a distance and thus, visual cues predominate in this species (Shine 2005). Even so, skin lipid pheromones are involved even in this group of highly specialized snakes, albeit only at close range and after the female has been located. There is no evidence that these pheromones play any role in trailing behavior.

The sexual attractiveness pheromone of *Thamnophis sirtalis parietalis* is nonvolatile in nature, and as discussed earlier, events during the breeding season can affect the persistence of these cues in nature and necessitate reliance on other cues, such as visual information, for mate location (Shine *et al.* 2005b). One recent study has suggested that airborne volatile cues may be used in the mate location behavior of a colubrid relative of garter snakes, the Northern Water Snake (*Nerodia sipedon*). That study found that males preferentially oriented toward aerial cages containing females (Aldridge *et al.* 2005). This finding is interesting because it suggests that olfaction, the other primary chemical sense system in snakes, is playing a role in orchestrating reproductive behavior. This work warrants further testing of the hypothesis on the role of volatile chemical cues and olfaction, as opposed to vomerolfaction, in snake reproduction. This line of research is especially compelling given that there is an increasing realization that airborne chemical cues play a more important role in snake prey trailing behavior than was previously appreciated (e.g., Begun *et al.* 1988; Waters 1993; Parker and Kardong 2005).

13.4 BROAD THEMES IN SNAKE REPRODUCTIVE BEHAVIOR AND THE ROLE OF PHEROMONES

There is a wealth of natural history observations on courtship behavior and general characteristics of male/female interactions in snakes. The most useful resources are all reviews (Davis 1936; Carpenter and Ferguson 1977; Gillingham 1987; Mason 1992; Mason *et al.* 1998). Since the last publication, several authors have described similar observations on field reproductive behaviors in the same species or in closely related species to those previously described. Thus, in this section, we will focus on publications that have described courtship behavior in distantly related groups that could reveal the role of pheromones in snake reproductive behavior.

In Davis's review (1936), he defines two general classifications of snake reproductive behavior: *Natrix* type and *Coluber* type. The former is typified by males maintaining close contact with the female throughout courtship until copulation and "requires" chin-pressing/rubbing to facilitate female receptivity. The latter resembles male-male combat behavior of snakes and is often referred to as a "dance," with darting and weaving of the female following *Natrix* type courtship where the male rapidly tongue-flicks and slowly progresses along the female's body. Once the male catches up with the evasive female, he rapidly "restrains her" by wrapping his body around hers and often results in neck-biting preceding copulation.

Both types of snake courtship behavior are initiated by tongue-flicking chemical cues produced by the female in her integument. Snakes exhibiting *Natrix* type mating require chemical information for the duration of courtship whereas conspecific chemical cues are important for mate location and recognition prior to engaging in courtship behavior in *Coluber* type mating. In instances where the female breaks from the male during *Coluber* type mating, a combination of visual and chemical cues likely facilitates relocation.

Carpenter and Ferguson (1977) wrote an excellent review of the behaviors involved in snake courtship and mating for a volume of *Biology of the Reptilia*. In that review, the authors surveyed the literature and noted the groups and species where tongue-flicking and chin-rubbing/pressing was part of the reported courtship behavior. More recently, Gillingham (1987) advanced our understanding of snake courtship behavior and divided it into three distinct phases: precourtship, courtship, and postmating. Male trailing behavior of female chemical trails would fall under precourtship, and it is the type of behavior for which the most accurate, consistent field and laboratory data have been gathered. This is due largely to the cryptic nature of snakes that prevents field observation of courtship, as noted by Gillingham, Carpenter, and others who have spent a great deal of time and effort to study snake reproductive behavior in nature. As noted by Gillingham (1987), there are grave inconsistencies in field accounts of "courtship behavior" because most observations lack consistency in terminology and the necessary detail for these observations to be useful in interspecific comparisons. We have chosen instead to highlight the overlying themes from the literature that relate directly to the role of pheromones in reproductive behavior in snakes. We have placed these hypotheses in the current phylogenetic framework for the major groups within the Ophidia (Lee *et al.* 2007; Vidal *et al.* 2007).

13.4.1 Chemical Cue Use during Trailing

The Pythonidae, Caenophidia (Colubroidea and Acrochordidae), and Boidae—thus, the bulk of Macrostomata—all use chemical information during trailing (precourtship behavior) to identify sex and species of the individual or trail encountered and induce appropriate behavioral responses. Therefore, there are likely sex- and species-specific chemical signals or pheromones on the skin of snakes in these groups that enable such discrimination. To date, the biochemistry of such critical compounds has only been rigorously studied in two species: *Thamnophis sirtalis* and *Boiga irregularis*.

13.4.2 Chemical Cue Use during Courtship

Only certain groups (Viperidae, Natricinae, Elapidae) utilize chemical cues throughout the duration of courtship until copulation, specifically during chin-rubbing/pressing and body alignment. The Pythonidae and Boidae appear to use primarily tactile information following mate identification until copulation. Therefore, the reliance on chemical cues in Viperidae, Natricinae, and Elapidae throughout courtship may be a derived state.

13.4.3 Chemical Cue Use by Basal Snake Groups

To date, little is known about the reproductive chemical ecology of Scolecophidia and the Uropeltidae. Only one study, to our knowledge, has examined the conspecific trailing behavior of Scolecophidians (Gehlbach *et al.* 1971), and no detailed account of the courtship behavior in either group of snakes exists. Given the position of these groups within the Ophidia, detailed studies on both the courtship behavior and chemical ecology of these species are needed. Although the Boidae and Pythonidae are thought of as exhibiting ancestral behavioral states and thus warrant significantly more study in the future, it may be even more insightful to examine the behavioral ecology of the fossorial Scolecophidia. Because of the paucity of information on the reproductive biology and chemical ecology of these groups, any studies in this regard would be most valuable.

13.5 FUTURE DIRECTIONS

The lack of detailed studies on both the chemical ecology and reproductive behavior in many representative groups of snakes is substantial. However, there are interesting aspects that can be pursued in systems where pheromones are either known to be central in reproduction (e.g., *Thamnophis sirtalis parietalis*) or directly involved in the entirety of the male-female interaction (e.g., pit vipers). One area of study that has been almost completely ignored in reptiles is the relationship between internal (hormones) and external (pheromones) chemical signals. Much work has been done recently in the Mason lab to describe the interaction between sex steroid hormones and pheromone production, but those studies have been conducted solely on one species (*T. s. parietalis*; Parker, unpubl. data). Although mechanisms for steroid action on the pheromone producing capacity of the skin are likely conserved in reptiles, nuances in the chemical constituents affected by treatment with the same steroid hormones are likely to exist.

One area of future research that seems especially promising in a number of snake species is the observation that there is undoubtedly a link between the process of ecdysis and the production of pheromones. Indeed, females of many snake species have been observed to elicit courtship from males, sometimes multiple males, at or near the time of ecdysis (e.g., Greene and Mason 1998; 2000; Panger and Greene 1998; Marques *et al.* 2009; pit vipers, Brown 1995; McCartney and Gregory 1988; Coupe 2002; Hill and Beaupre 2008). Recent work from the Mason lab has demonstrated that the most attractive components of the female sex pheromone of *Thamnophis sirtalis parietalis* (unsaturated methyl ketones weighing 504, 518, and 532 Da) are extremely abundant in female sheds, regardless of season. These lines of evidence enable us to suppose that pheromonal compounds have arisen and are derived from the processes occurring in the developing skin and eventually expressed during shedding. Application of common, modern molecular and microscopy techniques could readily reveal the physiological mechanisms involved in the synthesis of those chemical compounds that act as potent sexual signals in snakes.

Finally, the chemical ecology of unusual reptile species is starting to come into focus. A recent study by López and Martín (2009) demonstrated that male amphisbaenians react with aggression to precloacal secretions from other males. The authors isolated the major constituent in these secretions and found it to be squalene (López and Martín 2009; see section 13.3.1.2). Thus, the most basic chemical signal enabling sex recognition in one form of reptile may have been discovered and tested using a simple and appropriate bioassay. This is the key component of any pheromone study in any reptile, not just snakes. The bioassay is the most difficult part of any study because the identification of active semiochemicals or pheromones is crucially tied to the bioassay. Most investigators use a response-guided strategy wherein the behavioral response of the organisms to isolated chemical constituents helps to guide the identification of any putative pheromones. Thus, non-specific behaviors or generic behaviors are of little use in helping to identify pheromones or elucidate the chemical ecology of an organism. In the study of the reproductive biology of species relying heavily on pheromones, identification of those pheromones is the necessary first step before research integrating reproductive physiology, morphology and behavior can be effectively conducted. The field of reproductive chemical ecology has been slowly expanding in reptiles, and we hope that the data and observations in the groups we have highlighted in this chapter will pique the interest of researchers new and seasoned alike to further investigate the role that chemical cues and pheromones play in orchestrating snake reproduction.

13.6 ACKNOWLEDGMENTS

Funding for the writing of this review was partially provided by an NSF grant to RTM (0620135-IOB). Our work on the Canadian populations of Red-sided Garter Snakes, which has comprised much of this review, was made possible by the generosity and support of Manitoba Conservation and several people in Manitoba, Canada: William Watkins, David Roberts, Al and Gerry Johnson.

13.7 LITERATURE CITED

Aldridge, R., Bufalino, A. and Reeves, A. 2005. Pheromone communication in the watersnake, *Nerodia sipedon*: a mechanistic difference between semi-aquatic and terrestrial species. American Midland Naturalist 154: 412-422.

Aleksiuk, M. and Gregory, P. 1974. Regulation of seasonal mating behavior in *Thamnophis sirtalis parietalis*. Copeia 1974: 681-689.

Andrén, C. 1982. The role of the vomeronasal organs in the reproductive behavior of the adder *Vipera berus*. Copeia 1982: 148-157.

Andrén, C. 1986. Courtship, mating and agonistic behaviour in a free-living population of adders, *Vipera berus* (L.). Amphibia-Reptilia 7: 353-383.

Baumann, F. 1929. Experimente über den Geruchssinn und den Beuteerwerb der Viper (*Vipera aspis*). Journal of Comparative Physiology A: Neuroethology, Sensory, Neural, and Behavioral Physiology 10: 36-119.

Baxi, K., Dorries, K. and Eisthen, H. 2006. Is the vomeronasal system really specialized for detecting pheromones? Trends in Neurosciences 29: 1-7.

Begun, D., Kubie, J., O'Keefe, M. and Halpern, M. 1988. Conditioned discrimination of airborne odorants by garter snakes (*Thamnophis radix* and *T. sirtalis sirtalis*). Journal of Comparative Psychology 102: 35-43.

Brown, W. 1995. Heterosexual groups and the mating season in a northern population of timber rattlesnakes, *Crotalus horridus*. Herpetological Natural History 3: 127-133.

Burghardt, G. 1967. Chemical-cue preferences of inexperienced snakes: comparative aspects. Science 157: 718-720.

Burghardt, G. 1969. Comparative prey-attack studies in newborn snakes of the genus *Thamnophis*. Behaviour 33: 77-114.

Cardwell, M. 2008. The reproductive ecology of Mohave rattlesnakes. Journal of Zoology 274: 65-76.

Carpenter, C. and Ferguson, G. 1977. Variation and evolution of stereotyped behavior in repiles. Pp. 335–554. In C. Gans and D. Tinkle (eds), *Biology of the Reptilia*, vol. 7. Academic Press, London.

Cooper, W. 1998. Evaluation of swab and related tests as a bioassay for assessing responses by squamate reptiles to chemical stimuli. Journal of Chemical Ecology 24: 841-866.

Cooper, W. and Burghardt, G. 1990. Vomerolfaction and vomodor. Journal of Chemical Ecology 16: 103-105.

Coupe, B. 2002. Pheromones, search patterns, and old haunts: how do male timber rattlesnakes (*Crotalus horridus*) locate mates? Pp. 139-148. In G. W. Schuett, M. Höggren, M. E. Douglas and H. W. Greene (eds), *Biology of the Vipers*. Eagle Mountain Publishing, Eagle Mountain, Utah.

Crews, D., Camazine, B., Diamond, M., Mason, R., Tokarz, R. and Garstka, W. 1984. Hormonal independence of courtship behavior in the male garter snake. Hormones and Behavior 10: 29-41.

Davis, D. 1936. Courtship and mating behavior in snakes. Field Museum of Natural History Publications in Zoology 20: 257-290.

Devine, M. 1975. Copulatory plugs in snakes: enforced chastity. Science 187: 844-845.

Devine, M. 1977. Copulatory plugs, restricted mating opportunities and reproductive competition among male garter snakes. Nature 267: 345-346.

Duvall, D., King, M. and Gutzwiller, K. 1985. Behavioral ecology and ethology of the prairie rattlesnake. National Geographic Research 1: 80-111.

Ford, N. 1981. Seasonality of pheromone trailing behavior in two species of garter snake, *Thamnophis* (Colubridae). The Southwestern Naturalist 26: 385-388.

Ford, N. 1982. Species specificity of sex pheromone trails of sympatric and allopatric garter snakes (*Thamnophis*). Copeia 1982: 10-13.

Ford, N. 1986. The role of pheromone trails in the sociobiology of snakes. Pp. 261-278. In D. Duvall, D. Mueller-Schwarze and R. Silverstein (eds), *Chemical Signals in Vertebrates 4*. Plenum Press, New York.

Ford, N. 1995. Experimental design in studies of snake behavior. Herpetological Monographs 9: 130-139.

Ford, N. and Low, J. 1984. Sex pheromone source location by garter snakes. Journal of Chemical Ecology 10: 1193-1199.
Ford, N. and O'Bleness, M. 1986. Species and sexual specificity of pheromone trails of the garter snake, *Thamnophis marcianus*. Journal of Herpetology 20: 259-262.
Fornasiero, S., Bresciani, E., Dendi, F. and Zuffi, M. 2007. Pheromone trailing in male European whip snake, *Hierophis viridiflavus*. Amphibia-Reptilia 28: 555-559.
Gehlbach, F., Watkins, J. and Kroll, J. 1971. Pheromone trail-following studies of typhlopid, leptotyphlopid, and colubrid snakes. Behaviour 40: 282-294.
Gillingham, J. 1987. Social behavior. Pp. 184–209. In R. Seigel, J. Collins and S. Novak (eds), *Snakes: Ecology and Evolutionary Biology*. McGraw-Hill, New York.
Gillingham, J. and Dickinson, J. 1980. Postural orientation during courtship in the eastern garter snake, *Thamnophis s. sirtalis*. Behavioral and Neural Biology 28: 211-217.
Gove, D. 1979. A comparative study of snake and lizard tongue-flicking, with an evolutionary hypothesis. Zeitschrift fur Tierpsychologie 51: 58-76.
Greene, M. and Mason, R. 1998. Chemically mediated sexual behavior of the brown tree snake, *Boiga irregularis*. Ecoscience 5: 405-409.
Greene, M. and Mason, R. 2000. Courtship, mating, and male combat of the brown tree snake, *Boiga irregularis*. Herpetologica 56: 166-175.
Greene, M., Stark, S. and Mason, R. 2001. Pheromone trailing behavior of the brown tree snake, *Boiga irregularis*. Journal of Chemical Ecology 27: 2193-2201.
Halpern, M. 1992. Nasal chemical senses in reptiles: structure and function. Pp. 423-523. In C. Gans and D. Crews (eds), *Biology of the Reptilia*, vol. 18. University Of Chicago Press, Chicago, Illinois.
Hill, J., III and Beaupre, S. 2008. Body size, growth, and reproduction in a population of western cottonmouths (*Agkistrodon piscivorus leucostoma*) in the Ozark Mountains of northwest Arkansas. Copeia 2008: 105-114.
Houck, L., Palmer, C., Watts, R., Arnold, S., Feldhoff, P. and Feldhoff, R. 2007. A new vertebrate courtship pheromone, PMF, affects female receptivity in a terrestrial salamander. Animal Behaviour 73: 315-320.
Inouchi, J., Wang, D., Jiang, X., Kubie, J. and Halpern, M. 1993. Electrophysiological analysis of the nasal chemical senses in garter snakes. Brain, Behavior and Evolution 41: 171-182.
Jiang, X., Inouchi, J., Wang, D. and Halpern, M. 1990. Purification and characterization of a chemoattractant from electric shock-induced earthworm secretion, its receptor binding, and signal transduction through the vomeronasal system of garter snakes. Journal of Biological Chemistry 265: 8736-8744.
Kahmann, H. 1932. Sinnesphysiologische studien an Reptilien I: Exprimentelle untersuchungen über das Jakobson'sche Organ der eidechsen und schlangen. Zoologische Jahrbücher, Abteilungen Allgemeine Zoologische Physiologie Tiere 51: 173-238.
Karlson, P. and Lüscher, M. 1959. 'Pheromones': a new term for a class of biologically active substances. Nature 183: 55-56.
Kubie, J., Vagvolgyi, A. and Halpern, M. 1978. Roles of the vomeronasal and olfactory systems in courtship behavior of male garter snakes. Journal of Comparative Physiology and Psychology 92: 627-641.
Lee, M., Hugall, A., Lawson, R. and Scanlon, J. 2007. Phylogeny of snakes (Serpentes): combining morphological and molecular data in likelihood, Bayesian and parsimony analyses. Systematics and Biodiversity 5: 371-389.
LeMaster, M. and Mason, R. 2001a. Evidence for a female sex pheromone mediating male trailing behavior in the red-sided garter snake, *Thamnophis sirtalis parietalis*. Chemoecology 11: 149-152.

LeMaster, M. and Mason, R. 2001b. Annual and seasonal variation in the female sexual attractiveness pheromone of the red-sided garter snake, *Thamnophis sirtalis parietalis*. Pp. 369-376. In A. Marchlewska-Koj, J. Lepri and D. Muller-Schwarze (eds), *Chemical Signals in Vertebrates 9*. Kluwer Academic/Plenum Publishers, New York.

LeMaster, M. and Mason, R. 2002. Variation in a female sexual attractiveness pheromone controls male mate choice in garter snakes. Journal of Chemical Ecology 28: 1269-1285.

LeMaster, M. and Mason, R. 2003. Pheromonally mediated sexual isolation among denning populations of red-sided garter snakes, *Thamnophis sirtalis parietalis*. Journal of Chemical Ecology 29: 1027-1043.

LeMaster, M., Moore, I. and Mason, R. 2001. Conspecific trailing behaviour of red-sided garter snakes, *Thamnophis sirtalis parietalis*, in the natural environment. Animal Behaviour 61: 827-833.

LeMaster, M., Stefani, A., Shine, R. and Mason, R. 2008. Cross-dressing in chemical cues: exploring "she-maleness" in newly emerged male garter snakes. Pp. 223-231. In J. Hurst, R. Beynon, S. Roberts and T. Wyatt (eds), *Chemical Signals in Vertebrates 11*. Springer, London.

López, P. and Martín, J. 2009. Potential chemosignals associated with male identity in the amphisbaenian *Blanus cinereus*. Chemical Senses 34: 479-486.

Marques, O., Almeida-Santos, S., Rodrigues, M. and Camargo, R. 2009. Mating and reproductive cycle in the neotropical colubrid snake *Chironius bicarinatus*. South American Journal of Herpetology 4: 76-80.

Martin, D. 1984. An instance of sexual defense in the cottonmouth, *Agkistrodon piscivorus*. Copeia 1984: 772-774.

Martín, J., Moreira, P. and López, P. 2007. Status-signalling chemical badges in male Iberian rock lizards. Functional Ecology 21: 568-576.

Mason, R. 1992. Reptilian pheromones. Pp. 114-228. In C. Gans and D. Crews (eds), *Biology of the Reptilia*, vol. 18. The University of Chicago Press, Chicago, Illinois.

Mason, R. and Crews, D. 1985. Female mimicry in garter snakes. Nature **316:** 59-60

Mason, R. and Crews, D. 1986. Pheromone mimicry in snakes. Pp. 279-283. In D. Duvall, D. Muller-Schwarze and R. Silverstein (eds), *Chemical Signals in Vertebrates 4*. Plenum Press, New York.

Mason, R., Chinn, J. and Crews, D. 1987. Sex and seasonal differences in the skin lipids of garter snakes. Comparative Biochemistry and Physiology B 87: 999-1003.

Mason, R., Chivers, D., Mathis, A. and Blaustein, A. 1998. Bioassays with Amphibians and Reptiles. Pp. 271-325. In J. Millar and K. Haynes (eds), *Methods in Chemical Ecology*. Chapman and Hall, New York.

Mason, R., Fales, H., Jones, T., Pannell, L., Chinn, J. and Crews, D. 1989. Sex pheromones in snakes. Science 245: 290-293.

Mason, R., Jones, T., Fales, H., Pannell, L. and Crews, D. 1990. Characterization, synthesis, and behavioral responses to sex attractiveness pheromones of red-sided garter snakes (*Thamnophis sirtalis parietalis*). Journal of Chemical Ecology 16: 2353-2369.

McCartney, J. and Gregory, P. 1988. Reproductive biology of female rattlesnakes (*Crotalus viridus*) in British Columbia. Copeia 1988: 47-57.

Mendonca, M. and Crews, D. 2001. Control of attractivity and receptivity in female red-sided garter snakes. Hormones and Behavior 40: 43-50.

Moore, I., LeMaster, M. and Mason, R. 2000. Behavioural and hormonal responses to capture stress in the male red-sided garter snake, *Thamnophis sirtalis parietalis*. Animal Behaviour 59: 529-534.

Noble, G. K. 1937. The sense organs involved in the courtship of *Storeria, Thamnophis* and other snakes. Bulletin of the American Museum of Natural History 73: 673-725.
Noble, G. K and Clausen, H. 1936. The aggregation behavior of *Storeria dekayi* and other snakes, with special reference to the sense organs involved. Ecological Monographs 6: 269-316.
O'Donnell, R., Ford, N., Shine, R. and Mason, R. 2004. Male red-sided garter snakes, *Thamnophis sirtalis parietalis,* determine female mating status from pheromone trails. Animal Behaviour 68: 677-683.
Panger, M. and Greene, H. 1998. *Micrurus nigrocinctus* (coral snake): reproduction. Herpetological Review 29: 46.
Parker, M. and Kardong, K. 2005. Rattlesnakes can use airborne cues during post-strike prey relocation. Pp. 397-402. In R. Mason, M. LeMaster and D. Müller-Schwarze (eds), *Chemical Signals in Vertebrates 10*. Springer, New York.
Parker, M. and Mason, R. 2009. Low temperature dormancy affects the quantity and quality of the female sexual attractiveness pheromone in red-sided garter snakes. Journal of Chemical Ecology 35: 1234-1241.
Phoenix, C., Goy, R., Gerall, A. and Young, W. 1959. Organizing action of prenatally administered testosterone propionate on the tissues mediating mating behavior in the female guinea pig. Endocrinology 65: 369-382.
Rollmann, S., Houck, L. and Feldhoff, R. 1999. Proteinaceous pheromone affecting female receptivity in a terrestrial salamander. Science 285: 1907-1909.
Ross, P. and Crews, D. 1977. Influence of the seminal plug on mating behaviour in the garter snake. Nature 267: 344-345.
Schuett, G. and Gillingham, J. 1988. Courtship and mating of the copperhead, *Agkistrodon contortrix*. Copeia 1988: 374-381.
Shine, R. 2005. All at sea: aquatic life modifies mate-recognition modalities in sea snakes (*Emydocephalus annulatus*, Hydrophiidae). Behavioral Ecology and Sociobiology 57: 591-598.
Shine, R., O'Connor, D. and Mason, R. 2000a. Female mimicry in garter snakes: behavioural tactics of "she-males" and the males that court them. Canadian Journal of Zoology 78: 1391-1396.
Shine, R., Olsson, M. and Mason, R. 2000b. Chastity belts in gartersnakes: the functional significance of mating plugs. Biological Journal of the Linnean Society 70: 377-390.
Shine, R., Phillips, B., Waye, H., LeMaster, M. and Mason, R. 2001. Benefits of female mimicry in snakes. Nature 414: 267.
Shine, R., Reed, R., Shetty, S., LeMaster, M. and Mason, R. 2002. Reproductive isolating mechanisms between two sympatric sibling species of sea snakes. Evolution 56: 1655-1662.
Shine, R., Phillips, B., Waye, H., LeMaster, M. and Mason, R. 2003a. Chemosensory cues allow courting male garter snakes to assess body length and body condition of potential mates. Behavioral Ecology and Sociobiology 54: 162-166.
Shine, R., Langkilde, T. and Mason, R. 2003b. Confusion within 'mating balls' of garter snakes: does misdirected courtship impose selection on male tactics? Animal Behaviour 66: 1011-1017.
Shine, R., Langkilde, T., Wall, M. and Mason, R. 2005a. Do female garter snakes evade males to avoid harassment or to enhance mate quality? American Naturalist 165: 660-668.
Shine, R., Webb, J., Lane, A. and Mason, R. 2005b. Mate location tactics in garter snakes: effects of rival males, interrupted trails and non-pheromonal cues. Ecology 19: 1017-1024.

Slip, D. and Shine, R. 1988. The reproductive biology and mating system of diamond pythons, *Morelia spilota* (Serpentes: Boidae). Herpetologica 44: 396-404.
Smith, C., Schwenk, K., Earley, R. and Schuett, G. 2008. Sexual size dimorphism of the tongue in a North American pitviper. Journal of Zoology 274: 367-374.
Vagvolgyi, A. and Halpern, M. 1983. Courtship behavior in garter snakes: effects of artificial hibernation. Canadian Journal of Zoology 61: 1171-1174.
Vidal, N., Delmas, A., David, P., Cruaud, C., Couloux, A. and Hedges, S. 2007. The phylogeny and classification of caenophidian snakes inferred from seven nuclear protein-coding genes. Comptes rendus-Biologies 330: 182-187.
Wang, D., Chen, P., Liu, W., Li, C. and Halpern, M. 1997. Chemosignal transduction in the vomeronasal organ of garter snakes: Ca^{2+}-dependent regulation of adenylate cyclase. Archives of Biochemistry and Biophysics 348: 96-106.
Waters, R. 1993. Odorized air current trailing by garter snakes, *Thamnophis sirtalis*. Brain Behavior and Evolution 41: 219-219.
Whittier, J., Mason, R. and Crews, D. 1985. Mating in the red-sided garter snake, *Thamnophis sirtalis parietalis*: differential effects on male and female sexual behavior. Behavioral Ecology and Sociobiology 16: 257-261.
Young, B. 1993. Evaluating hypotheses for the transfer of stimulus particles to Jacobson's organ in snakes. Brain, Behavior and Evolution 41: 203-209.

CHAPTER 14

Offspring Size Variation in Snakes

Neil B. Ford[1] *and Richard A. Seigel*[2]

14.1 INTRODUCTION

Understanding the factors that cause variation in offspring size is fundamental to resolving important theoretical and conceptual issues in studies of life-history evolution. Models of selection of offspring size usually consider how (or if) offspring size correlates with offspring survival or performance (e.g., Jayne and Bennett 1990; Congdon *et al.* 1999; Janzen and Warner 2009) and/or how reproductive resources are partitioned by females into either larger offspring size or larger numbers of offspring (Smith and Fretwell 1974; Stearns 1992; Bernardo 1996; Charnov 1997; Marshall and Uller 2007; Brown and Shine 2009; Janzen and Warner 2009).

Although studies of snake reproductive biology and life history evolution have made major strides in the past 20 years (see reviews in Seigel and Ford 1987; Shine 1992, 2003, 2005), a synthesis examining patterns of offspring size variation has not been available. Part of the reason for this absence is historical; studies of the reproductive biology of snakes are often descriptive in nature and have not always attempted to place results in a context to test predications from life-history theory. Other syntheses of snake life history evolution (e.g., Shine 2003, 2005) have examined broad questions such as the evolution of viviparity and phylogenetic patterns of life-history variation, which we will not attempt here. In this chapter we will instead concentrate on factors affecting offspring size at the individual, species, and population level, with emphasis on the roles of maternal size, geographical variation, and effects of environmental factors such as prey availability. Our specific goals are as follows: (1) provide a concise overview of the importance of basic life-history theory regarding offspring size, (2) review what is known about the basic sources of variation in offspring

[1]Department of Biology, Univ. of Texas at Tyler, Tyler, Texas, 75799 USA
[2]Department of Biological Sciences, Towson Univ., Towson, Maryland, 21252 USA

size in snakes, (3) provide "case histories" of studies that have provided particularly useful insights into this life-history trait, and (4) point out unanswered questions and fruitful areas of research for future studies.

14.2 A BRIEF REVIEW OF MODELS OF OPTIMAL OFFSPRING SIZE

Our starting point for discussing patterns of offspring size variation in snakes is to first briefly review the basic concepts involved in the relationship between maternal fitness and offspring fitness. Most life history models assume (1) that mothers have control over how their energy resources are allocated between offspring number and size (e.g., Smith and Fretwell 1974; Brockelman 1975; Brown and Shine 2009; Uller and Olsson 2009) and (2) that selection occurs for an optimal offspring size that will produce the highest fitness for the parent (Roff 1992; Stearns 1992; Charnov 1997). However, this size may not necessarily maximize offspring fitness, thus resulting in the potential for parent-offspring conflict (e.g., Bernardo 1996; Janzen and Warner 2009). Because models generally assume energy devoted to reproduction is fixed, females must trade-off offspring size with number to produce the highest number of the fittest offspring over her lifetime (Smith and Fretwell 1974; Brockelman 1975; Uller and Olsson 2007). In addition to the fitness consequences of offspring size, knowledge of the costs to the female as a function of reproduction is essential to developing realistic models (e.g., Shine 1980; Reznick 1992; Niewiarowski and Dunham 1994). For snakes and lizards, the costs of reproduction include reduction in locomotor ability (Shine 1980; Seigel *et al.* 1987; Brodie 1989) reduced growth rates, (Madsen and Shine 1993; Gregory *et al.* 1999; Lourdais *et al.* 2002), increased risk of death (Madsen and Shine 1992, 1994; Luiselli *et al.* 1996, 1997; Plummer 1997; Gregory *et al.* 1999; Brown and Weatherhead 2004) and reductions in future fecundity (Bell 1980; Shine 1980).

In addition to maternal and offspring fitness, trade-offs, and reproductive costs, two other issues important in understanding offspring size variation in snakes are morphological constraints imposed by the size of the mother and the degree to which offspring size is heritable versus environmentally induced. Both of these will be discussed in detail below.

14.3 OFFSPRING SIZE AND OFFSPRING FITNESS IN SNAKES

The first point we address is whether larger offspring have higher fitness than smaller offspring. Intuitively, we might expect that larger offspring have an advantage in survival due to an increased locomotor ability which may enhance their ability to escape predators, shorter time to sexual maturity due to larger initial size, and, especially in gape-limited predators such as snakes, enhanced foraging success since they can subdue and swallow a wider variety of prey (see Smith and Fretwell 1974; Arnold 1993; Kissner and Weatherhead 2005; Janzen and Warner 2009). The hypothesis that larger offspring have higher survival rates has received considerable attention in lizards (e.g, Ferguson and Fox 1984; Ferguson *et al.* 1990) and is a point of some contention in turtles (Congdon *et al.* 1999; Janzen *et al.* 2000). However, until recently, relatively little actual information existed on survival of neonate snakes, let alone how variation in size is correlated with survival. To date, results of both field and laboratory or mesocosm studies have provided mixed results. In a common-garden experiment, Bronikowski (2000) showed that larger neonate mass was correlated with pre-hibernation survival in lab-reared Western Terrestrial Garter Snakes (*Thamnophis elegans*). Conversely, Ji *et al.* (2009) produced small or large eggs in different female Cobras (*Naja atra*) by either giving them exogenous FSH or by ablating some yolking follicles. In these lab-manipulated clutches, the size of the eggs had no effect on hatching rate or growth of the hatchlings for 240 days following emergence. Kissner and Weatherhead (2005) used outdoor enclosures to examine overwinter survival in neonate Northern Water Snakes (*Nerodia sipedon*) and found a strong association between neonate mass and survival.

At least two field studies suggested that larger offspring have higher survival than smaller snakes. First, Jayne and Bennett (1990) found directional selection favoring larger individuals in at least one cohort of the Common Garter Snake (*Thamnophis sirtalis*) in California. More recently, Brown and Shine (2005) found that there was a positive correlation between offspring SVL and probability of recapture rate in Keelback Snakes (*Tropidonophis mairii*) in Australia. However, no differences were found in overwinter survival for Western Rattlesnakes (*Crotalus viridis*) of different sizes (Charland 1989), and Madsen and Shine (1998) found no correlation between neonate size and probability of recapture in the tropical Water Python (*Liasis fuscus*). Finally, even for the correlation between offspring size and probability of recapture rates in *Tropidonophis mairii*, the majority of variation in offspring survival was left unexplained (Brown and Shine 2005). Obviously, more empirical data are required before any assumptions can be made about the advantage to the individual offspring of being born large. Given the difficulty in obtaining recapture data on many neonate snakes in the wild, we suggest that an increased use of radiotelemetry on neonates (e.g., Jellen and Kowalski 2007), increased use of mesocosm experiments (Kissner and Weatherhead 2005), and increased emphasis of laboratory tests on how initial neonate size affects both age at maturity and foraging success will help amass data on this subject more rapidly.

14.4 MAJOR FACTORS AFFECTING OFFSPRING SIZE IN SNAKES

14.4.1 Maternal Size and Clutch Size

The importance of maternal size on reproductive traits has long been recognized for squamates (e.g., Fitch 1970, 1985; Seigel and Ford 1987; Shine 1992; King 1993; Brown and Shine 2005). As outlined by Shine

(1992) and Brown and Shine (2005), larger females have expanded room in the abdominal cavity and can therefore accommodate larger numbers of similar-sized offspring than can smaller females. Although data allowing tests of the relationship between clutch size and female size have been reported routinely for some time (see summaries in Seigel and Ford 1987; Bonnet *et al.* 2000), data on the offspring size/maternal size relationship are less abundant. Information on the lengths and masses of individual eggs and/or offspring is especially lacking. To some extent, this represents a methodological constraint, as finding gravid snakes and then obtaining their offspring is sometimes difficult for snakes. Increasingly, investigators have recognized that holding gravid females in semi-natural outdoor enclosures or appropriate laboratory conditions for modest amounts of time allows for the collection of large amounts of reproductive data, thus allowing for much more quantitative tests of the relationships between maternal size and offspring size (e.g., Ford and Seigel 1989a; King 1993; Farrell *et al.* 1995; Gregory and Skebo 1998; Weatherhead *et al.* 1999; Brown and Shine 2005; Sparkman *et al.* 2007).

The simplest way of approaching the effects of maternal size is to use a linear regression or correlation analysis between female length or mass and offspring length or mass. Such studies often show that larger or older females have larger offspring (see example in Fig. 14.1), although there is considerable variation in this relationship (Lemen and Voris 1981; Ford and Seigel 1989b; King 1993; Farrell *et al.* 1995; Bonnet *et al.* 2000; Brown and Shine 2005; Sparkman *et al.* 2007), and the relationship is not seen at all

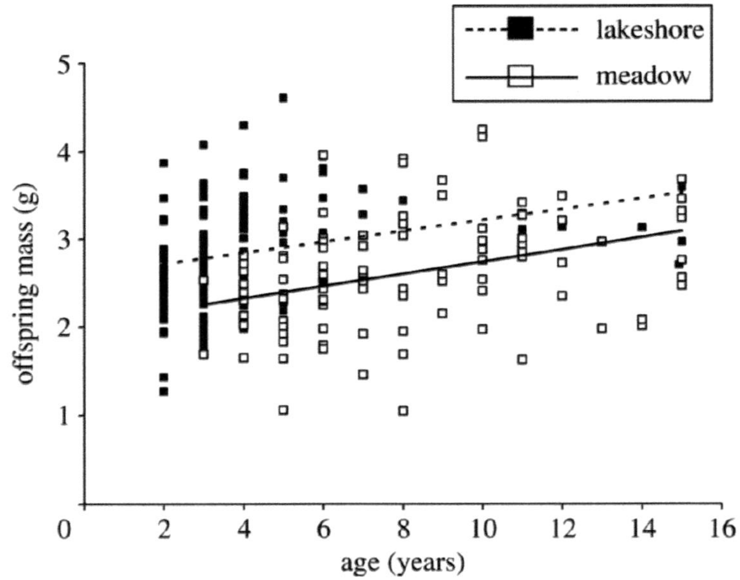

Fig. 14.1 Mean live offspring mass as a function of maternal age in two ecotypes of Wandering Garter Snakes (*Thamnophis elegans vagrans*). From Sparkman *et al.* 2007, Proceedings of the Royal Society B, 274: 943-950, Fig. 3.

in some species (e.g., *Nerodia sipedon*; Weatherhead *et al.* 1999). However, such an analysis misses the fact that other factors simultaneously affect offspring size, including maternal condition and, especially, clutch size. Simply put, larger females not only produce larger offspring, but more offspring as well. However, since producing more offspring should result in a reduction in offspring size (see references above), understanding the size of offspring that a female actually produces requires an analysis that examines the residuals of the female size-clutch size relationship and the residuals from the female size/offspring size relationship in a so-called partial correlation analysis (e.g., Ford and Seigel 1989b; Bonnet *et al.* 2001; see Fig. 14.2 for an illustration).

Although a partial correlation analysis is more useful than the simple correlations described above, some authors have attempted to model multiple variables at the same time, using a path analysis (e.g., King 1993; Shine 1996; Weatherhead *et al.* 1999; Brown and Shine 2005). Such an analysis (see Fig. 14.3 for an example) illustrates how dependent variables change (in standardized units) as independent variables change by the same amount (Brown and Shine 2005). Data from a very large sample of *Tropidonophis mairii* (Brown and Shine 2005) show that maternal SVL and maternal condition both affect egg mass, and that egg mass, incubation period, and the mass gained by eggs during incubation all affect offspring length (Fig. 14.3). Although an analysis requires a large sample size to be

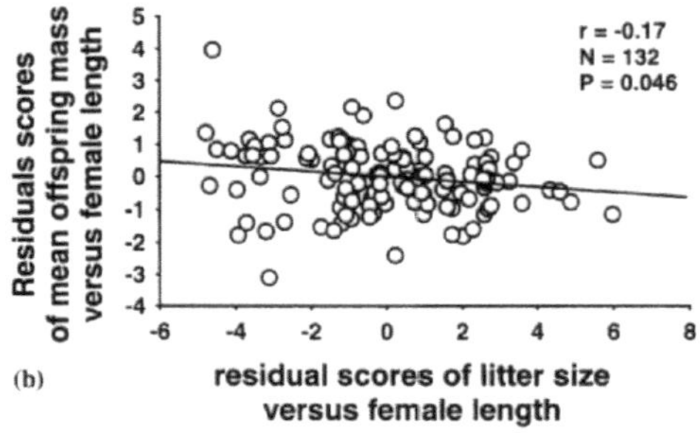

Fig. 14.2 Relationship between litter size and offspring size in *Vipera aspis* after the effect of maternal size is removed by partial correlation analysis. From Bonnet *et al.* 2001, Oikos 92: 297-308, Fig. 4.

valid and has some statistical drawbacks (see Kingsolver and Schemske 1991; Weatherhead *et al.* 1999), studies that have incorporated path analysis have been extremely useful in understanding the interrelationships between variables affecting offspring size in snakes (discussed below).

14.4.2 Environmentally-induced Variation in Offspring Size

Even after the effects of maternal size are accounted for, a fundamental question in understanding offspring size variation in snakes is whether

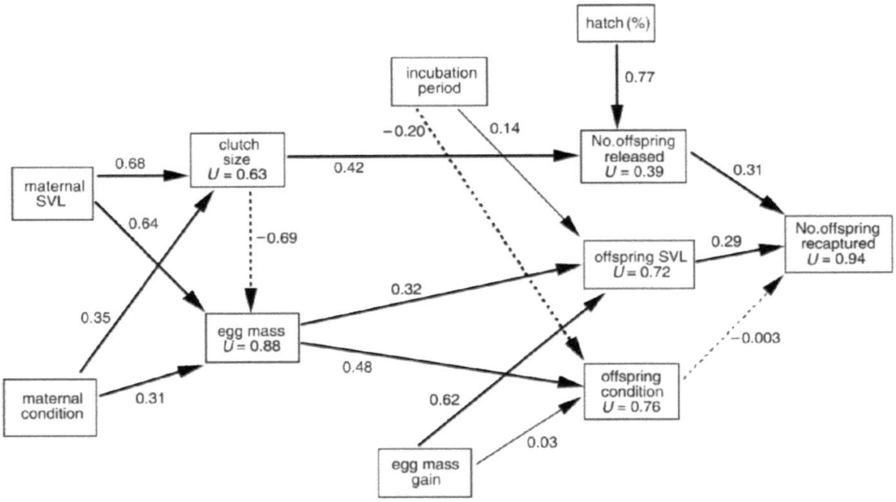

Fig. 14.3 Path diagram for factors that affect offspring size in Keelback Snakes (*Tropidonophis mairii*). Numbers beside arrows are coefficients based on the total data set and *U* is the proportion of unexplained variance. From Brown and Shine 2005. Ecology 86: 2763-2770, Fig. 1.

observed variation in offspring size represents genetic-based differences among females or whether such variation is induced by differences in the environment that those females experienced. For this question, differences in reproductive mode (oviparity versus viviparity) become critical, in that "offspring size" "neonate size" and "egg size" are not necessarily equivalent terms. For example, Brown and Shine's (2005) path analysis (Fig. 14.3) showed that egg size (mass) was correlated directly with both maternal SVL and maternal condition, but that the offspring size (neonate SVL) was correlated not only with initial egg mass, but also incubation period and egg mass gain from water uptake during incubation. Clearly then, if "offspring size" means neonate SVL or mass (as opposed to initial egg size), then environmentally induced variation is likely to be common (if not ubiquitous) in egg-laying snakes. Incubation duration and timing of parturition has also been shown to affect offspring size in viviparous snakes (King 1993; Weatherhead *et al.* 1999), but whether this is an artifact of captivity is unclear.

Incubation effects not withstanding, we still need to understand to what degree differences in initial egg mass in oviparous snakes and neonate size in newborn viviparous snakes, are determined environmentally. Much of the focus on this topic has been on the effects of differences in prey availability. The concept that increased prey availability affects reproductive traits was first documented by Andren and Nilson (1983) when they correlated higher vole densities with larger clutch sizes and increased offspring mass in Adders (*Vipera berus*) in Sweden. Numerous studies have now tested whether there is a link between food availability and offspring size in snakes, both under experimental laboratory (e.g., Ford and Seigel 1989a, 1994, 2006; Seigel and Ford 1991; Gregory and Skebo 1998) and field conditions (e.g., Weatherhead *et al.* 1999; Shine and Madsen 1997; Brown and Shine 2002; Reading 2004). Most of the experimental data fail to show any correlation between changes in female diet and offspring size, in both viviparous Checkered and Common Garter Snakes (*Thamnophis marcianus* and *T. sirtalis*, respectively) and oviparous Corn Snakes (*Pantherophis guttatus [=Elaphe guttata]*) (see references immediately above), though one population of *T. marcianus* showed a marginal difference in neonate mass from females with access to higher prey availability (Seigel and Ford 2001).

Field data on annual differences in offspring size have provided generally mixed results. Populations of *Vipera berus* (Andren and Nilson 1983) and *Nerodia sipedon* (Weatherhead *et al.* 1999) showed significant differences among years in neonate size, but no differences were seen among years in such diverse taxa as Smooth Snakes (*Coronella austriaca;* Luiselli *et al.* 1997), *Liasis fuscus* (Shine and Madsen 1997; Madsen and Shine, 2001), Cottonmouths (*Agkistrodon piscivorus;* Ford *et al.* 2004), and for a second population of *V. berus* (Madsen and Shine 1992).

When sufficient "longitudinal" data are available, a more elegant and perhaps informative way of examining environmental effects on offspring size is to examine variation among successive clutches for the same female. This has now been conducted at least five times for snakes, four times using field data (Luiselli *et al.* 1997; Bronikowski and Arnold 1999; Brown and Shine 2002, 2006, 2007; Farrell *et al.* 2009), and once in the laboratory (Ford and Seigel 2006). In general, these data indicate a high degree of repeatability in offspring size among successive clutches, indicating that individual females produce similar sized-eggs or neonates among successive clutches, especially when adjusted for body size (see Table 14.1). In addition, the repeatability scores for snakes maintained in the laboratory (where prey availability was regulated) were considerably higher than those from the field (Table 14.1), suggesting that environmental variation is playing at least some role in affecting variation in offspring size (Farrell *et al.* 2009).

Table 14.1 Summary of repeatability of reproductive traits of snakes. Values indicate R, the coefficient of intraclass correlation. Values of R close to 1 indicate that most variation is among rather than within females. Values are size-adjusted where possible. A "*" indicates $P < 0.05$; "ns" indicates a non-significant value. Table taken partially from Farrell *et al.* (2009)

Species	*Litter size*	*Offspring mass* (g)	*Source*
Lamprophis fuliginosus	0.65*	0.67*	Ford and Seigel 2006
Tropidonophis mairii	0.28*	0.26*	Brown and Shine 2007
Thamnophis elegans	0.20 ns	0.43*	Bronikowski and Arnold 1999
Sistrurus miliarius	0.42*	0.33*	Farrell *et al.* 2009

14.4.3 Geographic Variation in Offspring Size

Tests of geographic differences in life history traits have a long history for reptiles, especially lizards and turtles (Tinkle 1961; Fitch 1970, 1985; Tinkle *et al.* 1970; Moll and Legler 1971). The concept that different populations of the same species may produce different sized clutches and offspring is based on the idea that local selection pressures drive reproductive traits to maximize fitness for that geographic area. For example, the size of available prey for neonates might influence the minimum size that offspring are born. However, such geographic comparisons are confounded by several factors, especially (1) the statistical complications arising from examining multiple related traits (maternal size, clutch size, and offspring size; see above) and (2) determining whether observed differences among localities is due to local adaptation or phenotypic plasticity (Stearns 1976). For example, if clutch size or offspring size are affected by differences in prey availability, then much of the observed geographic variation in these traits may have little to do with local adaptation and may simply represent phenotypic plasticity (Seigel and Ford 1991). Thus, informative tests for geographic differences in offspring size must use appropriate statistical methods to correct for effects of maternal size and clutch size and must take the effects of phenotypic plasticity into account.

Despite these constraints, there have been several recent studies that have examined both broad-scale and micro-geographic variation in offspring size (Ji and Wang 2005; Zuffi *et al.* 2007, 2009). In two papers, Gregory and Larsen (1993, 1996) attempted to analyze the factors that might be the cause of geographic variation in offspring size found in the widely-distributed Common Garter Snake (*Thamnophis sirtalis*). Although their final conclusion was that more work is necessary, there was evidence that populations from eastern Canada produced large broods of small offspring when compared with snakes from western Canada, where larger offspring and smaller brood sizes were found. This pattern may be linked to differences in the primary prey species, with smaller offspring more common in areas where earthworms were the primary prey for neonates.

Three recent studies have taken advantage of the ease of maintaining gravid snakes in the laboratory to examine geographic differences in offspring size. On a microgeographic scale, Bronikowski (2000) used a "common-garden" design to show significant differences in offspring size and growth rates among local populations of Western Terrestrial Garter Snakes in California. On a broader geographic scale, Ji and Wang (2005) also used a common garden experiment to compare reproductive traits in island versus mainland populations of *Naja atra*. Their results confirmed that geographic differences in clutch size and clutch mass were not the result of phenotypic plasticity but that there were no geographic differences in egg mass between populations. Conversely, Seigel *et al.* (2000) showed that offspring of Checkered Garter Snakes from Arizona were significantly larger than offspring from southern Texas, and that these differences were not the result of phenotypic plasticity. However, other experiments (Seigel and Ford 2001) suggested that even the presence of phenotypic plasticity was geographically variable in these garter snakes, further showing the utility of, and the need for, additional common garden experiments to better understand life-history patterns in snakes.

14.5 CONCLUSIONS

Although snakes were once considered less than suitable subjects for ecological studies (Turner 1977), some authors now consider snakes to be "model organisms", especially for studies of reproductive biology (e.g., Seigel 1993; Shine and Bonnet 2000). Even though data on offspring size for snakes were not collected in many early studies, the combination of well-designed laboratory experiments, increased use of outdoor enclosures and radio-telemetry, and intensive mark-recapture studies has allowed a considerable recent increase in the quantity and quality of data available to test some fundamental questions in life-history theory for snakes.

Despite this greatly enlarged data set, much remains to be done. In addition to some of the areas we have already drawn attention to above (most notably offspring size versus offspring survival), we here point out some other areas for future research. For example, much remains to be learned about how offspring size is affected by the interaction between the timing of ovulation and subsequent intake of energy in viviparous snakes. After ovulation has occurred for species without a placenta, any intake of food after ovulation cannot (theoretically) impact offspring size, but exactly when the "decision" is made about how much energy to allocate to each embryo is poorly known (Lourdais *et al.* 2003). In addition, how the available energy is allocated to the individual embryos during vitellogenesis is not known (but see Stewart *et al.* 1990; Stewart and Thompson 2000).

Another poorly studied aspect that affects offspring size is how space allocation within females is involved in the trade-off with clutch size (Brown and Shine 2002; Ji *et al.* 2006). Small species and younger (smaller) females of any species have different allometric constraints in terms of adding additional eggs or embryos even if more resources are available. For example, Ji *et al.* (2006) used follicle ablation to show that there is a fixed upper limit on egg size in King Ratsnakes (*Elaphe carinata*). How this relates to offspring size and possible trade-offs with clutch size and growth has not been well-studied in snakes.

Finally, the variance (often expressed as coefficient of variation [CV]) among and within litters in offspring size has been examined only rarely (Ford *et al.* 1990; Farrell *et al.* 1995; Weatherhead *et al.* 1999; Seigel and Ford 2001; Blouin-Demers and Weatherhead 2007). That is, are all offspring for an individual female of the same size or is the mean size made up of offspring of greatly different sizes? Ford *et al.* (1990) examined reproduction in 17 species of snakes in eastern Texas and found CVs of up to 25% in offspring from a viviparous species. Seigel and Ford (2001) showed that snakes receiving a high-energy diet had higher variation in offspring size than females from a low-energy diet, indicating that diet affects the degree of variability of offspring within a clutch. How (and if) such variation is related to offspring fitness, remains unknown.

14.6 ACKNOWLEDGMENTS

We start by thanking our colleagues whose efforts have made reviewing the literature on snake reproductive biology so enjoyable as well as time consuming. We also thank the organizers of this volume (Robert Aldridge and Dave Sever) for inviting us to participate. Neil Ford was supported by Research Opportunity Award grant DEB 0713969 from NSF and a Texas Parks and Wildlife State Wildlife grant. Rich Seigel was supported by grants from Dynamac Corporation and the Maryland State Highway Authority during the writing of this chapter.

14.7 LITERATURE CITED

Andren, C. and Nilson, G. 1983. Reproductive tactics in an island population of adders, *Vipera berus* (L.), with a fluctuating food resource. Amphibia-Reptilia 4: 63-79.

Arnold, S. J. 1993. Foraging theory and prey-size-predator-size relations in snakes. Pp. 87-115. In R. A. Seigel and J. T. Collins (eds), *Snakes: Ecology and Behavior*. McGraw-Hill, New York.

Bell, G. 1980. The costs of reproduction and their consequences. American Naturalist 116: 45-76.

Bernardo, J. 1996. Maternal effects in animal ecology. American Zoologist 36: 83-105.

Blouin-Demers, G. and Weatherhead, P. J. 2007. Allocation of offspring size and sex by female black ratsnakes. Oikos 10: 1759-1767.

Bonnet, X., Naulleau, G., Shine, R. and Lourdais, O. 2000. Reproductive versus ecological advantages to larger body size in female snakes, *Vipera aspis*. Oikos 89: 509-518.

Bonnet, X., Naulleau, G., Shine, R. and Lourdais, O. 2001. Short-term versus long-term effects of food intake on reproductive output in a viviparous snake, *Vipera aspis*. Oikos 92: 297-308.

Brockelman, W. Y. 1975. Competition, the fitness of offspring, and optimal clutch size. American Naturalist 109: 677-699.

Brodie, E. D. III. 1989. Behavioral modification as a means of reducing the cost of reproduction. American Naturalist 134: 225-238.

Bronikowski, A.M. 2000. Experimental evidence for the adaptive evolution of growth rate in the garter snake *Thamnophis elegans*. Evolution 64: 1760-1767.

Bronikowski, A. M. and Arnold, S. J. 1999. Evolutionary ecology of life history variation in the garter snake *Thamnophis elegans*. Ecology 80: 2314-2325.

Brown, G. P. and Shine, R. 2002. Reproductive ecology of a tropical natricine snake, *Tropidonophis mairii* (Colubridae). Journal of Zoology (London) 258: 63-72.

Brown, G. P. and Weatherhead, P. J. 2004. Sexual abstinence and the cost of reproduction in adult male water snakes, *Nerodia sipedon*. Oikos 104: 269-276.

Brown, G. P. and Shine, R. 2005. Links between female phenotype, life-history and reproductive success in free-ranging snakes (*Tropidonophis mairii*, Colubridae). Ecology 86: 2763-2770.
Brown, G. P. and Shine, R. 2006. Why do most tropical animals reproduce seasonally? Testing hypotheses on an Australian snake. Ecology 87: 133-143.
Brown, G. P. and Shine, R. 2007. Repeatability and heritability of reproductive traits in free-ranging snakes. Journal of Evolutionary Biology 20: 588-596.
Brown, G. P. and Shine, R. 2009. Beyond size-number tradeoffs: clutch size as a maternal effect. Philosophical Transactions of the Royal Society B 364: 1097-1106.
Charland, M. B. 1989. Size and winter survivorship in neonatal western rattlesnakes (*Crotalus viridis*). Canadian Journal of Zoology 67: 1620-1625.
Charnov, E. L. 1989. Phenotypic evolution under Fisher's fundamental theorem of natural selection. Heredity 62: 113-116.
Congdon, J. D., Nagle, R. D., Dunham, A. E., Beck, C. W., Kinney, O. M. and Yeoman, S. R. 1999. The relationship of body size to survivorship of hatchling snapping turtles (*Chelydra serpentina*): an evaluation of the "bigger is better" hypothesis. Oecologia 121: 224-235.
Farrell, T. M., May, P. G. and Pilgrim, M. A. 1995. Reproduction in the rattlesnake, *Sistrurus miliarius barbouri*, in central Florida. Journal of Herpetology 29: 21-27.
Farrell, T. M., May, P. G. and Pilgrim, M. A. 2009. Repeatability of female reproductive traits in Pigmy Rattlesnakes (*Sistrurus miliarius*). Journal of Herpetology 43: 332–335.
Ferguson, G. W. and Fox, S. F. 1984. Annual variation of survival advantage of large juvenile side-blotched lizards, *Uta stansburiana*: its causes and evolutionary significance. Evolution 38: 342-349.
Ferguson, G. W., Snell, H. L. and Landwer, A. J. 1990. Proximate control of variation of clutch, egg and body size in a west-Texas population of *Uta stansburiana stejnegeri* (Sauria: Iguanidae). Herpetologica 46: 227-238.
Fitch, H. S. 1970. Reproductive cycles in lizards and snakes. University of Kansas Museum of Natural History Miscellaneous Publications 52: 1-247.
Fitch, H. S. 1985. Variation in clutch and litter size in New World reptiles. University of Kansas Museum Natural History, Miscellaneous Publication 52: 1–76.
Ford, N. B. and Seigel, R. A. 1989a. Phenotypic plasticity in reproductive characteristics: Evidence from a viviparous snake. Ecology 70: 1768-1774.
Ford, N. B. and Seigel, R. A. 1989b. Relationships among body size, clutch size, and egg size in three species of oviparous snakes. Herpetologica 45: 75-83.
Ford, N. B. and Seigel, R. A. 1994. An experimental study of the trade-offs between age and size at maturity: effects of energy availability. Functional Ecology 8: 91-96.
Ford, N. B. and Seigel, R. A. 2006. Intra-individual variation in clutch and offspring size in an oviparous snake. Journal of Zoology (London) 268: 171-176.
Ford, N. B., Cobb V. A. and Lamar, W. 1990. Reproductive data on snakes of Northeastern Texas. Texas Journal of Science 42: 355-368.
Ford, N. B., Brischoux, F. and Lancaster, D. 2004. Reproduction in the Western Cottonmouth, *Agkistrodon piscivorus leucostoma*, in a floodplain forest. Southwestern Naturalist 49: 465-471.
Gregory, P. T. 2006. Influence of income and capital on reproduction in a viviparous snake: direct and indirect effects. Journal of Zoology: 270: 414-419.
Gregory, P. T. and Larsen, K. W. 1993. Geographic variation in reproductive characteristics among Canadian populations of the common garter snake (*Thamnophis sirtalis*). Copeia 1993: 946-958.

Gregory, P. T. and Larsen, K. W. 1996. Are there any meaningful correlates of geographic life-history variation in the garter snake, *Thamnophis sirtalis*? Copeia 1996: 183-189.

Gregory, P. T. and Skebo, K. M. 1998. Trade-offs between reproductive traits and the influence of food intake during pregnancy in the garter snake, *Thamnophis elegans*. American Naturalist 151: 477-486.

Gregory, P. T., Crampton, L. H. and Skebo, K. M. 1999. Conflicts and interactions among reproduction, thermoregulation and feeding in viviparous reptiles: are gravid snakes anorexic? Journal of Zoology (London) 248: 231-241.

Janzen, F. J. and Warner, D. A. 2009. Parent–offspring conflict and selection on egg size in turtles. Journal of Evolutionary Biology 22: 2222-2230.

Janzen, F. J., Tucker, J. K. and Paukstis, G. L. 2000. Experimental analysis of an early life-history stage: avian predation selects for larger body size of hatchling turtles. Journal of Evolutionary Biology 13: 947-954.

Jayne B. C. and Bennett, A. F. 1990. Selection on locomotor performance capacity in a natural population of garter snakes. Evolution 44: 1204-1229.

Jellen, B. C. and Kowalski, M. J. 2007. Movement and growth of neonate eastern Massasaugas (*Sistrurus catenatus*). Copeia 2007: 994-1000.

Ji, X. and Wang, Z-W. 2005. Geographic variation in reproductive traits and trade-offs between size and number of eggs of the Chinese cobra (*Naja atra*). Biological Journal of the Linnaean Society 85: 27-40.

Ji, X., Du, W. G., Li, H. and Lin, L. H. 2006. Experimentally reducing clutch size reveals a fixed upper limit to egg size in snakes, evidence from the king ratsnake, *Elaphe carinata*. Comparative Biochemistry and Physiology, Part A 144: 474-478.

Ji, X., Du, W. G., Li, H. and Lin, L. H. 2009. Nonlinear continuum of egg size-number trade-offs in a snake: is egg-size variation fitness related? Oecologia 159: 689-696.

King, R. B. 1993. Determinants of offspring number and size in the brown snake, *Storeria dekayi*. Journal of Herpetology 27: 175-185.

Kingsolver, J. G. and Schemske, D. W. 1991. Path analysis of selection. Trends in Ecology and Evolution 6: 276-280.

Kissner, K. J. and Weatherhead, P. J. 2005. Phenotypic effects on survival of neonatal northern watersnakes *Nerodia sipedon*. Ecology 74: 259–265.

Lemen, C. A. and Voris, H. K. 1981. A comparison of reproductive strategies among marine snakes. Journal of Animal Ecology 50: 89-101.

Lourdais O., Bonnet X. and Doughty, P. 2002. Costs of anorexia during pregnancy in a viviparous snake (*Vipera aspis*). Journal of Experimental Zoology 292: 487-493.

Lourdais, O., Bonnet, X., Shine, R. and Taylor, E. N. 2003. When does a reproducing female viper (*Vipera berus*) "decide" on her litter size? Journal of Zoology (London) 259: 123-129.

Luiselli, L., Capula, M. and Shine, R. 1996. Food habits, growth rates, and reproductive biology of grass snakes, *Natrix natrix* (Colubridae) in the Italian Alps. Journal of Zoology (London) 241: 371-380.

Luiselli, L., Capula, M. and Shine, R. 1997. Reproductive output, costs of reproduction, and ecology of the smooth snake, *Coronella austriaca*, in the eastern Italian Alps. Oecologia 106: 100-110.

MacArthur, R. H. and Wilson, E. O. 1967. *The Theory of Island Biogeography*. Princeton University Press, Princeton, New Jersey. Pp. 228

Madsen, T. 1987. Cost of reproduction and female life-history tactics in a population of grass snakes, *Natrix natrix*, in southern Sweden. Oikos 49: 129-132.

Madsen, T. and Shine, R., 1992. Determinants of reproductive success in female adders, *Vipera berus*. Oecologia 92: 40-47.

Madsen, T. and Shine, R. 1993. Costs of reproduction in a population of European adders. Oecologia 94: 488-495.
Madsen, T. and Shine, R. 1994. Costs of reproduction influence the evolution of sexual size dimorphism in snakes. Evolution 48: 1389-1397.
Madsen, T. and Shine, R. 1996. Determinants of reproductive output in female water pythons (*Liasis fuscus*: Pythonidae). Herpetologica 52: 146-159.
Madsen, T. and Shine, R. 1998. Quantity or quality? Determinants of maternal reproductive success in tropical pythons (*Liasis fuscus*). Proceedings Royal Society London B 265: 1521-1525.
Madsen, T. and Shine, R. 2000. Energy versus risk: costs of reproduction in free-ranging pythons in tropical Australia. Australian Ecology 25: 67-675.
Madsen, T. and Shine, R. 2001. Conflicting conclusions from long-term versus short-term studies on growth and reproduction of a tropical snake. Herpetologica 57: 147-156.
Marshall, D. J. and Uller, T. 2007. When is a maternal effect adaptive? Oikos 116: 1957-1963.
Miles, D. B., Sinervo, B. and Frankino, W. A. 2000. Reproductive burden, locomotor performance, and the cost of reproduction in free-ranging lizards. Evolution 54: 1386-1395.
Moll, E. O. and Legler, J. M. 1971. The life history of a neotropical slider turtle, *Pseudemys scripta* (Schoepff), in Panama. Bulletin Los Angeles County Museum Natural History 11: 1-102.
Niewiarowski, P. H. and Dunham, A. E. 1994. The evolution of reproductive effort in squamate reptiles: Costs, trade-offs, and assumptions reconsidered. Evolution 48: 137-145.
Olsson, M. and Shine, R. 1997. The limits to reproductive output: offspring size versus number in the sand lizard (*Lacerta agilis*). American Naturalist 149: 179-188.
Plummer, M. V. 1997. Speed and endurance of gravid and non-gravid green snakes, *Opheodrys aestivus*. Copeia 1997: 191-194.
Reading, C. J. 2004. The influence of body condition and prey availability on female breeding success in the smooth snake (*Coronella austriaca* Laurenti). Journal of Zoology 264: 61-67
Reznick, D. 1992. Measuring the costs of reproduction. Trends in Ecology and Evolution 7: 42-45.
Roff, D. A. 1992. *Life History Evolution*. Chapman and Hall, New York. Pp. 465.
Seigel, R. A. 1993. Summary: Future research on snakes, or how to combat "lizard envy". Pp. 395-402 In: R. A. Seigel and J. T. Collins (eds), *Snakes: Ecology and Behavior*. McGraw-Hill, New York.
Seigel, R. A. and Ford, N. B. 1987. Reproductive biology. Pp. 210-252. In: R. A. Seigel, J. T. Collins and S. S. Novak (eds), *Snakes: Ecology and Evolutionary Biology*. Macmillan Publ. Co, New York.
Seigel, R. A. and Ford, N. B. 1991. Phenotypic plasticity in the reproductive characteristics of an oviparous snake, *Elaphe guttata*: Implications for life history studies. Herpetologica 47: 301-307.
Seigel, R. A. and Ford, N. B. 2001. Phenotypic plasticity in reproductive traits: Geographic variation in plasticity in a viviparous snake. Functional Ecology 15: 36-42.
Seigel, R. A., Huggins, M. M. and Ford, N. B. 1987. Reduction in locomotor ability as a cost of reproduction in gravid snakes. Oecologia 73: 481-485.
Seigel, R. A., Ford, N. B. and. Mahrt, L. A. 2000. Ecology of an aquatic snake (*Thamnophis marcianus*) in a desert environment: implications of early timing of

birth and geographic variation in reproduction. American Midland Naturalist. 143: 453-462.
Shine, R. 1980. "Costs" of reproduction in reptiles". Oecologia 46: 92-100.
Shine, R. 1992. Relative clutch mass and body shape in lizards and snakes: Is reproductive investment constrained or optimized? Evolution 46: 828-833.
Shine, R. 1996. Life-history evolution in Australian snakes: a path analysis. Oecologia 107: 484-489.
Shine, R. 2003. Reproductive strategies in snakes. Proceedings Royal Society London B 270: 995-1004.
Shine, R. 2005. Life-history evolution in reptiles. Annual Review Ecology, Evolution and Systematics. 36: 23-46.
Shine, R. and Madsen, T. 1997. Prey abundance and predator reproduction: rats and pythons on a tropical Australian floodplain. Ecology 78: 1078-1086.
Shine, R. and Bonnet, X. 2000. Snakes: a new "model" organism in ecological research? Trends in Ecology and Evolution 15: 221-222.
Smith, C. C. and Fretwell, S. D. 1974. The optimal balance between size and number of offspring. American Naturalist 108: 499-506.
Sparkman, A. M., Arnold, S. J. and Bronikowski, A. M. 2007. An empirical test of evolutionary theories for reproductive senescence and reproductive effort in the garter snake *Thamnophis elegans*. Proceedings Royal Society London B 274: 943-950.
Stearns, S. C. 1976. Life-history tactics: A review of the ideas. Quarterly Review of Biology 51: 3-47.
Stearns, S. C. 1992. *The Evolution of Life Histories*. Oxford University Press, Oxford, Pp. 248. UK.
Stewart, J. R. and Thompson, M. B. 2000. Evolution of placentation among squamate reptiles: recent research and future directions. Comparative Biochemistry and Physiology 127: 411-431.
Stewart, J. R., Blackburn, D. G., Baxter, D. C. and Hoffman, L. H. 1990. Nutritional provision to embryos in a predominantly lecithotrophic placental reptile, *Thamnophis ordinoides* (Squamata: Serpentes). Physiological Zoology 63: 722-734.
Tinkle, D. W. 1961. Geographic variation in reproduction, size, sex ratio, and maturity of *Sternotherus odoratus* (Testudinata, Chelydridae). Ecology 42: 68-76.
Tinkle, D. W., Wilbur, H. M. and Tilley, S. G. 1970. Evolutionary strategies in lizard reproduction. Evolution 24: 55-74.
Turner, F. B. 1977. The dynamics of populations of squamates, crocodilians and rhynchocephalians. Pp. 157–264. In C. Gans and D. W. Tinkle (eds), *Biology of the Reptilia*, vol. 7. Academic Press, New York.
Uller, T. and Olsson, M. 2009. Offspring size-number trade-off in a lizard with small clutch sizes: tests of invariants and potential implications. Evolutionary Ecology 23: 363-372.
Weatherhead, P. J., Brown, G. P., Prosser, M. R. and Kissner, K. J. 1999. Factors affecting neonate size variation in northern water snakes, *Nerodia sipedon*. Journal of Herpetology 33: 577-589.
Zuffi, M., Fornasiero, S. and Bonnet, X. 2007. Geographic variation in reproductive output of female European whip snakes (*Hierophis viridiflavus*). Herpetological Journal 17: 219-224.
Zuffi, M., Gentilli, A., Cecchinelli E., Pupin F., Bonnet X., Fillippi, E., Luiselli, L. M., Barbanera, F., Dini, F. and Fasola, M. 2009. Geographic variation of body size and reproductive patterns in Continental versus Mediterranean asp vipers, *Vipera aspis*. Biological Journal of the Linnaean Society 96: 383-391.

CHAPTER 15

IGF-1 and Reproduction in Snakes

A.M. Sparkman[1], *A.M. Bronikowski*[2], *and N.B. Ford*[3]

15.1 INTRODUCTION

The endocrine system comprises a diverse array of hormonal signaling pathways that regulate the expression of key life-history traits, such as developmental and growth rates, and timing and magnitude of reproductive effort. The role of hormones in the expression of life-history traits has become increasingly of interest in the field of evolutionary ecology, which seeks to understand the genetic and physiological mechanisms underlying the evolution of life-history strategies (Ketterson and Nolan 1992; Finch and Rose 1995; Zera and Bottsford 2001; Ricklefs and Wikelski 2002). The hormone insulin-like growth factor-1 (IGF-1) is particularly of interest in this regard, as it can exert far-reaching pleiotropic effects on organismal growth, reproduction, and lifespan. As a member of the insulin-like family of signaling molecules found throughout the vertebrate and invertebrate phylogeny, IGF-1 has been the subject of intense research in domesticated and model species, including those of great commercial interest. Little, however, is known regarding the physiology of IGF-1 in reptiles, or in wild species in general. Reptiles display various combinations of terrestrial ectothermy, determinate or indeterminate growth, diversity in reproductive mode, metabolic flexibility, and remarkable plasticity in life-history strategy, which make them a particularly intriguing focus for IGF-1 study. Nevertheless, IGF-1 function has been examined in only five reptile species to date: the Loggerhead Sea Turtle (*Caretta caretta*), Pond Slider (*Trachemys picta*), American Alligator (*Alligator mississippiensis),* Terrestrial

[1]Department of Biology, Trent University, Peterborough, ON, Canada
[2]Department of Ecology, Evolution and Organismal Biology, Iowa State University, Ames, IA, USA
[3]Department of Biology, University of Texas at Tyler, Tyler, Texas, USA

Gartersnake (*Thamnophis elegans*), and Brown House Snake (*Lamprophis fuliginosus*) (Crain *et al.* 1995a,b; Guillette *et al.* 1996; Sparkman *et al.* 2009; Sparkman *et al.* 2010). Though these few studies have done much to show both consistencies and disparities in activity of IGF-1 with that in other taxa, many questions remain.

To begin to understand how IGF-1 is involved in snake reproduction, we must bear in mind the complexity of interrelated genetic, physiological, ecological, and evolutionary factors that determine how IGF-1 is regulated, and the functions that it performs. The aim of this chapter is to describe both what is known regarding IGF-1 in snake reproduction, what might be inferred from studies in other taxa, and what remains to be explored. The discussion will delve into four main lines of research that need to be more fully developed to gain a comprehensive understanding of how IGF-1 is involved in reproductive success. First, it will examine what is known regarding the basic components of IGF-1 system, such as the conservation of IGF-1 sequence and structure among taxa, the existence of distinct insulin, IGF-1, and IGF-2 receptors, as the well as the number and character of IGF-1 binding proteins. Second, it will explore the ways in which IGF-1 can respond plastically to a variety of environmental variables, which may have important ramifications for the timing of sexual development and reproductive frequency and success. Third, it will situate the role of IGF-1 in reproduction firmly in the context of its role in the regulation of a suite of life-history traits, such as growth rate and body size, which may influence the size and number of offspring produced in each reproductive event, and lifespan, which may influence the number of reproductive events. Finally, it will evaluate how the pleiotropic involvement of IGF-1 in such traits may serve as a mechanism for life-history trade-offs, and thereby mediate the evolution of correlated traits.

The discussion will center on the role of IGF-1 in snake reproduction on two colubrid snake species, *Thamnophis elegans* and *Lamprophis fuliginosus*, which constitute the first and only squamate reptiles in which plasma IGF-1 levels have been examined to date. We have studied plasma IGF-1 in two ecotypes of *T. elegans* that show variation in several physiological and life-history parameters (Bronikowski and Arnold 1999; Sparkman *et al.* 2007; Bronikowski 2008; Robert *et al.* 2009; Sparkman and Palacios 2009; Robert and Bronikowski 2010). One ecotype, which lives along the lakeshore of Eagle Lake, a large natural lake in northeastern California, exhibits fast growth to large body sizes, early maturation, high reproduction, low adult survivorship, and short median lifespan. Another ecotype, which lives in montane meadows within a few kilometers of the lake, exhibits slow growth to small body sizes, late maturation, low annual reproduction, and high adult survivorship. This naturally occurring variation in life-history traits presents an ideal opportunity, therefore, to examine the ecological and evolutionary importance of IGF-1 in the expression of these traits, both in the field and in the laboratory (Sparkman *et al.* 2009; A. Sparkman and A. Bronikowski, unpublished data).

Our second study system comprised lab-reared offspring of wild-caught adult *Lamprophis fuliginosus* from Tanzania (Sparkman *et al.* 2010). Since *L. fuliginosus* is a tropical non-seasonal breeder and a rapidly developing species with sexual maturity occurring as early as six months of age (Ford and Seigel 2005; Byars 2008), we were able to raise these snakes through a full developmental and reproductive cycle within one year and test for a relationship between plasma IGF-1 and nutrition, growth, sexual maturation and reproduction.

Although limited to two study systems, our findings reveal that certain aspects of IGF-1 function are conserved across taxa and thus comparable to the wealth of data on captive and laboratory study models. Our studies further reveal that IGF-1 function is remarkably plastic and suggests a complementary role of the comparative biology of IGF-1 in informing and enhancing our understanding of IGF-1 in the animal kingdom.

15.2 COMPARATIVE EVOLUTION AND REGULATION OF IGF-1 SYSTEM

15.2.1 IGF-1 Evolution

In vertebrates, the insulin/insulin-like growth factor/relaxin (In/IGF/RLN) family comprises a single insulin, two IGFs (IGF-1 and IGF-2), and varying numbers of insulin-like factors and relaxin genes. Invertebrates also carry an array of insulin homologs, called insulin-like peptides, situating the root of the insulin family tree deep in the animal kingdom (Wu *et al.* 2006). Comparative studies of the vertebrate In/IGF/RLN family have suggested that it originated from a single ancestral preproinsulin-like gene, still retained in the extant cephalochordate amphioxus (*Branchiostoma californiensis*). It is thought to have diverged following several gene duplication events prior to and during the vertebrate radiation (Chan *et al.* 1990; Olinski *et al.* 2006). During this time, the family also split into two more or less distinct groups, where insulin and the IGF proteins retained more functional and structural similarity to each other than to the relaxin and insulin-like factors (Olinski *et al.* 2006). The IGFs are composed of A, B, C, and D domains, as well an E-domain that is proteolytically removed in the mature polypeptide. Insulin is composed of A, B, and D domains homologous to those of the IGFs, with the C-domain proteolytically removed in the mature polypeptide. The IGF-1 coding sequence has been highly conserved among vertebrates, particularly in the A and B domains, and sequences from snake, lizards, and turtle IGF-1 show reptiles to be no exception (Sparkman *et al.* in prep; Table 15.1). Thus in future studies of IGF-1 in snakes, we may expect to find a large degree of conservation in IGF-1 function with other taxa, even (as we shall explore) as there may be uniquely evolved differences.

The insulin/IGF-1 receptors are homologous tyrosine kinase receptors with a heterotetrameric structure, and are thought to have originated via duplication of a single receptor-coding gene, still found in amphioxus

Table 15.1 Alignment of IGF-1 amino acid sequences for outgroup and reptile species. *Homo sapiens* (GenBank accession # NM_000618), two representative fish (Fish1=*Siganus guttatus*, AY198184.1; Fish2=*Salmo salar*, NM_001123623.1), a frog (*Xenopus laevis*, M29857.1), and a conserved bird sequence (identical in several species of birds, e.g., EU031044.1), followed by original sequences from the Bronikowski lab: a turtle sequence from two turtles species, Snapping Turtle (*Chelydra serpentina*) and Pond Slider (*Trachemys scripta*); a sequence from the Ground Skink (*Scincella lateralis*), and a sequence from the Terrestrial Gartersnake (*Thamnophis elegans*). Dots represent identity corresponding to *H. sapiens* residues. Amino acid residues 1-9 and 70 were not sequenced in reptiles. Domains A-D are indicated.

```
                10        20        30        40        50        60        70
        ....|....|....|....|....|....|....|....|....|....|....|....|....|....|
Human   GPETLCGAELVDALQFVCGDRGFYFNKPTGYGSSSRRAPQTGIVDECCFRSCDLRRLEMYCAPLKPAKSA
Fish1   ............T......E.....S......PN...P--R........Q..E..........A.TS.A.
Fish2   ............T......E.....S......P....SHNR........Q..E....... ..V.SG.A.
Frog    ............T............S.......NN..SHHR........Q...F.........A......
Bird    .........................S...........LHHK........Q.............I..P...
Turtle  ~~~~~~~~~................S...........LHHK........Q.............I..P..~
Lizard  ~~~~~~~~~..........E.....S..A...GGR.PLPTK........Q.............S..P..~
Snake   ~~~~~~~~~..........E.....S..A...G.R.SSSTR........Q...I.........V..P..~
        <------------------------------> <-----------> <---------------------> <----->
                       B                        C                 A              D
```

(Pashmforoush *et al.* 1996). In mammals, IGF-2 binds to a monomeric transmembrane receptor called the mannose-6-phosphate receptor. There has also been at least one report of IGF-2 binding to this receptor in rainbow trout (Mendez *et al.* 2001). However, while this receptor is also present in amphibians and birds, it does not appear to bind IGF-2 due to a lack of binding site-specificity (Kelley *et al.* 2002). Whether IGF-2 binds to a separate receptor than IGF-1 and insulin in non-avian reptiles is as yet unknown. Due to their structural similarity, insulin, IGF-1, and IGF-2 have some degree of cross-reactivity with each other's receptors. Nevertheless, they do carry out distinguishable functions, with insulin predominately involved in the regulation of intermediary metabolic processes (such as glucose homeostasis), and the IGF proteins specializing in the regulation of cell growth and development—IGF-2 tending to specialize on prenatal growth in mammals, IGF-1 on postnatal functions. There appear to be evolutionary differences in IGF-1 and IGF-2 function, however, as specific binding of IGF-1 in skeletal muscles is higher than insulin binding in ectothermic vertebrates, while the reverse is true for endotherms (Parrizas *et al.* 1995). Furthermore in fish, IGF-2 has been shown to be high in adults as well as juveniles, and involved in regulating postnatal growth rate (reviewed in Reinecke and Collet 1998; Peterson *et al.* 2004).

15.2.2 IGF-1 Regulation

IGF-1 can be produced both locally in diverse tissues, and centrally by the liver. The majority of circulating plasma IGF-1 is produced by the liver and acts in an endocrine capacity, binding to receptors on target tissues throughout the body. Locally produced IGF-1 tends to work in an autocrine or paracrine fashion (i.e., affecting either the source cell or neighboring cells). Plasma production of IGF-1 is regulated primarily by growth hormone (GH), which is released by the pituitary, and binds to its receptors in the liver to stimulate IGF-1 synthesis and secretion. The growth-promoting effects of GH are largely carried out by IGF-1 (though it can also stimulate IGF-2 activity in some taxa—see Ponce *et al.* 2008), which stimulates cells in tissues such as cartilage, muscle, and adipose to increase DNA, RNA and protein synthesis, and expand cell size and number. Ultimately this results in expansion of organ size and function throughout the body, as well as linear growth in bones.

Binding proteins are also important factors in IGF-1 regulation. Six IGF-1 binding proteins are known to exist in mammals and fish, at least three in birds, and an unknown number in non-avian reptiles (though at least one has been identified in the Iguana, *Iguana iguana*) (Kelley *et al.* 2002; Kamangar 2006). Binding proteins, which can be produced both locally and centrally by the liver, allow IGF-1 to have a much longer half-life in the blood than GH. They regulate IGF-1 transport, turnover, and distribution, and can both inhibit and facilitate IGF-1 synthesis and activity in target tissues. For instance, IGFBP-4 and -6 have been shown to consistently inhibit IGF actions, while IGFBP-1, -2, -3 and -5 can either inhibit or facilitate IGF actions, depending on cell type and/or method or level of administration (reviewed in Duan *et al.* 2005).

15.3 ENVIRONMENTAL PLASTICITY OF IGF-1

15.3.1 IGF-1 and Seasonality

Seasonal changes in IGF-1 activity have been reported in fish, amphibians, reptiles, and mammals, with plasma levels generally elevated during peak growth/reproductive seasons (Pancak-Roessler and Lee 1990; Crain *et al.* 1995b; Webster *et al.* 1996; Mingarro *et al.* 2002; Sparkman *et al.* 2009). In hibernating species, the most dramatic seasonal decline in IGF-1 activity most likely occurs during the winter—in the only study testing for a hibernation effect to date, plasma IGF-1 and IGFBP-3 were shown to decrease by 75% in hibernating ground squirrels (*Spermophilus lateralis*; Schmidt and Kelley 2001). Untangling the relative influence of different environmental factors on IGF-1 activity, however, is no small challenge, as changes in reproductive activity, feeding, temperature, or photoperiod, can all play significant (and simultaneous) roles. For instance, loggerhead sea turtles (*Caretta caretta*) show a seasonal fluctuation in plasma IGF-1, with higher levels during the spring and summer months, but as this pattern is only present in large individuals, it may be driven primarily by reproductive activity rather than increased feeding/growth rates per se (Crain *et al.* 1995b). Nevertheless, environmental factors such as nutrition, temperature, and photoperiod have been shown to exert both independent and interacting effects on IGF-1 regulation.

15.3.2 IGF-1 and Nutrition

It has been well established that higher levels of feeding are accompanied by higher levels of plasma IGF-1 in mammals, fish, and birds (Clemmons and Underwood 1991; Duan 1998; McMurtry 1998; Webster *et al.* 1999). This relationship has also been confirmed in reptiles: Painted Turtles (*Trachemys picta)* maintained on a high diet show higher levels of circulating IGF-1 than those on a low diet (Crain *et al.* 1995a), and there is a positive correlation between feeding rate and IGF-1 in lab-reared juvenile *Lamprophis fuliginosus* and *Thamnophis elegans* (Sparkman *et al.* 2010, A. Sparkman and A. Bronikowski unpublished data). In addition to the amount of food consumed, diet protein content has also been shown to be positively correlated to IGF-1 in fish and poultry (Lauterio and Scanes 1987; Rosebrough *et al.* 1999; Perez-Sanchez *et al.* 1995).

No studies to date have directly tested the relationship between annual and/or seasonal prey availability (or protein content) and IGF-1 levels in wild populations. Our work with free-ranging *Thamnophis elegans*, however, suggests that such a relationship is indeed present, and may serve as a mechanism driving differences in growth rate in populations exposed to varying levels of food availability (Sparkman *et al.* 2009). In *T. elegans* populations residing along the lakeshore of Eagle Lake, where fish prey is abundant and consistently available from year to year, we found no seasonal changes with IGF-1 from May to July over two years of study (Fig. 15.1A,B). However, in populations residing in meadow habitats, where prey availability is much more variable, two different patterns were manifest over the two years (Fig. 15.1C). In the first year, as in the lakeshore populations, there was no relationship between IGF-1 and Julian date in meadow populations. However, during the second year, there was a negative relationship between the two, with IGF-1 levels declining over time. The most compelling interpretation of these findings is based on dramatic differences in rainfall between the two years of study, and the known positive relationship between annual rainfall and abundance of amphibian prey in the meadows (Bronikowski and Arnold 1999). The first year was a wet year, providing ideal conditions for amphibian breeding. Date-independent IGF-1 levels during this year, therefore, suggest that meadow snakes likely continued feeding and growing under optimal conditions throughout the sampling period. The second year, however, was a dry year, in which amphibian breeding (primarily *Pseudacris regilla*) was likely minimal, with tadpole and metamorph abundance declining as the meadow pools dried. It is therefore possible that declining IGF-1 levels between May and July is linked to increasing difficulty in obtaining amphibian prey. In light of these promising results, further study is warranted to more concretely establish the relationship between annual and seasonal *prey availability* and IGF-1 levels in the wild in *T. elegans* and in other species. Furthermore, an experimental design involving repeated blood sampling and measuring of free-ranging individuals

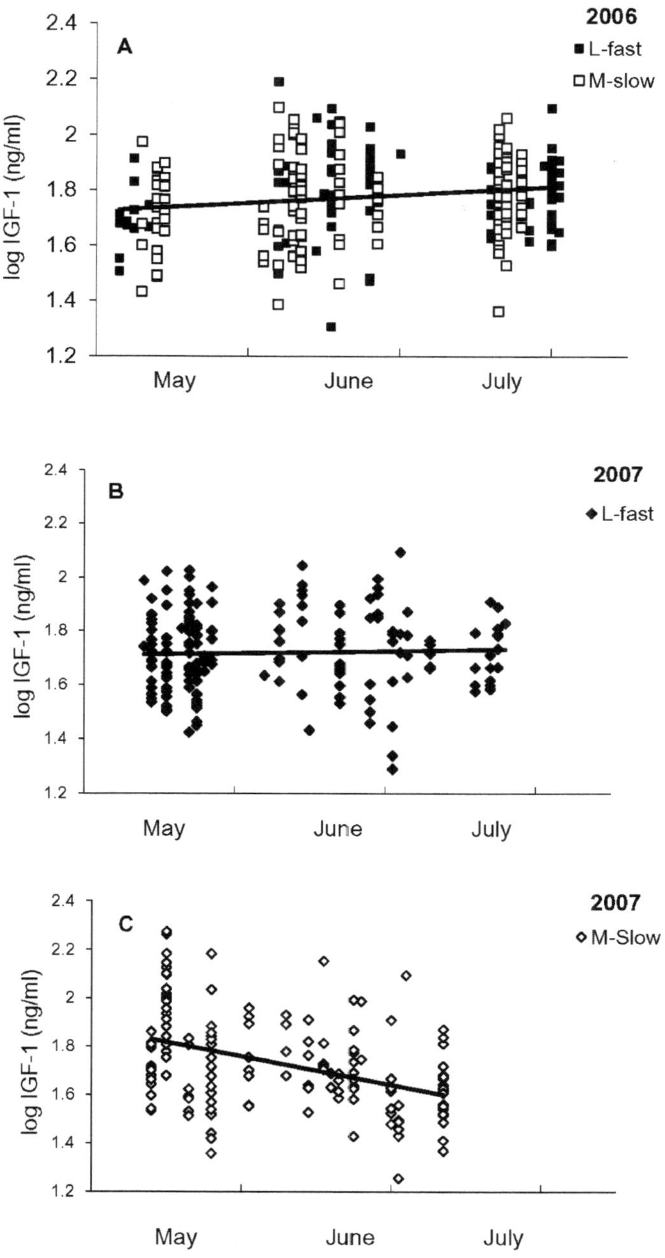

Fig. 15.1 Curve of log IGF-1 (ng/ml) against Julian days in 2006 in free-ranging *Thamnophis elegans* (**A**) for gravid and non-gravid snakes in both ecotypes, and 2007 (**B,C**) for both fast-growth lakeshore and slow-growth meadow individually. Months are denoted on the x-axis for the sake of clarity. Asterisks denote significant slopes; NS=non-significant. (From Sparkman *et al.* 2009. Ecology 90: 720-728, Fig. 2, with permission of Ecological Society of America.)

via mark-recapture over a single season could be used to test for a relationship between prey availability and IGF-1 that translates directly into modifications of individual growth rate.

Since IGF-1 can act as an index of nutritional status it has been put forward as a candidate for signaling somatic readiness to mature to the reproductive system, triggering the onset of sexual maturation. We will explore this in further detail in a subsequent section (see *Maturation*). However, it is important to bear in mind that any causal relationship between feeding levels and IGF-1 activity can be a two-way street: even as high nutrition can result in higher IGF-1 levels, and exerts permissive effects on IGF-1 stimulation by other environmental cues, there is evidence that other cues, such as photoperiod, can result in higher IGF-1 levels even without a concomitant increase in food intake (Webster *et al.* 2001; see next section).

15.3.3 IGF-1, Photoperiod and Temperature

In temperate species, a variety of seasonal changes in physiology are stimulated by changes in photoperiod, temperature, and/or precipitation, signaling the onset of the growth/reproductive season. Studies in fish have shown that exposure to long days results in both faster growth rates and higher IGF-1 levels (Boeuf and Le Bail 1999; Taylor *et al.* 2005; Imsland *et al.* 2008). On-going study continues to examine whether this effect is induced either directly via photoperiod-dependent GH production by the pituitary, resulting in increased levels of IGF-1, or indirectly via stimulation of the GH/IGF-1 axis by photoperiod-dependent hormones such as melatonin. The most thorough studies of photoperiod regulation of IGF-1 have been conducted in male Red Deer (*Cervus elaphus*) which exhibit (even in captivity) natural declines in IGF-1 in the autumn, subsequent to the fall/summer growing season (Webster *et al.* 1996; Webster *et al.* 1999; Webster *et al.* 2001). Experimental extension of long day lengths into the autumn has been shown to delay this decline relative to controls subject to naturally declining photoperiods (Webster *et al.* 1999). Furthermore, when juvenile red deer are exposed to long day lengths in the winter, they are able to elevate GH/IGF-1 levels even when prevented from increasing food consumption (Webster *et al.* 2001).

The effects of photoperiod can be temperature-dependent. In one study in fish, exposure to increased day length for seven days at 2°C resulted in much smaller and short-lived increases in plasma GH, IGF-1 and thyroxine than fish maintained at 10°C (McCormick *et al.* 2000). This is to be expected, as temperature itself can exert a powerful influence in IGF-1 activity. Higher environmental temperatures tend to be associated with more rapid development and growth rates in a variety of taxa, albeit different species/populations exhibit different optimal temperatures (e.g., O'Steen and Janzen 1999; Bronikowski 2000). In fish, the positive relationship between water temperature and individual growth rate has been shown to be at least partially mediated by IGF-1 levels (Beckman *et al.* 1998; Taylor *et al.* 2005; Davis and Peterson 2006; Imsland *et al.* 2007). This effect may vary according to stage/age and body temperature. In fish, IGF-2 appears to be the most important temperature-mediated factor in embryonic growth rate; however, in late-stage embryos and juveniles, increases in environmental temperature can lead to stimulation of the GH/IGF-1 axis (Gabillard *et al.* 2005; Li and Leatherland 2008). Furthermore, though several fish studies have shown water temperatures between 25-30°C to result in higher IGF-1 levels than 20°C, the optimal temperature for growth may vary among species. For juvenile Flounder (*Paralichthys lethostigma*) raised at 23°C and 28°C (well within the 5-30°C range at which they can be found in the wild), growth rates are similar until 100 mm, at which length they diverge, with fish from the lower-temperature environment growing 65-83% larger than those from the higher temperature.

The effects of photoperiod and temperature on IGF-1 levels in snakes and other reptiles, either in captivity or in the wild, remains entirely unexplored, but may provide insight into mechanisms signaling the onset of growth and reproduction following an inactive season.

15.3.4 IGF-1 and Stress

The effects of stress on IGF-1 activity are of interest for two main reasons: (1) naturally-occurring environmental stressors, such as frequency of predator encounters, may affect growth rates via effects on the GH/IGF-1 axis, and (2) anthropogenically-induced stress, such as capture, handling, and confinement, may significantly alter IGF-1 expression.

The effects of stress on IGF-1 activity have been most directly tested in the context of anthropogenically-induced stress. A handful of studies in fish have shown stress to have a negative effect on IGF-1 activity, either via decreased plasma IGF-1, or modulation of IGF-1 binding protein levels (e.g., Dyer *et al.* 2004; Davis and Peterson 2006; but see McCormick *et al.* 1998). In salmonids, stunted growth in fish subject to the stress of premature transfer from salt to freshwater conditions is associated with both reduced feeding and reduced IGF-1 levels (Dyer *et al.* 2004). The timescale in which the effects of stress on IGF-1 levels is manifest, however, is variable among studies. Wild-caught Southern Bluefin Tuna (*Thunnus maccoyii*), exhibited a decrease in IGF-1 from 48 ng/ml at capture to 28 ng/ml after three weeks in confinement, and did not recover normal levels until the sixth week of confinement (Dyer *et al.* 2004). In domesticated Sunshine Bass (*Morone chrysops* ♀ ×*Morone saxatilis* ♂ hybrid), plasma IGF-1 levels decreased by two hours, and return to normal by twenty-four hours (Davis and Peterson 2006). Handling and isolation of hatchery-reared Silver Perch (*Bidyanus bidyanus*) reduced plasma IGF-1 from 36 ng/ml to 19.5 ng/ml after twelve hours, with recovery commencing after twenty-four hours (Dyer *et al.* 2004). In contrast, wild-caught Black Bream (*Acanthopagrus butcherii*) subject to the same experimental design, showed significantly reduced plasma

IGF-1 levels twenty-four hours after handling, and retained lower levels than unstressed controls for a full week, in association with higher cortisol concentrations. Dyer *et al.* (2004) note that while the differences in recover time between perch and bream may be due to species-specific differences, it is also possible that wild fish recover less rapidly from stressors than domesticated fish.

No study has directly tested for the effects of stress on IGF-1 in any reptile. However, a small sample of free-ranging *Thamnophis elegans* bled hours after capture showed significantly lower plasma IGF-1 levels than those bled immediately upon capture (A. Sparkman, unpublished data; Fig. 15.2). Comparison with blood samples taken after several days of confinement suggests, however, that *T. elegans* IGF-1 levels are able to recover in the short-term. Interestingly, these differences in plasma IGF-1 levels with sampling time are inversely related to changes in corticosterone levels within the same individuals (Fig. 15.2). Furthermore, while repeated measures across individual *T. elegans*, bled immediately upon capture (A) and again three hours later (B) are positively correlated (n = 41; P = 0.0035), there is a significant time by SVL interaction in a repeated measures analysis (n = 41; sampling time*SVL: F = 4.66; P = 0.0371). The interaction is the result of a significant positive relationship between IGF-1 and SVL for A samples, in contrast to a lack of relationship between IGF-1 and SVL in B samples (A. Sparkman, unpublished data). Thus larger *T. elegans* individuals may experience a more pronounced response to capture stress than smaller individuals, perhaps indicating that small individuals prioritize growth over stress response relative to large individuals.

These data suggest that in snakes, as in fish, stress is capable of suppressing IGF-1 activity, and it may do so in a size/age-specific manner. Whether acute or chronic stress is capable of exerting discernable effects on long-term growth rate via IGF-1 suppression in either natural or experimental conditions will likely depend on the type, strength, duration, and frequency of the stressor, and the ability of a particular species to adjust to new and/or stressful conditions. The effects of stress on IGF-1 activity, and its importance for growth rate in snakes would itself be an interesting line of research, both in the laboratory and in the wild. For instance, if predator presence and/or attack is capable of inducing stress-related declines in IGF-1 activity on a regular basis, this—especially in combination with any predator-mediated reduction in foraging and thermoregulatory behavior—could act as a mechanism for reduced growth rates, particularly in food-limited systems. It would also be of interest to test whether there is latitudinal or seasonal variation in the response of IGF-1 to stress, with adaptive ramifications for growth and/or reproduction, even as there is evidence of adaptive variation in the ability to suppress the corticosterone mediated-stress response in the context of reproductive status and the cost to overall reproductive success (Silverin and Wingfield 1998; Moore and Mason 2001).

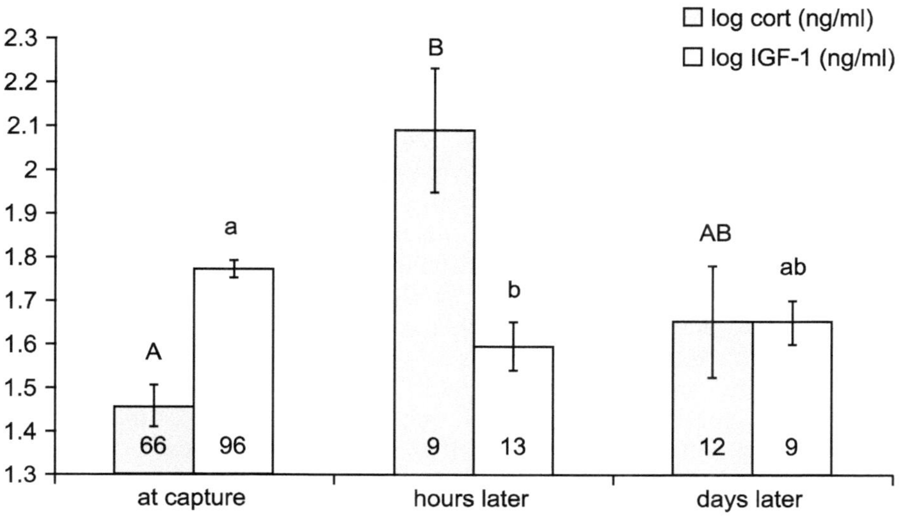

Fig. 15.2 Field levels of corticosterone and IGF-1 in *T. elegans* immediately after capture, within hours, and within days (A. Sparkman, unpublished data). Sample sizes for each group are indicated at the base of columns. Standard errors of the means are indicated. Capital letters show significant differences between groups for corticosterone ($P < 0.0001$), small-case letters for IGF-1 ($P = 0.0018$) (Sparkman and Bronikowski, unpublished data).

The effects of stress on IGF-1 are also of methodological importance as study of IGF-1 is pursued with regard to other research questions. The preponderance of studies of IGF-1 activity have thus far been conducted in domesticated species and laboratory model systems, and very few studies have focused on the effects of stress. As we begin to examine IGF-1 in free-ranging, non-model species, it is imperative that we understand how stress affects our ability to track true patterns and test hypotheses regarding IGF-1 activity. In the field, for instance, care should be taken to procure blood samples shortly after capture, and in the laboratory, care should be taken to ascertain that blood samples are taken only when and if snakes have become acclimated to their new environment.

15.4 IGF-1 AND LIFE-HISTORY TRAITS

15.4.1 IGF-1, Growth Rate and Body Size

Growth rate has direct ramifications for the expression of other life history traits, which develop as an organism grows—for instance, it is intimately involved in the determination of adult body size and timing of sexual maturation. In turn, adult size can play a pivotal role in determining the volume of reproductive output, particularly in species with indeterminate growth, where reproductive output may increase with size far past maturity (e.g., Sparkman *et al.* 2007).

While several hormones are known to influence growth and metabolism, including insulin, thyroid hormones, and corticosteroids, IGF-1 has been shown to be one of the major players in regulating growth rate and adult body size across vertebrates. For instance, chickens selected for fast growth show higher levels of IGF-1 than slow-growth lines, and larger breeds of dogs and pigs show higher levels of IGF-1 than smaller breeds (Eigenmann *et al.* 1984; Buonomo *et al.* 1987; Scanes *et al.* 1989; Beccavin *et al.* 2001). Circulating levels of IGF-1 are also correlated with adult body size in mice and humans (e.g., Binoux and Gourmelen 1987; Naar 1991). Furthermore, one recent study has shown that a single nucleotide polymorphism (SNP) in the IGF-1 gene is linked to body size differentiation among dog breeds (Sutter *et al.* 2007).

As in other species, plasma IGF-1 in *Lamprophis fuliginosus* is positively correlated to prematuration growth rate (Sparkman *et al.* 2010). However, when reproduction commences, the relationship between IGF-1 and growth rate disappears, presumably because at this time it is involved in a variety of reproduction-related functions as well. IGF-1 levels have been shown to be elevated in larger relative to smaller *Caretta caretta*, but this trend may be driven by a large representation of reproductive females in the large-individual sample, suggesting the difference between large/small, or young/old turtles may be due to reproductive status rather than size *per se* (Crain *et al.* 1995a). Similarly, IGF-1 increases with size in male and female *Thamnophis elegans*, but only after maturity when growth rates have begun to slow, suggesting that this pattern is driven at least in part by increasing reproductive output with size (Sparkman *et al.* 2009; Fig. 15.3; see section 15.4.2). Nevertheless, it is likely that even when any direct relationship between IGF-1 and growth rate is masked by simultaneous involvement in reproductive activities, it continues to be involved in regulating growth rate throughout the lives of indeterminately growing organisms. The extent to which it is involved in generating life histories characterized by determinate vs. indeterminate growth is as yet unknown, but ripe for investigation.

Interestingly, plasma IGF-1 in free-ranging *Thamnophis elegans* showed a significant relationship to shedding, or ecdysis, a condition that is by nature strongly associated with active growth (Sparkman *et al.* 2009). It is possible that higher IGF-1 levels in ecdytic individuals is simply indicative of a positive relationship between IGF-1 and growth rate. However, one would expect that many non-ecdytic snakes were nevertheless growing at comparable rates at the time of capture, making it unlikely that ecdytic individuals would have significantly higher levels of IGF-1 for this reason alone. An alternative possibility is that IGF-1 is itself actively involved in the growth of new skin during ecdysis. IGF-1 is known to act in an autocrine manner to promote epithelial development, suggesting plasma IGF-1 may also support this function in snakes (Haase *et al.* 2003).

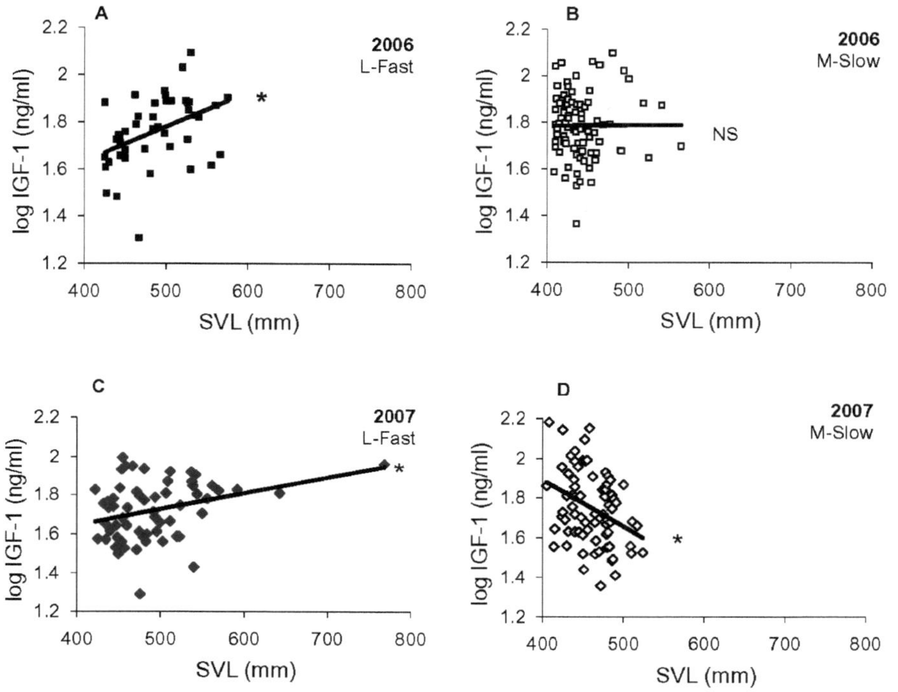

Fig. 15.3 Curves of log IGF-1 (ng/ml) against snout-vent length of free-ranging *Thamnophis elegans* for two years. **A.** 2006 fast-growth lakeshore, **B.** 2006 slow-growth meadow, **C.** 2007 lakeshore, and **D.** 2007 meadow snakes. Asterisks denote significant effects. Note: significant effect remains even if large snakes are excluded from C. (From Sparkman *et al.* 2009. Ecology 90: 720-728, Fig. 3, with permission of Ecological Society of America.)

15.4.2 IGF-1 and Maturation

IGF-1 is known to stimulate the production of gonadotropin-releasing hormone (GnRH) and luteinizing hormone (LH) by the hypothalamus and pituitary, respectively, which are responsible for the initiation of sexual maturation in vertebrates (reviewed in Daftary and Gore 2005). Consequently, since elevated IGF-1 is associated with high nutrition, large body sizes, and high growth rates, it is thought to provide a communication link between the somatic and reproductive systems, signaling an organism's readiness to mature. In ocean Chum Salmon (*Oncorhynchus keta*) for instance, plasma levels of maturing adults are two to three times that of immature fish, and in rodents and primates, administration of exogenous IGF-1 can significantly advance onset of maturation (Hiney *et al.* 1996; Wilson 1998; Onuma *et al.* 2009).

Cross-sectional analysis of free-ranging *Thamnophis elegans* blood samples reveals no significant variation in IGF-1 with size prior to

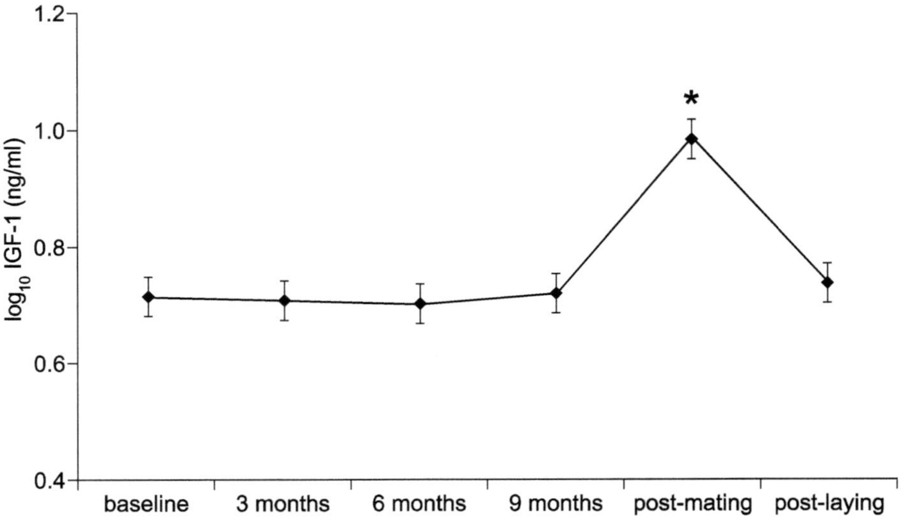

Fig. 15.4 Least square means and standard errors of the means of *Lamprophis fuliginosus* $\log_{10}$ plasma IGF-1 according to stage from a repeated measures analysis. (Sparkman *et al.* 2010.)

maturation (Sparkman *et al.* 2009). Similarly, there is no clear of evidence of a strong "spike" in IGF-1 prior to maturation in repeated measures of lab-reared *Lamprophis fuliginosus* (Fig. 15.4). Nevertheless, it is possible that we would have picked up a signal in free-ranging snakes had we been able to obtain repeated measures of individuals, and in the case of lab snakes, if we had taken blood samples at shorter time intervals. In *L. fuliginosus*, IGF-1 levels at six months of age were negatively correlated to age at first mating (Sparkman *et al.* 2010). Since each female in this study was given the opportunity to mate twice a month from the age of four months, age at first mating serves as a robust index of age at maturation. This suggests that individuals that matured earlier had higher levels of IGF-1 at six months (the earliest age at maturation) than those that matured late. This finding is consistent with the prediction that IGF-1 would rise prior to maturation in snakes as in other taxa. However, this pattern may also be driven simply by the fact that those that had the highest IGF-1 and were growing fastest at six months were the first to mature simply because they were first to reach a threshold size for maturity. The two possibilities are in fact complementary, if at a sufficiently advanced developmental stage high IGF-1 levels both facilitate continued growth and signal that an organism is ready to mature. Nevertheless, direct mechanistic relationships between IGF-1, gonadotropins, and maturation have yet to be established in any reptile. Snakes and lizards will be particularly interesting to investigate in this regard, as they appear to carry only a single gonadotropin and receptor, in contrast to other vertebrate groups, which have two (Licht 1983; Bluhm *et al.* 2004). The ramifications of this unique evolutionary development in squamates for the underlying mechanisms of maturation, and the role of IGF-1 therein, is a promising avenue for future study.

15.4.3 IGF and Reproduction

IGF-1 is involved in reproductive activities at almost every level. In addition to any role in maturation via gonadotropins, the effects of IGF-1 also extend directly to stimulation of the gonads, where it can facilitate sex steroid secretion, gamete (sperm and egg) development, follicle selection, luteal maintenance, embryonic implantation, and maternal transfer of nutrients into the embryo (e.g., Kagawa *et al.* 1994; Maestro *et al.* 1999; Macpherson *et al.* 2002; Ginther *et al.* 2004; Wuertz *et al.* 2007; Silva *et al.* 2009). In viviparous species such as humans, horses, and cattle, IGF-1 has been shown to rise during gestation, and peak during mid-gestation (e.g., Baxter and Martin 1989; Hess-Dudan *et al.* 1994; Moyes *et al.* 2003).

While the mechanistic details of IGF-1 involvement in reproduction have not yet been investigated in detail in reptiles, a role of IGF-1 in reproduction has so far received strong support. IGF-1 immunoreactivity has been found in the reproductive tracts of female *Alligator mississippiensis*, during vitellogenesis, and plasma IGF-1 is elevated in both reproductive *A. mississippiensis* and, *Caretta caretta* (Cox and Guillette 1993; Crain *et al.* 1995b; Guillette *et al.* 1996). Furthermore, gravid free-ranging *Thamnophis elegans* characterized by high reproductive output have higher levels of IGF-1 than those from populations with low reproductive output (Sparkman *et al.* 2009). Thus IGF-1 clearly appears to be involved in one or more aspects of reptile reproduction. What is less clear, however, is how IGF-1 is involved in phenotypes characterized by determinate vs. indeterminate reproductive output, and possibly spontaneous vs. mating-induced ovulation.

As previously mentioned, we found a consistent positive relationship between IGF-1 and body size in adult, non-food limited wild populations of *Thamnophis elegans* living along the lakeshore of Eagle Lake and sampled over three consecutive years (Sparkman *et al.* 2009; Fig. 15.3). This pattern is distinct from that found in mammalian species, where IGF-1 peaks during maturation, declines to relatively stable levels during adulthood, and exhibits senescent decline at later ages (e.g., Yamamoto *et al.* 1991; Hess and Roser 2001). It is possible that the ability to continue to increase reproductive output with size/age in *T. elegans* past maturity is linked at least in part to the ability to secrete increasingly higher levels of IGF-1, to facilitate increasingly higher reproductive output. This explanation could be applicable to both males and females, as IGF-1 is involved in the development of both eggs and sperm.

Interestingly, in populations of *Thamnophis elegans* living in meadows surrounding Eagle Lake, we found a plastic relationship between IGF-1 and body size. Meadow populations sampled from May through July exhibited no relationship between IGF-1 and adult size in one year, and a significant negative relationship the next. Furthermore, populations sampled in May alone during a third year showed a positive relationship between IGF-1 and adult size. As meadow populations experience dramatic fluctuations in prey availability from year to year, and across the summer months, we believe that this plasticity may reflect a plastic response of IGF-1 to prey availability. The optimal relationship between IGF-1 and adult size is likely to be positive, therefore, as seen both in snakes from lakeshore populations (where prey are consistently abundant) and meadow populations in early summer (when amphibian prey are more likely to be abundant, even in a dry year—see section 15.3.2). Whether IGF-1 provides a mechanistic link between environmental food availability and diverse aspects of reproduction in the wild should be the subject of future research.

While we did not find clear evidence of a peak in IGF-1 prior to maturation in lab-reared *Lamprophis fuliginosus*, we did find a surprising peak in IGF-1 post-mating (Fig. 15.4). It is possible that this peak is actually indicative of a peak in IGF-1 associated with maturation prior to mating, since snakes were bled at three-month intervals prior to mating. Alternatively, it may be that these snakes exhibit a mating-induced surge in plasma IGF-1. Repeated measures of two individuals within a short time interval support this latter interpretation: one individual bled at six months and successfully mated and bled three days later showed an increase in IGF-1 from 7.5 to 45 ng/ml, while another individual also bled at six months and mated and bled seven days later showed an increase from 6.38 to 34.6 ng/ml. Finer-scale sampling should be conducted to establish whether this peak in IGF-1 occurs just prior to mating, or is actually induced by mating. Nevertheless, these results are intriguing, and suggest that in some oviparous snake species, IGF-1 may play a unique and previously undescribed role in the initiation of reproduction. It is not currently known whether *L. fuliginosus* exhibits mating-induced ovulation or vitellogenesis, as reported in some species of mammals, birds, and reptiles (reviewed in Bakker and Baum 2000; DeNardo and Autumn 2001, Mathies *et al.* 2004; Manire *et al.* 2008). However, should this be the case, it may be that IGF-1 supports a rapid initiation of reproductive activity subsequent to mating.

One other unique pattern, also consistent with a mating-induced peak in IGF-1, was also found in *Lamprophis fuliginosus*. As individuals were bled anywhere from 0-16 days subsequent to mating, we were able to examine the relationship between plasma IGF-1 and days elapsed after mating. We found a strong negative relationship, particularly between days 0-9 (Sparkman *et al.* 2010). This pattern is in contrast to what has been found in mammals, where IGF-1 is highest mid-gestation, as well as in another oviparous reptile, *Alligator mississippiensis*, where IGF-1 peaks mid-gravidity, when luteal activity and progesterone secretion—known to be stimulated by IGF-1—are also elevated (Guillette *et al.* 1996). Since

A. mississippiensis and *L. fuliginosus* are gravid for approximately the same length of time (20-25 days), one would expect these species to show similar patterns in IGF-1 levels during gravidity. However, plasma IGF-1 shows a decline in *L. fuliginosus* during the same gestational period that it is rising in *A. mississippiensis*. Whether this indicates a lesser involvement of IGF-1 mid-reproduction physiology (such as luteal activity and progesterone secretion), but a greater involvement in early-reproductive events (such as ovulation) in snakes remains to be determined.

15.4.4 IGF-1, Aging and Lifespan

There has been a dramatic increase in the study of IGF-1 over the past decade in large part due to its involvement in controlling the rate of aging—potentially through stress resistance—and ultimately, lifespan. A number of studies, encompassing a diversity of laboratory organisms, have found that up- or down-regulation of insulin/IGF signaling impacts lifespan, shortening or lengthening it, respectively (reviewed in Piper *et al.* 2008; Greer and Brunet 2009). Relevant to the topic of this volume, much of the lifespan extending effects of IGF-1 are brought about by modifying stress resistance and decreasing reproduction. Our interest, therefore, in IGF-1 vis-à-vis lifespan is as a pleiotropic hormone that may underpin the universal trade-off between lifespan and reproduction. With regards to snakes in particular, a further consideration is that within their iteroparous reproductive strategy, lifespan itself will determine the length of the reproductive lifespan, and thereby Darwinian fitness. Thus, IGF-1 may impact reproduction relative to lifespan, either directly through altered cell signaling, or indirectly, through lifespan effects on reproductive opportunity.

Our understanding of IGF-1 function has been fostered by its pedigree as a member of the insulin/insulin-like family of signaling molecules that have been highly conserved across invertebrates and vertebrates. Insulin/IGF signaling was first shown to be an integral regulator of lifespan in *Caenorhabditis elegans*, where mutations decreasing the activity of the DAF-2 (insulin-like) receptor extended lifespan of worms by over 100% (Kenyon *et al.* 1993; Kimura *et al.* 1997). Similarly, mutations of the insulin-like receptor, which causes lowered signaling levels in *Drosophila melanogaster*, increased female lifespan by as much as 85% (Tatar *et al.* 2001). In mammals, a landmark study in heterozygous IGF-1R knockout mice demonstrated that a reduction in IGF-1 signaling increased female lifespan by 26% (Holzenberger *et al.* 2003). Furthermore, there is some evidence from genetic association studies in humans that polymorphisms in the IGF-1 signaling cascade are involved in exceptional longevity (Suh *et al.* 2008; Willcox *et al.* 2008; Flachsbart *et al.* 2009). The life-extending effects of reduced insulin/IGF signaling are thought to be mediated by increased activity of a downstream forkhead transcription factor (DAF-16/FOXO) that moves from the cytosol to the nucleus and upregulates key cellular stress resistance genes (reviewed in Tater *et al.* 2003, Schwartz and Bronikowski 2010). The oxidative damage theory of aging postulates that lifespan is determined by a combination of rates of reactive oxygen species (ROS) production (which inflicts damage on cellular protein, lipids, and nucleic acids), antioxidant capacity, and repair mechanisms (reviewed in Monaghan *et al.* 2008). Thus, insulin/IGF signaling constitutes an endocrine mechanism for understanding rates of aging via modulation of oxidative stress resistance.

Although the study of IGF-1 as it relates to aging, and the trade-off between lifespan and reproduction, is in its initial stages in reptiles, our recent findings are relevant to this area of inquiry. In *Thamnophis elegans* from representative populations of the two ecotypes discussed earlier (short- and long-lived with concomitant high and low reproductive effort respectively), we have found that long-lived and highly iteroparous meadow individuals show enhanced plasticity across season, but less effect of body size on IGF-1 blood plasma levels (discussed above). Neonates from these long-lived/low-reproductive effort populations are found to have smaller body sizes, higher mitochondrial conversion of ADP to ATP with lower amounts of ROS generated during electron transport, and higher efficiency of repair of inducible DNA damage (Robert and Bronikowski 2010). This suite of physiological variables combined with lower corticosterone reactivity (discussed above and in Robert *et al.* 2009), lower size-mediated IGF-1 levels, and potentially lower IGF-1 levels in years without food suggest a complex network of physiological traits underlying life-history traits (and trade-offs), many potentially controlled by IGF-1 signaling.

Elsewhere we have argued that the indeterminate growth and indeterminate fecundity in the garter snake study system may lead to negligible senescence despite lifespan differentiation (Sparkman *et al.* 2007; Robert and Bronikowski 2010). However, it is an interesting hypothesis to entertain that overall lifetime exposure or load of IGF-1 is lower in the meadow ecotype (see section 15.4.5), through both environmental effects and genetic differentiation of the two ecotypes, and that IGF-1 may also be the main determinant of life-history differentiation in this system.

15.4.5 IGF-1 and Life-History Evolution

Little is yet known regarding how the endocrine system is involved in the evolution of life-history traits in wild populations—what kinds of changes in hormonal structure, regulation, or cellular environment, for instance, are involved in the transition of a lineage from slow to fast growth, or from low to high reproductive output (Ketterson and Nolan 1992; Finch and Rose 1995; Zera and Bottsford 2001; Ricklefs and Wikelski 2002). The potential of the endocrine system as a chief player in the evolution of life-history trade-offs is supported by three general characteristics: (1) Individual hormones may exert widespread pleiotropic effects. Each hormone may bind to multiple target tissues, and initiate multiple signal transduction pathways and changes in gene transcription within each tissue. Thus, adjustments in the regulation of a single hormone and or/its receptors and binding proteins have the potential to either constrain or facilitate the evolution of correlated traits (Ketterson and Nolan 1992, 1999; Finch and Rose 1995; Zera and Bottsford 2001). (2) Selection experiments indicate that hormone levels can have a quantitative genetic basis. For example, chickens selected for low juvenile body weight exhibit lower titers of insulin-like growth factors-1 and -2 at early ages than those selected for high juvenile body weight (Scanes *et al.* 1989). Thus selection on hormonally-mediated traits can create sustained cross-generational change in hormonal expression. (3) The endocrine system can be plastic in response to external stimuli, such as temperature or food availability, and is thus continually mediating adaptive responses to the environment.

A number of recent studies suggest that looking at diverse hormones such as follicle-stimulating hormone (FSH), corticosterone and testosterone to understand the mechanisms underlying life-history trade-offs is a productive approach (e.g., Sinervo and Licht 1991; Clark *et al.* 1997; Wingfield *et al.* 1998; Lancaster *et al.* 2007; Mills *et al.* 2008). Most notable have been investigations demonstrating the role of juvenile hormone in a trade-off between reproduction (ovary mass) and dispersal ability (wing size) in the Field Cricket (*Gryllus firmus*) (Zera 1999; Zera *et al.* 2007), and the role of testosterone in trade-offs among parental care, sexual displays, and condition in Juncos (*Junco hyemalis*) (Ketterson and Nolan 1999; Reed *et al.* 2006; McGlothlin *et al.* 2007). Recent studies in insects have also uncovered a trade-off between growth/reproduction and lifespan mediated by juvenile hormone (Tatar and Yin 2001; Flatt *et al.* 2005; Flatt and Kawecki 2007). In vertebrates, IGF-1 is also a prime candidate in this regard. IGF-1 exemplifies each of the three general characteristics of the endocrine system listed above: (1) IGF-1 has well-established pleiotropic effects, and is highly involved in growth, gamete development, and lifespan (reviewed in Tatar *et al.* 2003; Bartke 2005, 2008). (2) Selection experiments in domestic species show that levels of IGF-1 have a quantitative genetic basis and that lines selected for fast growth exhibit higher levels (e.g., Scanes *et al.* 1989; Beccavin *et al.* 2001), (3) IGF-1 responds plastically to environmental factors such as nutrition, photoperiod, temperature, and stress (e.g., Crain *et al.* 1995a; Davis and Peterson 2006; Shimizu *et al.* 2006).

We have presented evidence that IGF-1 in snakes also shares these three general characteristics, as it has been shown to be involved in both growth and reproduction (though its effect on lifespan remains unknown) and respond plastically to nutritional cues; furthermore, we found a maternal effect on IGF-1 levels in *Lamphrophis fuliginosus* that was significant at every major life stage (juvenile, post-mating, post-laying), suggesting that IGF-1 levels in snakes are indeed heritable (Sparkman *et al.* 2009; Sparkman *et al.* 2010).

In the only study to test for a role of IGF-1 in the correlated evolution of life-history traits in the wild, we found significant distinctions in lifetime IGF-1 profiles between fast-living lakeshore and slow-living meadow *Thamnophis elegans* ecotypes (Sparkman *et al.* 2009). In all three years of study, plasma IGF-1 was higher in gravid females from the high-reproduction lakeshore ecotype than those from the low-reproduction meadow ecotype, suggesting that evolution in IGF-1 regulation may be facilitating the evolution of reproductive output in this system. Furthermore, lakeshore non-gravid adults consistently exhibited a positive correlation between IGF-1 and body size, while in non-gravid meadow adults this relationship was different each year—there was no relationship the first (wet) year, a negative relationship the second (dry) year, and a positive relationship during the third (restricted to a brief, early wet period) year. In general, both lakeshore and meadow snakes showed no relationship to sampling date in a wet year, but meadow snakes showed a decline in IGF-1 over the summer months in a dry year. We suggest that the variability in the relationship between IGF-1 and body size and season in meadow snakes is due to variability in amphibian prey availability in the meadows in years with different amounts of rainfall, since IGF-1 responds plastically to changes in nutrition. We also found significantly higher levels of IGF-1 in shedding snakes. As lakeshore snakes grow faster, and therefore shed more frequently, they may have more frequent peaks in IGF-1 than meadow snakes. Overall, we conclude that IGF-1 may have facilitated the evolution of high-reproductive performance in lakeshore snakes, and that these animals (whether primarily due to genetic or environmentally-based differences in plasma IGF-1) are likely to have cumulatively higher lifetime levels of IGF-1 than meadow snakes. Thus we provide a foundation for determining whether higher lifetime levels of IGF-1 activity in lakeshore snakes makes them less resistant to oxidative stress over the long term, and thus produces a trade-off with lifespan.

This predicted higher lifetime level of IGF-1 in lakeshore snakes is a promising mechanism for a trade-off between early and late-life traits in these snakes—though that trade-off has yet to be established (see Sparkman *et al.* 2007). But even so, it wouldn't necessarily be an evolved trade-off. The significant difference in IGF-1 levels between gravid lakeshore females and gravid meadow females is the most promising in that regard, because it is present in both wet and dry years. But a trade-off with lifespan resulting from fluctuating levels IGF-1 in response to variable prey availability, while certainly interesting in its own right, would not be an evolutionary explanation for differences in lifespan between the two ecotypes. Whether there is a genetic link between early-life growth rate and IGF-1, that is present regardless of prey availability, remains to be established.

Of course, to fully understand the activity of any one hormone, it is important to consider not only the hormone itself, but also its binding proteins and receptors, as well as key molecular agents in the signaling pathways downstream. Selection may result in changes in one or all of these components to influence the phenotypic end-result of their cumulative activity. Thus, though plasma IGF-1 has been associated with higher growth rate in other study systems, this measure should not stand alone. Not only should the IGF-1 receptor be considered, but also the multiple binding proteins that serve to increase its half-life in the blood and may either facilitate or inhibit IGF-1 movement and receptor binding (Duan 2005). Members of the IGF-1 signal transduction pathway, such as $p66^{shc}$, which has known effects on stress resistance and lifespan in mammals, are also of interest here (Pelicci *et al.* 1992; Migliaccio *et al.* 1999). Other growth and metabolic hormones, such as the thyroid hormones, triiodothyronine (T_3) and thyroxine (T_4), corticosterone, insulin, and IGF-2 may also contribute to our understanding of the physiological trade-offs that mould life-history strategies. While we found no significant difference in T_3 between the two ecotypes (A. Sparkman, unpublished data), T_4 should also be examined. Furthermore, there are plastic ecotype differences in stress physiology, including corticosterone activity, that may well factor into how life-histories have evolved in these snakes (Robert *et al.* 2009; Robert and Bronikowski 2010; Sparkman *et al.* in prep).

There are clear differences in plasma IGF-1 levels associated with differences in adult size, age at maturation and longevity between wild-derived and laboratory mice, with wild mice exhibiting lower IGF-1 levels, maturing later at smaller sizes, and living longer (Miller 2002). Thus the likelihood that plasma IGF-1 has been, at least in some instances, involved in diversification in life-history traits among closely related species is strong. Only one study, conducted in captive papionin primates (baboons, mangabeys, macaques, etc.), has tested for a role of IGF-1 in the evolution of body size across species. Contrary to expectations, they found no relationship between juvenile or adult body size and plasma IGF-1 levels across species (Bernstein *et al.* 2007). Whether the IGF-1 system is not in fact a key player in body size differentiation across papionid species, or whether this differentiation is manifest on other levels, such as receptor regulation, remains to be seen. With their tremendous diversity in growth rate, body size, maturation, and lifespan, we contend that snakes are an ideal group on which to focus future work on the role of the IGF-1 system in life-history strategy, both through studies of plasticity within species and evolution across species.

15.5 MEASURING IGF-1 IN SNAKES

Among snakes, plasma IGF-1 has been measured in only two species, *Thamnophis elegans* and *Lamprophis fuliginosus.* A heterologous (i.e., cross-species) radioimmunoassay using an anti-human IGF-1 antibody that is commonly used in a variety of taxa was successfully validated for both of these species—in both cases, native IGF-1 bound the anti-human antibody in a consistent manner. In the interests of stimulating further study of IGF-1 in reptiles, we have attempted to validate the assay for a range of snake species. We attempted to validate the IGF-1 assay for 11 additional snake species, as well as re-validate (for purposes of comparison) the *Chrysemys picta* (originally validated in Crain *et al.* 1995b). Interestingly, we were only able to validate the assay for the painted turtle and seven out of the eleven snake species examined (Sparkman, unpublished data). Of those seven snake species, the assay was validated for five natricine species and a single Xenodontinae species; however we were unsuccessful with four out of five Colubrine species examined and a Pseudoxyrophiinae species validation was also unsuccessful (taxonomic categorization based on Lawson *et al.* 2005). In these latter species, native IGF-1 did not consistently bind to the anti-human antibody in a predictable manner, making the assay inappropriate for accurate measures of plasma IGF-1. This finding is surprising, given that the anti-human IGF-1 antibody has consistently bound IGF-1 in diverse vertebrate groups—mammals, birds, crocodilians and turtles alike (e.g., Wilson and Hintz 1982; Crain *et al.* 1995b; Webster *et al.* 1996; Schmidt and Kelley 2001). We are currently testing the possibility that the IGF-1 coding sequence may have diverged in some snake species so as to alter protein shape in ways that prevent heterologous antibody binding. In the meantime, appropriate precautions should be taken to validate the IGF-1 assay using the human antibody for all new snake species studied.

Two other practical considerations should be kept in mind. First, we have observed that in taking blood samples from snakes in the tail region, it is quite common to tap not only blood vessels, but also lymphatic fluid. Any sample containing a mixture of lymphatic fluid and blood will yield inaccurate results when the concentration of IGF-1 (or any other hormone) is assayed. Reptile blood tends to be lighter and more dilute than mammalian or bird blood; nevertheless, it is quite possible to evaluate whether a sample is primarily composed of blood alone, and care should be taken to obtain samples of a dark, concentrated red coloration.

Second, as we have emphasized in the section on the relation between stress and IGF-1, it is important to ascertain that the results in any given experimental design are not compromised by stress. High levels of physiological stress can interfere not only with the expression of various physiological actors, such as IGF-1, but can also induce unanticipated plasticity in life-history traits. We ourselves have had difficulty establishing a relationship between IGF-1 and growth rate in lab-reared *Thamnophis elegans* due to unforeseen stressful events that resulted in highly plastic changes in growth rate among individuals during the intervals between length and mass measurements (A. Sparkman and A. Bronikowski, unpublished data). In the field, samples should be taken immediately upon capture (unless, of course, the effect of stress itself is of interest). In the laboratory, the acclimation of snakes to the captive environment should be assessed by appropriate physiological (such as corticosterone or heterophil:lymphocyte ratios in the blood) and/or behavioral measures.

15.6 SUMMARY AND FUTURE RESEARCH

We have reviewed our research on IGF-1 and life-history traits, including reproduction, and the relevant literature on IGF-1 in reptiles, with an emphasis on snakes. Several themes have emerged: (1) as in model laboratory systems, IGF-1 appears to be important in determining the life-history and likely reproductive schedule of snakes. (2) IGF-1 blood plasma levels are related either directly or indirectly to environmental stimuli such as food availability and temperature, as well as ecdysis. And (3) IGF-1 may be necessary for maturation, reproduction, and perhaps even lifespan determination in snakes. The details and specifics of IGF-1 involvement in reproduction and other life-history traits is in exciting area of research, made even more so by recent discoveries of links to aging rates, yet we are only at the beginning of understanding the impacts of IGF-1 expression in snakes. Future research questions in snakes are sorely needed to determine: (1) The role of IGF-1 in maturation, including whether it responds to mating or initiates receptivity to mating. (2) Detailed studies of both direct and indirect effects of IGF-1 on growth and reproduction. (3) The overall plasticity of IGF-1 in relation to specific ecological factors such as stress, temperature, nutrition, and hibernation, all of which have been shown to impact IGF-1 levels in these and other systems. (4) Whether IGF-1 modulates aging as it does in laboratory systems, through increased resistance to damaging ROS and efficiency of repair mechanisms, and (5) How receptors and binding proteins are ultimately involved in regulating IGF-1 activity.

15.7 ACKNOWLEDGMENTS

Thanks to Carol Vleck for helping develop the IGF-1 RIA for snakes and to Dawn Byars for her work on the house snake project. Thanks also to Jill Madden and Tonia Schwartz for their on-going collaboration on the genetics of IGF-1 in snakes.

15.8 LITERATURE CITED

Bakker, J. and Baum, M. J. 2000. Neuroendocrine regulation of GnRH release in induced ovulators. Frontiers in Neuroendocrinology 21: 220-262.

Bartke, A. 2005. Minireview: Role of the growth hormone/insulin-like growth factor system in mammalian aging. Endocrinology 146: 3718-3723.

Bartke, A. 2008. Impact of reduced insulin-like growth factor-1/insulin signaling on aging in mammals: novel findings. Aging Cell 7: 285-290.

Baxter, R. C. and Martin, J. L. 1989. Binding proteins for the insulin-like growth factors: structure, regulation and function. Progress in Growth Factor Research 1: 49-68.

Beccavin, C., Chevalier, B., Cogburn, L. A., Simon, J. and Duclos, M. J. 2001. Insulin-like growth factors and body growth in chickens divergently selected for high or low growth rate. Journal of Endocrinology 168: 297-306.

Beckman, B. R., Larsen, D. A., Moriyama, S., Lee-Pawlak, B. and Dickhoff, W. W. 1998. Insulin-like growth factor-I and environmental modulation of growth during smoltification of Spring Chinook Salmon (*Oncorhynchus tshawytscha*). General and Comparative Endocrinology 109(3): 325-335.

Bernstein, R. M., Leigh, S. R., Donovan, S. M. and Monaco, M. H. 2007. Hormones and body size evolution in papionin primates. American Journal of Physical Anthropology 132: 247-260.

Binoux, M. and Gourmelen, M. 1987. Statural development parallels IGF I levels in subjects of constitutionally variant stature. Acta Endocrinologica 114: 524-530.

Bluhm, A. P. C., Toledo, R. A., Mesquita, F. M., Pimenta, M. T., Fernandes, F. M. C., Ribelac, M. T. C. P., Fátima, M. and Lazari, M. 2004. Molecular cloning, sequence analysis and expression of the snake follicle-stimulating hormone receptor. General and Comparative Endocrinology 137: 300-311.

Boeuf, G. and Le Bail, P.Y. 1999. Does light have an influence on fish growth? Aquaculture 144: 129-152.

Bronikowski, A. M. 2000. Experimental evidence for the adaptive evolution of growth rate in the garter snake *Thamnophis elegans*. Evolution 54: 1760-1767.

Bronikowski, A. M. 2008. The evolution of aging phenotypes in snakes: a review and synthesis with new data. Age 30: 169-176.

Bronikowski, A. M. and Arnold, S. J. 1999. The evolutionary ecology of life-history variation in the garter snake *Thamnophis elegans*. Ecology 80: 2314-2325.

Buonomo, F. C., Lauterio, T. J., Baile, C. A. and Campion, D. R. 1987. Determination of insulin-like growth factor-I (IGF1) and IGF binding-protein levels in swine. Domestic Animal Endocrinology 4: 23-31.

Byars, D. J. 2008. Influences of diet and dam on age of maturation in African house snakes, *Lamprophis fuliginosus*. M.S. thesis, University of Texas at Tyler, Tyler, Texas.

Chan, S. J., Cao, Q. P. and Steiner, D. F. 1990. Evolution of the insulin superfamily: cloning of a hybrid insulin/insulin-like growth factor cDNA from amphioxus. Proceedings of the National Academy of Sciences USA 87: 9319-9323.

Clark, M. M., Desousa, D., Vonk, J. and Galef, B. G., Jr. 1997. Parenting and potency: alternative routes to reproductive success in male Mongolian gerbils. Animal Behaviour 54: 635-642.

Clemmons, D. R. and Underwood, L. E. 1991. Nutritional regulation of IGF-I and IGF-II binding proteins. Annual Review of Nutrition 11: 393-412.

Cox, C. and Guillette, L. J. 1993. Localization of insulin-like growth factor-I-like immunoreactivity in the reproductive tract of the vitellogenic female American alligator, *Alligator mississippiensis*. The Anatomical Record 236: 635-640.

Crain, A. D., Bolten, A. B., Bjorndal, K. A., Guillette, L. J. and Gross, T. S. 1995a. Size-dependent, sex-dependent, and seasonal changes in insulin-like growth factor I in the Loggerhead Sea Turtle (*Caretta caretta)*. General and Comparative Endocrinology 98: 219-226.

Crain, A. D., Gross, T. S., Cox, M. C. and Guillette, L.J. 1995b. Insulin-like growth factor-I in the plasma of two reptiles: assay development and validations. General and Comparative Endocrinology 98: 26-34.

Daftary, S. S. and Gore, A. C. 2005. IGF-1 in the brain as a regulator of reproductive neuroendocrine function. Experimental Biology and Medicine 230(5): 292-306.

Davis, K. B. and Peterson, B. C. 2006. The effect of temperature, stress and cortisol on plasma IGF-1 and IGFBPs in sunshine bass. General and Comparative Endocrinology 149: 219-225.

DeNardo, D. F. and Autumn, K. 2001. Effect of male presence on reproductive activity in captive female blood pythons, *Python curtus*. Copeia 2001: 1138-1141.

Duan, C. 1998. Nutritional and developmental regulation of insulin-like growth factors in fish. Journal of Nutrition 128: 306S.

Duan, C. and Xu, Q. 2005. Roles of insulin-like growth factor (IGF) binding proteins in regulating IGF actions. General and Comparative Endocrinology 142: 44-52.

Dyer, A. R., Upton, Z., Stone, D., Thomas, P. M., Sool, K. L., Higgs, N., Quinn, K. and Carragher, J. F. 2004. Development and validation of a radioimmunoassay for fish insulin-like growth factor I (IGF-I) and the effect of aquaculture related stressors on circulating IGF-I levels. General and Comparative Endocrinology 135: 268-275.

Eigenmann, J. E., Patterson, D. F., Zapf, J. and Froesch, E. R. 1984. Insulin-like growth factor I in the dog: a study in different dog breeds and in dogs with growth hormone elevation. Acta Endocrinologica 105: 294-301.

Finch, C. E. and Rose, M. R. 1995. Hormones and the physiological architecture of life history evolution. The Quarterly Review of Biology 70: 1-52.

Flachsbart, F., Caliebe, A., Kleindorp, R., Blanche, H., von Eller-Eberstein, H., Nikolaus, S., Schreiber, S. and Nebel, A. 2009. Association of FOXO3A variation with human longevity confirmed in German centenarians. Proceedings of the National Academy of Sciences USA 106: 2700-2705.

Flatt, T. and Kawecki, T. J. 2007. Juvenile hormone as a regulator of the trade-off between reproduction and life span in *Drosophila melanogaster*. Evolution 61: 1980-1991.

Flatt, T., Tu, M.-P. and Tatar, M. 2005. Hormonal pleiotropy and the juvenile hormone regulation of *Drosophila* development and life history. BioEssays 27: 999-1010.

Ford, N. B. and Seigel, R. A. 2006. Intra-individual variation in clutch and offspring size in an oviparous snake. Journal of Zoology 268: 171-176.

Gabillard, J.C., Weil, C., Rescan, P.Y., Navarro, I., Gutierrez, J. and Le Bail, P.Y. 2005. Does the GH/IGF system mediate the effect of water temperature on fish growth? Cybium 29: 107-117.

Ginther, O. J., Gastal, E. L., Gastal, M. O. and Beg, M. A. 2004. Critical role of insulin-like growth factor system in follicle selection and dominance in mares. Biology of Reproduction 70: 1374-1379.

Greer, E. L. and Brunet, A. 2009. Different dietary restriction regimens extend lifespan by both independent and overlapping genetic pathways in *C. elegans*. Aging Cell 8(2): 113-127.

Guillette, L. J., Cox, M. C. and Crain, D. A. 1996. Plasma insulin-like growth factor-1 concentration during the reproductive cycle of the American alligator (*Alligator mississippiensis*). General and Comparative Endocrinology 104: 116-122.

Haase, I., Evans, R., Pofahl, R. and Watt, F. M. 2003. Regulation of keratinocyte shape, migration and wound epithelialization by IGF-1- and EGF-dependent signalling pathways. Journal of Cell Science 116: 3227-3238.

Hess, M. F. and Roser, J. F. 2001. The effects of age, season and fertility status on plasma and intratesticular insulin-like growth factor I concentration in stallions. Theriogenology 56: 723-733.

Hess-Dudan, F., Vacher, P. Y., Bruckmaier, R. M., Weishaupt, M. A., Burger, D. and Blum, J. W. 1994. Immunoreactive insulin-like growth factor I and insulin in blood plasma and milk of mares and in blood plasma of foals. Equine Veterinary Journal 26: 134-139.

Hiney, J. K., Srivastava, V., Nyberg, C. L., Ojeda, S. R. and Dees, W. L. 1996. Insulin-like growth factor-1 of peripheral origin acts centrally to accelerate the initiation of female puberty. Endocrinology 137: 3717-3728.

Holzenberger, M., Dupont, J., Ducos, B., Leneuve, P., Geloen, A., Even, P. C., Cervera, P. and Bouc, Y. L. 2003. IGF-1 receptor regulates lifespan and resistance to oxidative stress in mice. Nature 421: 182-187.

Imsland, A. K., Björnsson, B. T., Gunnarsson, S., Foss, A. and Stefansson, S. O. 2007. Temperature and salinity effects on plasma insulin-like growth factor-I concentrations and growth in juvenile turbot (*Scophthalmus maximus*). Aquaculture 271: 546-552.

Imsland, A. K., Foss, A., Roth, B., Stefansson, S. O., Vikingstad, E., Pedersen, S., Sandvik, T. and Norberg, B. 2008. Plasma insulin-like growth factor-I concentrations and growth in juvenile halibut (*Hippoglossus hippoglossus*): Effects of photoperiods and feeding regimes. Comparative Biochemistry and Physiology—Part A: Molecular and Integrative Physiology 151: 66-70.

Kagawa, H., Kobayashi, M., Hasegawa, Y. and Aida, K. 1994. Insulin and insulin-like growth factors I and II induce final maturation of oocytes of red seabream, *Pagrus major*, in vitro. General and Comparative Endocrinology 95: 293-300.

Kamangar, B. B., Gabillard, J.-C. and Bobe, J. 2006. Insulin-like growth factor-binding protein (IGFBP)-1, -2, -3, -4, -5, and -6 and IGFBP-related protein 1 during rainbow trout postvitellogenesis and oocyte maturation: Molecular characterization, expression profiles, and hormonal regulation. Endocrinology 147: 2399-2410.

Kelley, K. M., Schmidt, K. E., Berg, L., Sak, K., Galima, M. M., Gillespie, C., Balogh, L., Hawayek, A., Reyes, J. A. and Jamison, M. 2002. Comparative endocrinology of the insulin-like growth factor-binding protein. Journal of Endocrinology 175: 3-18.

Kenyon, C., Chang, J., Gensch, E., Rudner, A. and Tabtlang, R. 1993. A *C. elegans* mutant that lives twice as long as wild type. Nature 366: 461-463.

Ketterson, E. D. and Nolan, V. J. 1992. Hormones and life histories: an integrative approach. The American Naturalist 140: S33-S62.

Ketterson, E. D. and Nolan, V. J. 1999. Adaptation, exaptation, and constraint: a hormonal perspective. The American Naturalist 154: S4-S25.

Kimura, K. D., Tissenbaum, H. A., Liu, Y. and Ruvkun, G. 1997. daf-2, an insulin receptor-like gene that regulates longevity and diapause in *Caenorhabditis elegans*. Science 277: 942-946.

Lancaster, L. T., McAdam, A. G., Wingfield, J. C. and Sinervo, B. R. 2007. Adaptive social and maternal induction of antipredator dorsal patterns in a lizard with alternative social strategies. Ecology Letters 10: 798-808.

Lauterio, T. J. and Scanes, C. G. 1987. Time course of changes in plasma concentrations of the growth related hormones during protein restriction in the domestic fowl (*Gallus domesticus*). Proceedings of the Society of Experimental Biology and Medicine 185: 420-426.

Lawson, R., Slowinski, J. B., Crother, B. I. and Burbrink, F. T. 2005. Phylogeny of the Colubroidea (Serpentes): New evidence from mitochondrial and nuclear genes. Molecular Phylogenetics and Evolution 37: 581-601.

Li, M. and Leatherland, J. 2008. Temperature and ration effects on components of the IGF system and growth performance of rainbow trout (*Oncorhynchus mykiss*)

during the transition from late stage embryos to early stage juveniles. General and Comparative Endocrinology 155: 668-679.
Licht, P. 1983. Evolutionary divergence in the structure and function of pituitary gonadotropins of tetrapod vertebrates. American Zoologist 23: 673-783.
Macpherson, M. L., Simmen, R. C. M., Simmen, F. A., Hernandez, J., Sheerin, B. R., Varner, D. D., Loomis, P., Cadario, M. E., Miller, C. D., Brinsko, S. P., Rigby, S. and Blanchard, T. L. 2002. Insulin-Like growth factor-I and insulin-like growth factor binding protein-2 and -5 in equine seminal plasma: Association with sperm characteristics and fertility. Biology of Reproduction 67: 648-654.
Maestro, M. A., Mendez, E., Planas, J. V. and Gutierrez, J. 1999. Dynamics of insulin and insulin-like growth factor-I (IGF-I) ovarian receptors during maturation in the brown trout (*Salmo trutta*). Fish Physiology and Biochemistry 20: 341-349.
Manire, C. A., Byrd, L., Therrien, C. L. and Martin, K. 2008. Mating-induced ovulation in loggerhead sea turtles, *Caretta caretta*. Zoo Biology 27: 213-225.
Mathies, T., Franklin, E. A. and Miller, L. A. 2004. Proximate cues for ovarian recrudescence and ovulation in the brown treesnake (*Boiga irregularis*) under laboratory conditions. Herpetological Review 35: 46-49.
McCormick, S. D., Moriyama, S. and Bjornsson, B. T. 2000. Low temperature limits photoperiod control of smolting in Atlantic salmon through endocrine mechanisms. American Journal of Physiology—Regulatory, Integrative, and Comparative Physiology 278(5): R1352-1361.
McCormick, S. D., Shrimpton, J. M., Carey, J. B., O'Dea, M. F., Sloan, K. E., Moriyama, S. and Bjornsson, B. T. 1998. Repeated acute stress reduces growth rate of Atlantic salmon parr and alters plasma levels of growth hormone, insulin-like growth factor I and cortisol. Aquaculture 168: 2201-2235.
McGlothlin, J. W., Jawor, J. M. and Ketterson, E. D. 2007. Natural variation in a testosterone-mediated trade-off between mating effort and parental effort. The American Naturalist 170: 864-875.
McMurtry, J. P. 1998. Nutritional and developmental roles of insulin-like growth factors in poultry. Annual Poultry Nutrition Conference: Nutritional and Developmental Roles of Insulin-like Growth Factors between Species 62: 302S-305S.
Mendez, E., Planas, J. V., Castillo, J., Navarro, I. and Gutierrez, J. 2001. Identification of a type II insulin-Like growth factor receptor in fish embryos. Endocrinology 142(3): 1090-1097.
Migliaccio, E., Giorgio, M., Mele, S., Pelicci, G., Reboldi, P., Pandolfi, P. P., Lanfrancone, L. and Peliici, P. G. 1999. The p66shc adaptor protein controls oxidative stress response and life span in mammals. Nature 402: 309-313.
Miller, R. A., Harper, J. M., Dysko, R. C., Durkee, S. J. and Austad, S. N. 2002. Longer lifespans and delayed maturation in wild-derived mice. Experimental Biology and Medicine 227: 500-508.
Mills, S. C., Hazard, L., Lancaster, L., Mappes, T., Miles, D., Oksanen, T. A. and Sinervo, B. 2008. Gonadotropin hormone modulation of testosterone, immune function, performance, and behavioral trade-offs among male morphs of the lizard *Uta stansburiana*. American Naturalist 171: 339-357.
Mingarro, M., Vega-Rubin de Celis, S., Astola, A., Pendón, C., Valdiviab, M. M. and Pérez-Sánchez, J. 2002. Endocrine mediators of seasonal growth in gilthead sea bream (*Sparus aurata*): the growth hormone and somatolactin paradigm. General and Comparative Endocrinology 128: 102-111.
Monaghan, P., Charmantier, A., Nussey, D. H. and Ricklefs, R. E. 2008. The evolutionary ecology of senescence. Functional Ecology 22: 371-378.

Moore, I. T. and Mason, R. T. 2001. Behavioral and hormonal responses to corticosterone in the male red-sided garter snake, *Thamnophis sirtalis parietalis*. Physiology and Behavior 72: 669-674.

Moyes, T. E., Stockdale, C. R., Humphrys, S. and Macmillan, K. L. 2003. Differences in plasma concentration of insulin-like growth factor-1 between pregnant and non-pregnant dairy cows. Reproduction, Fertility and Development 15S: 22.

Naar, E. M., Bartke, A., Majumdar, S. S., Buonomo, F. C., Yun, J. S. and Wagner, T. E. 1991. Fertility of transgenic female mice expressing bovine growth hormone or human growth hormone variant genes. Biology of Reproduction 45: 178-187.

O'Steen, S. and Janzen, F. J. 1999. Embryonic temperature affects metabolic compensation and thyroid hormones in hatchling snapping turtles. Physiological and Biochemical Zoology 72: 520-533.

Olinski, R. P., Lundin, L. G. and Hallbook, F. 2006. Conserved synteny between the *Ciona* genome and human paralogons identifies large duplication events in the molecular evolution of the insulin-relaxin gene family. Molecular Biology and Evolution 23: 10-22.

Onuma, T. A., Makino, K., Ban, M., Ando, H., Masa-aki, F., Azumaya, T., Swanson, P. and Urano, A. 2009. Elevation of the plasma level of insulin-like growth factor-I with reproductive maturation prior to initiation of spawning migration of chum salmon. Annals of the New York Academy of Sciences 1163: 497-500.

Pancak-Roessler, M. K. and Lee, P. D. 1990. Insulin-like growth factor-I and insulin-like growth factor-binding protein in the toad, *Bufo woodhousei*. General and Comparative Endocrinology 78: 263-272.

Parrizas, M., Maestro, M. A., Banos, N., Navarro, I., Planas, J. and Gutierrez, J. 1995. Insulin/IGF-I binding ratio in skeletal and cardiac muscles of vertebrates: A phylogenetic approach. American Journal of Physiology-Regulatory Integrative and Comparative Physiology 269: R1370-R1377.

Pashmforoush, M., Chan, S. J. and Steiner, D. F. 1996. Structure and expression of the insulin-like peptide receptor from amphioxus. Molecular Endocrinology 10: 857-866.

Pelicci, G., Lanfrancone, L., Grignani, F., McGlade, J., Cavallo, F., Forni, G., Nicoletti, I., Grignani, F., Pawson, T. and Pelicci, P. G. 1992. A novel transforming protein (SHC) with an SH2 domain is implicated in mitogenic signal transduction. Cell 70: 93-104.

Perez-Sanchez, J., Martipalanca, H. and Kaushik, S. J. 1995. Ration size and protein-intake affect circulating growth-hormone concentration, hepatic growth-hormone binding and plasma insulin-like growth-factor-I immunoreactivity in a marine teleost, the gilthead sea bream (*Sparus aurata*). Journal of Nutrition 125: 546-552.

Peterson, B. C., Waldbieser, G. C. and Bilodeau, L. 2004. IGF-I and IGF-II mRNA expression in slow and fast growing families of USDA103 channel catfish (*Ictalurus punctatus*). Comparative Biochemistry and Physiology—Part A: Molecular and Integrative Physiology 139: 317-323.

Piper, M. D. W., Selman, C., McElwee, J. J. and Partridge, L. 2008. Separating cause from effect: how does insulin/IGF signalling control lifespan in worms, flies and mice? Journal of Internal Medicine 263(2): 179-191.

Ponce, M., Infante, C., Funes, V. and Manchado, M. 2008. Molecular characterization and gene expression analysis of insulin-like growth factors I and II in the redbanded seabream, *Pagrus auriga*: transcriptional regulation by growth hormone. Comparative Biochemistry and Physiology, Part B 150: 418-426.

Reed, W. L., Clark, M. E., Parker, P. G., Raouf, S. A., Arguedas, N., Monk, D. S., Snajdr, E., Nolan Jr., V. and Ketterson, E. D. 2006. Physiological effects on

demography: A long-term experimental study of testosterone's effects on fitness. The American Naturalist 167: 667-683.
Reinecke, M. and Collet, C. 1998. The phylogeny of the insulin-like growth factors. International Review of Cytology 183: 1-94.
Ricklefs, R. E. and Wikelski, M. 2002. The physiology/life history nexus. Trends in Ecology and Evolution 17: 462-468.
Robert, K. A. and Bronikowski A. M. 2010 Evolution of senescence in nature: physiological evolution in populations of garter snake with divergent life histories. The American Naturalist 175(2): 147-159.
Robert, K. A., Vleck, C. M. and Bronikowski, A. M. 2009. The effects of maternal corticosterone levels on offspring behavior in fast- and slow-growth garter snakes (*Thamnophis elegans*). Hormones and Behavior 55: 24-32.
Rosebrough, R. W., McMurtry, J. P. and Vasilatos-Younken, R. 1999. Dietary fat and protein interactions in the broiler. Poultry Science 78: 992-998.
Scanes, C. G., Dunnington, E. A., Buonomo, F. C., Donoghue, D. J. and Siegel, P. B. 1989. Plasma concentrations of insulin-like growth factors IGF-I and IGF-II in dwarf and normal chickens of high and low weight selected lines. Growth, Development, and Aging 53: 151-157.
Schmidt, K. E. and Kelley, K. M. 2001. Down-regulation in the insulin-like growth factor (IGF) axis during hibernation in the golden-mantled ground squirrel, *Spermophilus lateralis*: IGF-I and the IGF-binding proteins (IGFBPs). Journal of Experimental Zoology 289: 66-73.
Schwartz, T. and Bronikowski, A. M. 2010. Molecular stress pathways and the evolution of reproduction and aging in reptiles. Pp. In Press. In T. Flatt and A. Heyland (eds), *Molecular Mechanisms of Life History Evolution*. Oxford Univ. Press. UK.
Shimizu, M., Beckman, B. R., Hara, A. and Dickhoff, W. W. 2006. Measurement of circulating salmon IGF binding protein-1: assay development, response to feeding ration and temperature, and relation to growth parameters. Journal of Endocrinology 188: 101-110.
Silva, J. R. V., Figueiredo, J. R. and van den Hurk, R. 2009. Involvement of growth hormone (GH) and insulin-like growth factor (IGF) system in ovarian folliculogenesis. Theriogenology 71: 1193-1208.
Silverin, B. and Wingfield, J. C. 1998. Adrenocortical responses to stress in breeding pied flycatchers *Ficedula hypoleuca*: Relation to latitude, sex and mating status. Journal of Avian Biology 29: 228-234.
Sinervo, B. and Licht, P. 1991. Hormonal and physiological control of clutch size, egg size, and egg shape in side-blotched lizards (*Uta stansburiana*): constraints on evolution of lizard life histories. Journal of Experimental Zoology 257: 252-264.
Sparkman, A. M. and Palacios, M. G. 2009. A test of life-history theories of immune defence in two ecotypes of the garter snake, *Thamnophis elegans*. Journal of Animal Ecology 78: 1242 - 1248.
Sparkman, A. M., Arnold, S. J. and Bronikowski, A. M. 2007. An empirical test of evolutionary theories for reproductive senescence and reproductive effort in the garter snake *Thamnophis elegans*. Proceedings of the Royal Society B—Biological Sciences 274 (1612): 943-950.
Sparkman, A. M., Vleck, C. M. and Bronikowski, A. M. 2009. Evolutionary ecology of endocrine-mediated life-history variation in the garter snake *Thamnophis elegans*. Ecology 90: 720-728.
Sparkman, A. M., Byars, D., Ford, N. B. and Bronikowski, A. M. 2010. The role of insulin-like growth factor-1 (IGF-1) in growth and reproduction in the brown house snake (*Lamprophis fuliginosus*). General and Comparative Endocrinology: 168(3): 401-414.

Suh, Y., Atzmon, G., Cho, M.-O., Hwang, D., Liu, B., Leahy, D. J., Barzilai, N. and Cohen, P. 2008. Functionally significant insulin-like growth factor I receptor mutations in centenarians. Proceedings of the National Academy of Sciences USA 105: 3438-3442.

Sutter, N. B., Bustamante, C. D., Chase, K., Gray, M. M., Zhao, K., Zhu, L., Padhukasahasram, B., Karlins, E., Davis, S., Jones, P. G., Quignon, P., Johnson, G. S., Parker, H. G., Fretwell, N., Mosher, D. S., Lawler, D. F., Satyaraj, E., Nordborg, M., Lark, K. G., Wayne, R. K. and Ostrander, E. A. 2007. A single IGF1 allele is a major determinant of small size in dogs. Science 316 (5821): 112-115.

Tatar, M. and Yin, C. M. 2001. Slow aging during insect reproductive diapause: why butterflies, grasshoppers and flies are like worms. Experimental Gerontology 36(4-6): 723-738.

Tatar, M., Bartke, A. and Antebi, A. 2003. The endocrine regulation of aging by insulin-like signals. Science 299: 1346-1350.

Tatar, M., Kopelman, A., Epstein, D., Tu, M. P., Yin, C. M. and Garofalo, R. S. 2001. A mutant *Drosophila* insulin receptor homolog that extends lifespan and impairs neuroendocrine function. Science 292 (5514): 107-110.

Taylor, J. F., Migaud, H., Porter, M. J. R. and Bromage, N. R. 2005. Photoperiod influences growth rate and plasma insulin-like growth factor-I levels in juvenile rainbow trout, *Oncorhynchus mykiss*. General and Comparative Endocrinology 142: 169-185.

Webster, J. R., Corson, I. D., Littlejohn, R. P., Stuart, S. K. and Suttie, J. M. 1996. Effects of season and nutrition on growth hormone and insulin-like growth factor-I in male red deer. Endocrinology 137: 698-704.

Webster, J. R., Corson, I. D., Littlejohn, R. P., Stuart, S. K. and Suttie, J. M. 1999. Effects of photoperiod on the cessation of growth during autumn in male red deer and growth hormone and insulin-like growth factor-I secretion. General and Comparative Endocrinology 113: 464-477.

Webster, J. R., Corson, I. D., Littlejohn, R. P., Martin, S. K. and Suttie, J. M. 2001. The roles of photoperiod and nutrition in the seasonal increases in growth and insulin-like growth factor-1 secretion in male red deer. Animal Science 73: 305-311.

Willcox, B. J., Donlon, T. A., He, Q., Chen, R., Grove, J. S., Yano, K., Masaki, K. H., Willcox, D. C., Rodriguez, B. and Curb, J. D. 2008. FOXO3A genotype is strongly associated with human longevity. Proceedings of the National Academy of Sciences USA 105: 13987-13992.

Wilson, D. M. and Hintz, R. L. 1982. Inter-species comparison of somatomedin structure using immunological probes. Journal of Endocrinology 95(1): 59-64.

Wilson, M. E. 1998. Premature elevation in serum insulin-like growth factor-I advances first ovulation in rhesus monkeys. Journal of Endocrinology 158: 247-257.

Wingfield, J. C., Maney, D. L., Breuner, C. W., Jacobs, J. D., Lynn, S., Ramenofsky, M. and Richardson, R. D. 1998. Ecological bases of hormone-behavior interactions: the "Emergency Life History Stage". American Zoologist 38: 191-206.

Wu, Q. and Brown, M. R. 2006. Signaling and function of insulin-like peptides in insects. Annual Review of Entomology 51: 1-24.

Wuertz, S., Nitsche, A., Jastroch, M., Gessner, J., Klingenspor, M., Kirschbaum, F. and Kloas, W. 2007. The role of the IGF-I system for vitellogenesis in maturing female sterlet, *Acipenser ruthenus* Linnaeus, 1758. General and Comparative Endocrinology 150: 140-150.

Yamamoto, H., Sohmiya, M., Oka, N. and Kato, Y. 1991. Effects of aging and sex on plasma insulin-like growth factor I (IGF-1) levels in normal adults. Acta Endocrinologica 124: 497-500.

Zera, A. J. 1999. The endocrine genetics of wing polymorphism in *Gryllus*: Critique of recent studies and state of affairs. Evolution 53: 973.

Zera, A. J. and Bottsford, J. 2001. The endocrine-genetic basis of life-history variation: the relationship between the ecdysteroid titer and morph-specific reproduction in the wing-Polymorphic cricket *Gryllus firmus*. Evolution 55: 539-549.

Zera, A. J., Zhao, Z. and Kaliseck, K. 2007. Hormones in the field: Evolutionary endocrinology of juvenile hormone and ecdysteroids in field populations of the wing-dimorphic cricket *Gryllus firmus*. Physiological and Biochemical Zoology 80: 592-606.

CHAPTER 16

Paternity Patterns

Benjamin C. Jellen and Robert D. Aldridge

16.1 A BRIEF HISTORY OF MULTIPLE PATERNITY

Natural selection theory predicts that individuals/sexes will engage in behaviors that maximize their reproductive success. Intersexual differences in reproductive investment and order of gamete release have led to a dichotomy of factors structuring male and female mating strategies in many taxa (Trivers 1972). Because male reproductive success is directly and positively associated with the number of females inseminated (Parker 1984; Prosser *et al.* 2002), it is advantageous for males to inseminate as many females as possible throughout their lifetime. Conversely, additional mating by females is generally not associated with increased fecundity, and the limited number of ova available for fertilization, coupled with the large energetic investment inherent in vitellogenesis, embryogenesis, and parental care, suggests that females should be highly selective with respect to mate choice.

Despite the limited advantage to multiple matings by females, females routinely mate with multiple males during a single mating period. Consequently, multiple paternity (two or more males siring offspring within a single litter) is a common occurrence in the animal kingdom and has been reported in many internally-fertilizing taxa including amphibians (Halliday 1998), arachnids (Austad 1984), birds (Birkhead and Møller 1992), fish (Avise *et al.* 2002), insects (Parker 1970), placental mammals (Schwagmeyer *et al.* 1998), marsupials (Kraaijeveld-Smit *et al.* 2002), and reptiles (Uller and Olsson 2008). However, multiple mating incurs increased (and potentially severe) costs to the female such as: 1) increased potential mortality (Shine *et al.* 2001b); 2) increased exposure to predators (Wing 1988); 3) increased energy usage to avoid undesired males (Watson *et al.* 1998); 4) increased exposure to sexually transmitted diseases (Sheldon 1993); and, 5) decreased foraging opportunities (Madsen and Shine 1993a).

A number of theories have been proposed to explain multiple matings in females, these include: 1) guarding against mating with a sterile male

Biology Department, Saint Louis University, St. Louis, Missouri 63103, USA

(Gibson and Jewell 1982); 2) inbreeding depression avoidance (Stockley *et al.* 1993); 3) genetic incompatibility avoidance (Zeh and Zeh 1996); 4) "trading up" to a higher quality male (Halliday 1983; Keller and Reeve 1995); 5) infanticide reduction (Hrdy 1979); 6) increased reception of nuptial gifts (Thornhill and Alcock 1983); and, 7) transfer of nutrients in seminal fluids (Leopold 1976). Though empirical evidence exists to support each hypothesis within its own specific framework, none has been successfully extrapolated to other mating systems. The ramifications of multiple mating on the resultant litter (i.e. proximate mechanisms of multiple paternity), however, have received considerably less attention and are the focus of this review.

Molecular paternity analysis techniques (e.g., protein electrophoresis and microsatellite DNA analysis) have revolutionized our understanding of animal mating systems by demonstrating that even females of species once believed to be strictly monogamous (e.g., many birds) routinely mate with additional males during a single mating period (Smith 1988; Westneat and Sherman 1997; Bennett and Owens 2002). These technological advances have also allowed for the detection of differential paternity, the occurrence of males siring unequal numbers of offspring within a multiply-sired litter. Parker (1990) likened the allocation of sperm to ova to a raffle in which the patrons with more tickets (males contributing more sperm) were more likely to win the raffle (fertilize ova). However, assuming a random distribution of sperm to available ova, discrepancies are commonly observed between the expected and observed paternity ratios in multiply-sired litters even when controlling for ejaculate quantity (Newcomer *et al.* 1999). Differential paternity is believed to arise through a manifestation of sperm competition (the competition between the sperm of two or more males to fertilize the ova of a single female; Parker 1970), cryptic female choice (biases in the use of stored sperm within the female reproductive tract; Pitnick and Brown 2000), as a function of mating order with respect to sperm storage and sperm precedence (Fox 1956), or a combination thereof. For internally fertilizing species, it is difficult to distinguish between different potential post-copulatory sexual selection processes (Stockley 1999); therefore, sperm competition and cryptic female choice will be discussed concurrently as "genotypic effects".

16.2 MULTIPLE PATERNITY IN THE OPHIDIA

The association between the Ophidia and paternity investigations has been a long and fruitful one as multiple paternity was first documented in the Ophidia in 1941 by Blanchard and Blanchard using genetically-based morphological markers (e.g., pigmentation, pattern), and the Ophidia represent the first taxon for which empirical evidence demonstrating a direct and positive relationship between female fitness and multiple mating was documented (the Adder, *Vipera berus* (Madsen *et al.* 1992)). By systematically crossing melanistic and regularly-pigmented Common

Gartersnakes (*Thamnophis sirtalis*) and noting the ratio of melanistic to regularly-pigmented offspring, Blanchard and Blanchard (1941) determined melanism was an autosomal recessive trait which followed Mendel's law of segregation. They then mated two differently-pigmented males with a single female and noted a deviation from the expected offspring phenotypic ratio which was best explained by each male siring offspring in the litter (Blanchard and Blanchard 1941). Gibson and Falls (1975) provided quantitative support for multiply paternity in *T. sirtalis* and Schuett and Gillingham (1986) observed multiply-sired Copperhead (*Agkistrodon contortrix*) litters by crossing different subspecies (each characterized by a unique banding pattern) and noting the banding pattern of the subsequent offspring. The advent of molecular techniques allowed for paternity investigations in populations and taxa not exhibiting multiple phenotypes and for the examination into genotypic effects as a possible mechanism by which differential paternity may arise.

By 2009, 23 studies examining multiple paternity had been conducted across 12 ophidian species with the natricines comprising half of these studies (Table 16.1). Multiple paternity was detected in all studies, with its incidence in a given population ranging from to 37.5% in *Thamnophis sirtalis* (Garner *et al.* 2002) to 100% in *Vipera berus* and the Mekong Mud Snake (*Enhydris subtaeniata*) (Höggren and Teggelström 1995; Voris *et al.* 2008, respectively) (minimum of 4 litters examined). Linear regression analysis of these 23 studies revealed that the detection of multiple paternity was not a function of the number of loci examined (R^2 = 0.054, $F_{1,13}$ = 0.69, P = 0.42; range 2–13 loci) (Zweifel and Dessauer 1983; Ursenbacher *et al.* 2009, respectively), average litter size (R^2 = 0.06, $F_{1,13}$ = 0.75, P = 0.40; range 6–32.5 offspring per litter) (Zweifel and Dessauer 1983; Voris *et al.* 2008, respectively), or the number of litters examined (R^2 = 0.01, $F_{1,14}$ = 0.11, P = 0.74; range 1–53 litters) (Zweifel and Dessauer 1983; Kissner *et al.* 2005, respectively) suggesting that multiple paternity is fairly ubiquitous among the Ophidia. The largest number of sires thus far detected in a litter (5) has been observed in *E. subtaeniata* (Voris *et al.* 2008), the Northern Watersnake (*Nerodia sipedon*) (Prosser *et al.* 2002), and the Slatey-Grey Snake (*Stegonotus cucullatus*) (Dubey *et al.* 2009).

It is not surprising that multiple paternity is so prevalent within the Ophidia. Rivas and Burghardt (2005) contend that the ophidian mating system is best described as polygynandrous, as both sexes mate with multiple partners during a single mating period resulting in males siring offspring with multiple females and the litters of females being comprised of the offspring of multiple males. Body size plays an important role in the mating success (propensity of an individual to secure a copulation) of both sexes. For species in which male-male combat is present, larger males win combat bouts over smaller males thereby gaining priority access to females (Schuett and Gillingham 1989; Madsen *et al.* 1993; Schuett 1997; Fearn *et al.* 2005); whereas, in females, snout-vent length (SVL) is directly and positively associated with fecundity (Madsen 1983; 1988;

Table 16.1 Taxon, reference, percent multiple paternity in sample, number of litters/clutches multiple paternity present in, type of effect (genotypic [G], phenotypic [P], mating order [M]) examined and whether this effect was observed (yes [Y], or no [N]).

Taxon *authors*	*% Multiple paternity (number of litters)*	*Effect examined (observed?)*
Nerodia sipedon		
Barry *et al.* 1992	85.7% (12/14)	NA
Kissner *et al.* 2005	54% (25/46)	G (N, Y)[1]; P (N, Y)[2]
Prosser *et al.* 2002	58% (26/45)	G (N)
Weatherhead *et al.* 2002	58% (26/45)	G (N); P (N)
Thamnophis elegans		
Garner and Larsen 2005	50% (3/6)	G(N); P (N, Y)[3]
Thamnophis radix		
Wusterbarth 2009	80.9% (17/21)	G (N); P (N)
Thamnophis sirtalis		
Blanchard and Blanchard 1941	62.5% (5/8)	NA
Garner *et al.* 2002	37.5% (6/16)	G (N); P (N)
Gibson and Falls 1975	43% (6/14)	NA
King *et al.* 2001	100% (4/4)	NA
McCracken *et al.* 1999	75% (6/8)	NA
Schwartz *et al.* 1989	50% (16/32)	NA
Enhydris enhydris		
Voris *et al.* 2008	100% (2/2)	NA
Enhydris subtaeniata		
Voris *et al.* 2008	100% (4/4)	NA
Lampropeltis getula		
Zweifel and Dessauer 1983	100% (1/1)	NA
Liasis fuscus		
Madsen *et al.* 2005	85.7% (12/14)	G (Y)
Pantherophis obsoletus[4]		
Blouin-Demers *et al.* 2005	88% (30/34)	G (N); P (N, Y)[5]
Stegonotus cucullatus		
Dubey *et al.* 2009	70% (16/23)	P(Y)[6]
Agkistrodon contortrix[4]		
Schuett and Gillingham 1986	66.7% (2/3)	NA
Vipera berus[4]		
Höggren and Teggelström 1995	100% (6/6)	NA
Höggren and Teggelström 2002	75% (6/8)	M (Y); P (N)
Madsen *et al.* 1992	NA	G (N, Y)[7]
Stille *et al.* 1986	80% (4/5)	NA
Ursenbacher *et al.* 2009	69% (9/13)	G (N); P (Y)[8]

[1]Multiply-sired litters comprised more offspring than singly-sired litters but number of offspring sired did not vary with genetic similarity of dyad; [2]Males in poor body condition sired more offspring than males in better body condition; male size and tail length did not affect number of offspring sired; mating order could be a potential confound; [3]Average neonate SVL and mass did not differ between singly- and multiply-sired litters but multiply-sired litters comprised more offspring than singly-sired litters; [4]Male-male combat reported in taxon; [5]Larger males sired more offspring than smaller males but males with longer tails or in better body condition did not; [6]Larger males sired more offspring within a litter than smaller males; mating order could be a potential confound (male-male combat may be present); [7]Study population suffered from inbreeding depression; multiple matings by female did not reduce number of unfertilized ova but did reduce number of stillborn offspring; [8]Larger males sired more offspring than smaller males (male-male combat present).

Aldridge *et al.* 1995; Weatherhead *et al.* 1995; Garner *et al.* 2002) and larger females are more likely to produce multiply-sired litters than smaller females (Kissner *et al.* 2005). Therefore, female SVL has implications both for potential suitors and the incidence of multiple paternity.

In their review of multiple paternity in the reptiles, Uller and Olsson (2008) did not observe an association between litter size and the incidence of multiple paternity in the squamates. However, because squamate average clutch size varies greatly (from 1 in the amphisbaenids to over 100 in the boids (Fitch 1970)), this occurrence may have been masked in relatively small taxonomic groups (Eccard and Wolf 2009), such as the natricines. Intra-litter paternity patterns have been observed in some ophidian taxa as the number of males siring offspring has been reported to be directly and positively associated with litter size in *Nerodia sipedon* (Prosser *et al.* 2002) and in the Plains Gartersnake (*Thamnophis radix*; Wusterbarth 2009). Additionally, upon linearly regressing the data presented by McCracken *et al.* (1999) and King *et al.* (2001), the relationship between multiple matings and litter size also occurs in *T. sirtalis* ($R^2 = 0.38$, $F_{1,11} = 6.26$, $P = 0.03$). Further support within *T. sirtalis* is provided by: 1) Barry *et al.* (1992) who reported that the two singly-sired litters were the third and fourth smallest of 12 litters; 2) McCracken *et al.* (1999) who reported that the two singly-sired litters were the two smallest of 8 litters; and, 3) Schwartz *et al.* (1989) who reported multiple paternity was present in litters ranging from 11 to 40 offspring but not present in litters of 5 to 11 offspring. This trend was not observed in the Terrestrial Gartersnake (*T. elegans*; Garner and Larsen 2005); however, this result may be an artifact of small sample size ($N = 6$ litters).

Examining all other published studies, the relationship of increasing number of sires with respect to litter size has not been readily observed outside the natricines. For example, Blouin-Demers *et al.* (2005) failed to detect this trend in the Black Ratsnake (*Pantherophis obsoletus*). Neither was this trend observed in *Vipera berus* ($R^2 = 0.001$, $F_{1,26} = 0.04$, $P = 0.85$; data combined from Höggren and Teggelström 1995, 2002; Ursenbacher 2009) nor in *Enhydris subtaeniata* ($R^2 = 0.069$, $F_{1,3} = 0.15$, $P = 0.74$; Voris *et al.* 2008). With the relatively small number of studies ($N = 5$) conducted on non-natricine taxa, and with only one of these studies directly assessing this relationship (Blouin-Demers *et al.* 2005), it is not surprising this trend has not yet been detected in non-natricines. Clearly, studies of additional taxa are warranted.

Prosser *et al.* (2002) reported that the increased number of sires associated with increased litter size in *Nerodia sipedon* was likely due to two factors. First, that the greater number of ova possessed by larger females afforded more paternity opportunities, and second, simply that more males mated with larger females (Prosser *et al.* 2002). Male Red-sided Gartersnakes (*Thamnophis sirtalis parietalis*), Grass Snakes (*Natrix natrix*), and Yellow-Lipped Sea Kraits (*Laticauda colubrina*) have also been shown to preferentially court large females over small ones (Luiselli 1996; LeMaster and Mason 2002; Shetty and Shine 2002; Shine *et al.* 2006). Because female fecundity increases with SVL, males may increase their reproductive success by selectively mating with large females. Male *T. s. parietalis* have shown the ability to discriminate among differently-sized females based solely on olfactory cues (LeMaster and Mason 2002; Shine *et al.* 2003) as the sexual attractiveness pheromone profile of large females has been reported to contain a greater percentage of unsaturated methyl ketones compared to that of small females (LeMaster and Mason 2002). Therefore, it is not surprising that size assortative mating (large males tending to mate with large females leaving small males to mate with small females) is present in some ophidian taxa (Luiselli 1996; Shine *et al.* 2001c; Blouin-Demers *et al.* 2005; Kissner *et al.* 2005; Shine *et al.* 2006).

Populations of *Thamnophis sirtalis parietalis* exhibit a primarily explosive mating system in which females are immediately courted by myriad males upon emerging from large communal over-wintering sites (Joy and Crews 1988; Shine *et al.* 2000b), thus allowing males to potentially select from multiple receptive females of varying SVLs. However, males also have been reported to mate indiscriminately in this population (Shine *et al.* 2000a). Discriminating between females of varying SVLs and size assortative mating likely plays a lesser role in the mating system of species which are wide-ranging during the mating season thereby making intersexual encounters infrequent and the probability of locating an aggregation of females of varying SVLs unlikely. It would also likely play a diminished role for species in which females do not exhibit a spatially clumped distribution during the mating period (Weatherhead *et al.* 2002).

The coarse methods (e.g., pigmentation, protein electrophoresis) available to early researchers of ophidian paternity analysis allowed solely for the detection of the presence/absence of multiple paternity in a given litter or population. Thus, the detection of differential paternity and examinations into its underlying proximate mechanisms have largely coincided with technological advances in paternity assignment. That being said, several studies have examined possible mechanisms by which differential paternity may arise in the Ophidia but have reported inconsistent results within and between populations and taxa. Here, we review these studies to surmise what is currently known regarding ophidian mating success, reproductive success, and differential paternity.

16.3 POTENTIAL MECHANISMS UNDERLYING PATERNITY PATTERNS

This chapter focuses on three potential mechanisms which may influence paternity patterns and differential paternity: 1) genotypic effects (sperm competition and cryptic female choice due to the genotypic profiles of mothers and sires); 2) phenotypic effects (size attributes of mating males and size effects to resultant offspring); and, 3) mating order effects.

While the number of polymorphic loci examined does not seem to influence the detection of multiple paternity, it does influence the probability of detecting fine-scale intra-litter paternity patterns. Therefore, not all

23 studies conducted on ophidian multiple paternity were conducted in such a capacity as to examine differential paternity or investigate its potential proximate mechanisms. Additionally, because many studies collected gravid females from the wild after the mating period ended, information regarding the phenotypic attributes of mating males and mating order in these studies is unknown, thus preventing any investigation of genotypic and mating order effects on differential paternity. Of the 23 ophidian multiple paternity studies examined, only 12 directly examined potential mechanisms underlying differential paternity (Table 16.1).

16.3.1 Genotypic Effects

Genotypic effects are frequently measured by juxtaposing attributes of the genotypic profile (such as relatedness to female, heterozygosity, and genetic variability) of each male siring offspring in a litter with the genotypic profile of the mother while quantifying the number and viability of offspring produced by each respective dyad. A caveat to this method is that only the aspect of being born is quantified. Male haplotypes which are incompatible with those of the female would be terminated early in embryogenesis (Birkhead *et al.* 2004) and therefore not counted thus skewing paternity ratios and providing an inaccurate depiction of genotypic effects and male reproductive success. The difficulty arises when trying to determine an unfertilized ova from one in which embryogenesis was terminated early in the gestation period. Genotypic effects are theorized to occur due to the specific combination of parental haplotypes during fertilization which may have ramifications on fertilization success (proportion of unfertilized ova in the litter/clutch) and/or offspring quantity and viability (number of stillborn offspring in the litter in viviparous and ovoviviparous species and the number of fertilized eggs failing to hatch in oviparous species). A second caveat is that no intrinsic measure of offspring quality (e.g., fitness, growth, survival) is quantified following birth/hatching. Therefore, whether paternal (or maternal) effects influence offspring life history characteristics remains unknown. However, offspring viability has been anecdotally reported in a few studies. For example, Kissner *et al.* (2005) reported that offspring from multiply-sired *Nerodia sipedon* litters overwintered more successfully than offspring from singly-sired litters, and Madsen and Shine (1998) reported that offspring from broods with greater hatching success demonstrated greater survival in the Water Python (*Liasis fuscus*). Hatching success in *L. fuscus* was later positively related to the total number of paternal alleles in the brood (discussed below; Madsen *et al.* 2005). However, Ursenbacher *et al.* (2009) reported that the number of sires per litter had no effect on neonate survival in *Vipera berus*. Studies monitoring individuals throughout their ontogeny are clearly needed to adequately address the potential long-term ramifications of genotypic effects on offspring fitness and to provide an accurate depiction of male and female lifetime reproductive success.

Data assessing potential genotypic effects in the Ophidia are available from 10 studies (Table 16.1), including investigations in *Liasis fuscus* (Madsen *et al.* 2005), *Nerodia sipedon* (Prosser *et al.* 2002; Weatherhead *et al.* 2002; Kissner *et al.* 2005), *Pantherophis obsoletus* (Blouin-Demers *et al.* 2005), *Thamnophis elegans* (Garner and Larson 2005), *T. radix* (Wusterbarth 2009), *T. sirtalis* (Garner *et al.* 2002), and *Vipera berus* (Madsen *et al.* 1992; Ursenbacher *et al.* 2009). Seven of these studies reported that genotypic effects were not associated with offspring number or viability (Table 16.1).

Prosser *et al.* (2002) observed no difference in the number of stillborn offspring or unfertilized ova between multiply- and singly-sired *Nerodia sipedon* litters. Similarly, the proportion of stillborn offspring did not vary between multiply- and singly-sired litters in *Thamnophis elegans* (Garner and Larson 2005), *T. radix* (Wusterbarth 2009), and *T. sirtalis* (Garner *et al.* 2002). However, Kissner *et al.* (2005) observed that multiply-sired litters had a higher proportion of stillborn offspring than singly-sired litters in captive *N. sipedon*.

Prosser *et al.* (2002) reported that female reproductive success appeared to increase as a function of the number of sires present in *Nerodia sipedon* litters (i.e. increased paternal genotypic effects); however, variation in female SVL (and its influence on fecundity) accounted for this difference. Variation in female SVL also accounted for the apparently larger size of multiply-sired litters in *Thamnophis radix* (Wusterbarth 2009). However, Kissner *et al.* (2005) reported that the number of offspring in multiply-sired litters was greater than that of singly-sired litters and that variation in female SVL did not account for this difference. Kissner *et al.* (2005) also reported that neither the proportion of offspring nor absolute number of offspring sired varied with respect to the genetic similarity of each dyad comprising offspring within *N. sipedon* litters, thus confounding interpretation of the underlying cause of this observation.

Using the same data set (811 offspring from 45 litters) as Prosser *et al.* (2002), Weatherhead *et al.* (2002) failed to observe genotypic effects acting on the number of *Nerodia sipedon* offspring due to the genetic similarity of each respective dyad comprising offspring in a litter, heterozygosity, or genetic variation. Blouin-Demers *et al.* (2005) reported that hatching success was not associated with dyad relatedness in a population of *Pantherophis obsoletus*. Examining 164 eggs from 14 clutches of *Liasis fuscus*, Madsen *et al.* (2005) reported that clutch size did not differ between multiply- and singly-sired *L. fuscus* litters; however, the proportion of eggs that hatched was positively related to the number of paternal alleles observed in the clutch. Madsen *et al.* (2002) reported that female *Vipera berus* which mated with multiple males experienced a reduced number of stillborn offspring in their litters which was attributed to genotypic effects. It is important to note, however, the small population in southern Sweden that Madsen *et al.* (2002) studied suffered from inbreeding depression (Madsen *et al.* 1996). These results support findings by Olsson *et al.* (1996) in which a negative relationship was observed between dyad relatedness and the number of offspring produced in a population of the Sand Lizard (*Lacerta agilis*). A study on a non inbred *V. berus* population reported no relationship between multiple matings and the number of stillborn offspring (Ursenbacher *et al.* 2009).

The role of genotypic effects as a potential mechanism by which differential paternity may arise in the Ophidia remains unclear. The majority of studies show no association between genotypic effects and the number and/or viability of offspring in the resultant litter. Multiple matings, and the increased genetic diversity associated with it, appears to benefit in bred populations. One possibility is females cryptically select male gametes less genetically similar to their own thereby minimizing the deleterious effects of inbreeding depression. However, an increase in paternal alleles was also found to be beneficial in two studies of non inbred populations (Kissner *et al.* 2005; Madsen *et al.* 2005). Female birds preferentially use sperm of non-related males to fertilize their ova (Pizzari *et al.* 2004) but the mechanism by which females assess sperm compatibility remains unknown. This mechanism may involve the major histocompatibility complex (MHC) which contains multiple closely linked and highly polymorphic genes integral in self/non-self recognition (Klein 1986). Compatibility at the MHC locus has been shown to be a pre-copulatory deterrent to consanguineous mating in House Mice (*Mus musculus domesticus*) and *Lacerta agilis* as females preferentially mate with males whose MHC genes are dissimilar to their own in these taxa (Penn and Potts 1998; Olsson *et al.* 2003, respectively). However, recent studies suggest complementarity between MHC haplotypes may also provide a post-copulatory mechanism to recognize offspring from consanguineous matings as MHC haplotypes have been shown to be expressed on the surface of male gametes (Ziegler *et al.* 2002) and on the female oviduct following fertilization (Zheng *et al.* 2001), thus possibly allowing for detection of an intrinsic measure of inter-gamete relatedness and/or compatibility subsequent to fertilization.

16.3.2 Phenotypic Effects

The term sperm competition is often used erroneously to describe both pre-copulatory as well as and post-copulatory male-male interactions. The inclusion of pre-copulatory competition stems from the proposition that sperm from a more physically adept male are themselves more fit than sperm from a less physically adept male and will therefore result in more offspring sired in a multiply-sired litter. Due to the haploid nature of gametes as a result of meiosis and the ensuing recombinational load, no direct evidence supports the contention that sperm quality covaries with male quality. Because sperm competition is defined as "the competition between the sperm of two or more males to fertilize the ova of a single female" (Parker 1970), it is strictly a post-copulatory process in which ejaculate size, sperm swimming velocity, and longevity are possible sources of inter-male variation leading to variances in reproductive success. It has been proposed that females will cryptically select sperm to fertilize their ova from certain males over others based on the genotypic and/or phenotypic attributes of the mating males (Eberhard 1996; Penn and Potts 1998; Pitnick and Brown 2000; Olsson *et al.* 2003). Genotypic effects were previously discussed and potential phenotypic effects are discussed below.

Pre-copulatory forms of intra-sexual competition, e.g., male-male combat (Schuett 1997), tail wrestling (Weatherhead *et al.* 1995; Luiselli 1996), and movement tactics (Duvall and Schuett 1997; Jellen *et al.* 2007) have been shown to be important determinants of male mating success and likely set up a mechanism by which differential paternity may arise either by: 1) limiting sperm of rival males from entering the female reproductive tract [i.e. actively guarding the female from rival males (female defense polygyny)]; 2) occluding the female cloaca with a copulatory plug to prevent insemination by rival males (Devine 1975); 3) decreasing the attractiveness of a recently-mated female via compounds transferred in male mating fluids (Shine *et al.* 2000b); and/or, 4) by securing a particular position in the mating order relative to other males mating with that female during a single mating period (i.e. first mating male).

Two types of potential phenotypic effects have been examined: 1) the phenotypic attributes of mating males, e.g., SVL, mass, body condition, tail length in relation to number of offspring sired; and, 2) phenotypic effects to resultant offspring (i.e., intra-litter paternal effects on offspring phenotypic variation). Additionally, we discuss the relationship between male phenotypic attributes and mating success; however, because male mating success is often a poor predictor of reproductive success in some ophidian taxa (Prosser *et al.* 2002), we are careful to distinguish between the two.

16.3.2.1 Phenotypic attributes of mating males

The operational sex ratio (OSR) (Emlen and Oring 1977) of most temperate ophidian mating systems are male-biased thus intensifying male-male competition for access to mates. This, coupled with the propensity for male reproductive success to increase with each additional mate, suggests that males should experience a high level of sexual selection compared to females. Prosser *et al.* (2002) measured the intensity of sexual selection in *Nerodia sipedon* and determined the opportunity for sexual selection was over five times greater for males than females. Sexual selection may influence body size for species in which males physically compete for access to mates (Darwin 1871). Therefore, variations in male size should equate to variations in male reproductive success in these taxa. This is exactly the case in the Ophidia where greater male size is directly and positively associated with mating success in species exhibiting pre-copulatory intrasexual male interactions. An obvious example is the larger male size observed in species exhibiting male-male combat (Shine 1978; 1994); however, because males of some taxa also physically compete with one another in intense mating aggregations (i.e. mating balls) by attempting to push aside other males who are concurrently attempting to mate with a particular female, we also investigate this trend in the natricines. Little data exists on the mating and reproductive success of ophidian species not exhibiting pre-copulatory intrasexual physical interactions.

Of the 23 ophidian paternity studies, six have examined a possible association between the phenotypic attributes of mating males and their subsequent reproductive success (Table 16.1) including investigations in *Nerodia sipedon* (Weatherhead *et al.* 2002; Kissner *et al.* 2005), *Pantherophis obsoletus* (Blouin-Demers *et al.* 2005), *Stegonotus cucullatus* (Dubey *et al.* 2009), and *Vipera berus* (Höggren and Tegelström 2002; Ursenbacher *et al.* 2009). Male-male combat is present in the mating system of *P. obsoletus* (Rigley 1971; Stickley *et al.* 1980) and, in accordance with theory, Blouin-Demers *et al.* (2005) observed that longer (SVL) males sired more offspring per clutch than smaller males but that tail length and body condition did not affect male reproductive success. Similarly, in a free-ranging *V. berus* population (a taxon also exhibiting male-male combat (Madsen *et al.* 1993)), Ursenbacher *et al.* (2009) observed that: 1) males which sired offspring in multiple clutches were longer than other sires; 2) longer males sired more offspring than shorter males; and, 3) singly-sired litters were sired by the longest males in the population. Madsen *et al.* (1993) also reported that male size and mobility factored greatly in the mating success of *V. berus*. They also reported that male size strongly influenced combat success and that larger males fought more frequently and won more bouts than smaller males (Madsen *et al.* 1993). Male *S. cucullatus* grow larger than female conspecifics and male-male combat is suspected to occur in this taxon (Dubey *et al.* 2009). Larger (SVL) male *S. cucullatus* sired more offspring than smaller males over a 10 year period (Dubey *et al.* 2009). Interestingly, the increased reproductive success of large males was not due to mating with more females than smaller males or to size assortative mating, but rather because large males sired more offspring within a given litter compared to smaller males (Dubey *et al.* 2009). While the mechanism underlying this result is unknown, it may have arisen due to: 1) larger males being able to successfully defend mates from rival males in agonistic bouts; 2) larger males contributing larger total ejaculates, including more total sperm, than smaller males (e.g., sperm competition); or, 3) mating order effects which were not investigated (Dubey *et al.* 2009).

Male length has also been shown to be an important determinant of combat, and thus, mating success in *Agkistrodon contortrix* (Schuett and Gillingham 1989; Schuett 1997) and the Australian Scrub Python (*Morelia kinghorni*; Fearn *et al.* 2005) and winners of these agonistic bouts gain priority access to females (Schuett and Gillingham 1989). Therefore, it is not surprising that larger males sire more offspring than smaller males in mating systems in which male-male combat is present. Males may utilize additional tactics to influence mating and/or reproductive success. Males of some species exhibit mate guarding (a post-copulatory form of female defense polygyny) in which a male actively defends one or more females from other potential male suitors. Mate guarding transpires over a period of multiple days (Jellen *et al.* 2007) and likely corresponds to an abbreviated window of time during the mating period in which female sex pheromone concentration likely peaks (BCJ unpublished data). Males that successfully defend females from the sexual advances of rival males during this critical time period potentially increase their reproductive success by preventing the sperm of rival males from entering the reproductive tract of the female. Additionally, the timing of copulation with respect to other mating males during the mating period may also be important as Jellen *et al.* (2007) observed that larger male Massasaugas (*Sistrurus catenatus*) courted females as the mating season progressed. Because larger males displace smaller ones in agonistic bouts, a potential advantage may exist for matings later in the mating period in this taxon. This tactic could potentially correspond with a period of peak female attractivity, receptivity, or with a juncture which otherwise maximizes male reproductive success.

Male-male combat is generally absent in the ophidian taxa lacking male-biased sexual size dimorphism (Shine 1978, 1994), such as the natricines (Rossman *et al.* 1996; Gibbons and Dorcas 2004). In *Nerodia sipedon*, females grow larger than male conspecifics (Weatherhead *et al.* 1995; Gibbons and Dorcas 2004), and Weatherhead *et al.* (2002) observed no association between male SVL, tail length, mass, or body condition and the number of offspring sired in a free-ranging population of *N. sipedon*. However, in a previous study examining mating success in *N. sipedon*, Weatherhead *et al.* (1995) observed that longer (SVL) males had greater mating success than smaller males in mating aggregations in one of two field seasons (the season in which the OSR was male-biased). No relationship was observed between male tail length or body condition and mating success however (Weatherhead *et al.* 1995). Kissner *et al.* (2005) devised experimental enclosures to investigate the effects of varying OSRs and male phenotypic attributes in staged mating aggregations of *N. sipedon* and observed that longer (SVL) males were more successful in siring offspring than shorter males when the inter-male SVL variation was relatively large (22 cm; 85% of the total SVL range for the population); however, this effect was not observed when inter-male SVL variation was considerably smaller (8 cm; 30% of the total SVL range in the population). Additionally, in a male-biased OSR (3:1), Kissner *et al.* (2005) observed that male SVL was positively associated with mating success and that males in poor body condition sired more offspring than those in good body condition.

The effects of male size on mating success have also been investigated in other natricines including *Thamnophis sirtalis* (Joy and Crews 1988; Shine *et al.* 2000c) and *Natrix natrix* (Luiselli 1996). Joy and Crews (1988) observed that neither male SVL nor mass affected mating success in *T. sirtalis*. Shine *et al.* (2000c) expanded the study conducted by Joy and Crews (1988; N = 123) and reported that SVL was a factor in male mating success. Though statistically significant, the small difference in the inter-male SVL range observed by Shine *et al.* (2000c) of successful males (average SVL = 45.6 cm) compared to unsuccessful males (average SVL = 45.2 cm) may represent a statistical artifact (N>5000 individuals) rather than a biologically meaningful result. However, male size has been shown to be an important determinant of mating success in *N. natrix* in which longer (total length) and heavier males secured more copulations than smaller and lighter males (Luiselli 1996). Madsen and Shine (1993b) also reported that larger male size (SVL) was positively related to mating success in large mating aggregations of *N. natrix*.

Though male-male combat is absent in the natricines, direct male intrasexual pre-copulatory physical competition is often intense. Natricine mating balls are comprised several males jockeying for position to mate with a single female. Luiselli (1996) reported a personal communication from T. Madsen of a *Natrix natrix* mating ball comprised of 22 males and one female and, in *Thamnophis sirtalis*, and Shine *et al.* (2001a) reported mating balls consisting of up to 62 males per female. For successful intromission, one male must physically out-maneuver and push aside multiple rival males to align his cloaca with that of the female. In these pre-copulatory male-male physical interactions, it is plausible that larger male SVL confers a benefit allowing these larger (and potentially more powerful) males to push aside the bodies and/or tails of smaller and presumably weaker males thus increasing the mating success of larger males and perpetuating the sexual selection of male size. Large male size is also positively associated with mating success in the Green Anaconda (*Eunectes murinus*; Rivas 2000). However, in ophidian taxa where female-biased SSD is present, stabilizing selection may act on male size. Larger male size may be selected for due to the benefit it confers in pre-copulatory male-male interactions; however, an unusually large male may be mistaken for a female by other males and thus have to allocate time and energy fighting off these male suitors (Rivas and Burghardt 2001). Studies directly examining for a potential relationship between male SVL and strength are lacking; however, larger males have been reported to push aside the tails of smaller males (Luiselli 1996; Shine *et al.* 2000c; Shine and Shetty 2001) and Shine *et al.* (2004a) reported that heavier males had greater mating success than lighter males.

Male Ophidia have longer tails relative to body length than females (Shine 1993) and three hypotheses (Morphological Constraint, Female Reproductive Output, and Male Mating Ability) have been proposed to explain this sexual size dimorphism (King 1989). While data exist to support each of the first two hypotheses (see King 1989 for review), the association between male tail length and mating success (i.e. Male Mating Ability Hypothesis) is largely unknown. To confound the issue of examining the association between male tail length and mating success, multiple selection processes concurrently act on male tail length. Sexual selection theory predicts longer male tail length has arisen due to a potential benefit in courtship activities (males with longer tails may physically out compete males with shorter, and potentially weaker, tails in "tail-wrestling" events) which has been demonstrated in

Thamnophis sirtalis (Shine *et al.* 1999) and *Laticauda colubrina* (Shine and Shetty 2001) whereas natural selection theory proposes longer tails may facilitate locomotion, as has been demonstrated in *L. colubrina* (Shine and Shetty 2001) and *T. elegans* (Arnold 1988). Similar to male SVL, male tail length may be a good predictor of mating success, which in turn, may influence reproductive success. Clearly, further studies are needed.

Male SVL and tail length appear to be important determinants of mating and possibly reproductive success in ophidian species in which male pre-copulatory intrasexual physical interactions (e.g., combat, tail wrestling) are present in the mating system; particularly when: 1) the OSR is strongly male-biased; 2) when a large inter-male SVL variance exists; and/or 3) when male-male competition is otherwise intense. Male mass and body condition, however, do not appear to be associated with mating or reproductive success. To better understand the role of phenotypic attributes on male reproductive success, long-term field studies examining paternity, mating strategy, and OSR are needed. The OSR may vary greatly between mating periods (Madsen and Shine (1993c) report the OSR ranged from 0.04 to 0.79 during a seven-year period) which may influence inter-year male mating tactics (e.g., time allocated to searching for mates vs. time allocated to defending mates) between mating periods. Additionally, in a given mating period, individual male snakes experience a great deal of variance in reproductive success with some males attaining a high degree of mating and/or reproductive success while other males attain little to none, with this relative success fluctuating temporally as well (Madsen and Shine 1993c; Weatherhead *et al.* 2002; Jellen *et al.* 2007; Ursenbacher *et al.* 2009). Therefore, long-term field studies documenting male lifetime reproductive success across mating periods are needed. Finally, studies are needed for taxa whose mating systems are devoid of male intrasexual pre-copulatory physical interactions.

16.3.2.2 Phenotypic effects to offspring

Offspring size is not a direct measure of reproductive success; however, because offspring size is associated with survival (Ferguson and Fox 1984; Ford and Seigel 1989; Jayne and Bennett 1990; Brown and Shine 2005), offspring size may be an indirect measure of parental reproductive success. Studies investigating offspring phenotypic attributes (length, mass, and body condition) have largely focused on maternal effects to the offspring and have been primarily couched under optimal offspring size theory examining for a trade-off between offspring number and size within a given litter (for review see Ford and Seigel, **Chapter 14**). However, because offspring size has been shown to vary within litters (Weatherhead *et al.* 1999) and at the intra- (Blouin-Demers and Weatherhead 2007) and inter-populational (Sinervo 1990) level, optimal offspring size theory has been largely abandoned in favor of examining the causes of this observed variation. Paternal phenotypic effects represent one possible source of this variation, but has not yet been investigated.

Four ophidian paternity studies have examined potential phenotypic effects of multiply- vs. singly-sired litters on offspring morphological attributes (Table 16.1). No association between offspring SVL or mass was observed between multiply- and singly-sired litters in *Thamnophis elegans* (Garner and Larsen 2005), *T. radix* (Wusterbarth 2009), or *T. sirtalis* (Garner *et al.* 2002). Additionally, the number of matings did not affect offspring mass or total litter mass in *Vipera berus* (Madsen *et al.* 1992). A direct relationship between paternal effects and offspring size was not directly examined in any study and may account for some of the intra-litter offspring size variation in the Ophidia.

16.3.3 Sperm Precedence and Mating Order Effects

Because mating and fertilization are decoupled in the Ophidia (for review see Aldridge and Duvall 2002; Aldridge *et al.* 2009), sperm are stored in specialized receptacles (sperm storage tubules; SSTs) in the female infundibulum prior to fertilization (Fox 1956). Depending on the type (I or II) of secondary vitellogenesis experienced (Aldridge 1979), sperm may necessarily be stored for as long as eight months in some species (Aldridge *et al.* 2008), thus providing an ideal arena for sperm competition and/or cryptic female choice.

Alternatively, it is plausible that the location of the SST in the female infundibulum in which a particular sperm is stored may influence its fertilization probability (i.e. sperm precedence). Conceivably, sperm could be stored in the first vacant SSTs encountered; thus, sperm from the first mating male would be stored in the caudal-most SSTs (Fox 1956; Hoffman and Wimsatt 1972) and sperm from later-mating males would correspondingly be stored in more anterior SSTs. During ovulation, ova move caudally through the female infundibulum, thus passing the anterior SSTs first. Therefore, it is plausible that sperm stored in the anterior SSTs would have more time to penetrate the cell membrane of the ova than more caudally-stored sperm, thus conferring a fertilization advantage to anteriorly-stored sperm. Alternatively, sperm from all mating males could be equally distributed throughout the SSTs (Briskie and Montgomery 1993) thus allowing sperm from multiple males to be stored together allowing ample opportunity for sperm-sperm interactions. It is plausible that sperm from the last mating male could be layered upon the sperm from earlier mating males conferring a fertilization advantage to those later arriving sperm. The relationship between fertilization success and SST location is unknown. Additionally, whether the relative position of a sperm within a SST influences its fertilization probability is unknown as well as any association between mating order, SST location, and reproductive success.

Mating order effects have been intensively studied in insects (Parker 1970), arachnids (Zeh and Zeh 1994), and birds (Birkhead and Møller 1992), but have also been examined in other taxa including mammals

(Schwagmeyer *et al.* 1998), salamanders (Jones *et al.* 2002), turtles (Pearse *et al.* 2002), lizards (Olsson *et al.* 1994), and snakes (Höggren and Tegglestöm 2002). Typically, either a first male (Austad 1984; Schwagmeyer *et al.* 1998; Höggren and Tegglestöm 2002; Jones *et al.* 2002) or last male (Parker 1970; Birkhead and Møller 1992; Pearse *et al.* 2002) mating advantage is observed, with a last male mating advantage commonly observed in birds and insects.

The majority of studies demonstrating a last male reproductive advantage have been restricted to laboratory mating trials using *only* two males and this scenario may not accurately reflect the incidence of multiple matings in the wild as up to seven sires have been reported in marsupials (Kraaijeveld-Smit *et al.* 2002), five in the Ophidia (Prosser *et al.* 2002; Voris *et al.* 2008; Dubey *et al.* 2009) and placental mammals (Say *et al.* 1999), and three in turtles (Valenzuela 2000) and salamanders (Adams *et al.* 2005). To test whether the last male reproductive advantage was simply due to a two-mating-male-experimental-design or indicative of animal mating systems in general, Zeh and Zeh (1994) devised a study using the Harlequin Beetle-Riding Pseudoscorpion (*Cordylochernes scorpioides*) to examine paternity ratios in litters of females mated to only two males compared to the paternity ratio in litters of females mated to three males. Their results show that a last male reproductive advantage was observed when the female was mated with only two males; however, the last male reproductive advantage was not observed when three males sired offspring in the litter (Zeh and Zeh 1994).

Potential mating order effects in the Ophidia were anecdotally reported by Blanchard and Blanchard (1941) upon the observation of a skewed paternity ratio in *Thamnophis sirtalis* litters. They theorized that since the first mating male sired fewer offspring than the second in one particular litter, perhaps the female "did not receive the usual quantity or quality of seminal fluid" from that male (Blanchard and Blanchard 1941). Schuett and Gillingham (1986) provided raw data sufficient for paternity assignment in three multiply-sired *Agkistrodon contortrix* litters. Their data suggest a first male mating advantage as, over three litters, the first mating male sired 12 of 19 (63%) offspring whereas the second mating male sired 7 of 19 (37%) offspring (Schuett and Gillingham 1986). However, it is important to note that different *A. contortrix* "subspecies" were crossed in this study thus genotypic effects potentially confound these results.

Only one study directly examining mating order effects in the Ophidia has been conducted. Höggren and Tegglestöm (2002) reported a first male reproductive advantage in *Vipera berus*. Female *V. berus* were collected from a non inbred population and mated with three males on three consecutive days in the laboratory. Each female was mated to three different males, thus controlling for possible male sperm depletion between matings. Recently-mated male *Thamnophis sirtalis* produced smaller copulatory plugs during their second mating event within the mating period than during their first (Shine *et al.* 2000b). This is likely due to a smaller amount of total ejaculate available to be transferred in temporally frequent matings; therefore the confound of potential sperm depletion must be controlled for when examining mating order effects. Additionally, in the Höggren and Teggleström (2002) study, only one male was allowed access to the female at a time thus controlling for the potential phenotypic effects of larger male size on combat, mating, and (potentially) reproductive success. Six of eight litters were multiply-sired with only the largest of these litters (N = 10 offspring) being sired by all three mating males (Höggren and Teggleström 2002). It is interesting to note in the triply-sired litter that, similar to Zeh and Zeh (1994), the second male sired the fewest number of offspring (first male: 6 offspring; second male: 1 offspring; third male: 3 offspring; Höggren and Teggleström 2002). Also, within doubly-sired litters, it is interesting to note that, similar to the paternity ratio observed by Schuett and Gillingham (1986), the first mating male sired 32 of 47 (68%) offspring whereas the second mating male sired 15 of 47 (32%) offspring showing a clear first mating male reproductive advantage in *V. berus* (Höggren and Teggleström 2002).

16.4 SUMMARY

The vast majority of our knowledge of ophidian mating systems stems from a small number of species; primarily the natricines and *Vipera berus*. Clearly, studies of other ophidian taxa (particularly non-temperate species and those not exhibiting intrasexual pre-copulatory physical interactions) are needed in order to provide a more accurate depiction of ophidian mating systems. The majority of studies on non-natricine and *V. berus* taxa have examined factors influencing male mate acquisition and mating success without directly assessing the impacts of those behaviors on reproductive success; which is often poorly predicted from strictly behavioral studies (i.e. avian extra-pair copulations and paternity). Additionally, offspring survival or some measure of fitness must be quantified throughout the ontogeny to accurately measure lifetime reproductive success.

Advances in technology have allowed us to shift from simply documenting the presence/absence of multiple paternity in a given population to investigating the potential proximate mechanisms of intra-litter paternity patterns. Differential paternity is a common occurrence in the animal kingdom, but the mechanism(s) underlying intra-litter paternity patterns are not clear, as this review illustrates. Genotypic effects are likely beneficial to populations suffering from low levels of genetic diversity; however, if they are present in non-inbred populations, the post-copulatory mechanism (e.g., genetic compatibility) by which they operate remains uncertain. For species exhibiting pre-copulatory male-male physical interactions, larger SVL and tail length (and the benefit these attributes confer in physical interactions) appear to influence mating success; however, these attributes alone do not influence the number of offspring sired. Neither male body condition nor mass appears to affect reproductive success, and paternal phenotypic attributes do not appear to influence offspring morphology. However, the one study examining mating order effects has shown a clear mechanism by which differential paternity may occur in the Ophidia (Höggren and Tegglestrӧm 2002) but how mating order relates to sperm precedence and storage in the female infundibulum is unknown. Taken together, all three mechanisms may potentially and concurrently influence differential paternity and teasing apart these effects will be tedious.

A controlled setting allowing for the solitary and sequential mating of multiple males (thus controlling for male size as a potential factor) with a virgin female (to eliminate the possibility of inter-season sperm storage) is necessary to assess potential mating order effects. Additionally, the genotypes of the mating males should be known (preferably using littermates from the same sire) thus controlling for potential paternal genotypic effects on differential paternity. For most ophidian taxa, mating order effects cannot be accurately determined for females not under constant mating surveillance, thus precluding studies on radio-equipped, free-ranging individuals. Snakes are highly secretive and often retire to seclusion (i.e. burrows, foundation cracks, brush piles) out of sight for lengthy periods thus preventing observations of any mating activities during that period (Jellen and Aldridge unpubl.). Additionally, the successful capture of each mating male in the wild may not be possible, thus resulting in an incomplete data set for the examination of potential genotypic and mating order effects.

Currently, little is known regarding the relationship between mating order, sperm precedence, and sperm storage location in the female infundibulum. No data exist on whether sperm from multiple males are stored in different SSTs throughout the female infundibulum, whether they are co-mingled within SSTs, or if SST location itself influences fertilization success. Because sperm from different males can be differently fluorescently labeled (Parrish and Foote 1985), this methodology could be used in conjunction with staged mating trials to observe the storage location in the female infundibulum of the sperm from each sequential mating male while examining for a relationship between mating order and SST location. This information could then be used in conjunction with information relating mating order to reproductive success to look for a possible association between mating order, SST location, and differential paternity.

Why females mate with multiple males within mating periods remains largely unknown. Because female reproductive success scarcely increases due to mating with multiple males and the costs associated with additional matings are relatively high, this behavior is perplexing. However, growing evidence from a variety of taxa (including crustaceans (Thiel and Hinojosa 2003), sharks (DiBattista *et al.* 2008), bony fishes (Head and Brooks 2006), insects (Trontti *et al.* 2007), snakes (Shine *et al.* 2004b), and turtles (Lee and Hayes 2004)) suggests that females may engage in multiple matings simply as a function to minimize these costs; particularly the costs of male harassment. Convenience polyandry (Wilcox 1984), a form of sexual coercion (Clutton-Brock and Parker 1995), is believed to occur when the costs of refusal to mate (e.g., physical harm or mortality (Shine *et al.* 2001b), disrupted foraging (Madsen and Shine 1993a), increased exposure to predators (Wing 1988)) outweigh the benefits of additional matings (e.g., increased reproductive success, increased reception of nuptial gifts (Thornhill and Alcock 1983)). Because female Ophidia experience little to no benefit by mating with multiple males (Prosser *et al.* 2002) but do experience substantial costs (Madsen and Shine 1993a; Shine *et al.* 2001b), females may elect to engage in multiple matings in accordance with convenience polyandry theory. These multiple matings coupled with the relatively large litter sizes experienced by many ophidian taxa provide an ideal setting for multiple paternity which, in turn, allows differential paternity to arise through mating order effects, genotypic effects, phenotypic effects, or a combination thereof.

Snakes have played a seminal role in reproductive studies as they were the first taxa for which it was shown that increased matings by females led to increased fecundity and one of the first taxa in which multiple paternity was observed. Snakes represent an ideal taxonomic group for multiple and differential paternity studies due to their polygynandrous mating system, large litter sizes, intense sexual selection pressures, large variances in phenotypic attributes, and ease of captive maintenance. Field and laboratory studies coupling behavioral and genetically-based approaches should continue to foster exciting results in differential paternity and factors influencing reproductive success far into the future.

16.5 ACKNOWLEDGMENTS

We thank the Herpetologists' League for sponsoring the Reproductive Biology and Phylogeny of Ophidia Symposium at the 2009 Joint Meeting of Ichthyologists and Herpetologists and for their invitation to submit this manuscript. We wish to thank R. Shine and P. Weatherhead for improving earlier drafts and Saint Louis University for their continued support of our research.

16.6 LITERATURE CITED

Adams, E. M., Jones, A. G. and Arnold, S. J. 2005. Multiple paternity in a natural population of a salamander with long-term sperm storage. Molecular Ecology 14: 1803-1810.

Aldridge, R. D. 1979. Female reproductive cycles of the snakes *Arizona elegans* and *Crotalus viridis*. Herpetologica 35: 256-261.

Aldridge, R. D. and Duvall, D. 2002. The evolution of the mating season in the pitvipers of North America. Herpetological Monographs 16: 1-25.

Aldridge, R. D., Flanagan, W. P. and Swarthout, J. T. 1995. Reproductive biology of the water snake *Nerodia rhombifer* from Veracruz, Mexico, with comparisons of tropical and temperate snakes. Herpetologica 51: 182-192.

Aldridge, R. D., Goldberg, S. R., Wisniewski, S. and Bufalino, A. P. 2009. The reproductive cycle and estrus in the colubrid snakes of temperate North America. Contemporary Herpetology 2009: 1-30.

Aldridge, R. D., Jellen, B. C., Allender, M. C., Dreslik, M. J., Shepard, D. B., Cox, J. M. and Phillips, C. A. 2008. Reproductive biology of the massasauga. Pp. 403-412. In W. K. Hayes, K. R. Beaman, M. D. Cardwell and S. P. Bush (eds), *The Biology of Rattlesnakes*. Loma Linda University Press, Loma Linda.

Arnold, S. J. 1988. Quantitative genetics and selection in natural populations: microevolution of vertebral numbers in the garter snake *Thamnophis elegans*, Pp. 619-636. In B. S. Weir, M. M. Goodman, E. J. Eisen and G. Namkong (eds), *Proceedings of the Second International Conference on Quantitative Genetics*. Sinauer Associates, Sunderland, U.K.

Austad, S. N. 1984. Evolution of sperm priority patterns in spiders. Pp 223-249. In R. L. Smith (ed.), *Sperm Competition and the Evolution of Animal Mating Systems.* Academic Press. London, U.K.

Avise, J. C., Jones, A. G., Walker, D. and DeWoody, J. A. 2002. Genetic mating systems and reproductive natural histories of fishes: Lessons for ecology and evolution. Annual Review of Genetics 36: 19-45.

Barry, F. E., Weatherhead, P. J. and Philipp, D. P. 1992. Multiple paternity in a wild population of northern water snakes, *Nerodia sipedon*. Behavioral Ecology and Sociobiology 30: 193-199.

Bennett, P. M. and Owens, I. P. F. 2002. *Evolutionary Ecology of Birds: Life History, Mating Systems and Extinction*. Oxford University Press, Oxford, U.K. Pp. 296.

Birkhead, T. R. and Møller, A. P. 1992. *Sperm Competition in Birds: Evolutionary Causes and Consequences*. Academic Press, London, U.K. Pp. 288.

Birkhead, T. R., Chaline, N., Biggins, J. D., Burke, T. and Pizzari, T. 2004. Non-transitivity of male reproductive success in a bird. Evolution 58: 416-420.

Blanchard, F. N. and Blanchard, F. C. 1941. The inheritance of melanism in the garter snake *Thamnophis sirtalis sirtalis* (Linnaeus), and some evidence of effective autumn mating. Papers of the Michigan Academy of Science, Arts and Letters 26: 177-193.

Blouin-Demers, G. and Weatherhead, P. J. 2007. Allocation of offspring size and sex by female black ratsnakes. Oikos 116: 1759-1767.

Blouin-Demers, G., Gibbs, H. L. and Weatherhead, P. J. 2005. Genetic evidence for sexual selection in black ratsnakes (*Elaphe obsoleta*). Animal Behaviour 69: 225-234.

Briskie, J. V. and Montgomery, R. 1993. Patterns of sperm storage in relation to sperm competition in passerine birds. The Condor 95: 442-454.

Brown, G. P. and Shine, R. 2005. Links between female phenotype, life-history and reproductive success in free-ranging snakes (*Tropidonophis mairii*, Colubridae). Ecology 86: 2763-2770.

Clutton-Brock, T. H. and Parker, G. A. 1995. Sexual coercion in animal societies. Animal Behaviour 49: 1345-1365.

Darwin, C. 1871. *The Descent of Man and Selection in Relation to Sex*, 2nd ed. John Murray, London, U.K. Pp. 528.

Devine, M. C. 1975. Copulatory plugs in snakes: enforced chastity. Science 187: 844-845.

DiBattista, J. D., Feldheim, K. A., Gruber, S. H. and Hendry, A. P. 2008. Are indirect genetic benefits associated with polyandry? Testing predictions in a natural population of lemon sharks. Molecular Ecology 17: 783-795.

Dubey, S., Brown, G. P., Madsen, T. and Shine, R. 2009. Sexual selection favours large body size in males of a tropical snake (*Stegonotus cucullatus*, Colubridae). Animal Behaviour 77: 177-182.
Duvall, D. and Schuett, G. W. 1997. Straight-line movement and competitive mate searching in prairie rattlesnakes, *Crotalus viridis viridis*. Animal Behaviour 54: 329-334.
Eberhard, W. G. 1996. *Female Control: Sexual Selection by Cryptic Female Choice*. Princeton University Press, New Jersey. Pp. 501.
Eccard, J. A. and Wolf, J. B. W. 2009. Effects of brood size on multiple paternity rates: a case for "paternity share" as an offspring-based estimate. Animal Behaviour 78: 563-571.
Emlen, S. T. and Oring, L. W. 1977. Ecology, sexual selection, and the evolution of mating systems. Science 197: 215-223.
Fearn, S., Schwarzkopf, L. and Shine, R. 2005. Giant snakes in tropical forests: a field study of the Australian scrub python, *Morelia kinghorni*. Wildlife Research 32: 193-201.
Ferguson, G. W. and Fox, S. F. 1984. Annual variation and survival advantage of large juvenile side-blotched lizards, *Uta stansburiana*: Its causes and evolutionary significance. Evolution 38: 342-349.
Fitch, H. S. 1970. Reproductive cycles of lizards and snakes. Miscellaneous Publication 52, University of Kansas Museum of Natural History, Lawrence, Kansas.
Ford, N. B. and Seigel, R. A. 1989. Relationships among body size, clutch size, and egg size in three species of oviparous snakes. Herpetologica 45: 75-83.
Fox, W. 1956. Seminal receptacles of snakes. Anatomical Record 124: 519-539.
Garner, W. J. and Larsen, K. W. 2005. Multiple paternity in the western terrestrial garter snake, *Thamnophis elegans*. Canadian Journal of Zoology 83: 656-663.
Garner, T. W., Gregory, P. T., McCracken, G. F., Burghardt, G. M. Koop, B. F., McLain, S. E. and Nelson, R. J. 2002. Geographic variation of multiple paternity in the common garter snake (*Thamnophis sirtalis*). Copeia 2002: 15-23.
Gibbons, J. W. and Dorcas, M. E. 2004. *North American Watersnakes: A Natural History*. University of Oklahoma Press, Norman, Oklahoma. Pp. 438.
Gibson, A. R. and Falls, J. B. 1975. Evidence for multiple insemination in the common garter snake, *Thamnophis sirtalis*. Canadian Journal of Zoology 53: 1362-1368.
Gibson, R. M. and Jewell, P. A. 1982. Semen quality, female choice, and multiple mating in domestic sheep: a test of Trivers' sexual competence hypothesis. Behaviour 80: 9-31.
Halliday, T. R. 1983. The study of mate choice. Pp. 3-22. In P. Bateson (ed.), *Mate Choice*. Cambridge University Press, Cambridge, U.K.
Halliday, T. R. 1998. Sperm competition in amphibians. Pp. 465-502. In T. R. Birkhead and A. P. Møller (eds), *Sperm Competition and Sexual Selection*. Academic Press, London, U.K.
Head, M. L. and Brooks, R. 2006. Sexual coercion and the opportunity for sexual selection in guppies. Animal Behaviour 71: 515-522.
Hoffman, L. H. and Wimsatt, W. A. 1972. Histochemical and electron microscopic observations on the sperm receptacles in the garter snake oviduct. Journal of Anatomy 134: 71-96.
Höggren, M. and Tegelström, H. 1995. DNA fingerprinting shows within-season multiple paternity in the adder (*Vipera berus*). Copeia 1995: 271-277.
Höggren, M. and Tegelström, H. 2002. Genetic evidence for first-male mating advantage in the adder (*Vipera berus*). Pp. 235-242. In G. W. Schuett, M. Höggren,

M. E. Douglas and H. W. Greene (eds), *Biology of the Vipers*. Eagle Mountain Publishing, Eagle Mountain, Utah.
Hrdy, S. B. 1979. Infanticide among animals: a review, classification, and examination of the implications for the reproductive strategies of females. Ethology and Sociobiology 1: 13-40.
Jayne, B. C. and Bennett, A. F. 1990. Selection on locomotor performance capacity in a natural population of garter snakes. Evolution 44: 1204-1229.
Jellen, B. C., Shepard, D. B., Dreslik, M. J. and Phillips, C. A. 2007. Male movement and body size affect mate acquisition in the eastern massasauga (*Sistrurus catenatus*). Journal of Herpetology 41: 451-457.
Jones, A. G., Adams, E. M. and Arnold, S. J. 2002. Topping off: a mechanism of first-male sperm precedence in a vertebrate. Proceedings of the National Academy of Sciences of the United States of America 99: 2078-2081.
Joy, J. E. and Crews, D. 1988. Male mating success in red-sided garter snakes: size is not important. Animal Behaviour 36: 1839-1841.
Keller, L. and Reeve, H. K. 1995. Why do females mate with multiple males? The sexually selected sperm hypothesis. Advances in the Study of Behavior 24: 291-315.
King, R. B. 1989. Sexual dimorphism in snake tail length: sexual selection, natural selection, or morphological constraint? Biological Journal of the Linnean Society 38: 133-154.
King, R. B., Milstead, W. B., Gibbs, H. L., Prosser, M. R., Burghardt, G. M. and McCracken, G. F. 2001. Application of microsatellite DNA markers to discriminate between maternal and genetic effects on scalation and behaviour in multiply-sired garter snake litters. Canadian Journal of Zoology 79: 121-128.
Kissner, K. J., Weatherhead, P. J. and Gibbs, H. L. 2005. Experimental assessment of ecological and phenotypic factors affecting male mating success and polyandry in northern watersnakes, *Nerodia sipedon*. Behavioral Ecology and Sociobiology 59: 207-214.
Klein, J. 1986. *Natural History of the Histocompatibility Complex*. Wiley, New York. Pp. 775.
Kraaijeveld-Smit, F. J. L., Ward, S. J. and Temple-Smith, P. D. 2002. Multiple paternity in a field population of a small carnivorous marsupial, the agile antechinus, *Antechinus agilis*. Behavioral Ecology and Sociobiology 52: 84-91.
Lee, P. L. M. and Hays, G. C. 2004. Polyandry in a marine turtle: females make the best of a bad job. Proceedings of the National Academy of Sciences USA 101: 6530-6535.
LeMaster, M. P. and Mason, R. T. 2002. Variation in a female sexual attractiveness pheromone controls male mate choice in garter snakes. Journal of Chemical Ecology 28: 1269-1285.
Leopold, R. A. 1976. The role of male accessory glands in insect reproduction. Annual Review of Entomology 21: 199-221.
Luiselli, L. 1996. Individual success in mating balls of the grass snake, *Natrix natrix*: size is important. Journal of Zoology 239: 731-740.
Madsen, T. 1983. Growth rates, maturation and sexual size dimorphism in a population of grass snakes, *Natrix natrix*, in southern Sweden. Oikos 40: 277-282.
Madsen, T. 1988. Reproductive success, mortality and sexual size dimorphism in the adder, *Vipera berus*. Holarctic Ecology 11: 77-80.
Madsen, T. and Shine, R. 1993a. Costs of reproduction in a population of European adders. Acta Oecologia 94:488-495.
Madsen, T. and Shine, R. 1993b. Male mating success and body size in European grass snakes. Copeia 1993: 561-564.

Madsen, T. and Shine, R. 1993c. Temporal variability in sexual selection acting on reproductive tactics and body size in male snakes. The American Naturalist 141: 167-171
Madsen, T. and Shine, R. 1998. Quality or quantity? Natural selection on female reproductive output in pythons (*Liasis fuscus*). Proceedings of the Royal Society of London Series B 265: 1521-1525.
Madsen, T., Stille, B. and Shine, R. 1996. Inbreeding depression in an isolated population of adders (*Vipera berus*). Biological Conservation 75: 113-118.
Madsen, T., Shine, R., Loman, J. and Håkansson, T. 1992. Why do female adders copulate so frequently? Nature 355: 440-441.
Madsen, T., Shine, R., Loman, J. and Håkansson, T. 1993. Determinants of mating success in male adders, *Vipera berus*. Animal Behaviour 45: 491-499.
Madsen, T., Ujvari, B., Olsson, M. and Shine, R. 2005. Paternal alleles enhance female reproductive success in tropical pythons. Molecular Ecology 14: 1783-1787.
McCracken, G. F., Burghardt, G. M. and Houts, S. E. 1999. Microsatellite markers and multiple paternity in the garter snake *Thamnophis sirtalis*. Molecular Ecology 8: 1475-1479.
Newcomer, S. D., Zeh, J. A. and Zeh, D. W. 1999. Genetic benefits enhance the reproductive success of polyandrous females. Proceedings of the National Academy of Sciences of the United States of America 96: 10236-10241.
Olsson, M., Gullberg, A. and Tegelström, H. 1994. Sperm competition in the sand lizard (*Lacerta agilis*). Animal Behaviour 48: 193-200.
Olsson, M., Shine, R., Madsen, T., Gullberg, A. and Tegelström, H. 1996. Sperm selection by females. Nature 383: 585.
Olsson, M., Madsen, T., Nordby, J., Wapstra, E., Ujvari, B. and Wittsell, H. 2003. Major histocompatibility complex and mate choice in sand lizards. Proceedings of the Royal Society of London Series B (Supplement) 270: S254-S256.
Parrish, J. J. and Foote, R. H. 1985. Fertility differences among male rabbits determined by heterospermic insemination of fluorochrome-labeled spermatozoa. Biology of Reproduction 33: 940-949.
Parker, G. A. 1970. Sperm competition and its evolutionary consequences in the insects. Biological Reviews of the Cambridge Philosophical Society 45: 525-567.
Parker, G. A. 1984. Sperm competition and the evolution of animal mating strategies, Pp. 1-60. In R. L. Smith (ed.), *Sperm Competition and the Evolution of Animal Mating Systems*. Academic Press, London, U.K.
Parker, G. A. 1990. Sperm competition games: raffles and roles. Proceedings: Biological Sciences 242: 120-126.
Pearse, D. E., Janzen, F. J. and Avise, J. C. 2002. Multiple paternity, sperm storage, and reproductive success of female and male painted turtles (*Chrysemys picta*) in nature. Behavioral Ecology and Sociobiology 51: 164-171.
Penn, D. and Potts, W. 1998. MHC-disassortative mating preferences reversed by cross-fostering. Proceedings of the Royal Society of London Series B 265: 1299-1306.
Pitnick, S. and Brown, W. D. 2000. Criteria for demonstrating female sperm choice. Evolution 54: 1052-1056.
Pizzari, T., Løvlie, H. and Cornwallis, C. K. 2004. Sex-specific, counteracting responses to inbreeding in a bird. Proceedings of the Royal Society of London Series B 271: 2115-2121.
Prosser, M. R., Weatherhead, P. J., Gibbs, H. L. and Brown, G. P. 2002. Genetic analysis of the mating system and opportunity for sexual selection in northern water snakes (*Nerodia sipedon*). Behavioral Ecology 13: 800-807.

Rigley, L. 1971. "Combat dance" of the black rat snake, *Elaphe o. obsoleta*. Journal of Herpetology 5: 65-66.
Rivas, J. A. 2000. Life history of the green anaconda (*Eunectes murinus*) with emphasis on its reproductive biology. Ph.D. thesis, University of Tennessee. Knoxville, Tennessee.
Rivas, J. A. and G. M. Burghardt. 2001. Understanding sexual size dimorphism in snakes: wearing the snake's shoes. Animal Behaviour 62: F1-F6.
Rivas, J. A. and G. M. Burghardt. 2005. Snake mating systems, behavior, and evolution: the revisionary implications of recent findings. Journal of Comparative Psychology 119: 447-454.
Rossman, D. A., Ford, N. B. and Seigel, R. A. 1996. *The Garter Snakes: Evolution and Ecology*. University of Oklahoma Press, Norman, Oklahoma. Pp. 332.
Say, L., Pontier, D. and Natoli, E. 1999. High variation in multiple paternity of domestic cats (*Felis catus* L.) in relation to environmental conditions. Proceedings of the Royal Society of London Series B 266: 2071-2074.
Schuett, G. W. 1997. Body size and agonistic experience affect dominance and mating success in male copperheads. Animal Behaviour 54: 213-224.
Schuett, G. W. and Gillingham, J. C. 1986. Sperm storage and multiple paternity in the copperhead, *Agkistrodon contortrix*. Copeia 1986: 807-811.
Schuett, G. W. and Gillingham, J. C. 1989. Male-male agonistic behaviour of the copperhead, *Agkistrodon contortrix*. Amphibia-Reptilia 10: 243-266.
Schwagmeyer, P. C., Parker, G. A. and Mock, D. W. 1998. Information asymmetries among males: implications for fertilization success in the thirteen-lined ground squirrel. Proceedings of the Royal Society of London Series B 265: 1861-1865.
Schwartz, J. M., McCracken, G. F. and Burghardt, G. M. 1989. Multiple paternity in wild populations of the garter snake, *Thamnophis sirtalis*. Behavioral Ecology and Sociobiology 25: 269-273.
Sheldon, B. C. 1993. Sexually transmitted disease in birds: occurrence and evolutionary implications. Philosophical Transactions of the Royal Society of London Series B 39: 491-497.
Shetty, S. and Shine, R. 2002. The mating system of yellow-lipped sea kraits (*Laticauda colubrina*: Laticaudidae). Herpetologica 58: 170-180.
Shine, R. 1978. Sexual size dimorphism and male combat in snakes. Oecologia 33: 269-277.
Shine, R. 1993. Sexual dimorphism, Pp. 49-86. In R. Seigel and J. Collins (eds), *Snakes: Ecology and Behavior*. McGraw-Hill, New York.
Shine, R. 1994. Sexual size dimorphism in snakes revisited. Copeia 1994: 326-346.
Shine, R. and Shetty, S. 2001. The influence of natural selection and sexual selection on the tails of sea-snakes (*Laticauda colubrina*). Biological Journal of the Linnean Society 74: 121-129.
Shine, R., O'Connor, D. and Mason, R. T. 2000a. Sexual conflict in the snake den. Behavioral Ecology and Sociobiology 48: 392-401.
Shine, R., Olsson, M. M. and Mason, R. T. 2000b. Chastity belts in gartersnakes: the functional significance of mating plugs. Biological Journal of the Linnean Society 70: 377-390.
Shine, R., Olsson, M. M., Moore, I. T., LeMaster, M. P., Greene, M., and Mason, R. T. 2000c. Body size enhances mating success in male garter snakes. Animal Behaviour 59: F4-F11.
Shine, R. Langkilde, T. and Mason, R. T. 2004a. Courtship tactics in garter snakes: how do a male's morphology and behaviour influence his mating success? Animal Behaviour 67: 477-483.

Shine, R., Phillips, B., Langkilde, T., Lutterschmidt, D. I., Waye, H. and Mason, R. T. 2004b. Mechanisms and consequences of sexual conflict in garter snakes (*Thamnophis sirtalis*, Colubridae). Behavioral Ecology 15: 654-660.
Shine, R., Phillips, B., Waye, LeMaster, M. and Mason, R. T. 2003. Chemosensory cues allow courting male garter snakes to assess body length and body condition of potential mates. Behavioral Ecology and Sociobiology 54: 162-166.
Shine, R., Webb, J. K., Lane, A. and Mason, R. T. 2006. Flexible mate choice: a male snake's preference for larger female is modified by the sizes of the females encountered. Animal Behaviour 71: 203-209.
Shine, R., Olsson, M. M., Moore, I. T., LeMaster, M. P. and Mason, R. T. 1999. Why do male snakes have longer tails than females? Proceedings of the Royal Society of London Series B 266: 2147-2151.
Shine, R., Elphick, M. J., Harlow, P. S., Moore, I. T., LeMaster, M. P. and Mason, R. T. 2001a. Movements, mating, and dispersal of red-sided gartersnakes (*Thamnophis sirtalis parietalis*) from a communal den in Manitoba. Copeia 2001: 82-91.
Shine, R., LeMaster, M. P., Moore, I. T., Olsson, M. M. and Mason, R. T. 2001b. Bumpus in the snake den: effects of sex, size, and body condition on mortality of red-sided garter snakes. Evolution 55: 598-604.
Shine, R., O'Connor, D., LeMaster, M. P. and Mason, R. T. 2001c. Pick on someone your own size: ontogenetic shifts in mate choice by male garter snakes results in size assortative mating. Animal Behaviour 61: 1133-1141.
Sinervo, B. 1990. The evolution of maternal investment in lizards: an experimental and comparative analysis of egg size and its effect on offspring performance. Evolution 44: 279-294.
Smith, S. M. 1988. Extra-pair copulations in black-capped chickadees: the role of the female. Behaviour 107: 15-23.
Stickel, L. F., Stickel, W. H. and Schmid, F. C. 1980. Ecology of a Maryland population of black rat snakes (*Elaphe o. obsoleta*). American Midland Naturalist 103: 1-14.
Stille, B., Madsen, T. and Niklasson, M. 1986. Multiple paternity in the adder, *Vipera berus*. Oikos 47: 173-175.
Stockley, P. 1999. Sperm selection and genetic incompatibility: does relatedness of mates affect male success in sperm competition? Proceedings of the Royal Society of London Series B 266: 1663-1669.
Stockley, P., Searle, J. B., MacDonald, D. W. and Jones, C. S. 1993. Female multiple mating behaviour in the common shrew as a strategy to reduce inbreeding. Proceedings of the Royal Society of London Series B 254: 173-179.
Thiel, M. and Hinojosa I. A. 2003. Mating behaviour of female rock shrimp *Rhynchocinetes typus* (Decapoda: Caridea)—indication for convenience polyandry and cryptic female choice. Behavioral Ecology and Sociobiology 55: 113-121.
Thornhill, R. and Alcock, J. 1983. *The Evolution of Insect Mating Systems*. Harvard University Press, Cambridge, Pp. 547.
Trivers, R. L. 1972. Parental investment and sexual selection. Pp. 136-172. In B. Campbell (ed.), *Sexual Selection and the Descent of Man 1871-1971*. Aldine Publishing Company, Chicago, Illinois.
Trontti, K., Thurin, N., Sundström, L. and Aron, S. 2007. Mating for convenience or genetic diversity? Mating patterns in the polygynous ant *Plagiolepis pygmaea*. Behavioral Ecology 18: 298-303.
Uller, T. and Olsson, M. 2008. Multiple paternity in reptiles: patterns and processes. Molecular Ecology 17: 2566-2580.
Ursenbacher, S., Erny, C. and Fumagalli, L. 2009. Male reproductive success and multiple paternity in wild, low-density populations of the adder (*Vipera berus*). Journal of Heredity 100: 365-370.

Valenzuela, N. 2000. Multiple paternity in side-necked turtles *Podocnemis expansa*: evidence from microsatellite DNA data. Molecular Ecology 9: 99-105.
Voris, H. K., Karns, D. R., Feldheim, K. A., Kechavarzi, B. and Rinehart, M. 2008. Multiple paternity in the Oriental-Australian rear fanged watersnakes (Homolopsidae). Herpetological Conservation and Biology 3: 88-102.
Watson, P. J., Arnqvist, G. and Stallmann, R. R. 1998. Sexual conflict and the energetic costs of mating and mate choice in water striders. American Naturalist 151: 46-58.
Weatherhead, P. J., Barry, F. E., Brown, G. P. and M. R. L. Forbes. 1995. Sex ratios, mating behaviour and sexual size dimorphism of the northern water snake, *Nerodia sipedon*. Behavioral Ecology and Sociobiology 36: 401-311.
Weatherhead, P. J., Brown, G. P., Prosser, M. E. and Kissner, K. J. 1999. Factors affecting neonate size variation in northern water snakes, *Nerodia sipedon*. Journal of Herpetology 33: 577-589.
Weatherhead, P. J., Prosser, M. R., Gibbs, H. L. and Brown, G. P. 2002. Male reproductive success and sexual selection in northern water snakes determined by microsatellite DNA analysis. Behavioral Ecology 13: 808-815.
Westneat, D. F. and P. W. Sherman. 1997. Density and extra-pair fertilizations in birds: a comparative analysis. Behavioral Ecology and Sociobiology 41: 205-215.
Wilcox, R. S. 1984. Male copulatory guarding enhances female foraging success in a water strider. Behavioral Ecology and Sociobiology 15: 171-174.
Wing, S. R. 1988. Cost of mating for female insects: risk of predation in *Photinus collustrans* (Coleoptera: Lampyridae). American Naturalist 131: 139-142.
Wusterbarth, T. 2009. Sexual selection and the mating strategies of new world Natricine snakes. unpubl. Ph.D. Northern Illinois University. Dekalb, Ill.
Zeh, J. A. and Zeh, D. W. 1994. Last-male sperm precedence breaks down when females mate with three males. Proceedings of the Royal Society of London Series B 257: 287-292.
Zeh, J. A. and Zeh, D. W. 1996. The evolution of polyandry. I. Intragenomic conflict and genetic incompatibility. Proceedings of the Royal Society of London Series B 263: 1711-1717.
Zheng, W. M., Nishibori, M., Isobe, N. and Yoshimura, Y. 2001. An *in situ* hybridization study of the effects of artificial insemination on the localization of cells expressing MHC class II mRNA in the chicken oviduct. Reproduction 122: 581-586.
Ziegler, A., Dohr, G. and Uchanska-Ziegler, B. 2002. Possible roles for products of polymorphic MHC and linked olfactory receptor genes during selection processes in reproduction. American Journal of Reproductive Immunology 48: 34-42.
Zwiefel, R. G. and Dessauer, H. C. 1983. Multiple insemination demonstrated experimentally in the kingsnake (*Lampropeltis getulus*). Experientia 39: 317-319.

CHAPTER 17

The Evolution of Semelparity

Xavier Bonnet

17.1 OVERVIEW

Mathematical modeling has been the major source of progress in the understanding of the evolution of two contrasting reproductive strategies: semelparity (death following a single reproduction) versus iteroparity (iterative reproduction). However, current models do not allow us to understand why some animal groups (e.g., insects and fish) are more oriented towards semelparity compared to others (e.g., birds and mammals), in which this strategy is under-represented. In addition, the putative links between allelic combinations and their associated respective reproductive strategies (semelparity versus iteroparity) rely on the personal choice and convenience of the modeler, and hence are subject to speculation. Based on field and laboratory research on the reproductive traits of the viviparous Aspic Viper (*Vipera aspis*), this chapter proposes a different approach and a novel scenario for the tendency toward semelparity observed in a snake population monitored in the field. The main purpose of this scenario is to provide rational links between physiological requirements for reproduction, lifetime reproductive success (a proxy of Darwinian fitness) and demographic consequences. This scenario is testable both in the field and in the laboratory and consequently it also opens a door for modeling, criticisms and generalization.

17.2 WHAT DOES SEMELPARITY MEAN?

17.2.1 From the Myth

Sémélé, the daughter of the king of Thebes named Cadmos, was strikingly beautiful (Hesiode 700BC). She was also a mistress of Zeus who commanded every mortal and the Gods. Era (Zeus' wife) became aware of the love affair and she hatched a cruel revenge. Using a subterfuge,

Centre d'Etudes Biologiques de Chizé, UPR 1934-CNRS, 79360 Villiers-en-Bois, France.

Era instilled a fatal doubt in the heart of Sémélé. Tormented, the princess asked Zeus to take an oath to fulfill her most desired vow. Because nothing was supposed to be impossible to him, Zeus imprudently swore the oath. Sémélé then asked Zeus to present himself in his full glory to prove he was Zeus and not a usurper. Despite desperate attempts to get free from the oath in order to not kill his most beloved Sémélé, Zeus complied. As no one, divine or mortal, could sustain the manifestation of Zeus in his full power, Sémélé died under raging thunderbolts. At the time Sémélé was pregnant and Zeus rescued the unborn child from Sémélé by sewing it into his thigh (and thereby was consequently a successful precursor to Rick Shine in manipulating viviparity). A few months later, Dionysus was born, and became reputed for his own reproductive career (Ovidius Naso 8).

17.2.2 To the Concept

Likely, the term semelparity derived from this myth (in Latin "pario" means giving birth). Death following reproduction has been documented in various species from a wide diversity of taxa, across bacteria, plants, and almost all animal classes (Cole 1954; Wodinsky 1977; Finch 1990; Young and Augspurger 1991; Stearns 1992; Rodhouse 1998; Hautekèete *et al.* 2001; Karsten *et al.* 2008; Mayor *et al.* 2009). Several authors even employed unbridled terminology such as "suicidal reproduction" to describe this strategy (e.g., Smith and Charnov 2001). However, all researchers agree that death is a consequence of an intense reproductive effort, and not a consequence of a voluntary sacrificial behavior; therefore research has focused on the costs versus benefits of various reproductive investments per reproductive bout (Stearns 1992; Hautekèete *et al.* 2001; Crespi and Teo 2002). In fact, semelparity represents one extremity of a reproductive strategy gradient, with long life species that can reproduce many times representing the other extreme. Although Cole (1954) was probably not the first to coin the term, he published a seminal article comparing population consequences, notably lifetime reproductive success, associated with the respective life history traits (such as maturity and fecundity), of semelparous and iteroparous organisms. Cole (1954) produced the first significant mathematical model that provided a solid background to further developments. From this pivotal publication, major improvements and multiple refinements permitted the resolution of early paradoxical conclusions, notably via incorporating various key factors such as survival rates of different age classes, competition, stability of population dynamics, density dependence and resource availability (Bryant 1971; Charnov and Schaffer 1973; Young 1981; Bulmer 1985; Ranta *et al.* 2000a, b; 2002; Davydova *et al.* 2005; Zeineddine and Jansen 2009). One of the main outcomes of the numerous developments is a shift away from a simple dichotomized selection scheme (e.g., the r–K selection paradigm) towards more realistic multi-polar systems (Stearns 1992; Benton and Grant 1999; Reznick *et al.* 2002). There is no doubt that this theoretical progress represents the major contribution to our understanding of the evolution of semelparity as an extreme reproductive strategy.

However, whatever the level of sophistication of theoretical developments, modeling relies heavily on demographic parameters (e.g., maturity schedule and density) in relation to environmental factors (e.g., predation and resource availability). Therefore, an important step from genetic make-up to population make-up was bypassed: the genetic and physiological regulatory mechanisms that underlie phenotypes (Franklin 1989; Stern 1998; Zera and Harshman 2001; Nachman *et al.* 2003; Abzhanov *et al.* 2004; Burggren and Warburton 2005; Carroll *et al.* 2005; Breuker *et al.* 2006; Davidson and Erwin 2006). Various studies clearly reveal that endocrine systems effectively play a central role in the negative consequences of reproduction for parental survival, thereby establishing functional links between the physiology of reproduction and semelparity (Wodinsky 1977; Bradley *et al.* 1980; Carruth *et al.* 2002; Barry *et al.* 2010). Other physiological parameters such as metabolic mode and hormonal regulation of follicle recruitment, although probably largely involved, have not yet been integrated.

17.2.3 Sex and Metabolic Modes

Strictly speaking, semelparity refers to species in which individuals die after a unique reproductive bout (Cole 1954; Zeineddine and Jansen 2009). Implicitly, this notion applies to species that reproduce sexually; and thus excludes vegetative reproduction where individuals cannot die from reproduction. For simplicity, I also discard social species (e.g., social ants) where the distinction between the individual's and the colony's life history traits, notably the reproductive strategy, is somewhat tricky (i.e., is it the queen or the colony that can be semelparous? If so what is the appropriate time scale to consider?). From this starting point, an important distinction in terminology should be raised when sex and lineage are taken into account. In plants, hermaphroditic or dioecious species classified as semelparous effectively die shortly after reproduction (e.g., annuals) and the terminology is strictly respected. In the animal kingdom, whatever the phylum, researchers classify females as typically semelparous when they die after their first and thus unique reproductive episode; this implies a single reproductive event before death (Morse and Stephens 1996; Bilde and Lubin 2001; Crespi and Teo 2002; Morse 2009; but see Futami and Akimoto 2005 for an interesting complication). In mammals however, males are considered as semelparous if they die after a single reproductive season, irrespective of the number of copulations or the number of different partners; and in fact multiple copulations with different females seems common (Bradley *et al.* 1980; Oakwood *et al.* 2001; Smith and Charnov 2001; Holleley *et al.* 2006; Martins *et al.* 2006). Such a flexible definition does not apply equally for other lineages where semelparity is generally considered under its absolute definition. For instance, male spiders are usually classified as semelparous if they die after a unique copulation (generally killed by the female), with extreme cases where the transfer of sperm necessitates the perforation of the male by the fangs of the female to secure the mating (Andrade 1996); but usually not if they die after a reproductive season with multiple mating events. Apparently trivial, this neglected distinction has an immediate broad consequence. Strict (or absolute) semelparity has been observed in ectotherms solely. Indeed, semelparity remains undocumented for endotherm females, and the application of the definition remains unclear for males. I therefore suspect that metabolic mode is a key factor for the expression of semelparity.

In the following paragraphs, I report a case study that identifies the proximal physiological and environmental factors, along with their interactions, that favor semelparity in a free ranging ectothermic vertebrate, a viviparous snake. Long-term studies of *Vipera aspis* in France suggest that under certain environmental conditions this species exhibits a marked trend toward semelparity (Bonnet *et al.* 2002). As males are clearly iteroparous, the focus is on females. Most of the results discussed below have been published in previous articles. Importantly, the results presented have never been assembled to form a broader picture and the scenario currently proposed was only slightly evoked in a paper published a decade ago (Bonnet *et al.* 1998). The synthetic approach is therefore the main contribution of this chapter. I fully realize that proposing a novel scenario for the evolution of a reproductive strategy is somewhat perilous, especially using a single species, and I therefore emphasize that my goal is essentially to attract criticisms and to stimulate further discussion.

17.3 A CASE STUDY FROM REPRODUCTIVE PHYSIOLOGY TO POPULATION DYNAMICS

17.3.1 Overview

In western central France, reproductive females (more than 500 individuals monitored in the field as well as laboratory studies) exhibit a low survival rate and consequently tended to breed only once during their lifespan. Prior to reproduction females store large amounts of body reserves in well-developed fat bodies, the liver, locomotor muscles and a dense vertebral skeleton. Immediately after emergence from hibernation in spring, a sharp elevation of plasma levels of oestradiol (oestradiol-17β) provokes the mobilization of maternal body reserves (e.g., lipids, amino-acids, minerals etc.) required for yolk formation. Such a physiological shift is associated with a marked exposure to predators because reproductive females must often bask in the sun to meet the high temperature (metabolic) requirements for vitellogenesis. Similarly, gestation imposes strong metabolic and predatory costs notably through important sun basking behaviors required to elevate body temperature (hence oxygen consumption of the entire organism). Importantly, a substantial proportion of the costs of reproduction is independent of fecundity. Indeed, the optimal body temperatures (close to 30°C) for vitellogenesis and embryogenesis are not influenced by offspring number; hence maternal metabolism and the risk of being killed by a predator are determined by reproductive status rather than fecundity. To minimize such costs, selection favors a maximal increase of fecundity per reproductive bout instead of iteroparity. Capital breeding offers a solution to increase offspring number. Under constraining climatic or environmental conditions this strategy can evolve towards semelparity.

17.3.2 Reproductive Phenology of Females in Western Central France

Vipera aspis is a medium sized polymorphic viviparous snake (~50 cm snout-vent length for adult females) distributed in France, Switzerland and Italy. Current taxonomy is still imprecise and subject to discussion. The species "*Vipera aspis*" probably includes a complex of species and/or sub-species (Golay *et al.* 2008; Barbanera *et al* 2009). This snake has been the subject of numerous ecological and physiological studies (Saint Girons 1957a,b; Naulleau 1970, 1973; Naulleau and Bidaut 1981; Naulleau and Fleury 1990; Saint Girons *et al.* 1993; Bonnet *et al.* 1994, 2001; Naulleau *et al.* 1999; Zuffi *et al.* 1999; 2009). Consequently, the reproductive biology and ecology of *V. aspis* is relatively well studied.

The phenology of reproduction is variable within and among populations; with climatic factors playing a major role and generating strong geographic and inter-annual fluctuations. In western central France, females usually emerge from hibernation in March after a succession of sunny and windless days. Mating is observed in March and April. Roughly, one third of the adult females undergo vitellogenesis immediately after emergence, and the two other thirds skip reproduction for one year or more. Vitellogenesis is a physiologically demanding process that broadly lasts three months with ovulation occurring in early June, followed by fertilization. Gestation takes place in summer, and parturition occurs from late August to late September. With the cooling of the ambient temperatures in October, the snakes progressively return to their winter shelter for a four-month hibernation period on average.

17.3.2.1 Methods

The results used in the current manuscript were obtained in the field and in captivity, several individuals being transported from one field site to the laboratory for various (usually short: days or weeks) periods of time, and often released at their initial place of capture. The main field site, Les Moutiers-en-Retz (Loire Atlantique, 47°03′N, 02°00′W), is situated at the western-northern limit of the distribution area of the species. Captive studies were conducted in the CEBC-CNRS also situated in western central France. The field and laboratory sites are separated by 120 km and are characterized by similar climatic oceanic conditions. Several females that were involved in the laboratory observations were captured in other locations. These additional field sites were located in a radius of 50 km around the CEBC-CNRS. Overall, all the snakes involved originated from previously interconnected populations living in a relatively homogenous oceanic climate characterized by a low occurrence of prolonged hot sunny periods. This means that the snakes must often bask in the sun to reach elevated body temperatures, especially in spring and in autumn.

The main population of *Vipera aspis aspis*, monitored in the field between 1992 and 2008 in Les Moutiers-en-Retz, was large and well isolated by roads and a village. Guy Naulleau initiated this population study with Xavier Bonnet, and over time more than 500 females have been marked individually and regularly recaptured. Broadly 150 gravid females were captured shortly before parturition, brought to the lab and maintained in individual cages until they gave birth. This technique allowed us to collect precise data on litters and post-parturient females. Newborns, stillborns and undeveloped eggs were counted and measured. Soon after the birth, females were released with their progeny at the last place of capture. Each year, field sessions were organized from hibernation emergence until the return to the winter hibernacula in October. The snakes were sexed, measured (±1 cm) weighed (±1 g), individually marked with PIT tags and precisely (±2 m) localized on a map. During successive recaptures, females were palpated to detect the presence of prey, follicles and embryos. Each year, ten to thirty females were fitted with temperature sensitive radio-transmitters (forced ingurgitation) and radio-located at least once a day for one to five consecutive months. Since 2002, the population declined sharply following pine tree plantation that entailed a rapid closing of the habitat, and the results presented here are limited to the field research undertaken between 1992 and 2002.

In parallel studies, we investigated different aspects of the reproductive physiology of captive females maintained under natural climatic conditions in outdoor enclosures. Females were weighed regularly, at least once a week, and fed regularly with dead mice. Reproductive individuals were placed in individual cages before parturition in order to obtain data on their litter. Palpation determined reproductive status around mid-vitellogenesis (follicles larger than 2 cm) and estimated litter size with almost no error. Another set of females was maintained in individual cages fitted with artificial grass substratum, a water dish, a shelter and a source of heat (halogen lamp). Several females were examined using ultra-sonography and nuclear magnetic imaging (Bonnet *et al.* 2008). These two techniques were useful to monitor developing follicles during vitellogenesis and embryos over the entire pregnancy. Blood of both reproductive and non-reproductive females (along with males) was sampled via intra-cardiac punctures. Several steroid hormones (oestradiol-17 , progesterone and testosterone) and plasma metabolites (total calcium, phosphorus, triglycerides, glucose, total protein, and albumin) were assayed over the complete annual cycle, thereby encompassing the reproductive cycle and hibernation (see Bonnet *et al.* 1994 for description of the methods).

We also studied individuals kept under controlled laboratory conditions. This methodology was essential to identify and better describe the key phase of the mobilization of maternal resources during vitellogenesis. Using indirect calorimetry (oxygen consumption), we assessed the metabolic rate of adult reproductive and non-reproductive females placed under different ambient temperatures. Individuals found dead in the field (killed by predators and road kills) or in the lab (emaciated and/or anorexic snakes) were dissected in order to determine their body composition. More details on the techniques and methods briefly presented above are available in previous publications (Bonnet *et al.* 1994, 1999, 2000, 2001a,b, 2003a,b, 2008; Bonnet and Naulleau 1996; Naulleau and Bonnet 1996; Naulleau *et al.* 1999; Vacher-Vallas *et al.* 1999; Aubret *et al.* 2002; Lourdais *et al* 2002a,b,c, 2003; Ladyman *et al.* 2003).

17.3.2.2 Maternal reproductive physiology: Plasma metabolites and hormones

Vitellogenesis involves a succession of physiological stages from the initial recruitment of the follicles that will then undergo full development to ovulation. Although all the successive phases are essential, I will focus on the aspects that are the most demanding in terms of resource investment following early recruitment steps. Broadly speaking, the results presented concern follicles larger than 0.5 cm in diameter.

The analyses below combine data collected directly in the field, in captive snakes maintained in outdoor enclosures and in laboratory-maintained individuals. Our empirical and experimental data clearly show that during the spring phase of vitellogenesis, immediately after emergence from hibernation until ovulation in June, the maternal organism is heavily oriented to fuel the demands of the rapidly developing follicles. Over an average of three months, reproductive females display distinctive physiology and ecology compared to non-reproductive females. Such divergence is so pronounced that for many plasma parameters non-reproductive females are more similar to males rather than to reproductive females (Fig. 17.1). The main maternal body reserve components are involved (Fig. 17.1). Importantly, the plasma concentration values of the metabolites presented below were recorded in non-digesting snakes because of the confounding effects of postprandial physiology. The plasma metabolite concentration values result from the mobilization of maternal reserves from different storage tissues that are directed to the follicles, with a fundamental detour for most of them via the liver that synthesizes yolk precursors; consequently they mix various types of elements. For example the total plasma calcium incorporates the portion bound to the vitellogenin. Below, the main types of maternal reserves are described in a concise manner.

1) Mineral reserves (Fig. 17.1A): the maternal skeleton is heavily mobilized during vitellogenesis; the vertebrae are subject to a deep demineralization (osteoporosis) that results in extremely high plasma

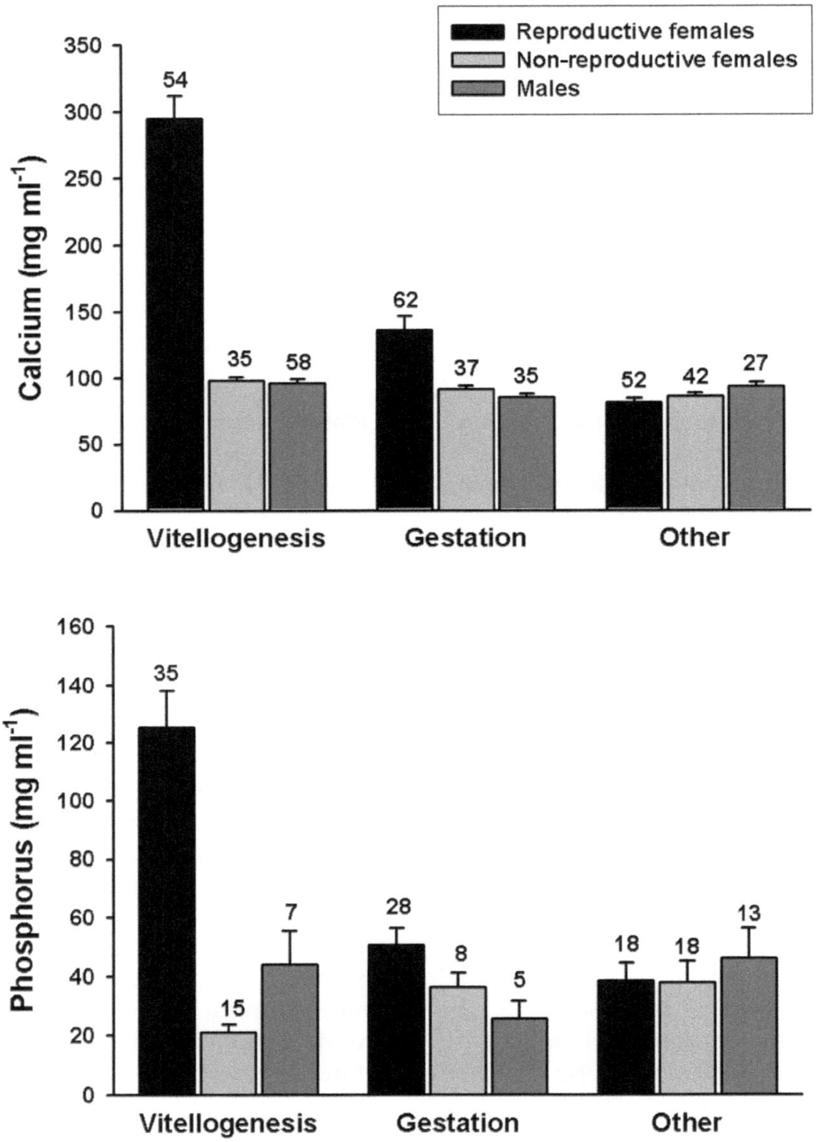

A

Fig. 17.1 Comparison of plasma metabolite levels of reproductive female *Vipera aspis* (black bars) relative to non-reproductive females (light grey bars) and males (dark grey bars) during vitellogenesis (~3 months), gestation (~3 months), and after parturitions (other, ~2 months before hibernation). **A.** The very high calcium and phosphorus plasma values observed in

Figure 17.1 Contd. ...

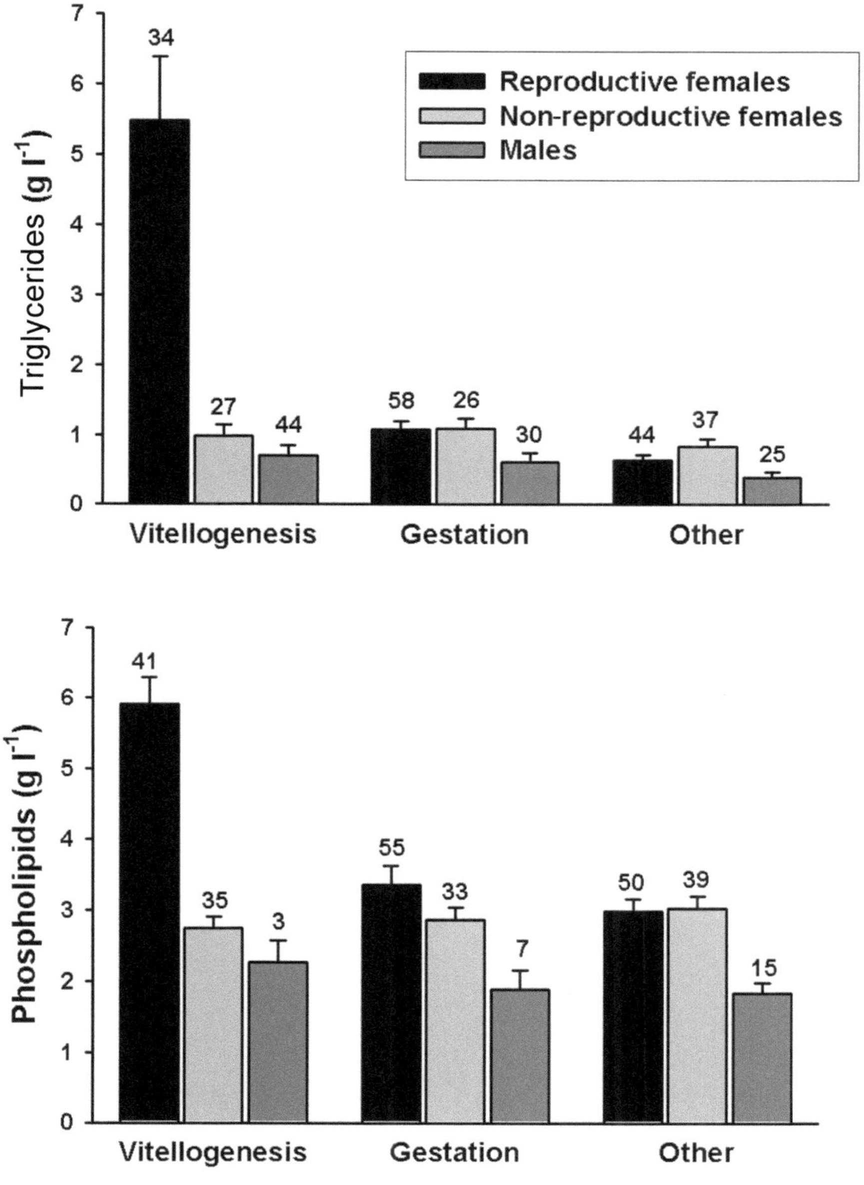

B

... *Figure 17.1 Contd.*

vitellogenic females are caused by the strong mobilization of maternal skeleton. **B.** Similar patterns are observed for the metabolites involved in the mobilization of maternal lipid body

Figure 17.1 Contd. ...

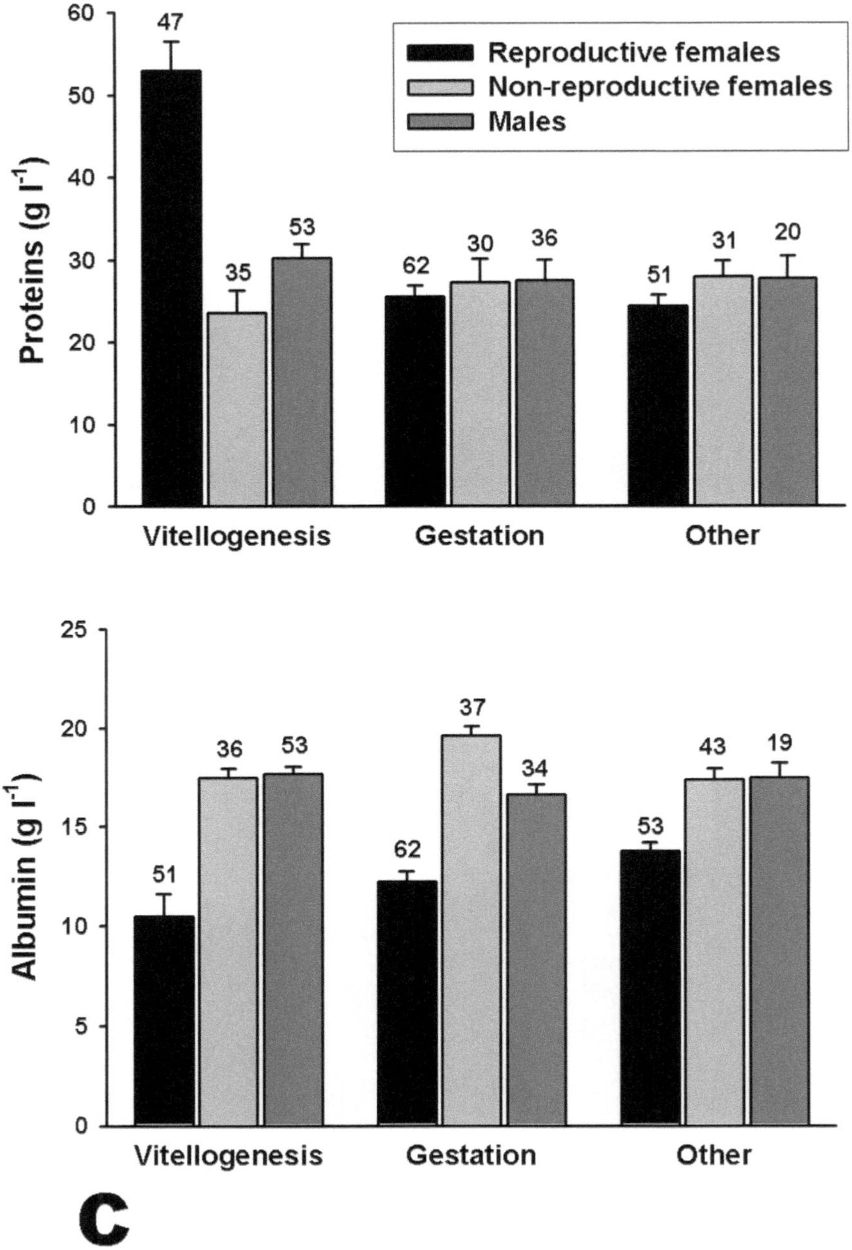

... Figure 17.1 Contd.

reserves (tryglycerides, phospholipids). **C.** The massive synthesis of vitellogenin (yolk precursor) by the liver entails both an elevation of plasma protein concentration and a decrease of albumin levels due to a trade-off between vitellogenin *versus* albumin syntheses by the hepatocytes. Means ± SE, sample size above each bar.

values of total calcium and phosphorus (Alcobendas 1989; Alcobendas *et al.* 1991; Bonnet *et al.* 1994). Both elements are essential for the formation of the yolk. The mean plasma value calculated over three months, close to 300 mg per ml (maximal value >500 mg/ml), is amongst the highest ever recorded in vertebrates (including pathologies). After ovulation, the plasma values return to normal concentrations as observed in non-reproductive females and males.

2) Lipid reserves (Fig. 17.1B): similarly, the rapid mobilization of the lipids stored in the form of fat bodies and in the liver provokes an extreme elevation of the plasma values. In fact, the plasma becomes very thick, with values culminating above 20 g/l for triglycerides and above 10 g/l for the phospholipids.

3) Vitellogenin synthesis: The liver produces and releases this macromolecule into the general circulation and this entails a strong elevation of plasma proteins (indeed vitellogenin is a large phospho-lipo-protein that binds calcium; Fig. 17.1C). Such production requires the mobilization of large amounts of amino acids, and consequently the locomotor muscles are subjected to a marked catabolism (also observed in other snakes species; Lourdais *et al.* 2004). Interestingly, the production of albumin drops during vitellogenesis (this plasma protein is not incorporated into vitellogenin), suggesting that the liver cannot simultaneously increase drastically the production of vitellogenin and maintain the production of albumin. Such a trade-off between competing functions has been observed at the cellular level: the experimental hormonal stimulation of vitellogenesis by the hepatocytes massively recruits the intra-cellular machinery (e.g., mRNA synthesis), and other protein biosyntheses are markedly slowed (Ho *et al.* 1982; Callard *et al.* 1990).

Overall, over three consecutive months, the massive changes in plasma metabolite concentrations observed in reproductive females reveal a pronounced and prolonged physiological shift. In terms of mobilization of maternal resources, vitellogenesis is a more extreme process than gestation and there are major metabolic and behavioral consequences associated with the development of the follicles. Surprisingly, almost all field ecologists ignore this crucial phase whilst gestation has received considerable attention. I do not dispute the importance of gestation, but I suggest that neglecting vitellogenesis may lead to erroneous, or at least incomplete, understanding of the reproductive biology and ecology of many animal species that produce large and/or numerous follicles. For the purpose of the current manuscript, the drastic shift imposed by vitellogenesis over three months on the maternal organism in terms of investment in body reserves, metabolism and behaviors (see below) is a major element that favors semelparity as explained below. Information on the mechanisms that control vitellogenesis is therefore interesting.

Our data clearly established that oestradiol is the key hormone that initiates the massive mobilization of all the maternal reserves during the spring phase of vitellogenesis (Fig. 17.2). An artificial increase in plasma oestradiol provokes identical changes in the plasma composition compared to naturally vitellogenic females, including the marked decline in the plasma levels of albumin (Bonnet *et al.* 1994). In all vertebrate species investigated, oestradiol has been identified as the primary stimulus for vitellogenesis in the hepatocytes (Ho *et al.* 1982). However, in snakes this role has been questioned (Saint Girons *et al.* 1993). I suspect that this disagreement was simply the result of the misclassification of the females (reproductive versus not), especially as our experimental data are particularly clear regarding the influence of elevated circulating levels of oestradiol on the mobilization of body reserves. The other important steroid hormone involved in the regulation of female reproduction, progesterone, is not (or perhaps indirectly through complex mechanisms?) involved in the mobilization of maternal reserves; an elevation of plasma values was observed in pregnant females only, hence after the completion of vitellogenesis. In addition, experimental elevation of the plasma concentration of progesterone was not accompanied by the characteristic oestradiol induced changes in the plasma metabolite composition (Bonnet *et al.* 2001a). Overall, oestradiol is the key hormone for the drastic mobilization of the maternal body reserves required to fuel vitellogenesis, which induces (directly or indirectly) a rapid catabolism of the vertebrae, fat stores, locomotor muscles, and liver, and that also stimulates the synthesis of vitellogenin by the liver. Because the concentrations we observed are particularly elevated, I suspect that the variations in plasma levels of this steroid are an important determinant of reproductive effort (although the complex but not-investigated systems that involve the receptors and numerous intra-cellular regulations also play

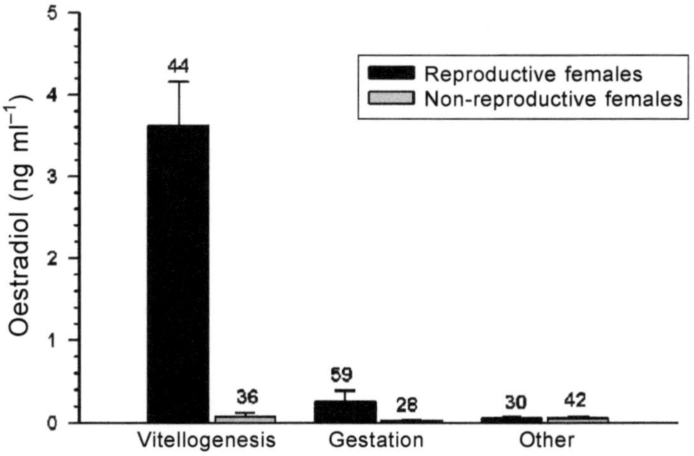

Fig. 17.2 Oestradiol (E2) is the key hormone in the extremely intense mobilization of maternal reserves during vitellogenesis in *Vipera aspis*. Three main periods of female cycle are considered: vitellogenesis (~3 months), gestation (~3 months), and post-parturition (~2 months before hibernation). Means ± SE, sample size above each bar.

central roles). The other major hormones that likely control the number of follicles at an early stage of vitellogenesis (gonadotrophins and leptins; Gobbetti *et al.* 1994; Schneider *et al.* 2000) remain virtually unexplored in snakes in general.

17.3.2.3 Changes in body condition over time

Considering the prolonged and intensive phase of spring vitellogenesis, we may expect significant consequences for maternal condition with an increasing depletion of reserves during the course of reproduction. Using different techniques (e.g., dissection of accidentally killed females and NMI, Nuclear Magnetic resonance Imaging) we observed that at the end of vitellogenesis females are very emaciated, despite an apparent external high body condition, in fact slightly higher than at the onset of vitellogenesis (Bonnet *et al.* 2003b). Comparison of the main body reserves of adult females dissected (death was always accidental, hence sample sizes follow availability of dead snakes) shows that at the onset of vitellogenesis, reproductive females possess large body reserves (fat bodies, liver and muscles), but that almost all the reserves have been transferred to the follicles by ovulation, and that post-parturient females are even more emaciated (Table 17.1). Other (non-reproductive) adult females are in intermediate position. In the course of vitellogenesis, almost all the fat bodies have been degraded, two thirds of the liver also, and a large proportion of the body (essentially represented by locomotor muscles) as well. Gestation generates further degradation of maternal reserves, and after birth, the females have almost no body reserves. Non-reproductive females are not emaciated, but their body reserves are significantly reduced compared to reproductive females at the onset of vitellogenesis; they can be considered as being in an intermediate stage.

Table 17.1 Dissections of adult female *Vipera aspis* provided the mass of the main body reserve tissues at different stages (1-4). 1: "Vitello" means at the onset of vitellogenesis (N=14), 2: "Ovulation" means shortly after ovulation (N=5), 3: "Parturition" means after parturition and before hibernation (N=31), and 3: "Other" means females not reproductive during the year they died (N=27). The last columns refer to comparisons between the stages, e.g., 1/2 is the comparison (%) between the masses of the body reserves observed in ovulated versus early-vitellogenic females (Bonnet *et al.* 2003b). The mean body size (SVL) was not significantly different among four categories.

Tissue (g)	1 *Vitello*	2 *Ovulation*	3 *Parturition*	4 *Other*	1/2	2/3	1/3	1/4
Fat	15.5±6.1	2.6±1.8	1.7±1.5	7.8±4.5	83%	34%	89%	50%
Liver	10.8±6.7	3.2±1.5	3.1±1.4	5.2±2.1	70%	2%	71%	52%
Carcass	77.5±23.9	46.1±8.9	41.6±10.2	58.2±12.3	41%	10%	46%	25%

These results show that at ovulation reproductive females are already in poor body condition, although large ovulated eggs distend their abdomen. During the course of gestation they use the small amounts of reserves left, and their extremely poor condition is clearly revealed after parturition:

large abdominal skin folds and very poor stamina characterize this category of females.

Given the marked physiological differences between reproductive and non-reproductive females, we may expect parallel divergences in terms of their behavioral ecology in the field.

17.3.2.4 Field behavior

Reproductive females are considerably more often exposed, basking in the sun, compared to non-reproductive females (Bonnet and Naulleau 1996). On average the annual capture frequency of individual reproductive females was 2.64±1.74 (mean±SD, range 1-8 per year) captures per year, and 1.76±1.02 (range 1-5 per year) captures per year in non-reproductive females (Bonnet and Naulleau 1996). As a result, most of the females captured in the field are reproductive, although non-reproductive females represent an average two thirds of the adult female population. Part of the greater visibility of reproductive females is explained by mating behavior in March and April, but the trend for a higher capture rate of reproductive females increases over time and is not associated with copulation: the proportion of reproductive females shifts from 45% to more than 70% between March and late August (Bonnet and Naulleau 1996). This is largely due to the progressive increase of ambient temperatures. In early spring when air temperatures often remain below 15°C, non-reproductive females are forced to bask in the sun for digestion, sloughing and to forage; in summer they can reach a satisfactory body temperature (25-30°C) with limited sun basking, indeed air temperature is often above 25°C. Consequently, non-reproductive females become progressively more and more secretive. By contrast, reproductive females are relatively easily observed in the field while the air temperature is below 25°C whilst they still need to bask in the sun basking to warm up their body to meet their specific reproductive metabolic requirements. This is typically the case after rainfall or during the first half-hour after sunrise, even in summer. These results show that during vitellogenesis and gestation reproductive females are under strong pressure to maintain their body temperature close to the maximal plateau body temperatures recorded in the field, 30-31°C, with brief peaks at 32-35°C. The difference in visibility is associated with a divergence in the risk of predation: at our field site, raptors, the main predators for *Vipera aspis* (Naulleau *et al.* 1997), patrol the area intensively during the entire activity period of snakes.

Body temperatures (N>2500) recorded from April to November in the field using internal temperature-sensitive transmitters revealed that reproductive females (N=18) maintain their body temperature an average of 3 to 4 degrees above the body temperature recorded in non-reproductive females (N=9). Interestingly, this difference is significant at any time of the day, even at night, suggesting that reproductive females select warmer night refuges than non-reproductive females. In summer, we also observed that non-reproductive females select cool (below 20°C) refuges when they are not involved in physiologically constraining processes such as skin sloughing or digestion. This means that not only do reproductive females select elevated and stable body temperatures in order to process and rapidly metabolize resources required for vitellogenesis and for precise thermoregulation to provide optimal temperatures to the developing embryos during gestation, but also that non-reproductive females select low body temperatures likely to minimize their metabolic costs when possible. Interestingly, in reproductive females, the mean distance traveled per day and home range decreased rapidly after ovulation (from 5.1 m/day to 2.7 m/day and from 3,187 m^2 to 670 m^2, during vitellogenesis [3 months] and gestation [3 months] respectively; Naulleau *et al.* 1996). This behavioral shift corresponds to two contrasting phases in terms of physiological priorities: during vitellogenesis reproductive females are foraging actively in search of prey (voles) to be invested into the yolk (Bonnet *et al.* 2001b), but during gestation they are essentially involved in careful thermoregulation to minimize thermal instability of the developing embryo, an activity somewhat antagonistic to foraging.

Overall, capture rate and body temperature data are complementary; they both indicate that reproductive females spend considerably greater amounts of time basking in the sun to reach elevated temperatures and/ or to maintain precise thermoregulation profiles, whilst non-reproductive females minimize their exposure to avian predation and their energy expenditure.

Considering the physiological and behavioral information presented above, it is obvious that reproduction generates cumulative survival risks: indeed, the metabolic requirements of vitellogenesis and gestation result in an almost complete depletion of body reserves, but they also expose reproductive females to an increased predation risk. Therefore, population dynamics of adult females should be largely influenced by reproductive status.

17.3.2.5 Population consequences

In our study population, each year an average of two thirds of the adult females skip reproduction. The very emaciated post-parturient females (from the previous year) represent part of the non-reproductive females, but apparently healthy females that are in an intermediate body condition are the most abundant snakes in this non-reproductive category. These proportions have been calculated using the capture-recapture technique and individual monitoring over several years: adult female *V. aspis* cannot reproduce every year in our study site, and they require an average of two years to restore their body reserves after parturition (Naulleau and Bonnet 1996; Bonnet *et al.* 1999, 2002). This extended recovery process means that most of the females cannot reproduce twice in their life. Indeed, roughly fifty percent of reproductive females do not survive the long, exhausting and perilous period of vitellogenesis + gestation + parturition (see above). The surviving post-parturient females that can safely hibernate are in very poor body condition most of the time (Table 17.1); about half of them die from extreme emaciation the next active season following spring emergence (Bonnet *et al.* 1999; 2002). The remaining females that are able to capture prey rapidly after hibernation (and hence escape lethal emaciation) must nonetheless survive for two years on average to restore their body condition in order to be able to reproduce again. For that, they must also evade predation during their second period of vitellogenesis and gestation.

From a simplistic perspective, the chances for a female *V. aspis* to breed more than once are limited to 13% in our study area. This value results from the combined survival rates calculated during the first reproductive year (survival of reproductive females is ≈0.50), during the recovery year after birth (survival of the very lean post-parturient females is ≈0.50), during the following body year of restoring body reserves (survival of intermediate non-reproductive females is ≈0.80), and during the second reproduction (survival rate is ≈0.65 to parturition; the immediate post-parturition low survival is ignored here). We marked and monitored more than 500 females in the field. Of these, we captured 148 shortly before parturition in order to keep them briefly in captivity to collect precise data on their litters. Some of these females were captured more than once, indicating that they were iteroparous. Interestingly, this number of iteroparous females (N = 19) matches almost perfectly the expected number based on our calculated survival rate: 19.2 (148*0.13). Therefore, the average survival rate estimated for the different categories of adult females using the CMR technique (capture-mark-recapture) is somewhat validated by the average respective proportions of semelparous (90%) and iteroparous (10%) females that we have been able to capture and bring to the laboratory for precise examination of their reproductive output.

On average, female *Vipera aspis* exhibits a strong trend toward semelparity, with roughly 90% of the individuals reproducing only once, and the breeding frequency of the few remaining iteroparous females is very low. In fact, these proportions are derived from averaging long-term monitoring data, and significant annual fluctuations have been observed (this important aspect will not be discussed in the current ms; but see Bonnet *et al.* 2000). In terms of the strict definition, *V. aspis* is not semelparous, but our study population is clearly oriented toward this extreme strategy. Strictly speaking, the *V. aspis* is not semelparous, complicating any generalization about this extreme strategy. However, the flexibility of post-reproduction survival that is apparent from our field data provides an opportunity to compare iteroparous versus semelparous individuals, otherwise impossible in species where post-reproduction death is obligate.

17.3.2.6 Lifetime reproductive success

Although poorly represented in our data set, the few iteroparous females that we monitored provide an opportunity to assess important questions. Do semelparous and iteroparous females partition their reproductive effort differently? Does the total number of offspring produced (Lifetime Reproductive Success, LRS, a proxy of Darwinian fitness) also differ? Are there morphological differences between semelparous and iteroparous females that will impart a survival advantage to iteroparous females?

We found no evidence of partitioning of reproductive effort between reproductive events ($P > 0.5$). On average semelparous females produced 6.0 ± 0.2 (mean litter size adjusted by maternal SVL ± SE) offspring ($N = 127$ litters), females that reproduced twice produced first 6.1 ± 0.5 and then 7.1 ± 0.5 ($N = 17$ litters) offspring; finally, females that reproduced three times produced successively 5.8 ± 1.0, 5.9 ± 1.0, and 6.3 ± 1.0 ($N = 4$ litters). The cumulated mean number of offspring increased linearly ($N = 3$) with the number of successful reproduction events. Clearly, the LRS of iteroparous females increases with reproductive events.

We found no difference between the morphological characteristics of semelparous ($N = 19$) and iteroparous ($N = 129$) females. Both groups were indistinguishable for body size (SVL) and early body condition (recorded at the onset of vitellogenesis) (unpublished). We extended the analyses by incorporating females for which the reproductive strategy could be determined but for which detailed information on reproductive output was not collected (to ensure that capture rate did not falsify our results, we waited at least two years to classify a given female as dead or not; Bonnet *et al.* 1999, 2002). The results remained unchanged. Consequently we found no difference between the two categories of females at the onset of reproduction: body size and/or body reserves did not influence the probability of reproducing more than once.

Perhaps divergent reproductive investment between the two categories of females was involved. Our data do not support this hypothesis either; females that lost less body mass over the reproductive period were not more likely to breed again (Bonnet *et al.* 2002). Similarly, iteroparous females that lost more mass when producing their first litter did not delay subsequent reproduction for a longer period compared to females that invested relatively less resources (Bonnet *et al.* 2002).

However, iteroparous females produced significantly heavier offspring during their first reproduction compared to semelparous ones. Taking into account two important factors that influence reproductive output, maternal SVL and initial body condition (Bonnet *et al.* 2001b) the mean offspring mass of the females that bred only once was 6.2 ± 0.1 g (SE, $N = 52$ litters) versus 6.9 ± 0.3 g (SE, $N = 14$ litters) for the first litter of females that bred more than once. This difference was essentially due to a differential food intake between the two categories of females (Bonnet *et al.* 2001b); iteroparous females consumed more prey (Lourdais *et al.* 20002b, 2003). Food availability varied greatly between years in our study area (Bonnet *et al.* 2000). Pregnant females that were able to capture three or more voles during gestation were also in better condition after parturition (Lourdais *et al.* 2003) and were more likely to survive the following year (Bonnet *et al.* 1999).

Overall, semelparous and iteroparous females were indistinguishable in terms of body size and body reserves at the onset of vitellogenesis and in the number of offspring produced per reproductive event. We nonetheless observed a significant effect of food intake on the mean mass of the offspring and on the post-parturition body condition of the mother. Costs of reproduction represent the second fundamental aspect that must be examined.

17.3.2.7 Fecundity independent costs of reproduction (FIC) and optimal reproductive effort

The notion that low frequencies of reproduction are most likely to evolve in taxa that display a high fecundity-independent cost of reproduction was proposed 30 years ago (Bull and Shine 1979). Our data provide some evidence in support of this idea.

In the field, we found no relationship between fecundity and survival rate. The number of follicles palpated (an accurate estimate of fecundity in this species) varied from one to thirteen, and did not influence the probability of recapturing a female during the following three years ($P > 0.6$; Bonnet *et al.* 1999, 2002). In the laboratory, using three different ambient temperatures (17.5°C, 25.0°C and 32.5°C), we found no (or a very weak) correlation between fecundity and metabolism measured as oxygen consumption (VO_2 calculated in ml/g/h; Ladyman *et al.* 2003). By contrast, ambient temperature had a strong effect on oxygen consumption (Ladyman *et al.* 2003). Therefore, at least two types of potentially important costs of reproduction, survival during reproduction and temperature dependent energy consumption during pregnancy, were independent of fecundity.

These counterintuitive results have a strong consequence in terms of optimal reproductive effort. Female *Vipera aspis* are under strong selection to amortize the fecundity independent costs of reproduction: the greater number of offspring they produce per reproduction, the better the level of amortization. However, a significant amount of reproductive effort is also represented by the considerable resources invested in yolk synthesis. The balance between the two types of costs, fecundity dependent versus fecundity independent, is likely a strong determinant of optimal reproductive effort. Because relative litter mass (the ratio of total litter mass to maternal post-partum mass; often been used as a measure of relative reproductive investment) is high in *V. aspis*, close to 50% on average with extreme values above 100% (Bonnet *et al.* 2003a), and because breeding frequency is particularly low, I suggest that FIC are elevated. Further results support this notion.

17.3.2.8 Thermoregulation: A key factor

Regardless of the number of eggs (fecundity), the optimal body temperature required for a female to rapidly complete vitellogenesis (i.e., in early June to leave enough time for gestation after ovulation and before winter) is likely to be the same. Indeed, metabolic rate (the sum of the catabolism for reserve mobilization, blood transport, anabolism in the liver and ovaries) is highly dependent on maternal body temperature but not on fecundity, and high body temperatures accelerate this physiological process. During vitellogenesis, it is expected that reproductive females would bask with the same intensity irrespective of the number of follicles they carry. Unpublished data gathered by Olivier Lourdais (CEBC-CNRS) on females fitted with temperature data loggers suggested that, during gestation, embryos are sensitive to thermal perturbations: all the gravid females maintained an identical high and stable body temperature (31°C) in order to optimize offspring phenotypes. During the entire reproductive period (6 months on average), all reproductive females tend to have identical thermoregulatory profiles (depending upon ambient conditions) irrespective of their offspring number. Consequently they take identical sun exposure risks with regard to avian predation, and the elevation of their body temperature generates an equal overall energy expenditure of the whole organism. Improving the reproductive process (vitellogenesis, gestation) via selection of high body temperatures automatically results in "parasitic" energy expenditure from the rest of the body not directly involved in reproduction (e.g., skin and muscles during gestation). For instance, an elevation of maternal body temperature from 25°C to 30°C generates a significant increase in oxygen consumption (>40%).

Thus high predation risk and significant parasitic energy expenditure represent typical FIC. Such FIC automatically decrease per offspring with increasing litter size. Our calculations (unpublished) suggest that a female should produce at least four to five offspring per reproduction. Importantly, females undergo vitellogenesis (at least the resource demanding part) immediately after winter emergence, and huge amounts of resources are required for the growth of more than four follicles. This means that the resources for the development of a large litter must be available very early during reproduction. A physiological solution is to store large body reserves prior to reproduction, and to not utilize them unless they are sufficient for the production of a large litter. In other words, this is a definition of the capital breeding strategy in which a body condition threshold is expected to dictate reproduction. This is also exactly what we observed in the field: female *Vipera aspis* do not undergo vitellogenesis unless they reach a high and precise body condition threshold (Naulleau and Bonnet 1996).

A corollary of these assumptions and observations is that we should not observe small litter sizes. However, on a few occasions, we recorded litter sizes of one to three offspring only (Bonnet *et al.* 2001b, 2003a). Repeated magnetic nuclear imaging sessions on the same individuals showed cases of "egg disappearance" during gestation. Combined with the more classical follicle atresia, these physiological processes offer an alternative explanation (Bonnet *et al.* 2008). Unexpected small litter sizes (<4) result from a reduction of fecundity after an initial "normal" recruitment (≥4 follicles). Repeated palpations of females maintained in captivity suggest that this process can be common in stressed females (e.g., involved in experiments that require frequent manipulations).

17.4 A NOVEL SCENARIO

By assembling the above conclusions in a single scenario I propose the following chain of causes and effects (Fig. 17.3):

A) If the costs independent of fecundity (FIC) represent a major proportion of the total costs of reproduction compared to the costs that increase with offspring number, then it is profitable to maximize offspring number (fecundity) per reproductive event. Indeed, the invariable total FIC would be divided per clutch size, and thus the total cost per offspring equally divided. Offspring production would be automatically more efficient. Therefore, selection should favor physiological processes that enable individuals to maximize offspring number. Among others, key endocrine systems such as those involving FSH and or LH in vertebrates should represent potential targets of selection (Gobbetti *et al.* 1994).

B) However, increasing offspring number is not an easy task because large amounts of resources are required along with space to accommodate the growing follicles. In addition, egg production and embryo development

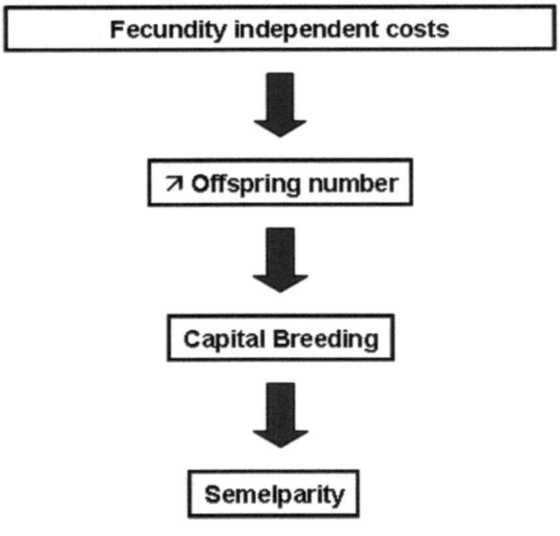

Fig. 17.3 Cascade of causalities from costs of reproduction to extreme physiological exhaustion in the course of a reproductive event. To amortize heavy fecundity independent costs of reproduction (FIC, see text) it is profitable to maximize offspring number per reproductive event. One of the best options to secure a large clutch size is to rely on a capital breeding strategy, and to make a massive reproductive effort. As a consequence, the probability of reproducing again is degraded (e.g., owing to the long time needed to restore body reverses). If FIC become predominant, selection should favor an extreme mobilization of the resources to maximize offspring number. The death of individuals after a single reproduction (semelparity) would be merely a consequence of such intense reproductive investment, not necessarily linked with other demographic traits (e.g., maturity schedule, growth rates), but strongly related to environmental factors (e.g., food availability).

might be strongly constrained by time and climatic factors. Therefore, in many cases the best option to secure a large clutch size is to rely on the accumulation of large maternal body reserves prior to reproduction. Body reserve storage necessitates both time and space, and potentially can handicap individuals (to escape predators for instance). The advantage of body reserve storage in terms of reproductive efficiency is therefore balanced by inevitable costs. A physiological system locking reproduction until sufficient body reserves have been accumulated is theoretically required to prevent inefficient investments. Storage of large body reserves prior to reproduction along with the existence of a body condition threshold to meet optimal reproductive investment requirements corresponds to a definition of a capital breeding reproductive strategy. Therefore, proportionally high FIC favors capital breeding.

C) For species with low breeding frequency, delayed maturity, or where reproduction entails a large shift in term of survival risk (e.g., long migration observed in European eels or in different species of salmonid fishes; Crespi and Teo 2002), the probability of reproducing twice is low. In fact, most of these traits can be interpreted with respect to the FIC concept (Bull and Shine 1979). In other words, when FIC become especially survival costs over energy costs, selection should favor an extreme mobilization of the resources for reproduction in order to maximize offspring number. The death of individuals would be merely a consequence of an intense reproductive effort.

This scenario allows us to answer one of the questions raised earlier: why are some ectotherms more oriented towards semelparity compared to edotherms?

In an eco-physiological perspective, capital breeding occupies a central position for the evolution of semelparity, but this strategy is more likely to evolve in ectotherms than in endotherms (see Bonnet *et al.* 1998 for further discussion of this issue). In fact, there is as yet no endotherm species in which reproduction is entirely fueled by capital whilst abundant examples have been documented among ectotherms. If we consider that semelparity is an extreme example of capital breeding, then the virtual absence of strict semelparity observed among endotherms is expected.

Secondly, in this scenario, evolution toward semelparity does not rely on the sudden occurrence of a mutation in an iteroparous population. However, one of the basic assumptions of most of the mathematic models in the literature is that the change between semelparity versus iteroparity rests on a single mutation that provokes a rigid all-or-nothing shift between two extremes (McNamara 1997; Ranta, *et al.* 2000). The eco-physiological (bottom up?) approach developed above seriously challenges this theoretical view. It is indeed possible to envisage an uncoupling of the various traits that are usually invoked to differentiate these two strategies (Baird *et al.* 1986). For example, a species simultaneously exhibiting late maturity (Saint Girons

1957b), relatively low fecundity and semelparity would be considered an impossible chimera by currently available demographic models as mixing typical and incompatible iteroparous and semelparous traits. But this is just what *Vipera aspis* does. Such annoying examples have been simply ranked as paradoxes (Stearns 1992). In the scenario above, I propose that the targets for selection are both genetic and hormonal mechanisms that control follicle recruitment, offspring size and the physiological investment underpinning reproduction. The physiological and behavioral systems involved can be gradually modified. The case of *V. aspis* shows that the transformation from iteroparity to semelparity does not involve a rearrangement of alleles with the creation of a genetic barrier between these two extremes, but rather differential environmental constraints. Other studies have reached similar conclusions for both ectotherms and endotherms (Schmidt *et al.* 2006; Mayor *et al.* 2009). For instance different capelin populations exhibit absolute semelparity versus iteroparity strategies; such facultative semelparity results from interactions between spawning habitat, physical forcing, and predatory pressure (Christiansen *et al.* 2008). In a copepod, the probability of death following a massive reproductive investment depends upon food availability (Mayor *et al.* 2009).

17.5 FUTURE DIRECTIONS

17.5.1 Field and Experimental Studies

The notion that FIC can represent a substantial proportion of the total costs of reproduction is not familiar to ecologists, and such costs have been neglected. I suggest devoting more effort in characterizing and estimating FIC versus costs that increase with fecundity in various systems. This can be done both in the field and in the laboratory. For instance, in many ectotherms, survival and metabolic consequences of the shift from non-reproductive to reproductive status could be measured independently from fecundity, for example by estimating the higher metabolic expenditure related to the specific thermoregulatory requirements of reproduction.

Comparative studies with other species represent another important future direction. For instance, although similar data to those presented above are lacking on other snake species, there is strong evidence that the more northerly distributed Adder (*Vipera berus*) exhibits even more semelparous traits than the aspic viper (Madsen and Shine 1993; pers. obs.). Post parturient females of *V. berus* are almost systematically extremely emaciated, and many of them cannot survive whatever environmental conditions (unpublished). Comparative studies would offer an opportunity to better take into account the impact of environmental constraints. As advocated in the current paper, because *V. berus* faces cold climatic constraints, we would expect its reproductive frequency of to be even lower and closer to semelparity than in *V. aspis*. Similarly, intra-specific comparisons of populations distributed over a wide range of climatic situations (altitude, latitude...) would be particularly useful (e.g., comparing aspic vipers from northern distribution areas with those living in relatively warm areas in several places in Italy).

17.5.2 Mathematical Modelling

Mathematical modeling could incorporate physiological constraints, for instance by estimating the fitness consequences of various combinations of FIC and fecundity dependent costs, both in terms of energy and survival. The potential links between capital breeding and semelparity could help to propose novel scenario for evolution toward this extreme strategy, a possibility clearly outlined in a salmon: "Adaptation-by-time may play an important role in life history evolution within some species, particularly those with breeding systems characterized by semelparity, capital breeding, and heritable breeding times" (Hendry *et al.* 1999). More generally, incorporating physiological constraints in relation to the environment rather than focusing too heavily on demographic parameters might be a fruitful approach (e.g., page 22-23 in Willson 1997). Finally, considering that some flexibility between iteroparous and semelparous individuals can occur would help to better understand genetic correlates between semelparous versus iteroparous phenotypes, for instance to better quantify the respective importance of gradual versus discrete selection, and therefore the extent of the limitation of gene flow to allow divergence (Mueller 1987; Hendry *et al.* 2003).

17.5.3 Conservation Issues

The low number of offspring produced during lifetime by European vipers in northern or alpine areas suggests that these populations are particularly sensitive to perturbations (climatic changes, habitat loss...). Therefore, it is essential to devote important conservation efforts (monitoring, habitat management...) to the species that exhibit a marked trend toward semelparity, and that simultaneously display reproductive trait typical of long live species, notably low fecundity, low breeding frequency, late maturity. Indeed, this particular combination of traits has been neglected in a conservation perspective, possibly because it was previously considered as improbable (or not easily integrated into current demographic models).

17.6 LITERATURE CITED

Abzhanov, A. M., Protas, B. R., Grant, P. R. and Tabin, C. J. 2004. Bmp4 and morphological variation of beaks in Darwin's finches. Science 305: 1462-1465.

Alcobendas, M. 1989. Recherche sur le métabolisme phosphocalcique au cour du cycle annuel et du cycle de la reproduction chez un reptile, *Vipera aspis*. Doctoral thesis, Université Paris VII, France.

Alcobendas, M., Baud, C. A. and Castanet, J. 1991. Structural changes of the periosteocytic area in *Vipera aspis* L. Ophidia, Viperidae bone tissue in various physiological conditions. Calcified Tissue International 49: 53-57.

Andrade, M. C. B. 1996. Sexual selection for male sacrifice in the Australian redback spider. Science 271: 70-72.

Aubret, F., Bonnet X., Shine, R. and Lourdais, O. 2002. Fat is sexy for females but not males: the influence of body reserves on reproduction in snakes *Vipera aspis*. Hormones and Behavior 42: 135-147.

Baird, D. J., Linton, L. R. and Davies, R. W. 1986. Life-history evolution and post-reproductive mortality risk. Journal of Animal Ecology 55: 295-302.

Barbanera, F., Zuffi, M. A. L., Guerri, M., Gentilli, A., Tofanelli, S., Fasola, M. and Dini, F. 2009. Molecular phylogeography of the asp viper *Vipera aspis* Linnaeus, 1758 in Italy: Evidence for introgressive hybridization and mitochondrial DNA capture. Molecular Phylogenetics and Evolution 52: 103-114.

Barry, T. P., Marwah, A. and Nunez, S. 2010. Inhibition of cortisol metabolism by 17a,20b-P: Mechanism mediating semelparity in salmon? General and Comparative Endocrinology 165: 53-59.

Benton, T. G. and Grant, A. 1999. Optimal reproductive effort in stochastic, density-dependent environments. Evolution 53: 677-688.

Bilde, T. and Lubin, Y. 2001. Kin recognition and cannibalism in a subsocial spider. Journal of Evolutionary Biology 14: 959-966.

Bonnet, X. and Naulleau, G. 1996. Catchability in snakes: consequences on breeding frequency estimates. Canadian Journal of Zoology 74: 233-239.

Bonnet, X., Naulleau, G. and Mauget, R. 1994. The influence of body condition on 17-β estradiol levels in relation to vitellogenesis in female *Vipera aspis* Reptilia, Viperidae. General and Comparative Endocrinology 93: 424-437.

Bonnet, X., Bradshaw, S. D. and Shine, R. 1998. Capital versus income breeding: an ectothermic perspective. Oikos 82: 333-342.

Bonnet, X., Naulleau G., Shine, R. and Lourdais, O. 1999. What is the appropriate time scale for measuring costs of reproduction in a capital breeder such as the aspic viper. Evolutionary Ecology 13: 485-497.

Bonnet, X., Naulleau, G., Shine, R. and Lourdais, O. 2000. Reproductive versus ecological advantages to larger body size in female *Vipera aspis*. Oikos 89: 509-518.

Bonnet, X., Naulleau, G., Bradshaw, S. D. and Shine, R. 2001a. Changes in plasma progesterone in relation to vitellogenesis and gestation in the viviparous snake *Vipera aspis*. General and Comparative Endocrinology 121: 84-94.

Bonnet, X., Naulleau, G., Shine, R. and Lourdais, O. 2001b. Short-term versus long-term effects of food intake on reproductive output in a viviparous snake, *Vipera aspis*. Oikos 92: 297-308.

Bonnet, X., Lourdais, O., Shine, R. and Naulleau, G. 2002. Reproduction in snakes *Vipera aspis*: costs, currencies and complications. Ecology 83: 2124-2135.

Bonnet, X., Shine, R., Lourdais, O. and Naulleau, G. 2003a. Measures of reproductive allometry are sensitive to sampling bias. Functional Ecology 17: 39-49.

Bonnet, X., Naulleau, G. and Lourdais, O. 2003b. The benefits of complementary techniques: using capture-recapture and physiological approaches to understand costs of reproduction in the asp viper. Pp. 483-495. In G. W. Schuett, M. Höggren, M. E. Douglas and H. W. Greene (eds), *Biology of the Vipers*. Eagle Mountain Publishing, Eagle Mountain, Utah.

Bonnet, X., Akoka, S., Shine, R. and Pourcelot, L. 2008. Disappearance of eggs during gestation in a viviparous snake *Vipera aspis* detected using non-invasive techniques. Acta Herpetologica 3: 129-137.

Bradley, A. J., McDonald, I. R. and Lee, A. K. 1980. Stress and mortality in a small marsupial *Antechinus stuartii*, Macleay. General and Comparative Endocrinology 40: 188-200.

Breuker, C. J., Debat, V. and Klingenberg, C. P. 2006. Functional evo-devo. Trends in Ecology and Evolution 21: 488-492.

Bull, J. J. and Shine, R. 1979. Iteroparous animals that skip opportunities for reproduction. American Naturalist 14: 296-316.

Bulmer, M. G. 1985. Selection of iteroparity in a variable environment. American Naturalist 126: 63-71.

Burggren, W. and Warburton, S. 2005. Comparative developmental physiology: an interdisciplinary convergence. Annual Review of Physiology 67: 203-223.

Bryant, E. H. 1971. Life history consequences of natural selection: Cole's result. The American Naturalist 104: 75-76.

Callard, I. P., Riley, D. and Perez, L. 1990. Vertebrate vitellogenesis: molecular model for multihormonal control of gene regulation. Progress in Comparative Endocrinology 1990: 343-348.

Carroll, S. B., Grenier, J. K. and Weatherbee, S. D. 2005a. Evolution at two levels: on genes and form. PLoS Biology 3: 1159-1166.

Carruth, L. L., Jones, R. E. and Norris, D. O. 2002. Cortisol and pacific salmon: a new look at the role of stress hormones in olfaction and home-stream migration. Integrative and Comparative Biology 42: 574-581.

Charnov, E. L. and Schaffer, W. M. 1973. Life-history consequences of natural selection: Cole's result revisited. American Naturalist 107: 791-793.

Christiansen, J. S., Præbel, K., Siikavuopio, S. I. and Carscadden, J. E. 2008. Facultative semelparity in capelin *Mallotus villosus* (Osmeridae)—an experimental test of a life history phenomenon in a sub-arctic fish. Journal of Experimental Marine Biology and Ecology 360: 47-55.

Cole, L. C. 1954. The population consequences of life history phenomena. Quarterly Review of Biology 29: 103-137.

Crespi, B. J. and Teo, R. 2002. Comparative phylogenetic analysis of the evolution of semelparity and life history in salmonid fishes. Evolution 56: 1008-1020.

Davidson, E. H. and Erwin, D. H. 2006. Gene regulatory networks and the evolution of animal body plans. Science 311: 796-800.

Davydova, N. V., Diekmann, O. and Van Gils, S. A. 2005. On circulant populations. I. The algebra of semelparity. Linear Algebra and its Applications 398: 185-243.

Finch, C. E. 1990. *Longevity, Senescence and the Genome.* Chicago University Press, Chicago, Illinois. Pp. 938.

Franklin, G. B. 1989. Androgen-regulated gene expression. Annual Review of Physiology 51: 51-65.

Futami, K. and Akimoto, S. I. 2005. Facultative second oviposition as an adaptation to egg loss in a semelparous crab spider. Ethology 111: 1126-1138.

Gobbetti, A., Zerani, M. and Di Fiore, M. M. 1994. GnRH and substance P regulate prostaglandins and sex steroids from reptilian *Podarcis sicula sicula* ovarian follicles and corpora lutea. General and Comparative Endocrinology 93: 153-162.

Golay, P., Monney, J. C., Conelli, A., Durand, T., Thiery, G., Zuffi, M. A. L. and Ursenbacher, S. 2008. Systematics of the Swiss asp vipers: some implications for the European *Vipera aspis* Linnaeus, 1758 complex Serpentes: Viperidae—A tribute to Eugen Kramer. Amphibia-Reptilia 29: 71-83.

Hautetèete, N. C., Piquot, Y. and Van Dijk, H. 2001. Investment in survival and reproduction along a semelparity-iteroparity gradient in Beta species complex. Journal of Evolutionary Biology 14: 795-804.

Hesiode ≈700BC. *La Théogonie, les Travaux et les Jours et Autres Poèmes*. Classique de Poche, Paris, France. Pp. 350.

Hendry, A. P., Berg, O. K. and Quinn, T. P. 1999. Condition dependence and adaptation-by-time: breeding date, life history and energy allocation within a population of salmon. Oikos 85: 499-514.

Hendry, A. P., Morbey, Y. E., Berg, O. K. and Wenburg, J. 2003. Adaptive variation in senescence: reproductive lifespan in a wild salmon population. Proceedings of the Royal Society London B 271: 259-266.

Ho, S. M., Kleis-San Francisco, S., McPherson, R., Heiserman, G. J. and Callard, I. P. 1982. Regulation of vitellogenesis in reptiles. Herpetologica 38: 40-50.

Holleley, C. E., Dickman, C. R., Crowther, M. S. and Oldroyd, B. P. 2006. Size breeds success: multiple paternity, multivariate selection and male semelparity in a small marsupial, *Anthechinus stuartii*. Molecular Ecology 15: 3439-3448.

Karsten, K. B., Andriamandimbiarisoa L. N., Fox S. F. and Raxworthy C. J. 2008. A unique life history among tetrapods: An annual chameleon living mostly as an egg. Proceedings of the National Academy of Sciences USA 105: 8980-8984.

Ladyman, M., Bonnet, X., Lourdais, O., Bradshaw, S. D. and Naulleau, G. 2003. Gestation, thermoregulation and metabolism in a viviparous snake, *Vipera aspis*: evidence for fecundity-independent costs. Physiological and Biochemical Zoology 76: 497-510.

Lourdais, O., Bonnet, X., Shine, R. and Taylor, E. 2003. When does a reproducing female viper "decide" on her litter size? Journal of Zoology 259: 123-129.

Lourdais, O., Brischoux, F., DeNardo, D. and Shine, R. 2004. Protein catabolism in pregnant snakes *Epicrates cenchria maurus* Boidae compromises musculature and performance after reproduction. Journal of Comparative Physiology B 174: 383-391.

Lourdais, O., Bonnet, X., Shine, R., DeNardo, D., Naulleau, G. and Guillon, M. 2002a. Capital-breeding and reproductive effort in a variable environment: a longitudinal study of a viviparous snake. Journal of Animal Ecology 71: 470-479.

Lourdais, O., Bonnet, X. and Doughty, P. 2002b. Costs of anorexia during pregnancy in a viviparous snake *Vipera aspis*. Journal of Experimental Zoology 292: 487-493.

Lourdais, O., Bonnet, X., DeNardo, D. and Naulleau, G. 2002c. Do sex divergences in reproductive eco-physiology translate into dimorphic demographic patterns? Population Ecology 44: 241-249.

Madsen, T. and Shine, R. 1993. Costs of reproduction in a population of European adders. Oecologia 94: 488-495.

MacNamara, J. M. 1997. Optimal life histories for structured populations in fluctuating environments. Theoretical Population Biology 51: 94-108.

Martins, E. G., Bonato, V., da-Silva, C. Q. and dos Reis, S. F. 2006. Partial semelparity in the Neotropical didelphid marsupial *Gracilinanus microtarsus*. Journal of Mammalogy 87: 915-920.

Mayor, D. J., Anderson, T. R., Pond, D. W. and Irigoien, X. 2009. Egg production and associated losses of carbon, nitrogen and fatty acids from maternal biomass in *Calanus finmarchicus* before the spring bloom. Journal of Marine Systems 78: 505-510.

Morse, D. H. 2009. Post-reproductive changes in female crab spiders *Misumena vatia* exposed to a rich prey source. Journal of Arachnology 37: 72-77.

Morse, D. H. and Stephens, E. G. 1996. The consequences of adult foraging success on the components of lifetime fitness in a semelparous, sit and wait predator. Evolutionary Ecology 10: 1573-8477.

Mueller, L. D. 1987. Evolution of accelerated senescence in laboratory populations of *Drosophila*. Proceedings of the National Academy of Sciences, USA 84: 1974-1977.

Nachman, M. W., Hoekstra, H. E. and D'Agostino, S. L. 2003. The genetic basis of adaptive melanism in pocket mice. Proceedings of the National Academy of Sciences, USA 100: 5268–5273.
Naulleau, G. 1970. La reproduction de *Vipera aspis* en captivité dans des conditions artificielles. Journal of Herpetolology 4: 113-121.
Naulleau, G. 1973. Reproduction twice in one year in a captive viper *Vipera aspis*. British Journal of Herpetology 5: 353-357.
Naulleau, G. and Bidaut, C. 1981. Intervalle entre l'accouplement, l'ovulation et la parturition chez *Vipera aspis* L. Reptiles, Ophidiens, Vipéridés, dans différentes conditions expérimentales, étudié par radiographie. Bulletin de la Société Zoologique de France 106: 137-143.
Naulleau, G. and Fleury, F. 1990. Changes ion plasma progesterone in females *Vipera aspis* L. Reptilia, Viperidae during the sexual cycle in pregnant and non-pregnant females. General and Comparative Endocrinology 78: 433-443.
Naulleau, G. and Bonnet, X. 1996. Body condition threshold for breeding in a viviparous snake. Oecologia 107: 301-306.
Naulleau, G., Bonnet, X. and Duret, S. 1996. Déplacements et domaines vitaux des femelles reproductrices de vipères aspic *Vipera aspis* Reptilia, Viperidae dans le centre ouest de la France. Société Herpétologique de France 78: 5-18.
Naulleau, G., Verheyden, C. and Bonnet, X. 1997. Prédation spécialisée sur la Vipère aspic, *Vipera aspis*, par un couple de buses variables *Buteo buteo*. Alauda. 65: 155-160.
Naulleau, G., Bonnet, X., Vacher-Vallas, M., Shine, R. and Lourdais, O. 1999. Does less-than-annual production of offspring by female vipers *Vipera aspis* mean less-than-annual mating? Journal of Herpetology 33: 688-691.
Oakwood, M., Bradley, A. J. and Cockburn, A. 2001. Semelparity in a large marsupial. Proceedings of the Royal Society of London, B 268: 407-411.
Ovidius Naso 1996. *Metamorphoses*. Garnier Frères, Paris, France. Pp. 504.
Ranta, E., Tesar, D. and Kaitala, V. 2002. Environmental variability and semelparity vs. iteroparity as life histories. Journal of Theoretical Biology 217: 391-396.
Ranta, E., Kaitala, V., Alaja, S. and Tesar, D. 2000a. Nonlinear dynamics and the evolution of semelparous and iteroparous reproductive strategies. American Naturalist 155: 294-300.
Ranta, E., Tesar, D., Alaja, S. and Kaitala, V. 2000b. Does evolution of iteroparous and semelparous reproduction call for spatially structured systems? Evolution 54: 145-150.
Reznick, D. N., Bryant, M. J. and Bashey, S. 2002. R- and K-selection revisited: the role of population regulation in life-history evolution. Ecology 83: 1509-1520.
Rodhouse, P. G. 1998. Physiological progenesis in cephalopod molluscs. Biological Bulletin 195: 17-20.
Saint Girons, H. 1957a. Le cycle sexuel chez *Vipera aspis* (L.) dans l'ouest de la France. Bulletin Biologique de la France et de la Belgique 91: 284-350.
Saint Girons, H. 1957b. Croissance et fécondité de *Vipera aspis* L. Vie et Milieu 8: 265-286.
Saint Girons, H., Bradshaw, S. D. and Bradshaw, F. J. 1993. Sexual activity and plasma levels of sex steroids in the Aspic viper, *Vipera aspis* L. Reptilia, Viperidae. General and Comparative Endocrinology 91: 287-297.
Schmidt, A. L., Taggart, D. A., Holz, P., Temple-Smith, P. D. and Bradley, A. J. 2006. Plasma steroids and steroid-binding capacity in male semelparous dasyurid marsupials *Phascogale tapoatafa* that survive beyond the breeding season in captivity. General and Comparative Endocrinology 149: 236-243.

Schneider, J. E., Zhou, D. and Blum, R. B. 2000. Leptin and metabolic control of reproduction. Hormones and Behavior 37: 306-326.

Sinervo, B. and Licht, P. 1991. Hormonal and physiological control of clutch size, egg size and egg shape in side-blotched lizards *Uta stansburiana*: constraints on the evolution of lizard life histories. Journal of Experimental Zoology 257: 252-264.

Smith, F. A. and Charnov, E. L. 2001. Fitness trade-offs select for semelparous reproduction in an extreme environment. Evolutionary Ecology Research 3: 595-602.

Stearns, S. C. 1992. *The Evolution of Life Histories*. Oxford University Press, Oxford, U. K. Pp. 249.

Stern, D. L. 1998. A role of ultrabithorax in morphological differences between *Drosophila* species. Nature 396: 463-466.

Vacher-Vallas, M., Bonnet, X. and Naulleau, G. 1999. Relations entre les comportements sexuels et les accouplements chez *Vipera aspis* : étude en milieu naturel. Revue d'Ecologie, Terre et Vie 54: 375-391.

Willson, M. F. 1997. Variation in salmonid life histories: patterns and perspectives. Research Paper PNW-RP-498. Portland, OR: U.S. Department of Agriculture, Forest Service, Pacific Northwest Research Station. Pp. 50.

Wodinsky, J. 1977. Hormonal inhibition of feeding and death in Octopus: control by optic gland secretion. Science 198: 948-951.

Young, T. P. 1981. A general model of comparative fecundity for semelparous and iteroparous life histories. American Naturalist 118: 27-36.

Young, T. P. and Augspurger, C. K. 1991. Ecology and evolution of long-lived semelparous plants. Annual Review of Ecology, Evolution and Systematics 6: 285-289.

Zeineddine, M. and Janse, V. A. A. 2009. To age, to die: parity, evolutionary tracking and Cole's paradox. Evolution 63: 1498-1507.

Zera, A. J. and Harshman, L. G. 2001. The physiology of life history trade-offs in animals. Annual Review of Ecology, Evolution and Systematics 32: 95-126.

Zuffi, M., Giudici, F. and Ioalé, P. 1999. Frequency and effort of reproduction in female *Vipera aspis* from a southern population. Acta Oecologica 20: 633-638.

Zuffi, M. A. L., Gentilli, A., Cecchinelli, E., Pupin, F., Bonnet, X., Filippi, E., Luiselli, L. M., Barbanera, F., Dini, F. and Fasola, M. 2009. Geographic variation of life-history traits and reproductive patterns in Continental and Mediterranean asp vipers, *Vipera aspis*. Biological Journal of the Linnean Society 96: 383-391.

CHAPTER 18

Parental Care in Snakes

Zachary R. Stahlschmidt and Dale F. DeNardo

18.1 INTRODUCTION

Broadly defined, parental care refers to any non-genetic contribution by a parent that appears likely to increase the fitness of its offspring (modified from Clutton-Brock 1991). This includes behavioral and physiological traits that occur before, during, and after parition (used to collectively define oviposition and parturition, Smith 1975). Similar to other components of life history, parental care is characterized by tradeoffs, including the classic concept of parent-offspring tradeoffs, as well intra-offspring tradeoffs, which result from different offspring needs requiring conflicting parental behaviors.

Despite these tradeoffs, parental care can provide considerable selective advantages. Thus, parental care, particularly parental attendance of eggs or offspring, is remarkably widespread across the animal kingdom (Clutton-Brock 1991). Despite its costs to future reproductive success, nest-attending parents can increase their current reproductive success by reducing embryonic predation (frogs, Townsend 1986), improving egg water balance (skinks, Somma and Fawcett 1989), thermoregulating embryos (bumblebees, Heinrich 1979), promoting embryonic respiration (fish, Lissaker and Kvarnemo 2006), reducing pathogen infiltration of eggs (crickets, West and Alexander 1963), and provisioning offspring with food (birds, Clutton-Brock 1991). Further, female-only parental care is the predominant mode of care in internally fertilizing vertebrates (e.g., reptiles and mammals, Clutton-Brock 1991) including species within major taxa in which external fertilization predominates (i.e., fish and amphibians, Gross and Shine 1981), as well as terrestrial arthropods (Zeh and Smith 1985). In this broad context, post-paritive parental care (i.e., care provided after oviposition or parturition) in snakes may be particularly relevant as it is overwhelmingly represented by female-only parental attendance. While this chapter emphasizes post-paritive parental care in snakes, it also summarize

School of Life Sciences, Arizona State University, Tempe, Arizona, U.S.A.

pre-paritive parental care for completeness. This chapter concludes also highlights python egg brooding, because of the preponderance of data in this taxon relative to other snakes groups. We conclude with suggestions for future directions in the study of snake parental care.

18.2 PRE-PARITIVE PARENTAL CARE

Non-genetic parental contributions to offspring fitness begin well before parition, and such early forms of parental care are widespread among vertebrates, including snakes. The influence of pre-paritive parental care in snakes is often profound and encompasses multiple variables important to offspring development and quality. Thus, for simplicity, we review the dynamics of snake pre-paritive care in relation to two well-studied developmental variables, energy balance and thermoregulation. We then follow with a discussion on the oviparity-viviparity continuum before summarizing nest site selection in snakes.

18.2.1 Energy Balance

While a ubiquitous requirement regardless of taxon or reproductive mode, energy investment for embryonic development is a significant form of parental care as the extent of investment can vary and affect future reproductive success (Clutton-Brock 1991). Most notably, females must provide their offspring with sufficient energy reserves to allow for development at least until parition. However, the source and timing of these nutrients can vary. In lecithotrophy, nutrients are invested prior to fertilization in the form of follicular yolk, whereas matrotrophy refers to the provisioning of nutrients post-fertilization via another mechanism (e.g., placenta or uterine secretions) (Wourms 1981). While all energy provisioning to offspring by reptiles is pre-paritive, the majority of nutrient utilization by oviparous offspring occurs post-paritively. Moreover, energy provided pre-paritively typically exceeds that required to support development until hatching (e.g., Ji *et al.* 1997a; Speake *et al.* 2003). Prior to hatching, offspring absorb excess nutrients to support neonatal maintenance and growth, which enhances neonate survival (Ji *et al.* 1997a, b, 1999).

Energy stores in neonatal reptiles consist of two forms, residual yolk and stored lipids in adipose tissue. The partitioning of energy reserves between residual yolk and adipose tissue probably reflects the relative advantages of these two sites for the neonate. Mobilization of fats from adipose tissue is very rapid, whereas yolk provides a more versatile store as it contains lipids, protein, vitamins, and minerals (Speake *et al.* 2003). In general, hatchling snakes contain more residual yolk than do hatchling lizards (snakes: 11–25% and lizards: 0–12% of the total hatchling dry mass, Cai *et al.* 2007), and this energy resource may confer an adaptive advantage. For example, residual yolk and fat bodies in hatchling Water Pythons (*Liasis fuscus*) make up approximately 10% and 30%, respectively, of the initial egg lipids (Speake *et al.* 2003), and these substantial reserves enable these neonates to survive for several weeks without feeding (Bedford and Christian 2001). As with other hatchling phenotype variables, abiotic factors can affect the amounts of residual yolk and stored lipid (e.g., incubation temperature in the Bullsnake (*Pituophis catenifer sayi*, Gutzke and Packard 1987). However, in this same study, hydric conditions of the incubation environment did not have an effect on energy reserves at hatching. Less is known about residual yolk content of viviparous offspring. However, neonatal Red-backed Ratsnakes (*Elaphe rufodorsata*) had a total lipid content of 22% (Xiang 1995), which is similar to that of oviparous snakes.

While macronutrients receive the vast majority of scientific interest, yolk contains other components that affect offspring fitness among reptiles. A few of these components including minerals (Ji *et al.* 1997a, b, 1999), immunoglobins (Hassl 2005a, b), antioxidants (reviewed in Thompson and Speake 2004), and defensive toxins (Hutchinson *et al.* 2008) have received some attention in snakes. However, their adaptive significance, provisioning dynamics, and phylogenetic context are poorly understood.

18.2.2 Thermoregulation

Thermal sensitivity is a nearly universal feature of biochemical processes; thus, temperature influences many facets of an organism's life (Huey and Kingsolver 1989; Hochachka and Somero 2002). Effects related to temperature begin during embryogenesis. In fact, temperature has been shown to affect snake developmental rate, hatching success, and offspring phenotype (Burger and Zappalorti 1988; Deeming and Ferguson 1991; O'Donnell and Arnold 2005; Webb *et al.* 2006). Furthermore, the effects of developmental temperature can have long-lasting consequences on individual growth, developmental stability, and fitness (Shine 2004; Webb *et al.* 2006). Thus, snakes have developed multiple parental strategies to reduce potentially deleterious environmental impacts on their progeny. Although ectothermic vertebrates produce negligible metabolic heat, they can promote thermal regulation of their developing offspring. Notable pre-paritive parental thermoregulatory strategies include behavioral thermoregulation and temperature-based selection of nest sites (Tu and Hutchison 1994; Charland 1995; Shine 1995; Shine and Harlow 1996; Chiaraviglio 2006). Despite their benefits to offspring, reproduction-related shifts in behavioral thermoregulation likely incur fitness-related costs to mothers similar to other forms of parental care (e.g., increased energy requirements due to temperature-induced elevations in metabolism result in increased risk of maternal predation, Bonnet *et al.* 2002; Ladyman *et al.* 2003).

Although the discussion of maternal thermoregulation and its resulting benefits to the offspring is often restricted to viviparous species, oviparous squamates (lizards and snakes) provide similar pre-paritive parental care to their embryos for at least part of development. Unlike what is seen in other reptiles, at least one-fourth of embryonic development in oviparous squamates is completed in the oviduct prior to oviposition (Shine 1983; Blackburn 1985; Andrews 2004). Such intra-oviductal development allows oviparous squamates to behaviorally regulate the thermal conditions experienced by the embryos during a significant proportion of early development. Shine (2006) recently proposed a pre-adaptation hypothesis, according to which modified maternal thermoregulation observed in viviparous species would be an extension of a pre-existing behavior in oviparous species. An experimental study on the Eastern Three-lined Skink (*Bassiana duperreyi*) suggests that maternal thermophily before oviposition influences the developmental rate as well as the phenotypic traits of offspring (Shine 2006). This hypothesis is likely applicable to snakes, as basking is more frequent, body temperature higher, and/or body temperature is less variable in gravid female Montpellier Snakes (*Malpolon monspessulanus*, Blazquez 1995), *Liasis fuscus* (Shine and Madsen 1996), Ratsnakes (*Pantherophis obsoletus* complex, Blouin-Demers and Weatherhead 2001), and Children's Pythons (*Antaresia childreni*, Lourdais *et al.* 2008), but see Isaac and Gregory (2004) for an instance of lower and more variable body temperature in gravid Grass Snakes (*Natrix natrix*).

18.2.3 Costs and Benefits along the Oviparity-Viviparity Continuum

Reproductive mode influences the degree to which females regulate the developmental environment of their offspring. Viviparity within squamates has evolved on more than 100 occasions (Blackburn 1985; Shine 1985), and the selective force responsible for these repeated transitions is probably linked to improved developmental conditions (Shine 1995). Comparative analyses by Shine (1985) have demonstrated that the transitions to viviparity have primarily occurred in cold climates where maternal thermoregulation benefits incubation temperature. This supports a "cold climate hypothesis" for the evolution of viviparity (Mell 1929; Weekes 1933; Sergeev 1940). However, to address the presence of large numbers of viviparous species in the tropics, Shine (1995) proposed a more broad "maternal manipulation hypothesis" whereby females enhance fitness-relevant phenotypic traits of their offspring by manipulating thermal conditions of the developmental environment (Shine 1995). This latter hypothesis incorporates the ability of females to maintain a different and or more stable developmental temperature regardless of climate, and thus explains the existence of viviparity in tropical climates. In support of the maternal manipulation hypothesis, female body temperature has been shown to be higher, less variable, or both during reproduction in a diverse group of viviparous snakes, including the Northern Pacific Rattlesnake (*Crotalus oreganus oreganus*, Gier *et al.* 1989), Prairie Rattlesnake (*Crotalus viridis*, Charland and Gregory 1990; Graves and Duvall 1995), Diamond-backed Watersnake (*Nerodia rhombifer*, Tu and Hutchison 1994), Terrestrial and Common Gartersnake (*Thamnophis elegans* and *Thamnophis sirtalis*, Charland 1995),

Northern Watersnake (*Nerodia sipedon*, Brown and Weatherhead 2000), Aspic Viper (*Vipera aspis*, Ladyman *et al.* 2003), Northern Death Adder (*Acanthophis praelongus*, Webb *et al.* 2006), and Argentine Boa Constrictor (*Boa constrictor occidentalis*, Chiaraviglio 2006).

While improved developmental conditions were likely a common driving force for the evolution of viviparity, viviparity itself should be considered a reproductive mode rather than a form of pre-paritive parental care (contra Clutton-Brock 1991; Gans 1996). This is not to discount the importance that reproductive mode has on pre-paritive parental care. As discussed previously, maternal thermoregulation is an important pre-paritive parental behavior of snakes. Although this behavior occurs in both oviparous and viviparous species, the duration of this behavior is greatly influenced by reproductive mode. While viviparous species can use behavioral thermoregulation to regulate the developmental environment throughout development, this parental behavior is only available for the first one-quarter to one-third of development in oviparous species.

Similarly, viviparity also influences female provisioning to offspring. In species where some degree of matrotrophy exists, viviparity provides females with an extended opportunity to provide energy and other nutrients to offspring. While not discussed earlier, water is also a critical material provided to offspring (reviewed in Belinsky *et al.* 2004). While all females must make some degree of water investment into their offspring, the parchment or elastic, ectohydric shell typical of all oviparous snakes allows for water absorption from the environment and thus may reduce water investment demands on the female (at least in moist nest environments) (reviewed in Belinsky *et al.* 2004). In contrast, viviparity requires females to invest the full allotment of water required for offspring development. While likely more hydrically demanding, viviparity provides the offspring greater independence from environmental hydric conditions. Thus, hydroregulation is likely another form of pre-paritive parental care that is affected by mode of reproduction.

Beyond its effects on provisioning offspring and regulating the developmental environment, viviparity also impacts other aspects of life history. Viviparity may reduce predatory risk to the developing offspring but may exacerbate the predatory risk of the female as a result (Bonnet *et al.* 2002; Ladyman *et al.* 2003). Lastly, because viviparity reduces the demands for the environment to provide a suitable developmental environment, choice of parition site may be less critical in viviparous species, although see the following discussion on nest site selection.

18.2.4 Nest Site Selection

Nest site selection is an important form of pre-paritive parental care as post-oviposition nest conditions influence the developmental environment in the majority of snakes studied. For example, the magnitude or variability of nest temperature has been shown to affect offspring phenotypic traits associated with fitness in a diverse group of snakes including the Pinesnake (*Pituophis melanoleucus*, Burger 1998), Bullsnake (*Pituophis catenifer sayi*, Gutzke and Packard 1987), Eastern Racer (*Coluber constrictor*, Burger 1990), Eastern Kingsnake (*Lampropeltis getula*, Burger 1990), *Liasis fuscus* (Shine *et al.* 1997), Chinese Cobra (*Naja atra*, Ji and Du 2001), Five-pace Pit Viper (*Deinagkistrodon acutus*, Lin *et al.* 2005), Keelback Snake (*Tropidonophis mairii*, Brown and Shine 2006), and Ratsnakes (*Pantherophis obsoletus* complex, Patterson and Blouin-Demers 2008). Further, the hydric characteristics of the nest (i.e., nest humidity or the water content/potential of nest substrate) affect offspring phenotype in *P. catenifer sayi* (Gutzke and Packard 1987), the Rough Greensnakes (*Opheodrys aestivus*, Plummer and Snell 1988), *T. mairii* (Brown and Shine 2006), and Children's Python (*Antaresia childreni*, Lourdais *et al.* 2007).

Not unexpectedly, some snakes demonstrate adaptive nest site selection in response to abiotic factors, particularly hydric conditions (e.g., *Opheodrys aestivus*, Plummer and Snell 1988; *Tropidonophis mairii*, Brown and Shine 2006). The physical characteristics of the nest may also influence adaptive nest site selection as gravid *Pituophis melanoleucus* choose nest sites with soft soil in minimally shaded locations (potentially due to these sites' favorable thermal or hydric properties) (Burger and Zappalorti 1986). However, "perfect" nest site selection may not be possible in some instances. For example, free-ranging *Liasis fuscus* choose between nest sites that are thermally favorable but vulnerable to predation, or nest sites that are thermally poor but minimally vulnerable to predation (Madsen and Shine 1999). Also, the co-variation between soil temperature and water content due to depth (i.e., deeper nests are cooler but moister) may force female snakes to navigate a temperature-hydration tradeoff in some systems.

Biotic factors may also influence nest site selection decisions. Many snakes reportedly nest communally, including the Texas Threadsnake (*Leptotyphlops* [=*Rena*] *dulcis*, Hibbard 1964) Copperheads (*Agkistrodon contortrix*, Palmer and Braswell 1995), and at least 13 species of colubrids (reviewed in Swain and Smith 1978; Doody *et al.* 2009). To our knowledge, its significance has not been tested in snakes; yet, communal nesting reduces nest-clutch water flux and results in larger, faster offspring in scincid lizards (Radder and Shine 2007). Physical remains of successful prior incubation (e.g., eggshells) also influences nest site selection in some snakes (i.e., *Opheodrys aestivus*, Plummer 1981; *Pituophis melanoleucus*, Burger and Zappalorti 1992; *Tropodinophis mairii*, Brown and Shine 2005). In particular, some female snakes have a tendency to choose nest sites that they themselves had either hatched from or used in the past, including *O. aestivus* (Plummer 1981), *P. melanoleucus* (Burger and Zappalorti 1992), *T. mairii* (Brown and Shine 2007), *Liasis fuscus* (Madsen and Shine 1999), and the Western Whip Snake (*Hierophis viridiflavus*, Filippi *et al.* 2007). Although nest site selection related to chemosensory cues may be generally adaptive, it does not seem to be "perfect" in all cases. For example, *T. mairii* are just as likely to oviposit in the nest of an egg predator (Slatey-grey Snake,

Stegonotus cucullatus) as they are to oviposit in the nest of a conspecific (Brown and Shine 2005). Clearly, the roles of biotic (e.g., chemosensory) cues in snake nest site selection need to be investigated in more depth.

Additionally, some snakes reportedly construct or excavate sites prior to oviposition, and such nest site manipulation is often related to other types of parental behavior. For example, gravid *Pituophis melanoleucus* excavate nest substrate prior to oviposition and choose soft, sandy nest sites that are easier to excavate than sites obstructed by roots (Burger and Zappalorti 1986). Further, biparental nest excavation has been reported in cobras (reviewed in Somma 2003a). Notably, the notion of bi-parental care in snakes has traditionally been viewed with skepticism due to conflicting results (Shine 1988). However, if the reports are accurate, this combination may persist because parents are more likely to invest in additional parental behaviors if the parent-specific costs of care are reduced, or parents may be under selection to enhance the developmental environment (e.g., thermal or hydric properties of the nest) as much as possible if they choose to invest in nest-attending behaviors. The latter hypothesis appears to be the case for female Diamond Pythons (*Morelia spilota spilota*) as they construct insulated incubation mounds before investing heavily into thermogenic egg brooding behavior (Slip and Shine 1988). Thus, nest site manipulation in snakes may be due to physical constraints (e.g., the feasibility of substrate excavation), the mediation of parent-offspring tradeoffs (e.g., substantial increase of offspring benefits at minimal cost to the parent[s]), or both.

18.3 POST-PARITIVE PARENTAL CARE

Oftentimes, discussion of parental care is limited to events that occur after parturition or oviposition (e.g., Greene *et al.* 2002; Somma 2003a). Such post-paritive parental care is a major component of reproductive strategies in both birds and mammals. In fact, it has been proposed that the ability to provide parental care was the primary driving force for the independent evolution of endothermy in both birds and mammals (Farmer 2000; Koteja 2000). Post-paritive parental care is also common in some ectotherm lineages, including fish, salamanders, and crocodilians. While post-paritive parental care is not as widespread among snakes as pre-paritive care, it has been documented in a relatively diverse number of snakes and even appears to be ubiquitous in some groups. This section provides an overview of parental care among snakes, but in no way attempts to be an all inclusive review. Extremely comprehensive works on post-paritive parental care in non-crocodilian reptiles and viperid snakes are provided in the exceptional works by Somma (2003a, b, c) and Greene *et al.* (2002), respectively.

In most snake families, post-paritive parental care has been documented only sporadically or not at all. Generally, details of parental provisioning are limited and, although sometimes speculated, the functional costs and benefits of parental care mostly remain unexplored. Most commonly, post-paritive parental care entails the female remaining with her clutch or neonates for some period of time after parition, and the duration of attendance greatly varies among species and even within species. Such behavior is reportedly uncommon in Boidae, Colubridae, and Leptotyphlopidae, common but not ubiquitous in Elapidae, but appears to be ubiquitous or nearly so in the Pythonidae and Crotalinae (Fitch 1970; Somma 2003a, b, c, d).

18.3.1 Care of Eggs

Variable terminology has been devoted to describing reptilian egg care, including (but not limited to): "brood defended", "coil around brood", "egg attendance", "egg brooding", "egg guarding", "false brooding", "nest attendance", "nest guarding", "passive protection" (Shine 1988; Somma 1988, 2003a). After reviewing the literature, Somma (1988) and Blackburn (1994) conclude that it is best to follow Peters (1964) as well as Carpenter and Ferguson (1977) in defining "brooding" as any situation where the female remains with the eggs after oviposition. However, "brooding" has also been used as a subset of such situations to identify only those cases where the egg attendance provides thermoregulatory benefits (Campbell and Quinn 1975; Lillywhite 2008). Because the vast majority of the reports of snakes remaining with their eggs after oviposition provide little to no empirical evidence of function, we opt to describe this behaviorial trait as "egg attendance" because this term has no history in the literature of suggesting a functional role. We limit our use of "brooding" to those instances of egg attendance where physiological benefits have been documented. In doing so, our use of brooding is in line with both the broader (e.g., Peters 1964; Carpenter and Ferguson 1977) and more specific (e.g., Campbell and Quinn 1975) definitions, and thus avoids confusion.

Reports of egg attendance are numerous, particularly within the Colubridae, Elapidae, Pythonidae, and Crotalinae, and have been thoroughly reviewed by Fitch (1970), Shine (1988), and Somma (2003a, b, c). However, as noted by these authors, many of the reports within the Colubridae and Elapidae are anecdotal observations based on small sample sizes (e.g., Forest Cobra [*Naja melanoleuca*], Haagner 1990, $N = 1$; Blue Beauty Snake [*Orthriophis taeniura*], Humphrey 2000, $N = 1$), found in nonrefereed publications (e.g., Eastern Milksnake [*Lampropeltis triangulum triangulum*], Winstel 1991), not primary observations (e.g., Rainbow Snake [*Farancia erytrogramma*], Mitchell 1994; *Naja atra*, Spitting Cobra [*Naja siamensis*], Monocled Cobra [*Naja kaouthia*], Lim Leong Keng and Lee Tat-Mong 1989), or some combination thereof. Further, the details of primary accounts are often insufficient. As one of many examples, Winstel (1991) wrote the following regarding *L. t. triangulum*: "Females typically remained with the eggs." Such wording may have referred to the proportion of females who remained with their eggs or the amount of time females spent with their eggs. Further, the spatial relationship between the attending female and her eggs is unclear.

However, some reports within the Colubridae and Elapidae are more reliable and insightful. Female Skaapstekers (*Psammophylax rhombeatus* and *Psammophylax variabilis*) reportedly coil around their eggs for up to six weeks (reviewed in Broadley 1977). Similarly, despite being in a non-refereed publication, picture evidence supports claims that *Orthriophis taeniura* may remain with their clutch until hatching, only leaving the clutch periodically to drink (Humphrey 2000). Furthermore, a single account of egg attendance within the Leptotyphlopidae is of special interest because this family is ancestral among snakes (Greene 1997; Slowinski and Lawson 2002), and the egg attendance was associated with communal nesting (i.e., multiple egg-attending females were found in close proximity of each other) (Texas Threadsnake, [*Leptotyphlops* {=Rena} dulcis], Hibbard 1964).

It is unclear whether the sporadic reports of egg attendance within the Colubridae, Elapidae, and Leptotyphlopidae accurately reflect that the behavior is relatively uncommon in these lineages or whether it reflects a need for a much more expansive exploration into this phenomenon among these and other lineages. Despite the relatively secretive nature of snakes, especially during parition, egg attendance has been documented widely among the Pythonidae, oviparous Afro-Asian Elapidae, and the Crotalinae (Wall 1921; Pope 1935; Smith 1943; Fitch 1970; Shine 1988; Greene *et al.* 2002; Somma 2003a, b, c).

Egg attendance has been documented in every species of python (Somma 2003a, b; Weigel pers. com.), and the behavior is atypical from that of most other reports of snake egg attendance in that the females tightly coil around their clutch. In fact, female pythons coil so effectively around their clutches that oftentimes the eggs are barely visible, if at all (Fig. 18.1). As there is a considerable and growing body of literature on post-paritive parental care in pythons, we will discuss python parental care more in-depth in a later section.

Fig. 18.1 A female Children's Python (*Antaresia childreni*) in a tightly coiled egg brooding posture.

Although not documented in all oviparous species of the diverse Crotalinae, egg attendance has been widely documented in this group including in the few existing oviparous New World pit vipers and numerous Old World pit vipers (reviewed in Greene *et al.* 2002). In some cases, the egg attendance was atypically tight, much like that seen in pythons (Ripa 1994). As a result of its wide occurrence within the Crotalinae, egg attendance likely originated at or prior to the origin of the ancestral oviparous crotaline at least 23 mya (Greene *et al.* 2002). Interestingly, while pit vipers are members of the Viperidae, egg attendance in the Viperinae (the other subfamily of the viperids) is extremely rare. The only report of egg attendance in a viperine snake is in the primitive Rhombic Night Adder (*Causus rhombeatus*), which coils around its eggs throughout incubation (FitzSimons 1912). Why egg attendance is so prevalent in Pythonidae and Crotalinae is unclear, as these groups are both phylogenetically and ecological quite distinct from each other. It has been theorized that parental care should characterize organisms that are especially capable of defense (Shine 1988), and most pythons and pit vipers clearly have such abilities. However, such theory does not explain the prevalence of post-paritive parental care in the more diminutive python species, and, more importantly, the lack of post-paritive parental care in the viperines (Greene *et al.* 2002).

Of interest but yet unknown relevance is the fact that both the Pythonidae and Crotalinae possess heat-sensing organs, an otherwise uncommon trait among snakes (but also present in some Boidae). While it has been postulated that female pythons use their heat-sensing organs to monitor clutch temperature (G. Schuett in Kend 1997), there is no evidence that elucidates whether these organs enabled the evolution of this relatively complex post-paritive parental care, evolved as a result of existing post-paritive parental care, or function completely independent of post-paritive parental.

18.3.2 Care of Neonates

Akin to egg attendance in oviparous species, neonate attendance has been reported in some viviparous snakes. While there are sporadic reports of neonate attendance with limited detail in other snake taxa (see Somma 2003a for review), neonate attendance is most prevalent and best described in the Crotalinae (see Greene *et al.* 2002 for a thorough review). This distinction may in some part reflect a greater attention of scientists toward pit vipers. Yet, it is worth noting that neonate attendance has been documented in 17 species of North American crotalines, but there are no documented accounts of attendance in any of the 36 species of viviparous natricine snakes in the United States (Greene *et al.* 2002).

Neonate attendance appears nearly or completely ubiquitous among temperate viviparous crotalines, but it is absent in tropical viviparous crotalines. Accordingly, the Neotropical Rattlesnake (*Crotalus durissus*) is the lone rattlesnake species with a lowland tropical distribution and the only rattlesnake known to lack neonate attendance. Similar to the pattern in the prevalence of egg attendance, offspring attendance has not been convincingly documented in the Viperinae, despite being common in the Crotalinae. While neonate attendance has been alleged in the Adder (*Vipera berus*), it has not been observed by numerous researchers who are highly experienced with this species (Greene *et al.* 2002).

In viviparous pit vipers, neonate attendance seems to consistently last from parturition until the first ecdysis (i.e., skin shedding) of the neonates, which is about one to two weeks post-parturition for most species. Coincidently, neonates of *Crotalus durissus*, the lone rattlesnake species where neonate attendance has not been documented, shed within 24 hours after birth (Reinert, reported in Greene *et al.* 2002). While the female and neonates might move around the oviposition site (e.g., emerge from and retreat into a subterranean refuge), females and their litters do not depart from the oviposition site until after neonatal ecdysis wherein each individual departs on its own (reviewed in Greene *et al.* 2002).

Post-paritive care is apparently a common component of reproductive behavior among pit vipers regardless of reproductive mode. As hatchling snakes also show a neonatal ecdysis typically several days after emergence, it is interesting to consider whether egg-attending female snakes show pre-ecdysis neonatal attendance. Field observations of egg-attending females also attending newly hatched offspring has been recorded in the Southern African Rock Python (*Python natalensis*, Alexander 2004; Alexander and Marais 2007) and the Malayan Pit Viper (*Calloselasma rhodostoma*, Hill *et al.* 2006). Whether or not the existence of a short-term interaction with newly hatched neonates is common in these species, let alone other egg-attending species, is unknown.

18.3.3 Oophagy

Oophagy and the consumption of stillborn neonates is reportedly widespread among viviparous snakes, including Bimini Island Boas (*Epicrates striatus*, Huff 1980), Rainbow Tree Boas (*E. cenchria*, Groves 1981; Lourdais *et al.* 2005), Green Anacondas (*Eunectes murinus*, Holmstrom and Behler 1981), Amazon Tree Boas (*Corallus hortulanus*, Miller 1983; Jes 1984), Yellow Anacondas (*E. notaeus*, Townson 1985), Egyptian Sand Boas (*Gongylophis colubrinus*, Ross and Marzec 1990), Cantils (*Agkistrodon bilineatus*, Mitchell and Groves 1993), Abaco Island Boas (*Epicrates exsul*, Tolson and Henderson 1993), and Mexican Lance-headed Rattlesnakes (*Crotalus polystictus*, Deloya *et al.* 2009). Oophagy has been hypothesized to reduce chemical cues that could attract predators (Groves 1981; Polis 1981; Shine 1988) or reduce fungal infestation of healthy eggs or neonates (Polis 1981). An alternative hypothesis posits that it may not be a parental care behavior but, rather, function as a means to facilitate post-parturition energy recovery (Lourdais *et al.* 2005). In Iberian Rock Lizards (*Iberolacerta monticola*), the presence of unfertilized eggs did not affect mold proliferation or clutch predation of viable eggs (Moreira and Barata 2005). Alternatively, research in *E. cenchria* and *C. polystictus* demonstrated that oophagy does help post-parturient females recover otherwise wasted energy (Lourdais *et al.* 2005; Deloya *et al.* 2009). In sum, while there is some evidence that oophagy and the consumption of stillborn neonates in snakes benefits female energy balance, there is not yet any evidence for a benefit to the offspring. Thus, oophagy should likely not be considered a parental care behavior until offspring benefits are verified by further research.

18.3.4 Function of Post-paritive Attendance

Contrary to some literature (e.g., Clutton-Brock 1991; Gardiner 2002), post-paritive parental attendance occurs sporadically across diverse lineages of snakes (Pope 1935; Fitch 1970; Greene *et al.* 2002; Somma 2003a). In fact, post-paritive parental care is the rule, and not the exception, in some lineages (e.g., pythons and pit vipers). Because parental behaviors are generally viewed as adaptive (Clutton-Brock 1991), the benefits to the offspring are assumed to outweigh the costs to the parent. As in Trivers (1974), the dynamics of a parent-offspring tradeoff may be resolved in the context of a benefit:cost model (Fig. 18.2). Parental attendance may be relatively beneficial and, thus, persist for a certain period of time. However, over time the costs may exceed the benefits; thus, attendance should desist at some optimal point in time ($T_{optimal}$) (Fig. 18.2). While the significance of adaptations can be tested in a present context (Reeve and Sherman 1993), data regarding offspring benefits and parental costs derived from post-paritive attendance in snakes are only beginning to accumulate. Unfortunately, the vast majority of reports of post-paritive attendance in non-pythonid snakes are limited to descriptive observations. When focusing

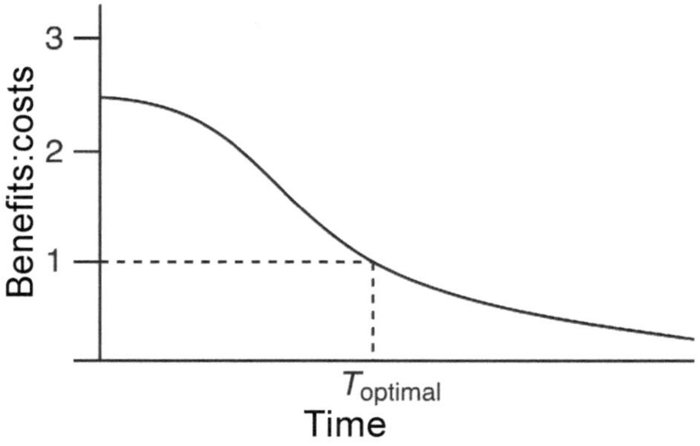

Fig. 18.2 Benefit:cost ratio of offspring attendance as a function of time. Benefits are measured in units of reproductive success to offspring while costs are measured in comparable units of reproductive success to parent's future reproductive success. At some point in time ($T_{optimal}$), the costs begin to outweigh the benefits and offspring attendance should desist. Unfortunately, benefit and cost determination in snake parental care is in its infancy.

on field data, parental attendance has been most extensively described in crotalines. However, despite widespread documentation, there is little empirical data that provides insight into the functional role of attendance in this group of snakes. Yet, several benefits of parental attendance have been postulated.

Neonatal attendance by pit vipers has frequently been referred to as "guarding". While there have been several reports of female pit vipers aggressively defending their offspring, there is an equal number of reports where the female did not appear to be aggressive toward intruders (reviewed in Green *et al.* 2002). However, many potential predators are likely familiar with pit vipers and may choose to avoid them upon detection, regardless of the snake's response to their presence (Greene 1988). Therefore, the mere presence of a female, let alone any anti-predator behavior, might provide substantial protection to the neonates (Greene *et al.* 2002). Oviparous females may deter predation of their eggs not only through protection, but also by using their cryptic coloration to better conceal their white eggs (York and Burghardt 1988). While protection is likely a critical function of parental attendance in pit vipers, empirical data are needed to ascertain the effectiveness of this protection and its importance to reproductive success.

Physiological benefits of parental attendance in pit vipers are even more poorly understood. In the lone assessment of a physiological benefit associated with pit viper egg attendance, there was no evidence that *Calloselasma rhodostoma* provide any thermal benefit to their eggs (York and Burghardt 1988). However, in the same study, females adjusted their coiling behavior in response to laboratory manipulations of humidity. When humidity dropped below 70%, females increased their coverage of the eggs, oftentimes covering them completely. However, field observations of an egg-attending female *C. rhodostoma* revealed only slight variation in the extent to which the female covered her eggs (Hill *et al.* 2006). In that instance, relative humidity never dropped below 70% and the authors estimated egg exposure to never exceed 20%, even during rainfall (Hill *et al.* 2006). Relatedly, the activity patterns and body posture of non-reproductive *C. rhodostoma* were strongly influenced by ambient humidity in a separate field study, suggesting acute hydrosensory abilities (Daltry *et al.* 1998).

Physiological benefits of mother-neonate aggregation in viviparous pit vipers is also unsubstantiated. Transcutaneous water loss is greater prior to a neonate's first ecdysis (Duvall *et al.* 1985; Tu *et al.* 2002). Because attendance predominantly occurs pre-ecdysis, some have suggested a hydroregulatory, and possibly a thermoregulatory, advantage to parental attendance (Finneran 1953; Graves and Duvall 1995). However, quantitative experiments examining the hydroregulatory benefits of offspring attendance are needed before any conclusions can be made. Another possible, but completely untested, benefit of neonatal attendance is that tongue-flicking among the mother and neonates might facilitate chemically-mediated social mechanisms (Greene *et al.* 2002).

18.3.5 Parental Costs of Post-paritive Care

Parental costs are equally important in understanding the adaptive significance of parental attendance in non-pythonid snakes, but they too are poorly known. Unlike the high energy demands associated with energy provisioning in both birds and mammals (Clutton-Brock 1991), parental attendance in most snakes entails no energy provisioning and minimal activity. However, parental attendance in snakes may lead to a negative energy balance due to lost foraging time (Shine 1988). Attending females have little to no opportunity to actively forage or even select optimal ambush sites. Nonetheless, attending females under captive conditions have been reported to leave their clutch to feed (Gloyd and Conant 1990). In the wild, while attendance and foraging is likely in much greater conflict than it is in captivity, some evidence of feeding in egg-attending females exists (Hill *et al.* 2006). Despite these occurrences of feeding during egg attendance, opportunities to feed are likely considerably reduced; thus, a negative energy balance during attendance is probable. In viviparous pit vipers, where attendance is brief (i.e., 1-2 weeks) and reproduction infrequent (i.e., super-annual), the negative energy balance which might occur during attendance may not significantly affect future fecundity (Price 1988; Butler *et al.* 1995). However, prolonged negative energy balance potentially associated with egg attendance might not be inconsequential.

18.4 PYTHON PARENTAL CARE

Parental care in snakes has been most studied in the Pythonidae. Pre-paritively, female pythons invest heavily into yolk stores that provide vital energy to the offspring during development and post-hatching (Bedford and Christian 2001; Speake *et al.* 2003). Additionally, gravid pythons also alter thermoregulatory behavior to optimize early offspring development (Shine and Madsen 1996; Lourdais *et al.* 2008). While pre-paritive parental care in pythons is similar to that of other snakes, python parental care is deserving of separate attention in this chapter because researchers have gained considerable insight into the functional significance of its post-paritive parental care. As mentioned earlier, egg attendance is ubiquitous among pythons wherein females coil tightly around their eggs as to oftentimes encompass the clutch completely or nearly so (Fig. 18.1).

18.4.1 Thermoregulatory Benefits of Egg Attendance

Python egg attendance can clearly be called brooding as multiple physiological benefits have been documented. Most notably, facultative thermogenesis by egg brooding pythons can raise clutch temperature as much as 7°C above ambient temperature (Vinegar *et al.* 1970). This phenomenon is frequently mentioned in introductory science textbooks and parental care reviews (e.g., Shine 1988; Farmer 2000; Somma 2003a).

However, it has thus far been convincingly documented (i.e., increased metabolic rate or the maintenance of clutch temperature with decreasing nest temperature) in only three species (Burmese Python [*Python bivittatus*], Indian Python [*Python molurus*], Hutchison *et al.* 1966; Vinegar *et al.* 1970; *Morelia spilota spilota*, Harlow and Grigg 1984; Slip and Shine 1988). Facultative thermogenesis has been either disputed or convincingly disproven in at least 10 other python species (Table 18.1). Discrepancies in published observations could be attributed to population variation, individual variation, variation in data collection, or experimental technique. For example, because they are concurrent with endogenous heat production in thermogenic species, coordinated large-scale muscular contraction (i.e., shivering) has been implicated as the mechanism of facultative thermogenesis (Vinegar *et al.* 1970) and used as a proxy for facultative thermogenesis in egg brooding pythons. However, poor correlations between contraction frequency and both oxygen consumption and the clutch-nest temperature gradient suggest that simple observation of muscular contraction during brooding does not provide definitive evidence that a female is sufficiently thermogenic to influence incubation temperature. Furthermore, these poor correlations suggest the possibility that thermogenic pythons may also incorporate non-shivering heat generating processes, such as those seen in birds and mammals (Van Mierop and Barnard 1976, 1978).

Table 18.1 Studies refuting facultative thermogenesis during maternal egg brooding within the Pythonidae

Species	*Common name*	*References*
Antaresia childreni	Children's Python	Stahlschmidt and DeNardo 2009a
Aspidites melanocephalus	Black-headed Python	Murphy *et al.* 1981; DeNardo unpublished
A. ramsayi	Woma	DeNardo unpublished
Morelia kinghorni	Scrub Python	Charles *et al.* 1985
M. spilota cheynei	Jungle Carpet Python	DeNardo unpublished
M. viridis	Green Tree Python	DeNardo unpublished
Python brongersmai	Blood Python	Noble 1935
P. regius	Ball Python	Van Mierop and Bessette 1981; Ellis and Chappell 1987
P. reticulatus	Reticulated Python	Honegger 1970; Vinegar *et al.* 1970; La Panouse and Pellier 1973; Pitman 1974
P. natalensis	Southern African Rock Python	Pitman 1974
P. sebae	African Rock Python	Vinegar *et al.* 1970

Although many species of pythons are incapable of significant heat production, they may use other behaviors to enhance the thermal environment around their eggs. For example, free-ranging Black-headed Pythons (*Aspidites melanocephalus*) and *Python natalensis* use heat radiated from the sun or conducted from substrate to warm their clutches (Johnson *et al.* 1975; Alexander 2007). Additionally, egg brooding behavior by free-ranging pythons is dynamic in that females adjust their posture (i.e., increase tight coiling behavior) to prevent egg cooling (Johnson *et al.* 1975).

Similarly, and in a more controlled environment, *Antaresia childreni* have been documented to spend more time tightly coiled around their eggs during cooling than during warming (Stahlschmidt and DeNardo 2009a, 2010a). Further, the amount of time that females spent tightly coiled during warming significantly affected the temperature gradient between the nest and clutch environment (Tnest-Tclutch gradient) (Stahlschmidt and DeNardo 2009a). Thus, although most female pythons are not facultatively thermogenic, they are likely capable of assessing the Tnest-Tclutch gradient and making behavioral adjustments to enhance the thermal microenvironment of their developing offspring.

The selection for physiological and behavioral traits that enhance egg temperature regulation is likely related to the thermal sensitivity of python embryos. Specifically, python embryos require a relatively high, stable incubation temperature (i.e., 30-33°C) for normal development, and deviations from this narrow range results in a combination of reduced hatching success, developmental rate, growth rate, body size, escape behavior, and willingness to feed (*Python bivittatus*, Vinegar 1973; *Python natalensis*, Branch and Patterson 1975; *Morelia spilota spilota*, Harlow and Grigg 1984; *Liasis fuscus*, Shine *et al.* 1997; *Antaresia childreni*, DeNardo unpublished).

Given the thermal sensitivity of python embryos, the question remains: Why are only two python species facultatively thermogenic? Phylogenetically, the three thermogenic pythons, *Python bivittatus, Python molurus,* and *Morelia spilota spilota*, are members of two different python clades (i.e., Afro-Asian and Indo-Australian, respectively) that diverged over 45 million years ago (Rawlings *et al.* 2008). Thus, *M. s. spilota* is more closely related to 90% of non-endothermic python species than it is to *P. bivittatus* and *P. molurus*. Alternatively, the variation in thermogenic capability may be explained by morphological variation among python species. Specifically, large snakes have lower surface area to volume ratios, and thus they lose heat at a slower rate relative to smaller snakes. Additionally, larger snakes can store far greater quantities of energy. In support of a morphological basis for the appearance of thermogenesis in pythons, *P. molurus* and *P. bivittatus* are two of five giant python species and can grow to over 5.8 m in length and 90 kg in mass (Mattison 2007). Yet, female *M. s. spilota* are not particularly large (i.e., 182-230 cm snout-vent length, Slip and Shine 1988), and the other three giant species (*Python sebae, P. natalensis,* and *P. reticulatus*) are not thermogenic (Vinegar *et al.* 1970). What might drive the existence of thermogenesis in pythons is an interaction between biogeography and morphology. The distributional ranges of *P. bivittatus, P. molurus,* and *M. s. spilota* extend to among the highest known latitudes for any python (Jacobs *et al.* 2009), and definitely higher than all species thus far shown not to be thermogenic. In fact, the range of the moderate sized *M. s. spilota* extends as far south as 37.5°S, the most polar of any python (Rawlinson 1969). In addition to being thermogenic, this species, unlike most other pythons, is known to build insulated incubation mounds (Slip and Shine 1988). Furthermore, the higher latitudinal distribution of *P. bivittatus* and *P. molurus* relative to the congeneric and similarly sized *P. reticulatus* has been attributed to the presence of thermogenic ability in the former but not the latter (Vinegar *et al.* 1970). In sum, python thermogenesis may only occur in species which live in more thermally challenging environments and can minimize the appreciable costs of endogenous heat production (e.g., large body size in *P. bivittatus* and *P. molurus*, and nest construction in *M. s. spilota*).

18.4.2 The Role of Tradeoffs in the Adaptive Significance of Python Parental Care

Like other parental care systems, python egg brooding represents parent-offspring tradeoffs where the costs to egg brooding females are offset by the benefits to the developing offspring. For example, facultative thermogenesis represents a substantial portion of female energy expenditure during egg brooding at cool temperatures (*Python bivittatus*: 92%, *Morelia spilota spilota*: 95%) (Vinegar *et al.* 1970; Harlow and Grigg 1984). Although the results of one study suggest that maternal costs of egg brooding are minimal (Aubret *et al.* 2005a), other studies demonstrate that egg brooding independent of thermogenesis obligates substantial maternal costs. Egg brooding necessitates lost foraging time and is generally accompanied by anorexia (Madsen and Shine 1999; Aubret *et al.* 2005a). Accordingly, in female *Antaresia childreni* under laboratory conditions, it obligates significant epaxial muscle atrophy and reduces contraction strength (Lourdais and DeNardo unpublished; Stahlschmidt *et al.* unpublished), as well as increases susceptibility to oxidative stress (Stahlschmidt *et al.* unpublished). Under natural conditions, the duration of egg brooding is negatively related to reproductive frequency and post-reproductive maternal body condition in free-ranging *Liasis fuscus* (Madsen and Shine 1999).

While likely to infer some cost to the parent, egg brooding conveys benefits to offspring beyond those associated with thermoregulation. For example, egg brooding duration is negatively related to egg predation in free-ranging *Liasis fuscus* (Madsen and Shine 1999). Further, python eggshells have extremely high water vapor conductance, and eggs can desiccate in conditions as moist as 75% relative humidity when not maternally brooded (Lourdais *et al.* 2007; Stahlschmidt *et al.* 2008). Thus, removal of females from their eggs results in reduced hatching success and altered hatchling phenotypes in Ball Pythons (*Python regius*, Aubret *et al.* 2005b) and 0% hatching success in *Antaresia childreni* under some conditions (Lourdais *et al.* 2007). Using a less extreme manipulation, Aubret *et al.* (2003) showed that experimentally increasing clutch size by 50% prohibited female *P. regius* from fully encompassing their clutch, leading to desiccation of the eggs. As a result, embryos in these "enlarged" clutches were more likely to die before hatching or hatch later.

Although less studied, intra-offspring tradeoffs (i.e., those that reflect competing offspring needs) may be important to parental care dynamics (Stahlschmidt and DeNardo 2009b). For example, *Liasis fuscus* females that nest in thermally superior sites abandon their clutches shortly after oviposition ($\overline{X}$ = 6.5 d), while females that nest in thermally poorer sites brood their clutches for the entire incubation period ($\overline{X}$ = 53.8 d) (Madsen and Shine 1999). Although *L. fuscus* nest site selection mitigates a parent-offspring tradeoff (i.e., thermal benefits to offspring vs. lost foraging time for females), an intra-offspring tradeoff between embryonic temperature and predation also exists. That is, thermally favorable nests are more prone to predation than are thermally poor nests (Madsen and Shine 1999).

Intra-offspring tradeoffs associated with python egg brooding have also been detected at a finer scale of brooding behavior. As stated earlier, egg brooding behavior is dynamic and entails shifts in coiling posture. Broadly, dynamics of egg brooding behavior can be divided between (1) a tight coiling state wherein the female is still and encompassing the clutch completely or nearly so, and (2) a postural adjustment state wherein the female makes small but significant movements and exposure of the eggs to the nest environment increases. These distinct brooding behaviors mediate several intra-offspring tradeoffs. During brooding, females predominately adopt the tightly coiled posture, and this positioning serves as an effective barrier to the exchange of gas and heat between the clutch environment (i.e., the space within the females coils) and the nest environment (i.e., the space immediately outside of the female's coils) (Stahlschmidt and DeNardo 2008, 2009a; Stahlschmidt *et al.* 2008).

While an effective barrier to gas exchange dramatically reduces water loss from the eggs, it also limits the transfer of oxygen and carbon dioxide (Stahlschmidt and DeNardo 2008, Stahlschmidt *et al.* 2008). Female *Antaresia childreni* mediate this intra-offspring tradeoff between embryonic water balance and respiration by periodically performing postural adjustments to facilitate nest-clutch gas (O_2 and H_2O vapor) exchange to benefit respiration at the cost of embryonic water balance (Stahlschmidt and DeNardo 2008, 2009b; Stahlschmidt *et al.* 2008). However, while embryonic oxygen consumption increases dramatically over the course of development, female *A. childreni* do not alter the relative frequency or duration of their postural adjustments, and this results in late-incubation developmental hypoxia that reduces offspring size, speed, and strength (Stahlschmidt and DeNardo 2008, 2009b; Stahlschmidt *et al.* 2008).

While oxygen depletion does not impact the relative frequency which female *Antaresia childreni* utilize tight coiling and postural adjustments, females do respond to other environmental conditions. As described above, females enhance the thermal environment of the eggs by performing more postural adjustments when ambient temperature is increasing than when it is cooling (Stahlschmidt and DeNardo 2009a, 2010a). However, this response to environmental temperature is dependent on the nest's hydric condition. During moderate and high nest humidity treatments (23 and

32 g·m^{-3} H_2O, respectively), females show the previously described reduction in tight coiling during nest warming. However, brooding females in low humidity nest environments (13 g·m^{-3} H_2O) maintain a high frequency of tight coiling even when the nest is warming. Thus, females "choose" embryonic thermoregulation over embryonic water balance in relatively humid nest conditions, but *vice versa* during relatively dry nest conditions (Stahlschmidt and DeNardo 2010a).

In sum, python egg brooding significantly impacts several critical developmental variables including embryonic predation (Madsen and Shine 1999), thermoregulation (Vinegar *et al.* 1970; Stahlschmidt and DeNardo 2009a), water balance (Aubret *et al.* 2005; Lourdais *et al.* 2007; Stahlschmidt *et al.* 2008), and respiration (Stahlschmidt and DeNardo 2008). Further, python egg brooding behavior is dynamic with females using assessments of environmental conditions to adjust brooding behavior on both large scales (i.e., choosing to abandon or continue brooding the clutch) and fine scales (i.e., altering the frequency and duration of postural shifts during brooding). Whether such parental care complexity is unique to pythons or remaining to be discovered in other taxa remains uncertain.

18.5 FUTURE DIRECTIONS

It has been well established that both pre- and post-paritive parental care are widespread among snakes. Many interesting questions remain regarding pre-paritive care, such as "how is the amount of post-hatching yolk determined" and "how is the yolk's energy allocated to various needs". However, the greatest voids to be filled involve post-paritive parental care, in particular egg and neonate attendance.

18.5.1 Documentation of the Occurrence of Parental Care

Currently, the existence of maternal attendance has been well documented in both Crotalinae and Pythonidae, and to a lesser extent Afro-Asian Elapidae, but other taxa have been poorly explored. By better understanding the true occurrence of parental care among snakes, we might better tease out any phylogenetic, ecological, and perhaps even morphological influences on the evolution of parental behaviors. It is critical that further exploration cover diverse taxa and that documentation be sound and thorough. Documenting parental behavior in the field is extremely challenging, especially when it involves extended observations over time. However, such work will be valuable in understanding traits and conditions that influence parental decisions.

18.5.2 The Ecophysiology of Parental Attendance

Our understanding of the function of parental care and its regulation is in its infancy. While recent work using manipulative experiments has greatly expanded our understanding of both costs and benefits of parental care in pythons, many unknowns still exist. We are only now beginning to explore the regulation of parental care. What environmental cues do females use to adjust both large- and fine-scale parental decisions? What neural, hormonal, or other chemical cues are used to initiate, maintain, and terminate parental behavior?

To explore these questions and others, it is critical that we continue examining species for which data are most abundant (e.g., *Antaresia childreni, Liasis fuscus, Python bivittatus, Python molurus, Python regius*), but also begin to explore other lineages which would provide exceptional comparative data. One such lineage is the pit vipers, where parental attendance in nearly ubiquitous but divided between oviparous species that invest long periods of time in attending their eggs and viviparous species that limit attendance predominantly to the relatively short pre-ecdysis periods. These two parental behaviors, while both falling into the category of parental attendance, likely present considerably different costs to the parents and benefits to the offspring. Regardless, it is of interest to examine whether these behaviors are similarly regulated.

While the lineages where post-paritive parental care is commonplace likely provide the best target for studies, we should not discount the value of comparative assessments of costs, benefits, and regulators of maternal attendance. The Colubridae are a diverse (and likely polyphyletic) group and parental attendance either occurs throughout egg incubation, is limited to the immediate post-paritive period, or is non-existent. What factors drive the selection for these different strategies, and are the regulators of parental care in those species that provide it similar to the regulators in the lineages where parental attendance is the norm? We specifically recommend investigating parental care tradeoffs in species whose parental behaviors have been most frequently observed but mostly unstudied (Somma 2003a, b), including both colubrids (e.g., Taiwan Beauty Snake [*Orthriophis taeniurus friesi*], Mud Snake [*Farancia abacura*], Skaapstekers [*Psammophylax* spp.], Keelbacks [*Xenochrophis* spp.]) and elapids (e.g., Shield-nosed Cobras [*Aspidelaps* spp.], kraits [*Bungarus* spp.], cobras [*Naja* spp.], and King Cobra [*Ophiophagus hannah*]).

18.5.3 Why Do We Care?

Without an accurate understanding of reproductive investment, it is impossible to fully understand reproductive strategies and success. Thus, understanding parental behavior more thoroughly will enable us to better understand the evolution and ecology of snakes. However, the value of a thorough documentation and understanding of parental care and its functions in snakes has much broader significance.

Python egg brooding is already demonstrating its value as a useful and relevant parental care model for a number of reasons. First, python egg brooding represents female-only nest attendance, which is the most widespread mode of parental care across taxa (Gross and Shine 1981; Zeh and Smith 1985; Clutton-Brock 1991). Additionally, it significantly impacts several critical developmental variables (e.g., embryonic predation, thermoregulation, water balance, and respiration) and obligates maternal costs (e.g., fecundity, body condition, oxidative stress susceptibility). Thus, python egg brooding can be used to examine cost-benefit parent-offspring and intra-offspring tradeoffs of parental care that can elucidate the evolution of parental care across taxa.

Expanding detailed investigations beyond pythons will only increase the value of snakes as valuable contributors to understanding the evolution or parental care. Snakes provide an amazing diversity of parental behaviors. Yet, even in its most complex form, snake parental care is quite amenable to study. While post-paritive parental care by snakes serves to balance multiple tradeoffs, the factors involved (e.g., temperature, water balance, respiration) are easily quantifiable. Furthermore, while attendance behavior is dynamic, behaviors are spatially limited, making uninterrupted monitoring and assessment possible. In addition, several advances in snake captive propagation and the development of technical equipment needed to quantify functional benefits are emerging. Together, there is little reason to allow such a valuable information trove to lie virtually unexplored.

18.6 SUMMARY

Parental care (i.e., any non-genetic contribution by a parent that appears likely to increase the fitness of its offspring) is remarkably widespread across the animal kingdom, including snakes. Here, we summarize parental behavioral and physiological traits in snakes that occur before and after parition (used to collectively define oviposition and parturition).

Pre-paritive parental care is ubiquitous among snakes (and likely all vertebrates), and can include investment of resources into the offspring, behavioral thermoregulation, and nest site selection. Energy, whether provided lecithotrophically or matritrophically, is the resource that has received greatest attention from researchers, but other resources that have been shown to be transferred from female snakes to their offspring include water, minerals, immunoglobulins, antioxidants, and defensive toxins.

Since development is highly sensitive to temperature, numerous snakes have been shown to increase and/or more tightly regulate body temperature pre-parition. Increased female thermoregulatory activity, while beneficial to the developing offspring, can impose metabolic and survival costs on the female. Although reproduction-induced changes in thermoregulation have been most studied in viviparous species, they have also been documented in oviparous specie where up to one third of development can occur prior to oviposition. However, oviparous species must also assure a suitable developmental environment post-parition, and thus nest site selection is a vital component of reproduction in these species.

Post-paritive parental care is much less common among snakes but has been documented in a variety of taxa. Most notable are the ubiquitous egg brooding behavior in pythons and nearly ubiquitous egg/neonate attendance in pit vipers. While the functional significance of post-paritive parental care is mostly untested in most snakes, there is a growing body of data regarding the functional significance of python egg brooding. Egg brooding by female pythons has been shown to impact embryonic predation, water balance, thermoregulation, and respiration. It is unknown whether the implications of post-paritive parental care in other species are equally or less complex. Clearly, more work is needed to explore the occurrence and functional significance of parental care among a greater diversity of snakes.

18.7 ACKNOWLEDGMENTS

We thank Bob Aldridge and Dave Sever for inviting our participation in the *Reproduction of the Ophidia* symposium at the 2009 Joint Meeting of Ichthyologists and Herpetologists. We are grateful to Louis Somma for his insightful comments on the chapter. We also appreciate the National Science Foundation (IOS-0543979 to DFD and a Graduate Research Fellowship to ZRS) for funding our research in this area.

18.8 LITERATURE CITED

Alexander, G. J. 2004. Breeding records of *Python natalensis* in the wild non-shivering incubation and maternal care of young. Herpetological Association of Africa Symposium (Port Elizabeth) 7: 8.

Alexander, G. J. 2007. Thermal biology of the Southern African Python (*Python natalensis*): Does temperature limit its distribution? Pp. 51-75. In R. W. Henderson and R. Powell (eds), *Biology of the Boas and Pythons.* Eagle Mountain Publishing, Eagle Mountain, Utah.

Alexander, G. J. and Marais, J. 2007. *A Guide to the Reptiles of Southern Africa.* Struik Publishers, Cape Town. P. 408.

Andrews, R. M. 2004. Patterns of embryonic development. Pp. 75-102. In D. C. Deeming (ed.), *Reptilian Incubation: Environment, Evolution, and Behaviour.* Nottingham University Press, Nottingham, U.K.

Aubret, F., Bonnet, X., Shine, R. and Maumelat, S. 2003. Clutch size manipulation, hatching success and offspring phenotype in the ball python (*Python regius).* Biological Journal of the Linnean Society 78: 263-272.

Aubret, F., Bonnet, X., Shine, R. and Maumelat, S. 2005a. Energy expenditure for parental care may be trivial for brooding pythons, *Python regius.* Animal Behaviour 69: 1043-1053.

Aubret, F., Bonnet, X., Shine, R. and Maumelat, S. 2005b. Why do female ball pythons (*Python regius*) coil so tightly around their eggs? Evolutionary Ecology Research 7: 743-758.

Bedford, G. and Christian, K. A. 2001. Metabolic response to feeding and fasting in the water python (*Liasis fuscus*). Australian Journal of Zoology 49: 379-387.

Belinsky, A., Ackerman, R. A., Dmi'el, R. and Ar, A. 2004. Water in reptilian eggs and hatchlings. Pp. 125-142. In D. C. Deeming (ed.), *Reptilian Incubation: Environment, Evolution, and Behaviour.* Nottingham University Press, Nottingham, U.K.

Blackburn, D. G. 1985. Evolutionary origins of viviparity in the Reptilia. II. Serpentes, Amphisbaenia, and Ichthyosauria. Amphibia-Reptilia 5: 259-291.
Blackburn, D. G. 1994. Discrepant usage of the term 'ovoviviparity' in the herpetological literature. Herpetological Journal 4: 65-72.
Blazquez, M. C. 1995. Body temperature, activity patterns and movements by gravid and non-gravid females of *Malpolon monspessulanus*. Journal of Herpetology 29: 264-266.
Blouin-Demers, G. and Weatherhead, P. J. 2001. Habitat use by black rat snakes (*Elaphe obsoleta obsoleta*) in fragmented forests. Ecology 82: 2882-2896.
Bonnet, X., Naulleau, G. and Lourdais, O. 2002. Benefits of complementary techniques: Using capture-recapture and physiological approaches to understand costs of reproduction in the aspic viper. Pp. 483-495. In G. W. Schuett, M. Höggren, M. E. Douglas and H. W. Greene (eds), *Biology of the Vipers*. Eagle Mountain Publishing, Eagle Mountain, Utah.
Branch, W. R. and Patterson, R. W. 1975. Notes on the development of embryos of the African rock python, *Python sebae* (Serpentes: Boidae). Journal of Herpetology 9: 242-248.
Broadley, D. G. 1977. A revision of the African snakes of the genus *Psammophylax* Fitzinger (Colubridae). Occasional Papers of the National Museums and Monuments of Rhodesia 1976B: 1-44.
Brown, G. P. and Weatherhead, P. J. 2000. Thermal ecology and sexual size dimorphism in northern water snakes, *Nerodia sipedon*. Ecological Monographs 70: 311-330.
Brown, G. P. and Shine, R. 2005. Nesting snakes (*Tropidonophis mairii*, Colubridae) selectively oviposit in sites that provide evidence of previous successful hatching. Canadian Journal of Zoology 83: 1134-1137.
Brown, G. and Shine, R. 2006. Effects of nest temperature and moisture on phenotypic traits of hatchling snakes (*Tropidonophis mairii*, Colubridae) from tropical Australia. Biological Journal of the Linnean Society 89: 159-168.
Brown, G. P. and Shine, R. 2007. Like mother, like daughter: Inheritance of nest-site location in snakes. Biology Letters 3: 131-133.
Burger, J. 1990. Effects of incubation temperature on behavior of young black racers (*Coluber constrictor*) and kingsnakes (*Lampropeltis getulus*). Journal of Herpetology 24: 158-163.
Burger, J. 1998. Effects of incubation temperature on hatchling pine snakes: implications for survival. Behavioral Ecology and Sociobiology 43: 11-18.
Burger, J. and Zappalorti, R. T. 1986. Nest site selection by pine snakes, *Pituophis melanoleucus*, in the New Jersey pine barrens. Copeia 1986: 116-121.
Burger, J. and Zappalorti, R. T. 1988. Effects of incubation temperature on sex-ratios in pine snakes—differential vulnerability of males and females. American Naturalist 132: 492-505.
Burger, J. and Zappalorti, R. T. 1992. Philopatry and nesting phenology of pine snakes *Pituophis melanoleucus* in the New Jersey pine barrens. Behavioral Ecology and Sociobiology 30: 331-336.
Butler, J. A., Hull, T. W. and Franz, R. 1995. Neonate aggregations and maternal attendance of young in the eastern diamondback rattlesnake, *Crotallus adamanteus*. Copeia 1995: 196-198.
Cai, Y., Zhou, T. and Ji, X. 2007. Embryonic growth and mobilization of energy and material in oviposited eggs of the red-necked keelback, *Rhabdophis tigrinus lateralis*. Comparative Biochemistry and Physiology A 147: 57-63.
Campbell, J. A. and Quinn, H. R. 1975. Reproduction in a pair of Asiatic cobras, *Naja naja* (Serpentes, Elapidae). Journal of Herpetology 9: 229-233.

Carpenter, C. C. and Ferguson, G. W. 1977. Variation and evolution of stereotyped behavior in reptiles. Pp. 335-554. In C. Gans and D. W. Tinkle (eds), *Biology of the Reptilia*. Volume 7. Academic Press, London, U.K.

Charland, M. B. 1995. Thermal consequences of reptilian viviparity—thermoregulation in gravid and nongravid garter snakes (*Thamnophis*). Journal of Herpetology 29: 383-390.

Charland, M. B. and Gregory, P. T. 1990. The influence of female reproductive status on thermoregulation in a viviparous snakes, *Crotalus viridis*. Copeia 1990: 1089-1098.

Charles, N., Field, R. and Shine, R. 1985. Notes on the reproductive biology of Australian pythons, genera *Aspidites*, *Liasis*, and *Morelia*. Herpetological Review 16: 45-48.

Chiaraviglio, M. 2006. The effects of reproductive condition on thermoregulation in the Argentina boa constrictor (*Boa constrictor occidentalis*) (Boidae). Herpetological Monographs 20: 172-177.

Clutton-Brock, T. H. 1991. *The Evolution of Parental Care*. Princeton University Press, Princeton, New Jersey. Pp. 368.

Daltry, J. C., Ross, T., Thorpe, R. S. and Wuster, W. 1998. Evidence that humidity influences snake activity patterns: A field study of the Malayan pit viper *Calloselasma rhodostoma*. Ecography 21: 25-34.

Deeming, D. C. and Ferguson, M. W. J. 1991. *Egg Incubation: Its Effects on Embryonic Development in Birds and Reptiles*. Cambridge University Press, Cambridge, U.K. Pp. 462.

Deloya, E. M., Setser, K., Pleguezuelos, J. M., Kardon, A. and Lazcano, D. 2009. Cannibalism of nonviable offspring by postparturient Mexican lance-headed rattlesnakes, *Crotalus polystictus*. Animal Behaviour 77: 145-150.

Doody, J. S., Freedberg, S. and Keogh, J. S. 2009. Communal egg-laying in reptiles and amphibians: Evolutionary patterns and hypotheses. Quarterly Review of Biology 84: 229-252.

Duvall, D., King, M. B. and Gutzwiller, K. J. 1985. Behavioral ecology and ethology of the prairie rattlesnake. National Geographic Research 1: 80-111.

Ellis, T. M. and Chappell, M. A. 1987. Metabolism, temperature relations, maternal behavior, and reproductive energetics in the ball python (*Python regius*). Journal of Comparative Physiology B 157: 393-402.

Farmer, C. G. 2000. Parental care: The key to understanding endothermy and other convergent features in birds and mammals. American Naturalist 155: 326-334.

Filippi, E., Anibaldi, C., Capizzi, D., Arianna, C. I., Capula, M. and Luiselli, L. 2007. Long-term fidelity to communal oviposition sites in *Hierophis viridiflavus*. Herpetological Journal 17: 7-13.

Finneran, L. C. 1953. Aggregation behavior of the female copperhead, *Agkistrodon contortrix*, during gestation. Copeia 1953: 61-62.

Fitch, H. S. 1970. Reproductive cycles in lizards and snakes. University of Kansas Miscellaneous Publications, Museum of Natural History 52: 1-247.

FitzSimons, F. W. 1912. *The Snakes of South Africa. Their Venom and the Treatment of Snake Bite*. T. Maskew Miller, Cape Town. Pp. 298.

Gans, C. 1996. An overview of parental care among the Reptilia. Advances in the Study of Behavior 25: 145-157.

Gardiner, B. G. 2002. Crocodile relationships. The Linnean 18(2): 33-40.

Gier, P. J., Wallace, R. L. and Ingermann, R. L. 1989. Influence of pregnancy on behavioral thermoregulation in the northern pacific rattlesnakes *Crotalus viridis oreganus*. Journal of Experimental Biology 145: 465-469.

Gloyd, H. K. and Conant, R. 1990. *Snakes of the Agkistrodon Complex: A Monographic Review. Contributions to Herpetology 6*. Society for the Study of Amphibian and Reptiles, Saint Louis. Pp. 614.
Graves, B. M. and Duvall, D. 1995. Aggregation of squamate reptiles associate with gestation, oviposition, and parturition. Herpetological Monographs 9: 102-119.
Greene, H. W. 1988. Antipredator mechanisms in reptiles. Pp. 1-152. In C. Gans and R. B. Huey (eds), *Biology of the Reptilia*, Volume 16. Alan R. Liss, New York.
Greene, H. W. 1997. *Snakes. The Evolution of Mystery in Nature*. University of California Press, Berkeley and Los Angeles. Pp. 351.
Greene, H. W., May, P. G., Hardy, D. L., Sr., Sciturro, J. M. and Farrell, T. M. 2002. Parental behavior by vipers. Pp. 179-205 In G. W. Schuett, M. Höggren, M. E. Douglas and H. W. Greene (eds), *Biology of the Vipers*. Eagle Mountain Publishing, Eagle Mountain, Utah.
Gross, M. R. and Shine, R. 1981. Parental care and mode of fertilization in ectothermic vertebrates. Evolution 35: 775-793.
Groves, J. D. 1981. Observations and comments on the post-parturient behaviour of some tropical boas of the genus *Epicrates*. British Journal of Herpetology. 6: 89-91.
Gutzke, W. H. N. and Packard, G. C. 1987. Influence of the hydric and thermal environments on eggs and hatchlings of bull snakes *Pituophis melanoleucus*. Physiological Zoology 60: 9-17.
Haagner, G. V. 1990. The husbandry and captive propagation of forest cobras, *Naja melanoleuca*. Journal of the Herpetological Association of Africa 38: 57-60.
Harlow, P. and Grigg, G. 1984. Shivering thermogenesis in a brooding diamond python, *Morelia spilotes spilotes*. Copeia 1984: 959-965.
Hassl, A. 2005a. Functional egg immunoglobulins in the snake *Elaphe guttata*. Amphibia-Reptilia 26: 109-112.
Hassl, A. 2005b. Snake egg immunoglobulins: Biochemical characteristics and adjusted isolation procedure. Journal of Immunological Methods 297: 253-257.
Heinrich, B. 1979. *Bumblebee Economics*. Harvard University Press, Cambridge, Massachusetts. Pp. 245.
Hibbard, C. W. 1964. A brooding colony of the blind snake, *Leptotyphlops dulcis dissecta* Cope. Copeia 1964: 222.
Hill, J. G. III, Chanhome, L., Artchawakom, T., Thirakhupt, K. and Voris, H. K. 2006. Nest attendance by a female Malayan pit viper (*Calloselasma rhodostoma*) in northeast Thailand. The Natural History Journal of Chulalongkorn University 6: 57-66.
Hochachka, P. W. and Somero, G. N. 2002. *Biochemical Adaptation: Mechanism and Process in Physiological Evolution*. Oxford University Press, Oxford, U.K. Pp. 480.
Holmstrom, W. F., Jr. and Behler, J. L. 1981. Post-parturient behavior of the common anaconda, *Eunectes murinus*. Zoologische Garten N. F. (Jena) 51: 353-356.
Honegger, R. E. 1970. Beitrag zur Fortpflanzungsbiologie von *Boa constrictor* und *Python reticulatus* (Reptilia, Boidae). Salamandra 6: 73-79.
Huey, R. B. and Kingsolver, J. G. 1989. Evolution of thermal sensitivity of ectotherm performance. Trends in Ecology and Evolution 4: 131-135.
Huff, T. A. 1980. Captive propagation of the subfamily Boinae with emphasis on the genus *Epicrates*. Pp. 125-134. In J. B. Murphy and J. T. Collins (eds), *Reproductive Biology and Diseases of Captive Reptiles. Contributions in Herpetology.* Society for the Study of Amphibians and Reptiles, Athens, Ohio.
Humphrey, M. 2000. Breeding the blue beauty snake. Reptiles 8: 74-80.

Hutchinson, D. A., Savitzky, A. H., Mori, A., Meinwald, J. and Schroeder, F. C. 2008. Maternal provisioning of sequestered defensive steroids by the Asian snake *Rhabdophis tigrinus*. Chemoecology 18: 181-190.

Hutchison, V. H., Dowling, H. G. and Vinegar, A. 1966. Thermoregulation in a brooding female Indian python, *Python molurus bivittatus*. Science 151: 694-696.

Isaac, L. A. and Gregory, P. T. 2004. Thermoregulatory behaviour of gravid and non-gravid female grass snakes (*Natrix natrix*) in a thermally limiting high-latitude environment. Journal of Zoology (London) 264: 403-409.

Jacobs, H. J., Auliya, M. and Böhme, W. 2009. Zur Taxonomie des Dunklen Tigerpythons, *Python molurus bivittatus* Kuhl, 1820, speziell der Population von Sulawesi. Sauria 31(3): 5-16.

Jes, H. 1984. Beobachtungen an einigen ovovivipare gebärenden Riesenschlangen (Boidae). Salamandra 20: 268-269.

Ji, X. and Du, W.-G. 2001. The effects of thermal and hydric environments on hatching success, embryonic use of energy and hatchling traits in a colubrid snake, *Elaphe carinata*. Comparative Biochemistry and Physiology A 129: 461-471.

Ji, X., Sun, P.-Y., Fu, S.-Y. and Zhang, H.-S. 1997a. Utilization of energy and nutrients in incubating eggs and post-hatching yolk in a colubrid snake, *Elaphe carinata*. Herpetological Journal 7: 7-12.

Ji, X., Sun, P.-Y., Zhang, H.-S. and Fu, S.-Y., 1997b. Incubation and utilization of energy and material during embryonic development in eggs of *Naja naja atra*. Journal of Herpetology 31: 302-306.

Ji, X., Sun, P.-Y., Fu, S.-Y. and Zhang, H.-S. 1999. Utilization of energy and material in eggs and post-hatching yolk in an oviparous snake, *Elaphe taeniura*. Asiatic Herpetological Research 8: 53-59.

Johnson, C. R., Webb, G. J. W. and Johnson, C. 1975. Thermoregulation in pythons: III. Thermal ecology and behavior of the black-headed rock python, *Aspidites melanocephalus*. Herpetologica 31: 326-332.

Kend, B. A. 1997. *Pythons of Australia*. Canyonlands Publishing Group, LC, Provo, Utah. Pp. 206.

Koteja, P. 2000. Energy assimilation, parental care and the evolution of endothermy. Proceedings of the Royal Society of London B: Biological Sciences 267: 479-484.

Ladyman, M., Bonnet, X., Lourdais, O., Bradshaw, D. and Naulleau, G. 2003. Gestation, thermoregulation, and metabolism in a viviparous Snake, *Vipera aspis*: Evidence for fecundity-independent costs. Physiological and Biochemical Zoology 76: 497-510.

La Panouse, R. de and Pellier, C. 1973. Ponte d'un python réticulé (*Python reticulatus*) élevé en terrarium, et incubation des œufs. Bulletin du Muséum National d'Histoire Naturelle Serie 3, Zoologie (Paris) 79(105): 37-48.

Lim Leong Keng, F. and Lee Tat-Mong, M. 1989. *Fascinating snakes of Southeast Asia—An Introduction.* Tropical Press Sdn. Bhd., Kuala Lumpur, Malaysia. Pp. 124.

Lin, Z. H., Ji, X., Luo, L. G. and Ma, X. M. 2005. Incubation temperature affects hatching success, embryonic expenditure of energy and hatchling phenotypes of a prolonged egg-retaining snake, *Deinagkistrodon acutus* (Viperidae). Journal of Thermal Biology 30: 289-297.

Lissaker, M. and Kvarnemo, C. 2006. Ventilation or nest defense–parental care trade-offs in a fish with male care. Behavioral Ecology and Sociobiology 60: 864-873.

Lourdais, O., Brischoux, F., Shine, R. and Bonnet, X. 2005. Adaptive maternal cannibalism in snakes (*Epicrates cenchria maurus*, Boidae). Biological Journal of the Linnean Society 84: 767-774.

Lourdais, O., Hoffman, T. and DeNardo, D. F. 2007. Maternal brooding in the Children's python (*Antaresia childreni*) promotes egg water balance. Journal of Comparative Physiology B 177: 569-577.

Lourdais, O., Heulin, B. and DeNardo, D. F. 2008. Thermoregulation during gravidity in the Children's python (*Antaresia childreni*): A test of the pre-adaptation hypothesis for maternal thermophily in snakes. Biological Journal of the Linnaean Society 93: 499-508.

Madsen, T. and Shine, R. 1999. Life history consequences of nest-site variation in tropical pythons (*Liasis fuscus*). Ecology 80: 989-997.

Mattison, C. 2007. *The New Encyclopedia of Snakes*. Revised and Updated Edition. Princeton University Press, Princeton, New Jersey. Pp. 272.

Mell, R. 1929. *Beiträge zur Fauna Sinica. IV. Grundzüge Einer Ökologie der Chinesischen Reptilien und Einer Herpetologischen Tiergeographie Chinas.* Walter de Gruyter, Berlin, Germany. Pp. 282.

Miller, T. J. 1983. Corallus enhydris enhydris (garden tree boa). Food. Herpetological Review 14: 46-47.

Mitchell, J. C. 1994. *The Reptiles of Virginia*. Smithsonian Institution Press, Washington, D. C. Pp. 352.

Mitchell, J. C. and Groves, J. D. 1993. Intraspecific oophagy in reptiles. Herpetological Review 24: 126-130.

Moreira, P. L. and Barata, M. 2005. Egg mortality and early embryo hatching caused by fungal infection of Iberian rock lizard (*Larcerta monticola*) clutches. Herpetological Journal 15: 265-272.

Murphy, J. B., Lamoreaux, W. L. and Barker, D. G. 1981. Miscellaneous notes on the reproductive biology of reptiles. 4. Eight species of the family Boidae, genera *Acrantophis, Aspidites, Candoia, Liasis*, and *Python*. Transactions of the Kansas Academy of Science 84: 39-49.

Noble, G. K. 1935. The brooding habit of the blood python and of other snakes. Copeia 1935: 1-3.

O'Donnell, R. P. and Arnold, S. J. 2005. Evidence for selection on thermoregulation: Effects of temperature on embryo mortality in the garter snake *Thamnophis elegans*. Copeia 2005: 930-934.

Palmer, W. M. and Braswell, A. L. 1995. *Reptiles of North Carolina*. The University of North Carolina Press, Chapel Hill, North Carolina. Pp. 448.

Patterson, L. D. and Blouin-Demers, G. 2008. The effect of constant and fluctuating incubation temperatures on the phenotype of black ratsnakes (*Elaphe obsoleta*) Canadian Journal of Zoology 86: 882-889.

Peters, J. A. 1964. *A Dictionary of Herpetology*. Hafner Publishing Company, New York. Pp. 392.

Pitman, C. R. S. 1974. *A Guide to the Snakes of Uganda*. Revised Edition. Wheldon and Wesley Ltd., London, U.K. Pp. 290.

Plummer, M. V. 1981. Communal nesting of *Opheodrys aestivus* in the laboratory. Copeia 1981: 243-246.

Plummer, M. V. and Snell, H. L. 1988. Nest site selection and water relations of eggs in the snake, *Opheodrys aestivus*. Copeia 1988: 58-64.

Polis, G. A. 1981. The evolution and dynamics of intraspecific predation. Annual Review of Ecology and Systematics 12: 225-251.

Pope, C. H. 1935. *The Reptiles of China. Turtles, Crocodilians, Snakes, Lizards. Natural History of Central Asia.* Volume X. The American Museum of Natural History, New York. Pp. 604.

Price, A. H. 1988. Observations on maternal behavior and neonate aggregation in the western diamondback rattlesnake, *Crotalus atrox* (Crotalidae). Southwestern Naturalist 33: 370-373.
Radder, R. and Shine, R. 2007. Why do female lizards lay their eggs in communal nests? Journal of Animal Ecology 76: 881-887.
Rawlings, L. H., Rabosky, D. L., Donnellan, S. C. and Hutchinson, M. N. 2008. Python phylogenetics: Inference from morphology and mitochondrial DNA. Biological Journal of the Linnean Society 93: 603-619.
Rawlinson, P. A. 1969. The reptiles of East Gippsland. Proceedings of the Royal Society of Victoria 82: 113-128.
Reeve, H. K. and Sherman, P. W. 1993. Adaptation and the goals of evolutionary research. Quarterly Review of Biology 68: 1-32.
Ripa, D. 1994. Reproduction of the Central American bushmaster (*Luchesis muta stenophrys*) and the black-headed bushmaster (*Lachesis muta melanocephala*) for the first time in captivity. Bulletin of the Chicago Herpetological Society 29: 165-183.
Ross, R. A. and Marzec, G. 1990. The Reproductive Husbandry of Pythons and Boas. The Institute for Herpetological Research, Stanford, California. Pp. 270.
Sergeev, A. M. 1940. Researches in the viviparity of reptiles. Moscow Society of Naturalists Jubilee Issue: 1-34.
Shine, R. 1983. Reptilian reproductive modes: The oviparity-viviparity continuum. Herpetologica 39: 1-8.
Shine, R. 1985. The evolution of viviparity in reptiles: An ecological analysis. Pp. 605-694. In C. Gans and F. Billet (eds), *Biology of the Reptilia,* Volume 15. John Wiley and Sons, New York.
Shine, R. 1988. Parental care in reptiles. Pp. 275-330. In C. Gans and R. B. Huey (eds), *Biology of the Reptilia,* Volume 16. Alan R. Liss, New York.
Shine, R. 1995. A new hypothesis for the evolution of viviparity in reptiles. American Naturalist 145: 809-823.
Shine, R. 2004. Adaptive consequences of developmental plasticity. Pp. 187-210. In D. C. Deeming (ed.), *Reptilian Incubation: Environment, Evolution and Behaviour.* Nottingham University Press, Nottingham, U.K.
Shine, R. 2006. Is increased maternal basking an adaptation or a pre-adaptation to viviparity in lizards? Journal of Experimental Zoology 305A: 524-535.
Shine, R. and Harlow, P. S. 1996. Maternal manipulation of offspring phenotypes via nest-site selection in an oviparous lizard. Ecology 77: 1808-1817.
Shine, R. and Madsen, T. 1996. Is thermoregulation unimportant for most reptiles? An example using water pythons (*Liasis fuscus*) in tropical Australia. Physiological Zoology 69: 252-269.
Shine, R., Madsen, T. R. L., Elphick, M. J. and Harlow, P. S. 1997. The influence of nest temperatures and maternal brooding on hatchling phenotypes of water pythons. Ecology 78: 1713-1721.
Slip, D. J. and Shine, R. 1988. Reptilian endothermy: A field study of thermoregulation by brooding pythons. Journal of Zoology (London) 216: 367-378.
Slowinski, J. B. and Lawson, R. 2002. Snake phylogeny: Evidence from nuclear and mitochondrial genes. Molecular Phylogenetics and Evolution 24: 194-202.
Smith, M. A. 1943. *The Fauna of British India Ceylon and Burma, including the Whole of the Indo-Chinese Sub-Region. Reptila and Amphibia. Volume III.—Serpentes.* Taylor and Francis, London, U.K. Pp. 583.
Smith, H. M. 1975. Grist for the mills of herpetophiles in Mexico. Bulletin of the Maryland Herpetological Society 11: 40-44.

Somma, L. A. 1988. Comments on the use of the term "brooding" to describe parental behaviour in squamate reptiles. Amphibia-Reptilia 9: 90-91.
Somma, L. A. 2003a. *Parental Behavior in Lepidosaurian and Testudinian Reptiles: A Literature Survey*. Krieger Publishing Company, Malabar, Florida. Pp. 174.
Somma, L. A. 2003b. Parental behavior in lepidosaurs and turtles: Source addendum. Bulletin of the Chicago Herpetological Society 28: 65-76.
Somma, L. A. 2003c. A few more additions to the literature on parental behavior in lizards and snakes. Bulletin of the Chicago Herpetological Society 38: 217-220.
Somma, L. A. 2003d. Reptilian parental behaviour. The Linnean 19: 42-44.
Somma, L. A. and Fawcett, J. D. 1989. Brooding behaviour of the prairie skink, *Eumeces septentrionalis*, and its relationship to the hydric environment of the nest. Zoological Journal of the Linnean Society 95: 245-256.
Speake, B. K., Thompson, M. B., Thacker, F. E. and Bedford, G. S. 2003. Distribution of lipids from the yolk to the tissues during development of the water python (*Liasis fuscus*). Journal of Comparative Physiology B—Biochemical, Systemic and Environmental Physiology 173: 541-547.
Stahlschmidt, Z. R. and DeNardo, D. F. 2008. Alternating egg brooding behaviors create and modulate a hypoxic developmental micro-environment in Children's pythons (*Antaresia childreni*). Journal of Experimental Biology 211: 1535-1540.
Stahlschmidt, Z. R. and DeNardo, D. F. 2009a. Effect of nest temperature on egg brooding dynamics in Children's pythons. Physiology and Behavior 98: 302-306.
Stahlschmidt, Z. R. and DeNardo, D. F. 2009b. Obligate costs of parental care to offspring: Egg brooding induced hypoxia creates smaller, slower, and weaker python offspring. Biological Journal of the Linnean Society 98: 414-421.
Stahlschmidt, Z. R. and DeNardo, D. F. 2010. Parental behavior in pythons is responsive to both the hydric and thermal dynamics of the nest. Journal of Experimental Biology. In press.
Stahlschmidt, Z. R., Hoffman, T. C. M. and DeNardo, D. F. 2008. Postural shifts during egg brooding and their impact on egg water balance in Children's pythons (*Antaresia childreni*). Ethology 114: 1113-1121.
Swain, T. and Smith, H. M. 1978. Communal nesting in *Coluber constrictor* in Colorado (Reptilia: Serpentes). Herpetologica 34: 175-177.
Thompson, M. B. and Speake, B. K. 2004. Egg morphology and composition. Pp. 45-74. In D. C. Deeming (ed.), *Reptilian Incubation: Environment, Evolution and Behaviour*. Nottingham University Press, Nottingham, U.K.
Tolson, P. J. and Henderson, R. W. 1993. The Natural History of West Indian Boas. R. & A. Publishing, Somerset, England. Pp. 125.
Townsend, D. S. 1986. The costs of male parental care and its evolution in a Neotropical frog. Behavioral Ecology and Sociobiology 19: 187-195.
Townson, S. 1985. The captive reproduction and growth of the yellow anaconda (*Eunectes notaeus*). Pp. 33-43. In S. Townson and K. Lawrence (eds), *Reptiles: Breeding, Behaviour and Veterinary Aspects*. British Herpetological Society, London, U.K.
Trivers, R. L. 1974. Parent-offspring conflict. American Zoologist 11: 249-264.
Tu, M. C. and Hutchison, V. H. 1994. Influence of pregnancy on thermoregulation of water snakes (*Nerodia rhombifera*). Journal of Thermal Biology 19: 255-259.
Tu, M. C., Lillywhite, H. B., Menon, J. G. and Menon, G. K. 2002. Postnatal ecdysis establishes the permeability barrier in snake skin: New insights into barrier lipid structures. Journal of Experimental Biology 205: 3019-3030.
Van Mierop, L. H. S. and Barnard, S. M. 1976. Thermoregulation in a brooding female *Python molurus bivittatus* (Serpentes: Boidae). Copeia 1976: 398-401.

Van Mierop, L. H. S. and Barnard, S. M. 1978. Further observations on thermoregulation in the brooding female *Python molurus bivittatus* (Serpentes: Boidae). Copeia 1978: 615-621.

Van Mierop, L. H. S. and Besette, E. L. 1981. Reproduction of the ball python, *Python regius*, in captivity. Herpetological Review 12: 20-22.

Vinegar, A. 1973. The effects of temperature on the growth and development of embryos of the Indian python, *Python molurus* (Reptilia: Serpentes: Boidae). Copeia 1973: 171-173.

Vinegar, A., Hutchison, V. H. and Dowling, H. G. 1970. Metabolism, energetics, and thermoregulation during brooding of snakes of genus *Python* (Reptilia, Boidae). Zoologica 55: 19-48.

Wall, F. 1921. *Phidia Taprobanica or the Snakes of Ceylon.* H. R. Cottle, Government Printer, New Dehli, India. Pp. 581.

Webb, J. K., Shine, R. and Christian, K. A. 2006. The adaptive significance of reptilian viviparity in the tropics: Testing the maternal manipulation hypothesis. Evolution 60: 115-122.

Weekes, H. C. 1933. On the distribution, habitat and reproductive habits of certain European and Australian snakes and lizards, with particular regard to their adoption of viviparity. Proceedings of the Linnean Society of New South Wales 58: 270-274.

West, K. J. and Alexander, R. D. 1963. Sub-social behavior in a burrowing cricket *Anurogryllus muticus* (De Geer). Orthoptera: Gryllidae. Ohio Journal of Science 63: 19-24.

Winstel, A. 1991. Captive husbandry of the eastern milk snake (*Lampropeltis t. triangulum*). Vivarium 2: 16-17, 28.

Wourms, J. P. 1981. Viviparity: The maternal-fetal relationship in fishes. American Zoologist 21: 473-515.

Xiang, J. 1995. Egg and hatchling components of a viviparous snake, *Elaphe rufodorsata*. Journal of Herpetology 29: 298-300.

York, D. S. and Burghardt, G. M. 1988. Brooding in the Malayan pit viper, Calloselasma rhodostoma: Temperature, relative humidity, and defensive behaviour. Herpetological Journal 1: 210-214.

Zeh, D. W. and Smith, R. L. 1985. Paternal investment by terrestrial arthropods. American Zoologist 25: 785-805.

Index

B

C

D

R

S

T

U

X

Y

Z

About the Editors

VOLUME EDITORS

Dr. Robert D. Aldridge is a Professor of Biology at Saint Louis University, St. Louis, Missouri. He has been at Saint Louis University since receiving his Ph.D. at the University of New Mexico in 1973. He has over 50 publications on the reproductive biology of snakes. These studies have focused on the anatomical, physiological and behavioral adaptations of a variety of temperate and tropical zone species of snakes. He has served as Treasurer and Publications Secretary for the Society for the Study of Amphibians and Reptiles and as an Associate Editor for the *American Midland Naturalist*, and as Chairman of the Biology Department, Director of the Reis Biological Station, and Director of the Lay Center for Education and the Arts at Saint Louis University.

Dr. David M. Sever is Professor and Department Head of Biological Sciences, Southeastern Louisiana University, Hammond, Louisiana, where he has been since 2004 after 30 years on the Biology faculty at Saint Mary's College, Notre Dame, Indiana. He received his Ph.D. at Tulane University in 1974. He has over 130 publications covering a broad range of topics on fish, amphibians, and reptiles, but most concern the anatomy of reproduction at the tissue and cellular levels. He was volume editor for *Reproductive Biology and Phylogeny of Urodela* and has served long terms as Associate Editor for *Journal of Herpetology* and *Journal of Morphology*.

SERIES EDITOR

Dr. Barrie Jamieson is Emeritus Professor of Zoology in the School of Integrative Biology, University of Queensland. He holds a Ph.D. from the University of Bristol, England, a D.Sc. from the University of Queensland, and is a former Visiting Fellow of, and member of the Association of, Corpus Christi College, Cambridge. In 1990 he was awarded the Clarke Medal for Research in Natural Sciences, early recipients of which were Thomas Henry Huxley and Richard Owen. His chief field of research is spermatozoal ultrastructure and its relevance to phylogeny but he is also an authority on taxonomy of earthworms and has published on bioluminescence, trematode taxonomy and life cycles, and DNA-based phylogenetics. He has named some 170 species, has published more than 200 scientific papers and is the author, coauthor or editor of twenty books.

Color Plate Section

CHAPTER 2

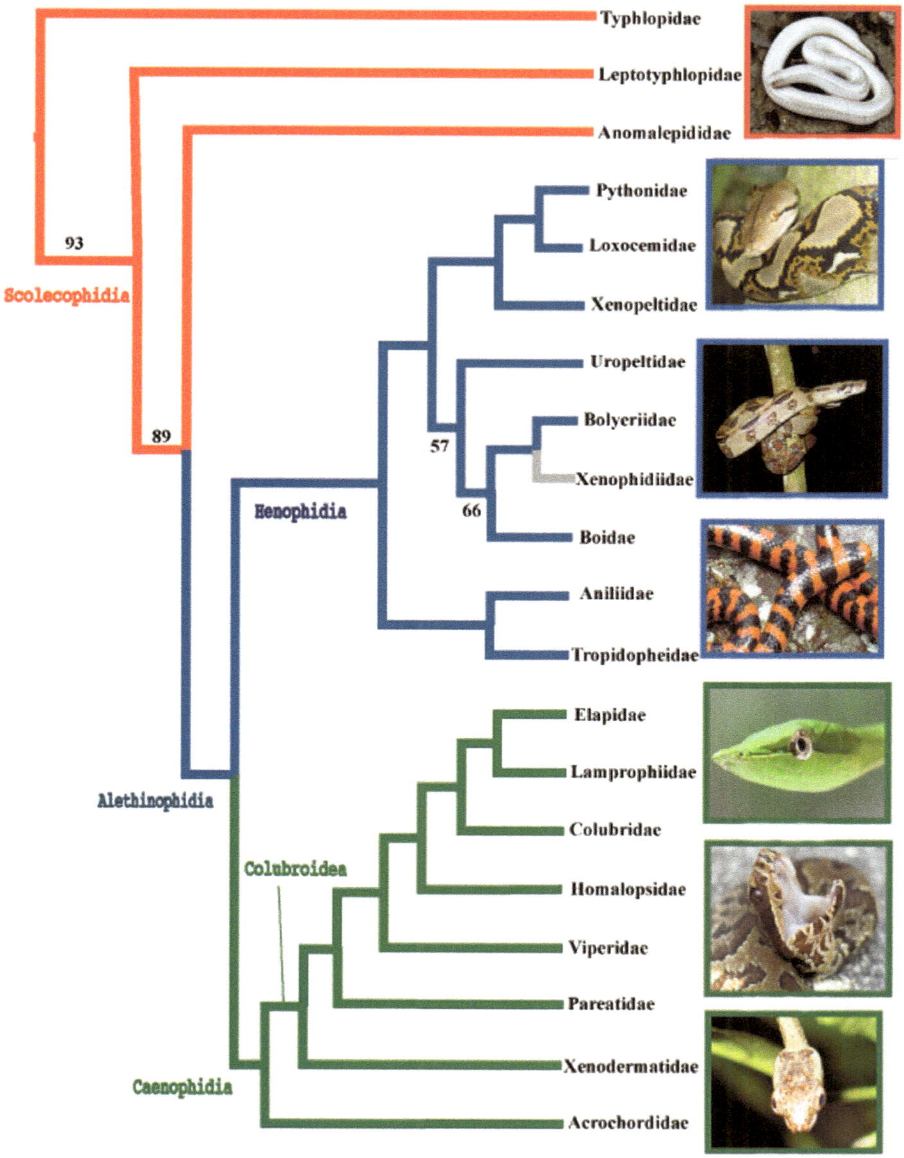

Figure. 2.1 See text page 27 for caption.

CHAPTER 3

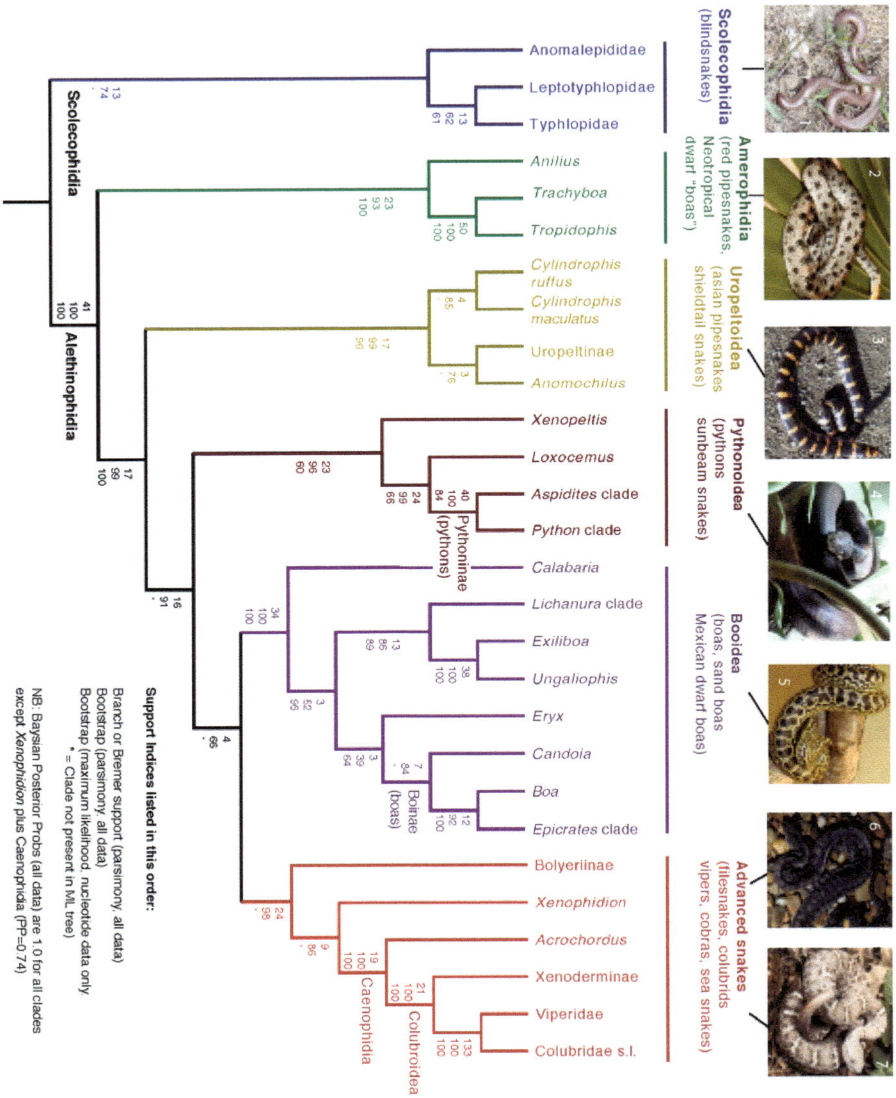

Figure 3.1 See text page **56** for caption.

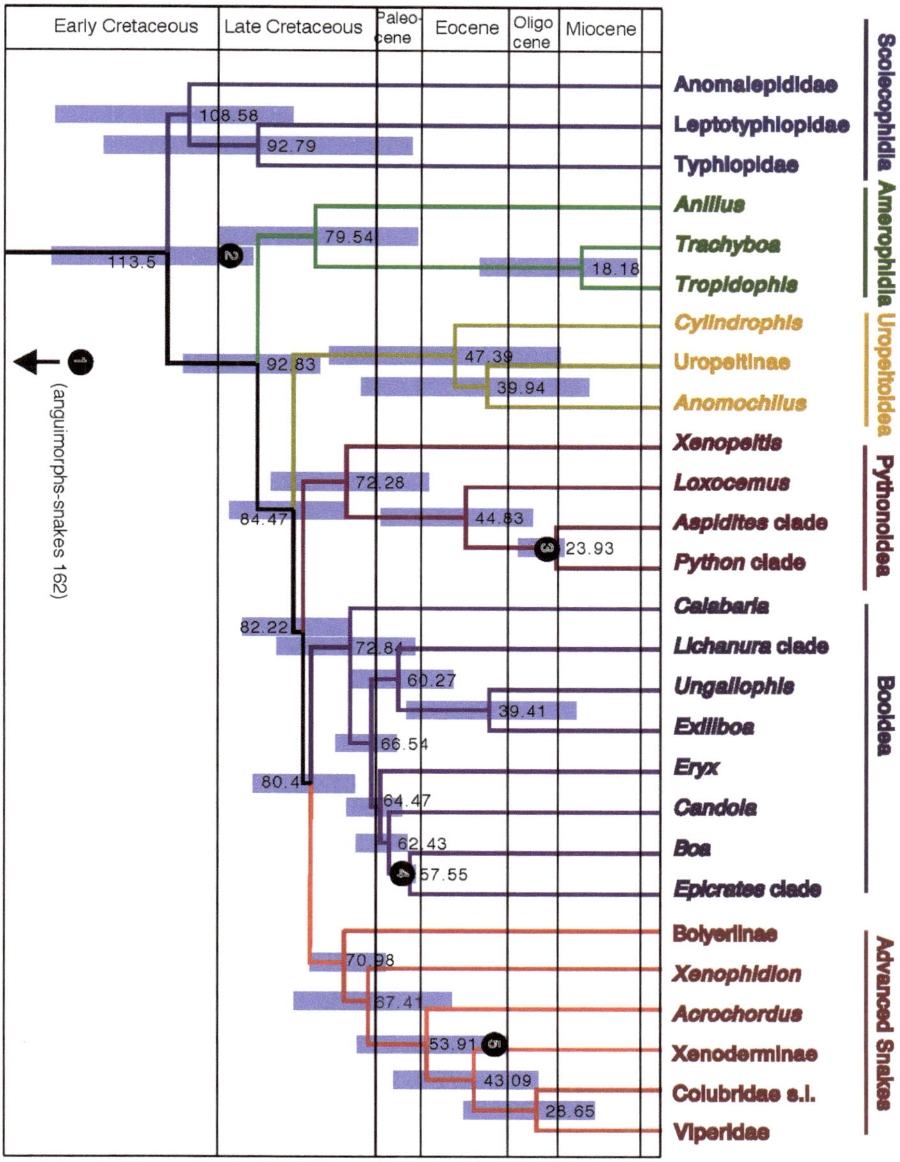

Figure 3.2 See text page **66** for caption.

CHAPTER 5

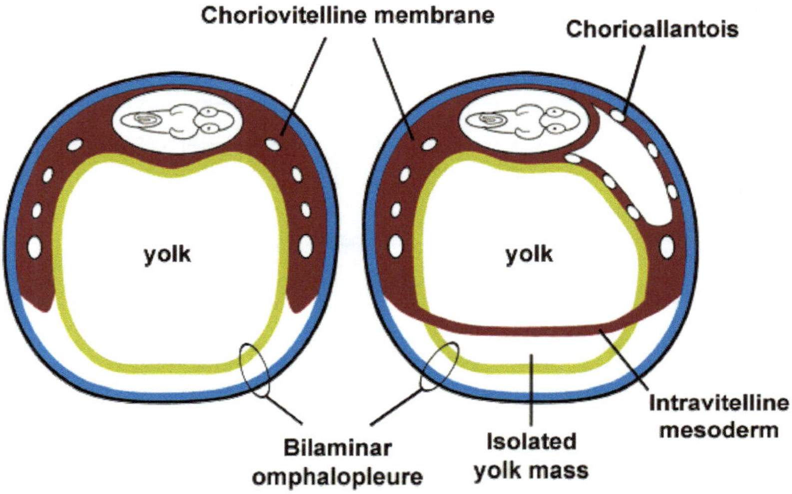

Figure 5.2 See text page 145 for caption.

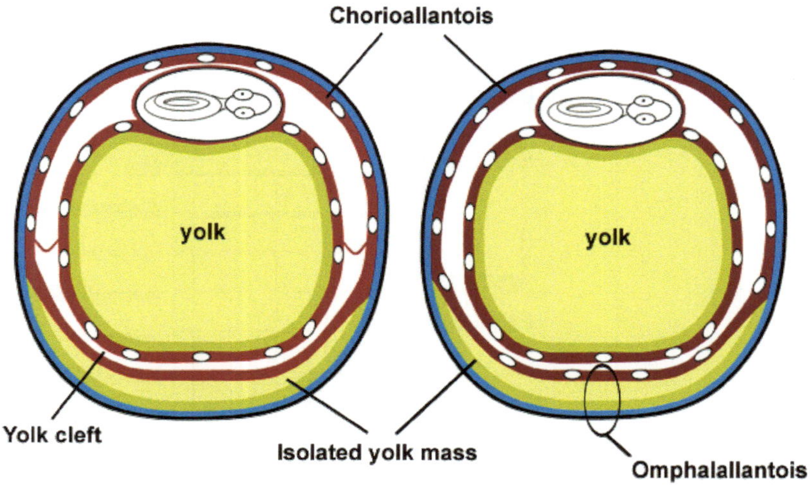

Figure 5.3 See text page 145 for caption.

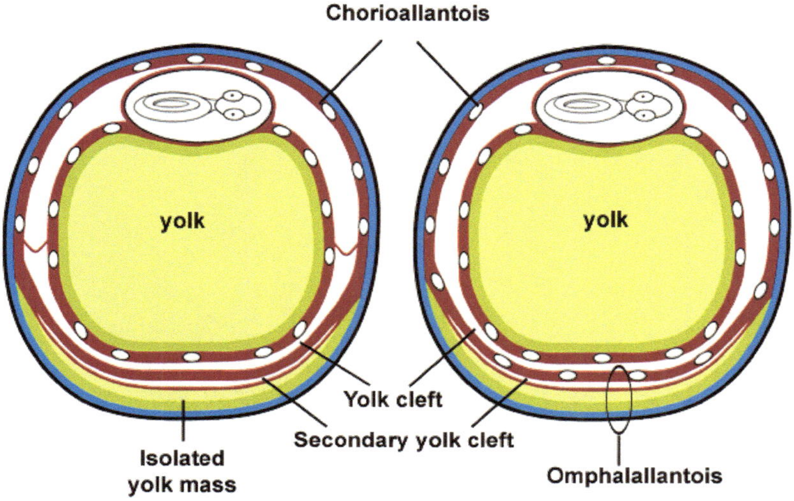

Figure 5.4 See text page **146** for caption.

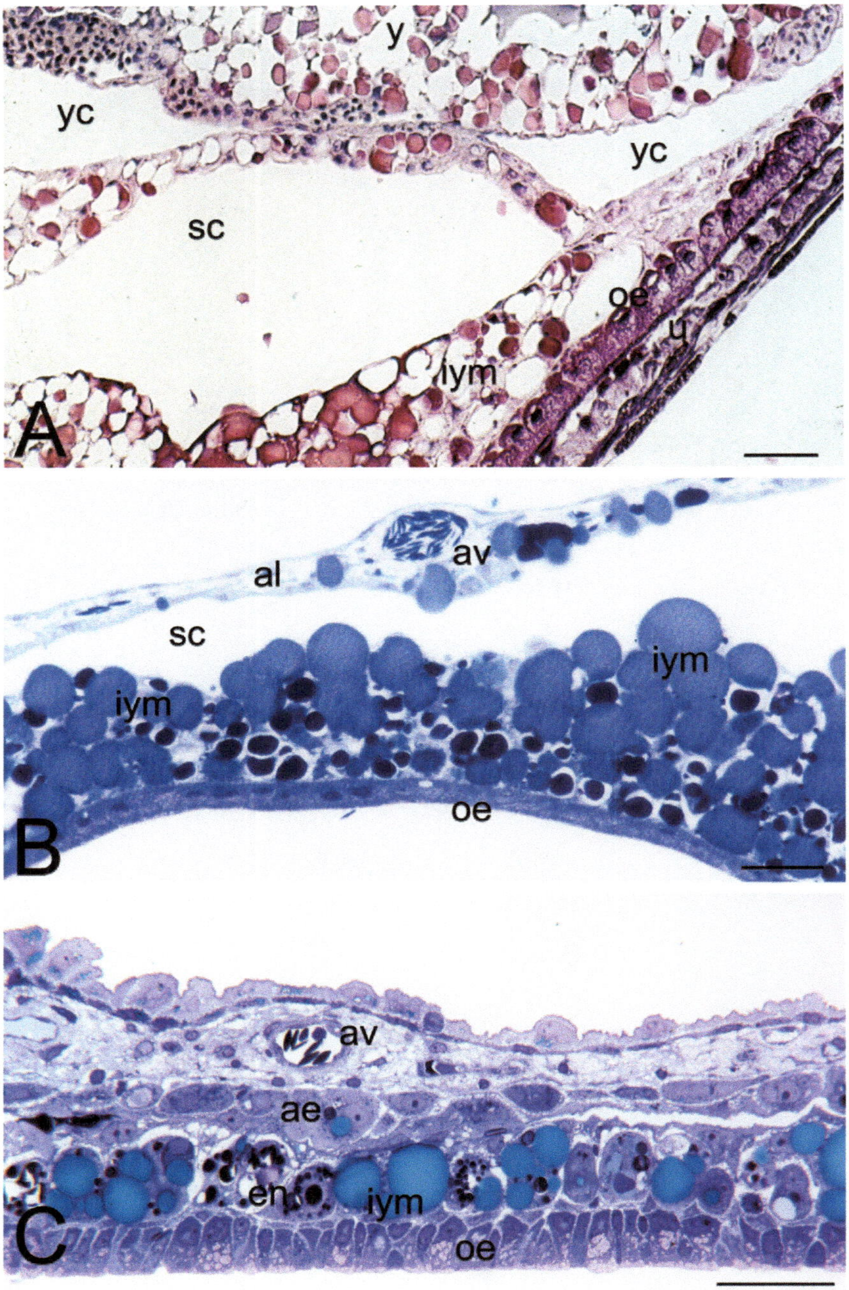

Figure 5.5 See text page 147 for caption.

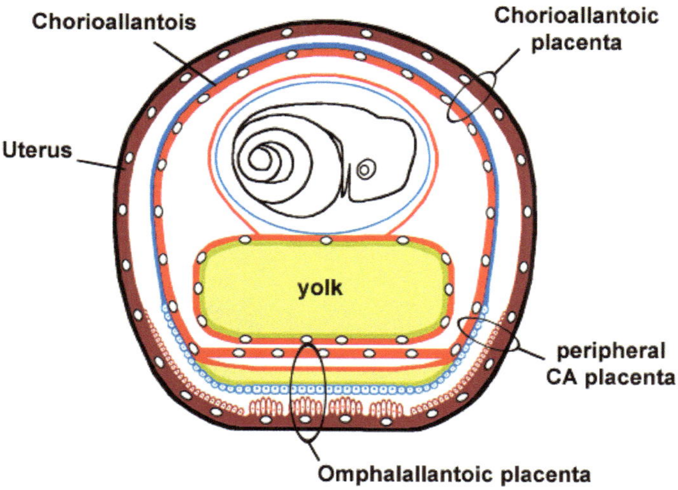

Figure 5.6 See text page **149** for caption.

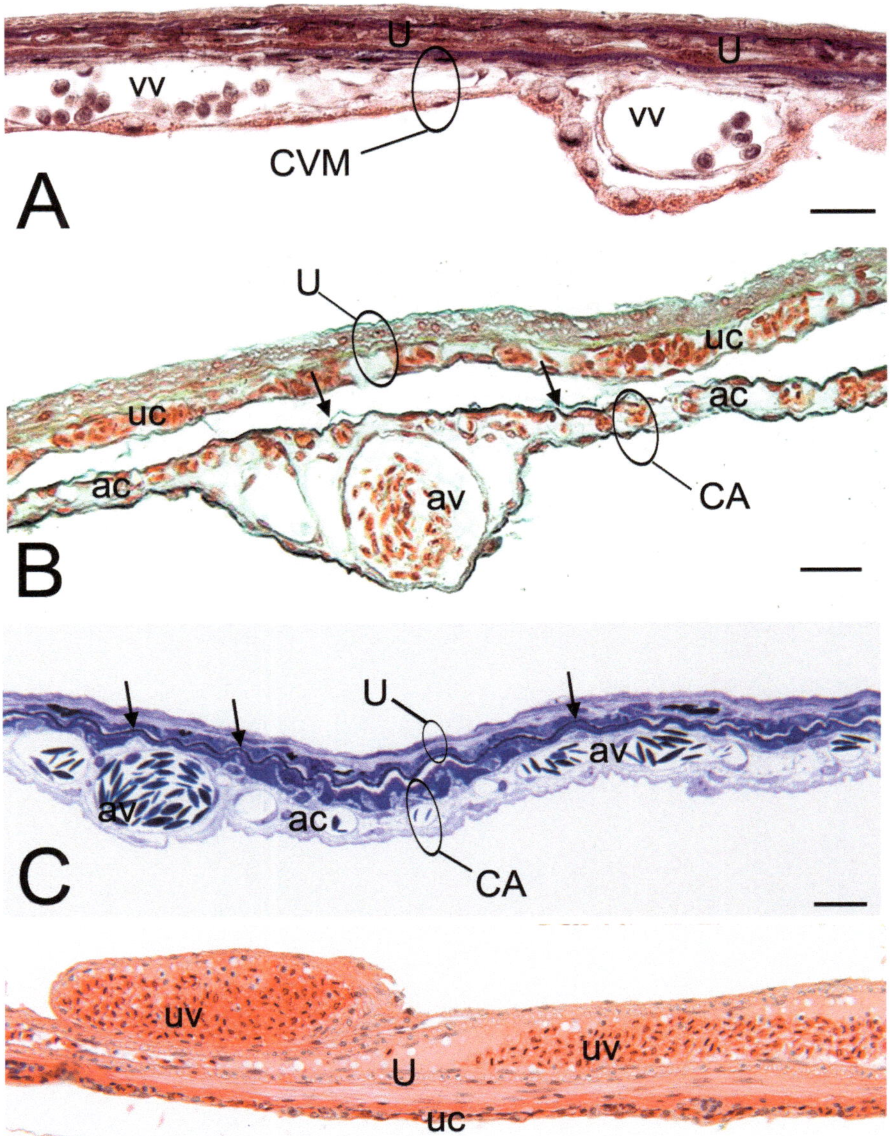

Figure 5.7 See text page 151 for caption.

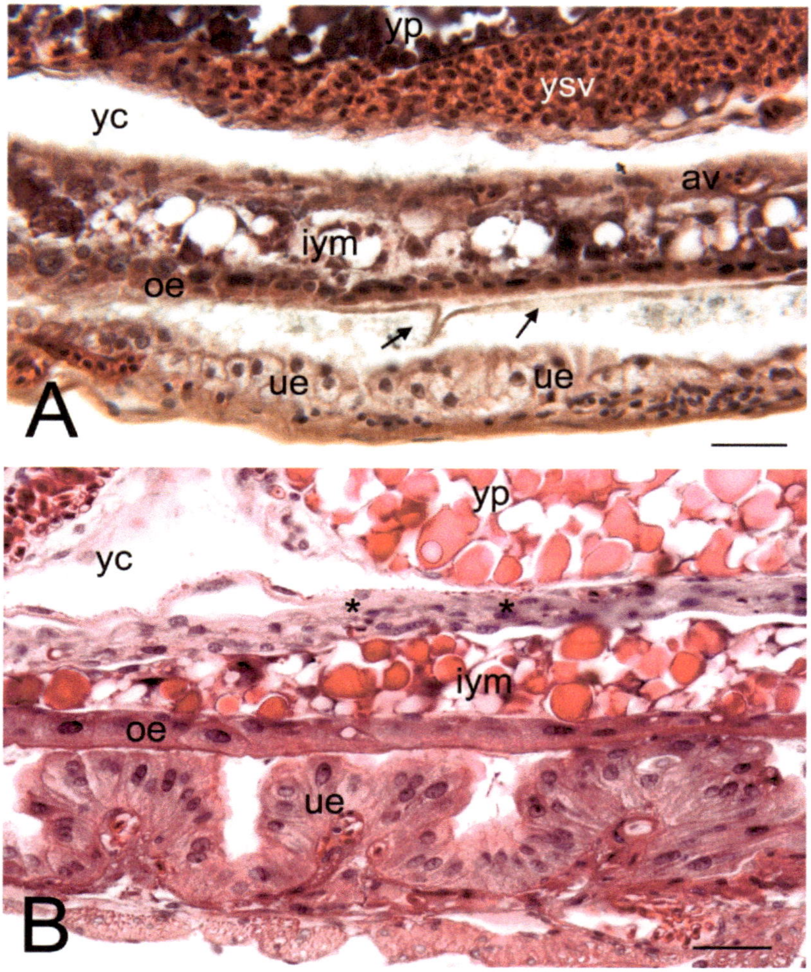

Figure 5.11 See text page **155** for caption.

CHAPTER 6

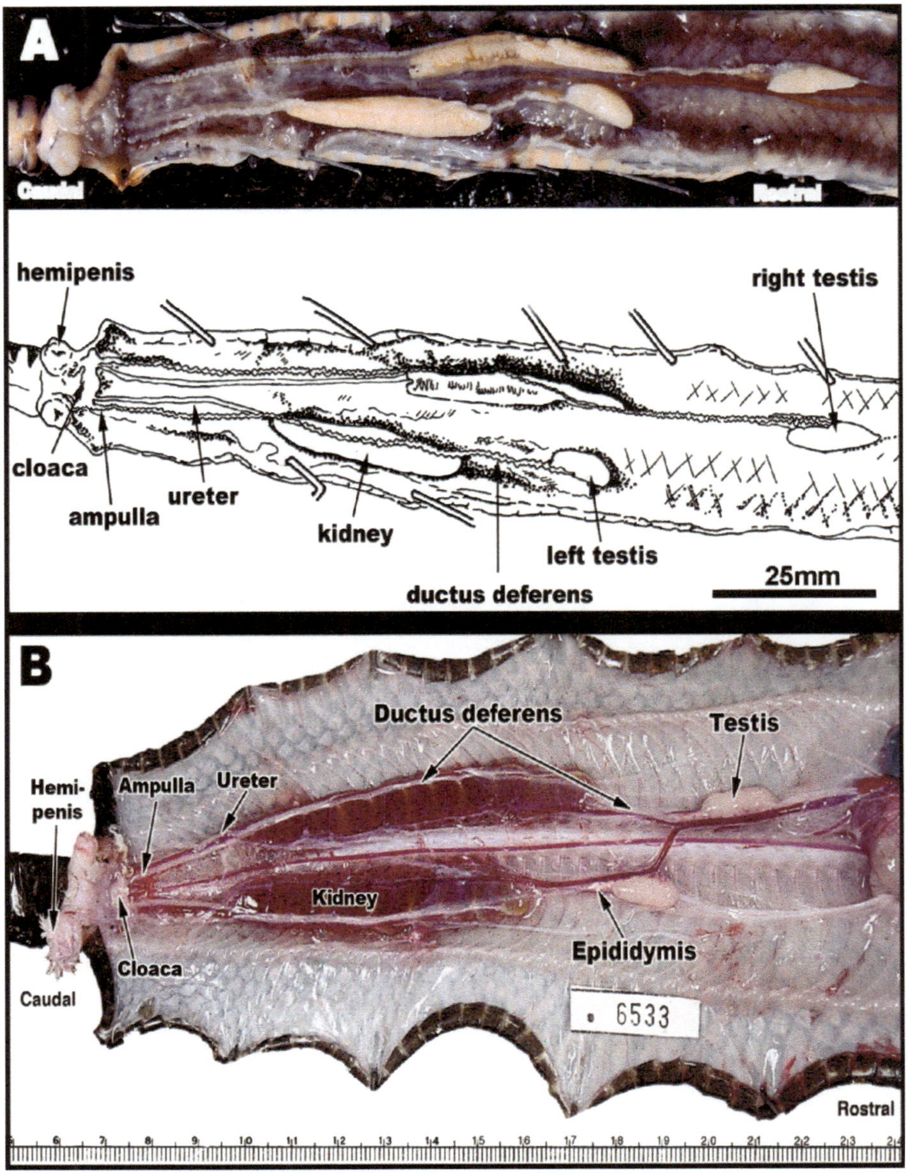

Figure 6.1 See text page **185** for caption.

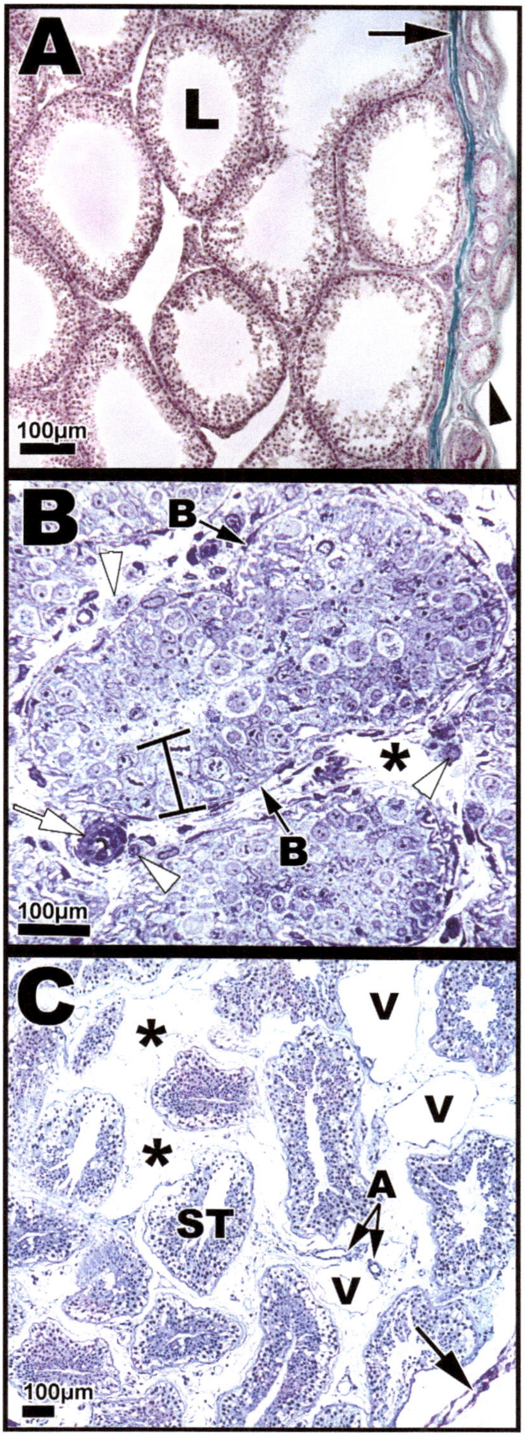

Figure 6.3 See text page 187 for caption.

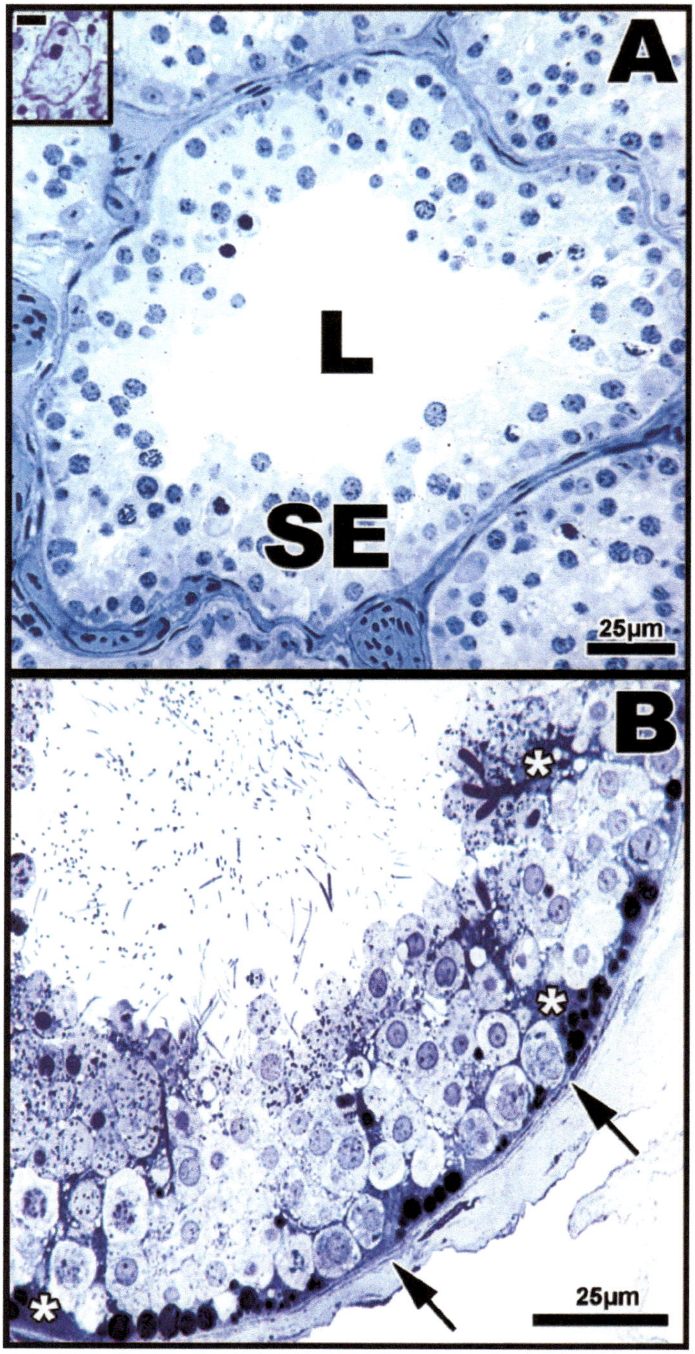

Figure 6.4 See text page 189 for caption.

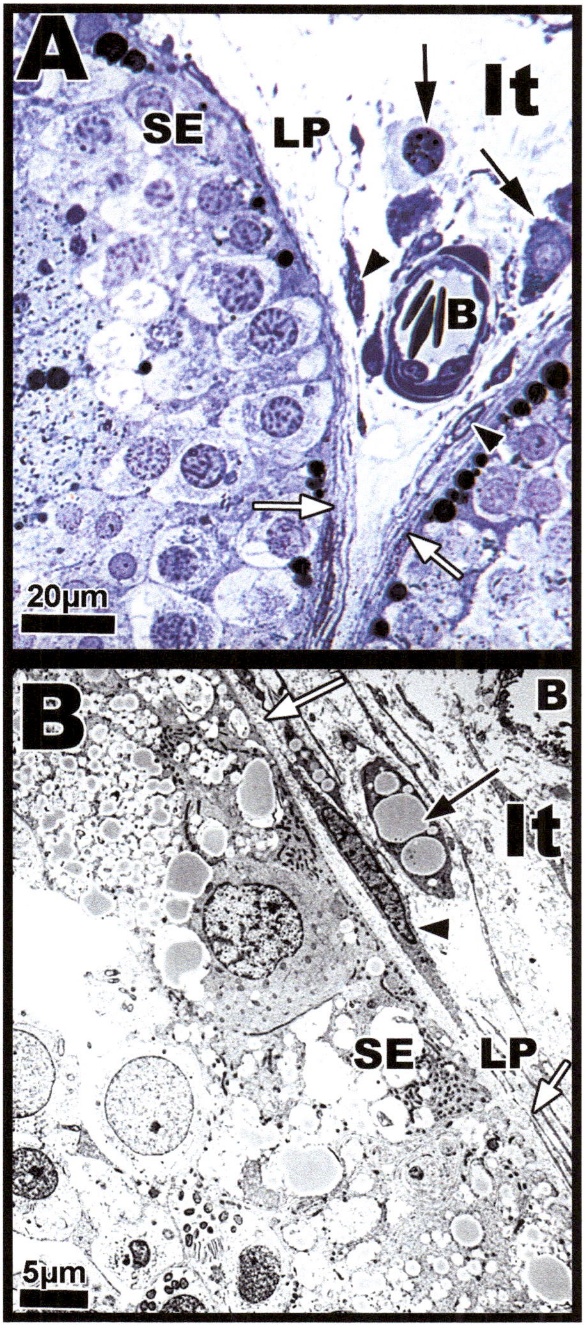

Figure 6.14 See text page 202 for caption.

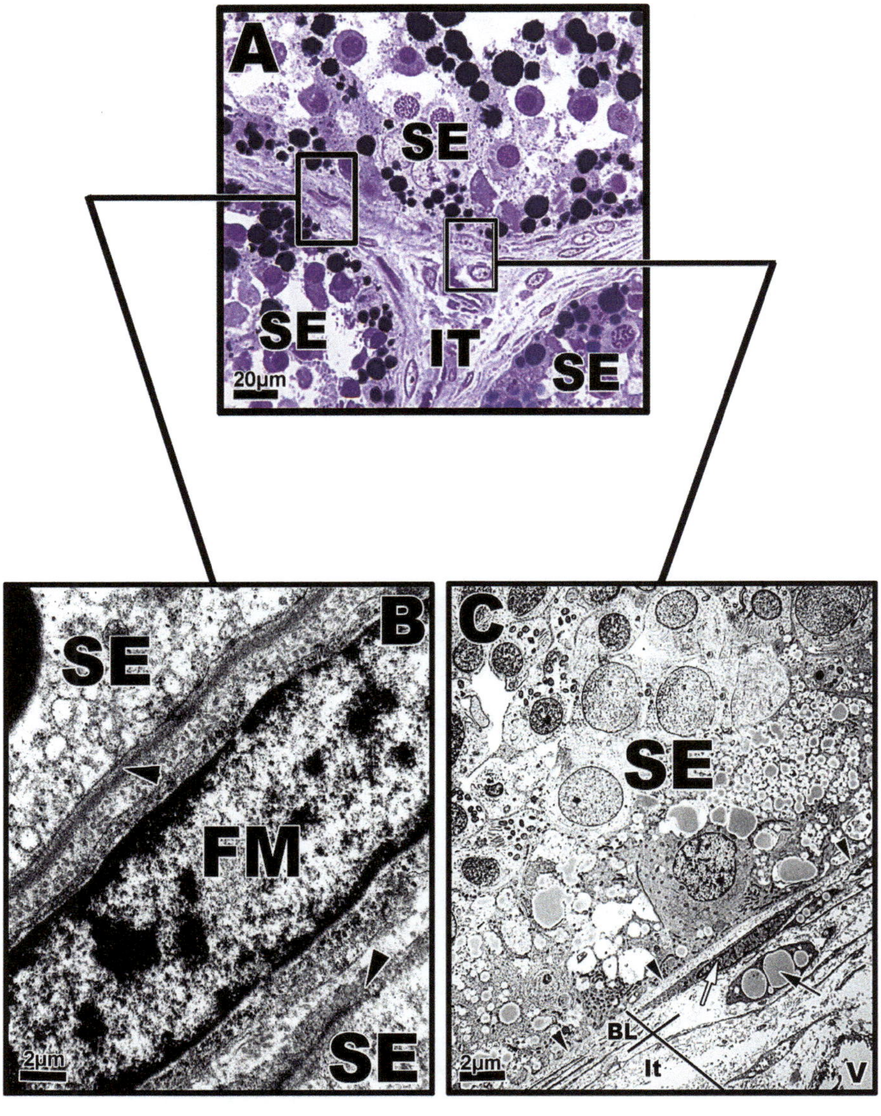

Figure 6.20 See text page 210 for caption.

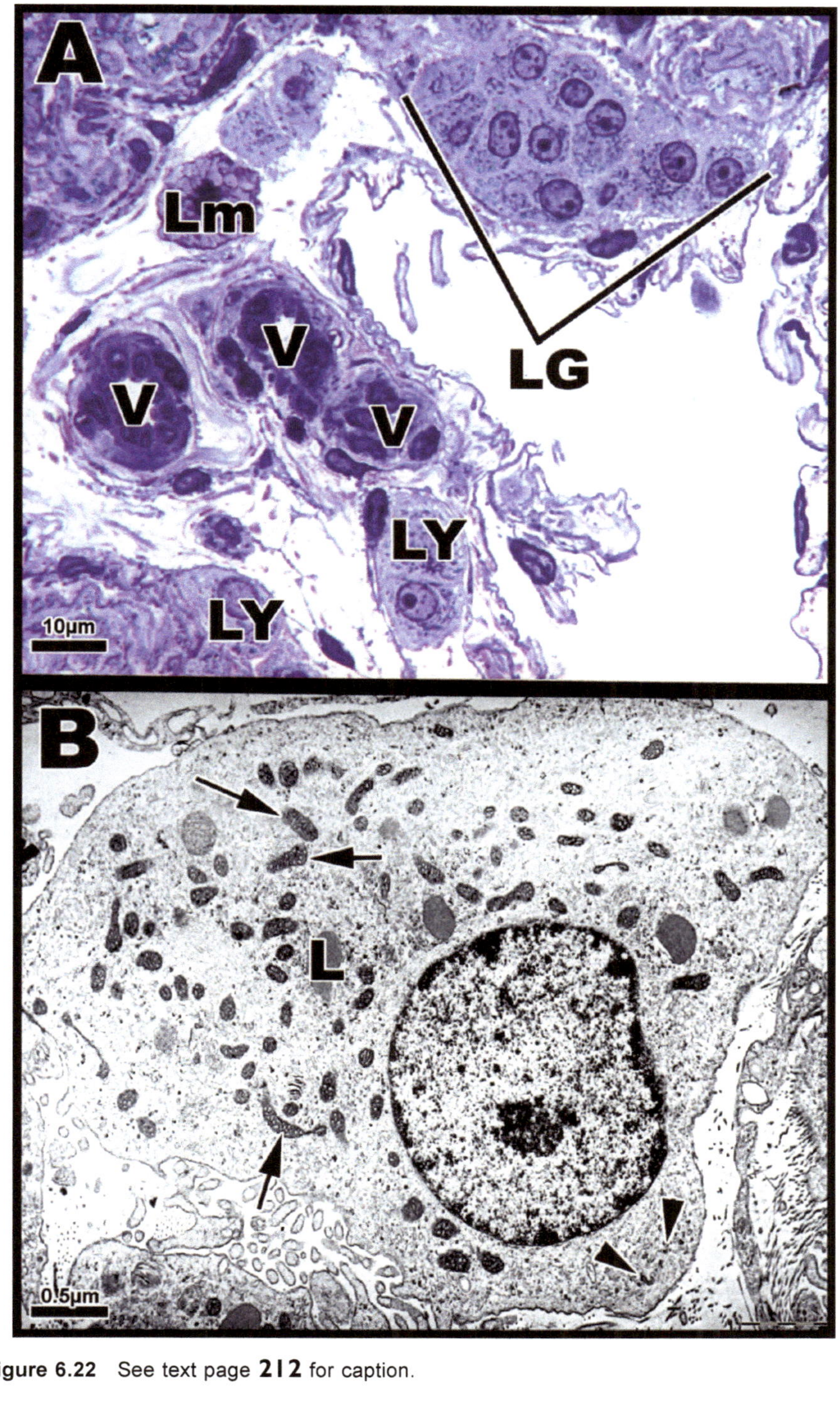

Figure 6.22 See text page 212 for caption.

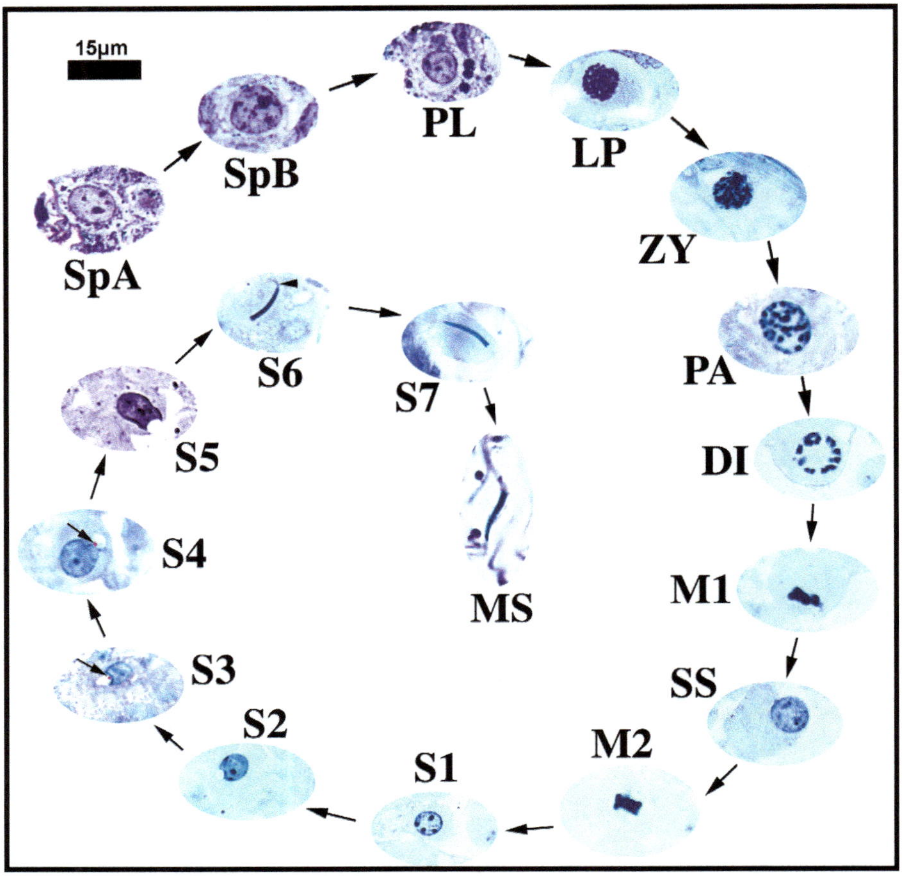

Figure 6.26 See text page 217 for caption.

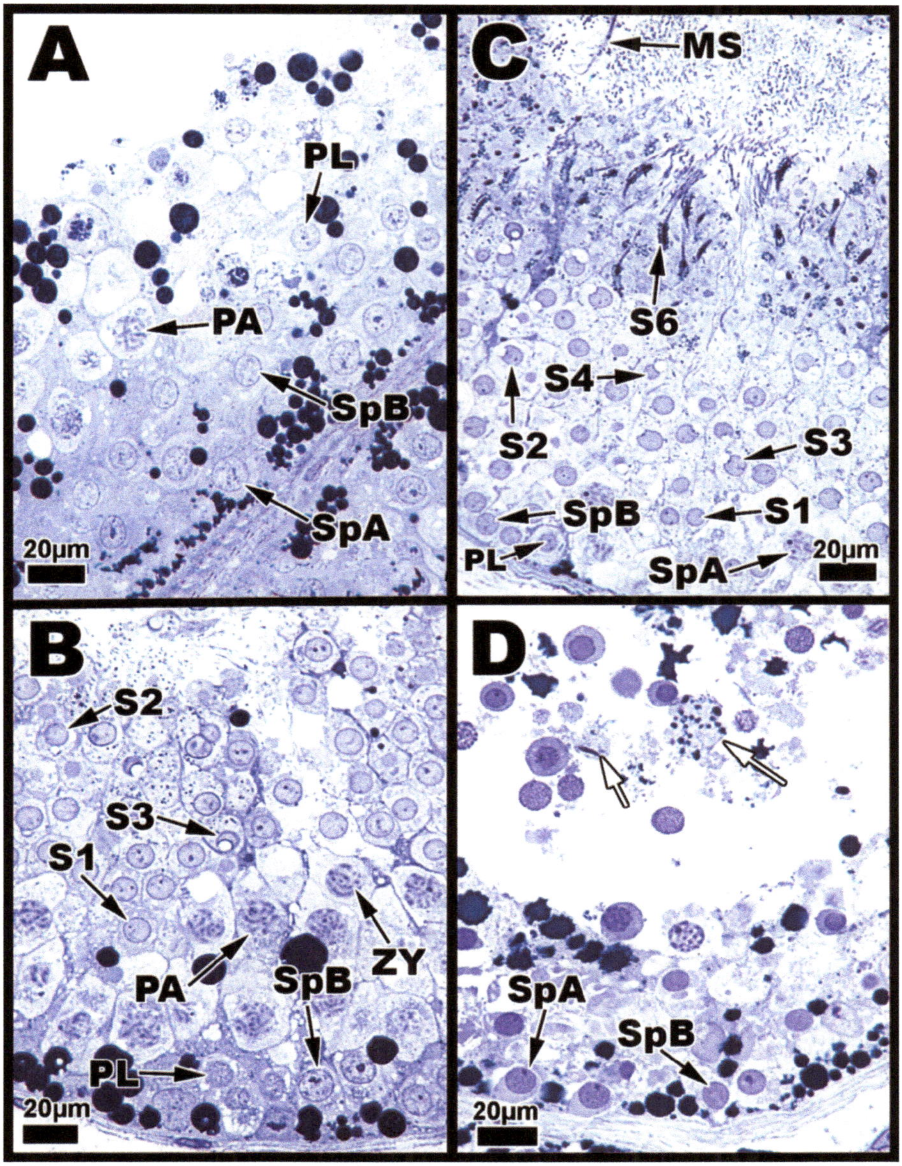

Figure 6.27 See text page **220** for caption.

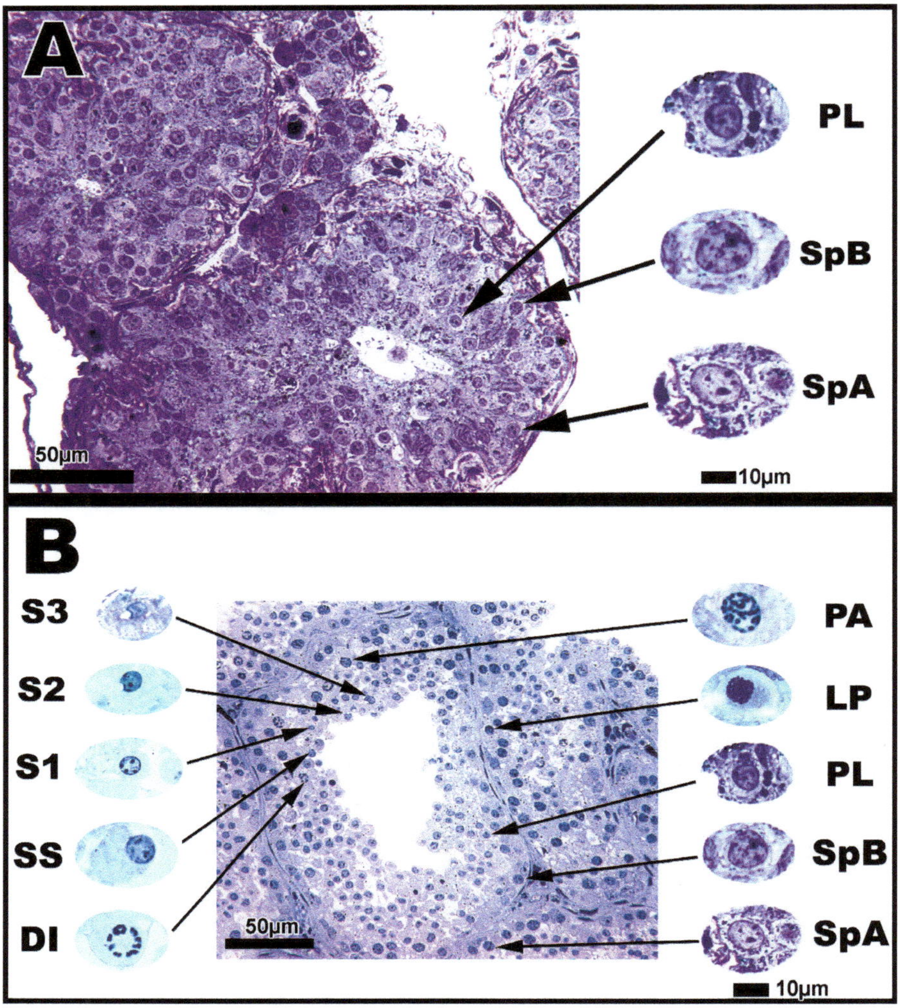

Figure 6.28 See text page **222** for caption.

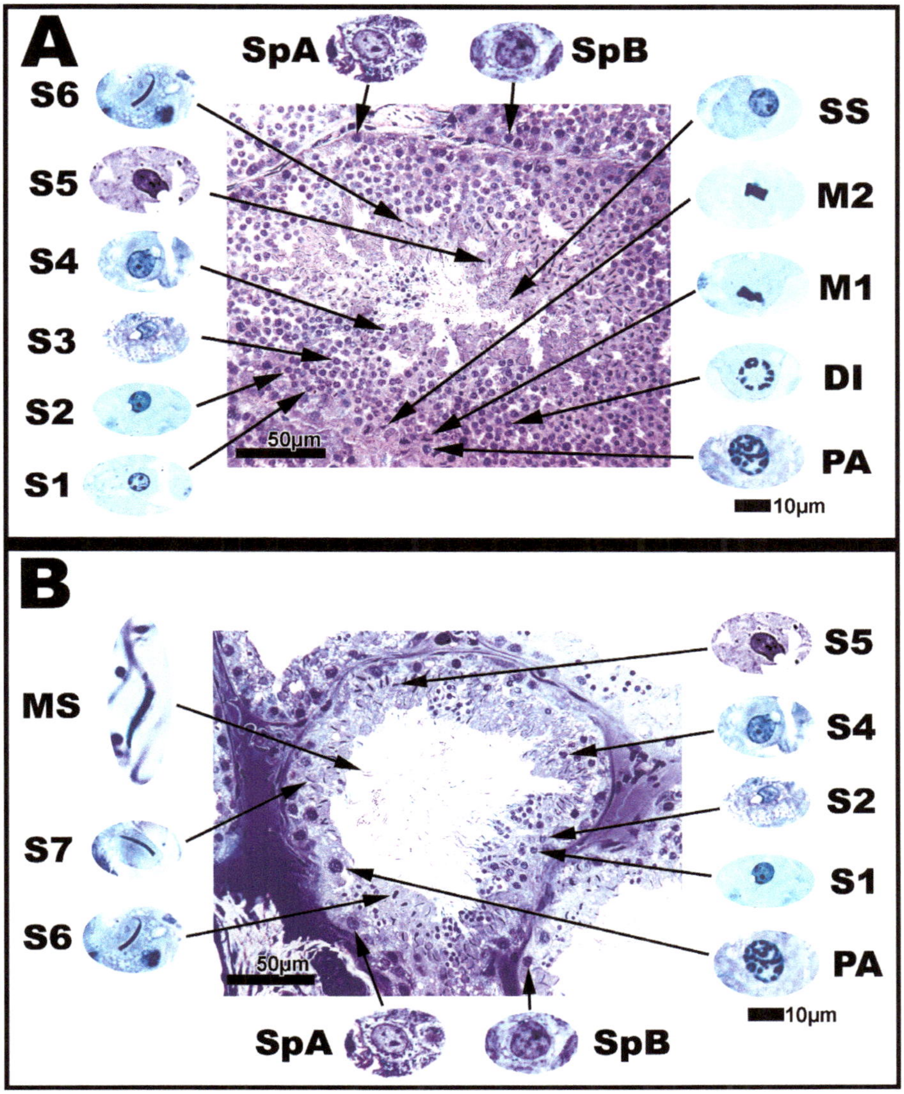

Figure 6.29 See text page 223 for caption.

CHAPTER 8

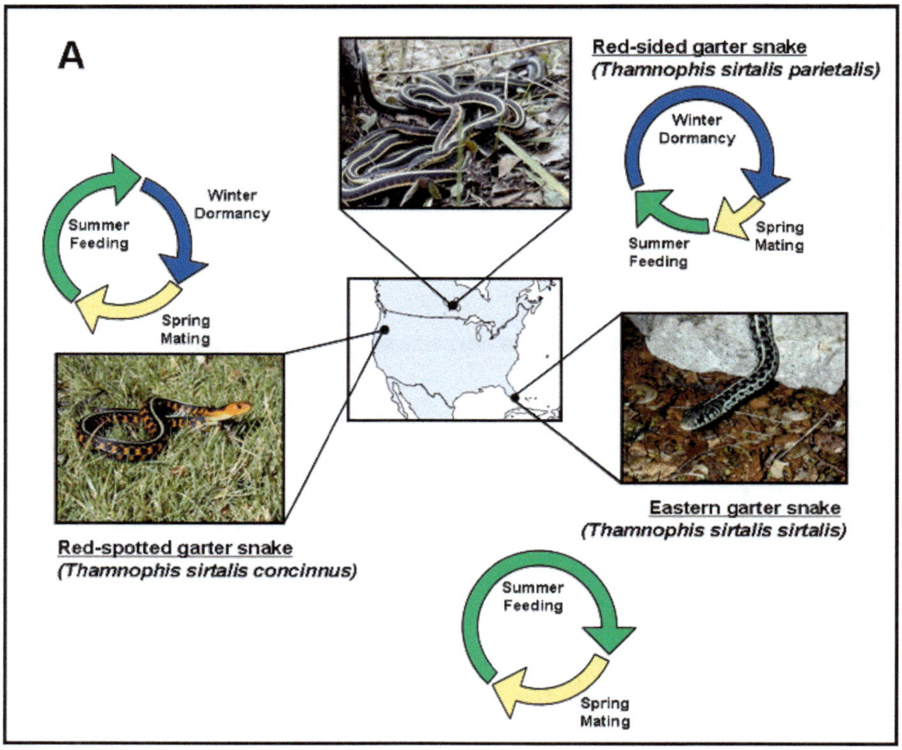

Figure 8.9A See text page 309 for caption.

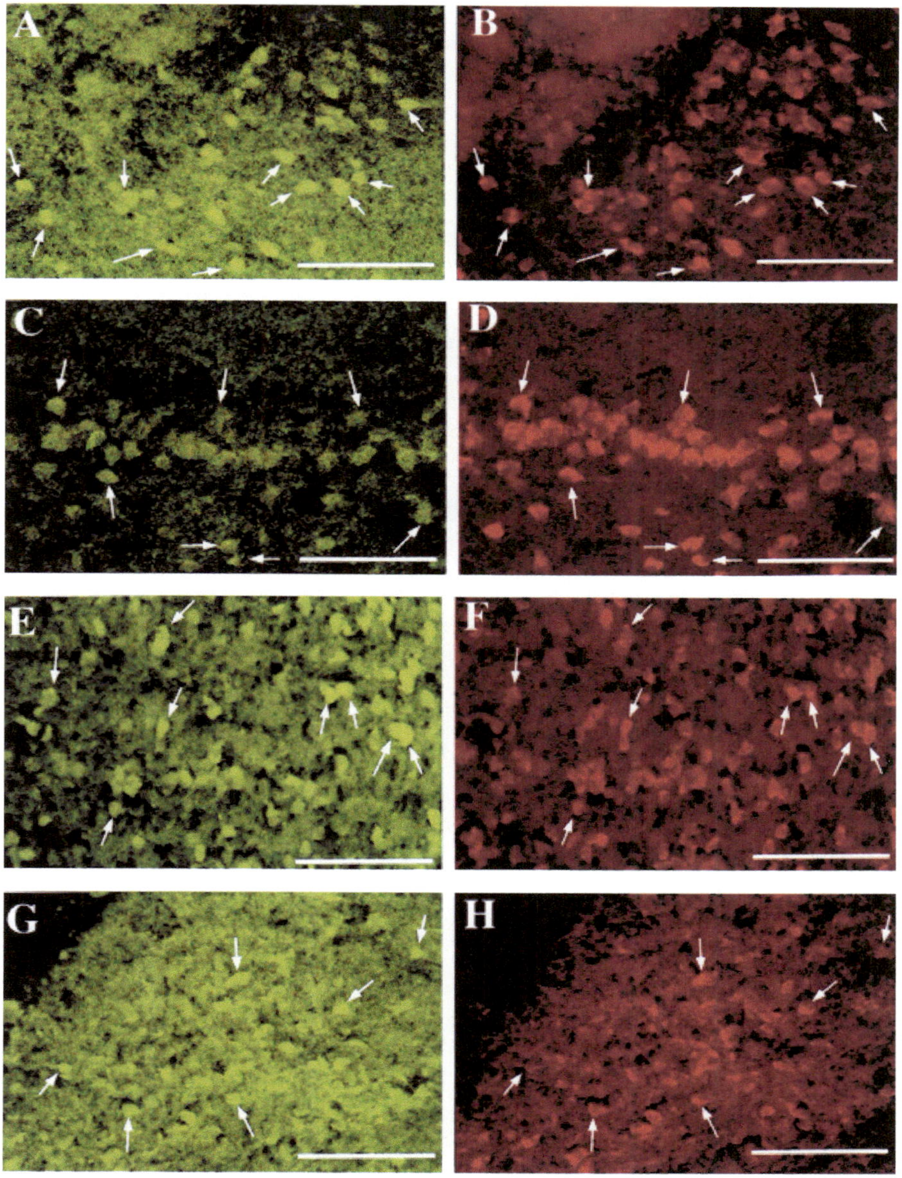

Figure 8.17 See text page 326 for caption.

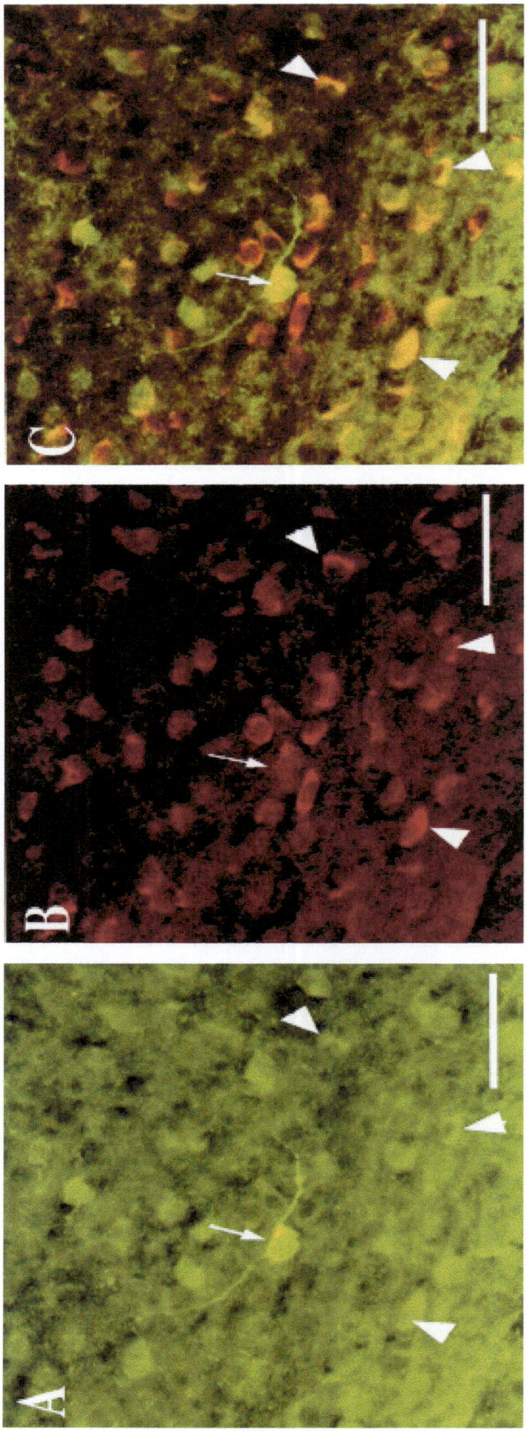

Figure 8.18 See text page **327** for caption.

CHAPTER 9

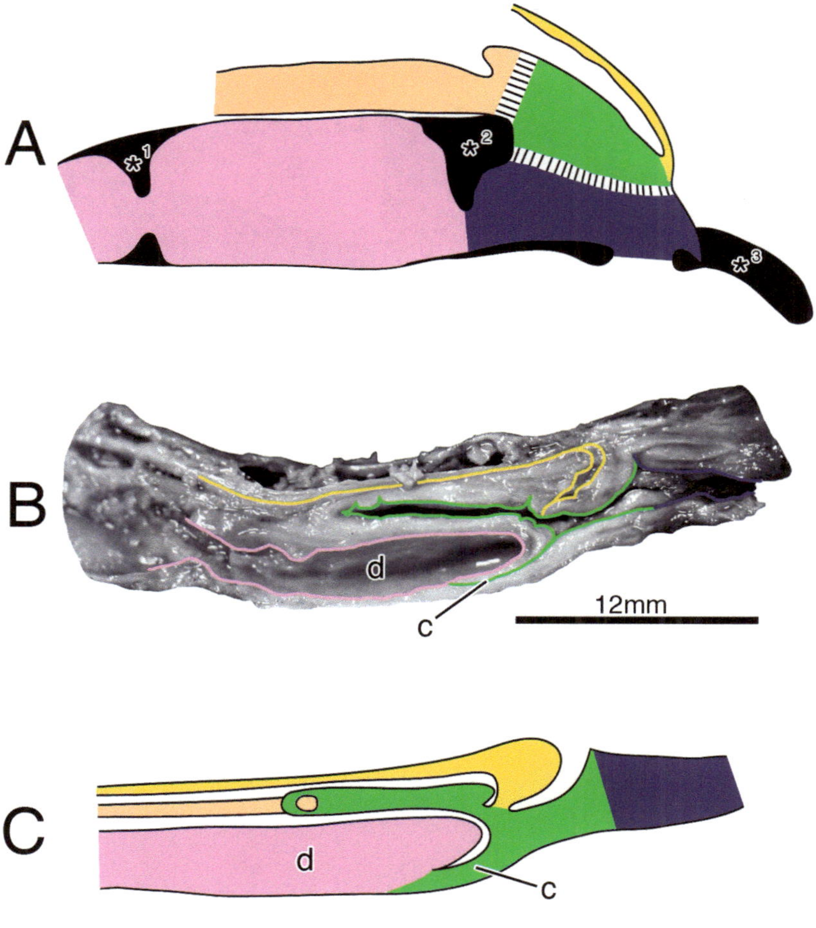

Figure 9.5 See text page **368** for caption.

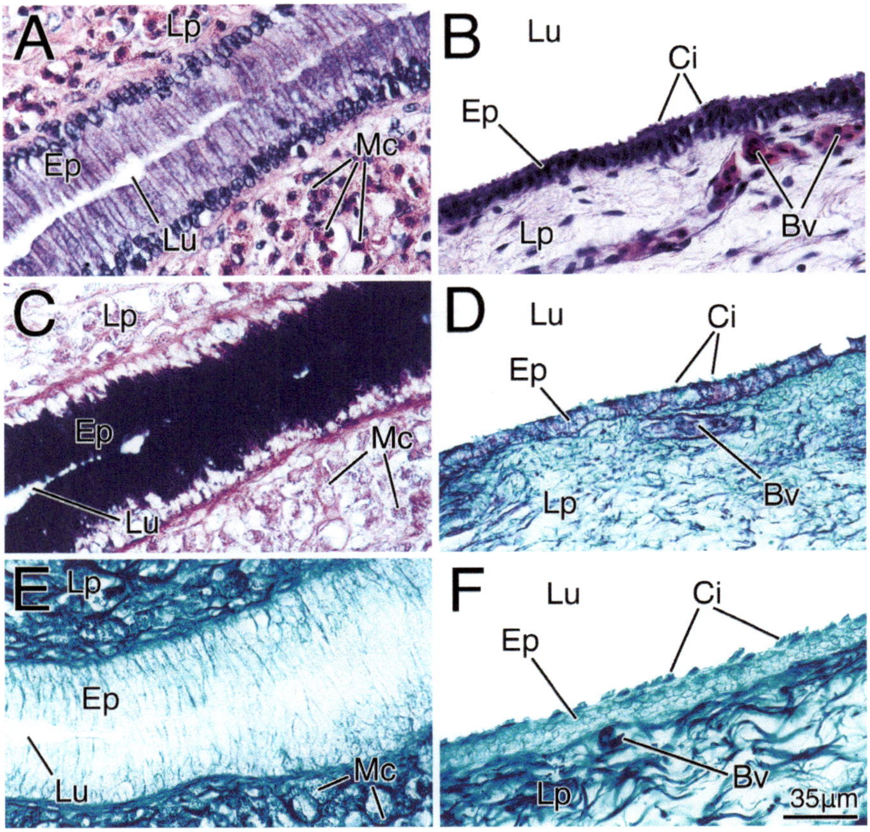

Figure 9.7 See text page **384** for caption.

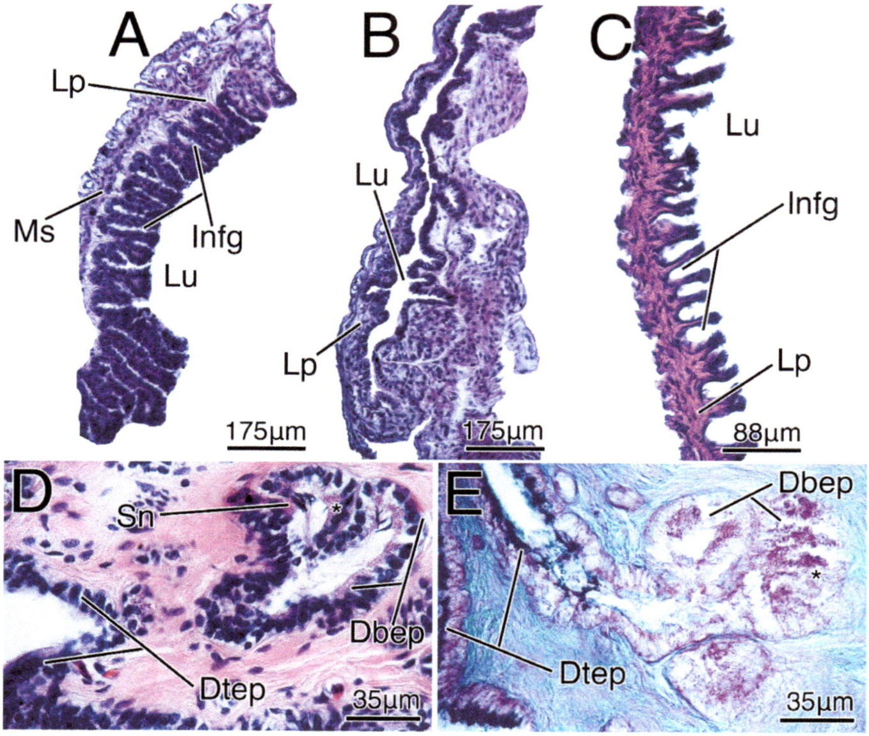

Figure 9.9 See text page **392** for caption.

CHAPTER 10

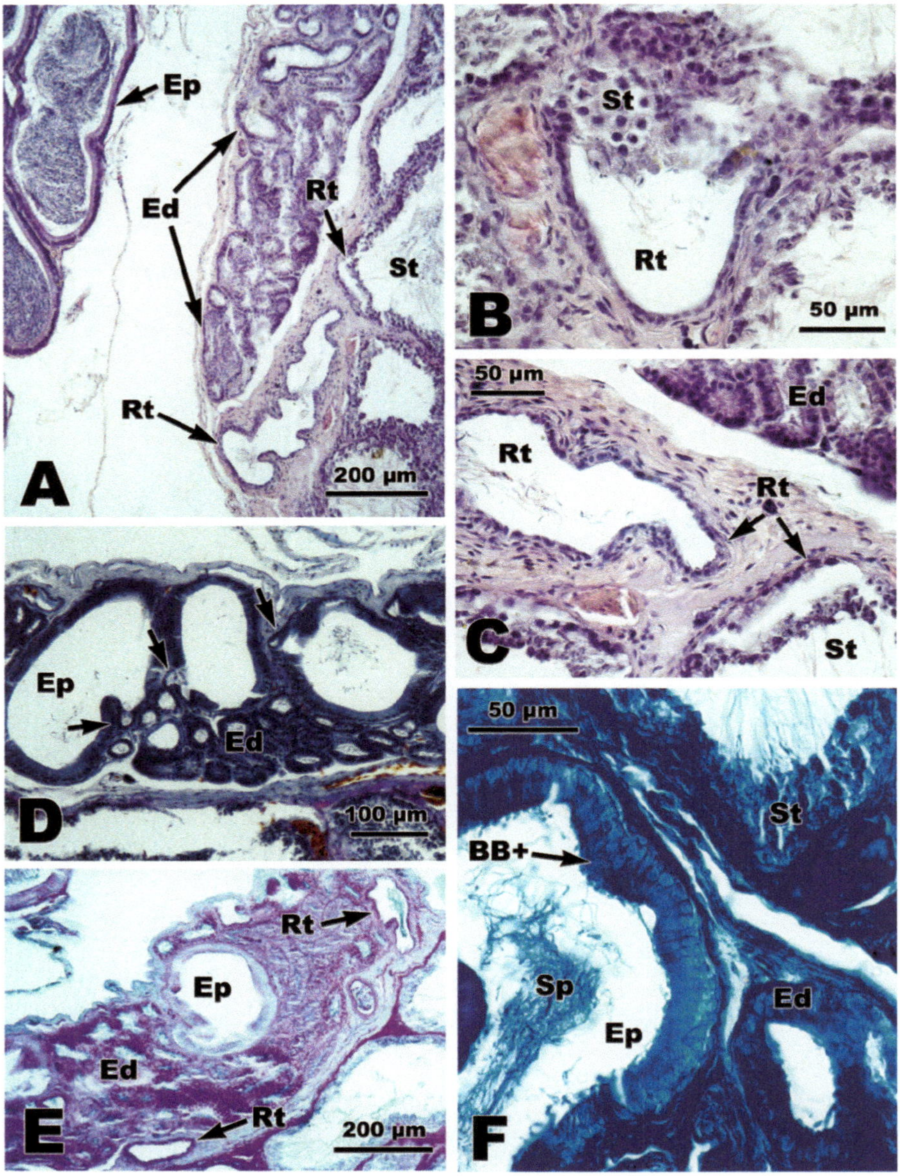

Figure 10.1 See text page 414 for caption.

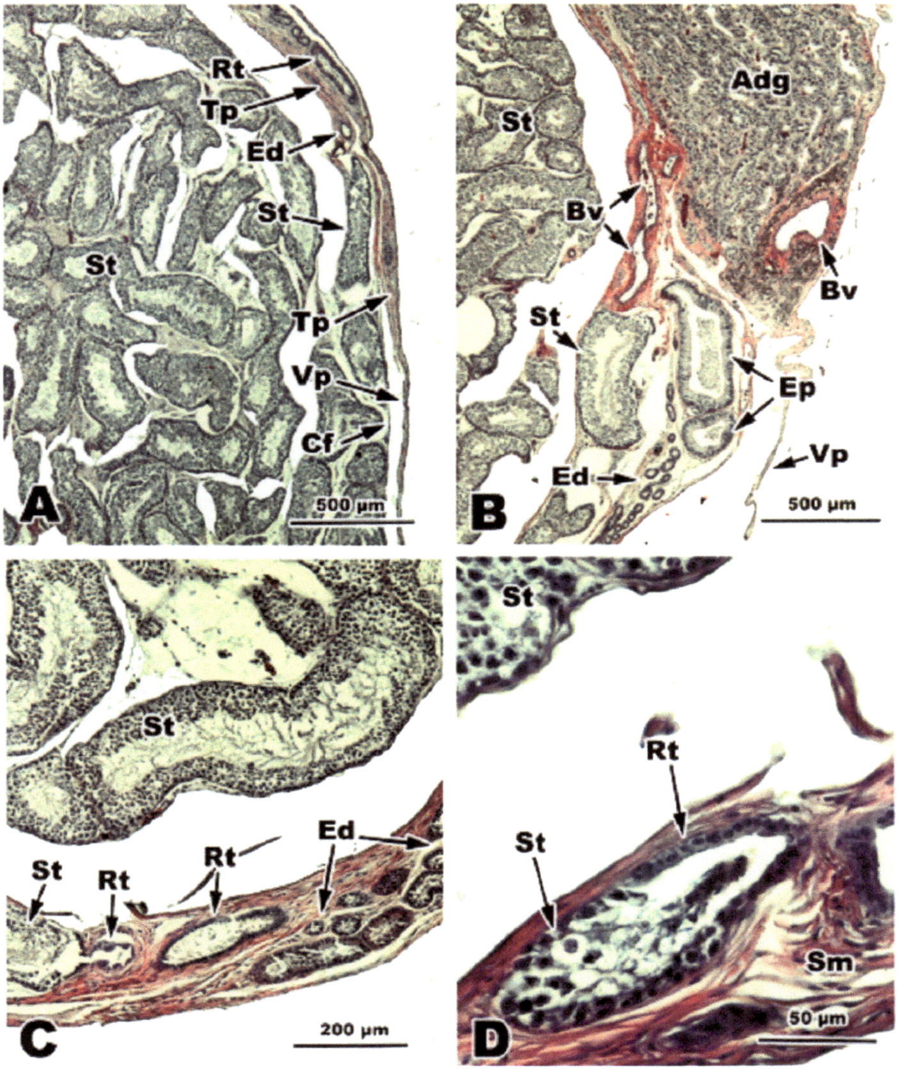

Figure 10.2 See text page **415** for caption.

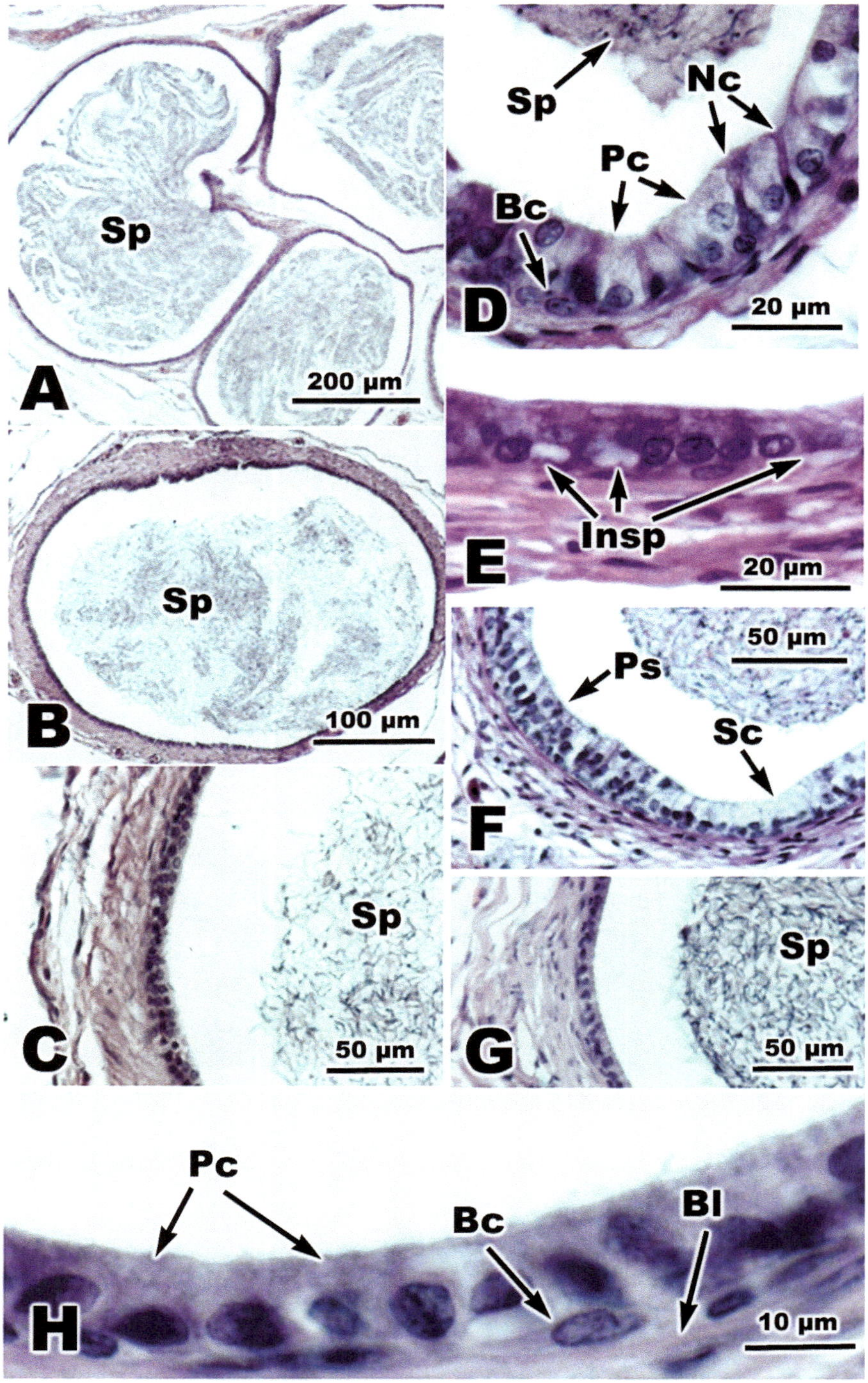

Figure 10.14 See text page **426–427** for caption.

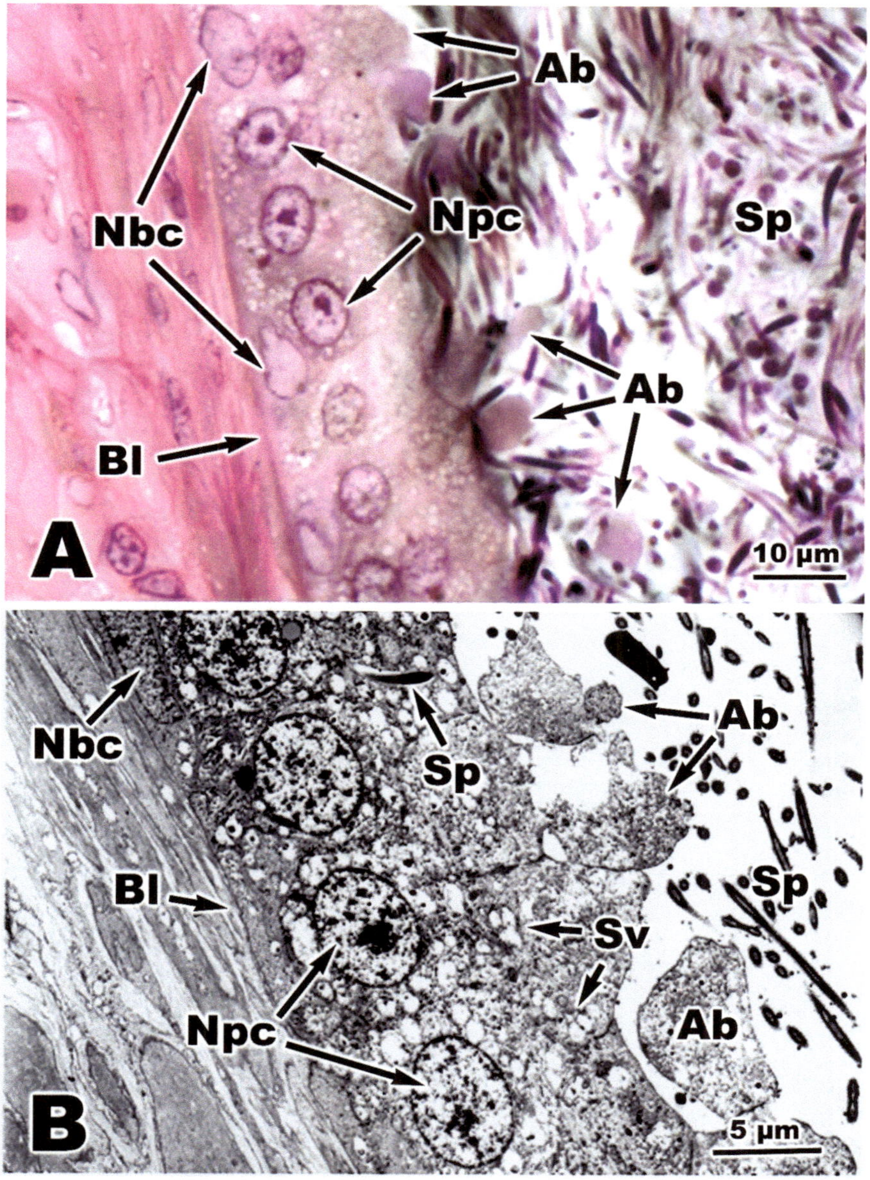

Figure 10.15 See text page **427** for caption.

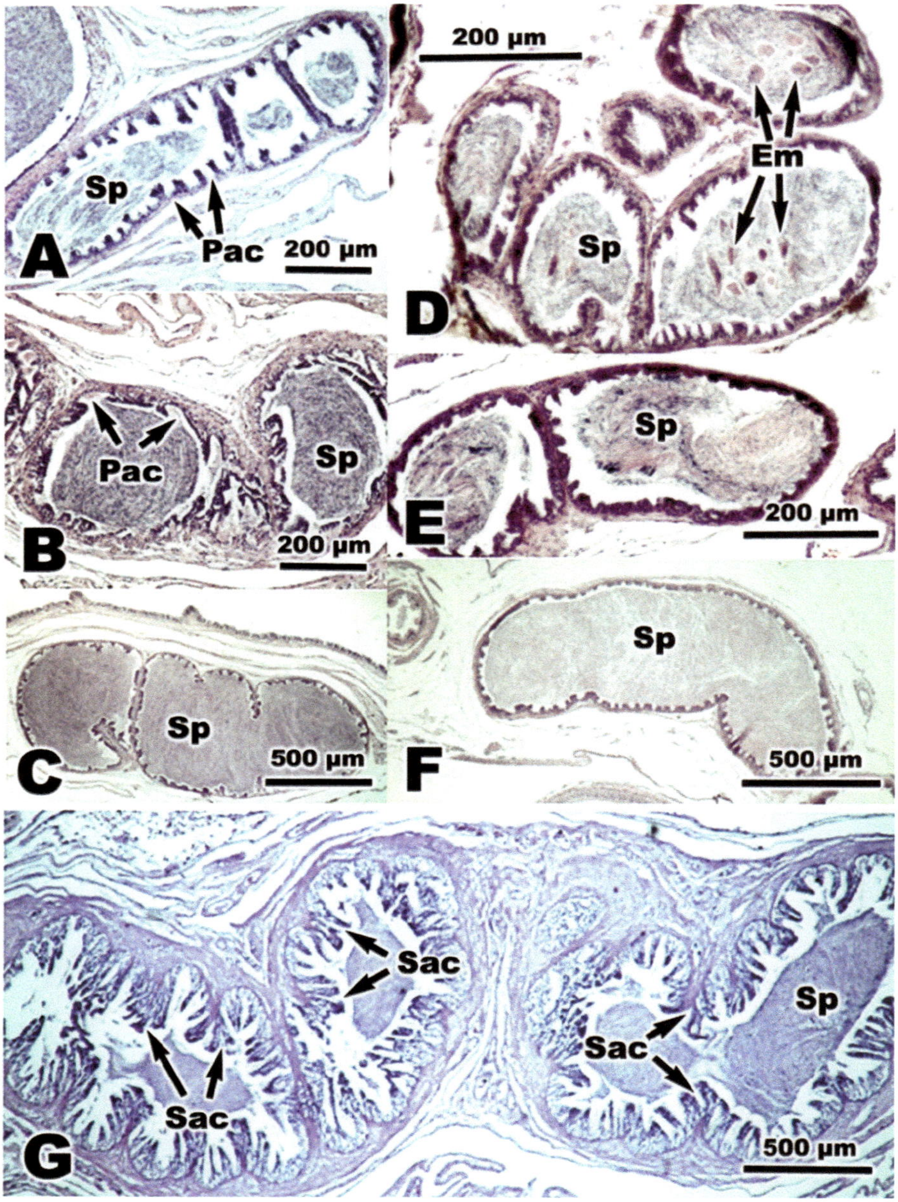

Figure 10.16 See text page **430** for caption.

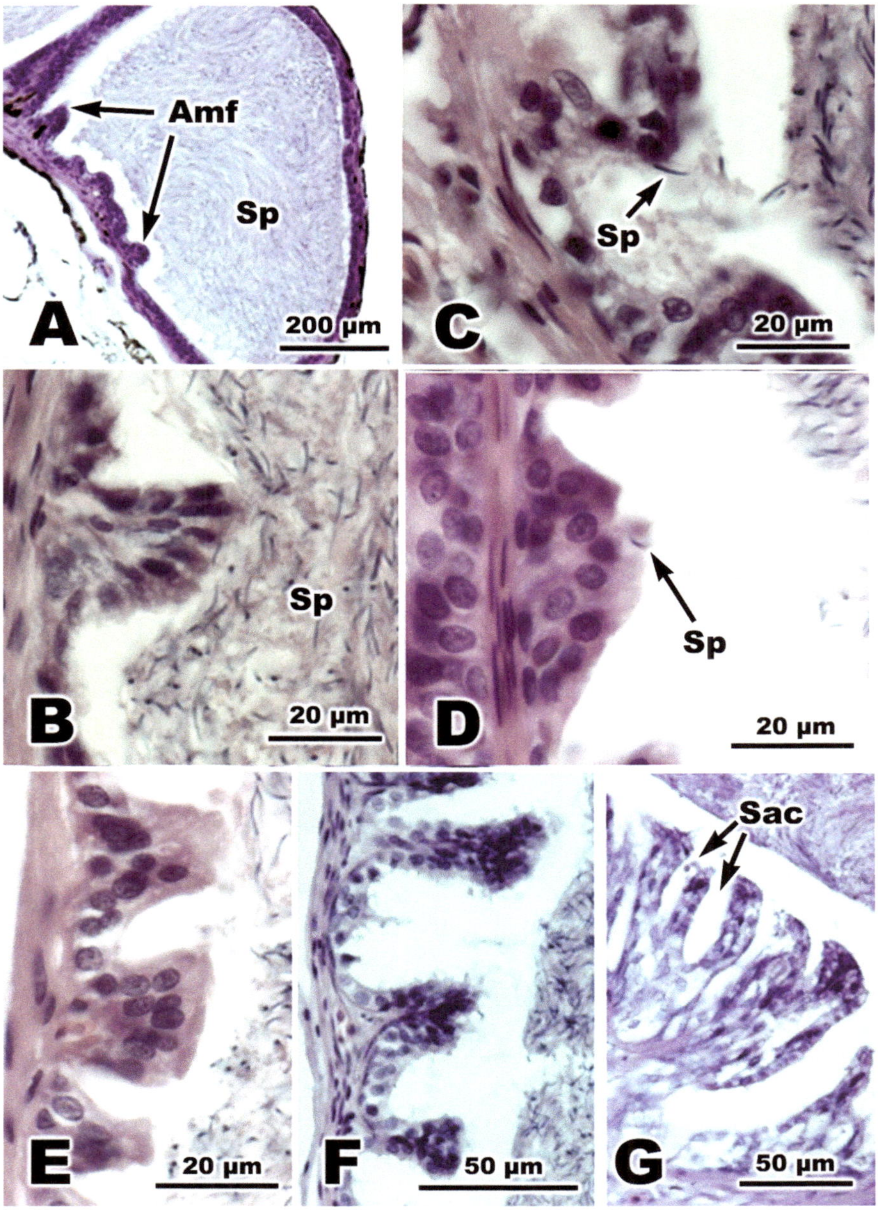

Figure 10.17 See text page **431** for caption.

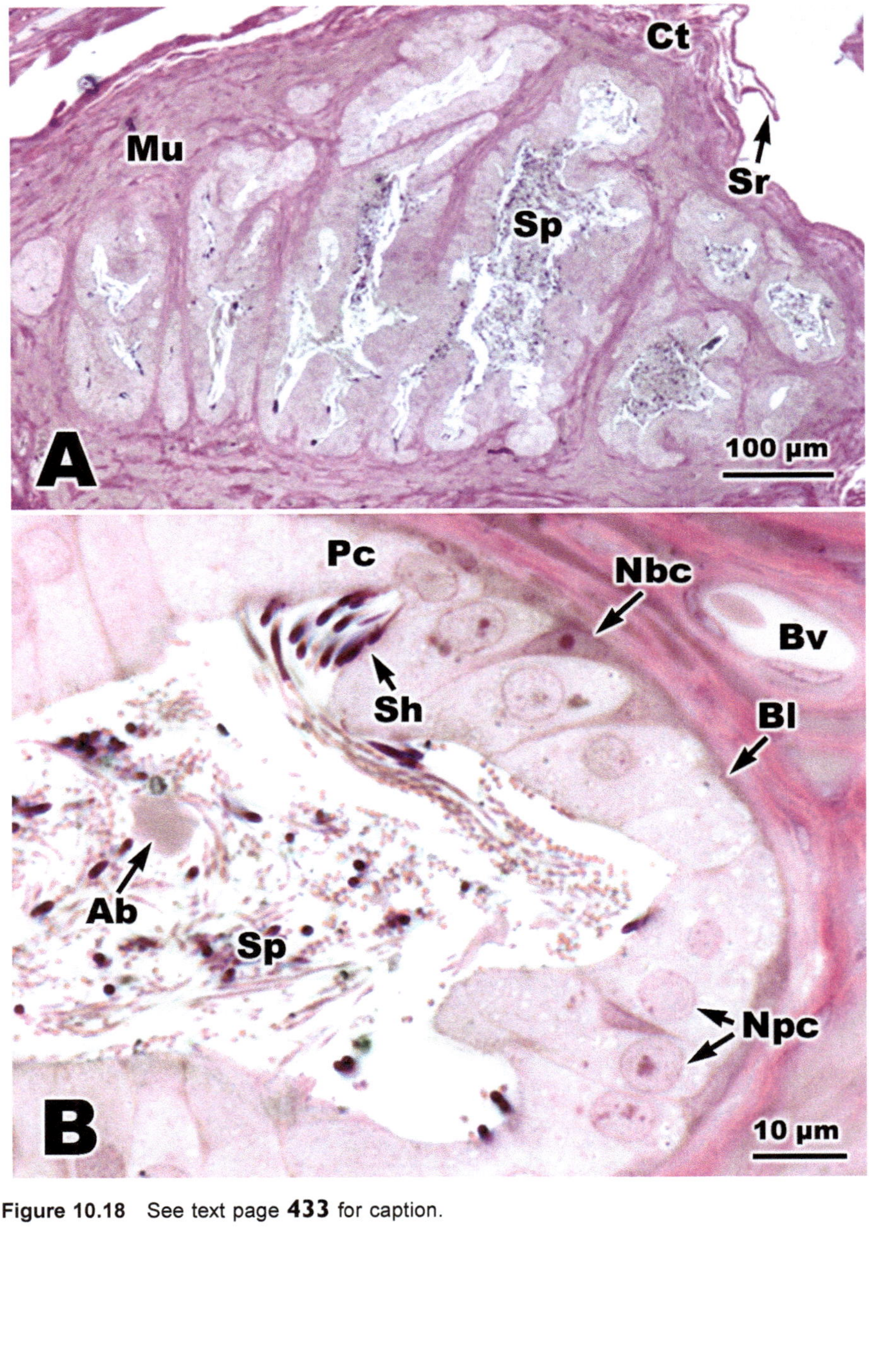

Figure 10.18 See text page **433** for caption.

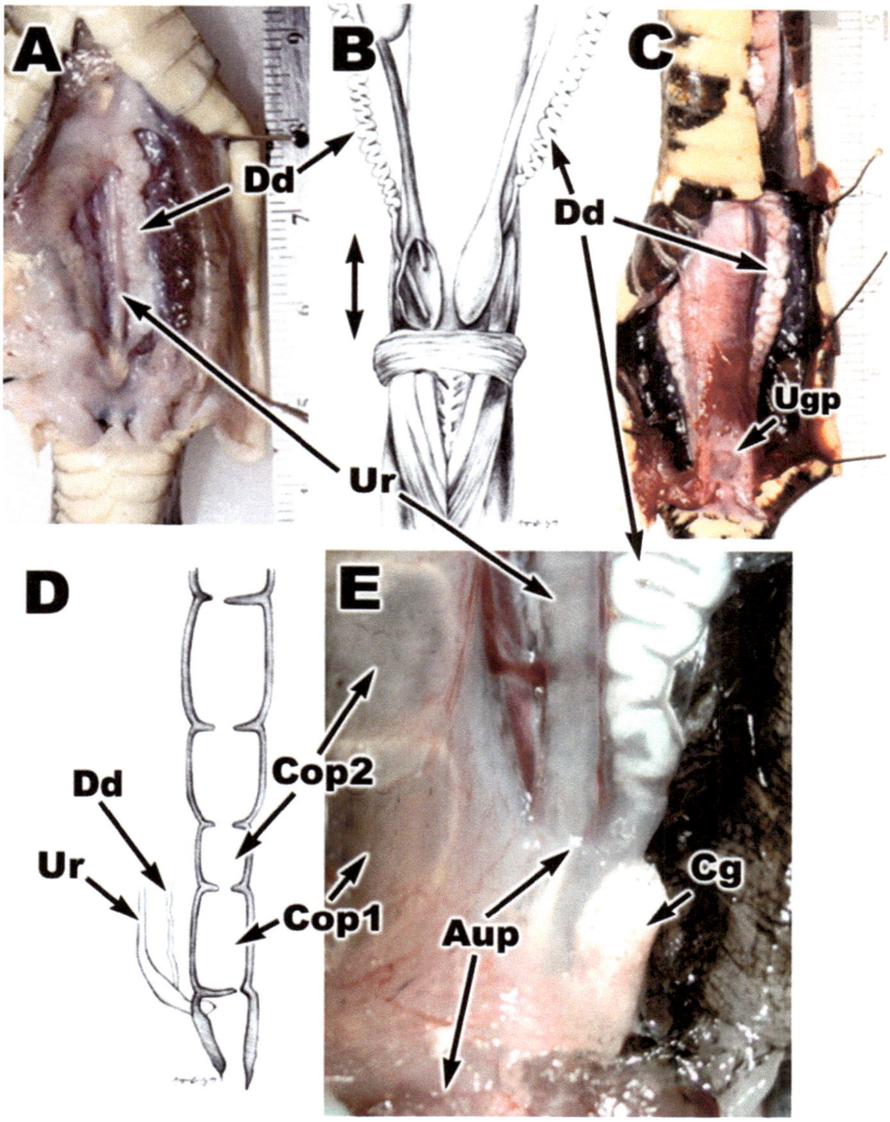

Figure 10.20 See text page **436** for caption.

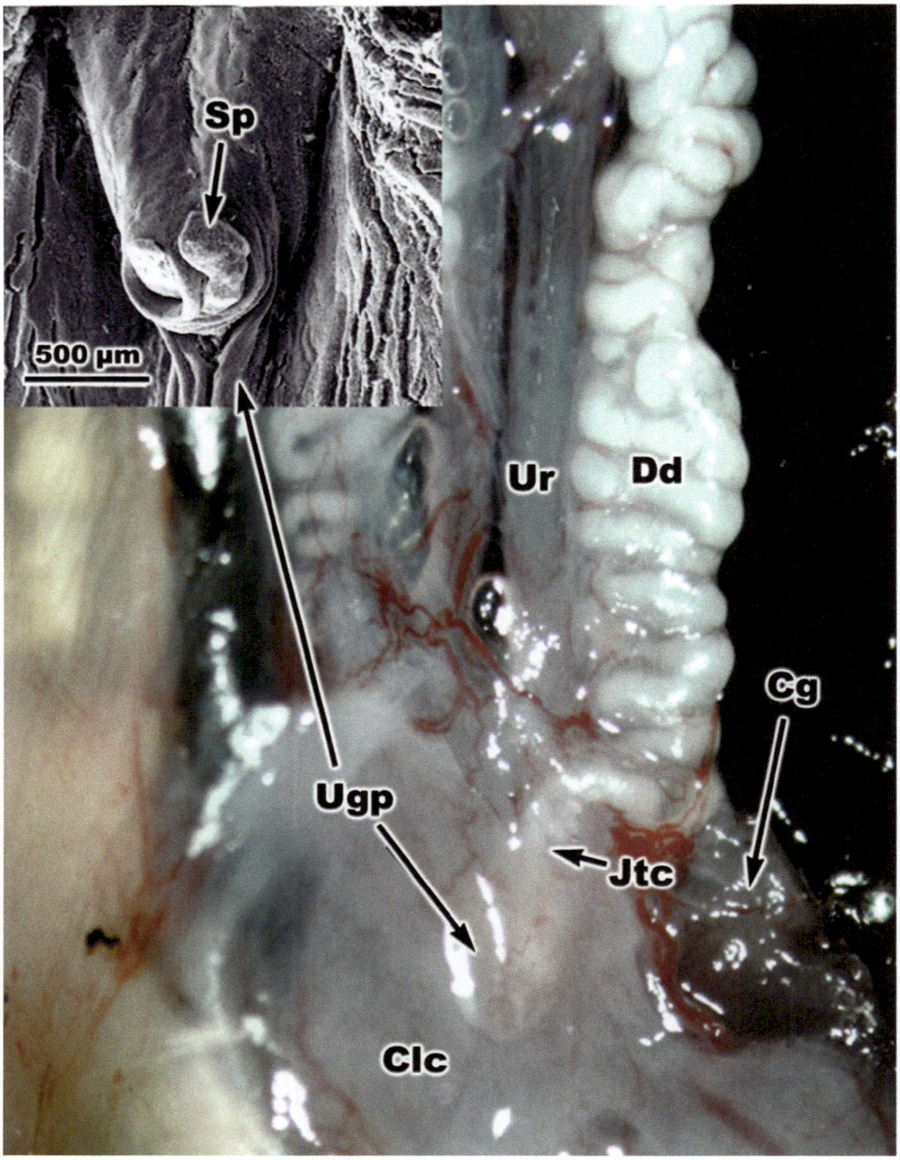

Figure 10.21 See text page **437** for caption.

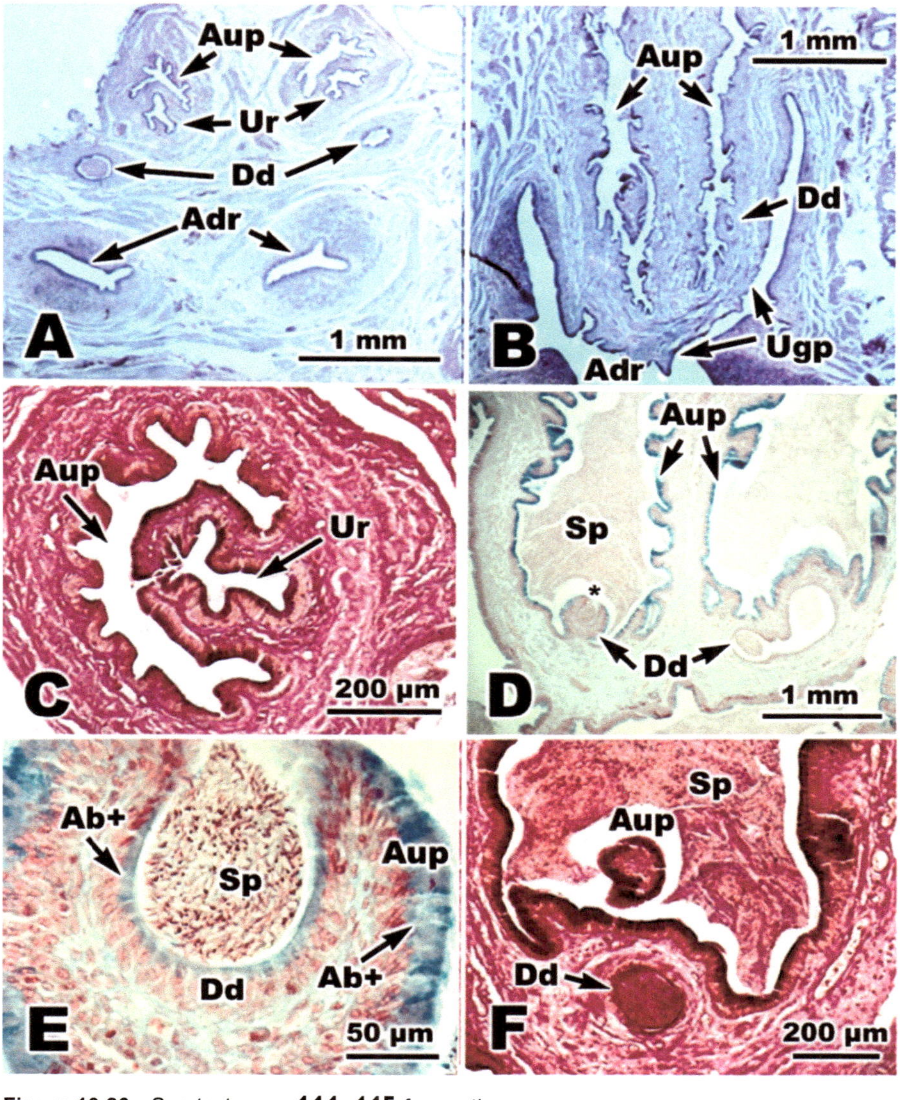

Figure 10.26 See text page **444–445** for caption.

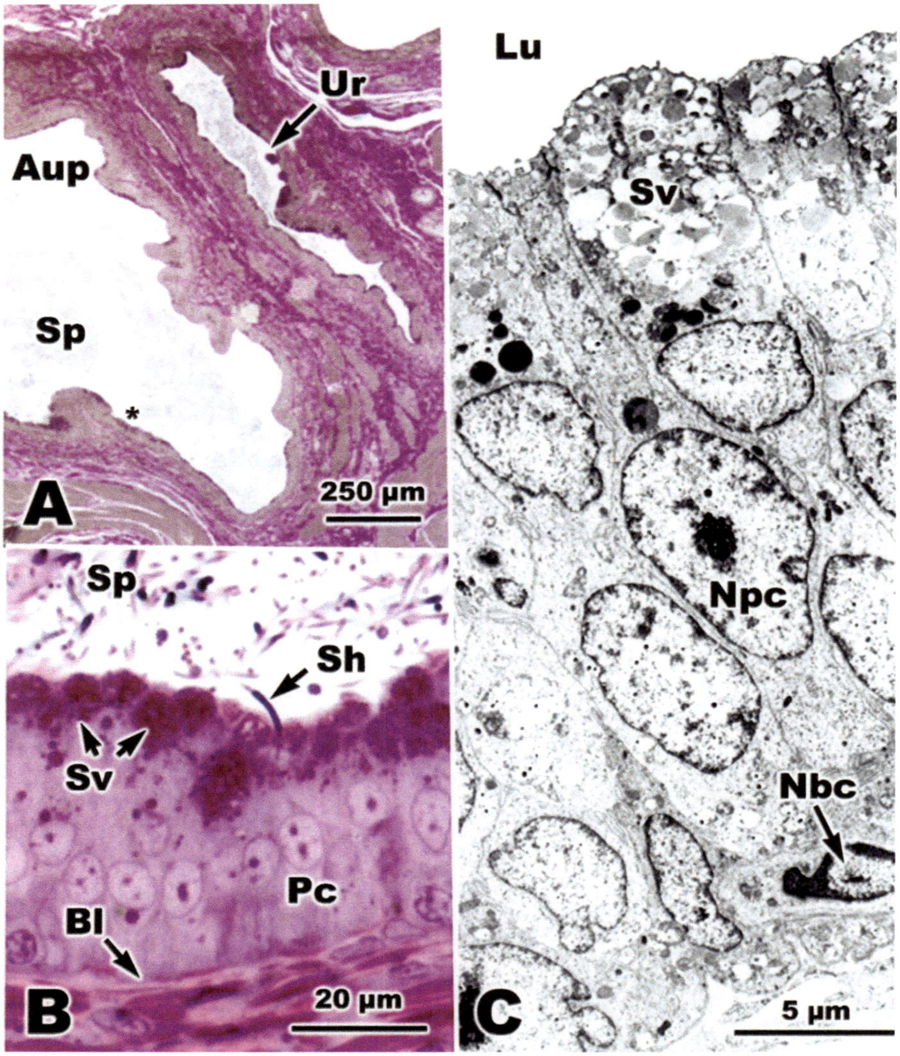

Figure 10.27 See text page **446** for caption.

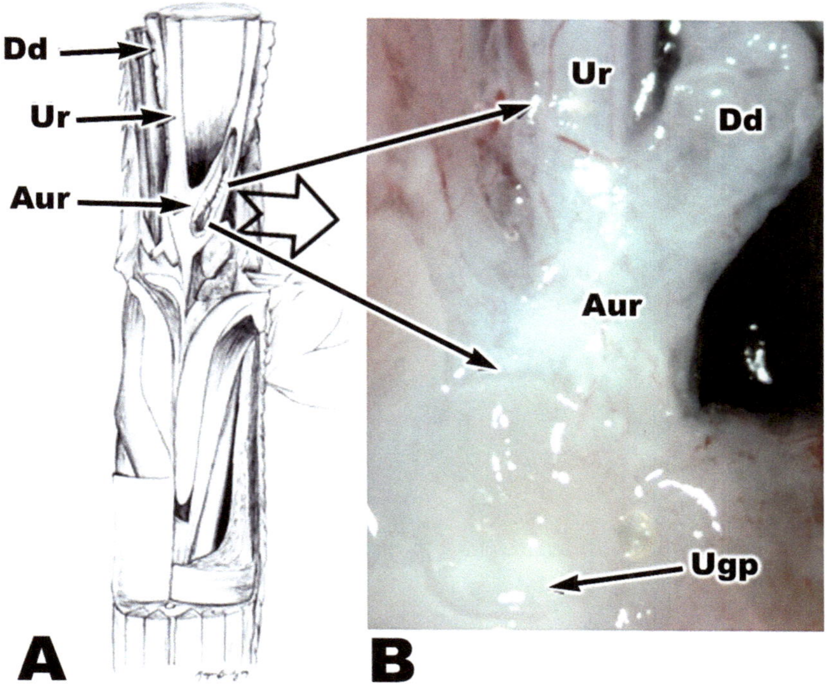

Figure 10.28 See text page **447** for caption.

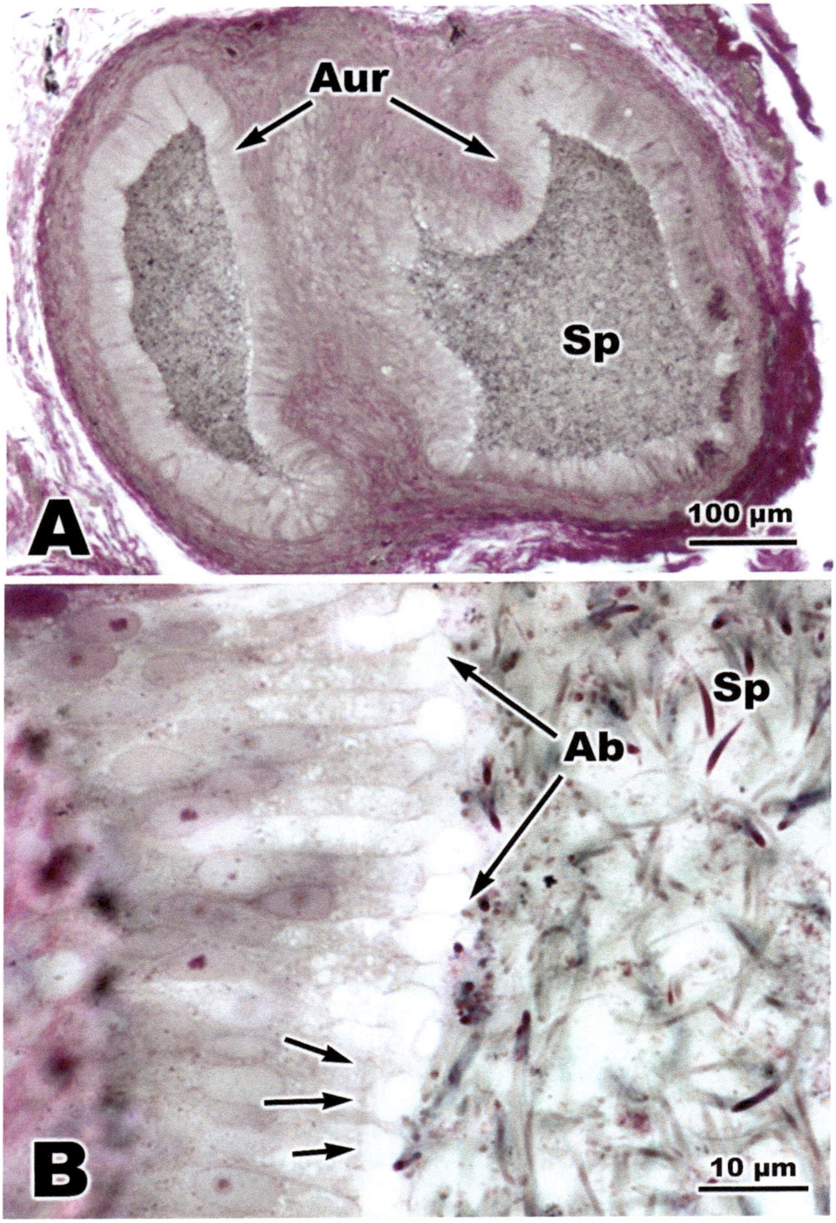

Figure 10.30 See text page 450 for caption.

CHAPTER 11

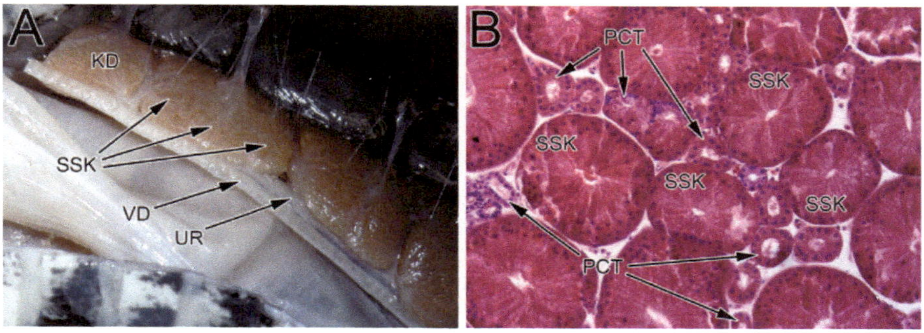

Figure 11.1 See text page **480** for caption.

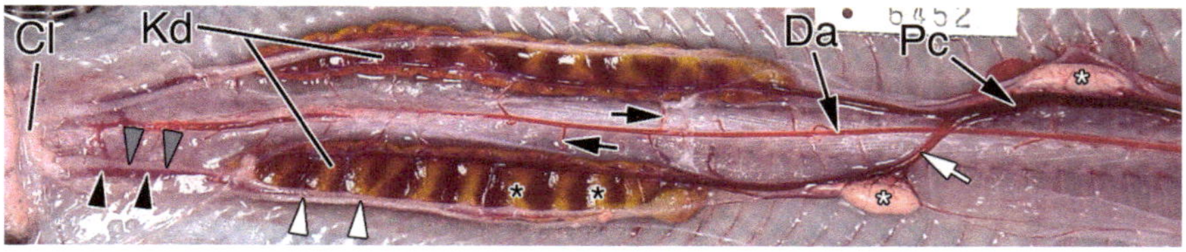

Figure 11.2 See text page **482** for caption.

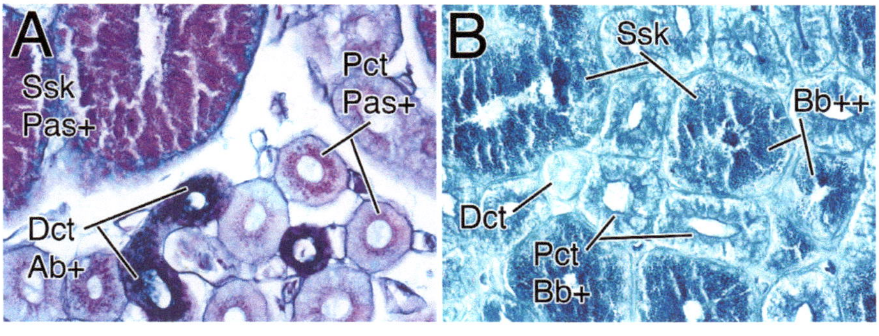

Figure 11.4 See text page **485** for caption.